Lecture Notes in Computer Science   10993

Commenced Publication in 1973
Founding and Former Series Editors:
Gerhard Goos, Juris Hartmanis, and Jan van Leeuwen

More information about this series at http://www.springer.com/series/7410

Hovav Shacham · Alexandra Boldyreva (Eds.)

Advances in Cryptology – CRYPTO 2018

38th Annual International Cryptology Conference
Santa Barbara, CA, USA, August 19–23, 2018
Proceedings, Part III

 Springer

Editors
Hovav Shacham
The University of Texas at Austin
Austin, TX
USA

Alexandra Boldyreva
Georgia Institute of Technology
Atlanta, GA
USA

ISSN 0302-9743   ISSN 1611-3349 (electronic)
Lecture Notes in Computer Science
ISBN 978-3-319-96877-3   ISBN 978-3-319-96878-0 (eBook)
https://doi.org/10.1007/978-3-319-96878-0

Library of Congress Control Number: 2018949031

LNCS Sublibrary: SL4 – Security and Cryptology

This Springer imprint is published by the registered company Springer Nature Switzerland AG
The registered company address is: Gewerbestrasse 11, 6330 Cham, Switzerland

Preface

The 38th International Cryptology Conference (Crypto 2018) was held at the University of California, Santa Barbara, California, USA, during August 19–23, 2018. It was sponsored by the International Association for Cryptologic Research (IACR). For 2018, the conference was preceded by three days of workshops on various topics. And, of course, there was the awesome Beach BBQ at Goleta Beach.

Crypto continues to grow, year after year, and Crypto 2018 was no exception. The conference set new records for both submissions and publications, with a whopping 351 papers submitted for consideration. It took a Program Committee of 46 cryptography experts working with 272 external reviewers almost 2.5 months to select the 79 papers which were accepted for the conference. It also took one program chair about 30 minutes to dig up all those stats.

In order to minimize intentional and/or subconscious bias, papers were reviewed in the usual double-blind fashion. Program Committee members were limited to two submissions, and their submissions were scrutinized more closely and held to higher standards. The two program chairs were not allowed to submit papers. Of course, they were fine with that restriction since they were way too busy to actually write any papers.

The Program Committee recognized two papers and their authors for standing out among the rest. "Yes, There Is an Oblivious RAM Lower Bound!", by Kasper Green Larsen and Jesper Buus Nielsen, was voted best paper of the conference. Additionally, "Multi-Theorem Preprocessing NIZKs from Lattices," by Sam Kim and David J. Wu, was voted Best Paper Authored Exclusively By Young Researchers. There was no award for Best Paper Authored Exclusively by Old Researchers.

Crypto 2018 played host for the IACR Distinguished Lecture, delivered by Shafi Goldwasser. Crypto also welcomed Lea Kissner as an invited speaker from Google.

We would like to express our sincere gratitude to all the reviewers for volunteering their time and knowledge in order to select a great program for 2018. Additionally, we are very appreciative of the following individuals and organizations for helping make Crypto 2018 a success:

Tal Rabin - Crypto 2018 General Chair and Workshops Organizer
Elette Boyle - Workshops Chair
Fabrice Benhamouda - Workshops Organizer
Shafi Goldwasser - IACR Distinguished Lecturer
Lea Kissner - Invited Speaker from Google
Shai Halevi - Author of the IACR Web Submission and Review System
Anna Kramer and her colleagues at Springer
Sally Vito and UCSB Conference Services

We would also like to say thank you to our numerous sponsors, everyone who submitted papers, the session chairs, the rump session chair, and the presenters.

Lastly, a big thanks to everyone who attended the conference at UCSB. Without you, we would have had a lot of leftover potato salad at the Beach BBQ.

August 2018   Alexandra Boldyreva
                                                    Hovav Shacham

Crypto 2018

The 38th IACR International Cryptology Conference

University of California, Santa Barbara, CA, USA
August 19–23, 2018

Sponsored by the *International Association for Cryptologic Research*

General Chair

Tal Rabin — IBM T.J. Watson Research Center, USA

Program Chairs

Hovav Shacham — University of Texas at Austin, USA
Alexandra Boldyreva — Georgia Institute of Technology, USA

Program Committee

Shweta Agrawal — Indian Institute of Technology, Madras, India
Benny Applebaum — Tel Aviv University, Israel
Foteini Baldimtsi — George Mason University, USA
Gilles Barthe — IMDEA Software Institute, Spain
Fabrice Benhamouda — IBM Research, USA
Alex Biryukov — University of Luxembourg, Luxembourg
Jeremiah Blocki — Purdue University, USA
Anne Broadbent — University of Ottawa, Canada
Chris Brzuska — Aalto University, Finland
Chitchanok Chuengsatiansup — Inria and ENS de Lyon, France
Dana Dachman-Soled — University of Maryland, USA
Léo Ducas — Centrum Wiskunde & Informatica, The Netherlands
Pooya Farshim — CNRS and ENS, France
Dario Fiore — IMDEA Software Institute, Spain
Marc Fischlin — Darmstadt University of Technology, Germany
Georg Fuchsbauer — Inria and ENS, France
Steven D. Galbraith — University of Auckland, New Zealand
Christina Garman — Purdue University, USA
Daniel Genkin — University of Pennsylvania and University
of Maryland, USA
Dov Gordon — George Mason University, USA
Viet Tung Hoang — Florida State University, USA

Tetsu Iwata   Nagoya University, Japan
Stanislaw Jarecki   University of California, Irvine, USA
Seny Kamara   Brown University, USA
Markulf Kohlweiss   University of Edinburgh, UK
Farinaz Koushanfar   University of California, San Diego, USA
Xuejia Lai   Shanghai Jiao Tong University, China
Tancrède Lepoint   SRI International, USA
Anna Lysyanskaya   Brown University, USA
Alex J. Malozemoff   Galois, USA
Sarah Meiklejohn   University College London, UK
Daniele Micciancio   University of California, San Diego, USA
María Naya-Plasencia   Inria, France
Kenneth G. Paterson   Royal Holloway, University of London, UK
Ananth Raghunathan   Google, USA
Mike Rosulek   Oregon State University, USA
Ron Rothblum   MIT and Northeastern University, USA
Alessandra Scafuro   North Carolina State University, USA
abhi shelat   Northeastern University, USA
Nigel P. Smart   Katholieke Universiteit Leuven, Belgium
Martijn Stam   University of Bristol, UK
Noah Stephens-Davidowitz   Princeton University, USA
Aishwarya Thiruvengadam   University of California, Santa Barbara, USA
Hoeteck Wee   CNRS and ENS, France
Daniel Wichs   Northeastern University, USA
Mark Zhandry   Princeton University, USA

Additional Reviewers

Aydin Abadi
Archita Agarwal
Divesh Aggarwal
Shashank Agrawal
Adi Akavia
Navid Alamati
Martin Albrecht
Miguel Ambrona
Ghous Amjad
Megumi Ando
Ralph Ankele
Gilad Asharov
Achiya Bar-On
Manuel Barbosa
Paulo Barreto
James Bartusek
Guy Barwell

Balthazar Bauer
Carsten Baum
Amos Beimel
Itay Berman
Marc Beunardeau
Sai Lakshmi Bhavana
Simon Blackburn
Estuardo Alpirez Bock
Andrej Bogdanov
André Schrottenloher
Xavier Bonnetain
Charlotte Bonte
Carl Bootland
Jonathan Bootle
Christina Boura
Florian Bourse
Elette Boyle

Zvika Brakerski
Jacqueline Brendel
David Butler
Matteo Campanelli
Brent Carmer
Ignacio Cascudo
Wouter Castryck
Andrea Cerulli
André Chailloux
Nishanth Chandran
Panagiotis Chatzigiannis
Stephen Checkoway
Binyi Chen
Michele Ciampi
Benoit Cogliati
Gil Cohen
Ran Cohen

Aisling Connolly
Sandro Coretti
Henry Corrigan-Gibbs
Geoffroy Couteau
Shujie Cui
Ting Cui
Joan Daemen
Wei Dai
Yuanxi Dai
Alex Davidson
Jean Paul Degabriele
Akshay Degwekar
Ioannis Demertzis
Itai Dinur
Jack Doerner
Nico Döttling
Benjamin Dowling
Tuyet Thi Anh Duong
Frédéric Dupuis
Betul Durak
Lior Eldar
Karim Eldefrawy
Lucas Enloe
Andre Esser
Antonio Faonio
Prastudy Fauzi
Daniel Feher
Serge Fehr
Nils Fleischhacker
Benjamin Fuller
Tommaso Gagliardoni
Martin Gagné
Adria Gascon
Pierrick Gaudry
Romain Gay
Nicholas Genise
Marilyn George
Ethan Gertler
Vlad Gheorghiu
Esha Ghosh
Brian Goncalves
Junqing Gong
Adam Groce
Johann Großschädl
Paul Grubbs
Jiaxin Guan

Jian Guo
Siyao Guo
Joanne Hall
Ariel Hamlin
Abida Haque
Patrick Harasser
Gottfried Herold
Naofumi Homma
Akinori Hosoyamada
Jialin Huang
Siam Umar Hussain
Chloé Hébant
Yuval Ishai
Ilia Iliashenko
Yuval Ishai
Håkon Jacobsen
Christian Janson
Ashwin Jha
Thomas Johansson
Chethan Kamath
Bhavana Kanukurthi
Marc Kaplan
Pierre Karpman
Sriram Keelveedhi
Dmitry Khovratovich
Franziskus Kiefer
Eike Kiltz
Sam Kim
Elena Kirshanova
Konrad Kohbrok
Lisa Maria Kohl
Ilan Komargodski
Yashvanth Kondi
Venkata Koppula
Lucas Kowalczyk
Hugo Krawczyk
Thijs Laarhoven
Marie-Sarah Lacharite
Virginie Lallemand
Esteban Landerreche
Phi Hung Le
Eysa Lee
Jooyoung Lee
Gaëtan Leurent
Baiyu Li
Benoit Libert

Fuchun Lin
Huijia Lin
Tingting Lin
Feng-Hao Liu
Qipeng Liu
Tianren Liu
Zhiqiang Liu
Alex Lombardi
Sébastien Lord
Steve Lu
Yiyuan Luo
Atul Luykx
Vadim Lyubashevsky
Fermi Ma
Varun Madathil
Mohammad Mahmoody
Mary Maller
Giorgia Azzurra Marson
Daniel P. Martin
Samiha Marwan
Christian Matt
Alexander May
Sogol Mazaheri
Bart Mennink
Carl Alexander Miller
Brice Minaud
Ilya Mironov
Tarik Moataz
Nicky Mouha
Fabrice Mouhartem
Pratyay Mukherjee
Mridul Nandi
Samuel Neves
Anca Nitulescu
Kaisa Nyberg
Adam O'Neill
Maciej Obremski
Olya Ohrimenko
Igor Carboni Oliveira
Claudio Orlandi
Michele Orrù
Emmanuela Orsini
Dag Arne Osvald
Elisabeth Oswald
Elena Pagnin
Chris Peikert

Sponsors

Contents – Part III

Zero Knowledge

Obfuscation

Efficient MPC

TinyKeys: A New Approach to Efficient Multi-Party Computation

Carmit Hazay[1], Emmanuela Orsini[2], Peter Scholl[3(✉)],
and Eduardo Soria-Vazquez[4]

[1] Bar-Ilan University, Ramat Gan, Israel
carmit.hazay@biu.ac.il
[2] KU Leuven ESAT/COSIC, Leuven, Belgium
emmanuela.orsini@kuleuven.be
[3] Aarhus University, Aarhus, Denmark
peter.scholl@cs.au.dk
[4] University of Bristol, Bristol, UK
eduardo.soria-vazquez@bristol.ac.uk

Abstract. We present a new approach to designing concretely efficient MPC protocols with semi-honest security in the dishonest majority setting. Motivated by the fact that within the dishonest majority setting the efficiency of most practical protocols *does not depend on the number of honest parties*, we investigate how to construct protocols which improve in efficiency as the number of honest parties increases. Our central idea is to take a protocol which is secure for $n - 1$ corruptions and modify it to use short symmetric keys, with the aim of basing security on the concatenation of all honest parties' keys. This results in a more efficient protocol tolerating fewer corruptions, whilst also introducing an LPN-style syndrome decoding assumption.

We first apply this technique to a modified version of the semi-honest GMW protocol, using OT extension with short keys, to improve the efficiency of standard GMW with fewer corruptions. We also obtain more efficient constant-round MPC, using BMR-style garbled circuits with short keys, and present an implementation of the online phase of this protocol. Our techniques start to improve upon existing protocols when there are around $n = 20$ parties with $h = 6$ honest parties, and as these increase we obtain up to a 13 times reduction (for $n = 400, h = 120$)

C. Hazay—Supported by the European Research Council under the ERC consolidators grant agreement No. 615172 (HIPS), and by the BIU Center for Research in Applied Cryptography and Cyber Security in conjunction with the Israel National Cyber Bureau in the Prime Minister's Office.

E. Orsini—Supported in part by ERC Advanced Grant ERC-2015-AdG-IMPaCT.

P. Scholl—Supported by the European Union's Horizon 2020 research and innovation programme under grant agreement No. 731583 (SODA), and the Danish Independent Research Council under Grant-ID DFF-6108-00169 (FoCC).

E. Soria-Vazquez—Supported by the European Union's Horizon 2020 research and innovation programme under the Marie Skłodowska-Curie grant agreement No. 643161, and by ERC Advanced Grant ERC-2015-AdG-IMPaCT.

H. Shacham and A. Boldyreva (Eds.): CRYPTO 2018, LNCS 10993, pp. 3–33, 2018.
https://doi.org/10.1007/978-3-319-96878-0_1

in communication complexity for our GMW variant, compared with the best-known GMW-based protocol modified to use the same threshold.

1 Introduction

Secure multi-party computation (MPC) protocols allow a group of n parties to compute some function f on the parties' private inputs, while preserving a number of security properties such as *privacy* and *correctness*. The former property implies data confidentiality, namely, nothing leaks from the protocol execution but the computed output. The latter requirement implies that the protocol enforces the integrity of the computations made by the parties, namely, honest parties are not lead to accept a wrong output. Security is proven either in the presence of an honest-but-curious adversary that follows the protocol specification but tries to learn more than allowed from its view of the protocol, or a malicious adversary that can arbitrarily deviate from the protocol specification in order to compromise the security of the other parties in the protocol.

The efficiency of a protocol typically also depends on how many corrupted parties can be tolerated before security breaks down, a quantity known as the *threshold*, t. With semi-honest security, most protocols either require $t < n/2$ (where n is the number of parties), in which case unconditionally secure protocols [BOGW88, CCD88] based on Shamir secret-sharing can be used, or support any choice of t up to $n - 1$, as in computationally secure protocols based on oblivious transfer [GMW87, Gol04]. Interestingly, within these two ranges, the efficiency of most practical semi-honest protocols *does not depend on t*. For instance, the GMW [GMW87] protocol (and its many variants) is *full-threshold*, so supports any $t < n$ corruptions. However, we *do not know* of any practical protocols with threshold, say, $t = \frac{2}{3}n$, or even $t = n/2 + 1$, that are more efficient than full-threshold GMW-style protocols. One exception to this is when the number of parties becomes very large, in which case protocols based on *committees* can be used. In this approach, due to an idea of Bracha [Bra85], first a random committee of size $n' \ll n$ is chosen. Then every party secret-shares its input to the parties in the committee, who runs a secure computation protocol for $t < n'$ to obtain the result. The committee size n' must be chosen to ensure (with high probability) that not the whole committee is corrupted, so clearly a lower threshold t allows for smaller committees, giving significant efficiency savings. However, this technique is only really useful when n is very large, at least in the hundreds or thousands.

In this paper we investigate designing MPC protocols where *an arbitrary threshold for the number of corrupted parties can be chosen*, which are practical both when n is very large, and also for small to medium sizes of n. Specifically, we ask the question:

Can we design concretely efficient MPC protocols where the performance improves gracefully as the number of honest parties increases?

Note that the performance of an MPC protocol can be measured both in terms of *communication overhead* and *computational overhead*. Using fully homomorphic encryption [Gen09], it is possible to achieve very low communication overhead that is independent of the circuit size [AJL+12] even in the malicious setting, but for reasonably complex functions FHE is impractical due to very high computational costs. On the other hand, practical MPC protocols typically communicate for every AND gate in the circuit, and use *oblivious transfer* (OT) to carry out the computation. Fast OT extension techniques allow a large number of secret-shared bit multiplications[1] to be performed using only symmetric primitives and an amortized communication complexity of $O(\kappa)$ [IKNP03] or $O(\kappa/\log \kappa)$ [KK13, DKS+17] bits, where κ is a computational security parameter. This leads to an overall communication complexity which grows with $O(n^2\kappa/\log \kappa)$ bits per AND gate in protocols based on secret-sharing following the [GMW87] style, and $O(n^2\kappa)$ in those based on garbled circuits in the style of [Yao86, BMR90, BLO16].

Short keys for secure computation. Our main idea towards achieving the above goal is to build a secure multi-party protocol with h honest parties, by distributing secret key material so that each party only holds a *small part of the key*. Instead of basing security on secret keys held by each party individually, we then base security on the *concatenation of all honest parties' keys*.

As a toy example, consider the following simple distributed encryption of a message m under n keys:

$$\mathsf{E}_k(m) = \bigoplus_{i=1}^{n} \mathsf{H}(i, k_i) \oplus m$$

where H is a suitable hash function and each key $k_i \in \{0,1\}^\ell$ belongs to party P_i. In the full-threshold setting with up to $n-1$ corruptions, to hide the message we need each party's key to be of length $\ell = 128$ to achieve 128-bit computational security. However, if only $t < n-1$ parties are corrupted, it seems that, intuitively, an adversary needs to guess all $h := n-t$ honest parties' keys to recover the message, and potentially each key k_i can be *much less than 128 bits* long when h is large enough. This is because the "obvious" way to try to guess m would be to brute force all h keys until decrypting "successfully".

In fact, recovering m when there are h unknown keys corresponds to solving an instance of the *regular syndrome decoding problem* [AFS03], which is related to the well-known *learning parity with noise* (LPN) problem, and believed to be hard for suitable choices of parameters.

1.1 Our Contribution

In this work we use the above idea of short secret keys to design new MPC protocols in both the constant round and non-constant round settings, which

---

[1] Note that OT is equivalent to secret-shared bit multiplication, and when constructing MPC it is more convenient to use the latter definition.

improve in efficiency as the number of honest parties increases. We consider security against a static, honest-but-curious adversary, and leave it for future work to extend our techniques to the malicious case based on, e.g. message authentication codes. Our contribution is captured by the following:

GMW-STYLE MPC WITH SHORT KEYS (SECT. 3). We present a GMW-style MPC protocol for binary circuits, where multiplications are done with OT extension using short symmetric keys. This reduces the communication complexity of OT extension-based GMW from $O(n^2\kappa/\log\kappa)$ [KK13] to $O(nt\ell)$, where the key length ℓ decreases as the number of honest parties, $h = n - t$, increases. When h is large enough, we can even have ℓ *as small as* 1.

To construct this protocol, we first analyse the security of the IKNP OT extension protocol [IKNP03] when using short keys, and formalise the leakage obtained by a corrupt receiver in this case. We then show how to use this version of "leaky OT" to generate multiplication triples using a modified version of the GMW method, where pairs of parties use OT to multiply their shares of random values. We also optimize our protocol by reducing the number of communication channels using two different-sized committees, improving upon the standard approach of choosing one committee to do all the work.

MULTI-PARTY GARBLED CIRCUITS WITH SHORT KEYS (SECT. 4). Our second contribution is the design of a constant round, BMR-style [BMR90] protocol based on garbled circuits with short keys. Our offline phase uses the multiplication protocol from the previous result in order to generate the garbled circuit, using secret-shared bit and bit/string multiplications as done in previous works [BLO16,HSS17], with the exception that the keys are shorter. In the online phase, we then use the LPN-style assumption to show that the combination of all honest parties' ℓ-bit keys suffices to obtain a secure garbling protocol. This allows us to save on the key length as a function of the number of honest parties. As well as reducing communication with a smaller garbled circuit, we also reduce computation when evaluating the circuit, since each garbled gate can be evaluated with only $O(n^2\ell/\kappa)$ block cipher calls (assuming the ideal cipher model), instead of $O(n^2)$ when using κ-bit keys. For this protocol, ℓ *can be as small as* 8, giving a significant saving over 128-bit keys used previously.

Concrete Efficiency Improvements. The efficiency of our protocols depends on the total number of parties, n, and the number of honest parties, h, so there is a large range of parameters to explore when comparing with other works. We discuss this in more detail in Sect. 5. Our protocols seem most significant in the *dishonest majority* setting, since when there is an honest majority there are unconditionally secure protocols with $O(n\log n)$ communication overhead and reasonable computational complexity e.g. [DN07], whilst our protocols have $\Omega(nt)$ communication overhead.

Our GMW-style protocol starts to improve upon previous protocols when we reach $n = 20$ parties and $t = 14$ corruptions: here, our triple generation method requires less than *half the communication cost* of the fastest GMW-style protocol based on OT extension [DKS+17] tolerating up to $n - 1$ corruptions. When the number of honest parties is large enough, we can use *1-bit keys*, giving a *25-fold reduction* in communication over previous protocols when $n = 400$ and $t = 280$. In addition, we describe a simple threshold-t variant of GMW-style protocols, which our protocol still outperforms by 1.1x and 13x, respectively, in these two scenarios.

For our constant round protocol, with $n = 20, t = 10$ we can use 32-bit keys, so the size of each garbled AND gate is 1/4 the size of [BLO16]. As n increases the improvements become greater, with a *16-fold reduction* in garbled AND gate size for $n = 400, t = 280$. We also reduce the communication cost of *creating* the garbled circuit. Here, the improvement starts at around 50 parties, and goes up to a 7 times reduction in communication when $n = 400, t = 280$. Note that our protocol does incur a slight additional overhead, since we need to use extra "splitter gates", but this cost is relatively small.

To demonstrate the practicality of our approach, we also present an implementation of the online evaluation phase of our constant-round protocol for key lengths ranging between 1–4 bytes, and with an overall number of parties ranging from 15–1000; more details can be found in Sect. 5.

Applications. Our techniques seem most useful for large-scale MPC with around 70% corruptions, where we obtain the greatest concrete efficiency improvements. An important motivation for this setting is privacy-preserving statistical analysis of data collected from a large network with potentially thousands of nodes. In scenarios where the nodes are not always online and connected, our protocols can also be used with the "random committee" approach discussed earlier, so only a small subset of, say, a hundred nodes need to be online and interacting during the protocol.

An interesting example is safely measuring the Tor network [DMS04] which is among the most popular tools for digital privacy, consisting of more than 6000 relays that can opt-in for providing statistics about the use of the network. Nowadays and due to privacy risks, the statistics collected over Tor are generally poor: There is a reduced list of computed functions and only a minority of the relays provide data, which has to be obfuscated before publishing [DMS04]. Hence, the statistics provide an incomplete picture which is affected by a noise that scales with the number of relays. Running MPC in this setting would enable for more complex, accurate and private data processing, for example through anomaly detection and more sophisticated censorship detection. Moreover, our protocols are particularly well-suited to this setting since all relays in the network must be connected to one another already, by design.

Another possible application is for securely computing the interdomain routing within the Border Gateway Protocol (BGP), which is performed at a large scale of thousands of nodes. A recent solution in the dishonest majority setting [ADS+17] centralizes BGP so that two parties run this computation for all Autonomous Systems. Our techniques allow scaling to a large number of systems computing the interdomain routing themselves using MPC, hence further reducing the trust requirements.

Decisional Regular Syndrome Decoding problem. The security of our protocols relies on the *Decisional Regular Syndrome Decoding (DRSD)* problem, which, given a random binary matrix **H**, is to distinguish between the syndrome obtained by multiplying **H** with an error vector $e = (e_1 \| \cdots \| e_h)$ where each $e_i \in \{0,1\}^{2^\ell}$ has Hamming weight one, and the uniform distribution. This can equivalently be described as distinguishing $\bigoplus_{i=1}^h \mathsf{H}(i, k_i)$ from the uniform distribution, where H is a random function and each k_i is a random ℓ-bit key (as in the toy example described earlier).

We remark that when h is large enough, the problem is *unconditionally hard* even for $\ell = 1$, which means for certain parameter choices in our GMW-based protocol we can use 1-bit keys *without introducing any additional assumptions*. This introduces a significant saving in our triple generation protocol.

Additional related work. Another work which applies a similar assumption to secure computation is that of Applebaum [App16], who built garbled circuits with the free-XOR technique in the standard model under the LPN assumption. Conceptually, our work differs from Applebaum's since our focus is to improve the efficiency of multi-party protocols with fewer corruptions, whereas in [App16], LPN is used in a more modular way in order to achieve encryption with stronger properties and under a more standard assumption.

In a recent work [NR17], Nielsen and Ranellucci designed a protocol in the dishonest majority setting with malicious, adaptive security in the presence of $t < cn$ corruption for $t \in [0, 1)$. Their protocol is aimed to work with a large number of parties and uses committees to obtain a protocol with poly-logarithmic overhead. This protocol introduces high constants and is not useful for practical applications.

Finally, in a concurrent work [BO17], Ben-Efraim and Omri also explore how to optimize garbled circuits in the presence of non-full-threshold adversaries. By using deterministic committees they achieve AND gates of size $4(t+1)\kappa$, where κ is the computational security parameter. By using the same technique we achieve a size of $4(t + h)\ell$, where $\ell \ll \kappa$ depends on h, a parameter for the minimum number of honest parties in the committee. The rest of their results apply only to the honest majority setting.

1.2   Technical Overview

In what follows we explain the technical side of our results in more detail.

Leaky oblivious transfer (OT). We first present a two-party secret-shared bit multiplication protocol, based on a variant of the IKNP OT extension protocol [IKNP03] with short keys. Our protocol performs a batch of r multiplications at once. Namely, the parties create r correlated OTs on ℓ-bit strings using the OT extension technique of [IKNP03], by transposing a matrix of ℓ OTs on r-bit strings and swapping the roles of sender and receiver. In contrast to the IKNP OT extension and followups, that use κ 'base' OTs for computational security parameter κ, we use $\ell = O(\log \kappa)$ base OTs.

This protocol leaks some information on the global secret $\Delta \leftarrow \{0,1\}^\ell$ picked by the receiver, as well as the inputs of the receiver. Roughly speaking, the leakage is of the form $\mathsf{H}(i, \Delta) + x_i$, where $x_i \in \{0,1\}$ is an input of the receiver and H is a hash function with 1-bit output. Clearly, when ℓ is short this is not secure to use on its own, since all of the receiver's inputs only have ℓ bits of min-entropy (based on the choice of Δ).

MPC from leaky OT. We then show how to apply this leaky two-party protocol to the multi-party setting, whilst preventing any leakage on the parties shares. The main observation is that, when using additive secret-sharing, we only need to ensure that the *sum* of all honest parties' shares is unpredictable; if the adversary learns just a few shares, they can easily be rerandomized by adding pseudorandom shares of zero, which can be done non-interactively using a PRF. However, we still have a problem, which is that in the standard GMW approach, each party P_i uses OT to multiply their share x^i with every other party P_j's share y^j. Now, there is leakage on the *same share* x^i from each of the OT instances between all other parties, which seems much harder to prevent than leakage from just a single OT instance.

To work around this problem, we have the parties add shares of zero to their x^i inputs *before* multiplying them. So, every pair (P_i, P_j) will use leaky OT to multiply $x^i \oplus s^{i,j}$ with y^j, where $s^{i,j}$ is a random share of zero satisfying $\bigoplus_{i=1}^n s^{i,j} = 0$. This preserves correctness of the protocol, because the parties end up computing an additive sharing of:

$$\bigoplus_{i=1}^n \bigoplus_{j=1}^n (x^i \oplus s^{i,j}) y^j = \bigoplus_{j=1}^n y^j \bigoplus_{i=1}^n (x^i \oplus s^{i,j}) = xy.$$

This also effectively removes leakage on the individual shares, so we only need to be concerned with the *sum* of the leakage on all honest parties' shares, and this turns out to be of the form: $\bigoplus_{i=1}^n (\mathsf{H}(i, \Delta_i) + x^i)$ which is pseudorandom under the decisional regular syndrome decoding assumption.

We realize our protocol using a hash function with a polynomial-sized domain, so that is can be implemented using a CRS which simply outputs a random lookup-table. This means that, unlike when using the IKNP protocol, we do not need to rely on a random oracle or a correlation robustness assumption.

When the number of parties is large enough, we can improve our triple generation protocol using *random committees*. In this case the amortized communication cost is $\leq n_h n_1 (\ell + \ell \kappa / r + 1)$ bits per multiplication where we need to choose two committees of sizes n_h and n_1 which have at least h and 1 honest parties, respectively.

Garbled circuits with short keys. We next revisit the multi-party garbled circuits technique by Beaver, Micali and Rogaway, known as BMR, that extends the classic Yao garbling [Yao86] to an arbitrary number of parties, where essentially all the parties jointly garble using one set of keys each. This method was recently improved in a sequence of works [LPSY15, LSS16, BLO16, HSS17], where the two latter works further support the Free-XOR property.

Our garbling method uses an expansion function $\mathsf{H} : [n] \times \{0,1\} \times \{0,1\}^\ell \to \{0,1\}^{n\ell+1}$, where ℓ is the length of each parties' keys used as wire labels in the garbled circuit. To garble a gate, the hash values of the input wire keys $k^i_{u,b}$ and $k^i_{v,b}$ are XORed over i and used to mask the output wire keys.

Specifically, for an AND gate g with input wires u, v and output wire w, the 4 garbled rows $\tilde{g}_{a,b}$, for each $(a,b) \in \{0,1\}^2$, are computed as:

$$\tilde{g}_{a,b} = \left(\bigoplus_{i=1}^{n} \mathsf{H}(i, b, k^i_{u,a}) \oplus \mathsf{H}(i, a, k^i_{v,b}) \right) \oplus (c, k^1_{w,c}, \ldots, k^n_{w,c}).$$

Security then relies on the DRSD assumption, which implies that the sum of h hash values on short keys is pseudorandom, which suffices to construct a secure garbling method with h honest parties.

Using this assumption instead of a PRF (as in recent works) comes with difficulties, as we can no longer garble gates with arbitrary fan-out, or use the free-XOR technique, without degrading the DRSD parameters. To allow for arbitrary fan-out circuits with our protocol we use *splitter gates*, which take as input one wire w and provide two outputs wires u, v, representing the same wire value. Splitter gates were previously introduced as a fix for an error in the original BMR paper in [TX03]. We stress that transforming a general circuit description into a circuit with only fan-out-1 gates requires adding at most a single splitter gate per AND or XOR gate.

The restriction to fan-out-1 gates and the use of splitter gates additionally allows us to garble XOR gates for free in BMR without relying on circular security assumptions or correlation-robust hash functions, based on the FlexOR technique [KMR14] where each XOR gate uses a unique offset. Furthermore, the overhead of splitter gates is very low, since garbling a splitter gate does not use the underlying MPC protocol: shares of the garbled gate can be generated non-interactively. We note that this observation also applies to Yao's garbled circuits, but the overhead of adding splitter gates there is more significant; this is because in most 2-party protocols, the *size* of the garbled circuit is the dominant cost factor, whereas in multi-party protocols the main cost is *creating* the garbled circuit in a distributed manner.

2   Preliminaries

We denote the security parameter by κ. We say that a function $\mu : \mathbb{N} \to \mathbb{N}$ is *negligible* if for every positive polynomial $p(\cdot)$ and all sufficiently large κ it holds that $\mu(\kappa) < \frac{1}{p(\kappa)}$. The function μ is *noticeable* (or non-negligible) if there exists a positive polynomial $p(\cdot)$ such that for all sufficiently large κ it holds that $\mu(\kappa) \geq \frac{1}{p(\kappa)}$. We use the abbreviation PPT to denote probabilistic polynomial-time. We further denote by $a \leftarrow A$ the uniform sampling of a from a set A, and by $[d]$ the set of elements $\{1, \ldots, d\}$. We often view bit-strings in $\{0,1\}^k$

as vectors in $\mathbb{F}_2^k$, depending on the context, and denote exclusive-or by "$\oplus$" or "+". If $a, b \in \mathbb{F}_2$ then $a \cdot b$ denotes multiplication (or AND), and if $c \in \mathbb{F}_2^\kappa$ then $a \cdot c \in \mathbb{F}_2^\kappa$ denotes the product of a with every component of c.

Security and Communication Models. We prove security of our protocols in the universal composability (UC) framework [Can01]. See We assume all parties are connected via secure, authenticated point-to-point channels, which is the default method of communication in our protocols. The adversary model we consider is a static, honest-but-curious adversary who corrupts a subset $A \subset [n]$ of parties at the beginning of the protocol. We denote by $\bar{A}$ the subset of honest parties, and define $h = |\bar{A}| = n - t$.

Functionality $\mathcal{F}_{\text{Zero}}^r(\mathcal{P})$

On receiving (zero) from all parties in $\mathcal{P} = \{P_1, \ldots, P_n\}$:

1. Sample random shares $s^2, \ldots, s^n \leftarrow \{0,1\}^r$ and let $s^1 = s^2 \oplus \cdots \oplus s^n$
2. Send s^i to party P_i

Fig. 1. Random zero sharing functionality.

Random Zero-Sharing. Our protocols require the parties to generate random additive sharings of zero, as in the $\mathcal{F}_{\text{Zero}}$ functionality in Fig. 1. This can be done efficiently using a PRF F, with interaction *only* during a setup phase, as in [AFL+16].

2.1 Regular Syndrome Decoding Problem

We now describe the Regular Syndrome Decoding (RSD) problem and some of its properties.

Definition 2.1. *A vector* $e \in \mathbb{F}_2^m$ *is* (m, h)-*regular if* $e = (e_1 \| \cdots \| e_h)$ *where each* $e_i \in \{0,1\}^{m/h}$ *has Hamming weight one. We denote by* $R_{m,h}$ *the set of all the* (m, h)-*regular vectors in* $\mathbb{F}_2^m$.

Definition 2.2 (Regular Syndrome Decoding (RSD)). *Let* $r, h, \ell \in \mathbb{N}$ *with* $m = h \cdot 2^\ell$, $\mathbf{H} \leftarrow \mathbb{F}_2^{r \times m}$ *and* $e \leftarrow R_{m,h}$. *Given* $(\mathbf{H}, \mathbf{H}e)$, *the* $\text{RSD}_{r,h,\ell}$ *problem is to recover* e *with noticeable probability.*

The decisional version of the problem, given below, is to distinguish the syndrome $\mathbf{H}e$ from uniform.

Definition 2.3 (Decisional Regular Syndrome Decoding (DRSD)). *Let* $\mathbf{H} \leftarrow \mathbb{F}_2^{r \times m}$ *and* $e \leftarrow R_{m,h}$, *and let* U_r *be the uniform distribution on* r *bits. The* $\text{DRSD}_{r,h,\ell}$ *problem is to distinguish between* $(\mathbf{H}, \mathbf{H}e)$ *and* $(\mathbf{H}, U_r)$ *with noticeable advantage.*

Hash function formulation. The DRSD problem can be equivalently described as distinguishing from uniform $\bigoplus_{i=1}^{h} \mathsf{H}(i, k_i)$ where $\mathsf{H} : [h] \times \{0,1\}^{\ell} \to \{0,1\}^{r}$ is a random hash function, and each $k_i \leftarrow \{0,1\}^{\ell}$. With this formulation, it is easier to see how the DRSD problem arises when using our protocols with short keys, since this appears when summing up a hash function applied to h honest parties' secret keys.

To see the equivalence, we can define a matrix $\mathbf{H} \in \mathbb{F}_2^{r \times h \cdot 2^{\ell}}$, where for each $i \in \{0, \ldots, h-1\}$ and $k \in [2^{\ell}]$, column $i \cdot 2^{\ell} + k$ of $\mathbf{H}$ contains $\mathsf{H}(i, k)$. Then, multiplying $\mathbf{H}$ with a random (m, h)-regular vector $\mathbf{e}$ is equivalent to taking the sum of H over h random inputs, as above.

Statistical hardness of DRSD. We next observe that for certain parameters where the output size of $\mathbf{H}$ is sufficiently smaller than the min-entropy of the error vector $\mathbf{e}$, the distribution in the decisional problem is *statistically close to uniform*. Proofs and the general case of ℓ-bit keys are given in [HOSS18].

Lemma 2.1. *If $\ell = 1$ and $h \geq r + s$ then $\mathsf{DRSD}_{r,h,\ell}$ is statistically hard, with distinguishing probability 2^{-s}.*

Search-to-decision reduction. For *all* parameter choices of DRSD, there is a simple reduction to the search version of the regular syndrome decoding problem with the same parameters.

Lemma 2.2. *Any efficient distinguisher for the $\mathsf{DRSD}_{r,h,\ell}$ problem can be used to efficiently solve $\mathsf{RSD}_{r,h,\ell}$.*

3   GMW-Style MPC with Short Keys

In this section we design a protocol for generating multiplication triples over $\mathbb{F}_2$ using short secret keys, with reduced communication complexity as the number of honest parties increases. More concretely, we first design a leaky protocol for secret-shared two-party bit multiplication, based on correlated OT and OT extension techniques with short keys. This protocol is not fully secure and we precisely define the leakage obtained by the receiver. We next show how to use the leaky protocol to produce multiplication triples, removing the leakage by rerandomizing the parties' shares with shares of zero, and using the DRSD assumption. Finally, this protocol can be used with Beaver's multiplication triple technique [Bea92] to obtain MPC for binary circuits with an amortized communication complexity of $O(nt\ell)$ bits per triple, where t is the threshold and ℓ is the secret key length. When the number of honest parties is large enough we can even use $\ell = 1$ and avoid relying on DRSD.

3.1   Leaky Two-Party Secret-Shared Multiplication

We first present our protocol for two-party secret-shared bit multiplication, based on a variant of the [IKNP03] OT extension protocol, modified to use short keys.

Functionality $\mathcal{F}_{\Delta\text{-ROT}}^{r,\ell}$

After receiving $\Delta \in \{0,1\}^\ell$ from P_S and $(x_1, \ldots, x_r) \in \{0,1\}^r$ from P_R, do the following:

1. Sample $\boldsymbol{q}_i \leftarrow \{0,1\}^\ell$, for $i \in [r]$, and let $\boldsymbol{t}_i = \boldsymbol{q}_i \oplus x_i \cdot \Delta$.
2. Output $\boldsymbol{q}_i$ to P_S and $\boldsymbol{t}_i$ to P_R, for $i \in [r]$.

Fig. 2. Functionality for oblivious transfer on random, correlated strings.

With short keys we cannot hope for computational security based on standard symmetric primitives, because an adversary can search every possible key in polynomial time. Our goal, therefore, is to define the precise *leakage* that occurs when using short keys, in order to remove this leakage at a later stage.

OT extension and correlated OT. Recall that the main observation of the IKNP protocol for extending oblivious transfer [IKNP03] is that *correlated OT is symmetric*, so that κ correlated OTs on r-bit strings can be locally converted into r correlated OTs on κ-bit strings. Secondly, a κ-bit correlated OT can be used to obtain an OT on chosen strings with computational security. The first stage of this process is abstracted away by the functionality $\mathcal{F}_{\Delta\text{-ROT}}$ in Fig. 2.

Using IKNP to multiply an input bit x_k from the sender, P_A, with an input bit y_k from P_B, the receiver, P_B sends y_k as its choice bit to $\mathcal{F}_{\Delta\text{-ROT}}$ and learns $\boldsymbol{t}_k = \boldsymbol{q}_k \oplus y_k \cdot \Delta$. The sender P_A obtains $\boldsymbol{q}_k$, and then sends

$$d_k = \mathsf{H}(\boldsymbol{q}_k) \oplus \mathsf{H}(\boldsymbol{q}_k \oplus \Delta) \oplus x_k,$$

where H is a 1-bit output hash function. This allows the parties to compute an additive sharing of $x_k \cdot y_k$ as follows: P_A defines the share $\mathsf{H}(\boldsymbol{q}_k)$, and P_B computes $\mathsf{H}(\boldsymbol{t}_k) \oplus y_k \cdot d_k$. This can be repeated many times with the same Δ to perform a large batch of $\mathsf{poly}(\kappa)$ secret-shared multiplications, because the randomness in Δ serves to computationally mask each x with the hash values (under a suitable correlation robustness assumption for H). The downside of this is that for $\Delta \in \{0,1\}^\kappa$, the communication cost is $O(\kappa)$ bits per two-party bit multiplication, to perform the correlated OTs.

Variant with short keys. We adapt this protocol to use short keys by performing the correlated OTs on ℓ-bit strings, instead of κ-bit, for some small key length $\ell = O(\log \kappa)$ (we could have ℓ as small as 1). This allows $\mathcal{F}_{\Delta\text{-ROT}}$ to be implemented with only $O(\ell)$ bits of communication per OT instead of $O(\kappa)$.

Our protocol, shown in Fig. 4, performs a batch of r multiplications at once. First the parties create r correlated OTs on ℓ-bit strings using $\mathcal{F}_{\Delta\text{-ROT}}$. Next, the parties hash the output strings of the correlated OTs, and P_A sends over the correction values d_k, which are used by P_B to convert the random OTs into a secret-shared bit multiplication. Finally, we require the parties to add a random value (from $\mathcal{F}_{\mathrm{Zero}}$, shown in Fig. 1) to their outputs, which ensures that they have a uniform distribution.

Note that if $\ell \in O(\log \kappa)$ then the hash function H_{AB} has a polynomial-sized domain, so can be described as a lookup table provided as a common input to the protocol by both parties. At this stage we do not make any assumptions about H_{AB}; this means that the leakage in the protocol will depend on the hash function, so its description is also passed to the functionality $\mathcal{F}_{\text{Leaky-2-Mult}}$ (Fig. 3). We require H_{AB} to take as additional input an index $k \in [r]$ and a bit in $\{0, 1\}$, to provide independence between different uses, and our later protocols require the function to be different in protocol instances between different pairs of parties (we use the notation H_{AB} to emphasize this).

Functionality $\mathcal{F}_{\text{Leaky-2-Mult}}^{r, \ell}$

INPUT: $(x_1, \ldots, x_r) \in \mathbb{F}_2^r$ from P_A and $(y_1, \ldots, y_r) \in \mathbb{F}_2^r$ from P_B.
COMMON INPUT: A hash function $\mathsf{H}_{AB} : [r] \times \{0, 1\} \times \{0, 1\}^\ell \to \{0, 1\}$.

1. Sample $\boldsymbol{z}^A, \boldsymbol{z}^B \leftarrow \mathbb{F}_2^r$ such that $\boldsymbol{z}^A + \boldsymbol{z}^B = \boldsymbol{x} * \boldsymbol{y}$ (where $*$ denotes component-wise product).
2. Output $\boldsymbol{z}^A$ to P_A and $\boldsymbol{z}^B$ to P_B.

Leakage: If P_B is corrupt:

1. Let $\mathbf{H} \in \mathbb{F}_2^{r \times 2^\ell}$ be defined so that entry (k, k') of $\mathbf{H}$ is $\mathsf{H}_{AB}(k, 1 \oplus y_k, \boldsymbol{t}_k \oplus k')$, where $\boldsymbol{t}_k \leftarrow \{0, 1\}^\ell$.
2. Sample a random unit vector $\boldsymbol{e} \in \mathbb{F}_2^{2^\ell}$ and send $(\mathbf{H}, \boldsymbol{u} = \mathbf{H}\boldsymbol{e} + \boldsymbol{x})$ to $\mathcal{A}$.

Fig. 3. Ideal functionality for leaky secret-shared two-party bit multiplication

Leakage. We now analyse the exact security of the protocol in Fig. 4 when using short keys, and explain how this is specified in the functionality $\mathcal{F}_{\text{Leaky-2-Mult}}$ (Fig. 3). Since a random share of zero is added to the outputs, note that the output distribution is uniformly random. Also, like IKNP, the protocol is *perfectly secure* against a corrupt P_A (or sender), so we only need to be concerned with leakage to a corrupt P_B who also sees the intermediate values of the protocol.

The leakage is different for each k, depending on whether $y_k = 0$ or $y_k = 1$, so we consider the two cases separately. Within each case, there are two potential sources of leakage: firstly, the corrupt P_B's knowledge of $\boldsymbol{t}_k$ and ρ_k may cause leakage (where ρ_k is a random share of zero), since these values are used to define P_A's output. Secondly, the d_k values seen by P_B, which equal

$$d_k = \mathsf{H}_{AB}(k, y_k, \boldsymbol{t}_k) \oplus \mathsf{H}_{AB}(k, 1 \oplus y_k, \boldsymbol{t}_k \oplus \Delta) \oplus x_k, \tag{1}$$

may leak information on P_A's inputs x_k.

Case 1 $(y_k = 1)$. In this case there is only leakage from the values $\boldsymbol{t}_k$ and ρ_k, which are used to define P_A's output. Since $z_k^A = \mathsf{H}_{AB}(k, 0, \boldsymbol{t}_k \oplus \Delta) \oplus \rho_k$, all of P_A's outputs (and hence, also inputs) where $y_k = 1$ effectively have only ℓ bits of min-entropy in the view of P_B, corresponding to the random choice of Δ. In this case P_B's output is $z_k^B = z_k^A \oplus x_k = \mathsf{H}_{AB}(k, 0, \boldsymbol{t}_k \oplus \Delta) \oplus \rho_k \oplus x_k$. To ensure that P_B's view is simulable the functionality needs to sample a random string $\Delta \leftarrow \{0,1\}^\ell$ and leak $\mathsf{H}_{AB}(k, 0, \boldsymbol{t}_k \oplus \Delta) \oplus x_k$ to a corrupt P_B.

Concerning the d_k values, notice that when $y_k = 1$ P_B can compute $\mathsf{H}_{AB}(k, 1, \boldsymbol{t}_k)$ and use (1) to recover $\mathsf{H}_{AB}(k, 0, \boldsymbol{q}_k) + x_k$, which equals $z_k^A + \rho_k + x_k$. However, this is not a problem, because in this case we have $z_k^B = z_k^A + x_k$, so d_k can be simulated given P_B's output.

Case 2 ($y_k = 0$). Here the d_k values seen by P_B causes leakage on P_A's inputs, because Δ is short. Looking at (1), d_k leaks information on x_k because $\Delta \leftarrow \{0,1\}^\ell$ is the only unknown in the equation, and is fixed for every k. Similarly to the previous case, this means that all of P_A's inputs where $y_k = 0$ have only ℓ bits of min-entropy in the view of an adversary who corrupts P_B. We can again handle this leakage, by defining $\mathcal{F}_{\text{Leaky-2-Mult}}$ to leak $\mathsf{H}_{AB}(k, 1, \boldsymbol{t}_k \oplus \Delta) + x_k$ to a corrupt P_B.

Note that there is no leakage from the $\boldsymbol{t}_k$ values when $y_k = 0$, because then $\boldsymbol{t}_k = \boldsymbol{q}_k$, so these messages are independent of Δ and the inputs of P_A.

In the functionality $\mathcal{F}_{\text{Leaky-2-Mult}}$, we actually modify the above slightly so that the leakage is defined in terms of linear algebra, instead of the hash function H_{AB}, to simplify the translation to the DRSD problem later on. Therefore, $\mathcal{F}_{\text{Leaky-2-Mult}}$ defines a matrix $\mathbf{H} \in \mathbb{F}_2^{r \times 2^\ell}$, which contains the 2^ℓ values $\{\mathsf{H}_{AB}(k, 1 \oplus y_k, \boldsymbol{t}_k \oplus \Delta)\}_{\Delta \in \{0,1\}^\ell}$ in row k, where each $\boldsymbol{t}_k$ is uniformly random. Given $\mathbf{H}$, the leakage from the protocol can then be described by sampling a random unit vector $\boldsymbol{e} \in \mathbb{F}_2^{2^\ell}$ (which corresponds to $\Delta \in \{0,1\}^\ell$ in the protocol) and leaking $\boldsymbol{u} = \mathbf{H}\boldsymbol{e} + \boldsymbol{x}$ to a corrupt P_B.

Communication complexity. The cost of computing r secret-shared products is that of ℓ random, correlated OTs on r-bit strings, and a further r bits of communication. Using OT extension [IKNP03, ALSZ13] to implement the correlated OTs the amortized cost is $\ell(r + \kappa)$ bits, for computational security κ. This gives a total cost of $\ell(r + \kappa) + r$ bits.

In [HOSS18] we prove the following.

Theorem 3.1. *Protocol* $\Pi_{\text{Leaky-2-Mult}}^{r,\ell}$ *securely implements the functionality* $\mathcal{F}_{\text{Leaky-2-Mult}}^{r,\ell}$ *with perfect security in the* $(\mathcal{F}_{\Delta\text{-ROT}}, \mathcal{F}_{\text{Zero}})$*-hybrid model in the presence of static honest-but-curious adversaries.*

3.2 MPC for Binary Circuits From Leaky OT

We now show how to use the leaky OT protocol to compute multiplication triples over $\mathbb{F}_2$, using a GMW-style protocol [GMW87, Gol04] optimized for the case of at least h honest parties. This can then be used to obtain a general MPC protocol for binary circuits using Beaver's method [Bea92].

Triple generation. We implement the triple generation functionality over $\mathbb{F}_2$, shown in Fig. 5. Recall that to create a triple using the GMW method, first each party locally samples shares $x^i, y^i \leftarrow \mathbb{F}_2$. Next, the parties compute shares of the product based on the fact that:

Protocol $\Pi_{\text{Leaky-2-Mult}}^{r,\ell}$

PARAMETERS: r, number of multiplications; ℓ, key length.
INPUT: $\boldsymbol{x} = (x_1, \ldots, x_r) \in \mathbb{F}_2^r$ from P_A and $\boldsymbol{y} = (y_1, \ldots, y_r) \in \mathbb{F}_2^r$ from P_B.
COMMON INPUT: A hash function $\mathsf{H}_{AB} : [r] \times \{0,1\} \times \{0,1\}^\ell \to \{0,1\}$.

1. P_A and P_B invoke $\mathcal{F}_{\Delta\text{-ROT}}^{r,\ell}$ where P_A is sender with a random input $\Delta \leftarrow \{0,1\}^\ell$, and P_B is receiver with inputs $(y_1, \ldots, y_r)$. P_A receives random strings $\boldsymbol{q}_k \in \{0,1\}^\ell$ and P_B receives $\boldsymbol{t}_k = \boldsymbol{q}_k \oplus y_k \cdot \Delta$, for $k \in [r]$.
2. Call $\mathcal{F}_{\text{Zero}}^r$ so that P_A and P_B obtain the same random $\rho_k \in \{0,1\}$ for every $k \in [r]$.
3. For each $k \in [r]$, P_A privately sends to P_B:
$$d_k = \mathsf{H}_{AB}(k, 0, \boldsymbol{q}_k) + \mathsf{H}_{AB}(k, 1, \boldsymbol{q}_k + \Delta) + x_k.$$

4. P_B outputs
$$z_k^B = \mathsf{H}_{AB}(k, y_k, \boldsymbol{t}_k) + y_k \cdot d_k + \rho_k, \quad \text{for } k \in [r].$$

5. P_A outputs
$$z_k^A = \mathsf{H}_{AB}(k, 0, \boldsymbol{q}_k) + \rho_k, \quad \text{for } k \in [r].$$

Fig. 4. Leaky secret-shared two-party bit multiplication protocol

$$\left(\sum_{i=1}^n x^i\right) \cdot \left(\sum_{i=1}^n y^i\right) = \sum_{i=1}^n x^i y^i + \sum_{i=1}^n \sum_{j \neq i} x^i y^j.$$

where x^i denotes P_i's share of $x = \sum_i x^i$.

Since each party can compute $x^i y^i$ on its own, in order to obtain additive shares of $z = xy$ it suffices for the parties to obtain additive shares of $x^i y^j$ for every pair $i \neq j$. This is done using oblivious transfer between P_i and P_j, since a 1-out-of-2 OT implies two-party secret-shared bit multiplication.

Functionality $\mathcal{F}_{\text{Triple}}^r$

1. Sample $(x_j^i, y_j^i, z_j^i) \leftarrow \mathbb{F}_2^3$, for $i \in [n]$ and $j \in [r]$, subject to the constraint that
$$\sum_i z_j^i = \left(\sum_i x_j^i\right) \cdot \left(\sum_i y_j^i\right)$$

2. Output (x_j^i, y_j^i, z_j^i) to party P_i, for $j \in [r]$.

Fig. 5. Multiplication triple generation functionality.

If we use the *leaky* two-party batch multiplication protocol from the previous section, this approach fails to give a secure protocol because the leakage in

$\mathcal{F}_{\text{Leaky-2-Mult}}$ allows a corrupt P_B to guess P_A's inputs with probability $2^{-\ell}$. When using this naively, P_A carries out a secret-shared multiplication using the *same input shares* with every other party, which allows every corrupt party to attempt to guess P_A's shares, increasing the success probability further. If the number of corrupted parties is not too small then this gives the adversary a significant chance of successfully guessing the shares of *every honest party*, completely breaking security.

To avoid this issue, we require P_A to *randomize* the shares used as input to $\mathcal{F}_{\text{Leaky-2-Mult}}$, in such a way that we still preserve correctness of the protocol. To do this, the parties will use $\mathcal{F}_{\text{Zero}}$ to generate random zero shares $s^{i,j} \in \mathbb{F}_2$ (held by P_i), satisfying $\sum_i s^{i,j} = 0$ for all $j \in [n]$, and then P_i and P_j will multiply $x^i + s^{i,j}$ and y^j. This means that all parties end up computing shares of:

$$\sum_{i=1}^{n}\sum_{j=1}^{n}(x^i + s^{i,j})y^j = \sum_{j=1}^{n} y^j \sum_{i=1}^{n}(x^i + s^{i,j}) = xy,$$

so still obtain a correct triple.

Finally, to ensure that the output shares are uniformly random, fresh shares of zero will be added to each party's share of xy. Note that masking each x^i input to $\mathcal{F}_{\text{Leaky-2-Mult}}$ means that it doesn't matter if the individual shares are leaked to the adversary, as long as it is still hard to guess the *sum of all shares*. This means that we only need to be concerned with the *sum of the leakage* from $\mathcal{F}_{\text{Leaky-2-Mult}}$. Recall that each individual instance leaks the input of an honest party P_i masked by $\mathbf{H}_i e_i$, where $\mathbf{H}_i$ is a random matrix and $e_i \in \mathbb{F}_2^{2^\ell}$ is a random unit vector. Summing up all the leakage from h honest parties, we get

$$\sum_{i=1}^{h}\mathbf{H}_i e_i = (\mathbf{H}_1\| \cdots \|\mathbf{H}_h)\begin{pmatrix}e_1\\ \vdots \\ e_h\end{pmatrix}$$

This is exactly an instance of the $\text{DRSD}_{r,h,\ell}$ problem, so is pseudorandom for an appropriate choice of parameters.

We remark that the number of triples generated, r, affects the hardness of DRSD. However, we can create an arbitrary number of triples without changing the assumption by repeating the protocol for a fixed r.

Reducing the number of OT channels. The above approach reduces communication of GMW by a factor κ/ℓ, for ℓ-bit keys, but still requires a complete network of $n(n-1)$ OT and communication channels between the parties. We can reduce this further by again taking advantage of the fact that there are at least h honest parties. We observe that when using our two-party secret-shared multiplication protocol to generate triples, information is only leaked on the x^i shares, and not the y^i shares of each triple. This means that $h-1$ parties can choose their shares of y to be zero, and y will still be uniformly random to an adversary who corrupts up to $t = n - h$ parties. This reduces the number of OT channels needed from $n(n-1)$ to $(t+1)(n-1)$.

When the number of parties is large enough, we can do even better using *random committees*. We randomly choose two committees, $\mathcal{P}_{(h)}$ and $\mathcal{P}_{(1)}$, such that except with negligible probability, $\mathcal{P}_{(h)}$ has at least h honest parties and $\mathcal{P}_{(1)}$ has at least one honest party. Only the parties in $\mathcal{P}_{(h)}$ choose non-zero shares of x, and parties in $\mathcal{P}_{(1)}$ choose non-zero shares of y; all other parties do not take part in any OT instances, and just output random sharings of zero. We remark that it can be useful to choose the parameter h *lower than* the actual number of honest parties, to enable a smaller committee size (at the cost of potentially larger keys). When the total number of parties, n, is large enough, this means the number of interacting parties can be independent of n. The complete protocol, described for two fixed committees satisfying our requirements, is shown in Fig. 6.

Protocol Π_{Triple}^r

CRS: Random hash functions $\mathsf{H}_i : [r] \times \{0,1\} \times \{0,1\}^\ell \to \{0,1\}$, for $i \in [n]$.

The protocol runs between a set of parties $\mathcal{P} = \{P_1, \ldots, P_n\}$, containing two (possibly overlapping) subsets $\mathcal{P}_{(h)}, \mathcal{P}_{(1)}$, such that $\mathcal{P}_{(h)}$ has at least h honest parties and $\mathcal{P}_{(1)}$ has at least one honest party.

1. Each party $P_i \in \mathcal{P}_{(h)}$ samples $x_k^i \leftarrow \mathbb{F}_2$, and each $P_j \in \mathcal{P}_{(1)}$ samples $y_k^j \leftarrow \mathbb{F}_2$, for $k \in [r]$.
2. Call $\mathcal{F}_{\text{Zero}}^{(n+1)r}$ so that each $P_i \in \mathcal{P}$ obtains shares $(\rho_1^i, \ldots, \rho_r^i), (s_1^{i,j}, \ldots, s_r^{i,j})_{j \in [n]}$, such that $\bigoplus_i \rho_k^i = 0$ and $\bigoplus_i s_k^{i,j} = 0$.
3. Every pair $(P_i, P_j) \in \mathcal{P}_{(h)} \times \mathcal{P}_{(1)}$ runs $\mathcal{F}_{\text{Leaky-2-Mult}}^{r,\ell}(\mathsf{H}_i)$ on input $\{x_k^i + s_k^{i,j}\}_{k \in [r]}$ from P_i and $\{y_k^j\}_{k \in [r]}$ from P_j. For $k \in [r]$, P_i receives $a_k^{i,j}$ and P_j receives $b_k^{j,i}$ such that $a_k^{i,j} + b_k^{j,i} = (x_k^i + s_k^{i,j}) \cdot y_k^j$.
4. Each $P_i \in \mathcal{P}$ computes, for $k \in [r]$:

$$z_k^i = (x_k^i + s_k^{i,i}) \cdot y_k^i + \sum_{j \neq i} (a_k^{i,j} + b_k^{i,j}) + \rho_k^i$$

   where if any value $x_k^i, y_k^i, a_k^{i,j}, b_k^{i,j}$ has not been defined by P_i, it is set to zero.
5. P_i outputs the shares $(x_k^i, y_k^i, z_k^i)_{k \in [r]}$.

Fig. 6. Secret-shared triple generation using leaky two-party multiplication.

Communication complexity. Recall from the analysis in Sect. 3.1 that when using protocol $\Pi_{\text{Leaky-2-Mult}}$ with Π_{Triple}, the cost of computing r secret-shared triples is that of ℓ random, correlated OTs on r-bit strings, and a further r bits of communication between every pair of parties. This gives a total cost of $\ell(r+\kappa)+r$ bits between every pair of parties who has an OT channel (ignoring $\mathcal{F}_{\text{Zero}}$ and the seed OTs for OT extension, since their communication cost is independent of the number of triples). If the two committees $\mathcal{P}_{(h)}, \mathcal{P}_{(1)}$ have sizes $n_h \leq n$ and $n_1 \leq t + 1$ then we have the following theorem (proven in [HOSS18]).

Theorem 3.2. *Protocol* Π_{Triple} *securely realizes* $\mathcal{F}_{\text{Triple}}^r$ *in the* $(\mathcal{F}_{\text{Leaky-2-Mult}}^{r,\ell},$ $\mathcal{F}_{\text{Zero}}^{(n+1)r})$-*hybrid model, based on the* $\text{DRSD}_{r,h,\ell}$ *assumption, where* h *is the number of honest parties in* $\mathcal{P}_{(h)}$. *The amortized communication cost is* $\leq$ $n_h n_1(\ell + \ell\kappa/r + 1)$ *bits per triple.*

Parameters for unconditional security. Recall from Lemma 2.1 that if $\ell = 1$ and $h \geq r + s$ for any ℓ, then $\text{DRSD}_{r,h,\ell}$ is *statistically hard*, with statistical security 2^{-s}. This means when h is large enough we can use 1-bit keys, and every pair of parties who communicates only needs to send $2 + \kappa/r$ bits over the network.[2]

MPC using multiplication triples. Our protocol for multiplication triples can be used to construct a semi-honest MPC protocol for binary circuits using Beaver's approach [Bea92]. The parties first secret-share their inputs between all other parties. Then, XOR gates can be evaluated locally on the shares, whilst an AND gate requires consuming a multiplication triple, and two openings with Beaver's method. Each opening can be done with $2(n-1)$ bits of communication as follows: all parties send their shares to P_1, who sums the shares together and sends the result back to every other party.

In the 1-bit key case mentioned above, using two (deterministic) committees of sizes n and $t+1$ and setting, for instance, $r = \kappa$ implies the following corollary. Note that the number of communication channels is $(t+1)(n-1)$ and not $(t+1)n$, because in the deterministic case $\mathcal{P}_{(1)}$ is contained in $\mathcal{P}_{(h)}$, so $t+1$ sets of the shared cross-products can be computed locally.

Corollary 3.3. *Assuming OT and OWF, there is a semi-honest MPC protocol for binary circuits with an amortized communication complexity of no more than* $3(t+1)(n-1) + 4(n-1)$ *bits per AND gate, if there are at least* $\kappa + s$ *honest parties.*

4   Multi-Party Garbled Circuits with Short Keys

In this section we present our second contribution: a constant-round MPC protocol based on garbled circuits with short keys. The protocol has two phases, a preprocessing phase independent of the parties' actual inputs where the garbled circuit is mutually generated by all parties, and an online phase where the computation is performed. We first abstractly discuss the details of our garbling method, and then turn to the two protocols for generating and evaluating the garbled circuit.

4.1   The Multi-Party Garbling Scheme

Our garbling method is defined by the functionality $\mathcal{F}_{\text{Preprocessing}}^{\ell_{\text{BMR}}}$ (Fig. 7), which creates a garbled circuit that is given to all the parties. It can be seen as a variant

---

[2] Note that we still need computational assumptions for OT and zero sharing in order to implement $\mathcal{F}_{\text{Leaky-2-Mult}}$ and $\mathcal{F}_{\text{Zero}}$.

Functionality $\mathcal{F}_{\text{Preprocessing}}^{\ell_{\text{BMR}}}$

COMMON INPUT: A function $\mathsf{H} : [n] \times \{0,1\} \times \{0,1\}^{\ell_{\text{BMR}}} \to \{0,1\}^{n\ell_{\text{BMR}}+1}$. Let H' denote the same function excluding the least significant bit of the output.

Let C_f be a boolean circuit with fan-out-one gates. Denote by $\mathsf{AND}, \mathsf{XOR}$ and SPLIT its sets of AND, XOR and Splitter gates, respectively. Given a gate, let I and O be the set of its input and output wires, respectively. If $g \in \mathsf{SPLIT}$, then $I = \{w\}$ and $O = \{u,v\}$, otherwise $O = \{w\}$.
The functionality proceeds as follows $\forall i \in [n]$:

1. $\forall g \in \mathsf{XOR}$, sample $\Delta_g^i \leftarrow \{0,1\}^{\ell_{\text{BMR}}}$.
2. For each circuit-input wire u, sample $\lambda_u \leftarrow \{0,1\}$ and $k_{u,0}^i \leftarrow \{0,1\}^{\ell_{\text{BMR}}}$. If u is input to a XOR gate g, set $k_{u,1}^i = k_{u,0}^i \oplus \Delta_g^i$, otherwise $k_{u,1}^i \leftarrow \{0,1\}^{\ell_{\text{BMR}}}$.
3. Passing topologically through all the gates $g \in \{\mathsf{AND} \cup \mathsf{XOR} \cup \mathsf{SPLIT}\}$ of the circuit:
   - If $g \in \mathsf{XOR}$:
     • Set $\lambda_w = \bigoplus_{x \in I} \lambda_x$
     • Set $k_{w,0}^i = \bigoplus_{x \in I} k_{x,0}^i$ and $k_{w,1}^i = k_{w,0}^i \oplus \Delta_g^i$
   - If $g \in \mathsf{AND}$:
     • Sample $\lambda_w \leftarrow \{0,1\}$.
     • $k_{w,0}^i \leftarrow \{0,1\}^{\ell_{\text{BMR}}}$. If w is input to a XOR gate g' set $k_{w,1}^i = k_{w,0}^i \oplus \Delta_{g'}^i$, else $k_{w,1}^i \leftarrow \{0,1\}^{\ell_{\text{BMR}}}$.
     • For $a,b \in \{0,1\}$, representing the public values of wires u and v, let $c = (a \oplus \lambda_u) \cdot (b \oplus \lambda_v) \oplus \lambda_w$. Store the four entries of the garbled version of g as:
$$\tilde{\boldsymbol{g}}_{a,b} = \left(\bigoplus_{i=1}^{n} \mathsf{H}(i,b,k_{u,a}^i) \oplus \mathsf{H}(i,a,k_{v,b}^i) \right)$$
$$\oplus (c, k_{w,c}^1, \dots, k_{w,c}^n), \quad (a,b) \in \{0,1\}^2.$$
   - If $g \in \mathsf{SPLIT}$:
     • Set $\lambda_x = \lambda_w$ for every $x \in O$.
     • $\forall x \in O$, sample $k_{x,0}^i \leftarrow \{0,1\}^{\ell_{\text{BMR}}}$. If $x \in O$ is input to a XOR gate g', set $k_{x,1}^i = k_{x,0}^i \oplus \Delta_{g'}^i$, otherwise $k_{x,1}^i \leftarrow \{0,1\}^{\ell_{\text{BMR}}}$.
     • For $c \in \{0,1\}$, the public value on w, store the two entries of the garbled version of g as:
$$\tilde{g}_c = \left(\bigoplus_{i=1}^{n} \mathsf{H}'(i,0,k_{w,c}^i), \bigoplus_{i=1}^{n} \mathsf{H}'(i,1,k_{w,c}^i) \right)$$
$$\oplus (k_{u,c}^1, \dots, k_{u,c}^n, k_{v,c}^1, \dots, k_{v,c}^n), \ c \in \{0,1\}$$

4. **Output:** For each circuit-input wire u, send λ_u to the party providing inputs to C_f on u. For every circuit wire v and $i \in [n]$, send $k_{v,0}^i, k_{v,1}^i$ to P_i. Finally, send to all parties $\tilde{g}$ for each $g \in \mathsf{AND} \cup \mathsf{SPLIT}$ and λ_w for each circuit-output wire w.

Fig. 7. Multi-party garbling functionality

of the multi-party garbling technique by Beaver, Micali and Rogaway [BMR90], known as BMR, which has been used and improved in a recent sequence of works [LPSY15,LSS16,BLO16,HSS17].

The main idea behind BMR is that every party P_i contributes a pair of keys $k_{w,0}^i, k_{w,1}^i \in \{0,1\}^\kappa$ and a share of a wire mask $\lambda_w^i \in \{0,1\}$ for each wire w in the circuit. To garble a gate, the corresponding output wire key from every party is encrypted under the combination of all parties' input wire keys, using a PRF or PRG, so that no single party knows all the keys for a gate. In addition, the free-XOR property can be supported by having each party choose their keys such that $k_{w,0}^i \oplus k_{w,1}^i = \Delta^i$, where Δ^i is a global fixed random string known to P_i.

The main difference between our work and recent related protocols is that we use short keys of length ℓ_{BMR} instead of κ, and then garble gates using a random, expanding function $\mathsf{H} : [n] \times \{0,1\} \times \{0,1\}^{\ell_{\mathrm{BMR}}} \to \{0,1\}^{n\ell_{\mathrm{BMR}}+1}$. Instead of basing security on a PRF or PRG, we then reduce the security of the protocol to the pseudorandomness of the *sum* of H when applied to each of the honest parties' keys, which is implied by the DRSD problem from Sect. 2.1. We also use H' to denote H with the least significant output bit dropped, which we use for garbling splitter gates.

To garble an AND gate g with input wires u, v and output wire w, each of the 4 garbled rows $\tilde{g}_{a,b}$, for $(a, b) \in \{0,1\}^2$, is computed as:

$$\tilde{g}_{a,b} = \left(\bigoplus_{i=1}^{n} \mathsf{H}(i, b, k_{u,a}^i) \oplus \mathsf{H}(i, a, k_{v,b}^i) \right) \oplus (c, k_{w,c}^1, \dots, k_{w,c}^n), \qquad (2)$$

where $c = (a \oplus \lambda_u) \cdot (b \oplus \lambda_v) \oplus \lambda_w$ and $\lambda_u, \lambda_v, \lambda_w$ are the secret-shared wire masks. Each row can be seen as an encryption of the correct n output wire keys under the corresponding input wire keys of all parties. Note that, for each wire, P_i holds the keys $k_{u,0}^i, k_{u,1}^i$ and an additive share λ_u^i of the wire mask. The extra bit value that H takes as input is added to securely increase the stretch of H when using the same input key twice, preventing a 'mix-and-match' attack on the rows of a garbled gate. The output of H is also extended by an extra bit, to allow encryption of the output wire mask c.[3]

Splitter gates. When relying on the DRSD problem, the reuse of a key in multiple gates degrades parameters and makes the problem easier (as the parameter r grows, the key length must be increased), so we cannot handle circuits with arbitrary fan-out. For this reason, we restrict our exposition of the garbling to fan-out-one circuits with so-called *splitter gates*. This type of gate takes as input a single wire w and provides two output wires u, v, each of them with fresh, independent keys representing the same value carried by the input wire. Converting an arbitrary circuit to use splitter gates incurs a cost of roughly a factor of two in the circuit size (see [HOSS18]).

---

[3] This only becomes necessary when using short keys — in BMR with full-length keys the parties can recover the wire mask by comparing the output with their own two keys, but this does not work if collisions are possible.

Splitter gates were previously introduced in [TX03] as a fix for a similar issue in the original BMR paper [BMR90], where the wire "keys" were used as seeds for a PRG in order to garble the gates, so that when a wire was used as input to multiple gates, their garbled versions did not use independent pseudorandom masks. Other recent BMR-style papers avoid this issue by applying the PRF over the gate identifier as well, which produces distinct, independent PRF evaluations for each gate.

Free-XOR. The Free-XOR [KS08] optimization results in an improvement in both computation and communication for XOR gates where a global fixed random Δ_i is chosen by each party P_i and the input keys are locally XORed, yielding the output key of this gate. We cannot use the standard free-XOR technique [KS08,BLO16] for the same reason discussed above: reusing a single offset across multiple gates would make the DRSD problem easier and not be secure. We therefore introduce a new free-XOR technique (inspired by FleXOR [KMR14]) which, combined with our use of splitter gates, allows garbling XOR gates for free without additional assumptions. For each arbitrary fan-in XOR gate g, each party chooses a different offset Δ_g^i, allowing for a free-XOR computation for wires using keys with that offset. For general circuits, this would normally introduce the problem that the input wires may not have the correct offset, requiring some 'translation' to Δ_g. However, because we restrict to gates with fan-out-one and splitter gates, we know that each input wire to g is not an input wire to any other gate, so we can always ensure the keys use the correct offset without any further changes.

Compiling to fan-out-one circuits with splitter gates. Let C_f be an arbitrary fan-out circuit, with A AND gates and X XOR gates, both with fan-in-two. Let I_{C_f} and O_{C_f} be the number of circuit-input and circuit-output wires, respectively. We will now compute the number S of splitter gates that the compiled circuit needs. First, note that each time a wire w is used as input to another gate or as a circuit-output wire, w's fan-out is increased by one. Each of the AND, XOR gates in the pre-compiled circuit provides a fresh output wire to be used in C_f, while using for its inputs two pre-existing wires in the circuit. Output wires also use one pre-existing wire each, while input wires use no pre-existing wires. This means that, to compile C_f to be a fan-out-one circuit, we need to add up to $(2 \cdot X + 2 \cdot A + O_{C_f}) - (A + X + I_{C_f})$ wires. Each of these missing wires, however, can be created by using a splitter gate in the compiled circuit, since each of these gates uses one wire to generate two fresh new wires. So, putting all the pieces together, the compiled circuit requires $S \leq X + A + O_{C_f} - I_{C_f}$ splitter gates. This gives a close upper bound, as if w is a circuit output wire *and* an input wire of another gate then it is being counted twice rather than once in the formula.

Functionality $\mathcal{F}_{\text{Bit} \times \text{Bit}}$

After receiving $(x^i, y^i) \in \mathbb{F}_2^2$ from each party P_i, sample $z_i \leftarrow \mathbb{F}_2$ such that $\sum_i z^i = (\sum_i x^i) \cdot (\sum_i y^i)$, and send z^i to party P_i.

Fig. 8. Secret-shared bit multiplication functionality

Functionality $\mathcal{F}_{\text{BitString}}^{\ell_{\text{BMR}}}(P_j)$

After receiving $(x^i, y^i) \in \mathbb{F}_2^2$ from each party P_i, as well as $\Delta \in \mathbb{F}_2^{\ell_{\text{BMR}}}$ from P_j, sample $Z_i \leftarrow \mathbb{F}_2$ such that $\sum_i Z^i = (\sum_i x^i) \cdot \Delta$, and send Z^i to party P_i.

Fig. 9. Secret-shared bit/string multiplication functionality

4.2 Protocol and Functionalities for Bit and Bit/String Multiplication

Even though we could implement both $\mathcal{F}_{\text{Bit} \times \text{Bit}}$ and $\mathcal{F}_{\text{BitString}}^{\ell_{\text{BMR}}}(P_j)$ using $\mathcal{F}_{\text{Triple}}$, there are more efficient ways to implement the latter: One by building directly from $\mathcal{F}_{\text{Leaky-2-Mult}}$, and another using [ALSZ13].

- $\mathcal{F}_{\text{Leaky-2-Mult}}$-hybrid implementation (Fig. 11): As the length-ℓ_{BMR} string R_g^j is not secret-shared and just known to one party, we only need to perform $n - 1$ invocations of $\mathcal{F}_{\text{Leaky-2-Mult}}$ in order to multiply it with a secret-shared bit $x = x^1 + \cdots + x^n$. The protocol uses random shares of zero to mask the inputs and outputs of $\mathcal{F}_{\text{Leaky-2-Mult}}$, similarly to the Π_{Triple} protocol. Note that this does not directly implement the functionality shown in Fig. 9, because $\Pi_{\text{Bit} \times \text{String}}^{r, \ell_{\text{BMR}}}$ performs a batch of r independent multiplications in parallel. However, in the protocol $\Pi_{\text{Preprocessing}}^{\ell_{\text{BMR}}}$ all the gates can be garbled in parallel, so a batch version of the functionality (as described in Fig. 10) suffices. The amortized communication complexity obtained is $\ell_{\text{BMR}}(1 + \ell_{\text{OT}} + \ell_{\text{OT}}\kappa/r)$ bits.
- [ALSZ13] implementation: The amortized communication complexity is $\kappa + \ell_{\text{BMR}}$ bits.

Functionality $\mathcal{F}_{\text{BitString}}^{r,\ell_{\text{BMR}}}$

After receiving input $(P_j, x_1^i, \ldots, x_m^i)$ from every party P_i, and additional inputs $\Delta_1, \ldots, \Delta_r$ from P_j, where each $x_k^i \in \{0,1\}$ and $\Delta_k \in \{0,1\}^{\ell_{\text{BMR}}}$:

1. Sample $Z_k^i \leftarrow \{0,1\}^{\ell_{\text{BMR}}}$, for $i \in [n]$ and $k \in [r]$, subject to the constraint that

$$\bigoplus_i Z_k^i = \Delta_k \cdot \bigoplus_i x_k^i, \quad \text{for } k \in [r]$$

2. Output $Z_1^i, \ldots, Z_r^i$ to party P_i

Fig. 10. Batch secret-shared bit/string multiplication between P_j and all parties

---

Protocol $\Pi_{\text{Bit}\times\text{String}}^{r,\ell_{\text{BMR}}}$, n-**party Bit/String-Mult**

To multiply the strings $\Delta_1, \ldots, \Delta_r \in \{0,1\}^{\ell_{\text{BMR}}}$ held by P_j with secret-shared bits $(x_1^i, \ldots, x_r^i)_{i\in[n]}$:

1. Denote the v-th bit of Δ_k by $\Delta_{k,v}$. For $v \in [\ell_{\text{BMR}}]$:
   (a) Call $\mathcal{F}_{\text{Zero}}^{2r}$ so that each P_i obtains fresh shares $(\rho_{1,v}^i, \ldots, \rho_{m,v}^i, \sigma_{1,v}^i, \ldots, \sigma_{m,v}^i)$, such that $\bigoplus_i \rho_{k,v}^i = 0$ and $\bigoplus_i \sigma_{k,v}^i = 0$
   (b) For each $i \neq j$, P_i and P_j run $\mathcal{F}_{\text{Leaky-2-Mult}}^{r,\ell_{\text{OT}}}$ on input $(x_k^i \oplus \sigma_{k,v}^i)_{k\in[r]}$ from P_i and $(\Delta_k[v])_{k\in[r]}$ from P_j. P_i receives $a_{k,v}^i$ and P_j receives $b_{k,v}^i$ such that $a_{k,v}^i \oplus b_{k,v}^i = \Delta_k[v] \cdot (x_k^i \oplus \sigma_k^i)$.
2. Each P_i, for $i \neq j$, outputs the ℓ_{BMR}-bit strings $Z_k^i := (a_{k,1}^i \oplus \rho_{k,1}^i, \ldots, a_{k,\ell_{\text{BMR}}}^i \oplus \rho_{k,\ell_{\text{BMR}}}^i)$, for $k \in [r]$.
3. P_j outputs the ℓ_{BMR}-bit strings $Z_k^j := \bigoplus_{i\neq j} (b_{k,1}^i, \ldots, b_{k,\ell_{\text{BMR}}}^i) \oplus (\rho_{k,1}^j, \ldots, \rho_{k,\ell_{\text{BMR}}}^j)$, for $k \in [r]$.

Fig. 11. n-party secret-shared bit/string multiplication using leaky 2-party multiplication

Communication complexity. The communication complexity of $\Pi_{\text{Bit}\times\text{String}}^{r,\ell_{\text{BMR}}}$ is exactly that of $(n-1)\ell_{\text{BMR}}$ instances of $\mathcal{F}_{\text{Leaky-2-Mult}}^{r,\ell_{\text{OT}}}$, where ℓ_{OT} is the leakage parameter used in the protocol $\Pi_{\text{Leaky-2-Mult}}^{r,\ell_{\text{OT}}}$. Note that ℓ_{OT} is independent of ℓ_{BMR} used in the bit/string protocol, but affects the security and cost of realising $\mathcal{F}_{\text{Leaky-2-Mult}}$. The total complexity is then $(n-1)\ell_{\text{BMR}}(\ell_{\text{OT}}(r+\kappa)+r)$ bits, or an amortized cost of $(n-1)\ell_{\text{BMR}}(\ell_{\text{OT}} + \ell_{\text{OT}}\kappa/r + 1)$ bits per multiplication.

Theorem 4.1. *Protocol* $\Pi_{\text{Bit}\times\text{String}}^{r,\ell_{\text{BMR}}}$ *UC-securely realizes* $\mathcal{F}_{\text{BitString}}^{r,\ell_{\text{BMR}}}$ *in the* $\mathcal{F}_{\text{Zero}}^{2r}$-*hybrid in the presence of static honest-but-curious adversaries, under the* $\text{DRSD}_{r,h,\ell_{\text{OT}}}$ *assumption.*

The proof is a direct extension of the proof of Theorem 3.2.

4.3   The Preprocessing Protocol

Our protocol for generating the garbled circuit is shown in Fig. 12. We use two functionalities $\mathcal{F}_{\text{Bit}\times\text{Bit}}$ (Fig. 8) and $\mathcal{F}_{\text{BitString}}(P_j)$ (Fig. 9) for multiplying two additively shared bits, and multiplying an additively shared bit with a string held by P_j, respectively. $\mathcal{F}_{\text{Bit}\times\text{Bit}}$ can be easily implemented using a multiplication triple from $\mathcal{F}_{\text{Triple}}$ in the previous section, whilst $\mathcal{F}_{\text{BitString}}$ uses a variant of the Π_{Triple} protocol optimized for this task.

Most of the preprocessing protocol is similar to previous works [BLO16, HSS17], where first each party samples their sets of wire keys and shares of wire masks, and then the parties interact to obtain shares of the garbled gates.

The Preprocessing Protocol $\Pi_{\text{Preprocessing}}^{\ell_{\text{BMR}}}$

COMMON INPUT: $H : [n] \times \{0,1\} \times \{0,1\}^{\ell_{\text{BMR}}} \to \{0,1\}^{n\ell_{\text{BMR}}+1}$, a uniformly random sampled function and H' defined from H excluding the least significant bit of the output. A boolean circuit C_f with fan-out 1. Let AND, XOR and SPLIT be the sets of AND, XOR and splitter gates, respectively. Given a gate, let I and O be the set of its input and output wires, respectively. If $g \in$ SPLIT, then $I = \{w\}$ and $O = \{u, v\}$, otherwise $O = \{w\}$.

For each $i \in [n]$, the protocol proceeds as follows:

1. **Free-XOR offsets:** For every $g \in$ XOR, P_i samples a random value $\Delta_g^i \leftarrow \{0,1\}^{\ell_{\text{BMR}}}$

2. **Circuit-input wires' masks and keys:** If w is a *circuit-input* wire:
   (a) P_i samples a key $k_{w,0}^i \leftarrow \{0,1\}^{\ell_{\text{BMR}}}$ and a wire mask share $\lambda_w^i \leftarrow \{0,1\}$.
   (b) If w is input to a XOR gate g', P_i sets $k_{w,1}^i = k_{w,0}^i \oplus \Delta_{g'}^i$, otherwise $k_{w,1}^i \leftarrow \{0,1\}^{\ell_{\text{BMR}}}$.

3. **Intermediate wires' masks and keys:** Passing topologically through all the gates $g \in G = \{\text{AND} \cup \text{XOR} \cup \text{SPLIT}\}$ of the circuit:
   (a) If $g \in$ XOR, P_i computes:
       - $\lambda_w^i = \bigoplus_{x \in I} \lambda_x^i$.
       - $k_{w,0}^i = \bigoplus_{x \in I} k_{x,0}^i$ and $k_{w,1}^i = k_{w,0}^i \oplus \Delta_g^i$.
   (b) If $g \notin$ XOR, P_i does as follows:
       - If $g \in$ AND, $\lambda_w^i \leftarrow \{0,1\}$. Else if $g \in$ SPLIT, sets $\lambda_x^i = \lambda_w^i$ for every $x \in O$.
       - For every $x \in O$, $k_{x,0}^i \leftarrow \{0,1\}^{\ell_{\text{BMR}}}$. If $x \in O$ is input to a XOR gate g', set $k_{x,1}^i = k_{x,0}^i \oplus \Delta_{g'}^i$, otherwise sample $k_{x,1}^i \leftarrow \{0,1\}^{\ell_{\text{BMR}}}$.

4. **Garble gates:** For each gate $g \in \{\text{AND} \cup \text{SPLIT}\}$, the parties run the subprotocol $\Pi_{\text{GateGarbling}}^{\ell_{\text{BMR}}}$, obtaining back shares $\tilde{g}^i$ of each garbled gate.

5. **Reveal input/output wires' masks:** For every *circuit-output* wire w, P_i broadcasts λ_w^i. For every *circuit-input* wire w, P_i sends λ_w^i to the party P_j who provides input on it. Each party reconstructs the wire masks from her received values as $\lambda_w = \bigoplus_{i=1}^n \lambda_w^i$.

6. **Open Garbling** For each $g \in \{\text{AND} \cup \text{SPLIT}\}$, P_i sends $\tilde{g}^i$ to P_1. P_1 reconstructs every garbled gate, $\tilde{g} = \bigoplus_{i=1}^n \tilde{g}^i$, and broadcasts it.

Fig. 12. The preprocessing protocol that realizes $\mathcal{F}_{\text{Preprocessing}}^{\ell_{\text{BMR}}}$

It is the second stage where our protocol differs, so we focus here on the details of the gate garbling procedures.

The Gate Garbling Protocol. We describe the details of the sub-protocol $\Pi_{\text{GateGarbling}}^{\ell_{\text{BMR}}}$ (Fig. 13), implementing the gate garbling phase of $\mathcal{F}_{\text{Preprocessing}}^{\ell_{\text{BMR}}}$. Creating garbled AND gates is done similarly to the OT-based protocol [BLO16], with the exception that we use short wire keys of length ℓ_{BMR} instead of κ. We also show how to create sharings of garbled splitter gates *without any interaction*, so these are much cheaper than AND gates.

Suppose that for an AND gate g, each P_i holds the wire mask share λ_v^i and keys $k_{v,0}^i, k_{v,1}^i \leftarrow \{0,1\}^{\ell_{\text{BMR}}}$. P_i defines $R_g^i = k_{w,0}^i \oplus k_{w,1}^i$. After that all parties call $\mathcal{F}_{\text{Bit}\times\text{Bit}}$ once to compute additive shares of $\lambda_{uv} = \lambda_u \cdot \lambda_v \in \{0,1\}$, which are then used to locally compute shares of $\chi_{g,a,b} = (a \oplus \lambda_u) \cdot (b \oplus \lambda_v) \oplus \lambda_w$, for each $(a,b) \in \{0,1\}^2$. Each P_i obtains $\chi_{g,a,b}^i$ such that $\chi_{g,a,b} = \oplus_{i \in [n]} \chi_{g,a,b}^i$. To compute shares of the products $\chi_{g,a,b} \cdot R_g^i$, the parties call $\mathcal{F}_{\text{BitString}}^{\ell_{\text{BMR}}}(P_i)$ three times, for each $i \in [n]$, to multiply R_g^i with each of the bits $\lambda_u, \lambda_v, (\lambda_{uv} \oplus \lambda_w)$. These can then be used for each P_j to locally obtain the shares $(\chi_{g,a,b} \cdot R_g^i)^j$, for all $(a,b) \in \{0,1\}^2$ (just as in [BLO16]).

The Gate Garbling Sub-protocol $\Pi_{\text{GateGarbling}}^{\ell_{\text{BMR}}}$

COMMON INPUT: a function $\mathsf{H} : [n] \times \{0,1\} \times \{0,1\}^{\ell_{\text{BMR}}} \to \{0,1\}^{n\ell_{\text{BMR}}+1}$; H' defined as H excluding the least significant output bit; the gate g to be garbled.
PRIVATE INPUT: Each P_i, $i \in [n]$, holds λ_v^i and $k_{v,0}^i, k_{v,1}^i$, for each wire v.

1. If $g \in$ AND with input wires $\{u,v\}$ and output wire w:
   (a) Each party P_i defines $R_g^i = k_{w,0}^i \oplus k_{w,1}^i$, for each $i \in [n]$
   (b) Call $\mathcal{F}_{\text{Bit}\times\text{Bit}}$ to compute shares of $\lambda_u \cdot \lambda_v$, and use these to locally obtain shares of

   $$\chi_{g,a,b} = (a \oplus \lambda_u) \cdot (b \oplus \lambda_v) \oplus \lambda_w, \quad \text{for } (a,b) \in \{0,1\}^2$$

   (c) Call $\mathcal{F}_{\text{BitString}}^{\ell_{\text{BMR}}}(P_i)$ to get shares of $\chi_{g,a,b} \cdot R_g^i$, for each $i \in [n]$ and $(a,b) \in \{0,1\}^2$. P_i then sets $\rho_{i,a,b}^i = k_{w,0}^i \oplus (\chi_{g,a,b} \cdot R_g^i)^i$, and $\forall j \neq i$, P_j sets $\rho_{i,a,b}^j = (\chi_{g,a,b} \cdot R_g^i)^j$.
   (d) Each P_i sets $\tilde{g}_{a,b}^i = \mathsf{H}(i, b, k_{u,a}^i) \oplus \mathsf{H}(i, a, k_{v,b}^i) \oplus (\chi_{g,a,b}^i, \rho_{1,a,b}^i, \ldots, \rho_{n,a,b}^i)$, for $a,b \in \{0,1\}$.

2. If $g \in$ SPLIT with input wire w and output wires $\{u,v\}$:
   (a) Call $\mathcal{F}_{\text{Zero}}^{2n\ell_{\text{BMR}}}$ twice, so that each P_i receives shares $s_0^i, s_1^i \in \{0,1\}^{2n\ell_{\text{BMR}}}$.
   (b) P_i sets $\rho_c^i = s_c^i \oplus (0, \ldots, k_{u,c}^i, 0, \ldots, k_{v,c}^i, \ldots, 0)$ for $c \in \{0,1\}$.
   (c) Set $\tilde{g}_c^i = \left(\mathsf{H}'(i, 0, k_{w,c}^i), \mathsf{H}'(i, 1, k_{w,c}^i) \right) \oplus \rho_c^i$, for $c \in \{0,1\}$.

Fig. 13. The gate garbling sub-protocol

After computing the bit/string products, P_j then computes for each $(a, b) \in \{0,1\}^2$:

$$\rho^j_{i,a,b} = \begin{cases} (\chi_{g,a,b} \cdot R^i_g)^j & j \neq i \\ k^i_{w,0} \oplus (\chi_{g,a,b} \cdot R^i_g)^i & j = i. \end{cases}$$

These values define shares of $\chi_{g,a,b} \cdot R^i_g \oplus k^i_{w,0}$. Finally, each party's share of the garbled AND gate is obtained as:

$$\tilde{g}^i_{a,b} = \mathsf{H}(i, b, k^i_{u,a}) \oplus \mathsf{H}(i, a, k^i_{v,b}) \oplus (\chi^i_{g,a,b}, \rho^i_{1,a,b}, \ldots, \rho^i_{n,a,b}), \quad a, b \in \{0,1\}$$

Summing up these values we obtain:

$$\bigoplus_i \tilde{g}^i_{a,b} = \bigoplus_i \mathsf{H}(i, b, k^i_{u,a}) \oplus \mathsf{H}(i, a, k^i_{v,b}) \oplus (\chi^i_{g,a,b}, \rho^i_{1,a,b}, \ldots, \rho^i_{n,a,b})$$

$$= \bigoplus_{i=1}^{n} (\mathsf{H}(i, b, k^i_{u,a}) \oplus \mathsf{H}(i, a, k^i_{v,b})) \oplus (c, k^1_{w,c}, \ldots, k^n_{w,c}),$$

where $c = \chi_{g,a,b}$, as required.

To garble a splitter gate, we observe that here there is no need for any secure multiplications within MPC, and the parties can produce shares of the garbled gate *without any interaction*. This is because the two output wire values are the same as the input wire value, so to obtain a share of the encryption of the two output keys on wires u, v with input wire w, party P_i just computes:

$$(\mathsf{H}'(i, 0, k^i_{w,c}), \mathsf{H}'(i, 1, k^i_{w,c})) \oplus (0, \ldots, k^i_{u,c}, 0, \ldots, k^i_{v,c}, 0, \ldots, 0)$$

for $c \in \{0, 1\}$, where the right-hand vector contains P_i's keys in positions i and $n+i$. The parties then re-randomize this sharing with a share of zero from $\mathcal{F}_{\text{Zero}}$, so that opening the shares does not leak information on the individual keys.[4]

4.4 Security and Complexity

The above approach reduces size of the garbled circuit by a factor κ/ℓ_{BMR}, for ℓ_{BMR}-bit keys, but still requires n keys for every row in the garbled gates. Similarly to Sect. 3, when n is large we can reduce this by using a (random) committee $\mathcal{P}_{(h)}$ of size n_h that has at least h honest parties. $\Pi^{\ell_{\text{BMR}}}_{\text{Preprocessing}}$ and $\Pi^{\ell_{\text{BMR}}}_{\text{BMR}}$ are then run as if called only by the parties in $\mathcal{P}_{(h)}$. For circuit-input wires w where parties in $\mathcal{P} \setminus \mathcal{P}_{(h)}$ provide input, they are sent the masks λ_w in $\Pi^{\ell_{\text{BMR}}}_{\text{Preprocessing}}$, so in the online phase they can then broadcast $\Lambda_w = \rho^i_w \oplus \lambda_w$ in the same way as parties in $\mathcal{P}_{(h)}$.

---

[4] For AND gates, the shares output by $\mathcal{F}^{\ell_{\text{BMR}}}_{\text{BitString}}$ are uniformly random, so do not need re-randomizing with sharings of zero.

This reduces the size of the garbled circuit by an additional factor of n/n_h. Finally, the same committee $\mathcal{P}_{(h)}$ can be combined with a (random) committee $\mathcal{P}_{(1)}$ with a single honest party in order to optimize the bit multiplications needed to compute the $\chi_{g,a,b}$ values, as was described in Sect. 3.

In Sect. 5, we give some examples of committee sizes and key lengths that ensure security, and compare this with the naive approach of running the preprocessing phase of BMR in $\mathcal{P}_{(1)}$ only. The following theorem is proved in [HOSS18].

Theorem 4.2. *Protocol* $\Pi_{\text{Preprocessing}}^{\ell_{\text{BMR}}}$ *UC-securely realizes the functionality* $\mathcal{F}_{\text{Preprocessing}}^{\ell_{\text{BMR}}}$ *with perfect security in the* $(\mathcal{F}_{\text{Bit}\times\text{Bit}}, \mathcal{F}_{\text{BitString}}^{\ell_{\text{BMR}}}, \mathcal{F}_{\text{Zero}}^{2n\ell_{\text{BMR}}})$*-hybrid model in the presence of static honest-but-curious adversaries.*

4.5 The Online Phase

Given the previous description of the garbling phase, the online phase is quite straightforward, where upon reconstructing the garbled circuit and obtaining all input keys, the evaluation process is similar to [BMR90]. As in that work, all parties run the evaluation algorithm, which in our case involves each party computing just $2n$ hash evaluations per gate. During evaluation, the parties only see the randomly masked wire values, which we call "public values", and cannot determine the actual values being computed. Upon completion, the parties obtain the actual output using the output wire masks revealed from $\mathcal{F}_{\text{Preprocessing}}^{\ell_{\text{BMR}}}$. The security of the protocol reduces to the $\mathsf{DRSD}_{r,h,\ell_{\text{BMR}}}$ problem, where ℓ_{BMR} is the key length, h is the number of honest parties, and r is twice the output length of the function H (sampled by the CRS).

Table 1. Amortized communication cost (in kbit) of producing a single triple in GMW. We consider [DKS+17] for 1-out-of-4 OT extension in the GMW protocols, and the protocol from Sect. 3 in our work.

# parties n (honest)	20 (6)	50 (15)	60 (20)	80 (30)	150 (40)	200 (50)	400 (120)
(ℓ_{OT}, r)	(31, 300)	(14, 300)	(11, 300)	(8, 300)	(7, 400)	(6, 450)	(1, 80)
GMW ($t = n - 1$)	25.46	164.15	237.18	423.44	1497.5	2666.6	10693.2
GMW ($t = n - h$)	14.07	84.42	109.88	170.85	818.07	1517.55	5271.56
Ours	**12.89**	**37**	**40.38**	**50.01**	**169.36**	**261.6**	**403.63**

We remark that in practice, we may want to implement the random function H in the CRS using fixed-key AES in the ideal cipher model, as is common for garbling schemes based on free-XOR. In [HOSS18], we show that this reduces the number of AES calls from $O(n^2)$ in previous BMR protocols to $O(n^2 \ell_{\text{BMR}}/\kappa)$. The protocol and the complete proof can be found in [HOSS18].

5 Complexity Analysis and Implementation Results

We now compare the complexity of the most relevant aspects of our approach to the state-of-the-art prior results in semi-honest MPC protocols with dishonest majority. To demonstrate the practicality of our approach, we also present implementation results for the online evaluation phase of our BMR-based protocol. Further details can be found on [HOSS18].

5.1 Threshold Variants of Full-Threshold Protocols

Since the standard GMW and BMR-based protocols allow for up to $n-1$ corruptions, we also show how to modify previous protocols to support some threshold t, and compare our protocols with these variants. The method is very simple (and similar to the use of committees in our protocols), but does not seem to have been explicitly mentioned in previous literature. To evaluate a circuit C, all parties first secret-share their inputs to an arbitrarily chosen committee $\mathcal{P}'$, of size $t+1$. Committee $\mathcal{P}'$ runs the full-threshold protocol for a modified circuit C', which takes all the shares as input, and first XORs them together so that it computes the same function as C. The committee $\mathcal{P}'$ then sends the output to all parties in $\mathcal{P}$. The complexity of the threshold-t variant of a full-threshold protocol, Π, is then essentially the same as running Π between $t+1$ parties instead of n.

5.2 GMW-Style Protocol

We now compare the communication cost of our triple generation protocol with the best-known instantiation of GMW, namely a variant based on 1-out-of-4 OT to generate triples, recently optimized by [DKS+17] in the 2-party setting. This easily extends to the multi-party case with communication complexity $O(n^2\kappa/\log\kappa)$ bits per AND gate, so we consider both full-threshold and threshold-t (Sect. 5.1) variants. Note that our protocol from Sect. 3 has complexity $O(nt\ell)$ when using deterministic committees.

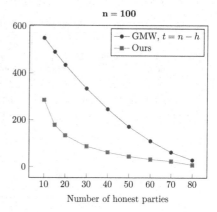

Fig. 14. Amortized communication cost (in kbit) for producing triples in GMW for $n=100$ parties

As can be seen in Table 1 and Fig. 14, for a fixed number of honest parties h, the improvement of our protocol over GMW (threshold t) becomes greater as the total number of parties increases. Our protocol starts to beat the best-known GMW protocol for producing multiplication triples when there are just 6 honest parties. For example, with 20 parties and 14 corruptions, the communication cost of our protocol is roughly 10% lower than threshold-14 GMW, and only 2 times lower than the cost of standard, full threshold GMW. As the number of parties (and honest parties) grows, our improvements become even greater, and when the number of honest parties is more than 80, we can use 1-bit keys and improve upon the threshold variant of GMW by *more than* 13 *times*.

In [HOSS18], we also analyse the complexity of our protocol when using random committees, and compare this with the standard approach of running full-threshold GMW in a single random committee.

5.3 BMR-Style Protocol

Communication Complexity. To show the efficiency of our constant-round garbling protocol from Sect. 4.5, we provide Table 2, which has two parts. First, it compares the amortized communication complexity incurred for garbling an AND gate with [BLO16]. We recall that this is the dominating cost for BMR-style protocols using Free-XOR, and that we incur no communication for creating shares of garbled splitter gates. Note that in the first setting of $n = 20, t = 10$, although our communication costs are around 3 times lower than [BLO16], we do not improve upon the threshold-t variant of that protocol, described earlier. Once we get to 50 parties, though, we start to improve upon [BLO16], with a reduction in communication going up to 7x for 400 parties and 10x for 1000 parties.

Table 2. Communication complexity for garbling, and size of garbled gates, in BMR-style protocols in kbit. A = #AND gates, S = #Splitter gates, X = #XOR gates.

# parties (honest)	20 (10)	50 (20)	80 (32)	100 (40)	200 (60)	400 (120)	1000 (160)
$(\ell_{BMR}, \ell_{OT}, r)$	$(32, 23, 530)$	$(27, 13, 450)$	$(17, 8, 380)$	$(15, 7, 400)$	$(8, 5, 370)$	$(8, 1, 80)$	$(8, 1, 120)$
[BLO16] (Gb $\mathcal{P}$)	341.24	2200.1	5675.36	8890	35740	143320.8	897102
[BLO16] (Gb $\mathcal{P}_{(1)}$)	**98.78**	835.14	2112.1	3286.7	17726.45	70654.7	634383.12
Ours (Garbling)	111.7	**747.63**	**1750.48**	**2678.74**	**5448.36**	**10114.99**	**64474.1**
[BLO16] ($|GC|$ $\mathcal{P}$)	10.24A	25.6A	40.96A	51.2A	102.4A	204.8A	512A
[BLO16] ($|GC|$ $\mathcal{P}_{(1)}$)	5.632A	15.88A	25.1A	31.23A	72.19A	143.9A	430.6A
[BLO17] ($|GC|$)	12.29$(A+X)$	12.29$(A+X)$	12.29$(A+X)$	12.29$(A+X)$	12.29$(A+X)$	12.29$(A+X)$	12.29$(A+X)$
Ours ($|GC|$)	2.56$(A+S)$	5.4$(A+S)$	5.45$(A+S)$	6$(A+S)$	6.4$(A+S)$	12.8$(A+S)$	32$(A+S)$

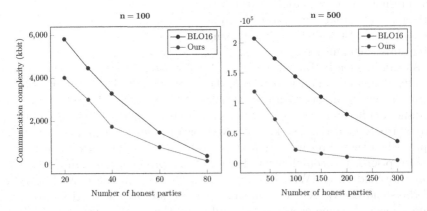

Fig. 15. Communication complexity cost (in kbit) for garbling when $n = 100$ and $n = 500$

The second half of the table shows the size of the garbled circuit in terms of the total number of AND, XOR and splitter gates. Garbled circuit size only has a slight impact on communication complexity, when opening the garbled circuit, which is much lower than the communication in the rest of the garbling phase. However, if an implementation needs to store the entire garbled circuit in memory (either for evaluation, or storage for later use) then it is also important to optimize its size. Here we also compare with [BLO17], which recently showed how to construct a compact multi-party garbled circuit based on key-homomorphic PRFs. The size of their garbled circuit is constant and grows with $O(\kappa)$ per gate, with security proven in the presence of $n - 1$ corrupted parties. On the other hand, their construction has much larger key sizes, does not support free-XOR, and has a more expensive preprocessing phase needing $O(n)$ secret-shared finite field multiplications per gate. In Fig. 15 we show the communication complexity of garbling when $n = 100, 500$ and for different number of honest parties.

References

[ADS+17] Asharov, G., Demmler, D., Schapira, M., Schneider, T., Segev, G., Shenker, S., Zohner, M.: Privacy-preserving interdomain routing at internet scale. PoPETs **2017**(3), 147 (2017)

[AFL+16] Araki, T., Furukawa, J., Lindell, Y., Nof, A., Ohara, K.: High-throughput semi-honest secure three-party computation with an honest majority. In: Weippl, E.R., Katzenbeisser, S., Kruegel, C., Myers, A.C., Halevi, S. (eds.) ACM CCS 2016, pp. 805–817. ACM Press, October 2016

[AFS03] Augot, D., Finiasz, M., Sendrier, N.: A fast provably secure cryptographic hash function. IACR Cryptology ePrint Archive 2003:230 (2003)

[AJL+12] Asharov, G., Jain, A., López-Alt, A., Tromer, E., Vaikuntanathan, V., Wichs, D.: Multiparty computation with low communication, computation and interaction via threshold FHE. In: Pointcheval, D., Johansson, T. (eds.) EUROCRYPT 2012. LNCS, vol. 7237, pp. 483–501. Springer, Heidelberg (2012). https://doi.org/10.1007/978-3-642-29011-4_29

[ALSZ13] Asharov, G., Lindell, Y., Schneider, T., Zohner, M.: More efficient oblivious transfer and extensions for faster secure computation. In: Sadeghi, A.-R., Gligor, V.D., Yung, M. (eds.) ACM CCS 2013, pp. 535–548. ACM Press, November 2013

[App16] Applebaum, B.: Garbling XOR gates "for free" in the standard model. J. Cryptol. **29**(3), 552–576 (2016)

[Bea92] Beaver, D.: Efficient multiparty protocols using circuit randomization. In: Feigenbaum, J. (ed.) CRYPTO 1991. LNCS, vol. 576, pp. 420–432. Springer, Heidelberg (1992). https://doi.org/10.1007/3-540-46766-1_34

[BLO16] Ben-Efraim, A., Lindell, Y., Omri, E.: Optimizing semi-honest secure multiparty computation for the internet. In: Weippl, E.R., Katzenbeisser, S., Kruegel, C., Myers, A.C., Halevi, S. (eds.) ACM CCS 2016, pp. 578–590. ACM Press, October 2016

[BLO17] Ben-Efraim, A., Lindell, Y., Omri, E.: Efficient scalable constant-round MPC via garbled circuits. In: Takagi, T., Peyrin, T. (eds.) ASIACRYPT 2017. LNCS, vol. 10625, pp. 471–498. Springer, Cham (2017). https://doi.org/10.1007/978-3-319-70697-9_17

[BMR90]   Beaver, D., Micali, S., Rogaway, P.: The round complexity of secure protocols (extended abstract). In: 22nd ACM STOC, pp. 503–513. ACM Press,
May 1990

[BO17]   Ben-Efraim, A., Omri, E.: Concrete efficiency improvements for multiparty
garbling with an honest majority. In: Latincrypt 2017 (2017)

[BOGW88]   Ben-Or, M., Goldwasser, S., Wigderson, A.: Completeness theorems
for non-cryptographic fault-tolerant distributed computation (extended
abstract). In: 20th ACM STOC, pp. 1–10. ACM Press, May 1988

[Bra85]   Bracha, G.: An $O(\lg n)$ expected rounds randomized byzantine generals
protocol. In: 17th ACM STOC, pp. 316–326. ACM Press, May 1985

[Can01]   Canetti, R.: Universally composable security: a new paradigm for cryptographic protocols. In: 42nd FOCS, pp. 136–145. IEEE Computer Society
Press, October 2001

[CCD88]   Chaum, D., Crépeau, C., Damgård, I.: Multiparty unconditionally secure
protocols (extended abstract). In: 20th ACM STOC, pp. 11–19. ACM Press,
May 1988

[DKS+17]   Dessouky, G., Koushanfar, F., Sadeghi, A.-R., Schneider, T., Zeitouni, S.,
Zohner, M.: Pushing the communication barrier in secure computation
using lookup tables. In: NDSS (2017)

[DMS04]   Dingledine, R., Mathewson, N., Syverson, P.F.: Tor: the second-generation
onion router. In: USENIX, pp. 303–320 (2004)

[DN07]   Damgård, I., Nielsen, J.B.: Scalable and unconditionally secure multiparty
computation. In: Menezes, A. (ed.) CRYPTO 2007. LNCS, vol. 4622, pp.
572–590. Springer, Heidelberg (2007). https://doi.org/10.1007/978-3-540-
74143-5_32

[Gen09]   Gentry, C.: Fully homomorphic encryption using ideal lattices. In: Mitzenmacher, M. (ed.) 41st ACM STOC, pp. 169–178. ACM Press, May/June
2009

[GMW87]   Goldreich, O., Micali, S., Wigderson, A.: How to play any mental game or a
completeness theorem for protocols with honest majority. In: Aho, A. (ed.)
19th ACM STOC, pp. 218–229. ACM Press, May 1987

[Gol04]   Goldreich, O.: The Foundations of Cryptography - Volume 2, Basic Applications. Cambridge University Press, Cambridge (2004)

[HOSS18]   Hazay, C., Orsini, E., Scholl, P., Soria-Vazquez, E.: Efficient MPC from
syndrome decoding (or: Honey, I shrunk the keys) (2018). https://eprint.
iacr.org/2018/208

[HSS17]   Hazay, C., Scholl, P., Soria-Vazquez, E.: Low cost constant round MPC
combining BMR and oblivious transfer. In: Takagi, T., Peyrin, T. (eds.)
ASIACRYPT 2017. LNCS, vol. 10624, pp. 598–628. Springer, Cham (2017).
https://doi.org/10.1007/978-3-319-70694-8_21

[IKNP03]   Ishai, Y., Kilian, J., Nissim, K., Petrank, E.: Extending oblivious transfers efficiently. In: Boneh, D. (ed.) CRYPTO 2003. LNCS, vol. 2729, pp.
145–161. Springer, Heidelberg (2003). https://doi.org/10.1007/978-3-540-
45146-4_9

[KK13]   Kolesnikov, V., Kumaresan, R.: Improved OT extension for transferring
short secrets. In: Canetti, R., Garay, J.A. (eds.) CRYPTO 2013, Part II.
LNCS, vol. 8043, pp. 54–70. Springer, Heidelberg (2013). https://doi.org/
10.1007/978-3-642-40084-1_4

[KMR14] Kolesnikov, V., Mohassel, P., Rosulek, M.: FleXOR: flexible garbling for XOR gates that beats free-XOR. In: Garay, J.A., Gennaro, R. (eds.) CRYPTO 2014, Part II. LNCS, vol. 8617, pp. 440–457. Springer, Heidelberg (2014). https://doi.org/10.1007/978-3-662-44381-1_25

[KS08] Kolesnikov, V., Schneider, T.: Improved garbled circuit: free XOR gates and applications. In: Aceto, L., Damgård, I., Goldberg, L.A., Halldórsson, M.M., Ingólfsdóttir, A., Walukiewicz, I. (eds.) ICALP 2008, Part II. LNCS, vol. 5126, pp. 486–498. Springer, Heidelberg (2008). https://doi.org/10.1007/978-3-540-70583-3_40

[LPSY15] Lindell, Y., Pinkas, B., Smart, N.P., Yanai, A.: Efficient constant round multi-party computation combining BMR and SPDZ. In: Gennaro, R., Robshaw, M. (eds.) CRYPTO 2015, Part II. LNCS, vol. 9216, pp. 319–338. Springer, Heidelberg (2015). https://doi.org/10.1007/978-3-662-48000-7_16

[LSS16] Lindell, Y., Smart, N.P., Soria-Vazquez, E.: More efficient constant-round multi-party computation from BMR and SHE. In: Hirt, M., Smith, A. (eds.) TCC 2016, Part I. LNCS, vol. 9985, pp. 554–581. Springer, Heidelberg (2016). https://doi.org/10.1007/978-3-662-53641-4_21

[NR17] Nielsen, J.B., Ranellucci, S.: On the computational overhead of MPC with dishonest majority. In: Fehr, S. (ed.) PKC 2017, Part II. LNCS, vol. 10175, pp. 369–395. Springer, Heidelberg (2017). https://doi.org/10.1007/978-3-662-54388-7_13

[TX03] Tate, S.R., Xu, K.: On garbled circuits and constant round secure function evaluation. CoPS Lab, University of North Texas, Technical report 2:2003 (2003)

[Yao86] Yao, A.C.-C.: How to generate and exchange secrets (extended abstract). In: 27th FOCS, pp. 162–167. IEEE Computer Society Press, October 1986

Fast Large-Scale Honest-Majority MPC for Malicious Adversaries

Koji Chida[1], Daniel Genkin[2], Koki Hamada[1], Dai Ikarashi[1], Ryo Kikuchi[1], Yehuda Lindell[3(✉)], and Ariel Nof[3]

[1] NTT Secure Platform Laboratories, Tokyo, Japan
{chida.koji,hamada.koki,ikarashi.dai,kikuchi.ryo}@lab.ntt.co.jp
[2] University of Michigan, Ann Arbor, USA
genkin@umich.edu
[3] Bar-Ilan University, Ramat Gan, Israel
{yehuda.lindell,ariel.nof}@biu.ac.il

Abstract. Protocols for secure multiparty computation enable a set of parties to compute a function of their inputs without revealing anything but the output. The security properties of the protocol must be preserved in the presence of adversarial behavior. The two classic adversary models considered are *semi-honest* (where the adversary follows the protocol specification but tries to learn more than allowed by examining the protocol transcript) and *malicious* (where the adversary may follow any arbitrary attack strategy). Protocols for semi-honest adversaries are often far more efficient, but in many cases the security guarantees are not strong enough.

In this paper, we present new protocols for securely computing any functionality represented by an arithmetic circuit. We utilize a new method for verifying that the adversary does not cheat, that yields a cost of just *twice* that of semi-honest protocols in some settings. Our protocols are information-theoretically secure in the presence of a malicious adversaries, assuming an honest majority. We present protocol variants for small and large fields, and show how to efficiently instantiate them based on replicated secret sharing and Shamir sharing. As with previous works in this area aiming to achieve high efficiency, our protocol is *secure with abort* and does not achieve fairness, meaning that the adversary may receive output while the honest parties do not.

We implemented our protocol and ran experiments for different numbers of parties, different network configurations and different circuit depths. Our protocol significantly outperforms the previous best for this setting (Lindell and Nof, CCS 2017); for a large number of parties, our implementation runs almost an order of magnitude faster than theirs.

Supported by the European Research Council under the ERC consolidators grant agreement no. 615172 (HIPS) and by the BIU Center for Research in Applied Cryptography and Cyber Security in conjunction with the Israel National Cyber Bureau in the Prime Minister's Office.

H. Shacham and A. Boldyreva (Eds.): CRYPTO 2018, LNCS 10993, pp. 34–64, 2018.
https://doi.org/10.1007/978-3-319-96878-0_2

1   Introduction

1.1   Background

Protocols for secure computation enable a set of parties with private inputs to compute a joint function of their inputs while revealing nothing but the output. The security properties typically required from secure computation protocols include *privacy* (meaning that nothing but the output is revealed), *correctness* (meaning that the output is correctly computed), *independence of inputs* (meaning that a party cannot choose its input as a function of the other parties' inputs), *fairness* (meaning that if one party gets output then so do all), and *guaranteed output delivery* (meaning that all parties always receive output). Formally, the security of a protocol is proven by showing that it behaves like an ideal execution with an incorruptible trusted party who computes the function for the parties [3,7,16,17]. In some cases, fairness and guaranteed output delivery are not required, in which case we say that the protocol is secure with abort. This is standard in the case of no honest majority since not all functions can be computed fairly without an honest majority [9], but security with abort can also in order to aid the construction of highly efficient protocols (e.g., as in [2,22]).

Despite the stringent requirements on secure computation protocols, in the late 1980s it was shown that any probabilistic polynomial-time functionality can be securely computed. This was demonstrated in the computational setting for any $t < n$ [16,19,26] (with security with abort for the case of $t \geq n/2$), in the information-theoretic setting with $t < n/3$ [5,8], and in the information-theoretic setting with $t < n/2$ assuming a broadcast channel [24]. These feasibility results demonstrate that secure computation is possible. However, significant work is needed to construct protocols that are efficient enough to use in practice.

1.2   Our Contributions

In this paper, we consider the problem of constructing highly efficient protocols that are secure in the presence of *static malicious adversaries* who control at most $t < n/2$ corrupted parties. Our protocol is fundamentally *information-theoretic*, but some efficient instantiations are computational (e.g., in order to generate correlated randomness). In the aim of achieving high efficiency, our protocols do not achieve fairness (even though this is fundamentally possible in our setting where $t < n/2$).

Our constructions work by securely computing an *arithmetic circuit* representation of the functionality over a finite field $\mathbb{F}$. This representation is very efficient for computations where many additions and multiplications are needed, like secure statistics. The starting point of our protocols utilizes the significant observation made by [14,15] that many protocols for *semi-honest multiplication* are actual secure in the presence of *malicious adversaries up to an additive attack*. This means that the only way an adversary can cheat is to make the result of the multiplication of shares of x and y be shares of $x \cdot y + d$, where d is an explicit value that the adversary knows (in this paper, we formalize this property via an ideal functionality definition). Since d is known by the adversary, it is independent of the values being multiplied, unless the adversary has some prior knowledge of these values. This property was utilized by [14,15] by having the multiplication be over randomized values, and making any cheating be detected unless the adversary is lucky enough to make the additive attack value match between different randomizations.

Our protocol works by running multiple circuit computations in parallel: one that computes the circuit over the real inputs and others which compute the circuit over randomized inputs. The outputs of the randomized circuits are then used to verify the correctness of the "original" circuit computation, thereby constituting a SPDZ-like MAC [13]. Security is achieved by the fact that the randomness is kept secret throughout the computation, and so any cheating by the adversary will be detected. All multiplications of shares are carried out using semi-honest protocols (that are actually secure for malicious adversaries up to an additive attack). Since this dominates the cost of the secure computation overall, the resulting protocol is highly-efficient. We present different protocols for the case of small and large fields, where a field is "large" if it is bigger than 2^σ where σ is the statistical security parameter. Our protocol for large fields requires computing one randomized circuit only, and the protocol for small fields requires δ randomized circuits where δ is such that $(|\mathbb{F}|/3)^\delta \geq 2^\sigma$. We note that our protocol for small fields can be run with $\delta = 1$ in the case of a large field, but in this case has about 10% more communication than the protocol that is dedicated to large fields. Both protocols have overall communication complexity that grows *linearly* with the number of parties (specifically, each party sends a constant number of field elements for each multiplication gate).

Based on the above, the running time of our protocol over large fields is just *twice* the cost of a semi-honest protocol, and the running time of our protocol over small fields is just $\delta + 1$ times the cost of a semi-honest protocol. As we discuss in the related work below, this is far more efficient than the protocols of [14,15] and the more recent protocol of [22]. The exact efficiency of our protocols depends on the specific instantiations of the secret sharing method, multiplication protocol and more. As in [22], we consider two main instantiations: one based on Shamir secret sharing for any number of parties, and one based on replicated secret sharing for the specific case of three parties. With our protocol we show that it is possible to compute any arithmetic circuit over large fields in the presence of malicious adversaries and an honest majority, at the cost of each party sending only *12 field elements* per multiplication gate. For 3-party computation, we show that using replicated sharing, this cost can be reduced to only *2 field elements* sent by each party per multiplication gate.

1.3 Experimental Results

We implemented our protocol for large fields, using replicated secret sharing for 3 parties and Shamir sharing for any number of parties. We then ran our implementation on AWS in two configurations: a *LAN network configuration* in a single AWS region (specifically, North Virginia), and a *WAN network configuration* with parties spread over three AWS regions (specifically, North Virginia, Germany and India). Each party was run in an independent AWS C4.large instance (2-core Intel Xeon E5-2666 v3 with 2.9 GHz clock speed and 3.75 GB RAM). We ran extensive experiments to analyze the efficiency of our protocols for different numbers of parties on a series of circuits of different depths, each with 1,000,000 multiplication gates, 1,000 inputs wires, and 50 output wires. The field we used for all our experiments was the 61-bit Mersenne field (and so security is approximately 2^{-60}). Our experiments show that our protocols have very good performance for all ranges of numbers of parties, especially in the LAN configuration (due to the protocol not being constant round). In particular, this 1 million gates large circuit with depth-20 can be computed in the LAN configuration in about 300 ms with three parties, 4 s with 50 parties, and 8 s with 110 parties. In the WAN configuration, the running time for this circuit is about 20 s for 3 parties, and about 2 min for 50 parties (at depth 100, the running time ranges from 45 s for 3 parties to 3.25 min for 50 parties). Thus, our protocols can be used in practice to compute arithmetic computations (like joint statistics) between many parties, while providing malicious security.

The previous best result in this setting was recently achieved in [22]. A circuit of the same size and depth-20 was computed by them in half a second with three parties, 29 s with 50 parties and 70 s with 100 parties. Our protocols run much faster than theirs, from approximately *twice as fast* for a small number of parties and up to *10 times faster* for a large number of parties.

1.4 Related Work

There is a large body of research focused on improving the efficiency of secure computation protocols. This work is roughly divided up into constructions of *concretely efficient* and *asymptotically efficient* protocols. Concretely efficient protocols are often implemented and aim to obtain the best overall running time, even if the protocol is not asymptotically optimal (e.g., it may have quadratic complexity and not linear complexity, but for a small number of parties the constants are such that the quadratic protocol is faster). Asymptotically efficient protocols aim to reduce the cost of certain parts of the protocols (rounds, communication complexity, etc.), and are often not concretely very efficient. However, in many cases, asymptotically efficient protocols provide techniques that inform the construction of concretely efficient protocols.

In the case of multiparty computation (with more than two parties) with a dishonest majority, concretely efficient protocols were given in [6,11,13,20]. This setting is much harder than that of an honest majority, and the results are therefore orders of magnitude slower (the state-of-the art SPDZ protocol [21] achieves a throughput of around 30,000 multiplication gates per second with 2 parties in some settings whereas we achieve a throughput of more than 1 million gates per second). For the case of an honest majority and arithmetic circuits, the previous best protocol is that of [22], and they include an in-depth comparison of their protocol to previous work, both concretely and asymptotically. Our protocol is fundamentally different from [22]. In their protocol, they use Beaver triples to verify correctness. Their main observation is that it is much more efficient to replace expensive opening operations with multiplication operation, since multiplication can be done with constant communication cost per party in the honest majority setting. We also use this observation but do not use Beaver triples at all. Thus, in our protocol, the parties are not required to generate and store these triples. As a result, our protocol has half the communication cost of [22] for 3 parties using replicated secret sharing (with each party sending 2 field elements here versus 4 field elements in [22] per multiplication gate), and less than a third of the communication cost of [22] for many parties using Shamir sharing (with each party sending 12 field elements here versus 42 field elements in [22] per multiplication gate). Experimentally, our protocol way outperforms [22], as shown in Sect. 6.3, running up to almost 10 times faster for a large number of parties.

The setting of $t < n/2$ and malicious adversaries was also studied in [1,2, 23], including implementations. However, they consider only three parties and Boolean circuits.

2   Preliminaries and Definitions

Notation. Let $P_1, ..., P_n$ denote the n parties participating in the computation, and let t denote the number of corrupted parties. In this work, we assume an honest majority, and thus $t < \frac{n}{2}$. Throughout the paper, we use H to denote the subset of honest parties and $\mathcal{C}$ to denote the subset of corrupted parties. Finally, we denote by $\mathbb{F}$ a finite field and by $|\mathbb{F}|$ its size.

2.1   Threshold Secret Sharing

A t-out-of-n secret sharing scheme enables n parties to share a secret $v \in \mathbb{F}$ so that no subset of t parties can learn any information about it, while any subset of $t + 1$ parties can reconstruct it. We require that the secret sharing scheme used in our protocol supports the following procedures:

– share(v): In this procedure, a dealer shares a value $v \in \mathbb{F}$. For simplicity, we consider *non-interactive* secret sharing, where there exists a probabilistic dealer D that receives v (and some randomness) and outputs shares $v_1, \ldots, v_n$, where v_i is the share intended for party P_i. We denote the sharing of a value v by $[v]$. We use the notation $[v]_J$ to denote the shares held by a subset of parties $J \subset \{P_1, \ldots, P_n\}$. We stress that if the dealer is corrupted, then the shares received by the parties may not be correct. Nevertheless, we abuse notation and say that the parties hold shares $[v]$ even if these are not correct. We will define correctness of a sharing formally below.

– share($v, [v]_J$): This non-interactive procedure is similar to the previous procedure, except that here the shares of a subset J of parties with $|J| \leq t$ are fixed in advance. We assume that there exists a probabilistic algorithm $\tilde{D}$ that receives $v, [v]_J = \{v'_j\}_{j|P_j \in J}$ (and some randomness) and outputs shares $v_1, \ldots, v_n$ where v_i is party P_i's share, and $v_j = v'_j$ for every $P_j \in J$.

We also assume that if $|J| = t$, then $[v]_J$ together with v *fully determine* all shares $v_1, \ldots, v_n$. This also means that any $t + 1$ shares fully determine all shares. (This follows since with $t+1$ shares one can always obtain v. However, for the secret sharing schemes we use, this holds directly as well.)

- reconstruct($[v], i$): Given a sharing of v and an index i held by the parties, this interactive protocol guarantees that if $[v]$ is not correct (see formal definition below), then P_i will output $\perp$ and abort. Otherwise, if $[v]$ is correct, then P_i will either output v or will abort.
- open($[v]$): Given a sharing of v held by the parties, this procedure guarantees that at the end of the execution, if $[v]$ is not correct, then *all* the honest parties will abort. Otherwise, if $[v]$ is correct, then each party will either output v or will abort. Clearly, open can be run by any subset of $t + 1$ or more parties. We require that if any subset J of $t + 1$ honest parties output a value v, then any superset of J will output either v or $\perp$ (but no other value).
- *local operations*: Given correct sharings $[u]$ and $[v]$ and a scalar $\alpha \in \mathbb{F}$, the parties can generate correct sharings of $[u + v]$, $[\alpha \cdot v]$ and $[v + \alpha]$ using local operations only (i.e., without any interaction). We denote these local operations by $[u] + [v]$, $\alpha \cdot [v]$, and $[v] + \alpha$, respectively.

Standard secret sharing schemes like Shamir and replicated secret sharing support all of these procedures (with their required properties). Throughout the entire paper, we set the threshold for the secret sharing scheme to be $\lfloor \frac{n-1}{2} \rfloor$, and we denote by t the number of corrupted parties. Since we assume an honest majority, it holds that $t < n/2$ and so the corrupted parties can learn nothing about a shared secret.

We now define correctness for secret sharing. Let J be a subset of honest parties of size $t + 1$, and denote by $\mathsf{val}([v])_J$ the value obtained by these parties after running the open procedure, where no corrupted parties or additional honest parties participate. Note that $\mathsf{val}([v])_J$ may equal $\perp$ if the shares held by the honest parties are not valid. Informally, a secret sharing is correct if every subset of $t + 1$ honest parties reconstruct the same value (which is not $\perp$). Formally:

Definition 2.1. *Let $H \subseteq \{P_1, \ldots, P_n\}$ denote the set of honest parties. A sharing $[v]$ is* correct *if there exists a value $v' \in \mathbb{F}$ ($v' \neq \perp$) such that for every $J \subseteq H$ with $|J| = t + 1$ it holds that $\mathsf{val}([v])_J = v'$.*

In the full version of the paper we show how to efficiently verify that a series of m shares are correct. Although not required in our general protocol, in some of our instantiations it is needed to verify that sharing of secrets is carried out correctly.

2.2   Security Definition

We use the standard definition of security based on the ideal/real model paradigm [7,19], with security formalized for non-unanimous abort. This means that the adversary first receives the output, and then determines for each honest party whether they will receive abort or receive their correct output.

3   Building Blocks and Sub-Protocols

In this section, we define a series of building blocks that we need for our protocol. The presentation here is general, and each basic protocol can be efficiently realized using standard secret sharing schemes. We describe these instantiations in Sect. 6.2.

3.1 Generating Random Shares

We define the ideal functionality $\mathcal{F}_{\text{rand}}$ to generate a sharing of a random value unknown to the parties. A formal description appears in Functionality 3.1. The functionality lets the adversary choose the corrupted parties' shares, which together with the random secret chosen by the functionality, are used to compute the shares of the honest parties.

FUNCTIONALITY 3.1 ($\mathcal{F}_{\text{rand}}$ - **Generating Random Shares**)

Upon receiving α_i for each i with $P_i \in \mathcal{C}$ from the ideal adversary $\mathcal{S}$, the ideal functionality $\mathcal{F}_{\text{rand}}$ chooses a random $r \in \mathbb{F}$, sets $[r]_\mathcal{C} = \{\alpha_i\}_{i|P_i \in \mathcal{C}}$ and runs share$(r, [r]_\mathcal{C})$ to receive a share r_i for each party P_i. Then, it hands each honest party P_j its share r_j.

As we have mentioned, the way we compute this functionality depends on the specific secret sharing scheme that is being used, and will be described in the instantiations in Sect. 6.2.

3.2 Generating Random Coins

$\mathcal{F}_{\text{coin}}$ is an ideal functionality that chooses a random element from $\mathbb{F}$ and hands it to all parties. A simple way to compute $\mathcal{F}_{\text{coin}}$ is to use $\mathcal{F}_{\text{rand}}$ to generate a random sharing and then open it. We formally describe and prove this in the full version of the paper.

3.3 $\mathcal{F}_{\text{input}}$ – Secure Sharing of Inputs

In this section, we present our protocol for secure sharing of the parties' inputs. The protocol is very simple: for each input x belonging to a party P_j, the parties call $\mathcal{F}_{\text{rand}}$ to generate a random sharing $[r]$; denote the share held by P_i by r_i. Then, r is reconstructed to P_j, who echo/broadcasts $x - r$ to all parties. Finally, each P_i outputs the share $[r + (x - r)] = [x]$. This is secure since $\mathcal{F}_{\text{rand}}$ guarantees that the sharing of r is correct, which in turn guarantees that the sharing of x is correct (since adding $x - r$ is a local operation only). In order to ensure that P_j sends the same value $x - r$ to all parties, a basic echo-broadcast is used. This is efficient since all inputs can be shared in parallel, utilizing a single echo broadcast. The formal definition of the ideal functionality for input sharing appears in Functionality 3.2.

FUNCTIONALITY 3.2 ($\mathcal{F}_{\text{input}}$ - **Sharing of Inputs**)

1. Functionality $\mathcal{F}_{\text{input}}$ receives inputs $v_1, \ldots, v_M \in \mathbb{F}$ from the parties. For every $i = 1, \ldots, M$, $\mathcal{F}_{\text{input}}$ also receives from $\mathcal{S}$ the shares $[v_i]$ of the corrupted parties for the ith input.
2. For every $i = 1, \ldots, M$, $\mathcal{F}_{\text{input}}$ computes all shares $(v_i^1, \ldots, v_i^n) = $ share$(v_i, [v_i]_\mathcal{C})$.
   For every $j = 1, \ldots, n$, $\mathcal{F}_{\text{input}}$ sends P_j its output shares $(v_1^j, \ldots, v_M^j)$.

A formal description appears in Protocol 3.3.

PROTOCOL 3.3 (Secure Sharing of Inputs)

- **Inputs:** Let $v_1, \ldots, v_M \in \mathbb{F}$ be the series of inputs; each v_i is held by some P_j.
- **The protocol:**
  1. The parties call $\mathcal{F}_{\text{rand}}$ M times to obtain sharings $[r_1], \ldots, [r_M]$.
  2. For $i = 1, \ldots, M$, the parties run reconstruct($[r_i], j$) for P_j to receive r_i, where P_j is the owner of the ith input. If P_j receives $\perp$, then it sends $\perp$ to all parties, outputs abort and halts.
  3. For $i = 1, \ldots, M$, party P_j sends $w_i = v_i - r_i$ to all parties.
  4. All parties send $\vec{w} = (w_1, \ldots, w_M)$, or a collision-resistant hash of the vector, to all other parties. If any party receives a different vector to its own, then it outputs $\perp$ and halts.
  5. For each $i = 1, \ldots, M$, the parties compute $[v_i] = [r_i] + w_i$.
- **Outputs:** The parties output $[v_1], \ldots, [v_M]$.

We now prove that Protocol 3.3 securely computes $\mathcal{F}_{\text{input}}$ specified in Functionality 3.2.

Proposition 3.4. *Protocol 3.3 securely computes Functionality 3.2 with abort in the presence of malicious adversaries controlling $t < n/2$ parties.*

Proof: Let $\mathcal{A}$ be the real adversary. We construct a simulator $\mathcal{S}$ as follows:

1. $\mathcal{S}$ receives $[r_i]_C$ (for $i = 1, \ldots, M$) that $\mathcal{A}$ sends to $\mathcal{F}_{\text{rand}}$ in the protocol.
2. For every $i \in \{1, \ldots, M\}$, $\mathcal{S}$ chooses a random r_i and computes $(r_i^1, \ldots, r_i^n) =$ share($r_i, [r_i]_C$). (This computation may be probabilistic or deterministic, depending on how many parties are corrupted.)
3. $\mathcal{S}$ simulates the honest parties in all reconstruct executions. If an honest party P_j receives $\perp$ in the reconstruction, then $\mathcal{S}$ simulates it sending $\perp$ to all parties. Then, $\mathcal{S}$ simulates all honest parties aborting.
4. $\mathcal{S}$ simulates the remainder of the execution, obtaining all w_i values from $\mathcal{A}$ associated with corrupted parties' inputs, and sending random w_j values for inputs associated with honest parties' inputs.
5. For every i for which the ith input is that of a corrupted party P_j, simulator $\mathcal{S}$ sends the trusted party computing $\mathcal{F}_{\text{input}}$ the input value $v_i = w_i + r_i$.
6. For every $i = 1, \ldots, M$, $\mathcal{S}$ defines the corrupted parties' shares $[v_i]_C$ to be $[r_i + w_i]_C$. (Observe that $\mathcal{S}$ has $[v_i]_C$ and merely needs to add the scalar w_i to each corrupted party's share.) Then, $\mathcal{S}$ sends $[v_1]_C, \ldots, [v_M]_C$ to the trusted party computing $\mathcal{F}_{\text{input}}$.
7. For every honest party P_j, if it aborted in the simulation, then $\mathcal{S}$ sends abort$_j$ to the trusted party computing $\mathcal{F}_{\text{input}}$; else, it sends continue$_j$.
8. $\mathcal{S}$ outputs whatever $\mathcal{A}$ outputs.

The only difference between the simulation by $\mathcal{S}$ and a real execution is that $\mathcal{S}$ sends random values w_j for inputs associated with honest parties' inputs. However, by the perfect secrecy of secret sharing, this is distributed identically to a real execution. ∎

3.4   Secure Multiplication up to Additive Attacks [14,15]

Our construction works by running a multiplication protocol (for multiplying two values that are shared among the parties) that is *not* fully secure in the presence of a malicious adversary and then running a verification step that enables the honest parties to detect cheating. In order to do this, we start with a multiplication protocols with the property that the adversary's ability to cheat is limited to carrying a so-called "additive attack" on the output. Formally, we say that a multiplication protocol is secure up to an additive attack if it realizes $\mathcal{F}_{\mathsf{mult}}$ defined in Functionality 3.5. This functionality receives input sharings $[x]$ and $[y]$ from the honest parties and an additive value d from the adversary, and outputs a sharing of $x \cdot y + d$. (Since the corrupted parties can determine their own shares in the protocol, the functionality allows the adversary to provide the shares of the corrupted parties, but this reveals nothing about the shared value.)

FUNCTIONALITY 3.5 ($\mathcal{F}_{\mathsf{mult}}$ - **Secure Mult. Up To Additive Attack**)

1. Upon receiving $[x]_H$ and $[y]_H$ from the honest parties, the ideal functionality $\mathcal{F}_{\mathsf{mult}}$ computes x, y and the corrupted parties shares $[x]_C$ and $[y]_C$.
2. $\mathcal{F}_{\mathsf{mult}}$ hands $[x]_C$ and $[y]_C$ to the ideal-model adversary/simulator $\mathcal{S}$.
3. Upon receiving d and $\{\alpha_i\}_{i|P_i \in C}$ from $\mathcal{S}$, functionality $\mathcal{F}_{\mathsf{mult}}$ defines $z = x \cdot y + d$ and $[z]_C = \{\alpha_i\}_{i|P_i \in C}$. Then, it runs $\mathsf{share}(z, [z]_C)$ to obtain a share z_j for each party P_j.
4. The ideal functionality $\mathcal{F}_{\mathsf{mult}}$ hands each honest party P_j its share z_j.

As we will discuss in the instantiations section (Sect. 6.2), the requirements defined by this functionality can be met by several semi-honest multiplication protocols. This will allow us to compute this functionality in a very efficient way.

3.5   Checking Equality to 0

In this section, we present a protocol to check whether a given sharing is a sharing of the value 0, without revealing any further information on the shared value. The idea behind the protocol is simple. Holding a sharing $[v]$, the parties generate a random sharing $[r]$ and multiply it with $[v]$. Then, the parties open the obtained sharing and check equality to 0. This works since if $v = 0$, then multiplying it with a random r will still yield 0. In contrast, if $v \neq 0$, then the multiplication will result with 0 only when $r = 0$, which happens with probability $\frac{1}{\mathbb{F}}$ only. The protocol is formally described in Protocol 3.7. For multiplying the sharings, the parties use the $\mathcal{F}_{\mathsf{mult}}$ functionality, which allows the adversary to change the output value via an additive attack. However, since the actual value is kept unknown, the adversary does not know which value should be added in order to achieve a sharing of 0.

We prove that Protocol 3.7 realizes the ideal functionality $\mathcal{F}_{\mathsf{checkZero}}$, which is defined in Functionality 3.6.

FUNCTIONALITY 3.6 ($\mathcal{F}_{\text{checkZero}}$ – **Checking Equality to 0**)

The ideal functionality $\mathcal{F}_{\text{checkZero}}$ receives $[v]_H$ from the honest parties and uses them to compute v. Then:

- If $v = 0$, then $\mathcal{F}_{\text{checkZero}}$ sends 0 to the simulator $\mathcal{S}$. If $\mathcal{S}$ sends reject (resp., accept), then $\mathcal{F}_{\text{checkZero}}$ sends reject (resp., accept) to the honest parties.
- If $v \neq 0$, then $\mathcal{F}_{\text{checkZero}}$ proceeds as follows:
  - With probability $\frac{1}{|\mathbb{F}|}$ it sends accept to the honest parties and $\mathcal{S}$.
  - With probability $1 - \frac{1}{|\mathbb{F}|}$ it sends reject to the honest parties and $\mathcal{S}$.

$\mathcal{F}_{\text{checkZero}}$ receives the honest parties shares, and use them to reconstruct the shared value. Then, if it is 0, then the simulator decides whether to send accept to the honest parties or reject. Otherwise, $\mathcal{F}_{\text{checkZero}}$ tosses a coin to decide what to send to the parties (i.e., in this case, the simulator is not given the opportunity to modify the output). In particular, when the checked value does not equal to 0, the output will still be accept with probability $\frac{1}{|\mathbb{F}|}$. This captures the event in Protocol 3.7 where $v \neq 0$ but $T = 0$, which happens also with probability $\frac{1}{\mathbb{F}}$ since r is uniformly distributed over $\mathbb{F}$.

PROTOCOL 3.7 (Checking Equality to 0 in the $(\mathcal{F}_{\text{rand}}, \mathcal{F}_{\text{mult}})$-Hybrid Model)

- **Inputs:** The parties hold a sharing $[v]$.
- **The protocol:**
  1. The parties call $\mathcal{F}_{\text{rand}}$ to obtain a sharing $[r]$.
  2. The parties call $\mathcal{F}_{\text{mult}}$ on $[r]$ and $[v]$ to obtain $[T] = [r \cdot v]$.
  3. The parties run $\text{open}([T])$. If a party receives $\perp$, then it outputs $\perp$. Else, it continues.
  4. Each party checks that $T = 0$. If yes, it outputs accept; else, it outputs reject.

Proposition 3.8. *Protocol 3.7 securely computes $\mathcal{F}_{\text{checkZero}}$ with abort in the $(\mathcal{F}_{\text{rand}}, \mathcal{F}_{\text{mult}})$-hybrid model in the presence of malicious adversaries who control $t < n/2$ parties.*

Proof: Let $\mathcal{A}$ be the real adversary. We construct the simulator $\mathcal{S}$ as follows. The ideal execution begins with $\mathcal{S}$ receiving the value 0 or 1 from $\mathcal{F}_{\text{checkZero}}$ (depending on if $v = 0$ or if $v \neq 0$, respectively). It then proceeds according to the following cases:

Case 1 – $v = 0$: $\mathcal{S}$ plays the role of $\mathcal{F}_{\text{rand}}$ by receiving the corrupted parties' shares. Then, $\mathcal{S}$ plays the role of $\mathcal{F}_{\text{mult}}$: it receives the corrupted parties' shares of T (i.e., $[T]_C$) and the value d to add to the output. Finally, $\mathcal{S}$ simulates the opening of $[T]$ by playing the role of the honest parties. In this case, $\mathcal{S}$ can simulate the real world execution precisely, since it knows that $T = d$ (regardless of the value of r), and thus it can define the honest parties' shares of T by running $\text{share}(d, [T]_C)$. Then, if $d = 0$ and $\mathcal{A}$ sent the correct shares when opening, then it sends accept to $\mathcal{F}_{\text{checkZero}}$. Otherwise, the parties open to a value that is not 0, or the opening fails (if $\mathcal{A}$ sends incorrect shares). In the first case, $\mathcal{S}$ sends reject to

$\mathcal{F}_{\text{checkZero}}$, whereas in the latter it sends abort and simulates the honest parties aborting in the protocol. Finally, $\mathcal{S}$ outputs whatever $\mathcal{A}$ outputs.

Observe that in this case, $\mathcal{S}$ simulates the real world execution exactly, and thus the view of $\mathcal{A}$ in the simulation is identical to its view in a real execution. Therefore, the output of the honest parties, which is determined in this case by $\mathcal{A}$, is the same in both the real and simulated execution.

Case 2 – $v \neq 0$: In this case, $\mathcal{S}$ receives the final output from $\mathcal{F}_{\text{checkZero}}$ without being able to influence it (beyond aborting). As in the previous case, $\mathcal{S}$ receives from $\mathcal{A}$ the corrupted parties' shares of r and T and the value d to add to T. In this case, it holds that $T = r \cdot v + d$, where r and v are unknown to $\mathcal{A}$. To simulate the opening of $[T]$, the simulator $\mathcal{S}$ either sets $T = 0$ (in the case where accept was received from $\mathcal{F}_{\text{checkZero}}$) or chooses a random $T \in \mathbb{F} \setminus \{0\}$ (in the case where reject was received) and defines the honest parties' shares by running $\text{share}(T, [T]_C)$. Then, $\mathcal{S}$ simulates the opening by playing the role of the honest parties. If $\mathcal{A}$ sends incorrect shares, causing the opening to fail, then $\mathcal{S}$ simulated the honest parties aborting in the real world execution and sends abort to $\mathcal{F}_{\text{checkZero}}$. Finally, $\mathcal{S}$ outputs whatever $\mathcal{A}$ outputs.

We claim that the view of $\mathcal{A}$ is identically distributed in the simulation and in the real world execution. This holds since the only difference is in the opening of T. In particular, in the real world, a random r is chosen and then the value of T is set to be $r \cdot v + d$, whereas in the simulation T is chosen randomly from $\mathbb{F}$ (this follows from the fact that T is set to be 0 with probability $\frac{1}{|\mathbb{F}|}$ and to any value other than 0 with probability $\frac{1}{|\mathbb{F}-1|}$). In both cases, therefore, T is distributed uniformly over $\mathbb{F}$. Thus, the view of the adversary $\mathcal{A}$ is distributed identically in both executions. We now proceed to show that the output of the honest parties is also distributed identically in both executions. In the simulation, the output is accept with probability $\frac{1}{|\mathbb{F}|}$, as determined by the trusted party. In the real execution, the honest parties output accept when $T = 0$. This happens when $r \cdot v + d = 0$, i.e., when $r = -d \cdot v^{-1}$ (recall that $v \neq 0$ and so it has an inverse). Since r is distributed uniformly over $|\mathbb{F}|$, then $T = 0$ with probability $\frac{1}{|\mathbb{F}|}$, and so the honest parties' output is accept with the same probability as in the simulation. This concludes the proof. ∎

4   The Protocol for Large Fields

With all the building blocks described in the previous section, we are now ready to present our protocol that computes an arithmetic circuit over a large field on the private inputs of the parties. We stress that by a large field, we mean that $2/|\mathbb{F}| \leq 2^{-\sigma}$, where σ is the statistical security parameter determining the allowed error. The protocol works by computing the circuit using $\mathcal{F}_{\text{mult}}$ (i.e., a multiplication protocol that is secure up to additive attacks), and then running a verification step where the computations of all multiplication gates are verified.

The idea behind the protocol is for the parties to generate a random sharing $[r]$, and then evaluate the arithmetic circuit while preserving the invariant that on every wire, the parties hold shares of the value $[x]$ on the wire and shares of the randomized value $[r \cdot x]$. This is achieved by generating $[r]$ using $\mathcal{F}_{\text{rand}}$ (Sect. 3.1) and then multiplying each shared input with $[r]$ using $\mathcal{F}_{\text{mult}}$ (Sect. 3.4). For addition and multiplication-by-a-constant gates, each party can locally compute the sharings of both the output and the randomized output (since $[r \cdot x] + [r \cdot y] = [r \cdot (x + y)]$). For multiplication gates, the parties interact to compute shares of $[x \cdot y]$ and $[r \cdot x \cdot y]$ (given shares of $x, r \cdot x, y, r \cdot y$). This is achieving by running $\mathcal{F}_{\text{mult}}$ on $[x]$ and $[y]$ to obtain a sharing $[z] = [x \cdot y]$, and running $\mathcal{F}_{\text{mult}}$ on $[r \cdot x]$ and $[y]$ to obtain a sharing $[r \cdot z] = [r \cdot x \cdot y]$ (equivalently, this latter sharing could be generated by multiplying $[x]$ with $[r \cdot y]$).

As we have described in Sect. 3.4, the multiplication subprotocol that we use – and is modeled in $\mathcal{F}_{\text{mult}}$ – is only secure up to additive attacks. The circuit randomization technique described above is aimed at preventing the adversary from carrying out such an attack without getting caught. In order to see why, consider an attacker who carries out an additive attack when multiplying $[x]$ and $[y]$ so that the result is $[x \cdot y + d]$ for some $d \neq 0$. Then, in order for the invariant to be maintained, the adversary needs to cheat in the other multiplication for this gate so that the result is $[r \cdot (x \cdot y + d)]$. Now, since the adversary can only carry out an *additive* attack, it must make the result be $[r \cdot x \cdot y + d']$, where $d' = d \cdot r$. However, r is not known, and thus the attacker can only succeed in this with probability $1/|\mathbb{F}|$. Thus, in order to prevent cheating in the multiplication gates, it suffices to check that the invariant is preserved over all wires.

This verification is carried out as follows. After the entire circuit has been evaluated, the parties compute a *random* linear combination of the sharings of the values and the sharings of the randomized values on each multiplication gate's output wire; denote the former by $[w]$ and the latter by $[u]$. Then, $[r]$ is opened, and the parties locally multiply it with $[w]$. Clearly, if there was no cheating, then $r \cdot [w]$ equals $[u]$, and thus $[u] - r \cdot [w]$ equals 0, which can be checked using $\mathcal{F}_{\text{checkZero}}$ (Sect. 3.5). In contrast, if the adversary did cheat, then as we have mentioned above, the invariant will not hold except with probability $1/|\mathbb{F}|$. In this case, as we will show below, $[u] = r \cdot [w]$ with probability only $1/|\mathbb{F}|$ (since they are generated via a random linear combination). When $[u - r \cdot w] \neq 0$, then $\mathcal{F}_{\text{checkZero}}$ outputs reject except with probability $1/|\mathbb{F}|$. Overall, we therefore have that the adversary can cheat with probability at most $2/|\mathbb{F}|$. We prove this formally in Lemma 4.2. A full specification appears in Protocol 4.1.

We will provide an exact complexity analysis below (in Sect. 6.2), but for now observe that the cost is dominated by just 2 invocations of $\mathcal{F}_{\text{mult}}$ per multiplication gate. As we have mentioned in Sect. 3.4, $\mathcal{F}_{\text{mult}}$ is securely realized in the presence of malicious adversaries by several *semi-honest* multiplication protocols. Thus, the overall cost of achieving malicious security here is very close to the semi-honest cost.

We begin by proving that the verification step has the property that if the adversary cheats in a multiplication gate, then the $T = 0$ with probability $\frac{2}{|\mathbb{F}|} \leq 2^{-\sigma}$ (where σ is the statistical security parameter). We call a multiplication triple of a multiplication gate, the triple of values $([x], [y], [z])$, where $[x], [y]$ are the shares on the inputs wires, and $[z]$ is the shares on the output wire, after the multiplication. (Note that z may not equal $x \cdot y$, if the adversary cheated in the multiplication.)

PROTOCOL 4.1 (Computing Arithmetic Circuits Over Large Fields)

Inputs: Each party P_j ($j \in \{1, \ldots, n\}$) holds an input $x_j \in \mathbb{F}^\ell$.

Auxiliary Input: The parties hold the description of a finite field $\mathbb{F}$ (with $3/|\mathbb{F}| \leq 2^{-\sigma}$) and an arithmetic circuit C over $\mathbb{F}$ that computes f on inputs of length $M = \ell \cdot n$. Let N be the number of multiplication gates in C.

The protocol (throughout, if any party receives $\perp$ as output from a call to a sub-functionality, then it sends $\perp$ to all other parties, outputs $\perp$ and halts):

1. *Secret sharing the inputs:*
   (a) For each input v_i held by party P_j, party P_j sends v_i to $\mathcal{F}_{\text{input}}$.
   (b) Each party P_j records its vector of shares $(v_1^j, \ldots, v_M^j)$ of all inputs, as received from $\mathcal{F}_{\text{input}}$. If a party received $\perp$ from $\mathcal{F}_{\text{input}}$, then it sends abort to the other parties and halts.

2. *Generate randomizing share:* The parties call $\mathcal{F}_{\text{rand}}$ to receive a sharing $[r]$.

3. *Randomization of inputs:* For each input wire sharing $[v_m]$ (where $m \in \{1, \ldots, M\}$), the parties call $\mathcal{F}_{\text{mult}}$ on $[r]$ and $[v_m]$ to receive $[r \cdot v_m]$.

4. *Circuit emulation:* Let $G_1, \ldots, G_N$ be a predetermined topological ordering of the gates of the circuit. For $k = 1, \ldots, N$ the parties work as follows:
   - G_k *is an addition gate:* Given pairs $([x], [r \cdot x])$ and $([y], [r \cdot y])$ on the *left* and *right* input wires respectively, the parties locally compute $([x + y], [r \cdot x] + [r \cdot y]) = ([x + y], [r \cdot (x + y)])$.
   - G_k *is a multiplication-by-constant gate:* Given $([x], [r \cdot x])$ on the input wire and constant $a \in \mathbb{F}$, the parties locally compute $([a \cdot x], [r \cdot (a \cdot x)])$.
   - G_k *is a multiplication gate:* Given pair $([x], [r \cdot x])$ and $([y], [r \cdot y])$ on the *left* and *right* input wires respectively:
     (a) The parties call $\mathcal{F}_{\text{mult}}$ on $[x]$ and $[y]$ to receive $[x \cdot y]$.
     (b) The parties call $\mathcal{F}_{\text{mult}}$ on $[r_i \cdot x]$ and $[y]$ to receive $[r_i \cdot x \cdot y]$.

5. *Verification stage:* Before the secrets on the output wires are reconstructed, the parties verify that all the multiplications were carried out correctly, as follows. Let $\left\{ ([z_k], [r \cdot z_k]) \right\}_{k=1}^{N}$ be the pairs on the output wires of all multiplication gates and let $\left\{ ([v_m], [r \cdot v_m]) \right\}_{m=1}^{M}$ be the pairs on the input wires of the circuit.
   (a) The parties call $\mathcal{F}_{\text{coin}}$ to receive $\alpha_1, \ldots, \alpha_N, \beta_1, \ldots, \beta_M \in \mathbb{F}$.
   (b) The parties locally compute
   $$[u] = \sum_{k=1}^{N} \alpha_k \cdot [r \cdot z_k] + \sum_{m=1}^{M} \beta_m \cdot [r \cdot v_m] \quad \text{and} \quad [w] = \sum_{k=1}^{N} \alpha_k \cdot [z_k] + \sum_{m=1}^{M} \beta_m \cdot [v_m].$$
   (c) The parties run $\text{open}([r])$ to receive r.
   (d) Each party locally computes $[T] = [u] - r \cdot [w]$.
   (e) The parties call $\mathcal{F}_{\text{checkZero}}$ on $[T]$. If $\mathcal{F}_{\text{checkZero}}$ outputs reject, the parties output $\perp$ and abort. Else, if it outputs accept, the parties proceed to the next step.

6. *Output reconstruction:* For each output wire of the circuit, the parties run $\text{reconstruct}([v], j)$, where $[v]$ is the sharing of the value on the output wire, and P_j is the party whose output is on the wire.
   If a party received $\perp$ in any call to the reconstruct procedure, then it sends $\perp$ to the other parties, outputs $\perp$ and halts.

Output: If a party has not output $\perp$, then it outputs the values it received on its output wires.

Lemma 4.2. *If $\mathcal{A}$ sends an additive value $d \neq 0$ in any of the calls to $\mathcal{F}_{\mathsf{mult}}$ in the execution of Protocol 4.1, then the value $[T]$ computed in the verification stage of Step 5 in Protocol 4.1 equals 0 with probability less than $2/|\mathbb{F}|$.*

Proof: The intuition has been discussed above, and we therefore proceed directly to the proof.

Consider the multiplication triple $([x_k], [y_k], [z_k])$ for the kth multiplication gate. We stress that the values on the input wires $[x_k], [y_k]$ may not actually be the appropriate values as when the circuit is computed by honest parties. However, in order to prove the lemma, we consider each gate separately, and all that is important is whether the invariant described above holds on the output wire (i.e., the randomized result is $[r \cdot z_k]$ for whatever z_k is here). By the definition of $\mathcal{F}_{\mathsf{mult}}$, a malicious adversary is able to carry out an additive attack, meaning that it can add a value to the output of each multiplication gate. Thus, it holds that $\mathsf{val}([z_k])_H = x_k \cdot y_k + d_k$ and $\mathsf{val}([r \cdot z_k])_H = (r \cdot x_k + e_k) \cdot y_k + f_k$, where $d_k, e_k, f_k \in \mathbb{F}$ are the added values in the additive attacks, as follows. The value d_k is the value added by the adversary when $\mathcal{F}_{\mathsf{mult}}$ is called with $[x_k]$ and $[y_k]$. The value e_k is such that the input to $\mathcal{F}_{\mathsf{mult}}$ for the randomized multiplication is $[y_k]$ and $[r \cdot x_k + e_k]$. This is an accumulated error on the randomized value from previous gates. Finally, f_k is the value added by the adversary when $\mathcal{F}_{\mathsf{mult}}$ is called with the shares $[y_k]$ and $[r \cdot x_k + e_k]$. Similarly, for each input wire with sharing $[v_m]$, it holds that $\mathsf{val}([r \cdot v_m])_H = r \cdot v_m + g_m$, where $g_m \in \mathbb{F}$ is the value added by the adversary when $\mathcal{F}_{\mathsf{mult}}$ is called with $[r]$ and the shared input $[v_m]$. Thus, we have that

$$\mathsf{val}([u])_H = \sum_{k=1}^{N} \alpha_k \cdot ((r \cdot x_k + e_k) \cdot y_k + f_k) + \sum_{m=1}^{M} \beta_m \cdot (r \cdot v_m + g_m)$$

$$\mathsf{val}([w])_H = \sum_{k=1}^{N} \alpha_k \cdot (x_k \cdot y_k + d_k) + \sum_{m=1}^{M} \beta_m \cdot v_m$$

and so

$$\mathsf{val}([T])_H = \mathsf{val}([u])_H - r \cdot \mathsf{val}([w])_H =$$

$$= \sum_{k=1}^{N} \alpha_k \cdot ((r \cdot x_k + e_k) \cdot y_k + f_k) + \sum_{m=1}^{M} \beta_m \cdot (r \cdot v_m + g_m)$$

$$- r \cdot \left(\sum_{k=1}^{N} \alpha_k \cdot (x_k \cdot y_k + d_k) + \sum_{m=1}^{M} \beta_m \cdot v_m \right)$$

$$= \sum_{k=1}^{N} \alpha_k \cdot (e_k \cdot y_k + f_k - r \cdot d_k) + \sum_{m=1}^{M} \beta_m \cdot g_m. \qquad (1)$$

where the second equality holds because r is opened and so the multiplication $r \cdot [w]$ always yields $[r \cdot w]$. Our aim is to show that $\mathsf{val}([T])_H$, as shown in Eq. (1), equals 0 with probability at most $1/|\mathbb{F}|$. We have the following cases.

- *Case 1 – there exists some* $m \in \{1, \ldots, M\}$ *such that* $g_m \neq 0$: Let m_0 be the smallest such m for which this holds. Then, $\mathsf{val}([T])_H = 0$ if and only if

$$\beta_{m_0} = \left(-\sum_{k=1}^{N} \alpha_k \cdot (e_k \cdot y_k + f_k - r \cdot d_k) - \sum_{\substack{m=1 \\ m \neq m_0}}^{M} \beta_m \cdot g_m \right) \cdot g_{m_0}^{-1}$$

which holds with probability $\frac{1}{|\mathbb{F}|}$ since β_{m_0} is distributed uniformly over $\mathbb{F}$, and chosen independently of all other values.
- *Case 2 – all* $g_m = 0$: By the assumption in the lemma, some additive value $d \neq 0$ was sent to $\mathcal{F}_{\mathsf{mult}}$. Since none was sent for the input randomization, there exists some $k \in \{1, \ldots, N\}$ such that $d_k \neq 0$ or $f_k \neq 0$. Let k_0 be the smallest such k for which this holds. Note that since this is the first error added, it holds that $e_{k_0} = 0$. Thus, in this case, $\mathsf{val}([T])_H = 0$ if and only if

$$\alpha_{k_0} \cdot (f_{k_0} - r \cdot d_{k_0}) = -\sum_{\substack{k=1 \\ k \neq k_0}}^{N} \alpha_k \cdot (e_k \cdot y_k + f_k - r \cdot d_k). \tag{2}$$

If $f_{k_0} - r \cdot d_{k_0} \neq 0$, then the above equality holds with probability $1/|\mathbb{F}|$ since α_{k_0} is distributed uniformly over $\mathbb{F}$, and chosen independently of all other values. However, if $f_{k_0} - r \cdot d_{k_0} = 0$, then equality may hold. (Indeed, the best strategy of an adversary is to cheat in both multiplications of a single gate, and hope that the additive values cancel each other out.) Nevertheless, the probability that $f_{k_0} - r \cdot d_{k_0} = 0$ is at most $1/|\mathbb{F}|$, since r is not known to the adversary when the k_0'th gate is computed (and by the security of the secret sharing scheme, it is completely random). Thus, the probability that Eq. (2) holds is at most $\frac{1}{|\mathbb{F}|} + \left(1 - \frac{1}{|\mathbb{F}|}\right) \cdot \frac{1}{|\mathbb{F}|} < \frac{2}{|\mathbb{F}|}$.

In both cases, the probability of equality is upper bounded by $2/|\mathbb{F}|$ and this completes the proof. ∎

We are now ready to prove the security of Protocol 4.1.

Theorem 4.3. *Let* σ *be a statistical security parameter, and let* $\mathbb{F}$ *be a finite field such that* $3/|\mathbb{F}| \leq 2^{-\sigma}$. *Let* f *be an n-party functionality over* $\mathbb{F}$. *Then, Protocol 4.1 securely computes* f *with abort in the* $(\mathcal{F}_{\mathsf{input}}, \mathcal{F}_{\mathsf{mult}}, \mathcal{F}_{\mathsf{coin}}, \mathcal{F}_{\mathsf{rand}}, \mathcal{F}_{\mathsf{checkZero}})$-*hybrid model with statistical error* $2^{-\sigma}$, *in the presence of a malicious adversary controlling* $t < \frac{n}{2}$ *parties.*

Proof: Intuitively, the protocol is secure since if the adversary cheats in any multiplication, then the value T computed in the verification stage will equal zero with probability at most $2/|\mathbb{F}|$, as shown in Lemma 4.2. Then, if indeed $T \neq 0$, this will be detected in the call to $\mathcal{F}_{\mathsf{checkZero}}$, except with probability $1/|\mathbb{F}|$. Thus, overall, the adversary can avoid detection with probability at most $3/|\mathbb{F}| \leq 2^{-\sigma}$.

Let $\mathcal{A}$ be the real adversary who controls the set of corrupted parties $\mathcal{C}$; the simulator $\mathcal{S}$ works as follows:

1. *Secret sharing the inputs:* S receives from A the set of corrupted parties inputs (values v_j associated with parties $P_i \in C$) and the corrupted parties' shares $\{[v_i]_C\}_{i=1}^M$ that A sends to $\mathcal{F}_{\text{input}}$ in the protocol. For each honest party's input v_j, S computes $(v_j^1, \ldots, v_j^n) = \text{share}(0, [v_i]_C)$ (i.e., uses 0 as the input on the wire). Then, S hands A the shares of the corrupted parties for all inputs.

2. *Generate the randomizing share:* Simulator S receives the share $[r]_C$ of the corrupted parties that A sends to $\mathcal{F}_{\text{rand}}$.

3. *Randomization of inputs:* For every input wire $m = 1, \ldots, M$, simulator S plays the role of $\mathcal{F}_{\text{mult}}$ in the multiplication of the mth input $[v_m]$ with r. Specifically, S hands A the corrupted parties shares in $[v_m]$ and $[r]$ (it has these shares from the previous steps). Next, S receives the additive value $d = g_m$ and the corrupted parties' shares $[z]_C$ of the result that A sends to $\mathcal{F}_{\text{mult}}$. Simulator S stores all of these corrupted parties shares.

4. *Circuit emulation:* Throughout the emulation, S will use the fact that it knows the corrupted parties' shares on the input wires of the gate being computed. This holds initially from the steps above, and we will show it computes the output wires of each gate below. For each gate G_k in the circuit,
   - If G_k *is an addition gate:* Given the shares of the corrupted parties on the input wires, S locally adds them as specified by the protocol, and stores them.
   - If G_k *is a multiplication-by-a-constant gate:* Given the shares of the corrupted parties on the input wire, S locally multiplies them by the constant, them as specified by the protocol, and stores them.
   - If G_k *is a multiplication gate:* S plays the role of $\mathcal{F}_{\text{mult}}$ in this step (as in the randomization of inputs above). Specifically, simulator S hands A the corrupted parties' shares on the input wires as it expects to receive from $\mathcal{F}_{\text{mult}}$ (it has these shares by the invariant), and receives from A the additive value as well as the corrupted parties' shares for the output. These additive values are d_k (for the multiplication of the actual values) and f_k (for the multiplication of the randomized value), as defined in the proof of Lemma 4.2. S stores the corrupted parties' shares.

5. *Verification stage:* Simulator S works as follows. S chooses random $\alpha_1, \ldots, \alpha_N, \beta_1, \ldots, \beta_M \in \mathbb{F}$ and hands them to A, as it expects to receive from $\mathcal{F}_{\text{coin}}$. Then, S chooses a random $r \in \mathbb{F}$ and computes the shares of r by $(r_1, \ldots, r_n) = \text{share}(r, [r]_C)$, using the shares $[r]_C$ provided by A in the "generate randomizing share" step above. Next, S simulates the honest parties sending their shares in $\text{open}([r])$ to A, and receives the shares that A sends to the honest parties in this open. If any honest party would abort (it knows whether this would happen since it has all the honest parties' shares), then S simulates it sending $\perp$ to all parties, externally sends abort_j for every $P_j \in H$ to the trusted party computing f, and halts.

   Finally, S simulates $\mathcal{F}_{\text{checkZero}}$, as follows. If any non-zero g_m, d_k, f_k was provided to $\mathcal{F}_{\text{mult}}$ by A in the simulation, then S simulates $\mathcal{F}_{\text{checkZero}}$ sending reject, and then all honest parties sending $\perp$. Then, S externally sends abort_j for every $P_j \in H$ to the trusted party computing f. Otherwise, S proceeds to the next step.

6. *Output reconstruction:* If no abort had occurred, S externally sends the trusted party computing f the corrupted parties' inputs that it received in the "secret sharing the inputs" step above. S receives back the output values for each output wire associated with a corrupted party. Then, S simulates the honest parties in the reconstruction of the corrupted parties' outputs. It does this by computing the shares of the honest parties on this wire using the corrupted parties' shares on the wire (which it has by the invariant) and the actual output value it received from the trusted party.

   In addition, S receives the messages from A for the reconstructions to the honest parties. If any of the messages in the reconstruction of an output wire associated with an honest P_j are incorrect (i.e., the shares sent by A are not the correct shares it holds), then S sends abort_j to instruct the trusted party to not send the output to P_j. Otherwise, S sends $\mathsf{continue}_j$ to the trusted party, instructing it to send P_j its output.

We claim that the view of the adversary in the simulation is identical to its view in the real execution, except with probability $3/|\mathbb{F}|$. In order to see this, observe first that if all g_m, d_k, f_k values equal 0, then the simulation is perfect. The only difference is that the input shares of the honest parties are to 0. However, by the perfect secrecy of secret sharing, this has the same distribution as in a real execution.

Next, consider the case that some g_m, d_k, f_k value does not equal 0. In this case, the simulator S *always* simulates $\mathcal{F}_{\mathsf{checkZero}}$ outputting reject. However, in a real execution where some g_m, d_k, f_k value does not equal 0, functionality $\mathcal{F}_{\mathsf{checkZero}}$ may return accept either if $T = 0$, or if $T \neq 0$ but it chose accept with probability $1/|\mathbb{F}|$ in the computation of the functionality output. By Lemma 4.2, the probability that $T = 0$ in such a real execution is less than $2/|\mathbb{F}|$, and thus $\mathcal{F}_{\mathsf{checkZero}}$ outputs accept with probability less than $\frac{2}{|\mathbb{F}|} + \left(1 - \frac{2}{|\mathbb{F}|}\right) \cdot \frac{1}{|\mathbb{F}|} < \frac{3}{|\mathbb{F}|}$. Since this is the only difference between the real execution and the ideal-model simulation, we have that the statistical difference between these distributions is less than $\frac{3}{|\mathbb{F}|} \leq 2^{-\sigma}$, and so the protocol is secure with statistical error $2^{-\sigma}$. ∎

Using pseudo-randomness to reduce the number of calls to $\mathcal{F}_{\mathsf{coin}}$. Observe that in the verification phase, we need to call $\mathcal{F}_{\mathsf{coin}}$ many times; once for each input wire and multiplication gate, to be exact. Instead of calling $\mathcal{F}_{\mathsf{coin}}$ for every value (since this would be expensive), it suffices to call it once to obtain a seed for a pseudorandom generator, and then each party locally uses the seed to obtain as much randomness as needed. (Practically, the key would be an AES key, and randomness is obtained by running AES in counter mode.) It is not difficult to show that by the pseudorandomness assumption, the probability that the adversary can cheat is only negligibly different.[1]

---

[1] Note that this is not as immediate as it seems since the adversary has the seed/key as well, and so at this point the pseudorandom property is actually lost. However, the checks work by generating the randomness after everything else is finished and then verifying that some equality holds, or that the results are correct. These properties are actually determined *before* the key is revealed, and thus security is maintained even after the key is revealed.

Concrete efficiency. We analyze the performance of our protocol. The functionality $\mathcal{F}_{\text{mult}}$ is called once for every input wire and twice for every multiplication gate (once for multiplying $[x]$ and $[y]$, and another time for multiplying $[r \cdot x]$ with $[y]$). Thus, the overall number of multiplications is $(M + 2N) \cdot \mathcal{F}_{\text{mult}}$, where M denotes the number of inputs and N the number of multiplication gates. In addition, there are M calls to $\mathcal{F}_{\text{input}}$, which by Protocol 3.3 reduces to M invocations of $\mathcal{F}_{\text{rand}}$ and M reconstructions. Furthermore, there is one call to $\mathcal{F}_{\text{rand}}$ for generating $[r]$, one call to $\mathcal{F}_{\text{coin}}$ for generating all the α_k, β_k values (which reduces to one $\mathcal{F}_{\text{rand}}$ and one open), one call to open for $[r]$, and one call to $\mathcal{F}_{\text{checkZero}}$ (which reduces to one call to $\mathcal{F}_{\text{rand}}$, one $\mathcal{F}_{\text{mult}}$ and one opening). Finally, let L denote the number of output values, and so the number of reconstruct operations equals L in order to obtain output. We have that the overall exact cost of the protocol is

$$(M + 2N + 1) \cdot \mathcal{F}_{\text{mult}} + (M + 3) \cdot \mathcal{F}_{\text{rand}} + (M + L) \cdot \text{reconstruct} + 3 \cdot \text{open}.$$

Clearly, amortizing over the size of the circuit, we have that the average cost is $2 \cdot \mathcal{F}_{\text{mult}}$ per multiplication gate.

Reducing memory. One issue that can arise in the implementation of Protocol 4.1 is due to the fact that the parties need to store all of the shares used throughout the computation in order to run the verification stage. If the circuit is huge (e.g., has billions of gates), then this can be problematic. However, in such cases, it is possible to run the verification multiple times. For example, one can determine that the verification is run after every million gates processed. Since this involves opening the randomizing share $[r]$, a new randomizing share $[r']$ is chosen by running $\mathcal{F}_{\text{rand}}$, and the shares on all wires that are still "active" (meaning that they are input into later gates) are randomized using $[r']$ (in the same way that the input wires are randomized). The protocol then proceeds as before. The additional cost is calling $\mathcal{F}_{\text{rand}}$ and $\mathcal{F}_{\text{checkZero}}$ once every million gates (or whatever is determined) instead of just once, and multiplying $[r']$ by all of the active wires using $\mathcal{F}_{\text{mult}}$ at each such iteration instead of just for the inputs. This will typically only be worthwhile for extremely large circuits.

Small fields. Protocol 4.1 works for fields that are large enough so that $3/|\mathbb{F}|$ is an acceptable probability of an adversary cheating. In cases where it is desired to work in a smaller field, one could consider the following strategy. Instead of having a single randomizing share $[r]$, generate δ such random shares $[r_1], \ldots, [r_\delta]$ (where $(3/|\mathbb{F}|)^\delta$ is small enough). Then, run the same circuit emulation and verification steps using each r_i separately. Since each verification is independent, this will yield a cheating probability of at most $(3/|\mathbb{F}|)^\delta$, as required. The problem with such a strategy is that the simulator must be able to simulate the $[T]$ values for each verification. Unlike the case of a large field, in this case, there is a good probability that some of the $[T]$ values will equal 0, even if the adversary cheated (the only guarantee is that *not all* $[T]$ values will equal 0). Looking at Eq. (1), observe that the value of $[T]$ is dependent on the values $\alpha_k, e_k, f_k, r, d_k, \beta_m, g_m$ known to the simulator, and an unknown value y_k (this is the actual value on the wire). However, *all* of these values are known to the distinguisher (since it knows the actual inputs, and also has the adversary's view) and thus it can know for certain which $[T]$ values should equal 0 and which should not. Thus, a simulation strategy where the simulator determines whether $[T]$ equals 0 with probability $1/|\mathbb{F}|$ if the adversary cheated will fail (since the distinguisher can verify if the value should actually be zero, depending on the given values). In the next section, we present a different strategy that solves this problem. In short, the strategy involves generating the linear combinations in Step 5b of Protocol 4.1 using *shared and secret* α_k and β_m values. Since these values are never revealed, the distinguisher cannot know if an actual $[T]$ should be 0 or not, and it suffices to simulate by choosing $[T]$ to equal 0 with probability $1/|\mathbb{F}|$ in the case that the adversary cheats.

5   A Protocol for Small Fields

Motivation. As discussed at the end of Sect. 4, Protocol 4.1 only works for large fields. In this section, we describe a protocol variant that works for any field size. The protocol is similar to Protocol 4.1 except that multiple randomizing shares and verifications are carried out. In particular, the parties generate δ random shares $[r_1], \ldots, [r_\delta]$ and then verify the correctness of all multiplications by generating δ independent random linear combinations as in Step 5b of Protocol 4.1 (each with a different r_i, and with independent α_k, β_m values). The main difference is that instead of α_k, β_m being *public* values generated by calls to $\mathcal{F}_{\text{coin}}$, they are random shares generated by calling $\mathcal{F}_{\text{rand}}$. Furthermore, they are kept secret and not opened. We show that this yields a cheating probability of at most $(3/|\mathbb{F}|)^\delta$, which can be made arbitrarily small by increasing δ. Since $\mathcal{F}_{\text{rand}}$ is somewhat more expensive than $\mathcal{F}_{\text{coin}}$ (see Sect. 6.2), Protocol 4.1 is better for large fields.

Secure sum of products. In order to implement the verification step with shared and secret α_k, β_m, it is necessary to compute the following linear combinations efficiently:

$$[u] = \sum_{k=1}^{N}[\alpha_k] \cdot [r \cdot z_k] + \sum_{m=1}^{M}[\beta_m] \cdot [r \cdot v_m] \quad \text{and} \quad [w] = \sum_{k=1}^{N}[\alpha_k] \cdot [z_k] + \sum_{m=1}^{M}[\beta_m] \cdot [v_m].$$

This seems to require an additional *four multiplications* (e.g., calls to $\mathcal{F}_{\text{mult}}$) per multiplication gate. Given that Protocol 4.1 requires only two calls to $\mathcal{F}_{\text{mult}}$ per multiplication gate overall, this seems to be considerably more expensive. In Sect. 6.1, we show how to compute a sum of products, for any number of terms, essentially at the cost of just a *single multiplication*. Our construction works for Shamir and replicated secret sharing, as we use in this paper. This subprotocol is of independent interest, and can be useful in many other scenarios. For example, in statistical computations, a sum-of-squares is often needed, and our method can be used to compute the sum-of-square of millions of values at the cost of just one multiplication. We formally define the sum-of-products functionality, denoted $\mathcal{F}_{\text{product}}$, in Functionality 5.1. It is very similar to $\mathcal{F}_{\text{mult}}$, with the exception that it receives two lists of values instead of a single pair. As with $\mathcal{F}_{\text{mult}}$, security is defined up to additive attacks.

FUNCTIONALITY 5.1 ($\mathcal{F}_{\text{product}}$ - **Sum-of-Products Up To Additive Attacks**)

1. Upon receiving $\{[x_i]_H\}_{i=1}^{\ell}$ and $\{[y_i]_H\}_{i=1}^{\ell}$ from the honest parties, the ideal functionality $\mathcal{F}_{\text{product}}$ computes x_i and y_i and the corrupted parties shares $[x_i]_C$ and $[y_i]_C$, for each $i \in \{1, \ldots, \ell\}$.
2. $\mathcal{F}_{\text{product}}$ hands $\{[x_i]_C\}_{i=1}^{\ell}$ and $\{[y_i]_C\}_{i=1}^{\ell}$ to the ideal-model adversary $\mathcal{S}$.
3. Upon receiving d and $\{\alpha_i\}_{i|P_i \in C}$ from $\mathcal{S}$, functionality $\mathcal{F}_{\text{product}}$ defines $z = \sum_{i=1}^{\ell} x_i \cdot y_i + d$ and $[z]_C = \{\alpha_i\}_{i|P_i \in C}$. Then, it runs $\mathsf{share}(z, [z]_C)$ to obtain a share z_j for each party P_j.
4. The ideal functionality $\mathcal{F}_{\text{product}}$ hands each honest party P_j its share z_j.

The protocol. We now proceed to describe the protocol. As we have described above, the protocol is very similar to Protocol 4.1 with the exception that the share randomization and verification are run δ times, and the linear combinations are computed using secret and shared α_k, β_m. The formal description of the protocol appears in Protocol 5.3. Observe that the computation of $[u_i]$ and $[w_i]$ in order to compute $[T_i]$ in Steps 6(c)i–6(c)iv in Protocol 5.3 is exactly the same as the computation of T in Step 5b Protocol 4.1. Namely, we obtain

$$[u_i] = \sum_{k=1}^{N} [\alpha_{k,i}] \cdot [r_i \cdot z_k] + \sum_{m=1}^{M} [\beta_{m,i}] \cdot [r_i \cdot v_m] \text{ and } [w_i] = \sum_{k=1}^{N} [\alpha_{k,i}] \cdot [z_k] + \sum_{m=1}^{M} [\beta_{m,i}] \cdot [v_m].$$

Thus, the intuition as to why $[T_i] = [u_i] - r_i \cdot [w_i]$ equals 0 with probability $3/|\mathbb{F}|$ is the same as in Protocol 4.1. Despite this, the proof is different since here $3/|\mathbb{F}|$ is noticeable, and this affects the simulation. As such, the proof of the protocol is similar to that of Protocol 4.1 with the exception that the simulator needs to compute the *exact probability* that each $T_i = 0$, depending on the different cases of possible additive attacks. This is due to the fact that some T_i may equal 0 with probability $1/|\mathbb{F}|$ even when an additive attack does take place. Unlike the case of large fields, $1/|\mathbb{F}|$ may be noticeable and thus the simulation cannot afford to just fail in such cases. As we will see in the proof, if the adversary cheats in a multiplication gate, then each $T_i = 0$ with probability at most $3/|\mathbb{F}|$, and so all $T_i = 0$ with probability at most $(3/|\mathbb{F}|)^\delta \leq 2^{-\sigma}$, as required. Thus, the adversary cannot cheat undetected with probability greater than $2^{-\sigma}$. Nevertheless, the simulation of when $T_i = 0$ and when $T_i \neq 0$ is needed to show that revealing this fact does not leak any information about the real input.

Theorem 5.2. *Let σ be a statistical security parameter, let $\mathbb{F}$ be a finite field, and let f be a n-party functionality over $\mathbb{F}$. Then, Protocol 5.3 securely computes f with abort in the $(\mathcal{F}_{\text{input}}, \mathcal{F}_{\text{mult}}, \mathcal{F}_{\text{coin}}, \mathcal{F}_{\text{rand}}, \mathcal{F}_{\text{checkZero}}, \mathcal{F}_{\text{product}})$-hybrid model with statistical error $2^{-\sigma}$, in the presence of a malicious adversary controlling $t < \frac{n}{2}$ parties.*

Proof: We have already described the intuition behind the proof, and so proceed directly. Let $\mathcal{A}$ be the real adversary; we construct the ideal adversary/simulator $\mathcal{S}$ as follows. The simulation up to the verification stage is almost identical to the simulator in the proof of Theorem 4.3 for Protocol 4.1, with appropriate differences for the fact that the randomization is carried out δ times.

PROTOCOL 5.3 (Computing Arithmetic Circuits Over Any Finite $\mathbb{F}$)

Inputs: Each party P_j ($j \in \{1, \ldots, n\}$) holds an input $x_j \in \mathbb{F}^\ell$.

Auxiliary Input: The parties hold the description of a finite field $\mathbb{F}$ and an arithmetic circuit C over $\mathbb{F}$ that computes f on inputs of length $M = \ell \cdot n$. Let N be the number of multiplication gates in C.

The protocol:

1. *Parameter computation:* Set δ to be the smallest value for which $\delta \geq \frac{\sigma}{\log(|\mathbb{F}|/3)}$.

2. *Secret sharing the inputs:*
   (a) For each input v_i held by party P_j, party P_j sends v_i to $\mathcal{F}_{\text{input}}$.
   (b) Each party P_j records its vector of shares $(v_1^j, \ldots, v_M^j)$ of all inputs, as received from $\mathcal{F}_{\text{input}}$. If a party received $\perp$ from $\mathcal{F}_{\text{input}}$, then it sends **abort** to the other parties and halts.

3. *Generate randomizing shares:* For $i = 1$ to δ, the parties call $\mathcal{F}_{\text{rand}}$ to receive a sharing $[r_i]$.

4. *Randomization of inputs:* For each input wire sharing $[v_m]$ (where $m \in \{1, \ldots, M\}$) and for every $i = 1, \ldots, \delta$, the parties call $\mathcal{F}_{\text{mult}}$ on $[r_i]$ and $[v_m]$ to receive $[r_i \cdot v_m]$.

5. *Circuit emulation:* Let $G_1, \ldots, G_N$ be a predetermined topological ordering of the gates of the circuit. For $k = 1, \ldots, N$ the parties work as follows:
   - G_k *is an addition gate:* Given tuples $([x], [r_1 \cdot x], \ldots, [r_\delta \cdot x])$ and $([y], [r_1 \cdot y], \ldots, [r_\delta \cdot y])$ on the *left* and *right* input wires respectively, the parties locally compute $([x+y], [r_1 \cdot (x+y)], \ldots, [r_\delta \cdot (x+y)])$.
   - G_k *is a multiplication-by-a-constant gate:* Given a constant $a \in \mathbb{F}$ and tuple $([x], [r_1 \cdot x], \ldots, [r_\delta \cdot x])$ on the input wire, the parties locally compute $([a \cdot x], [r_1 \cdot (a \cdot x)], \ldots, [r_\delta \cdot (a \cdot x)])$.
   - G_k *is a multiplication gate:* Given tuples $([x], [r_1 \cdot x], \ldots, [r_\delta \cdot x])$ and $([y], [r_1 \cdot y], \ldots, [r_\delta \cdot y])$ on the *left* and *right* input wires respectively:
     (a) The parties call $\mathcal{F}_{\text{mult}}$ on $[x]$ and $[y]$ to receive $[x \cdot y]$.
     (b) For $i = 1$ to δ, the parties call $\mathcal{F}_{\text{mult}}$ on $[r_i \cdot x]$ and $[y]$ to receive $[r_i \cdot x \cdot y]$.

6. *Verification stage:* Let $\{([z_k], [r_1 \cdot z_k], \ldots, [r_\delta \cdot z_k])\}_{k=1}^N$ be the tuples on the output wires of all multiplication gates and let $\{([\beta_{m,1}], \ldots, [\beta_{m,\delta}])\}_{m=1}^M$ be the tuples on the input wires of the circuit.
   (a) For $m = 1, \ldots, M$, the parties call $\mathcal{F}_{\text{rand}}$ to receive $[\beta_{m,1}], \ldots, [\beta_{m,\delta}]$.
   (b) For $k = 1, \ldots, N$, the parties call $\mathcal{F}_{\text{rand}}$ to receive $[\alpha_{k,1}], \ldots, [\alpha_{k,\delta}]$.
   (c) *Compute linear combinations:* For $i = 1, \ldots, \delta$:
      i. The parties call $\mathcal{F}_{\text{product}}$ on vectors $([\alpha_{1,i}], \ldots, [\alpha_{N,i}], [\beta_{1,i}], \ldots, [\beta_{M,i}])$ and $([r_i \cdot z_1], \ldots, [r_i \cdot z_N], [r_i \cdot v_1], \ldots, [r_i \cdot v_M])$ to receive $[u_i]$.
      ii. The parties call $\mathcal{F}_{\text{product}}$ on vectors $([\alpha_{1,i}], \ldots, [\alpha_{N,i}], [\beta_{1,i}], \ldots, [\beta_{M,i}])$ and $([z_1], \ldots, [z_N], [v_1], \ldots, [v_M])$ to receive $[w_i]$.
      iii. The parties run $\text{open}([r_i])$ to receive r_i.
      iv. Each party locally computes $[T_i] = [u_i] - r_i \cdot [w_i]$.
      v. The parties call $\mathcal{F}_{\text{checkZero}}$ on $[T_i]$. If $\mathcal{F}_{\text{checkZero}}$ outputs **reject**, the parties output $\perp$ and abort. Else, if it outputs **accept**, they proceed.

7. *Output reconstruction:* For each output wire of the circuit, the parties run $\text{reconstruct}([v], j)$, where $[v]$ is the sharing of the value on the output wire, and P_j is the party whose output is on the wire.
   If a party received $\perp$ in any call to the reconstruct procedure, then it sends $\perp$ to the other parties, outputs $\perp$ and halts.

Output: If a party has not aborted, it outputs the values received on its output wires.

We now show how to simulate the verification step. As in the proof of Theorem 4.3, the simulator $\mathcal{S}$ chooses $r_1, \ldots, r_\delta \in \mathbb{F}$ at random, and generates all shares by computing $(r_i^1, \ldots, r_i^n) = \mathsf{share}(r_i, [r_i]_C)$, for every $i = 1, \ldots, \delta$. Next, $\mathcal{S}$ simulates $\delta \cdot (N + M)$ calls to $\mathcal{F}_{\mathsf{rand}}$ used to obtain all of the $\beta_{m,i}$ and $\alpha_{k,i}$ values. Now, for every $i = 1, \ldots, \delta$, $\mathcal{S}$ works as follows:

1. $\mathcal{S}$ simulates two invocations of $\mathcal{F}_{\mathsf{product}}$ with $\mathcal{A}$, receiving $d_{i,1}$ and $d_{i,2}$, respectively, as the additive attack of $\mathcal{A}$ in these invocations.
2. $\mathcal{S}$ simulates the opening of r_i by handing $\mathcal{A}$ all of the honest parties' shares as computed above. If any honest party would abort due to the opening values sent by $\mathcal{A}$ ($\mathcal{S}$ knows whether this would happen since it has all the honest parties' shares), then $\mathcal{S}$ simulates the honest party sending $\perp$ to all parties, externally sends abort_j for every $P_j \in H$ to the trusted party computing f, and halts.
3. $\mathcal{S}$ simulates $\mathcal{F}_{\mathsf{checkZero}}$, determining the value of T_i to be equal or not equal to zero, based on the process described below. If $T_i \neq 0$, then $\mathcal{S}$ simulates an abort, as in the proof of Theorem 4.3. Else, $\mathcal{S}$ proceeds with the simulation.

If $\mathcal{A}$ carried out an additive attack when calling $\mathcal{F}_{\mathsf{mult}}$ with $[x]$ and $[y]$ on a wire (i.e., the actual value multiplication and not the randomization), and yet all $\mathcal{F}_{\mathsf{checkZero}}$ simulations return accept (either because $T_i = 0$ or because $\mathcal{F}_{\mathsf{checkZero}}$ returns accept with probability $1/|\mathbb{F}|$ even when $T_i \neq 0$), then $\mathcal{S}$ outputs fail.

If $\mathcal{S}$ did not halt, then it concludes the output reconstruction as in the proof of Theorem 4.3.

It remains to show how $\mathcal{S}$ determines the value of T_i as equal or not equal to zero, for each $i = 1, \ldots, \delta$, and to show that this is the same distribution as in a real execution. Fix i, and let $d_k, e_{k,i}, f_{k,i}, g_{m_i}$ be as in the proof of Lemma 4.2 (the additional subscript of i for $e_{k,i}, f_{k,i}, g_{m,i}$ is due to the fact that there are separate $\mathcal{F}_{\mathsf{mult}}$ calls for each randomization multiplication; i.e., for each $i = 1, \ldots, \delta$ and the associated r_i). $\mathcal{S}$ determines the probability that $T_i = 0$ based on the following mutually-exclusive cases:

1. *Case 1 – there exists an $m \in \{1, \ldots, M\}$ such that $g_{m,i} \neq 0$:* In this case, $\mathcal{S}$ sets $T_i = 0$ with probability $1/|\mathbb{F}|$ exactly.
2. *Case 2 – $g_{m,i} = 0$ for all $m \in \{1, \ldots, M\}$ and $d_k = 0$ for all $k \in \{1, \ldots, N\}$, but there exists some $k \in \{1, \ldots, N\}$ for which $f_{k,i} \neq 0$:* As in the previous case, in this case $\mathcal{S}$ sets $T_i = 0$ with probability $1/|\mathbb{F}|$ exactly.
3. *Case 3 – $g_{m,i} = 0$ for all $m \in \{1, \ldots, M\}$ and for all $k \in \{1, \ldots, N\}$ it holds that $f_{k,i} - r_i \cdot d_k = 0$:* In this case, $\mathcal{S}$ sets $T_i = d_{i,1} - r_i \cdot d_{i,2}$ with probability 1. (Note that this case includes cases that some $d_k, f_{k,i} \neq 0$ and it happens that $f_{k,i} - r_i \cdot d_k = 0$, as well as the case that all $d_k, f_{k,i}$ equal 0 and so $\mathcal{A}$ did not cheat.)
4. *Case 4 – $g_{m,i} = 0$ for all $m \in \{1, \ldots, M\}$ and there exists a $k \in \{1, \ldots, N\}$ such that $d_k \neq 0$ and $f_{k,i} - r_i \cdot d_k \neq 0$:* In this case, $\mathcal{S}$ sets $T_i = 0$ with probability $1/|\mathbb{F}|$ exactly.

Observe that $\mathcal{S}$ knows all the additive values, and uses the random choice of r_i above, and so can determine all of the above cases. In addition, observe that this covers all possible cases.

We now analyze all of the above cases and show that the distribution over the zero/non-zero value of T_i generated by $\mathcal{S}$ is identical to that of a real execution. As in Eq. (1) in the proof of Lemma 4.2, we have that

$$\mathsf{val}([T_i])_H = \sum_{k=1}^{N} \alpha_{k,i} \cdot (e_{k,i} \cdot y_k + f_{k,i} - r_i \cdot d_k) + \sum_{m=1}^{M} \beta_{m,i} \cdot g_{m,i} + d_{i,1} - r_i \cdot d_{i,2}. \quad (3)$$

We use this to analyze the cases:

1. *Case 1:* Let $m_0 \in \{1, \ldots, M\}$ be such that $g_{m_0,i} \neq 0$. By Eq. (3) we have $\mathsf{val}([T_i])_H = 0$ if and only if $\beta_{m_0,i} = \big(-\sum_{k=1}^{N} \alpha_{k,i} \cdot (e_{k,i} \cdot y_k + f_{k,i} - r_i \cdot d_k) - \sum_{\substack{m=1 \\ m \neq m_0}}^{M} \beta_{m,i} \cdot g_{m,i} - d_{i,1} + r_i \cdot d_{i,2} \big) \cdot g_{m_0,i}^{-1}$. By the uniform choice of $\beta_{m_0,i}$, this holds in a real execution with probability $1/|\mathbb{F}|$ exactly.

2. *Case 2:* Let $k_0 \in \{1, \ldots, N\}$ be such that $f_{k_0,i} \neq 0$. As above, $\mathsf{val}([T_i])_H = 0$ if and only if $\alpha_{k_0,i} \cdot (f_{k_0,i} - r_i \cdot d_{k_0}) = -\sum_{\substack{k=1 \\ k \neq k_0}}^{N} \alpha_{k,i} \cdot (e_{k,i} \cdot y_k + f_{k,i} - r_i \cdot d_k) - d_{i,1} + r_i \cdot d_{i,2}$, but since all $d_k = 0$ we have $\alpha_{k_0,i} \cdot (f_{k_0,i} - r_i \cdot d_{k_0}) = \alpha_{k_0,i} \cdot f_{k_0,i}$ and so $\mathsf{val}([T_i])_H = 0$ if and only if $\alpha_{k_0,i} = \big(-\sum_{\substack{k=1 \\ k \neq k_0}}^{N} \alpha_{k,i} \cdot (e_{k,i} \cdot y_k + f_{k,i} - r_i \cdot d_k) - d_{i,1} + r_i \cdot d_{i,2} \big) \cdot f_{k_0,i}^{-1}$. As in the previous case, by the uniform choice of $\alpha_{k_0,i}$, this holds in a real execution with probability $1/|\mathbb{F}|$ exactly.

3. *Case 3:* In this case, all $g_{m,i} = 0$, and all $f_{k,i} - r_i \cdot d_k = 0$. If this occurs since all $f_{k,i} = 0$ and all $d_k = 0$, then clearly $T_i = d_{i,1} - r_i \cdot d_{i,2}$ since $\mathcal{A}$ did not cheat during the circuit emulation step. Otherwise, assume that for all $f_{k,i}, d_k \neq 0$ it holds that $f_{k,i} - r_i \cdot d_k = 0$. The computation of multiplication gate G_k involves two calls to $\mathcal{F}_{\mathsf{mult}}$: one with x_k and y_k, and the other with $r_i \cdot x_k$ and y_k. By the definition of $\mathcal{F}_{\mathsf{mult}}$ and the values $d_k, f_{k,i}$, the output of the first call to $\mathcal{F}_{\mathsf{mult}}$ is $z_k = x_k \cdot y_k + d_k$, and the output of the second call to $\mathcal{F}_{\mathsf{mult}}$ is $z_k' = r_i \cdot x_k \cdot y_k + f_{k,i}$. Writing $x_k \cdot y_k = z_k - d_k$, we have that $z_k' = r_i \cdot (z_k - d_k) + f_{k,i} = r_i \cdot z_k - r_i \cdot d_k + f_{k,i}$. However, by this case assumption, $f_{k,i} - r_i \cdot d_k = 0$ and so $z_k' = r_i \cdot z_k$. This means that the invariant of the relation between the real and randomized values on the wires is maintained, and formally that the kth term in the sum for T_i equals zero. Since this holds for all $k \in \{1, \ldots, N\}$, we have that $T_i = d_{i,1} - r_i \cdot d_{i,2}$ with probability 1, as determined by the simulator. (We remark that there is no accumulated error $e_{k,i}$ in this case, since $e_{k,i}$ appears when the invariant on the wires is *not* preserved.)

4. *Case 4:* Let k_0 be the first $k \in \{1, \ldots, N\}$ for which $f_{k,i} - r_i \cdot d_k \neq 0$. Since this is the first such k, it holds that $e_{k_0,i} = 0$ (note that some previous $d_k, f_{k,i}$ may be non-zero, but as we saw in the previous case, if $f_{k,i} - r_i \cdot d_k = 0$ then there is no accumulated error). As in case 2, we have that $\mathsf{val}([T_i])_H = 0$ if and only if $\alpha_{k_0,i} = \big(-\sum_{\substack{k=1 \\ k \neq k_0}}^{N} \alpha_{k,i} \cdot (e_{k,i} \cdot y_k + f_{k,i} - r_i \cdot d_k) - d_{i,1} + r_i \cdot d_{i,2} \big) \cdot (f_{k_0,i} - r_i \cdot d_{k_0})^{-1}$. where division by $f_{k_0,i} - r_i \cdot d_{k_0}$ is possible since this value is non-zero. As above, this equality holds with probability exactly $1/|\mathbb{F}|$, by the uniform choice of $\alpha_{k_0,i}$.

The above demonstrates that the simulation by $\mathcal{S}$ of the zero/non-zero value of T_i is identical to a real execution. Furthermore, since the actual values of $\alpha_{k,i}, \beta_{m,i}$ are never revealed in this protocol, the simulation only requires that the probability that T_i is zero/non-zero be the same as in a real execution.[2]

It remains to show that $\mathcal{S}$ outputs fail with probability at most $\left(\frac{3}{|\mathbb{F}|}\right)^\delta$, which is at most $2^{-\sigma}$ by the choice of δ in the protocol. Recall that $\mathcal{S}$ outputs fail if and only if there exists some $d_k \neq 0$ and yet all $\mathcal{F}_{\text{checkZero}}$ invocations return accept in the simulation. This is indeed a fail, since the outputs received by the honest parties in the real and ideal executions in this case would be different. Now, assume that $d_k \neq 0$ for some $k \in \{1, \ldots, N\}$. Then, for *every* i, the simulation case is either Case 3 or Case 4, where the actual case depends on the value of r_i chosen. $\mathcal{F}_{\text{checkZero}}$ returns accept in the ith invocation in the simulation if either **(a)** case 3 occurs, meaning that $f_{k,i} - r_i \cdot d_k = 0$ which is equivalent to $r_i = f_{k,i}/d_k$, or **(b)** case 4 occurs and $\alpha_{k,i}$ results in $T_i = 0$, or **(c)** Case 4 occurs and $T_i \neq 0$ but $\mathcal{F}_{\text{checkZero}}$ returns accept nevertheless. The probability that accept is received from $\mathcal{F}_{\text{checkZero}}$ for any given i equals the probability that one of **(a)**, **(b)** or **(c)** occur. Each one independently occurs with probability $1/|\mathbb{F}|$: **(a)** because of the random choice of r_i, **(b)** because of the random choice of $\alpha_{k,i}$, and **(c)** because of the $1/|\mathbb{F}|$ probability that $\mathcal{F}_{\text{checkZero}}$ returns accept on non-zero input. By the union bound, the probability that *one* of these occur is therefore upper bound by $3/|\mathbb{F}|$. We conclude by noting that the above holds *independently* for each $i \in \{1, \ldots, \delta\}$, and thus the probability that $\mathcal{F}_{\text{checkZero}}$ returns accept for *all* $i \in \{1, \ldots, \delta\}$ is upper bound by $(3/|\mathbb{F}|)^\delta$, as required. ∎

Concrete efficiency. We analyze the performance of our protocol. The main difference compared to Protocol 4.1 is that functionality $\mathcal{F}_{\text{mult}}$ is called δ times for every input wire and $1 + \delta$ for every multiplication gate (once for multiplying $[x]$ and $[y]$, and δ additional times for multiplying $[r_i \cdot x]$ with $[y]$). Thus, the overall number of multiplications is $(\delta \cdot M + (1+\delta) \cdot N) \cdot \mathcal{F}_{\text{mult}}$, where M denotes the number of inputs and N the number of multiplication gates. Another difference is that now there are δ calls to $\mathcal{F}_{\text{rand}}$ for generating $[r_i]$, and $\delta \cdot (M + N)$ calls for generating all the $\alpha_{k,i}, \beta_{m,i}$ values (which are secret here, unlike in Protocol 4.1). In addition, there are 2δ calls to $\mathcal{F}_{\text{product}}$, δ calls to open for $[r_i]$, δ calls to $\mathcal{F}_{\text{checkZero}}$ (each of which reduces to one call to $\mathcal{F}_{\text{rand}}$, one $\mathcal{F}_{\text{mult}}$ and one opening), and $M + L$ calls to reconstruct as part of $\mathcal{F}_{\text{input}}$ and obtaining output (where L equals the number of output wires). Assuming that $\mathcal{F}_{\text{product}}$ is equivalent to $\mathcal{F}_{\text{mult}}$ (as will be shown in Sect. 6.2), we have that the overall exact cost of the protocol is

$$(\delta \cdot M + (1 + \delta) \cdot N + 3\delta) \cdot \mathcal{F}_{\text{mult}} + (\delta \cdot (M + N) + 2\delta) \cdot \mathcal{F}_{\text{rand}} + (M + L) \cdot \text{reconstruct} + 2\delta \cdot \text{open}.$$

---

[2] If $\beta_{m_0,i}$ were to be revealed, as in Protocol 4.1 for large fields, then the question of whether the equation holds is something that the distinguisher could determine (since it knows all of the y_k values from the input, and it can receive all of the $d_k, e_{k,i}, f_{k,i}, g_{m,i}$ values from the adversary). Thus, it would *not* suffice to set $T_i = 0$ with the correct probability but as a function of the actual values. However, $\mathcal{S}$ does not know the y_k values and so could not determine this.

Amortizing over the size of the circuit, we have that the average cost is $(1 + \delta) \cdot \mathcal{F}_{\text{mult}} + \delta \cdot \mathcal{F}_{\text{rand}}$ per multiplication gate.

We compare now the cost of running Protocol 5.3 with $\delta = 1$ to the cost of running Protocol 4.1 for large fields. The amortized cost of Protocol 4.1 is $2 \cdot \mathcal{F}_{\text{mult}}$ per multiplication gate, whereas the cost of Protocol 5.3 with $\delta = 1$ is $2 \cdot \mathcal{F}_{\text{mult}} + 1 \cdot \mathcal{F}_{\text{rand}}$. Thus, the difference between these protocols depends on the cost of $\mathcal{F}_{\text{rand}}$. As we will see in Sect. 6.2, the cost of $\mathcal{F}_{\text{rand}}$ for our specific instantiation for a not-small number of parties is about a third of the cost of $\mathcal{F}_{\text{mult}}$, making Protocol 5.3 about 17% slower.

It is also instructive to compare the cost of running Protocol 4.1 with a large field versus running Protocol 5.3 with a smaller field. Concretely, assume that the computation being carried out is over the integers, and that all values are smaller than 2^{30}, and that security 2^{-60} is desired. Then, the question that may arise is whether one should run Protocol 4.1 over a 60-bit field, or whether one should run Protocol 5.3 with $\delta = 2$ over a 30-bit field. The amortized cost is $2 \cdot \mathcal{F}_{\text{mult}}$ for Protocol 4.1 versus $3 \cdot \mathcal{F}_{\text{mult}} + 1 \cdot 2 \cdot \mathcal{F}_{\text{rand}} \approx 3.66 \cdot \mathcal{F}_{\text{mult}}$ for Protocol 5.3 (assuming the cost of $\mathcal{F}_{\text{rand}}$ to be one-third of $\mathcal{F}_{\text{mult}}$). Clearly, the communication cost is double for a 60-bit field, and so the expected communication using Protocol 5.3 is lower in such a case. Regarding computation, empirical experimentation is needed to make a comparison.

Reducing memory. As in Protocol 4.1, when the circuit is huge, it is highly undesirable to store all values until completion in order to carry out the verification. Thus, in such cases, it is preferable to compute the verification while evaluating the circuit. However, Protocol 4.1 required running a full verification at intermediate steps to do this, and this incurred additional work to rerandomize the active wires for the next phase, and so on (see the discussion at the end of Sect. 4). In contrast, Protocol 5.3 is much more amenable to verification-on-the-fly because the α, β values are never revealed. Thus, it is possible to call $\mathcal{F}_{\text{rand}}$ to obtain the $[\beta_{m,i}]$ shares at the input phase, and to call $\mathcal{F}_{\text{rand}}$ to obtain $[\alpha_{k,i}]$ shares during multiplications. Then, the parties can locally store the partial sums for u_i and w_i, and all previous shares that are no longer needed for the circuit evaluation can be discarded. This method for verification-on-the-fly is also very easy to implement.

Reactive computation. In Protocol 4.1 where the α, β values are public, it is necessary to open $[r]$ in order to compute $[T] = [u] - r \cdot [w]$. This is because otherwise the adversary can input an additive value in the multiplication $[r] \cdot [w]$ that can cancel out a previous error (note that at this stage, α, β are already known and so the adversary has enough information to make the errors cancel). In contrast, in Protocol 5.3, the α, β values are never revealed. Thus, it is not necessary to open the $[r_i]$ shares, and the parties can compute $[T_i] = [u_i] - [r_i] \cdot [w_i]$, using $\mathcal{F}_{\text{mult}}$ with $[r_i]$ and $[w_i]$. In a regular one-off computation, this makes no real difference. However, in the case of reactive computation, where outputs are revealed, and the computation continues, it is undesirable to open the $[r_i]$ shares, since new randomization is necessary. Thus, in such cases, one can leave the $[r_i]$ shares secret, and compute $[T_i]$ using $\mathcal{F}_{\text{mult}}$ as described.

6 Instantiations and Experimental Results

Our protocol is generic and can be instantiated in many ways (with different secret sharing schemes, multiplication protocols, and more). Clearly, the efficiency of our protocol depends significantly on the instantiations. In order to demonstrate the efficiency of our protocol, we plug in the instantiations presented in [22], which meet all of our requirements. We consider two secret sharing schemes: replicated secret sharing for 3 parties, and Shamir sharing [25] for any number of parties. Recall that our protocol requires instantiations for functionalities $\mathcal{F}_{\mathsf{mult}}$ and $\mathcal{F}_{\mathsf{rand}}$, and for procedures open and reconstruct ($\mathcal{F}_{\mathsf{input}}$, $\mathcal{F}_{\mathsf{coin}}$ and $\mathcal{F}_{\mathsf{checkZero}}$ are constructed generically using these functionalities and procedures). We also need to show how to securely realize $\mathcal{F}_{\mathsf{product}}$ for Protocol 5.3; we begin by showing this in Sect. 6.1. Then, in Sect. 6.2 we present the concrete costs of the instantiations from [22] along with $\mathcal{F}_{\mathsf{product}}$ from Sect. 6.1. Finally, in Sect. 6.3, we present experimental results of the implementation of our protocol and compare it to prior work. In the full version of this paper, we describe the protocols for the instantiation based on the Shamir sharing, including proofs that the protocols securely compute $\mathcal{F}_{\mathsf{rand}}$ and $\mathcal{F}_{\mathsf{mult}}$.

6.1 Securely Realizing Functionality 5.1 – $\mathcal{F}_{\mathsf{product}}$

$\mathcal{F}_{\mathsf{product}}$ **with Shamir secret sharing.** We begin by describing how to securely realize $\mathcal{F}_{\mathsf{product}}$ when Shamir sharing is used. Let $[x_1], \ldots, [x_\ell]$ and $[y_1], \ldots, [y_\ell]$ be two vectors of inputs, where the parties wish to compute shares of $\sum_{i=1}^{\ell} x_i \cdot y_i$. The key observation here is that most (if not all) protocols for multiplication based on Shamir sharing have two phases:

1. *Local multiplication:* In this phase, each party locally multiplies its shares on the two values. This yields a sharing of the product of the two values on a degree-$2t$ polynomial. Since $t < n/2$, there is enough "information" to reconstruct the polynomial (since $2t < n$).
2. *Degree reduction:* In the second phase, the parties run an interactive protocol that reduces the degree of the polynomial generated in the previous step back to degree-t, without changing its constant term.

Observe that the protocols of [5,12,18] and others all follow this framework.

The crucial observation regarding how to compute $\mathcal{F}_{\mathsf{product}}$ is that the parties can begin by locally computing the sum of the products of their input shares. Specifically, denote P_j share of x_i and y_i by x_i^j and y_i^j, respectively. Then, each P_j can locally compute $z_j = \sum_{i=1}^{\ell} x_i^j \cdot y_i^j$, and the shares $z_1, \ldots, z_n$ constitute a sharing of degree-$2t$ polynomial with constant-term $\sum_{i=1}^{\ell} x_i \cdot y_i$. All that therefore remains is for the parties to run the *degree reduction* on these shares, and they obtain a good Shamir sharing of the sum of products.

The above strategy securely computes $\mathcal{F}_{\mathsf{product}}$ if the degree reduction phase of the protocol has the property that the only attack possible by the adversary is an additive attack. That is, if the input shares define a degree-$2t$ polynomial hiding the secret z, then the adversary can cause the parties to output a degree-t

sharing of $z + d$ where d can be extracted by a simulator (exactly as in $\mathcal{F}_{\text{mult}}$). In the full version of this paper, we show that this property holds for the semi-honest multiplication protocol of [12].

$\mathcal{F}_{\text{product}}$ **with replicated secret sharing.** In the multiplication protocol of [2] which is also shown to be secure up to an additive attack in [22]), the parties first locally compute a sum of 3 products of their local shares (given replicated shares (s_i, s_{i+1}) and (t_i, t_{i+1}) of two values s and t held by P_i, each party computes $u_i = s_i \cdot t_i + s_{i+1} \cdot t_i + s_i \cdot t_{i+1}$). Then, in the next step, each party sends its share u_i – randomized using correlated randomness – to party P_{i+1}, who defines the pair (u_{i+1}, u_i) as its share of the output. The simple observation here is that if each party computes many u_i's for each product in the vector and then sums them all together, the result will be a replicated secret sharing of the entire sum of products.

Efficiency. Using the above method, the cost of a sum of products for *any number of terms* is local operations on the vector (similar to addition gates in the circuit) and interaction equivalent to a *single multiplication*. Thus, $\mathcal{F}_{\text{product}}$ essentially costs the same as $\mathcal{F}_{\text{mult}}$.

Applications. Beyond the use of $\mathcal{F}_{\text{product}}$ in Protocol 5.3 for computing the random linear combinations, this subprotocol can be used to significantly speed up many secure statistical operations. For example, in order to securely computed the standard deviation over a large list, the main cost is computing the sum of squares (of the difference between each item and the mean), and then dividing by the length of the vector. Using our method, this can be carried out on millions of data items at the cost of a *single multiplication* followed by a *single division* (and if the number of data items is known, then the division can be carried out on the result).

6.2 Instantiations from [22] and their Cost

As we have discussed above, we present the cost of $\mathcal{F}_{\text{mult}}$, $\mathcal{F}_{\text{rand}}$, open and reconstruct for Shamir sharing (for any number of parties n) and for replicated secret sharing (for 3 parties), as described in [22]. The communication costs are presented in Table 1. In the two instantiations we consider for Shamir sharing, $\mathcal{F}_{\text{rand}}$ can be instantiated using PRSS [10] which has zero communication cost but has computation that is exponential in the number of parties and so is only good for up to 7 or so parties (as shown in [22]). In addition, $\mathcal{F}_{\text{rand}}$ can be instantiated using VAN, which is the hyper-invertible matrices method of [4]

Table 1. The communication cost per party for instantiations in [22], written as the number of *field elements* sent.

	$\mathcal{F}_{\text{mult}}$	$\mathcal{F}_{\text{rand}}$	open	reconstruct
Replicated secret sharing (three parties)	1	0	4	2
Shamir sharing (few parties), $\mathcal{F}_{\text{rand}}$ with PRSS	6	0	$n-1$	1
Shamir sharing (many parties), $\mathcal{F}_{\text{rand}}$ with VAN	6	2	$n-1$	1

that utilizes Vandermonde matrices. In both cases, Shamir sharing uses the DN multiplication protocol of [12].

Table 1 counts the communication costs of each protocol instantiation. The computational costs are low overall (since we use secret-sharing based primitives), except for PRSS which is exponential in the number of parties and thus only suitable for a small number. In Sect. 6.3, we show concrete running times for the replicated secret sharing and Shamir-sharing with VAN instantiations.

Overall protocol costs. As shown in Sects. 4 and 5, the cost per multiplication gate of Protocol 4.1 is $2 \cdot \mathcal{F}_{\mathsf{mult}}$, and the cost per multiplication gate of Protocol 5.3 is $(1+\delta) \cdot \mathcal{F}_{\mathsf{mult}} + \delta \cdot \mathcal{F}_{\mathsf{rand}}$. Plugging these into the above instantiations, we obtain a maliciously secured protocol for three-parties that requires each party to send only *2 field elements per multiplication gate* when the filed is large. For the multi-party setting, we obtain a protocol with a communication cost of only *12 field elements per multiplication gate* for each party when the field is large. This is shown in Table 2, including a comparison to the cost of the protocol of [22].

Table 2. The communication cost per party for the instantiations in Table 1 and the protocol of [22], written as the number of *field elements* sent *per multiplication gate*. (Note that Protocol 5.3 with $\delta = 2$ has smaller field elements and thus more elements sent could actually mean less bandwidth.)

	Protocol of [22] with $\delta = 1$	Protocol 4.1 (large field)	Protocol 5.3 with $\delta = 1$	Protocol 5.3 with $\delta = 2$
Replicated secret sharing (three parties)	4	2	2	3
Shamir (few parties), $\mathcal{F}_{\mathsf{rand}}$ with PRSS	36	12	12	18
Shamir (many parties), $\mathcal{F}_{\mathsf{rand}}$ with VAN	42	12	14	22

6.3   Experimental Results

We implemented Protocol 4.1 with two instantiations: replicated secret sharing for 3 parties and Shamir sharing using VAN for $\mathcal{F}_{\mathsf{rand}}$ and DN [12] for $\mathcal{F}_{\mathsf{mult}}$ (see Sect. 6.2). The field we used for all our experiments was the 61-bit Mersenne field (and so security is approximately 2^{-60}). We ran our protocols for different numbers of parties on a series of circuits of different depths, each with 1,000,000 multiplication gates, 1,000 inputs wires, 50 output wires. The circuits had 4 different depths: 20, 100, 1,000 and 10,000. The experiment was run on AWS in two configurations: a *LAN network configuration* in a single AWS region (specifically, North Virginia), and a *WAN network configuration* in three AWS regions (specifcally, North Virginia, Germany and India). Each party was run in an independent AWS C4.large instance (2-core Intel Xeon E5-2666 v3 with 2.9 GHz clock speed and 3.75 GB RAM). Each execution (configuration, number of parties, circuit) was run 5 times, and the result reported is the average run-time (Table 3).

Table 3. *LAN configuration* execution times in milliseconds of a circuit with 1,000,000 multiplication gates, for different depths. The first column gives the running time for the replicated secret sharing version; all other columns are the Shamir sharing for different numbers of parties.

Circuit Depth	3 (replicated)	3	5	7	9	11	30	50	70	90	110
20	319	826	844	1,058	1,311	1,377	2,769	4,053	5,295	6,586	8,281
100	323	842	989	1,154	1,410	1,477	3,760	6,052	8,106	11,457	15,431
1,000	424	1,340	1,704	1,851	2,243	2,887	12,144	26,310	33,294	48,927	79,728
10,000	1,631	6,883	7,424	8,504	12,238	16,394	61,856	132,160	296,047	411,195	544,525

Table 4. *LAN configuration* execution times in milliseconds of a circuit with 1,000,000 multiplication gates and depth 20. The times for [22] are for the best protocol for the number of parties.

	3 (replicated)	3	5	7	9	11	30	50	70	90	110
Protocol 4.1	319	826	844	1,058	1,311	1,377	2,769	4,053	5,295	6,586	8,281
Protocol of [22]	513	1,229	1,890	3,056	4,009	5,187	15,954	28,978	44,599	58,966	72,096
Speedup	161%	149%	224%	289%	306%	377%	576%	715%	842%	895%	871%

In order to compare our protocol to that of [22], we compare the running times in a LAN configuration for depth 20 (this is because that is the only configuration run by them); see Table 4.

As can be seen, our protocol outperforms the best protocol of [22] significantly, even for a small number of parties. However, as the number of parties increases, the gap widens. Observe that the communication difference between the protocols, as shown in Table 2 would only predict that our protocol would run 3 times faster than that of [22], whereas experiment yield an almost 10 times faster result for a large number of parties. This may be due to additional computational work involved in generating the Beaver triples in [22].

Finally, in Table 5, we present the experimental results of running our protocol in the WAN configuration. Due to the many rounds of communication, the results are significantly slower, but demonstrate that it is even possible to run for quite a large number of parties (e.g., 50 parties) with reasonable time.

Table 5. *WAN configuration* (North Virginia, Germany and India) execution times in milliseconds of a circuit with 1,000,000 multiplication gates, for different depths.

Circuit Depth	3 (replicated)	3	5	7	9	11	30	50
20	3502	20,492	27,772	28,955	24,482	24,729	87,355	128,366
100	10,712	45,250	53,872	50,719	55,716	56,482	134,860	197,321

References

1. Araki, T., Barak, A., Furukawa, J., Lichter, T., Lindell, Y., Nof, A., Ohara, K., Watzman, A., Weinstein, O.: Optimized honest-majority MPC for malicious adversaries - breaking the 1 billion-gate per second barrier. In: The IEEE S&P (2017)
2. Araki, T., Furukawa, J., Lindell, Y., Nof, A., Ohara, K.: High-throughput semi-honest secure three-party computation with an honest majority. In: The 23rd ACM CCS, pp. 805–817 (2016)
3. Beaver, D.: Foundations of secure interactive computing. In: Feigenbaum, J. (ed.) CRYPTO 1991. LNCS, vol. 576, pp. 377–391. Springer, Heidelberg (1992). https://doi.org/10.1007/3-540-46766-1_31
4. Beerliová-Trubíniová, Z., Hirt, M.: Perfectly-secure MPC with linear communication complexity. In: Canetti, R. (ed.) TCC 2008. LNCS, vol. 4948, pp. 213–230. Springer, Heidelberg (2008). https://doi.org/10.1007/978-3-540-78524-8_13
5. Ben-Or, M., Goldwasser, S., Wigderson, A.: Completeness theorems for non-cryptographic fault-tolerant distributed computation. In: 20th STOC (1988)
6. Burra, S.S., Larraia, E., Nielsen, J.B., Nordholt, P.S., Orlandi, C., Orsini, E., Scholl, P., Smart, N.P.: High performance multi-party computation for binary circuits based on oblivious transfer. ePrint Cryptology Archive, 2015/472 (2015)
7. Canetti, R.: Security and composition of multiparty cryptographic protocols. J. Cryptol. **13**(1), 143–202 (2000)
8. Chaum, D., Crépeau, C., Damgård, I.: Multi-party unconditionally secure protocols. In: 20th STOC, pp. 11–19 (1988)
9. Cleve, R.: Limits on the security of coin flips when half the processors are faulty. In: 18th STOC, pp. 364–369 (1986)
10. Cramer, R., Damgård, I., Ishai, Y.: Share conversion, pseudorandom secret-sharing and applications to secure computation. In: Kilian, J. (ed.) TCC 2005. LNCS, vol. 3378, pp. 342–362. Springer, Heidelberg (2005). https://doi.org/10.1007/978-3-540-30576-7_19
11. Damgård, I., Keller, M., Larraia, E., Pastro, V., Scholl, P., Smart, N.P.: Practical covertly secure MPC for dishonest majority – or: breaking the SPDZ limits. In: Crampton, J., Jajodia, S., Mayes, K. (eds.) ESORICS 2013. LNCS, vol. 8134, pp. 1–18. Springer, Heidelberg (2013). https://doi.org/10.1007/978-3-642-40203-6_1
12. Damgård, I., Nielsen, J.B.: Scalable and unconditionally secure multiparty computation. In: Menezes, A. (ed.) CRYPTO 2007. LNCS, vol. 4622, pp. 572–590. Springer, Heidelberg (2007). https://doi.org/10.1007/978-3-540-74143-5_32
13. Damgård, I., Pastro, V., Smart, N.P., Zakarias, S.: Multiparty computation from somewhat homomorphic encryption. In: Safavi-Naini, R., Canetti, R. (eds.) CRYPTO 2012. LNCS, vol. 7417, pp. 643–662. Springer, Heidelberg (2012). https://doi.org/10.1007/978-3-642-32009-5_38
14. Genkin, D., Ishai, Y., Prabhakaran, M., Sahai, A., Tromer, E.: Circuits resilient to additive attacks with applications to secure computation. In: STOC 2014 (2014)
15. Genkin, D., Ishai, Y., Polychroniadou, A.: Efficient multi-party computation: from passive to active security via secure SIMD circuits. In: Gennaro, R., Robshaw, M. (eds.) CRYPTO 2015. LNCS, vol. 9216, pp. 721–741. Springer, Heidelberg (2015). https://doi.org/10.1007/978-3-662-48000-7_35
16. Goldreich, O., Micali, S., Wigderson, A.: How to play any mental game. In: 19th STOC, pp. 218–229 (1987)

17. Goldwasser, S., Levin, L.: Fair computation of general functions in presence of immoral majority. In: Menezes, A.J., Vanstone, S.A. (eds.) CRYPTO 1990. LNCS, vol. 537, pp. 77–93. Springer, Heidelberg (1991). https://doi.org/10.1007/3-540-38424-3_6

18. Gennaro, R., Rabin, M., Rabin, T.: Simplified VSS and fact-track multiparty computations with applications to threshold cryptography. In: 17th PODC (1998)

19. Goldreich, O.: Foundations of Cryptography: Basic Applications, vol. 2 (2004)

20. Keller, M., Orsini, E., Scholl, P.: MASCOT: faster malicious arithmetic secure computation with oblivious transfer. In: 23rd ACM CCS, pp. 830–842 (2016)

21. Keller, M., Pastro, V., Rotaru, D.: Overdrive: making SPDZ great again. In: Nielsen, J.B., Rijmen, V. (eds.) EUROCRYPT 2018. LNCS, vol. 10822, pp. 158–189. Springer, Cham (2018). https://doi.org/10.1007/978-3-319-78372-7_6

22. Lindell, Y., Nof, A.: A framework for constructing fast MPC over arithmetic circuits with malicious adversaries and an honest-majority. In: ACM CCS (2017)

23. Mohassel, P., Rosulek, M., Zhang, Y.: Fast and secure three-party computation: the garbled circuit approach. In: ACM CCS, pp. 591–602 (2015)

24. Rabin, T., Ben-Or, M.: Verifiable secret sharing and multi-party protocols with honest majority. In: 21st STOC, pp. 73–85 (1989)

25. Shamir, A.: How to share a secret. CACM 22(11), 612–613 (1979)

26. Yao, A.: How to generate and exchange secrets. In: 27th FOCS, pp. 162–167 (1986)

Quantum Cryptography

Quantum FHE (Almost) As Secure As Classical

Zvika Brakerski$^{(\boxtimes)}$

Weizmann Institute of Science, Rehovot, Israel
`zvika.brakerski@weizmann.ac.il`

Abstract. Fully homomorphic encryption schemes (FHE) allow to apply arbitrary efficient computation to encrypted data without decrypting it first. In Quantum FHE (QFHE) we may want to apply an arbitrary *quantumly* efficient computation to (classical or quantum) encrypted data.

We present a QFHE scheme with classical key generation (and classical encryption and decryption if the encrypted message is itself classical) with comparable properties to classical FHE. Security relies on the hardness of the learning with errors (LWE) problem with polynomial modulus, which translates to the worst case hardness of approximating short vector problems in lattices to within a *polynomial* factor. Up to polynomial factors, this matches the best known assumption for classical FHE. Similarly to the classical setting, relying on LWE alone only implies *leveled* QFHE (where the public key length depends linearly on the maximal allowed evaluation depth). An additional *circular security* assumption is required to support completely unbounded depth. Interestingly, our circular security assumption is the same assumption that is made to achieve unbounded depth *multi-key* classical FHE.

Technically, we rely on the outline of Mahadev (arXiv 2017) which achieves this functionality by relying on super-polynomial LWE modulus and on a new circular security assumption. We observe a connection between the *functionality* of evaluating quantum gates and the *circuit privacy* property of classical homomorphic encryption. While this connection is not sufficient to imply QFHE by itself, it leads us to a path that ultimately allows using classical FHE schemes with polynomial modulus towards constructing QFHE with the same modulus.

1 Introduction

A fully homomorphic encryption (FHE) scheme [17,32] is one where the transformation $\mathsf{Enc}(x) \to \mathsf{Enc}(f(x))$ can be performed efficiently for any efficiently

The full version of this work is available at https://eprint.iacr.org/2018/338.

Supported by the Israel Science Foundation (Grant No. 468/14), Binational Science Foundation (Grants No. 2016726, 2014276), and by the European Union Horizon 2020 Research and Innovation Program via ERC Project REACT (Grant 756482) and via Project PROMETHEUS (Grant 780701).

H. Shacham and A. Boldyreva (Eds.): CRYPTO 2018, LNCS 10993, pp. 67–95, 2018.
https://doi.org/10.1007/978-3-319-96878-0_3

computable f, without violating the security of the scheme. This primitive is very useful for cryptographic applications, and in particular it allows *private outsourcing of computation*. That is, using the resources of a powerful third party to perform a computation without giving up privacy. In recent years it was shown how to construct FHE based on standard cryptographic assumptions (mostly lattice related), including ones that are assumed to be secure against quantum adversaries. In particular, it was shown [1,5,6,9,10,20] that FHE can be based on the hardness of the learning with errors (LWE) problem introduced by Regev [29]. LWE was proven to be as hard to solve as the hardness of finding approximate shortest vectors in arbitrary worst-case lattices, a task for which no significant quantum speedup is known. The approximation factor directly relates to a parameter of the LWE problem known as the *noise ratio*, expressed as a function of the dimension of the problem.[1] Initial schemes [9] relied on LWE with *sub-exponential* noise ratio, and thus the hardness of sub-exponential approximation for lattice problems. Extensive research effort improved the schemes all the way down to only requiring a *polynomial* noise ratio [10], which is the gold standard for LWE-based security.

Understanding the capabilities and boundaries of FHE in various computational models is a fundamental question in cryptographic study. In this work, we focus on extending the set of supported functions f to the set of functions computable in *quantum* polynomial time, at the necessary cost of the evaluation process itself becoming quantum as well. This extension is called Quantum FHE (QFHE).

With developments in quantum computing occurring at an increasing rate, one could anticipate outsourcing of *quantum computation* becoming a quite common. Specifically it is quite likely that the first scalable quantum computers will be very expensive and require specialized maintenance and thus will not be directly available to the public. Rather, users will need to send their inputs to be processed by third party providers. If privacy is desired in this scenario, then QFHE could become a useful tool. While current research on QFHE, including this work, is well within the theoretical regime, developing theoretical tools and techniques could serve as basis for the development of concrete systems in due time.

Previous Works. Broadbent and Jeffery [11] showed that any classical FHE scheme can be translated into a quantum one that supports a limited set of gates (specifically, the evaluation of Clifford gates). Their idea is quite natural and elegant, and while not explicitly stated in this way, is related to the well established cryptographic notion of key encapsulation mechanisms (KEM). They rely on the notion of quantum one time pad (QOTP) that allows to information theoretically encrypt a quantum state using a single-use *classical* random pad. They propose to encrypt a quantum state using a QOTP, and then encrypt the pad itself using a classical homomorphic encryption scheme. They then show that Clifford operations in the quantum regime translate into applying a public

---

[1] To the informed reader we clarify that the noise ratio is the inverse of the Gaussian parameter of the *relative* noise, i.e. $1/\alpha$ in the common notation.

operation on the quantum part of the QOTP ciphertext, and applying public classical operations on the classical secret bits of the pad. The latter can be applied homomorphically since the secret bits of the pad are encrypted using a classical FHE scheme. They also show that evaluating an a-priori bounded number of non-Clifford gates is possible at the cost of the ciphertext size blowing up polynomially with the number of supported non-Clifford gates.

Dulek, Schaffner and Speelman [14] showed how to transfer the dependence on the number of non-Clifford gates from the ciphertext to the key. Specifically, their key generation involves generating a quantum gadget for every non-Clifford gate to be evaluated throughout the lifetime of the scheme, and transferring these gadgets to the homomorphic evaluator. The gadgets are consumed after a single use and their quantum nature prevents them from being duplicated or shared. This allowed for the first time to outsource quantum computation privately and compactly, but at the cost of quantum preprocessing. The [14] solution used the KEM approach as well, but required the decryption complexity of the classical FHE scheme to be bounded (roughly logarithmic space). They instantiate their scheme with the [9] FHE scheme, thus inheriting its unfavorable properties, but we believe it can also be instantiated using newer schemes such as [5,6,20], but it is not clear whether it applies to schemes based on the hardness of polynomial lattice approximation due to the sequentialization technique of [10] used in these schemes.

Mahadev [21] very recently presented a scheme whose key generation process is completely classical. This immediately implies that the keys can be duplicated and there is no longer a global bound on the total homomorphic capacity of the system. This scheme also uses key encapsulation, and requires specific properties of the underlying classical homomorphic encryption. An important property of the [21] scheme is that the homomorphic evaluation of each quantum gate is not necessarily perfectly correct, but rather it is only guaranteed to be within small trace distance of the correct state. These errors accumulate so in the worst case they are multiplied by the total circuit size. Thus, in order to achieve correctness up to a negligible trace distance, the per-gate error needs to be negligible as well. In the [21] solution, the per-gate error is (inversely) related to the noise rate of the underlying LWE assumption, so in order to achieve correctness for all polynomial size circuits, it is required to rely on the hardness of super-polynomial approximation to lattice problems (or even larger, depending on the type of computation and the user's desired level of confidence).

Another unusual requirement of [21] from the underlying classical FHE scheme is randomness recoverability. Namely, that using the secret key it is possible to recover the randomness of a ciphertext. This is achieved using the dual scheme to the [1,10,20] scheme, but requires changing the secret key from being a single vector to a *trapdoor* to the lattice corresponding to the public key. This would all be in the realm of low order technicalities, except for the issue of *circular security*, which we explain next. Even in the classical setting, relying on LWE alone only allows to construct *leveled* FHE, where an a-priori bound on the *depth* (but not on the size) of evaluated circuits needs to be known. Overcoming this issue to obtain a scheme that is secure for any depth requires encrypting the scheme's own secret key, and explicitly assuming that this does not adversely impact the security of the scheme. Making this assumption for standard LWE-based encryption is by now the norm, but one might be less confident about making this assumption for new distributions of secret keys.

To conclude this overview, we note that there is a distinction in the literature between QFHE for classical vs. for quantum inputs. The former requires that the encryption and key generation process are completely classical, so that quantum computation on classical inputs can be outsourced by a classical entity. This distinction could suggest that the two notions are incomparable, however we believe that it is instructive to aspire to achieve a notion that generalizes both. Specifically, we propose to aspire for QFHE with classical keys, that can encrypt classical messages using a classical encryption process, and can encrypt quantum messages using a quantum process, and likewise if the output of homomorphic evaluation is classical then it should be decryptable by a classical decryption process. This stronger notion is in fact achieved by [21], although this property is not highlighted.

Our Results and Approach. We present a QFHE scheme using the high level outline of [21], but with per-gate error that decays *exponentially* with the noise rate of the underlying LWE assumption. Thus, using polynomial noise rate we are able to achieve exponentially small per-gate error, which means that we can securely evaluate any polynomial (or even super-polynomial) quantum circuit while incurring only an exponentially small skew between the output of homomorphic evaluation and the desired result. We do this by (again) relying on key encapsulation, this time using the (primal) [1,10,20] scheme as the KEM component. As for the distribution of secret keys, we do not require to use a lattice trapdoor as secret key, but our scheme requires publishing an encryption of the secret key of a [20]-style scheme, and keeping the randomness used to generate this encryption as a part of its own secret key.

Therefore, if we wish to create a scheme that works for a-priori unbounded depth, we need to assume circular security respective to a key containing a standard LWE key as well as randomness that was used to generate encryptions of this key. Interestingly, this exact assumption is required in order to construct unbounded depth classical *multi-key* FHE from [20]-style encryption [8,12,25, 28].[2]

In terms of our approach, we observe that the [21] method is implicitly intimately connected to the *circuit privacy* property of the underlying classical homomorphic scheme. Circuit privacy is the property that after homomorphically evaluating a function f, the resulting ciphertext $\mathsf{Enc}(f(x))$ does not contain any information about f except the value $f(x)$ (even statistically). While circuit privacy is not a sufficient condition, it appears to be necessary for ensuring functionality in the [21] method.

Circuit private homomorphic encryption schemes are useful for various applications and this property has been extensively studied in the FHE literature, e.g.

---

[2] Curiously, there is a syntactic resemblance between the randomness of a [1,10,20] ciphertext and lattice trapdoors generated using the method of [22].

in [4,13,16,18]. However, this property is usually considered to be a security feature, and we find it quite curious that in the quantum setting it turns out to be related to the *correctness* of homomorphic evaluation.

Through the circuit privacy lens, the [21] scheme can be viewed as applying the most rudimentary method for achieving function privacy, known as *noise flooding* [16]. This method guarantees privacy that is roughly relative to the noise rate of the underlying LWE assumption, hence super-polynomial rate is required to achieve privacy with all but negligible probability. It is not immediately clear how to apply more modern circuit privacy approaches in the QFHE setting (due to the additional properties required for quantum homomorphic evaluation), and the bulk of our technical work goes towards developing techniques to allow this application. We elaborate more on our techniques below.

1.1 Technical Overview

Our basic approach, traced back to [11], is to rely on key encapsulation. The ciphertext is encrypted using a quantum one time pad (QOPT), and the (classical) secret pad is encrypted using a classical FHE. QOTP encryption of a qubit can be expressed as applying a random Pauli operation, namely a random bit flip and a random phase flip. This allows to easily evaluate Clifford gates. As observed in previous works [14,21], a missing piece that would imply QFHE is being able to homomorphically evaluate the CNOT operation on a given quantum state, but given a classical control bit in encrypted form. To be more explicit, given a 2-qubit superposition $\sum_{a,b} \alpha_{a,b}|a,b\rangle$ and an encrypted control bit x, output an encapsulated encryption of $\sum_{a,b} \alpha_{a,b}|a, b \oplus ax\rangle$, i.e. a two-qubit register and a classically encrypted pad that would decrypt the quantum register to the aforementioned superposition. The encapsulated version we produce will be a superposition of the form $\sum_{a,b} (-1)^{a\gamma_{\text{phase}}} \alpha_{a,b}|a, b \oplus ax \oplus \gamma_{\text{flip}}\rangle$ for some bits $\gamma_{\text{flip}}, \gamma_{\text{phase}}$, together with encryptions of the bits $\gamma_{\text{flip}}, \gamma_{\text{phase}}$. One can verify that indeed $\sum_{a,b} (-1)^{a\gamma_{\text{phase}}} \alpha_{a,b}|a, b \oplus ax \oplus \gamma_{\text{flip}}\rangle$ can be corrected to the prescribed output using a proper bit flip and phase flip. We start by describing at a high level the [21] approach and its relation to circuit privacy.

The [21] Approach and Circuit Privacy. Given $\sum_{a,b} \alpha_{a,b}|a,b\rangle$ and $\text{Enc}(x)$, the idea is to apply classical homomorphic evaluation to generate a superposition of the form

$$\sum_{a,b,\mu} \alpha_{a,b}|a, b \oplus \mu\rangle |\text{Enc}(ax \oplus \mu)\rangle |\mu\rangle$$

(we ignore normalization factors). This can be done using the properties of the classical FHE by applying to $\text{Enc}(x)$ the function $f_{a,\mu}(x) = ax \oplus \mu$. Now, measure the register containing $|\text{Enc}(ax \oplus \mu)\rangle$ to obtain some ciphertext c', let γ_{flip} denote the bit that it encrypts and note that $\mu = ax \oplus \gamma_{\text{flip}}$. Then the remainder superposition is: $\sum_{a,b} \alpha_{a,b}|a, b \oplus ax \oplus \gamma_{\text{flip}}\rangle |ax \oplus \gamma_{\text{flip}}\rangle$. So far we used the homomorphic ciphertext to introduce an added ax term into the $|b\rangle$ register. Finally, to remove the last register $|ax \oplus \gamma_{\text{flip}}\rangle$, measure it in the Hadamard basis, or alternatively, apply Fourier Transform and measure the result. We get a measured bit w and the state $\sum_{a,b}(-1)^{(wx)a}\alpha_{a,b}|a, b \oplus ax \oplus \gamma_{\text{flip}}\rangle$ (with a global factor $(-1)^{w\gamma_{\text{flip}}}$ that can be ignored). Therefore, setting $\gamma_{\text{phase}} = wx$ should complete the proof.

Unfortunately, this outline is too simplistic. We ignored the fact that there are many possible ciphertexts of the form $\mathsf{Enc}(ax \oplus \mu)$, and the specific ciphertext output by homomorphically evaluating $f_{a,\mu}$ might depend on a, μ, which means that measuring it might collapse the superposition completely. This is why *circuit privacy* seems useful, since it will ensure that regardless of a, μ the distribution of $\mathsf{Enc}(ax \oplus \mu)$ depends only on the bit it encrypts. However, making a ciphertext private necessarily requires randomness, and we cannot use classical randomness since it will cause the superposition to collapse just as before. Therefore, the randomness is taken in superposition, and after measuring c' we are left with an additional register containing the randomness conditioned on c'. In a sense the privacy transformation transferred the information about the applied circuit from the ciphertext to the randomness register. We are thus left with $\sum_{a,b}(-1)^{a\gamma_{\text{phase}}}\alpha_{a,b}|a, b \oplus ax \oplus \gamma_{\text{flip}}\rangle|r_a\rangle$ and we need to find a way to get rid of this additional randomness register.

In [21] it is shown that using their specific scheme, it is possible to express r_a as $r_0 \oplus (ar_1)$ where r_0, r_1 are binary vectors, and thus again measuring this register in the Hadamard basis will be effective. This crucially relies on having a one-to-one mapping between the randomness in the privacy transformation and the ciphertext c'. This property indeed holds for noise flooding, but not for later privacy techniques.

To complete this description, we note that after the Hadamard measurement, the value of r_1 now contributes to γ_{phase}, and an additional process involving the lattice trapdoor is introduced in order to show that a classical encryption of the new γ_{phase} can be recovered.

Our Solution. We are inspired by the circuit privacy argument of Bourse et al. [4] which is applicable to encryption schemes of the type introduced in [20] (henceforth referred to as GSW) and shows how to achieve circuit privacy with polynomial noise rate. In GSW an encryption of a bit x is represented by a matrix over $\mathbb{Z}_q$ for some modulus q of the form $\mathbf{C} = \mathbf{AR}_c + x\mathbf{G}$, where $\mathbf{A}$ is the public key of the scheme, $\mathbf{R}_c$ is a matrix of low norm (say all entries are $\ll q$) and $\mathbf{G}$ is a special "gadget" matrix. For our purposes it will be useful to choose the modulus q to be even (this does not have an effect on the resulting hardness assumption). The circuit privacy argument of [4] implies that if we sample a integer vector $\mathbf{r}$ from a *discrete Gaussian* distribution over the set $\{\mathbf{r} : \mathbf{Gr} = a\frac{q}{2}\boldsymbol{\Delta} \pmod q\}$ (for some vector $\boldsymbol{\Delta}$), and compute the vector $\mathbf{c}' = \mathbf{Cr} + (\frac{q}{2}\mu + y)\boldsymbol{\Delta}$, where y is a discrete Gaussian over $\mathbb{Z}$, then $\mathbf{c}'$ is a circuit private representation of $ax \oplus \mu$, i.e. $\mathbf{c}'$ does not reveal information about a, μ beyond the value $ax \oplus \mu$.[3]

Let us now see how this method fits into the [21] outline. Specifically, every $\mathbf{c}'$ in this setting will have multiple randomness values associated with it, so there is no longer a single r_a associated with each $\mathbf{c}'$. We will therefore try to find

---

[3] Indeed, $\mathbf{c}'$ does not have the same form as the original ciphertext $\mathbf{C}$, but it can be correctly decrypted, which is the property we care about.

an alternative structural property of the randomness register that will allow us to remove it without collapsing the superposition. Looking closely, we see that the randomness consistent with $\mathbf{c}'$ is a discrete Gaussian over variables $\mathbf{r}, y, \mu$ s.t. $\{\mathbf{r} : \mathbf{Gr} = a\frac{q}{2}\mathbf{\Delta} \pmod{q}\}$ and $\mathbf{c}' = \mathbf{Cr} + (\frac{q}{2}\mu + y)\mathbf{\Delta} = \mathbf{AR}_c\mathbf{r} + (\frac{q}{2}(ax \oplus \mu) + y)\mathbf{\Delta}$. Indeed we observe that this is a Gaussian superposition over the solutions of a set of linear equations modulo q. In other words over *a coset of a q-ary lattice*, where the coset is determined by $\mathbf{c}'$ and by $a\frac{q}{2}\mathbf{\Delta}$. This suggests a way out, if we are willing to replace the binary Fourier Transform with q-ary Fourier Transform $\left(\mathrm{FT}_q : |\mathbf{x}\rangle \to \sum |\mathbf{w}\rangle e^{-\frac{2\pi i}{q}\langle \mathbf{w}, \mathbf{x}\rangle}\right)$. As a rule of thumb, applying FT_q on different cosets of the same lattice, results in the same output, up to a phase that depends on the difference between the cosets. In our case, the difference is a multiple of a, just like we wanted.

Unfortunately, things are not so simple. First of all, indeed the phase is a multiple of a, but since we applied FT_q, this phase might be relative to a q-ary root of unity, and not to (-1) as we require for our key encapsulation.[4] Luckily, in our case the difference between the cosets is a multiple of $\frac{q}{2}$, which translates to a phase relative to (-1). A greater difficulty comes from the fact that we are not actually uniform over a the coset, but rather Gaussian, which makes the transference between the pre-FT_q and post-FT_q regimes more messy. In particular, instead of all points having the same phase shift, each measured value receives phase contributions from many sources which can interfere with each other. It is known that if the Gaussian parameter is large enough (larger than the so called "smoothing parameter" of the lattice), then the interference is negligible. Unfortunately this is *not* the case here, and we need to explicitly analyze the post-FT_q superposition in order to show that the effect of the interference only amounts to exponentially small trace distance.

Finally, we note that in order to make the analysis go through, we add an additional component to the privacy transformation and actually set $\mathbf{c}' = \mathbf{Cr} + \mathbf{A}\hat{\mathbf{r}} + (\frac{q}{2}\mu + y)\mathbf{\Delta}$, with $\hat{\mathbf{r}}$ being an additional Gaussian parameter. This allows us to prove useful properties for the resulting lattice, as well as provides us with a way to recover the new γ_{phase} without requiring lattice trapdoors, but rather using only an encrypted form of $\mathbf{R}_c$ and the LWE secret key.

1.2 Paper Organization

The main technical contribution of this paper is the homomorphic evaluation of classically controlled CNOT, which is outlined above in Sect. 1.1 and formally analyzed in Sect. 5.

General preliminaries appear in Sect. 2, preliminaries related to the definition of homomorphic encryption and results from previous works that we use appear in Sect. 3. In Sect. 4 we describe how to put together the components from previous works together with our classically controlled CNOT to create the QFHE scheme.

---

[4] One could consider using q-ary QOTP, but this introduces other difficulties since it changes the class of circuits that are "easy", analogous to Clifford in the binary setting.

2 Preliminaries

We denote the unit ball by $\mathcal{B}_m = \{\mathbf{x} \in \mathbb{R}^m : \|\mathbf{x}\|_2 \le 1\}$, we omit the subscript when m is clear from the context. Similarly we denote the unit cube by $\mathcal{H}_m = \{\mathbf{x} \in \mathbb{R}^m : \forall i.\ \mathbf{x}[i] \in (-1, 1]\}$. We will sometimes use the shorthand $\mathcal{B}_m^t$, $\mathcal{H}_m^t$ to denote $t \cdot \mathcal{B}_m$, $t \cdot \mathcal{H}_m$ respectively.

Let $F : X \to \mathbb{C}$, and let $W \subseteq X$, then we denote $F(W) = \sum_{x \in W} F(x)$. For all $q \in \mathbb{N}$ we let $\mathbb{Z}_q$ denote the ring of integers modulo q. We represent elements in $\mathbb{Z}_q$ using numbers in the range $(-\frac{q}{2}, \frac{q}{2}] \cap \mathbb{Z}$. We denote by $[x]_q$ the value y s.t. $y = x \pmod q$ and $y \in (-\frac{q}{2}, \frac{q}{2}]$. We let $[\mathbb{Z}]_q$ denote the set $\mathbb{Z} \cap (-\frac{q}{2}, \frac{q}{2}]$.

We say that we δ-compute a quantum state if we compute a superposition that is within trace distance $O(\delta)$ of that state.

Quantum Rejection Sampling. We recall that quantum rejection sampling allows to take a superposition $\sum_{x \in X} \alpha_x |x\rangle$ and any sequence $\{\alpha_x'\}_x$ s.t. $|\alpha_x'| \le 1$ for all x, and produce a superposition $\frac{1}{A} \sum_{x \in X} \alpha_x \alpha_x' |x\rangle$, where $A = \sum_{x \in X} |\alpha_x \alpha_x'|^2$. The success probability of this procedure (i.e. the probability of not rejecting) is A. If it is efficient to generate the original superposition then the process can be repeated until successful, $1/A$ times in expectation.

Log-Infinity Uniformity. It will be convenient for us to consider a measure we call *log-infinity variance*.[5]

Definition 2.1. *The log-infinity variance of a vector $\mathbf{v} \in (\mathbb{R}^+)^m$ is defined as*

$$\mathsf{loginf}(\mathbf{v}) = \ln\left(\frac{\max_i \mathbf{v}[i]}{\min_i \mathbf{v}[i]}\right). \tag{1}$$

If $\mathsf{loginf}(\mathbf{v}) \le \epsilon$, we say that $\mathbf{v}$ is ϵ-loginf uniform.

We will often use loginf-uniformity for general indexed sets $V = \{v_z \in \mathbb{R}^+\}_{z \in M}$, where M is some set of indices.

The following properties are easy to verify by definition.

Lemma 2.2. *Let $V = \{v_z\}_{z \in M}$ be ϵ-loginf uniform. Then the following hold:*

1. ***Conditioning.*** *$\forall M' \subseteq M$ the sequence $V' = \{v_z\}_{z \in M'}$ is ϵ-loginf uniform.*
2. ***Aggregation.*** *$\forall a_1, \ldots, a_k \in \mathbb{R}^+$ the sequence $\{a_1 v_{z_1} + \cdots + a_k v_{z_k}\}_{z_1, \ldots, z_k \in M}$ is ϵ-loginf uniform.*
3. ***ℓ_p^p-Uniformity.*** *Let $p \in \mathbb{R}^+$. The distribution defined on M by assigning probabilities $\Pr[z] \propto v_z^p$ is within statistical distance $O(p\epsilon)$ of uniform.*

2.1 Quantum One Time Pad

The quantum one time pad (QOTP) allows to encrypt a qubit in an information theoretically secure manner using two random classical bits as symmetric key. Encrypting a multi-qubit state can be done in a bit by bit manner (using an independently sampled symmetric key for each qubit in the state).

---

[5] We suspect that this measure has been considered before, but were not able to find any reference or a well established name for it.

- QOTP.Keygen(). Sample two classical bits $x, z \xleftarrow{\$} \{0,1\}$ and outputs (x, z).
- QOTP.Enc$((x, z), \phi)$. Given a qubit ϕ apply the Pauli transformation $X^x Z^z$ to ϕ and output the resulting $\hat{\phi}$. More explicitly, the applied transformation is: $(\alpha_0 |0\rangle + \alpha_1 |1\rangle) \rightarrow (\alpha_0 |x\rangle + (-1)^z \alpha_1 |\bar{x}\rangle)$.
- QOTP.Dec$((x, z), \hat{\phi})$. Apply the reverse transformation $Z^z X^x$ to $\hat{\phi}$.

We note that if the message to be encrypted ϕ is classical, then it is possible to generate a syntactically correct and unconditionally secure QOTP of ϕ using a classical algorithm by simply applying a classical one time pad using randomness x, and setting $z = 0$. Furthermore, given any QOTP encryption of a classical value, it is possible to measure $\hat{\phi}$ and the resulting classical value can be correctly decrypted using the key (x, z) (or even $(x, 0)$) by the standard classical one time pad decryption.

2.2 Discrete and Periodic Gaussians

For $s > 0$ we define the Gaussian density function $\rho_s(\mathbf{x}) := e^{-\pi(\|\mathbf{x}\|/s)^2}$, where $\mathbf{x} \in \mathbb{R}^n$. For a set of points $X \subseteq \mathbb{R}^n$ we denote $\rho_s(X) = \sum_{\mathbf{x} \in X} \rho_s(\mathbf{x})$. The discrete Gaussian distribution $D_{\mathbb{Z}^n, s}$ is one that is supported only over $\mathbf{x} \in \mathbb{Z}^n$ and such that $\Pr[D_{\mathbb{Z}^n, s} = \mathbf{x}] \propto \rho_s(\mathbf{x})$.

Definition 2.3 (Periodic Gaussian). *The q-periodic Gaussian function $\rho_{s,q}$ is the periodic continuation of ρ_s. Namely $\rho_{s,q}(\mathbf{x}) = \rho_s(\mathbf{x} + q\mathbb{Z}^m)$.*

We show next that when s is sufficiently smaller than q, $\rho_{s,q}(\mathbf{x})$ is close to the non-periodic (but truncated) Gaussian.

Lemma 2.4. *Let $s > 0$, $q \in \mathbb{N}$, $\mathbf{x} \in \mathbb{Z}^m$ be such that $\|[\mathbf{x}]_q\| < q/4$. Then*

$$1 \leq \frac{\rho_{s,q}(\mathbf{x})}{\rho_s([\mathbf{x}]_q)} < 1 + 2^{-(\frac{1}{2}(q/s)^2 - m)} \tag{2}$$

Proof. The lower bound holds by definition. For the upper bound,

$$\frac{\rho_{s,q}(\mathbf{x})}{\rho_s([\mathbf{x}]_q)} = \frac{\sum_{\mathbf{v} \in \mathbb{Z}^m} (\rho_s([\mathbf{x}]_q + q\mathbf{v}))}{\rho_s([\mathbf{x}]_q)} \tag{3}$$

$$= \sum_{\mathbf{v} \in \mathbb{Z}^m} \exp\left(-\pi \left(\|[\mathbf{x}]_q + q\mathbf{v}\|^2 - \|[\mathbf{x}]_q\|^2\right)/s^2\right) \tag{4}$$

$$= 1 + \underbrace{\sum_{\mathbf{v} \in \mathbb{Z}^m \setminus \{\mathbf{0}\}} \exp\left(-\pi \left(\|[\mathbf{x}]_q + q\mathbf{v}\|^2 - \|[\mathbf{x}]_q\|^2\right)/s^2\right)}_{\text{denote by } \delta} \tag{5}$$

However, since $\|[\mathbf{x}]_q\| < q/4$, it holds that for all $\mathbf{v} \in \mathbb{Z}^m \setminus \{\mathbf{0}\}$

$$\|[\mathbf{x}]_q + q\mathbf{v}\|^2 - \|[\mathbf{x}]_q\|^2 \geq \|q\mathbf{v}\| \cdot (\|q\mathbf{v}\| - 2\|[\mathbf{x}]_q\|) \tag{6}$$

$$> \|q\mathbf{v}\| \cdot (\|q\mathbf{v}\| - q/2) \tag{7}$$

$$\geq \|q\mathbf{v}\| \cdot (\|q\mathbf{v}\|/2) \tag{8}$$

$$= \|q\mathbf{v}\|^2/2 \tag{9}$$

Therefore

$$\delta \leq \rho\left(\left(\frac{q}{s\sqrt{2}}\mathbb{Z}^m\right) \setminus \{\mathbf{0}\}\right) \tag{10}$$

$$\leq 2^{m-\frac{1}{2}(q/s)^2}, \tag{11}$$

where the last inequality follows by Lemma 2.10, with $t = \frac{q}{s\sqrt{2}}$.   □

For one dimensional Gaussians, another bound can be achieved.

Lemma 2.5. *Let $q \in \mathbb{N}$, $s > 0$ and $x \in [\mathbb{Z}]_q$. Then*

$$\rho_{s,q}(x) \leq 2\rho_s(x)/(1 - \rho_s(q)) \tag{12}$$

Proof. We expand the expression:

$$\rho_{s,q}(x) = \sum_{j \in \mathbb{Z}} e^{-\pi\left(\frac{x+jq}{s}\right)^2} \tag{13}$$

$$= \sum_{j \in \mathbb{N}} e^{-\pi\left(\frac{|x|+jq}{s}\right)^2} + \sum_{j \in \mathbb{N}} e^{-\pi\left(\frac{(q-|x|)+jq}{s}\right)^2} \tag{14}$$

$$\leq \sum_{j \in \mathbb{N}} e^{-\pi\left(\frac{x}{s}\right)^2} \cdot e^{-\pi j\left(\frac{q}{s}\right)^2} + \sum_{j \in \mathbb{N}} e^{-\pi\left(\frac{(q-|x|)}{s}\right)^2} \cdot e^{-\pi j\left(\frac{q}{s}\right)^2}. \tag{15}$$

Since $e^{-\pi\left(\frac{(q-|x|)}{s}\right)^2} \leq e^{-\pi\left(\frac{x}{s}\right)^2}$, and $\sum_{j \in \mathbb{N}} e^{-\pi j\left(\frac{q}{s}\right)^2} = 1/(1 - e^{-\pi\left(\frac{q}{s}\right)^2})$, the lemma follows.   □

Corollary 2.6. *Let $s > 0$, $q \in \mathbb{N}$, $\mathbf{x} \in \mathbb{Z}^m$ be such that $\|[\mathbf{x}]_q\| \geq t$. Then*

$$\rho_{s,q}(\mathbf{x}) \leq \frac{2^m \rho_s(t)}{1 - m\rho_s(q)}, \tag{16}$$

Proof. We will use Lemma 2.5 as follows:

$$\rho_{s,q}(\mathbf{x}) \leq \prod_{i=1}^m \rho_{s,q}(x_i) \leq \prod_{i=1}^m \frac{2\rho_s(x)}{1 - \rho_s(q)} \leq \frac{2^m}{1 - m\rho_s(q)} \cdot \rho_s(\mathbf{x}) \leq \frac{2^m \rho_s(t)}{1 - m\rho_s(q)}.$$

2.3 Lattices

A lattice, formally, is a discrete subgroup of $\mathbb{R}^m$. In this work we focus on integer lattices, which are subgroups of $\mathbb{Z}^m$. Any lattice can be represented as the $\mathbb{Z}$-span of a set of *basis vectors*. The basis is usually represented as a matrix $\mathbf{B}$ whose columns are the elements of the basis. The lattice spanned by the basis $\mathbf{B} \in \mathbb{Z}^{m \times k}$ is denoted $\mathcal{L}(\mathbf{B}) = \{\mathbf{Bt} : \mathbf{t} \in \mathbb{Z}^k\}$. We will usually consider *full rank* lattices where $\mathbf{B}$ is a square matrix. A *coset* of a lattice is defined by a vector $\mathbf{c} \in \mathbb{R}^m$ and denoted as $\mathbf{c} + \Lambda = \{\mathbf{x} : \mathbf{x} - \mathbf{v} \in \Lambda\}$ (note that many different $\mathbf{c}$ vectors can define the same coset). The *dual* of Λ is the set $\Lambda^* = \{\mathbf{y} : \forall \mathbf{x} \in \Lambda. \langle \mathbf{y}, \mathbf{x} \rangle \in \mathbb{Z}\}$.

The following is an immediate corollary from Banaszczyk's transference theorems [2].

Corollary 2.7. *Let Λ be a rank n lattice, and assume that Λ contains k linearly independent vectors of length $\leq \ell$. Then any set of $(n-k+1)$ linearly independent vectors in Λ^* contains a vector of length $\geq 1/\ell$.*

Specifically, if Λ contains $(n-1)$ linearly independent vectors of length $\leq \ell$, then all vectors in Λ^* of length $< 1/\ell$ are on the same line.

Given a lattice $\Lambda \subseteq \mathbb{R}^m$, we say that $\mathbf{T} \in \mathbb{Z}^{m \times m'}$ is a σ-*trapdoor* for Λ if it has the same rank as Λ and its *orthogonalized norm* $\left\| \widetilde{\mathbf{T}} \right\|$ is at most σ. The orthogonalized norm is the maximal norm of the columns of $\widetilde{\mathbf{T}}$, which is in turn the Gram-Schmidt orthogonalization of the columns of $\mathbf{T}$. An upper bound on the norm of the columns of $\mathbf{T}$ itself is also an upper bound for its trapdoor quality.

The ϵ-*smoothing parameter* of the lattice Λ, denoted $\eta_\epsilon(\Lambda)$ is defined as the maximal Gaussian measure over Λ whose Fourier Transform is concentrated around $\mathbf{0}$. For our purposes we will only require the following two properties proven in [19,23,29].

Lemma 2.8. *If Λ is of rank m and has a σ-trapdoor then for all $\epsilon < 1/2$ it holds that $\eta_\epsilon(\Lambda) \leq \sigma \cdot \sqrt{\frac{1}{\pi} \log(4m/\epsilon)}$.*

Lemma 2.9. *If $\eta_\epsilon(\Lambda) \leq s$ then the sequence $\{\rho_s(\Lambda + \mathbf{d})\}_{\mathbf{d} \in \mathbb{R}^m}$ is $O(\epsilon)$-loginf uniform.*

We also use the following lemma, a parameterized version of [31, Lemma 7], which is in turn a simplified version of [2], and follows by an identical proof.

Lemma 2.10. *For any m dimensional lattice Λ, for all $\mathbf{d} \in \mathbb{R}^m$ and for all s, t it holds that*

$$\rho_s((\Lambda + \mathbf{d}) \setminus \mathcal{B}_m^t) \leq 2^{m-(t/s)^2} \rho_s(\Lambda). \tag{17}$$

2.4 The Class of q-Ary Lattices

This class of lattices that is very useful in cryptography, and plays a prominent role in this work as well. A lattice is q-ary, for a modulus $q \in \mathbb{N}$, if it contains all of the vectors in $q\mathbf{I}$ (where $\mathbf{I}$ is the identity matrix). All such lattices have full rank.

Every matrix of the form $\mathbf{L} \in \mathbb{Z}_q^{n \times m}$ defines two useful q-ary lattices. The "perp lattice" $\Lambda_q^\perp(\mathbf{L}) = \{\mathbf{x} : \mathbf{L}\mathbf{x} = 0 \pmod{q}\}$, and the row span $\mathrm{Span}_q(\mathbf{L}) = \{\mathbf{y} \in \mathbb{Z}^m : \exists \mathbf{s} \in \mathbb{Z}_q^n. \; \mathbf{y} = \mathbf{s}\mathbf{L} \pmod{q}\}$, which contrary to our usual convention will be considered as a lattice of row vectors. The dual of $\mathrm{Span}_q(\mathbf{L})$ is $\frac{1}{q}\Lambda_q^\perp(\mathbf{L})$. For all $\mathbf{v} \in \mathbb{Z}_q^n$ define $\Lambda_q^\perp(\mathbf{L}, \mathbf{v}) = \{\mathbf{x} : \mathbf{L}\mathbf{x} = \mathbf{v} \pmod{q}\}$ and note that these are cosets of $\Lambda_q^\perp(\mathbf{L})$.

Translating Corollary 2.7, we get the following.

Corollary 2.11. *If $\Lambda_q^\perp(\mathbf{L})$ contains $(n-1)$ linearly independent vectors of length $\leq \ell$, then all vectors in $\mathrm{Span}_q(\mathbf{L})$ of length $< q/\ell$ are on the same line.*

For all n, we define the *gadget matrix* $\mathbf{G} \in \mathbb{Z}_q^{n \times n \lceil \log q \rceil}$ as the block matrix $\mathbf{G} = [\mathbf{I} \| 2\mathbf{I} \| \cdots \| 2^{\lceil \log q \rceil - 1} \mathbf{I}]$ (where $\mathbf{I}$ is the $n \times n$ identity matrix). For all $\mathbf{V} \in \{0,1\}^{n \times k}$ we define $\mathbf{G}^{-1}(\mathbf{V}) \in \{0,1\}^{n \lceil \log q \rceil \times k}$ to be the binary matrix s.t. $\mathbf{G}\mathbf{G}^{-1}(\mathbf{V}) = \mathbf{V} \pmod{q}$. The matrix $\mathbf{G}$ has a $\sqrt{5}$-trapdoor (for any values of n, q).

By the leftover hash lemma, for all $m > (n \log q + 2)$, all but 2^{-n} fraction of the matrices $\mathbf{L} \in \mathbb{Z}_q^{n \times m}$ have a $\sqrt{m}$-trapdoor. The matrix $\mathbf{G}$ also has a $\sqrt{m}$-trapdoor (which is efficiently computable, but we will not require it for the purpose of this work).

Lastly, the following is a direct corollary of the fact that $\frac{1}{q}\mathrm{Span}_q(\mathbf{D})$ is the dual of $\Lambda_q^{\perp}(\mathbf{D})$, the Poisson summation formula and basic properties of the Fourier Transform (see, e.g., [30]).

Corollary 2.12. *For any full rank* $\mathbf{D} \in \mathbb{Z}_q^{n \times m}$, *for all* $\mathbf{v} \in \mathbb{Z}_q^n$, $\mathbf{w} \in \mathbb{Z}_q^m$ *and any* $\sigma \in \mathbb{R}^+$ *it holds that*

$$\sum_{\mathbf{x} \in \Lambda_q^{\perp}(\mathbf{D}, \mathbf{v})} \rho_\sigma(\mathbf{x}) e^{-\frac{2\pi i}{q}\langle \mathbf{w}, \mathbf{x} \rangle} = \frac{\sigma^m}{q^n} \cdot \sum_{\mathbf{t} \in \mathbb{Z}^n} \rho_{q/\sigma}(\mathbf{w} + \mathbf{t}\mathbf{D}) \cdot e^{\frac{2\pi i}{q}\langle \mathbf{t}, \mathbf{v} \rangle} \tag{18}$$

$$= \frac{\sigma^m}{q^n} \cdot \sum_{\mathbf{t} \in \mathbb{Z}_q^n} \rho_{q/\sigma, q}(\mathbf{w} + \mathbf{t}\mathbf{D}) \cdot e^{\frac{2\pi i}{q}\langle \mathbf{t}, \mathbf{v} \rangle}. \tag{19}$$

2.5   Learning with Errors

The learning with errors (LWE) problem was defined by Regev [29]. In this work we exclusively use the decisional version. The $\mathrm{LWE}_{n,m,q,\chi}$ problem, for $n, m, q \in \mathbb{N}$ and for a distribution χ supported over $\mathbb{Z}$ is to distinguish between the distributions $(\mathbf{A}, \mathbf{s}\mathbf{A} + \mathbf{e} \pmod{q})$ and $(\mathbf{A}, \mathbf{u})$, where $\mathbf{A}$ is uniform in $\mathbb{Z}_q^{n \times m}$, $\mathbf{s}$ is a uniform row vector in $\mathbb{Z}_q^n$, $\mathbf{e}$ is a uniform row vector drawn from χ^m, and $\mathbf{u}$ is a uniform vector in $\mathbb{Z}_q^m$. Often we consider the hardness of solving LWE for *any* $m = \mathrm{poly}(n \log q)$. This problem is denoted $\mathrm{LWE}_{n,q,\chi}$.

As shown in [27,29], the $\mathrm{LWE}_{n,q,\chi}$ problem with χ being the discrete Gaussian distribution with parameter $\sigma = \alpha q \geq 2\sqrt{n}$ (i.e. the distribution over $\mathbb{Z}$ where the probability of x is proportional to $e^{-\pi(|x|/\sigma)^2}$, see more details below), is at least as hard as approximating the shortest independent vector problem (SIVP) to within a factor of $\gamma = \tilde{O}(n/\alpha)$ in *worst case* dimension n lattices. This is proven using a quantum reduction. Classical reductions (to a slightly different problem) exist as well [7,26] but with somewhat worse parameters. The best known (classical or quantum) algorithm for these problems run in time $2^{\tilde{O}(n/\log \gamma)}$, and in particular are conjectured to be intractable for $\gamma = \mathrm{poly}(n)$.

2.6   The q-Ary Fourier Transform

We will use the following flavor of Fourier Transform over the ring $\mathbb{Z}_q$ for $q \in \mathbb{N}$ (this is sometimes called discrete Fourier Transform) which maps functions f:

$\mathbb{Z}^n \to \mathbb{C}$ to $\hat{f} : \mathbb{Z}_q^n \to \mathbb{C}$ as

$$\hat{f}_q(\mathbf{w}) = \sum_{\mathbf{x} \in \mathbb{Z}^n} f(\mathbf{x}) \cdot e^{-\frac{2\pi i}{q} \langle \mathbf{w}, \mathbf{x} \rangle}. \tag{20}$$

We note that if f is only supported over the cube modulo q, i.e. over $\mathcal{H}_n^{q/2} \cap \mathbb{Z}_q^n$, then the q-ary Fourier Transform operator is unitary (up to a global normalization factor).

2.7 Generating Gaussian Superpositions over Lattices

It has been shown in previous works [7,19] how to sample from a Gaussian superposition over a lattice, or a coset of a lattice, given a good enough basis. We observe that these methods can be extended to generating a Gaussian superposition by carefully repeating the argument from [7, Sect. 5], replacing rejection sampling with quantum rejection sampling, and neglecting the far tail of the Gaussian distribution. We state the result only for integer lattices to avoid handling matters of precision.

Lemma 2.13 (Lattice Superposition Generation). *Let $\Lambda = \mathcal{L}(\mathbf{B}) \subseteq \mathbb{Z}^m$ be an m-dimensional lattice, let $\mathbf{c} \in \mathbb{Z}^m$ and let $r \geq \sqrt{\ln(2m+4)/\pi} \cdot \|\widetilde{\mathbf{B}}\|$. Let $\delta \in (0,1)$. Then there exists a quantum expected polynomial time algorithm* GenGauss *s.t.* GenGauss$(\mathbf{B}, \mathbf{c}, r, 1/\delta)$ *outputs a quantum state which is within $O(\delta)$ trace distance of*

$$\frac{1}{\sqrt{\rho_r(\Lambda + \mathbf{c})}} \sum_{\mathbf{x} \in \Lambda + \mathbf{c}} \rho_{\sqrt{2}r}(\mathbf{x}) |\mathbf{x}\rangle. \tag{21}$$

Furthermore if $r \geq \sqrt{\log(4m/\delta)/\pi} \cdot \|\widetilde{\mathbf{B}}\|$ then the resulting quantum state is supported only over $\mathbb{Z}^m \cap \mathcal{B}_m^{r\sqrt{m + \log(1/\delta)}}$.

A proof is provided in the full version for the sake of completeness.

3   Homomorphic Encryption Tools and Techniques

3.1   Classical Homomorphic Encryption and Bootstrapping

We now define fully homomorphic encryption in the classical and quantum setting, and introduce Gentry's bootstrapping theorem.

A homomorphic (public-key) encryption scheme HE = (HE.Keygen, HE.Enc, HE.Dec, HE.Eval) is a tuple of PPT algorithms as follows (λ is the security parameter):

- **Key generation** (pk, sk)←HE.Keygen(1^λ): Outputs a public encryption key pk and a secret decryption key sk.
- **Encryption** c←HE.Enc(pk, x): Using the public key pk, encrypts a single bit message $x \in \{0, 1\}$ into a ciphertext c.

- **Decryption** $x \leftarrow \mathsf{HE.Dec}(\mathsf{sk}, c)$: Using the secret key sk, decrypts a ciphertext c to recover the message $x \in \{0, 1\}$.
- **Homomorphic evaluation** $\widehat{c} \leftarrow \mathsf{HE.Eval}(\mathcal{C}, (c_1, \ldots, c_\ell), \mathsf{pk})$: Using the public key pk, applies a circuit $\mathcal{C} : \{0,1\}^\ell \to \{0,1\}^{\ell'}$ to $c_1, \ldots, c_\ell$, and outputs ciphertexts $\widehat{c}_1, \ldots, \widehat{c}_{\ell'}$.

We overload the functionality of the encryption and decryption procedures by allowing the encryption to take multi-bit messages as input, and produce a sequence of ciphertexts corresponding to a bit-by-bit encryption. Similarly we allow the decryption to take as input a sequence of ciphertexts, decrypt them one after the other and output the result. We note that when we refer to the "decryption complexity" of the scheme, we refer to the single ciphertext procedure (although we will mostly be concerned with computation depth which remains the same in the overloaded version).

A homomorphic encryption scheme is said to be secure if it is semantically secure.

Full homomorphism and leveled full homomorphism is defined next.[6]

Definition 3.1 (compactness and full homomorphism). *A scheme* HE *is fully homomorphic, if for any efficiently computable circuit* $\mathcal{C}$ *and any set of inputs* $x_1, \ldots, x_\ell$, *letting* $(\mathsf{pk}, \mathsf{sk}) \leftarrow \mathsf{HE.Keygen}(1^\lambda)$ *and* $c_i \leftarrow \mathsf{HE.Enc}(\mathsf{pk}, x_i)$, *it holds that*

$$\Pr\left[\mathsf{HE.Dec}(\mathsf{sk}, \mathsf{HE.Eval}(\mathcal{C}, (c_1, \ldots, c_\ell), \mathsf{pk})) \neq \mathcal{C}(x_1, \ldots, x_\ell)\right] = \mathrm{negl}(\lambda).$$

A fully homomorphic encryption scheme is compact if its decryption circuit is independent of the evaluated function. The scheme is leveled fully homomorphic if it takes 1^L *as additional input in key generation, and can only evaluate depth* L *Boolean circuits.*

Gentry's bootstrapping theorem shows how to go from limited amount of homomorphism to full homomorphism. This method has to do with the *augmented decryption circuit* and, in the case of pure fully homomorphism, relies on the *weak circular security* property of the scheme.

Definition 3.2 (Bootstrappable Homomorphic Encryption). *Consider a homomorphic encryption scheme* HE. *Let* $(\mathsf{sk}, \mathsf{pk})$ *be properly generated keys and let* $\mathcal{C}$ *be the set of properly decryptable ciphertexts. Then the set of augmented decryption functions,* $\{f_{c_1, c_2}\}_{c_1, c_2 \in \mathcal{C}}$ *is defined by*

$$f_{c_1, c_2}(x) = \overline{\mathsf{HE.Dec}_x(c_1) \wedge \mathsf{HE.Dec}_x(c_2)}.$$

Namely, the function that uses its input as secret key, decrypts c_1, c_2 *and returns the* NAND *of the results.*

The scheme HE *is* bootstrappable *if it can homomorphically evaluate its family of augmented decryption circuits.*

---

[6] An informed reader will notice that we define *single-hop* homomorphism. However this notion is sufficient and implies the multi-hop version via bootstrapping.

Definition 3.3. *A public key encryption scheme* PKE *is said to be* weakly circular secure *if it is secure even against an adversary who gets encryptions of the bits of the secret key.*

The bootstrapping theorem is thus as follows.

Theorem 3.4 (bootstrapping [16,17]**).** *A bootstrappable homomorphic encryption scheme can be transformed into a leveled fully homomorphic encryption scheme with the same decryption circuit, ciphertext space and public key.*

Furthermore, if the aforementioned scheme is also weakly circular secure, then it can be made into a (non-leveled) fully homomorphic encryption scheme.

3.2 Quantum Fully Homomorphic Encryption

A quantum fully homomorphic encryption (QFHE) is one that can encrypt qubit registers and apply quantum circuits to encrypted data. For the purpose of this paper we will only consider QFHE schemes with classical keys.

We start by considering quantum homomorphic encryption. This is a scheme with similar syntax to the classical setting described above, and is likewise defined as a sequence of algorithms (HE.Keygen, QHE.Enc, QHE.Dec, QHE.Eval). The syntactic differences are as follows.

1. HE.Keygen remains a classical probabilistic algorithm.
2. QHE.Enc takes as input a qubit x rather than a bit, and outputs a ciphertext represented in qubits.
3. QHE.Dec takes as input a ciphertext represented as a quantum register and outputs the plaintext as a qubit.
4. QHE.Eval takes as input a classical description of a quantum circuit with ℓ input qubits and ℓ' output qubits, and a sequence of ℓ quantum ciphertexts. Its output is a sequence of ℓ' quantum ciphertexts.

A quantum homomorphic encryption scheme is secure if it is semantically secure. For the definition of quantum semantic security see [11].

Definition 3.5 (compactness and full homomorphism). *A scheme* QHE *is fully homomorphic, if for any BQP circuit C and any ℓ-qubit state $x_1, \dots, x_\ell$, the states ρ_1, ρ_2 defined henceforth are within negligible trace distance.*

We define ρ_1 to be the ℓ'-qubit state of the output of $C(x_1, \dots, x_\ell)$. We define ρ_2 to be the ℓ'-qubit state produced as follows. Generate $(\mathsf{pk}, \mathsf{sk}) \leftarrow \mathsf{HE.Keygen}(1^\lambda)$ and $c_i \leftarrow \mathsf{HE.Enc}(\mathsf{pk}, x_i)$, and output $\mathsf{QHE.Dec}(\mathsf{sk}, \mathsf{QHE.Eval}(C, (c_1, \dots, c_\ell), \mathsf{pk}))$. As in the classical case, a fully homomorphic encryption scheme is compact if its decryption circuit is independent of the evaluated function. The scheme is leveled fully homomorphic if it takes 1^L as additional input in key generation, and can only evaluate depth L Boolean circuits.

3.3 GSW-Style Classical FHE with Polynomial Modulus

We consider the LWE based fully homomorphic encryption scheme of Gentry, Sahai and Waters [20]. Specifically we use a result due to Brakerski and Vaikuntanathan [10] showing that it is possible to achieve secure FHE using polynomial modulus.

Theorem 3.6 ([10]). *There exist polynomials $q_0(n), B_r(n), B_e(n)$, and a (classical) bootstrappable fully homomorphic encryption scheme parameterized by any function $q(n)$ s.t. $\forall n.\ q(n) \in [q_0(n), 2^n]$, with the following properties.*

1. *The scheme is secure based on the $\mathsf{LWE}_{n,q,\chi}$ assumption, with $\chi = D_{\mathbb{Z}, 2\sqrt{n}}$, and thus on the hardness of SIVP_γ for $\gamma = \tilde{O}(\sqrt{n} \cdot q)$. Specifically if q is polynomial then so is γ.*
2. *The public key of the scheme is a matrix $\mathbf{A} \in \mathbb{Z}_q^{n \times m}$ for some $m = O(n \log q)$, for $m > n(\log q + 2)$, of the form $\mathbf{A} = \left[\begin{smallmatrix} \mathbf{B} \\ \mathbf{sB+e} \end{smallmatrix} \right] \pmod{q}$, where $\mathbf{B} \in \mathbb{Z}_q^{(n-1) \times m}$ is a random matrix, $\mathbf{s} \xleftarrow{\$} \mathbb{Z}_q^{n-1}$, and $\|\mathbf{e}\| \leq B_e(n)$. The secret key is the vector $\mathbf{s}$.*
3. *When the output of a homomorphic evaluation is a ciphertext encrypting a bit $x \in \{0, 1\}$, this ciphertext is a matrix $\mathbf{C} \in \mathbb{Z}_q^{n \times n \lceil \log q \rceil}$ of the form $\mathbf{C} = \mathbf{AR}_c + x\mathbf{G} \pmod{q}$ where $\mathbf{R}_c \in \mathbb{Z}^{m \times n \lceil \log q \rceil}$. Furthermore, the maximum Euclidean norm of any column in $\mathbf{R}_c$ is at most $B_r(n)$ (note that this bound is independent on q, so long as q is in the aforementioned regime).*
4. *There exists a deterministic polynomial time computable function*

$$\mathsf{TrackRand}((\mathcal{C}, (c_1, \ldots, c_\ell), \mathsf{pk}), (r_1, \ldots, r_\ell), (x_1, \ldots, x_\ell))$$

whose input consists of $(\mathcal{C}, (c_1, \ldots, c_\ell), \mathsf{pk})$ which is an input to the homomorphic evaluation function, as well as the random tapes and messages r_i, x_i used to generate each of the ciphertexts c_i. Its output is the matrix $\mathbf{R}_c$ (where $\mathbf{C} = \mathbf{AR}_c + x\mathbf{G}$ is the output of the original homomorphic evaluation). Furthermore, the depth of $\mathsf{TrackRand}$ is only dependent on the depth of $\mathcal{C}$.

We note that property 4 was not proven directly in [10] but follows from analysis of the GSW method in followups [1,3].

3.4 A Randomness Propagating Classical FHE Scheme

We show that using the scheme from Theorem 3.6 it is possible to generate a cryptosystem with the same properties, but that in addition produces, as the output of Eval an encryption of the randomness $\mathbf{R}_c$ of the output ciphertext. We call such a scheme *randomness propagating*.

Corollary 3.7 (Randomness Propagating Classical FHE). *There exists a (parameterized) scheme with the exact same properties as that of the scheme from Theorem 3.6, but with an additional property:*

5. *The output of homomorphic evaluation is a ciphertext* $\mathbf{C}$ *as above, in addition to an encryption of* $\mathbf{R}_c$ *(in bit representation).*

The idea for constructing the scheme relies on bootstrapping and is similar to the construction of fully-dynamic multi-key FHE by [8] via bootstrapping the schemes of [12, 25].

Proof. Since the scheme from Theorem 3.6 is bootstrappable, it can be extended to one that supports homomorphic evaluation of depth L circuits, for any a-priori polynomial L. In the new scheme, we change the encryption procedure to first encrypt the message and then encrypt the randomness that was used to generate that first ciphertext. Then, to perform the new homomorphic evaluation, first produce $\mathbf{C}$ using the homomorphic evaluation of the original scheme, and then homomorphically evaluate TrackRand on the encryption of the randomness in order to produce the encryption of $\mathbf{R}_c$. Since the decryption function did not change, it is possible to choose L large enough so that the scheme remains bootstrappable.   □

4 Our Quantum FHE Scheme

Our scheme follows an outline going back to Broadbent and Jeffery [11] and used also in [14, 21]. The idea is to encrypt messages using a quantum one-time pad (QOTP), and then encrypt the secret pad using a classical FHE scheme (this is often called *key encapsulation* or *hybrid encryption* in cryptographic literature). It is shown in [11] that applying Clifford gates on the encrypted message can be carried out by applying it to the QOTP encrypted state, and applying an appropriate classical operation on the encapsulated key. Since the encapsulated key is encrypted using a classical FHE, this classical operation can be carried out thus completing the homomorphic evaluation.

However, to allow evaluating general BQP functionality, it is required to evaluate gates beyond the Clifford family, in particular it is sufficient to evaluate the Toffoli gate. It has been shown (see, e.g., [21, Appendix A.3]) that in order to carry out this operation, it is sufficient to be able to evaluate a CNOT operation on a quantum input with an encrypted classical control bit. Specifically, it is sufficient to support the operation that takes as input a register encoding a general 2-qubit superposition $\sum_{a,b} \alpha_{a,b} |a, b\rangle$ and an encrypted control bit x, and output an encapsulated encryption of $\sum_{a,b} \alpha_{a,b} |a, b \oplus ax\rangle$. Namely a QOTP encrypted state together with a classical encryption of the QOTP key.

Our encryption scheme will be based on the key encapsulation methodology, using the randomness propagating scheme from Corollary 3.7 as the key encapsulation scheme (this is sometimes called a "key encapsulation mechanism", or KEM). To show that this scheme can indeed evaluate a CNOT with a classical control bit we prove the following theorem which constitutes the main technical contribution of this work. We present the theorem here and explain how to use it to construct our quantum FHE scheme. The theorem is then proven in Sect. 5 below.

Theorem 4.1. *For all δ and an appropriately set value of $q = \text{poly}(n, \log(1/\delta))$, let $\mathbf{A} = \begin{bmatrix} \mathbf{B} \\ \mathbf{sB+e} \end{bmatrix} \pmod{q}$ and $\mathbf{C} = \mathbf{AR}_c + x\mathbf{G} \pmod{q}$ be such that there exist global $\text{poly}(n \log q)$ bounds on the norms of $\mathbf{e}, \mathbf{R}_c$ and such that $\mathbf{B}$ has a $\sqrt{m}$-trapdoor (which does not need to be known to any entity).*

There exists a quantum polynomial time algorithm taking as input $\mathbf{A}, \mathbf{C}$ and a general superposition over two qubits $\sum_{a,b} \alpha_{a,b}|a, b\rangle$. Its output, with probability $1 - O(\delta)$, is a superposition over two qubits of the form

$$\sum_{a,b} (-1)^{a \cdot \gamma_{\text{phase}}} \alpha_{a,b}|a, b \oplus ax \oplus \gamma_{\text{flip}}\rangle, \tag{22}$$

as well as two vectors $\mathbf{c}_{\text{flip}}, \mathbf{c}_{\text{phase}}$ and two implicit vectors $\mathbf{s}_{\text{flip}}, \mathbf{s}_{\text{phase}}$, defined as a function of $\mathbf{s}, \mathbf{e}, \mathbf{R}_c, x$, s.t.

$$\left| \left[\langle \mathbf{c}_{\text{flip}}, \mathbf{s}_{\text{flip}} \rangle - \tfrac{q}{2} \gamma_{\text{flip}} \right]_q \right| \leq q/10, \tag{23}$$

and likewise for $\langle \mathbf{c}_{\text{phase}}, \mathbf{s}_{\text{phase}} \rangle$.

We note that we purposely provide a theorem with parameterized dependence on δ, even though it would have been sufficient to just show that there *exists* a negligible δ for which the theorem holds. We do this to emphasize the robustness of our techniques that allow taking the error to be even exponentially small in the security parameter while still keeping q polynomial.

Putting the Components Together. We follow a similar outline to [21], with the required changes from our different method of evaluating classically controlled CNOT. Security follows immediately from the KEM mechanism by combining the security of the quantum one time pad and the security of the classical homomorphic encryption. This argument is identical to previous works.

Let $\delta > 2^{-\text{poly}(n)}$ be some negligible function. We start with instantiating the randomness propagating scheme from Corollary 3.7. We let q be the (polynomial in n) value implied by Theorem 4.1, when instantiated with the bounds $B_e(n), B_r(n)$ from Corollary 3.7 (note that these bounds are independent of q so there is no circularity here), and instantiate the randomness propagating scheme accordingly. We furthermore notice that since the matrix $\mathbf{B}$ in the public key is uniformly sampled, it has a $\sqrt{m}$-trapdoor with all but negligible probability. Since the scheme is bootstrappable, it can be extended to support depth L computation for any predefined polynomial L. We will set a proper value for L later.

As explained, we use this scheme as KEM (key encapsulator) for a QOTP. As in previous works, homomorphically evaluating a BQP circuit is done gate by gate (or rather layer by layer). Clifford gates are evaluated as in [11]. To evaluate CNOT with classical control, we recall that by Corollary 3.7, and our definition of q, the structure of the matrices $\mathbf{A}, \mathbf{C}$ allows to apply Theorem 4.1 to obtain an output 2-bit register, along with the values $\mathbf{c}_{\text{flip}}, \mathbf{c}_{\text{phase}}$.

From this point and on, our outline is again similar to [21]. We note that the values $\gamma_{\text{flip}}, \gamma_{\text{phase}}$ can be recovered via a (classical) polynomial time process out of $\mathbf{c}_{\text{flip}}, \mathbf{c}_{\text{phase}}$ using $(\mathbf{s}, \mathbf{e}, \mathbf{R}_c, x)$ by computing the vectors $\mathbf{s}_{\text{flip}}, \mathbf{s}_{\text{phase}}$, evaluating the respective inner product and rounding to the nearest multiple of $q/2$. Since we have encryptions of these values, we can set L to be large enough to allow us to apply this process homomorphically, followed by bootstrapping the resulting value, thus getting a bootstrapped KEM encryption of $\gamma_{\text{flip}}, \gamma_{\text{phase}}$. In other words, we set L to be large enough so that the resulting scheme is bootstrappable even after evaluating the quantum circuit.

This completes the proof. We can use Theorem 3.4 to bootstrap the resulting scheme to a leveled FHE of any desired depth, while still relying on the same LWE assumption as the original scheme. Recalling Theorem 3.6, the LWE parameters used imply hardness under the hardness of approximating SIVP to within a factor of $\widetilde{O}(\sqrt{n}q) = \text{poly}(n)$. Alternatively, if we assume circular security, we get a (non-leveled) FHE scheme. We will need to assume the circular security of the randomness propagating scheme, i.e. of a scheme that also encrypts the randomness used to generate ciphertexts. Interestingly, as we mention above, this assumption was already proposed in the literature for bootstrapping LWE-based *multi-key* FHE schemes [8,12,25].

5   Evaluating a Classically Controlled CNOT

In this section we prove Theorem 4.1 by providing a BQP algorithm, setting parameters and a value for q and proving that the requirements of the theorem are met.

5.1   The Algorithm

We define $m' = m + n\lceil \log q \rceil + 2$. The choice of parameters for the values σ, q is described in Sect. 5.2 below. We recall that we use the term "δ-computing a quantum state" to refer to computing a state that is within $O(\delta)$ trace distance of the prescribed state.

1. We start with a superposition $\sum_{a,b} \alpha_{a,b} |a, b\rangle$ stored in a register we denote by INP.
2. Use the algorithm from Sect. 2.7 to δ-compute the superposition

$$|\psi\rangle = \frac{1}{\sqrt{\rho_{\frac{\sigma}{\sqrt{2}}}(\mathbb{Z}^{m+2})}} \sum_{\substack{\hat{\mathbf{r}} \in \mathbb{Z}^m \\ y, \mu \in \mathbb{Z}}} \rho_\sigma(\hat{\mathbf{r}}, y, \mu) |\hat{\mathbf{r}}, y, \mu\rangle. \tag{24}$$

Specifically, our choice of parameters will ensure that we generate a quantum state which is supported only over $\mathbb{Z}^{m+2} \cap \mathcal{H}_{m+2}^{q/2}$ but is within trace distance $O(\delta)$ from the above.

3. We note that it is possible to δ-compute, for any vector $\mathbf{v} \in \mathbb{Z}_q^n$, the superposition

$$|\psi_{\mathbf{v}}\rangle = \frac{1}{\sqrt{\rho_{\frac{\sigma}{\sqrt{2}}}(\Lambda_q^\perp(\mathbf{G}, \mathbf{v}))}} \sum_{\mathbf{r} \in \Lambda_q^\perp(\mathbf{G}, \mathbf{v})} \rho_\sigma(\mathbf{r}) |\mathbf{r}\rangle, \tag{25}$$

again we will show that we generate a superposition supported only over $\mathbb{Z}^{n\lceil \log q \rceil} \cap \mathcal{H}_{n\lceil \log q \rceil}^{q/2}$ which is within trace distance $O(\delta)$ from the above.

For all $a \in \{0, 1\}$ we define $\mathbf{v}_a = a \cdot \begin{bmatrix} \mathbf{0} \\ q/2 \end{bmatrix} \in \mathbb{Z}_q^n$, and using the above we δ-compute the superposition

$$\sum_{a,b} \alpha_{a,b} |a, b\rangle \underbrace{|\psi_{\mathbf{v}_a}\rangle |\psi\rangle}_{\text{register } \Psi} . \tag{26}$$

4. Let μ_0 denote the least significant bit of μ (the last coordinate in the Ψ register), we apply the transformation $|a, b\rangle \to |a, b \oplus \mu_0\rangle$ to the INP register.
5. Consider the (classical, deterministic) ciphertext randomization function $\mathsf{RandCT}_{\mathbf{A},\mathbf{C}}(\tilde{\mathbf{r}}) : \mathbb{Z}^{m'} \to \mathbb{Z}_q^{n\lceil \log q \rceil}$ which is defined as follows. Parse $\tilde{\mathbf{r}}$ as a concatenation of $\mathbf{r} \in \mathbb{Z}^{n\lceil \log q \rceil}$, $\hat{\mathbf{r}} \in \mathbb{Z}^m$, $y, \mu \in \mathbb{Z}$ and compute

$$\mathsf{RandCT}_{\mathbf{A},\mathbf{C}}(\tilde{\mathbf{r}}) = \mathbf{C}\mathbf{r} + \mathbf{A}\hat{\mathbf{r}} + \begin{bmatrix} \mathbf{0} \\ 1 \end{bmatrix} y + \begin{bmatrix} \mathbf{0} \\ q/2 \end{bmatrix} \mu \pmod{q}. \tag{27}$$

Apply RandCT to the register Ψ, and add the output to a new $|0\rangle$ register. Measure the new register to obtain a value $\mathbf{c}'$.
6. Apply q-ary Fourier Transform (see Sect. 2.6) over $\mathbb{Z}_q$ to the register Ψ, and measure the result to obtain a value $\mathbf{w}$. We note that since Ψ contains a superposition which is supported over $\mathbb{Z}^{m'} \cap \mathcal{H}_{m'}^{q/2}$, the q-ary Fourier Transform is indeed a unitary transformation.
7. Output the register INP, and the vectors $\mathbf{c}_{\text{flip}} = \mathbf{c}'$ and $\mathbf{c}_{\text{phase}} = \mathbf{w}$, relative to $\mathbf{s}_{\text{flip}} = [-\mathbf{s}, 1]$ and

$$\mathbf{s}_{\text{phase}} = \boldsymbol{v} = \begin{bmatrix} \mathbf{G}^{-1}(\frac{q}{2}\boldsymbol{\Delta}) \\ -\mathbf{R}_c \cdot \mathbf{G}^{-1}(\frac{q}{2}\boldsymbol{\Delta}) \\ 0 \\ -x \end{bmatrix}.$$

5.2   Parameters and Definitions

The following matrix $\mathbf{D} \in \mathbb{Z}_q^{2n \times m'}$, where $m' = m + n\lceil \log q \rceil + 2$, and the lattices induced by it will play a central role in our analysis. This matrix is defined as follows.

$$\mathbf{D} = \begin{bmatrix} \mathbf{G} & 0 & 0 & 0 \\ \mathbf{C} & \mathbf{A} & 0 & 0 \\ & & 1 & q/2 \end{bmatrix}. \tag{28}$$

The $m' - 1$ columns of the following matrix are all in the lattice $\Lambda_q^\perp(\mathbf{D})$:

$$\mathbf{T}' = \begin{bmatrix} \mathbf{T_G} & 0 & 0 \\ -\mathbf{R}_c \mathbf{T_G} & \mathbf{T_B} & 0 \\ 0 & -\mathbf{e}\mathbf{T_B} & 0 \\ 0 & 0 & 2 \end{bmatrix} \in \mathbb{Z}^{m' \times (m'-1)}, \tag{29}$$

where $\mathbf{T_G} \in \mathbb{Z}^{n\lceil \log q \rceil \times n\lceil \log q \rceil}$ is a $\sqrt{n\lceil \log q \rceil}$-trapdoor for $\mathbf{G}$ and $\mathbf{T_B} \in \mathbb{Z}^{m \times m}$ is a $\sqrt{m}$-trapdoor for $\mathbf{B}$. Note that we will never need to explicitly compute $\mathbf{T}'$. We furthermore notice that the columns of $\mathbf{T}'$ are vectors in $\Lambda_q^\perp(\mathbf{D})$ since

$$\mathbf{D}\mathbf{T}' = \mathbf{0} \pmod{q}. \tag{30}$$

An additional important vector is the offset vector:

$$v = \begin{bmatrix} \mathbf{G}^{-1}(\frac{q}{2}\boldsymbol{\Delta}) \\ -\mathbf{R}_c \cdot \mathbf{G}^{-1}(\frac{q}{2}\boldsymbol{\Delta}) \\ 0 \\ -x \end{bmatrix}, \tag{31}$$

where $\boldsymbol{\Delta} = \begin{bmatrix} 0 \\ 1 \end{bmatrix} \in \{0,1\}^n$ (i.e. all zeros except the last coordinate). We note that

$$\mathbf{D} \cdot v = \begin{bmatrix} \mathbf{G} & 0 & 0 & 0 \\ \mathbf{C} & \mathbf{A} & 1 & q/2 \end{bmatrix} \cdot \begin{bmatrix} \mathbf{G}^{-1}(\frac{q}{2}\boldsymbol{\Delta}) \\ -\mathbf{R}_c \cdot \mathbf{G}^{-1}(\frac{q}{2}\boldsymbol{\Delta}) \\ 0 \\ -x \end{bmatrix} = \begin{bmatrix} \frac{q}{2}\boldsymbol{\Delta} \\ 0 \end{bmatrix}. \tag{32}$$

Finally we consider the row vector $\mathbf{d}^* = [2\mathbf{eR}_c \| 2\mathbf{e} \| 2 \| 0]$ (which we prove below is the shortest vector in $\mathrm{Span}_q(\mathbf{D})$).

Setting the Parameters. We let $p = \mathrm{poly}(n \log q)$ denote a polynomial upper bound on $\max \{ \|\mathbf{T}'\|, 10 \cdot \|v\|, \|\mathbf{d}^*\| \}$ (where $\|\mathbf{T}'\|$ refers to the maximal column norm), and set

$$\sigma = p \cdot \sqrt{2n \log q + m'(\log q + 1) + 2\log(4m'/\delta) + 1}, \tag{33}$$

finally we set $q = 2 \cdot \left\lceil 10 \cdot p \cdot \sigma \cdot \sqrt{m' + \log(1/\delta)} \right\rceil$, it will be useful for us that q is even.

One might be worried about circularity of this definition, since p, σ are used to determine the value of q but depend themselves on $\log q$. Indeed this situation frequently occurs when choosing parameters for LWE-based constructions, but it is easily resolved since the dependence of p, σ on q is logarithmic. Specifically, upper bound $\log q$ in the expressions for p, σ by, e.g., $\log^2 n$, and compute the value of q that is implied by these values of p, σ. The result will be $q = \mathrm{poly}(n)$ which indeed justifies the bound $\log q < \log^2 n$.

Properties of Lattices Induced by D. We prove a few properties that will be useful down the line.

We let p denote an upper bound on the ℓ_2 norm of the columns of $\mathbf{T}'$, note that $p = \mathrm{poly}(n \log q)$ for a suitable polynomial. We now invoke Corollary 2.11 to conclude that $\mathrm{Span}_q(\mathbf{D})$ has at most a single nonzero vector of norm $< q/p$ (up to multiplication by scalar). The next claim identifies the shortest vector in $\mathrm{Span}_q(\mathbf{D})$.

Claim 5.1. *The shortest vector in $\mathrm{Span}_q(\mathbf{D})$ is the vector $\mathbf{d}^* = [2\mathbf{eR}_c \| 2\mathbf{e} \| 2 \| 0]$ (where $\mathbf{d}^* = \mathbf{t}^* \mathbf{D} \pmod{q}$ for $\mathbf{t}^* = 2 \cdot [-x(\mathbf{s}, -1) \| (\mathbf{s}, -1)]$). All vectors in $\mathrm{Span}_q(\mathbf{D})$ that are not integer multiples of $\mathbf{d}^*$ are of length at least q/p.*

Proof. Since $\mathbf{T}'$ contains $(m' - 1)$ vectors in $\Lambda_q^{\perp}(\mathbf{D})$ of length at most p, Corollary 2.11 guarantees that $\mathrm{Span}_q(\mathbf{D})$ has at most a single nonzero vector of norm $< q/p$ (up to integer multiplications). We next verify that the shortest of these vectors is $\mathbf{d}^*$.

We can verify that indeed $\mathbf{d}^* \in \mathrm{Span}_q(\mathbf{D})$ since $\mathbf{d}^* = \mathbf{t}^*\mathbf{D} \pmod{q}$. Furthermore, $\|\mathbf{d}^*\| \le p < q/p$, and therefore either $\mathbf{d}^*$ is the shortest vector, or is an integer multiple of a shorter vector. However, $\mathbf{d}^*$ is only divisible by 2 (recall that $\mathrm{Span}_q(\mathbf{D})$ is an integer lattice), and the vector $\mathbf{d}^*/2 = [\mathbf{eR}_c\|\mathbf{e}\|1\|0]$ is not in $\mathrm{Span}_q(\mathbf{D})$ since $q|2$. ∎

For the next claim we recall the definition of loginf-uniformity in Definition 2.1 and its properties from Lemma 2.2.

Claim 5.2. *The sequence* $\left\{\rho_{\tilde{\sigma}}(\Lambda_q^\perp(\mathbf{D}, \hat{\mathbf{v}}))\right\}_{\hat{\mathbf{v}} \in \mathbb{Z}_q^{2n}}$ *is* $O(\delta)$-loginf *uniform for any* $\tilde{\sigma} \ge p \cdot \sqrt{\frac{1}{\pi}\log(4m'/\delta)}$.

Proof. Denote $\mathbf{h} = [\mathbf{eR}_c\|\mathbf{e}\|1\|0]$ and notice that $\mathbf{h}$ is orthogonal to all columns of $\mathbf{T}'$. By definition it holds that $\rho_{\tilde{\sigma}}(\Lambda_q^\perp(\mathbf{D}, \hat{\mathbf{v}})) = \sum_{\tilde{\mathbf{r}} \in \Lambda_q^\perp(\mathbf{D}, \hat{\mathbf{v}})} \rho_{\tilde{\sigma}}(\tilde{\mathbf{r}})$, and we can decompose each element in this sum to a component parallel to $\mathbf{h}$ and one orthogonal to $\mathbf{h}$:

$$
\begin{aligned}
\rho_{\tilde{\sigma}}(\Lambda_q^\perp(\mathbf{D}, \hat{\mathbf{v}})) &= \sum_{\tilde{\mathbf{r}} \in \Lambda_q^\perp(\mathbf{D}, \hat{\mathbf{v}})} \rho_{\tilde{\sigma}}(\tilde{\mathbf{r}}) \\
&= \sum_{k \in \mathbb{Z}} \rho_{\tilde{\sigma}}(k/\|\mathbf{h}\|) \sum_{\substack{\tilde{\mathbf{r}} \in \Lambda_q^\perp(\mathbf{D}, \hat{\mathbf{v}}) \\ \mathbf{h}\tilde{\mathbf{r}} = k}} \rho_{\tilde{\sigma}}(\tilde{\mathbf{r}} - k\mathbf{h}/\|\mathbf{h}\|^2).
\end{aligned}
$$

Fix a value of $k \in \mathbb{Z}$ and consider the sum $\sum \rho_{\tilde{\sigma}}(\tilde{\mathbf{r}} - k\mathbf{h}/\|\mathbf{h}\|^2)$ ranging over all $\tilde{\mathbf{r}} \in \Lambda_q^\perp(\mathbf{D}, \hat{\mathbf{v}})$ for which $\mathbf{h}\tilde{\mathbf{r}} = k$. Consider the lattice $\widehat{\Lambda}_\mathbf{D}$ containing all vectors in $\Lambda_q^\perp(\mathbf{D})$ which are orthogonal to $\mathbf{h}$. Then the set of vectors $S = \{(\tilde{\mathbf{r}} - k\mathbf{h}/\|\mathbf{h}\|^2) : \tilde{\mathbf{r}} \in \Lambda_q^\perp(\mathbf{D}, \hat{\mathbf{v}}), \mathbf{h}\tilde{\mathbf{r}} = k\}$ is exactly a coset of $\widehat{\Lambda}_\mathbf{D}$, and furthermore is supported only over the hyperplane that is orthogonal to $\mathbf{h}$.

Since $\mathbf{T}'$ is an p-trapdoor for $\widehat{\Lambda}_\mathbf{D}$ (for p defined above), then $\eta_\delta(\widehat{\Lambda}_\mathbf{D}) \le p \cdot \sqrt{\frac{1}{\pi}\log(4(m'-1)/\delta)} \le \tilde{\sigma}$. Lemma 2.9 implies therefore that the sequence $\{\rho_{\tilde{\sigma}}(\widehat{\Lambda}_\mathbf{D} + \mathbf{d})\}_{\mathbf{d} \perp \mathbf{h}}$ is $O(\delta)$-loginf uniform. Since the decomposition above shows that $\rho_{\tilde{\sigma}}(\Lambda_q^\perp(\mathbf{D}, \hat{\mathbf{v}}))$ is a linear combination of elements from the above sequence, applying Lemma 2.2 concludes the proof. ∎

5.3   Analysis

We now prove that the algorithm described above indeed has the properties required in the theorem statement.

Before Ciphertext Randomization. Recall that in the end of Step 3 of the algorithm, we δ-compute the superposition

$$
\sum_{a,b} \alpha_{a,b}|a,b\rangle|\psi_{\mathbf{v}_a}\rangle|\psi\rangle, \tag{34}
$$

which can also be written as

$$\sum_{a,b} \alpha_{a,b}|a,b\rangle \frac{1}{\sqrt{\rho_{\frac{\sigma}{\sqrt{2}}}(\mathbb{Z}_q^{m+2})\rho_{\frac{\sigma}{\sqrt{2}}}(\Lambda_q^{\perp}(\mathbf{G},\mathbf{v}_a))}} \sum_{\mathbf{r},\hat{\mathbf{r}},y,\mu} \rho_{\sigma}(\mathbf{r},\hat{\mathbf{r}},y,\mu)|\mathbf{r},\hat{\mathbf{r}},y,\mu\rangle, \qquad (35)$$

where the sum is over all $\mathbf{r} \in \Lambda_q^{\perp}(\mathbf{G},\mathbf{v}_a)$, $\hat{\mathbf{r}} \in \mathbb{Z}^m$, $y, \mu \in \mathbb{Z}$.

Recall that by Lemma 2.8, since $\mathbf{G}$ has a $O(1)$-trapdoor then $\eta_{\delta}(\Lambda_q^{\perp}(\mathbf{G})) \leq O(\log(n\log q/\delta))$. It follows by Lemma 2.9 that the set $\{\rho_{\frac{\sigma}{\sqrt{2}}}(\Lambda_q^{\perp}(\mathbf{G},\mathbf{v}))\}_{\mathbf{v}\in\mathbb{Z}_q^n}$ is δ-loginf uniform. Therefore, the above is within $O(\delta)$ trace distance of the superposition

$$\frac{1}{\sqrt{\rho_{\frac{\sigma}{\sqrt{2}}}(\Lambda_q^{\perp}(\mathbf{G})\times\mathbb{Z}^{m+2})}} \sum_{a,b} \alpha_{a,b}|a,b\rangle \sum_{\mathbf{r},\hat{\mathbf{r}},y,\mu} \rho_{\sigma}(\mathbf{r},\hat{\mathbf{r}},y,\mu)|\mathbf{r},\hat{\mathbf{r}},y,\mu\rangle, \qquad (36)$$

with $\mathbf{r}, \hat{\mathbf{r}}, y, \mu$ as before.

After applying Step 4, the resulting superposition is thus (ignoring global normalization)

$$\sum_{a,b} \alpha_{a,b} \sum_{\mathbf{r},\hat{\mathbf{r}},y,\mu} \rho_{\sigma}(\mathbf{r},\hat{\mathbf{r}},y,\mu)|a,b\oplus\mu_0\rangle|\mathbf{r},\hat{\mathbf{r}},y,\mu\rangle. \qquad (37)$$

Ciphertext Randomization. In Step 5 we compute

$$\sum_{a,b} \alpha_{a,b} \sum_{\mathbf{r},\hat{\mathbf{r}},y,\mu} \rho_{\sigma}(\mathbf{r},\hat{\mathbf{r}},y,\mu)|a,b\oplus\mu_0\rangle|\mathbf{r},\hat{\mathbf{r}},y,\mu\rangle \underbrace{|\mathsf{RandCT}_{\mathbf{A},\mathbf{C}}(\mathbf{r},\hat{\mathbf{r}},y,\mu)\rangle}_{\mathbf{c}'}, \qquad (38)$$

and measure $\mathbf{c}'$. We prove next that with all but $O(\delta)$ probability, $\mathbf{c}'$ is a ciphertext that decrypts to the value $\mu' = \mu_0 \oplus ax$.

Claim 5.3. *It holds that*

$$\left| \left[(-\mathbf{s},1)\cdot\mathbf{c}' - \tfrac{q}{2}\mu' \right]_q \right| < q/10 \qquad (39)$$

with probability $1 - O(\delta)$.

Proof. Consider the register holding $\mathbf{c}'$ before it is measured, we have (recalling that $\mathbf{r}$ is only supported over values where $\mathbf{G}\mathbf{r} = \mathbf{v}_a \pmod q$ and that q is even)

$$\begin{aligned}
\mathbf{c}' &= \mathbf{C}\mathbf{r} + \mathbf{A}\hat{\mathbf{r}} + \begin{bmatrix}0\\1\end{bmatrix}y + \begin{bmatrix}0\\q/2\end{bmatrix}\mu \pmod q \\
&= \mathbf{A}\mathbf{R}_c\mathbf{r} + \mathbf{A}\hat{\mathbf{r}} + \begin{bmatrix}0\\1\end{bmatrix}y + \begin{bmatrix}0\\q/2\end{bmatrix}(\mu+ax) \pmod q \\
&= \mathbf{A}(\mathbf{R}_c\mathbf{r} + \hat{\mathbf{r}}) + \begin{bmatrix}0\\1\end{bmatrix}y + \begin{bmatrix}0\\q/2\end{bmatrix}(\mu \oplus ax) \pmod q.
\end{aligned}$$

Recalling that $(-\mathbf{s},1)\mathbf{A} = \mathbf{e}$, we get that for $\mathbf{c}'$ as above

$$\begin{aligned}
(-\mathbf{s},1)\mathbf{c}' &= (\mathbf{e}\mathbf{R}_c\mathbf{r} + \mathbf{e}\hat{\mathbf{r}} + y) + \tfrac{q}{2}\mu' \pmod q \\
&= \mathbf{h}\tilde{\mathbf{r}} + \tfrac{q}{2}\mu' \pmod q,
\end{aligned}$$

where the vector $\mathbf{h} = [\mathbf{e}\mathbf{R}_c\|\mathbf{e}\|1\|0]$ (which also equals $\mathbf{d}^*/2$) is as defined in Claim 5.2. By definition of p we have that $\|\mathbf{h}\| \leq p/2$.

Therefore, it holds that in order for $\mathbf{c}'$ to not comply with Eq. (39), it must be the case that $|\mathbf{h}\tilde{\mathbf{r}}| > q/10$. Due to the bound on the norm of $\mathbf{h}$, this means that it must be the case that $\|\tilde{\mathbf{r}}\| > q/(5p) \geq \sigma\sqrt{m' + \log(1/\delta)}$. The probability that this happens, by Lemma 2.10, is at most

$$\frac{\rho_{\frac{\sigma}{\sqrt{2}}}\left(\Lambda_q^{\perp}(\mathbf{G}) \times \mathbb{Z}^{m+2} \setminus \mathcal{B}_{m'}^{q/(5p)}\right)}{\rho_{\frac{\sigma}{\sqrt{2}}}\left(\Lambda_q^{\perp}(\mathbf{G}) \times \mathbb{Z}^{m+2}\right)} \leq \delta, \tag{40}$$

and the claim follows.   ∎

We note that by definition after measuring $\mathbf{c}'$, it holds that $\mathbf{r}, \hat{\mathbf{r}}, y, \mu$ are only supported over values for which

$$\mathbf{D} \cdot \underbrace{\begin{bmatrix} \mathbf{r} \\ \hat{\mathbf{r}} \\ y \\ \mu \end{bmatrix}}_{\text{denote } \tilde{\mathbf{r}}} = \hat{\mathbf{v}}_a = \begin{bmatrix} \mathbf{v}_a \\ \mathbf{c}' \end{bmatrix} \pmod{q}, \tag{41}$$

where $\mathbf{D}$ is as defined in Eq. (28).

Namely, up to this point, we δ-computed the superposition

$$\sum_{a,b} \frac{\alpha_{a,b}}{\sqrt{\rho_{\frac{\sigma}{\sqrt{2}}}(\Lambda_q^{\perp}(\mathbf{D}, \hat{\mathbf{v}}_a))}} |a, b \oplus ax \oplus \mu'\rangle \sum_{\tilde{\mathbf{r}} \in \Lambda_q^{\perp}(\mathbf{D}, \hat{\mathbf{v}}_a)} \rho_{\sigma}(\tilde{\mathbf{r}})|\tilde{\mathbf{r}}\rangle, \tag{42}$$

where we note that since we defined $\mu' = ax \oplus \mu_0$ then it holds that $b \oplus \mu_0 = b \oplus ax \oplus \mu'$.

Fourier Transform and Measurement. From Claim 5.2 we deduce that we can remove the $\hat{\mathbf{v}}_a$-dependent normalization factor from Eq. (42) at the cost of $O(\delta)$ trace distance, so we conclude that at this point, before Step 6 of the algorithm, we δ-computed

$$\sum_{a,b} \alpha_{a,b}|a, b \oplus ax \oplus \mu'\rangle \underbrace{\sum_{\tilde{\mathbf{r}} \in \Lambda_q^{\perp}(\mathbf{D}, \hat{\mathbf{v}}_a)} \rho_{\sigma}(\tilde{\mathbf{r}})|\tilde{\mathbf{r}}\rangle}_{\text{denote } |\phi_a\rangle}. \tag{43}$$

In Step 6, we apply a q-ary Fourier transform on the register holding $|\tilde{\mathbf{r}}\rangle$. We recall that this register is actually supported only over $\mathbb{Z}^{m'} \cap \mathcal{H}_{m'}^{q/2}$, and therefore we can perform q-ary Fourier Transform as a unitary operation. Since the state of the register is $O(\delta)$-close in trace distance to the superposition in Eq. (43), the output of this operation will be $O(\delta)$ close in trace distance to the q-ary Fourier transform of Eq. (43). Formally, the q-ary Fourier transform of $|\phi_a\rangle$ is

$$|\hat{\phi}_a\rangle = \sum_{\mathbf{w} \in \mathbb{Z}_q^{m'}} |\mathbf{w}\rangle \sum_{\tilde{\mathbf{r}} \in \Lambda_q^{\perp}(\mathbf{D}, \hat{\mathbf{v}}_a)} \rho_{\sigma}(\tilde{\mathbf{r}}) e^{-\frac{2\pi i}{q}\langle \mathbf{w}, \tilde{\mathbf{r}}\rangle}. \tag{44}$$

By Corollary 2.12 it holds that

$$\sum_{\tilde{\mathbf{r}} \in \Lambda_q^{\perp}(\mathbf{D}, \hat{\mathbf{v}}_a)} \rho_{\sigma}(\tilde{\mathbf{r}}) e^{-\frac{2\pi i}{q}\langle \mathbf{w}, \tilde{\mathbf{r}}\rangle} = \frac{\sigma^{m'}}{q^{2n}} \cdot \sum_{\mathbf{t} \in \mathbb{Z}_q^{2n}} \rho_{q/\sigma, q}(\mathbf{w} + \mathbf{tD}) \cdot e^{\frac{2\pi i}{q}\langle \mathbf{t}, \hat{\mathbf{v}}_a\rangle}, \tag{45}$$

where we recall the definition of periodic Gaussian from Sect. 2.2: $\rho_{\sigma',q}(x) = \rho_{\sigma'}(x + q\mathbb{Z})$. Therefore it holds that

$$|\hat{\phi}_a\rangle = \frac{\sigma^{m'}}{q^{2n}} \cdot \sum_{\mathbf{w} \in \mathbb{Z}_q^{m'}} |\mathbf{w}\rangle \sum_{\mathbf{t} \in \mathbb{Z}_q^{2n}} \rho_{q/\sigma,q}(\mathbf{w} + \mathbf{tD}) \cdot e^{\frac{2\pi i}{q}\langle \mathbf{t}, \hat{\mathbf{v}}_a \rangle} \tag{46}$$

$$= \frac{\sigma^{m'}}{q^{2n}} \cdot \sum_{\mathbf{w} \in \mathbb{Z}_q^{m'}} |\mathbf{w}\rangle \sum_{\mathbf{t} \in \mathbb{Z}_q^{2n}} \rho_{q/\sigma,q}((\mathbf{w} - \mathbf{d}_w) + \mathbf{tD}) \cdot e^{\frac{2\pi i}{q}\langle \mathbf{t} - \mathbf{t}_w, \hat{\mathbf{v}}_a \rangle}. \tag{47}$$

For all $\mathbf{w}$, let $\mathbf{d}_w$ denote the vector in $\mathrm{Span}_q(\mathbf{D})$ that is closest to $\mathbf{w}$ and let $\mathbf{t}_w \in \mathbb{Z}_q^{2n}$ be s.t. $\mathbf{t}_w \mathbf{D} = \mathbf{d}_w \pmod{q}$. We let W denote the set of vectors that are close to $\mathrm{Span}_q(\mathbf{D})$

$$W = \{\mathbf{w} \in \mathbb{Z}_q^{m'} : \|\mathbf{w} - \mathbf{d}_w\| \le q/p\}. \tag{48}$$

We define

$$|\hat{\phi}_a'\rangle = \frac{\sigma^{m'}}{q^{2n}} \sum_{\mathbf{w} \in W} |\mathbf{w}\rangle \sum_{k \in \mathbb{Z}_q} \rho_{q/\sigma,q}((\mathbf{w} - \mathbf{d}_w) + k\mathbf{d}^*) \cdot e^{\frac{2\pi i}{q}\langle k\mathbf{t}^* - \mathbf{t}_w, \hat{\mathbf{v}}_a \rangle}. \tag{49}$$

Claim 5.4. *The trace distance between (the normalized versions of) the superpositions $|\hat{\phi}_a\rangle$ and $|\hat{\phi}_a'\rangle$ is $O(\delta)$.*

Proof. We start by bounding the norm of the difference $\|\hat{\phi}_a - \hat{\phi}_a'\|^2$. We first consider $\mathbf{w} \in \mathbb{Z}_q^{m'} \setminus W$. Then in particular it holds that $\left\|[\mathbf{w} + \mathbf{d}]_q\right\| \ge q/p$ for all $\mathbf{d} \in \mathrm{Span}_q(\mathbf{D})$ and therefore

$$\left| \sum_{\mathbf{t} \in \mathbb{Z}_q^{2n}} \rho_{q/\sigma,q}(\mathbf{w} + \mathbf{tD}) \cdot e^{\frac{2\pi i}{q}\langle \mathbf{t}, \hat{\mathbf{v}}_a \rangle} \right| \le \frac{q^{2n} \cdot 2^{m'} \cdot \rho_{q/\sigma}(q/p)}{(1 - m'\rho_{q/\sigma}(q))} \tag{50}$$

$$= 2^{2n \log q + m'} \cdot \frac{e^{-\pi(\sigma/p)^2}}{1 - m'e^{-\pi\sigma^2}}. \tag{51}$$

Since we chose $\sigma = p \cdot \sqrt{2n \log q + m'(\log q + 1) + 2\log(4m'/\delta) + 1}$ then in particular $m'e^{-\pi\sigma^2} < 1/2$ and $2^{2n \log q + m'} \cdot e^{-\pi(\sigma/p)^2} < \delta \cdot q^{-m'}/2$ which implies that the above is bounded by $\delta \cdot q^{-m'}$.

Now let us consider $w \in W$, the absolute value of the difference between $\hat{\phi}_a$, $\hat{\phi}_a'$ at $\mathbf{w}$ is at most

$$\sum_{\substack{\{\mathbf{t} \in \mathbb{Z}_q^{2n} : \\ \mathbf{t} \ne k\mathbf{t}^* \bmod q\}}} \rho_{q/\sigma,q}((\mathbf{w} - \mathbf{d}_w) + \mathbf{tD}). \tag{52}$$

If $\mathbf{t} \ne k\mathbf{t}^* \pmod{q}$ then $[\mathbf{tD}]_q \ge q/p$. This is since $\mathbf{d}^* = [\mathbf{t}^*\mathbf{D}]_q$ is the only vector in $\mathrm{Span}_q(\mathbf{D})$ of length $< q/p$, up to integer multiples. Since $\left\|[\mathbf{x}]_q\right\| \le \|\mathbf{x}\|$ it follows that for all $\mathbf{x}$, if $\|[\mathbf{x}]_q\| < q/p$ then $[\mathbf{x}]_q = [k\mathbf{d}^*]_q$ for some $k \in \mathbb{Z}_q$.

By definition of W we have $\|\mathbf{w} - \mathbf{d}_w\| \leq q/p$ and therefore by triangle inequality $\|[(\mathbf{w} - \mathbf{d}_w) + \mathbf{t}\mathbf{D}]_q\| \leq 2q/p$. Using a similar argument to above we get

$$\sum_{\substack{\{\mathbf{t} \in \mathbb{Z}_q^{2n}: \\ \mathbf{t} \neq k\mathbf{t}^* \bmod q\}}} \rho_{q/\sigma,q}((\mathbf{w} - \mathbf{d}_w) + \mathbf{t}\mathbf{D}) \leq \frac{q^{2n} \cdot 2^{m'} \cdot \rho_{q/\sigma}(2q/p)}{(1 - m'\rho_{q/\sigma}(q))} < \delta \cdot q^{-m'}. \quad (53)$$

It follows that:

$$\|\hat{\phi}_a - \hat{\phi}_a'\|^2 \leq \left(\frac{\sigma^{m'}}{q^{2n}}\right)^2 q^{m'} \cdot (\delta \cdot q^{-m'})^2 < \left(\frac{\sigma^{m'}}{q^{2n}}\right)^2 \cdot \delta. \quad (54)$$

We now lower bound $\|\hat{\phi}_a\|$ by simply looking at $\mathbf{w} = \mathbf{0}$:

$$\|\hat{\phi}_a\| \geq \sum_{\tilde{\mathbf{r}} \in \Lambda_q^\perp(\mathbf{D}, \hat{\mathbf{v}}_a) \cap \mathcal{H}_{m'}^{q/2}} \rho_{\sigma,q}(\tilde{\mathbf{r}}) = \sum_{\tilde{\mathbf{r}} \in \Lambda_q^\perp(\mathbf{D}, \hat{\mathbf{v}}_a)} \rho_\sigma(\tilde{\mathbf{r}}), \quad (55)$$

however by Claim 5.2, this is lower bounded by

$$(1 - O(\delta))\rho_\sigma(\Lambda_q^\perp(\mathbf{D})) = (1 - O(\delta))\frac{\sigma^{m'}}{q^{2n}}\rho_{q/\sigma}(\mathrm{Span}_q(\mathbf{D})) \geq (1 - O(\delta))\frac{\sigma^{m'}}{q^{2n}}.$$

Where the first equality follows from Corollary 2.12. The claim thus follows. ∎

We conclude that up to this point we δ-computed the superposition

$$\sum_{a,b} \alpha_{a,b}|a, b \oplus ax \oplus \mu'\rangle \sum_{\mathbf{w} \in W} |\mathbf{w}\rangle \sum_{k \in \mathbb{Z}_q} \rho_{q/\sigma,q}((\mathbf{w} - \mathbf{d}_w) + k\mathbf{d}^*) \cdot e^{\frac{2\pi i}{q}\langle k\mathbf{t}^* - \mathbf{t}_w, \hat{\mathbf{v}}_a\rangle}. \quad (56)$$

The next step is to measure the register $|\mathbf{w}\rangle$. Since $\mathbf{w} \in W$ it holds that $\|\mathbf{w} - \mathbf{d}_w\| < q/p$. We are left with the superposition

$$\sum_{a,b} \alpha_{a,b}|a, b \oplus ax \oplus \mu'\rangle \sum_{k \in \mathbb{Z}_q} \rho_{q/\sigma,q}((\mathbf{w} - \mathbf{d}_w) + k\mathbf{d}^*) \cdot e^{\frac{2\pi i}{q}\langle k\mathbf{t}^* - \mathbf{t}_w, \hat{\mathbf{v}}_a\rangle}. \quad (57)$$

We recall that $\hat{\mathbf{v}}_a$ can be written as $\hat{\mathbf{v}}_a = \hat{\mathbf{v}}_0 + a \cdot \frac{q}{2} \cdot \left[\begin{smallmatrix}\boldsymbol{\Delta}\\\mathbf{0}\end{smallmatrix}\right]$ for $\boldsymbol{\Delta} = \left[\begin{smallmatrix}\mathbf{0}\\1\end{smallmatrix}\right] \in \{0,1\}^n$. Let us now analyze the term $\langle k\mathbf{t}^* - \mathbf{t}_w, \hat{\mathbf{v}}_a\rangle \pmod{q}$ that is the exponent of the above expression (the $\bmod q$ comes from this term being in the exponent of the q-th root of unity). We recall that $\mathbf{t}^* = 2 \cdot [-x(\mathbf{s}, -1)\|(\mathbf{s}, -1)]$ is a multiple of 2, and therefore $\frac{q}{2}\mathbf{t}^* = \mathbf{0} \pmod{q}$. Let us also denote $\mathbf{t}_w = [\mathbf{t}_1\|\mathbf{t}_2]$, where $\mathbf{t}_1, \mathbf{t}_2 \in \mathbb{Z}_q^n$. We get that

$$\langle k\mathbf{t}^* - \mathbf{t}_w, \hat{\mathbf{v}}_a\rangle = \langle k\mathbf{t}^* - \mathbf{t}_w, \hat{\mathbf{v}}_0\rangle + a \cdot \frac{q}{2}\langle k\mathbf{t}^* - \mathbf{t}_w, \left[\begin{smallmatrix}\boldsymbol{\Delta}\\\mathbf{0}\end{smallmatrix}\right]\rangle \quad (58)$$

$$= \langle k\mathbf{t}^* - \mathbf{t}_w, \hat{\mathbf{v}}_0\rangle - a \cdot \frac{q}{2}\langle \mathbf{t}_1, \boldsymbol{\Delta}\rangle \pmod{q} \quad (59)$$

and plugging into the superposition above we have

$$\sum_{a,b} \alpha_{a,b} |a, b \oplus ax \oplus \mu'\rangle \sum_{k \in \mathbb{Z}_q} \rho_{q/\sigma,q}((\mathbf{w} - \mathbf{d}_w) + k\mathbf{d}^*) \cdot e^{\frac{2\pi i}{q}\langle k\mathbf{t}^* - \mathbf{t}_w, \hat{\mathbf{v}}_0\rangle} \cdot (-1)^{a \cdot \langle \mathbf{t}_1, \mathbf{\Delta}\rangle}.$$

Rearranging, we get that the above is equal to

$$\sum_{a,b} \alpha_{a,b} (-1)^{a \cdot \langle \mathbf{t}_1, \mathbf{\Delta}\rangle} |a, b \oplus ax \oplus \mu'\rangle \cdot \underbrace{\left(\sum_{k \in \mathbb{Z}_q} \rho_{q/\sigma,q}((\mathbf{w} - \mathbf{d}_w) + k\mathbf{d}^*) \cdot e^{\frac{2\pi i}{q}\langle k\mathbf{t}^* - \mathbf{t}_w, \hat{\mathbf{v}}_0\rangle} \right)}_{\text{Constant scaling factor, independent of } a, b.}.$$

We can thus remove the constant scaling factor and remain with

$$\sum_{a,b} \alpha_{a,b} (-1)^{a \cdot \langle \mathbf{t}_1, \mathbf{\Delta}\rangle} |a, b \oplus ax \oplus \mu'\rangle. \tag{60}$$

It is left to be shown that $\langle \mathbf{t}_1, \mathbf{\Delta}\rangle \pmod 2$ is efficiently recoverable given $\mathbf{R}_c, x$. We recall that we can write $\mathbf{w} = \mathbf{t}_w \mathbf{D} + \mathbf{e}_w \pmod q$ with $\|\mathbf{e}_w\| \leq q/p$. Next, we consider the vector

$$\mathbf{v} = \begin{bmatrix} \mathbf{G}^{-1}(\frac{q}{2}\mathbf{\Delta}) \\ -\mathbf{R}_c \cdot \mathbf{G}^{-1}(\frac{q}{2}\mathbf{\Delta}) \\ 0 \\ -x \end{bmatrix}, \tag{61}$$

and note that

$$\mathbf{D} \cdot \mathbf{v} = \begin{bmatrix} \mathbf{G} & 0 & 0 & 0 \\ \mathbf{C} & \mathbf{A} & 1 & q/2 \end{bmatrix} \cdot \begin{bmatrix} \mathbf{G}^{-1}(\frac{q}{2}\mathbf{\Delta}) \\ -\mathbf{R}_c \cdot \mathbf{G}^{-1}(\frac{q}{2}\mathbf{\Delta}) \\ 0 \\ -x \end{bmatrix} = \begin{bmatrix} \frac{q}{2}\mathbf{\Delta} \\ 0 \end{bmatrix}, \tag{62}$$

which implies that

$$\mathbf{w} \cdot \mathbf{v} = (\mathbf{t}_w \mathbf{D} + \mathbf{e}_w) \cdot \mathbf{v} = \mathbf{t}_w \mathbf{D} \mathbf{v} + \mathbf{e}_w \mathbf{v} = \frac{q}{2}\langle \mathbf{t}_1, \mathbf{\Delta}\rangle + \mathbf{e}_w \mathbf{v} \pmod q, \tag{63}$$

and since $|\mathbf{e}_w \mathbf{v}| \leq \|\mathbf{e}_w\| \cdot \|\mathbf{v}\| \leq q/p \cdot (p/10) \leq q/10$, the theorem follows.

Acknowledgments. The author wishes to thank Urmila Mahadev for numerous insightful discussions.

References

1. Alperin-Sheriff, J., Peikert, C.: Faster bootstrapping with polynomial error. In: Garay, J.A., Gennaro, R. (eds.) CRYPTO 2014, Part I. LNCS, vol. 8616, pp. 297–314. Springer, Heidelberg (2014). https://doi.org/10.1007/978-3-662-44371-2_17
2. Banaszczyk, W.: New bounds in some transference theorems in the geometry of numbers. Math. Ann. **296**(1), 625–635 (1993)

3. Boneh, D., et al.: Fully key-homomorphic encryption, arithmetic circuit ABE and compact garbled circuits. In: Nguyen, P.Q., Oswald, E. (eds.) EUROCRYPT 2014. LNCS, vol. 8441, pp. 533–556. Springer, Heidelberg (2014). https://doi.org/10.1007/978-3-642-55220-5_30

4. Bourse, F., Del Pino, R., Minelli, M., Wee, H.: FHE circuit privacy almost for free. In: Robshaw, M., Katz, J. (eds.) CRYPTO 2016, Part II. LNCS, vol. 9815, pp. 62–89. Springer, Heidelberg (2016). https://doi.org/10.1007/978-3-662-53008-5_3

5. Brakerski, Z.: Fully homomorphic encryption without modulus switching from classical GapSVP. In: Safavi-Naini, R., Canetti, R. (eds.) CRYPTO 2012. LNCS, vol. 7417, pp. 868–886. Springer, Heidelberg (2012). https://doi.org/10.1007/978-3-642-32009-5_50

6. Brakerski, Z., Gentry, C., Vaikuntanathan, V.: (Leveled) fully homomorphic encryption without bootstrapping. In: Goldwasser, S. (ed.) ITCS, pp. 309–325. ACM (2012)

7. Brakerski, Z., Langlois, A., Peikert, C., Regev, O., Stehlé, D.: Classical hardness of learning with errors. In: Boneh, D., Roughgarden, T., Feigenbaum, J. (eds.) Symposium on Theory of Computing Conference, STOC 2013, Palo Alto, CA, USA, 1–4 June 2013, pp. 575–584. ACM (2013)

8. Brakerski, Z., Perlman, R.: Lattice-based fully dynamic multi-key FHE with short ciphertexts. In: Robshaw, M., Katz, J. (eds.) CRYPTO 2016, Part I. LNCS, vol. 9814, pp. 190–213. Springer, Heidelberg (2016). https://doi.org/10.1007/978-3-662-53018-4_8

9. Brakerski, Z., Vaikuntanathan, V.: Efficient fully homomorphic encryption from (standard) LWE. In: Ostrovsky, R. (ed.) FOCS, pp. 97–106. IEEE (2011). https://eprint.iacr.org/2011/344.pdf

10. Brakerski, Z., Vaikuntanathan, V.: Lattice-based FHE as secure as PKE. In: Naor, M. (ed.) Innovations in Theoretical Computer Science, ITCS 2014, Princeton, NJ, USA, 12–14 January 2014, pp. 1–12. ACM (2014)

11. Broadbent, A., Jeffery, S.: Quantum homomorphic encryption for circuits of low t-gate complexity. In: Gennaro and Robshaw [15], pp. 609–629 (2015)

12. Clear, M., McGoldrick, C.: Multi-identity and multi-key leveled FHE from learning with errors. In: Gennaro and Robshaw [15], pp. 630–656 (2015)

13. Ducas, L., Stehlé, D.: Sanitization of FHE ciphertexts. In: Fischlin, M., Coron, J.-S. (eds.) EUROCRYPT 2016, Part I. LNCS, vol. 9665, pp. 294–310. Springer, Heidelberg (2016). https://doi.org/10.1007/978-3-662-49890-3_12

14. Dulek, Y., Schaffner, C., Speelman, F.: Quantum homomorphic encryption for polynomial-sized circuits. In: Robshaw, M., Katz, J. (eds.) CRYPTO 2016, Part III. LNCS, vol. 9816, pp. 3–32. Springer, Heidelberg (2016). https://doi.org/10.1007/978-3-662-53015-3_1

15. Gennaro, R., Robshaw, M. (eds.): CRYPTO 2015, Part II. 9216. Springer, Heidelberg (2015). https://doi.org/10.1007/978-3-662-48000-7

16. Gentry, C.: A fully homomorphic encryption scheme. Ph.D. thesis, Stanford University (2009)

17. Gentry, C.: Fully homomorphic encryption using ideal lattices. In: Mitzenmacher [24], pp. 169–178 (2009)

18. Gentry, C., Halevi, S., Vaikuntanathan, V.: i-hop homomorphic encryption and rerandomizable yao circuits. In: Rabin, T. (ed.) CRYPTO 2010. LNCS, vol. 6223, pp. 155–172. Springer, Heidelberg (2010). https://doi.org/10.1007/978-3-642-14623-7_9

19. Gentry, C., Peikert, C., Vaikuntanathan, V.: Trapdoors for hard lattices and new cryptographic constructions. In: Dwork, C. (ed.) STOC, pp. 197–206. ACM (2008)

20. Gentry, C., Sahai, A., Waters, B.: Homomorphic encryption from learning with errors: conceptually-simpler, asymptotically-faster, attribute-based. In: Canetti, R., Garay, J.A. (eds.) CRYPTO 2013, Part I. LNCS, vol. 8042, pp. 75–92. Springer, Heidelberg (2013). https://doi.org/10.1007/978-3-642-40041-4_5
21. Mahadev, U.: Classical homomorphic encryption for quantum circuits. CoRR, abs/1708.02130 (2017)
22. Micciancio, D., Peikert, C.: Trapdoors for lattices: simpler, tighter, faster, smaller. In: Pointcheval, D., Johansson, T. (eds.) EUROCRYPT 2012. LNCS, vol. 7237, pp. 700–718. Springer, Heidelberg (2012). https://doi.org/10.1007/978-3-642-29011-4_41
23. Micciancio, D., Regev, O.: Worst-case to average-case reductions based on Gaussian measures. In: Proceedings of the 45th Symposium on Foundations of Computer Science (FOCS 2004), Rome, Italy, 17–19 October 2004, pp. 372–381 (2004)
24. Mitzenmacher, M. (ed.): Proceedings of the 41st Annual ACM Symposium on Theory of Computing, STOC 2009, Bethesda, MD, USA, 31 May–2 June 2009. ACM (2009)
25. Mukherjee, P., Wichs, D.: Two round multiparty computation via multi-key FHE. In: Fischlin, M., Coron, J.-S. (eds.) EUROCRYPT 2016, Part II. LNCS, vol. 9666, pp. 735–763. Springer, Heidelberg (2016). https://doi.org/10.1007/978-3-662-49896-5_26
26. Peikert, C.: Public-key cryptosystems from the worst-case shortest vector problem: extended abstract. In: Mitzenmacher [24], pp. 333–342 (2009)
27. Peikert, C., Regev, O., Stephens-Davidowitz, N.: Pseudorandomness of ring-LWE for any ring and modulus. In: Hatami, H., McKenzie, P., King, V. (eds.) Proceedings of the 49th Annual ACM SIGACT Symposium on Theory of Computing, STOC 2017, Montreal, QC, Canada, 19–23 June 2017, pp. 461–473. ACM (2017)
28. Peikert, C., Shiehian, S.: Multi-key FHE from LWE, revisited. In: Hirt, M., Smith, A. (eds.) TCC 2016, Part II. LNCS, vol. 9986, pp. 217–238. Springer, Heidelberg (2016). https://doi.org/10.1007/978-3-662-53644-5_9
29. Regev, O.: On lattices, learning with errors, random linear codes, and cryptography. In: Gabow, H.N., Fagin, R. (eds.) STOC, pp. 84–93. ACM (2005). Full version in J. ACM **56**(6) (2009)
30. Regev, O., Kol, G.: Lattices in computer science lecture notes - lecture 9 - fourier transform (2004). https://cims.nyu.edu/~regev/teaching/lattices_fall_2004/ln/FourierTransform.pdf
31. Regev, O., Verbin, E.: Lattices in computer science lecture notes - lecture 11 - transference theorems (2004). https://cims.nyu.edu/~regev/teaching/lattices_fall_2004/ln/transference.pdf
32. Rivest, R., Adleman, L., Dertouzos, M.: On data banks and privacy homomorphisms. In: Foundations of Secure Computation, pp. 169–177. Academic Press (1978)

IND-CCA-Secure Key Encapsulation Mechanism in the Quantum Random Oracle Model, Revisited

Haodong Jiang[1,2], Zhenfeng Zhang[2,3(✉)], Long Chen[2,3], Hong Wang[1], and Zhi Ma[1,4(✉)]

[1] State Key Laboratory of Mathematical Engineering and Advanced Computing, Zhengzhou, Henan, China
hdjiang13@gmail.com, wfallmoon@163.com, ma_zhi@163.com
[2] TCA Laboratory, State Key Laboratory of Computer Science, Institute of Software, Chinese Academy of Sciences, Beijing, China
{zfzhang,chenlong}@tca.iscas.ac.cn
[3] University of Chinese Academy of Sciences, Beijing, China
[4] CAS Center for Excellence and Synergetic Innovation Center in Quantum Information and Quantum Physics, USTC, Hefei, Anhui, China

Abstract. With the gradual progress of NIST's post-quantum cryptography standardization, the Round-1 KEM proposals have been posted for public to discuss and evaluate. Among the IND-CCA-secure KEM constructions, mostly, an IND-CPA-secure (or OW-CPA-secure) public-key encryption (PKE) scheme is first introduced, then some generic transformations are applied to it. All these generic transformations are constructed in the random oracle model (ROM). To fully assess the post-quantum security, security analysis in the quantum random oracle model (QROM) is preferred. However, current works either lacked a QROM security proof or just followed Targhi and Unruh's proof technique (TCC-B 2016) and modified the original transformations by adding an additional hash to the ciphertext to achieve the QROM security.

In this paper, by using a novel proof technique, we present QROM security reductions for two widely used generic transformations without suffering any ciphertext overhead. Meanwhile, the security bounds are much tighter than the ones derived by utilizing Targhi and Unruh's proof technique. Thus, our QROM security proofs not only provide a solid post-quantum security guarantee for NIST Round-1 KEM schemes, but also simplify the constructions and reduce the ciphertext sizes. We also provide QROM security reductions for Hofheinz-Hövelmanns-Kiltz modular transformations (TCC 2017), which can help to obtain a variety of combined transformations with different requirements and properties.

Keywords: Quantum random oracle model
Key encapsulation mechanism · IND-CCA security
Generic transformation

© International Association for Cryptologic Research 2018
H. Shacham and A. Boldyreva (Eds.): CRYPTO 2018, LNCS 10993, pp. 96–125, 2018.
https://doi.org/10.1007/978-3-319-96878-0_4

1   Introduction

As a foundational cryptography primitive, key encapsulation mechanism (KEM) is efficient and versatile. It can be used to construct, in a black-box manner, PKE (the KEM-DEM paradigm [1]), key exchange and authenticated key exchange [2,3]. Compared with designing a full PKE scheme, the KEM construction is usually somewhat easier or more efficient. In December 2016, National Institute of Standards and Technology (NIST) announced a competition with the goal to standardize post-quantum cryptographic (PQC) algorithms including digital-signature, public-key encryption (PKE), and KEM (or key exchange) with security against quantum adversaries [4]. Among the 69 Round-1 algorithm submissions, posted in December 2017 by NIST for public to discuss and evaluate [4], there are 39 proposals for KEM constructions.

Indistinguishability against chosen-ciphertext attacks (IND-CCA) [5] is widely accepted as a standard security notion for many cryptography applications. However, the security is usually much more difficult to prove than IND-CPA (and OW-CPA) security, i.e., indistinguishability (and one-way) against chosen-plaintext attacks. Mostly, generic transformations [6,7] are used to create an IND-CCA-secure KEM from some weakly secure (OW-CPA or IND-CPA) PKEs.

Recently, considering the drawbacks of previous analysis of Fujisaki-Okamoto (FO) transformation [8,9], such as a non-tight security reduction and the need for a perfectly correct scheme, Hofheinz, Hövelmanns and Kiltz [7] revisited the KEM version of FO transformation [6] and provided a fine-grained and modular toolkit of transformations $U^{\not\perp}$, $U^{\perp}$, $U_m^{\not\perp}$, $U_m^{\perp}$, $QU_m^{\not\perp}$ and $QU_m^{\perp}$ (In what follows, these transformations will be categorized as modular FO transformations for brevity), where m (without m) means $K = H(m)$ ($K = H(m,c)$), $\not\perp$ ($\perp$) means implicit (explicit) rejection[1] and Q means adding an additional hash to the ciphertext. Combing these modular transformations, they obtained several variants of FO transformation $FO^{\not\perp}$, $FO^{\perp}$, $FO_m^{\not\perp}$, $FO_m^{\perp}$, $QFO_m^{\not\perp}$ and $QFO_m^{\perp}$ (These transformations will be categorized as FO transformations in the following).

All the (modular) FO transformations are in the random oracle model (ROM) [10]. When the KEM scheme is instantiated, the random oracle is usually replaced by a hash function, which a quantum adversary may evaluate on a quantum superposition of inputs. As a result, to fully assess post-quantum security, we should analyze security in the quantum random oracle model (QROM), as introduced in [11]. However, proving security in the QROM is quite challenging, as many classical ROM proof techniques will be invalid [11].

In [7], Hofheinz et al. presented QROM security reductions for $QU_m^{\not\perp}$, $QU_m^{\perp}$, $QFO_m^{\not\perp}$ and $QFO_m^{\perp}$. For these transformations, there is an additional hash in the ciphertext, which plays an important role in their reductions. The security reductions for $U^{\not\perp}$, $U^{\perp}$, $U_m^{\not\perp}$, $U_m^{\perp}$, $FO^{\not\perp}$, $FO^{\perp}$, $FO_m^{\not\perp}$ and $FO_m^{\perp}$ are just presented in the ROM.

---

[1] In implicit (explicit) rejection, a pseudorandom key (an abnormal symbol $\perp$) is returned for an invalid ciphertext.

Among the 39 KEM submissions, there are 35 schemes that take IND-CCA as the security goal. Particularly, 25 IND-CCA-secure KEM schemes are constructed by utilizing above transformations (see Table 1) from different PKE schemes, with different security notions (e.g., IND-CPA vs OW-CPA), and underlying hardness of certain problems over lattice, code theory and isogeny. In the submissions of LAC, Odd Manhattan, LEDAkem and SIKE, the QROM security is not considered. In the 16 submissions including FrodoKEM etc., $QFO^{\not\perp}_m$, $QFO^\perp$, $QFO^{\not\perp}_m$ and $QFO^\perp_m$ are used, where an additional hash is appended to the ciphertext. In the other 5 submissions including CRYSTALS-Kyber, LIMA, SABER, ThreeBears and Classic McEliece, the additional hash is removed according to recent works [12,13].

For the (modular) FO transformations, the underlying PKE schemes differ in the following aspects including additional hash, correctness, determinacy, and security.

- **Additional hash.** Additional hash here is a length-preserving hash function (that has the same domain and range size) appended to the ciphertext, which was first introduced by Targhi and Unruh [14] to prove the QROM security of the variants of FO transformation [8,9] and OAEP transformation [15,16]. Following Targhi and Unruh's trick, Hofheinz et al. gave the transformations $QU^{\not\perp}_m$, $QU^\perp_m$, $QFO^{\not\perp}_m$ and $QFO^\perp_m$ by adding an additional hash to the corresponding ROM constructions, and presented the QROM security reductions for them.

  Among NIST Round-1 submissions of an IND-CCA-secure KEM, 16 proposals use this trick to achieve QROM security. Intuitively, for 128-bit post-quantum security, this additional hash merely increases the ciphertext size by 256 bits [17]. However, we note that the QROM security proof in [7,14] requires the additional hash to be length-preserving. Thus, for some schemes where the message space is strictly larger than the output space of the hash function, the increasement of ciphertext size is significant. Hülsing et al. [18] tried several ways to circumvent this issue, unfortunately all straight forward approaches failed. For their specific NTRU-based KEM, additional 1128 bits are needed, which accounts for 11% of the final encapsulation size.

  In the ROM, this additional hash is clearly redundant for the constructions of an IND-CCA-secure KEM [6,7]. Some proposals, e.g., ThreeBears [19], believe this additional hash adds no security. To accomplish the QROM security proof, this additional hash was deliberately introduced, which increased the ciphertext size and complicated the implementation. Thus, a natural question is that: can we improve the QROM security proofs without suffering any ciphertext overhead for these constructions?

- **Correctness error.** For many practical post-quantum PKE schemes, e.g., DXL [20], Peikert [21], BCNS [22], New hope [23], Frodo [24], Lizard [25], Kyber [26], NTRUEncrypt [27], NTRU Prime [28], and QC-MDPC [29], there

---

[2] $QFO^\perp$ ($QFO^{\not\perp}$) is the same as $QFO^\perp_m$ ($QFO^{\not\perp}_m$) except that $K = H(m,c)$. Its security proof can be easily obtained from the one for $QFO^\perp_m$ ($QFO^{\not\perp}_m$) in [7].

exists a small correctness error δ, i.e., the probability of decryption failure in a legitimate execution of the scheme. Specially, among the KEM submissions in Table 1, there are 18 proposals that have a correctness error issue.

From a security point of view, it turns out that correctness errors not only influence the validity of a security proof, but also leak information on the private key [30]. Particularly, the chosen-ciphertext attacks by exploiting the gathered correctness errors [30,31] were demonstrated for CCA versions of NTRUEncrypt and QC-MDPC obtained by using generic transformations, whose securities were proved assuming the underlying PKEs perfectly correct. Additionally, recently, Bernstein et al. [32] showed that the HILA5 KEM [33] does not provide IND-CCA security by demonstrating a key-recovery attack in the standard IND-CCA attack model using the information obtained from the correctness errors.

To date, it is not clear how highly these correctness errors can affect the CCA security of these KEM schemes and how high these correctness errors should be to achieve a fixed security strength. To the best of our knowledge, for all previous security analyses about (modular) FO transformations except the work [7], perfect correctness, i.e., $\delta = 0$, is assumed. Therefore, QROM security analyses of above (modular) FO transformations with correctness errors into consideration are preferred.

- **Determinacy.** According to the work [7], an IND-CCA-secure KEM in the ROM can be easily constructed by applying the transformation $U_m^\perp$ (or $U_m^{\not\perp}$) to a deterministic PKE (DPKE). Saito et al. [12] showed that a DPKE can be constructed based on the concepts of the GPV trapdoor function for LWE [34], NTRU [27], the McEliece PKE [35], and the Niederreiter PKE [36]. However, the popular LWE cryptosystem and variants [37–40] are probabilistic encryption, which are referred by CRYSTALS-Kyber, EMBLEM and R.EMBLEM, FrodoKEM, KINDI, LAC, Lepton, LIMA, Lizard, NewHope, Round2, SABER and ThreeBears [4]. Particularly, of the underlying PKEs in the KEM proposals in Table 1, DPKEs just account for 28%.

- **Security notion.** IND-CPA security and OW-CPA security are widely accepted as standard security notions for PKE. In the KEM submissions in Table 1, all the underlying PKE schemes satisfy the OW-CPA security. The IND-CPA security is taken as a security goal of a PKE/KEM scheme during NIST's PQC standardization, and satisfied for most latticed-based and isogeny-based PKE schemes. FO transformations are widely used as they just require the PKE schemes to have the standard CPA security.

There are also some non-standard security notions, e.g., one-way against plaintext checking attacks (OW-PCA), one-way against validity checking attacks (OW-VA), one-way against plaintext and validity checking attacks (OW-PVCA) for PKE [6,7] and disjoint simulatability (DS) for DPKE [12]. According to [7,12], if the underlying PKE satisfies these non-standard securities, modular FO transformations can be used to construct an IND-CCA-secure KEM with a tighter security reduction. Particularly, Saito et al. [12] presented a tight security proof for $U_m^{\not\perp}$ with stronger assumptions for underlying DPKE scheme, DS security and perfect correctness, which are satisfied by Classical McEliece in Table 1.

To accurately evaluate the CCA security of the KEM proposals in Table 1 in the QROM, taking correctness error into account, we revisit the QROM security of above (modular) FO transformations without additional hash and with different assumptions for the underlying PKE scheme in terms of determinacy and security.

1.1   Our Contributions

1. For any correctness error δ ($0 \le \delta < 1$), we prove the QROM security of two generic transformations, $FO^{\not{\perp}}$ and $FO_m^{\not{\perp}}$ in [7], by reducing the standard OW-CPA security of the underlying PKE to the IND-CCA security of KEM, see Table 2.

   The obtained security bounds are both $\epsilon' \approx q\sqrt{\delta} + q\sqrt{\epsilon}$, where ϵ' is the success probability of an adversary against the IND-CCA security of the resulting KEM, ϵ is the success probability of another adversary against the OW-CPA security of the underlying PKE, and q is the total number of $\mathcal{B}$'s queries to various oracles. Our security bounds are much better than $\epsilon' \approx q\sqrt{q^2\delta + q\sqrt{\epsilon}}$, achieved by [7]. Meanwhile, the additional hash is not required as it is redundant for our security proofs. In [12], Saito et al. also obtained a same tight security bound $\epsilon' \approx q\sqrt{\epsilon}$ for a variant of $FO_m^{\not{\perp}}$, $FO_m^{\not{\perp}} = \text{TPunc} \circ U_m^{\not{\perp}}$ [3], by assuming the underlying PKE scheme IND-CPA-secure and perfectly correct (i.e., $\delta = 0$).

   With our tighter QROM security proofs, 16 KEM constructions including FrodoKEM etc., where $QFO^{\not{\perp}}$, $QFO^{\perp}$, $QFO_m^{\not{\perp}}$ and $QFO_m^{\perp}$ are used, can be simplified by cutting off the additional hash and improved in performance with respect to speed and sizes. Additionally, although LAC and SIKE are constructed by using $FO^{\not{\perp}}$ without the additional hash, the QROM security proof is not considered in their proposals. Thus, our proofs also provide a solid post-quantum security guarantee for these two KEM schemes without any additional ciphertext overhead.

2. For modular FO transformations including $U^{\not{\perp}}$, $U^{\perp}$, $U_m^{\not{\perp}}$ and $U_m^{\perp}$ in [7], we provide QROM security reductions without additional hash for any correctness error δ ($0 \le \delta < 1$), see Table 3.

   Specifically, we first define the quantum version of OW-PCA and OW-PVCA by one-way against quantum plaintext checking attacks (OW-qPCA) and one-way against quantum plaintext and (classical) validity checking attacks (OW-qPVCA) (quantum plaintext checking attacks mean that the adversary can make quantum queries to the plaintext checking oracle). For any correctness error δ ($0 \le \delta < 1$), we provide QROM security reductions for, $U^{\not{\perp}}$ from

---

[3] TPunc is a variant of T in [7].

Table 1. List of KEM submissions based on (modular) FO transformations.

Proposals	Transformations	Correctness error	DPKE?	QROM consideration?
CRYSTALS-Kyber	$\mathrm{FO}^{\not\perp}$	Y	N	Y
EMBLEM and R.EMBLEM	$\mathrm{QFO}^{\perp}$	Y	N	Y
FrodoKEM	$\mathrm{QFO}^{\not\perp}$	Y	N	Y
KINDI	$\mathrm{QFO}_m^{\not\perp}$	Y	N	Y
LAC	$\mathrm{FO}^{\not\perp}$	Y	N	N
Lepton	$\mathrm{QFO}^{\perp}$	Y	N	Y
LIMA	$\mathrm{FO}_m^{\perp}$	N^a	N	Y
Lizard	$\mathrm{QFO}^{\not\perp}$	Y	N	Y
NewHope	$\mathrm{QFO}^{\not\perp}$	Y	N	Y
NTRU-HRSS-KEM	$\mathrm{QFO}_m^{\perp}$	N	N	Y
Odd Manhattan	$\mathrm{U}_m^{\perp}$	N	N	N
OKCN-AKCN-CNKE	$\mathrm{QFO}^{\not\perp}$	Y	N	Y
Round2	$\mathrm{QFO}^{\not\perp}$	Y	N	Y
SABER	$\mathrm{FO}^{\not\perp}$	Y	N	Y
ThreeBears	$\mathrm{FO}_m^{\perp}$	Y	N	Y
Titanium	$\mathrm{QFO}^{\not\perp}$	Y	N	Y
BIG QUAKE	$\mathrm{QFO}^{\perp}$	N	N	Y
Classic McEliece	$\mathrm{U}^{\not\perp}$	N	Y	Y
DAGS	$\mathrm{QFO}_m^{\perp}$	N	N	Y
HQC	$\mathrm{QFO}^{\perp}$	Y	N	Y
LEDAkem	$\mathrm{U}_m^{\not\perp}$	Y	Y	N
LOCKER	$\mathrm{QFO}^{\perp}$	Y	N	Y
QC-MDPC	$\mathrm{QFO}_m^{\perp}$	Y	N	Y
RQC	$\mathrm{QFO}^{\perp}$	N	N	Y
SIKE	$\mathrm{FO}^{\not\perp}$	N	N	N

[a] In the round-1 submission, the LIMA team uses rejection sampling in encryption to avoid correctness errors. But they claim that they will replace the rejection sampling in encryption with a "standard" analysis of correctness errors to fix a mistake in previous analysis if LIMA survives until the second round [41].

Table 2. FO transformations from standard security assumptions.

Transformation	Underlying security	Security bound	Additional hash	Perfectly correct?
QFO$_m^{\not\perp}$ and QFO$_m^\perp$ [7]	OW-CPA	$q\sqrt{q^2\delta + q\sqrt{\epsilon}}$	Y	N
FO$'^{\not\perp}_m$ [12]	IND-CPA	$q\sqrt{\epsilon}$	N	Y
FO$^{\not\perp}$ and FO$_m^{\not\perp}$ Our work	OW-CPA	$q\sqrt{\delta} + q\sqrt{\epsilon}$	N	N

Table 3. Modular FO transformations from non-standard security assumptions.

Transformation	Underlying security	Security bound	Additional hash	DPKE	Perfectly correct?
QU$_m^\perp$ [7]	OW-PCA	$q\sqrt{\epsilon}$	Y	N	N
QU$_m^{\not\perp}$ [7]	OW-PCA	$q\sqrt{\epsilon}$	Y	N	N
U$_m^{\not\perp}$ [12]	DS	ϵ	N	Y	Y
U$^{\not\perp}$ Our work	OW-qPCA	$q\sqrt{\epsilon}$	N	N	N
U$^\perp$ Our work	OW-qPVCA	$q\sqrt{\epsilon}$	N	N	N
U$_m^{\not\perp}$ Our work	OW-CPA	$q\sqrt{\delta} + q\sqrt{\epsilon}$	N	Y	N
U$_m^{\not\perp}$ Our work	DS	$q\sqrt{\delta} + \epsilon$	N	Y	N
U$_m^\perp$ Our work	OW-VA	$q\sqrt{\delta} + q\sqrt{\epsilon}$	N	Y	N

OW-qPCA, U$^\perp$ from OW-qPVCA, U$_m^{\not\perp}$ from OW-CPA (and DS), U$_m^\perp$ from OW-VA, to IND-CCA without additional hash.

OW-qPCA (OW-qPVCA) security is just a proof artefact for simulating H. Compared with the DS security notion introduced by [12], the OW-qPCA security is less restrained and weaker. We note that the DS security notion is defined for the DPKE scheme which satisfies (1) statistical disjointness and (2) ciphertext-indistinguishability. Actually, all the DPKE schemes satisfy the OW-qPCA security as the plaintext checking oracle can be simulated by re-encryption in a quantum computer. Therefore, all the instantiations of DS-secure DPKE in [12] are also OW-qPCA-secure. Particularly, the OW-qPCA security is not restrained to the DPKE scheme. Many post-quantum PKE schemes satisfy OW-qPCA security, e.g., NTRU [27], McEliece [35], and Niederreiter [36]. Additionally, we show that the resulting PKE scheme achieved by applying the transformation T to a OW-CPA-secure PKE [7] is also OW-qPCA-secure.

Our security reductions preserve the tightness of the ones in [7,12] without additional hash for any correctness error δ ($0 \leq \delta < 1$), see Table 3. Our QROM security analyses not only provide post-quantum security guarantees for the KEM schemes constructed by using these modular FO transformations, e.g., Odd Manhattan, Classic McEliece and LEDAkem, but also can help to obtain a variety of combined transformations with different requirements and properties.

1.2   Techniques

Remove the additional hash. As explained by Targhi and Unruh [14], their proof technique strongly relies on the additional hash. In their paper, they discussed the QROM security of a variant of FO transformation from a OW-CPA-secure PKE to an IND-CCA-secure PKE. To implement the security reduction, one needs to simulate the decryption oracle without possessing the secret key. In classical proof, a RO-query list is used to simulate such an oracle. In the QROM, the simulator has no way to learn the actual content of adversarial RO queries, therefore such a RO-query list does not exist. Targhi and Unruh circumvented this issue by adding an additional length-preserving hash (modeled as a RO) to the ciphertext. In the security reduction, this additional RO is simulated by a k-wise independent function. For every output of this RO, the simulator can recover the corresponding input by inverting this function. Thereby, the simulator can answer the decryption queries without a secret key.

When considering the generic transformations from a weakly secure PKE to an IND-CCA-secure KEM, one needs to simulate the decapsulation oracle DECAPS without the secret key. Indeed, obviously, we can modify the transformations by adding an additional length-preserving hash to the ciphertext so that the simulator can carry out the decryption. Thus, using the key-derivation-function (KDF, modeled as a random oracle H), he can easily simulate the DECAPS oracle.

In [11, Theorem 6], Boneh et al. proved the QROM security of a generic hybrid encryption scheme [10], built from an injective trapdoor function and symmetric key encryption scheme. Inspired by their proof idea, we present a novel approach to simulate the DECAPS oracle[4].

The high level idea is that we associate the random oracle H (KDF in the KEM) with a secret random function H' by setting $H = H' \circ g$ such that $H'(\cdot) = $ DECAPS$(sk, \cdot)$. We demand that the function g should be indistinguishable from an injective function for any efficient quantum adversary. Thus, in the view of the adversary against the IND-CCA security of KEM, H is indeed a random oracle. Meanwhile, we can simulate the DECAPS oracle just by using H'. Note that in our simulation of the DECAPS oracle, we circumvent the decryption computation. Thereby, there is no need to read the content of adversarial RO queries, which makes it unnecessary to add an additional length-preserving hash to the ciphertext.

Tighten the security bound. When proving the IND-CCA security of KEM from the OW-CPA security of underlying PKE for $FO^{\not\perp}$ and $FO^{\not\perp}_m$, reprogramming the random oracles G and H is a natural approach. In quantum setting, the

---

[4] This method is also used by a concurrent and independent work [12].

one-way to hiding (OW2H) lemma [42, Lemma 6.2] is a practical tool to argue the indistinguishability between games where the random oracles are reprogrammed. However, the OW2H lemma inherently incurs a quadratic security loss.

To tighten the security bounds, we have to decrease the times of the usage of the OW2H lemma. [7] analyzed the QROM security of $QFO_m^{\not\perp}$ (and $QFO_m^{\perp}$) by two steps. First, they presented a QROM security reduction from the OW-CPA security of the underlying PKE to the OW-PCA security of an intermediate scheme PKE'. In this step, the random oracle G was reprogrammed, thus by using the OW2H lemma they obtained that $\epsilon'' \leq q^2\delta + q\sqrt{\epsilon}$, where ϵ'' is the success probability of an adversary against the OW-PCA security of PKE'. In the second step, they reduced the OW-PCA security of PKE' to the IND-CCA security of KEM, where the random oracles H and H'' (the additional hash) were reprogrammed. Again, by using the OW2H lemma, they gained $\epsilon' \leq q\sqrt{\epsilon''}$. Finally, combing above two bounds, they obtained the security bound of KEM, $\epsilon' \leq q\sqrt{q^2\delta + q\sqrt{\epsilon}}$. Direct combination of the modular analyses leads to twice utilization of the OW2H lemma, which makes the security bound highly non-tight.

When considering the QROM security of $FO^{\not\perp}$ and $FO_m^{\not\perp}$, instead of modular analysis, we choose to reduce the OW-CPA security of underlying PKE to the IND-CCA security of KEM directly without introducing an intermediate scheme PKE'. In this way, G and H are reprogrammed simultaneously, thus the OW2H lemma is used only once in our reductions.

We also find that the order of the games can highly affect the tightness of the security bound. If we reprogram G and H before simulating the DECAPS oracle with the secret random function H', the obtained security bound will be $q\sqrt{\epsilon} + q\sqrt{\delta}$, where the ϵ term has quadratic loss and the δ term has quartic loss. Therefore, we choose to simulate the DECAPS oracle with H' before reprogramming G and H. But, in this way, when using the OW2H lemma to argue the indistinguishability between games where G and H are reprogrammed, one has to guarantee the consistency of H and H'. We solve this by generalizing the OW2H lemma to the case where the reprogrammed oracle and other redundant oracle can be sampled simultaneously according to some joint distribution (for complete description of the generalized OW2H lemma, see Lemma 3).

Finally, our derived security bound is $q\sqrt{\delta} + q\sqrt{\epsilon}$, which is much tighter than the bound $q\sqrt{q^2\delta + q\sqrt{\epsilon}}$ obtained by [7].

1.3 Discussion

Tightness. Having a tight security reduction is a desirable property for practice cryptography, especially in large-scale scenarios. In the ROM, if we assume that the underlying PKE scheme in $FO^{\not\perp}$ and $FO_m^{\not\perp}$ is IND-CPA-secure, we can obtain a tight reduction from the IND-CPA security of underlying PKE to IND-CCA security of resulting KEM [7]. Specially, if the PKE scheme in $FO_m^{\not\perp}$ is instantiated with a Ring-LWE-based PKE scheme [39], the security of the underlying Ring-LWE problem can be reduced to the IND-CCA security of KEM [43].

In [12], Saito et al. presented a tight security reduction for $U_m^{\not\perp}$ by assuming a stronger underlying DPKE, which is only satisfied by Classic McEliece in Table 1. For the widely used $FO^{\not\perp}$ and $FO_m^{\not\perp}$, quadratic security loss still exists even assuming the IND-CPA security of the underlying PKE scheme, see Table 2. For the tight ROM security reductions in [7,43], the simulators need to make an elaborate analysis of the RO-query inputs and determine which one of the query inputs can be used to break the IND-CPA security of the underlying PKE scheme [7] or solve a decision Ring-LWE problem [43]. However, in the QROM, such a proof technique will be invalid for the reason that there is no way for the simulators to learn the RO-query inputs [44,45]. Thus, in the QROM, it is still an important open problem that whether one can develop a novel proof technique to obtain a tight reduction for $FO^{\not\perp}$ and $FO_m^{\not\perp}$ assuming standard IND-CPA security of the underlying PKE.

Implicit rejection. For most of the previous generic transformations from a OW-CPA-secure (or IND-CPA-secure) PKE to an IND-CCA-secure KEM, explicit rejection is adopted. In [7], Hofheinz et al. presented several transformations with implicit rejection. These two different versions (explicit rejection and implicit rejection) have their own merits. The transformation with implicit rejection [7] does not require the underlying PKE scheme to be γ-spread [8,9] (meaning that the ciphertexts generated by the probabilistic encryption algorithm have sufficiently large entropy), which may allow choosing better system parameters for the same security level. Whereas, the ones with explicit rejection have a relatively simple decapsulation algorithm.

In our paper, we just give QROM security reductions for the transformations with implicit rejection. It is not obvious how to extend our QROM security proofs for the transformations with explicit rejection, since the simulator has no way to tell if the submitted ciphertext is valid. In classical ROM, we usually assume the underlying PKE is γ-spread. Then, we can recognize invalid ciphertexts just by testing if they are in the RO-query list, as the probability that the adversary makes queries to the decapsulation oracle with a valid ciphertext which is not in the RO-query list is negligible [7–9,43]. Unfortunately, in the QROM, the adversary makes quantum queries to the RO, above RO-query list does not exist. Thus, the ROM proof technique for the recognition of invalid ciphertexts is invalid in the QROM. Here, we leave it as an open problem to prove the QROM security of the transformations $FO^{\not\perp}$ and $FO_m^{\not\perp}$ with explicit rejection.

2 Preliminaries

Symbol description. Denote $\mathcal{K}$, $\mathcal{M}$, $\mathcal{C}$ and $\mathcal{R}$ as key space, message space, ciphertext space and randomness space, respectively. For a finite set X, we denote the sampling of a uniform random element x by $x \xleftarrow{\$} X$, and we denote the sampling according to some distribution D by $x \leftarrow D$. By $x =?y$ we denote the integer that is 1 if $x = y$, and otherwise 0. $\Pr[P : G]$ is the probability that the predicate P holds true where free variables in P are assigned according to the program in G. Denote deterministic (probabilistic) computation of an algorithm A on input x by $y := A(x)$ ($y \leftarrow A(x)$). A^H means that the algorithm A gets access to the oracle H.

2.1 Quantum Random Oracle Model

In the ROM [10], we assume the existence of a random function H, and give all parties oracle access to this function. The algorithms comprising any cryptographic protocol can use H, as can the adversary. Thus we modify the security games for all cryptographic systems to allow the adversary to make random oracle queries.

When a random oracle scheme is implemented, some suitable hash function H is included in the specification. Any algorithm (including the adversary) replaces oracle queries with evaluations of this hash function. In quantum setting, because a quantum algorithm can evaluate H on an arbitrary superposition of inputs, we must allow the quantum adversary to make quantum queries to the random oracle. We call this the quantum random oracle model [11]. Unless otherwise specified, the queries to random oracles are quantum in our paper.

Tools. Next we state four lemmas that we will use throughout the paper. The first two lemmas have been proved in other works, and the complete proofs of last two are presented in the full version [13]. We refer the reader to [46] for basic of quantum computation. Here, we just recall two facts about quantum computation.

- Fact 1. Any classical computation can be implemented on a quantum computer.

- Fact 2. Any function that has an efficient classical algorithm computing it can be implemented efficiently as a quantum-accessible oracle.

Lemma 1 (Simulating the random oracle [47, Theorem 6.1]). *Let H be an oracle drawn from the set of $2q$-wise independent functions uniformly at random. Then the advantage any quantum algorithm making at most q queries to H has in distinguishing H from a truly random function is identically 0.*

Lemma 2 (Generic search problem [48,49]). *Let $\gamma \in [0,1]$. Let Z be a finite set. $N_1 : Z \rightarrow \{0,1\}$ is the following function: For each z, $N_1(z) = 1$ with probability p_z ($p_z \leq \gamma$), and $N_1(z) = 0$ else. Let N_2 be the function with $\forall z : N_2(z) = 0$. If an oracle algorithm A makes at most q quantum queries to N_1 (or N_2), then*

$$\left| \Pr[b = 1 : b \leftarrow A^{N_1}] - \Pr[b = 1 : b \leftarrow A^{N_2}] \right| \leq 2q\sqrt{\gamma}.$$

Particularly, the probability of A finding a z such that $N_1(z) = 1$ is at most $2q\sqrt{\gamma}$, i.e., $\Pr[N_1(z) = 1 : z \leftarrow A^{N_1}] \leq 2q\sqrt{\gamma}$.

Note. [48, Lemma 37] and [49, Theorem 1] just consider the specific case where all p_zs are equal to γ. But in our security proof, we need to consider the case where $p_z \leq \gamma$ and p_zs are in general different from each other. Fortunately, it is not difficult to verify that the proof of [48, Lemma 37] can be extended to this generic case.

The one-way to hiding (OW2H) lemma [42, Lemma 6.2] is a useful tool for reducing a hiding (i.e., indistinguishability) property to a guessing (i.e., one-wayness) property in the security proof. Roughly speaking, the lemma states that if there exists an oracle algorithm A who issuing at most q_1 queries to random oracle $\mathcal{O}_1$ can distinguish $(x, \mathcal{O}_1(x))$ from (x, y), where y is chosen uniformly at random, we can construct another oracle algorithm B who can find x by running A and measuring one of A's query. However, in our security proof, the oracle $\mathcal{O}_1$ is not a perfect random function and A can have access to other oracle $\mathcal{O}_2$ associated to $\mathcal{O}_1$. Therefore, we generalize the OW2H lemma.

Lemma 3 (One-way to hiding, with redundant oracle). *Let oracles $\mathcal{O}_1$, $\mathcal{O}_2$, input parameter inp and x be sampled from some joint distribution D, where $x \in \{0,1\}^n$ (the domain of $\mathcal{O}_1$) and $\mathcal{O}_1(x)$ is uniformly distributed on $\{0,1\}^m$ (the codomain of $\mathcal{O}_1$) conditioned on any fixed $\mathcal{O}_1(x')$ for all $x' \neq x$, $\mathcal{O}_2$, inp and x, and independent from $\mathcal{O}_2$.*

Consider an oracle algorithm $A^{\mathcal{O}_1,\mathcal{O}_2}$ that makes at most q_1 queries to $\mathcal{O}_1$ and q_2 queries to $\mathcal{O}_2$. Denote E_1 as the event that $A^{\mathcal{O}_1,\mathcal{O}_2}$ on input $(inp, x, \mathcal{O}_1(x))$ outputs 1. Reprogram $\mathcal{O}_1$ at x and replace $\mathcal{O}_1(x)$ by a uniformly random y from $\{0,1\}^m$. Denote E_2 as the event that $A^{\mathcal{O}'_1,\mathcal{O}_2}$ on input (inp, x, y) outputs 1 after $\mathcal{O}_1$ is reprogrammed, where $\mathcal{O}'_1$ is denoted as the reprogrammed $\mathcal{O}_1$. Let $B^{\mathcal{O}_1,\mathcal{O}_2}$ be an oracle algorithm that on input (inp, x) does the following: pick $i \xleftarrow{\$} \{1, \ldots, q_1\}$ and $y \xleftarrow{\$} \{0,1\}^m$, run $A^{\mathcal{O}'_1,\mathcal{O}_2}(inp, x, y)$ until the i-th query to $\mathcal{O}'_1$, measure the argument of the query in the computational basis, and output the measurement outcome. (When A makes less than i queries, B outputs $\perp \notin \{0,1\}^n$.) Let

$$\Pr[E_1] = \Pr[b' = 1 : (\mathcal{O}_1, \mathcal{O}_2, inp, x) \leftarrow D, b' \leftarrow A^{\mathcal{O}_1,\mathcal{O}_2}(inp, x, \mathcal{O}_1(x))]$$
$$\Pr[E_2] = \Pr[b' = 1 : (\mathcal{O}_1, \mathcal{O}_2, inp, x) \leftarrow D, y \xleftarrow{\$} \{0,1\}^m, b' \leftarrow A^{\mathcal{O}'_1,\mathcal{O}_2}(inp, x, y)]$$
$$P_B := \Pr[x' = x : (\mathcal{O}_1, \mathcal{O}_2, inp, x) \leftarrow D, x' \leftarrow B^{\mathcal{O}_1,\mathcal{O}_2}(inp, x)].$$

Then

$$|\Pr[E_1] - \Pr[E_2]| \leq 2q_1 \sqrt{P_B}.$$

Note that $\mathcal{O}_2$ is unchanged during the reprogramming of $\mathcal{O}_1$ at x. Thus, intuitively, $\mathcal{O}_2$ is redundant and unhelpful for A distinguishing $(x, \mathcal{O}_1(x))$ from (x, y). The complete proof of Lemma 3 is similar to the proof of the OW2H lemma [42, Lemma 6.2] and we present it in the full version [13].

Lemma 4. *Let Ω_H ($\Omega_{H'}$) be the set of all functions $H : \{0,1\}^{n_1} \times \{0,1\}^{n_2} \to \{0,1\}^m$ ($H' : \{0,1\}^{n_2} \to \{0,1\}^m$). Let $H \xleftarrow{\$} \Omega_H$, $H' \xleftarrow{\$} \Omega_{H'}$, $x \xleftarrow{\$} \{0,1\}^{n_1}$. Let $F_0 = H(x, \cdot)$, $F_1 = H'(\cdot)$ Consider an oracle algorithm A^{H,F_i} that makes at* *most q queries to H and F_i ($i \in \{0,1\}$). If x is independent from the A^{H,F_i}'s view,*

$$\left| \Pr[1 \leftarrow A^{H,F_0}] - \Pr[1 \leftarrow A^{H,F_1}] \right| \leq 2q \frac{1}{\sqrt{2^{n_1}}}.$$

We now sketch the proof of Lemma 4. For the complete proof, please refer to the full version [13].

Proof sketch. In classical setting, it is obvious that $\left| \Pr[1 \leftarrow A^{H,F_0}] - \Pr[1 \leftarrow A^{H,F_1}] \right|$ can be bounded by the probability that A performs an H-query with input $(x, *)$. As x is independent from A^{H,F_i}'s view, $\left| \Pr[1 \leftarrow A^{H,F_0}] - \Pr[1 \leftarrow A^{H,F_1}] \right| \leq q \frac{1}{2^{n_1}}$. In quantum setting, it is not well-defined that A queries $(x, *)$ from H, since H can be queried in superposition. To circumvent this problem, we follow Unruh's proof technique in [42, Lemma 6.2] and define a new adversary B who runs A, but at some random query stops and measures the query input. Let P_B be the probability that B measures x. Similarly to [42, Lemma 6.2], we can bound $\left| \Pr[1 \leftarrow A^{H,F_0}] - \Pr[1 \leftarrow A^{H,F_1}] \right|$ by $2q\sqrt{P_B}$. Since x is independent from the A^{H,F_i}'s view, $P_B = \frac{1}{2^{n_1}}$. Thus, $\left| \Pr[1 \leftarrow A^{H,F_0}] - \Pr[1 \leftarrow A^{H,F_1}] \right| \leq 2q \frac{1}{\sqrt{2^{n_1}}}$.

2.2 Cryptographic Primitives

Definition 1 (Public-key encryption). *A public-key encryption scheme* PKE = (Gen, Enc, Dec) *consists of a triple of polynomial time (in the security parameter λ) algorithms and a finite message space $\mathcal{M}$. Gen, the key generation algorithm, is a probabilistic algorithm which on input 1^λ outputs a public/secret key-pair (pk, sk). The encryption algorithm Enc, on input pk and a message $m \in \mathcal{M}$, outputs a ciphertext $c \leftarrow Enc(pk, m)$. If necessary, we make the used randomness of encryption explicit by writing $c := Enc(pk, m; r)$, where $r \xleftarrow{\$} \mathcal{R}$ ($\mathcal{R}$ is the randomness space). Dec, the decryption algorithm, is a deterministic algorithm which on input sk and a ciphertext c outputs a message $m := Dec(sk, c)$ or a special symbol $\perp \notin \mathcal{M}$ to indicate that c is not a valid ciphertext.*

Definition 2 (Correctness [7]). *A PKE is δ-correct if*

$$E[\max_{m \in \mathcal{M}} \Pr[Dec(sk, c) \neq m : c \leftarrow Enc(pk, m)]] \leq \delta,$$

where the expectation is taken over $(pk, sk) \leftarrow Gen$.

We now define four security notions for public-key encryption: one-way against chosen plaintext attacks (OW-CPA), one-way against validity checking attacks (OW-VA), one-way against quantum plaintext checking attacks (OW-qPCA) and one-way against quantum plaintext and (classical) validity checking attacks (OW-qPVCA).

Definition 3 (OW-ATK-secure PKE). *Let* PKE $= (Gen, Enc, Dec)$ *be a public-key encryption scheme with message space* $\mathcal{M}$. *For* ATK $\in$ $\{$CPA, VA, qPCA, qPVCA$\}$, *we define OW-ATK games as in Fig. 1, where*

$$O_{ATK} := \begin{cases} \bot & \text{ATK} = \text{CPA} \\ \text{VAL}(\cdot) & \text{ATK} = \text{VA} \\ \text{PCO}(\cdot, \cdot) & \text{ATK} = \text{qPCA} \\ \text{PCO}(\cdot, \cdot), \text{VAL}(\cdot) & \text{ATK} = \text{qPVCA}. \end{cases}$$

Define the OW-ATK advantage function of an adversary $\mathcal{A}$ *against PKE as* $\text{Adv}_{\text{PKE}}^{\text{OW-ATK}}(\mathcal{A}) := \Pr[\text{OW-ATK}_{\text{PKE}}^{\mathcal{A}} = 1]$.

Game OW-ATK	PCO(m,c)	VAL(c)
1 : $(pk, sk) \leftarrow Gen$	1 : **if** $m \notin \mathcal{M}$	1 : $m := Dec(sk, c)$
2 : $m^* \xleftarrow{\$} \mathcal{M}$	2 : **return** $\bot$	2 : **if** $m \in \mathcal{M}$
3 : $c^* \leftarrow Enc(pk, m^*)$	3 : **else return**	3 : **return** 1
4 : $m' \leftarrow \mathcal{A}^{O_{\text{ATK}}}(pk, c^*)$	4 : $Dec(sk, c) =?m$	4 : **else return** 0
5 : **return** $m' =?m^*$		

Fig. 1. Games OW-ATK (ATK $\in \{$CPA, VA, qPCA, qPVCA$\}$) for PKE, where O_{ATK} is defined in Definition 3. In games qPCA and qPVCA, the adversary $\mathcal{A}$ can query the PCO oracle with quantum state.

Remark. We note that the security game OW-qPCA (OW-qPVCA) is the same as OW-PCA (OW-PVCA) except the adversary $\mathcal{A}$'s queries to the PCO oracle. In OW-qPCA (OW-qPVCA) game, $\mathcal{A}$ can make quantum queries to the PCO oracle, while in OW-PCA (OW-PVCA) game only the classical queries are allowed. These two new security notions will be used in the security analysis of modular FO transformations in Sect. 4.

Definition 4 (DS-secure DPKE [12]). *Let* $D_{\mathcal{M}}$ *denote an efficiently sampleable distribution on* $\mathcal{M}$. *A DPKE scheme (Gen,Enc,Dec) with plaintext and ciphertext spaces* $\mathcal{M}$ *and* $\mathcal{C}$ *is* $D_{\mathcal{M}}$-*disjoint simulatable if there exists a PPT algorithm* S *that satisfies (1) Statistical disjointness:* $\text{DISJ}_{\text{PKE},S} := \max_{pk} \Pr[c \in Enc(pk, \mathcal{M}) : c \leftarrow S(pk)]$ *is negligible. (2) Ciphertext-indistinguishability: For any PPT adversary* $\mathcal{A}$, $\text{Adv}_{\text{PKE},D_{\mathcal{M}},S}^{\text{DS-IND}}(\mathcal{A}) := |\Pr[\mathcal{A}(pk, c^*) \to 1 : (pk, sk) \leftarrow Gen; m^* \leftarrow D_{\mathcal{M}}; c^* := Enc(pk, m^*)] - \Pr[\mathcal{A}(pk, c^*) \to 1 : (pk, sk) \leftarrow Gen; c^* \leftarrow S(pk)]|$ *is negligible.*

Definition 5 (Key encapsulation). *A key encapsulation mechanism KEM consists of three algorithms Gen, Encaps and Decaps. The key generation algorithm Gen outputs a key pair* (pk, sk). *The encapsulation algorithm Encaps, on input pk, outputs a tuple* (K, c) *where c is said to be an encapsulation of* *the key K which is contained in key space $\mathcal{K}$. The deterministic decapsulation algorithm Decaps, on input sk and an encapsulation c, outputs either a key $K := Decaps(sk, c) \in \mathcal{K}$ or a special symbol $\perp \notin \mathcal{K}$ to indicate that c is not a valid encapsulation.*

Game IND-CCA	$\mathrm{DECAPS}(sk, c)$
1 : $(pk, sk) \leftarrow Gen$	1 : **if** $c = c^*$
2 : $b \xleftarrow{\$} \{0, 1\}$	2 : **return** $\perp$
3 : $(K_0^*, c^*) \leftarrow Encaps(pk)$	3 : **else return**
4 : $K_1^* \xleftarrow{\$} \mathcal{K}$	4 : $K := Decaps(sk, c)$
5 : $b' \leftarrow \mathcal{A}^{\mathrm{DECAPS}}(pk, c^*, K_b^*)$	
6 : **return** $b' =?b$	

Fig. 2. IND-CCA game for KEM.

We now define a security notion for KEM: indistinguishability against chosen ciphertext attacks (IND-CCA).

Definition 6 (IND-CCA-secure KEM). *We define the IND-CCA game as in Fig. 2 and the IND-CCA advantage function of an adversary $\mathcal{A}$ against KEM as* $\mathrm{Adv}_{\mathrm{KEM}}^{\mathrm{IND\text{-}CCA}}(\mathcal{A}) := \left| \Pr[\mathrm{IND\text{-}CCA}_{\mathrm{KEM}}^{\mathcal{A}} = 1] - \frac{1}{2} \right|.$

We also define OW-ATK security of PKE, DS security of DPKE and IND-CCA security of KEM in the QROM, where adversary $\mathcal{A}$ can make quantum queries to random oracles. Following the work [7], we also make the convention that the number q_H of adversarial queries to a random oracle H counts the total number of times H is executed in the experiment. That is, the number of $\mathcal{A}$'s explicit queries to H plus the number of implicit queries to H made by the experiment.

3 Security Proofs for Two Generic KEM Constructions in the QROM

In this section, we revisit two generic transformations, $\mathrm{FO}^{\not\perp}$ and $\mathrm{FO}_m^{\not\perp}$, see Figs. 3 and 4. These two transformations are widely used in the post-quantum IND-CCA-secure KEM constructions, see Table 1. But, there are no QROM security proofs for them. To achieve QROM security, some proposals, e.g., FrodoKEM, followed Hofheinz et al.'s work [7] and modified $\mathrm{FO}^{\not\perp}$ and $\mathrm{FO}_m^{\not\perp}$ by adding an additional length-preserving hash function to the ciphertext. Here, we present two QROM security proofs for $\mathrm{FO}^{\not\perp}$ and $\mathrm{FO}_m^{\not\perp}$ respectively without suffering any ciphertext overhead.

Gen'	Encaps(pk)	Decaps(sk', c)
1: $(pk, sk) \leftarrow Gen$	1: $m \xleftarrow{\$} \mathcal{M}$	1: Parse $sk' = (sk, s)$
2: $s \xleftarrow{\$} \mathcal{M}$	2: $c = Enc(pk, m; G(m))$	2: $m' := Dec(sk, c)$
3: $sk' := (sk, s)$	3: $K := H(m, c)$	3: if $Enc(pk, m'; G(m')) = c$
4: return (pk, sk')	4: return (K, c)	4: return $K := H(m', c)$
		5: else return
		6: $K := H(s, c)$

Fig. 3. IND-CCA-secure KEM-I=FO$^{\not\perp}$[PKE,G,H]

Gen'	Encaps(pk)	Decaps(sk', c)
1: $(pk, sk) \leftarrow Gen$	1: $m \xleftarrow{\$} \mathcal{M}$	1: Parse $sk' = (sk, k)$
2: $k \xleftarrow{\$} \mathcal{K}^{prf}$	2: $c = Enc(pk, m; G(m))$	2: $m' := Dec(sk, c)$
3: $sk' := (sk, k)$	3: $K := H(m)$	3: if $Enc(pk, m'; G(m')) = c$
4: return (pk, sk')	4: return (K, c)	4: return $K := H(m')$
		5: else return
		6: $K := f(k, c)$

Fig. 4. IND-CCA-secure KEM-II=FO$^{\not\perp}_m$[PKE,G,H,f]

To a public-key encryption scheme PKE $= (Gen, Enc, Dec)$ with message space $\mathcal{M}$ and randomness space $\mathcal{R}$, hash functions $G : \mathcal{M} \rightarrow \mathcal{R}$, $H : \{0,1\}^* \rightarrow \{0,1\}^n$ and a pseudorandom function (PRF) f with key space $\mathcal{K}^{prf}$, we associate KEM-I=FO$^{\not\perp}$[PKE,G,H] and KEM-II= FO$^{\not\perp}_m$[PKE,G,H,f][5] shown in Figs. 3 and 4, respectively. The following two theorems establish that IND-CCA securities of KEM-I and KEM-II can both reduce to the OW-CPA security of PKE, in the QROM.

Theorem 1 (PKE OW-CPA $\overset{QROM}{\Rightarrow}$ KEM-I IND-CCA). *If PKE is δ-correct, for any IND-CCA $\mathcal{B}$ against KEM-I, issuing at most q_D queries to the decapsulation oracle* DECAPS, *at most q_G queries to the random oracle G and at most q_H queries to the random oracle H, there exists a OW-CPA adversary $\mathcal{A}$ against PKE such that* $\mathrm{Adv}^{\mathrm{IND\text{-}CCA}}_{\mathrm{KEM\text{-}I}}(\mathcal{B}) \leq 2q_H \frac{1}{\sqrt{|\mathcal{M}|}} + 4q_G\sqrt{\delta} + 2(q_G + q_H) \cdot \sqrt{\mathrm{Adv}^{\mathrm{OW\text{-}CPA}}_{\mathrm{PKE}}(\mathcal{A})}$ *and the running time of $\mathcal{A}$ is about that of $\mathcal{B}$.*

---

[5] FO$^{\not\perp}_m$ here is the generic version of FO$^{\not\perp}_m$ in [7]. In their work, such a pseudorandom function f is instantiated with $H(s, \cdot)$ (s is a random seed and contained in the secret key sk').

Proof. Let $\mathcal{B}$ be an adversary against the IND-CCA security of KEM-I, issuing at most q_D queries to DECAPS, at most q_G queries to G and at most q_H queries to H. Denote Ω_G, Ω_H and $\Omega_{H'}$ as the sets of all functions $G : \mathcal{M} \to \mathcal{R}$, $H : \mathcal{M} \times \mathcal{C} \to \mathcal{K}$ and $H' : \mathcal{C} \to \mathcal{K}$, respectively. Consider the games in Figs. 5 and 9.

GAME G_0. Since game G_0 is exactly the IND-CCA game,

$$\left| \Pr[G_0^{\mathcal{B}} \Rightarrow 1] - \frac{1}{2} \right| = \mathsf{Adv}_{\mathrm{KEM\text{-}I}}^{\mathrm{IND\text{-}CCA}}(\mathcal{B}).$$

GAME G_1. In game G_1, we change the DECAPS oracle that $H_2(c)$ is returned instead of $H(s, c)$ for an invalid encapsulation c. Define an oracle algorithm A^{H, F_i} ($i \in \{0, 1\}$), see Fig. 6. Let $H = H_3$, $F_0(\cdot) = H_3(s, \cdot)$ ($s \xleftarrow{\$} \mathcal{M}$) and $F_1 = H_2$, where H_2 and H_3 are chosen in the same way as G_0 and G_1. Then, $\Pr[G_i^{\mathcal{B}} \Rightarrow 1] = \Pr[1 \leftarrow A^{H, F_i}]$. Since the uniform secret s is chosen independently from A^{H, F_i}'s view, we can use Lemma 4 to obtain

$$\left| \Pr[G_0^{\mathcal{B}} \Rightarrow 1] - \Pr[G_1^{\mathcal{B}} \Rightarrow 1] \right| \le 2q_H \cdot \frac{1}{\sqrt{|\mathcal{M}|}}.$$

GAME G_2. Note that in game G_1, $H(m, c) = H_3(m, c)$. In game G_2, if H-query input (m, c) satisfies $g(m) = c$, the response is replaced by $H_1^g(m) = H_1 \circ g(m) = H_1(g(m)) = H_1(c)$, where

$$g(\cdot) = Enc(pk, \cdot; G(\cdot)).$$

GAMES $G_0 - G_4$	$H(m, c)$
1: $(pk, sk') \leftarrow Gen'; G \xleftarrow{\$} \Omega_G$	1: **if** $Enc(pk, m; G(m)) = c$ $//G_2 - G_4$
2: $H_1, H_2 \xleftarrow{\$} \Omega_{H'}; H_3 \xleftarrow{\$} \Omega_H$	2: **return** $H_1(c)$ $//G_2 - G_4$
3: $m^* \xleftarrow{\$} \mathcal{M}$	3: **return** $H_3(m, c)$
4: $r^* := G(m^*)$ $//G0 - G_3$	DECAPS $(c \neq c^*)$ $//G_0 - G_2$
5: $r^* \xleftarrow{\$} \mathcal{R}$ $//G_4$	1: Parse $sk' = (sk, s)$
6: $c^* := Enc(pk, m^*; r^*)$	2: $m' := Dec(sk, c)$
7: $k_0^* := H(m^*, c^*)$	3: **if** $Enc(pk, m'; G(m')) = c$
8: $k_0^* \xleftarrow{\$} \mathcal{K}$ $//G_4$	4: **return** $K := H(m', c)$
9: $k_1^* \xleftarrow{\$} \mathcal{K}$	5: **else return**
10: $b \xleftarrow{\$} \{0, 1\}$	6: $K := H(s, c)$ $//G_0$
11: $b' \leftarrow \mathcal{B}^{G, H, \mathrm{DECAPS}}(pk, c^*, k_b^*)$	7: $K := H_2(c)$ $//G_1 - G_2$
12: **return** $b' =?b$	DECAPS $(c \neq c^*)$ $//G_3 - G_4$
	1: **return** $K := H_1(c)$

Fig. 5. Games G_0-G_4 for the proof of Theorem 1

A^{H,F_i}	DECAPS $(c \neq c^*)$
1 : $(pk, sk) \leftarrow Gen; G \xleftarrow{\$} \Omega_G$	1 : $m' := Dec(sk, c)$
2 : $m^* \xleftarrow{\$} \mathcal{M}$	2 : **if** $Enc(pk, m'; G(m')) = c$
3 : $r^* := G(m^*)$	3 : **return** $K := H(m', c)$
4 : $c^* := Enc(pk, m^*; r^*)$	4 : **else return**
5 : $k_0^* := H(m^*, c^*); k_1^* \xleftarrow{\$} \mathcal{K}$	5 : $K := F_i(c)$
6 : $b \xleftarrow{\$} \{0, 1\}$	
7 : $b' \leftarrow \mathcal{B}^{G,H,\text{DECAPS}}(pk, c^*, k_b^*)$	
8 : **return** $b' =?b$	

Fig. 6. A^{H,F_i} for the proof of Theorem 1.

$A^N(pk, sk)$	$\widetilde{G}(m)$
1 : Pick $2q_G$-wise function f	1 : **if** $N(m) = 0$
2 : $b'' \leftarrow B^{\widetilde{G}}(pk, sk)$	2 : $\widetilde{G}(m) = Sample(\mathcal{R} \setminus \mathcal{R}_{\text{bad}}(pk, sk, m); f(m))$
3 : **return** b''	3 : **else**
	4 : $\widetilde{G}(m) = Sample(\mathcal{R}_{\text{bad}}(pk, sk, m); f(m))$
	5 : **return** $\widetilde{G}(m)$

Fig. 7. A^N for the proof of Theorem 1

Given (pk, sk) and $m \in \mathcal{M}$, let

$$\mathcal{R}_{\text{bad}}(pk, sk, m) := \{r \in \mathcal{R} : Dec(sk, Enc(pk, m; r)) \neq m\}$$

denote the set of "bad" randomness. Define

$$\delta(pk, sk, m) = \frac{|\mathcal{R}_{\text{bad}}(pk, sk, m)|}{|\mathcal{R}|}$$

as the fraction of bad randomness and $\delta(pk, sk) = \max_{m \in \mathcal{M}} \delta(pk, sk, m)$. With this notation $\delta = \mathbf{E}[\delta(pk, sk)]$, where the expectation is taken over $(pk, sk) \leftarrow Gen$.

Let G' be a random function such that $G'(m)$ is sampled from the uniform distribution in $\mathcal{R} \setminus \mathcal{R}_{\text{bad}}(pk, sk, m)$. Let

$$g'(\cdot) = Enc(pk, \cdot; G'(\cdot)).$$

Distinctly, g' is an injective function. $H_1 \circ g'$ has the same output distribution as H in G_1. Thus, distinguishing G_2 from G_1 is equivalent to distinguishing g from g', which is essentially the distinguishing problem between G and G'.

Let N_1 be the function such that $N_1(m)$ is sampled from the Bernoulli distribution $B_{\delta(pk,sk,m)}$, i.e., $\Pr[N_1(m) = 1] = \delta(pk, sk, m)$ and $\Pr[N_1(m) = 0] = 1 - \delta(pk, sk, m)$. Let N_2 be a constant function that always outputs 0 for any input. Next, we will show that any algorithm that distinguishes G from G' can be converted into an algorithm that distinguishes N_1 from N_2.

For any efficient quantum adversary $B^{\widetilde{G}}(pk, sk)$, we can construct an adversary $A^N(pk, sk)$ as in Fig. 7. $Sample(\mathcal{Y})$ is a probabilistic algorithm that returns a uniformly distributed $y \overset{\$}{\leftarrow} \mathcal{Y}$. $Sample(\mathcal{Y}; f(m))$ denotes the deterministic execution of $Sample(\mathcal{Y})$ using explicitly given randomness $f(m)$.

Note that $\widetilde{G} = G$ if $N = N_1$ and $\widetilde{G} = G'$ if $N = N_2$. Thus, for any fixed (pk, sk) that is generated by Gen, $\Pr[1 \leftarrow A^{N_1} : (pk, sk)] = \Pr[1 \leftarrow B^G : (pk, sk)]$ and $\Pr[1 \leftarrow A^{N_2} : (pk, sk)] = \Pr[1 \leftarrow B^{G'} : (pk, sk)]$. Conditioned on a fixed (pk, sk) we obtain by Lemma 2

$$\left| \Pr[1 \leftarrow B^G : (pk, sk)] - \Pr[1 \leftarrow B^{G'} : (pk, sk)] \right|$$
$$= \left| \Pr[1 \leftarrow A^{N_1} : (pk, sk)] - \Pr[1 \leftarrow A^{N_2} : (pk, sk)] \right| \leq 2q_G \sqrt{\delta(pk, sk)}.$$

Note that $\left| \Pr[G_1^B \Rightarrow 1 : (pk, sk)] - \Pr[G_2^B \Rightarrow 1 : (pk, sk)] \right|$ can be bounded by the maximum distinguishing probability between G and G' for $B^{\widetilde{G}}(pk, sk)$. Thus,

$$\left| \Pr[G_1^B \Rightarrow 1 : (pk, sk)] - \Pr[G_2^B \Rightarrow 1 : (pk, sk)] \right| \leq 2q_G \sqrt{\delta(pk, sk)}.$$

By averaging over $(pk, sk) \leftarrow Gen$ we finally obtain

$$\left| \Pr[G_1^B \Rightarrow 1] - \Pr[G_2^B \Rightarrow 1] \right| \leq 2q_G \sqrt{\delta}.$$

GAME G_3. In game G_3, the DECAPS oracle is changed that it makes no use of the secret key sk' any more. When B queries the DECAPS oracle on c ($c \neq c^*$), $K := H_1(c)$ is returned as the response. Let $m' := Dec(sk, c)$ and consider the following two cases.

Case 1: $Enc(pk, m'; G(m')) = c$. In this case, $H(m', c) = H_1(c)$. Thus, both DECAPS oracles in G_2 and G_3 return the same value.

Case 2: $Enc(pk, m'; G(m')) \neq c$. Random values $H_2(c)$ and $H_1(c)$ are returned in G_2 and G_3 respectively. In G_2, H_2 is a random function independent of the oracles G and H, thus $H_2(c)$ is uniform at random in B's view. In G_3, B's queries to H can only help him get access to H_1 at $\hat{c}$ such that $g(\hat{m}) = \hat{c}$ for some $\hat{m}$. Consequently, if B can not find a m'' such that $g(m'') = c$, $H_1(c)$ is also a fresh random key just like $H_2(c)$ in his view. Since $m'' \neq m'$, finding such an m'' is exactly the event E that B finds a plaintext m'' such that $Dec(sk, g(m'')) \neq m''$. That is, in this case, if E does not happen, the output distributions of the DECAPS oracles in G_2 and G_3 are same in B's view.

As a result, G_2 and G_3 only differ when E happens. By [7, Lemma 4.3], we know that if B can find a plaintext m'' such that $Dec(sk, g(m'')) \neq m''$ with at most

q_G quantum queries to g, we can easily construct another adversary $\mathcal{B}'$ who can find a plaintext m'' such that $N_1(m'') = 1$ with at most q_G quantum queries to N_1. Considering that the PKE scheme is δ-correct, we can derive the upper bound of $\Pr[E]$ by utilizing Lemma 2, $\Pr[E] \leq \Pr[N_1(m'') = 1 : (pk, sk) \leftarrow Gen, m'' \leftarrow \mathcal{B}'^{N_1}] \leq 2q_G\sqrt{\delta}$. Therefore,

$$\left|\Pr[G_2^{\mathcal{B}} \Rightarrow 1] - \Pr[G_3^{\mathcal{B}} \Rightarrow 1]\right| \leq \Pr[E] \leq 2q_G\sqrt{\delta}.$$

GAME G_4. In game G_4, r^* and k_0^* are chosen uniformly at random from $\mathcal{R}$ and $\mathcal{K}$, respectively. In this game, bit b is independent from $\mathcal{B}$'s view. Hence,

$$\Pr[G_4^{\mathcal{B}} \Rightarrow 1] = \frac{1}{2}.$$

Note that in this game we reprogram the oracles G and H on inputs m^* and (m^*, c^*) respectively. In classical setting, this will be unnoticed unless the event QUERY that $\mathcal{B}$ queries G on m^* or H on (m^*, c^*) happens. Then we can argue that G_3 and G_4 are indistinguishable until QUERY happens. In quantum setting, due to the quantum queries to G and H, the case is complicated and we will use Lemma 3 to bound $\left|\Pr[G_3^{\mathcal{B}} \Rightarrow 1] - \Pr[G_4^{\mathcal{B}} \Rightarrow 1]\right|$. Note that (m^*, c^*) is a valid plaintext-ciphertext pair, i.e., $g(m^*) = c^*$. Therefore, $H(m^*, c^*) = H_1(c^*) = H_1^g(m^*)$. Actually, we just reprogram G and H_1^g at m^*.

Let $(G \times H_1^g)(x) := (G(x), H_1^g(x))^6$. H_1^g and H_3 are internal random oracles that $\mathcal{B}$ can have access to only by querying the oracle H. Then, the number of total queries to $G \times H_1^g$ is at most $q_G + q_H$. Let H_1' be the function such that $H_1'(g(m^*)) = \perp$ and $H_1' = H_1$ everywhere else. H_1' is exactly the DECAPS oracle in G_3 and G_4 and unchanged during the reprogramming of $G \times H_1^g$.

Let $A^{G \times H_1^g, H_1'}$ be an oracle algorithm that has quantum access to $G \times H_1^g$ and H_1', see Fig. 8. Sample G, H_1, H_1^g and pk in the same way as G_3 and G_4, i.e., $(pk, sk') \leftarrow Gen', G \xleftarrow{\$} \Omega_G, H_1 \xleftarrow{\$} \Omega_{H'}, H_1^g := H_1 \circ g$. Let $m^* \xleftarrow{\$} \mathcal{M}$.

Then, if $r^* := G(m^*)$ and $k_0^* := H_1^g(m^*)$, $A^{G \times H_1^g, H_1'}$ on input $(pk, m^*, (r^*, k_0^*))$ perfectly simulates G_3. And, if $r^* \xleftarrow{\$} \mathcal{R}$ and $k_0^* \xleftarrow{\$} \mathcal{K}$, $A^{G \times H_1^g, H_1'}$ on input $(pk, m^*, (r^*, k_0^*))$ perfectly simulates G_4. Let $B^{G \times H_1^g, H_1'}$ be an oracle algorithm that on input (pk, m^*) does the following: pick $i \xleftarrow{\$} \{1, \ldots, q_G + q_H\}$, $r^* \xleftarrow{\$} \mathcal{R}$ and $k_0^* \xleftarrow{\$} \mathcal{K}$, run $A^{G \times H_1^g, H_1'}(pk, m^*, (r^*, k_0^*))$ until the i-th query to $G \times H_1^g$, measure the argument of the query in the computational basis, output the measurement outcome (when $A^{G \times H_1^g, H_1'}$ makes less than i queries, output $\perp$). Define game G_5 as in Fig. 9. Then, $\Pr[B^{G \times H_1^g, H_1'} \Rightarrow m^*] = \Pr[G_5^{\mathcal{B}} \Rightarrow 1]$.

Applying Lemma 3 with $\mathcal{O}_1 = G \times H_1^g$, $\mathcal{O}_2 = H_1'$, $inp = pk$, $x = m^*$ and $y = (r^*, k_0^*)$, we have

$$\left|\Pr[G_3^{\mathcal{B}} \Rightarrow 1] - \Pr[G_4^{\mathcal{B}} \Rightarrow 1]\right| \leq 2(q_G + q_H)\sqrt{\Pr[G_5^{\mathcal{B}} \Rightarrow 1]}.$$

---

[6] Note that if one wants to make queries to G (or H_1^g) by accessing to $G \times H_1^g$, he just needs to prepare a uniform superposition of all states in the output register responding to H_1^g (or G). This trick [14,50,51] has been used to ignore part of the output of an oracle.

$$A^{G \times H_1^g, H_1'}(pk, m^*, (r^*, k_0^*)) \qquad H(m, c)$$

1 :	$H_3 \xleftarrow{\$} \Omega_H$
2 :	$c^* := Enc(pk, m^*; r^*)$
3 :	$k_1^* \xleftarrow{\$} \mathcal{K}$
4 :	$b \xleftarrow{\$} \{0, 1\}$
5 :	$b' \leftarrow \mathcal{B}^{G, H, \text{DECAPS}}(pk, c^*, k_b^*)$
6 :	**return** $b' =?b$

$H(m, c)$

1 :   **if** $g(m) = c$
2 :   **return** $H_1^g(m)$
3 :   **else return** $H_3(m, c)$

DECAPS $(c \neq c^*)$

1 :   **return** $K := H_1'(c)$

Fig. 8. $A^{G \times H_1^g, H_1'}$ for the proof of Theorem 1.

GAMES G_5

1 : $i \xleftarrow{\$} \{1, \ldots, q_G + q_H\}, (pk, sk') \leftarrow Gen', G \xleftarrow{\$} \Omega_G$
2 : $H_1 \xleftarrow{\$} \Omega_{H'}, H_3 \xleftarrow{\$} \Omega_H$
3 : $m^* \xleftarrow{\$} \mathcal{M}, r^* \xleftarrow{\$} \mathcal{R}$
4 : $c^* := Enc(pk, m^*; r^*)$
5 : $k_0^*, k_1^* \xleftarrow{\$} \mathcal{K}$
6 : $b \xleftarrow{\$} \{0, 1\}$
7 :   run $\mathcal{B}^{G, H, \text{DECAPS}}(pk, c^*, k_b^*)$ until the i–th query to $G \times H_1^g$
8 :   measure the argument $\hat{m}$
9 :   **return** $\hat{m} =?m^*$

$H(m, c)$ | DECAPS $(c \neq c^*)$

1 :   **if** $Enc(pk, m; G(m)) = c$ 1 :   **return** $K := H_1(c)$
2 :   **return** $H_1(c)$
3 :   **else return** $H_3(m, c)$

Fig. 9. Game G_5 for the proof of Theorem 1

Next, we construct an adversary $\mathcal{A}$ against the OW-CPA security of the PKE scheme such that $\text{Adv}_{\text{PKE}}^{\text{OW-CPA}}(\mathcal{A}) = \Pr[G_5^{\mathcal{B}} \Rightarrow 1]$. The adversary $\mathcal{A}$ on input $(1^\lambda, pk, c)$ does the following:

1. Run the adversary $\mathcal{B}$ in Game G_5.
2. Use a $2q_G$-wise independent function and two different $2q_H$-wise independent functions to simulate the random oracles G, H_1 and H_3 respectively. The random oracle H is simulated in the same way as the one in game G_5.
3. Answer the decapsulation queries by using the DECAPS oracle in Fig. 9.

4. Select $k^* \xleftarrow{\$} \mathcal{K}$ and respond to $\mathcal{B}$'s challenge query with (c, k^*).

5. Select $i \xleftarrow{\$} \{1, \dots, q_G + q_H\}$, measure the argument $\hat{m}$ of i-th query to $G \times H_1^g$ and output $\hat{m}$.

According to Lemma 1, $\mathrm{Adv}_{\mathrm{PKE}}^{\mathrm{OW\text{-}CPA}}(\mathcal{A}) = \Pr[G_5^{\mathcal{B}} \Rightarrow 1]$. Finally, combing this with the bounds derived above, we can conclude that

$$\mathrm{Adv}_{\mathrm{KEM\text{-}I}}^{\mathrm{IND\text{-}CCA}}(\mathcal{B}) \leq 2q_H \frac{1}{\sqrt{|\mathcal{M}|}} + 4q_G\sqrt{\delta} + 2(q_G + q_H) \cdot \sqrt{\mathrm{Adv}_{\mathrm{PKE}}^{\mathrm{OW\text{-}CPA}}(\mathcal{A})}.$$

$\square$

Theorem 2 (PKE OW-CPA $\overset{QROM}{\Rightarrow}$ KEM-II IND-CCA). *If PKE is δ-correct, for any IND-CCA $\mathcal{B}$ against KEM-II, issuing at most q_D classical queries to the decapsulation oracle DECAPS and at most q_G (q_H) queries to random oracle G (H), there exist a quantum OW-CPA adversary $\mathcal{A}$ against PKE and an adversary $\mathcal{A}'$ against the security of PRF with at most q_D classical queries such that $\mathrm{Adv}_{\mathrm{KEM\text{-}II}}^{\mathrm{IND\text{-}CCA}}(\mathcal{B}) \leq \mathrm{Adv}_{\mathrm{PRF}}(\mathcal{A}') + 4q_G\sqrt{\delta} + 2(q_H + q_G) \cdot \sqrt{\mathrm{Adv}_{\mathrm{PKE}}^{\mathrm{OW\text{-}CPA}}(\mathcal{A})}$ and the running time of $\mathcal{A}$ is about that of $\mathcal{B}$.*

The only difference between KEM-I and KEM-II is the KDF function. In KEM-I, $K = H(m, c)$, while $K = H(m)$ in KEM-II. Note that given pk and random oracle G, c is determined by m. The proof of Theorem 2 is similar to the one of Theorem 1 and we present it in the full version [13].

4   Modular Analysis of FO Transformation in the QROM

In [7], Hofheinz et al. introduced seven modular transformations T, $\mathrm{U}^{\not{\perp}}$, $\mathrm{U}^{\perp}$, $\mathrm{U}_m^{\not{\perp}}$, $\mathrm{U}_m^{\perp}$, $\mathrm{QU}_m^{\not{\perp}}$ and $\mathrm{QU}_m^{\perp}$. But, they just presented QROM security reductions for the transformations T, $\mathrm{QU}_m^{\not{\perp}}$ and $\mathrm{QU}_m^{\perp}$. Different from the transformations $\mathrm{U}^{\not{\perp}}$, $\mathrm{U}^{\perp}$, $\mathrm{U}_m^{\not{\perp}}$ and $\mathrm{U}_m^{\perp}$, the transformations $\mathrm{QU}_m^{\not{\perp}}$ and $\mathrm{QU}_m^{\perp}$ have an additional length-preserving hash in the ciphertext, thus they can follow the proof technique in [14,52] to give QROM security reductions for them. As they pointed [14], their QROM security reductions quite rely on this additional hash. And, QROM security reductions for $\mathrm{U}^{\not{\perp}}$, $\mathrm{U}^{\perp}$, $\mathrm{U}_m^{\not{\perp}}$ and $\mathrm{U}_m^{\perp}$ are missing in [7]. In [12], Saito et al. presented a tight QROM security reduction for $\mathrm{U}_m^{\not{\perp}}$ with stronger assumptions for underlying DPKE scheme, DS-security and perfect correctness.

In this section, we revisit the transformations $\mathrm{U}^{\not{\perp}}$, $\mathrm{U}^{\perp}$, $\mathrm{U}_m^{\not{\perp}}$ and $\mathrm{U}_m^{\perp}$, and argue their QROM security without any modification to the constructions and with correctness error into consideration. [7] has shown that the transformation T can turn a OW-CPA-secure PKE into a OW-PCA-secure PKE in the QROM. In Sect. 4.1, we first show that the resulting PKE scheme by applying T to a OW-CPA-secure PKE is also OW-qPCA-secure. The QROM security reduction for $\mathrm{U}^{\not{\perp}}$ ($\mathrm{U}^{\perp}$) from the OW-qPCA (OW-qPVCA) security of PKE to the IND-CCA security of KEM is given in Sect. 4.2 (4.3). In Sect. 4.4, we show that $\mathrm{U}_m^{\not{\perp}}$ ($\mathrm{U}_m^{\perp}$) transforms any OW-CPA-secure or DS-secure (OW-VA-secure) DPKE into an IND-CCA-secure KEM in the QROM.

4.1   T: from OW-CPA to OW-qPCA in the QROM

To a public-key encryption PKE = (Gen, Enc, Dec) with message space $\mathcal{M}$ and randomness space R, and a hash function $G : \mathcal{M} \rightarrow \mathcal{R}$, we associate PKE$'$ = $T[\text{PKE}, G]$. The algorithms of PKE$'$ = (Gen, Enc', Dec') are defined in Fig. 10.

Theorem 3 (PKE OW-CPA $\overset{QROM}{\Rightarrow}$ PKE$'$ OW-qPCA). *If PKE is δ-correct, for any OW-qPCA $\mathcal{B}$ against PKE$'$, issuing at most q_G quantum queries to the random oracle G and at most q_P quantum queries to the plaintext checking oracle* PCO, *there exists a OW-CPA adversary $\mathcal{A}$ against PKE such that* $\text{Adv}_{\text{PKE}'}^{\text{OW-qPCA}}(\mathcal{B}) \leq 2q_G \cdot \sqrt{\delta} + (1 + 2q_G) \cdot \sqrt{\text{Adv}_{\text{PKE}}^{\text{OW-CPA}}(\mathcal{A})}$ *and the running time of $\mathcal{A}$ is about that of $\mathcal{B}$.*

The proof is essentially the same as the one of [7, Theorem 4.4] except the argument about the difference in $\mathcal{B}$'s success probability between game G_0 and game G_1. Game G_0 is exactly the original OW-qPCA game. In game G_1, the PCO oracle is replaced by a simulation that $Enc(pk, m; G(m)) = ?c$ is returned for the query input (m, c). As pk is public and G is a quantum random oracle, such a PCO simulation can be queried on a quantum superposition of inputs. Note that G_0 and G_1 are indistinguishable unless there exits an adversary who issuing at most q_G queries to G can distinguish N_1 from a constant function N_2 that always outputs 0 for any input, where $N_1(m) = 0$ if $Dec(sk, Enc(pk, m; G(m))) = m$, and otherwise $N_1(m) = 1$. Thus, using Lemma 2, we can obtain that $\left| \Pr[G_0^{\mathcal{B}} \Rightarrow 1] - \Pr[G_1^{\mathcal{B}} \Rightarrow 1] \right| \leq 2q_G \cdot \sqrt{\delta}$. Then, following the security proof of [7, Theorem 4.4], we can easily prove Theorem 3.

$Enc'(pk, m)$	$Dec'(sk, c)$
1 : $c = Enc(pk, m; G(m))$	1 : $m' := Dec(sk, c)$
2 : **return** c	2 : **if** $Enc(pk, m'; G(m')) = c$
	3 : **return** m'
	4 : **else return** $\perp$

Fig. 10. OW-qPCA-secure PKE$'$ = $T[\text{PKE}, G]$

4.2   U$^{\not\perp}$: from OW-qPCA to IND-CCA in the QROM

To a public-key encryption PKE$'$ = (Gen', Enc', Dec') and a hash function H, we associate KEM-III = $U^{\not\perp}[\text{PKE}', H]$. The algorithms of KEM-III = $(Gen, Encaps, Decaps)$ are defined in Fig. 11.

Gen	Encaps(pk)	Decaps(sk', c)
1 : $(pk, sk) \leftarrow Gen'$	1 : $m \xleftarrow{\$} \mathcal{M}$	1 : Parse $sk' = (sk, s)$
2 : $s \xleftarrow{\$} \mathcal{M}$	2 : $c \leftarrow Enc'(pk, m)$	2 : $m' := Dec'(sk, c)$
3 : $sk' := (sk, s)$	3 : $K := H(m, c)$	3 : if $m' = \perp$
4 : **return** (pk, sk')	4 : **return** (K, c)	4 : **return** $K := H(s, c)$
		5 : **else return**
		6 : $K := H(m', c)$

Fig. 11. IND-CCA-secure KEM-III $= U^{\not=}[\text{PKE}', H]$

Theorem 4 (PKE' OW-qPCA $\overset{QROM}{\Rightarrow}$ KEM-III IND-CCA). *If* PKE' *is δ-correct, for any IND-CCA $\mathcal{B}$ against KEM-III, issuing at most q_D (classical) queries to the decapsulation oracle* DECAPS *and at most q_H queries to the quantum random oracle H, there exists a quantum OW-qPCA adversary $\mathcal{A}$ against PKE' that makes at most q_H queries to the* PCO *oracle such that* $\text{Adv}_{\text{KEM-III}}^{\text{IND-CCA}}(\mathcal{B}) \leq 2q_H \frac{1}{\sqrt{|\mathcal{M}|}} + 2q_H \cdot \sqrt{\text{Adv}_{\text{PKE}'}^{\text{OW-qPCA}}(\mathcal{A})}$ *and the running time of $\mathcal{A}$ is about that of $\mathcal{B}$.*

The proof skeleton of Theorem 4 is essentially the same as the one of Theorem 1. Here, we briefly state the main differences. The complete proof is presented in the full version [13].

In KEM-I, the randomness used in the encryption algorithm is determined by the random oracle G. Given a plaintext m, we can deterministically evaluate the ciphertext $c = Enc(pk, m; G(m))$. Thus, we can divide H-query inputs (m, c) into two categories by judging if (m, c) is a matching plaintex-ciphertext pair (i.e., $c = Enc(pk, m; G(m))$) or not. In KEM-III, the encryption algorithm may be probabilistic, thus the above method will be invalid. Instead, we can query the PCO oracle to judge whether (m, c) is a matching plaintex-ciphertext pair. If $\text{PCO}(m, c) = 1$, the random oracle H returns $H_1(c)$, otherwise $H_3(m, c)$. To simulate the random oracle H, we make quantum queries to PCO (this is the reason why we require the scheme PKE' to be OW-qPCA-secure). Note that it is impossible that $\text{PCO}(m_1, c) = \text{PCO}(m_2, c) = 1$ for $m_1 \neq m_2$. Thus, H is perfectly simulated without introducing the δ term. As $\mathcal{B}$'s queries to H can only help him get access to H_1 at c such that $Dec'(sk, c) = \hat{m}$ for some $\hat{m} \neq \perp$, the DECAPS oracle can be perfectly simulated by H_1. Therefore, different from the security bounds obtained in Theorems 1 and 2, the δ term is removed with the OW-qPCA security of underlying PKE.

Gen	Encaps(pk)	Decaps$^{\perp}$(sk, c)
1 : $(pk, sk) \leftarrow Gen'$	1 : $m \xleftarrow{\$} \mathcal{M}$	1 : $m' := Dec'(sk, c)$
2 : **return** (pk, sk)	2 : $c \leftarrow Enc'(pk, m)$	2 : **if** $m' = \perp$
	3 : $K := H(m, c)$	3 : **return** $\perp$
	4 : **return** (K, c)	4 : **else return**
		5 : $K := H(m', c)$

Fig. 12. IND-CCA-secure KEM-IV = $U^{\perp}[\text{PKE}', H]$

4.3 $U^{\perp}$: from OW-qPVCA to IND-CCA in the QROM

To a public-key encryption PKE' = (Gen', Enc', Dec') and a hash function H, we associate KEM-IV = $U^{\perp}[\text{PKE}', H]$. We remark that $U^{\perp}$ is essentially the transformation [6, Table 2], a KEM variant of the REACT/GEM transformations [53,54]. The algorithms of KEM-IV = $(Gen, Encaps, Decaps^{\perp})$ are defined in Fig. 12.

Theorem 5 (PKE' OW-qPVCA $\overset{QROM}{\Rightarrow}$ KEM-IV IND-CCA). *If* PKE' *is* δ-correct, for any IND-CCA $\mathcal{B}$ against KEM-IV, issuing at most q_D (classical) queries to the decapsulation oracle DECAPS and at most q_H queries to the quantum random oracle H, there exists a OW-qPVCA adversary $\mathcal{A}$ against PKE' that makes at most q_H queries to the PCO oracle and at most q_D queries to the VAL oracle such that $\text{Adv}_{\text{KEM-IV}}^{\text{IND-CCA}}(\mathcal{B}) \leq 2q_H \cdot \sqrt{\text{Adv}_{\text{PKE}'}^{\text{OW-qPVCA}}(\mathcal{A})}$ and the running time of $\mathcal{A}$ is about that of $\mathcal{B}$.*

The only difference between KEM-III and KEM-IV is the response to the invalid ciphertext in the decapsulation algorithm. When the ciphertext c is invalid, the decapsulation algorithm in KEM-III returns a pseudorandom key related to c. In this way, whatever the ciphertext (valid or invalid) is submitted, the return values have the same distribution. As a result, $\mathcal{A}$ can easily simulate the decapsulation oracle DECAPS without recognition of the invalid ciphertexts. While the decapsulation algorithm in KEM-IV returns $\perp$ when the submitted c is invalid. Thus, in order to simulate DECAPS, $\mathcal{A}$ needs to judge if the ciphertext c is valid. As we assume that the scheme PKE' is OW-qPVCA-secure, $\mathcal{A}$ can query the VAL oracle to fulfill such a judgement. Then, it is easy to verify that by using the same proof method in Theorem 4 we can obtain the desired security bound.

4.4 $U_m^{\not\perp}/U_m^{\perp}$: from OW-CPA/OW-VA to IND-CCA for Deterministic Encryption in the QROM

The transformation $U_m^{\not\perp}$ ($U_m^{\perp}$) is a variant of $U^{\not\perp}$ ($U^{\perp}$) that derives the KEM key as $K = H(m)$, instead of $K = H(m, c)$. To a deterministic public-key encryption scheme $PKE' = (Gen', Enc', Dec')$ with message space $\mathcal{M}$, a hash function $H : \mathcal{M} \rightarrow \mathcal{K}$, and a pseudorandom function f with key space $\mathcal{K}^{prf}$, we associate KEM-V $= U_m^{\not\perp}[PKE', H, f]$ and KEM-VI $= U_m^{\perp}[PKE', H]$ shown in Figs. 13 and 14, respectively.

Gen	$Encaps(pk)$	$Decaps(sk', c)$
1 : $(pk, sk) \leftarrow Gen'$	1 : $m \xleftarrow{\$} \mathcal{M}$	1 : Parse $sk' = (sk, k)$
2 : $k \xleftarrow{\$} \mathcal{K}^{prf}$	2 : $c := Enc'(pk, m)$	2 : $m' := Dec'(sk, c)$
3 : $sk' := (sk, k)$	3 : $K := H(m)$	3 : if $Enc'(pk, m') = c$
4 : return (pk, sk')	4 : return (K, c)	4 : return $K := H(m')$
		5 : else return
		6 : $K := f(k, c)$

Fig. 13. IND-CCA-secure KEM-V $= U_m^{\not\perp}[PKE', H, f]$

Gen	$Encaps(pk)$	$Decaps(sk, c)$
1 : $(pk, sk) \leftarrow Gen'$	1 : $m \xleftarrow{\$} \mathcal{M}$	1 : $m' := Dec(sk, c)$
2 : return (pk, sk)	2 : $c := Enc'(pk, m)$	2 : if $Enc'(pk, m') = c$
	3 : $K := H(m)$	3 : return $K := H(m')$
	4 : return (K, c)	4 : else return $\perp$

Fig. 14. IND-CCA-secure KEM-VI $= U_m^{\perp}[PKE', H]$

We note that for a deterministic PKE scheme the OW-PCA security is equivalent to the OW-CPA security as we can simulate the Pco oracle via re-encryption during the proof. Thus, combing the proofs of Theorem 2, Theorem 4, Theorem 5 and [12, Theorem 4.1], we can easily obtain the following two theorems.

Theorem 6 (PKE' OW-CPA $\overset{QROM}{\Rightarrow}$ KEM-V IND-CCA). *If* PKE' *is* δ-*correct and deterministic, for any IND-CCA $\mathcal{B}$ against KEM-V, issuing at most q_E quantum queries to the encryption oracle*[7]*, at most q_D (classical) queries to the decapsulation oracle* Decaps *and at most q_H quantum queries to the random oracle H, there exist a quantum OW-CPA adversary $\mathcal{A}$ against* PKE'*, an adversary $\mathcal{A}'$ against the security of* PRF *with at most q_D classical queries and an adversary $\mathcal{C}$ against the $U_\mathcal{M}$-DS security with a simulator S of* PKE'

---

[7] For the deterministic scheme PKE', given public key pk, quantum adversary $\mathcal{B}$ can execute the encryption algorithm Enc' in a quantum computer.

($U_{\mathcal{M}}$ is the uniform distribution in $\mathcal{M}$) such that $\mathrm{Adv}_{\mathrm{KEM\text{-}V}}^{\mathrm{IND\text{-}CCA}}(\mathcal{B}) \leq \mathrm{Adv}_{\mathrm{PRF}}(\mathcal{A}') +$
$4q_E\sqrt{\delta} + 2q_H \cdot \sqrt{\mathrm{Adv}_{\mathrm{PKE}'}^{\mathrm{OW\text{-}CPA}}(\mathcal{A})}$ *and* $\mathrm{Adv}_{\mathrm{KEM\text{-}V}}^{\mathrm{IND\text{-}CCA}}(\mathcal{B}) \leq \mathrm{Adv}_{\mathrm{PRF}}(\mathcal{A}') + 4q_E\sqrt{\delta} +$
$\mathrm{Adv}_{\mathrm{PKE}',U_{\mathcal{M}},S}^{\mathrm{DS\text{-}IND}}(\mathcal{C}) + \mathrm{DISJ}_{\mathrm{PKE}',S}$, *and the running time of* $\mathcal{A}$ *($\mathcal{C}$) is about that of* $\mathcal{B}$.

Theorem 7 (PKE′ OW-VA $\overset{QROM}{\Rightarrow}$ KEM-VI IND-CCA). *If* PKE′ *is δ-correct and deterministic, for any IND-CCA $\mathcal{B}$ against KEM-VI, issuing at most q_E quantum queries to the encryption oracle, at most q_D (classical) queries to the decapsulation oracle* DECAPS *and at most q_H quantum queries to the random oracle H, there exists a quantum OW-VA adversary $\mathcal{A}$ against* PKE′ *who makes at most q_D queries to the* VAL *oracle such that* $\mathrm{Adv}_{\mathrm{KEM\text{-}VI}}^{\mathrm{IND\text{-}CCA}}(\mathcal{B}) \leq 2q_E\sqrt{\delta} + 2q_H \cdot \sqrt{\mathrm{Adv}_{\mathrm{PKE}'}^{\mathrm{OW\text{-}VA}}(\mathcal{A})}$ *and the running time of $\mathcal{A}$ is about that of $\mathcal{B}$.*

Acknowledgements. We would like to thank anonymous reviews of Crypto 2018, Keita Xagawa, Takashi Yamakawa, Jiang Zhang, and Edoardo Persichetti for their helpful comments and suggestions. This work is supported by the National Key Research and Development Program of China (No. 2017YFB0802000), the National Natural Science Foundation of China (No. U1536205, 61472446, 61701539, 61501514), and the Open Project Program of the State Key Laboratory of Mathematical Engineering and Advanced Computing (No. 2016A01).

References

1. Cramer, R., Shoup, V.: Design and analysis of practical public-key encryption schemes secure against adaptive chosen ciphertext attack. SIAM J. Comput. **33**(1), 167–226 (2003)
2. Boyd, C., Cliff, Y., Gonzalez Nieto, J., Paterson, K.G.: Efficient one-round key exchange in the standard model. In: Mu, Y., Susilo, W., Seberry, J. (eds.) ACISP 2008. LNCS, vol. 5107, pp. 69–83. Springer, Heidelberg (2008). https://doi.org/10.1007/978-3-540-70500-0_6
3. Fujioka, A., Suzuki, K., Xagawa, K., Yoneyama, K.: Strongly secure authenticated key exchange from factoring, codes, and lattices. Des. Codes Crypt. **76**(3), 469–504 (2015)
4. NIST: National institute for standards and technology. Post quantum crypto project (2017). https://csrc.nist.gov/projects/post-quantum-cryptography/round-1-submissions
5. Rackoff, C., Simon, D.R.: Non-interactive zero-knowledge proof of knowledge and chosen ciphertext attack. In: Feigenbaum, J. (ed.) CRYPTO 1991. LNCS, vol. 576, pp. 433–444. Springer, Heidelberg (1992). https://doi.org/10.1007/3-540-46766-1_35
6. Dent, A.W.: A designer's guide to KEMs. In: Paterson, K.G. (ed.) Cryptography and Coding 2003. LNCS, vol. 2898, pp. 133–151. Springer, Heidelberg (2003). https://doi.org/10.1007/978-3-540-40974-8_12
7. Hofheinz, D., Hövelmanns, K., Kiltz, E.: A modular analysis of the Fujisaki-Okamoto transformation. In: Kalai, Y., Reyzin, L. (eds.) TCC 2017. LNCS, vol. 10677, pp. 341–371. Springer, Cham (2017). https://doi.org/10.1007/978-3-319-70500-2_12

8. Fujisaki, E., Okamoto, T.: Secure integration of asymmetric and symmetric encryption schemes. In: Wiener, M.J. (ed.) CRYPTO 1999. LNCS, vol. 1666, pp. 537–554. Springer, Heidelberg (1999). https://doi.org/10.1007/3-540-48405-1_34
9. Fujisaki, E., Okamoto, T.: Secure integration of asymmetric and symmetric encryption schemes. J. Cryptol. **26**(1), 1–22 (2013)
10. Bellare, M., Rogaway, P.: Random oracles are practical: a paradigm for designing efficient protocols. In: Denning, D.E., Pyle, R., Ganesan, R., Sandhu, R.S., Ashby, V. (eds.) Proceedings of the 1st ACM Conference on Computer and Communications Security - CCS 1993, pp. 62–73. ACM (1993)
11. Boneh, D., Dagdelen, Ö., Fischlin, M., Lehmann, A., Schaffner, C., Zhandry, M.: Random oracles in a quantum world. In: Lee, D.H., Wang, X. (eds.) ASIACRYPT 2011. LNCS, vol. 7073, pp. 41–69. Springer, Heidelberg (2011). https://doi.org/10.1007/978-3-642-25385-0_3
12. Saito, T., Xagawa, K., Yamakawa, T.: Tightly-secure key-encapsulation mechanism in the quantum random oracle model. In: Nielsen, J.B., Rijmen, V. (eds.) EUROCRYPT 2018. LNCS, vol. 10822, pp. 520–551. Springer, Cham (2018). https://doi.org/10.1007/978-3-319-78372-7_17
13. Jiang, H., Zhang, Z., Chen, L., Wang, H., Ma, Z.: IND-CCA-secure key encapsulation mechanism in the quantum random oracle model, revisited. Technical report, Cryptology ePrint Archive, Report 2017/1096 (2017). https://eprint.iacr.org/2017/1096
14. Targhi, E.E., Unruh, D.: Post-quantum security of the Fujisaki-Okamoto and OAEP transforms. In: Hirt, M., Smith, A.D. (eds.) TCC 2016-B. LNCS, vol. 9986, pp. 192–216. Springer, Heidelberg (2016). https://doi.org/10.1007/978-3-662-53644-5_8
15. Bellare, M., Rogaway, P.: Optimal asymmetric encryption. In: De Santis, A. (ed.) EUROCRYPT 1994. LNCS, vol. 950, pp. 92–111. Springer, Heidelberg (1995). https://doi.org/10.1007/BFb0053428
16. Fujisaki, E., Okamoto, T., Pointcheval, D., Stern, J.: RSA-OAEP is secure under the RSA assumption. In: Kilian, J. (ed.) CRYPTO 2001. LNCS, vol. 2139, pp. 260–274. Springer, Heidelberg (2001). https://doi.org/10.1007/3-540-44647-8_16
17. Grover, L.K.: A fast quantum mechanical algorithm for database search. In: Miller, G.L. (ed.) Proceedings of the Twenty-Eighth Annual ACM Symposium on Theory of Computing - STOC 1996, pp. 212–219. ACM (1996)
18. Hülsing, A., Rijneveld, J., Schanck, J.M., Schwabe, P.: High-speed key encapsulation from NTRU. In: Fischer, W., Homma, N. (eds.) CHES 2017. LNCS, vol. 10529, pp. 232–252. Springer, Cham (2017). https://doi.org/10.1007/978-3-319-66787-4_12
19. Hamburg, M.: Module-LWE: the three bears. Technical report. https://www.shiftleft.org/papers/threebears/
20. Ding, J.: A simple provably secure key exchange scheme based on the learning with errors problem. IACR Cryptology ePrint Archive 2012/688 (2012)
21. Peikert, C.: Lattice cryptography for the internet. In: Mosca, M. (ed.) PQCrypto 2014. LNCS, vol. 8772, pp. 197–219. Springer, Cham (2014). https://doi.org/10.1007/978-3-319-11659-4_12
22. Bos, J.W., Costello, C., Naehrig, M., Stebila, D.: Post-quantum key exchange for the TLS protocol from the ring learning with errors problem. In: 2015 IEEE Symposium on Security and Privacy - SP 2015, pp. 553–570 (2015)
23. Alkim, E., Ducas, L., Pöppelmann, T., Schwabe, P.: Post-quantum key exchange - a new hope. In: Holz, T., Savage, S. (eds.) 25th USENIX Security Symposium - USENIX Security 2016, pp. 327–343. USENIX Association (2016)

24. Bos, J.W., Costello, C., Ducas, L., Mironov, I., Naehrig, M., Nikolaenko, V., Raghunathan, A., Stebila, D.: Frodo: take off the ring! Practical, quantum-secure key exchange from LWE. In: Weippl, E.R., Katzenbeisser, S., Kruegel, C., Myers, A.C., Halevi, S. (eds.) Proceedings of the 2016 ACM SIGSAC Conference on Computer and Communications Security - CCS 2016, pp. 1006–1018. ACM (2016)
25. Cheon, J.H., Kim, D., Lee, J., Song, Y.S.: Lizard: cut off the tail! practical post-quantum public-key encryption from LWE and LWR. Technical report, Cryptology ePrint Archive, Report 2016/1126 (2016). http://eprint.iacr.org/2016/1126
26. Bos, J., Ducas, L., Kiltz, E., Lepoint, T., Lyubashevsky, V., Schanck, J.M., Schwabe, P., Stehlé, D.: Crystals-kyber: a CCA-secure module-lattice-based KEM. In: 2018 IEEE European Symposium on Security and Privacy - EuroSP 2018 (2018, to appear)
27. Hoffstein, J., Pipher, J., Silverman, J.H.: NTRU: a ring-based public key cryptosystem. In: Buhler, J.P. (ed.) ANTS-III 1998. LNCS, vol. 1423, pp. 267–288. Springer, Heidelberg (1998). https://doi.org/10.1007/BFb0054868
28. Bernstein, D.J., Chuengsatiansup, C., Lange, T., van Vredendaal, C.: NTRU prime: reducing attack surface at low cost. In: Adams, C., Camenisch, J. (eds.) SAC 2017. LNCS, vol. 10719, pp. 235–260. Springer, Cham (2018). https://doi.org/10.1007/978-3-319-72565-9_12
29. Misoczki, R., Tillich, J.P., Sendrier, N., Barreto, P.S.: MDPC-McEliece: new McEliece variants from moderate density parity-check codes. In: Proceedings of the 2013 IEEE International Symposium on Information Theory (ISIT), pp. 2069–2073. IEEE (2013)
30. Howgrave-Graham, N., et al.: The impact of decryption failures on the security of NTRU encryption. In: Boneh, D. (ed.) CRYPTO 2003. LNCS, vol. 2729, pp. 226–246. Springer, Heidelberg (2003). https://doi.org/10.1007/978-3-540-45146-4_14
31. Guo, Q., Johansson, T., Stankovski, P.: A key recovery attack on MDPC with CCA security using decoding errors. In: Cheon, J.H., Takagi, T. (eds.) ASIACRYPT 2016. LNCS, vol. 10031, pp. 789–815. Springer, Heidelberg (2016). https://doi.org/10.1007/978-3-662-53887-6_29
32. Bernstein, D.J., Groot Bruinderink, L., Lange, T., Panny, L.: HILA5 pindakaas: on the CCA security of lattice-based encryption with error correction. In: Joux, A., Nitaj, A., Rachidi, T. (eds.) AFRICACRYPT 2018. LNCS, vol. 10831, pp. 203–216. Springer, Cham (2018). https://doi.org/10.1007/978-3-319-89339-6_12
33. Saarinen, M.-J.O.: HILA5: on reliability, reconciliation, and error correction for ring-LWE encryption. In: Adams, C., Camenisch, J. (eds.) SAC 2017. LNCS, vol. 10719, pp. 192–212. Springer, Cham (2018). https://doi.org/10.1007/978-3-319-72565-9_10
34. Gentry, C., Peikert, C., Vaikuntanathan, V.: Trapdoors for hard lattices and new cryptographic constructions. In: Dwork, C. (ed.) Proceedings of the 40th Annual ACM Symposium on Theory of Computing - STOC 2008, pp. 197–206. ACM (2008)
35. Mceliece, R.J.: A public-key cryptosystem based on algebraic. DSN progress report 42-44, pp. 114–116 (1978)
36. Niederreiter, H.: Knapsack-type cryptosystems and algebraic coding theory. Probl. Control Inf. Theory 15(2), 159–166 (1986)
37. Regev, O.: On lattices, learning with errors, random linear codes, and cryptography. J. ACM (JACM) 56(6), 34 (2009)
38. Lindner, R., Peikert, C.: Better key sizes (and attacks) for LWE-based encryption. In: Kiayias, A. (ed.) CT-RSA 2011. LNCS, vol. 6558, pp. 319–339. Springer, Heidelberg (2011). https://doi.org/10.1007/978-3-642-19074-2_21

39. Lyubashevsky, V., Peikert, C., Regev, O.: On ideal lattices and learning with errors over rings. In: Gilbert, H. (ed.) EUROCRYPT 2010. LNCS, vol. 6110, pp. 1–23. Springer, Heidelberg (2010). https://doi.org/10.1007/978-3-642-13190-5_1

40. Lyubashevsky, V., Peikert, C., Regev, O.: A toolkit for ring-LWE cryptography. In: Johansson, T., Nguyen, P.Q. (eds.) EUROCRYPT 2013. LNCS, vol. 7881, pp. 35–54. Springer, Heidelberg (2013). https://doi.org/10.1007/978-3-642-38348-9_3

41. Google: PQC-forum. LIMA (2018). https://groups.google.com/a/list.nist.gov/forum/#!topic/pqc-forum/6khIivE2KE0

42. Unruh, D.: Revocable quantum timed-release encryption. J. ACM 62(6), 49:1–49:76 (2015)

43. Albrecht, M.R., Orsini, E., Paterson, K.G., Peer, G., Smart, N.P.: Tightly secure ring-LWE based key encapsulation with short ciphertexts. In: Foley, S.N., Gollmann, D., Snekkenes, E. (eds.) ESORICS 2017. LNCS, vol. 10492, pp. 29–46. Springer, Cham (2017). https://doi.org/10.1007/978-3-319-66402-6_4

44. Giovannetti, V., Lloyd, S., Maccone, L.: Quantum private queries. Phys. Rev. Lett. 100(23), 230502 (2008)

45. De Martini, F., Giovannetti, V., Lloyd, S., Maccone, L., Nagali, E., Sansoni, L., Sciarrino, F.: Experimental quantum private queries with linear optics. Phys. Rev. A 80(1), 010302 (2009)

46. Nielsen, M.A., Chuang, I.L.: Quantum Computation and Quantum Information, no 2. Cambridge University Press, Cambridge (2000)

47. Zhandry, M.: Secure identity-based encryption in the quantum random oracle model. In: Safavi-Naini, R., Canetti, R. (eds.) CRYPTO 2012. LNCS, vol. 7417, pp. 758–775. Springer, Heidelberg (2012). https://doi.org/10.1007/978-3-642-32009-5_44

48. Ambainis, A., Rosmanis, A., Unruh, D.: Quantum attacks on classical proof systems: the hardness of quantum rewinding. In: 55th IEEE Annual Symposium on Foundations of Computer Science - FOCS 2014, pp. 474–483. IEEE (2014)

49. Hülsing, A., Rijneveld, J., Song, F.: Mitigating multi-target attacks in hash-based signatures. In: Cheng, C.-M., Chung, K.-M., Persiano, G., Yang, B.-Y. (eds.) PKC 2016. LNCS, vol. 9614, pp. 387–416. Springer, Heidelberg (2016). https://doi.org/10.1007/978-3-662-49384-7_15

50. Boneh, D., Zhandry, M.: Secure signatures and chosen ciphertext security in a quantum computing world. In: Canetti, R., Garay, J.A. (eds.) CRYPTO 2013. LNCS, vol. 8043, pp. 361–379. Springer, Heidelberg (2013). https://doi.org/10.1007/978-3-642-40084-1_21

51. Zhandry, M.: A note on the quantum collision and set equality problems. Quant. Inf. Comput. 15(7–8), 557–567 (2015)

52. Unruh, D.: Non-interactive zero-knowledge proofs in the quantum random oracle model. In: Oswald, E., Fischlin, M. (eds.) EUROCRYPT 2015. LNCS, vol. 9057, pp. 755–784. Springer, Heidelberg (2015). https://doi.org/10.1007/978-3-662-46803-6_25

53. Okamoto, T., Pointcheval, D.: REACT: rapid enhanced-security asymmetric cryptosystem transform. In: Naccache, D. (ed.) CT-RSA 2001. LNCS, vol. 2020, pp. 159–174. Springer, Heidelberg (2000). https://doi.org/10.1007/3-540-45353-9_13

54. Jean-Sébastien, C., Handschuh, H., Joye, M., Paillier, P., Pointcheval, D., Tymen, C.: GEM: a generic chosen-ciphertext secure encryption method. In: Preneel, B. (ed.) CT-RSA 2002. LNCS, vol. 2271, pp. 263–276. Springer, Heidelberg (2002). https://doi.org/10.1007/3-540-45760-7_18

Pseudorandom Quantum States

Zhengfeng Ji[1(✉)], Yi-Kai Liu[2,3(✉)], and Fang Song[4(✉)]

[1] Centre for Quantum Software and Information, School of Software,
Faculty of Engineering and Information Technology,
University of Technology Sydney, Ultimo, NSW, Australia
Zhengfeng.Ji@uts.edu.au
[2] Applied and Computational Mathematics Division,
National Institute of Standards and Technology (NIST), Gaithersburg, MD, USA
yi-kai.liu@nist.gov
[3] Joint Center for Quantum Information and Computer Science (QuICS),
University of Maryland, College Park, MD, USA
[4] Computer Science Department, Portland State University, Portland, OR, USA
fang.song@pdx.edu

Abstract. We propose the concept of pseudorandom quantum states, which appear random to any quantum polynomial-time adversary. It offers a *computational* approximation to perfectly random quantum states analogous in spirit to cryptographic pseudorandom generators, as opposed to *statistical* notions of quantum pseudorandomness that have been studied previously, such as quantum t-designs analogous to t-wise independent distributions.

Under the assumption that quantum-secure one-way functions exist, we present efficient constructions of pseudorandom states, showing that our definition is achievable. We then prove several basic properties of pseudorandom states, which show the utility of our definition. First, we show a cryptographic no-cloning theorem: no efficient quantum algorithm can create additional copies of a pseudorandom state, when given polynomially-many copies as input. Second, as expected for random quantum states, we show that pseudorandom quantum states are highly entangled on average. Finally, as a main application, we prove that any family of pseudorandom states naturally gives rise to a private-key quantum money scheme.

1 Introduction

Pseudorandomness is a foundational concept in modern cryptography and theoretical computer science. A distribution $\mathcal{D}$, e.g., over a set of strings or functions, is called *pseudorandom* if no computationally-efficient observer can distinguish between an object sampled from $\mathcal{D}$, and a truly random object sampled from the uniform distribution [10,56,63]. Pseudorandom objects, such as pseudorandom generators (PRGs), pseudorandom functions (PRFs) and pseudorandom permutations (PRPs) are fundamental cryptographic building blocks, such as in the design of stream ciphers, block ciphers and message authentication

H. Shacham and A. Boldyreva (Eds.): CRYPTO 2018, LNCS 10993, pp. 126–152, 2018.
https://doi.org/10.1007/978-3-319-96878-0_5

codes [23,24,27,37,53]. Pseudorandomness is also essential in algorithm design and complexity theory such as derandomization [32,47].

The law of quantum physics asserts that truly random bits can be generated easily even with untrusted quantum devices [15,41]. Is pseudorandomness, a seemingly weaker notion of randomness, still relevant in the context of quantum information processing? The answer is yes. By a simple counting argument, one needs exponentially many bits even to specify a truly random function on n-bit strings. Hence, in the *computational* realm, pseudorandom objects that offer efficiency as well as other unique characteristics and strengths are indispensable.

A fruitful line of work on pseudorandomness in the context of quantum information science has been about quantum t-designs and unitary t-designs [4,11,12,16,17,26,33,40,43–45,59,69]. However, while these objects are often called "pseudorandom" in the mathematical physics literature, they are actually analogous to t-wise independent random variables in theoretical computer science. Our focus in this work is a notion of *computational* pseudorandomness, and in particular suits (complexity-theoretical) cryptography.

The major difference between t-wise independence and cryptographic pseudorandomness is the following. In the case of t-wise independence, the observer who receives the random-looking object may be computationally unbounded, but only *a priori* (when the random-looking object is constructed) fixed number t samples are given. Thus, quantum t-designs satisfy an "information-theoretic" or "statistical" notion of security. In contrast, in the case of cryptographic pseudorandomness, the observer who receives the random-looking object is assumed to be computationally efficient, in that it runs in probabilistic polynomial time for an arbitrary polynomial that is chosen by the observer, *after* the random-looking object has been constructed. This leads to a "computational" notion of security, which typically relies on some complexity-theoretic assumption, such as the existence of one-way functions.

In general, these two notions, t-wise independence and cryptographic pseudorandomness, are incomparable. In some ways, the setting of cryptographic pseudorandomness imposes stronger restrictions on the observer, since it assumes a bound on the observer's total computational effort (say, running in probabilistic polynomial time). In other ways, the setting of t-wise independence imposes stronger restrictions on the observer, since it forces the observer to make a limited number of non-adaptive "queries," specified by the parameter t, which is usually a constant or a fixed polynomial. In addition, different distance measures are often used, e.g., trace distance or diamond norm, versus computational distinguishability.

Cryptographic pseudorandomness in quantum information, which has received relatively less study, mostly connects with quantum money and post-quantum cryptography. Pseudorandomness is used more-or-less implicitly in quantum money, to construct quantum states that look complicated to a dishonest party, but have some hidden structure that allows them to be verified by the bank [1–3,39,68]. In post-quantum cryptography, one natural question is whether the classical constructions such as PRFs and PRPs remain secure against quantum attacks. This is a challenging task as, for example, a quantum adversary may query the underlying function or permutation in *superposition*. Fortunately, people have so far restored several positive results. Assuming a one-way function that is hard to invert for polynomial-time quantum algorithms, we can attain quantum-secure PRGs as well as PRFs [27,65]. Furthermore, one can construct quantum-secure PRPs from quantum-secure PRFs using various *shuffling* constructions [57,67].

In this work, we study pseudorandom *quantum* objects such as quantum states and unitary operators. Quantum states (in analogy to strings) and unitary operations (in analogy to functions) form continuous spaces, and the Haar measure is considered the perfect randomness on the spaces of quantum states and unitary operators. A basic question is:

How to define and construct computational pseudorandom *approximations of Haar randomness, and what are their applications?*

Our contributions. We propose definitions of pseudorandom quantum states (PRS's) and pseudorandom unitary operators (PRUs), present efficient constructions of PRS's, demonstrate basic properties such as no-cloning and high entanglement of pseudorandom states, and showcase the construction of private-key quantum money schemes as one of the applications.

1. We propose a suitable definition of *quantum pseudorandom states.*
   We employ the notion of quantum *computational indistinguishability* to define quantum pseudorandom states. Loosely speaking, we consider a collection of quantum states $\{|\phi_k\rangle\}$ indexed by $k \in \mathcal{K}$, and require that no efficient quantum algorithm can distinguish between $|\phi_k\rangle$ for a random k and a state drawn according to the Haar measure. However, as a unique consideration in the quantum setting, we need to be cautious about *how many copies* of the input state are available to an adversary.
   Classically, this is a vacuous concern for defining a pseudorandom distribution on strings, since one can freely produce many copies of the input string. The quantum no-cloning theorem, however, forbids copying an unknown quantum state in general. Pseudorandom states in terms of *single-copy* indistinguishability have been discussed in the literature (see for example [13] and a recent study [14]). Though this single-copy definition may be suitable for certain cryptographic applications, it also loses many properties of Haar random states as a purely classical distributions already satisfies the definition[1].
   Therefore we require that no adversary can tell a difference even given any *polynomially many* copies of the state. This subsumes the single-copy version and is strictly stronger. We gain from it many interesting properties, such as the no-cloning property and entanglement property for pseudorandom states as discussed later in the paper.

---

[1] For example, a uniform distribution over the computational basis state $\{|k\rangle\}$ has an identical density matrix as a Haar random state and satisfy the single-shot definition of PRS. But distinguishing them becomes easy as soon as we have more than one copies. These states also do not appear to be hard to clone or possess entanglement.

2. We present concrete efficient constructions of PRS's with the minimal assumption that quantum-secure one-way functions exist.

Our construction uses any quantum-secure $\mathsf{PRF} = \{\mathsf{PRF}_k\}_{k \in \mathcal{K}}$ and computes it into the phases of a uniform superposition state (see Eq. (8)). We call such family of PRS the *random phase states*. This family of states can be efficiently generated using the quantum Fourier transform and a phase kick-back trick. We prove that this family of state is pseudorandom by a hybrid argument. By the quantum security of PRF, the family is computationally indistinguishable from a similar state family defined by truly random functions.

We then prove that, this state family corresponding to truly random functions is statistically indistinguishable from Haar random states. Finally, by the fact that PRF exists assuming quantum-secure one-way functions, we can base our PRS construction on quantum-secure one-way functions.

We note that Aaronson [1, Theorem 3] has described a similar family of states, which uses some polynomial function instead of a PRF in the phases. In that construction, however, the size of the state family depends on (i.e., has to grow with) the adversary's number of queries that the family wants to tolerate. It therefore fails to satisfy our definition, in which any polynomial number queries independent of the family are permitted.

3. We prove *cryptographic no-cloning theorems* for PRS's, and they give a simple and generic construction of private-key quantum money schemes based on any PRS.

We prove that a PRS remains pseudorandom, even if we additionally give the distinguisher an oracle that reflects about the given state (i.e., $O_\phi := \mathbb{1} - 2|\phi\rangle\langle\phi|$). This establishes the equivalence between the standard and a strong definition of PRS's. Technically, this is proved using the fact that with polynomially many copies of the state, one can approximately simulate the reflection oracle O_ϕ.

We obtain general *cryptographic no-cloning theorems* of PRS's both with and without the reflection oracle. The theorems roughly state that given any polynomially many copies of pseudorandom states, no polynomial-time quantum algorithm can produce even one more copy of the state. We call them cryptographic no-cloning theorems due to the computational nature of our PRS. The proofs of these theorems use SWAP tests in the reduction from a hypothetical cloning algorithm to an efficient distinguishing algorithm violating the definition of PRS's.

Using the strong pseudorandomness and the cryptographic no-cloning theorem with reflection oracle, we show that any PRS immediately gives a *private-key quantum money scheme*. While much attention has been focused on public-key quantum money [1–3,39,68], we emphasize that private-key quantum money is already non-trivial. Early schemes for private-key quantum money due to Wiesner and others were not *query secure*, and could be broken by online attacks [9,20,38,61]. Aaronson and Christiano finally showed a query-secure scheme in 2012, which achieves information-theoretic security in the random oracle model, and computational security in the standard model [2]. They used a specific construction based on hidden subspace states, whereas our construction (which is also query-secure) is more generic and can be based on any PRS. The freedom to choose and tweak the underlying pseudorandom functions or permutations in the PRS may motivate and facilitate the construction of public-key quantum money schemes in future work.

4. We show that pseudorandom states are highly entangled.

It is known that a Haar random state is entangled with high probability. We establish a similar result for any family of pseudorandom states. Namely, the states in any PRS family are entangled on average. It is shown that the expected Schmidt rank for any PRS is superpolynomial in κ and that the expected min entropy and von Neumann entropy are of the order $\omega(\log \kappa)$ where κ is the security parameter. This is yet another evidence of the suitability of our definition.

The proof again rests critically on that our definition grants multiple copies to the distinguisher—if the expected entanglement is low, then SWAP test with respect to the corresponding subsystems of two copies of the state will serve as a distinguisher that violates the definition.

5. We propose a definition of *quantum pseudorandom unitary operators* (PRUs). We also present candidate constructions of PRUs (without a proof of security), by extending our techniques for constructing PRS's.

Loosely speaking, these candidate PRUs resemble unitary t-designs that are constructed by interleaving random permutations with the quantum Fourier transform [26], or by interleaving random diagonal unitaries with the Hadamard transform [43,44], and iterating this construction several times. We conjecture that a PRU can be obtained in this way, using only a constant number of iterations. This is in contrast to unitary t-designs, where a parameter counting argument suggests that the number of iterations must grow with t. This conjecture is motivated by examples such as the Luby-Rackoff construction of a pseudorandom permutation using multi-round Feistel network built using a PRF.

Table 1. Summary of various notions that approximate true randomness

	Classical	Quantum
True randomness	Uniform distribution	Haar measure
t-wise independence	*t*-wise independent random variables	Quantum *t*-designs
Pseudorandomness	PRGs PRFs, PRPs	*(this work)* PRS's PRUs

Discussion. We summarize the mentioned variants of randomness in Table 1. The focus of this work is mostly about PRS's and we briefly touch upon PRUs. We view our work as an initial step and anticipate further fundamental investigation inspired by our notion of pseudorandom states and unitary operators.

We mention some immediate open problems. First, can we prove the security of our candidate PRU constructions? The techniques developed in quantum unitary designs [12,26] seem helpful. Second, are quantum-secure one-way functions necessary for the construction of PRS's? Third, can we establish security proofs for more candidate constructions of PRS's? Different constructions may have their own special properties that may be useful in different settings. It is also interesting to explore whether our quantum money construction may be adapted to a public-key money scheme under reasonable cryptographic assumptions. Finally, the entanglement property we prove here refers to the standard definitions of entanglement. If we approach the concept of pseudo-entanglement as a quantum analogue of pseudo-entropy for a distribution [7], can we improve the quantitative bounds?

We point out a possible application in physics. PRS's may be used in place of high-order quantum t-designs, giving a performance improvement in certain applications. For example, pseudorandom states can be used to construct toy models of quantum *thermalization*, where one is interested in quantum states that can be prepared efficiently via some dynamical process, yet have "generic" or "typical" properties as exemplified by Haar-random pure states, for instance [51]. Using t-designs with polynomially large t, one can construct states that are "generic" in a information-theoretic sense [35]. Using PRS, one can construct states that satisfy a weaker property: they are computationally indistinguishable from "generic" states, for a polynomial-time observer.

In these applications, PRS states may be more physically plausible than high-order quantum t-designs, because PRS states can be prepared in a shorter time, e.g., using a polylogarithmic-depth quantum circuit, based on known constructions for low-depth PRFs [6,46].

2 Preliminaries

2.1 Notions

For a finite set $\mathcal{X}$, $|\mathcal{X}|$ denotes the number of elements in $\mathcal{X}$. We use the notion $\mathcal{Y}^{\mathcal{X}}$ to denote the set of all functions $f : \mathcal{X} \to \mathcal{Y}$. For finite set $\mathcal{X}$, we use $x \leftarrow \mathcal{X}$ to mean that x is drawn uniformly at random from $\mathcal{X}$. The permutation group over elements in $\mathcal{X}$ is denoted as $S_{\mathcal{X}}$. We use poly(κ) to denote the collection of polynomially bounded functions of the security parameter κ, and use negl(κ) to denote negligible functions in κ. A function $\epsilon(\kappa)$ is *negligible* if for all constant $c > 0$, $\epsilon(\kappa) < \kappa^{-c}$ for large enough κ.

In this paper, we use a *quantum register* to name a collection of qubits that we view as a single unit. Register names are represented by capital letters in a *sans serif* font. We use S($\mathcal{H}$), D($\mathcal{H}$), U($\mathcal{H}$) and L($\mathcal{H}$) to denote the set of pure quantum states, density operators, unitary operators and bounded linear operators on space $\mathcal{H}$ respectively. An ensemble of states $\{(p_i, \rho_i)\}$ represents a system prepared in ρ_i with probability p_i. If the distribution is uniform, we write the ensemble as $\{\rho_i\}$. The adjoint of matrix M is denoted as M^*. For matrix M, $|M|$ is defined to be $\sqrt{M^*M}$. The operator norm $\|M\|$ of matrix M is the largest eigenvalue of $|M|$. The trace norm $\|M\|_1$ of M is the trace of $|M|$. For two operators $M, N \in L(\mathcal{H})$, the Hilbert-Schmidt inner product is defined as

$$\langle M, N \rangle = \text{tr}(M^*N).$$

A quantum channel is a physically admissible transformation of quantum states. Mathematically, a quantum channel

$$\mathcal{E} : L(\mathcal{H}) \to L(\mathcal{K})$$

is a completely positive, trace-preserving linear map.

The trace distance of two quantum states $\rho_0, \rho_1 \in D(\mathcal{H})$ is

$$\text{TD}(\rho_0, \rho_1) \overset{\text{def}}{=} \frac{1}{2}\|\rho_0 - \rho_1\|_1. \tag{1}$$

It is known (Holevo-Helstrom theorem [29,30]) that for a state drawn uniformly at random from the set $\{\rho_0, \rho_1\}$, the optimal distinguish probability is given by

$$\frac{1 + \text{TD}(\rho_0, \rho_1)}{2}.$$

Define number $N = 2^n$ and set $\mathcal{X} = \{0, 1, \ldots, N-1\}$. The quantum Fourier transform on n qubits is defined as

$$F = \frac{1}{\sqrt{N}} \sum_{x,y \in \mathcal{X}} \omega_N^{xy}|x\rangle\langle y|. \tag{2}$$

It is a well-known fact in quantum computing that F can be implemented in time $\text{poly}(n)$.

For Hilbert space $\mathcal{H}$ and integer m, we use $\vee^m \mathcal{H}$ to denote the symmetric subspace of $\mathcal{H}^{\otimes m}$, the subspace of states that are invariant under permutations of the subsystems. Let N be the dimension of $\mathcal{H}$ and let $\mathcal{X}$ be the set $\{0, 1, \ldots, N-1\}$ such that $\mathcal{H}$ is the span of $\{|x\rangle\}_{x \in \mathcal{X}}$. For any $\mathbf{x} = (x_1, x_2, \ldots, x_m) \in \mathcal{X}^m$, let m_j be the number of j in $\mathbf{x}$ for $j \in \mathcal{X}$. Define state

$$|\mathbf{x}; \text{Sym}\rangle = \sqrt{\frac{\prod_{j \in \mathcal{X}} m_j!}{m!}} \sum_\sigma \left|x_{\sigma(1)}, x_{\sigma(2)}, \ldots, x_{\sigma(m)}\right\rangle. \tag{3}$$

The summation runs over all possible permutations σ that give different tuples $(x_{\sigma(1)}, x_{\sigma(2)}, \ldots, x_{\sigma(m)})$. Equivalently, we have

$$|\mathbf{x}; \text{Sym}\rangle = \frac{1}{\sqrt{m! \prod_{j \in \mathcal{X}} m_j!}} \sum_{\sigma \in S_m} \left|x_{\sigma(1)}, x_{\sigma(2)}, \ldots, x_{\sigma(m)}\right\rangle. \tag{4}$$

The coefficients in the front of the above two equations are normalization constants. The set of states

$$\left\{|\mathbf{x}; \text{Sym}\rangle\right\}_{\mathbf{x} \in \mathcal{X}^m} \tag{5}$$

forms an orthonormal basis of the symmetric subspace $\vee^m \mathcal{H}$ [58, Proposition 7.2]. This implies that the dimension of the symmetric subspace is

$$\binom{N + m - 1}{m}.$$

Let Π_m^{Sym} be the projection onto the symmetric subspace $\vee^m \mathcal{H}$. For a permutation $\sigma \in S_m$, define operator

$$W_\sigma = \sum_{x_1, x_2, \ldots, x_m \in \mathcal{X}} \big| x_{\sigma^{-1}(1)}, x_{\sigma^{-1}(2)}, \ldots, x_{\sigma^{-1}(m)} \big\rangle \big\langle x_1, x_2, \ldots, x_m \big|.$$

The following identity will be useful [58, Proposition 7.1]

$$\Pi_m^{\mathrm{Sym}} = \frac{1}{m!} \sum_{\sigma \in S_m} W_\sigma. \tag{6}$$

Let μ be the Haar measure on $S(\mathcal{H})$, it is known that [25, Proposition 6]

$$\int \big(|\psi\rangle\langle\psi| \big)^{\otimes m} \, \mathrm{d}\mu(\psi) = \binom{N + m - 1}{m}^{-1} \Pi_m^{\mathrm{Sym}}. \tag{7}$$

2.2 Cryptography

In this section, we recall several definitions and results from cryptography that is necessary for this work.

Pseudorandom functions (PRF) and pseudorandom permutations (PRP) are important constructions in classical cryptography. Intuitively, they are families of functions or permutations that looks like truly random functions or permutations to polynomial-time machines. In the quantum case, we need a strong requirement that they still look random even to polynomial-time quantum algorithms.

Definition 1 (Quantum-Secure Pseudorandom Functions and Permutations). *Let $\mathcal{K}, \mathcal{X}, \mathcal{Y}$ be the key space, the domain and range, all implicitly depending on the security parameter κ. A keyed family of functions $\{\mathsf{PRF}_k : \mathcal{X} \to \mathcal{Y}\}_{k \in \mathcal{K}}$ is a quantum-secure pseudorandom function (QPRF) if for any polynomial-time quantum oracle algorithm $\mathcal{A}$, PRF_k with a random $k \leftarrow \mathcal{K}$ is indistinguishable from a truly random function $f \leftarrow \mathcal{Y}^{\mathcal{X}}$ in the sense that:*

$$\left| \Pr_{k \leftarrow \mathcal{K}} \big[\mathcal{A}^{\mathsf{PRF}_k}(1^\kappa) = 1 \big] - \Pr_{f \leftarrow \mathcal{Y}^{\mathcal{X}}} \big[\mathcal{A}^f(1^\kappa) = 1 \big] \right| = \mathrm{negl}(\kappa).$$

Similarly, a keyed family of permutations $\{\mathsf{PRP}_k \in S_{\mathcal{X}}\}_{k \in \mathcal{K}}$ is a quantum-secure pseudorandom permutation (QPRP) if for any quantum algorithm $\mathcal{A}$ making at most polynomially many queries, PRP_k with a random $k \leftarrow \mathcal{K}$ is indistinguishable from a truly random permutation in the sense that:

$$\left| \Pr_{k \leftarrow \mathcal{K}} \big[\mathcal{A}^{\mathsf{PRP}_k}(1^\kappa) = 1 \big] - \Pr_{P \leftarrow S_{\mathcal{X}}} \big[\mathcal{A}^P(1^\kappa) = 1 \big] \right| = \mathrm{negl}(\kappa).$$

In addition, both PRF_k *and* PRP_k *are polynomial-time computable (on a classical computer).*

Fact 1. *QPRFs and QPRPs exist if quantum-secure one-way functions exist.*

Zhandry proved the existence of QPRFs assuming the existence of one-way functions that are hard to invert even for quantum algorithms [65]. Assuming QPRF, one can construct QPRP using various *shuffling* constructions [57,67]. Since a random permutation and a random function is indistinguishable by efficient quantum algorithms [64,66], existence of QPRP is hence equivalent to existence of QPRF.

3   Pseudorandom Quantum States

In this section, we will discuss the definition and constructions of pseudorandom quantum states.

3.1   Definition of Pseudorandom States

Intuitively speaking, a family pseudorandom quantum states are a set of random states $\{|\phi_k\rangle\}_{k\in\mathcal{K}}$ that is indistinguishable from Haar random quantum states.

The first idea on defining pseudorandom states can be the following. Without loss of generality, we consider states in $\mathrm{S}(\mathcal{H})$ where $\mathcal{H} = (\mathbb{C}^2)^{\otimes n}$ is the Hilbert space for n-qubit systems. We are given either a state randomly sampled from the set $\{|\phi_k\rangle \in \mathcal{H}\}_{k\in\mathcal{K}}$ or a state sampled according to the Haar measure on $\mathrm{S}(\mathcal{H})$, and we require that no efficient quantum algorithm will be able to tell the difference between the two cases.

However, this definition does not seem to grasp the quantum nature of the problem. First, the state family where each $|\phi_k\rangle$ is a uniform random bit string will satisfy the definition—in both cases, the mixed states representing the ensemble are $\mathbb{1}/2^n$. Second, many of the applications that we can find for PRS's will not hold for this definition.

Instead, we require that the family of states looks random even if polynomially many copies of the state are given to the distinguishing algorithm. We argue that this is the more natural way to define pseudorandom states. One can see that this definition also naturally generalizes the definition of pseudorandomness in the classical case to the quantum setting. In the classical case, asking for more copies of a string is always possible and one does not bother making this explicit in the definition. This of course also rules out the example of classical random bit strings we discussed before. Moreover, this strong definition, once established, is rather flexible to use when studying the properties and applications of pseudorandom states.

Definition 2 (Pseudorandom Quantum States (PRS's)). *Let* κ *be the security parameter. Let* $\mathcal{H}$ *be a Hilbert space and* $\mathcal{K}$ *the key space, both parameterized by* κ. *A keyed family of quantum states* $\{|\phi_k\rangle \in \mathrm{S}(\mathcal{H})\}_{k\in\mathcal{K}}$ *is* **pseudorandom**, *if the following two conditions hold:*

1. **(Efficient generation).** *There is a polynomial-time quantum algorithm G that generates state $|\phi_k\rangle$ on input k. That is, for all $k \in \mathcal{K}$, $G(k) = |\phi_k\rangle$.*
2. **(Pseudorandomness).** *Any polynomially many copies of $|\phi_k\rangle$ with the same random $k \in \mathcal{K}$ is **computationally indistinguishable** from the same number of copies of a Haar random state. More precisely, for any efficient quantum algorithm $\mathcal{A}$ and any $m \in \operatorname{poly}(\kappa)$,*

$$\left| \Pr_{k \leftarrow \mathcal{K}}\left[\mathcal{A}(|\phi_k\rangle^{\otimes m}) = 1\right] - \Pr_{|\psi\rangle \leftarrow \mu}\left[\mathcal{A}(|\psi\rangle^{\otimes m}) = 1\right] \right| = \operatorname{negl}(\kappa),$$

where μ is the Haar measure on $\mathrm{S}(\mathcal{H})$.

3.2 Constructions and Analysis

In this section, we give an efficient construction of pseudorandom states which we call random phase states. We will prove that this family of states satisfies our definition of PRS's. There are other interesting and simpler candidate constructions, but the family of random phase states is the easiest to analyze.

Let $\mathrm{PRF} : \mathcal{K} \times \mathcal{X} \to \mathcal{X}$ be a quantum-secure pseudorandom function with key space $\mathcal{K}$, $\mathcal{X} = \{0, 1, 2, \ldots, N-1\}$ and $N = 2^n$. $\mathcal{K}$ and N are implicitly functions of the security parameter κ. The family of pseudorandom states of n qubits is defined

$$|\phi_k\rangle = \frac{1}{\sqrt{N}} \sum_{x \in \mathcal{X}} \omega_N^{\mathrm{PRF}_k(x)} |x\rangle, \qquad (8)$$

for $k \in \mathcal{K}$ and $\omega_N = \exp(2\pi i/N)$.

Theorem 1. *For any QPRF $\mathrm{PRF} : \mathcal{K} \times \mathcal{X} \to \mathcal{X}$, the family of states $\{|\phi_k\rangle\}_{k \in \mathcal{K}}$ defined in Eq. (8) is a PRS.*

Proof. First, we prove that the state can be efficiently prepared with a single query to PRF_k. As PRF_k is efficient, this proves the efficient generation property.

The state generation algorithm works as follows. First, it prepares a state

$$\frac{1}{N} \sum_{x \in \mathcal{X}} |x\rangle \sum_{y \in \mathcal{X}} \omega_N^y |y\rangle.$$

This can be done by applying $H^{\otimes n}$ to the first register initialized in $|0\rangle$ and the quantum Fourier transform to the second register in state $|1\rangle$.

Then the algorithm calls PRF_k on the first register and subtract the result from the second register, giving state

$$\frac{1}{N} \sum_{x \in \mathcal{X}} |x\rangle \sum_{y \in \mathcal{X}} \omega_N^y |y - \mathrm{PRF}_k(x)\rangle.$$

The state can be rewritten as

$$\frac{1}{N} \sum_{x \in \mathcal{X}} \omega_N^{\mathrm{PRF}_k(x)} |x\rangle \sum_{y \in \mathcal{X}} \omega_N^y |y\rangle.$$

Therefore, the effect of this step is to transform the first register to the required form and leaving the second register intact.

Next, we prove the pseudorandomness property of the family. For this purpose, we consider three hybrids. In the first hybrid H_1, the state will be $|\phi_k\rangle^{\otimes m}$ for a uniform random $k \in \mathcal{K}$. In the second hybrid H_2, the state is $|f\rangle^{\otimes m}$ for truly random functions $f \in \mathcal{X}^{\mathcal{X}}$ where

$$|f\rangle = \frac{1}{\sqrt{N}} \sum_{x \in \mathcal{X}} \omega_N^{f(x)} |x\rangle.$$

In the third hybrid H_3, the state is $|\psi\rangle^{\otimes m}$ for $|\psi\rangle$ chosen according to the Haar measure.

By the definition of the quantum-secure pseudorandom functions for PRF, we have for any polynomial-time quantum algorithm $\mathcal{A}$ and any $m \in \text{poly}(\kappa)$,

$$\left| \Pr[\mathcal{A}(H_1) = 1] - \Pr[\mathcal{A}(H_2) = 1] \right| = \text{negl}(\kappa).$$

By Lemma 1, we have for any algorithm $\mathcal{A}$ and $m \in \text{poly}(\kappa)$,

$$\left| \Pr[\mathcal{A}(H_2) = 1] - \Pr[\mathcal{A}(H_3) = 1] \right| = \text{negl}(\kappa).$$

This completes the proof by triangle inequality.

Lemma 1. *For function $f : \mathcal{X} \to \mathcal{X}$, define quantum state*

$$|f\rangle = \frac{1}{\sqrt{N}} \sum_{x \in \mathcal{X}} \omega_N^{f(x)} |x\rangle.$$

For $m \in \text{poly}(\kappa)$, the state ensemble $\{|f\rangle^{\otimes m}\}$ is statistically indistinguishable from $\{|\psi\rangle^{\otimes m}\}$ for Haar random $|\psi\rangle$.

Proof. Let $m \in \text{poly}(\kappa)$ be the number of copies of the state. We have

$$\mathop{\mathbb{E}}_{f}\left[\left(|f\rangle\langle f|\right)^{\otimes m}\right] = \frac{1}{N^m} \sum_{\mathbf{x} \in \mathcal{X}^m, \mathbf{y} \in \mathcal{X}^m} \mathop{\mathbb{E}}_{f} \omega_N^{f(x_1)+\cdots+f(x_m)-[f(y_1)+\cdots+f(y_m)]} |\mathbf{x}\rangle\langle \mathbf{y}|,$$

where $\mathbf{x} = (x_1, x_2, \ldots, x_m)$ and $\mathbf{y} = (y_1, y_2, \ldots, y_m)$. For later convenience, define density matrix

$$\rho^m = \mathop{\mathbb{E}}_{f}\left[\left(|f\rangle\langle f|\right)^{\otimes m}\right].$$

We will compute the entries of ρ^m explicitly.

For $\mathbf{x} = (x_1, x_2, \ldots, x_m) \in \mathcal{X}^m$, let m_j be the number of j in $\mathbf{x}$ for $j \in \mathcal{X}$. Obviously, one has $\sum_{j \in \mathcal{X}} m_j = m$. Note that we have omitted the dependence of m_j on $\mathbf{x}$ for simplicity. Recall the basis states defined in Eq. (4)

$$|\mathbf{x}; \text{Sym}\rangle = \frac{1}{\sqrt{\left(\prod_{j \in \mathcal{X}} m_j!\right)m!}} \sum_{\sigma \in S_m} \left| x_{\sigma(1)}, x_{\sigma(2)}, \ldots, x_{\sigma(m)} \right\rangle.$$

For $\mathbf{x}, \mathbf{y} \in \mathcal{X}^m$, let m_j be the number of j in $\mathbf{x}$ and m'_j be the number of j in $\mathbf{y}$. We can compute the entries of ρ^m as

$$\langle \mathbf{x}; \mathrm{Sym} | \rho^m | \mathbf{y}; \mathrm{Sym} \rangle$$

$$= \frac{m!}{N^m \sqrt{\left(\prod_{j \in \mathcal{X}} m_j!\right)\left(\prod_{j \in \mathcal{X}} m'_j!\right)}} \mathop{\mathbb{E}}_{f}\left[\exp\left(\frac{2\pi i}{N} \sum_{l=1}^{m}\left(f(x_l) - f(y_l)\right)\right)\right].$$

When $\mathbf{x}$ is not a permutation of $\mathbf{y}$, the summation $\sum_{l=1}^{m}\left(f(x_l) - f(y_l)\right)$ is a summation of terms $\pm f(z_j)$ for distinct values z_j. As f is a truly random function, $f(z_j)$ is uniformly random and independent of $f(z_{j'})$ for $z_j \neq z_{j'}$. So it is not hard to verify that the entry is nonzero only if $\mathbf{x}$ is a permutation of $\mathbf{y}$. These nonzero entries are on the diagonal of ρ^m in the basis of $\{|\mathbf{x}; \mathrm{Sym}\rangle\}$. These diagonal entries are

$$\langle \mathbf{x}; \mathrm{Sym} | \rho^m | \mathbf{x}; \mathrm{Sym} \rangle = \frac{m!}{N^m \prod_{j \in \mathcal{X}} m_j!}.$$

Let ρ_μ^m be the density matrix of a random state $|\psi\rangle^{\otimes m}$, for $|\psi\rangle$ chosen from the Haar measure μ. From Eqs. (5) and (7), we have that

$$\rho_\mu^m = \binom{N + m - 1}{m}^{-1} \sum_{\mathbf{x}; \mathrm{Sym}} |\mathbf{x}; \mathrm{Sym}\rangle\langle \mathbf{x}; \mathrm{Sym}|.$$

We need to prove

$$\mathrm{TD}\left(\rho^m, \rho_\mu^m\right) = \mathrm{negl}(\kappa).$$

Define

$$\delta_{\mathbf{x}; \mathrm{Sym}} = \frac{m!}{N^m \prod_{j \in \mathcal{X}} m_j!} - \binom{N + m - 1}{m}^{-1}.$$

Then

$$\mathrm{TD}(\rho^m, \rho_\mu^m) = \frac{1}{2} \sum_{\mathbf{x}; \mathrm{Sym}} |\delta_{\mathbf{x}; \mathrm{Sym}}|.$$

The ratio of the two terms in $\delta_{\mathbf{x}; \mathrm{Sym}}$ is

$$\frac{m!\binom{N + m - 1}{m}}{N^m \prod\limits_{j \in \mathcal{X}} m_j!} = \frac{\prod\limits_{l=0}^{m-1}\left(1 + \frac{l}{N}\right)}{\prod\limits_{j \in \mathcal{X}} m_j!}.$$

For sufficient large security parameter κ, the ratio is larger than 1 only if $\prod_{j \in \mathcal{X}} m_j! = 1$, which corresponds to $\mathbf{x}$'s whose entries are all distinct. As there are $\binom{N}{m}$ such $\mathbf{x}$'s, we can calculate the trace distance as

$$\mathrm{TD}\left(\rho^m, \rho_\mu^m\right) = \binom{N}{m}\left[\frac{m!}{N^m} - \binom{N + m - 1}{m}^{-1}\right]$$

$$= \frac{N(N-1)\cdots(N-m+1)}{N^m} - \frac{N(N-1)\cdots(N-m+1)}{(N+m-1)\cdots N}.$$

As first term is less than 1 and is at least

$$(1 - \frac{1}{N}) \cdots (1 - \frac{m-1}{N}) \geq 1 - \frac{1 + 2 + \cdots + (m-1)}{N}$$

For our choices of $m \in \text{poly}(\kappa)$ and $N \in 2^{\text{poly}(\kappa)}$, this term is $1 - \text{negl}(\kappa)$ for sufficiently large security parameter κ. Similar analysis applies to the second term and this completes the proof.

3.3 Comparison with Related Work

We remark that a similar family of states was considered in [1] (Theorem 3). However, the size of the state family there depends on a parameter d which should be larger than the sum of the number of state copies and the number of queries. In our construction, the key space is fixed for a given security parameter, which may be advantageous for various applications.

We mention several other candidate constructions of PRS's and leave detailed analysis of them to future work. A construction closely related to the random phase states in Eq. (8) uses random ± 1 phases,

$$|\phi_k\rangle = \frac{1}{\sqrt{N}} \sum_{x \in \mathcal{X}} (-1)^{\text{PRF}_k(x)} |x\rangle.$$

Intuitively, this family is less random than the random phase states in Eq. (8) and the corresponding density matrix ρ^m has small off-diagonal entries, making the proof more challenging. The other family of candidate states on $2n$ qubits takes the form

$$|\phi_k\rangle = \frac{1}{\sqrt{N}} \text{PRP}_k \left[\sum_{x \in \mathcal{X}} |x\rangle \otimes |0^n\rangle \right].$$

In this construction, the state is an equal superposition of a random subset of size 2^n of $\{0,1\}^{2n}$ and PRP is any pseudorandom permutation over the set $\{0,1\}^{2n}$. We call this the *random subset states* construction.

Finally, we remark that under plausible cryptographic assumptions our PRS constructions can be implemented using shallow quantum circuits of polylogarithmic depth. To see this, note that there exist PRFs that can be computed in polylogarithmic depth [6], which are based on lattice problems such as "learning with errors" (LWE) [52], and are believed to be secure against quantum computers. These PRFs can be used directly in our PRS construction. (Alternatively, one can use low-depth PRFs that are constructed from more general assumptions, such as the existence of trapdoor one-way permutations [46].)

This shows that PRS states can be prepared in surprisingly small depth, compared to quantum state t-designs, which generally require at least linear depth when t is a constant greater than 2, or polynomial depth when t grows polynomially with the number of qubits [4,12,40,43]. (Note, however, that for $t = 2$, quantum state 2-designs can be generated in logarithmic depth [16].) Moreover, PRS states are sufficient for many applications where high-order t-designs are used [35,51], provided that one only requires states to be *computationally* (not statistically) indistinguishable from Haar-random.

4 Cryptographic No-cloning Theorem and Quantum Money

A fundamental fact in quantum information theory is that unknown or random quantum states cannot be cloned [18,48,50,60,62]. The main topic of this section is to investigate the cloning problem for pseudorandom states. As we will see, even though pseudorandom states can be efficiently generated, they do share the no-cloning property of generic quantum states.

Let $\mathcal{H}$ be the Hilbert space of dimension N and $m < m'$ be two integers. The numbers N, m, m' depend implicitly on a security parameter κ. We will assume that N is exponential in κ and $m \in \text{poly}(\kappa)$ in the following discussion.

We first recall the fact that for Haar random state $|\psi\rangle \in S(\mathcal{H})$, the success probability of producing m' copies of the state given m copies is negligibly small. Let $\mathcal{C}$ be a cloning channel that on input $(|\psi\rangle\langle\psi|)^{\otimes m}$ tries to output a state that is close to $(|\psi\rangle\langle\psi|)^{\otimes m'}$ for $m' > m$. The expected success probability of $\mathcal{C}$ is measured by

$$\int \left\langle (|\psi\rangle\langle\psi|)^{\otimes m'}, \mathcal{C}((|\psi\rangle\langle\psi|)^{\otimes m}) \right\rangle d\mu(\psi).$$

It is known that [60], for all cloning channel $\mathcal{C}$, this success probability is bounded by

$$\binom{N+m-1}{m} \bigg/ \binom{N+m'-1}{m'},$$

which is $\text{negl}(\kappa)$ for our choices of N, m, m'.

We establish a no-cloning theorem for PRS's which says that no efficient quantum cloning procedure exists for a general PRS. The theorem is called the cryptographic no-cloning theorem because of its deep roots in pseudorandomness in cryptography.

Theorem 2 (Cryptographic No-cloning Theorem). *For any PRS family* $\{|\phi_k\rangle\}_{k\in\mathcal{K}}$, $m \in \text{poly}(\kappa)$, $m < m'$ *and any polynomial-time quantum algorithm* $\mathcal{C}$, *the success cloning probability*

$$\mathop{\mathbb{E}}_{k\in\mathcal{K}} \left\langle (|\phi_k\rangle\langle\phi_k|)^{\otimes m'}, \mathcal{C}((|\phi_k\rangle\langle\phi_k|)^{\otimes m}) \right\rangle = \text{negl}(\kappa).$$

Proof. Assume on the contrary that there is a polynomial-time quantum cloning algorithm $\mathcal{C}$ such that the success cloning probability of producing $m + 1$ from m copies is κ^{-c} for some constant $c > 0$. We will construct a polynomial-time distinguisher $\mathcal{D}$ that violates the definition of PRS's. Distinguisher $\mathcal{D}$ will draw $2m + 1$ copies of the state, call $\mathcal{C}$ on the first m copies, and perform the SWAP test on the output of $\mathcal{C}$ and the remaining $m + 1$ copies. It is easy to see that $\mathcal{D}$ outputs 1 with probability $(1 + \kappa^{-c})/2$ if the input is from PRS, while if the input is Haar random, it outputs 1 with probability $(1 + \text{negl}(\kappa))/2$. Since $\mathcal{C}$ is polynomial-time, it follows that $\mathcal{D}$ is also polynomial-time. This is a contradiction with the definition of PRS's and completes the proof.

4.1   A Strong Notion of PRS and Equivalence to PRS

In this section, we show that, somewhat surprisingly, PRS in fact implies a seemingly stronger notion, where indistinguishability needs to hold even if a distinguisher additionally has access to an oracle that reflects about the given state. There are at least a couple of motivations to consider an augmented notion. Firstly, unlike a classical string, a quantum state is inherently *hidden*. Give a quantum register prepared in some state (i.e., a physical system), we can only choose some observable to measure which just reveals partial information and will collapse the state in general. Therefore, it is meaningful to consider offering a distinguishing algorithm more information *describing* the given state, and the reflection oracle comes naturally. Secondly, this stronger notion is extremely useful in our application of quantum money schemes, and could be interesting elsewhere too.

More formally, for any state $|\phi\rangle \in \mathcal{H}$, define an oracle $O_\phi := \mathbb{1} - 2|\phi\rangle\langle\phi|$ that reflects about $|\phi\rangle$.

Definition 3 (Strongly Pseudorandom Quantum States). *Let $\mathcal{H}$ be a Hilbert space and $\mathcal{K}$ be the key space. $\mathcal{H}$ and $\mathcal{K}$ depend on the security parameter κ. A keyed family of quantum states $\{|\phi_k\rangle \in S(\mathcal{H})\}_{k\in\mathcal{K}}$ is **strongly pseudorandom**, if the following two conditions hold:*

1. *(**Efficient generation**). There is a polynomial-time quantum algorithm G that generates state $|\phi_k\rangle$ on input k. That is, for all $k \in \mathcal{K}$, $G(k) = |\phi_k\rangle$.*
2. *(**Strong Pseudorandomness**). Any polynomially many copies of $|\phi_k\rangle$ with the same random $k \in \mathcal{K}$ is **computationally indistinguishable** from the same number of copies of a Haar random state. More precisely, for any efficient quantum oracle algorithm $\mathcal{A}$ and any $m \in \mathrm{poly}(\kappa)$,*

$$\left| \Pr_{k\leftarrow\mathcal{K}}\left[\mathcal{A}^{O_{\phi_k}}(|\phi_k\rangle^{\otimes m}) = 1\right] - \Pr_{|\psi\rangle\leftarrow\mu}\left[\mathcal{A}^{O_\psi}(|\psi\rangle^{\otimes m}) = 1\right] \right| = \mathrm{negl}(\kappa),$$

where μ is the Haar measure on $S(\mathcal{H})$.

Note that since the distinguisher $\mathcal{A}$ is polynomial-time, the number of queries to the reflection oracle (O_{ϕ_k} or O_ψ) is also polynomially bounded.

We prove the advantage that a reflection oracle may give to a distinguisher is limited. In fact, standard PRS implies strong PRS, and hence they are equivalent.

Theorem 3. *A family of states $\{|\phi_k\rangle\}_{k\in\mathcal{K}}$ is strongly pseudorandom if and only if it is (standard) pseudorandom.*

Proof. Clearly a strong PRS is also a standard PRS by definition. It suffice to prove that any PRS is also strongly pseudorandom.

Suppose for contradiction that there is a distinguishing algorithm $\mathcal{A}$ that breaks the strongly pseudorandom condition. Namely, there exists $m \in \mathrm{poly}(\kappa)$ and constant $c > 0$ such that for sufficiently large κ,

$$\left| \Pr_{k\leftarrow\mathcal{K}}\left[\mathcal{A}^{O_{\phi_k}}(|\phi_k\rangle^{\otimes m}) = 1\right] - \Pr_{|\psi\rangle\leftarrow\mu}\left[\mathcal{A}^{O_\psi}(|\psi\rangle^{\otimes m}) = 1\right] \right| = \varepsilon(\kappa) \geq \kappa^{-c}.$$

We assume $\mathcal{A}$ makes $q \in \text{poly}(\kappa)$ queries to the reflection oracle. Then, by Theorem 4, there is an algorithm $\mathcal{B}$ such that for any l

$$\left| \Pr_{k \leftarrow \mathcal{K}} \left[\mathcal{A}^{O_{\phi_k}} (|\phi_k\rangle^{\otimes m}) \right] - \Pr_{k \leftarrow \mathcal{K}} \left[\mathcal{B}(|\phi_k\rangle^{\otimes(m+l)}) \right] \right| \leq \frac{2q}{\sqrt{l+1}},$$

and

$$\left| \Pr_{|\psi\rangle \leftarrow \mu} \left[\mathcal{A}^{O_\psi} (|\psi\rangle^{\otimes m}) \right] - \Pr_{|\psi\rangle \leftarrow \mu} \left[\mathcal{B}(|\psi\rangle^{\otimes(m+l)}) \right] \right| \leq \frac{2q}{\sqrt{l+1}}.$$

By triangle inequality, we have

$$\left| \Pr_{k \leftarrow \mathcal{K}} \left[\mathcal{B}(|\phi_k\rangle^{\otimes(m+l)}) \right] - \Pr_{|\psi\rangle \leftarrow \mu} \left[\mathcal{B}(|\psi\rangle^{\otimes(m+l)}) \right] \right| \geq \kappa^{-c} - \frac{4q}{\sqrt{l+1}}.$$

Choosing $l = 64q^2 \kappa^{2c} \in \text{poly}(\kappa)$, we have

$$\left| \Pr_{k \leftarrow \mathcal{K}} \left[\mathcal{B}(|\phi_k\rangle^{\otimes(m+l)}) \right] - \Pr_{|\psi\rangle \leftarrow \mu} \left[\mathcal{B}(|\psi\rangle^{\otimes(m+l)}) \right] \right| \geq \kappa^{-c}/2,$$

which is a contradiction with the definition of PRS for $\{|\phi_k\rangle\}$. Therefore, we conclude that PRS and strong PRS are equivalent.

We now show a technical ingredient that allows us to simulate the reflection oracle about a state by using multiple copies of the given state. This result is inspired by a similar theorem proved by Ambainis et al. [5, Lemma 42]. Our simulation applies the reflection about the standard symmetric subspace, as opposed to a reflection operation about a particular subspace in [5], on the multiple copies of the given state, which we know how to implement efficiently.

Theorem 4. *Let $|\psi\rangle \in \mathcal{H}$ be a quantum state. Define oracle $O_\psi = \mathbb{1} - 2|\psi\rangle\langle\psi|$ to be the reflection about $|\psi\rangle$. Let $|\xi\rangle$ be a state not necessarily independent of $|\psi\rangle$. Let $\mathcal{A}^{O_\psi}$ be an oracle algorithm that makes q queries to O_ψ. For any integer $l > 0$, there is a quantum algorithm $\mathcal{B}$ that makes no queries to O_ψ such that*

$$\text{TD}\left(\mathcal{A}^{O_\psi}(|\xi\rangle), \mathcal{B}(|\psi\rangle^{\otimes l} \otimes |\xi\rangle) \right) \leq \frac{q\sqrt{2}}{\sqrt{l+1}}.$$

Moreover, the running time of $\mathcal{B}$ is polynomial in that of $\mathcal{A}$ and l.

Proof. Consider a quantum register T, initialized in the state $|\Theta\rangle_\mathsf{T} = |\psi\rangle^{\otimes l} \in \mathcal{H}^{\otimes l}$. Let Π be the projection onto the symmetric subspace $\vee^{l+1}\mathcal{H} \subset \mathcal{H}^{\otimes(l+1)}$, and let $R = \mathbb{1} - 2\Pi$ be the reflection about the symmetric subspace.

Assume without loss of generality that algorithm $\mathcal{A}$ is unitary and only performs measurements at the end. We define algorithm $\mathcal{B}$ to be the same as $\mathcal{A}$, except that when $\mathcal{A}$ queries O_ψ on register D, $\mathcal{B}$ applies the reflection R on the collection of quantum registers D and T. We first analyze the corresponding states after the first oracle call to O_ψ in algorithms $\mathcal{A}$ and $\mathcal{B}$,

$$|\Psi_A\rangle = O_\psi(|\phi\rangle_\mathsf{D}) \otimes |\Theta\rangle_\mathsf{T}, \quad |\Psi_B\rangle = R(|\phi\rangle_\mathsf{D} \otimes |\Theta\rangle_\mathsf{T}).$$

For any two states $|x\rangle, |y\rangle \in \mathcal{H}$, we have

$$\big((\langle x| \otimes \langle \Theta|) R(|y\rangle \otimes |\Theta\rangle)\big) = \langle x|y\rangle - 2 \underset{\pi \in S_{l+1}}{\mathbb{E}} \big((\langle x| \otimes \langle \Theta|) W_\pi (|y\rangle \otimes |\Theta\rangle)\big)$$

$$= \langle x|y\rangle - \frac{2}{l+1} \langle x|y\rangle - \frac{2l}{l+1} \langle x|\psi\rangle \langle \psi|y\rangle$$

$$= \frac{l-1}{l+1} \langle x|y\rangle - \frac{2l}{l+1} \langle x|\psi\rangle \langle \psi|y\rangle,$$

where the first step uses the identity in Eq. (6) and the second step follows by observing that the probability of a random $\pi \in S_{l+1}$ mapping 1 to 1 is $1/(l+1)$. These calculations imply that,

$$\big(\mathbb{1} \otimes \langle \Theta|\big) R\big(\mathbb{1} \otimes |\Theta\rangle\big) = \frac{l-1}{l+1}\mathbb{1} - \frac{2l}{l+1}|\psi\rangle\langle\psi|.$$

We can compute the inner product of the two states $|\Psi_A\rangle$ and $|\Psi_B\rangle$ as

$$\langle \Psi_A | \Psi_B \rangle = \mathrm{tr}\Big((|\phi\rangle \otimes |\Theta\rangle)(\langle\phi| \otimes \langle\Theta|)(O_\psi \otimes \mathbb{1}) R\Big)$$

$$= \mathrm{tr}\Big(|\phi\rangle\langle\phi| O_\psi (\mathbb{1} \otimes \langle\Theta|) R (\mathbb{1} \otimes |\Theta\rangle)\Big)$$

$$= \mathrm{tr}\left(|\phi\rangle\langle\phi|(\mathbb{1} - 2|\psi\rangle\langle\psi|)\left(\frac{l-1}{l+1}\mathbb{1} - \frac{2l}{l+1}|\psi\rangle\langle\psi|\right)\right)$$

$$= \frac{l-1}{l+1} + \frac{2l}{l+1}|\langle\phi|\psi\rangle|^2 - \frac{2(l-1)}{l+1}|\langle\phi|\psi\rangle|^2$$

$$= \frac{l-1}{l+1} + \frac{2}{l+1}|\langle\phi|\psi\rangle|^2$$

$$\geq 1 - \frac{2}{l+1}.$$

This implies that

$$\big\| |\Psi_A\rangle - |\Psi_B\rangle \big\| \leq \frac{2}{\sqrt{l+1}}.$$

Let $|\Psi_A^q\rangle$ and $|\Psi_B^q\rangle$ be the final states of algorithm $\mathcal{A}$ and $\mathcal{B}$ before measurement respectively. Then by induction on the number of queries, we have

$$\big\| |\Psi_A^q\rangle - |\Psi_B^q\rangle \big\| \leq \frac{2q}{\sqrt{l+1}}.$$

This concludes the proof by noticing that

$$\mathrm{TD}(|\Psi_A^q\rangle, |\Psi_B^q\rangle) \leq \big\| |\Psi_A^q\rangle - |\Psi_B^q\rangle \big\|.$$

Finally, we show that if $\mathcal{A}$ is polynomial-time, then so is $\mathcal{B}$. Based on the construction of $\mathcal{B}$, it suffices to show that the reflection R is efficiently implementable for any polynomially large l. Here we use a result by Barenco et al. [8] which provides an efficient implementation for the projection Π onto $\vee^{l+1}\mathcal{H}$. More precisely, they design a quantum circuit of size $O(\mathrm{poly}(l, \log \dim \mathcal{H}))$ that implements a unitary U such that $U|\phi\rangle = \sum_j |\xi_j\rangle|j\rangle$ on $\mathcal{H}^{\otimes(l+1)} \otimes \mathcal{H}'$ for an auxiliary space $\mathcal{H}'$ of dimension $O(l!)$. Here $|\xi_0\rangle = \Pi|\phi\rangle$ corresponds to the projection of $|\phi\rangle$ on the symmetric subspace. With U, we can implement the reflection R as U^*SU where S is the unitary that introduces a minus sign conditioned on the second register being 0.

$$
S|\Psi\rangle|j\rangle = \begin{cases} -|\Psi\rangle|j\rangle & \text{if } j = 0, \\ |\Psi\rangle|j\rangle & \text{otherwise.} \end{cases}
$$

4.2 Quantum Money from PRS

Using Theorem 3, we can improve Theorem 2 to the following version. The proof is omitted as it is very similar to that for Theorem 2 and uses the complexity-theoretic no-cloning theorem [1,2] for Haar random states.

Theorem 5 (Cryptographic no-cloning Theorem with Oracle). *For any PRS* $\{|\phi_k\rangle\}_{k\in\mathcal{K}}$, $m \in \text{poly}(\kappa)$, $m < m'$ *and any polynomial-time quantum query algorithm* $\mathcal{C}$, *the success cloning probability*

$$
\mathop{\mathbb{E}}_{k\in\mathcal{K}} \left\langle \left(|\phi_k\rangle\langle\phi_k|\right)^{\otimes m'}, \mathcal{C}^{O_{\phi_k}}\left(\left(|\phi_k\rangle\langle\phi_k|\right)^{\otimes m}\right) \right\rangle = \text{negl}(\kappa).
$$

A direct application of this no-cloning theorem is that it gives rise to new constructions for private-key quantum money. As one of the earliest findings in quantum information [9,61], quantum money schemes have received revived interests in the past decade (see e.g. [1,3,20,21,39,42]). First, we recall the definition of quantum money scheme adapted from [2].

Definition 4 (Quantum Money Scheme). *A private-key quantum money scheme* $\mathcal{S}$ *consists of three algorithms:*

- *KeyGen, which takes as input the security parameter* 1^κ *and randomly samples a private key* k.
- *Bank, which takes as input the private key* k *and generates a quantum state* $|\$\rangle$ *called a* **banknote**.
- *Ver, which takes as input the private key* k *and an alleged banknote* $|\mathbb{c}\rangle$, *and either accepts or rejects.*

The money scheme $\mathcal{S}$ *has* **completeness error** ε *if* $Ver(k, |\$\rangle)$ *accepts with probability at least* $1 - \varepsilon$ *for all valid banknote* $|\$\rangle$.

Let Count be the money counter that output the number of valid banknotes when given a collection of (possibly entangled) alleged banknotes $|\mathbb{c}_1, \mathbb{c}_2, \ldots, \mathbb{c}_r\rangle$. *Namely, Count will call Ver on each banknotes and return the number of times that Ver accepts. The money scheme* $\mathcal{S}$ *has* **soundness error** δ *if for any polynomial-time counterfeiter* C *that maps* q *valid banknotes* $|\$_1\rangle, \ldots, |\$_q\rangle$ *to* r *alleged banknotes* $|\mathbb{c}_1, \ldots, \mathbb{c}_r\rangle$ *satisfies*

$$
\Pr\left[Count\left(k, C(|\$_1\rangle, \ldots, |\$_q\rangle)\right) > q\right] \le \delta.
$$

The scheme $\mathcal{S}$ *is* **secure** *if it has completeness error* $\le 1/3$ *and negligible soundness error.*

For any $\mathsf{PRS} = \{|\phi_k\rangle\}_{k \in \mathcal{K}}$ with key space $\mathcal{K}$, we can define a private-key quantum money scheme $\mathcal{S}_{\mathsf{PRS}}$ as follows:

- $\mathsf{KeyGen}(1^\kappa)$ randomly outputs $k \in \mathcal{K}$.
- $\mathsf{Bank}(k)$ generates the banknote $|\$\rangle = |\phi_k\rangle$.
- $\mathsf{Ver}(k, \rho)$ applies the projective measurement that accepts ρ with probability $\langle \phi_k | \rho | \phi_k \rangle$.

We remark that usually the money state $|\$\rangle$ takes the form $|\$\rangle = |s, \psi_s\rangle$ where the first register contains a classical serial number. Our scheme, however, does not require the use of the serial numbers. This simplification is brought to us by the strong requirement that any polynomial copies of $|\phi_k\rangle$ are indistinguishable from Haar random states.

Theorem 6. *The private-key quantum money scheme $\mathcal{S}_{\mathsf{PRS}}$ is secure for all PRS.*

Proof. It suffices to prove the soundness of $\mathcal{S}_{\mathsf{PRS}}$ is negligible. Assume to the contrary that there is a counterfeiter C such that

$$\Pr\left[\mathsf{Count}(k, C(|\phi_k\rangle^{\otimes q})) > q\right] \geq \kappa^{-c}$$

for some constant $c > 0$ and sufficiently large κ. From the counterfeiter C, we will construct an oracle algorithm $\mathcal{A}^{O_{\phi_k}}$ that maps q copies of $|\phi_k\rangle$ to $q + 1$ copies with noticeable probability and this leads to a contradiction with Theorem 5.

The oracle algorithm $\mathcal{A}$ first runs C and implement the measurement

$$\left\{M^0 = \mathbb{1} - |\phi_k\rangle\langle\phi_k|, M^1 = |\phi_k\rangle\langle\phi_k|\right\}$$

on each copy of the money state C outputs. This measurement can be implemented by attaching an auxiliary qubit initialized in $(|0\rangle + |1\rangle)/\sqrt{2}$ and call the reflection oracle O_ϕ conditioned on the qubit being at 1 and performs the X measurement on this auxiliary qubit. This gives r-bit of outcome $\mathbf{x} \in \{0, 1\}^r$. If $\mathbf{x}$ has Hamming weight at least $q + 1$, algorithm $\mathcal{A}$ outputs any $q + 1$ registers that corresponds to outcome 1; otherwise, it outputs $|0\rangle^{\otimes(q+1)}$. By the construction of $\mathcal{A}$, it succeeds in cloning $q + 1$ money states from q copies with probability at least κ^{-c}.

Our security proof of the quantum money scheme is arguably simpler than that in [2]. In [2], to prove their hidden subspace money scheme is secure, one needs to develop the so called inner-product adversary method to show the worst-case query complexity for the hidden subspace states and use a random self-reducible argument to establish the average-case query complexity. In our case, it follows almost directly from the cryptographic no-cloning theorem with oracles. The quantum money schemes derived from PRS's enjoy many nice features of the hidden subspace scheme. Most importantly, they are also *query-secure* [2], meaning that the bank can simply return the money state back to the user after verification.

It is also interesting to point out that quantum money states are not necessarily pseudorandom states. The hidden subspace state [2], for example, do not satisfy our definition of PRS as one can measure polynomially many copies of the state in the computational basis and recover a basis for the hidden subspace with high probability.

5   Entanglement of Pseudorandom Quantum States

In this section, we study the entanglement property of pseudorandom quantum states. Our result shows that any PRS consists of states that have high entanglement on average.

The entanglement property of a bipartite pure quantum state is well understood and is completely determined by the Schmidt coefficients of a bipartite state (see e.g. [31]). Any state $|\psi\rangle \in \mathcal{H}_A \otimes \mathcal{H}_B$ on system A and B can be written as

$$|\psi\rangle = \sum_{j=1}^{R} \sqrt{\lambda_j} \, |\psi_A^j\rangle \otimes |\psi_B^j\rangle,$$

where $\lambda_j > 0$ for all $1 \le j \le R$ and the states $|\psi_A^j\rangle$ (and $|\psi_B^j\rangle$) form a set of orthonormal states on A (and B respectively). Here, the positive real numbers λ_j's are the Schmidt coefficients and R is the Schmidt rank of state $|\psi\rangle$. Let ρ_A be the reduced density matrix of $|\psi\rangle$ on system A, then λ_j is the nonzero eigenvalues of ρ_A. Entanglement can be measured by the Schmidt rank R or entropy-like quantities derived from the Schmidt coefficients. We consider the quantum α-Rényi entropy of ρ_A

$$S_\alpha(\rho_A) := \frac{1}{1-\alpha} \log\left(\sum_{j=1}^{R} \lambda_j^\alpha \right).$$

When $\alpha \to 1$, S_α coincides with the von Neumann entropy of ρ_A

$$S(\rho_A) = -\sum_{j=1}^{R} \lambda_j \log \lambda_j.$$

When $\alpha \to \infty$, S_α coincides with the quantum min entropy of ρ_A

$$S_{\min}(\rho_A) = -\log \|\rho_A\| = -\log \lambda_{\max},$$

where $\lambda_{\max}$ is the largest eigenvalue of ρ_A. For $\alpha = 2$, the entropy S_2 is the quantum analogue of the collision entropy.

For Haar random state $|\psi\rangle \sim \mu(\mathcal{H}_A \otimes \mathcal{H}_B)$ where the dimensions of $\mathcal{H}_A$ and $\mathcal{H}_B$ are d_A and d_B respectively, the Page conjecture [49] proved in [22,54,55] states that for $d_A \le d_B$, the average entanglement entropy is explicitly given as

$$\mathbb{E}\, S(\rho_A) = \frac{1}{\ln 2} \left[\left(\sum_{j=d_B+1}^{d_A d_B} \frac{1}{j} \right) - \frac{d_B - 1}{2 d_A} \right] > \log d_A - O(1).$$

That is, the Haar random states are highly entangled on average and, in fact, a typical Haar random state is almost maximumly entangled. A more detailed discussion on this phenomena is give in [28,34]. The following theorem and its corollary tell us that pseudorandom states are also entangled on average though the quantitative bound is much weaker.

Theorem 7. *Let* $\{|\phi_k\rangle\}_{k \in \mathcal{K}}$ *be a family of PRS with security parameter* κ. *Consider partitions of the state* $|\phi_k\rangle$ *into systems A and B consisting of* n_A *and* n_B *qubits each where both* n_A *and* n_B *are polynomial in the security parameter. Let* ρ_k *be the reduced density matrix on system A. Then,*

$$\mathbb{E}_k \operatorname{tr}(\rho_k^2) = \operatorname{negl}(\kappa).$$

Proof. Assume to the contrary that

$$\mathbb{E}_k \operatorname{tr}(\rho_k^2) \geq \kappa^{-c}$$

for some constant $c > 0$ and sufficiently large κ. We will construct a distinguisher $\mathcal{A}$ that tells the family of state $\{|\phi_k\rangle\}$ apart from the Haar random states.

Consider the SWAP test performed on the system A of two copies of $|\phi_k\rangle$. The test accepts with probability

$$\frac{1 + \operatorname{tr}(\rho_k^2)}{2}.$$

Let distinguisher $\mathcal{A}$ be the above SWAP test, we have

$$\left| \Pr_{k \leftarrow \mathcal{K}} \left[\mathcal{A}(|\phi_k\rangle^{\otimes 2}) = 1 \right] - \Pr_{|\psi\rangle \leftarrow \mu} \left[\mathcal{A}(|\psi\rangle^{\otimes 2}) = 1 \right] \right|$$

$$= \frac{1}{2} \left| \mathbb{E}_k \operatorname{tr}(\rho_k^2) - \mathbb{E}_\mu \operatorname{tr}(\rho_\psi^2) \right| \geq \kappa^{-c}/4,$$

for sufficiently large κ. The last step follows by a formula of Lubkin [36]

$$\mathbb{E}_{|\psi\rangle \leftarrow \mu} \operatorname{tr}(\rho_\psi^2) = \frac{d_A + d_B}{d_A d_B + 1} = \frac{2^{n_A} + 2^{n_B}}{2^{n_A + n_B} + 1} = \operatorname{negl}(\kappa).$$

Corollary 1. *Let* $\{|\phi_k\rangle\}_{k \in \mathcal{K}}$ *be a family of PRS with security parameter* κ. *Consider partitions of the state* $|\phi_k\rangle$ *into systems A and B consisting of* n_A *and* n_B *qubits each where both* n_A *and* n_B *are polynomial in the security parameter. We have*

1. *Let* R_k *be the Schmidt rank of state* $|\phi_k\rangle$ *under the A, B partition, then* $\mathbb{E}_k R_k \geq \kappa^c$ *for all constant* $c > 0$ *and sufficiently large* κ.
2. $\mathbb{E}_k S_{\min}(\rho_k) = \omega(\log \kappa)$ *and* $\mathbb{E}_k S(\rho_k) = \omega(\log \kappa)$.

Proof. The first item follows from the fact that

$$\mathrm{tr}(\rho_k^2) \geq \frac{1}{R_k}.$$

where R_k is the Schmidt rank of state $|\phi_k\rangle$. The second item for the min entropy follows by Jensen's inequality and

$$\mathrm{tr}(\rho_k^2) \geq \lambda_{\max}^2.$$

Finally, the bound on the expected entanglement entropy follows by the fact that min entropy is the smallest α-Rényi entropy for all $\alpha > 0$.

6   Pseudorandom Unitary Operators (PRUs)

6.1   Definitions

Our notion of pseudorandom states readily extends to distributions over unitary operators. Let $\mathcal{H}$ be a Hilbert space and let $\mathcal{K}$ a key space, both of which depend on a security parameter κ. Let μ be the Haar measure on the unitary group $\mathrm{U}(\mathcal{H})$.

Definition 5. *A family of unitary operators* $\{U_k \in \mathrm{U}(\mathcal{H})\}_{k \in \mathcal{K}}$ *is **pseudorandom**, if two conditions hold:*

1. *(**Efficient computation**). There is an efficient quantum algorithm Q, such that for all k and any $|\psi\rangle \in \mathrm{S}(\mathcal{H})$, $Q(k, |\psi\rangle) = U_k|\psi\rangle$.*
2. *(**Pseudorandomness**). U_k with a random key k is **computationally indistinguishable** from a Haar random unitary operator. More precisely, for any efficient quantum algorithm $\mathcal{A}$ that makes at most polynomially many queries to the oracle,*

$$\left| \Pr_{k \leftarrow \mathcal{K}}[\mathcal{A}^{U_k}(1^\kappa) = 1] - \Pr_{U \leftarrow \mu}[\mathcal{A}^{U}(1^\kappa) = 1] \right| = \mathrm{negl}(\kappa).$$

The extensive literature on approximation of Haar randomness on unitary groups concerns with unitary *designs* [12,19], which are statistical approximations to the Haar random distribution up to a fixed t-th moment. Our notion of pseudorandom unitary operators in terms of computational indistinguishability, in addition to independent interest, supplements and could substitute for unitary designs in various applications.

6.2   Candidate Constructions

Clearly, given a pseudorandom unitary family $\{U_k\}$, it immediately gives pseudorandom states as well (e.g., $\{U_k|0\rangle\}$). On the other hand, our techniques for constructing pseudorandom states can be extended to give candidate constructions for pseudorandom unitary operators (PRUs) in the following way. Let

$\mathcal{H} = (\mathbb{C}^2)^{\otimes n}$. Assume we have a pseudorandom function PRF : $\mathcal{K} \times \mathcal{X} \to \mathcal{X}$, with domain $\mathcal{X} = \{0, 1, 2, \ldots, N - 1\}$ and $N = 2^n$. Using the phase kick-back technique, we can implement the unitary transformation $T_k \in U(\mathcal{H})$ that maps

$$T_k : |x\rangle \mapsto \omega_N^{\mathsf{PRF}_k(x)}|x\rangle, \quad \omega_N = \exp(2\pi i/N). \tag{9}$$

Our pseudorandom states were given by $|\phi_k\rangle = T_k H^{\otimes n}|0\rangle$, where $H^{\otimes n}$ denotes the n-qubit Hadamard transform. We conjecture that by repeating the operation $T_k H^{\otimes n}$ a constant number of times (with different keys k), we get a PRU. This is resembles the construction of unitary t-designs in [43,44].

Alternatively, one can give a candidate construction for PRUs based on pseudorandom permutations (PRPs) as follows. First, let PRP_k be a pseudorandom permutation (with key $k \in \mathcal{K}$) acting on $\{0, 1\}^n$, and suppose we have efficient quantum circuits that compute the permutation $P_k : |x\rangle|y\rangle \mapsto |x\rangle|y \oplus \mathsf{PRP}_k(x)\rangle$ as well as its inverse $R_k : |x\rangle|y\rangle \mapsto |x\rangle|y \oplus \mathsf{PRP}_k^{-1}(x)\rangle$ (where $\oplus$ denotes the bitwise xor operation). Then we can compute the permutation in-place by applying the following sequence of operations:

$$
\begin{aligned}
|x\rangle|0\rangle &\xrightarrow{P_k} |x\rangle|\mathsf{PRP}_k(x)\rangle \\
&\xrightarrow{SWAP} |\mathsf{PRP}_k(x)\rangle|x\rangle \\
&\xrightarrow{R_k} |\mathsf{PRP}_k(x)\rangle|0\rangle.
\end{aligned}
\tag{10}
$$

For simplicity, let us denote this operation by $S_k : |x\rangle \mapsto |\mathsf{PRP}_k(x)\rangle$ (ignoring the second register, which stays in the state $|0\rangle$). Now we can consider repeating the operation $S_k H^{\otimes n}$ several times (with different keys k), as a candidate for a PRU. Note that this resembles the construction of unitary t-designs in [26].

It is an interesting challenge to prove that these constructions actually yield PRUs. For the special case of non-adaptive adversaries, one could try to use the proof techniques of [26,43,44] for unitary t-designs. For the general case, where the adversary can make adaptive queries to the pseudorandom unitary, new proof techniques seem to be needed. Finally, we can consider combining all of these ingredients (the pseudorandom operations S_k and T_k, and the Hadamard transform) to try to obtain more efficient constructions of PRUs.

References

1. Aaronson, S.: Quantum copy-protection and quantum money. In: Proceedings of the Twenty-Fourth Annual IEEE Conference on Computational Complexity (CCC 2009), pp. 229–242. IEEE Computer Society (2009). https://doi.org/10.1109/CCC.2009.42
2. Aaronson, S., Christiano, P.: Quantum money from hidden subspaces. In: Proceedings of the Forty-Fourth Annual ACM Symposium on Theory of Computing, STOC 2012, pp. 41–60. ACM, New York (2012). https://doi.org/10.1145/2213977.2213983

3. Aaronson, S., Farhi, E., Gosset, D., Hassidim, A., Kelner, J., Lutomirski, A.: Quantum money. Commun. ACM **55**(8), 84–92 (2012). https://doi.org/10.1145/2240236. 2240258
4. Ambainis, A., Emerson, J.: Quantum t-designs: t-wise independence in the quantum world. In: Proceedings of the Twenty-Second Annual IEEE Conference on Computational Complexity (CCC 2007), pp. 129–140, June 2007
5. Ambainis, A., Rosmanis, A., Unruh, D.: Quantum attacks on classical proof systems: the hardness of quantum rewinding. In: Proceedings of the 2014 IEEE 55th Annual Symposium on Foundations of Computer Science, pp. 474–483. IEEE Computer Society (2014). https://doi.org/10.1109/FOCS.2014.57. Full version at https://arxiv.org/abs/1404.6898
6. Banerjee, A., Peikert, C., Rosen, A.: Pseudorandom functions and lattices. In: Pointcheval, D., Johansson, T. (eds.) EUROCRYPT 2012. LNCS, vol. 7237, pp. 719–737. Springer, Heidelberg (2012). https://doi.org/10.1007/978-3-642-29011-4_42
7. Barak, B., Shaltiel, R., Wigderson, A.: Computational analogues of entropy. In: Arora, S., Jansen, K., Rolim, J.D.P., Sahai, A. (eds.) APPROX/RANDOM-2003. LNCS, vol. 2764, pp. 200–215. Springer, Heidelberg (2003). https://doi.org/10.1007/978-3-540-45198-3_18
8. Barenco, A., Berthiaume, A., Deutsch, D., Ekert, A., Jozsa, R., Macchiavello, C.: Stabilization of quantum computations by symmetrization. SIAM J. Comput. **26**(5), 1541–1557 (1997). https://doi.org/10.1137/S0097539796302452
9. Bennett, C.H., Brassard, G., Breidbart, S., Wiesner, S.: Quantum cryptography, or unforgeable subway tokens. In: Chaum, D., Rivest, R.L., Sherman, A.T. (eds.) Advances in Cryptology, pp. 267–275. Springer, Boston, MA (1983). https://doi.org/10.1007/978-1-4757-0602-4_26
10. Blum, M., Micali, S.: How to generate cryptographically strong sequences of pseudorandom bits. SIAM J. Comput. **13**(4), 850–864 (1984). https://doi.org/10.1137/0213053
11. Brandão, F.G.S.L., Harrow, A.W., Horodecki, M.: Efficient quantum pseudorandomness. Phys. Rev. Lett. **116**, 170502 (2016). https://doi.org/10.1103/PhysRevLett.116.170502
12. Brandão, F.G.S.L., Harrow, A.W., Horodecki, M.: Local random quantum circuits are approximate polynomial-designs. Commun. Math. Phys. **346**(2), 397–434 (2016). https://doi.org/10.1007/s00220-016-2706-8
13. Bremner, M.J., Mora, C., Winter, A.: Are random pure states useful for quantum computation? Phys. Rev. Lett. **102**, 190502 (2009). https://doi.org/10.1103/PhysRevLett.102.190502
14. Chen, Y.H., Chung, K.M., Lai, C.Y., Vadhan, S.P., Wu, X.: Computational notions of quantum min-entropy. arXiv:1704.07309 (2017)
15. Chung, K.M., Shi, Y., Wu, X.: Physical randomness extractors: generating random numbers with minimal assumptions. arXiv preprint arXiv:1402.4797 (2014)
16. Cleve, R., Leung, D., Liu, L., Wang, C.: Near-linear constructions of exact unitary 2-designs. Quantum Inf. Comput. **16**(9&10), 721–756 (2016). http://www.rintonpress.com/xxqic16/qic-16-910/0721-0756.pdf
17. Dankert, C., Cleve, R., Emerson, J., Livine, E.: Exact and approximate unitary 2-designs and their application to fidelity estimation. Phys. Rev. A **80**, 012304 (2009). https://doi.org/10.1103/PhysRevA.80.012304
18. Dieks, D.: Communication by EPR devices. Phys. Lett. A **92**(6), 271–272 (1982)

19. Emerson, J., Weinstein, Y.S., Saraceno, M., Lloyd, S., Cory, D.G.: Pseudo-random unitary operators for quantum information processing. Science **302**(5653), 2098–2100 (2003)
20. Farhi, E., Gosset, D., Hassidim, A., Lutomirski, A., Nagaj, D., Shor, P.: Quantum state restoration and single-copy tomography for ground states of hamiltonians. Phys. Rev. Lett. **105**, 190503 (2010). https://doi.org/10.1103/PhysRevLett.105.190503
21. Farhi, E., Gosset, D., Hassidim, A., Lutomirski, A., Shor, P.: Quantum money from knots. In: Proceedings of the 3rd Innovations in Theoretical Computer Science Conference, ITCS 2012, pp. 276–289. ACM, New York (2012). https://doi.org/10.1145/2090236.2090260
22. Foong, S.K., Kanno, S.: Proof of Page's conjecture on the average entropy of a subsystem. Phys. Rev. Lett. **72**, 1148–1151 (1994). https://doi.org/10.1103/PhysRevLett.72.1148
23. Goldreich, O., Goldwasser, S., Micali, S.: On the cryptographic applications of random functions (extended abstract). In: Blakley, G.R., Chaum, D. (eds.) CRYPTO 1984. LNCS, vol. 196, pp. 276–288. Springer, Heidelberg (1985). https://doi.org/10.1007/3-540-39568-7_22
24. Goldreich, O., Goldwasser, S., Micali, S.: How to construct random functions. J. ACM **33**(4), 792–807 (1986). https://doi.org/10.1145/6490.6503
25. Harrow, A.W.: The church of the symmetric subspace. arXiv:1308.6595 (2013)
26. Harrow, A.W., Low, R.A.: Efficient quantum tensor product expanders and k-designs. In: Dinur, I., Jansen, K., Naor, J., Rolim, J. (eds.) APPROX/RANDOM-2009. LNCS, vol. 5687, pp. 548–561. Springer, Heidelberg (2009). https://doi.org/10.1007/978-3-642-03685-9_41
27. Håstad, J., Impagliazzo, R., Levin, L.A., Luby, M.: A pseudorandom generator from any one-way function. SIAM J. Comput. **28**(4), 1364–1396 (1999)
28. Hayden, P., Leung, D.W., Winter, A.: Aspects of generic entanglement. Commun. Math. Phys. **265**(1), 95–117 (2006). https://doi.org/10.1007/s00220-006-1535-6
29. Helstrom, C.W.: Detection theory and quantum mechanics. Inf. Control **10**(3), 254–291 (1967)
30. Holevo, A.S.: An analogue of statistical decision theory and noncommutative probability theory. Tr. Mosk. Matematicheskogo Obshchestva **26**, 133–149 (1972)
31. Horodecki, R., Horodecki, P., Horodecki, M., Horodecki, K.: Quantum entanglement. Rev. Mod. Phys. **81**, 865–942 (2009). https://doi.org/10.1103/RevModPhys.81.865
32. Impagliazzo, R., Wigderson, A.: P = BPP if E requires exponential circuits: derandomizing the XOR lemma. In: Proceedings of the Twenty-Ninth Annual ACM Symposium on Theory of Computing, STOC 1997, pp. 220–229. ACM, New York (1997). https://doi.org/10.1145/258533.258590
33. Kueng, R., Gross, D.: Qubit stabilizer states are complex projective 3-designs. arXiv:1510.02767 (2015)
34. Liu, Z.W., Lloyd, S., Zhu, E.Y., Zhu, H.: Entropic scrambling complexities. arXiv:1703.08104 (2017)
35. Low, R.A.: Large deviation bounds for k-designs. Proc. R. Soc. Lond. A: Math. Phys. Eng. Sci. **465**(2111), 3289–3308 (2009). http://rspa.royalsocietypublishing.org/content/465/2111/3289
36. Lubkin, E.: Entropy of an n-system from its correlation with a k-reservoir. J. Math. Phys. **19**(5), 1028–1031 (1978)
37. Luby, M., Rackoff, C.: How to construct pseudorandom permutations from pseudorandom functions. SIAM J. Comput. **17**(2), 373–386 (1988)

38. Lutomirski, A.: An online attack against Wiesner's quantum money. arXiv:1010.0256 (2010)
39. Lutomirski, A., Aaronson, S., Farhi, E., Gosset, D., Hassidim, A., Kelner, J., Shor, P.: Breaking and making quantum money: toward a new quantum cryptographic protocol. In: Proceedings of the Innovations in Theoretical Computer Science Conference, ITCS 2010, pp. 20–31. Tsinghua University Press (2010)
40. Mezher, R., Ghalbouni, J., Dgheim, J., Markham, D.: Efficient quantum pseudorandomness with simple graph states. arXiv:1709.08091 (2017)
41. Miller, C.A., Shi, Y.: Robust protocols for securely expanding randomness and distributing keys using untrusted quantum devices. J. ACM (JACM) 63(4), 33 (2016)
42. Mosca, M., Stebila, D.: Quantum coins. In: Bruen, A.A., Wehlau, D.L. (eds.) Error-Correcting Codes, Finite Geometries and Cryptography. Contemporary Mathematics, vol. 523, pp. 35–47. American Mathematical Society, Providence (2010). http://www.ams.org/bookstore?fn=20&arg1=conmseries&ikey=CONM-523
43. Nakata, Y., Hirche, C., Koashi, M., Winter, A.: Efficient quantum pseudorandomness with nearly time-independent Hamiltonian dynamics. Phys. Rev. X 7, 021006 (2017). https://doi.org/10.1103/PhysRevX.7.021006
44. Nakata, Y., Hirche, C., Morgan, C., Winter, A.: Unitary 2-designs from random X- and Z-diagonal unitaries. J. Math. Phys. 58(5), 052203 (2017). https://doi.org/10.1063/1.4983266
45. Nakata, Y., Koashi, M., Murao, M.: Generating a state t-design by diagonal quantum circuits. New J. Phys. 16(5), 053043 (2014). http://stacks.iop.org/1367-2630/16/i=5/a=053043
46. Naor, M., Reingold, O.: Synthesizers and their application to the parallel construction of pseudo-random functions. J. Comput. Syst. Sci. 58(2), 336–375 (1999). https://doi.org/10.1006/jcss.1998.1618
47. Nisan, N., Wigderson, A.: Hardness vs randomness. J. Comput. Syst. Sci. 49(2), 149–167 (1994). https://doi.org/10.1016/S0022-0000(05)80043-1
48. Ortigoso, J.: Twelve years before the quantum no-cloning theorem. arXiv:1707.06910 (2017)
49. Page, D.N.: Average entropy of a subsystem. Phys. Rev. Lett. 71, 1291–1294 (1993). https://doi.org/10.1103/PhysRevLett.71.1291
50. Park, J.L.: The concept of transition in quantum mechanics. Found. Phys. 1, 23–33 (1970)
51. Popescu, S., Short, A.J., Winter, A.: Entanglement and the foundations of statistical mechanics. Nat. Phys. 2(11), 754 (2006)
52. Regev, O.: On lattices, learning with errors, random linear codes, and cryptography. J. ACM (JACM) 56(6), 34 (2009)
53. Rompel, J.: One-way functions are necessary and sufficient for secure signatures. In: Proceedings of the Twenty-Second Annual ACM Symposium on Theory of Computing, pp. 387–394. ACM (1990)
54. Sánchez-Ruiz, J.: Simple proof of Page's conjecture on the average entropy of a subsystem. Phys. Rev. E 52, 5653–5655 (1995). https://doi.org/10.1103/PhysRevE.52.5653
55. Sen, S.: Average entropy of a quantum subsystem. Phys. Rev. Lett. 77, 1–3 (1996). https://doi.org/10.1103/PhysRevLett.77.1
56. Shamir, A.: On the generation of cryptographically strong pseudorandom sequences. ACM Trans. Comput. Syst. 1(1), 38–44 (1983). https://doi.org/10.1145/357353.357357

57. Song, F.: Quantum-secure pseudorandom permutations, June 2017. Blog post. http://qcc.fangsong.info/2017-06-quantumprp/
58. Watrous, J.: The Theory of Quantum Information. Cambridge University Press, Cambridge (2018, to be published). A draft copy is available at https://cs.uwaterloo.ca/~watrous/TQI/
59. Webb, Z.: The Clifford group forms a unitary 3-design. Quantum Inf. Comput. **16**(15&16), 1379–1400 (2016). http://www.rintonpress.com/xxqic16/qic-16-1516/1379-1400.pdf
60. Werner, R.F.: Optimal cloning of pure states. Phys. Rev. A **58**, 1827–1832 (1998). https://doi.org/10.1103/PhysRevA.58.1827
61. Wiesner, S.: Conjugate coding. SIGACT News **15**(1), 78–88 (1983). Original manuscript written Circa 1970
62. Wootters, W.K., Zurek, W.H.: A single quantum cannot be cloned. Nature **299**, 802–803 (1982)
63. Yao, A.C.: Theory and application of trapdoor functions. In: 23rd Annual Symposium on Foundations of Computer Science (SFCS 1982), pp. 80–91, November 1982
64. Yuen, H.: A quantum lower bound for distinguishing random functions from random permutations. Quantum Inf. Comput. **14**(13–14), 1089–1097 (2014). http://dl.acm.org/citation.cfm?id=2685166
65. Zhandry, M.: How to construct quantum random functions. In: FOCS 2012, pp. 679–687. IEEE (2012). http://eprint.iacr.org/2012/182
66. Zhandry, M.: A note on the quantum collision and set equality problems. Quantum Inf. Comput. **15**(7&8) (2015). http://arxiv.org/abs/1312.1027
67. Zhandry, M.: A note on quantum-secure PRPs (2016). https://eprint.iacr.org/2016/1076
68. Zhandry, M.: Quantum lightning never strikes the same state twice. iACR eprint 2017/1080 (2017)
69. Zhu, H.: Multiqubit Clifford groups are unitary 3-designs. arXiv:1510.02619 (2015)

Quantum Attacks Against Indistinguishablility Obfuscators Proved Secure in the Weak Multilinear Map Model

Alice Pellet-Mary[(✉)]

Univ Lyon, CNRS, ENS de Lyon, Inria, UCBL, LIP, Lyon, France
alice.pellet__mary@ens-lyon.fr

Abstract. We present a quantum polynomial time attack against the GMMSSZ branching program obfuscator of Garg et al. (TCC'16), when instantiated with the GGH13 multilinear map of Garg et al. (EURO-CRYPT'13). This candidate obfuscator was proved secure in the weak multilinear map model introduced by Miles et al. (CRYPTO'16).

Our attack uses the short principal ideal solver of Cramer et al. (EUROCRYPT'16), to recover a secret element of the GGH13 multilinear map in quantum polynomial time. We then use this secret element to mount a (classical) polynomial time mixed-input attack against the GMMSSZ obfuscator. The main result of this article can hence be seen as a classical reduction from the security of the GMMSSZ obfuscator to the short principal ideal problem (the quantum setting is then only used to solve this problem in polynomial time).

As an additional contribution, we explain how the same ideas can be adapted to mount a quantum polynomial time attack against the DGGMM obfuscator of Döttling et al. (ePrint 2016), which was also proved secure in the weak multilinear map model.

1 Introduction

An obfuscator is a cryptographic primitive that should enable a user to compute a function, without revealing anything about it, except its input-output behaviour. Unfortunately, such a security notion for obfuscators, called Virtual Black Box (or VBB) security, has been shown to be impossible to achieve for all circuits [7]. To circumvent this impossibility result, two directions have been explored. The first direction is to build a VBB obfuscator for a restricted class of functions. Recently, the authors of [36] and [25] managed to prove VBB security of their obfuscator, for the restricted class of compute-and-compare functions,[1] under the LWE assumption. The second direction is to consider weaker security notions, and try to build obfuscators for all circuits under these weaker security

---

[1] A compute-and-compare function CC[f,α] on input x outputs 1 if $f(x) = \alpha$ and 0 otherwise.

© International Association for Cryptologic Research 2018
H. Shacham and A. Boldyreva (Eds.): CRYPTO 2018, LNCS 10993, pp. 153–183, 2018.
https://doi.org/10.1007/978-3-319-96878-0_6

notions. In addition to their impossibility result, the authors of [7] proposed such a weaker security notion, called indistinguishablility obfuscation (or iO).

Indistinguishability obfuscation requires that it should be hard to distinguish between the obfuscation of two equivalent circuits, i.e., circuits that compute the same function. Even if iO security is weaker than VBB security, achieving iO for all circuits would have a lot of applications (see, e.g., [22,34]). The first candidate obfuscator for iO security was proposed in 2013 by Garg, Gentry, Halevi, Raykova, Sahai and Waters [22], based on the GGH13 approximate multilinear map [21]. They showed that iO for the class of polynomial-size branching programs[2] could be bootstrapped to iO for all polynomial-size circuits,[3] and they then described a candidate iO obfuscator for polynomial-size branching programs (without a security proof). Since 2013, numerous candidate obfuscators for polynomial-size branching programs have been proposed, all relying on one of the three candidate cryptographic multilinear map constructions [17,21,24].[4] However, none of these candidate obfuscators could be proven secure under classical hardness assumptions.

The main security weakness of these candidate obfuscators stems from the underlying candidate multilinear maps. Indeed, all candidate multilinear maps have been shown to suffer from so-called zeroizing attacks [15,26], and these zeroizing attacks and their generalizations have made it difficult to design potentially secure iO obfuscators. In the following, we will instantiate all the obfuscators with the GGH13 [21] multilinear map,[5] as our attack exploits a weakness of this specific multilinear map.

In order to improve security confidence, recent obfuscator constructions carefully instantiate the underlying multilinear map (to try to avoid zeroizing attacks) and prove VBB security of their obfuscator in some idealised model. First, the authors of [2,6,12] proved VBB security of their obfuscators in the so-called ideal graded encoding model, introduced in [11]. But zeroizing attacks against multilinear maps and the resulting annihilation attacks against obfuscators [3,14,31] showed that this model was not adapted to capture potential attacks against obfuscators. Another model was then proposed in [31]: the weak multilinear map model. This model captures all the attacks mentioned above, and two candidate obfuscators were proved secure in this model [19,23].

Previous work. The annihilation attack of Miles, Sahai and Zhandry [31] already impacted many obfuscators: [2,5,6,12,30,32]. One limitation of this attack is that it is captured by the weak multilinear map model and so cannot apply against the recent obfuscators of [19,23]. A formalisation and generalisation of

---

[2] See Sect. 2.3 for the definition of a matrix branching program.

[3] The proof relies on Barrington's theorem [8], and on a bootstrapping procedure enabled by fully homomorphic encryption.

[4] The GGH15 multilinear map is a restricted multilinear map that cannot be used for all obfuscator constructions.

[5] Some obfuscators, like [19] are specifically designed to work with the GGH13 multilinear map. Some others can be instantiated with either GGH13 or CLT13 multilinear map. For those, we only consider the GGH13 instantiation.

this attack was then proposed by [3]. This attack enables to distinguish a larger class of circuits than the one of [31], but applies to the same candidate obfuscators. Moreover, it only works for single-input branching programs. In a parallel work, Chen, Gentry and Halevi [14], proposed an attack against the original obfuscator of [22], and a quantum attack against the GGH15 construction [24], that were both unbroken so far. These attacks rely on specific branching programs, namely input partitionable branching programs. Since then, Fernando, Rasmussen and Sahai [20] proposed a technique to transform any branching program into an equivalent branching program which is not input partitionable. This transformation can be used either with the GGH13 map or with the CLT map. Hence, using the [22] obfuscator combined with the technique of [20] prevents the attack of [14].

Our contribution. In this work, we propose quantum polynomial time attacks against the branching program obfuscators of [19,23], when instantiated with the GGH13 multilinear map. These candidate obfuscators were not broken yet, and were proven secure in the weak multilinear map model (the current strongest ideal model for obfuscators). As a secondary contribution, our attack also applies to the obfuscators of [2,5,6,30,32], which were already broken in classical polynomial time by [31]. Our attack is still interesting for these obfuscators, as it uses different techniques than those of [31], and in particular, techniques that are not captured by the weak multilinear map model. Note that our attack does not work against the obfuscator of [12], while [31] does. Finally, as a last contribution, our attack also applies to the circuit obfuscators of [4,37], when instantiated with the GGH13 multilinear map.[6] Overall, we prove the following theorem (informally stated for the moment).

Theorem 1 (Informal, heuristic). *Let $\mathcal{O}$ be any of the branching program obfuscators in [2,5,6,23,30,32], on single or dual input branching programs (respectively, let $\mathcal{O}$ be any of the circuit obfuscators in [4,19,37]), instantiated with the GGH13 multilinear map [21]. There exist two explicit equivalent branching programs (respectively, two equivalent circuits) A and A' such that $\mathcal{O}(A)$ and $\mathcal{O}(A')$ can be distinguished in quantum polynomial time, under some conjecture and heuristic (see Theorem 3 for a formal statement).*

We note that the only part of our attack which is quantum is the principal ideal solver of Biasse and Song [10]. All the other steps of our attack are classical. Hence, our attack can also be viewed as a (classical) reduction from the iO security of the candidate obfuscators mentioned in Theorem 1 to the principal ideal problem. One might then want to use the classical sub-exponential principal ideal solver of Biasse, Espitau, Fouque, Gélin and Kirchner [9] to obtain a classical sub-exponential attack against the above obfuscators. However, the dimension of the cyclotomic ring used in current instantiations on the GGH multilinear map is chosen to be at least λ^2 where λ is the security parameter.

---

[6] These obfuscators need composite-order multilinear maps, and hence were originally instantiated with the CLT multilinear map. However, as observed in [19], the GGH13 multilinear map can also be used with composite-order.

This is done to thwart the attacks of [1,16,27] over the GGH13 multilinear map, but it also means that the classical variant of the attack described in this article is exponential in the security parameter, even when using the sub-exponential principal ideal solver of [9]. It is still interesting to note that any future improvement for solving the principal ideal problem will directly imply an improvement for the attack described in this article.

Technical overview. Recent branching program obfuscators, starting with the one of [6], use the underlying multilinear map to prevent mixed-input attacks, using so-called straddling set systems. A mixed-input attack is an attack in which the attacker does not evaluate honestly the obfuscated circuit, but changes the value of one bit along the computation: for example, if the same bit of the entry is used twice during the computation, the attacker puts it to 1 the first time and to 0 the second time. By choosing good levels for the encodings of the multilinear map, the authors of [6] proved that one could prevent such dishonest computations: an attacker that tries to mix the bits of the input will obtain a final encoding which does not have the good level to be zero-tested and provide a useful output. Following this idea, the obfuscators of [2,5,23,30,32] also used straddling set systems to prevent mixed-input attacks.

However, straddling set systems only ensures that an attacker cannot mixed the inputs of the obfuscated program to obtain a dishonest top level encoding of zero. But it does not prevent an attacker to create a dishonest encoding of zero at a level higher than the top level. In the case where the multilinear map is ideal, this is not a security threat, because the attacker should not be able to test at a level higher than the top level whether it has created an encoding of zero or not. However, this is not the case of the GGH13 multilinear map. Indeed, using recent improvements on the short Principal Ideal Problem [10,13,18] (abbreviated as sPIP), it has been shown that it is possible to recover in quantum polynomial time some secret zero-testing element h of the GGH13 map (see Sect. 2.2 for more details on the GGH13 map). Recovering this secret element will then allow us to zero-test at a higher level than the one initially authorised.[7] This is the starting point of our mixed-input attack against the iO security of [2,5,6,23,30,32].

As said above, all these candidate obfuscators use straddling set systems, meaning that performing a dishonest evaluation of the branching program outputs an encoding at a forbidden level. However, if we perform two well-chosen dishonest evaluations and take the product of the resulting encodings, we can obtain an encoding whose level is twice the maximal level of the multilinear map. The idea to construct well-chosen dishonest evaluations is to take complementary ones. For instance, assume the first bit of the input is used three times during the evaluation of the branching program. A first illegal computation could be to take this first bit to be equal to 0 the first time it is used, and then to 1 for the other two times. The complementary illegal computation will then be to take the first bit to be equal to 1 the first time, and to 0 the other two times. These

---

[7] To be correct, we cannot really test whether we have an encoding of 0, but rather whether we have an encoding which is a product of two encodings of 0. More details can be found in Sect. 4.

two illegal computation will result in encodings that are not at the top level, but there levels will be complementary in the sense that taking the product of them gives an encoding whose level is twice the top-level. We can then use the new zero-test parameter obtained above to determine whether this product of illegal encodings is an encoding of zero or not. It then remains to find a pair of equivalent branching programs such that the illegal encoding obtained above is an encoding of zero for one of the two branching programs only. We exhibit such a pair of branching programs in Sect. 4.3. While we just exhibit one pair, it should be possible to find many other pairs that can also be distinguished. We do not pursue this, as finding one such pair suffices to violate the iO property.

All the branching program obfuscators described above have a similar structure. In order to simplify the description of the attack, and to highlight which characteristics of these obfuscators are needed for the attack, we describe in Sect. 3 an abstract obfuscator, that captures the obfuscators of [2,5,23,30,32]. This abstract obfuscator is elementary, and it suffices to describe our attack against it, in order to attack all the obfuscators of [2,5,23,30,32]. The obfuscator of [6] does not completely fit in this abstract obfuscator and is discussed later.

We finally handle the case of the [19] obfuscator. This obfuscator is different from the ones presented above, as it encodes a circuit rather than a branching program. However, it also uses straddling set system to prevent mixed-input attacks. The same ideas as above can then be adapted to mount a mixed-input attack against the obfuscator of [19], in quantum polynomial time. Here, a new difficulty arises, as a dishonest evaluation of the circuit may not always be possible (for example it can lead to impossible additions, between encodings which are not at the same level). We handle this difficulty by choosing a specific universal circuit, for which we know that some dishonest evaluations are possible. As in the case of the branching program obfuscators, we then give an explicit example of two circuits whose obfuscated versions can be efficiently distinguished by a quantum attacker. Also, as for the the branching program obfuscators, we describe our attack against a simple circuit obfuscator, which captures the circuit obfuscator of [19]. This simple circuit also captures the circuits obfuscators of [4,37], hence the attack also applies to these obfuscators, when they are instantiated with the GGH13 multilinear map.

Impact and open problems. To our knowledge, the only GGH13-based branching program or circuit obfuscator still standing against quantum attackers is the [22] branching program obfuscator, when combined with the technique of [20] to prevent input partitioning. We summarize in Fig. 1 the current state of the art attacks against branching program or circuit obfuscators based on the GGH13 multilinear map. The obfuscators relying on the CLT multilinear map are already known to be insecure against quantum attackers, as the CLT multilinear map is known to be broken if we can factor some public modulus, and we have a quantum polynomial time algorithm for factoring integers [35]. Finally, the obfuscator of [24], based on the GGH15 multilinear map, has been proven insecure against quantum attackers in [14]. In light of this, an interesting question could be to assess the post-quantum security of the obfuscator of [22] when combined with [20].

Obfuscator (instantiated with the GGH13 map)	Quantum attack	Classical attack
[22] without [20]	[14]	[14]
[22] combined with [20]	none	none
[2, 6, 32] [5, 30]	[3, 31] and this work	[3, 31]
[12]	[3, 31]	[3, 31]
[4, 19, 23, 37]	this work	none

Fig. 1. Attacks against GGH13-based branching program and circuit obfuscators

Also, we show that solving the short Principal Ideal Problem enables us to mount a classical attack against the candidate obfuscators of [19,23]. We could wonder whether the opposite is true: can we base the security of these candidate obfuscators or variants thereof on the short Principal Ideal Problem?

Finally, it is interesting to note that the mixed-input attack described in this article crucially relies on the use of straddling set systems. This may seem paradoxical, as straddling set systems were introduced to build obfuscators secure in idealized models, hence supposedly more secure than the first candidates. The first candidate obfuscators [12,22] tried to prevent mixed-input attacks by using so-called bundling scalars, but it was heuristic and came with no proof. On the contrary, the use of straddling set systems allows us to *prove* that the schemes are resistant to mixed-input attacks if the underlying multilinear map is somehow ideal, hence giving us a security proof in some idealized model. However, this comes at the cost of relying more on the security of the underlying multilinear map. So when the obfuscators are instantiated with the GGH13 multilinear map, which is known to have some weaknesses, this gives more possibilities to an attacker to transform these weaknesses of the multilinear map into weaknesses of the obfuscators. This is what we do is this article, by transforming a weakness of the GGH13 map into a concrete attack against obfuscators using straddling set systems. It also explains why our attack does not apply to the obfuscators of [12,22], which did not use straddling set systems.

Roadmap. In Sect. 2, we recall the GGH13 multilinear map, and the notion of matrix branching programs. In Sect. 3, we define an abstract obfuscator, which captures all the obfuscators of [2,5,23,30,32], with both single input and dual input variants. We will then use this abstract obfuscator to present our attack in Sect. 4. This will prove Theorem 1, except for the obfuscators of [6,19]. We then discuss in Sect. 4.4 how to adapt the attack to the obfuscator of [6]. Finally, we describe in Sect. 5 the obfuscator of [19] and explain how to adapt the mixed-input attack to this obfuscator, hence completing the proof of Theorem 1.

2   Preliminaries

In this section, we first recall some mathematical background and define some notations. We then recall the settings of the GGH13 multilinear map and the definition of matrix branching programs. Finally, we recall recent results for the Principal Ideal Problem, that we will use in our attack.

2.1   Mathematical Background

Rings. Let R be the ring $\mathbb{Z}[X]/(X^n + 1)$ for n a power of two, and $K = \mathbb{Q}[X]/(X^n + 1)$ be its fraction field. We let $R^\times$ denote the set of invertible elements of R. For an element $x \in K$, we let x_i denote its coefficients when seen as a polynomial of degree less than n, that is $x = \sum_{i=0}^{n-1} x_i X^i$. An ideal of R is a subset $I \subseteq R$ which is stable by addition and by multiplication by an element of R. If $I = gR = \{gr | r \in R\}$ for some element $g \in R$, we say that I is a principal ideal generated by g, and we denote it by gR or $\langle g \rangle$. We denote by $\{\sigma_j\}_{j \in [n]}$ the complex embeddings of K in $\mathbb{C}$. We can write these embeddings as $\sigma_1, \cdots, \sigma_{n/2}, \overline{\sigma_1}, \cdots, \overline{\sigma_{n/2}}$, where $\bar{\cdot}$ denotes the complex conjugation. The (algebraic) norm of an element $x \in K$ is $\mathcal{N}(x) = \prod_{j \in [n]} \sigma_j(x) \in \mathbb{R}$. The norm of an ideal $I \subseteq R$ is $\mathcal{N}(I) = |R/I|$. If $I = gR$ is a principal ideal, then $\mathcal{N}(I) = \mathcal{N}(g)$. The product of two ideals $I, J \subseteq R$, denoted by $I \cdot J$, is the smallest ideal containing $\{ab \mid a \in I, b \in J\}$. We say that an ideal $I \subseteq R$ is prime if $I \neq R$ and if for all ideals $J_1, J_2 \subseteq R$ such that $I = J_1 \cdot J_2$, then we have either $J_1 = R$ or $J_2 = R$.

Lattices. We view the ring R as an n-dimensional lattice, where the elements of R are mapped to the vectors of their coefficients, when seen as polynomials of degree $n-1$. For $x, y \in K$, the inner product of x and y is $\langle x, y \rangle = \sum_i x_i y_i$. We also define the ℓ_2 norm (or Euclidean norm) of $x \in K$ by $\|x\| = \sqrt{\sum_i x_i^2}$ and the infinite norm of x by $\|x\|_\infty = \max_i(x_i)$. Recall the following properties, for any $x, y \in K$

$$\|x \cdot y\| \leq \sqrt{n} \cdot \|x\| \cdot \|y\| \tag{1}$$

$$\|x\|_\infty \leq \|x\| \leq \sqrt{n} \cdot \|x\|_\infty. \tag{2}$$

For $x \in K$, the Minkowski embeddings of x is $\sigma(x) := (\mathrm{Re}(\sigma_1(x)), \mathrm{Im}(\sigma_1(x)), \cdots, \mathrm{Re}(\sigma_{n/2}(x)), \mathrm{Im}(\sigma_{n/2}(x))) \in \mathbb{R}^n$. We define the inner product of the Minkowski embeddings of two elements $x, y \in K$ by the usual inner product over $\mathbb{R}^n$ of $\sigma(x)$ and $\sigma(y)$. As we are in a cyclotomic ring of order a power of two, the geometry induced by the coefficient embeddings is the same, up to scaling, as the one induced by the Minkowski embeddings. This means that for any $x, y \in K$, we have

$$\langle \sigma(x), \sigma(y) \rangle = n/2 \cdot \langle x, y \rangle.$$

In particular, for all $x \in K$, we have

$$\|\sigma(x)\|_2 = \sqrt{n/2} \cdot \|x\|, \tag{3}$$

where $\|\sigma(x)\|_2 = \sqrt{\langle \sigma(x), \sigma(x)\rangle}$.

An ideal I can be seen as a sub-lattice of R, and hence described by a $\mathbb{Z}$-basis. The Principal Ideal Problem (PIP) is, given a basis of a principal ideal I, to recover a generator of I, that is an element $g \in R$ such that $I = \langle g \rangle$.

For any lattice L, real $\sigma > 0$ and point $c \in L$, we define the Gaussian weight function over L by

$$\rho_{L,\sigma,c}(x) = \exp\left(\frac{-\|x - c\|^2}{2\sigma^2}\right).$$

We define the discrete (spherical) Gaussian distribution over L of parameter σ and centered in c by

$$\forall x \in L, \ D_{L,\sigma,c}(x) = \frac{\rho_{L,\sigma,c}(x)}{\rho_{L,\sigma,c}(L)},$$

where $\rho_{L,\sigma,c}(L) = \sum_{x\in L} \rho_{L,\sigma,c}(x)$. We simplify $\rho_{L,\sigma,0}$ and $D_{L,\sigma,0}$ into $\rho_{L,\sigma}$ and $D_{L,\sigma}$, and say in that case that the distribution is centered.

2.2 The GGH13 Multilinear Map

We recall in this section the GGH13 multilinear map (or shortly GGH map) of [21], in its asymmetric setting. The GGH multilinear map allows to encode elements of ring. We can then homomorphically perform additions and multiplications on these elements, under some constraints. It also allows to publicly test if an encoding encodes zero. Let q be a large integer (usually taken exponential in n) and define $R_q = R/qR$. Let g be some small element of $R^\times$ chosen such that the ideal $\langle g \rangle$ is prime and has a prime norm. The plaintext space will be R/gR and the encoding space will be R_q.

Encodings. Let κ be some positive integer and $z_1, \cdots, z_\kappa$ be chosen randomly in $R_q^\times$.[8] These z_i's are chosen during the initialisation phase of the GGH map. Let S be a subset of $[\kappa]$ and $a + gR$ be an element of R/gR. An encoding of $a + gR$ at level S is an element of the form

$$u = c \cdot \prod_{i\in S} z_i^{-1} \bmod q,$$

where c is a small representative of $a + gR$ in R. We sometimes abuse notation by saying that u is an encoding of $a \in R$ instead of $a + gR \in R/gR$. We use the notation $[a]_S$ to denote an encoding of $a + gR$ at level S. When there is no ambiguity on the level of the encoding, we just write it $[a]$, with no subscript. We say that c is the numerator of the encoding u and $\prod_{i\in S} z_i$ is its denominator.

---

[8] The distribution of the z_i's does not matter here.

Operations on encodings. Let u_1 and u_2 be the encodings of two elements a_1 and a_2 at the same level S. Then $u_1 + u_2$ is an encoding of $a_1 + a_2$ at level S. Let u_1 and u_2 be the encodings of two elements u_1 and u_2 at level S_1 and S_2 respectively, with $S_1 \cap S_2 = \emptyset$. Then $u_1 \cdot u_2$ is an encoding of $a_1 \cdot a_2$ at level $S_1 \cup S_2$.[9]

Zero-testing. Let S_{zt} denote the set $[\kappa]$ and $z^* = \prod_{i \in S_{zt}} z_i$. Let h be some element in R of ℓ_2-norm approximately $\sqrt{q}$. We define $p_{zt} = h z^* g^{-1} \bmod q$ and call it the zero-testing parameter. To test if an encoding u at level S_{zt} is an encoding of zero or not (i.e., to test if the numerator of u is a multiple of g or not), compute $w = u \cdot p_{zt} \bmod q$. If this is smaller than $q^{3/4}$,[10] then u is an encoding of zero, otherwise it is not. Indeed, if $u = bg(z^*)^{-1} \bmod q$ (i.e., u is an encoding of zero), then $w = bh \bmod q$ and the parameters are set such that $||bh|| \leq q^{3/4}$ for a correct level-S_{zt} encoding. On the other hand, if u is not an encoding of zero, then the g^{-1} in the zero-testing parameter does not cancel out, and $g^{-1} \bmod q$ is very unlikely to be small compared to q. We can prove that in this case, w will never be smaller than $q^{3/4}$ (see [21] for more details).

The elements (n, q, κ, p_{zt}) of the multilinear map are public, while the parameters $(h, g, \{z_i\}_{i \in [\kappa]})$ are secret. In our case, the obfuscator generates the multilinear maps and retains these secret elements. Note that to encode an element, we need to know the secret parameters g and $\{z_i\}_i$. This means that only the obfuscator will be able to create encodings from scratch. An encoding generated by the obfuscator, using the secret parameters, is called a fresh encoding, by opposition to the encodings obtained by adding or multiplying other encodings.

Size of the parameters. The size of the parameters of the GGH multilinear map may vary depending on the obfuscator. We present here the size recommended in the original article [21], with a small change for the size of q, due to the fact that we use the multilinear map in a different way for obfuscators than what was described in [21].

- The dimension n of R should be taken such that $n = \Omega(\kappa\lambda^2)$, where λ is the security parameter of the scheme. Taking a lower bound in λ^2 was the original choice of [21] to avoid some lattice attacks. It was reduced to $n = \Omega(\kappa\lambda \log(\lambda))$ in [28]. However, with the recent sub-exponential algorithm of [9] to solve PIP, it should be increased back to $\Omega(\kappa\lambda^2)$. Looking ahead, the attack we describe in Sect. 4 has a classical variant which is sub-exponential in the dimension n of the lattice (it has a complexity $O(2^{\sqrt{n}+o(1)})$). However, as $n \geq \Omega(\lambda^2)$, this remains exponential in the security parameter λ.
- The secret element g is sampled using a Gaussian distribution, with rejection, such that $||g|| = O(n)$ and $||1/g|| = O(n^2)$.
- The modulus q is chosen such that $q \geq n^{O(\kappa)}$. In the original GGH scheme, the modulus q was chosen greater than $2^{8\kappa\lambda} \cdot n^{O(\kappa)}$. This extra factor $2^{8\kappa\lambda}$

---

[9] Even if $S_1 \cap S_2 \neq \emptyset$, we can still see $u_1 \cdot u_2$ as an encoding of $a_1 \cdot a_2$ at level $S_1 \cup S_2$, where $S_1 \cup S_2$ is a multiset, that is we keep multiple copies of elements that appear both in S_1 and S_2.

[10] This bound is the one chosen in [21], but it is flexible.

came from the re-randomisation procedure used originally to publicly generate level-1 encodings. In the case of obfuscators, as the one that generates
encodings knows the secret parameters, it can generates the fresh encodings
with a numerator of size $O(\text{poly}(n))$ instead of $O(2^\lambda \text{poly}(n))$, and hence get
ride of this factor $2^{8\kappa\lambda}$. In all the obfuscators described here, except [19], the
modulus q is exponential in λ. In [19], the obfuscator is built such that q
remains polynomial in λ (even if κ is polynomial in λ, the authors managed
to obtain a polynomial modulus q).

- The secret element h is sampled using a centered Gaussian distribution of
  parameter $\sqrt{q}$, so that $\|h\| = \Theta(\sqrt{n} \cdot \sqrt{q})$. In [21, Sect. 6.4], the authors
  suggest to sample h according to a non spherical Gaussian distribution instead
  of a spherical one. In the following we will always assume that h is sampled
  according to a spherical Gaussian distribution. We discuss the case of non
  spherical distributions in the full version [33].

2.3   Matrix Branching Programs

We recall in this section the definition of matrix branching programs, and we
introduce some notation that will be used throughout the article. A branching
program is defined over a ring $\mathcal{R}$.

Definition 1 (d-ary Matrix Branching Program [3]). *A d-ary matrix
branching program $\boldsymbol{A}$ of length ℓ and width w over m-bit inputs is given by
a sequence of square matrices*

$$\{A_{i,b}\}_{i\in[\ell],b\in\{0,1\}^d} \in \mathcal{R}^{w\times w},$$

two bookend vectors

$$A_0 \in \mathcal{R}^{1\times w} \text{ and } A_{\ell+1} \in \mathcal{R}^{w\times 1},$$

and an input function $\boldsymbol{inp} : [\ell] \rightarrow [m]^d$.

*Let $x \in \{0,1\}^m$ and let x_i denote the i-th bit of x, for i in $[m]$. We will use
the notation $x[\boldsymbol{inp}(i)] = (x_{\boldsymbol{inp}(i)_1}, x_{\boldsymbol{inp}(i)_2}, \cdots, x_{\boldsymbol{inp}(i)_d}) \in \{0,1\}^d$, where $\boldsymbol{inp}(i) =
(\boldsymbol{inp}(i)_1, \cdots, \boldsymbol{inp}(i)_d) \in [m]^d$.*

The output of the matrix branching program on input $x \in \{0,1\}^m$ is given by

$$\boldsymbol{A}(x) = \begin{cases} 0 \text{ if } A_0 \cdot \left(\prod_{i\in[\ell]} A_{i,x[\boldsymbol{inp}(i)]}\right) \cdot A_{\ell+1} = 0 \\ 1 \text{ otherwise.} \end{cases}$$

Remark. A branching program with $d = 1$ (respectively with $d = 2$) is also called
a single input (respectively dual input) branching program. In the following, we
will not distinguish between the single input and dual input cases, as our attack
works in the same way in both cases (and even for higher arity d).

We say that two branching programs are equivalent if they compute the same
function. We also introduce a notion of strong equivalence between branching
programs, which will be useful later for the description of the abstract obfuscator
and our attack.

Definition 2 (Strongly equivalent branching programs). *We say that two d-ary matrix branching programs* $A = (A_0, \{A_{i,b}\}_{i\in[\ell], b\in\{0,1\}^d}, A_{\ell+1})$ *and* $A' = (A'_0, \{A'_{i,b}\}_{i\in[\ell], b\in\{0,1\}^d}, A'_{\ell+1})$, *with the same length ℓ and the same input function $\mathbf{inp}$ (but not necessarily defined over the same rings) are strongly equivalent if, for all $\{\mathbf{b}_i\}_{i\in[\ell]} \in (\{0,1\}^d)^\ell$, we have*

$$A_0 \cdot \prod_{i\in[\ell]} A_{i,b_i} \cdot A_{\ell+1} = 0 \iff A'_0 \cdot \prod_{i\in[\ell]} A'_{i,b_i} \cdot A'_{\ell+1} = 0. \qquad (4)$$

Remark. This notion is stronger than simple equivalence between branching programs, because we ask that (4) holds for all possible choices of $\{\mathbf{b}_i\}_{i\in[\ell]}$, and not only for the ones of the form $\{x[\mathbf{inp}(i)]\}_{i\in[\ell]}$ for some input x (corresponding to an honest evaluation of the branching program on x). The pair of branching programs described in Sect. 4.3 gives an example of equivalent branching programs that are not strongly equivalent.

2.4 The Short Principal Ideal Problem

We define the short Principal Ideal Problem in the following way.

Definition 3 (Short Principal Ideal Problem). *Let $h \in R$ be sampled according to some distribution D. The short Principal Ideal Problem is, given any basis of the ideal $\langle h \rangle$ (when seen as a sub-lattice of R), to recover $\pm X^i \cdot h$ for some $i \in [n]$.*

For cyclotomic fields of order a power of two, when D is a discrete Gaussian distribution, this problem can be solved in quantum polynomial time, using the results of [10,13,18]. In [10], the authors show that given any basis of $\langle h \rangle$, an attacker can recover a generator $\widetilde{h}$ of the ideal $\langle h \rangle$ in quantum polynomial time.[11] Then, the authors of [18], based on an observation of [13], proved that from any generator $\widetilde{h}$ of $\langle h \rangle$, if h has been sampled using a discrete Gaussian distribution, then an attacker can recover $\pm X^i \cdot h$, for some $i \in [n]$, in (classical) polynomial time. This second part (recovering $\pm X^i \cdot h$ from $\widetilde{h}$) relies on the conjecture that the set of cyclotomic units of R is equal to $R^\times$ for power-of-two cyclotomic fields. We summarise this in the following theorem.

Theorem 2 (adapted from [10,18]). *Let $h \in R$ be sampled according to a discrete spherical Gaussian distribution of parameter larger than $200 \cdot n^{1.5}$. Then, under Conjecture 1, there is a quantum polynomial time algorithm such that, given any basis of the ideal $\langle h \rangle$, it recovers $\pm X^i \cdot h$ for some $i \in [n]$, with constant probability close to 1 over the choice of h.*

Conjecture 1. The set of cyclotomic units of R is equal to $R^\times$ (see [18] for a definition of cyclotomic units and a discussion of this conjecture).

---

[11] Note that there also exists a classical sub-exponential time algorithm to recover $\widetilde{h}$, due to [9]. However, their algorithm runs in time $O(2^{\sqrt{n}+o(1)})$, but we chose $n \geq \Omega(\lambda^2)$, so this algorithm is exponential in the security parameter λ.

3   An Abstract Obfuscator

Following an idea of Miles, Sahai and Zhandry in [31], we define here an abstract obfuscation scheme. This abstract obfuscator is inspired by the one of [31] but is a bit simpler and more general. In particular, it captures all the obfuscators of Theorem 1, except the ones of [6] and [19]. We will then show in Sect. 4 how to apply our quantum attack to this abstract obfuscator, resulting in an attack against the obfuscators of [2,5,23,30,32] and we will explain how to adapt the attack to the branching program obfuscator of [6] (which is just slightly different from the abstract obfuscator defined in this section). The case of the [19] obfuscator is postponed in Sect. 5 as it is not a branching program obfuscator, and so the formalism of the abstract branching program obfuscator does not apply to it.

The abstract obfuscator takes as input a polynomial size d-ary matrix branching program $\mathbf{A}$ (for some integer $d > 0$), over the ring of integers $\mathbb{Z}$,[12] with a fixed input function $\mathtt{inp}$ and with coefficients in $\{0,1\}$. Usually, the obfuscators pad the branching program with identity matrices, to ensure that the input function has the desired structure. Here, to simplify the obfuscator, we will assume that the obfuscator only accepts branching programs with the desired $\mathtt{inp}$ function (the user has to pad the branching program himself before giving it to the obfuscator). For the attack to work, we ask that there exist two different integers j_1 and j_2 such that $\mathtt{inp}(j_1) \cap \mathtt{inp}(j_2) \neq \emptyset$ (meaning that there is a bit of the input which is inspected at least twice during the evaluation of the branching program). This can be assumed for all the obfuscators of Theorem 1.[13] Let w be the width of $\mathbf{A}$, ℓ be its length, A_0, $A_{\ell+1}$ be its bookend vectors and $\{A_{i,\mathbf{b}}\}_{i\in[\ell],\mathbf{b}\in\{0,1\}^d} \in \{0,1\}^{w\times w}$ be its square matrices. Recall that the function computed by the branching program $\mathbf{A}$ is defined by

$$\mathbf{A}(x) = \begin{cases} 0 \text{ if } A_0 \cdot \left(\prod_{i\in[\ell]} A_{i,x[\mathtt{inp}(i)]}\right) \cdot A_{\ell+1} = 0 \\ 1 \text{ otherwise.} \end{cases}$$

The abstract obfuscator then proceeds as follows.

- It instantiates the GGH multilinear map and retains its secret parameters $(g, h, \{z_i\}_{i\in[\kappa]})$ and its public parameters (n, q, κ, p_{zt}). The choice of the parameters of the GGH map depends on the parameters ℓ, w and d of the branching program $\mathbf{A}$.
- It transforms the matrices of branching program $\mathbf{A}$ to obtain a new branching program $\hat{\mathbf{A}}$, with the same parameters w, d, ℓ, the same input function $\mathtt{inp}$, and which is strongly equivalent to $\mathbf{A}$. We denote by $\{\hat{A}_{i,\mathbf{b}}\}_{i\in[\ell],\mathbf{b}\in\{0,1\}^d} \in (R/gR)^{w\times w}$ and $\hat{A}_0 \in (R/gR)^{1\times w}, \hat{A}_{\ell+1} \in (R/gR)^{w\times 1}$ the matrices and

---

[12] Most of the time, the matrices of the branching program will be permutation matrices, and the underlying ring will have no importance.

[13] This is even mandatory for the dual input version of the obfuscators, as it is usually required that all pairs (s,t) (or (t,s)) appear in the $\mathtt{inp}$ function, for any $s,t \in [m]$ with $s \neq t$.

bookend vectors of $\hat{\mathbf{A}}$. Note that this new matrix branching programs has its coefficients in the ring R/gR and not in $\{0,1\}$. Recall that strong equivalence means that

$$A_0 \cdot \prod_{i\in[\ell]} A_{i,\mathbf{b}_i} \cdot A_{\ell+1} = 0 \iff \hat{A}_0 \cdot \prod_{i\in[\ell]} \hat{A}_{i,\mathbf{b}_i} \cdot \hat{A}_{\ell+1} = 0 \text{ (in } R/gR) \quad (5)$$

for all choices of $\mathbf{b}_i \in \{0,1\}^d$, with $i \in [\ell]$. This condition is required for our attack to work, and is satisfied by all the obfuscators of [2,5,23,30,32]. To transform the initial branching program $\mathbf{A}$ into this new branching program $\hat{\mathbf{A}}$, the obfuscators of [2,5,23,30,32] first embed the matrices of $\mathbf{A}$ into the ring R/gR (this is possible since the coefficients of the matrices are 0 and 1). Then, they use various tools, taken among the following.[14]

1. Transform the matrices $A_{i,\mathbf{b}}$ into block-diagonal matrices $\begin{pmatrix} A_{i,\mathbf{b}} & \\ & B_{i,\mathbf{b}} \end{pmatrix}$, were $B_{i,\mathbf{b}}$ are square $w' \times w'$ matrices in R/gR, chosen arbitrarily (they can be fixed, or chosen at random, this will have no importance for us), with w' polynomial in the security parameter λ. In order to cancel the extra diagonal block, the vector A_0 is transformed into $\begin{pmatrix} A_0 & 0 \end{pmatrix}$, with a block of zeros of size $1 \times w'$. The vector $A_{\ell+1}$ is transformed into $\begin{pmatrix} A_{\ell+1} \\ B_{\ell+1} \end{pmatrix}$, with $B_{\ell+1}$ an arbitrary $w' \times 1$ vector.

2. Use Killian randomisation, that is, choose $\ell + 1$ non singular matrices $\{R_i\}_{i\in[\ell+1]} \in (R/gR)^{w\times w}$ and transform $A_{i,\mathbf{b}}$ into $R_i \cdot A_{i,\mathbf{b}} \cdot R_{i+1}^{\mathrm{adj}}$, where R_{i+1}^{adj} is the adjugate matrix of R_{i+1}, i.e., $R_{i+1} \cdot R_{i+1}^{\mathrm{adj}} = \det(R_{i+1}) \cdot I_n$. Transform also A_0 into $A_0 \cdot R_1^{\mathrm{adj}}$ and $A_{\ell+1}$ into $R_{\ell+1} \cdot A_{\ell+1}$.

3. Multiply by random scalars, i.e., multiply each matrix $A_{i,\mathbf{b}}$ by some random scalar $\alpha_{i,\mathbf{b}} \in (R/gR)^\times$. Also multiply A_0 and $A_{\ell+1}$ by α_0 and $\alpha_{\ell+1}$ respectively.

We can check that all the transformations described above output a branching program which is strongly equivalent to the one given in input, so the final branching program $\hat{\mathbf{A}}$ is also strongly equivalent to $\mathbf{A}$ (as in (5)). In the following, we will only be interested in (5), not in the details of the transformation.

- Finally, the obfuscator encodes the matrices $\{\hat{A}_{i,\mathbf{b}}\}_{i,\mathbf{b}}$, $\hat{A}_0$ and $\hat{A}_{\ell+1}$ at some level $\{S_{i,\mathbf{b}}\}_{i,\mathbf{b}}$, S_0 and $S_{\ell+1}$ respectively, using the GGH multilinear map. The choice of these levels (called a straddling set system) depends on the obfuscators, but will have no importance in the following. The only property that we need, and that is fulfilled by the above obfuscators, is that for any entry x, the sets $S_0, S_{\ell+1}$ and $S_{i,x[inp(i)]}$ for $i \in [l]$ are disjoint and we have

$$S_0 \cup \left(\cup_{i\in[l]} S_{i,x[inp(i)]} \right) \cup S_{\ell+1} = S_{zt}. \quad (6)$$

This means that every honest evaluation of the encoded branching program outputs an element at level S_{zt}, that can be zero-tested. This condition is

---

[14] The obfuscators of [23,32] use the three tools while the ones of [2,5,30] use Tools 2 and 3 only.

necessary for the above obfuscators to be correct (otherwise we cannot evaluate the obfuscated branching program).

- The obfuscator then outputs the elements $[\hat{A}_0]_{S_0}$, $\{[\hat{A}_{i,\mathbf{b}}]_{S_{i,\mathbf{b}}}\}_{i\in[\ell],\mathbf{b}\in\{0,1\}^d}$, $[\hat{A}_{\ell+1}]_{S_{\ell+1}}$ and the public parameters of the GGH map (n, q, κ, p_{zt}).

To evaluate the obfuscated branching program on input x, compute

$$u_x = [\hat{A}_0]_{S_0} \times \prod_{i\in[\ell]} [\hat{A}_{i,x[\mathtt{inp}(i)]}]_{S_{i,x[\mathtt{inp}(i)]}} \times [\hat{A}_{\ell+1}]_{S_{\ell+1}}.$$

By Property (5), this is an encoding of zero if and only if the output of the original branching program was zero. And by Property (6), this encoding is at level S_{zt}. So using p_{zt}, we can perform a zero-test and output 0 if this is an encoding of 0 and 1 otherwise. In the following, we will sometimes simplify notations and forget about the subscripts $S_{i,\mathbf{b}}$, as the levels of the encodings are entirely determined by the encoded matrices $A_{i,\mathbf{b}}$.

For our attack to work, we will need to assume that if we evaluate the obfuscated branching program on enough inputs for which the output is zero, then we can recover a basis of the ideal $\langle h \rangle$ (where h is a secret element of the GGH13 map, as described in Sect. 2.2). More formally, we make the following heuristic assumption.

Heuristic 1. Let X_0 be the set of inputs on which the branching program evaluates to 0 and let $x \in X_0$. If we evaluate the obfuscated branching program on x and zero-test the final encoding, we obtain a ring element of the form $r_x \cdot h \in R$. We assume that the set of all $r_x \cdot h$ for $x \in X_0$ spans the ideal $\langle h \rangle$ (and not a smaller ideal contained in $\langle h \rangle$). We also assume that if x is chosen uniformly in X_0, then we can obtain a basis of $\langle h \rangle$ with a polynomial number of samples.

Discussion about Heuristic 1. We make the heuristic assumption above to simplify the description of our attack. This heuristic assumption is coherent with the numerical experiments we made (see the full version [33] for a description of the experimental results). Moreover, we also observe that, even if we recover an ideal $J \subseteq \langle h \rangle$ instead of the ideal $\langle h \rangle$, we can still handle it if $\langle h \rangle$ has a constant number of prime factors (see the full version for more details).

This completes the definition of our abstract obfuscator, which captures the obfuscators of [2,5,23,30,32]. In the next section, we describe a mixed-input attack against this abstract obfuscator, where all we use is that it satisfies Properties (5) and (6).

4   The Main Attack

We will now prove our main theorem.

Theorem 3. *Let $\mathcal{O}$ be any of the obfuscators in [2, 5, 6, 23, 30, 32], on single or dual input branching programs, instantiated with the GGH13 multilinear map [21] (respectively, let $\mathcal{O}$ be any of the circuit obfuscators in [4, 19, 37]). Assume the secret parameter h of the GGH13 multilinear map is sampled using a spherical Gaussian distribution (as in Sect. 2.2). Then, there exist two explicit equivalent branching programs (respectively, two equivalent circuits) $\boldsymbol{A}$ and $\boldsymbol{A'}$ such that $\mathcal{O}(\boldsymbol{A})$ and $\mathcal{O}(\boldsymbol{A'})$ can be distinguished in quantum polynomial time, under Conjecture 1 and Heuristic 1.*

The limitation to the case where h is sampled according to a spherical Gaussian distribution is discussed in the full version. We show that if q is large enough, or if h is a product of a small number of spherical Gaussian distributions, then our result still holds. We leave as an open problem to show that the attack goes through for every efficient way of sampling h, or to find a way that allows to thwart the attack (although we lean towards the former rather than the latter). The necessity for h being sampled according to a spherical Gaussian distribution appears in Theorem 2, to solve the short Principal Ideal Problem and recover the secret element h. It is not used anywhere else in the attack, in particular, it is not used in the mixed-input part of the attack (see Sect. 4.2).

To prove Theorem 3, we present a quantum polynomial time attack against the abstract obfuscator described in Sect. 3. This results into an attack against the iO security of the branching program obfuscators of [2, 5, 23, 30, 32]. We then explain how to slightly modify this attack to use it against the obfuscator of [6], whose structure is very close to the one of the abstract obfuscator. Finally, adapting the attack to the circuit obfuscator of [19] will require more work, because its structure is further away from the abstract obfuscator.

The attack works in two steps. We first recover the secret element h of the GGH multilinear map. Using the results of [10, 13, 18], recalled in Sect. 2.4, this can be done in quantum polynomial time. Knowing this secret element h, we are able to construct a zero-testing parameter p'_{zt} at a higher level than S_{zt}. We can then use this new parameter p'_{zt} to mount a (classical) polynomial time mixed-input attack against the abstract obfuscator.

4.1   Creating a New Zero-Testing Parameter in Quantum Polynomial Time

We first explain in this section how we can recover the secret parameter h of the multilinear map in quantum polynomial time. We then describe how to construct a new zero-testing parameter at a level higher than S_{zt}, using h. Note that the following is folklore, we recall it for the sake of completeness.

The first step is to recover sufficiently many multiples of h, to obtain a basis of the ideal $\langle h \rangle$ (when seen as a sub-lattice of R). This part of the attack was already described in the original article [21], and can be done in classical polynomial time, under Heuristic 1. Observe that for each top-level encoding that pass the zero-test, we obtain a multiple of h. We make the heuristic Assumption 1 to ensure that we indeed recover a basis of the ideal $\langle h \rangle$, by zero-testing sufficiently many top-level encodings of zero. For this step to work, we need that the branching program evaluates sufficiently often to 0, to obtain sufficiently many encodings of 0. In the following, we will choose branching programs that compute the always zero function, hence the condition on the number of encodings that pass the zero-test will be satisfied.

We then recover $\pm X^i h$ from the basis of the ideal $\langle h \rangle$, using Theorem 2. This can be done in quantum polynomial time, under Conjecture 1, as h is sampled according to a Gaussian distribution of parameter larger than $200 \cdot n^{1.5}$. The fact that we recover $\pm X^j h$ instead of h will have no importance for our attack,[15] so in the following we will assume that we recovered h exactly. In [21, Sect. 6.4], the authors propose another distribution for the secret parameter h (the element h is sampled according to a non spherical Gaussian distribution). Theorem 2 does not apply as it in this case, but we show in the full version that our attack can be extended to some other distributions of h.

We now explain how to use h to create a new zero-testing parameter p'_{zt} at a higher level than S_{zt}. A close variant of this step was already mentioned in [21, Sect. 6.3.3]. The authors explained how to use a small multiple of $1/h$ and a low level encoding of zero to create a new zero-testing parameter that enabled to test at a higher level whether the numerator of an encoding was a multiple of g or not (i.e., if the encoding was an encoding of zero or not). In our case, the situation is a little different, as we do not know any low level encoding of zero. Hence, we only manage to create a new zero-testing parameter that enables us to determine whether the numerator of an encoding is a multiple of g^2 or not. In the following, we will say that an encoding is an encoding at level $2S_{zt}$ if its denominator is $(z^*)^2$. For instance, such an encoding can be obtained by multiplying two level S_{zt} encodings. We see the level $2S_{zt}$ as a multiset containing all the elements of S_{zt} twice. We use the secret h to compute a new zero-testing parameter p'_{zt} at level $2S_{zt}$. Recall that $p_{zt} = h z^* g^{-1} \bmod q$. We then define

$$p'_{zt} = p_{zt}^2 h^{-2} \bmod q = (z^*)^2 \cdot g^{-2} \bmod q.$$

Again, note that even if we call it a new zero-testing parameter, p'_{zt} only enables us to test whether the numerator of a level $2S_{zt}$ encoding is a multiple of g^2, and not g, as our original zero-test parameter p_{zt} did. But still, being able to test at a level higher than S_{zt} if the numerator is a multiple of g^2 will enable us to mount a mixed-input attack against the abstract obfuscator of Sect. 3. We describe this mixed-input attack in the next subsection.

4.2   The Mixed-Input Attack

We now assume that we have built a new pseudo-zero-test parameter p'_{zt}, as in Subsect. 4.1 (in quantum polynomial time), and that we are given an obfuscated branching program $([\hat{A}_0]_{S_0}, \{[\hat{A}_{i,\mathbf{b}}]_{S_{i,\mathbf{b}}}\}_{i \in [l], \mathbf{b} \in \{0,1\}^d}, [\hat{A}_{\ell+1}]_{S_{\ell+1}})$, obtained by using our abstract obfuscator defined in Sect. 3.

---

[15] This is because both X^j and its inverse $-X^{n-j}$ have euclidean norm 1.

Let x and y be two different inputs of the branching program. A mixed-input attack consists in changing the value of some bits of the input during the evaluation of the obfuscated branching program. For instance, the way we will do it is by taking some matrix $[\hat{A}_{i,y[\text{inp}(i)]}]_{S_{i,y[\text{inp}(i)]}}$, instead of $[\hat{A}_{i,x[\text{inp}(i)]}]_{S_{i,x[\text{inp}(i)]}}$, while evaluating the program on x. Such mixed-input attack can leak information on the program being obfuscated (see the specific choice of branching programs described in the next subsection). In order to prevent mixed-input attack, the abstract obfuscator uses a straddling set system. The intuition is that if the attacker tries to mix the matrices $[\hat{A}_{i,x[\text{inp}(i)]}]_{S_{i,x[\text{inp}(i)]}}$ and $[\hat{A}_{i,y[\text{inp}(i)]}]_{S_{i,y[\text{inp}(i)]}}$, it will not get an encoding at level S_{zt} at the end of the computation and hence it cannot zero-test it. However, we can use our new zero-testing parameter p'_{zt} to handle this difficulty.

Let $j \in [\ell]$ and compute

$$\tilde{u}_{x,j} = [\hat{A}_0] \cdot \prod_{i<j}[\hat{A}_{i,x[\text{inp}(i)]}] \cdot [\hat{A}_{j,y[\text{inp}(j)]}] \cdot \prod_{j<i\leq\ell}[\hat{A}_{i,x[\text{inp}(i)]}] \cdot [\hat{A}_{\ell+1}]$$

$$\tilde{u}_{y,j} = [\hat{A}_0] \cdot \prod_{i<j}[\hat{A}_{i,y[\text{inp}(i)]}] \cdot [\hat{A}_{j,x[\text{inp}(j)]}] \cdot \prod_{j<i\leq\ell}[\hat{A}_{i,y[\text{inp}(i)]}] \cdot [\hat{A}_{\ell+1}],$$

that is, we exchange $[\hat{A}_{j,x[\text{inp}(j)]}]_{S_{j,x[\text{inp}(j)]}}$ and $[\hat{A}_{j,y[\text{inp}(j)]}]_{S_{j,y[\text{inp}(j)]}}$ in the honest evaluations of the obfuscated branching program on x and y.

The encodings $\tilde{u}_{x,j}$ and $\tilde{u}_{y,j}$ will have illegal levels S_x and S_y that are different from S_{zt}. But as we only exchange two matrices between correct evaluations, we know that $\tilde{u}_{x,j} \cdot \tilde{u}_{y,j}$ will be encoded at the same level as $u_x \cdot u_y$ where u_x and u_y are the correct evaluations of the obfuscated branching program on x and y. As u_x and u_y are correct evaluations, using Property (6), we know that they are encoded at level S_{zt}. Hence $\tilde{u}_{x,j} \cdot \tilde{u}_{y,j}$ is encoded at level $2S_{zt}$, and we can zero-test $\tilde{u}_{x,j} \cdot \tilde{u}_{y,j}$ using p'_{zt}.

Remember that an encoding will pass this zero-test only if its numerator is a multiple of g^2 and not only g. A simple way to ensure that $\tilde{u}_{x,j} \cdot \tilde{u}_{y,j}$ has a numerator which is a multiple of g^2 is to choose x and y such that $\tilde{u}_{x,j}$ and $\tilde{u}_{y,j}$ are both encodings of 0 (i.e., their numerator are both multiples of g, and hence their product has a numerator which is a multiple of g^2). Using Property (5) of our abstract obfuscator, we know that $\tilde{u}_{x,j}$ is an encoding of 0 if and only if

$$A_0 \cdot \prod_{i<j} A_{i,x[\text{inp}(i)]} \cdot A_{j,y[\text{inp}(j)]} \cdot \prod_{j<i\leq\ell} A_{i,x[\text{inp}(i)]} \cdot A_{\ell+1} = 0.$$

We denote by $\tilde{a}_{x,j}$ the left hand side of this equation. In the same way, we define

$$\tilde{a}_{y,j} = A_0 \cdot \prod_{i<j} A_{i,y[\text{inp}(i)]} \cdot A_{j,x[\text{inp}(j)]} \cdot \prod_{j<i\leq\ell} A_{i,y[\text{inp}(i)]} \cdot A_{\ell+1},$$

and we have that $\tilde{u}_{y,j}$ is an encoding of 0 if and only if $\tilde{a}_{y,j} = 0$.

To conclude, if we manage to find two equivalent branching programs $\mathbf{A}$ and $\mathbf{A}'$, two inputs x and y and an integer $j \in [\ell]$ such that $\tilde{a}_{x,j} = \tilde{a}_{y,j} = 0$

for $\mathbf{A}$ but $\tilde{a}'_{x,j} \neq 0$ and $\tilde{a}'_{y,j} \neq 0$ for $\mathbf{A}'$, then we can distinguish between the obfuscation of $\mathbf{A}$ and the one of $\mathbf{A}'$. Indeed, the numerator of $\tilde{u}_{x,j} \cdot \tilde{u}_{y,j}$ will be a multiple of g^2 in the case of $\mathbf{A}$ but the numerator of $\tilde{u}'_{x,j} \cdot \tilde{u}'_{y,j}$ will not be a multiple of g in the case of $\mathbf{A}'$ (and therefore not a multiple of g^2 either). Hence, using p'_{zt}, we can determine which of the branching program $\mathbf{A}$ or $\mathbf{A}'$ has been obfuscated.

In the next subsection, we present two possible branching programs $\mathbf{A}$ and $\mathbf{A}'$ and inputs x and y that satisfy the condition above. We note that this condition is easily satisfied and it should be possible to find a lot of other branching programs satisfying it. We just propose here a simple example of such branching programs, in order to complete the proof of Theorem 1.

4.3   A Concrete Example of Branching Programs

In this section, we present an example of two branching programs $\mathbf{A}$ and $\mathbf{A}'$ that are equivalent, but such that their obfuscated versions, obtained using the abstract obfuscator, can be distinguished using the framework described above, hence attacking the iO security of the obfuscator.

Remember that for the first step of our attack (recovering h and creating p'_{zt}, see Sect. 4.1), we need to have a sufficient number of inputs x that evaluate to zero. Here, we choose branching programs that compute the always zero function. We now show how to satisfy the conditions for the second part of the attack (Sect. 4.2).

Let $I = I_w \in \{0,1\}^{w \times w}$ be the identity matrix and $J \in \{0,1\}^{w \times w}$ be a matrix of order two (i.e., $J \neq I$ and $J^2 = I$). One could for example take

$$J = \begin{pmatrix} 0 & 1 & \\ 1 & 0 & \\ & & I_{w-2} \end{pmatrix}.$$ Our first branching program will consist in identity matrices

only. We will build our second branching program such that when evaluating it on input x, we have a product of ℓ matrices I (when we forget about the bookend vectors), but on input y we have a product of $\ell - 2$ matrices I and 2 matrices J.[16] We will then exchange one of these J matrices with an I matrix in the evaluation on x. The resulting products will then be equal to matrix J instead of matrix I (as it is the case for the first branching program). We describe the two branching programs more precisely below.

Input selection function. Recall that the input selection function $\mathtt{inp}$ is fixed and is such that there are at least two distinct integers j_1 and j_2 such that $\mathtt{inp}(j_1) \cap \mathtt{inp}(j_2) \neq \emptyset$. Let s be such that $s \in \mathtt{inp}(j_1) \cap \mathtt{inp}(j_2)$. This means that when evaluating the branching program on some input, the j_1-th and the j_2-th matrices of the product both depend on the s-th bit of the input. Without loss of generality, we assume that $\mathtt{inp}(j_1) = (s, s_2, \cdots, s_d)$ and $\mathtt{inp}(j_2) = (s, t_2, \cdots, t_d)$ for some integers s_i and t_i in $[m]$.

---

[16] As J has order 2, the resulting product will still be the identity matrix.

Matrices. Our first branching program **A** consists in identity matrices only, i.e., $A_{i,\mathbf{b}} = I$ for all $i \in [\ell]$ and $\mathbf{b} \in \{0,1\}^d$. For our second branching program **A**$'$, we take

$$A'_{i,\mathbf{b}} = \begin{cases} I \text{ if } i \notin \{j_1, j_2\} \text{ or } b_1 = 0 \\ J \text{ if } i \in \{j_1, j_2\} \text{ and } b_1 = 1, \end{cases}$$

where $\mathbf{b} = (b_1, \cdots, b_d)$. This means that when evaluating the branching program **A**$'$ on some input x, if $x_s = 0$, then all the matrices of the product are identity matrices. And if $x_s = 1$, then the j_1-th and j_2-th matrices of the product are J matrices and the others are I matrices. As J has order two, the product will always be the identity.

Bookend vectors. We take A_0 and $A_{\ell+1}$ to be two vectors such that $A_0 I A_{\ell+1} = 0$ but $A_0 J A_{\ell+1} \neq 0$. For instance, with the choice of $J = \begin{pmatrix} 0 & 1 \\ 1 & 0 \\ & & I_{w-2} \end{pmatrix}$, we can take $A_0 = \begin{pmatrix} 1 & 0 \ldots 0 \end{pmatrix}$ and $A_{\ell+1} = \begin{pmatrix} 0 & 1 & 0 \ldots 0 \end{pmatrix}^T$, where A^T denotes the transpose of A for any matrix A. These bookend vectors are the same for both branching programs, i.e., $A'_0 = A_0$ and $A'_{\ell+1} = A_{\ell+1}$.

These two branching programs **A** and **A**$'$ are equivalent as they both compute the always zero function. Now, take $x = 0 \ldots 0$ and $y = 0 \ldots 010 \ldots 0$ where the 1 is at the s-th position, and let $j = j_1$. Let us compute $\tilde{a}_{x,j}, \tilde{a}_{y,j}$ for branching program **A** and $\tilde{a}'_{x,j}, \tilde{a}'_{y,j}$ for branching program **A**$'$.

Branching program A. As all matrices are identity matrices in **A**, exchanging two matrices does not change the product and we still have

$$\tilde{a}_{x,j} = A_0 \cdot \prod_{i<j} A_{i,x[\mathtt{inp}(i)]} \cdot A_{j,y[\mathtt{inp}(j)]} \cdot \prod_{j<i\leq\ell} A_{i,x[\mathtt{inp}(i)]} \cdot A_{\ell+1} = A_0 \cdot I \cdot A_{\ell+1} = 0,$$

$$\tilde{a}_{y,j} = A_0 \cdot \prod_{i<j} A_{i,y[\mathtt{inp}(i)]} \cdot A_{j,x[\mathtt{inp}(j)]} \cdot \prod_{j<i\leq\ell} A_{i,y[\mathtt{inp}(i)]} \cdot A_{\ell+1} = A_0 \cdot I \cdot A_{\ell+1} = 0.$$

Branching program A$'$. Here, we chose our parameters so that an honest evaluation of **A**$'$ on x leads to a product of only I matrices and an honest evaluation of **A**$'$ on y leads to a product of $\ell - 2$ matrices I and 2 matrices J. We also chose j so that we exchange a J matrix with a I matrix. Hence, we have

$$\tilde{a}'_{x,j} = A'_0 \cdot \prod_{i<j} A'_{i,x[\mathtt{inp}(i)]} \cdot A'_{j,y[\mathtt{inp}(j)]} \cdot \prod_{j<i\leq\ell} A'_{i,x[\mathtt{inp}(i)]} \cdot A'_{\ell+1} = A_0 \cdot J \cdot A_{\ell+1} \neq 0,$$

$$\tilde{a}'_{y,j} = A'_0 \cdot \prod_{i<j} A'_{i,y[\mathtt{inp}(i)]} \cdot A'_{j,x[\mathtt{inp}(j)]} \cdot \prod_{j<i\leq\ell} A'_{i,y[\mathtt{inp}(i)]} \cdot A'_{\ell+1} = A_0 \cdot J \cdot A_{\ell+1} \neq 0.$$

To conclude, this gives us the desired condition of Sect. 4.2. Indeed, for the branching program **A**, the numerator of $\tilde{u}_{x,j} \cdot \tilde{u}_{y,j}$ is a multiple of g^2, hence zero-testing it with the parameter p'_{zt} gives a positive result. Oppositely, for the branching program $\mathbf{A}'$, the numerator of $\tilde{u}_{x,j} \cdot \tilde{u}_{y,j}$ is not a multiple of g,[17] hence zero-testing it with the parameter p'_{zt} gives a negative result. We can then distinguish between the obfuscations of $\mathbf{A}$ and $\mathbf{A}'$. This completes the proof of Theorem 1 for the obfuscators of $[2,5,23,30,32]$.

4.4   Other Branching Program Obfuscators

We now discuss the possible extension of this attack to other branching program obfuscators that are not captured by the abstract obfuscator of Sect. 3.

Obfuscator of [6]. This obfuscator is close to the one described in the abstract model, except that it obfuscates a slightly different definition of branching programs. In [6], a branching program $\mathbf{A}$ comes with an additional value q_{acc}, and we have $\mathbf{A}(x) = 0$ if and only if $A_0 \cdot \prod_{i \in [\ell]} A_{i,x[\text{inp}(i)]} \cdot A_{\ell+1} = q_{acc}$. The only difference with the definition of branching programs given in Sect. 2.3 is that q_{acc} may be non-zero. Hence, when multiplying by the scalars $\alpha_{i,\mathbf{b}}$ in the obfuscator (see Tool 3), we may change the output of the function. To enable correct evaluation of the obfuscated branching program, the obfuscator of [6] also publishes encodings of the scalars $\alpha_{i,\mathbf{b}}$ at level $S_{i,\mathbf{b}}$.

More formally, the obfuscator of [6] uses Tools 2 and 3 of Sect. 3. In Tool 2, the authors use R_{i+1}^{-1} instead of R_{i+1}^{adj}, in order to keep the same product (otherwise the product would be multiplied by the determinants of the R_i matrices). Let $\hat{A}_{i,\mathbf{b}} = \alpha_{i,\mathbf{b}} R_i A_{i,\mathbf{b}} R_{i+1}^{-1}$ be the matrices obtained after re-randomization (using Tools 2 and 3). Let $\hat{A}_0 = A_0 R_1^{-1}$ and $\hat{A}_{\ell+1} = R_{\ell+1} A_{\ell+1}$. The obfuscator provides encodings of the matrices $\hat{A}_0$, $\{\hat{A}_{i,\mathbf{b}}\}_{i,\mathbf{b}}$ and $\hat{A}_{\ell+1}$ at levels S_0, $\{S_{i,\mathbf{b}}\}_{i,\mathbf{b}}$ and $S_{\ell+1}$, respectively. It also provides encodings of the $\{\alpha_{i,\mathbf{b}}\}_{i,\mathbf{b}}$ at levels $\{S_{i,\mathbf{b}}\}_{i,\mathbf{b}}$ and an encoding of q_{acc} at level $S_0 \cup S_{\ell+1}$. Then, to evaluate the obfuscated branching program on input x, one computes

$$[\hat{A}_0]_{S_0} \cdot \prod_{i \in [\ell]} [\hat{A}_{i,x[\text{inp}(i)]}]_{S_{i,x[\text{inp}(i)]}} \cdot [\hat{A}_{\ell+1}]_{S_{\ell+1}} - [q_{acc}]_{S_0 \cup S_{\ell+1}} \cdot \prod_{i \in [\ell]} [\alpha_{i,x[\text{inp}(i)]}]_{S_{i,x[\text{inp}(i)]}},$$

and tests whether this is an encoding of 0 or not. By construction, this will be an encoding of 0 at level S_{zt} if and only if $\mathbf{A}(x) = 0$.

The first part of our attack (recovering h and p'_{zt}) still goes through. We slightly modify the mixed-input part. Instead of exchanging only the j-th matrix between the evaluations of x and y, we will also exchange the corresponding $\alpha_{j,\mathbf{b}}$ in the second product. Doing so, we ensure that the product of the $\alpha_{i,\mathbf{b}}$'s remains the same in both sides of the difference. This also ensures that the level of both sides will be the same after the exchange, and hence we can still subtract them. The same example as in Sect. 4.3 will then work also for this obfuscator. This gives us a way to distinguish in quantum polynomial time between the obfuscated versions of two equivalent branching programs, hence attacking the iO security of the obfuscator of [6].

---

[17] The ideal $\langle g \rangle$ is chosen prime in GGH so the product of two elements that are not divisible by g is also not divisible by g.

Obfuscators of [12,22]. Our attack does not seem to extend to the obfuscators of [12,22]. The obstacle is that the security of these obfuscators against mixed-input attacks does not rely on the GGH map but on the scalars $\alpha_{i,\mathbf{b}}$, which are chosen with a specific structure to ensure that the branching program is correctly evaluated.

More precisely, these obfuscators use (single input) branching programs with a slightly different definition, where the product of matrices (with the bookend vectors) is never 0. For instance, the branching programs are chosen such that the product of the matrices (on honest evaluations) is either 1 or 2, in which cases we say that the output of the branching program is respectively 0 or 1. Hence, when evaluating the obfuscated branching program on input x, the user obtains a top-level encoding of either $\prod_i \alpha_{i,x_i}$ or $2\prod_i \alpha_{i,x_i}$ depending on the output of the branching program. In order for the user to determine which one of the two encodings it has obtained, the obfuscated branching program also provide him (via a so-called dummy branching program) with a top-level encoding $\prod_i \alpha_{i,x_i}$. The user then only has to subtract the two top-level encodings and zero-test to determine whether $\mathbf{A}(x) = 0$ or 1. Now, if the user tries to mix the inputs, it can obtain a top-level encoding of $(\alpha_{j,y_j} \cdot \prod_{i \neq j} \alpha_{i,x_i}) \cdot a_{x,j}$ for instance (where $a_{x,j} = 1$ or 2 is the product of the corresponding matrices). But, as it is not an honest evaluation, it will not have a top-level encoding of $\alpha_{j,y_j} \cdot \prod_{i \neq j} \alpha_{i,x_i}$ to compare it with.

Following the same idea as for the mixed-input attack described above, the attacker could compute two top-level encodings of $(\alpha_{j,y_j} \cdot \prod_{i \neq j} \alpha_{i,x_i}) \cdot a_{x,j}$ and $(\alpha_{j,x_j} \cdot \prod_{i \neq j} \alpha_{i,y_i}) \cdot a_{y,j}$ and then multiply them to obtain an encoding of $(\prod_i \alpha_{i,x_i} \cdot \prod_i \alpha_{i,y_i}) \cdot a_{x,j} \cdot a_{y,j}$ at level $2S_{zt}$. Now, using the top-level encodings of $\prod_i \alpha_{i,x_i}$ and $\prod_i \alpha_{i,y_i}$ that are provided by the obfuscated branching program, one can also obtain an encoding of $(\prod_i \alpha_{i,x_i} \cdot \prod_i \alpha_{i,y_i})$ at level $2S_{zt}$. So if we could zero-test at level $2S_{zt}$, then we could distinguish between a branching program where $a_{x,j} \cdot a_{y,j} = 1$ and one where $a_{x,j} \cdot a_{y,j} \neq 1$. But we cannot zero-test at level $2S_{zt}$: our new zero-testing parameter p'_{zt} only enables us to determine whether the numerator of an encoding is a multiple of g^2 or not. Here, we subtract two level-$2S_{zt}$ encodings of the same value, so the numerator of the result will be a multiple of g, but it is very unlikely to be a multiple of g^2. Hence, we do not learn anything by using p'_{zt}. Because of the final subtraction, we did not manage to obtain an encoding at level $2S_{zt}$ whose numerator was a multiple of g^2, and so we did not manage to adapt the mixed-input attack described above to the obfuscators of [12,22].

5   Adapting the Attack to the Obfuscator of [19]

Unlike the abstract branching program described in Sect. 3, the obfuscator of [19] does not obfuscate branching programs, but it obfuscates circuits directly. The structure of this obfuscator is very different from the abstract obfuscator described in Sect. 3 and so the attack described in Sect. 4 cannot be directly applied to it. However, similarly to the other obfuscators described above, the obfuscator of [19] also uses the levels of the GGH multilinear map to prevent mixed-input attacks. This is the weakness we exploited to mount a mixed-input attack against the abstract obfuscator, and here again, this will enable us to attack the [19] obfuscator, by attacking the underlying GGH multilinear map. In this section, we first describe in a simplified way the obfuscator of [19] (this simplified version also captures the obfuscators of [4,37]). We then show how to adapt our attack to mount a quantum polynomial-time mixed-input attack against this candidate obfuscator.

5.1   The Obfuscator

The obfuscator of [19] uses the GGH multilinear map [21] in its asymmetric version, but with a composite g. More concretely, sample three elements $g_1, g_2, g_3 \in R$ as for the original g in the GGH map, that is $\|g_i\| = O(n)$, $\|1/g_i\| = O(n^2)$ and such that $\mathcal{N}(g_i)$ is a prime integer, for all $i \in [3]$. Then, let $g = g_1 g_2 g_3$. If we denote by $R_i = R/g_i R$ the quotient rings for $i \in [3]$, then using the Chinese reminder theorem we know that the encoding space R/gR is isomorphic to $R_1 \times R_2 \times R_3$. In the following, it will be useful to choose this point of view, as we will encode triplets of elements $(a_1, a_2, a_3) \in R_1 \times R_2 \times R_3$, using the GGH map.

Let Σ be some subset of $\{0,1\}^l$ with both l and $|\Sigma|$ that are polynomial in the security parameter λ. We will be interested into arithmetic circuits $C : \Sigma \to \{0,1\}$. By arithmetic circuits, we mean that C performs addition, multiplication and subtraction over the bits of the element of Σ (i.e., C is an arithmetic circuit from $\{0,1\}^l$ to $\{0,1\}$, but we are only interested in its restriction to $\Sigma \subseteq \{0,1\}^l$). The operations over the bits are performed over $\mathbb{Z}$ but we only consider circuits whose output is in $\{0,1\}$. Let $\mathscr{C}$ be a class of such circuits, whose size is bounded by some polynomial (the properties of this class of circuit will not be interesting for our attack) and let U be a universal circuit for the class $\mathscr{C}$. The size of U is also bounded by some polynomial in the security parameter. We abuse notation by denoting by C both a circuit of $\mathscr{C}$ and its bit representation, that is we have $U(\sigma, C) = C(\sigma)$ for any $\sigma \in \Sigma$ (the first C denotes the bit representation of the circuit while the second one represent the function computed by the circuit).

To obfuscate a circuit C of the class $\mathscr{C}$, the main idea of [19] is that the obfuscator will produce GGH encodings of the bits of C and of the bits of all the possible inputs $\sigma \in \Sigma$. Then, to evaluate the obfuscated circuit, it suffices to homomorphically evaluate the universal circuit U on these encodings and to test whether the result is 0 or not. In order to prove the security of their obfuscators, the authors of [19] added other gadgets to their obfuscator. The first idea is to encode the useful information only in the second slot of the GGH map (in the ring R_2) and to use the two other slots to prevent some mixed-input attack (where we mix the bits of two circuits). They also use straddling set systems, like the abstract obfuscator defined in Sect. 3, to prevent other kind of mixed input attacks (where we mix the bits of two inputs). We describe below in more details how the obfuscator of [19] obfuscates a circuit $C \in \mathscr{C}$. In order to help

understanding what is happening, we also describe in parallel how to evaluate
the obfuscated circuit.

1. First, we encode each bit of all the possible inputs $\sigma \in \Sigma$ (recall that we chose
$|\Sigma|$ to be polynomial in the security parameter, so it is possible to enumerate
all the elements of Σ). For each symbol $\sigma \in \Sigma$ and each bit position $i \in [l]$,
define $W_{i,\sigma}^{(1)} = [r_\sigma^{(1)} \cdot w_{i,\sigma}^{(1)}]_{S_\sigma^{(1)}}$ and $R_\sigma^{(1)} = [r_\sigma^{(1)}]_{S_\sigma^{(1)}}$, where $r_\sigma^{(1)}$ is sampled
uniformly in $R/gR^\times$ (and only depends on σ) and

$$w_{i,\sigma}^{(1)} = (y_i^{(1)}, \sigma_i, \rho_{i,\sigma}^{(1)}) \in R_1 \times R_2 \times R_3,$$

for σ_i the i-th bit of σ and $y_i^{(1)}$ and $\rho_{i,\sigma}^{(1)}$ sampled uniformly in R_1 and R_3
respectively. The level $S_\sigma^{(1)}$ of the encoding will be chosen to prevent mixed-
input attacks. We will go into more details about the levels of the encodings
later. These encodings $W_{i,\sigma}^{(1)}$ and $R_\sigma^{(1)}$ are made public, for $i \in [l]$ and $\sigma \in \Sigma$.
Note that $y_i^{(1)}$ is the same for all symbols σ, this will be necessary for correctness.

2. Second, we encode the bits of the representation of the circuit $C \in \mathscr{C}$. We
denote by $|C|$ the size of the bit representation of C. For each $1 \leq j \leq |C|$,
define $W_j^{(2)} = [r^{(2)} \cdot w_j^{(2)}]_{S^{(2)}}$ and $R^{(2)} = [r^{(2)}]_{S^{(2)}}$, where $r^{(2)}$ is sampled
uniformly in $R/gR^\times$ and

$$w_j^{(2)} = (y_j^{(2)}, C_j, \rho_j^{(2)}) \in R_1 \times R_2 \times R_3,$$

for C_j the j-th bit of the representation of C and $y_j^{(2)}$ and $\rho_j^{(2)}$ sampled
uniformly in R_1 and R_3 respectively. Again, the level $S^{(2)}$ of the encoding
will be described later. These encodings $W_j^{(2)}$ and $R^{(2)}$ are made public, for
$1 \leq j \leq |C|$.

   Once we have encodings for the bits of C and for all the possible input values
$\sigma \in \Sigma$, as the universal circuit U only performs additions, subtractions and
multiplications, we can homomorphically evaluate it on the encodings. We can
always perform multiplications of encodings, it will only increase the level of the
encodings. However, there is a subtlety for addition and subtraction, as we can
only add and subtract encodings at the same level. To circumvent this difficulty,
the authors of [19] use the encodings $R^{(2)}$ and $R_\sigma^{(1)}$. During the evaluation of the
universal circuit U on the encodings, we will perform computations so that for all
intermediate encodings we compute, we always have encodings of the form $[r \cdot w]_S$
and $[r]_S$, with the same level S. At the beginning, all the encodings described
above have the desired form $[r \cdot w]_S$ and $[r]_S$. If we want to multiply $[r_1 \cdot w_1]_{S_1}$ and
$[r_2 \cdot w_2]_{S_2}$, we just compute the product of the encodings to get $[r_1 r_2 \cdot w_1 w_2]_{S_1 \cup S_2}$
and we also compute the product of the r part to obtain $[r_1 r_2]_{S_1 \cup S_2}$. Note that
here, the union of the two sets $S_1 \cup S_2$ keeps multiple copies of the elements
that appear both in S_1 and in S_2 (i.e., $S_1 \cup S_2$ is a multiset). If we want to add
$[r_1 \cdot w_1]_{S_1}$ and $[r_2 \cdot w_2]_{S_2}$, then two cases appear. If $r_1 = r_2$ and $S_1 = S_2$, then

add both encodings to get $[r_1 \cdot (w_1 + w_2)]_{S_1}$ and keep $[r_1]_{S_1}$. Otherwise, compute $[r_1]_{S_1} \cdot [r_2 \cdot w_2]_{S_2} + [r_2]_{S_2} \cdot [r_1 \cdot w_1]_{S_1} = [r_1 r_2 \cdot (w_1 + w_2)]_{S_1 \cup S_2}$ and compute the product $[r_1 r_2]_{S_1 \cup S_2}$. We proceed similarly for subtraction.

With this technique, we can evaluate the circuit U on the encodings provided by the obfuscator, independently of the levels used to encode them. Assume we evaluate it honestly on the encodings of C and of some input $\sigma \in \Sigma$, we then obtain encodings $W_\sigma = [r_\sigma \cdot w_\sigma]_{S_\sigma}$ and $R_\sigma = [r_\sigma]_{S_\sigma}$ at some level S_σ, for some $r_\sigma \in R/gR$, where

$$w_\sigma = (y^*, C(\sigma), \rho_\sigma) \in R_1 \times R_2 \times R_3,$$

for some $y^* \in R_1$ and $\rho_\sigma \in R_3$. Note that, as the $y_i^{(1)}$'s do not depend on the input σ, the value y^* is the same for all σ's. We then want to annihilate the values in the extra slots (that is y^* and ρ_σ) to recover the value of $C(\sigma)$ by zero-testing. To do that, the obfuscator provides two more encodings.

3. To annihilate the value in the third slot, the obfuscator output encodings $\widehat{W_\sigma} = [\widehat{r_\sigma} \cdot \widehat{w}]_{\widehat{S_\sigma}}$ and $\widehat{R_\sigma} = [\widehat{r_\sigma}]_{\widehat{S_\sigma}}$, for all $\sigma \in \Sigma$, where $\widehat{r_\sigma}$ is sampled uniformly in $R/gR^\times$ and

$$\widehat{w} = (\widehat{y}, \widehat{a}, 0),$$

for $\widehat{y}$ and $\widehat{a}$ uniformly chosen in R_1 and $R_2^\times$, respectively.

Multiplying the encoding of $w_\sigma = (y^*, C(\sigma), \rho_\sigma)$ obtained above, by this encoding of $\widehat{w} = (\widehat{y}, \widehat{a}, 0)$ enables us to cancel the last slot and to obtain an encoding of $\widehat{w_\sigma} := (\widehat{y} \cdot y^*, \widehat{a} \cdot C(\sigma), 0)$. We also multiply the r parts, as described above. Note that to cancel this third slot, the obfuscator outputs one pair of encodings for each symbol $\sigma \in \Sigma$. While this may seem useless because each encoding encodes the same $\widehat{w}$, this is in fact required to standardise the levels of the encodings. Indeed, after evaluating the universal circuit on the encodings of C and σ, we obtain an encoding whose level depends on σ. By multiplying with an encoding at a complementary level at this step, we can then ensure that the level of the product is independent of σ. This property will be important, because to zero-test the final encoding, we need it to be at the maximal level S_{zt}, independently of the input σ.

4. Finally, to cancel the first slot, the obfuscator provides two encodings $\bar{W} = [\bar{r} \cdot \bar{w}]_{\bar{S}}$ and $\bar{R} = [\bar{r}]_{\bar{S}}$, where $\bar{r}$ is sampled uniformly in $R/gR^\times$ and

$$\bar{w} = (\widehat{y} \cdot y^*, 0, 0).$$

Note that $\widehat{w_\sigma} - \bar{w} = 0$ if and only if $C(\sigma) = 0$. Hence, it suffices to subtract the corresponding encodings (using the r part, because the levels of the encodings will not match) and to zero-test the obtained encoding to determine whether $C(\sigma) = 0$ or 1.

This completes the description of the obfuscator, together with the correctness proof of the evaluation of the obfuscated program. Before describing the mixed-input attack, we would like to insist on some properties of the obfuscator described above.

- The levels of the encodings output by the obfuscator are chosen such that all honest evaluations of the obfuscated circuit on some input $\sigma \in \Sigma$ produce encodings with the same level. This level is then chosen to be the maximal level of the GGH map, and will be denoted by S_{zt}. The obfuscator also provides a zero-test parameter p_{zt} to enable zero-test at level S_{zt}. In the following, the only thing that will be interesting for our attack is that a honest evaluation of the obfuscated circuit on any input $\sigma \in \Sigma$ outputs an encoding at level S_{zt}, so we do not go into more details about the levels of the encodings.

- As we already noted, the value y^* obtained in the first slot after evaluating the universal circuit on the encodings of C and σ does not depend on σ. This is needed for the last step, where we subtract $\widehat{y} \cdot y^*$. As we want this to output 0 for any input (to cancel out the first slot), the value y^* has to be independent of σ. This first slot prevents us from mixing the bits of the circuit C, but does not prevent us from mixing the bits of the input σ (i.e., changing the value of some bit during the evaluation). Mixing the bits of the input is only prevented by the GGH map and the straddling set system (recall that the levels of the encodings depend on the input σ). This is the kind on mixed-input attack we will be able to perform after recovering the secret element h of the GGH map.

Differences between the DGGMM obfuscator and our simplification above. The obfuscator of [19] obfuscates circuits from Σ^c to $\{0,1\}$ for some constant c, instead of circuits from Σ to $\{0,1\}$ as described above. However, for our attack, we can take the constant c to be equal to 1, so we simplified a bit the description of the obfuscator and forgot about this constant c. If needed, the attack can be easily adapted to the case where c is a constant different from 1.

Also, the obfuscator of [19] uses an extra slot where it computes a PRF, and which is cancelled out before zero-testing by multiplying by an encoding of 0 in this slot (the principle is the same as for cancelling the third slot of the obfuscator described here). This extra slot is used only in the proof of security and does not interfere with our mixed-input attack, so we removed it from the description above.[18]

Finally, in the obfuscator of [19], we have $\bar{w} = (\widehat{y} \cdot y^*, \widehat{a}, 0)$ instead of $\bar{w} = (\widehat{y} \cdot y^*, 0, 0)$. So when subtracting, we obtain at the end an encoding of $(0, \hat{a}(1 - C(\sigma)), 0)$, which is 0 if and only if $C(\sigma) = 1$, instead of 0 if and only if $C(\sigma) = 1$ as in our simplification. However, both versions are equivalent, as we can always negate the output of the circuit. In order to be consistent with the other obfuscators described in this article, we decided to stick with the fact that obtaining an encoding of 0 means that the circuit outputs 0.

The DGGMM obfuscator was designed to obtain a candidate iO obfuscator from low noise multilinear maps. To do so, the class of circuit $\mathscr{C}$ targeted by

---

[18] This extra slot can be captured by the simplification above by taking g_3 to be a product of two prime elements and changing the distribution of the elements ρ in the third slot of the encodings. This has no impact on our attack.

the obfuscator described above is a very restrictive one (among other things, it requires that the circuits have a constant depth and a polynomial number of inputs). The authors then use a theorem from [29] to bootstrap their construction for this restricted class of circuit $\mathscr{C}$ to an obfuscator for all circuits in P/poly.

Remark. The DGGMM obfuscator is very similar to the previous circuit obfuscators of [4,37], and the simple circuit obfuscator described above also captures these obfuscators. Hence, the attack described below also applies to the obfuscators of [4,37], when instantiated with the GGH13 multilinear map (these obfuscators were originally instantiated with the CLT multilinear map, as they require composite-order multilinear maps, but they can also be instantiated with a modified version of the GGH13 map, as observed in [19]).

5.2   The Mixed-Input Attack

As mentioned above, the attack will consist in modifying a bit of the input σ during the computation. The idea is the same as for the attack of Sect. 4. We start by recovering the secret element h of the GGH map in quantum polynomial time, using the works of [10,13,18]. As above, we can obtain top level encodings of 0 each time the circuit evaluates to 0, so by choosing a circuit that evaluates to 0 sufficiently often, we can recover a basis of the ideal $\langle h \rangle$ (under Heuristic 1) and then recover h exactly (under Conjecture 1). We then construct a new zero-testing parameter p'_{zt} at level $2S_{zt}$ (testing whether the numerator of an encoding is a multiple of g^2, and not only g). This first step of the attack works exactly as described in Sect. 4.1 and we do not re-explain it here.

The second part of the attack (using p'_{zt} to mount a mixed input attack) will differ from the one for the abstract branching program obfuscator. The first difference is that in the abstract branching program obfuscator, we only computed products of matrices. So by changing a matrix, we just changed the final level of the encodings but all the operations remained possible (products of encodings are always possible, whatever their levels are). Here, as we evaluate a circuit with additions and multiplications, we must be careful. Indeed, if we change the level of one encoding of a sum but not the other one, we will not be able to perform the sum anymore. To circumvent this difficulty, we will use a specific universal circuit, which ends up by a multiplication. Let U be a universal circuit for the class of circuit $\mathscr{C}$. We define a new circuit $\tilde{U}$, which takes as input a concatenation of the description of two circuits in $\mathscr{C}$ and an input $\sigma \in \Sigma$ and computes the product of the evaluations of the two circuits on input σ. More formally, we define

$$\tilde{U}(\sigma, C_1 \cdot C_2) = U(\sigma, C_1) \cdot U(\sigma, C_2).$$

The circuit $\tilde{U}$ is a universal circuit for the class $\mathscr{C} \cdot \mathscr{C}$. Note that when evaluating the circuit $\tilde{U}$, we finish the evaluation with a multiplication. To perform our mixed input attack, we will evaluate $U(\cdot, C_1)$ and $U(\cdot, C_2)$ honestly on different inputs σ_1 and σ_2. As each partial evaluation is honest, we can perform all the required operations on the encodings. The dishonest computation will be the last multiplication only.

Let σ_1 and σ_2 be two distinct elements of Σ. Let C_{00} be a circuit that evaluates always to 0 on Σ. We also let C_{10} be a circuit that evaluates to 1 on σ_1 and to 0 otherwise and C_{01} be a circuit that evaluates to 1 on σ_2 and to 0 otherwise. The functions computed by $C_{00} \cdot C_{00}$ and by $C_{01} \cdot C_{10}$ are the same, so these circuits are equivalent. We will now show how to distinguish the obfuscated versions of $C_{00} \cdot C_{00}$ and $C_{01} \cdot C_{10}$, when using the universal circuit $\tilde{U}$. As both circuits are equivalent, this will result into an attack against the iO security of the obfuscator.

Objective: The obfuscator obfuscates the circuit $C_1 \cdot C_2 \in \{C_{00} \cdot C_{00}, C_{01} \cdot C_{10}\}$, and we want to distinguish whether $C_1 \cdot C_2 = C_{00} \cdot C_{00}$ or $C_1 \cdot C_2 = C_{01} \cdot C_{10}$.

1. The obfuscator encodes the bits of C_1 and C_2 under the GGH map, as well as the bits of all possible inputs $\sigma \in \Sigma$. In particular, we have encodings for σ_1 and σ_2. We homomorphically evaluate U on the encodings of C_1 and σ_1, C_1 and σ_2, C_2 and σ_1 and C_2 and σ_2.[19] These are honest partial evaluations of the circuit $\tilde{U}$ on input σ_1 and σ_2, so we can perform these evaluations (in particular, there will not be incompatibilities of encodings levels). We obtain four pairs of encodings $(R_{b_1 b_2} = [r_{b_1 b_2}]_{S_{b_1 b_2}}, W_{b_1 b_2} = [r_{b_1 b_2} \cdot w_{b_1 b_2}]_{S_{b_1 b_2}})$, for $b_1, b_2 \in \{1, 2\}^2$, where

$$w_{b_1 b_2} = (y_{b_1}, C_{b_1}(\sigma_{b_2}), \rho_{b_1 b_2}).$$

Recall that the y part of the encoding does not depend on the input σ, so this is independent of b_2 for our notations.

2. A honest evaluator of the obfuscated program would then multiply the encodings W_{11} and W_{21} (of $C_1(\sigma_1)$ and $C_2(\sigma_1)$) and the encodings W_{12} and W_{22} (of $C_1(\sigma_2)$ and $C_2(\sigma_2)$). However, in order to distinguish which circuit has been obfuscated, we do not perform these honest computations. Instead, following the idea of the mixed input attack described in Sect. 4.2, we compute $W_{11} \cdot W_{22}$ and $W_{12} \cdot W_{21}$ (and we do the same for the r part). We then obtain two encodings $\widetilde{W}_1$ and $\widetilde{W}_2$ of

$$\tilde{w}_1 := (y^*, C_1(\sigma_1) \cdot C_2(\sigma_2), \rho_{11}\rho_{22})$$
$$\text{and } \tilde{w}_2 := (y^*, C_1(\sigma_2) \cdot C_2(\sigma_1), \rho_{12}\rho_{21})$$

at levels $S_{11} \cup S_{22}$ and $S_{12} \cup S_{21}$ respectively. Note that the first slot of the encodings contains y^*, as it would for a honest evaluation.

3. We then complete the computation as if $\widetilde{W}_1$ was an honest evaluation on σ_1 and $\widetilde{W}_2$ was an honest evaluation on σ_2. That is, we first multiply $\widetilde{W}_1$ by $\widehat{W}_{\sigma_1}$ and $\widetilde{W}_2$ by $\widehat{W}_{\sigma_2}$ to cancel the third slot. We obtain two encodings $\widehat{W}_1$ and $\widehat{W}_2$ of

---

[19] Recall that $U(\sigma, C) = C(\sigma)$ and the universal circuit we chose is $\tilde{U}(\sigma, C_1 \cdot C_2) = U(\sigma, C_1) \cdot U(\sigma, C_2)$.

$$\widehat{w}_1 := (y^* \cdot \widehat{y}, \widehat{\alpha} \cdot C_1(\sigma_1) \cdot C_2(\sigma_2), 0)$$
$$\text{and} \quad \widehat{w}_2 := (y^* \cdot \widehat{y}, \widehat{\alpha} \cdot C_1(\sigma_2) \cdot C_2(\sigma_1), 0)$$

at levels $S_{11} \cup S_{22} \cup \widehat{S_{\sigma_1}}$ and $S_{12} \cup S_{21} \cup \widehat{S_{\sigma_2}}$, respectively.

4. Finally, we cancel the first slot by subtracting $\bar{W}$ to the encodings $\widehat{W}_1$ and $\widehat{W}_2$ obtained above. Note that this subtraction is between encodings that are not at the same level (for both honest and dishonest evaluations), so the resulting level is the union of the levels of both parts of the subtraction. We obtain two encodings $\bar{W}_1$ and $\bar{W}_2$ of

$$\bar{w}_1 := (0, \widehat{\alpha} \cdot C_1(\sigma_1) \cdot C_2(\sigma_2), 0)$$
$$\text{and} \quad \bar{w}_2 := (0, \widehat{\alpha} \cdot C_1(\sigma_2) \cdot C_2(\sigma_1), 0)$$

at levels $S_{11} \cup S_{22} \cup \widehat{S_{\sigma_1}} \cup \bar{S}$ and $S_{12} \cup S_{21} \cup \widehat{S_{\sigma_2}} \cup \bar{S}$, respectively.

5. Now, we would like to zero-test the encodings $\bar{W}_1$ and $\bar{W}_2$ obtained above, but because we mixed the inputs, the levels of the encodings are unlikely to be S_{zt} and we are not able to zero-test. However, we know that $S_{11} \cup S_{21} \cup \widehat{S_{\sigma_1}} \cup \bar{S} = S_{zt}$, because the encoding obtained by honestly evaluating the obfuscated program on σ_1 has this level. In the same way, we know that $S_{12} \cup S_{22} \cup \widehat{S_{\sigma_2}} \cup \bar{S} = S_{zt}$. Hence, the level of the product $\bar{W}_1 \cdot \bar{W}_2$ is $2S_{zt}$. Using our p'_{zt} parameter, we can then test whether its numerator is a multiple of g^2 or not.

   - In the case where $C_1 \cdot C_2 = C_{00} \cdot C_{00}$, we have $\bar{w}_1 = 0 \bmod g$ and $\bar{w}_2 = 0 \bmod g$. Hence, their product is a multiple of g^2. So the numerator of $\bar{W}_1 \cdot \bar{W}_2$ is a multiple of g^2, and the zero-test using p'_{zt} answers positively.
   - In the case where $C_1 \cdot C_2 = C_{01} \cdot C_{10}$, we have $\bar{w}_1 = 0 \bmod g$ and $\bar{w}_2 \neq 0 \bmod g$. So the product is a multiple of g^2 if and only if $\bar{w}_1$ is a multiple of g^2, which is very unlikely ($\bar{w}_1$ is obtained by subtracting two values that are equal modulo g_1, so this is a multiple of g_1 but this is unlikely to be a multiple of g_1^2).[20] Hence, the numerator of $\bar{W}_1 \cdot \bar{W}_2$ will not be a multiple of g^2 (with high probability), and the zero-test using p'_{zt} will fail.

   We can then distinguish between the obfuscated versions of $C_{00} \cdot C_{00}$ and $C_{01} \cdot C_{10}$ in (classical) polynomial time, using our new zero-testing parameter p'_{zt} obtained in quantum polynomial time.

This completes our quantum attack against the obfuscators of [4,19,37] and the proof of Theorem 1.

Acknowledgments. The author is grateful to Damien Stehlé for helpful discussions and comments on the draft. The author was supported by an ERC Starting Grant ERC-2013-StG-335086-LATTAC.

---

[20] Note that even if $\bar{w}_1$ were a multiple of g^2, then, by taking $p'_{zt} = (z^* \cdot g^{-1})^3 \bmod q$, we could mount the same kind of attack, at level $3S_{zt}$ instead of $2S_{zt}$.

References

1. Albrecht, M.R., Bai, S., Ducas, L.: A subfield lattice attack on overstretched NTRU assumptions. In: Robshaw, M., Katz, J. (eds.) CRYPTO 2016. LNCS, vol. 9814, pp. 153–178. Springer, Heidelberg (2016). https://doi.org/10.1007/978-3-662-53018-4_6

2. Ananth, P.V., Gupta, D., Ishai, Y., Sahai, A.: Optimizing obfuscation: avoiding Barrington's theorem. In: ACM CCS 14: 21st Conference on Computer and Communications Security, pp. 646–658. ACM Press, November 2014

3. Apon, D., Döttling, N., Garg, S., Mukherjee, P.: Cryptanalysis of indistinguishability obfuscations of circuits over GGH13. In: International Colloquium on Automata, Languages, and Programming. Springer, Heidelberg (2017)

4. Applebaum, B., Brakerski, Z.: Obfuscating circuits via composite-order graded encoding. In: Dodis, Y., Nielsen, J.B. (eds.) TCC 2015. LNCS, vol. 9015, pp. 528–556. Springer, Heidelberg (2015). https://doi.org/10.1007/978-3-662-46497-7_21

5. Badrinarayanan, S., Miles, E., Sahai, A., Zhandry, M.: Post-zeroizing obfuscation: new mathematical tools, and the case of evasive circuits. In: Fischlin, M., Coron, J.-S. (eds.) EUROCRYPT 2016. LNCS, vol. 9666, pp. 764–791. Springer, Heidelberg (2016). https://doi.org/10.1007/978-3-662-49896-5_27

6. Barak, B., Garg, S., Kalai, Y.T., Paneth, O., Sahai, A.: Protecting obfuscation against algebraic attacks. In: Nguyen, P.Q., Oswald, E. (eds.) EUROCRYPT 2014. LNCS, vol. 8441, pp. 221–238. Springer, Heidelberg (2014). https://doi.org/10.1007/978-3-642-55220-5_13

7. Barak, B., et al.: On the (im)possibility of obfuscating programs. In: Kilian, J. (ed.) CRYPTO 2001. LNCS, vol. 2139, pp. 1–18. Springer, Heidelberg (2001). https://doi.org/10.1007/3-540-44647-8_1

8. Barrington, D.A.M.: Bounded-width polynomial-size branching programs recognize exactly those languages in NC^1. In: 18th Annual ACM Symposium on Theory of Computing, pp. 1–5. ACM Press, May 1986

9. Biasse, J.-F., Espitau, T., Fouque, P.-A., Gélin, A., Kirchner, P.: Computing generator in cyclotomic integer rings. In: Coron, J.-S., Nielsen, J.B. (eds.) EUROCRYPT 2017. LNCS, vol. 10210, pp. 60–88. Springer, Cham (2017). https://doi.org/10.1007/978-3-319-56620-7_3

10. Biasse, J.-F., Song, F.: Efficient quantum algorithms for computing class groups and solving the principal ideal problem in arbitrary degree number fields. In: Proceedings of the Twenty-Seventh Annual ACM-SIAM Symposium on Discrete Algorithms, pp. 893–902. Society for Industrial and Applied Mathematics (2016)

11. Brakerski, Z., Rothblum, G.N.: Obfuscating conjunctions. In: Canetti, R., Garay, J.A. (eds.) CRYPTO 2013. LNCS, vol. 8043, pp. 416–434. Springer, Heidelberg (2013). https://doi.org/10.1007/978-3-642-40084-1_24

12. Brakerski, Z., Rothblum, G.N.: Virtual black-box obfuscation for all circuits via generic graded encoding. In: Lindell, Y. (ed.) TCC 2014. LNCS, vol. 8349, pp. 1–25. Springer, Heidelberg (2014). https://doi.org/10.1007/978-3-642-54242-8_1

13. Campbell, P., Groves, M., Shepherd, D.: Soliloquy: a cautionary tale. In: ETSI 2nd Quantum-Safe Crypto Workshop, pp. 1–9 (2014)

14. Chen, Y., Gentry, C., Halevi, S.: Cryptanalyses of candidate branching program obfuscators. In: Coron, J.-S., Nielsen, J.B. (eds.) EUROCRYPT 2017. LNCS, vol. 10212, pp. 278–307. Springer, Cham (2017). https://doi.org/10.1007/978-3-319-56617-7_10

15. Cheon, J.H., Han, K., Lee, C., Ryu, H., Stehlé, D.: Cryptanalysis of the multilinear map over the integers. In: Oswald, E., Fischlin, M. (eds.) EUROCRYPT 2015. LNCS, vol. 9056, pp. 3–12. Springer, Heidelberg (2015). https://doi.org/10.1007/978-3-662-46800-5_1

16. Cheon, J.H., Jeong, J., Lee, C.: An algorithm for NTRU problems and cryptanalysis of the GGH multilinear map without a low-level encoding of zero. LMS J. Comput. Math. **19**(A), 255–266 (2016)

17. Coron, J.-S., Lepoint, T., Tibouchi, M.: Practical multilinear maps over the integers. In: Canetti, R., Garay, J.A. (eds.) CRYPTO 2013. LNCS, vol. 8042, pp. 476–493. Springer, Heidelberg (2013). https://doi.org/10.1007/978-3-642-40041-4_26

18. Cramer, R., Ducas, L., Peikert, C., Regev, O.: Recovering short generators of principal ideals in cyclotomic rings. In: Fischlin, M., Coron, J.-S. (eds.) EUROCRYPT 2016. LNCS, vol. 9666, pp. 559–585. Springer, Heidelberg (2016). https://doi.org/10.1007/978-3-662-49896-5_20

19. Döttling, N., Garg, S., Gupta, D., Miao, P., Mukherjee, P.: Obfuscation from low noise multilinear maps. Cryptology ePrint Archive, Report 2016/599 (2016). http://eprint.iacr.org/2016/599

20. Fernando, R., Rasmussen, P.M.R., Sahai, A.: Preventing CLT attacks on obfuscation with linear overhead. In: Takagi, T., Peyrin, T. (eds.) ASIACRYPT 2017. LNCS, vol. 10626, pp. 242–271. Springer, Cham (2017). https://doi.org/10.1007/978-3-319-70700-6_9

21. Garg, S., Gentry, C., Halevi, S.: Candidate multilinear maps from ideal lattices. In: Johansson, T., Nguyen, P.Q. (eds.) EUROCRYPT 2013. LNCS, vol. 7881, pp. 1–17. Springer, Heidelberg (2013). https://doi.org/10.1007/978-3-642-38348-9_1

22. Garg, S., Gentry, C., Halevi, S., Raykova, M., Sahai, A., Waters, B.: Candidate indistinguishability obfuscation and functional encryption for all circuits. In: 54th FOCS, pp. 40–49. IEEE, October 2013

23. Garg, S., Miles, E., Mukherjee, P., Sahai, A., Srinivasan, A., Zhandry, M.: Secure obfuscation in a weak multilinear map model. In: Hirt, M., Smith, A. (eds.) TCC 2016-B. LNCS, vol. 9986, pp. 241–268. Springer, Heidelberg (2016). https://doi.org/10.1007/978-3-662-53644-5_10

24. Gentry, C., Gorbunov, S., Halevi, S.: Graph-induced multilinear maps from lattices. In: Dodis, Y., Nielsen, J.B. (eds.) TCC 2015. LNCS, vol. 9015, pp. 498–527. Springer, Heidelberg (2015). https://doi.org/10.1007/978-3-662-46497-7_20

25. Goyal, R., Koppula, V., Waters, B.: Lockable obfuscation. In: FOCS 2017, pp. 612–621. IEEE (2017)

26. Hu, Y., Jia, H.: Cryptanalysis of GGH map. In: Fischlin, M., Coron, J.-S. (eds.) EUROCRYPT 2016. LNCS, vol. 9665, pp. 537–565. Springer, Heidelberg (2016). https://doi.org/10.1007/978-3-662-49890-3_21

27. Kirchner, P., Fouque, P.-A.: Revisiting lattice attacks on overstretched NTRU parameters. In: Coron, J.-S., Nielsen, J.B. (eds.) EUROCRYPT 2017. LNCS, vol. 10210, pp. 3–26. Springer, Cham (2017). https://doi.org/10.1007/978-3-319-56620-7_1

28. Langlois, A., Stehlé, D., Steinfeld, R.: GGHLite: more efficient multilinear maps from ideal lattices. In: Nguyen, P.Q., Oswald, E. (eds.) EUROCRYPT 2014. LNCS, vol. 8441, pp. 239–256. Springer, Heidelberg (2014). https://doi.org/10.1007/978-3-642-55220-5_14

29. Lin, H.: Indistinguishability obfuscation from constant-degree graded encoding schemes. In: Fischlin, M., Coron, J.-S. (eds.) EUROCRYPT 2016. LNCS, vol. 9665, pp. 28–57. Springer, Heidelberg (2016). https://doi.org/10.1007/978-3-662-49890-3_2

30. Miles, E., Sahai, A., Weiss, M.: Protecting obfuscation against arithmetic attacks. Cryptology ePrint Archive, Report 2014/878 (2014). http://eprint.iacr.org/2014/878

31. Miles, E., Sahai, A., Zhandry, M.: Annihilation attacks for multilinear maps: cryptanalysis of indistinguishability obfuscation over GGH13. In: Robshaw, M., Katz, J. (eds.) CRYPTO 2016. LNCS, vol. 9815, pp. 629–658. Springer, Heidelberg (2016). https://doi.org/10.1007/978-3-662-53008-5_22

32. Pass, R., Seth, K., Telang, S.: Indistinguishability obfuscation from semantically-secure multilinear encodings. In: Garay, J.A., Gennaro, R. (eds.) CRYPTO 2014. LNCS, vol. 8616, pp. 500–517. Springer, Heidelberg (2014). https://doi.org/10.1007/978-3-662-44371-2_28

33. Pellet-Mary, A.: Quantum attacks against indistinguishablility obfuscators proved secure in the weak multilinear map model. Cryptology ePrint Archive, Report 2018/533 (2018). http://eprint.iacr.org/2018/533

34. Sahai, A., Waters, B.: How to use indistinguishability obfuscation: deniable encryption, and more. In 46th Annual ACM Symposium on Theory of Computing, pp. 475–484. ACM Press, May/June 2014

35. Shor, P.W.: Algorithms for quantum computation: discrete logarithms and factoring. In: FOCS 1994, pp. 124–134. IEEE (1994)

36. Wichs, D., Zirdelis, G.: Obfuscating compute-and-compare programs under LWE. In: FOCS 2017, pp. 600–611. IEEE (2017)

37. Zimmerman, J.: How to obfuscate programs directly. In: Oswald, E., Fischlin, M. (eds.) EUROCRYPT 2015. LNCS, vol. 9057, pp. 439–467. Springer, Heidelberg (2015). https://doi.org/10.1007/978-3-662-46803-6_15

Cryptanalyses of Branching Program Obfuscations over GGH13 Multilinear Map from the NTRU Problem

Jung Hee Cheon, Minki Hhan[(✉)], Jiseung Kim, and Changmin Lee

Seoul National University, Seoul, Republic of Korea
{jhcheon,hhan_,tory154,cocomi11}@snu.ac.kr

Abstract. In this paper, we propose cryptanalyses of all existing indistinguishability obfuscation (iO) candidates based on branching programs (BP) over GGH13 multilinear map for all recommended parameter settings. To achieve this, we introduce two novel techniques, *program converting* using NTRU-solver and *matrix zeroizing*, which can be applied to a wide range of obfuscation constructions and BPs compared to previous attacks. We then prove that, for the suggested parameters, the existing general-purpose BP obfuscations over GGH13 do not have the desired security. Especially, the first candidate indistinguishability obfuscation with input-unpartitionable branching programs (FOCS 2013) and the recent BP obfuscation (TCC 2016) are not secure against our attack when they use the GGH13 with recommended parameters. Previously, there has been no known polynomial time attack for these cases.

Our attack shows that the lattice dimension of GGH13 must be set much larger than previous thought in order to maintain security. More precisely, the underlying lattice dimension of GGH13 should be set to $n = \tilde{\Theta}(\kappa^2 \lambda)$ to rule out attacks from the subfield algorithm for NTRU where κ is the multilinearity level and λ the security parameter.

Keywords: Obfuscations · Multilinear maps
Graded encoding schemes · NTRU

1 Introduction

Constructing a general-purpose program obfuscation has been a long standing coveted open problem [8,9] in spite of their fruitful applications. At FOCS 2013, Garg *et al.* suggested the first plausible candidate general-purpose indistinguishability obfuscation (GGHRSW) [23] using branching program (BP) representation of functions [10]. This first candidate of iO has ignited the various subsequent studies [3,5–7,15,24,30,32,34] on obfuscations, all of which stand on the cryptographic multilinear maps.

To date, there are three plausible candidates of multilinear map; the first is due to Garg, Gentry, and Halevi [22] (GGH13), the second is due to Coron, Lepoint, and Tibouchi [19] and the last is due to Gentry, Gorbunov, and

ⓒ International Association for Cryptologic Research 2018
H. Shacham and A. Boldyreva (Eds.): CRYPTO 2018, LNCS 10993, pp. 184–210, 2018.
https://doi.org/10.1007/978-3-319-96878-0_7

Halevi [25]. The security of three candidates are not well clarifed, whereas some works [3,7,15,30,34] claim the security under the idealized model, so-called the *generic multilinear map model*.

Recently several works try to overcome this gap [6,24,29]. In particular, Garg *et al.* proved the security of the slightly modified first candidate iO construction (GMMSSZ) under the *weak multilinear map model* of GGH13, which captures all existing polynomial time attacks on BP obfuscations over GGH13 multilinear map [24]. Despite the provable security under these models, the practical security of obfuscations over GGH13 is still in dubious nature.

Direct attack to GGH13. As a direct method of analyzing obfuscations over GGH13, we may consider attacks on the GGH13 encoding scheme. The latent hardness problems of GGH13 are the (overstretched) NTRU problem and the short generator of principal ideal generator problem (SPIP).

The subfield attacks, proposed by Albrecht *et al.* and Cheon *et al.* independently [1,18], are the most notable algorithms to solve the NTRU problem. These attacks shows that the underlying NTRU problem of GGH13-based obfuscation is solved in polynomial time whenever the multilinear level κ is larger than the security parameter λ. By combining this with the algorithms to solve SPIP [12–14,20], GGH13 is broken in classical subexponential time on security parameter λ for the instantiations in [2,27] or quantum polynomial time. This work shows that the parameters of GGH13 should be set to prevent either the algorithms for NTRU or PIP.[1]

Attacks on BP Obfuscations over GGH13. For obfuscations over GGH13 multilinear map, several cryptanalyses have also been suggested. The *annihilation attack* introduced by Miles *et al.* [31] showed that some constructions of single/dual input BP obfuscations [3,6,7,30] do not have the desired security when they are used for general-purpose and implemented with GGH13. The authors presented a very simple example of BPs which are threatened by annihilation attacks. Soon after, Apon *et al.* [4] extended the range of annihilation attacks to BPs generated by Barrington's theorem [10] which is the fundamental method to transform $\mathcal{NC}^1$ circuits into bounded width BPs.

Chen *et al.* [16] presented another attack on BP obfuscation over GGH13 multilinear map. They showed that there exist two functionally equivalent programs with a special property called *input-partitionable*, and their obfuscated programs by GGHRSW can be efficiently distinguished.

Limitations of Previous Works. Despite the diverse attacks on BP obfuscations over GGH13 multilinear map, GGHRSW remains secure against all known PPT attack when it only takes *input-unpartitionable* BPs as input, such as BPs generated by Barrington's theorem. Meanwhile, there is no known polynomial

---

[1] Indeed, the parameters of original paper [22] are already set to be $n = \tilde{\Theta}(\kappa\lambda^2)$ so that known classical algorithms for PIP require exponential time for λ. On the other hands, the improved parameters [2,27] allow the subexponential time attacks.

time attack for multi-input branching program obfuscations including GMMSSZ. We also remark that the direct approach [1], with the current best algorithm to solve SPIP [13,20], has the classical exponential running time with respect to security parameter λ when the dimension n of the base number field satisfies $n = \Omega(\lambda^2)$.

Our Contribution. We present distinguishing attacks on candidates BP iO over GGH13 multilinear map based on the algorithm to solve the NTRU problem. With the novel two techniques, *program converting* and *matrix zeroizing attack*, we show that existing general-purpose BP obfuscations cannot achieve the desired security when the obfuscations use GGH13 with proposed parameters in [2,22,27]. In other words, there are two functionally equivalent BPs with same length such that their obfuscations obtained by an existing BP obfuscations over GGH13 can be distinguished in polynomial time for the suggested parameters.

Our attack is applicable to wide range of obfuscations and BPs compared to the previous attacks. In particular, we show that multi-input BP obfuscations such as GMMSSZ construction are insecure in the NTRU-solvable parameter regime. Further, we show that the first candidate indistinguishability obfuscation GGHRSW based on GGH13 with current parameters also does not have the desired security even if it only obfuscates input-unpartitionable BPs including branching programs generated by Barrington's theorem. Although a new property of BPs called *linear relationally inequivalence* is exploited in our attack, we show that various pairs of BPs satisfy this property.

As a result, we show that the BP obfuscations based on GGH13 multilinear map with suggested parameters are broken using the algorithm for NTRU solely. Therefore the underlying lattice dimension n of GGH13 should be set to $n = \tilde{\Theta}(\kappa^2 \lambda)$ to maintain 2^λ security of obfuscation schemes. This implies the iO based on GGH13 is even much inefficient than the previous results [1,28].

1.1   Technical Overview

Here we briefly show how our attack is applied to simplified GGHRSW.

Simplified GGHRSW Obfuscation. Let $P = \{M_{i,b} \in \mathbb{Z}^{d \times d}\}_{b \in \{0,1\}, 1 \le i \le \ell}$ be a set of matrices corresponding to a single input BP such that

$$P(\boldsymbol{x}) := \begin{cases} 0 & \text{if } \prod_{i=1}^{\ell} M_{i,x_i} = I_d \\ 1 & \text{if } \prod_{i=1}^{\ell} M_{i,x_i} \ne I_d, \end{cases}$$

where x_i is the i-th bit of $\boldsymbol{x}$. The obfuscator randomizes the given BP over several steps.

1. Sample random and independent scalars $\{\alpha_{i,b}, \alpha'_{i,b}\}_{b \in \{0,1\}, 1 \le i \le \ell}$ such that $\prod_{i=1}^{\ell} \alpha_{i,x_i} = \prod_{i=1}^{\ell} \alpha'_{i,x_i}$ for all $\boldsymbol{x} \in \{0,1\}^{\ell}$.[2]

---

[2] In fact $\alpha_{i,b} = \alpha'_{i,b}$ should holds in this simplified setting, but we do not use this equality to give the idea of our attack.

2. Sample bookend vectors $\{s, t, s', t'\}$ such that $s \cdot t = s' \cdot t'$.
3. Sample invertible matrices $\{K_i, K_i' \in \mathbb{Z}^{d \times d}\}_{0 \le i \le \ell}$ and set

$$
\begin{aligned}
R_0 &= s \cdot K_0^{-1}, & R_0' &= s' \cdot K_0'^{-1} \\
R_{i,b} &= \alpha_{i,b} \cdot K_{i-1} \cdot M_{i,b} \cdot K_i^{-1}, & R_{i,b}' &= \alpha_{i,b}' \cdot K_{i-1}' \cdot I_d \cdot K_i'^{-1} \\
R_{\ell+1} &= K_\ell \cdot t, & R_{\ell+1}' &= K_\ell' \cdot t'.
\end{aligned}
$$

For the sake of simplicity, we write $R_{0,b}$, $R_{\ell+1,b}$, $R_{0,b}'$, and $R_{\ell+1,b}'$ to denote R_0, $R_{\ell+1}$, R_0', and $R_{\ell+1}'$, respectively. The randomized BP can then maintain the same functionality as the following evaluation, where $x_0, x_{\ell+1}$ are 0.

$$
P(x) = \begin{cases} 0 & \text{if } \prod_{i=0}^{\ell+1} R_{i,x_i} - \prod_{i=0}^{\ell+1} R_{i,x_i}' = 0 \\ 1 & \text{if } \prod_{i=0}^{\ell+1} R_{i,x_i} - \prod_{i=0}^{\ell+1} R_{i,x_i}' \ne 0. \end{cases}
$$

As a final step, each entry of the R_i and R_i' is encoded through the GGH13 multilinear map. Let $\mathcal{R} = \mathbb{Z}[X]/\langle X^n + 1 \rangle$. The plaintext space and encoding space of GGH13 multilinear map is specified by $\mathcal{R}_g = \mathcal{R}/\langle g \rangle$ with some small element $g \in \mathcal{R}$ and $\mathcal{R}_q = \mathcal{R}/\langle q \rangle$ with some large integer $q \in \mathbb{Z}$, respectively. In GGH13 multilinear map, a random and invertible element $z \in \mathcal{R}_q$ is sampled. Then the encoding of m is of the form $\mathsf{enc}(m) = [(r \cdot g + m)/z]_q$ for some small random element $r \in \mathcal{R}$. The smallness of g and r implies that the size of the numerator is quite smaller than q. We write $\mathsf{enc}(R_{i,b})$ to denote the matrix whose entries are encoding of entries of $R_{i,b}$.

Then, in the case of $P(x) = 0$, evaluation of the encoded BP over input x can be computed as follows:

$$
\prod_{i=0}^{\ell+1} \mathsf{enc}(R_{i,x_i}) - \prod_{i=0}^{\ell+1} \mathsf{enc}(R_{i,x_i}') = \left[\frac{e \cdot g}{z^{\ell+2}} \right]_q
$$

where the term e is the small noise element of $\mathcal{R}$. If it is evaluated for another input x, the numerator of the evaluated value cannot be a multiple of g.

In order to check whether the numerator of the evaluation value of the encoded BP is a zero or not, the GGH13 multilinear map provide a zerotesting parameter $p_{zt} = [(h \cdot z^{\ell+2})/g]_q$ for some element $h \in \mathcal{R}$ of size $\approx \sqrt{q}$. More precisely, when the p_{zt} is multiplied by the evaluated value, it is of the form $h \cdot r'$ and its size is much smaller than q if the numerator is a multiple of g. Otherwise it is a large value. Hence, one can publicly test that whether the plaintext of the encoding is zero or not and an encoded BP give the same functionality with the original BP by employing the zerotesting parameter p_{zt}.

In summary, the GGHRSW obfuscator outputs the following set as an obfuscated BP.

$$
\{\mathsf{enc}(R_{i,b}), \mathsf{enc}(R_{i,b}'), p_{zt}\}
$$

Goal of Cryptanalysis on Simplified GGHRSW Obfuscation. The simplified GGHRSW obfuscation given above is called *indistinguishability obfuscation* if the following statement holds: For every two BPs $P^0 = \{M_{i,b}^0\}$, and

$P^1 = \{M^1_{i,b}\}$ with the same size and the same functionality and randomly chosen $c \in \{0,1\}$, any PPT adversary cannot recover c from the given obfuscated program $\{\text{enc}(R^c_{i,b}), \text{enc}(R'^c_{i,b}), p_{zt}\}$.

In other words, our purpose of the cryptanalysis is to recover such c for appropriately given P^0, P^1 and its obfuscation.

Program Converting Technique. In the first step, we remove the modulus q using the algorithm for NTRU. The $(1,1)$ and $(1,2)$ components of the $\text{enc}(R_{1,1})$ are of the form $[(r_{1,1} \cdot g + m_{1,1})/z]_q$ and $[(r_{1,2} \cdot g + m_{1,2})/z]_q$, respectively. The ratio $[(r_{1,1} \cdot g + m_{1,1})/(r_{1,2} \cdot g + m_{1,2})]_q$ of two encodings can be understood as an instance of the NTRU problem.

By solving the NTRU problem, we can obtain multiples of the denominator and numerator

$$\beta \cdot (r_{1,1} \cdot g + m_{1,1}, r_{1,2} \cdot g + m_{1,2}) \in \mathcal{R}^2$$

for some small element $\beta \in \mathcal{R}$. Further, dividing $\beta \cdot (r_{1,1} \cdot g + m_{1,1})$ by a $[(r_{1,1} \cdot g + m_{1,1})/z]_q$, we can compute $[\beta \cdot z]_q$. By multiplying this value to all entries of $\text{enc}(R_{i,b})$ and $\text{enc}(R'_{i,b})$, we replace $1/z$ with a small element β. The obtained entries are of the form $\beta \cdot (r_{j,k} \cdot g + m_{j,k})$, which can be understood as an element defined in $\mathcal{R}$, not $\mathcal{R}_q$ due to its small size. We denote these new BP matrices with entries in $\mathcal{R}$ by $\{D_{i,b}\}$ and $\{D'_{i,b}\}$, respectively.

Next we consider an input x such that $P(x) = 0$.[3] The corresponding computation of matrices R is zero, thus the following equation holds over $\mathcal{R}$ for such input.

$$\prod_{i=0}^{\ell+1} D_{i,x_i} - \prod_{i=0}^{\ell+1} D'_{i,x_i} = e \cdot g \cdot \beta^{\ell+2}$$

Hence, the term is a multiple of g. Using the same procedure for other zeros of P, one can recover several multiples of g and then we can recover a basis of ideal $\langle g \rangle$ using lattice algorithms.

Then we can do a plain-like procedure using the above results. More precisely, the following equations hold.

$$Eval_D(x) := \prod_{i=0}^{\ell+1} D_{i,x_i} = \prod_{i=0}^{\ell+1} \alpha_{i,x_i} \cdot s \cdot \prod_{i=1}^{\ell} M^c_{i,x_i} \cdot t \pmod{g}$$

$$Eval'_D(x) := \prod_{i=0}^{\ell+1} D'_{i,x_i} = \prod_{i=0}^{\ell+1} \alpha'_{i,x_i} \cdot s' \cdot \prod_{i=1}^{\ell} I_d \cdot t' \pmod{g}$$

Removing Scalars. In the above step, we removed the modulus q using the solutions of the NTRU problem and obtained matrices $\{D_{i,b}, D'_{i,b}\}$ and a basis of ideal $\langle g \rangle$. We now remove the effects of scalars α. $Eval_D(x)$ and $Eval'_D(x)$

---

[3] Because of this step, our attack cannot be applied to BP obfusaction for evasive functions.

share the same scalar $\prod_{i=0}^{\ell+1} \alpha_{i,x_i} = \prod_{i=0}^{\ell+1} \alpha'_{i,x_i}$ due to its definition. Thus, we can compute

$$Eval_D(\boldsymbol{x})/Eval'_D(\boldsymbol{x}) = 1/(s' \cdot t') \cdot \left(s \cdot \prod_{i=1}^{\ell} \boldsymbol{M}^c_{i,x_i} \cdot t \right) \quad (\text{mod } \boldsymbol{g}).$$

We note that these values $Eval_D(\boldsymbol{x})/Eval'_D(\boldsymbol{x})$ all share the same scalar $1/(s' \cdot t')$ (mod $\boldsymbol{g}$).

Matrix Zeroizing Attack. At last we introduce the *matrix zeroizing attack*. We denote $Eval_{M^\circ}(\boldsymbol{x})$ and $\widetilde{Eval_D}(\boldsymbol{x})$ as $\prod_{i=1}^{\ell} \boldsymbol{M}^0_{i,x_i}$ and $Eval_D(\boldsymbol{x})/Eval'_D(\boldsymbol{x})$, respectively.

Then, for several $Eval_{M^\circ}(\boldsymbol{x}_j)$ for $1 \leq j \leq \tau$, we can find a vector $\boldsymbol{q} = (q_1, \cdots, q_\tau)$ such that $\sum_{j=1}^{\tau} q_j \cdot Eval_{M^\circ}(\boldsymbol{x}_j) = \boldsymbol{0}_d$, where $\boldsymbol{0}_d$ is a zero matrix. If $c = 1$ so that the obfuscated BP is derived from P^0, the following equation also holds.

$$\sum_{j=1}^{\tau} c_j \cdot \widetilde{Eval_D}(\boldsymbol{x}_j) = \boldsymbol{0}_d \quad (\text{mod } \boldsymbol{g})$$

Otherwise, it would not be zero (mod $\boldsymbol{g}$).

As a result, we can distinguish two obfuscated program efficiently when we know corresponding branching programs. We remark that the matrix zeroizing attack and removing scalars step are slightly different for the other BP obfuscations.

Organization. In Sect. 2, we introduce the indistinguishability obfuscation, matrix branching program and GGH13 multilinear map. In Sect. 3, we show main results of our cryptanalyses on BP obfuscations over GGH13 multilinear map. We describe the attackable BP obfuscation Model over GGH13 throughout the Sect. 4. In addition, we present the algorithm called program converting technique in Sect. 5. We last propose the matrix zeroizing attack in Sect. 6.

2   Preliminaries

Notations. The set $\{1, \cdots, n\}$ is denoted by $[n]$ for a positive integer n. The set of integers modulo p is denoted by $\mathbb{Z}_p := \mathbb{Z}/p\mathbb{Z}$. All elements in $\mathbb{Z}_p$ are considered as integers in $(-p/2, p/2]$. We use the bold letters to denote matrices, vectors and elements of ring. For $\boldsymbol{a} = a_0 + \cdots + a_{n-1} \cdot X^{n-1} \in \mathcal{R} = \mathbb{Z}[X]/\langle X^n + 1 \rangle$, the size of $\boldsymbol{a}$ means the Euclidean norm of the coefficient vector $(a_0, \cdots, a_{n-1})$. We denote (j, k)-th entry of matrix $\boldsymbol{M}$ by $\boldsymbol{M}[j, k]$.

2.1   Matrix Branching Program

A branching program consists of several matrix chains and input functions with indices of input bit. To evaluate a matrix branching program, we multiply all matrices and output 0 or 1 depending on whether the product of the matrices is the same as a given matrix or not. We briefly review matrix branching programs.

Definition 1 (w-ary Matrix Branching Programs). *Let A_0 be a $d_1 \times d_{\ell+1}$ matrix and w, ℓ, d, and N be natural numbers. A w-ary matrix branching program BP with length ℓ over N-bit inputs consists of the following data; a set of input functions $\{\text{inp}_\mu : [\ell] \to [N]\}_{\mu \in [w]}$, a set of matrices $\{M_{i,b} \in \mathbb{Z}^{d_i \times d_{i+1}}\}_{i \in [\ell], b \in \{0,1\}^w}$. It has a domain for evaluations $\{0,1\}^N$, and evaluation of BP at $x = (x^v)_{v \in [w]}$ is computed by*

$$BP(x) = BP_{(\text{inp}_\mu)_{\mu \in [w]}, M}(x) = \begin{cases} 0 & \text{if } \prod_{i=1}^{\ell} M_{i, (x^\mu_{\text{inp}_\mu(i)})_{\mu \in [w]}} = A_0 \\ 1 & \text{if } \prod_{i=1}^{\ell} M_{i, (x^\mu_{\text{inp}_\mu(i)})_{\mu \in [w]}} \neq A_0 \end{cases}.$$

When w is set to 1 and ≥ 2, the matrix branching program is called a single-input and a multi-input matrix branching program, respectively. Throughout this paper, a matrix A_0 is used as the zero matrix 0 or the identity matrix I_d if $d_i = d$ for all i. Moreover, we simplify the notation $(x^\mu_{\text{inp}_\mu(i)})_{\mu \in [w]}$ as $x_{\text{inp}(i)}$.

Barrington proved all boolean functions can be expressed in the form of matrix branching program with bounded width [10]. The first candidate for iO [23] and following obfuscations [7,15,30,32] exploit Barrington's theorem to transform circuits into BPs.

We also note that there are other methods to convert circuits into branching programs. Ben-Or and Cleve proved that the similar result to Barrington's theorem for arithmetic circuits [11]. Follow-up studies such as [3,6] suggest more efficient methods for transformation. Their methods bypass the Barrington's theorem and make a circuit into a branching program directly. However, they still preserve the length of program, in other words, the length of branching program is equal to or larger than the size of circuit (number of gates).

We assume a mild condition on the branching programs: The length of branching program is $\Omega(N)$ for the number of input bits N. This is plausible since all input bits may affect the program, and the existing methods give much longer lengths. On the other hand, we do not restrict that the width/properties of the matrices in branching programs and the input function (such as single or dual input).

2.2 Indistinguishability Obfuscation

Definition 2 (Indistinguishability Obfuscation (iO)). *A PPT algorithm iO is an indistinguishability obfuscation for a circuit class $\mathcal{C}$ if the following conditions are satisfied:*

- *For all security parameters $\lambda \in \mathbb{N}$, for all circuits $C \in \mathcal{C}$, for all inputs x, the following probability holds:*

$$\Pr\left[C'(x) = C(x) : C' \leftarrow iO(\lambda, C)\right] = 1.$$

- *For any PPT distinguisher $\mathcal{D}$, there exists a negligible function α satisfying the following statement: For all security parameters $\lambda \in \mathbb{N}$ and all pairs of circuits C_0, $C_1 \in \mathcal{C}$, $C_0(x) = C_1(x)$ for all inputs x implies*

$$\left|\Pr\left[D(iO(\lambda, C_0)) = 1\right] - \Pr\left[D(iO(\lambda, C_1)) = 1\right]\right| \leq \alpha(\lambda).$$

Hereafter, we denote $iO(P)$ by an obfuscated program or obfuscation of a program, or a branching program P.

2.3 GGH13 Multilinear Map

Garg *et al.* suggest a candidate of multilinear map based on ideal lattice [22]. It is used to realize the indistinguishable obfuscation [23]. In this section, we briefly describe the GGH13 multilinear map. For more details, we recommend readers to refer [22]. Any parameters of multilinear maps are induced by the multilinearity parameter κ and the security parameters λ. For the sake of simplicity, we denote the multilinear maps which has the previous mentioned parameter as (κ, λ)-GGH multilinear map.

The multilinear map is sometimes called the graded encoding scheme. *i.e.*, All encodings of message have corresponding levels. Let g be a secret element in $\mathcal{R} = \mathbb{Z}[X]/\langle X^n + 1 \rangle$ and q a large integer. Then, the message space and encoding space are set by $\mathcal{M} = \mathcal{R}/\langle g \rangle$ and $\mathcal{R}_q = \mathcal{R}/\langle q \rangle$, respectively. In order to represent a level of encodings, the set of secret invertible elements $\mathbb{L} = \{z_i\}_{1 \le i \le \kappa} \subset \mathcal{R}_q$ is chosen. We call a subset of $\mathbb{L}$ *level set* and elements in $\mathbb{L}$ *level parameters*.

For a small message $m \in \mathcal{M}$, level-$L(\subset \mathbb{L})$ encoding of m is:

$$\mathsf{enc}_L(m) = \left[\frac{r \cdot g + m}{\prod_{i \in L} z_i} \right]_q,$$

where $r \in \mathcal{R}$ is a small random element. We call $\mathsf{enc}_{\mathbb{L}}(m)$, $\mathsf{enc}_{\{z_i\}}(m)$ a top-level and level 1 encoding of m, respectively. In addition, for a matrix M, we denote a matrix whose entries are level-L encodings of corresponding entries of M by $\mathsf{enc}_L(M)$.

The arithmetic operations between encodings are defined as follows:

$$\mathsf{enc}_L(m_1) + \mathsf{enc}_L(m_2) = \mathsf{enc}_L(m_1 + m_2),$$
$$\mathsf{enc}_{L_1}(m_1) \cdot \mathsf{enc}_{L_2}(m_2) = \mathsf{enc}_{L_1 \sqcup L_2}(m_1 \cdot m_2).$$

Additionally, the (κ, λ)-GGH scheme provides a zerotesting parameter which can be used to determine whether a hidden message of a top-level encoding is zero or not. The zerotesting parameter p_{zt} is of the form:

$$p_{zt} = \left[h \cdot \frac{\prod_{i \in \mathbb{L}} z_i}{g} \right]_q,$$

where h is an $O(\sqrt{q})$-size element of $\mathcal{R}$. Given a top-level encoding of zero $\mathsf{enc}_{\mathbb{L}}(0) = [r \cdot g / \prod_{i \in \mathbb{L}} z_i]_q$, a zerotesting value is:

$$[p_{zt} \cdot \mathsf{enc}_{\mathbb{L}}(0)]_q = \left[h \cdot \frac{\prod_{i \in \mathbb{L}} z_i}{g} \cdot \frac{r \cdot g}{\prod_{i \in \mathbb{L}} z_i} \right]_q = [h \cdot r]_q = h \cdot r \in \mathcal{R}.$$

We remark that a zerotesting value for a top-level encoding of nonzero gives an element of the form $[h \cdot (r + m \cdot g^{-1})]_q$, which is not small by Lemma 4 in [22]. Thus one can decide whether a message is zero or not by the zerotesting value.

Several papers [2,22,27] proposed the parameters of (κ, λ)-GGH13 multilinear map. Here we introduce the minimum conditions that satisfy the three works.

- $\log q = \tilde{\Theta}(\kappa \cdot \log n)$
- $n = \tilde{\Theta}(\kappa^\epsilon \cdot \lambda^\delta)$ for constants δ, ϵ
- $M = \tilde{O}(n^{\Theta(1)})$

Here M is the size bound of numerators $r \cdot g + m$ of level 1 encodings.[4] We note that the suggested parameters in [2,27] choose $\delta = \epsilon = 1$, which enables the subexponential attack with respect to λ for small κ [1,13]. When $\delta \geq 2$, all known direct attacks on GGH13 multilinear map require exponential time for classical adversary.

3 Main Theorem

In this section, we present the results from our attacks. We denote the obfuscation within our attack range as *the attackable obfuscation*, which is formally defined by *the attackable model* in the next section. The attackable obfuscation model encompasses all suggested BP obfuscations based on GGH13 multilinear map.

Proposition 1 (Universality of the Attackable Model). *BP obfuscations [3, 6, 7, 23, 24, 30, 32] satisfy all the constraints of the attackable model.*[5]

As a result, we obtain the following main theorem.

Theorem 1. *Let $\mathcal{O}$ be an attackable obfuscator, κ, λ be the multilinearity level and the security parameter of underlying GGH13 multilinear map. Suppose that the modulus q, dimension n, size bound M of numerators of level 1 encoding of underlying GGH13 satisfy $\log q = \tilde{\Theta}(\kappa \cdot \log n), M = \tilde{O}(n^{\Theta(1)})$. Then the following propositions hold:*

1. *For $n = \tilde{\Theta}(\kappa \cdot \lambda^\delta)$ for a constant δ as in [2, 22, 27], there exist two functionally equivalent branching programs with $\Omega(\lambda^\delta)$-length such that their obfuscated programs by $\mathcal{O}$ can be distinguished with high probability in polynomial time with respect to λ.*
2. *Moreover, for new parameter constraints $n = \tilde{\Theta}(\kappa^\epsilon \cdot \lambda^\delta)$ for constants $\epsilon < 2, \delta$, there exist two functionally equivalent branching programs with $\Omega(\lambda^{\delta/(2-\epsilon)})$-length such that their obfuscated programs by $\mathcal{O}$ can be distinguished with high probability in polynomial time with respect to λ.*

---

[4] The coefficients of random values are usually sampled from the Gaussian distribution. This do not hurt the result of this paper because the coefficients are bounded with overwhelming probability.

[5] We deal with easier model in the main body for simplicity. We can extend the model to capture the construction in [15]. This extended model is placed in Appendix A.

The main theorem is proven by combining *converting program technique* and *matrix zeroizing attack* which are described in Sects. 5 and 6. The bottleneck of the attack is the algorithm for NTRU, which is exploited in the middle step of converting technique; the other process can be done in polynomial time, while the time complexity to solve the NTRU problem relies on the parameters. The detailed analysis for the time complexity will be discussed in Sect. 5.3.

4   Attackable BP Obfuscations

In this section, we present a new BP obfuscation model which is attackable by our attack, *the attackable model.* We call a BP obfuscation captured by our model an *attackable BP obfuscation.*

The attackable model is composed of two steps; for a given BP, randomize BP, and encode randomized BPs by GGH13 multilinear map. More precisely, for a given branching program BP of the form

$$P = \left\{ M_{i,b} \in \mathbb{Z}^{d_i \times d_{i+1}} \right\}_{i \in [\ell], b \in \{0,1\}^w},$$

we randomize P by several methods satisfying Definition 3 which will be described later. And then we encode each entries of randomized matrices and outputs the obfuscated program as the set

$$\mathcal{O}(P) = \left\{ \widetilde{S}, \widetilde{S}' \in \mathcal{R}_q^{d_0 \times (d_1 + e_1)} \right\}$$
$$\cup \left\{ \{\widetilde{M}_{i,b}, \widetilde{M}'_{i,b} \in \mathcal{R}_q^{(d_i + e_i) \times (d_{i+1} + e_{i+1})} \}_{i \in [\ell], b \in \{0,1\}^w}, \right\}$$
$$\cup \left\{ \widetilde{T}, \widetilde{T}' \in \mathcal{R}_q^{(d_{\ell+1} + e_{\ell+1}) \times d_{\ell+2}} \right\}$$

and the public parameters of GGH13 multilinear map. S, T denote bookend matrices, and matrices with apostrophe mean the matrices of dummy program. In the attackable model, we specify the following property instead of establishing how to evaluate the program exactly. To evaluate the input value, a new function $Eval_{\widetilde{M}} : \{0,1\}^N \to \mathcal{R}_q^{d_0 \times d_{\ell+2}}$ is computed as follows:

$$Eval_{\widetilde{M}}(x) = \widetilde{S} \cdot \prod_{i=1}^{\ell} \widetilde{M}_{i,x_{\mathrm{inp}(i)}} \cdot \widetilde{T} - \widetilde{S}' \cdot \prod_{i=1}^{\ell} \widetilde{M}'_{i,x_{\mathrm{inp}(i)}} \cdot \widetilde{T}' \in \mathcal{R}_q^{d_0 \times d_{\ell+2}}.$$

Proposition 2 (Evaluation of Obfuscation). *For a program P and program $\mathcal{O}(P)$ obfuscated by the attackable model, the evaluation of $\mathcal{O}(P)$ at a root x of P yields a top-level GGH13 encoding of zero in specific entry of the matrix $Eval_{\widetilde{M}}(x)$. In other words, there are two integers u, v such that $Eval_{\widetilde{M}}(x)[u, v]$ is an encoding of zero at level $\mathbb{L}$ for every input x satisfying $P(x) = 0$.*

In the rest of this section, we explain specified descriptions of the attackable model in Sects. 4.1 and 4.2, and present a constraint of BPs to execute our attack in Sect. 4.3.

4.1   Randomization for Attackable Obfuscation Model

We introduce the conditions for BP randomization of attackable obfuscation model. These conditions for randomization covers all of the BP randomization methods suggested in the first candidate iO [23] and its subsequent works [3,6,7, 24,30,32]. In other words, higher dimension embedding, scalar bundling, Kilian randomization, bookend matrices (vectors), and dummy programs are captured by the attackable conditions.

Definition 3 (Attackable Conditions for Randomization). *For a branching program* $P = \left\{ M_{i,b} \in \mathbb{Z}^{d_i \times d_{i+1}} \right\}_{i \in [\ell], b \in \{0,1\}^w}$, *the attackable randomized branching program is the set*

$$Rand(P) = \left\{ R_S, R_S' \in \mathbb{Z}^{d_0 \times (d_1 + e_1)} \right\}$$
$$\cup \left\{ \{ R_{i,b}, R_{i,b}' \in \mathbb{Z}^{(d_i + e_i) \times (d_{i+1} + e_{i+1})} \}_{i \in [\ell], b \in \{0,1\}^w}, \right\}$$
$$\cup \left\{ R_T, R_T' \in \mathbb{Z}^{(d_{\ell+1} + e_{\ell+1}) \times d_{\ell+2}} \right\}$$

satisfying the following properties, where $d_0, d_{\ell+2}, e_i$*'s are integers.*

1. *There exist matrices* $S_0, S_0' \in \mathbb{Z}^{d_0 \times d_1}, T_0, T_0' \in \mathbb{Z}^{d_\ell \times d_{\ell+1}}$ *and scalars* α_S, α_S', $\alpha_T, \alpha_T', \{ \alpha_{i,b}, \alpha_{i,b}' \}_{i \in [\ell], b \in \{0,1\}^w}$ *such that the following equations hold for all* $\{ b_i \in \{0,1\}^w \}_{i \in [\ell]}$:

$$R_S \cdot \prod_{i=1}^{\ell} R_{i,b_i} \cdot R_T = \alpha_S \cdot \prod_{i=1}^{\ell} \alpha_{i,b_i} \cdot \alpha_T \cdot \left(S_0 \cdot \prod_{i=1}^{\ell} M_{i,b_i} \cdot T_0 \right),$$

$$R_S' \cdot \prod_{i=1}^{\ell} R_{i,b_i}' \cdot R_T' = \alpha_S' \cdot \prod_{i=1}^{\ell} \alpha_{i,b_i}' \cdot \alpha_T' \cdot \left(S_0' \cdot \prod_{i=1}^{\ell} M_{i,b_i}' \cdot T_0' \right).$$

2. *The evaluation of randomized program is done by checking whether the fixed entries of* $RP(x) := R_S \cdot \prod_{i=1}^{\ell} R_{i,x_{\mathsf{inp}(i)}} \cdot R_T - R_S' \cdot \prod_{i=1}^{\ell} R_{i,x_{\mathsf{inp}(i)}}' \cdot R_T'$ *are zero or not. Especially, there are two integers* u, v *such that* $P(x) = 0 \Rightarrow RP(x)[u,v] = 0$.

Matrices with apostrophe are called *dummy matrices*, R_S, R_S', R_T, R_T' bookend matrices (vectors), and α's *bundling scalars*. When some elements of $Rand(P)$ (or bundling scalars) are trivial elements, we say that there is no such element.

4.2   Encoding by Multilinear Map

After the randomization, we encode the randomized matrix branching program by GGH13 multilinear map. We stress that we *do not encode* dummy/bookend matrices if there are no dummy/bookends, respectively.

For each randomized matrices, $R_{i,b}, R_{i,b}'$ and randomized bookend matrices R_S, R_S', R_T, R_T', we obtain the encoded matrices $\mathsf{enc}_{L_{i,b}}(R_{i,b})$ whose entries are encoding of corresponding entries of randomized matrix $R_{i,b}$. For brevity we write $\widetilde{\boldsymbol{M}}_{i,b}$ to denote $\text{enc}_{L_{i,b}}(\boldsymbol{R}_{i,b})$, and the other matrices $\widetilde{\boldsymbol{M}}'_{i,b}$, $\widetilde{\boldsymbol{S}}$, $\widetilde{\boldsymbol{S}}'$, $\widetilde{\boldsymbol{T}}$, $\widetilde{\boldsymbol{T}}'$ are defined in similar manner.

Two conditions should hold in the attackable model

1. the evaluation of valid input is top-level, in other words, for all input $\boldsymbol{x}$, $\left(\cup_{i=1}^{\ell} L_{i,x_{\text{inp}(i)}}\right) \cup L_{\boldsymbol{S}} \cup L_{\boldsymbol{T}} = \mathbb{L}$ where $\mathbb{L}$ denotes top-level set,
2. the sizes of set L's are all similar, that is, there is a constant C such that $|L_{i,b}|/|L_{j,b'}| \leq C$ for all $i, j, \boldsymbol{b}, \boldsymbol{b}'$ and similar inequalities hold for $L_{\boldsymbol{S}}, L_{\boldsymbol{T}}$.

In practice, the level L's is determined by *the straddling set system* introduced in [7,30], and these constructions satisfy our conditions. Using the condition 1 and Definition 3, Proposition 2 can be easily verified. We also note that the condition 2 implies $\ell = \Theta(\kappa)$, where κ is the level of underlying multilinear map.

4.3   Linear Relationally Inequivalent Branching Programs

At last, we explain the condition, *linear relationally inequivalence*, for branching programs of attackable BP obfuscation. This condition is used at the last section, but we note that there are several linear relationally inequivalence BPs as stated in Proposition 3.

To define the linear relationally inequivalence, we consider evaluations of invalid inputs of branching program and denote $\prod_{i=1}^{\ell} \boldsymbol{M}_{i,b_i}$ by $\boldsymbol{M}(\boldsymbol{b})$ for $\boldsymbol{b} = (\boldsymbol{b}_1, \cdots, \boldsymbol{b}_\ell)$. We define linear relations of two BPs and the *linear relationally inequivalence* of BPs as

Definition 4 (Linear Relations of Branching Program). *For a given branching program*

$$P_M = \left\{ \boldsymbol{M}_{i,b} \in \mathbb{Z}^{d_i \times d_{i+1}} \right\}_{i \in [\ell], b \in \{0,1\}^w},$$

the set of linear relations of P_M is

$$L_M := \left\{ (q_b)_{b \in \{0,1\}^{w \times \ell}} : \sum_{b \in \{0,1\}^{w \times \ell}} q_b \cdot \boldsymbol{M}(\boldsymbol{b}) = \boldsymbol{0}^{d_1 \times d_{\ell+1}} \right\}$$

Definition 5 (Linear Relationally Inequivalence). *We say that two branching programs P_M and P_N with the same length are linear relationally inequivalent if $L_M \neq L_N$.*

The set of linear relations of a given BP is easily computed by computing the kernel, considering BP matrices as vectors. It is clear that L_M is a lattice. We note that the set of linear relations of BP is not determined by the functionality of BP, and indeed it seems that they are irrelevant.

Further, one can observe that if P_M, P_N are linear relationally inequivalent BPs, then so do two extended BPs P'_M, P'_N which are obtained by concatenating some other (functionally equivalent) BPs on the right (or left) of P_M, P_N. Therefore we can show that there exist arbitrary large two functionally equivalent BPs which are linear relationally inequivalent.

We conclude this section by presenting a proposition that shows concrete examples of linear relationally inequivalent BPs, which are placed in Appendix C.

Proposition 3. *There are two functionally equivalent, but linear relationally inequivalent branching programs. Especially, there are examples satisfying the linear relationally inequivalence which are*

(1) generated by Barrington's theorem and input-unpartitionable or
(2) from non-deterministic finite automata and read-once, in other words, inp is a bijection.

5   Program Converting Technique

In this section, we describe the program converting technique, which remove the hindrance of modulus q and $\boldsymbol{g}$. We first define new notion $\boldsymbol{Y}$ *program (of P)* if all entries of branching program matrices corresponding a program P are in a space $\boldsymbol{Y}$ while preserving many properties. For example, the obfuscated program $\mathcal{O}(P)$ is $\mathcal{R}_q$ program. Suppose that the obfuscated program $\mathcal{O}(P)$ of program P is given.

We will convert given obfuscated program $\mathcal{O}(P)$ into $\mathcal{R}$ and $\mathcal{R}/\langle\boldsymbol{g}\rangle$ program using the algorithm to solve the NTRU problem, especially *subfield attacks* [1,18] which solves the problem with large modulus q.

Proposition 4 ([1,17,18,26]). *Let q be a large integer, n a power of two, M a constant much smaller than q, $\mathcal{R} = \mathbb{Z}[X]/\langle X^n + 1\rangle$ and $\mathcal{R}_q = \mathcal{R}/q\mathcal{R}$. For a given $[\boldsymbol{f}_1/\boldsymbol{f}_2]_q \in \mathcal{R}_q$ for $\boldsymbol{f}_1, \boldsymbol{f}_2 \in \mathcal{R}$ with size smaller than M, there is an algorithm to compute $(\boldsymbol{c}\cdot\boldsymbol{f}_2, \boldsymbol{c}\cdot\boldsymbol{f}_1) \in \mathcal{R}^2$ such that sizes of $\boldsymbol{c}$, $\boldsymbol{c}\cdot\boldsymbol{f}_1$ and $\boldsymbol{c}\cdot\boldsymbol{f}_2$ are much smaller than q in time $2^{O(\beta)}\cdot poly(n)$ for a constant β satisfying $\beta/\log\beta = \Theta(n\log M/\log^2 q)$.*

We note that the similar results hold for other non-cyclotomic ring [17,26] or for $\boldsymbol{f}_1, \boldsymbol{f}_2$ from certain distribution [1]. Throughout in this paper, we only consider the bounded coefficient $\boldsymbol{f}_1\boldsymbol{f}_2$ in cyclotomic ring for brevity.

For given obfuscated program in $\mathcal{R}_q$, we first make the NTRU instances and solve the problem, and then convert to $\mathcal{R}$ program by some computations on obfuscated matrices. This procedure replaces the level parameter z_i with a small element c_i. The $\mathcal{R}$ program preserves same functionality with the $\mathcal{R}_q$ program. Subsequently, we convert this $\mathcal{R}$ program to $\mathcal{R}/\langle\boldsymbol{g}\rangle$ program by recovering the ideal $\langle\boldsymbol{g}\rangle$.

5.1   Converting to $\mathcal{R}$ Program

In order to remove the modulus q, we employ the algorithm for solving NTRU problem. Let $\widetilde{\boldsymbol{M}}_{i,b}$ be the obfuscated matrix of $\boldsymbol{R}_{i,b}$. Then, each (j, k)-th entries of obfuscated matrix $\widetilde{\boldsymbol{M}}_{i,b}$ is of the form

$$\boldsymbol{d}_{j,k,b} = \left[\frac{\boldsymbol{r}_{j,k,b}\cdot\boldsymbol{g} + \boldsymbol{a}_{j,k,b}}{\boldsymbol{z}_i}\right]_q,$$

where $a_{j,k,b}$ is the (j,k)-th entry of the matrix $R_{i,b}$ and $r_{j,k,b} \in \mathcal{R}$ are random small elements. Consider an element $v = [d_{1,1,0}/d_{1,2,0}]_q = [(r_{1,1,0} \cdot g + a_{1,1,0})/(r_{1,2,0} \cdot g + a_{1,2,0})]_q$. Then, v is the instance of the NTRU problem since the size of denominator and numerator of v is much smaller than q in the parameter setup of GGH13 multilinear map.

Applying Proposition 4 to an instance v, one can find a pair $(c_i \cdot (r_{1,1,0} \cdot g + a_{1,1,0}), \; c_i \cdot (r_{1,2,0} \cdot g + a_{1,2,0})) \in \mathcal{R}^2$ with relatively small $c_i \in \mathcal{R}$. Further, for any element $d_{j,k,b} \in \widetilde{M}_{i,b}$, we can remove the modulus q by computing

$$c_i \cdot (r_{1,1,0} \cdot g + a_{1,1}, 0) \cdot [d_{j,k,b}/d_{1,1,0}]_q = c_i \cdot (r_{j,k,0} \cdot g + a_{j,k,0}) \in \mathcal{R}$$

because of the small size of c_i. Consequently, one can obtain a new matrix $D_{i,b}$ over $\mathcal{R}$ whose (j,k)-th entry is $c_i \cdot (r_{j,k,0} \cdot g + a_{j,k,0})$.

Similarly, a new dummy matrix $D'_{i,b}$ over $\mathcal{R}$ can be obtained because $\widetilde{M}'_{i,b}$ shares the level parameter z_i with $\widetilde{M}_{i,b}$ by multiplying $c_i \cdot (r_{j,k,0} \cdot g + a_{j,k,0})$ to $[d'_{j,k,b}/d_{1,1,0}]_q$ where $d'_{j,k,b}$ is a (j,k)-th entry of $\widetilde{S}'_{i,b}$. We easily observe that $2 \cdot 2^w$ matrices $D_{i,b}$ and $D'_{i,b}$ share the parameter c_i.

For all matrices $\widetilde{M}_{i,b}$ and $\widetilde{M}'_{i,b}$ with $i \in [\ell]$ and $b \in \{0,1\}^w$, we can obtain new matrices $D_{i,b}$ and $D'_{i,b}$ over $\mathcal{R}$. In the case of bookend matrices $\widetilde{S}$ and $\widetilde{T}$, they are converted into matrices over $\mathcal{R}$ with small constants c_S and c_T, respectively. Note that this step runs in polynomial time if κ is large [1,17,18,26]. Detailed analysis of this part is discussed in Sect. 5.3.

Therefore, we can convert $\mathcal{R}_q$-program $\mathcal{O}(P)$ into a new program, $\mathcal{R}$-program of P:

$$\mathcal{R}(P) = \{D_S, D_T, D'_S, D'_T, \{D_{i,b}, D'_{i,b}\}_{i \in [\ell], b \in \{0,1\}^w}\}.$$

Note that the matrix $D_{i,b}$ of $\mathcal{R}(P)$ is of the form $c_i \cdot R_{i,b} \pmod{\langle g \rangle}$ in $\mathcal{R}/\langle g \rangle$.

Dummy and bookend matrices satisfies similar relations. We denote $c_i \cdot \alpha_{i,b}$ and $c_i \cdot \alpha'_{i,b}$ by $\rho_{i,b}, \rho'_{i,b}$ for simplicity. The properties of Definition 3 is naturally extended to the following. The Proposition 5 means an evaluation of $\mathcal{R}(P)$ preserves the functionality up to constant on the valid input x.

Proposition 5 (Evaluation of $\mathcal{R}$ and $\mathcal{R}/\langle g \rangle$ Branching Program). *For a $\mathcal{R}$ program given in this section, the following propositions holds:*

1. *The higher dimension embedding matrices U's are eliminated in the product of randomized matrix branching program, that is, there are matrices $S_0, S'_0 \in \mathbb{Z}^{d_0 \times d_1}, T_0, T'_0 \in \mathbb{Z}^{d_{\ell+1} \times d_{\ell+2}}$ such that the following equations hold for all input x:*

$$D_S \cdot \prod_{i=1}^{\ell} D_{i,b_i} \cdot D_T = \rho_S \cdot \prod_{i=1}^{\ell} \rho_{i,b_i} \cdot \rho_T \cdot \left(S_0 \cdot \prod_{i=1}^{\ell} M_{i,b_i} \cdot T_0 \right) \pmod{\langle g \rangle},$$

$$D'_S \cdot \prod_{i=1}^{\ell} D'_{i,b_i} \cdot D'_T = \rho'_S \cdot \prod_{i=1}^{\ell} \rho'_{i,b_i} \cdot \rho'_T \cdot \left(S'_0 \cdot \prod_{i=1}^{\ell} M'_{i,b_i} \cdot T'_0 \right) \pmod{\langle g \rangle}.$$

2. *The evaluation of $\mathcal{R}$ program is done by checking whether the fixed entries of*
$$Eval_D(x) := D_S \cdot \prod_{i=1}^{\ell} D_{i,x_{\mathrm{inp}(i)}} \cdot D_T - D_S' \cdot \prod_{i=1}^{\ell} D_{i,x_{\mathrm{inp}(i)}}' \cdot D_T' \text{ is multiple}$$
of g or not. Especially, there are two integers u, v such that $P(x) = 0 \Rightarrow$
$Eval_D(x)[u, v] = 0 \pmod{\langle g \rangle}$

5.2   Recovering $\langle G \rangle$ and Converting to $\mathcal{R}/\langle g \rangle$ Program

Next, we will compute a basis of the plaintext space $\langle g \rangle$ to transform $\mathcal{R}$ program into $\mathcal{R}/\langle g \rangle$-program. Unlike other attacks, we do not use the assumption 'input partitionability'. We exploits the fact that $\mathcal{R}$ program which comes from $\mathcal{R}_q$ program has the same functionality up to constant. However, existing attacks with input partitionable assumption and our cryptanalysis cannot be applied to a BP program for an 'evasive function' since it does not output multiples of g. It consists of following two steps:

Finding a multiple of g. This step is done by computing $Eval_D$ at the zeros of program P. We compute $Eval_D(x)$ for $\mathcal{R}$ program $\mathcal{R}(P)$ at x satisfying $P(x) = 0$. Then, Proposition 5 implies that $Eval_D(x)[u, v]$ is a multiple of g. More precisely, $Eval_D(x)[u, v]$ is of the form

$$c_S \cdot c_T \cdot \prod_{i=1}^{\ell} c_i \cdot a \cdot g$$

when $p_{zt} \cdot Eval_{\widetilde{M}}(x)[u, v] = a \cdot h \pmod{q}$ for some $a \in \mathcal{R}$ such that $\|a \cdot h\|_2$ is less than $q^{3/4}$.

This procedure outputs the value which is not only multiple of g but also c_i's. However, we can generate several different $\mathcal{R}$ program from $\mathcal{O}(P)$ for different solutions of Proposition 4. We assume that the multiples of g from different $\mathcal{R}$ program are independent multiples of g, with the randomized lattice reduction algorithm as in [21].

Computing Hermite Normal Form of $\langle g \rangle$. For given several random multiples $f_i \cdot g$ of g, we can recover a basis of $\langle g \rangle$ by computing sum of sufficiently many ideal $\langle f \cdot g \rangle$ represented by a lattice with basis $\{f \cdot g, f \cdot g \cdot X, \cdots, f \cdot g \cdot X^{n-1}\}$ or computing the Hermite Normal Form of union of their generating sets by applying the lemma [1, Lemma 1].

Both computations are done in polynomial time in λ and κ, since the evaluations and computing the Hermite normal form has a polynomial time complexity. Eventually, we recover the basis of ideal lattice $\langle g \rangle$ and we can efficiently compute the arithmetics in $\mathcal{R}/\langle g \rangle$. In other words, we get a $\mathcal{R}/\langle g \rangle$ program corresponding to $\mathcal{O}(P)$ (or P), whose properties are characterized by Proposition 5. For convenience, we abuse the notation; from now, $\mathcal{R}(P)$ is the $\mathcal{R}/\langle g \rangle$ program and D_S, D_T and $D_{i,b}$ for all $i \in [\ell], b \in \{0,1\}^w$ are matrices over $\mathcal{R}/\langle g \rangle$.

5.3   Analysis of the Converting Technique

We discuss the time complexity of our program converting technique. The program converting consists of converting to $\mathcal{R}$ program, evaluating of $\mathcal{R}$ program, computing a Hermite Normal Form of an ideal lattice $\langle g \rangle$. The last two steps take polynomial time complexity, so the total cost is dominated by the first step. More precisely, solving the NTRU problem for each encoded matrix is the dominant part of the program converting.

To estimate the cost of solving the NTRU problem, we assume that each component of branching program is encoded by GGH13 multilinear map in level-1. The general cases are similar but a bit more complex when we assume that the size of level sets are not too different so that $\ell = \Theta(\kappa)$.

Suppose that an obfuscated branching program $\mathcal{O}(P)$ over (κ, λ)-GGH13 multilinear map is given. As we written in Sect. 2.3, for constants δ, e and security parameter λ, multilinearity level κ, n, M, and $\log q$ are set to be $\tilde{\Theta}(\kappa^e \cdot \lambda^\delta)$, $n^{\Theta(1)}$, and $\tilde{\Theta}(\kappa \cdot \log n)$, respectively. Proposition 4 implies that one can convert the program in $2^{O(\beta)} \cdot poly(\lambda, \kappa)$ time for $\frac{\beta}{\log \beta} = \Theta(\frac{n \log M}{\log^2 q}) = \tilde{\Theta}\left(\frac{\lambda^\delta}{\kappa^{2-e}}\right)$. Therefore, the program converting technique is done in polynomial time for $\kappa = \tilde{\Omega}(\lambda^{\delta/(2-e)})$. Alternatively, the program converting technique is done in polynomial time for obfuscated programs with length $\ell = \tilde{\Omega}(\lambda^{\delta/(2-e)})$.

We note that choosing large n to make the subfield attack work in exponential time rules out our attack as well. More concretely, if one chooses $n = \tilde{\Theta}(\kappa^2 \lambda)$ then the underlying NTRU problem is hard enough to block known subexponential time attacks.

6 Matrix Zeroizing Attack

In this section, we present a distinguishing attack on $\mathcal{R}$ programs to complete our cryptanalysis of attackable BP obfuscation model. We note that we can evaluate the $\mathcal{R}$ program at invalid inputs, or *mixed input*, since the multilinearity level which was the obstacle of mixed inputs is removed in the previous step. We recall that $M(b)$ denotes $\prod_{i=1}^{\ell} M_{i,b_i}$ for $b = (b_1, \cdots, b_\ell)$ and the set of linear relations

$$ L_M = \left\{ (q_b)_{b \in \{0,1\}^{w \times \ell}} : \sum_{b \in \{0,1\}^{w \times \ell}} q_b \cdot M(b) = \mathbf{0}^{d_1 \times d_{\ell+1}} \right\} $$

which was defined in Sect. 4.3. We also recall that the two program M and N are linear relationally inequivalent if $L_M \neq L_N$.

For two functionally equivalent but linear relationally inequivalent BPs P_M and P_N, we will zeroize the R program corresponding to P_M by exploiting the linear relation, whereas R program corresponding to P_N would not be a zero matrix. The result of the matrix zeroizing attack is as follows.

Proposition 6 (Matrix Zeroizing Attack). *For functionally equivalent but linear relationally inequivalent branching programs P_M, P_N, there is a PPT algorithm which can distinguish between two $\mathcal{R}$ programs $\mathcal{R}(P_M)$ and $\mathcal{R}(P_N)$ obtained by the method in Sect. 5 with non-negligible probability.*

Now we explain how to distinguish two $\mathcal{R}$ programs using linear relationally inequivalence. Despite the absence of multilinearity level, we still have obstacles to directly exploit linear relationally inequivalence: scalar bundlings. To explain the main idea of the attack, we assume that, for the time being, all scalar bundling are trivial in the obtained program in Sect. 5. We later explain how to deal the scalar bundlings.

Suppose that two BPs P_M, P_N and an $\mathbf{R}$ program

$$\mathcal{R}(P_X) = \{\mathbf{D_S}, \mathbf{D_T}, \mathbf{D_{S'}}, \mathbf{D_{T'}}, \{\mathbf{D_{i,b}}, \mathbf{D'_{i,b}}\}_{i\in[\ell],b\in\{0,1\}^w}\}$$

are given. Our goal is to determine $\mathbf{X} = \mathbf{N}$ or $\mathbf{X} = \mathbf{M}$. We can compute a linear relation (q_b) which is an element of $L_M \setminus L_N$ in polynomial time[6] by computing a basis of kernel, and solve the membership problems of lattice for each vector in the basis. Then the following equation holds

$$\sum_{b\in\{0,1\}^{w\times\ell}} \left(q_b \cdot \mathbf{D_S} \cdot \prod_{i=1}^{\ell} \mathbf{D_{i,b_i}} \cdot \mathbf{D_T} \right) = \sum_{b\in\{0,1\}^{w\times\ell}} \left(q_b \cdot \mathbf{S_0} \cdot \prod_{i=1}^{\ell} \mathbf{M_{i,b_i}} \cdot \mathbf{T_0} \right)$$

$$= \mathbf{S_0} \cdot \sum_{b\in\{0,1\}^{w\times\ell}} \left(q_b \cdot \prod_{i=1}^{\ell} \mathbf{M_{i,b_i}} \right) \cdot \mathbf{T_0} = \mathbf{S_0} \cdot \mathbf{0}^{d_1 \times d_{\ell+1}} \cdot \mathbf{T_0} = \mathbf{0}^{d_0 \times d_{\ell+2}} \pmod{\langle g \rangle}$$

when $\mathbf{X} = \mathbf{M}$ whereas this is not hold when $\mathbf{X} = \mathbf{N}$. Therefore, the matrix zeroizing attack works when the scalar bundlings are all trivial.

When the scalar bundlings are not trivial, we can do the similar computation after recovering ratios of bundling scalars. Assume that we know $\rho_{i,u}/\rho_{i,v}$ for every $1 \le i \le \ell$ and $u, v \in \{0,1\}^w$. Consequently, for $r(b) := \prod_{i\in[\ell]} \rho_{i,b_i}$ where $b = (b_1, \cdots, b_\ell)$, we can compute $r(b)/r(c)$ for $b, c \in \{0,1\}^{w\times\ell}$ by multiplying ratios of bundling scalars. Then, we can calculate

$$\sum_{b\in\{0,1\}^{w\times\ell}} \left(q_b \cdot \frac{r(0)}{r(b)} \cdot \mathbf{D_S} \cdot \prod_{i=1}^{\ell} \mathbf{D_{i,b_i}} \cdot \mathbf{D_T} \right)$$

$$= \sum_{b\in\{0,1\}^{w\times\ell}} \left(q_b \cdot \rho_S \cdot r(0) \cdot \rho_T \cdot \mathbf{S_0} \cdot \prod_{i=1}^{\ell} \mathbf{M_{i,b_i}} \cdot \mathbf{T_0} \right)$$

$$= \rho_S \cdot r(0) \cdot \rho_T \cdot \mathbf{S_0} \cdot \sum_{b\in\{0,1\}^{w\times\ell}} \left(q_b \cdot \prod_{i=1}^{\ell} \mathbf{M_{i,b_i}} \right) \cdot \mathbf{T_0} \pmod{\langle g \rangle},$$

which is a zero matrix if and only if $\mathbf{X} = \mathbf{M}$.

Accordingly, we should remove the scalar bundlings or recover ratios of scalar bundlings to execute the matrix zeroizing attack. In the rest of this section, we

---

[6] The dimension of $(q_b)_{b\in\{0,1\}^{w\times\ell}}$ is $2^{w\times\ell}$, which is exponentially large. However, we can reduce this exponential part by considering a polynomial number of b so that there are linear relations.

show how to recover or remove (ratios of) scalar bundlings in several cases. In Sect. 6.2, we explain how to recover all ratios in general cases by complex techniques.

6.1 Existing BP Obfuscations

In this section, we show how to apply the matrix zeroizing attack on two remarkable obfuscations, GGHRSW and GMMSSZ. The other examples on obfuscations [6,32] are placed in Appendix B.

GGHRSW. As the first case, we consider the first BP obfuscation, GGHRSW, which has the identity dummy program. We note that the attack for this case works for the attackable BP obfuscations with fixed dummy program as well. For this case, a constraint on the bundling scalars $\alpha_x = \alpha'_x$ for every input x is given where $\alpha_x = \alpha_S \cdot \prod_{i=1}^{\ell} \alpha_{i,x_{\mathrm{inp}(i)}} \cdot \alpha_T$, $\alpha'_x = \alpha'_S \cdot \prod_{i=1}^{\ell} \alpha'_{i,x_{\mathrm{inp}(i)}} \cdot \alpha'_T$. Suppose $\mathcal{R}$ *program of P* is given by

$$\mathcal{R}(P) = \{\boldsymbol{D_S}, \boldsymbol{D_T}, \boldsymbol{D_{S'}}, \boldsymbol{D_{T'}}, \{\boldsymbol{D_{i,b}}, \boldsymbol{D'_{i,b}}\}_{i \in [\ell], b \in \{0,1\}^w}\}.$$

By Proposition 5, the following equations hold

$$\boldsymbol{D_S} \cdot \prod_{i=1}^{\ell} \boldsymbol{D}_{i,x_{\mathrm{inp}(i)}} \cdot \boldsymbol{D_T} = \rho_S \cdot \prod_{i=1}^{\ell} \rho_{i,x_{\mathrm{inp}(i)}} \cdot \rho_T \cdot \left(\boldsymbol{S_0} \cdot \prod_{i=1}^{\ell} \boldsymbol{M}_{i,x_{\mathrm{inp}(i)}} \cdot \boldsymbol{T_0}\right) \bmod \langle g \rangle,$$

$$\boldsymbol{D'_S} \cdot \prod_{i=1}^{\ell} \boldsymbol{D'}_{i,x_{\mathrm{inp}(i)}} \cdot \boldsymbol{D'_T} = \rho'_S \cdot \prod_{i=1}^{\ell} \rho'_{i,x_{\mathrm{inp}(i)}} \cdot \rho'_T \cdot \left(\boldsymbol{S'_0} \cdot \prod_{i=1}^{\ell} \boldsymbol{M'}_{i,x_{\mathrm{inp}(i)}} \cdot \boldsymbol{T'_0}\right) \bmod \langle g \rangle.$$

Here we assume that each $\boldsymbol{M'}_{i,x_{\mathrm{inp}(i)}}$ are identity matrices. Now we consider the two quantity of evaluations $Plain_D(\boldsymbol{x}) := \boldsymbol{D_S} \cdot \prod_{i=1}^{\ell} \boldsymbol{D}_{i,x_{\mathrm{inp}(i)}} \cdot \boldsymbol{D_T}$ and $Dummy_D(\boldsymbol{x}) := \boldsymbol{D'_S} \cdot \prod_{i=1}^{\ell} \boldsymbol{D'}_{i,x_{\mathrm{inp}(i)}} \cdot \boldsymbol{D'_T}$.

According to the condition of scalar bundlings, $\rho_S \cdot \prod_{i=1}^{\ell} \rho_{i,x_{\mathrm{inp}(i)}} \cdot \rho_T = \rho'_S \cdot \prod_{i=1}^{\ell} \rho'_{i,x_{\mathrm{inp}(i)}} \cdot \rho'_T$ since the value c's are shared for plain and dummy program. It is possible to remove scalar bundlings by dividing $Plain_D(\boldsymbol{x})$ by $Dummy_D(\boldsymbol{x})$. In other words, we can get $d \cdot \boldsymbol{S_0} \cdot \prod_{i=1}^{\ell} \boldsymbol{M}_{i,x_{\mathrm{inp}(i)}} \cdot \boldsymbol{T_0}$ for some fixed d from the above division. Since we know all $\boldsymbol{M}$'s, the matrix zeroizing attack works well for the computed quantities.

We remark that the previous analysis [16] analyzed the first candidate iO [23]. Whereas the work in [16] heavily relies on the input partitionable property of the single input branching program, our algorithm do not need this property. Moreover, our algorithm can be applied to dual input branching program, so this attack can be applied to wider range of branching programs.

GMMSSZ. Most notable result for BP obfuscation, GMMSSZ, is suggested by Garg *et al.* in TCC 2016 [24]. The authors claim the security of their construction against all known attack. Nevertheless, the matrix zeroizing attack can be applied to their obfuscation.

GMMSSZ obfuscates low-rank matrix branching program, which is evaluated by checking whether the product $M_0 \cdot \prod_{i \in [\ell]} M_{i,b_i} \cdot M_{\ell+1}$ is zero or not. There are two distinctive property of the obfuscation; the uniform random higher dimension embedding and given bookend vectors as inputs. Let $M_0 = (\beta_1, \cdots, \beta_{d_1})$, $M_{\ell+1} = (\gamma_1, \cdots, \gamma_{d_{\ell+1}})^T$ are the given bookend vectors. The bookend vectors are also extended as $H_0 = (M_0 \| 0), H_{\ell+1} = (M_{\ell+1} \| U_{\ell+1})^T$ for randomly chosen $U_{\ell+1}$ in the higher dimension embedding step to remove the higher dimension embedding matrices. Note that the branching programs of this obfuscation are square, we do not restrict the shape of matrices in this section.

For the evaluation, one compute $\widetilde{M_0} \cdot \prod_{i \in [\ell]} \widetilde{M}_{i,b_i} \cdot \widetilde{M}_{\ell+1}$, which is corresponding to

$$D_S \cdot \prod_{i=1}^{\ell} D_{i,b_i} \cdot D_T = \rho_S \cdot \prod_{i=1}^{\ell} \rho_{i,b_i} \cdot \rho_T \cdot \left(M_0 \cdot \prod_{i=1}^{\ell} M_{i,b_i} \cdot M_{\ell+!} \right) \pmod{\langle g \rangle}$$

in $\mathcal{R}$ program by Proposition 5. Since we know all M's, we can compute the ratios of scalar bundlings by

$$\rho_{j,b_j} / \rho_{j,b'_j} = \frac{D_S \cdot \prod_{i \in [\ell]} D_{i,b_i} \cdot D_T / M_0 \prod_{i \in [\ell]} M_{i,b_i} \cdot M_{\ell+1}}{D_S \cdot \prod_{i \in [\ell]} D_{i,b'_i} \cdot D_T / M_0 \prod_{i \in [\ell]} M_{i,b'_i} \cdot M_{\ell+1}}$$

for b, b' which are same at all but j-th bit. Therefore, the matrix zeroizing attack well works for the construction of [24]. We remark that this method works for *unknown* bookend matrices with more complicated technique, see Sect. 6.2.

6.2   Attackable BP Obfuscation, General Case

Now we consider the attackable BP obfuscations in general. We note that an attackable obfuscation without bookends can be considered as the obfuscation with bookends by re-naming the matrices. For example, if we name $D_S := D_{1,0} = \rho_{1,0} \cdot D_1$, then we can regard that D_S is a left bookend matrix and $\rho_{1,0}$ the corresponding scalar bundling.

The case of obfuscation with bookend matrices is most complex, and requires complicated technique. We will recover the bookend matrices up to constant multiplication, and proceed the algorithm similar to the case of [24].

Recovering the Bookends. For the sake of simplicity, we only consider the case of *bookend vectors*. To tackle constructions using bookend matrices, it is suffice to consider a fixed (u, v)-entry of output matrix given in Proposition 2.

If the obfuscation has bookend vectors, then the evaluation of $\mathcal{R}$ program is computed by

$$D_S \cdot \prod_{i=1}^{\ell} D_{i,b_i} \cdot D_T = \rho_S \cdot \prod_{i=1}^{\ell} \rho_{i,b_i} \cdot \rho_T \cdot \left(S_0 \cdot \prod_{i=1}^{\ell} M_{i,b_i} \cdot T_0 \right) \pmod{\langle g \rangle}$$

for some vectors $S_0 \in (\mathcal{R}/\langle g \rangle)^{1 \times d_1}$ and $T_0 \in (\mathcal{R}/\langle g \rangle)^{d_{\ell+1} \times 1}$. Let $S_0 = (\beta_1, \cdots, \beta_{d_1})$, $T_0 = (\gamma_1, \cdots, \gamma_{d_{\ell+1}})$ and the evaluation $D_S \cdot \prod_{i=1}^{\ell} D_{i,b_i} \cdot D_T$ is denoted by $Eval_D(b_1, \cdots, b_\ell)$.

Our idea is removing ρ's to make equations over S_0, T_0. Let $b_{i,t} \in \{0,1\}^w$ for $1 \le i \le \ell$ and $t \in \{0,1\}$ and $t = (t_1, \cdots, t_\ell) \in \{0,1\}^w$. Then the following two values share the same ρ's, precisely $(\rho_S \rho_T)^2 \cdot \prod_{i \in [\ell]} \rho_{i,b_{i,0}} \rho_{i,b_{i,1}}$:

$$Eval_D(b_{1,0}, \cdots, b_{\ell,0}) \cdot Eval_D(b_{1,1}, \cdots, b_{\ell,1}),$$
$$Eval_D(b_{1,t_1}, \cdots, b_{\ell,t_\ell}) \cdot Eval_D(b_{1,1-t_1}, \cdots, b_{\ell,1-t_\ell}).$$

We denote $S_0 \cdot \prod_{i=1}^{\ell} M_{i,b_i} \cdot T_0$ by $Eqn_M(b_1, \cdots, b_\ell)$. Then, by the above relations, we get a equation for $\beta_1, \cdots, \beta_{d_1}, \gamma_1, \cdots, \gamma_{d_{\ell+1}}$:

$$\frac{Eqn_M(b_{1,0}, \cdots, b_{\ell,0}) \cdot Eqn_M(b_{1,1}, \cdots, b_{\ell,1})}{Eval_D(b_{1,0}, \cdots, b_{\ell,0}) \cdot Eval_D(b_{1,1}, \cdots, b_{\ell,1})}$$
$$= \frac{Eqn_M(b_{1,t_1}, \cdots, b_{\ell,t_\ell}) \cdot Eqn_M(b_{1,1-t_1}, \cdots, b_{\ell,1-t_\ell})}{Eval_D(b_{1,t_1}, \cdots, b_{\ell,t_\ell}) \cdot Eval_D(b_{1,1-t_1}, \cdots, b_{\ell,1-t_\ell})}.$$

Both side of the equation is homogeneous polynomial of degree 4. If we substitute each degree 4 monomials by another variables, this equation become a homogeneous linear equation of new variables. The number of new variable is $O(d_1^2 d_{\ell+1}^2)$.

Now we assume that we can obtain sufficient number of linearly independent equations generated by the explained way. Then, since the system of linear equations can be solved in $O(M^3)$ time by Gaussian elimination for the number of variable M, we can find all ratios of degree 4 monomials.[7] In other words, we can compute $\delta\beta_1, \cdots, \delta\beta_{d_1}, \delta\gamma_1, \cdots, \delta\gamma_{d_{\ell+1}}$ for some constant δ.

Matrix Zeroizing Attack. The remaining part of the attack is exactly same with the attack on GMMSSZ. Precisely, we can recover the ratios of scalar bundlings by computing

$$\rho_{j,b_j}/\rho_{j,b_j'} = \frac{D_S \cdot \prod_{i \in [\ell]} D_{i,b_i} \cdot D_T / S_0 \prod_{i \in [\ell]} M_{i,b_i} \cdot T_0}{D_S \cdot \prod_{i \in [\ell]} D_{i,b_i'} \cdot D_T / S_0 \prod_{i \in [\ell]} M_{i,b_i'} \cdot T_0}$$

for b, b' which are same at all but j-th bits. We note that we do not know exact values of S_0, T_0, but we recovered $\delta S_0, \delta T_0$ in the above step. Thus we can compute $\rho_{j,b_j}/\rho_{j,b_j'}$ by

$$\frac{D_S \cdot \prod_{i \in [\ell]} D_{i,b_i} \cdot D_T / (\delta S_0) \prod_{i \in [\ell]} M_{i,b_i} \cdot (\delta T_0)}{D_S \cdot \prod_{i \in [\ell]} D_{i,b_i'} \cdot D_T / (\delta S_0) \prod_{i \in [\ell]} M_{i,b_i'} \cdot (\delta T_0)}.$$

Therefore the matrix zeroizing attack can be applied to the attackable BP obfuscations, which include all existing BP obfuscations over GGH13.

---

[7] Here we assume that g is hard to factorize. If g is factorized in the Gaussian elimination procedure, we can proceed the algorithm for a factor of g.

Acknowledgement. We sincerely thank the anonymous reviewers of Crypto 2018 for their fruitful comments. This work was supported by Institute for Information & communication Technology Promotion (IITP) grant funded by the Korea government (MSIT) (No. 2016-6-00598, The mathematical structure of functional encryption and its analysis) and was based upon work supported by the ARO and DARPA under Contract No. W911NF-15-C-0227.

References

1. Albrecht, M., Bai, S., Ducas, L.: A subfield lattice attack on overstretched NTRU assumptions. In: Robshaw, M., Katz, J. (eds.) CRYPTO 2016. LNCS, vol. 9814, pp. 153–178. Springer, Heidelberg (2016). https://doi.org/10.1007/978-3-662-53018-4_6
2. Albrecht, M.R., Cocis, C., Laguillaumie, F., Langlois, A.: Implementing candidate graded encoding schemes from ideal lattices. In: Iwata, T., Cheon, J.H. (eds.) ASIACRYPT 2015. LNCS, vol. 9453, pp. 752–775. Springer, Heidelberg (2015). https://doi.org/10.1007/978-3-662-48800-3_31
3. Prabhanjan, A., Gupta, D., Ishai, Y., Sahai, A.: Optimizing obfuscation: avoiding Barrington's theorem. In: Proceedings of the 2014 ACM SIGSAC Conference on Computer and Communications Security, pp. 646–658. ACM (2014)
4. Apon, D., Döttling, N., Garg, S., Mukherjee, P.: Cryptanalysis of indistinguishability obfuscations of circuits over GGH13. In: LIPIcs-Leibniz International Proceedings in Informatics, vol. 80. Schloss Dagstuhl-Leibniz-Zentrum fuer Informatik (2017)
5. Applebaum, B., Brakerski, Z.: Obfuscating circuits via composite-order graded encoding. In: Dodis, Y., Nielsen, J.B. (eds.) TCC 2015. LNCS, vol. 9015, pp. 528–556. Springer, Heidelberg (2015). https://doi.org/10.1007/978-3-662-46497-7_21
6. Badrinarayanan, S., Miles, E., Sahai, A., Zhandry, M.: Post-zeroizing obfuscation: new mathematical tools, and the case of evasive circuits. In: Fischlin, M., Coron, J.-S. (eds.) EUROCRYPT 2016. LNCS, vol. 9666, pp. 764–791. Springer, Heidelberg (2016). https://doi.org/10.1007/978-3-662-49896-5_27
7. Barak, B., Garg, S., Kalai, Y.T., Paneth, O., Sahai, A.: Protecting obfuscation against algebraic attacks. In: Nguyen, P.Q., Oswald, E. (eds.) EUROCRYPT 2014. LNCS, vol. 8441, pp. 221–238. Springer, Heidelberg (2014). https://doi.org/10.1007/978-3-642-55220-5_13
8. Barak, B., Goldreich, O., Impagliazzo, R., Rudich, S., Sahai, A., Vadhan, S., Yang, K.: On the (im)possibility of obfuscating programs. In: Kilian, J. (ed.) CRYPTO 2001. LNCS, vol. 2139, pp. 1–18. Springer, Heidelberg (2001). https://doi.org/10.1007/3-540-44647-8_1
9. Barak, B., Goldreich, O., Impagliazzo, R., Rudich, S., Sahai, A., Vadhan, S., Yang, K.: On the (im)possibility of obfuscating programs. J. ACM (JACM) **59**(2), 6 (2012)
10. Barrington, D.A.: Bounded-width polynomial-size branching programs recognize exactly those languages in NC 1. In: Proceedings of the Eighteenth Annual ACM Symposium on Theory of Computing, pp. 1–5. ACM (1986)
11. Ben-Or, M., Cleve, R.: Computing algebraic formulas using a constant number of registers. In: Proceedings of the 20th Annual ACM Symposium on Theory of Computing, pp. 254–257 (1988)
12. Biasse, J.-F.: Subexponential time relations in the class group of large degree number fields. Adv. Math. Commun. **8**(4), 407–425 (2014)

13. Biasse, J.-F., Espitau, T., Fouque, P.-A., Gélin, A., Kirchner, P.: Computing generator in cyclotomic integer rings. In: Coron, J.-S., Nielsen, J.B. (eds.) EUROCRYPT 2017. LNCS, vol. 10210, pp. 60–88. Springer, Cham (2017). https://doi.org/10.1007/978-3-319-56620-7_3

14. Biasse, J.-F., Song, F.: Efficient quantum algorithms for computing class groups and solving the principal ideal problem in arbitrary degree number fields. In: Proceedings of the Twenty-Seventh Annual ACM-SIAM Symposium on Discrete Algorithms, pp. 893–902. SIAM (2016)

15. Brakerski, Z., Rothblum, G.N.: Virtual black-box obfuscation for all circuits via generic graded encoding. In: Lindell, Y. (ed.) TCC 2014. LNCS, vol. 8349, pp. 1–25. Springer, Heidelberg (2014). https://doi.org/10.1007/978-3-642-54242-8_1

16. Chen, Y., Gentry, C., Halevi, S.: Cryptanalyses of candidate branching program obfuscators. In: Coron, J.-S., Nielsen, J.B. (eds.) EUROCRYPT 2017. LNCS, vol. 10212, pp. 278–307. Springer, Cham (2017). https://doi.org/10.1007/978-3-319-56617-7_10

17. Cheon, J.H., Hhan, M., Lee, C.: Cryptanalysis of the overstretched NTRU problem for general modulus polynomial. IACR Cryptology ePrint Archive, 2017:484 (2017)

18. Cheon, J.H., Jeong, J., Lee, C.: An algorithm for NTRU problems and cryptanalysis of the GGH multilinear map without a low-level encoding of zero. LMS J. Comput. Math. 19(A), 255–266 (2016)

19. Coron, J.-S., Lepoint, T., Tibouchi, M.: Practical multilinear maps over the integers. In: Canetti, R., Garay, J.A. (eds.) CRYPTO 2013. LNCS, vol. 8042, pp. 476–493. Springer, Heidelberg (2013). https://doi.org/10.1007/978-3-642-40041-4_26

20. Cramer, R., Ducas, L., Peikert, C., Regev, O.: Recovering short generators of principal ideals in cyclotomic rings. In: Fischlin, M., Coron, J.-S. (eds.) EUROCRYPT 2016. LNCS, vol. 9666, pp. 559–585. Springer, Heidelberg (2016). https://doi.org/10.1007/978-3-662-49896-5_20

21. Gama, N., Nguyen, P.Q.: Predicting lattice reduction. In: Smart, N. (ed.) EUROCRYPT 2008. LNCS, vol. 4965, pp. 31–51. Springer, Heidelberg (2008). https://doi.org/10.1007/978-3-540-78967-3_3

22. Garg, S., Gentry, C., Halevi, S.: Candidate multilinear maps from ideal lattices. In: Johansson, T., Nguyen, P.Q. (eds.) EUROCRYPT 2013. LNCS, vol. 7881, pp. 1–17. Springer, Heidelberg (2013). https://doi.org/10.1007/978-3-642-38348-9_1

23. Garg, S., Gentry, C., Halevi, S., Raykova, M., Sahai, A., Waters, B.: Candidate indistinguishability obfuscation and functional encryption for all circuits. In: Proceedings of the 2013 IEEE 54th Annual Symposium on Foundations of Computer Science, pp. 40–49. IEEE Computer Society (2013)

24. Garg, S., Miles, E., Mukherjee, P., Sahai, A., Srinivasan, A., Zhandry, M.: Secure obfuscation in a weak multilinear map model. In: Hirt, M., Smith, A. (eds.) TCC 2016. LNCS, vol. 9986, pp. 241–268. Springer, Heidelberg (2016). https://doi.org/10.1007/978-3-662-53644-5_10

25. Gentry, C., Gorbunov, S., Halevi, S.: Graph-induced multilinear maps from lattices. In: Dodis, Y., Nielsen, J.B. (eds.) TCC 2015. LNCS, vol. 9015, pp. 498–527. Springer, Heidelberg (2015). https://doi.org/10.1007/978-3-662-46497-7_20

26. Kirchner, P., Fouque, P.-A.: Revisiting lattice attacks on overstretched NTRU parameters. In: Coron, J.-S., Nielsen, J.B. (eds.) EUROCRYPT 2017. LNCS, vol. 10210, pp. 3–26. Springer, Cham (2017). https://doi.org/10.1007/978-3-319-56620-7_1

27. Langlois, A., Stehlé, D., Steinfeld, R.: GGHLite: more efficient multilinear maps from ideal lattices. In: Nguyen, P.Q., Oswald, E. (eds.) EUROCRYPT 2014. LNCS, vol. 8441, pp. 239–256. Springer, Heidelberg (2014). https://doi.org/10.1007/978-3-642-55220-5_14

28. Lewi, K., Malozemoff, A.J., Apon, D., Carmer, B., Foltzer, A., Wagner, D., Archer, D.W., Boneh, D., Katz, J., Raykova, M.: 5Gen: a framework for prototyping applications using multilinear maps and matrix branching programs. In: Proceedings of the 2016 ACM SIGSAC Conference on Computer and Communications Security, pp. 981–992. ACM (2016)

29. Ma, F., Zhandry, M.: The MMAP strikes back: obfuscation and new multilinear maps immune to CLT13 Zeroizing attacks. Cryptology ePrint Archive, Report 2017/946 (2017). https://eprint.iacr.org/2017/946

30. Miles, E., Sahai, A., Weiss, M.: Protecting obfuscation against arithmetic attacks. IACR Cryptology ePrint Archive, 2014:878 (2014)

31. Miles, E., Sahai, A., Zhandry, M.: Annihilation attacks for multilinear maps: cryptanalysis of indistinguishability obfuscation over GGH13. In: Robshaw, M., Katz, J. (eds.) CRYPTO 2016. LNCS, vol. 9815, pp. 629–658. Springer, Heidelberg (2016). https://doi.org/10.1007/978-3-662-53008-5_22

32. Pass, R., Seth, K., Telang, S.: Indistinguishability obfuscation from semantically-secure multilinear encodings. In: Garay, J.A., Gennaro, R. (eds.) CRYPTO 2014. LNCS, vol. 8616, pp. 500–517. Springer, Heidelberg (2014). https://doi.org/10.1007/978-3-662-44371-2_28

33. Sahai, A., Zhandry, M.: Obfuscating low-rank matrix branching programs. IACR Cryptology ePrint Archive, 2014:773 (2014)

34. Zimmerman, J.: How to obfuscate programs directly. In: Oswald, E., Fischlin, M. (eds.) EUROCRYPT 2015. LNCS, vol. 9057, pp. 439–467. Springer, Heidelberg (2015). https://doi.org/10.1007/978-3-662-46803-6_15

A   Extended Attackable BP Obfuscation Model

In this section we introduce an extended model of attackable BP obfuscation by our attack. The extended attackable BP obfuscation is modified in the randomization step to embraces the obfuscation in [15]. The definition of extended attackable conditions for randomization is as follows, which is similar to Definition 3:

Definition 6 (Extended Attackable Conditions for Randomization). *For a branching program* $P = \left\{ M_{i,b} \in \mathbb{Z}^{d_i \times d_{i+1}} \right\}_{i \in [\ell], b \in \{0,1\}^w}$, *the extended attackable randomized branching program is the set*

$$Rand(P) = \left\{ R_{i,b}, R'_{i,b} \in \mathbb{Z}^{d_i \times d_{i+1}} \right\}_{i \in [\ell], b \in \{0,1\}^w}$$

$$\cup \left\{ R_S, R'_S \in \mathbb{Z}^{d_0 \times d_1}, R_T, R'_T \in \mathbb{Z}^{d_{\ell+1} \times d_{\ell+2}} \right\}$$

$$\cup \left\{ \mathsf{aux}_{J,b}, \mathsf{aux}'_{J,b} \right\}_{J \subset [N], b \in \{0,1\}^{w \times |J|}}$$

satisfying the following properties, where $d_0, d_{\ell+2}, e_i$ *'s are integers.*

1. *There exist matrices* $S_0, S_0' \in \mathbb{Z}^{d_0 \times d_1}, T_0, T_0' \in \mathbb{Z}^{d_\ell \times d_{\ell+1}}$ *and scalars* α_S, α_S', $\alpha_T, \alpha_T', \{\alpha_{i,b}, \alpha_{i,b}'\}_{i \in [\ell], b \in \{0,1\}^w}$ *such that the following equations hold for all* $\{b_i \in \{0,1\}^w\}_{i \in [\ell]}$:

$$R_S \cdot \prod_{i=1}^{\ell} R_{i,b_i} \cdot R_T = \alpha_S \cdot \prod_{i=1}^{\ell} \alpha_{i,b_i} \cdot \alpha_T \cdot \left(S_0 \cdot \prod_{i=1}^{\ell} M_{i,b_i} \cdot T_0 \right),$$

$$R_S' \cdot \prod_{i=1}^{\ell} R_{i,b_i}' \cdot R_T' = \alpha_S' \cdot \prod_{i=1}^{\ell} \alpha_{i,b_i}' \cdot \alpha_T' \cdot \left(S_0' \cdot \prod_{i=1}^{\ell} M_{i,b_i}' \cdot T_0' \right).$$

2. *The evaluation of randomized program is done by checking whether the fixed entries of*

$$RP(x) = \prod_{J \subset [N]} \mathsf{aux}_{J,x|_J} \cdot R_S \cdot \prod_{i=1}^{\ell} R_{i,x_{\mathsf{inp}(i)}} \cdot R_T - \prod_{J \subset [N]} \mathsf{aux}_{J,x|_J}' \cdot R_S' \cdot \prod_{i=1}^{\ell} R_{i,x_{\mathsf{inp}(i)}}' \cdot R_T'$$

is zero or not. Especially, there are two integers u, v *such that* $P(x) = 0 \Rightarrow RP(x)[u,v] = 0$.

After randomizing matrices, we encode every entries and scalars of $Rand(P)$ separately by GGH13 multilinear map with respect to the level corresponding to the first index of elements. We denote $\mathsf{enc}(\mathsf{aux}_{J,a})$ by $\widetilde{\mathsf{aux}}_{J,a}$ for each $J \subset [N]$ and $a \in \{0,1\}^{w \times |J|}$.

We note that aux's were not discussed in the main body of our paper. However, our program converting technique is applied with small modification for auxiliary scalars as well. More precisely, for each $\widetilde{\mathsf{aux}}_{J,a}, \widetilde{\mathsf{aux}}_{J,b}$, we compute $h = \widetilde{\mathsf{aux}}_{J,a}/\widetilde{\mathsf{aux}}_{J,b}$ and solve the NTRU problem for the instance h. Then we obtain $c_J \cdot (\mathsf{aux}_{J,a} + r_a \cdot g)$ for small c_J. For an auxiliary scalar $\widetilde{\mathsf{aux}}_{J,c}$ corresponding to J, we compute $c_J \cdot (\mathsf{aux}_{J,c} + r_c \cdot g) = c_J \cdot (\mathsf{aux}_{J,a} + r_a \cdot g) \cdot \widetilde{\mathsf{aux}}_{J,c}/\widetilde{\mathsf{aux}}_{J,a}$. We can recover dummy auxiliaries as well.

From this calculation, $\mathcal{R}$ program is obtained for extended model. the other step such as recovering the ideal $\langle g \rangle$ and the matrix zeroizing attack work correctly as well.

B   Examples of Matrix Zeroizing Attack

Obfuscation in [32]. In this section, we prove that obfuscation in [32] cannot be iO for general-purpose. This scheme is characterized by several special randomizations; converting to merged branching program which consists of permutation matrices, and choose the right bookend vector $T = e_1$ and no left bookend vector, and then choose identity Kilian matrix $K_0 = I$ at the first left position. It implies that, by Proposition 5, the evaluation of the program is of the form:

$$\prod_{i=1}^{\ell} D_{i,b_i} \cdot D_T = \rho_T \cdot \prod_{i=1}^{\ell} \rho_{i,b_i} \cdot \prod_{i=1}^{\ell} M_{i,b_i} \cdot e_1 = \rho_T \cdot \prod_{i=1}^{\ell} \rho_{i,b_i} \cdot e_k \pmod{\langle g \rangle},$$

where k is an integer computed by M's. Therefore, we can compute $\rho_T \cdot \prod_{i=1}^{\ell} \rho_{i,b_i}$ from the computed value. As a next step, we recover ratios of scalar bundlings $\rho_{j,b_j}/\rho_{j,b_j'}$ for b, b' which satisfies $b_i = b_i'$ for all $i \in [\ell]$ except j by computing the ratio $\rho_T \cdot \prod_{i=1}^{\ell} \rho_{i,b_i}/\rho_T \cdot \prod_{i=1}^{\ell} \rho_{i,b_i'}$. Finally, we can run the matrix zeroizing attack.

Obfuscation in [6]. Badrinarayanan *et al.* suggest a construction for obfuscation based on branching program, especially for *evasive functions* [6].[8]. In this section, we prove that obfuscation of Badrinarayanan *et al.* cannot be a general-purpose iO. This construction is for low-rank branching program, thus it do not have dummy matrices and also does not apply higher dimension embeddings.

The original method for their construction is in the bookend; the authors use no bookend matrices and use special form of Kilian randomization at the first and last matrices. The first and last Kilian matrices are given as follows:

$$\boldsymbol{K}_0 = diag(\beta_1, \cdots, \beta_{d_1}), \boldsymbol{K}_{\ell+1}^{-1} = diag(\gamma_1, \cdots, \gamma_{d_{\ell+1}}),$$

where β_u, γ_v are randomly chosen scalars.

To evaluate the obfuscated program, we see $\left(\prod_{i=1}^{\ell} \widetilde{M}_{i,b_i} \right) [u, v]$ for some u, v. This is corresponding to the following value, which is computed by Proposition 5,

$$\left(\prod_{i \in [\ell]} \boldsymbol{D}_{i,b_i} \right) [u, v] = \beta_u \cdot \gamma_v \cdot \prod_{i \in [\ell]} \rho_{i,b_i} \cdot \left(\prod_{i \in [\ell]} M_{i,b_i} \right) [u, v] \pmod{\langle g \rangle}$$

since $\boldsymbol{S}_0, \boldsymbol{T}_0$ are exactly $\boldsymbol{K}_0, \boldsymbol{K}_{\ell+1}^{-1}$. We then can recover the ratio of scalar bundlings by computing $\prod_{i \in [\ell]} \boldsymbol{D}_{i,b_i}[u, v]/\prod_{i \in [\ell]} \boldsymbol{D}_{i,b_i'}[u, v]$ for b, b' which satisfies $b_i = b_i'$ for all $i \in [\ell]$ except j. Since we computed ratios of scalar bundlings $\rho_{j,b_j}/\rho_{j,b_j'}$, we can run the matrix zeroizing attack.

C   Examples of Linear Relationally Inequivalent BPs

We exhibit two examples of two functionally equivalent but linear relationally inequivalent branching programs here. This examples also certify Proposition 3. The first simple example from nondeterministic finite automata is read-once BPs, and the second example comes from Barrington's theorem and thus input-unpartitionable.

C.1   Read-Once BPs from NFA

Two read-once BPs in Table 1 are from non-deterministic finite automata and linear relationally inequivalent.

---

[8] We remark that the construction of [6] is similar to the construction of [33], which is used as a foundation of recent implementation 5Gen [28] and our attack is also applied to [33] in the same manner.

These two BPs are the point function which output 1 only for input 01, but they are linear relationally inequivalent. For example,

$$M_{0,1} \cdot M_{1,0} - M_{0,1} \cdot M_{1,1} \neq 0,$$
$$N_{0,1} \cdot N_{1,0} - N_{0,1} \cdot N_{1,1} = 0.$$

We note that the matrix $M_{i,b}$ is the adjacent matrix between $\{A_{i,c}\}_{c \in \{0,1\}}$ and $\{A_{i+1,c}\}_{c \in \{0,1\}}$, and N's are defined similarly.

Table 1. BPs from NFA

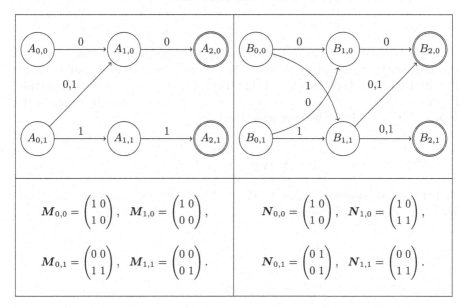

C.2   Input-Unpartionable BPs from Barrington's Theorem

In the case of Barrington's theorem, the linear relationally inequivalent matrix BPs are more complex. We consider the following two functionally equivalent circuits:

$$C_0 = (X_1 \wedge X_2) \wedge (\neg X_1 \wedge X_3),$$
$$C_1 = (\neg X_1 \wedge X_2) \wedge (X_1 \wedge X_3).$$

We transform two circuits into the following BPs by Barrington theorem as follow[9]:

---

[9] Barrington theorem can be implemented in various ways, but we only consider the first description in [10]. This description also can be found in [4].

$$P_{C_0} = \begin{array}{l} 0: \alpha_\rho \ \beta_\rho \ \alpha_\rho^{-1} \ \beta_\rho^{-1} \ e \ \beta_\delta \ e \ \beta_\delta^{-1} \cdots \\ 1: e \ \ e \ \ \ e \ \ \ \ e \ \ \alpha_\delta \ e \ \alpha_\delta^{-1} \ \ e \ \cdots \end{array}$$

$$P_{C_1} = \begin{array}{l} 0: e \ \beta_\rho \ e \ \ \beta_\rho^{-1} \ \alpha_\delta \ \beta_\delta \ \alpha_\delta^{-1} \ \beta_\delta^{-1} \cdots \\ 1: \alpha_\rho \ e \ \alpha_\rho^{-1} \ \ e \ \ \ e \ \ e \ \ \ e \ \ \ e \ \cdots \end{array}$$

$$\text{input bits} \quad 1 \ \ 2 \ \ 1 \ \ \ 2 \ \ \ 1 \ \ 3 \ \ 1 \ \ \ 3 \ \cdots$$

where τ_σ denotes $\sigma\tau\sigma^{-1}$ for permutations $\tau, \sigma \in S_5$. In the matrix representation, the permutations $\alpha, \beta, \gamma, \rho, \delta$ are of the form

$$\alpha = \begin{bmatrix} 0&1&0&0&0 \\ 0&0&1&0&0 \\ 0&0&0&1&0 \\ 0&0&0&0&1 \\ 1&0&0&0&0 \end{bmatrix}, \beta = \begin{bmatrix} 0&0&1&0&0 \\ 1&0&0&0&0 \\ 0&0&0&0&1 \\ 0&1&0&0&0 \\ 0&0&0&1&0 \end{bmatrix}, \gamma = \begin{bmatrix} 0&0&1&0&0 \\ 0&0&0&0&1 \\ 0&1&0&0&0 \\ 1&0&0&0&0 \\ 0&0&0&1&0 \end{bmatrix}, \rho = \begin{bmatrix} 1&0&0&0&0 \\ 0&0&1&0&0 \\ 0&1&0&0&0 \\ 0&0&0&0&1 \\ 0&0&0&1&0 \end{bmatrix}, \delta = \begin{bmatrix} 1&0&0&0&0 \\ 0&0&0&1&0 \\ 0&0&1&0&0 \\ 0&0&0&0&1 \\ 0&1&0&0&0 \end{bmatrix}.$$

We note that two functionally equivalent branching programs P_{C_0} and P_{C_1} are clearly input-unpartitionable. Now if we consider two (invalid) inputs $\boldsymbol{x} = 0110110111111111$ and $\boldsymbol{y} = 1111101011111111$. These yield, for example, $P_{C_0}(\boldsymbol{x}) = \alpha_\rho \cdot e \cdot e \cdot \beta_\rho^{-1} \cdot \alpha_\delta \cdot e \cdot e \cdot e \cdots = \alpha_\rho \cdot \beta_\rho^{-1} \cdot \alpha_\delta = \beta$. The terms in the right $\cdots$ are canceled. Then the equation

$$P_{C_0}(\boldsymbol{x}) - P_{C_0}(\boldsymbol{y}) = 0,$$
$$P_{C_1}(\boldsymbol{x}) - P_{C_1}(\boldsymbol{y}) \neq 0$$

hold. Thus two branching programs P_{C_0} and P_{C_1} are functionally equivalent but linear relationally inequivalent.

MPC

An Optimal Distributed Discrete Log Protocol with Applications to Homomorphic Secret Sharing

Itai Dinur[1]([✉]), Nathan Keller[2], and Ohad Klein[2]

[1] Department of Computer Science, Ben-Gurion University, Beersheba, Israel
dinuri@cs.bgu.ac.il
[2] Department of Mathematics, Bar-Ilan University, Ramat Gan, Israel

Abstract. The distributed discrete logarithm (DDL) problem was introduced by Boyle et al. at CRYPTO 2016. A protocol solving this problem was the main tool used in the share conversion procedure of their homomorphic secret sharing (HSS) scheme which allows non-interactive evaluation of branching programs among two parties over shares of secret inputs.

Let g be a generator of a multiplicative group $\mathbb{G}$. Given a random group element g^x and an unknown integer $b \in [-M, M]$ for a small M, two parties A and B (that cannot communicate) successfully solve DDL if $A(g^x) - B(g^{x+b}) = b$. Otherwise, the parties err. In the DDL protocol of Boyle et al., A and B run in time T and have error probability that is roughly linear in M/T. Since it has a significant impact on the HSS scheme's performance, a major open problem raised by Boyle et al. was to reduce the error probability as a function of T.

In this paper we devise a new DDL protocol that substantially reduces the error probability to $O(M \cdot T^{-2})$. Our new protocol improves the asymptotic evaluation time complexity of the HSS scheme by Boyle et al. on branching programs of size S from $O(S^2)$ to $O(S^{3/2})$. We further show that our protocol is optimal up to a constant factor for all relevant cryptographic group families, unless one can solve the discrete logarithm problem in a *short* interval of length R in time $o(\sqrt{R})$.

Our DDL protocol is based on a new type of random walk that is composed of several iterations in which the expected step length gradually increases. We believe that this random walk is of independent interest and will find additional applications.

Keywords: Homomorphic secret sharing · Share conversion
Fully homomorphic encryption · Discrete logarithm
Discrete logarithm in a short interval · Random walk

1 Introduction

Homomorphic Secret Sharing. Homomorphic secret sharing (HSS) is a practical alternative approach to fully homomorphic encryption (FHE) [12,17] that

© International Association for Cryptologic Research 2018
H. Shacham and A. Boldyreva (Eds.): CRYPTO 2018, LNCS 10993, pp. 213–242, 2018.
https://doi.org/10.1007/978-3-319-96878-0_8

provides some of its functionalities. It was introduced by Boyle et al. [5] at CRYPTO 2016 and further studied and extended in [4,6,7,10]. The main advantage of HSS over traditional secure multiparty computation protocols [1,8,20] is that, similarly to FHE, its communication complexity is smaller than the circuit size of the computed function.

HSS allows homomorphic evaluation to be distributed among two parties who do not interact with each other. A (2-party) HSS scheme randomly splits an input w into a pair of shares (w_0, w_1) such that: (1) each share w_i computationally hides w, and (2) there exists a polynomial-time local evaluation algorithm Eval such that for any program P from a given class (e.g., a boolean circuit or a branching program), the output $P(w)$ can be efficiently reconstructed from $\text{Eval}(w_0, P)$ and $\text{Eval}(w_1, P)$.

The main result of [5] is an HSS scheme for branching programs under the Decisional Diffie-Hellman (DDH) assumption that satisfies $P(w) = \text{Eval}(w_0, P) + \text{Eval}(w_1, P)$. It was later optimized in [4,6], where the security of the optimized variants relies on other discrete log style assumptions.

Let $\mathbb{G}$ be a multiplicative cyclic group of prime order N in which the discrete log problem is (presumably) hard and let g be a generator of this group. The scheme of [5] allows the parties to locally multiply an encrypted (small) input $w \in \mathbb{Z}$ with an additively secret-shared (small) value $y \in \mathbb{Z}$, such that the result $z = wy$ is shared between the parties. The problem is that at this stage g^z is multiplicatively shared by the parties, so they cannot multiply z with a new encrypted input w'. Perhaps the most innovative idea of [5] allows the parties to convert multiplicative shares of g^z into additive shares of z without any interaction via a share conversion procedure. Once the parties have an additive sharing of z, they can proceed to add it to other additive shares. These operations allow to evaluate restricted multiplication straight-line (RMS) programs which can emulate any branching program of size S using $O(S)$ instructions.

The share conversion procedure of [5] is not perfect in the sense that the parties may err. More specifically, the parties fail to compute correct additive shares of z with some error probability δ that depends on the running time T of the parties and on a small integer M that bounds the intermediate computation values. As share conversion is performed numerous times during the execution of Eval, its total error probability accumulates and becomes roughly $\delta \cdot S$, where S is the number of multiplications performed by the RMS program P. Thus, for the total error probability to be constant one has to set the running time T of the parties in the share conversion procedure such that $\delta \approx 1/S$. Consequently, the running time to error tradeoff has a significant impact on the performance of the HSS scheme.

Since the main motivation behind HSS is to provide a practical alternative to FHE, one of the main open problems posed in [5] was to improve the running time to error tradeoff of the share conversion procedure. Progress on this open problem was made in the followup works [4,6] which significantly improved the practicality of the HSS scheme. Despite this progress, the asymptotic running time to error tradeoff of the share conversion procedure was not substantially improved and the running time T in all the schemes grows (roughly) linearly with the inverse error probability $1/\delta$ (or as M/δ in general). Thus, to obtain $\delta \approx 1/S$, one has to set $T \approx S$, and since the total number of multiplications in P is S, the total running time becomes $O(S^2)$.

The Distributed Discrete Log Problem. The focus of this paper is on the "distributed discrete log" (DDL) problem which the parties collectively solve in the share conversion procedure. We now describe the DDL problem and abstract away the HSS details for simplicity. The DDL problem involves two parties A and B. The input of A consists of a group element g^x, were x is chosen uniformly at random from $\mathbb{Z}_N$. The input of B consists of g^{x+b}, where $b \in [-M, M]$ is an unknown uniformly chosen integer in the interval (for a small fixed integer parameter M). The algorithms A, B are restricted by a parameter T which bounds the number of group operations they are allowed to compute.[1] After executing its algorithm, each party outputs an integer. The parties successfully solve the DDL instance if $A(g^x) - B(g^{x+b}) = b$. We stress that A and B are not allowed to communicate.[2]

If g^z is multiplicatively shared by A (party 0) and B (party 1), then $g^{z_0} \cdot g^{z_1} = g^z$. In the share conversion procedure party A runs $A(g^{-z_0})$ while party B runs $B(g^{z_1})$. Assuming they correctly solve DDL for $|z| \leq M$, we have $A(g^{-z_0}) - B(g^{z_1}) = z_1 + z_0 = z$, namely, $A(g^{-z_0})$ and $-B(g^{z_1})$ are additive shares of z as required.

It is convenient to view the DDL problem as a synchronization problem: A and B try to agree or *synchronize* on a group element with a known offset in the exponent from their input. If they manage to do so, the parties solve DDL by outputting this offset. For example, if both parties synchronize on g^y, then $A(g^x) = y - x$ while $B(g^{x+b}) = y - (x+b)$, so $A(g^x) - B(g^{x+b}) = b$ as required. In particular, if both A and B can solve the discrete logarithm problem for their input (i.e., compute x and $x + b$, respectively), then they can synchronize on the generator g by outputting $A(g^x) = 1 - x$ and $B(g^{x+b}) = 1 - (x + b)$. Of course, this would violate the security of the HSS scheme, implying that the discrete logarithm problem in $\mathbb{G}$ should be hard and A, B have to find other means to succeed.

Our Goals. The goal of this paper is to devise algorithms for A and B (i.e., a DDL protocol) that maximize their success probability (taken over the randomness of x, b), or equivalently, minimize their error probability δ given T. Our

---

[1] In all algorithms presented in this paper, the bulk of computation involves performing group operations, hence this is a reasonable complexity measure. Alternatively, the parameter T may bound the complexity of A, B in some reasonable computational model.

[2] We note that in the applications of [4–6], the distribution of $b \in [-M, M]$ is arbitrary. However (as we show in Lemma 13), our choice to define and analyze DDL for the uniform distribution of $b \in [-M, M]$ is technically justified since the uniform distribution is the hardest for DDL: algorithms for A, B that solve DDL with an error probability δ for the uniform distribution, also solve DDL with with an error probability $O(\delta)$ for any distribution of $b \in [-M, M]$.

point of reference is the DDL protocol of [6] (which is a refined version of the original DDL protocol [5]) that achieves a linear tradeoff between the parameter T and error probability δ. More precisely, given that A, B are allowed T group operations, the DDL error probability is roughly M/T. In fact, there are several closely related protocols devised in [4–6] which give similar linear tradeoffs between the parameter T and error probability δ.

Yet another goal of this paper is to better understand the limitations of DDL protocols. More specifically, we aim to prove lower bounds on the error probability of DDL protocols by reducing a well-studied computational problem on groups to DDL. In particular, we are interested in the *discrete log in an interval (DLI)* problem, where the input consists of a group element in a known interval of length R and the goal is to compute its discrete log.

DLI has been the subject of intensive study in cryptanalysis and the best known algorithms for it are adaptations of the classical baby-step giant-step algorithm and the memory-efficient variant of Pollard [15] (see [11,16] for additional extensions). These algorithms are based on collision finding and have complexity of about $\sqrt{R}$. They are the best known in concrete prime-order group families (in which discrete log is hard) up to large values of the interval R. In particular, for elliptic curve groups, the best known DLI algorithm has complexity of about $\sqrt{R}$ where R is as large as the size of the group N (which gives the standard discrete logarithm problem). For some other groups (such as prime order subgroups of $\mathbb{Z}_p^*$), the best known complexity is about $\sqrt{R}$, where R can be up to subexponential in $\log N$ (as discrete log can be solved in subexponential complexity in these groups [13,14]). We note that besides its relevance in cryptanalysis, DLI is solved as part of the decryption process of some cryptosystems (notably in the cryptosystem by Boneh, Goh and Nissim [3]).

An alternative approach to establishing error probability lower bounds for DDL is to use the generic group model (GGM), introduced by Shoup [18]. In GGM, an algorithm is not allowed direct access to the bit representation of the group elements, but can only obtain randomized encodings of the elements, available via oracle queries. The generic group model is a standard model for proving computational lower bounds on certain (presumably hard) problems on groups and thus establishing confidence in their hardness. Although the bounds obtained in GGM are relevant to a restricted class of algorithms, it is essentially the only model in which meaningful lower bounds are known for some computational problems on groups (such as discrete log). Moreover, for several problems (such as discrete log computation in some elliptic curve groups), generic algorithms are essentially the best algorithms known. The downside of this alternative proof approach is that it does not directly relate DDL to any hard problem in a group family, but rather establishes a lower bound proof in an abstract model.

Our Contribution. The main result of this work is closing the gap for DDL in many concrete group families by presenting upper and lower bounds that are tight (within a constant factor) based on the hardness of DLI in these families.

We first develop an improved DDL protocol that is applicable in any group $\mathbb{G}$ and achieves a quadratic tradeoff between the parameter T and the error probability, namely $\delta = O(M/T^2)$. This is a substantial improvement over the linear tradeoff $\delta = O(M/T)$ obtained in [4–6]. Therefore, when executing Eval on an RMS program P with multiplicative complexity S, one can set $T = O(S^{1/2})$ to obtain $\delta = O(1/S)$ and the total running time is reduced from $O(S^2)$ in [4–6] to $O(S^{3/2})$. This result directly improves upon the computational complexity of some of the HSS applications given in [4–6]. For example, in private information retrieval [9] (PIR), a client privately searches a database distributed among several servers for the existence of a document satisfying a predicate P. The 1-round 2-server PIR scheme of [5] supports general searches expressed as branching programs of size S applied to each document. The computational complexity per document in the scheme of Boyle et al. is $O(S^2)$ and our result reduces this complexity to $O(S^{3/2})$.

On the practical side, we fully verified our protocol by extensive experiments. We hope that it will render HSS practical for new applications.

Our DDL protocol uses a new type of (pseudo) random walk composed of several iterations. Each one of these iterations resembles Pollard's "kangaroo" random walk algorithm for solving DLI using limited memory [15]. However DDL is different from DLI as the parties cannot communicate and seek to minimize their error probability (rather than make it constant). This leads to a more complex iterative algorithm, where the parties carefully distribute their time complexity T among several random walks iterations. These iterations use increasingly longer step lengths that gradually reduce the error probability towards $O(M/T^2)$.

The new random walk maximizes the probability that parties with close inputs agree (or synchronize) on a common output without communicating. We believe that this random walk is of independent interest and will find additional applications beyond homomorphic secret sharing schemes and cryptography in general.

After presenting our DDL protocol, we focus on lower bounds and show that any DDL protocol for a family of groups must have error probability of $\delta = \Omega(M/T^2)$, unless DLI (with interval of length R) can be solved in time $T' \approx T = o(\sqrt{R})$ in this family. This is currently not achievable for small (polynomial) T in standard cryptographic groups (for which the group-based HSS scheme is deemed to be secure).

Finally, we analyze DDL protocols in the generic group model. In this model, our DDL protocol is adaptive, as the oracle queries of A and B depend on the answers to their previous queries. This stands in contrast to the protocols of [4–6] in GGM, whose oracle queries are fixed in advance (or selected with high probability from a pre-fixed set of size $O(T)$). It is therefore natural to ask whether adaptivity is necessary to obtain optimal DDL protocols in GGM. Interestingly, we prove that the answer is positive. In fact, we show that the linear tradeoff obtained in [4–6] is essentially the best possible for non-adaptive DDL protocols in GGM.

Paper Organization. The rest of the paper is organized as follows. We describe preliminaries in Sect. 2 and present an overview of our new protocol and related work in Sect. 3. Our new DDL protocol is analyzed in Sect. 4. We prove lower bounds on the DDL error probability in concrete group families in Sect. 5 and finally prove lower bounds on non-adaptive algorithms in GGM in Sect. 6.

2 Preliminaries

In this section we describe the preliminaries required for this work. First we introduce notation that we use throughout the paper and then we present and analyze the DDL algorithm of [5], which will serve as a basis for our algorithms.

2.1 Notation for the Distributed Discrete Log Problem

Recall that the parties A and B successfully solve the DDL instance if $A(g^x) - B(g^{x+b}) = b$. To simplify our notation, we typically do not explicitly write the parameters $\mathbb{G}, g, N, M, T$ in the description of A, B, although some of them will appear in the analysis. We are interested in the success (or error) probability of A and B, taken over the randomness of x, b (and possibly over the randomness of A, B). We denote by $\mathbf{err}(A, B, x, b, T)$ the error event $A(g^x) - B(g^{x+b}) \neq b$, and by $\mathrm{Pr}_{\mathbf{err}}(A, B, [M_1, M_2], T)$ its probability $\Pr_{x,b}[\mathbf{err}(A, B, x, b, T)]$, where $x \in \mathbb{Z}_N$ and $b \in [M_1, M_2]$ are uniform (typically, we are interested in $M_2 = -M_1 = M$). We also denote by $\mathbf{suc}(A, B, x, b, T)$ the complementary success event $A(g^x) - B(g^{x+b}) = b$.

When both parties perform the same algorithm A, we shorten the notation into $\mathbf{err}(A, x, b, T)$, $\mathrm{Pr}_{\mathbf{err}}(A, [M_1, M_2], T)$, and $\mathbf{suc}(A, x, b, T)$, respectively. If the parameters A, B, x, b, T are apparent from the context, we sometimes use $\mathbf{err}$ and $\mathbf{suc}$ instead of $\mathbf{err}(A, B, x, b, T)$ and $\mathbf{suc}(A, B, x, b, T)$, respectively. As mentioned above, A and B can be randomized algorithms and in this case the success (and error) probabilities are taken over their randomness as well. However, to simplify our notation we will typically not refer to this randomness explicitly.

We note that the DDL problem considered in [4–6] is slightly different, as A, B are allowed to perform up to T group operations in expectation. In this alternative definition, one can construct DDL protocols that are more efficient than ours by a small constant factor, while our lower bounds remain the same (again, up to a constant factor).

In the description and analysis of the DDL algorithms, we make frequent use of group elements of the form g^{x+j}. For sake of simplicity, we denote $g_j := g^{x+j}$. In addition, we usually assume $b \geq 0$, as otherwise we can simply exchange the names of the parties A and B when they use the same algorithm. Finally, we refer to a group operation whose output is h as a query to h.

2.2   The Basic DDL Algorithm

Let $\phi : \mathbb{G} \rightarrow [0, N - 1]$ be a pseudo-random function (PRF) that maps group elements to integers. Our protocols evaluate ϕ on $O(T)$ group elements for $T \ll N^{1/2}$. We assume throughout the analysis that ϕ behaves as a truly random permutation on the evaluated group elements, and in particular, we do not encounter collisions in ϕ (i.e., for arbitrary $h \neq h'$, $\phi(h) \neq \phi(h')$). Our probabilistic calculations are taken over the choice of ϕ, even though we do not indicate this explicitly for simplicity.[3]

We describe the min-based DDL algorithm of [5] in Algorithm 1 and refer to it as the basic DDL algorithm. The algorithm is executed by both A and B. When applied to $g_0 = g^x$, the algorithm scans the T values $g_0, g_1, \ldots, g_{T-1}$ and chooses the index i_{min} for which $\phi(g_i)$ is minimal. The output of the algorithm is $\text{Basic}_T(g^x) = (i_{min}, g_{min})$. Note that the algorithm depends also on $\mathbb{G}, g$; however, we do not mention them explicitly in the notation. Furthermore, the output g_{min} will only be relevant later, when we use this algorithm as a subprocedure. For the sake of analysis, we slightly abuse notation below and refer to i_{min} as the (only) output of $\text{Basic}_T(g^x)$.

The motivation behind the algorithm is apparent: if party A applies $\text{Basic}_T(g^x)$ and party B applies $\text{Basic}_T(g^{x+b})$, where $b \ll T$, then the lists of values scanned by the two algorithms (i.e., $g_0, g_1, \ldots, g_{T-1}$ and $g_b, g_{b+1}, \ldots, g_{b+T-1}$) contain many common values, and thus, with a high probability the minimum is one of the common values, resulting in success of the algorithm.

2.3   Analysis of the Basic DDL Algorithm

Error Probability. The following lemma calculates the error probability of the basic DDL algorithm, as a function of $|b|$ and T.

Lemma 1. *The error probability of the basic DDL algorithm is*

$$\Pr_x[\text{err}(\text{Basic}_T, x, b, T)] = \Pr[(\text{Basic}_T(g^x) - \text{Basic}_T(g^{x+b})) \neq b)] = 2|b|/(|b| + T).$$

Proof. We assume $b \geq 0$, as otherwise we exchange the names of A and B. Since both A and B use Algorithm 1, then A computes the function ϕ on $g_0, g_1, \ldots, g_{T-1}$, while B computes this function on $g_b, g_{b+1}, \ldots, g_{b+T-1}$. If the minimum value of ϕ for each party is obtained on an element $g_{min} = g^x \cdot g^{i_{min}}$ which is queried by both, then we have $\text{Basic}_T(g^x) = i_{min}$ and $\text{Basic}_T(g^{x+b}) = i_{min} - b$, implying that $\text{Basic}_T(g^x) - \text{Basic}_T(g^{x+b}) = b$ and the parties are successful. Similarly, they fail when the minimal value of ϕ on the elements

---

[3] The function ϕ (and additional pseudo-random functions defined in this paper) can be implemented by a keyed MAC, where the key is pre-distributed to A and B. Thus, our probabilistic calculations should be formally taken over the choice of the key. They should include an error term that accounts for the distinguishing advantage of an efficient adversary (A or B in our case) that attempts to distinguish the PRF from a truly random permutation. However, for an appropriately chosen PRF, the distinguishing advantage is negligible and we ignore it for the sake of simplicity.

Algorithm 1. $\text{Basic}_T(g^x)$

```
 1 begin
 2 |     h' ← g^x, i ← 0, min ← ∞;
 3 |     while i < T do
 4 |         y ← φ(h');
 5 |         if y < min then
 6 |             g_min ← h';
 7 |             i_min ← i, min ← y;
 8 |         end
 9 |         h' ← h' · g;
10 |         i ← i + 1;
11 |     end
12 |     Output (i_min, g_min);
13 end
```

$g_0, g_1, \ldots, g_{b+T-1}$ is obtained on an element computed only by one party, namely on one of the $2b$ elements $g^x \cdot g^i$ for $0 \leq i < b$ or $T \leq i < b+T$. Assuming that the output of ϕ on each element is uniform and the outputs are distinct, this occurs with probability $2b/(b+T)$. Hence $\Pr_x[\mathbf{err}(\text{Basic}_T, x, b, T)] = 2|b|/(|b| + T)$, as asserted. ∎

The output difference in case of failure. An important quantity that plays a role in our improved protocol is the output difference of the parties in case they fail to synchronize on the same element g_{min} (i.e., their output difference is not b). The following lemma calculates the expectation of this difference, as function of $|b|$ and T.

Lemma 2.

$$\mathbb{E}\left[\left|\text{Basic}_T(g^x) - \text{Basic}_T(g^{x+b}) - b\right| \,\big|\, \mathbf{err}\right] = (|b| + T)/2.$$

Proof. We assume $b \geq 0$. As written above, A computes the function ϕ on $g_0, \ldots, g_{T-1}$, and B computes this function on $g_b, \ldots, g_{b+T-1}$. The ordering of the values $\phi(g_0), \ldots, \phi(g_{b+T-1})$ is uniform, and so the permutation π satisfying $\phi(g_{\pi(0)}) < \phi(g_{\pi(1)}) < \ldots < \phi(g_{\pi(b+T-1)})$ is uniformly random in the permutation group of $\{0, 1, \ldots, b + T - 1\}$. The event $\mathbf{err}$ is equivalent to the event $\pi(0) \notin [b, T - 1]$. Without loss of generality let us restrict ourselves to the event $\pi(0) < b$, i.e. A encounters the minimal group element, and B does not; the other possibility $\pi(0) \geq T$ is symmetric with respect to reflection. Clearly, $\pi(0)$, which equals $\text{Basic}_T(g^x)$, is uniformly random in $[0, b-1]$. Moreover, $\text{Basic}_T(g^{x+b}) + b$ is $\pi(\min\{i \mid \pi(i) > b\})$ which uniformly distributes in $[b, b + T - 1]$. Hence the expected final distance between the parties is $(2b+T+1)/2 - (b+1)/2 = (b+T)/2$. ∎

3 Overview of Our New Protocol and Related Work

3.1 The New DDL Protocol

For the sake of simplicity, we assume in this overview that $M = 1$, hence $|b| \leq 1$. The starting point of our new DDL protocol is Algorithm 1. It makes T queries (i.e., group operations) and fails with probability of roughly $2/T$ according to Lemma 1. Let us assume that we run this algorithm with only $T/2$ queries, which increases the error probability by a factor of 2 to about $4/T$. On the other hand, we still have a budget of $T/2$ queries and we can exploit them to reduce the error probability. Interestingly, simply proceeding to calculate more consecutive group elements is not an optimal way to exploit the remaining budget.

After the first $T/2$ queries, we say that A (or B) is placed at group element g^y if $\phi(g^y)$ is the minimal value in its computed set of size $T/2$. Assume that A and B fail to synchronize on the same group element after the first $T/2$ queries (which occurs with probability of roughly $4/T$). Then, by Lemma 2, A and B are placed at elements which are at distance of about $T/4$, i.e., if A is placed at g^y and B is placed at g^z, then $|y - z| \approx T/4$. Our main idea is to use a somewhat different procedure in order to try to synchronize A and B in case they fail to do so after the first $T/2$ queries, while keeping A and B synchronized if they already are.

The next procedure employed by both A and B is a (pseudo) random walk starting from their initial position, whose step length is uniformly distributed in $[1, L - 1]$, where $L \approx \sqrt{T}$. The step length at group element g^y is determined by $\psi_{L-1}(g^y)$, where ψ_{L-1} is a pseudo-random function independent of ϕ that outputs a uniform integer in $[1, L - 1]$.[4] Assume that after the first $T/2$ queries, B is placed at distance of about $T/4$ in front of A. Then A will pass B's initial position after about $\sqrt{T}/2$ steps and simple probabilistic analysis shows that A will land on one of B's steps after an additional expected number of about $\sqrt{T}/2$ steps. From this point, the walks coincide for the remaining $T/2 - \sqrt{T}$ steps. Similarly to Algorithm 1, each party outputs the offset of the minimal $\phi(g^y)$ value visited during its walk. Since both A and B use the same deterministic algorithm, they remain synchronized if they already are at the beginning of the walks. On the other hand, if they are not initially synchronized, their walks are expected to coincide on $T/2 - \sqrt{T}$ elements, and hence the probability that they remain unsynchronized is roughly $\sqrt{T}/(T/2) = 2 \cdot T^{-1/2}$. Thus, the error probability at this stage is about $4 \cdot T^{-1} \cdot 2 \cdot T^{-1/2} = 8 \cdot T^{-3/2}$, which already significantly improves upon the $2 \cdot T^{-1}$ error probability of Algorithm 1 for large T.

However, we can still do better. For the sake of simplicity, let us completely ignore constant factors in rest of this rough analysis. Note that we may reserve an additional number of $O(T)$ queries to be used in another random walk by shortening the first two random walks, without affecting the failure probability

---

[4] Our analysis assumes that ψ_{L-1} is a truly random function and our probabilistic calculations are taken over the choice of ψ_{L-1}.

significantly. Hence, assume that the parties fail to synchronize after the random walk (which occurs with probability of about $T^{-3/2}$) and that we still have enough available queries for another random walk with $O(T)$ steps. Since each party covers a distance of about $T^{3/2}$ during its walk, then the expected distance between the parties in case of failure is roughly $T^{3/2}$. We can now perform another random walk with expected step length of $T^{3/4}$ (hence the walks are expected to coincide after about $T^{3/4}$ steps), reducing the error probability to about $T^{-3/2} \cdot (T^{3/4} \cdot T^{-1}) = T^{-7/4}$. This further increases the expected distance between A and B in case of failure to approximately $T^{7/4}$. We continue executing random walk iterations with a carefully chosen step length (distributing a budget of $O(T)$ queries among them). After i random walk iterations, the error probability is reduced to about $T^{-2+2^{-i}}$ (and the expected distance between the parties is roughly $T^{2-2^{-i}}$). Choosing $i \approx \log \log T$ gives an optimal error probability of about $T^{-2+1/\log T} = O(T^{-2})$.

Our new DDL protocol is presented in Algorithms 2 and 3. Algorithm 2 describes a single iteration of the random walk, parameterized by (L, T) which determine the maximal step length and the number of steps, respectively.[5] Algorithm 3 describes the full protocol which is composed of application of the basic DDL algorithm (using $t_0 < T$ queries, reserving queries for the subsequent random walks), and then I additional random walks, where the i'th random walk is parameterized by (L_i, t_i) which determine its maximal step length and number of steps. Between each two iterations in Step 6, both parties are moved forward by a large (deterministic) number of steps, in order to guarantee independence between the iterations (the computation time used to perform these calculations is negligible compared to T). We are free to choose the parameters $I, \{L_i, t_i\}$, as long as $\sum_{i=0}^{I} t_i = T$ is satisfied.

The very rough analysis presented above assumes that we have about T queries in each of the $\log \log T$ iterations, whereas we are only allowed T queries overall. Moreover, it does not accurately calculate the error probability and the distance between the parties in case of failure in each iteration. Taking all of these into account in an accurate analysis results in an error probability of $\Omega(\log T \cdot T^{-2})$. Surprisingly, we can still achieve an error probability of $O(T^{-2})$. This is done by a fine tuning of the parameters which distribute the number of queries among the iterations and select the step length of each random walk. In particular, it is not optimal to independently optimize the step length of each iteration and one has to analyze the subtle dependencies between the iterations in order to achieve an error probability of $O(T^{-2})$.

As the fine tuning of the parameters is rather involved, in addition to the theoretical analysis we verified the failure probability by extensive experiments.

---

[5] We assume that the algorithm uses a table containing the pre-computed values $g, g^2, \ldots, g^{L-1}$. Otherwise, it has to compute $g^{z_{i+1}}$ on-the-fly in Step 10, which results in a multiplicative penalty of $O(\log(T))$ on the number of group operations. Of course, it is also possible to obtain a time-memory tradeoff here.

Algorithm 2. $\text{RandW}_{L,T}(h)$

---

1 **begin**
2 | $h' \leftarrow h$, $i \leftarrow 0$, $min \leftarrow \infty$, $d_0 \leftarrow 0$;
3 | **while** $i < T$ **do**
4 | | $y \leftarrow \phi(h')$;
5 | | **if** $y < min$ **then**
6 | | | $h_{min} \leftarrow h'$;
7 | | | $d_{min} \leftarrow d_i$, $min \leftarrow y$;
8 | | **end**
9 | | $z_{i+1} \leftarrow \psi_{L-1}(h')$;
10 | | $h' \leftarrow h' \cdot g^{z_{i+1}}$;
11 | | $d_{i+1} \leftarrow d_i + z_{i+1}$;
12 | | $i \leftarrow i + 1$;
13 | **end**
14 | Output (d_{min}, h_{min});
15 **end**

Algorithm 3. $\text{IteratedRandW}_{I,t_0,\{(L_i,t_i)_{i=1}^I\}}(h)$

---

1 **begin**
2 | $(c_0, h_0) \leftarrow \text{Basic}_{t_0}(h)$;
3 | $p_0 \leftarrow c_0$;
4 | $i \leftarrow 1$;
5 | **while** $i \leq I$ **do**
6 | | $h'_{i-1} \leftarrow h_{i-1} \cdot g^{\sum_{j<i} t_j L_j}$;
7 | | $(c_i, h_i) \leftarrow \text{RandW}_{L_i, t_i}(h'_{i-1})$;
8 | | $p_i \leftarrow p_{i-1} + c_i$;
9 | | $i \leftarrow i + 1$;
10 | **end**
11 | Output p_I;
12 **end**

3.2 Related Work

The most closely related work to our DDL algorithm is Pollard's "kangaroo" method for solving the discrete logarithm problem in an interval using limited memory (see [15] and [11,16] for further analysis and extensions). The kangaroo method launches two random walks (kangaroos), one from the input $h = g^x$ (where x the unknown discrete log) and one from g^y, where y is a known value in an interval of a fixed size R around x. The algorithm is optimized such that the walks meet at a "distinguished point", which reveals x. The kangaroo method thus resembles a single random walk iteration of our DDL algorithm.

On the other hand, there are fundamental differences between the standard DLI and DDL. These differences result in the iterative structure of our algorithm that differs from Pollard's method. First, in contrast to the DLI problem, in DDL

A and B cannot communicate and never know if they succeed to synchronize. Hence, the parties cannot abort the computation at any time. Second, the goal in DDL is to minimize the error probability, whereas achieving a constant error probability (as in standard DLI) is unsatisfactory. To demonstrate the effect of these differences, observe that solving the discrete log problem in an interval of size 3 can be trivially done with probability 1 using 3 group operations. On the other hand, our algorithm for solving DDL for $M = 1$ is much more complicated and achieves an error probability of about T^{-2} using T group operations (which is essentially optimal for many concrete group families).

Yet another difference between DLI and DDL is that in DLI the boundaries of the interval of the input h are known, whereas in DDL the input of each party is completely uniform. The knowledge of the interval boundaries in DLI allows to shift it to the origin (using the self-reducibility property of discrete log) and efficiently use preprocessing (with a limited amount of storage) to speed up the online computation [2]. On the other hand, it is not clear how to efficiently exploit preprocessing in DDL.

4   The New Distributed Discrete Log Protocol

In this section we study our new DDL protocol in more detail. In Sect. 4.1 we focus on a single iteration of our DDL protocol (i.e., a single random walk iteration) and analyze its failure probability and the expected distance between its outputs in case of a failure. In Sect. 4.2 we briefly analyze the complete protocol. Some parts of the analysis are quite involved and presented in the extended version of this paper. This includes the proofs of Lemmas 5, 8, 9 and Theorem 1.

The experimental verification of the protocol is presented in Sect. 4.3. We also describe some practical considerations regarding the protocol in the extended version of this paper.

4.1   A Single Iteration of Our DDL Protocol – The Random Walk DDL Algorithm

Recall that in Algorithm 2 applied with parameters (L, T), both parties perform a random walk of T steps of the form $g^y \rightarrow g^{y+a_i}$, where the length a_i of each step is determined by a (pseudo) random function $\psi_{L-1} : \mathbb{G} \rightarrow \{1, 2, \ldots, L-1\}$ which guarantees that the step length is uniformly distributed in the range $[1, L-1]$. Each party then chooses among the elements of $\mathbb{G}$ visited by its walk, the element h_{min} for which $\phi(h_{min})$ is minimal (as in the basic DDL algorithm).[6]

Once the parties synchronize in a given iteration, they remain synchronized in the subsequent ones as each iteration is deterministic. Thus, an application

---

[6] We assume in our analysis that during the application of the whole protocol by a single party, each function ϕ, ψ_{L-1} is not evaluated twice on the same input. These constrains are satisfied since $|\mathbb{G}| = N$ is much larger than T (e.g., $|\mathbb{G}| > cT^2$ for a sufficiently large constant c).

of Algorithm 2 is "relevant" only if the two parties failed to synchronize in the previous iterations. In this case, the initial distance b between the parties is the difference between their outputs in the previous iteration. We shall compute the probability of failure as a function of b (which allows us to treat b as a constant throughout the analysis), and then substitute the expectation of b – computed in the analysis of the previous step – into the computation. In particular, as the expected distance between the outputs in case of failure of the basic DDL algorithm was computed in Lemma 2, we will be able to substitute it as $\mathbb{E}[|b|]$ into the computation of the failure probability of the second iteration of our DDL protocol. At the same time, in addition to the failure probability we shall compute the expected distance between the outputs in case of failure of Algorithm 2 in order to be able to link the examined iteration to the subsequent one.

Additional Notation. In our analysis we use some auxiliary notation. Without loss of generality, we assume that $b \geq 0$ (namely, B is located at distance b in front of A). We let S_A be the number of steps of A until its walk lands on an element visited by B (i.e., the number of queries made by A strictly before the first element of A that is included in B's path). If this never occurs, we let $S_A = T$. Similarly, we define S_B as the number of steps of B until its walk lands on an element visited by A. Clearly, the walks of A and B coincide for $T - \max(S_A, S_B)$ steps.

We define U_A as the number of steps A performs until it is within reach of a single step from the starting point of B. Namely, $U_A = \min\{i \mid d_i^A > b - L\}$, where d_i^A is the variable d_i in Algorithm 2 applied by A. In addition, we let V_A, V_B denote the numbers of steps performed by A and B, respectively, starting from the point where A is within reach of a single step from the starting point of B, until the walks collide or one of them ends. Furthermore, we denote $V_m = \max\{V_A, V_B\}$. Notice that $S_A = U_A + V_A$ and $S_B = V_B$, and hence

$$\max(S_A, S_B) \leq U_A + V_m. \tag{1}$$

Below, we evaluate the expectations of the random variables U_A, V_m in order to bound the error probability of synchronization based on Algorithm 2.

Finally, while $\text{RandW}_{L,T}(h)$ has two outputs, we slightly abuse notation and refer to d_{min} as the (only) output of $\text{RandW}_{L,T}(h)$ since only this output is relevant for this analysis.

The Failure Probability of Algorithm 2. First, we bound the expected number of steps performed by A until it reaches the starting point of B.

Lemma 3. $\mathbb{E}[U_A] < 2b/L$.

Proof. By the definition of U_A, we have $d_{U_A}^A < b$. Consider the martingale $d_i' = d_i^A - iL/2$ (which is indeed a martingale, as $\psi_{L-1}(h')$ computed in the algorithm are independent and have expectation $L/2$). The classical *Doob's martingale theorem* yields

$$0 = d_0' = \mathbb{E}[d_{U_A}'] = \mathbb{E}[d_{U_A}^A] - L\,\mathbb{E}[U_A]/2.$$

As $d_{U_A} < b$, we deduce $\mathbb{E}[U_A] < 2b/L$. ∎

Our next lemma bounds the expectation of $S_A + S_B$, that is, the total number of steps performed by the two walks together before they meet.

Lemma 4. *Suppose the initial distance between the parties, b, satisfies $0 < b < L$. Then $\mathbb{E}[S_A + S_B] \leq L - 1$.*

Proof. For ease of computation, we do not trim S_A and S_B with T. Of course, this can only make the upper bound larger. One easily sees that $\mathbb{E}[S_A + S_B]$ is finite and depends only on b (and the parameter L). Write E_b for this expectation. We have $E_0 = 0$, and by dividing into cases according to the result of a single step of A, we obtain

$$E_b = 1 + \frac{1}{L-1} \left(E_{b-1} + E_{b-2} + \ldots + E_1 + E_0 + E_1 + \ldots + E_{L-1-b} \right).$$

A valid solution for this system of linear equations is $E_b = L - 1$ for all $0 < b < L$. This is actually the only solution, since the matrix corresponding to this system is strictly diagonally dominant, and thus is invertible by the *Levy-Desplanques theorem*. Therefore, $\mathbb{E}[S_A + S_B] = L - 1$, independently of b. ∎

The next lemma bounds the maximum between the numbers of steps performed by A and B between the time A "almost" reached the starting point of B and the meeting of the walks.

Lemma 5. $\mathbb{E}[V_m] \leq (L - 1)/2 + \sqrt{8(L - 1)}$.

The proof of the lemma is a lengthy technical argument, which mainly uses Lemma 4 and Doob's martingale theorem.

Now we are ready to estimate the failure probability of Algorithm 2.

Lemma 6. *Let $R = 2b/L + L/2 + \sqrt{8L}$ for $0 < b < L$. The error probability of the random walk DDL algorithm satisfies*

$$\Pr_x[\mathrm{err}(\mathrm{RandW}, x, b, T)] = \Pr[(\mathrm{RandW}_{L,T}(g^x) - \mathrm{RandW}_{L,T}(g^{x+b}) \neq b)] \leq 2R/(T + R).$$

Proof. The walks of A and B coincide for $T - \max(S_A, S_B)$ steps. Notice that we have,

$$\mathbb{E}[\max(S_A, S_B)] \leq \mathbb{E}[U_A + V_m] \leq L/2 + 2b/L + \sqrt{8L} = R, \qquad (2)$$

where the first inequality uses (1) and the second inequality uses Lemmas 3 and 5. Similarly to the basic DDL algorithm (Lemma 1), the error probability (assuming that the output of ϕ on each element is uniformly random) is

$$\Pr_x[\mathrm{err}(\mathrm{RandW}, x, b, T)] = \mathbb{E}[2\max(S_A, S_B)/(T + \max(S_A, S_B))]$$

$$\leq \frac{\mathbb{E}[2\max(S_A, S_B)]}{T + \mathbb{E}[\max(S_A, S_B)]} \leq \frac{2R}{T + R},$$

where the first inequality is Jensen's inequality applied to the increasing concave function $x \mapsto 2x/(T + x)$ in the domain $x > 0$, and the second inequality uses the monotonicity of the function $x \mapsto 2x/(T + x)$ and Eq. (2). ∎

The Output Difference in Case of Failure. Similarly to Lemma 2 which bounded the expected difference of outputs in case of failure for the basic DDL algorithm, we bound the analogous quantity for Algorithm 2. In order to achieve this result, we need a "conditional" version of the classical *Azuma martingale inequality.*

Lemma 7 (Azuma's inequality). *Let* $X_0, X_1, \ldots, X_n$ *be a martingale with* $|X_i - X_{i-1}| \leq V$. *Then for any* $t \geq 0$,

$$\Pr\left[|X_n - X_0| \geq V \cdot t\sqrt{n}\right] \leq 2\exp(-t^2/2).$$

Lemma 8. *Let* $X_1, \ldots, X_n$ *be independent random variables with* $|X_i - \mathbb{E}[X_i]| \leq V$ *and let* $\mathcal{E}$ *be an event. Then*

$$\mathbb{E}\left[\max_{k=1}^{n} \left|\sum_{i \leq k}(X_i - \mathbb{E}[X_i])\right| \,\middle|\, \mathcal{E}\right] \leq V\sqrt{8n\log(2/\Pr[\mathcal{E}])}.$$

To understand the intuition behind the lemma, consider the sum of independent $\{1, -1\}$ random variables $\{X_i\}_{i=1}^{n}$. By Chernoff's inequality, $\Pr[|\sum X_i| > t\sqrt{n}] < e^{-t^2/2}$. However, if we condition on the event $\mathcal{E} = \{\sum X_i = n\}$, then we have $\Pr[|\sum X_i| > t\sqrt{n}|\mathcal{E}] = 1$ for all $t < \sqrt{n}$. On the other hand, the probability of the event $\mathcal{E}$ is extremely small. We claim that if one is allowed to condition only on events with not-so-small probability, then all sums $|\sum_{i=1}^{k} X_i|$ are not much larger than the Chernoff bound, which applies without the conditioning. Our lemma is the martingale version of this intuition.

Finally, we show that either the error probability is "very small" (and so there is no need to continue the random walk iterations), or we can bound the distance between the outputs in case of failure.

Lemma 9. *If* $\Pr_{\mathbf{err}}(\mathrm{RandW}_{L,T}, [1,1], T) \geq \epsilon$, *then for* $0 < b < L$ *and* $h_1 \in \mathbb{G}$

$$\mathbb{E}\left[\left|\mathrm{RandW}_{L,T}(h_1) - \mathrm{RandW}_{L,T}(h_1 \cdot g^b) - b\right| \,\middle|\, \mathbf{err}\right] \\ \leq b + TL/4 + L\sqrt{32T\log(2/\epsilon)}. \tag{3}$$

The proof of the lemma is a somewhat lengthy technical argument which mainly uses Lemma 8.

4.2 The Iterated Random Walk DDL Algorithm

As described in Sect. 3, our full DDL protocol (i.e., Algorithm 3) runs iteratively several stages of Algorithm 2. It depends on a set of parameters: I, which is the number of iterations in the algorithm (on top of the basic DLL algorithm), $(t_i)_{i=0}^{I}$, which represent the number of queries in each of the $I+1$ iterations, and $(L_i)_{i=1}^{I}$, which determine the (maximal) sizes of steps performed in each random walk iteration.

Given a set of parameters, Lemmas 6 and 9 allow us to compute the failure probability of IteratedRandW under that set of parameters. A "naive" choice of parameters leads to a failure probability of $O(T^{-2} \log T)$. However, we show in the following theorem that the parameters can be chosen in such a way that the failure probability becomes $O(T^{-2})$.

Theorem 1. *There exists a parameter set PS for which the error probability of the iterated random walk DDL algorithm is*

$$\text{Pr}_{\mathbf{err}}(\text{IteratedRandW}_{PS}, [1, 1], T)$$
$$= \text{Pr}[\text{IteratedRandW}_{PS}(g^x) - \text{IteratedRandW}_{PS}(g^{x+1}) \neq 1]$$
$$\leq 2^{10.2+o(1)}/T^2.$$

We give concrete choices of parameters and experimentally obtained values of the error probability for various values of T (that take into consideration the low-order terms) in the extended version of this paper.

A simple *distance extension* argument (see Sect. 5.5, Lemma 14) allows us to obtain a similar result for larger distances between the starting points.

Corollary 1. *Consider Algorithm 3 with the parameter set PS chosen in Theorem 1. Then for any distribution of the initial distance b that has expectation $\mathbb{E}[|b|]$, the error probability of Algorithm 3 is at most $O(\mathbb{E}|b|/T^2)$. In particular,* $\text{Pr}_{\mathbf{err}}(\text{IteratedRandW}_{PS}, [-M, M], T) = O(M/T^2)$.

We note that when $\mathbb{E}[|b|] \gg 1$, it is more efficient to start the sequence of iterations directly with a random walk of expected step length of roughly $\sqrt{\mathbb{E}[|b|]}$ (instead of starting it with Algorithm 1). This reduces the error probability by a constant factor.

Dependency of Iterations and Parameter Selection. In Sect. 3, we noted that in order to minimize the error probability of the protocol it is necessary to consider the dependencies between the different iterations, rather than optimizing their error probability independently. We demonstrate this here by analyzing only two iterations of Algorithm 2. We first simplify the formulas in Lemmas 6 and 9 by ignoring low-order terms.

Lemma 6 asserts that the error probability of each iteration is $2R/(T + R)$, where $R \approx 2b/L + L/2$ (ignoring low-order terms). We can lower bound this probability by $2R/T \approx (4b + L^2)/2TL$. Lemma 9 asserts that the expected distance between the parties in case of error is at most $b + TL/4$ (ignoring low-order terms). Since L is roughly $\sqrt{b}$ and $b < T^2$ at any iteration (as noted in Sect. 3), then we estimate the expected distance as $TL/4$.

Let us assume that after iteration $i-1$ the parties are at distance b_i (which is a random variable). Assume we are allowed T' queries in the next two iterations, where in iteration i we use step length parameter L_i and query parameter t_i. After this iteration, the expected distance between the parties is b_{i+1}, which we estimated above as $t_i L_i/4$. Similarly, in iteration $i + 1$ we use parameters L_{i+1} and t_{i+1} (under the restriction that $t_i + t_{i+1} = T'$). Thus, the estimated error probability of the two iterations (assuming that the previous ones failed to synchronize) is

$$\frac{4b_i + L_i^2}{2t_i L_i} \cdot \frac{4b_{i+1} + L_{i+1}^2}{2t_{i+1}L_{i+1}}.$$

To simplify this expression, for parameters α, β write $L_i = \alpha\sqrt{b_i}$ and $L_{i+1} = \beta\sqrt{b_{i+1}} \approx \beta/2 \cdot \sqrt{t_i L_i}$. Then, the error probability expression simplifies to

$$\frac{b_i^{3/4}}{8} \cdot \frac{(4 + \alpha^2)(4 + \beta^2)}{\alpha^{1/2}\beta \cdot t_i^{1/2} t_{i+1}}.$$

It remains to minimize this expression by appropriately selecting $\alpha, \beta, t_i, t_{i+1}$ (such that $t_i + t_{i+1} = T'$). Taking partial derivatives, it is clear that the global minimum is obtained by selecting different parameters for the two iterations, namely, the values of α, t_i for iteration i are different from β, t_{i+1} for iteration $i+1$. In other words, in order to minimize the error probability it is necessary to solve the global optimization problem rather than optimize the error probability of each iteration independently.

4.3 Experimental Verification

In order to evaluate Algorithm 3 in practice, we programmed a simulator which simulates IteratedRandW$_{PS}(g^x)$ and IteratedRandW$_{PS}(g^{x+1})$, and empirically approximates Pr[**err**] as the percentage of pairs of simulations which disagree. The parameters for these simulations were chosen using a numerical optimizer based on the analysis above. The results are given in Table 1.[7] Since performing full simulations is too expensive for large values of T, we had to use three optimizations that do not affect the simulator's reliability. These are detailed in the extended version of this paper.

5 Error Probability Lower Bounds in Concrete Group Families

5.1 Overview of the Lower Bound Proof

We outline the main ideas of the lower bound proof in concrete family groups. For the sake of simplicity, we only consider DDL with $M = 1$ in this overview (whereas the proof considers general M).

---

[7] We note that according to the proof of Theorem 1 (given in the extended version of this paper), I can be taken to be any value between $\log\log(T)+\omega(1)$ and $o(\sqrt{\log(T)})$, without a significant effect on the provable performance of the algorithm. On the other hand, according to Table 1, I seems to grow more sharply. However, the restriction of $I = o(\sqrt{\log(T)})$ is merely an artifact of the proof and the optimal value of I could be asymptotically larger. Furthermore, the sharp increase of the values of I in Table 1 could be attributed to low-order terms that have a more noticeable effect for small T values.

Table 1. Experimental results

T	I	$\sim \log_2(t_k)$ $\sim \log_2(L_k)$	$T^2 \cdot \Pr[\text{err}]$ $\sigma(SD)$	$T^2 \cdot \Pr[\text{err}]$ $\sigma(SD)$
2^{13}	5	$6.0, 8.6, 9.6, 10.4, 11.1, 11.7$	334	336.6
		$1.6, 3.6, 5.6, 7.5, 9.4$	2	0.3
2^{16}	6	$7.1, 10.3, 11.6, 12.5, 13.3, 14.1, 14.7$	390	382.5
		$1.6, 3.9, 6.2, 8.3, 10.3, 12.2$	10	1
2^{19}	7	$7.0, 10.3, 12.6, 13.7, 14.7, 15.5, 16.3, 17.1, 17.7$	391	394
		$1.6, 2.8, 5.2, 7.4, 9.6, 11.5, 13.4, 15.2$	25	2
2^{22}	8	$8.2, 12.6, 15.1, 16.4, 17.5, 18.5, 19.3, 20.1, 20.7$	—	420
		$1.6, 3.7, 6.7, 9.4, 11.8, 14.1, 16.2, 18.1$	—	4
2^{25}	9	$8.4, 13.0, 16.5, 18.0, 19.3, 20.5, 21.5, 22.3, 23.1, 23.8$	—	427
		$1.6, 3.2, 6.4, 9.4, 12.2, 14.7, 17.0, 19.1, 21.1$	—	10

The fourth column gives the result of simulations without the third optimization (detailed in the extended version of this paper), and the last column uses that optimization.

We first prove in Lemma 10 that in a DDL protocol, using different algorithms for A, B cannot give a significant advantage in the error probability. As a result, we can assume that both A and B use A's algorithm, which simplifies the analysis.

Let us assume that we can solve DDL with error probability $\delta \ll 1$ in time T for $M = 1$. Our main reduction shows how to use A's algorithm to solve DLI in an interval of length about $R \approx 1/\delta$ in time less than $4T$ with probability $1/2$. If we assume that in a specific family of groups, DLI in an interval of length $c \cdot T^2$ (for a sufficiently large constant c) cannot be solved in complexity lower than $4T$ with probability $1/2$,[8] we must have $R \approx 1/\delta < c \cdot T^2$ or $\delta = \Omega(T^{-2})$, which gives our main lower bound for the case of $M = 1$. It is important to stress that A is a DDL algorithm for $M = 1$ that is not explicitly given the DLI interval length parameter R. Yet the reduction below will apply A's algorithm to solve DLI with parameter R in a black-box manner.

Recall that a DLI algorithm obtains as input a group element $h = g^x$, where x is in a known interval of length $R \approx 1/\delta$. By the self-reducibility of discrete log, we can assume that h is a uniform group element (i.e., we can multiply the input by a randomly chosen group element). Our reduction picks a point g^z in the interval (for a known z) and runs A on inputs g^x and g^z, where $|x - z| \leq R$. We hope that $A(g^x) - A(g^z) = z - x$ and thus we return $z - (A(g^x) - A(g^z)) = x$.

Clearly, the DLI algorithm runs in time less than $4T$ and it remains to upper bound its error probability by $1/2$. In other words, we need to upper bound

---

[8] We consider only uniform algorithms that can be applied to families of groups (such as elliptic curve groups) and not non-uniform algorithms that are specialized to a specific group $\mathbb{G}$. Indeed, in the non-uniform model, there exist algorithms that solve DLI in an interval of length R in time $o(\sqrt{R})$ for any specific group (see, e.g., [2]).

the probability of $A(g^x) - A(g^z) \neq z - x$ by $1/2$. We know that the DDL error probability is δ for $M = 1$, namely, if $|x - z| \leq 1$, then the required probability is[9] δ. Next, assume that $z = x + 2$. Then, if $A(g^x) - A(g^{x+2}) \neq 2$ this implies that either $A(g^x) - A(g^{x+1}) \neq 1$ or $A(g^{x+1}) - A(g^{x+2}) \neq 1$ (or both). Since the probability of each of these two events is δ, we can use a union bound to upper bound the probability that $A(g^x) - A(g^{x+2}) \neq 2$ by 2δ. Using a similar argument (which we refer to as *distance extension*, formalized in Lemma 14), we can upper bound the probability of the event $A(g^x) - A(g^z) \neq z - x$ for $|x - z| \leq R$ by $O(R \cdot \delta)$ and for $R = O(1/\delta)$, this gives error probability $1/2$, as required. Note that the same algorithm A is used for any distance $|x - z| \leq R$ (which is unknown in advance) and conditioning on this distance is only done for the sake of analysis.

5.2 The Single Algorithm Distributed Discrete Log Problem

We now define the *single algorithm* DDL problem, which is the same problem as general DDL with the restriction that the algorithms of the parties are the same (i.e., both parties use A's algorithm). Denote by $\mathbf{err}(A, x, b, T)$ the event $A(g^x) - A(g^{x+b}) \neq b$ and by $\mathrm{Pr}_{\mathbf{err}}(A, [M_1, M_2], T)$ its probability (over $x \in \mathbb{Z}_N, b \in [M_1, M_2]$). Obviously, the optimal 2-party DDL error probability is a lower bound on the optimal single algorithm DDL error probability. In this section, we prove that the bound in the other direction holds as well up to a constant factor in case $M_2 = -M_1 = M$.[10]

Lemma 10. $\mathrm{Pr}_{\mathbf{err}}(A, B, [-M, M], T) \geq 1/8 \cdot \mathrm{Pr}_{\mathbf{err}}(A, [-M, M], T)$.

Proof. Note that if $A(g^{x+b_1}) - A(g^{x+b_2}) \neq b_2 - b_1$, then $A(g^{x+b_1}) - B(g^x) \neq -b_1$ or $A(g^{x+b_2}) - B(g^x) \neq -b_2$ (or both). Therefore, for uniform $b_1, b_2 \in [-M, M]$,

$$\Pr_{x,b_1,b_2} [\mathbf{err}(A, x + b_1, b_2 - b_1, T)] = \Pr_{x,b_1,b_2} [A(g^{x+b_1}) - A(g^{x+b_2}) \neq b_2 - b_1]$$

$$\leq \Pr_{x,b_1,b_2} [(A(g^{x+b_1}) - B(g^x) \neq -b_1) \cup (A(g^{x+b_2}) - B(g^x) \neq -b_2)]$$

$$\leq \Pr_{x,b_1} [A(g^{x+b_1}) - B(g^x) \neq -b_1] + \Pr_{x,b_2} [A(g^{x+b_2}) - B(g^x) \neq -b_2]$$

$$= 2 \cdot \mathrm{Pr}_{\mathbf{err}}(A, B, [-M, M], T)$$

It remains to relate $\mathrm{Pr}_{\mathbf{err}}(A, [-M, M], T)$ to $\Pr_{x,b_1,b_2} [\mathbf{err}(A, x + b_1, b_2 - b_1, T)]$. Denote the event $|b_2 - b_1| \leq M$ by $\mathcal{E}$ and note that $\Pr_{b_1,b_2} [\mathcal{E}] \geq 1/2$. Conditioned on $\mathcal{E}$, if $b_2 - b_1$ was uniform in $[-M, M]$, then we would have $\Pr_{x,b_1,b_2} [\mathbf{err}(A, x + b_1, b_2 - b_1, T) | \mathcal{E}] = \mathrm{Pr}_{\mathbf{err}}(A, [-M, M], T)$. Although it is not uniform, $b_2 - b_1$ is almost uniform in the sense that for each $i \in [-M, M]$, we have $\Pr_{b_1,b_2} [b_2 - b_1 =$

---

[9] More accurately, it is $O(\delta)$, as δ is the average error probability in the interval $[-1, 1]$.

[10] It is also possible to prove a similar bound in case the interval $[M_1, M_2]$ is not symmetric around the origin.

$i] \geq \Pr_{b_1,b_2} [b_2 - b_1 = M] \geq (M+1)/(4M^2)$ and $\Pr_{b_1,b_2} [b_2 - b_1 = i] \leq \Pr_{b_1,b_2} [b_2 - b_1 = 0] \leq (2M+1)/(4M^2)$. As the minimal and maximal probabilities assigned to $|b_2 - b_1|$ in $[-M, M]$ are within a factor of 2,

$$\Pr_{x,b_1,b_2} [\mathrm{err}(A, x + b_1, b_2 - b_1, T) \,|\, \mathcal{E}] \geq 1/2 \cdot \mathrm{Pr}_{\mathrm{err}}(A, [-M, M], T)$$

and

$$\Pr_{x,b_1,b_2} [\mathrm{err}(A, x + b_1, b_2 - b_1, T)]$$

$$\geq \Pr_{x,b_1,b_2} [\mathrm{err}(A, x + b_1, b_2 - b_1, T) \,|\, \mathcal{E}] \cdot \Pr_{b_1,b_2} [\mathcal{E}] \geq 1/4 \cdot \mathrm{Pr}_{\mathrm{err}}(A, [-M, M], T).$$

Finally,

$$\mathrm{Pr}_{\mathrm{err}}(A, B, [-M, M], T) \geq 1/2 \cdot \Pr_{x,b_1,b_2} [\mathrm{err}(A, x + b_1, b_2 - b_1, T)]$$

$$\geq 1/8 \cdot \mathrm{Pr}_{\mathrm{err}}(A, [-M, M], T),$$

concluding the proof.   ∎

The consequence of the lemma is that for the sake of proving lower bounds on the error probability, we can restrict our attention to A's algorithm by analyzing $\mathrm{Pr}_{\mathrm{err}}(A, [-M, M], T)$. The lemma immediately gives us the same lower bound on $\mathrm{Pr}_{\mathrm{err}}(A, B, [-M, M], T)$, up to a constant factor.

Furthermore, note that by symmetry we have $\mathrm{Pr}_{\mathrm{err}}(A, B, [-M, M], T) \geq 1/8 \cdot \mathrm{Pr}_{\mathrm{err}}(B, [-M, M], T)$, hence the general DDL error probability is lower bounded by the maximal error probability of the (single) algorithms of the two parties (up to constant factors). Therefore, running different algorithms for the two parties cannot give a much better result than simply having both players run the best algorithm in the single algorithm setting.

5.3   Limitation on Randomness

The effect of the internal randomness of a DDL algorithm A on its outcome is quantified by $\mathrm{Pr}_{\mathrm{err}}(A, [0, 0], T)$. This quantity measures the probability that two different executions of A on the same input differ, where the probability is taken over A's input g^x and its internal randomness. We prove that A's internal randomness cannot significantly influence its outcome.

Lemma 11. *Assume* $\mathrm{Pr}_{\mathrm{err}}(A, [-M, M], T) = \delta$. *Then* $\mathrm{Pr}_{\mathrm{err}}(A, [0, 0], T) \leq 2\delta$.

Proof. To be more explicit, we denote by $A(r, g^x)$ the execution of A with a randomness string r. Assume we fix the output of $A(r, g^{x+b})$ for some $b \in [-M, M]$ and randomness string r. Then, if $A(r_1, g^x) \neq A(r_2, g^x)$ for r_1, r_2, either

$A(r_1, g^x) - A(r, g^{x+b}) \neq b$ or $A(r_2, g^x) - A(r, g^{x+b}) \neq b$ (or both). Hence,

$$
\begin{aligned}
&\mathrm{Pr}_{\mathbf{err}}(A, [0, 0], T) \\
&= \Pr_{r_1, r_2, x} [A(r_1, g^x) \neq A(r_2, g^x)] \\
&\leq \Pr_{r_1, r_2, r, x, b} [(A(r_1, g^x) - A(r, g^{x+b}) \neq b) \cup (A(r_2, g^x) - A(r, g^{x+b}) \neq b)] \\
&\leq \Pr_{r_1, r, x, b} [A(r_1, g^x) - A(r, g^{x+b}) \neq b] + \Pr_{r_2, r, x, b} [A(r_2, g^x) - A(r, g^{x+b}) \neq b] \\
&= 2\delta.
\end{aligned}
$$

∎

5.4 Symmetry

We prove the following symmetric property.

Lemma 12. *For $M_2 \geq M_1$, $\mathrm{Pr}_{\mathbf{err}}(A, [M_1, M_2], T) = \mathrm{Pr}_{\mathbf{err}}(A, [-M_2, -M_1], T)$.*

Proof. It is sufficient to prove that for any positive integer b, $\mathrm{Pr}_{\mathbf{err}}(A, [b, b], T) = \mathrm{Pr}_{\mathbf{err}}(A, [-b, -b], T)$. This indeed holds, since $\mathbf{err}(A, x, b, T)$ and $\mathbf{err}(A, x + b, -b, T)$ are identical events, $\mathrm{Pr}_{\mathbf{err}}(A, [b, b], T) = \Pr_x[\mathbf{err}(A, x, b, T)] = \Pr_x[\mathbf{err}(A, x + b, -b, T)] = \mathrm{Pr}_{\mathbf{err}}(A, [-b, -b], T)$. ∎

5.5 Distance Extension

In this section, we show that the distance parameter M of any DDL algorithm A can be extended at the expense of a linear loss in the error probability.

First, we prove the following lemma which reduces the error probability of $\mathrm{Pr}_{\mathbf{err}}(A, [-M, M], T)$ to each one of the indices in the interval. As mentioned in the Introduction (see Footnote 2), it proves that a DDL algorithm with error probability δ for uniform $b \in [-M, M]$ also solves DDL with an error probability $O(\delta)$ for any distribution on $b \in [-M, M]$.

Lemma 13. *Assume that $\mathrm{Pr}_{\mathbf{err}}(A, [-M, M], T) = \delta$. Then, for every $b \in [-M, M]$, $\mathrm{Pr}_{\mathbf{err}}(A, [b, b], T) \leq 4\delta$.*

Proof. We first assume that $b \in [1, M]$ and let $i \in [-M, M - b]$. Clearly, if g^x is a uniform group element, then so is g^{x+i}. Therefore, $\mathrm{Pr}_{\mathbf{err}}(A, [b, b], T) = \Pr_x[\mathbf{err}(A, x + i, b, T)]$. Furthermore, if the event $\mathbf{err}(A, x + i, b, T)$ occurs, then $A(g^{x+i}) - A(g^{x+i+b}) \neq b$ implying that at least one of the events $A(g^x) - A(g^{x+i+b}) \neq i + b$ and $A(g^x) - A(g^{x+i}) \neq i$ must occur. Consequently, $\mathrm{Pr}_{\mathbf{err}}(A, [b, b], T)$ (the error probability associated with index b) is upper bounded by $\mathrm{Pr_{err}}(A, [i+b, i+b], T) + \mathrm{Pr_{err}}(A, [i, i], T)$ (the sum of error probabilities associated with indices i and $i + b$). Formally,

$$\mathrm{Pr_{err}}(A, [b, b], T) = \Pr_x[\mathbf{err}(A, x + i, b, T)]$$
$$\leq \Pr_x[\mathbf{err}(A, x, i + b, T) \cup \mathbf{err}(A, x, i, T)]$$
$$\leq \Pr_x[\mathbf{err}(A, x, i + b, T)] + \Pr_x[\mathbf{err}(A, x, i, T)]$$
$$= \mathrm{Pr_{err}}(A, [i + b, i + b], T) + \mathrm{Pr_{err}}(A, [i, i], T).$$

We map the indices in $[-M, M]$ into disjoint pairs of the form $i, i + b$ (this implies that $i \in [-M, M - b]$). We can obtain at least $\lfloor (2M - b + 1)/2 \rfloor$ such pairs, which is at least $(M + 1)/2$ for $b \in [1, M - 1]$. On the other hand, for $b = M$, the number of pairs is $M \geq (M + 1)/2$. We apply the above inequality to each of the pairs:

$$\delta = \mathrm{Pr_{err}}(A, [-M, M], T)$$
$$= 1/(2M + 1) \cdot \sum_{i=-M}^{M} \mathrm{Pr_{err}}(A, [i, i], T)$$
$$\geq 1/(2M + 1) \cdot (M + 1)/2 \cdot \mathrm{Pr_{err}}(A, [b, b], T)$$
$$\geq \mathrm{Pr_{err}}(A, [b, b], T)/4.$$

Thus, $\mathrm{Pr_{err}}(A, [b, b], T) \leq 4\delta$ for $b \in [1, M]$.

The proof for $b \in [-M, -1]$ follows by symmetry (Lemma 12). Finally, for $b = 0$, we have $\mathrm{Pr_{err}}(A, [0, 0], T) \leq 2\delta$ by Lemma 11. ∎

Lemma 14. *Let* $\mathrm{Pr_{err}}(A, [-M, M], T) = \delta$. *Then for any* $\beta > 1$,

$$\mathrm{Pr_{err}}(A, [-\beta M, \beta M], T) \leq 8\beta \cdot \delta.$$

For the sake of simplicity we assume that βM is an integer (otherwise, we only consider integer values in $[-\beta M, \beta M]$).

Proof. First, we analyze $\mathrm{Pr_{err}}(A, [1, \beta M], T)$. Let $b \in [1, \beta M]$ and divide it by M, writing $b = b_1 \cdot M + b_2$, for integers $b_1 \leq \beta$ and $b_2 \in [0, M)$. We examine the following $b_1 + 1$ success events:

$$\mathcal{E}_1 : \mathbf{suc}(A, x, M, T)$$
$$\mathcal{E}_2 : \mathbf{suc}(A, x + M, M, T)$$
$$\cdots$$
$$\mathcal{E}_{b_1} : \mathbf{suc}(A, x + (b_1 - 1)M, M, T)$$
$$\mathcal{E}_{b_1+1} : \mathbf{suc}(A, x + b_1 M, b_2, T)$$

Observe that if all $b_1 + 1$ events hold (i.e. $\cap_{i=1}^{b_1+1} \mathcal{E}_i$), then $A(g^x) - A(g^{x+b_1 M + b_2}) = A(x) - A(g^{x+b}) = b$, i.e., $\mathbf{suc}(A, x, b, T)$ holds.

By Lemma 13, $\Pr_x[\bar{\mathcal{E}}_i] \le 4\delta$ holds for each $i \in 1, 2, \ldots, b_1$, while $\Pr_{x,b}[\bar{\mathcal{E}}_{b_1+1}] \le 4\delta$ holds as well by the same lemma. Hence,

$$1 - \Pr_{err}(A, [1, \beta M], T) \ge \Pr_{x,b}[\cap_{i=1}^{b_1+1}\mathcal{E}_i] = 1 - \Pr_{x,b}[\cup_{i=1}^{b_1+1}\bar{\mathcal{E}}_i]$$

$$\ge 1 - \sum_{i=1}^{b_1+1} \Pr_{x,b}[\bar{\mathcal{E}}_i] \ge 1 - (b_1 + 1)4\delta \ge 1 - (\beta + 1)4\delta \ge 1 - 8\beta \cdot \delta.$$

Therefore, $\Pr_{err}(A, [1, \beta M], T) \le 8\beta \cdot \delta$. By symmetry (Lemma 12), we have $\Pr_{err}(A, [-\beta M, -1], T) \le 8\beta \cdot \delta$ as well. Since $\Pr_{err}(A, [0, 0], T) \le 2\delta$ by Lemma 11, we conclude that $\Pr_{err}(A, [-\beta M, \beta M], T) \le 8\beta \cdot \delta$, as claimed. ∎

Remark 1. An open question of [5] asked whether the DDL error probability can be eliminated completely. If we apply the above lemma with no error (i.e., $\Pr_{err}(A, [-M, M], T) = \delta = 0$), we obtain $\Pr_{err}(A, [-\beta M, \beta M], T) = 0$, implying that the two parties (running A's algorithm) never err for any distance. This allows the parties to collectively solve the discrete log problem in $\mathbb{G}$ with probability 1 (a similar reduction will be formally presented in Algorithm 4), thus violating the security assumption of the underlying HSS scheme. Namely, the DDL error probability cannot be eliminated (in fact it is easy to show that it must be superpolynomial in $1/N$), answering negatively the open question of Boyle et al.

5.6 Reduction from Discrete Log in an Interval to Distributed Discrete Log

Recall that the discrete log problem in an interval (DLI) is parametrized by an interval length R for a cyclic multiplicative group $\mathbb{G}$ of size N with generator g. The input to the problem is a group element $h = g^x$, where[11] $x \in [0, R-1]$ and the goal is to recover x with high probability (which is at least a constant).

The following lemma reduces the DLI problem to DDL.

Lemma 15. *For a family of groups, assume that $\Pr_{err}(A, [-M, M], T) = \delta$, where $T \ge \log N$ and $\delta < 1/32$. Then discrete log in an interval of length $R = M/(32\delta)$ can be solved in complexity $4T$ with probability $1/2$.*

Proof. Consider Algorithm 4 for solving DLI on input $h = g^x$ for $x \in [0, R-1]$. For the sake of simplicity we assume that R is even.

The algorithm computes $h \cdot g^y$ and $g^{y+(R/2)}$, which can be carried out by performing $2 \log N$ group operations using the square-and-multiply algorithm. It further invokes A twice in complexity $2T$ and therefore its total complexity is $2T + 2 \log N \le 4T$ (since $T \ge \log N$).

---

[11] Alternatively, x could be in any fixed interval of length R. The exact interval is not important as one can easily reduce the problem in a given interval to any other interval.

Algorithm 4. DLI(h)

---

1 **begin**
2    $\quad y \xleftarrow{R} \mathbb{Z}_N$;
3    $\quad d_1 \leftarrow A(h \cdot g^y)$;
4    $\quad d_2 \leftarrow A(g^{y+(R/2)})$;
5    $\quad$ Output $(R/2) - (d_1 - d_2)$;
6 **end**

It remains to upper bound the error probability of the algorithm by $1/2$. The algorithm succeeds to return x if $(R/2)-(d_1-d_2) = x$, namely $d_1-d_2 = (R/2)-x$ or equivalently $A(g^{y+x}) - A(g^{y+(R/2)}) = (R/2) - x$. Since $y \in \mathbb{Z}_N$ is uniform, then g^{y+x} is a uniform group element. Moreover, since $x \in [0, R-1]$, then $(R/2) - x \in [-R/2, R/2]$. Therefore, by Lemma 13, the error probability of the algorithm is at most

$$4 \cdot \mathrm{Pr}_{\mathbf{err}}(A, [-R/2, R/2], T) \leq 4 \cdot 8 \cdot R/(2M) \cdot \mathrm{Pr}_{\mathbf{err}}(A, [-M, M], T)$$
$$= 16 \cdot R/M \cdot \delta = 1/2,$$

where the first inequality is due to Lemma 14. Note that we use Lemma 13 (and pay a factor of 4 in the error probability), as $x \in [0, R-1]$ may be selected by an adversary (whereas $\mathrm{Pr}_{\mathbf{err}}(A, [-R/2, R/2], T)$ averages the error probability). ∎

Theorem 2 is a simple corollary of Lemma 15.

Theorem 2. *For a specific family of groups, assume there exists a constant c such that for any group in the family of size N, DLI in an interval of length at least $c \cdot T^2$ cannot be solved in complexity $4T$ with probability at least $1/2$ (where $\log N \leq T < \mathcal{B}$ for a bound $\mathcal{B}$). Moreover, assume that there is a DDL protocol A for this family with time complexity parameter T, maximal distance parameter M and error probability $\mathrm{Pr}_{\mathbf{err}}(A, [-M, M], T) = \delta$ for $\delta < 1/32$. Then $\delta = \Omega(M \cdot T^{-2})$.*

We note that the bound $\mathcal{B}$ depends on N according to the concrete group family. For example, for some subgroups of $\mathbb{Z}_p^*$, $\mathcal{B}$ is subexponential in $\log N$.

Proof. By Lemma 15, discrete log in an interval of length $R = M/(32\delta)$ can be solved in complexity $4T$ with probability $1/2$. By our assumption, $R = M/(32\delta) < c \cdot T^2$ implying that $\delta = \Omega(M \cdot T^{-2})$ as claimed. ∎

6   Error Probability Lower Bounds for Non-Adaptive Algorithms in the Generic Group Model

In this section, we prove lower bounds on DDL algorithms in the generic group model (GGM), focusing on non-adaptive algorithms. We first review the generic group model (GGM) we consider (which is slightly different than the one proposed by Shoup [18]) and formulate DDL in this model. This formulation is given for the additive group $\mathbb{Z}_N$, which is isomorphic to the multiplicative group $\mathbb{G}$ of size N. We note that the proofs of most of the statements in this section are given in the extended version of this paper.

6.1   Distributed Discrete Log in the Generic Group Model

Let $\mathbb{Z}_N$ be the additive group of integers, and let S be a set of bit strings of cardinality at least N. An encoding function of $\mathbb{Z}_N$ on S is an injective map $\sigma : \mathbb{Z}_N \to S$.

A generic algorithm A for $\mathbb{Z}_N$ on S for the discrete logarithm problem is a probabilistic algorithm that takes as input an *encoding list* of size 2, $\sigma(1), \sigma(x)$, namely, the encodings of a generator of $\mathbb{Z}_N$ and a uniform $x \in \mathbb{Z}_N$, where σ is an encoding function of $\mathbb{Z}_N$ on S. Throughout its execution, A continues to maintain the encoding list, and is allowed to extend it using oracle queries. An oracle query in our model specifies two indices $i, j \in \mathbb{Z}_N$. The oracle computes $\sigma(i \cdot x + j)$ and the returned bit string is appended to the encoding list. A succeeds to solve the discrete log problem if $A(\sigma; 1, x) = x$, and its success probability is taken over the uniform choices of $\sigma : \mathbb{Z}_N \to S$ and $x \in \mathbb{Z}_N$ (and perhaps additional randomness of its own coin tosses). We measure the complexity of A according to the number of oracle queries it makes. The following success probability upper bound was proved in [18].

Theorem 3 ([18]). *If a generic discrete log algorithm A is allowed T oracle queries, then $\Pr_{\sigma,x}[A(\sigma; 1, x) = x] = O(T^2/N)$, assuming that N is prime.*

We note that our GGM formulation is slightly stronger than the one of [18], where the queries of A are limited to linear combinations with coefficients of ± 1 of elements in its encoding list. Since any query (i, j) can be issued in Shoup's original GGM after at most $O(\log N)$ queries using the double-and-add algorithm, a stronger GGM algorithm can be simulated by a standard one by increasing the query complexity by a multiplicative factor of $\log N$. However, by following its original proof in [18], it is easy to verify that Theorem 3 actually holds with no modification in our stronger GGM. Obviously, any algorithm in the original GGM is also an algorithm in the stronger GGM. Therefore, any lower bounds we obtain in the stronger GGM also apply in the original GGM.

We now describe the basic game of distributed discrete log in GGM. Obviously, all the results of Sect. 5 also hold in the generic group model. In particular, by Lemma 10 it is sufficient to consider single algorithm DDL to obtain general DDL lower bounds.

A party (algorithm) A is given as input $\sigma(1)$ and the encoding of an additional group element $\sigma(x)$ for $x \in \mathbb{Z}_N$, selected uniformly at random. Algorithm A is allowed to make T oracle queries. After obtaining the answers from the oracle, A returns an integer value. Two parties (both running A's algorithm) win the

DDL game in GGM if

$$A(\sigma; 1, x) - A(\sigma; 1, x + b) = b,$$

otherwise, they lose the game, or err.

We are interested in proving lower bounds on the DDL error probability as a function of T, namely

$$\Pr_{\sigma, x, b}[A(\sigma; 1, x) - A(\sigma; 1, x + b) \neq b].$$

Analogously to our notation for multiplicative groups, we denote by $\mathbf{err}(A, \sigma, x, b, T)$ the error event $A(\sigma; 1, x) - A(\sigma; 1, x + b) \neq b$, and by $\Pr_{\mathbf{err}}(A, \sigma, [M_1, M_2], T)$ its probability $\Pr_{\sigma, x, b}[\mathbf{err}(A, \sigma, x, b, T)]$, where $b \in [M_1, M_2]$ is a uniform integer. We further denote by $\mathbf{suc}(A, \sigma, x, b, T)$ the complementary success event.

6.2   An Error Probability Lower Bound for Arbitrary Generic Algorithms

The following theorem gives a DDL error probability lower bound in GGM. The theorem is a somewhat weaker statement than Theorem 2 (which has implications in concrete group families).

Theorem 4. *For any generic DDL algorithm* A, $\Pr_{\mathbf{err}}(A, \sigma, [-M, M], T) = \Omega(M \cdot T^{-2})$, *given that* $M = O(T^2)$, $T = o(\sqrt{N})$, *and* N *is prime.*

We omit the proof, as it is similar to the one of Theorem 2. It applies a reduction to discrete log, while using Theorem 3 to obtain the error probability lower bound. Alternatively, one could obtain a hardness result for DLI in GGM (extending Theorem 3 to smaller intervals) and apply Theorem 2 directly.

6.3   An Error Probability Lower Bound for Non-Adaptive Generic Algorithms

In this section, we prove a lower bound on the DDL error probability of non-adaptive generic algorithms, whose oracle queries $\{(i_1, j_1), (i_2, j_2), \ldots, (i_T, j_T)\}$ are fixed in advance and do not depend on previous answers.

We will prove the following lower bound:

Theorem 5. *Any non-adaptive DDL algorithm* A *satisfies* $\Pr_{\mathbf{err}} A, \sigma, [-1, 1], T = \Omega(1/T)$, *given that* $T = o(N^{1/2})$, *and* N *is prime.*

Overview of the Lower Bound Proof on Non-Adaptive Algorithms in the Generic Group Model. Let us first consider the class of algorithms that make T consecutive oracle queries to group elements (such as Algorithm 1 and the ones of [4–6] in general). Consider the executions $A(\sigma; 1, x)$ and $A(\sigma; 1, x + T)$, which query $2T$ disjoint group elements. In GGM, algorithm executions that query disjoint elements are essentially independent (as each group element is associated with a random string), which implies that the probability that $A(\sigma; 1, x) - A(\sigma; 1, x + T) \neq T$ is at least $1/2$. Recall that we are interested in the probability that $A(\sigma; 1, x) - A(\sigma; 1, x + 1) \neq 1$ and it can be lower bounded by $\Omega(T^{-1})$ using distance extension (Lemma 14). A similar lower bound applies if A only queries group elements in a short interval of length $O(T)$.[12]

Of course, we are interested in proving the $\Omega(T^{-1})$ lower bound for arbitrary non-adaptive algorithms. The main idea that allows us to achieve this is to define a transformation that takes an arbitrary non-adaptive algorithm A' and maps its T queries to a small interval of size $O(T)$, obtaining a new algorithm A (for which the error lower bound $\Omega(T^{-1})$ holds). We require that the query mapping preserves the error probability of A', thus proving that the error probability lower bound $\Omega(T^{-1})$ above also applies to non-adaptive algorithms in general. In order to preserve the error probability of A', the mapping will ensure that the joint input distribution of $A'(\sigma; 1, x)$ and $A'(\sigma; 1, x + 1)$ is equal to that of $A(\sigma; 1, x)$ and $A(\sigma; 1, x + 1)$. In the generic group model, this means that the mapping should preserve joint queries, namely, satisfy the condition that query i of $A'(\sigma; 1, x)$ and query j of $A'(\sigma; 1, x + 1)$ evaluate the same group element if and only if query i of $A(\sigma; 1, x)$ and query j of $A(\sigma; 1, x + 1)$ evaluate the same group element.[13] Based on this observation, it is possible to define an appropriate query mapping and complete the proof, since for non-adaptive algorithms we know in advance (independently of σ) if query i of $A'(\sigma; 1, x)$ and and query j of $A'(\sigma; 1, x + 1)$ evaluate the same group element.

Additional Notation. We begin by defining additional notation. Given a query (i, j), denote its evaluation on x as $(i, j)[x] = ix + j$. Thus, its oracle answer is $\sigma(ix + j)$. We denote by $Q(A(\sigma; 1, x))$ the query set of $A(\sigma; 1, x)$, excluding queries (i, j) for which $i = 0$ (which we call *constant queries*). Denote by $QE(A(\sigma; 1, x))$ the set of evaluations of all (non-constant) queries $Q(A(\sigma; 1, x))$.

We further denote by $Q(A)$ the set containing all of the potential (non-constant) queries of A on any input x and encoding σ. Note that for non-adaptive algorithms, $|Q(A)| \leq T$ and any adaptive algorithm A' can be simulated by a non-adaptive algorithm that makes $T' \stackrel{\text{def}}{=} |Q(A')|$ queries.

For the rest of this section, we focus on non-adaptive algorithms. For such algorithms, we can write $QE(A, x)$ (instead of $QE(A(\sigma; 1, x)))$, as the query evaluations are independent of σ.

Restricted Queries. We examine pairs of executions $A(\sigma; 1, x)$ and $A(\sigma; 1, x + b)$ for some $b \in [-M, M]$. For such a pair, we define a (non-trivial) collision as the event that two queries issued by these executions (i, j) and (i', j') with $i \neq i'$

---

[12] Our actual proof is slightly more general than outlined here and uses the notion of query chains.

[13] Our proof relaxes this strong condition, and requires that it holds unless a low-probability event (called a collision) occurs.

have the same evaluation. The actual evaluations depend on which algorithm issued the queries and there are 4 cases, e.g., $ix + j \bmod N = i'x + j' \bmod N$ if $A(\sigma; 1, x)$ issued both and $ix+j \bmod N = i'(x+b)+j' \bmod N$ if $A(\sigma; 1, x)$ issued (i, j) and $A(\sigma; 1, x+b)$ issued (i', j'), etc. In each of these 4 cases, both algorithms can exploit the collision to jointly solve the discrete logarithm problem using at most $2T$ queries (e.g., in the first case above, $x = (j' - j) \cdot (i - i')^{-1} \bmod N$). According to Theorem 3, the probability of this event is $O(T^2/N) = o(1)$ (by our assumption $T = o(N^{1/2})$), which is negligible. In the following we generally denote collision events by COL.

Most of the analysis below will be conditioned on the event $\overline{COL}$ (whose probability is $1 - o(1)$), but we will omit this explicit conditioning for simplicity, while ignoring a negligible factor in the probability calculation.

Lemma 16. *Assume that $T = o(N^{1/2})$. Then, any non-adaptive algorithm A' with $\mathrm{Pr_{err}}(A', \sigma, [-M, M], T) = \delta$ can be transformed into a non-adaptive query-restricted algorithm A with $\mathrm{Pr_{err}}(A, \sigma, [-M, M], T) \leq \delta \cdot (1 + o(1))$ such that A only issues restricted queries of the form (i, j) with $i \in \{0, 1\}$.*

Query Disjoint Indices. We say that a non-adaptive DDL algorithm A has a query disjoint index b if $QE(A, x) \cap QE(A, x + b) = \emptyset$ for any $x \in \mathbb{Z}_N$. We note that A can have many query disjoint indices. We prove the following error probability lower bound on algorithms with a (small) query disjoint index.

Lemma 17. *Any non-adaptive algorithm which is query disjoint on index $b \geq 1$ satisfies $\mathrm{Pr_{err}}(A, \sigma, [-1, 1], T) = \Omega(1/b)$.*

Query Chains. Given a query $(1, j)$, we refer to the value j as a query offset. For a non-adaptive algorithm A, we define a query chain of length c as a sequence of $c + 1$ query offsets $j, j + 1, j + 2, \ldots, j + c$ such that for each $k \in \{0, 1, \ldots, c\}$, $(1, j + k) \in Q(A)$, while $(1, j + c + 1) \notin Q(A)$ and $(1, j - 1) \notin Q(A)$ (i.e., the sequence is maximal).

Denote the length of the longest query chain of A by $C(A)$.

Lemma 18. *Any non-adaptive query-restricted algorithm A satisfies*

$$\mathrm{Pr_{err}}(A, \sigma, [-1, 1], T) \geq \Omega(1/C(A)).$$

Proof (of Theorem 5). The theorem is a simple corollary of Lemmas 16 and 18. Given a non-adaptive algorithm A, transform it into a query-restricted algorithm A' using Lemma 16, with a multiplicative loss of $1 + o(1)$ in error probability. Clearly, $C(A') \leq T$, hence by Lemma 18 we have $\mathrm{Pr_{err}}(A, \sigma, [-1, 1], T) \geq \mathrm{Pr_{err}}(A', \sigma, [-1, 1], T) \cdot 1/(1 + o(1)) = \Omega(1/C(A')) = \Omega(1/T)$, concluding the proof. ∎

A Generalization of Theorem 5. The theorem above does not completely render non-adaptive algorithms as inefficient since (for example) it does not rule out the possibility that $\Pr_{\mathbf{err}}(A, \sigma, [-T, T], T) = O(1/T)$ (which is optimal according to Theorem 4). However, the following theorem states the this is impossible and non-adaptive algorithms have a linear query-error tradeoff at best.

Theorem 6. *For all* $1 \leq M \leq T$, *any non-adaptive generic DDL algorithm* A *satisfies* $\Pr_{\mathbf{err}}(A, \sigma, [-M, M], T) = \Omega(M/T)$ *given that* $T = o(N^{1/2})$ *and* N *is prime. In particular, for* $M = T$, $\Pr_{\mathbf{err}}(A, \sigma, [-T, T], T) = \Omega(1)$.

Note that Theorem 5 is a special case of the one above, obtained for $M = 1$. The proof of Theorem 6 uses Fourier analysis and is given in the extended version of this paper.

Acknowledgements. The authors would like to thanks Elette Boyle, Niv Gilboa, Yuval Ishai and Yehuda Lindell for discussions and helpful suggestions regarding this work.

This research was supported by the European Research Council under the ERC starting grant agreement no. 757731 (LightCrypt) and by the BIU Center for Research in Applied Cryptography and Cyber Security in conjunction with the Israel National Cyber Bureau in the Prime Minister's Office.

The first author was additionally supported by the Israeli Science Foundation through grant No. 573/16.

References

1. Ben-Or, M., Goldwasser, S., Wigderson, A.: Completeness theorems for non-cryptographic fault-tolerant distributed computation (extended abstract). In: Simon [19], pp. 1–10 (1988)
2. Bernstein, D.J., Lange, T.: Computing small discrete logarithms faster. In: Galbraith, S., Nandi, M. (eds.) INDOCRYPT 2012. LNCS, vol. 7668, pp. 317–338. Springer, Heidelberg (2012). https://doi.org/10.1007/978-3-642-34931-7_19
3. Boneh, D., Goh, E.-J., Nissim, K.: Evaluating 2-DNF formulas on ciphertexts. In: Kilian, J. (ed.) TCC 2005. LNCS, vol. 3378, pp. 325–341. Springer, Heidelberg (2005). https://doi.org/10.1007/978-3-540-30576-7_18
4. Boyle, E., Couteau, G., Gilboa, N., Ishai, Y., Orrù, M.: Homomorphic secret sharing: optimizations and applications. In: Thuraisingham, B.M., Evans, D., Malkin, T., Xu, D. (eds.) Proceedings of the 2017 ACM SIGSAC Conference on Computer and Communications Security, CCS 2017, Dallas, TX, USA, 30 October–03 November 2017, pp. 2105–2122. ACM (2017)
5. Boyle, E., Gilboa, N., Ishai, Y.: Breaking the circuit size barrier for secure computation under DDH. In: Robshaw, M., Katz, J. (eds.) CRYPTO 2016, Part I. LNCS, vol. 9814, pp. 509–539. Springer, Heidelberg (2016). https://doi.org/10.1007/978-3-662-53018-4_19
6. Boyle, E., Gilboa, N., Ishai, Y.: Group-based secure computation: optimizing rounds, communication, and computation. In: Coron, J.-S., Nielsen, J.B. (eds.) EUROCRYPT 2017, Part II. LNCS, vol. 10211, pp. 163–193. Springer, Cham (2017). https://doi.org/10.1007/978-3-319-56614-6_6
7. Boyle, E., Gilboa, N., Ishai, Y., Lin, H., Tessaro, S.: Foundations of homomorphic secret sharing. In: Karlin, A.R. (ed.) 9th Innovations in Theoretical Computer Science Conference, ITCS 2018, Cambridge, MA, USA, 11–14 January 2018. LIPIcs, vol. 94, pp. 21:1–21:21. Schloss Dagstuhl - Leibniz-Zentrum fuer Informatik (2018)

8. Chaum, D., Crépeau, C., Damgård, I.: Multiparty unconditionally secure protocols (extended abstract). In: Simon [19], pp. 11–19 (1988)
9. Chor, B., Kushilevitz, E., Goldreich, O., Sudan, M.: Private information retrieval. J. ACM **45**(6), 965–981 (1998)
10. Fazio, N., Gennaro, R., Jafarikhah, T., Skeith III, W.E.: Homomorphic secret sharing from Paillier encryption. In: Okamoto, T., Yu, Y., Au, M.H., Li, Y. (eds.) ProvSec 2017. LNCS, vol. 10592, pp. 381–399. Springer, Cham (2017). https://doi.org/10.1007/978-3-319-68637-0_23
11. Galbraith, S.D., Pollard, J.M., Ruprai, R.S.: Computing discrete logarithms in an interval. Math. Comput. **82**(282), 1181–1195 (2013)
12. Gentry, C.: Fully homomorphic encryption using ideal lattices. In: Mitzenmacher, M. (ed.) Proceedings of the 41st Annual ACM Symposium on Theory of Computing, STOC 2009, Bethesda, MD, USA, 31 May–2 June 2009, pp. 169–178. ACM (2009)
13. Gordon, D.M.: Discrete logarithms in GF(P) using the number field sieve. SIAM J. Discret. Math. **6**(1), 124–138 (1993)
14. Lenstra, A.K., Lenstra, H.W. (eds.): The Development of the Number Field Sieve. LNM, vol. 1554. Springer, Heidelberg (1993). https://doi.org/10.1007/BFb0091534
15. Pollard, J.M.: Monte Carlo methods for index computation (mod p). Math. Comput. **32**(143), 918–924 (1978)
16. Pollard, J.M.: Kangaroos, monopoly and discrete logarithms. J. Cryptol. **13**(4), 437–447 (2000)
17. Rivest, R.L., Adleman, L., Dertouzos, M.L.: On data banks and privacy homomorphisms. In: Foundations of Secure Computation, pp. 169–179 (1978)
18. Shoup, V.: Lower bounds for discrete logarithms and related problems. In: Fumy, W. (ed.) EUROCRYPT 1997. LNCS, vol. 1233, pp. 256–266. Springer, Heidelberg (1997). https://doi.org/10.1007/3-540-69053-0_18
19. Simon, J. (ed.): Proceedings of the 20th Annual ACM Symposium on Theory of Computing, Chicago, Illinois, USA, 2–4 May 1988. ACM (1988)
20. Yao, A.C.: Protocols for secure computations (extended abstract). In: 23rd Annual Symposium on Foundations of Computer Science, Chicago, Illinois, USA, 3–5 November 1982, pp. 160–164. IEEE Computer Society (1982)

Must the Communication Graph of MPC Protocols be an Expander?

Elette Boyle[1], Ran Cohen[2,3(✉)], Deepesh Data[4], and Pavel Hubáček[5]

[1] IDC Herzliya, Herzliya, Israel
elette.boyle@idc.ac.il
[2] MIT, Cambridge, USA
rancohen@mit.edu
[3] Northeastern University, Boston, USA
[4] UCLA, Los Angeles, USA
deepeshdata@ucla.edu
[5] Computer Science Institute, Charles University, Prague, Czech Republic
hubacek@iuuk.mff.cuni.cz

Abstract. Secure multiparty computation (MPC) on incomplete communication networks has been studied within two primary models: (1) Where a partial network is fixed a priori, and thus corruptions can occur dependent on its structure, and (2) Where edges in the communication graph are determined dynamically as part of the protocol. Whereas a rich literature has succeeded in mapping out the feasibility and limitations of graph structures supporting secure computation in the fixed-graph model (including strong classical lower bounds), these bounds do not apply in the latter dynamic-graph setting, which has recently seen exciting new results, but remains relatively unexplored.

In this work, we initiate a similar foundational study of MPC within the dynamic-graph model. As a first step, we investigate the property of graph *expansion*. All existing protocols (implicitly or explicitly) yield communication graphs which are expanders, but it is not clear whether this is inherent. Our results consist of two types:

- Upper bounds: We demonstrate secure protocols whose induced communication graphs are *not* expanders, within a wide range of settings (computational, information theoretic, with low locality, and adaptive security), each assuming some form of input-independent setup.
- Lower bounds: In the setting without setup and adaptive corruptions, we demonstrate that for certain functionalities, *no* protocol can maintain a non-expanding communication graph against all

E. Boyle—Supported in part by ISF grant 1861/16, AFOSR Award FA9550-17-1-0069, and ERC Grant no. 307952.

R. Cohen—Supported in part by Alfred P. Sloan Foundation Award 996698, ISF grant 1861/16, ERC starting grant 638121, NEU Cybersecurity and Privacy Institute, and NSF TWC-1664445.

R. Cohen, D. Data and P. Hubáček—This work was done in part while visiting at the FACT Center at IDC Herzliya.

P. Hubáček—Supported by the project 17-09142S of GA ČR, Charles University project UNCE/SCI/004, and Charles University project PRIMUS/17/SCI/9. This work was done under financial support of the Neuron Fund for the support of science.

© International Association for Cryptologic Research 2018
H. Shacham and A. Boldyreva (Eds.): CRYPTO 2018, LNCS 10993, pp. 243–272, 2018.
https://doi.org/10.1007/978-3-319-96878-0_9

adversarial strategies. Our lower bound relies only on protocol correctness (not privacy), and requires a surprisingly delicate argument.

1   Introduction

The field of secure multiparty computation (MPC), and more broadly fault-tolerant distributed computation, constitutes a deep and rich literature, yielding a vast assortment of protocols providing strong robustness and even seemingly paradoxical privacy guarantees. A central setting is that of n parties who jointly compute a function of their inputs while maintaining correctness (and possibly privacy) facing adversarial behavior from a constant fraction of corruptions.

Since the original seminal results on secure multiparty computation [2,11, 25,36], the vast majority of MPC solutions to date assume that every party can (and will) communicate with every other party. That is, the underlying point-to-point communication network forms a complete graph. Indeed, many MPC protocols begin directly with every party secret sharing his input across all other parties (or simply sending his input, in the case of tasks without privacy such as Byzantine agreement [17,34,35]).

There are two classes of exceptions to this rule, which consider MPC on incomplete communication graphs.

Fixed-Graph Model. The first corresponds to an area of work investigating achievable security guarantees in the setting of a *fixed* partial communication network. In this model, communication is allowed only along edges of a fixed graph, known a priori, and hence where corruptions can take place as a function of its structure. This setting is commonly analyzed within the distributed computing community. In addition to positive results, this is the setting of many fundamental lower bounds: For example, to achieve secure Byzantine agreement against t corruptions, the graph must be $(t+1)$-connected [16,21].[1] For graphs with lower connectivity, the best one can hope for is a form of "almost-everywhere agreement," where some honest parties are not guaranteed to output correctly, as well as restricted notions of privacy [10,19,23,26,27]. Note that because of this, one cannot hope to achieve protocols with standard security in this model with $o(n^2)$ communication, even for simple functionalities such as Byzantine agreement.

Dynamic-Graph Model. The second, more recent approach addresses a model where all parties have the *ability* to initiate communication with one another, but make use of only a subset of these edges as determined dynamically during the protocol. We refer to this as the "dynamic-graph model." When allowing for negligible error (in the number of parties), the above lower bounds do not apply, opening the door for dramatically different approaches and improvements in complexity. Indeed, distributed protocols have been shown for Byzantine agreement in this model with as low as $\tilde{O}(n)$ bits of communication [6,33], and secure MPC protocols have been constructed whose communication graphs have degree

---

[1] If no setup assumptions are assumed, the connectivity bound increases to $2t + 1$.

$o(n)$—and as low as polylog(n) [3,5,9,15].[2] However, unlike the deep history of the model above, the current status is a sprinkling of positive results. Little is known about what types of communication graphs must be generated from a secure MPC protocol execution.

Gaining a better understanding of this regime is motivated not only to address fundamental questions, but also to provide guiding principles for future protocol design. In this work, we take a foundational look at the dynamic-graph model, asking:

*What properties of induced communication graphs
are necessary to support secure computation?*

On the necessity of graph expansion. Classical results tell us that the fully connected graph suffices for secure computation. Protocols achieving low locality indicate that a variety of significantly sparser graphs, with many low-weight cuts, can also be used [3,5,9,15]. We thus consider a natural extension of connectivity to the setting of low degree. Although the positive results in this setting take different approaches and result in different communication graph structures, we observe that in each case, the resulting sparse graph has high *expansion*.

Roughly, a graph is an expander if every subset of its nodes that is not "too large" has a "large" boundary. Expander graphs have good mixing properties and in a sense "mimic" a fully connected graph. There are various ways of formalizing expansion; in this work we consider a version of *edge* expansion, pertaining to the number of outgoing edges from any subset of nodes. We consider a variant of the expansion definition which is naturally monotonic: that is, expansion cannot decrease when extra edges are added (note that such monotonicity also holds for the capacity of the graph to support secure computation).

Indeed, expander graphs appear explicitly in some works [9,33], and implicitly in others (e.g., using random graphs [31], pseudorandom graphs [5], and averaging samplers [6], to convert from almost-everywhere to everywhere agreement). High connectivity and good mixing intuitively go hand-in-hand with robustness against corruptions, where adversarial entities may attempt to impede or misdirect information flow.

This raises the natural question: Is this merely an artifact of a convenient construction, or is high expansion *inherent*? That is, we investigate the question: Must the communication graph of a generic MPC protocol, tolerating a linear number of corruptions, be an expander graph?

1.1 Our Results

More explicitly, we consider the setting of secure multiparty computation with n parties in the face of a linear number of active corruptions. As common in the honest-majority setting, we consider protocols that guarantee output delivery. Communication is modeled via the dynamic-graph setting, where all parties have

---

[2] This metric is also called the communication *locality* of the protocol [5].

the ability to initiate communication with one another, and use a subset of edges as dictated by the protocol. We focus on the synchronous setting.

Our contributions are of the following three kinds:

Formal definitional framework. As a first contribution, we provide a formal framework for analyzing and studying the evolving communication graph of MPC protocols. The framework abstracts and refines previous approaches concerning specific properties of protocols implicitly related to the graph structure, such as the degree [5]. This gives a starting point for studying the relation between secure computation and further, more general, graph properties.

Upper bounds. We present secure protocols whose induced communication graphs are decidedly *not* expander graphs, within a range of settings. This includes with computational security, with information-theoretic security, with low locality, even with low locality *and* adaptive security (in a hidden-channels model [9]) — but all with the common assumption of some form of input-independent *setup* information. The resulting communication graph has a low-weight cut, splitting the n parties into two equal (linear) size sets with only poly-logarithmic edges connecting them.

Theorem 1 (MPC with non-expanding communication graph, informal). *For any efficient functionality f and any constant $\epsilon > 0$, there exists a protocol in the PKI model, assuming digital signatures, securely realizing f against $(1/4 - \epsilon) \cdot n$ static corruptions, such that with overwhelming probability the induced communication graph is non-expanding.*

Theorem 1 is stated in the computational setting with static corruptions; however, this approach extends to various other settings, albeit at the expense of a lower corruption threshold. (See Sect. 4 for more details.)

Theorem 2 (extensions of Theorem 1, informal). *For any efficient functionality f, there exists a protocol securely realizing f, in the settings listed below, against a linear number of corruptions, such that with overwhelming probability the induced communication graph is non-expanding:*

- *In the setting of Theorem 1 with poly-logarithmic locality.*
- *Unconditionally, in the information-theoretic PKI model (with or without low locality).*
- *Unconditionally, in the information-theoretic PKI model, facing adaptive adversaries.*
- *Under standard cryptographic assumptions, in the PKI model, facing adaptive adversaries, with poly-logarithmic locality.*

As an interesting special case, since our protocols are over point-to-point channels and do not require a broadcast channel, these results yield the first Byzantine agreement protocols whose underlying communication graphs are not expanders.

The results in Theorems 1 and 2 all follow from a central transformation converting existing secure protocols into ones with low expansion. At a high level, the first $n/2$ parties will run a secure computation to elect two representative committees of poly-logarithmic size: one amongst themselves and the other from the other $n/2$ parties. These committees will form a "communication bridge" across the two halves (see Fig. 1). The setup is used to certify the identities of the members of both committees to the receiving parties, either via public-key infrastructure for digital signatures (in the computational setting) or correlated randomness for information-theoretic signatures [37,38] (in the information-theoretic setting).

Interestingly, this committee-based approach can be extended to the adaptive setting (with setup), in the hidden-channels model considered by [9], where the adversary is not aware which communication channels are utilized between honest parties.[3] Here, care must be taken to not reveal more information than necessary about the identities of committee members to protect them from being corrupted.

As a side contribution, we prove the first instantiation of a protocol with polylogarithmic locality and information-theoretic security (with setup), by adjusting the protocol from [5] to the information-theoretic setting.

Theorem 3 (polylog-locality MPC with information-theoretic security, informal). *For any efficient functionality f and any constant $\epsilon > 0$, there exists a protocol with poly-logarithmic locality in the information-theoretic PKI model, securely realizing f against computationally unbounded adversaries statically corrupting $(1/6 - \epsilon) \cdot n$ parties.*

Lower bounds. On the other hand, we show that in some settings a weak form of expansion *is* a necessity. In fact, we prove a stronger statement, that in these settings the graph must have high connectivity.[4] Our lower bound is in the setting of adaptive corruptions, computational (or information-theoretic) security, and *without* setup assumptions. Our proof relies only on correctness of the protocol and not on any privacy guarantees; namely, we consider the *parallel broadcast* functionality (aka *interactive consistency* [35]), where every party distributes its input to all other parties. We construct an adversarial strategy in this setting such that no protocol can guarantee correctness against this adversary if its induced communication graph at the end of the protocol has any cut with sublinear many crossing edges (referred to as a "sublinear cut" from now on).

Theorem 4 (high connectivity is necessary for correct protocols, informal). *Let $t \in \Theta(n)$. Any $t(n)$-resilient protocol for parallel broadcast in the computational setting, tolerating an adaptive, malicious adversary cannot maintain an induced communication graph with a sublinear cut.*

Theorem 4 in particular implies that the resulting communication graph must have a form of expansion. We note that in a weaker communication model, a

---

[3] Sublinear locality is impossible in the adaptive setting if the adversary is aware of honest-to-honest communication, since it can simply isolate an honest party from the rest of the protocol.

[4] More concretely, the graph should be at least $\alpha(n)$-connected for every $\alpha(n) \in o(n)$.

weaker form of consensus, namely Byzantine agreement, can be computed in a way that the underlying graph (while still an expander) has low-weight cuts [32].

It is indeed quite intuitive that if a sublinear cut exists in the communication graph of the protocol, and the adversary can adaptively corrupt a linear number of parties $t(n)$, then he could corrupt the parties on the cut and block information flow. The challenge, however, stems from the fact that the cut is not known a priori but is only revealed over time, and by the point at which the cut is identifiable, all necessary information may have already been transmitted across the cut. In fact, even the identity of the cut and visible properties of the communication graph itself can convey information to honest parties about input values without actual bits being communicated. This results in a surprisingly intricate final attack, involving multiple indistinguishable adversaries, careful corruption strategies, and precise analysis of information flow. See below for more detail.

1.2   Our Techniques

We focus on the technical aspects of the lower bound result.

Overview of the attack. Consider an execution of the parallel broadcast protocol over *random* inputs. At a high level, our adversarial strategy, denoted $\mathcal{A}_n^{\text{honest-}i^*}$, will select a party P_{i^*} at random and attempt to block its input from being conveyed to honest parties. We are only guaranteed that somewhere in the graph will remain a sublinear cut. Because the identity of the eventual cut is unknown, it cannot be attacked directly. We take the following approach:

1. **Phase I.** Rather, our attack will first "buy time" by corrupting the neighbors of P_{i^*}, and blocking information flow of its input x_{i^*} to the remaining parties. Note that this can only continue up to a certain point, since the degree of P_{i^*} will eventually surpass the corruption threshold (as we prove). But, the benefit of this delay is that in the meantime, the communication graph starts to fill in, which provides more information about the locations of the potential cuts.

   For this to be the case, it must be that the parties cannot identify that P_{i^*} is under attack (otherwise, the protocol may instruct many parties to quickly communicate to/from P_{i^*}, forcing the adversary to run out of his "corruption budget" before the remaining graph fills in). The adversary thus needs to fool all honest parties and make each honest party believe that he participates in an honest execution of the protocol. This is done by maintaining two simulated executions: one pretending to be P_{i^*} running on a random input, and another pretending (to P_{i^*}) to be all other parties running on random inputs. Note that for this attack strategy to work it is essential that the parties do not have pre-computed correlated randomness such as PKI.

2. **Phase II.** We show that with noticeable probability, by the time we run out of the Phase I corruption threshold (which is a linear number of parties), *all parties* in the protocol have high (linear) degree. In turn, we prove that the current communication graph can have at most a constant number of sublinear cuts.

In the remainder of the execution, the adversary will simultaneously attack all of these cuts. Namely, he will block information flow from P_{i^*} across any of these cuts by corrupting the appropriate "bridge" party, giving up on each cut one by one when a certain threshold of edges have already crossed it.

If the protocol is guaranteed to maintain a sublinear cut, then necessarily there will remain at least one cut for which all Phase II communication across the cut has been blocked by the adversary. Morally, parties on the side of this cut opposite P_{i^*} should not have learned x_{i^*}, and thus the *correctness* of the protocol should be violated. Proving this, on the other hand, requires surmounting two notable challenges.

1. We must prove that there still remains an uncorrupted party P_{j^*} on the opposite side of the cut. It is not hard to show that each side of the cut is of linear size, that P_{i^*} has a sublinear number of neighbors across the cut (all of which are corrupted), and that a sublinear number of parties get corrupted in Phase II. Hence, there exists parties across the cut that are not neighbors of P_{i^*} and that are not corrupted in Phase II. However, by the attack strategy, all of the neighbors of the *virtual* P_{i^*} are corrupted in Phase I as well, and this is also a linear size set, which is independent of the real neighbors of P_{i^*}. Therefore, it is not clear that there will actually remain honest parties across the cut by the end of the protocol execution.
2. More importantly, even though we are guaranteed that no bits of communication have been passed along any path from P_{i^*} to P_{j^*}, this does not imply that no *information* about x_{i^*} has been conveyed. For example, since the graph develops as a function of parties' inputs, it might be the case that this situation of P_{j^*} being blocked from P_{i^*}, only occurs when x_{i^*} equals a certain value.

We now discuss how these two challenges are addressed.

Guaranteeing honest parties across the cut. Unexpectedly, we cannot guarantee existence of honest parties across the cut. Instead, we introduce a different adversarial strategy, which we prove *must* have honest parties blocked across a cut from P_{i^*}, and for which there exist honest parties who cannot distinguish which of the two attacks is taking place. More explicitly, we consider the "dual" version of the original attack, denoted $\mathcal{A}_n^{\mathsf{corrupt}\text{-}i^*}$, where party P_{i^*} is *corrupted* and instead pretends to be under attack as per $\mathcal{A}_n^{\mathsf{honest}\text{-}i^*}$ above.

Blocking honest parties from x_{i^*} in $\mathcal{A}_n^{\mathsf{corrupt}\text{-}i^*}$ does not contradict correctness explicitly on its own, as P_{i^*} is corrupted in this case. It is the combination of both of these attacks that will enable us to contradict correctness. Namely, we prove that:

- Under the attack $\mathcal{A}_n^{\mathsf{corrupt}\text{-}i^*}$, there exists a "blocked cut" $(S, \bar{S})$ with uncorrupted parties on both sides. By *agreement*, all uncorrupted parties output the same value y_{i^*} as the i^*'th coordinate of the output vector.
- The view of some of the uncorrupted parties under the attack $\mathcal{A}_n^{\mathsf{corrupt}\text{-}i^*}$ is identically distributed as that of uncorrupted parties in the original attack

$\mathcal{A}_n^{\text{honest-}i^*}$. Thus, their output distribution must be the same across the two attacks.

- Since under the attack $\mathcal{A}_n^{\text{honest-}i^*}$, the party P_{i^*} is honest, by *completeness*, all uncorrupted parties in $\mathcal{A}_n^{\text{honest-}i^*}$ must output the *correct* value $y_{i^*} = x_{i^*}$.
- Thus, uncorrupted parties in $\mathcal{A}_n^{\text{corrupt-}i^*}$ (who have the same view) must output the correct value x_{i^*} as well.

Altogether, this implies all honest parties in interaction with $\mathcal{A}_n^{\text{corrupt-}i^*}$, in particular P_{j^*} who is blocked across the cut from P_{i^*}, must output $y_{i^*} = x_{i^*}$.

Bounding information transmission about x_{i^*}. The final step is to show that this cannot be the case, since an uncorrupted party P_{j^*} across the cut in $\mathcal{A}_n^{\text{corrupt-}i^*}$ does not receive enough information about x_{i^*} to fully specify the input. This demands delicate treatment of the specific attack strategy and analysis, as many "side channel" signals within the protocol can leak information on x_{i^*}. Corruption patterns in Phase II, and their timing, can convey information "across" the isolated cut. In fact, even the event of successfully reaching Phase II may be correlated with the value of x_{i^*}.

For example, say the cut at the conclusion of the protocol is $(S_1, \bar{S}_1)$ with $i^* \in S_1$ and $j^* \in \bar{S}_1$, but at the beginning of Phase II there existed another cut $(S_2, \bar{S}_2)$, for which $S_1 \cap S_2 \neq \emptyset$, $S_1 \cap \bar{S}_2 \neq \emptyset$, $\bar{S}_1 \cap S_2 \neq \emptyset$, and $\bar{S}_1 \cap \bar{S}_2 \neq \emptyset$. Since any "bridge" party in $\bar{S}_2$ that receives a message from S_2, gets corrupted and discards the message, the view of honest parties in $\bar{S}_1$ might change as a result of the corruption related to the cut $(S_2, \bar{S}_2)$, which in turn could depend on x_{i^*}.

Ultimately, we ensure that the final view of P_{j^*} in the protocol can be simulated given only "Phase I" information, which is independent of x_{i^*}, in addition to the identity of the final cut in the graph, which reveals only a constant amount of additional entropy.

Additional subtleties. The actual attack and its analysis are even more delicate. E.g., it is important that the degree of the "simulated P_{i^*}," by the adversarial strategy $\mathcal{A}_n^{\text{honest-}i^*}$, will reach the threshold faster than the real P_{i^*}. In addition, in each of these cases, the threshold, and so the transition to the next phase, could possibly be reached in a middle of a round, requiring detailed treatment.

1.3   Open Questions

This work leaves open many interesting lines of future study.

- Bridging the gap between upper and lower bounds. This equates to identifying the core properties that necessitate graph expansion versus not. Natural candidates suggested by our work are existence of setup information and adaptive corruptions in the hidden or visible (yet private) channels model.
- What other graph properties are necessary (or not) to support secure computation? Our new definitional framework may aid in this direction.
- Our work connects graph theory and secure protocols, giving rise to further questions and design principles. For example, can good constructions of expanders give rise to new communication-efficient MPC? On the other

hand, can necessity of expansion (in certain settings) be used to argue new communication complexity lower bounds?

Paper Organization

Basic notations are presented in Sect. 2. In Sect. 3, we provide our formalization of the communication graph induced by a MPC protocol and related properties. In Sect. 4, we describe our upper bound results, constructing protocols with non-expanding graphs. In Sect. 5, we prove our lower bound.

2 Preliminaries

Graph-theoretic notations. Let $G = (V, E)$ be an undirected graph of size n, i.e., $|V| = n$. Given a set $S \subseteq V$, we denote its complement set by $\bar{S}$, i.e., $\bar{S} = V \setminus S$. Given two disjoint subsets $U_1, U_2 \subseteq V$ define the set of all the edges in G for which one end point is in U_1 and the other end point is in U_2 as

$$\mathsf{edges}_G(U_1, U_2) := \{(u_1, u_2) : u_1 \in U_1, u_2 \in U_2, \text{ and } (u_1, u_2) \in E\}.$$

We denote by $|\mathsf{edges}_G(U_1, U_2)|$ the total number of edges going across U_1 and U_2. For simplicity, we denote $\mathsf{edges}_G(S) = \mathsf{edges}_G(S, \bar{S})$. A cut in the graph G is a partition of the vertices V into two non-empty, disjoint sets $\{S, \bar{S}\}$. An α-cut is a cut $\{S, \bar{S}\}$ such that $|\mathsf{edges}_G(S)| \leq \alpha$.

Given a graph $G = (V, E)$ and a node $i \in V$, denote by $G \setminus \{i\} = (V', E')$ the graph obtained by removing node i and all its edges, i.e., $V' = V \setminus \{i\}$ and $E' = E \setminus \{(i, j) \mid j \in V'\}$.

MPC Model. We consider multiparty protocols in the stand-alone, synchronous model, and require security with guaranteed output delivery. We refer the reader to [7,24] for a precise definition of the model. Throughout the paper we assume malicious adversaries that can deviate from the protocol in an arbitrary manner. We will consider both *static* corruptions, where the set of corrupted parties is fixed at the onset of the protocol, and *adaptive* corruptions, where the adversary can dynamically corrupt parties during the protocol execution, In addition, we will consider both PPT adversaries and computationally unbounded adversaries.

Recall that in the synchronous model protocols proceed in rounds, where every round consists of a *send phase* followed by a *receive phase*. The adversary is assumed to be *rushing*, meaning that he can determine the messages for corrupted parties *after* seeing the messages sent by the honest parties. We assume a complete network of point-to-point channels (broadcast is not assumed), where every party has the ability to send a message to every other party. We will normally consider *secure* (private) channels where the adversary learns that a message has been sent between two honest parties, but not its content. If a public-key encryption is assumed, this assumption can be relaxed to *authenticated* channels, where the adversary can learn the content of all messages (but not change them). For our upper bound in the adaptive setting (Sect. 4.2) we consider *hidden* channels (as introduced in [9]), where the adversary does not even know whether two honest parties have communicated or not.

3 Communication Graphs Induced by MPC Protocols

In this section, we present formal definitions of properties induced by the communication graph of interactive protocols. These definitions are inspired by previous works in distributed computing [29,30,32,33] and multiparty computation [3,5,9] that constructed interactive protocols with *low locality*.

3.1 Ensembles of Protocols and Functionalities

In order to capture certain asymptotic properties of the communication graphs of generic n-party protocols, such as edge expansion and locality, it is useful to consider a family of protocols that are parametrized by the number of parties n. This is implicit in many distributed protocols and in generic multiparty protocols, for example [2,17,25,34,35]. We note that for many large-scale protocols, e.g., protocols with low locality [3,5,29,30,32,33], the security guarantees increase with the number of parties, and in fact, the number of parties is assumed to be polynomially related to the security parameter.

Definition 1 (protocol ensemble). *Let $f = \{f_n\}_{n \in \mathbb{N}}$ be an ensemble of functionalities, where f_n is an n-party functionality, let $\pi = \{\pi_n\}_{n \in \mathbb{N}}$ be an ensemble of protocols, and let $\mathcal{C} = \{\mathcal{C}_n\}_{n \in \mathbb{N}}$ be an ensemble of classes of adversaries (e.g., $\mathcal{C}_n$ is the class of PPT $t(n)$-adversaries). We say that π securely computes f tolerating adversaries in $\mathcal{C}$ if for every n that is polynomially related to the security parameter κ, it holds that π_n securely computes f_n tolerating adversaries in $\mathcal{C}_n$.*

In Sect. 4, we will consider several classes of adversaries. We use the following notation for clarity and brevity.

Definition 2. *Let $f = \{f_n\}_{n \in \mathbb{N}}$ be an ensemble of functionalities and let $\pi = \{\pi_n\}_{n \in \mathbb{N}}$ be an ensemble of protocols. We say that π securely computes f tolerating adversaries of the form* **type** *(e.g., static PPT $t(n)$-adversaries, adaptive $t(n)$-adversaries, etc.), if π securely computes f tolerating adversaries in $\mathcal{C} = \{\mathcal{C}_n\}_{n \in \mathbb{N}}$, where for every n, the set $\mathcal{C}_n$ is the class of adversaries of the form* **type**.

3.2 The Communication Graph of a Protocol's Execution

Intuitively, the communication graph induced by a protocol should include an edge (i, j) precisely if parties P_i and P_j exchange messages during the protocol execution. For instance, consider the property of *locality*, corresponding to the maximum degree of the communication graph. When considering malicious adversaries that can deviate from the protocol using an arbitrary strategy, it is important to consider only messages that are sent by honest parties and messages that are received by honest parties. Otherwise, every corrupted party can send a message to every other corrupted party, yielding a subgraph with degree $\Theta(n)$. We note that restricting the analysis to only consider honest parties is quite common in the analysis of protocols.

Another issue that must be taken under consideration is flooding by the adversary. Indeed, there is no way to prevent the adversary from sending messages from all corrupted parties to all honest parties; however, we wish to only count those message which are actually processed by honest parties. To model this, the *receive* phase of every communication round[5] is composed of two sub-phases:

1. *The filtering sub-phase:* Each party inspects the list of messages received in the previous round, according to specific filtering rules defined by the protocol, and discards the messages that do not pass the filter. The resulting list of messages is appended to the local transcript of the protocol.
2. *The processing sub-phase:* Based on its local transcript, each party computes the next-message function and obtains the list of messages to be sent in the current round along with the list of recipients, and sends them to the relevant parties.

In practice, the filtering procedure should be "lightweight," such as verifying validity of a signature. However, we assume only an abstraction and defer the actual choice of filtering procedure (as well as corresponding discussion) to specific protocol specifications.

We now turn to define the communication graph of a protocol's execution, by which we mean the deterministic instance of the protocol defined by fixing the adversary and all input values and random coins of the parties and the adversarial strategy. We consider protocols that are defined in the correlated-randomness model (e.g., for establishing PKI). This is without loss of generality since by defining the "empty distribution," where every party is given an empty string, we can model also protocols in the plain model. Initially, we focus on the *static* setting, where the set of corrupted parties is determined at the onset of the protocol. We discuss the *adaptive* setting in the full version [4].

Definition 3 (protocol execution instance). *For $n \in \mathbb{N}$, let π_n be an n-party protocol, let κ be the security parameter, let $x = (x_1, \ldots, x_n)$ be an input vector for the parties, let $\rho = (\rho_1, \ldots, \rho_n)$ be correlated randomness for the parties, let $\mathcal{A}$ be an adversary, let z be the auxiliary information of $\mathcal{A}$, let $\mathcal{I} \subseteq [n]$ be the set of indices of corrupted parties controlled by $\mathcal{A}$, and let $r = (r_1, \ldots, r_n, r_\mathcal{A})$ be the vector of random coins for the parties and for the adversary.*

Denote by $\mathsf{instance}(\pi_n) = (\pi_n, \mathcal{A}, \mathcal{I}, \kappa, x, \rho, z, r)$ the list of parameters that deterministically define an execution instance *of the protocol π_n.*

Note that $\mathsf{instance}(\pi_n)$ fully specifies the entire views and transcript of the protocol execution, including all messages sent to/from honest parties.

Definition 4 (communication graph of protocol execution). *For $n \in \mathbb{N}$, let $\mathsf{instance}(\pi_n) = (\pi_n, \mathcal{A}, \mathcal{I}, \kappa, x, \rho, z, r)$ be an execution instance of the protocol π_n. We now define the following communication graphs induced by this execution instance. Each graph is defined over the set of n vertices $[n]$.*

---

[5] Recall that in the synchronous model, every communication round is composed of a *send* phase and a *receive* phase.

- Outgoing communication graph. *The directed graph $G_{out}(instance(\pi_n)) = ([n], E_{out})$ captures all the communication lines that are used by honest parties to send messages. That is,*

$$E_{out}(instance(\pi_n)) = \{(i, j) \mid P_i \text{ is honest and sent a message to } P_j\}.$$

- Incoming communication graph. *The directed graph $G_{in}(instance(\pi_n)) = ([n], E_{in})$ captures all the communication lines in which honest parties received messages that were processed (i.e., excluding messages that were filtered out). That is,*

$$E_{in}(instance(\pi_n)) = \{(i, j) \mid P_j \text{ is honest and processed a message received from } P_i\}.$$

- Full communication graph. *The undirected graph $G_{full}(instance(\pi_n)) = ([n], E_{full})$ captures all the communication lines in which honest parties received messages that were processed, or used by honest parties to send messages.*
  That is,

$$E_{full}(instance(\pi_n)) = \{(i, j) \mid (i, j) \in E_{out} \text{ or } (i, j) \in E_{in}\}.$$

We will sometimes consider ensembles of protocol instances (for $n \in \mathbb{N}$) and the corresponding ensembles of graphs they induce.

Looking ahead, in subsequent sections we will consider the full communication graph G_{full}. Apart from making the presentation clear, the graphs G_{out} and G_{in} are used for defining G_{full} above, and the locality of a protocol in Definition 5. Note that G_{out} and G_{in} are interesting in their own right, and can be used for a fine-grained analysis of the communication graph of protocols in various settings, e.g., when transmitting messages is costly but receiving messages is cheap (or vice versa). We leave it open as an interesting problem to study various graph properties exhibited by these two graphs.

3.3   Locality of a Protocol

We now present a definition of communication locality, aligning with that of [5], with respect to the terminology introduced above.

Definition 5 (locality of a protocol instance). *Let $instance(\pi_n) = (\pi_n, \kappa, \boldsymbol{x}, \boldsymbol{\rho}, \mathcal{A}, z, \mathcal{I} \subseteq [n], \boldsymbol{r})$ be an execution instance as in Definition 4. For every honest party P_i we define the locality of party P_i to be the number of parties from which P_i received and processed messages, or sent message to; that is,*

$$\ell_i(instance(\pi_n)) = |\{j \mid (i, j) \in G_{out}\} \cup \{j \mid (j, i) \in G_{in}\}|.$$

The locality of $instance(\pi_n)$ is defined as the maximum locality of an honest party, i.e.,

$$\ell(instance(\pi_n)) = \max_{i \in [n] \setminus \mathcal{I}} \{\ell_i(instance(\pi_n))\}.$$

We proceed by defining locality as a property of a protocol ensemble. The protocol ensemble is parametrized by the number of parties n. To align with standard notions of security where asymptotic measurements are with respect to the security parameter κ, we consider the situation where the growth of n and κ are polynomially related.

Definition 6 (locality of a protocol). *Let $\pi = \{\pi_n\}_{n \in \mathbb{N}}$ be a family of protocols in the correlated-randomness model with distribution $D_\pi = \{D_{\pi_n}\}_{n \in \mathbb{N}}$, and let $\mathcal{C} = \{\mathcal{C}_n\}_{n \in \mathbb{N}}$ be a family of adversary classes. We say that π has locality $\ell(n)$ facing adversaries in $\mathcal{C}$ if for every n that is polynomially related to κ it holds that for every input vector $\boldsymbol{x} = (x_1, \ldots, x_n)$, every auxiliary information z, every adversary $\mathcal{A} \in \mathcal{C}_n$ running with z, and every set of corrupted parties $\mathcal{I} \subseteq [n]$, it holds that*

$$\Pr\left[\ell(\pi_n, \mathcal{A}, \mathcal{I}, \kappa, \boldsymbol{x}, z) > \ell(n)\right] \leq \mathsf{negl}(\kappa),$$

where $\ell(\pi_n, \mathcal{A}, \mathcal{I}, \kappa, \boldsymbol{x}, z)$ is the random variable corresponding to $\ell(\pi_n, \mathcal{A}, \mathcal{I}, \kappa, \boldsymbol{x}, \boldsymbol{\rho}, z, \boldsymbol{r})$ when $\boldsymbol{\rho}$ is distributed according to D_{π_n} and $\boldsymbol{r}$ is uniformly distributed.

3.4 Edge Expansion of a Protocol

The measure of complexity we study for the communication graph of interactive protocols will be that of *edge expansion* (see discussion below). We refer the reader to [18,28] for more background on expanders. We consider a definition of edge expansion which satisfies a natural monotonic property, where adding more edges cannot decrease the graph's measure of expansion.

Definition 7 *(edge expansion of a graph).* *Given an undirected graph $G = (V, E)$, the edge expansion ratio of G, denoted $h(G)$, is defined as*

$$h(G) = \min_{\{S \subseteq V : |S| \leq \frac{|V|}{2}\}} \frac{|edges(S)|}{|S|}, \tag{1}$$

where $edges(S)$ denotes the set of edges between S and its complement $\bar{S} = V \setminus S$.

Definition 8 *(family of expander graphs).* *A sequence $\{G_n\}_{n \in \mathbb{N}}$ of graphs is a family of expander graphs if there exists a constant $\epsilon > 0$ such that $h(G_n) \geq \epsilon$ for all n.*

We now consider the natural extension of graph expansion to the setting of protocol-induced communication graph.

Definition 9 *(bounds on edge expansion of a protocol).* *Let $\pi = \{\pi_n\}_{n \in \mathbb{N}}$, $D_\pi = \{D_{\pi_n}\}_{n \in \mathbb{N}}$, and $\mathcal{C} = \{\mathcal{C}_n\}_{n \in \mathbb{N}}$ be as in Definition 6.*

- *A function $f(n)$ is a lower bound of the edge expansion of π facing adversaries in $\mathcal{C}$, denoted $f(n) \leq h_{\pi, D_\pi, \mathcal{C}}(n)$, if for every n that is polynomially related* to κ, for every $\boldsymbol{x} = (x_1, \ldots, x_n)$, every $\mathcal{A} \in \mathcal{C}_n$ running with z, and every $\mathcal{I} \subseteq [n]$, it holds that

$$\Pr\left[h(G_{full}(\pi_n, \mathcal{A}, \mathcal{I}, \kappa, \boldsymbol{x}, z)) \leq f(n)\right] \leq negl(\kappa),$$

where $G_{full}(\pi_n, \mathcal{A}, \mathcal{I}, \kappa, \boldsymbol{x}, z)$ is the random variable $G_{full}(\pi_n, \mathcal{A}, \mathcal{I}, \kappa, \boldsymbol{x}, \boldsymbol{\rho}, z, \boldsymbol{r})$, when $\boldsymbol{\rho}$ is distributed according to D_{π_n} and $\boldsymbol{r}$ is uniformly distributed.

- A function $f(n)$ is a **upper bound of the edge expansion** of π facing adversaries in $\mathcal{C}$, denoted $f(n) \geq h_{\pi, D_\pi, \mathcal{C}}(n)$, if there exists a polynomial relation between n and κ such that for infinitely many n it holds that for every $\boldsymbol{x} = (x_1, \ldots, x_n)$, every $\mathcal{A} \in \mathcal{C}_n$ running with z, and every $\mathcal{I} \subseteq [n]$, it holds that

$$\Pr\left[h(G_{full}(\pi_n, \mathcal{A}, \mathcal{I}, \kappa, \boldsymbol{x}, z)) \geq f(n)\right] \leq negl(\kappa).$$

Definition 10 (expander protocol). *Let $\pi = \{\pi_n\}_{n \in \mathbb{N}}$, $D_\pi = \{D_{\pi_n}\}_{n \in \mathbb{N}}$, and $\mathcal{C} = \{\mathcal{C}_n\}_{n \in \mathbb{N}}$ be as in Definition 6. We say that the **communication graph of π is an expander**, facing adversaries in $\mathcal{C}$, if there exists a constant function $\epsilon(n) > 0$ such that $\epsilon(n) \leq h_{\pi, D_\pi, \mathcal{C}}(n)$.*

We note that most (if not all) secure protocols in the literature are expanders according to Definition 10, both in the realm of distributed computing [17, 20, 22, 29, 30, 32, 33] and in the realm of MPC [2, 3, 5, 9, 25]. Proving that a protocol is not an expander according to this definition requires showing an adversary for which the edge expansion is sub-constant. Looking ahead, both in our constructions of protocols that are not expanders (Sect. 4) and in our lower bound, showing that non-expander protocols can be attacked (Sect. 5), we use a stronger definition, that requires that the edge expansion is sub-constant facing *all* adversaries, see Definition 11 below. While it makes our positive results stronger, we leave it as an interesting open question to attack protocols that do not satisfy Definition 10.

Definition 11 (strongly non-expander protocol). *Let $\pi = \{\pi_n\}_{n \in \mathbb{N}}$, $D_\pi = \{D_{\pi_n}\}_{n \in \mathbb{N}}$, and $\mathcal{C} = \{\mathcal{C}_n\}_{n \in \mathbb{N}}$ be as in Definition 6. We say that the **communication graph of π is strongly not an expander**, facing adversaries in $\mathcal{C}$, if there exists a sub-constant function $\alpha(n) \in o(1)$ such that $\alpha(n) \geq h_{\pi, D_\pi, \mathcal{C}}(n)$.*

We next state a useful observation (proven in the full version [4]) that will come into play in Sect. 5, stating that if the communication graph of π is strongly not an expander, then there must exist a sublinear cut in the graph.

Lemma 1. *Let $\pi = \{\pi_n\}_{n \in \mathbb{N}}$ be a family of protocols in the correlated-randomness model with distribution $D_\pi = \{D_{\pi_n}\}_{n \in \mathbb{N}}$, and let $\mathcal{C} = \{\mathcal{C}_n\}_{n \in \mathbb{N}}$ be such that $\mathcal{C}_n$ is the class of adversaries corrupting at most $\beta \cdot n$ parties, for a constant $0 < \beta < 1$.*

Assuming the communication graph of π is strongly non-expanding facing adversaries in $\mathcal{C}$, there exists a sublinear function $\alpha(n) \in o(n)$ such that for infinitely many n's the full communication graph of π_n has an $\alpha(n)$-cut with overwhelming probability.

4   MPC with Non-Expanding Communication Graph

In this section, we show that in various standard settings, the communication graph of an MPC protocol is *not* required to be an expander graph, even when the communication locality is poly-logarithmic. In Sect. 4.1, we focus on static corruptions and computational security. In Sect. 4.2, we extend the construction to the information-theoretic setting and to the adaptive-corruption setting. The proof for these extensions can be found in the full version [4].

4.1   Computational Security with Static Corruptions

We start by considering the computational setting with static corruptions.

Theorem 5. *Let* $f = \{f_n\}_{n \in \mathbb{N}}$ *be an ensemble of functionalities, let* $\delta > 0$, *and assume that one-way functions exist. Then, the following holds in the PKI-hybrid model with secure channels:*

1. *Let* $\beta < 1/4 - \delta$ *and let* $t(n) = \beta \cdot n$. *Then,* f *can be securely computed by a protocol ensemble* π *tolerating static PPT* $t(n)$-*adversaries such that the communication graph of* π *is strongly not an expander.*
2. *Let* $\beta < 1/6 - \delta$ *and let* $t(n) = \beta \cdot n$. *Then,* f *can be securely computed by a protocol ensemble* π *tolerating static PPT* $t(n)$-*adversaries such that (1) the communication graph of* π *is strongly not an expander, and (2) the locality of* π *is poly-logarithmic in* n.
3. *Let* $\beta < 1/4 - \delta$, *let* $t(n) = \beta \cdot n$, *and assume in addition the* secret-key infrastructure (SKI) *model[6] and the existence of public-key encryption schemes. Then,* f *can be securely computed by a protocol ensemble* π *tolerating static PPT* $t(n)$-*adversaries such that (1) the communication graph of* π *is strongly not an expander, and (2) the locality of* π *is poly-logarithmic in* n.[7]

Proof. The theorem follows from Lemma 2 (below) by instantiating the hybrid functionalities using existing MPC protocols from the literature.

- The first part follows using honest-majority MPC protocols that exist assuming one-way functions in the secure-channels model, e.g., the protocol of Beaver et al. [1] or of Damgård and Ishai [14].
- The second part follows using the low-locality MPC protocol of Boyle et al. [5] that exists assuming one-way functions in the PKI model with secure channels and tolerates $t = (1/3 - \delta)n$ static corruptions.
- The third part follows using the low-locality MPC protocol of Chandran et al. [9] that exists assuming public-key encryption in the PKI and SKI model with authenticated channels and tolerates $t < n/2$ static corruptions.   □

---

[6] In the SKI model every pair of parties has a secret random string that is unknown to other parties.

[7] This item hold in the authenticated-channels model, since we assume PKE.

Ideal Functionalities used in the Construction. The proof to Theorem 5 relies on Lemma 2 (below). We start by defining the notations and the ideal functionalities that will be used in the protocol considered in Lemma 2.

Signature notations. Given a signature scheme (Gen, Sign, Verify) and m pairs of signing and verification keys $(\text{sk}_i, \text{vk}_i) \leftarrow \text{Gen}(1^\kappa)$ for $i \in [m]$, we use the following notations for signing and verifying with multiple keys:

- Given a message μ we denote by $\text{Sign}_{\text{sk}_1,\ldots,\text{sk}_m}(\mu)$ the vector of m signatures $\sigma = (\sigma_1,\ldots,\sigma_m)$, where $\sigma_i \leftarrow \text{Sign}_{\text{sk}_i}(\mu)$.
- Given a message μ and a signature $\sigma = (\sigma_1,\ldots,\sigma_m)$, we denote by $\text{Verify}_{\text{vk}_{m+1},\ldots,\text{vk}_{2m}}(\mu,\sigma)$ the verification algorithm that for every $i \in [m]$ computes $b_i \leftarrow \text{Verify}_{\text{vk}_{m+i}}(\mu,\sigma_i)$, and accepts the signature σ if and only if $\sum_{i=1}^{m} b_i \geq m - t$, i.e., even if up to t signatures are invalid.

We note that it is possible to use multi-signatures or aggregated signatures in order to obtain better communication complexity, however, we use the notation above both for simplicity and as a step towards the information-theoretic construction in the following section.

The Elect-and-Share functionality. In the Elect-and-Share m-party functionality, $f_{\text{elect-share}}^{(t',n')}$, every party P_i has a pair of inputs (x_i, sk_i), where $x_i \in \{0,1\}^*$ is the "actual input" and sk_i is a private signing key. The functionality starts by electing two random subsets $\mathcal{C}_1, \mathcal{C}_2 \subseteq [m]$ of size n', and signing each subset using all signing keys. In addition, every input value x_i is secret shared using a (t',n') error-correcting secret-sharing scheme. Every party receives as output the subset $\mathcal{C}_1$, whereas a party P_i, for $i \in \mathcal{C}_1$, receives an additional output consisting of a signature on $\mathcal{C}_1$, the signed subset $\mathcal{C}_2$, along with one share for each one of the m input values.

The Reconstruct-and-Compute functionality. The Reconstruct-and-Compute functionality, $f_{\text{recon-compute}}^{(\text{vk}_1,\ldots,\text{vk}_m)}$, is an m-party functionality. Denote the party-set by $\{P_{m+1},\ldots,P_{2m}\}$. Every party P_{m+i} has an input value $x_{m+i} \in \{0,1\}^*$, and a potential additional input value consisting of a signed subset $\mathcal{C}_2 \subseteq [m]$ and a vector of m shares. The functionality starts by verifying the signatures, where every invalid input is ignored. The signed inputs should define a single subset $\mathcal{C}_2 \subseteq [m]$ (otherwise the functionality aborts), and the functionality uses the additional inputs of parties P_{m+i}, for every $i \in \mathcal{C}_2$, in order to reconstruct the m-tuple $(x_1,\ldots,x_m)$. Finally, the functionality computes $y = f(x_1,\ldots,x_{2m})$ and hands y as the output for every party.

The Output-Distribution functionality. The m-party Output-Distribution functionality is parametrized by a subset $\mathcal{C}_1 \subseteq [m]$. Every party P_i, with $i \in \mathcal{C}_1$, hands in a value, and the functionality distributes the majority of these inputs to all the parties.

Constructing Non-Expander Protocols.

High-level overview of the protocol. Having defined the ideal functionalities, we are ready to present the main lemma. We start by describing the underlying idea behind the non-expanding MPC protocol π_n^{ne} (Fig. 2). At the onset of the protocol, the party-set is partitioned into two subsets of size $m = n/2$, a left subset and a right subset (see Fig. 1). The left subset will invoke the Elect-and-Share functionality, that elects two subsets $\mathcal{C}_1, \mathcal{C}_2 \subseteq [m]$ of size $n' = \log^2(n)$. The parties in the left subset corresponding to $\mathcal{C}_1$ and the parties in the right subset corresponding to $\mathcal{C}_2$ will form a "bridge". The parties in $\mathcal{C}_1$ will receive shares of all inputs values of parties in the left subset, and transfer them to $\mathcal{C}_2$. Next, the right subset of parties will invoke the Reconstruct-and-Compute functionality, where each party hands its input value, and parties in $\mathcal{C}_2$ additionally provide the shares they received from $\mathcal{C}_1$. The functionality reconstructs the left-subset's inputs, computes the function f and hands the output to the right subset. Finally, $\mathcal{C}_2$ transfers the output value to $\mathcal{C}_1$, and the left subset invoke the Output-Distribution functionality in order to distribute the output value to all the parties.

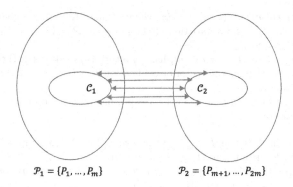

Fig. 1. The non-expanding subsets in the protocol π^{ne}. The sets $\mathcal{C}_1$ and $\mathcal{C}_2$ are of poly-logarithmic size and the sets $\mathcal{P}_1$ and $\mathcal{P}_2$ are of linear size. The number of edges between $\mathcal{P}_1$ and $\mathcal{P}_2$ is poly-logarithmic.

Lemma 2. *Let $f = \{f_n\}_{n \in 2\mathbb{N}}$,[8] where f_n is an n-party functionality for $n = 2m$, let $\delta > 0$, and assume that one-way functions exist. Then, in the PKI-hybrid model with secure channels, where a trusted party additionally computes the m-party functionality-ensembles $(f_{\mathsf{elect\text{-}share}}, f_{\mathsf{recon\text{-}compute}}, f_{\mathsf{out\text{-}dist}})$ tolerating $\gamma \cdot m$ corruptions, there exists a protocol ensemble π that securely computes f tolerating static PPT βn-adversaries, for $\beta < \min(1/4 - \delta, \gamma/2)$, with the following guarantees:*

1. *The communication graph of π is strongly not an expander.*

---

[8] For simplicity, we consider even n's. Extending the statement to any n is straightforward, however, adds more details.

2. *Denote by f_1, f_2, f_3 the functionality-ensembles $f_{\text{elect-share}}, f_{\text{recon-compute}}, f_{\text{out-dist}}$ (resp.). If protocol-ensembles ρ_1, ρ_2, ρ_3 securely compute f_1, f_2, f_3 (resp.) with locality $\ell_\rho = \ell_\rho(m)$, then $\pi^{f_i \to \rho_i}$ (where every call to f_i is replaced by an execution of ρ_i) has locality $\ell = 2 \cdot \ell_\rho + \log^2(n)$.*

Proof. For $m \in \mathbb{N}$ and $n = 2m$, we construct the n-party protocol π_n^{ne} (see Fig. 2) in the $(f_{\text{elect-share}}, f_{\text{recon-compute}}, f_{\text{out-dist}})$-hybrid model. The parameters for the protocol are $n' = \log^2(n)$ and $t' = (1/2 - \delta) \cdot n'$. We start by proving in Proposition 1 that the protocol π_n^{ne} securely computes f_n. Next, in Proposition 2 we prove that the communication graph of π^{ne} is strongly not an expander. Finally, in Proposition 3 we prove that by instantiating the functionalities $(f_{\text{elect-share}}, f_{\text{recon-compute}}, f_{\text{out-dist}})$ using low-locality protocols, the resulting protocol has low locality.

Protocol π_n^{ne}

- **Hybrid Model:** The protocol is defined in the $(f_{\text{elect-share}}, f_{\text{recon-compute}}, f_{\text{out-dist}})$-hybrid model.
- **Common Input:** A (t', n') ECSS scheme (Share, Recon), a signature scheme (Gen, Sign, Verify), and a partition of the party-set $\mathcal{P} = \{P_1, \ldots, P_n\}$ into $\mathcal{P}_1 = \{P_1, \ldots, P_m\}$ and $\mathcal{P}_2 = \mathcal{P} \setminus \mathcal{P}_1$.
- **PKI:** Every party P_i, for $i \in [n]$, has signature keys $(\text{sk}_i, \text{vk}_i)$; the signing key sk_i is private, whereas the vector of verification keys $(\text{vk}_1, \ldots, \text{vk}_n)$ is public and known to all parties.
- **Private Input:** Every party P_i, for $i \in [n]$, has private input $x_i \in \{0, 1\}^*$.
- **The Protocol:**

1. The parties in $\mathcal{P}_1$ invoke $f_{\text{elect-share}}^{(t', n')}$, where every $P_i \in \mathcal{P}_1$ sends input (x_i, sk_i), and receives back output consisting of a committee $\mathcal{C}_1 = \{i_{(1,1)}, \ldots, i_{(1,n')}\} \subseteq [m]$. Every party P_i with $i = i_{(1,j)} \in \mathcal{C}_1$, receives additional output consisting of a signature σ_1 on $\mathcal{C}_1$, a committee $\mathcal{C}_2 = \{i_{(2,1)}, \ldots, i_{(2,n')}\} \subseteq [m]$, a signature σ_2 on $\mathcal{C}_2$, and a vector $s_j = (s_1^j, \ldots, s_m^j)$.
2. For every $j \in [n']$, party $P_{i_{(1,j)}}$ sends $(\mathcal{C}_1, \sigma_1, \mathcal{C}_2, \sigma_2)$ to every party in $\mathcal{C}_2$, and s_j only to $P_{m+i_{(2,j)}}$.
   A party $P_{m+i} \in \mathcal{P}_2$ that receives a message $(\mathcal{C}_1, \sigma_1, \mathcal{C}_2, \sigma_2)$ from $P_j \in \mathcal{P}_1$ will discard the message in the following cases:
   (a) If $i \notin \mathcal{C}_2$ or $j \notin \mathcal{C}_1$.
   (b) If $\text{Verify}_{\text{vk}_1, \ldots, \text{vk}_m}(\mathcal{C}_1, \sigma_1) = 0$ or $\text{Verify}_{\text{vk}_1, \ldots, \text{vk}_m}(\mathcal{C}_2, \sigma_2) = 0$.
3. The parties in $\mathcal{P}_2$ invoke $f_{\text{recon-compute}}^{(\text{vk}_1, \ldots, \text{vk}_m)}$, where $P_{m+i} \in \mathcal{P}_2$ sends input (x_{m+i}, z_{m+i}) such that for $i \notin \mathcal{C}_2$, set $z_{m+i} = \epsilon$, and for $i = i_{(2,j)} \in \mathcal{C}_2$, set $z_{m+i} = (\mathcal{C}_2, \sigma_2, s_j)$. Every party in $\mathcal{P}_2$ receives back output y.
4. For every $j \in [n']$, party $P_{m+i_{(2,j)}}$ sends y to party $P_{i_{(1,j)}}$. In addition, every party in $\mathcal{P}_2$ outputs y and halts.
5. The parties in $\mathcal{P}_1$ invoke $f_{\text{out-dist}}^{\mathcal{C}_1}$, where party P_i, with $i \in \mathcal{C}_1$, has input y, and party P_i, with $i \notin \mathcal{C}_1$ has the empty input ϵ. Every party in $\mathcal{P}_1$ receives output y, outputs it, and halts.

Fig. 2. Non-expanding MPC in the $(f_{\text{elect-share}}, f_{\text{recon-compute}}, f_{\text{out-dist}})$-hybrid model

Proposition 1. *For sufficiently large* n, *the protocol* π_n^{ne} *securely computes the function* f_n, *tolerating static PPT* βn-*adversaries, in the* $(f_{\text{elect-share}}, f_{\text{recon-compute}}, f_{\text{out-dist}})$-*hybrid model.*

The proof of Proposition 1 can be found in the full version [4].

Proposition 2. *The communication graph of the Protocol* π^{ne} *is strongly not an expander, facing static PPT* βn-*adversaries.*

Proof. For $n = 2m$, consider the set $\mathcal{P}_1 = \{P_1, \ldots, P_m\}$ and its complement $\mathcal{P}_2 = \mathcal{P} \setminus \mathcal{P}_1$. For any input vector and for every static PPT βn-adversary it holds that with overwhelming probability that $|\mathcal{P}_1| = n/2$ and $\text{edges}(\mathcal{P}_1, \mathcal{P}_2) = \log^2(n) \cdot \log^2(n)$. Therefore, considering the function

$$f(n) = \frac{2 \log^4(n)}{n},$$

it holds that $f(n) \in o(1)$ and $f(n)$ is an upper bound of the edge expansion of π^{ne}. We conclude that the communication graph of π^{ne} is strongly not an expander. $\square$

Proposition 3. *Let* ρ_1, ρ_2, ρ_3, *and* $\pi^{f_i \to \rho_i}$ *be the protocols defined in Lemma 2, and let* $\ell_\rho = \ell_\rho(m)$ *be the upper bound of the locality of* ρ_1, ρ_2, ρ_3. *Then* $\pi^{f_i \to \rho_i}$ *has locality* $\ell = 2 \cdot \ell_\rho + \log^2(n)$.

Proof. Every party in $\mathcal{P}_1$ communicates with ℓ_ρ parties when executing ρ_1, and with at most another ℓ_ρ parties when executing ρ_3. In addition, every party in $\mathcal{C}_1$ communicates with all $n' = \log^2(n)$ parties in $\mathcal{C}_2$. Similarly, every party in $\mathcal{P}_2$ communicates with ℓ_ρ parties when executing ρ_2, and parties in $\mathcal{C}_2$ communicates with all n' parties in $\mathcal{C}_1$. It follows that maximal number of parties that a party communicates with during the protocol is $2 \cdot \ell_\rho + \log^2(n)$. $\square$

This concludes the proof of Lemma 2. $\square$

4.2 Additional Results

Information-Theoretic Security. The protocol in Sect. 4.1 relies on digital signatures, hence, security is guaranteed only in the presence of computationally bounded adversaries. Next, we gain security facing all-powerful adversaries by using information-theoretic signatures. We prove the following theorem in the full version [4].

Theorem 6. *Let* $f = \{f_n\}_{n \in \mathbb{N}}$ *be an ensemble of functionalities and let* $\delta > 0$. *Then, the following holds in the IT-PKI-hybrid model with secure channels:*

1. *Let* $\beta < 1/4 - \delta$ *and let* $t = \beta \cdot n$. *Then,* f *can be* t-*securely computed by a protocol ensemble* π *tolerating static* $t(n)$-*adversaries such that the communication graph of* π *is strongly not an expander.*
2. *Let* $\beta < 1/12 - \delta$ *and let* $t = \beta \cdot n$. *Then,* f *can be* t-*securely computed by a protocol ensemble* π *tolerating static* $t(n)$-*adversaries such that (1) the communication graph of* π *is strongly not an expander, and (2) the locality of* π *is poly-logarithmic in* n.

Adaptive Corruptions. In this section, we focus on the adaptive setting, where the adversary can corrupt parties dynamically, based on information gathered during the course of the protocol.

Adjusting Lemma 2 to the adaptive setting is not straightforward, since once the subsets C_1 and C_2 are known to the adversary he can completely corrupt them. A first attempt to get around this problem, is not to reveal the entire subsets in the output of the Elect-and-Share functionality, but rather, let each party in C_1 learn the identity of a single party in C_2 with which he will communicate. This way, if a party in C_1 (resp. C_2) gets corrupted, only one additional party in C_2 (resp. C_1) is revealed to the adversary. This solution comes with the price of tolerating a smaller fraction of corrupted parties, namely, $(1/8 - \delta)$ fraction.

This solution, however, is still problematic in the adaptive setting if the adversary can monitor the communication lines, even when they are completely private (as in the secure-channels setting). The reason is that once the adversary sees the communication that is sent between C_1 and C_2 he can completely corrupt both subsets. This problem is inherent when the communication lines are visible to the adversary, therefore, we turn to the hidden-channels setting that was used by Chandran et al. [9], where the adversary does not learn whether a message is sent between two honest parties.

Theorem 7. *Let $f = \{f_n\}_{n \in \mathbb{N}}$ be an ensemble of functionalities, let $\delta > 0$, let $\beta < 1/8 - \delta$, and let $t = \beta \cdot n$. Then, the following holds in the hidden-channels model:*

1. *Assuming the existence of one-way functions, f can be securely computed by a protocol ensemble π in the PKI model tolerating adaptive PPT $t(n)$-adversaries such that the communication graph of π is strongly not an expander.*
2. *Assume in addition the SKI model and non-committing encryption. Then, f can be securely computed by a protocol ensemble π in the PKI model tolerating adaptive PPT $t(n)$-adversaries such that (1) the communication graph of π is strongly not an expander, and (2) the locality of π is poly-logarithmic in n.*
3. *f can be securely computed by a protocol ensemble π in the IT-PKI model tolerating adaptive $t(n)$-adversaries such that the communication graph of π is strongly not an expander.*

5   Expansion is Necessary for Correct Computation

In this section, we show that in certain natural settings there exist functionalities such that the final communication graph of any MPC protocol that securely computes them *must* be an expander. In fact, we prove a stronger statement, that removing a sublinear number of edges from such graphs will not disconnect them. We consider the plain model, in which parties do not have any trusted setup assumptions, a PPT adaptive adversary, and focus on parallel multi-valued broadcast (also known as interactive consistency [35]), where every party has an input value, and all honest parties agree on a common output vector, such that if P_i is honest then the i'th coordinate equals P_i's input. In particular, our proof does not rely on any privacy guarantees of the protocol, merely its correctness.

For simplicity, and without loss of generality, we assume the security parameter is the number of parties n.

Definition 12 (parallel broadcast). *A protocol ensemble $\pi = \{\pi_n\}_{n \in \mathbb{N}}$ is a $t(n)$-resilient, parallel broadcast protocol with respect to input space $\{\{0,1\}^n\}_{n \in \mathbb{N}}$, if there exists a negligible function $\mu(n)$, such that for every $n \in \mathbb{N}$, every party P_i in π_n has input $x_i \in \{0,1\}^n$ and outputs a vector of n values $\boldsymbol{y}_i = (y_1^i, \ldots, y_n^i)$ such that the following is satisfied, except for probability $\mu(n)$. Facing any adaptive, malicious PPT adversary that dynamically corrupts and controls a subset of parties $\{P_j\}_{j \in \mathcal{I}}$, with $\mathcal{I} \subseteq [n]$ of size $|\mathcal{I}| \leq t(n)$, it holds that:*

- **Agreement.** *There exists a vector $\boldsymbol{y} = (y_1, \ldots, y_n)$ such that for every party P_i that is honest at the conclusion of the protocol it holds that $\boldsymbol{y}_i = \boldsymbol{y}$.*
- **Validity.** *For every party P_i that is honest at the conclusion of the protocol it holds that the i'th coordinate of the common output equals his input value, i.e., $y_i = x_i$.*

Recall that a connected graph is k-edge-connected if it remains connected whenever fewer than k edges are removed. We are now ready to state the main result of this section. We note that as opposed to Sect. 4.2, where we considered adaptive corruptions in the hidden-channels model, this section considers the *parallel secure message transmission (SMT)* model, formally defined in Sect. 5.1, where the adversary is aware of communication between honest parties, but not of the message content.

Theorem 8. *Let $\beta > 0$ be a fixed constant, let $t(n) = \beta \cdot n$, and let $\pi = \{\pi_n\}_{n \in \mathbb{N}}$ be a $t(n)$-resilient, parallel broadcast protocol with respect to input space $\{\{0,1\}^n\}_{n \in \mathbb{N}}$, in the parallel SMT hybrid model (in the computational setting, tolerating an adaptive, malicious PPT adversary). Then, the communication graph of π must be $\alpha(n)$-edge-connected, for every $\alpha(n) \in o(n)$.*

From Theorem 8 and Lemma 1 (stating that if π is strongly not an expander then there must exist a sublinear cut in the graph) we get the following corollary.

Corollary 1. *Consider the setting of Theorem 8. If the communication graph of π is strongly not an expander (as per Definition 11), then π is not a $t(n)$-resilient parallel broadcast protocol.*

The remainder of this section goes towards proving Theorem 8. We start by presenting the communication model in Sect. 5.1. In Sect. 5.2, we prove a graph-theoretic theorem that will be used in the core of our proof and may be of independent interest. Then, in Sect. 5.3 we present the proof of Theorem 8. Some of the proofs are deferred to the full version [4].

5.1   The Communication Model

We consider secure communication channels, where the adversary can see that a message has been sent but not its content (in contrast to the hidden-communication model, used in Sect. 4.2, where the communication between honest parties was hidden from the eyes of the adversary). A standard assumption when considering adaptive corruptions is that in addition to being notified that an honest party sent a message, the adversary can corrupt the sender *before* the receiver obtained the message, learn the content of the message, and replace it with another message of its choice that will be delivered to the receiver. Although the original modular composition framework [7] does not give the adversary such power, this ability became standard after the introduction of the *secure message transmission (SMT)* functionality in the UC framework [8]. As we consider *synchronous* protocols, we use the parallel SMT functionality that was formalized in [12,13].[9]

Definition 13 (parallel SMT). *The parallel secure message transmission functionality f_{psmt} is a two-phase functionality. For every $i, j \in [n]$, the functionality initializes a value x_j^i to be the empty string ϵ (the value x_j^i is the message to be sent from P_i to P_j).*

- *The input phase. Every party P_i sends a vector of n messages $(v_1^i, \ldots, v_n^i)$. The functionality sets $x_j^i = v_j^i$, and provides the adversary with leakage information on the input values. As we consider rushing adversaries, who can determine the messages to be sent by the corrupted parties after receiving the messages sent by the honest parties, the leakage function should leak the messages that are to be delivered from honest parties to corrupted parties. Therefore, the leakage function is*

$$l_{\mathsf{psmt}}\left((x_1^1, \ldots, x_n^1), \ldots, (x_1^n, \ldots, x_n^n)\right) = \left(y_1^1, y_2^1, \ldots, y_{n-1}^n, y_n^n\right),$$

where $y_j^i = |x_j^i|$ in case P_j is honest and $y_j^i = x_j^i$ in case P_j is corrupted.
We consider adaptive corruptions, and so, the adversary can corrupt an honest party during the input phase based on this leakage information, and send a new input on behalf of the corrupted party (note that the message are not delivered yet to the honest parties).
- *The output phase. In the second phase, the messages are delivered to the parties, i.e., party P_i receives the vector of messages $(x_i^1, \ldots, x_i^n)$.*

In addition, we assume that the parties do not have any trusted-setup assumption.

---

[9] We note that by considering secure channels, that hide the content of the messages from the adversary, we obtain a stronger lower bound than, for example, authenticated channels.

5.2   A Graph-Theoretic Theorem

Our lower-bound proof is based on the following graph-theoretic theorem, which we believe may be of independent interest. We show that every graph in which every node has a linear degree, can be partitioned into a constant number of linear-size sets that are pairwise connected by sublinear many edges. These subsets are "minimal cuts" in the sense that every sublinear cut in the graph is a union of some of these subsets. The proof of the theorem given in the full version [4].

Definition 14 ((α, d)-partition). *Let $G = (V, E)$ be a graph of size n. An (α, d)-partition of G is a partition $\Gamma = (U_1, \dots, U_\ell)$ of V that satisfies the following properties:*

1. *For every $i \in [\ell]$ it holds that $|U_i| \geq d$.*
2. *For every $i \neq j$, there are at most α edges between U_i and U_j, i.e., $|edges_G(U_i, U_j)| \leq \alpha$.*
3. *For every $S \subseteq V$ such that $\{S, \bar{S}\}$ is an α-cut, i.e., $|edges_G(S)| \leq \alpha$, it holds that there exists a subset $J \subsetneq [\ell]$ for which $S = \bigcup_{j \in J} U_j$ and $\bar{S} = \bigcup_{j \in [\ell] \setminus J} U_j$.*

In Theorem 9 we first show that if every node in the graph has a linear degree $d(n)$, and $\alpha(n)$ is sublinear, then for sufficiently large n there exists an $(\alpha(n), d(n))$-partition of the graph, and moreover, the partition can be found in polynomial time.

Theorem 9. *Let $c > 1$ be a constant integer, let $\alpha(n) \in o(n)$ be a fixed sublinear function in n, and let $\{G_n\}_{n \in \mathbb{N}}$ be a family of graphs, where $G_n = ([n], E_n)$ is defined on n vertices, and every vertex of G_n has degree at least $\frac{n}{c} - 1$. Then, for sufficiently large n it holds that:*

1. *There exists a $(\alpha(n), n/c)$-partition of G_n, denoted Γ; it holds that $|\Gamma| \leq c$.*
2. *A $(\alpha(n), n/c)$-partition Γ of G_n can be found in (deterministic) polynomial time, given the $n \times n$ adjacency matrix of G_n.*

Note that if for every n there exists an $\alpha(n)$-cut in G_n, then it immediately follows that $|\Gamma| > 1$, i.e., the partition is not the trivial partition of the set of all nodes.

5.3   Proof of Main Theorem (Theorem 8)

High-level overview of the attack. For $n \in \mathbb{N}$, consider an execution of the alleged parallel broadcast protocol π_n over uniformly distributed n-bit input values for the parties $(x_1, \dots, x_n) \in_R (\{0, 1\}^n)^n$. We define two ensembles of adversarial strategies $\{\mathcal{A}_n^{\text{honest-}i^*}\}_{n \in \mathbb{N}}$ and $\{\mathcal{A}_n^{\text{corrupt-}i^*}\}_{n \in \mathbb{N}}$ (described in full in Sect. 5.3).

The adversary $\mathcal{A}_n^{\text{corrupt-}i^*}$ corrupts a random party P_{i^*}, and simulates an honest execution on a random input $\tilde{x}_{i^*}$ until P_{i^*} has degree $\beta/4$. Next, $\mathcal{A}_n^{\text{corrupt-}i^*}$

switches the internal state of P_{i^*} with a view that is consistent with an honest execution over the initial input x_{i^*}, where all other parties have random inputs. The adversary $\mathcal{A}_n^{\text{corrupt-}i^*}$ continues by computing the $(\alpha(n), n/c)$-partition $\{U_1, \ldots, U_\ell\}$ of the communication graph, (where c is a constant depending only on β – this is possible due to Theorem 9), and blocking every message that is sent between every pair of U_i's. In Lemma 3, we show that there exist honest parties that at the conclusion of the protocol have received a bounded amount of information on the initial input value x_{i^*}.

The second adversary, $\mathcal{A}_n^{\text{honest-}i^*}$, is used for showing that under the previous attack, every honest party will eventually output the initial input value x_{i^*} (Lemma 4). This is done by having $\mathcal{A}_n^{\text{honest-}i^*}$ corrupt all the neighbors of P_{i^*}, while keeping P_{i^*} honest, and simulate the previous attack to the remaining honest parties.

We show that there exist honest parties whose view is identically distributed under both attacks, and since they output x_{i^*} in the latter, they must also output x_{i^*} in the former. By combining both of these lemmata, we then derive a contradiction.

Proof (Proof of Theorem 8). First, since we consider the plain model, without any trusted setup assumptions, known lower bounds [21,34,35] state that parallel broadcast cannot be computed for $t(n) \geq n/3$, therefore, we can focus on $0 < \beta < 1/3$, i.e., the case where $t(n) = \beta \cdot n < n/3$.

Assume toward a contradiction that π is $t(n)$-resilient parallel broadcast protocol in the above setting, and that there exists a sublinear function $\alpha(n) \in o(n)$ such that the communication graph of π is not $\alpha(n)$-edge-connected, i.e., for sufficiently large n there exists a cut $\{S_n, \bar{S}_n\}$ of weight at most $\alpha(n)$.

Notations. We start by defining a few notations. For a fixed n,[10] consider the following independently distributed random variables

$$\text{INPUTSANDCOINS} = \left(X_1, \ldots, X_n, R_1, \ldots, R_n, \tilde{X}_1, \ldots, \tilde{X}_n, \tilde{R}_1, \ldots, \tilde{R}_n, I^* \right),$$

where for every $i \in [n]$, each X_i and $\tilde{X}_i$ take values uniformly at random in the input space $\{0,1\}^n$, each R_i and $\tilde{R}_i$ take values uniformly at random in $\{0,1\}^*$, and I^* takes values uniformly at random in $[n]$. During the proof, (X_i, R_i) represent the pair of input and private randomness of party P_i, whereas $(\tilde{X}_1, \ldots, \tilde{X}_n, \tilde{R}_1, \ldots, \tilde{R}_n, I^*)$ correspond to the random coins of the adversary (used in simulating the two executions towards the honest parties). Unless stated otherwise, all probabilities are taken over these random variables.

Let REDEXEC be a random variable defined as

$$\text{REDEXEC} := \left(X_{-I^*}, \tilde{X}_{I^*}, R_{-I^*}, \tilde{R}_{I^*} \right).$$

That is, REDEXEC contains X_i and R_i for $i \in [n] \setminus \{I^*\}$, along with $\tilde{X}_{I^*}$ and $\tilde{R}_{I^*}$. We denote by the "red execution" an *honest* protocol execution when the inputs

---

[10] For clarity, we denote the random variables without the notation n.

and private randomness of the parties are $(X_{-I^*}, \tilde{X}_{I^*}, R_{-I^*}, \tilde{R}_{I^*})$. We denote by the "blue execution" an *honest* protocol execution when the inputs and private randomness of the parties are $(\tilde{X}_{-I^*}, X_{I^*}, \tilde{R}_{-I^*}, R_{I^*})$. Note that such a sample fully determines the view and transcript of all parties in an *honest* simulated execution of π_n.

Let $\text{FINALCUT}^{\text{corrupt}}$ be a random variable defined over $2^{[n]} \cup \{\bot\}$. The distribution of $\text{FINALCUT}^{\text{corrupt}}$ is defined by running protocol π until its conclusion with adversary $\mathcal{A}_n^{\text{corrupt-}i^*}$ (defined in Sect. 5.3) on inputs and coins sampled according to INPUTSANDCOINS. If at the conclusion of the protocol there is no $\alpha(n)$-cut in the graph, then set the value of $\text{FINALCUT}^{\text{corrupt}}$ to be $\bot$; otherwise, set the value to be the identity of the smallest $\alpha(n)$-cut $\{S, \bar{S}\}$ in the communication graph according to some canonical ordering on the $\alpha(n)$-cuts. We will prove that conditioned on the value of REDEXEC, then $\text{FINALCUT}^{\text{corrupt}}$ can only take one of a constant number of values depending only on β (and not on n).

Let $\mathcal{E}_1$ denote the event that P_{I^*} is the last among all the parties to reach degree $\beta n/4$ in both the red and the blue honest executions of the protocol. More precisely, the event that P_{I^*} reaches degree $\beta n/4$ in both executions, and if it has reached this degree in round ρ in the red (blue) execution, then all parties in the red (blue) execution have degree at least $\beta n/4$ in round ρ.

Let $\mathcal{E}_2$ denote the event that the degree of P_{I^*} reaches $\beta n/4$ in the red execution before, or at the same round as, in the blue execution. Note that $\mathcal{E}_1$ and $\mathcal{E}_2$ are events with respect to two honest executions of the protocol (the red execution and the blue execution) that are defined according to INPUTSANDCOINS. In the adversarial stategies that are used in the proof, the corrupted parties operate in a way that indeed induces the red and blue executions, and so, the events $\mathcal{E}_1$ and $\mathcal{E}_2$ are well defined in an execution of the protocol with those adversarial strategies.

In Sect. 5.3, we formally describe two adversarial strategies, $\mathcal{A}_n^{\text{honest-}i^*}$ and $\mathcal{A}_n^{\text{corrupt-}i^*}$. We denote by $Y_{I^*}^{\text{corrupt}}$, respectively $Y_{I^*}^{\text{honest}}$, the random variable that corresponds to the I^*'th coordinate of the common output of honest parties, when running over random inputs with adversarial strategy $\mathcal{A}_n^{\text{corrupt-}i^*}$, respectively $\mathcal{A}_n^{\text{honest-}i^*}$.

Proof structure. Our proof follows from two main steps. In Lemma 3, stated in Sect. 5.3, we show that in an execution of π_n on random inputs with adversary $\mathcal{A}_n^{\text{corrupt-}i^*}$, it holds that (1) $\Pr[\mathcal{E}_1 \cap \mathcal{E}_2] \geq 1/2n^2 - \text{negl}(n)$, and that (2) conditioned on the event $\mathcal{E}_1 \cap \mathcal{E}_2$, there exists an honest party P_{j^*} such that X_{I^*}, conditioned on $\mathcal{E}_1 \cap \mathcal{E}_2$ and on the view of P_{j^*} at the conclusion of the protocol, still retains at least $n/4$ bits of entropy. This means, in particular, that P_{j^*} will output the value X_{I^*} only with negligible probability. Hence, by *agreement*, the probability for any of the honest parties to output X_{I^*} in an execution with $\mathcal{A}_n^{\text{corrupt-}i^*}$ is negligible. In particular,

$$\Pr\left[Y_{I^*}^{\text{corrupt}} = X_{I^*} \mid \mathcal{E}_1 \cap \mathcal{E}_2\right] = \text{negl}(\kappa).$$

In Lemma 4, stated in Sect. 5.3, we show that in an execution of π on random inputs with adversary $\mathcal{A}_n^{\text{honest-}i^*}$, it holds that (1) with overwhelming probability all honest parties output X_{i^*} (this holds by correctness, since P_{I^*} remains honest), i.e.,

$$\Pr\left[Y_{I^*}^{\text{honest}} = X_{I^*}\right] \geq 1 - \text{negl}(\kappa),$$

and that (2) conditioned on the event $\mathcal{E}_1 \cap \mathcal{E}_2$, there exists an honest party whose view is identically distributed as in an execution with $\mathcal{A}_n^{\text{corrupt-}i^*}$, therefore,

$$\Pr\left[Y_{I^*}^{\text{corrupt}} = Y_{I^*}^{\text{honest}} \mid \mathcal{E}_1 \cap \mathcal{E}_2\right] \geq 1 - \text{negl}(\kappa).$$

From the combination of the two lemmata, we derive a contradiction.   □

Defining Adversarial Strategies. As discussed above, the main idea behind the proof is to construct two dual adversarial strategies that will show that on the one hand, the output of all honest parties must contain the initial value of a randomly chosen corrupted party, and on the other hand, there exist parties that only receive a bounded amount of information on this value during the coarse of the protocol.

We use the following notation for defining the adversarial strategies. Virtual parties that only exist in the head of the adversary are denoted with "tilde". In particular, for a random $i^* \in [n]$, we denote by $\tilde{P}_{i^*}$ a virtual party that emulates the role of P_{i^*} playing with the real parties using a random input in the so-called "red execution," and by $\{\tilde{Q}_i\}_{i \neq i^*}$ virtual parties that emulate an execution over random inputs towards P_{i^*}.[11]

The adversary $\mathcal{A}_n^{\text{honest-}i^*}$. At a high level, the adversary $\mathcal{A}_n^{\text{honest-}i^*}$ chooses a random $i^* \in [n]$ and isolates the honest party P_{i^*}. The adversary $\mathcal{A}_n^{\text{honest-}i^*}$ consists of three phases. In Phase I, $\mathcal{A}_n^{\text{honest-}i^*}$ induces two honestly distributed executions.

- The first (red) execution is set by simulating an honest execution of a virtual party $\tilde{P}_{i^*}$ over a random input $\tilde{x}_{i^*}$ towards all other parties. The adversary corrupts any party that sends a message to P_{i^*}, blocks its message, and simulates $\tilde{P}_{i^*}$ receiving this message. Whenever $\tilde{P}_{i^*}$ should send a message to some P_j, the adversary corrupts the P_j, and instructs him to proceed as if he received the intended message from $\tilde{P}_{i^*}$.
- For the second (blue) execution, $\mathcal{A}_n^{\text{honest-}i^*}$ emulates a virtual execution with virtual parties $(\tilde{Q}_1, \ldots, \tilde{Q}_n) \setminus \{\tilde{Q}_{i^*}\}$ on random inputs towards the honest party P_{i^*}. To do so, whenever P_{i^*} sends a message to P_j in the real execution, the adversary corrupts P_j, instructing him to ignore this message, and simulates this message from P_{i^*} to $\tilde{Q}_j$ in the virtual execution (that is running in the head of the adversary). Whenever a party $\tilde{Q}_j$ sends a message to P_{i^*} in the virtual execution, the adversary corrupts the real party P_j and instructs him to send this message to P_{i^*} in the real execution.

---

[11] Following the *red pill blue pill* paradigm, in the adversarial strategy $\mathcal{A}_n^{\text{honest-}i^*}$, the chosen party P_{i^*} is participating (without knowing it) in the *blue* execution, which is a fake execution that does not happen in the real world. The real honest parties participate in the *red* execution, where the adversary simulates P_{i^*} by running a virtual party.

Phase II begins when the degree of P_{i^*} in the *red* execution is at least $(\beta/4)\cdot n$; if P_{i^*} reaches this threshold faster in the *blue* execution, the attack fails. Phase III begins when the degree of P_{i^*} in the *real* execution is at least $(\beta/4) \cdot n$.

Ideally, Phase I will continue until all parties in the real execution have a linear degree, and before the adversary will use half of his "corruption budget", i.e., $(\beta/2) \cdot n$. This would be the case if we were to consider a single honest execution of the protocol, since we show that there always exists a party that will be the last to reach the linear-degree threshold with a noticeable probability. However, as the attack induces two *independent* executions, in which the degree of the parties can grow at different rates, care must be taken. We ensure that even though P_{i^*} runs in the blue execution, by the time P_{i^*} will reach the threshold, all other parties (that participate in the red execution) will already have reached the threshold, and can be partitioned into "minimal" $\alpha(n)$-cuts, as follows.

The adversary allocates $(\beta/4)\cdot n$ corruptions for the red execution and $(\beta/4)\cdot n$ corruptions for the blue execution. We show that with a noticeable probability, once $\tilde{P}_{i^*}$ has degree $(\beta/4) \cdot n$ in the red execution, all other parties in the red execution also have high degree. Consider the communication graph of the red execution without the virtual party $\tilde{P}_{i^*}$ (i.e., after removing the node i^* and its edges); by Theorem 9 there exists an $(\alpha(n), (\beta/4)n - 1)$ partition of this graph into a constant number of linear-size subsets that are connected with sublinear many edges, denoted $\Gamma = \{U_1, \ldots, U_\ell\}$ (in particular, this partition is independent of x_{i^*}). In Phase II, the adversary continues blocking outgoing messages from P_{i^*} towards the real honest parties, until the degree of P_{i^*} in the real execution is $\beta n/4$. In addition, $\mathcal{A}_n^{\mathsf{honest}\text{-}i^*}$ blocks any message that is sent between two subsets in the partition, by corrupting the recipient and instructing him to ignore messages from outside of his subset.

In Phase III, which begins when P_{i^*} has high degree in the real execution, the adversary adds P_{i^*} to one of the subsets in the partition, in which P_{i^*} has many neighbors, and continues to block messages between different subsets in the partition until the conclusion of the protocol.

We note that special care must be taken in the transition between the phases, since such a transition can happen in a middle of a round, after processing some of the messages, but not all. Indeed, if the transition to the next phase will happen at the end of the round, the adversary may need to corrupt too many parties. For this reason, in Phases I and II, we analyze the messages to and from P_{i^*} one by one, and check whether the threshold has been met after each such message.

The adversary $\mathcal{A}_n^{\mathsf{corrupt}\text{-}i^*}$. The adversary $\mathcal{A}_n^{\mathsf{corrupt}\text{-}i^*}$ corrupts the randomly chosen party P_{i^*}, and emulates the operations of an honest P_{i^*} that is being attacked by $\mathcal{A}_n^{\mathsf{honest}\text{-}i^*}$.

In Phase I, the adversary $\mathcal{A}_n^{\mathsf{corrupt}\text{-}i^*}$ induces two honestly distributed executions, by simulating an honest execution of a virtual party $\tilde{P}_{i^*}$ over a random input $\tilde{x}_{i^*}$ towards all other honest parties (the red execution), and furthermore, runs in its mind a virtual execution over the initial input x_{i^*} and random inputs $\tilde{x}_i$ for $i \neq i^*$ (the blue execution). This phase continues until $\tilde{P}_{i^*}$ has degree

$\beta n/4$ in the red execution (no parties other than P_{i^*} are being corrupted). If all other parties in the red execution have high degree, then the adversary finds the partition of the red graph as in the previous attack (the partition is guaranteed by Theorem 9).

In Phase II, the adversary continues simulating the corrupted P_{i^*} towards the real honest parties until the degree of P_{i^*} in the real execution is $\beta n/4$; however, his communication is based on the view in the blue execution at the end of Phase I (this is no longer an honest-looking execution). During this phase, $\mathcal{A}_n^{\text{corrupt-}i^*}$ blocks any message that is sent between two subsets in the partition.

In Phase III, that begins when P_{i^*} has high degree (in the real execution), $\mathcal{A}_n^{\text{corrupt-}i^*}$ adds P_{i^*} to one of the subsets in the partition, in which P_{i^*} has many neighbors, and continues to block messages between different subsets in the partition until the conclusion of the protocol.

The Core Lemmata. In the full version [4] we prove the following core lemmata that conclude the proof of the theorem.

Lemma 3. *Consider an execution of π_n on random inputs $(X_1, \ldots, X_n)$ for the parties with adversary $\mathcal{A}_n^{\text{corrupt-}i^*}$, and the events $\mathcal{E}_1$ and $\mathcal{E}_2$ as defined in Sect. 5.3. Then, it holds that:*

1. *$\Pr[\mathcal{E}_1 \cap \mathcal{E}_2] \geq 1/2n^2 - \text{negl}(n)$.*
2. *Conditioned on the event $\mathcal{E}_1 \cap \mathcal{E}_2$, there exists an honest party P_{J^*} such that*

$$H(X_{I^*} \mid \mathcal{E}_1 \cap \mathcal{E}_2, \text{VIEW}_{J^*}^{\text{corrupt}}) \geq n/4,$$

where $\text{VIEW}_{J^*}^{\text{corrupt}}$ is the random variable representing the view of P_{J^*} at the end of the protocol.

Lemma 4. *Consider an execution of π_n on random inputs $(X_1, \ldots, X_n)$ for the parties with adversary $\mathcal{A}_n^{\text{honest-}i^*}$. Then, conditioned on the event $\mathcal{E}_1 \cap \mathcal{E}_2$ it holds that:*

1. *The I^*'th coordinate of the common output $Y_{I^*}^{\text{honest}}$ equals the initial input X_{I^*} of P_{I^*}, except for negligible probability, i.e.,*

$$\Pr\left[Y_{I^*}^{\text{honest}} = X_{I^*} \mid \mathcal{E}_1 \cap \mathcal{E}_2\right] \geq 1 - \text{negl}(n).$$

2. *The I^*'th coordinate of the common output $Y_{I^*}^{\text{honest}}$ in an execution with $\mathcal{A}_n^{\text{honest-}i^*}$ equals the I^*'th coordinate of the common output $Y_{I^*}^{\text{corrupt}}$ in an execution with $\mathcal{A}_n^{\text{corrupt-}i^*}$, except for negligible probability, i.e.,*

$$\Pr\left[Y_{I^*}^{\text{honest}} = Y_{I^*}^{\text{corrupt}} \mid \mathcal{E}_1 \cap \mathcal{E}_2\right] \geq 1 - \text{negl}(n).$$

References

1. Beaver, D., Micali, S., Rogaway, P.: The round complexity of secure protocols (extended abstract). In: STOC, pp. 503–513 (1990)
2. Ben-Or, M., Goldwasser, S., Wigderson, A.: Completeness theorems for non-cryptographic fault-tolerant distributed computation (extended abstract). In: FOCS, pp. 1–10 (1988)
3. Boyle, E., Chung, K.-M., Pass, R.: Large-scale secure computation: multi-party computation for (parallel) RAM programs. In: Gennaro, R., Robshaw, M. (eds.) CRYPTO 2015, Part II. LNCS, vol. 9216, pp. 742–762. Springer, Heidelberg (2015). https://doi.org/10.1007/978-3-662-48000-7_36
4. Boyle, E., Cohen, R., Data, D., Hubáček, P.: Must the communication graph of MPC protocols be an expander? Cryptology ePrint Archive, Report 2018/540 (2018). https://eprint.iacr.org/2018/540
5. Boyle, E., Goldwasser, S., Tessaro, S.: Communication locality in secure multi-party computation. In: Sahai, A. (ed.) TCC 2013. LNCS, vol. 7785, pp. 356–376. Springer, Heidelberg (2013). https://doi.org/10.1007/978-3-642-36594-2_21
6. Braud-Santoni, N., Guerraoui, R., Huc, F.: Fast Byzantine agreement. In: PODC, pp. 57–64 (2013)
7. Canetti, R.: Security and composition of multiparty cryptographic protocols. JCRYPTOL 13(1), 143–202 (2000)
8. Canetti, R.: Universally composable security: a new paradigm for cryptographic protocols. In: FOCS, pp. 136–145 (2001)
9. Chandran, N., Chongchitmate, W., Garay, J.A., Goldwasser, S., Ostrovsky, R., Zikas, V.: The hidden graph model: communication locality and optimal resiliency with adaptive faults. In: ITCS, pp. 153–162 (2015)
10. Chandran, N., Garay, J.A., Ostrovsky, R.: Almost-everywhere secure computation with edge corruptions. JCRYPTOL 28(4), 745–768 (2015)
11. Chaum, D., Crépeau, C., Damgård, I.: Multiparty unconditionally secure protocols (extended abstract). In: STOC, pp. 11–19 (1988)
12. Cohen, R., Coretti, S., Garay, J.A., Zikas, V.: Round-preserving parallel composition of probabilistic-termination cryptographic protocols. In: ICALP, pp. 37:1–37:15 (2017)
13. Cohen, R., Coretti, S., Garay, J.A., Zikas, V.: Probabilistic termination and composability of cryptographic protocols. In: Robshaw, M., Katz, J. (eds.) CRYPTO 2016, Part III. LNCS, vol. 9816, pp. 240–269. Springer, Heidelberg (2016). https://doi.org/10.1007/978-3-662-53015-3_9
14. Damgård, I., Ishai, Y.: Constant-round multiparty computation using a black-box pseudorandom generator. In: Shoup, V. (ed.) CRYPTO 2005. LNCS, vol. 3621, pp. 378–394. Springer, Heidelberg (2005). https://doi.org/10.1007/11535218_23
15. Dani, V., King, V., Movahedi, M., Saia, J., Zamani, M.: Secure multi-party computation in large networks. Distrib. Comput. 30(3), 193–229 (2017)
16. Dolev, D.: The Byzantine generals strike again. J. Algorithms 3(1), 14–30 (1982)
17. Dolev, D., Strong, R.: Authenticated algorithms for Byzantine agreement. SICOMP 12(4), 656–666 (1983)
18. Dvir, Z., Wigderson, A.: Monotone expanders: constructions and applications. Theory Comput. 6(1), 291–308 (2010)
19. Dwork, C., Peleg, D., Pippenger, N., Upfal, E.: Fault tolerance in networks of bounded degree. SICOMP 17(5), 975–988 (1988)

20. Feldman, P., Micali, S.: An optimal probabilistic protocol for synchronous Byzantine agreement. SICOMP **26**(4), 873–933 (1997)
21. Fischer, M.J., Lynch, N.A., Merritt, M.: Easy impossibility proofs for distributed consensus problems. Distrib. Comput. **1**(1), 26–39 (1986)
22. Garay, J.A., Moses, Y.: Fully polynomial Byzantine agreement in t+1 rounds. In: STOC, pp. 31–41 (1993)
23. Garay, J.A., Ostrovsky, R.: Almost-everywhere secure computation. In: Smart, N. (ed.) EUROCRYPT 2008. LNCS, vol. 4965, pp. 307–323. Springer, Heidelberg (2008). https://doi.org/10.1007/978-3-540-78967-3_18
24. Goldreich, O.: Foundations of Cryptography - Volume 2: Basic Applications. Cambridge University Press, Cambridge (2004)
25. Goldreich, O., Micali, S., Wigderson, A.: How to play any mental game or a completeness theorem for protocols with honest majority. In: STOC, pp. 218–229 (1987)
26. Halevi, S., Ishai, Y., Jain, A., Kushilevitz, E., Rabin, T.: Secure multiparty computation with general interaction patterns. In: ITCS, pp. 157–168 (2016)
27. Halevi, S., Lindell, Y., Pinkas, B.: Secure computation on the web: computing without simultaneous interaction. In: Rogaway, P. (ed.) CRYPTO 2011. LNCS, vol. 6841, pp. 132–150. Springer, Heidelberg (2011). https://doi.org/10.1007/978-3-642-22792-9_8
28. Hoory, S., Linial, N., Wigderson, A.: Expander graphs and their applications. Bull. Am. Math. Soc. **43**(4), 439–561 (2006)
29. Kapron, B.M., Kempe, D., King, V., Saia, J., Sanwalani, V.: Fast asynchronous Byzantine agreement and leader election with full information. In: SODA, pp. 1038–1047 (2008)
30. King, V., Lonargan, S., Saia, J., Trehan, A.: Load balanced scalable Byzantine agreement through quorum building, with full information. In: Aguilera, M.K., Yu, H., Vaidya, N.H., Srinivasan, V., Choudhury, R.R. (eds.) ICDCN 2011. LNCS, vol. 6522, pp. 203–214. Springer, Heidelberg (2011). https://doi.org/10.1007/978-3-642-17679-1_18
31. King, V., Saia, J.: From almost everywhere to everywhere: Byzantine agreement with $\tilde{O}(n^{3/2})$ bits. In: Keidar, I. (ed.) DISC 2009. LNCS, vol. 5805, pp. 464–478. Springer, Heidelberg (2009). https://doi.org/10.1007/978-3-642-04355-0_47
32. King, V., Saia, J.: Breaking the $O(n^2)$ bit barrier: scalable Byzantine agreement with an adaptive adversary. In: PODC, pp. 420–429 (2010)
33. King, V., Saia, J., Sanwalani, V., Vee, E.: Scalable leader election. In: SODA, pp. 990–999 (2006)
34. Lamport, L., Shostak, R.E., Pease, M.C.: The Byzantine generals problem. ACM Trans. Program. Lang. Syst. **4**(3), 382–401 (1982)
35. Pease, M.C., Shostak, R.E., Lamport, L.: Reaching agreement in the presence of faults. J. ACM **27**(2), 228–234 (1980)
36. Rabin, T., Ben-Or, M.: Verifiable secret sharing and multiparty protocols with honest majority (extended abstract). In: FOCS, pp. 73–85 (1989)
37. Seito, T., Aikawa, T., Shikata, J., Matsumoto, T.: Information-theoretically secure key-insulated multireceiver authentication codes. In: Bernstein, D.J., Lange, T. (eds.) AFRICACRYPT 2010. LNCS, vol. 6055, pp. 148–165. Springer, Heidelberg (2010). https://doi.org/10.1007/978-3-642-12678-9_10
38. Shikata, J., Hanaoka, G., Zheng, Y., Imai, H.: Security notions for unconditionally secure signature schemes. In: Knudsen, L.R. (ed.) EUROCRYPT 2002. LNCS, vol. 2332, pp. 434–449. Springer, Heidelberg (2002). https://doi.org/10.1007/3-540-46035-7_29

Two-Round Multiparty Secure Computation Minimizing Public Key Operations

Sanjam Garg, Peihan Miao, and Akshayaram Srinivasan[(✉)]

University of California, Berkeley, Berkeley, USA
{sanjamg,peihan,akshayaram}@berkeley.edu

Abstract. We show new constructions of semi-honest and malicious two-round multiparty secure computation protocols using only (a fixed) $\mathsf{poly}(n, \lambda)$ invocations of a two-round oblivious transfer protocol (which use expensive public-key operations) and $\mathsf{poly}(\lambda, |C|)$ cheaper one-way function calls, where λ is the security parameter, n is the number of parties, and C is the circuit being computed. All previously known two-round multiparty secure computation protocols required $\mathsf{poly}(\lambda, |C|)$ expensive public-key operations.

1 Introduction

Secure multiparty computation (MPC) allows a set of mutually distrusting parties to compute a joint function on their private inputs with the guarantee that only the output of the function is revealed and everything else about the private inputs of the parties is hidden. This is a classic problem in cryptography and was originally studied by Yao [Yao82] for the case of two parties. Later, Goldreich, Micali and Wigderson [GMW87] considered the multiparty case and gave protocols for securely computing any multiparty functionality.

A key metric in determining the efficiency of a secure computation protocol is its *round complexity* or in other words, the number of sequential messages exchanged between the parties. Starting with the first constant round protocol by Beaver, Micali and Rogaway [BMR90], there has been a tremendous amount of research to reduce the round complexity to its *absolute minimum*. It was shown in [HLP11] that two rounds are necessary to securely compute certain functionalities and a sequence of works have tried to realize this goal. The first two-round construction was obtained by Garg, Gentry, Halevi and Raykova based on indistinguishability obfuscation [GGHR14, GGH+13]. Subsequently, a sequence of works improved the needed assumptions, first to witness encryption [GLS15, GGSW13], and then to learning with errors assumption

Research supported in part from DARPA/ARL SAFEWARE Award W911NF15C0210, AFOSR Award FA9550-15-1-0274, AFOSR YIP Award, DARPA and SPAWAR under contract N66001-15-C-4065, a Hellman Award and research grants by the Okawa Foundation, Visa Inc., and Center for Long-Term Cybersecurity (CLTC, UC Berkeley). The views expressed are those of the author and do not reflect the official policy or position of the funding agencies.

© International Association for Cryptologic Research 2018
H. Shacham and A. Boldyreva (Eds.): CRYPTO 2018, LNCS 10993, pp. 273–301, 2018.
https://doi.org/10.1007/978-3-319-96878-0_10

[MW16, BP16, PS16]. Improving these results, recent works obtained two-round constructions based on the DDH assumption [BGI16, BGI17b] (for the case of constant number of parties) or on bilinear maps [GS17] (in the general case). Finally, very recent results have also yielded constructions based on the minimal assumption of two-round oblivious transfer [BL18, GS18].

Apart from round complexity, another metric that is crucial for computational efficiency in MPC protocols is the *number of public-key operations* performed by each party. Typically, public key operations are orders of magnitude more expensive than symmetric key operations and minimizing them typically leads to more efficient protocols. The question of minimizing public key operations in secure computation was first considered by Beaver [Bea96] for the case of oblivious transfer. In particular, Beaver gave a construction for obtaining a large number $L \gg \lambda$ of oblivious transfers (OTs) using only a fixed number λ public key operations along with the use of $\mathsf{poly}(L)$ cheaper one-way function calls. This task of extending λ OTs to a larger L OTs using only one-way functions is referred to as oblivious transfer extension. Following Beaver's result, a rich line of work [IKNP03, Nie07, HIKN08, KK13] gave concretely efficient protocols for OT extension which have served as a crucial ingredient in the design of several concretely efficient secure computation protocols [HIK07, NNOB12, ALSZ17, KRS16].

In this work, we are interested in getting the best of both worlds, namely, constructing two-round MPC protocols while minimizing the number of public-key operations performed. Indeed, the number of public-key operations in the prior two-round MPC protocols grows with the size of the circuit computed. Given this state of affairs, we would like to address the following question.

Can we construct two-round, secure multiparty computation protocols where the number of public key operations performed by each party is independent of the size of the circuit being computed?

1.1   Our Results

We give a positive answer to the above question. We show new constructions of semi-honest and malicious two-round, multiparty computation protocols where the number of public key operations performed by each party is a *fixed polynomial* (in the security parameter and the number of participants) and is *independent* of the circuit size of the function being computed. Further, we prove the security of these protocols under the *minimal assumption* that two-round semi-honest/malicious oblivious transfer (OT) exists. More formally, our main theorem is:

Theorem 1 (Informal). *Let $\mathcal{X} \in \{$semi-honest in plain model, malicious in common random/reference sting model$\}$. Assuming the existence of a two-round $\mathcal{X}$ secure OT protocol, there exists a two-round, $\mathcal{X}$ secure, n-party protocol computing a function f (represented as a circuit C_f) where the number of public key operations performed by each party is $\mathsf{poly}(n, \lambda)$. Here, $\mathsf{poly}(\cdot)$ is a fixed polynomial independent of $|C_f|$ and λ is the security parameter.*

The focus of this work is theoretical feasibility rather than concrete optimization of the polynomial. We leave the goal of obtaining concretely efficient protocols for future work. Additionally, in the malicious case, this work focuses on obtaining protocols in the common random/reference string model. Obtaining round optimal MPC protocols in the plain model [GMPP16, ACJ17, BHP17, COSV17, HHPV17, BGJ+17, BL18] has been a problem of significant interest and we expect that our techniques will be useful in reducing the number of public-key operations needed in these protocols. We leave this as an open problem.

2   Technical Overview

In this section, we give a high-level overview of the main challenges and the techniques used to overcome them in our construction of two-round MPC protocols minimizing the number of public key operations.

Starting Point. The starting point of our work is the recent results of Benhamouda and Lin [BL18] and Garg and Srinivasan [GS18] that provide constructions of two-round, secure multiparty computation (MPC) protocol based on two-round oblivious transfer. These works provide a method of squishing the round complexity of an arbitrary round secure computation protocol to just two rounds. The key idea behind this method is the concept of "talking garbled circuits," i.e., garbled circuits that can interact with each other by sending and receiving messages. Let us briefly explain how this primitive helps in squishing the round complexity of a multi-round MPC protocol.

To squish the round complexity, each party generates "talking garbled circuits" that emulates its actions as per the specification of the multi-round MPC protocol. The parties then broadcast these "talking garbled circuits" so that every party has access to the "talking garbled circuits" of every other party. Finally, all parties evaluate these "talking garbled circuits" that internally executes the multi-round MPC protocol. This step does not involve any further interactions between the parties. Thus, the only overhead in the round complexity of this approach is the number of rounds needed for generating the "talking garbled circuits."

Let us give a very high level overview of how the "talking garbled circuits" are generated. In these two works, the "talking garbled circuits" are generated via a two-round protocol that makes use of (plain) garbled circuits and two-round oblivious transfer (OT).[1] At the end of the two rounds, every party has access to every other party's "talking garbled circuits" and can evaluate them without any further interaction. The first round of this two-round protocol can be visualized as setting up a channel for the garbled circuits to communicate. Without going into the actual details on how this is achieved, we note that this step involves generating several first round OT messages. Next, in the second

---

[1] Recall that in a two-round oblivious transfer, the first message is generated by the receiver and it encodes the receiver's choice bit and the second message is generated by the sender and it encodes its two messages.

round, the actual garbled circuits are sent which interact with each other via the channel set up in the first round. Again, without going into the details, a message sent from one party (the sender) to another party (the receiver) is communicated via the sender's garbled circuit outputting the randomness used in generating a subset of the first round OT messages and the receiver's garbled circuit outputting some second round OT messages.

Computational Overhead. One major source of inefficiency in the approaches of [BL18, GS18] is the number of expensive OT instances needed. In particular, these protocols use $\Omega(1)$ OTs in enabling the garbled circuits to communicate a single bit. Hence, the number of OTs needed for compiling an arbitrary secure computation protocol grows with the circuit size of the function being computed.[2] Our goal is to remove this dependency between the number of OTs needed and the circuit size of the function being computed.

Can We Use OT Extension? A natural first attempt to minimize the number of instances of oblivious transfer would to be use an OT extension protocol [Bea96, IKNP03]. We need this OT extension protocol to run in two-rounds, as otherwise the protocol for computing "talking garbled circuits" will run in more rounds. Further, we need the OT extension protocol to satisfy the following three properties for it to be useful in constructing "talking garbled circuits." We also explain why a general two-round OT satisfies each of these properties.

1. **Delegatability.** For every OT computed between a sender and a receiver, the receiver should be able to delegate its decryption capabilities for that OT to any party by revealing a decryption key. This key and the transcript could then be used to compute the message that the receiver would have obtained in the OT execution. A general two-round OT satisfies delegatability as revealing the receiver's random coins allows any party to obtain the receiver's message.
2. **Independence.** We require independence between multiple parallel invocations of the underlying OT protocol. More specifically, revealing the receiver's delegation key for one of the instances of an OT execution does not affect the receiver security for the other OTs. Again, a general two-round OT satisfies independence as each OT instance is generated using an independent random tape.
3. **Availability of Delegation Keys.** The keys for delegating the decryption must be available at the end of the first round i.e., after the receiver sends its message. This property is trivially satisfied by a two-round OT as the delegation key is in fact the receiver's random tape.

Let us first explain the intuition on why these three properties are required for the construction of "talking garbled circuits." The delegatability property is required since the garbled circuits sent in the second round reveal the delegation keys for a subset of the OT messages generated in the first round. Recall that this is required for one garbled circuit to send a message to another. The key

---

[2] In fact, the number of OTs grows with the computational complexity of the underlying multiparty protocol.

availability property is needed since the delegation keys are to be hardwired in the second round garbled circuits so that the appropriate delegation keys can be output by these circuits during evaluation. The independence property is needed since the second round garbled circuits reveal the delegation keys for only a subset of the first round OT messages. We need the other OT messages to still be secure.

We stress that even though the above three properties are trivially satisfied by every two-round OT, a two-round OT extension protocol need not satisfy all of them. To demonstrate this, let us first see why does the two-round version of Beaver's OT extension protocol [Bea96, GMMM17] not satisfy all the properties.

Why doesn't beaver's OT extension work? In order to understand why this does not work, we first recall a two-round version [GMMM17] of the OT extension protocol of Beaver that expands λ two-round, base OTs to $L = \text{poly}(\lambda)$ OTs. In the first round of the OT extension protocol, the receiver (having input $c \in \{0, 1\}^L$) samples a "short" seed s of a PRG $: \{0, 1\}^\lambda \rightarrow \{0, 1\}^L$ and computes $e = c \oplus \text{PRG}(s)$. Additionally, it computes λ first round OT messages using s as its choice bits. It sends these OT messages along with e to the sender. The sender garbles a circuit C that has its messages $\{\text{msg}_{i,0}, \text{msg}_{i,1}\}_{i \in [L]}$ hardwired along with the string e received in the first round. The circuit C takes as input the λ-bit string s, expands it to L bits using the PRG and uses it to unmask e to obtain c. Specifically, it computes $c := e \oplus \text{PRG}(s)$, and outputs $\{\text{msg}_{i,c[i]}\}_{i \in [L]}$. The sender sends this garbled circuit and uses the λ second round OT messages to communicate the labels of the garbled circuit to the receiver. The receiver decrypts the labels corresponding to the bits of its seed s and uses it to evaluate the garbled circuit to obtain $\{\text{msg}_{i,c[i]}\}_{i \in [L]}$.

The above OT extension protocol of Beaver is delegatable as revealing all the randomness used by the receiver allows any party to decrypt all the messages. However, the protocol does not satisfy the independence requirement as the randomness used for generating L different OTs is highly correlated. In fact, revealing all the random coins for generating the first round OT messages compromises the security of all the L OTs.

Delegatable and Independent Two-Round OT Extension. Towards constructing an OT extension that satisfies all the properties, we first construct a protocol that is both delegatable and independent. In the new protocol, the receiver's first round message is the same as before. However, the sender's message is generated differently. In particular, the sender samples a set of masks $M = \{\text{m}_{i,0}, \text{m}_{i,1}\}_{i \in [L]}$ where each mask $\text{m}_{i,b}$ is a random string with the same length as $\text{msg}_{i,b}$. It constructs the circuit C (described above) with the set of masks hardwired in place of the messages. It garbles this circuit. It additionally computes $ct_{i,b} = \text{msg}_{i,b} \oplus \text{m}_{i,b}$ for each $i \in [L]$ and $b \in \{0, 1\}$ and sends the garbled circuit, the set $\{ct_{i,b}\}_{i \in [L], b \in \{0,1\}}$ and λ second round OT messages to communicate the labels of the garbled circuit to the receiver. The receiver then recovers the labels corresponding to its seed s, evaluates the garbled circuit to obtain $\{\text{m}_{i,c[i]}\}_{i \in [L]}$, and computes $\text{msg}_{i,c[i]} = ct_{i,c[i]} \oplus \text{m}_{i,c[i]}$ for every $i \in [L]$.

This scheme is delegatable as the receiver can use $m_{i,c[i]}$ as the delegation key. It is also independent, as revealing $m_{i,c[i]}$ does not leak any information of $c[k]$ for $k \neq i$. However, this construction does not satisfy the third property, namely key availability. This is because $m_{i,c[i]}$ can be computed by the receiver only at the end of the second round and is not available at the end of the first round.

Weakening the Key Availability Property. We first observe that we can in fact, weaken the key availability property. Recall that the key availability property requires the delegation keys to be available at the end of the first round so that they can be hardwired inside the garbled circuits that performs the communication. However, for the construction to work, we just need the delegation keys to be given as inputs to these garbled circuits and need not be hardwired. We will now construct a two-round, OT extension that satisfies the weakened key availability property. For the ease of exposition, let us overload the notation and call the these communicating garbled circuits (sent in the second round) as "talking garbled circuits."

Satisfying All Properties. Recall that the problem with the previous approach was because the receiver could evaluate the sender's garbled circuit only at the end of the second round. Our solution to the key availability problem is in having the receiver "offload" its evaluation of this garbled circuit. This solution makes use of the fact that in the MPC setting the sender and the receiver are connected via a *simultaneous message exchange* model. At a high level, we require the sender to send its garbled circuit in the first round. The receiver now garbles a wrap-circuit, which has the sender's garbled circuit hardwired in it. This wrap-circuit evaluates the sender's circuit inside and translates its output to the labels of the "talking garbled circuits." In particular, the receiver "offloads" the evaluation of the sender's garbled circuit via the wrap-circuit which helps in achieving the weakened key availability property. Let us explain our idea in more detail.

Key Idea: "Offloading" Garbled Circuit Evaluation. We first give the description of the protocol and then explain why it satisfies all the three properties. The key steps in the protocol are depicted in Fig. 1.

In the new protocol, the receiver's first round message is unchanged. Additionally, in the first round, the sender samples the random set M as before and constructs a circuit C_B that has the set M hardwired in it. This circuit takes as input a seed s, expands it using the PRG and outputs $\{m_{i,\mathsf{PRG}(s)[i]}\}_{i \in [L]}$. The sender garbles C_B to obtain a garbled circuit $\widetilde{C}_B$ and sends this to the receiver.

In the second round, the sender computes $ct_{i,0} = \mathsf{msg}_{i,0} \oplus m_{i,e[i]}$ and $ct_{i,1} = \mathsf{msg}_{i,1} \oplus m_{i,1-e[i]}$ (where e is obtained from the receiver's first round message) and sends $\{ct_{i,b}\}_{i \in [L], b \in \{0,1\}}$ to the receiver. The receiver constructs a wrap-circuit C_{wrap} that has $\widetilde{C}_B$ and the input labels for the "talking garbled circuits" hardwired in it. C_{wrap} takes as input the labels for evaluating $\widetilde{C}_B$, evaluates it using these labels to obtain $\{m_{i,\mathsf{PRG}(s)[i]}\}_{i \in [L]}$, and outputs a set of labels corresponding to $\{m_{i,\mathsf{PRG}(s)[i]}\}_{i \in [L]}$. The output will later be treated as the input

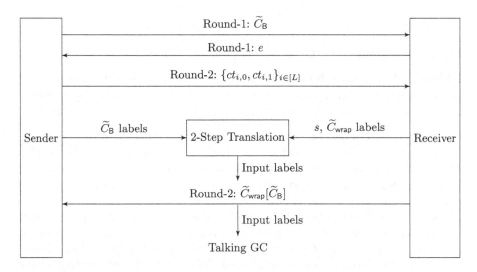

Fig. 1. Semi-honest OT extension satisfying delegatability, independence and weakened key availability

labels for evaluating the "talking garbled circuits." The receiver garbles C_{wrap} and sends the garbled circuit $\widetilde{C}_{\mathsf{wrap}}$ to the sender.

Notice that $m_{i,\mathsf{PRG}(s)[i]}$ can serve as the delegation keys as it can be used to unmask $ct_{i,c[i]}$ to obtain $\mathsf{msg}_{i,c[i]}$, and the other message $\mathsf{msg}_{i,1-c[i]}$ is hidden. This approach inherits the delegatability and independence from the previous approach. Now, this scheme also satisfies the weakened key availability property! In particular, the delegation keys are passed to the "talking garbled circuits" via the wrap circuit.

How to obtain labels for evaluating $\widetilde{C}_{\mathsf{wrap}}$? However, there is one question that we have not answered yet. In particular, how to obtain the labels for evaluating the garbled wrap-circuit $\widetilde{C}_{\mathsf{wrap}}$? Recall that the warp-circuit C_{wrap} takes as input the labels for evaluating $\widetilde{C}_{\mathsf{B}}$. Hence, to evaluate $\widetilde{C}_{\mathsf{wrap}}$ we need its input labels that correspond to the labels for evaluating $\widetilde{C}_{\mathsf{B}}$. We therefore need a two-step translation mechanism: one from the seed s to the labels for evaluating $\widetilde{C}_{\mathsf{B}}$ and then from these labels to the labels for evaluating $\widetilde{C}_{\mathsf{wrap}}$.

For this purpose, we use the two-round MPC protocol from [BL18,GS18] to securely compute the two-step translation functionality. This functionality takes as input the seed s and the set of labels for $\widetilde{C}_{\mathsf{wrap}}$ from the receiver and the set of labels for $\widetilde{C}_{\mathsf{B}}$ from the sender. It first chooses the labels of $\widetilde{C}_{\mathsf{B}}$ that correspond to the string s. It then outputs the labels of $\widetilde{C}_{\mathsf{wrap}}$ that correspond to those chosen labels of $\widetilde{C}_{\mathsf{B}}$. Given such a two-round MPC protocol, we can run this protocol in parallel of the aforementioned protocol to obtain the labels for evaluating $\widetilde{C}_{\mathsf{wrap}}$. We then evaluate $\widetilde{C}_{\mathsf{wrap}}$ to obtain the labels for evaluating the "talking garbled circuits." Note that the circuit size computing this two-step translation functionality is polynomially dependent on λ and is independent of L and hence we can use these two-round MPC results to securely compute this functionality. This helps in minimizing the number of public key operations.

Tackling Malicious Adversaries. Plugging the above OT extension protocol into the compilers of [BL18, GS18] gives us the desired result in the semi-honest setting. However, a couple of major challenges arise in the malicious setting.

1. **Adaptive Security.** The first issue arises because a malicious receiver might wait until it receives the garbled circuit $\widetilde{C}_\mathsf{B}$ before choosing its seed s. This leads to adaptive security issues [BHR12] in garbling C_B.
2. **Input Dependent Abort.** The second issue arises because a malicious sender might generate an ill-formed $\widetilde{C}_\mathsf{B}$ that may lead to an honest receiver to abort on specific choices of the receiver's input. This leaks information about the receiver's input to the sender. To give a concrete example, a corrupted sender might generate $\widetilde{C}_\mathsf{B}$ such that it outputs $\bot$ if the first bit of $\mathsf{PRG}(s)$ is 1 instead of outputting the valid mask. Thus, if the honest receiver aborts then the sender can recover $c[1]$ from $e[1]$.

Solving these two issues requires development of new tools and techniques which we now elaborate.

Solving Adaptive Security Issue. A tempting approach to solving this issue is use the recent constructions of adaptively secure garbling [HJO+16, JW16, JKK+17] to generate $\widetilde{C}_\mathsf{B}$. However, this does not work! Recall that the length of the garbled input of an adaptively secure garbling scheme must at least grow with the output length of the circuit [AIKW13]. In our case, the output length of C_B is L, hence the garbled input of $\widetilde{C}_\mathsf{B}$ grows with L. Therefore, the circuit size of the two-step translation functionality that first translates the seed s to the garbled input of $\widetilde{C}_\mathsf{B}$ must grow with L. This implies that the number of public key operations in the two-round protocol that securely computes this functionality grows with L. This kills the efficiency of the overall protocol.

On the one hand, we need our garbling scheme to satisfy the stronger notion of adaptive security and on the other hand, we need to minimize the number of public key operations. These two requirements seem contradictory to each other and it seems that we need to trade one requirement in order to achieve the other. We resolve this deadlock by observing that full blown adaptive security is not needed in garbling C_B. We note that it is sufficient for this garbling scheme to be *somewhere adaptive*. Let us explain this in more detail.

To understand our approach, the first step is to break the circuit C_B down to L individual circuits $C_1, \ldots, C_L$ where C_i has $\{\mathsf{m}_{i,0}, \mathsf{m}_{i,1}\}$ hardwired and outputs $\mathsf{m}_{i,\mathsf{PRG}(s)[i]}$ on input s. The garbled circuit $\widetilde{C}_\mathsf{B}$ comprises of garbled versions of each C_i, i.e., $\widetilde{C}_1, \ldots, \widetilde{C}_L$. The key trick we employ in garbling $C_1, \ldots, C_L$ is that we use the *same set of input labels* in generating each $\widetilde{C}_i$. Notice that even though we break C_B down to L circuits, the garbled input for $\widetilde{C}_\mathsf{B}$ only grows with the input length of C_B and is independent of L. To simulate $\widetilde{C}_\mathsf{B}$, we design a sequence of carefully chosen hybrids where in each hybrid, it is sufficient to simulate a single $\widetilde{C}_i$. But things get complicated as the simulation of this $\widetilde{C}_i$ requires knowledge of the adaptively chosen s. It seems that we again run into the adaptive security issue. However, notice that the output length of the circuit C_i is independent of L and thus the length of the garbled input for $\widetilde{C}_i$ (and hence all other $\widetilde{C}_j$, $j \neq i$) need not grow with L! Thus, we can now use the standard tricks in the adaptive garbling circuits literature to "adaptively garble" C_i. We now explain how this is done.

Instead of sending the garbled circuits $\{\widetilde{C}_i\}_{i\in[L]}$ in the clear, we encrypt them using a somewhere equivocal encryption scheme [HJO+16] and send the ciphertext as the garbled circuit $\widetilde{C}_\mathsf{B}$. The key for decrypting this ciphertext is revealed in the garbled input along with the labels for evaluating each $\widetilde{C}_i$. Recall that we use the same set of labels for evaluating each $\widetilde{C}_i$. Intuitively, a somewhere equivocal encryption allows to equivocate a bunch of positions of a ciphertext with arbitrary message values. What makes a somewhere equivocal encryption different from a fully equivocal encryption is that the size of the key only grows with the number of positions that are to be equivocated and is otherwise independent of the message size. Somewhere equivocal encryption allows us to solve the above adaptivity issue as we can equivocate the positions that correspond to $\widetilde{C}_i$ in the ciphertext to a simulated circuit (that can depend on the adaptively chosen s) by deriving a suitable key. Further, the size of the garbled input (that also includes the key) only grows with the size of $\widetilde{C}_i$ and is independent of L. This helps us in ensuring that the circuit size of the two-step translation functionality is independent of L.

Solving Input Dependent Aborts. Suppose the sender sends a proof that $\widetilde{C}_\mathsf{B}$ is correctly generated, then the problem of input dependent aborts does not arise. We additionally require this proof to be zero-knowledge so that it does not leak any information about the sender's secrets to the receiver. A natural approach would be to give a Non-Interactive Zero-Knowledge proof (NIZK). However, we only know constructions of NIZK based on public key assumptions such as trapdoor permutations or factoring. Furthermore, the number of public key operations in computing a NIZK proof grows with the instance size. Here, the instance size grows with the size of C_B which is at least L. This again kills the efficiency.

Our approach to solving this issue is to design a two-round, special purpose zero-knowledge proof (in the CRS model) where the number of public key operations is *independent* of the instance size. Indeed, given such a zero-knowledge proof, we can solve the problem of input dependent aborts and also ensure that the number of public key operations is independent of L. We now explain the main ideas behind this construction.

Let us first consider the simpler task of constructing a two-round, zero-knowledge proof with constant soundness error where the number of public key operations is independent of the instance size. We first observe that if we allow one more round of interaction then we know constructions (e.g., Blum's Hamiltonicity protocol) that completely avoid any public key operations. The main idea behind our construction is a method of compressing the round complexity of these protocols (in the simultaneous message exchange model) using a small number of public key operations (that is independent of the instance size). To explain the idea, let us take the example of compressing the Blum's Hamiltonicity protocol to two rounds using a two-round oblivious transfer (used in the recent works of [JKKR17,BGI+17a]). The Blum's protocol can be abstractly described using three messages: zk_1 sent by the prover in the first round, a random bit b sent by the verifier in the second round and $zk_{3,b}$ sent by the prover in the third round.

To compress the protocol to two rounds, we require the verifier to send a receiver OT message with b as its choice bit in the first round. In addition to sending zk_1 in the first round, the prover also sends commitment (c_0, c_1) to $zk_{3,0}$ and $zk_{3,1}$ respectively. In the second round, the sender sends a sender OT message with the randomness used to compute c_0 and c_1 as its messages.[3] The receiver obtains the randomness used in generating c_b and then uses it to check if $(zk_1, b, zk_{3,b})$ is a valid proof. Note that to minimize the number of public key operations, the length of the random string used to generate the commitment should be independent of the size of the message. This is indeed true when we use a pseudorandom generator to expand the length of the randomness to any desired length.

The above idea helps us in achieving constant soundness error but to be useful in solving the problem of input dependent aborts, we need the protocol to have negligible soundness error. One approach to achieve negligible soundness is to do a parallel repetition of the constant soundness protocol but it is well-known that parallel repetition is not guaranteed to preserve the zero-knowledge property. Fiege and Shamir [FS90] showed that parallel repetition preserves the weaker property of witness indistinguishability and we make use of this fact to to achieve the stronger property of zero-knowledge. In our actual construction, we incorporate a trapdoor (such as pre-image of a one-way function) in the CRS and the simulator uses this trapdoor while generating the zero-knowledge proof. Witness indistinguishability guarantees that no verifier can distinguish between the prover's messages that uses the real witness and the simulator's messages that uses the trapdoor witness. This helps us achieve *zero-knowledge* against malicious verifiers and parallel repetition helps us achieve *negligible* soundness error against cheating provers. Additionally, the number of public key operations is a fixed polynomial in the security parameter and is independent of the instance size. We believe that this primitive may be of independent interest.

3   Preliminaries

We recall some standard cryptographic definitions in this section. Let λ denote the security parameter. A function $\mu(\cdot) : \mathbb{N} \to \mathbb{R}^+$ is said to be negligible if for any polynomial $\mathsf{poly}(\cdot)$ there exists λ_0 such that for all $\lambda > \lambda_0$ we have $\mu(\lambda) < \frac{1}{\mathsf{poly}(\lambda)}$. We will use $\mathsf{negl}(\cdot)$ to denote an unspecified negligible function and $\mathsf{poly}(\cdot)$ to denote an unspecified polynomial function.

---

[3] We assume that given the randomness, we can obtain the message that is committed.

For a probabilistic algorithm A, we denote $A(x; r)$ to be the output of A on input x with the content of the random tape being r. When r is omitted, $A(x)$ denotes a distribution. For a finite set S, we denote $x \leftarrow S$ as the process of sampling x uniformly from the set S. We will use PPT to denote Probabilistic Polynomial Time algorithm.

For a binary string $x \in \{0,1\}^n$, we denote the ith bit of x by $x[i]$. Similarly, we denote the substring of x from the ith to j^{th} position for any $i \leq j$ by $x[i, j]$. For any $\overline{\mathsf{lab}} := \{\mathsf{lab}_{i,0}, \mathsf{lab}_{i,1}\}_{i \in [L]}$ where $\mathsf{lab}_{i,b} \in \{0,1\}^*$ and a string $c \in \{0,1\}^L$, we define $\mathsf{Projection}(c, \overline{\mathsf{lab}}) = \{\mathsf{lab}_{i,c[i]}\}_{i \in [L]}$. We treat the output of $\mathsf{Projection}$ as a string. That is, we treat the output as $\|_{i \in [L]}(\mathsf{lab}_{i,c[i]})$.

3.1 Selective Garbled Circuits

We recall the definition of selectively secure garbled circuits [Yao82] (see Lindell and Pinkas [LP09] and Bellare et al. [BHR12] for a detailed proof and further discussion). A garbling scheme for circuits is a tuple of PPT algorithms $(\mathsf{Garble}, \mathsf{Eval})$. Very roughly, Garble is the circuit garbling procedure and Eval the corresponding evaluation procedure. We use a formulation where input labels for a garbled circuit are provided as input to the garbling procedure rather than generated as output. This simplifies the presentation of our construction. We additionally model security wherein the simulator is provided with a set of labels corresponding to the input. This helps in simplifying the security proofs. More formally:

- $\widetilde{\mathsf{C}} \leftarrow \mathsf{Garble}\left(1^\lambda, \mathsf{C}, \{\mathsf{lab}_{w,b}\}_{w \in \mathsf{inp}(C), b \in \{0,1\}}\right)$: Garble takes as input a security parameter λ, a circuit C, and input labels $\mathsf{lab}_{w,b}$ where $w \in \mathsf{inp}(C)$ ($\mathsf{inp}(C)$ is the set of input wires to the circuit C) and $b \in \{0,1\}$. This procedure outputs a *garbled circuit* $\widetilde{\mathsf{C}}$. We assume that for each w, b, $\mathsf{lab}_{w,b}$ is chosen uniformly from $\{0,1\}^\lambda$.

- $y \leftarrow \mathsf{Eval}\left(\widetilde{\mathsf{C}}, \{\mathsf{lab}_{w,x_w}\}_{w \in \mathsf{inp}(C)}\right)$: Given a garbled circuit $\widetilde{\mathsf{C}}$ and a sequence of input labels $\{\mathsf{lab}_{w,x_w}\}_{w \in \mathsf{inp}(C)}$ (referred to as the garbled input), Eval outputs a string y.

Correctness. For correctness, we require that for any circuit C, input $x \in \{0,1\}^{|\mathsf{inp}(C)|}$ and input labels $\{\mathsf{lab}_{w,b}\}_{w \in \mathsf{inp}(C), b \in \{0,1\}}$ we have that:

$$\Pr\left[\mathsf{C}(x) = \mathsf{Eval}\left(\widetilde{\mathsf{C}}, \{\mathsf{lab}_{w,x_w}\}_{w \in \mathsf{inp}(C)}\right)\right] = 1$$

where $\widetilde{\mathsf{C}} \leftarrow \mathsf{Garble}\left(1^\lambda, \mathsf{C}, \{\mathsf{lab}_{w,b}\}_{w \in \mathsf{inp}(C), b \in \{0,1\}}\right)$.

Selective Security. For security, we require that there exists a PPT simulator $\mathsf{Sim}_{\mathsf{ckt}}$ such that for any circuit C, an input $x \in \{0,1\}^{|\mathsf{inp}(C)|}$ and $\{\mathsf{lab}_{w,x_w}\}_{w \in \mathsf{inp}(C)}$, we have that

$$\left\{\widetilde{\mathsf{C}}, \{\mathsf{lab}_{w,x_w}\}_{w \in \mathsf{inp}(C)}\right\} \overset{c}{\approx} \left\{\mathsf{Sim}_{\mathsf{ckt}}\left(1^\lambda, 1^{|C|}, \mathsf{C}(x), \{\mathsf{lab}_{w,x_w}\}_{w \in \mathsf{inp}(C)}\right), \{\mathsf{lab}_{w,x_w}\}_{w \in \mathsf{inp}(C)}\right\}$$

where $\widetilde{C} \leftarrow \mathsf{Garble}\left(1^\lambda, \mathsf{C}, \{\mathsf{lab}_{w,b}\}_{w\in\mathsf{inp}(C),b\in\{0,1\}}\right)$ and for each $w \in \mathsf{inp}(C)$ we have $\mathsf{lab}_{w,1-x_w} \leftarrow \{0,1\}^\lambda$. Here $\overset{c}{\approx}$ denotes that the two distributions are computationally indistinguishable.

3.2 Somewhere Adaptive Garbled Circuits

In this section, we define and construct somewhere adaptive garbled circuits. Intuitively, somewhere adaptive garbled circuits satisfy the stronger notion of adaptive security in the computation of a particular block of the output. Before we define this primitive, we give a notation to denote circuits.

Circuit Notation. We model a circuit $C : \{0,1\}^n \rightarrow \{0,1\}^{m\lambda}$ as a sequence of m circuits $C_1, C_2, \ldots, C_m$ where $C_i(x) = C(x)[(i-1)\lambda + 1, i\lambda]$ for every $x \in \{0,1\}^n$ and $i \in [m]$.

We now give the definition of somewhere adaptive garbled circuits.

Definition 1. *A somewhere adaptive garbling scheme for circuits is a tuple of PPT algorithms* (SAdpGarbleCkt, SAdpGarbleInp, SAdpEvalCkt) *such that:*

- *$(\widetilde{C}, \mathsf{state}) \leftarrow \mathsf{SAdpGarbleCkt}(1^\lambda, C)$: It is a PPT algorithm that takes as input the security parameter 1^λ (encoded in unary) and a circuit $C : \{0,1\}^n \rightarrow \{0,1\}^{m\lambda}$ as input and outputs a garbled circuit $\widetilde{C}$ and state information* state.
- *$\widetilde{x} \leftarrow \mathsf{SAdpGarbleInp}(\mathsf{state}, x)$: It is a PPT algorithm that takes as input the state information* state *and an input $x \in \{0,1\}^n$ and outputs the garbled input $\widetilde{x}$.*
- *$y = \mathsf{SAdpEvalCkt}(\widetilde{C}, \widetilde{x})$: Given a garbled circuit $\widetilde{C}$ and a garbled input $\widetilde{x}$, it outputs a value $y \in \{0,1\}^{m\lambda}$.*

Correctness. *For every $\lambda \in \mathbb{N}$, $C : \{0,1\}^n \rightarrow \{0,1\}^m$ and $x \in \{0,1\}^n$ it holds that:*

$$\Pr\left[(\widetilde{C}, \mathsf{state}) \leftarrow \mathsf{SAdpGarbleCkt}(1^\lambda, C); \widetilde{x} \leftarrow \mathsf{SAdpGarbleInp}(\mathsf{state}, x) : C(x) = \mathsf{SAdpEvalCkt}(\widetilde{C}, \widetilde{x})\right] = 1.$$

Security. *There exists a PPT simulator* Sim *such that for all non-uniform PPT adversary $\mathcal{A}$:*

$$\left| \Pr[\mathsf{Exp}_{\mathcal{A}}^{\mathsf{Adp}}(1^\lambda, 0) = 1] - \Pr[\mathsf{Exp}_{\mathcal{A}}^{\mathsf{Adp}}(1^\lambda, 1) = 1] \right| \leq \mathsf{negl}(\lambda)$$

where the experiment $\mathsf{Exp}_{\mathcal{A}}^{\mathsf{Adp}}(1^\lambda, b)$ is defined as follows:

1. *$(C, j) \leftarrow \mathcal{A}(1^\lambda)$ where $C : \{0,1\}^n \rightarrow \{0,1\}^{m\lambda}$ and $j \in [m]$. We assume that C is given as a sequence of m circuits $C_1, C_2, \ldots, C_m$.*
2. *The adversary obtains $\widetilde{C}$ where $\widetilde{C}$ is created as follows:*
   - *If $b = 0$: $(\widetilde{C}, \mathsf{state}) \leftarrow \mathsf{SAdpGarbleCkt}(1^\lambda, C)$.*
   - *If $b = 1$: $(\widetilde{C}, \mathsf{state}) \leftarrow \mathsf{Sim}(1^\lambda, C_1, \ldots, C_{j-1}, 1^{|C_j|}, C_{j+1}, \ldots, C_m)$.*
3. *The adversary $\mathcal{A}$ specifies the input x and gets $\widetilde{x}$ created as follows:*
   - *If $b = 0$: $\widetilde{x} \leftarrow \mathsf{SAdpGarbleInp}(\mathsf{state}, x)$.*
   - *If $b = 1$: $\widetilde{x} \leftarrow \mathsf{Sim}(\mathsf{state}, x, C_j(x))$.*

4. *Finally, the adversary outputs a bit b', which is the output of the experiment.*

Efficiency. *We require that the running time of* SAdpGarbleInp *to be* $\max_i |C_i| \cdot$ poly$(|x|, \lambda)$.

We give a construction of somewhere adaptive garbled circuits assuming the existence of one-way functions.

Lemma 1. *Assuming the existence of one-way functions, there exists a construction of somewhere adaptive garbled circuits.*

We give the proof of Lemma 1 in the full version [GMS18].

3.3 Universal Composability Framework

We work in the the Universal Composition (UC) framework [Can01] to formalize and analyze the security of our protocols. (Our protocols can also be analyzed in the stand-alone setting, using the composability framework of [Can00a]). We provide a brief overview of the framework in the full version of our paper [GMS18] and refer the reader to [Can00b] for details.

3.4 Prior MPC Results

We will use the two-round secure multiparty computation protocol from the work of [GS18] computing special functionalities that have small circuit size in our constructions. We could also use the protocol from [BL18] but their protocol against malicious adversaries additionally relies on non-interactive zero-knowledge proofs. Below we restate the result from [GS18]. The ideal functionality $\mathcal{F}_f$ for the MPC is defined in Fig. 2.

Theorem 2 ([GS18]). *For any polynomial-time function f computed by n parties, there exists a two-round UC-secure semi-honest/malicious multiparty computation protocol Π_f that realizes the ideal functionality $\mathcal{F}_f$, assuming the existence of semi-honest/malicious, two-round oblivious transfer. The number of total public key operations is bounded by $\mathsf{poly}(\lambda, |f|)$, where $|f|$ is the size of the Boolean circuit that computes f.*

$\mathcal{F}_f$ parameterized by a function f, running with n parties $P_1, P_2, \ldots, P_n$ (of which some may be corrupted) and an adversary $\mathcal{S}$, proceeds as follows:

- Every party P_i sends (sid, i, x_i) to the functionality.
- Upon receiving the inputs from all the parties, compute $y := f(x_1, \ldots, x_n)$, and output (sid, y) to every party and $\mathcal{S}$.

Fig. 2. Ideal functionality $\mathcal{F}_f$

4   Semi-Honest Protocol

In this section, we give a construction of two-round multiparty computation protocol with security against semi-honest adversaries that performs $\mathsf{poly}(n, \lambda)$ public key operations which is independent of the circuit size being computed. We start with the definition of conforming protocols which was a notion introduced in [GS18] in Subsect. 4.1 and then give our construction in Subsect. 4.2.

4.1   Conforming Protocols

This subsection is taken verbatim from [GS18]. Consider an n party deterministic[4] MPC protocol Φ between parties $P_1, \ldots, P_n$ with inputs $x_1, \ldots, x_n$, respectively. For each $i \in [n]$, we let $x_i \in \{0,1\}^m$ denote the input of party P_i. A conforming protocol Φ is defined by functions pre, post, and computation steps or what we call *actions* $\phi_1, \cdots \phi_T$. The protocol Φ proceeds in three stages: the pre-processing stage, the computation stage and the output stage.

- **Pre-processing phase:** For each $i \in [n]$, party P_i computes

$$(z_i, v_i) \leftarrow \mathsf{pre}(1^\lambda, i, x_i)$$

  where pre is a randomized algorithm. The algorithm pre takes as input the index i of the party, its input x_i and outputs $z_i \in \{0,1\}^{\ell/n}$ and $v_i \in \{0,1\}^\ell$ (where ℓ is a parameter of the protocol). Finally, P_i retains v_i as the secret information and broadcasts z_i to every other party. We require that $v_i[k] = 0$ for all $k \in [\ell] \backslash \{(i-1)\ell/n + 1, \ldots, i\ell/n\}$.
- **Computation phase:** For each $i \in [n]$, party P_i sets

$$\mathsf{st}_i := (z_1 \| \cdots \| z_n) \oplus v_i.$$

  Next, for each $t \in \{1 \cdots T\}$ parties proceed as follows:
  1. Parse action ϕ_t as (i, f, g, h) where $i \in [n]$ and $f, g, h \in [\ell]$.
  2. Party P_i computes *one* NAND gate as

$$\mathsf{st}_i[h] = \mathsf{NAND}(\mathsf{st}_i[f], \mathsf{st}_i[g])$$

     and broadcasts $\mathsf{st}_i[h] \oplus v_i[h]$ to every other party.
  3. Every party P_j for $j \neq i$ updates $\mathsf{st}_j[h]$ to the bit value received from P_i. We require that for all $t, t' \in [T]$ such that $t \neq t'$, we have that if $\phi_t = (\cdot, \cdot, \cdot, h)$ and $\phi_{t'} = (\cdot, \cdot, \cdot, h')$ then $h \neq h'$. Also, we denote $A_i \subset [T]$ to be the set of rounds in which party P_i sends a bit. Namely, $A_i = \{t \in T \mid \phi_t = (i, \cdot, \cdot, \cdot)\}$.
- **Output phase:** For each $i \in [n]$, party P_i outputs $\mathsf{post}(i, \mathsf{st}_i)$.

The following lemma was shown in [GS18]

Lemma 2 ([GS18]). *Any MPC protocol Π can be written as a conforming protocol Φ while inheriting the correctness and the security of the original protocol.*

---

[4] Randomized protocols can be handled by including the randomness used by a party as part of its input.

4.2   Construction

In this subsection, we describe our construction of two-round, n-party computation protocol computing a function f. Our construction uses the following primitives.

1. An n-party semi-honest secure conforming protocol Φ computing the function f.
2. (Garble, Eval) be a garbling scheme for circuits.
3. A pseudorandom generator $\mathsf{PRG} : \{0,1\}^\lambda \rightarrow \{0,1\}^{4T}$.
4. A UC-secure two-round MPC protocol computing the function g described in Fig. 3.

Notations. For a bit string c, we use $c[i]$ to denote the i-th bit of it. For each $t \in [T]$ and $\alpha, \beta \in \{0,1\}$, we use (t, α, β) to succinctly denote the integer $4t + 2\alpha + \beta - 3$. In particular, we use $c[(t, \alpha, \beta)]$ to denote $c[4t + 2\alpha + \beta - 3]$ for any $c \in \{0,1\}^{4T}$. We use $\overline{\mathsf{lab}}$ to denote the set of both labels per input wire of a garbled circuit, and $\widetilde{\mathsf{lab}}$ denotes the set of one label per input wire. Recall the definition of Projection from Sect. 3.

We give an overview of the construction below and describe the formal construction later.

Parties: $P_1, P_2, \ldots, P_n$.

Inputs:

- P_1 (also called as the receiver) inputs $s \in \{0,1\}^\lambda$ and $\overline{\mathsf{rlab}}_2, \ldots, \overline{\mathsf{rlab}}_n$ where each $\overline{\mathsf{rlab}}_i$ is a collection of labels $\{\mathsf{rlab}_{j,0}^{i\rightarrow 1}, \mathsf{rlab}_{j,1}^{i\rightarrow 1}\}_{j \in [\lambda^2]}$ with each label of length λ.
- For each $i \in [2, n]$, P_i (also called as the sender) inputs $\overline{\mathsf{slab}}_i$, where $\overline{\mathsf{slab}}_i$ is a collection of labels $\{\mathsf{slab}_{j,0}^{i\rightarrow 1}, \mathsf{slab}_{j,1}^{i\rightarrow 1}\}_{j \in [\lambda]}$ with each label having length λ.

Output: $\{\mathsf{Projection}(\mathsf{Projection}(s, \overline{\mathsf{slab}}_i), \overline{\mathsf{rlab}}_i)\}_{i \in [2,n]}$.

Fig. 3. The function g computed by the internal MPC where P_1 acts as the receiver

Overview. As explained in Sect. 2, our construction combines a special purpose OT extension protocol (which is delegatable, fine-grained secure and satisfies key availability) along with the two-round MPC protocols of [BL18, GS18] to obtain a protocol that minimizes the number of public key operations. Recall that the protocols of [BL18, GS18] used the concept of "talking garbled circuits" to squish the round complexity of a conforming protocol to two rounds. At a high level, in the first round, every pair of parties sets up a channel to enable their garbled circuits to interact, and then in the second round, they send "talking garbled circuits" that emulate the interactions in the conforming protocol. The interaction between the "talking garbled circuits" is done via oblivious transfer. In our new construction, we use a special purpose OT extension protocol that allows the parties to set-up the channel for interaction while minimizing the number of public key operations.

A major modification from the description given in Sect. 2 is in modeling the special oblivious transfer as a protocol between a single receiver and $n-1$ senders. We do this to ensure that the receiver uses the same choice bits in interactions with every sender. Even though this is not an issue in the semi-honest case, it causes issues in the malicious setting if the corrupted receiver uses different choice bits in two different interactions. For uniformity of treatment, we adopt an approach where the special oblivious transfer is a protocol between a single receiver and $n - 1$ senders.

Description of the Protocol. We give a formal description of our protocol below in the $\mathcal{F}_g$-hybrid model.

Round-1: Each party P_i does the following:
1. Compute $(z_i, v_i) \leftarrow \mathsf{pre}(1^\lambda, i, x_i)$.
2. For each $t \in [T]$ and for each $\alpha, \beta \in \{0,1\}$

$$c_i[(t, \alpha, \beta)] := v_i[h] \oplus \mathsf{NAND}(v_i[f] \oplus \alpha, v_i[g] \oplus \beta)$$

where $\phi_t = (\star, f, g, h)$.
3. Sample $s_i \leftarrow \{0,1\}^\lambda$ and compute $e_i := \mathsf{PRG}(s_i) \oplus c_i$.
4. For each $j \in [n] \setminus \{i\}$, sample

$$\mathsf{rlab}_{k,b}^{j \to i} \leftarrow \{0,1\}^\lambda \text{ for all } k \in [\lambda^2], b \in \{0,1\}$$
$$\mathsf{slab}_{k,b}^{i \to j} \leftarrow \{0,1\}^\lambda \text{ for all } k \in [\lambda], b \in \{0,1\}$$
$$\mathsf{m}_{k,b}^{i \to j} \leftarrow \{0,1\}^\lambda \text{ for all } k \in [4T], b \in \{0,1\}$$

5. For each $j \in [n] \setminus \{i\}$, compute

$$\widetilde{\mathsf{C}}_{\mathsf{B}}^{i \to j} \leftarrow \mathsf{Garble}\left(\mathsf{C}_{\mathsf{B}}\left[\left\{\mathsf{m}_{k,0}^{i \to j}, \mathsf{m}_{k,1}^{i \to j}\right\}_{k \in [4T], b \in \{0,1\}}\right], \left\{\mathsf{slab}_{k,b}^{i \to j}\right\}_{k \in [\lambda], b \in \{0,1\}}\right)$$

where C_{B} is described in Fig. 4.
6. Send $(\mathsf{ssid} = i, s_i, \{\mathsf{rlab}_{k,b}^{j \to i}\}_{j \in [n] \setminus \{i\}})$ to $\mathcal{F}_g$ acting as the receiver.
7. For each $j \in [n] \setminus \{i\}$, send $(\mathsf{ssid} = j, \{\mathsf{slab}_{k,b}^{i \to j}\})$ to $\mathcal{F}_g$ acting as the sender.
8. Send $\left(z_i, \{\widetilde{\mathsf{C}}_{\mathsf{B}}^{i \to j}\}_{j \in [n] \setminus \{i\}}, e_i\right)$ to every other party.

$$\mathsf{C}_{\mathsf{B}}\left[\{\mathsf{m}_{k,0}, \mathsf{m}_{k,1}\}_{k \in [4T]}\right]$$

Input: $s \in \{0,1\}^\lambda$.
1. $d := \mathsf{PRG}(s)$ where $d \in \{0,1\}^{4T}$.
2. Output $\{\mathsf{m}_{k,d[k]}\}_{k \in [4T]}$.

Fig. 4. Circuit C_{B}

Round-2: Each party P_i does the following:

1. Set $\mathsf{st}_i := (z_1 \| \dots \| z_n) \oplus v_i$.
2. Set $N = \ell + 4T\lambda(n-1)$.
3. Set $\overline{\mathsf{lab}}^{i,T+1} := \{\mathsf{lab}_{k,0}^{i,T+1}, \mathsf{lab}_{k,1}^{i,T+1}\}_{k \in [N]}$ where $\mathsf{lab}_{k,b}^{i,T+1} := 0^\lambda$ for each $k \in [N], b \in \{0,1\}$.
4. **for** each t from T down to 1 **do:**
   (a) Parse ϕ_t as (i^*, f, g, h).
   (b) If $i = i^*$ then compute (where P is described in Fig. 6)

   $$\left(\widetilde{\mathsf{P}}^{i,t}, \overline{\mathsf{lab}}^{i,t}\right) \leftarrow \mathsf{Garble}(1^\lambda, \mathsf{P}[i, \phi_t, v_i, \bot, \overline{\mathsf{lab}}^{i,t+1}]).$$

   (c) If $i \neq i^*$ then for every $\alpha, \beta \in \{0,1\}$, set $\mathsf{m}_{\alpha,\beta,0}' = \mathsf{m}_{(t,\alpha,\beta),e_{i^*}[(t,\alpha,\beta)]}^{i \to i^*}$ and $\mathsf{m}_{\alpha,\beta,1}' = \mathsf{m}_{(t,\alpha,\beta),1 \oplus e_{i^*}[(t,\alpha,\beta)]}^{i \to i^*}$.
   Compute $ct_{\alpha,\beta}^i := (\mathsf{m}_{\alpha,\beta,0}' \oplus \mathsf{lab}_{h,0}^{i,t+1}, \mathsf{m}_{\alpha,\beta,1}' \oplus \mathsf{lab}_{h,1}^{i,t+1})$ and compute

   $$\left(\widetilde{\mathsf{P}}^{i,t}, \overline{\mathsf{lab}}^{i,t}\right) \leftarrow \mathsf{Garble}(1^\lambda, \mathsf{P}[i, \phi_t, v_i, \{ct_{\alpha,\beta}^i\}, \overline{\mathsf{lab}}^{i,t+1}]).$$

5. Compute

   $$\widetilde{\mathsf{C}}_{\mathsf{wrap}}^i \leftarrow \mathsf{Garble}\left(\mathsf{C}_{\mathsf{wrap}}\left[\{\widetilde{\mathsf{C}}_{\mathsf{B}}^{j \to i}\}_{j \in [n] \setminus \{i\}}, \mathsf{st}_i, \overline{\mathsf{lab}}^{i,1}\right], \{\mathsf{rlab}_{k,b}^{j \to i}\}_{j \in [n] \setminus \{i\}, k \in [\lambda^2], b \in \{0,1\}}\right)$$

   where $\mathsf{C}_{\mathsf{wrap}}$ is described in Fig. 5.
6. Send $\left(\{\widetilde{\mathsf{P}}^{i,t}\}_{t \in [T]}, \widetilde{\mathsf{C}}_{\mathsf{wrap}}^i\right)$ to every other party.

$$\mathsf{C}_{\mathsf{wrap}}\left[\{\widetilde{\mathsf{C}}_{\mathsf{B}}^{j \to i}\}_{j \in [n] \setminus \{i\}}, \mathsf{st}_i, \overline{\mathsf{lab}}^{i,1}\right]$$

Input: $\{\widetilde{\mathsf{slab}}^{j \to i}\}_{j \in [n] \setminus \{i\}}$

1. For each $j \in [n] \setminus \{i\}$, compute $\{\mathsf{m}_k^{j \to i}\}_{k \in [4T]} \leftarrow \mathsf{Eval}\left(\widetilde{\mathsf{C}}_{\mathsf{B}}^{j \to i}, \widetilde{\mathsf{slab}}^{j \to i}\right)$.
2. Let $\mathsf{m} := \underset{j \in [n] \setminus \{i\}, k \in [4T]}{\|} (\mathsf{m}_k^{j \to i})$.
3. Output $\mathsf{Projection}(\mathsf{st}_i \| \mathsf{m}, \overline{\mathsf{lab}}^{i,1})$.

Fig. 5. Circuit $\mathsf{C}_{\mathsf{wrap}}$

Evaluation: Every party P_i does the following:

1. For each $j \in [n]$,
   (a) Obtain $(\mathsf{ssid} = j, \widetilde{\mathsf{rlab}}^j)$ from $\mathcal{F}_g$ where party P_j acts as the receiver.
   (b) $\widetilde{\mathsf{lab}}^{j,1} \leftarrow \mathsf{Eval}(\widetilde{\mathsf{C}}_{\mathsf{wrap}}^j, \widetilde{\mathsf{rlab}}^j)$
2. **for** each t from 1 to T **do:**
   (a) Parse ϕ_t as (i^*, f, g, h).

(b) Compute $((\alpha, \beta, \gamma), \{\omega^j\}_{j \in [n] \setminus \{i^*\}}, \widetilde{\mathsf{lab}}^{i^*, t+1}) := \mathsf{Eval}(\widetilde{\mathsf{P}}^{i^*, t}, \widetilde{\mathsf{lab}}^{i^*, t})$.

(c) Set $\mathsf{st}_i[h] := \gamma \oplus v_i[h]$.

(d) **for** each $j \neq i^*$ **do:**

    i. Compute $(ct = (\delta_0, \delta_1), \{\mathsf{lab}_k^{j, t+1}\}_{k \in [N] \setminus \{h\}}) := \mathsf{Eval}(\widetilde{\mathsf{P}}^{j, t}, \widetilde{\mathsf{lab}}^{j, t})$.

    ii. Recover $\mathsf{lab}_h^{j, t+1} := \delta_\gamma \oplus \omega^j$.

    iii. Set $\widetilde{\mathsf{lab}}^{j, t+1} := \{\mathsf{lab}_k^{j, t+1}\}_{k \in [N]}$.

3. Compute the output as $\mathsf{post}(i, \mathsf{st}_i)$.

Correctness. In order to prove correctness, it is sufficient to show that the label $\mathsf{lab}_h^{j, t+1}$ computed in Step 2(d)ii of the evaluation procedure corresponds to the bit $\mathsf{NAND}(\mathsf{st}_{i^*}[f], \mathsf{st}_{i^*}[g]) \oplus v_{i^*}[h]$. Notice that by the structure of v_{i^*} we have for every $j \neq i^*$, $\mathsf{st}_j[f] = \mathsf{st}_{i^*}[f] \oplus v_{i^*}[f]$.

First, ω^j is computed in Step 2b. Let $k := (t, \alpha, \beta)$, and we have $\omega^j = \mathsf{m}_k^{j \to i^*} = \mathsf{m}_{k, \mathsf{PRG}(s_{i^*})[k]}^{j \to i^*}$.

$P[i, \phi_t, v_i, \{ct_{\alpha, \beta}\}_{\alpha, \beta \in \{0,1\}}, \overline{\mathsf{lab}}]$

Input. $Z = (\mathsf{st}_i, \{\mathsf{m}_k^{j \to i}\}_{j \in [n] \setminus \{i\}, k \in [4T]})$.

Hardcoded. The index i of the party, the action $\phi_t = (i^*, f, g, h)$, the secret value v_i, the strings $\{ct_{\alpha, \beta}\}_{\alpha, \beta \in \{0,1\}}$, and a set of labels $\overline{\mathsf{lab}} = \{\mathsf{lab}_{k, 0}, \mathsf{lab}_{k, 1}\}_{k \in [N]}$.

1. **if** $i = i^*$ **then:**

    (a) Compute $\mathsf{st}_i[h] := \mathsf{NAND}(\mathsf{st}_i[f], \mathsf{st}_i[g])$, and update $Z[h]$ accordingly.

    (b) $\alpha := \mathsf{st}_i[f] \oplus v_i[f]$, $\beta := \mathsf{st}_i[g] \oplus v_i[g]$ and $\gamma := \mathsf{st}_i[h] \oplus v_i[h]$.

    (c) Output $\left((\alpha, \beta, \gamma), \{\mathsf{m}_{(t, \alpha, \beta)}^{j \to i}\}_{j \in [n] \setminus \{i\}}, \mathsf{Projection}(Z, \overline{\mathsf{lab}})\right)$.

2. **else:**

    (a) Output $(ct_{\mathsf{st}_i[f], \mathsf{st}_i[g]}, \{\mathsf{lab}_{k, Z[k]}\}_{k \in [N] \setminus \{h\}})$.

Fig. 6. The program P

Second, $ct = (\delta_0, \delta_1)$ is computed in Step 2(d)i. Note that $\alpha = \mathsf{st}_{i^*}[f] \oplus v_{i^*}[f] = \mathsf{st}_j[f]$, $\beta = \mathsf{st}_{i^*}[g] \oplus v_{i^*}[g] = \mathsf{st}_j[g]$. From the functionality of $\mathsf{P}^{j, t}$ we know that $ct = ct_{\mathsf{st}_j[f], \mathsf{st}_j[g]} = ct_{\alpha, \beta}^j = (\mathsf{m}_{\alpha, \beta, 0}' \oplus \mathsf{lab}_{h, 0}^{j, t+1}, \mathsf{m}_{\alpha, \beta, 1}' \oplus \mathsf{lab}_{h, 1}^{j, t+1}) = (\mathsf{m}_{k, e_{i^*}[k]}^{j \to i^*} \oplus \mathsf{lab}_{h, 0}^{j, t+1}, \mathsf{m}_{k, e_{i^*}[k] \oplus 1}^{j \to i^*} \oplus \mathsf{lab}_{h, 1}^{j, t+1})$.

Therefore, $\delta_\gamma \oplus \omega^j = \mathsf{m}_{k, e_{i^*}[k] \oplus \gamma}^{j \to i^*} \oplus \mathsf{lab}_{h, \gamma}^{j, t+1} \oplus \mathsf{m}_{k, \mathsf{PRG}(s_{i^*})[k]}^{j \to i^*}$. Recall that $c_{i^*}[k] = \mathsf{NAND}(\mathsf{st}_{i^*}[f], \mathsf{st}_{i^*}[g]) \oplus v_{i^*}[h] = \gamma$, thus $e_{i^*}[k] \oplus \gamma = e_{i^*}[k] \oplus c_{i^*}[k] = \mathsf{PRG}(s_{i^*})[k]$. Hence $\delta_\gamma \oplus \omega^j = \mathsf{lab}_{h, \gamma}^{j, t+1}$. This concludes the proof.

It is useful to keep in mind that for every $i, j \in [n]$ and $k \in [\ell]$, we have that $\mathsf{st}_i[k] \oplus v_i[k] = \mathsf{st}_j[k] \oplus v_j[k]$. Let us denote this shared value by st^*. Also, we denote the transcript of the interaction in the computation phase by Z.

Efficiency. Let the number of OT invocations in Φ be npk_Φ and in one execution of $\mathcal{F}_g$ be npk_g. Since we make non-black box use of the underlying conforming protocol Φ (but make black-box use of $\mathcal{F}_g$), we augment the circuit computing Π and $\mathcal{F}_g$ to have OT gates (this is similar in spirit to the works of [GMM17a, GMM17b]) to count the number of public-key operations. An OT gate enables one execution of one of the algorithms provided by the OT protocol. We choose the conforming protocol that performs OT extension between every pair of parties so that npk_Φ is bounded by $\mathcal{O}(n^2\lambda)$. Thus, the total number of public-key operations (including the non-black-box public-key operations) in our two-round construction is $\mathcal{O}(\mathsf{npk}_\Phi + n \cdot \mathsf{npk}_g)$. It follows from Theorem 2 that this number is bounded by $\mathsf{poly}(n, \lambda)$.

Security. The proof of security is given in the full version [GMS18].

5   Special Zero-Knowledge Protocol

In this section, we define and construct a special zero-knowledge protocol which will later be used in our construction against malicious adversaries. We give the formal definition below.

$\mathcal{F}_{\mathsf{ZK}}$ parameterized by an NP relation R, running with n parties $P_1, P_2, \ldots, P_n$ (of which some may be corrupted) and an adversary $\mathcal{S}$, proceeds as follows:

- P_1 sends $(\mathsf{prover}, \mathsf{sid}, x, w)$ to the functionality. The functionality sends $(\mathsf{request}, x, R(x, w))$ to $\mathcal{S}$. If $\mathcal{S}$ has corrupted P_2, then $\mathcal{S}$ sends $(\mathsf{response}, \mu)$ to the ideal functionality, and the ideal functionality broadcasts $(R(x, w), x, \mu)$ to every other party and goes offline. Else, P_2 sends $(\mathsf{verifier}, \mathsf{sid}, \mu_0, \mu_1)$ to the functionality, where $\mu_b \in \{0, 1\}^\lambda$.
- Upon receiving the inputs from both P_1 and P_2, functionality checks if $R(x, w) = 1$. If yes, it sends $(1, x, \mu_1)$ to every party. Otherwise, it sends $(0, x, \mu_0)$ to all parties.

Fig. 7. Special zero-knowledge functionality $\mathcal{F}_{\mathsf{ZK}}$

Definition 2. *A special zero-knowledge protocol is a two-round protocol that securely realizes the $\mathcal{F}_{\mathsf{ZK}}$ functionality given in Fig. 7. Further, we require the number of pubic key operations performed in the protocol to be bounded by $\mathsf{poly}(n, \lambda)$ independent of the size of x and w.*

We give a proof of the following theorem.

Theorem 3. *Assuming the existence of two-round UC secure oblivious transfer, there exists a construction of special zero-knowledge protocol.*

5.1   Construction

We first describe the tools used in the construction.

1. **Special Non-interactive Statistically Binding Commitment.** We use a special non-interactive, statistically binding commitment scheme

(com, decom) where the length of the randomness used to commit to arbitrary length messages is λ. We note that any standard commitment can be made to satisfy this property by using a pseudorandom generator to expand the random string to required length.

2. **Blum's Hamiltonicity Protocol.** We use the three-round, constant soundness zero-knowledge $(\mathsf{zk}_1, \mathsf{zk}_2, \mathsf{zk}_3)$ protocol of Blum. We note that in Blum's protocol $\mathsf{zk}_2 \in \{0,1\}$ and we let $\mathsf{zk}_{3,b}$ be the response when $\mathsf{zk}_2 = b$. We also assume without loss of generality that zk_1 includes the instance.

3. **Two-Round Secure Computation Protocol.** We make use of the two-round secure computation protocol of [GS18] (that can be based on any two-round UC secure oblivious transfer) computing the ideal functionality $\mathcal{F}_f$ described in Fig. 10.

4. **Length Doubling Pseudorandom Generator:** We use a pseudorandom generator $\mathsf{PRG} : \{0,1\}^\lambda \to \{0,1\}^{2\lambda}$.

Common Random String: Sample $\sigma \leftarrow \{0,1\}^{2\lambda}$ and set σ as the CRS.

Message from P_1: On input an instance x and a witness w, P_1 does the following:

1. If $R(x,w) = 0$, broadcast $(\mathsf{NotInL}, x, R(x,w))$ to every other party.
2. Else, **for each** $i \in [\lambda]$ **do:**
   (a) Prepare zk_1^i for the language $\mathcal{L}$ using the witness w where $\mathcal{L}$ is defined below.

$$\mathcal{L} := \{(x,\sigma) : \exists\ (w,s)\ \text{s.t.}\ R(x,w) = 1 \lor \mathsf{PRG}(s) = \sigma\}$$

   (b) Let $\mathsf{zk}_{3,b}^i$ be the third round message when $\mathsf{zk}_2^i = b$. Sample $r_b^i \leftarrow \{0,1\}^\lambda$ for each $b \in \{0,1\}$ and compute $c_b^i := \mathsf{com}(\mathsf{zk}_{3,b}^i; r_b^i)$.
   (c) Broadcast $\mathsf{zk}_1^i, c_0^i, c_1^i$ to every other party.

Message from P_2: On input the message from P_1 :

1. If P_1 has sent $(\mathsf{NotInL}, x, 0)$, broadcast μ_0 to every other party and every party outputs $(0, x, \mu_0)$. Else, do:
   (a) Sample $ch \leftarrow \{0,1\}^\lambda$.
   (b) Sample $\mathsf{lab}_{w,b}^i \leftarrow \{0,1\}^\lambda$ for each $i, w \in [\lambda]$ and $b \in \{0,1\}$.
   (c) Compute $\widetilde{C} \leftarrow \mathsf{Garble}(C[ch, \{\mathsf{zk}_1^i, c_0^i, c_1^i\}_{i \in [\lambda]}, \{\mu_b\}_{b \in \{0,1\}}], \{\mathsf{lab}_{w,b}^i\})$ where the C is described in Figure 9.
   (d) Broadcast $\widetilde{C}$ to every party.

Internal MPC: The parties in parallel call $\mathcal{F}_f$ to jointly compute the function f shown in Figure 10. More specifically, P_1 sends $\{r_0^i, r_1^i\}_{i \in [\lambda]}$ to $\mathcal{F}_f$; P_2 sends $ch, \{\mathsf{lab}_{w,b}^i\}_{i,w \in [\lambda], b \in \{0,1\}}$ to $\mathcal{F}_f$; and $P_3, P_4, \ldots, P_n$ send nothing. Every party then gets $\{\mathsf{lab}_w^i\}_{i,w \in [\lambda]}$ back from $\mathcal{F}_f$.

Evaluation: Every party does the following:

1. Compute $(b, x, \mu) \leftarrow \mathsf{Eval}\left(\widetilde{C}, \{\mathsf{lab}_w^i\}_{i,w \in [\lambda]}\right)$
2. Output (b, x, μ).

Fig. 8. Special zero-knowledge protocol Π_{ZK}

$C\left[ch, \{zk_1^i, c_0^i, c_1^i\}_{i\in[\lambda]}, \{\mu_b\}_{b\in\{0,1\}}\right]$

Input: $r^1, r^2, \ldots, r^\lambda$.

Hardcoded parameters: $ch, \{zk_1^i, c_0^i, c_1^i\}_{i\in[\lambda]}, \{\mu_b\}_{b\in\{0,1\}}$

1. Use the randomness r^i to obtain the message zk_3^i committed in $c_{ch[i]}^i$ for each $i \in [\lambda]$.
2. For each $i \in [\lambda]$, check if $(zk_1^i, ch[i], zk_3^i)$ is a valid proof for the membership in language $\mathcal{L}$.
3. If any of the checks fails, output $(0, x, \mu_0)$. Else, output $(1, x, \mu_1)$.

Fig. 9. Circuit C

Parties: $P_1, P_2, \ldots, P_n$.

Inputs:

- P_1 inputs $\{r_0^i, r_1^i\}_{i\in[\lambda]}$, where $r_i^b \in \{0,1\}^\lambda$.
- P_2 inputs $ch, \{lab_{w,b}^i\}_{i,w\in[\lambda], b\in\{0,1\}}$, where $lab_{w,b}^i \in \{0,1\}^\lambda$.
- $P_3, P_4, \ldots, P_n$ input nothing.

Output: $\{lab_{w, r_{ch[i]}^i[w]}^i\}_{i,w\in[\lambda]}$ (same for every party).

Fig. 10. The function f computed by the internal MPC

Overview. We present the formal construction in the $\mathcal{F}_f$ hybrid model in Fig. 8.

Correctness. To argue the correctness of the protocol, we only need to prove that in the evaluation step, μ is either μ_0 or μ_1 based on whether $R(x, w) = 0$ or $R(x, w) = 1$. We know that the output of $\mathcal{F}_f$ is $\{lab_w^i\}_{i,w\in[\lambda]}$, where $lab_w^i = lab_{w, r_{ch[i]}^i[w]}^i$. Notice that $lab_{w,b}^i$'s are the input keys of $\widetilde{C}$, hence lab_w^i is the label corresponding to the w-th bit of $r_{ch[i]}^i$. Using these input labels to evaluate $\widetilde{C}$ gives us $\mathsf{Eval}\left(\widetilde{C}, \{lab_w^i\}_{i,w\in[\lambda]}\right) = C\left(\{r_{ch[i]}^i\}_{i\in[\lambda]}\right)$.

In the circuit evaluation of C, $r_{ch[i]}^i$ is used to obtain $zk_{3,ch[i]}^i$ from $c_{ch[i]}^i$. It now follows from the completeness of $(zk_1^i, ch[i], zk_{3,ch[i]}^i)$ that μ is either μ_0 or μ_1 based on $R(x, w) = 0$ or $R(x, w) = 1$.

Efficiency. The number of public key operations performed in the protocol is $\mathsf{poly}(n, \lambda)$ which follows from Theorem 2 when applied to function f.

Security. We give the security proof in the full version [GMS18].

6 Malicious Secure Protocol

In this section, we give a construction of two-round, multiparty computation that is secure against malicious adversaries and minimizes the number of public key operations.

6.1   Construction

Our two-round protocol computing a function f uses the following primitives.

1. An n-party malicious secure conforming protocol Φ computing the function f.
2. A selective garbling scheme for circuits (Garble, Eval).
3. A pseudorandom generator $\mathsf{PRG_{mal}} : \{0,1\}^\lambda \to \{0,1\}^{4T}$ where each output bit can be computed by a circuit of size $\mathsf{poly}(\lambda, \log T)$.[5]
4. A somewhere adaptive garbling scheme for circuits (SAdpGarbleCkt, SAdpGarbleInp, SAdpEvalCkt) (defined in Sect. 3.2). We assume that the length of the garbled input when SAdpGarbleCkt is used to garbled $\mathsf{C_B}$ (described in Fig. 11) is M.
5. A maliciously secure two-round MPC protocol computing the function g described in Fig. 12.
6. A non-interactive statistically binding commitment scheme (Com, Decom).
7. The special ZK protocol parameterized by an NP relation R described below.

$$R := \left\{ \left(x = \left(\widetilde{\mathsf{C}}_\mathsf{B}, \mathsf{cm}\right), w = (\Omega, \mathsf{C_B}, \mathsf{state}, \omega)\right) : \right.$$

$$\left. (\mathsf{Decom}(\mathsf{cm}, \mathsf{state}, \omega) = 1) \wedge \left((\widetilde{\mathsf{C}}_\mathsf{B}, \mathsf{state}) = \mathsf{SAdpGarbleCkt}\,(\mathsf{C_B}; \Omega)\right) \right\}.$$

Description of the Protocol. We now give a formal description of our construction in below in the $\mathcal{F}_g$ and $\mathcal{F}_{\mathsf{zk}}$ hybrid model.

Round-1: Each party P_i does the following:

1. Compute $(z_i, v_i) \leftarrow \mathsf{pre}(1^\lambda, i, x_i)$.
2. For each $t \in [T]$ and for each $\alpha, \beta \in \{0,1\}$

$$c_i[(t, \alpha, \beta)] := v_i[h] \oplus \mathsf{NAND}(v_i[f] \oplus \alpha, v_i[g] \oplus \beta)$$

where $\phi_t = (\star, f, g, h)$.

3. Sample $s_i \leftarrow \{0,1\}^\lambda$ and compute $e_i := \mathsf{PRG_{mal}}(s_i) \oplus c_i$.
4. For each $j \in [n] \setminus \{i\}$, sample

$$\mu_0^{j \to i}, \mu_1^{j \to i} \leftarrow \{0,1\}^\lambda$$
$$\mathsf{rlab}_{k,b}^{j \to i} \leftarrow \{0,1\}^\lambda \text{ for all } k \in [M], b \in \{0,1\}$$
$$\mathsf{m}_{k,b}^{i \to j} \leftarrow \{0,1\}^\lambda \text{ for all } k \in [4T], b \in \{0,1\}$$

5. **Garbling $\mathsf{C_B}$:** For each $j \in [n] \setminus \{i\}$, compute

$$(\widetilde{\mathsf{C}}_\mathsf{B}^{i \to j}, \mathsf{state}^{i \to j}) := \mathsf{SAdpGarbleCkt} \left(\mathsf{C_B}\left[\left\{\mathsf{m}_{k,0}^{i \to j}, \mathsf{m}_{k,1}^{i \to j}\right\}_{k \in [4T], b \in \{0,1\}}\right]; \Omega\right)$$

$$\mathsf{cm}^{i \to j} := \mathsf{Com}(\mathsf{state}^{i \to j}; \omega^{i \to j})$$

where $\mathsf{C_B}$ is described in Fig. 11 and $\Omega, \omega^{i \to j}$ are sampled randomly.

---

[5] The GGM PRF [GGM86] can be easily modified to give such a PRG based on one-way functions.

6. **Messages to $\mathcal{F}_g$:** Send (ssid $= i, s_i, \{\mathsf{rlab}_{k,b}^{j\to i}\}_{j\in[n]\setminus\{i\},k\in[M],b\in\{0,1\}}$) to $\mathcal{F}_g$ acting as the receiver and for each $j \in [n] \setminus \{i\}$, send (ssid $= j, \{\mathsf{cm}^{i\to j}, \mathsf{state}^{i\to j}, \omega^{i\to j}\}$) to $\mathcal{F}_g$ acting as the sender.

7. **Messages to $\mathcal{F}_{\mathsf{zk}}$:** For each $j \in [n]\setminus\{i\}$, send (ssid $= (j \to i), \mu_0^{j\to i}, \mu_1^{j\to i}$) to $\mathcal{F}_{\mathsf{zk}}$ acting as the verifier, and send (ssid $= (i \to j), X^{i\to j}, W^{i\to j}$) to $\mathcal{F}_{\mathsf{zk}}$ acting as the prover where $X^{i\to j} = \left(\widetilde{C}_B^{i\to j}, \mathsf{cm}^{i\to j}\right)$ and $W^{i\to j} = \left(\Omega, C_B\left[\{m_{k,0}^{i\to j}, m_{k,1}^{i\to j}\}_{k\in[4T],b\in\{0,1\}}\right], \mathsf{state}^{i\to j}, \omega^{i\to j}\right)$.

8. Send $\left(z_i, \{\widetilde{C}_B^{i\to j}\}_{j\in[n]\setminus\{i\}}, e_i, \{\mathsf{cm}^{i\to j}\}_{j\in[n]\setminus\{i\}}\right)$ to every other party.

$$C_B\left[\{m_{k,0}, m_{k,1}\}_{k\in[4T]}\right]$$

Input: $s \in \{0,1\}^\lambda$.
1. $d := \mathsf{PRG}_{\mathsf{mal}}(s)$ where $d \in \{0,1\}^{4T}$.
2. Output $\{m_{k,d[k]}\}_{k\in[4T]}$.

Fig. 11. Circuit C_B

Round-2: Each party P_i does the following:
1. Set $\mathsf{st}_i := (z_1\|\ldots\|z_n) \oplus v_i$.
2. Set $N = \ell + 4T\lambda(n-1)$.
3. Set $\overline{\mathsf{lab}}^{i,T+1} := \{\mathsf{lab}_{k,0}^{i,T+1}, \mathsf{lab}_{k,1}^{i,T+1}\}_{k\in[N]}$ where $\mathsf{lab}_{k,b}^{i,T+1} := 0^\lambda$ for each $k \in [N], b \in \{0,1\}$.
4. **for** each t from T down to 1 **do:**
   (a) Parse ϕ_t as (i^*, f, g, h).
   (b) If $i = i^*$ then compute (where P is described in Fig. 6)
   $$\left(\widetilde{\mathsf{P}}^{i,t}, \overline{\mathsf{lab}}^{i,t}\right) \leftarrow \mathsf{Garble}(1^\lambda, \mathsf{P}[i, \phi_t, v_i, \bot, \overline{\mathsf{lab}}^{i,t+1}]).$$
   (c) If $i \neq i^*$ then for every $\alpha, \beta \in \{0,1\}$, set $m'_{\alpha,\beta,0} = m_{(t,\alpha,\beta),e_{i^*}[(t,\alpha,\beta)]}^{i\to i^*}$ and $m'_{\alpha,\beta,1} = m_{(t,\alpha,\beta),1\oplus e_{i^*}[(t,\alpha,\beta)]}^{i\to i^*}$.
   Compute $ct_{t,\alpha,\beta}^i := (m'_{\alpha,\beta,0} \oplus \mathsf{lab}_{h,0}^{i,t+1}, m'_{\alpha,\beta,1} \oplus \mathsf{lab}_{h,1}^{i,t+1})$ and compute
   $$\left(\widetilde{\mathsf{P}}^{i,t}, \overline{\mathsf{lab}}^{i,t}\right) \leftarrow \mathsf{Garble}(1^\lambda, \mathsf{P}[i, \phi_t, v_i, \{ct_{t,\alpha,\beta}^i\}, \overline{\mathsf{lab}}^{i,t+1}]).$$

Parties: $P_1, P_2, \ldots, P_n$.

Inputs:

- P_1 (also called as the receiver) inputs $s \in \{0,1\}^\lambda$ and $\overline{\text{rlab}}_2, \ldots, \overline{\text{rlab}}_n$ where each $\overline{\text{rlab}}_i$ is a collection of labels $\{\text{rlab}_{j,0}^{i\rightarrow 1}, \text{rlab}_{j,1}^{i\rightarrow 1}\}_{j \in [M]}$ with each label of length λ.

- For each $i \in [2,n]$, P_i (also called as the sender) inputs $(\text{cm}^{i\rightarrow 1}, \text{state}^{i\rightarrow 1}, \omega^{i\rightarrow 1})$, where $\text{cm}^{i\rightarrow 1}$ is a commitment and is a public input, $\text{state}^{i\rightarrow 1}$ is the secret state of the somewhere adaptive garbling scheme, and $\omega^{i\rightarrow 1}$ is a string.

Output: Check if for each $i \in [2,n]$, $\text{Decom}(\text{cm}^{i\rightarrow 1}, \text{state}^{i\rightarrow 1}, \omega^{i\rightarrow 1}) = 1$. If all the checks pass, output $\{\text{Projection}(\text{SAdpGarbleInp}(\text{state}^{i\rightarrow 1}, s), \overline{\text{rlab}}_i)\}_{i \in [2,n]}$ to every party.

Fig. 12. The function g computed by the internal MPC where P_1 acts as the receiver

5. **Garbling C_{wrap}:** Compute

$$\widetilde{C}_{\text{wrap}}^i \leftarrow \text{Garble}\Big(\mathsf{C}_{\text{wrap}}\Big[\{\widetilde{C}_{\mathsf{B}}^{j\rightarrow i}\}_{j \in [n]\setminus\{i\}}, \text{st}_i, \overline{\text{lab}}^{i,1}\Big],$$

$$\Big\{\{\mu_b^{j\rightarrow i}\}_{j \in [n]\setminus\{i\}}, \{\text{rlab}_{k,b}^{j\rightarrow i}\}_{j \in [n]\setminus\{i\}, k \in [M], b \in \{0,1\}}\Big\}\Big)$$

where C_{wrap} is described in Fig. 13.

6. Send $\Big(\{\widetilde{\mathsf{P}}^{i,t}\}_{t \in [T]}, \widetilde{C}_{\text{wrap}}^i\Big)$ to every other party.

$$\mathsf{C}_{\text{wrap}}\Big[\{\widetilde{C}_{\mathsf{B}}^{j\rightarrow i}\}_{j \in [n]\setminus\{i\}}, \text{st}_i, \overline{\text{lab}}^{i,1}\Big]$$

Input: $\{b^{j\rightarrow i}\}_{j \in [n]\setminus\{i\}}, \{\widetilde{s}^{j\rightarrow i}\}_{j \in [n]\setminus\{i\}}$

1. If $b^{j\rightarrow i} = 1$ for all $j \in [n] \setminus \{i\}$ do:

   (a) For each $j \in [n] \setminus \{i\}$, compute $\{m_k^{j\rightarrow i}\}_{k \in [4T]} \leftarrow$ $\text{SAdpEvalCkt}\left(\widetilde{C}_{\mathsf{B}}^{j\rightarrow i}, \widetilde{s}^{j\rightarrow i}\right)$.

   (b) Let $m := \underset{j \in [n]\setminus\{i\}, k \in [4T]}{\|} (m_k^{j\rightarrow i})$

   (c) Output $\text{Projection}(\text{st}_i \| m, \overline{\text{lab}}^{i,1})$.

2. Else, output $\bot$.

Fig. 13. Circuit C^{wrap}

Evaluation: Every party P_i does the following:

1. For each $j \in [n]$,

   (a) Obtain $(\text{ssid} = j, \{\widetilde{\text{rlab}}^j\})$ from $\mathcal{F}_g$ where party P_j acts as the receiver.

(b) For each $k \in [n] \setminus \{j\}$, obtain $(\mathsf{ssid} = (k \rightarrow j), b^{k \rightarrow j}, X^{k \rightarrow j}, \mu^{k \rightarrow j})$
from $\mathcal{F}_{\mathsf{zk}}$. Set $\widetilde{\mu}^j = \{\mu^{k \rightarrow j}\}_{k \in [n] \setminus \{j\}}$.

(c) $\widetilde{\mathsf{lab}}^{j,1} \leftarrow \mathsf{Eval}(\widetilde{\mathsf{C}}^j_{\mathsf{wrap}}, \widetilde{\mu}^j \| \widetilde{\mathsf{rlab}}^j)$.

2. **for** each t from 1 to T **do**:

(a) Parse ϕ_t as (i^*, f, g, h).

(b) Compute $((\alpha, \beta, \gamma), \{\omega^j\}_{j \in [n] \setminus \{i\}}, \widetilde{\mathsf{lab}}^{i^*, t+1}) := \mathsf{Eval}(\widetilde{\mathsf{P}}^{i^*, t}, \widetilde{\mathsf{lab}}^{i^*, t})$.

(c) Set $\mathsf{st}_i[h] := \gamma \oplus v_i[h]$.

(d) **for** each $j \neq i^*$ **do**:

  i. Compute $(ct = (\delta_0, \delta_1), \{\mathsf{lab}^{j,t+1}_k\}_{k \in [N] \setminus \{h\}}) := \mathsf{Eval}(\widetilde{\mathsf{P}}^{j,t}, \widetilde{\mathsf{lab}}^{j,t})$.

  ii. Recover $\mathsf{lab}^{j,t+1}_h := \delta_\gamma \oplus \omega^j$.

  iii. Set $\widetilde{\mathsf{lab}}^{j,t+1} := \{\mathsf{lab}^{j,t+1}_k\}_{k \in [N]}$.

3. Compute the output as $\mathsf{post}(i, \mathsf{st}_i)$.

Correctness. The correctness follows via a similar argument to the semi-honest case.

Efficiency. Let the number of public key operations in Φ be npk_Φ, in one execution of $\mathcal{F}_{\mathsf{zk}}$ be $\mathsf{npk}_{\mathsf{zk}}$, and in one execution of $\mathcal{F}_g$ be npk_g. We choose the conforming protocol that performs OT extension between every pair of parties so that npk_Φ is bounded by $\mathcal{O}(n^2 \lambda)$. The total number of public key operations in our two-round construction is $\mathcal{O}(\mathsf{npk}_\Phi + n^2 \cdot \mathsf{npk}_{\mathsf{zk}} + n \cdot \mathsf{npk}_g)$. It follows from Theorems 3, 2 that this number is bounded by $\mathsf{poly}(n, \lambda)$.

Security. The security proof will be given in the full version [GMS18].

References

[ACJ17] Ananth, P., Choudhuri, A.R., Jain, A.: A new approach to round-optimal secure multiparty computation. In: Katz, J., Shacham, H. (eds.) CRYPTO 2017, Part I. LNCS, vol. 10401, pp. 468–499. Springer, Cham (2017). https://doi.org/10.1007/978-3-319-63688-7_16

[AIKW13] Applebaum, B., Ishai, Y., Kushilevitz, E., Waters, B.: Encoding functions with constant online rate or how to compress garbled circuits keys. In: Canetti, R., Garay, J.A. (eds.) CRYPTO 2013, Part II. LNCS, vol. 8043, pp. 166–184. Springer, Heidelberg (2013). https://doi.org/10.1007/978-3-642-40084-1_10

[ALSZ17] Asharov, G., Lindell, Y., Schneider, T., Zohner, M.: More efficient oblivious transfer extensions. J. Cryptol. 30(3), 805–858 (2017)

[Bea96] Beaver, D.: Correlated pseudorandomness and the complexity of private computations. In: Proceedings of the Twenty-Eighth Annual ACM Symposium on the Theory of Computing, Philadelphia, Pennsylvania, 22–24 May 1996, pp. 479–488 (1996)

[BGI16] Boyle, E., Gilboa, N., Ishai, Y.: Breaking the circuit size barrier for secure computation under DDH. In: Robshaw, M., Katz, J. (eds.) CRYPTO 2016, Part I. LNCS, vol. 9814, pp. 509–539. Springer, Heidelberg (2016). https://doi.org/10.1007/978-3-662-53018-4_19

[BGI+17a] Badrinarayanan, S., Garg, S., Ishai, Y., Sahai, A., Wadia, A.: Two-message witness indistinguishability and secure computation in the plain model from new assumptions. In: Takagi, T., Peyrin, T. (eds.) ASI-ACRYPT 2017, Part III. LNCS, vol. 10626, pp. 275–303. Springer, Cham (2017). https://doi.org/10.1007/978-3-319-70700-6_10

[BGI17b] Boyle, E., Gilboa, N., Ishai, Y.: Group-based secure computation: optimizing rounds, communication, and computation. In: Coron, J.-S., Nielsen, J.B. (eds.) EUROCRYPT 2017, Part II. LNCS, vol. 10211, pp. 163–193. Springer, Cham (2017). https://doi.org/10.1007/978-3-319-56614-6_6

[BGJ+17] Badrinarayanan, S., Goyal, V., Jain, A., Kalai, Y.T., Khurana, D., Sahai, A.: Promise zero knowledge and its applications to round optimal MPC. Cryptology ePrint Archive, Report 2017/1088 (2017). https://eprint.iacr.org/2017/1088

[BHP17] Brakerski, Z., Halevi, S., Polychroniadou, A.: Four round secure computation without setup. In: Kalai, Y., Reyzin, L. (eds.) TCC 2017, Part I. LNCS, vol. 10677, pp. 645–677. Springer, Cham (2017). https://doi.org/10.1007/978-3-319-70500-2_22

[BHR12] Bellare, M., Hoang, V.T., Rogaway, P.: Foundations of garbled circuits. In: Yu, T., Danezis, G., Gligor, V.D., (eds.) ACM CCS 2012, pp. 784–796. ACM Press, October 2012

[BL18] Benhamouda, F., Lin, H.: k-round MPC from k-round OT via garbled interactive circuits. In: EUROCRYPT (2018, to appear). https://eprint.iacr.org/2017/1125

[BMR90] Beaver, D., Micali, S., Rogaway, P.: The round complexity of secure protocols (extended abstract). In: 22nd ACM STOC, pp. 503–513. ACM Press, May 1990

[BP16] Brakerski, Z., Perlman, R.: Lattice-based fully dynamic multi-key FHE with short ciphertexts. In: Robshaw, M., Katz, J. (eds.) CRYPTO 2016, Part I. LNCS, vol. 9814, pp. 190–213. Springer, Heidelberg (2016). https://doi.org/10.1007/978-3-662-53018-4_8

[Can00a] Canetti, R.: Security and composition of multiparty cryptographic protocols. J. Cryptol. 13(1), 143–202 (2000)

[Can00b] Canetti, R.: Universally composable security: a new paradigm for cryptographic protocols. Cryptology ePrint Archive, Report 2000/067 (2000). http://eprint.iacr.org/2000/067

[Can01] Canetti, R.: Universally composable security: a new paradigm for cryptographic protocols. In: 42nd FOCS, pp. 136–145. IEEE Computer Society Press, October 2001

[COSV17] Ciampi, M., Ostrovsky, R., Siniscalchi, L., Visconti, I.: Round-optimal secure two-party computation from trapdoor permutations. In: Kalai, Y., Reyzin, L. (eds.) TCC 2017, Part I. LNCS, vol. 10677, pp. 678–710. Springer, Cham (2017). https://doi.org/10.1007/978-3-319-70500-2_23

[FS90] Feige, U., Shamir, A.: Witness indistinguishable and witness hiding protocols. In: 22nd ACM STOC, pp. 416–426. ACM Press, May 1990

[GGH+13] Garg, S., Gentry, C., Halevi, S., Raykova, M., Sahai, A., Waters, B.: Candidate indistinguishability obfuscation and functional encryption for all circuits. In: 54th FOCS, pp. 40–49. IEEE Computer Society Press, October 2013

[GGHR14] Garg, S., Gentry, C., Halevi, S., Raykova, M.: Two-round secure MPC from indistinguishability obfuscation. In: Lindell, Y. (ed.) TCC 2014. LNCS, vol. 8349, pp. 74–94. Springer, Heidelberg (2014). https://doi.org/10.1007/978-3-642-54242-8_4

[GGM86] Goldreich, O., Goldwasser, S., Micali, S.: How to construct random functions. J. ACM **33**(4), 792–807 (1986)

[GGSW13] Garg, S., Gentry, C., Sahai, A., Waters, B.: Witness encryption and its applications. In: Boneh, D., Roughgarden, T., Feigenbaum, J. (eds.) 45th ACM STOC, pp. 467–476. ACM Press, June 2013

[GLS15] Dov Gordon, S., Liu, F.-H., Shi, E.: Constant-round MPC with fairness and guarantee of output delivery. In: Gennaro, R., Robshaw, M. (eds.) CRYPTO 2015, Part II. LNCS, vol. 9216, pp. 63–82. Springer, Heidelberg (2015). https://doi.org/10.1007/978-3-662-48000-7_4

[GMM17a] Garg, S., Mahmoody, M., Mohammed, A.: Lower bounds on obfuscation from all-or-nothing encryption primitives. In: Katz, J., Shacham, H. (eds.) CRYPTO 2017, Part I. LNCS, vol. 10401, pp. 661–695. Springer, Cham (2017). https://doi.org/10.1007/978-3-319-63688-7_22

[GMM17b] Garg, S., Mahmoody, M., Mohammed, A.: When does functional encryption imply obfuscation? In: Kalai, Y., Reyzin, L. (eds.) TCC 2017, Part I. LNCS, vol. 10677, pp. 82–115. Springer, Cham (2017). https://doi.org/10.1007/978-3-319-70500-2_4

[GMMM17] Garg, S., Mahmoody, M., Masny, D., Meckler, I.: On the round complexity of OT extension. Cryptology ePrint Archive, Report 2017/1187 (2017). https://eprint.iacr.org/2017/1187

[GMPP16] Garg, S., Mukherjee, P., Pandey, O., Polychroniadou, A.: The exact round complexity of secure computation. In: Fischlin, M., Coron, J.-S. (eds.) EUROCRYPT 2016, Part II. LNCS, vol. 9666, pp. 448–476. Springer, Heidelberg (2016). https://doi.org/10.1007/978-3-662-49896-5_16

[GMS18] Garg, S., Miao, P., Srinivasan, A.: Two-round multiparty secure computation minimizing public key operations. Cryptology ePrint Archive, Report 2018/180 (2018). https://eprint.iacr.org/2018/180

[GMW87] Goldreich, O., Micali, S., Wigderson, A.: How to play any mental game or a completeness theorem for protocols with honest majority. In: Aho, A. (ed.) 19th ACM STOC, pp. 218–229. ACM Press, May 1987

[GS17] Garg, S., Srinivasan, A.: Garbled protocols and two-round MPC from bilinear maps. In: 58th FOCS, pp. 588–599. IEEE Computer Society Press (2017)

[GS18] Garg, S., Srinivasan, A.: Two-round multiparty secure computation from minimal assumptions. In: Nielsen, J.B., Rijmen, V. (eds.) EUROCRYPT 2018. LNCS, vol. 10821, pp. 468–499. Springer, Cham (2018). https://doi.org/10.1007/978-3-319-78375-8_16. https://eprint.iacr.org/2017/1156

[HHPV17] Halevi, S., Hazay, C., Polychroniadou, A., Venkitasubramaniam, M.: Round-optimal secure multi-party computation. Cryptology ePrint Archive, Report 2017/1056 (2017). http://eprint.iacr.org/2017/1056

[HIK07] Harnik, D., Ishai, Y., Kushilevitz, E.: How many oblivious transfers are needed for secure multiparty computation? In: Menezes, A. (ed.) CRYPTO 2007. LNCS, vol. 4622, pp. 284–302. Springer, Heidelberg (2007). https://doi.org/10.1007/978-3-540-74143-5_16

[HIKN08] Harnik, D., Ishai, Y., Kushilevitz, E., Nielsen, J.B.: OT-combiners via secure computation. In: Canetti, R. (ed.) TCC 2008. LNCS, vol. 4948, pp. 393–411. Springer, Heidelberg (2008). https://doi.org/10.1007/978-3-540-78524-8_22

[HJO+16] Hemenway, B., Jafargholi, Z., Ostrovsky, R., Scafuro, A., Wichs, D.: Adaptively secure garbled circuits from one-way functions. In: Robshaw, M., Katz, J. (eds.) CRYPTO 2016, Part III. LNCS, vol. 9816, pp. 149–178. Springer, Heidelberg (2016). https://doi.org/10.1007/978-3-662-53015-3_6

[HLP11] Halevi, S., Lindell, Y., Pinkas, B.: Secure computation on the web: computing without simultaneous interaction. In: Rogaway, P. (ed.) CRYPTO 2011. LNCS, vol. 6841, pp. 132–150. Springer, Heidelberg (2011). https://doi.org/10.1007/978-3-642-22792-9_8

[IKNP03] Ishai, Y., Kilian, J., Nissim, K., Petrank, E.: Extending oblivious transfers efficiently. In: Boneh, D. (ed.) CRYPTO 2003. LNCS, vol. 2729, pp. 145–161. Springer, Heidelberg (2003). https://doi.org/10.1007/978-3-540-45146-4_9

[JKK+17] Jafargholi, Z., Kamath, C., Klein, K., Komargodski, I., Pietrzak, K., Wichs, D.: Be adaptive, avoid overcommitting. In: Katz, J., Shacham, H. (eds.) CRYPTO 2017, Part I. LNCS, vol. 10401, pp. 133–163. Springer, Cham (2017). https://doi.org/10.1007/978-3-319-63688-7_5

[JKKR17] Jain, A., Kalai, Y.T., Khurana, D., Rothblum, R.: Distinguisher-dependent simulation in two rounds and its applications. In: Katz, J., Shacham, H. (eds.) CRYPTO 2017, Part II. LNCS, vol. 10402, pp. 158–189. Springer, Cham (2017). https://doi.org/10.1007/978-3-319-63715-0_6

[JW16] Jafargholi, Z., Wichs, D.: Adaptive security of Yao's garbled circuits. In: Hirt, M., Smith, A. (eds.) TCC 2016, Part I. LNCS, vol. 9985, pp. 433–458. Springer, Heidelberg (2016). https://doi.org/10.1007/978-3-662-53641-4_17

[KK13] Kolesnikov, V., Kumaresan, R.: Improved OT extension for transferring short secrets. In: Canetti, R., Garay, J.A. (eds.) CRYPTO 2013, Part II. LNCS, vol. 8043, pp. 54–70. Springer, Heidelberg (2013). https://doi.org/10.1007/978-3-642-40084-1_4

[KRS16] Kumaresan, R., Raghuraman, S., Sealfon, A.: Network oblivious transfer. In: Robshaw, M., Katz, J. (eds.) CRYPTO 2016, Part II. LNCS, vol. 9815, pp. 366–396. Springer, Heidelberg (2016). https://doi.org/10.1007/978-3-662-53008-5_13

[LP09] Lindell, Y., Pinkas, B.: A proof of security of Yao's protocol for two-party computation. J. Cryptol. 22(2), 161–188 (2009)

[MW16] Mukherjee, P., Wichs, D.: Two round multiparty computation via multi-key FHE. In: Fischlin, M., Coron, J.-S. (eds.) EUROCRYPT 2016, Part II. LNCS, vol. 9666, pp. 735–763. Springer, Heidelberg (2016). https://doi.org/10.1007/978-3-662-49896-5_26

[Nie07] Nielsen, J.B.: Extending oblivious transfers efficiently - how to get robustness almost for free. Cryptology ePrint Archive, Report 2007/215 (2007). http://eprint.iacr.org/2007/215

[NNOB12] Nielsen, J.B., Nordholt, P.S., Orlandi, C., Burra, S.S.: A new approach to practical active-secure two-party computation. In: Safavi-Naini, R., Canetti, R. (eds.) CRYPTO 2012. LNCS, vol. 7417, pp. 681–700. Springer, Heidelberg (2012). https://doi.org/10.1007/978-3-642-32009-5_40

[PS16] Peikert, C., Shiehian, S.: Multi-key FHE from LWE, revisited. In: Hirt, M., Smith, A. (eds.) TCC 2016, Part II. LNCS, vol. 9986, pp. 217–238. Springer, Heidelberg (2016). https://doi.org/10.1007/978-3-662-53644-5_9

[Yao82] Yao, A.C.-C.: Protocols for secure computations (extended abstract). In: 23rd FOCS, pp. 160–164. IEEE Computer Society Press, November 1982

Limits of Practical Sublinear Secure Computation

Elette Boyle[1], Yuval Ishai[2], and Antigoni Polychroniadou[3(✉)]

[1] IDC Herzliya, Herzliya, Israel
eboyle@alum.mit.edu
[2] Technion, Haifa, Israel
yuvali@cs.technion.ac.il
[3] Cornell Tech and University of Rochester, New York, USA
antigoni@cornell.edu

Abstract. Secure computations on big data call for protocols that have sublinear communication complexity in the input length. While fully homomorphic encryption (FHE) provides a general solution to the problem, employing it on a large scale is currently quite far from being practical. This is also the case for secure computation tasks that reduce to weaker forms of FHE such as "somewhat homomorphic encryption" or single-server private information retrieval (PIR).

Quite unexpectedly, Aggarwal, Mishra, and Pinkas (Eurocrypt 2004), Brickell and Shmatikov (Asiacrypt 2005), and Shelat and Venkitasubramaniam (Asiacrypt 2015) have shown that in several natural instances of secure computation on big data, there are practical sublinear communication protocols that only require sublinear local computation and minimize the use of expensive public-key operations. This raises the question of whether similar protocols exist for other natural problems.

In this paper we put forward a framework for separating "practical" sublinear protocols from "impractical" ones, and establish a methodology for identifying "provably hard" big-data problems that do not admit practical protocols. This is akin to the use of NP-completeness to separate hard algorithmic problems from easy ones. We show that while the previous protocols of Aggarwal et al., Brickell and Shmatikov, and Shelat and Venkitasubramaniam are indeed classified as being "practical" in this framework, slight variations of the problems they solve and other natural computational problems on big data are hard.

Our negative results are established by showing that the problem at hand is "PIR-hard" in the sense that *any* secure protocol for the problem implies PIR on a large database. This imposes a barrier on the local computational cost of secure protocols for the problem. We also identify a new natural relaxation of PIR that we call *semi-PIR,* which is useful for establishing "intermediate hardness" of several practically motivated secure computation tasks. We show that semi-PIR implies *slightly* sublinear PIR via an adaptive black-box reduction and that ruling out a stronger black-box reduction would imply a major breakthrough in complexity theory. We also establish information-theoretic separations between semi-PIR and PIR, showing that some problems that we prove to be semi-PIR-hard are not PIR-hard.

© International Association for Cryptologic Research 2018
H. Shacham and A. Boldyreva (Eds.): CRYPTO 2018, LNCS 10993, pp. 302–332, 2018.
https://doi.org/10.1007/978-3-319-96878-0_11

Keywords: Secure computation · Private information retrieval
Sublinear communication · Locally decodable codes

1   Introduction

Protocols for *secure multi-party computation* (MPC) enable mutually distrusting parties to jointly evaluate a function on their private inputs, without revealing any information beyond the prescribed function outputs [Yao82, GMW87, BGW88, CCD88].

An important efficiency metric of MPC protocols is the required communication between parties. A great deal of research focus has gone towards minimizing the asymptotic communication complexity of MPC, as well as improving the practical efficiency of MPC. Our work proposes a theoretical framework for capturing the intersection. This framework can be used to provide a crude distinction between tasks that admit "practical" sublinear-communication protocols and ones that do not, akin to the use of NP-completeness to separate hard algorithmic problems from easy ones.

Secure computation on big data calls for MPC protocols that have sublinear communication complexity in the input size. The line of work on sublinear-communication MPC started with works on private information retrieval (PIR) [CKGS98, KO97] and related primitives, and culminated in the constructions of fully homomorphic encryption (FHE) [Gen09] schemes. FHE gives a general solution to the problem in that it essentially closes the gap between secure and insecure communication complexity.

The main concrete bottleneck of current FHE schemes, which makes them slow in practice, is their computational complexity. Even in the case of PIR, which can be viewed as the simplest instance of "somewhat-homomorphic encryption," local computation on the server side is by far the most significant cost. Indeed, PIR protocols without preprocessing provably require linear computational complexity, and all known PIR protocols on a database of length N require at least N "public-key operations" (comparable to the amortized cost of encrypting a bit in an underlying public-key encryption scheme). Consequently, the computational cost of such protocols is significantly higher than that of protocols that process a similar amount of information using only symmetric encryption. Moreover, unlike the case of OT-based protocols, little can be done for amortizing the cost of PIR or for pushing it to an input-independent preprocessing phase. Despite recent advances on the concrete cost of PIR [MBFK16] and the asymptotic cost of PIR with preprocessing [BIPW17, CHR17], performing a single instance of PIR on an N-bit database is more expensive in terms of local computation than, say, securely evaluating a boolean circuit of size N by relying on efficient OT extension techniques.

Notable exceptions to the above state of affairs are the work of Aggarwal et al. [AMP10] on medians, the work of Brickell and Shmatikov [BS05] on certain graph problems, and the work of Shelat and Venkitasubramaniam [SV15] that vastly generalizes them. These works show that for certain natural and practically motivated problems, including several central combinatorial optimization problems, one can enjoy the best of both worlds: sublinear communication complexity with low computational overhead, both asymptotically and concretely. These works leave several interesting open questions. In particular, it is not clear how robust the positive results are to natural variations of the functionality and whether they extend to other optimization problems.

1.1   Our Results

Towards addressing the above open questions in a systematic way, we propose a clean formal framework to capture the feature of the protocols of [AMP10,BS05,SV15] that distinguishes them from more generic alternatives based on FHE or PIR. Concretely, we consider a model of secure two-party computation with input-independent preprocessing in the form of correlated randomness. (The latter can be used to implement oblivious transfer unconditionally.) We distinguish between:

- "Easy" problems, namely ones admitting sublinear-communication secure protocols that may rely on input-independent correlated randomness and oblivious transfer, and
- "PIR-hard" problems, for which any sublinear-communication protocol implies a nontrivial PIR protocol on a large database.

Given the current state of the art, PIR-hard problems are unlikely to be shown "easy," and any protocol for such problems is likely to have poor concrete efficiency.

PIR-Hardness of Combinatorial Problems. We then revisit a class of combinatorial optimization problems for which "easy" protocols have been demonstrated. (In particular, all of the protocols from [AMP10,BS05,SV15] are easy in the above sense.) We show that, while the original formulation of the problem yields lightweight protocols, certain natural and useful variants of the same problems are in fact PIR-hard. We first demonstrate this for the case of *one-sided* variants—in which only one party learns the function output—for an assortment of combinatorial problems with different structures:

- **Median.** The *median* functionality accepts a list of numerical inputs from each party and outputs the median of the combined list.
- **Convex Hull.** The 2-dimensional *convex hull* functionality accepts a set of points in 2-space from each party and outputs the subset of those points on the convex hull of the combined set.
- **Single-Source Shortest Distance.** The *SSSD* functionality accepts a set of weighted edges from each party (on n fixed vertices) with distinguished vertex v^* and outputs the lengths of $n - 1$ shortest paths from v^* to each $v \neq v^*$ in the combined graph (taking parallel edges).

– **Approximate Set Cover.** The *approx set cover* functionality refers to the output of the polynomial-time greedy algorithm for polynomial time approximation of set cover (which iteratively selects the set that contains the largest number of uncovered elements).

We prove that the one-sided variant of each of the above problems is PIR-hard. This further implies that any *secret-shared* variant of the problems, in which the parties compute secret shares of the corresponding functionality output, is additionally PIR-hard. (Indeed, existence of an "easy" protocol for the latter directly yields an analogous protocol for the former, by having party 2 send his secret share to party 1). This may indicate that lightweight protocols for these problems cannot be effectively used "within" other larger MPC computations.

We remark that the previous "easy" protocol constructions for the above combinatorial problems frequently provide security within a promise setting, where certain restrictions are assumed to hold on parties' inputs (e.g., that parties' inputs are disjoint). Our negative results are each within the respective promise settings.

Our PIR-hardness results can be interpreted as imposing a barrier on the local computational cost of secure protocols for the problem. This barrier applies both asymptotically (linear computation is necessary without preprocessing) and concretely (achieving sublinear communication comes at a high computational cost given the current state of the art on PIR).

Semi-PIR-Hardness. We then identify a new natural relaxation of PIR that we call *semi-PIR,* which is useful for establishing "intermediate hardness" of several practically motivated secure computation tasks. Semi-PIR is defined analogously to PIR, except that the privacy requirement is relaxed to guarantee privacy of the client's query index $i \in [n]$ *only* if it holds that the corresponding database value z_i is equal to 1. This notion may be independently motivated by settings where there is natural asymmetry in privacy concerns for 0 versus 1 values (e.g., database of patients with and without some disease).

We show that semi-PIR with polylogarithmic communication complexity implies *slightly* sublinear PIR via an adaptive black-box reduction. Thus, semi-PIR-hardness can have a similar interpretation as PIR-hardness from a crude asymptotic perspective. Our reduction from PIR to semi-PIR makes use of query-efficient locally decodable codes (LDC). Correspondingly, ruling out a stronger black-box reduction would imply a major breakthrough in complexity theory, concerning existence of LDCs with polynomial rate and low query complexity.

Theorem 1 (Informal). *Suppose there is an efficient q-query LDC $C :$ $\{0,1\}^n \rightarrow \{0,1\}^N$. Then, there exists a protocol that implements PIR on a database $z \in \{0,1\}^n$ by using an expected $O(2^q)$ (adaptive) calls to semi-PIR on a database $z' \in \{0,1\}^N$ and no additional interaction.*

The reduction effectively attempts to reconstruct the desired database value z_i, $i \in [n]$ by accessing positions $j_1, \ldots, j_q$ in the encoded database $j_\ell \in [N]$, each time to either the direct bit value or a negated version (so that the read value will be 0 with probability $1/2$). At any point in which the queried location stores a 0, this query index is no longer hidden, and the reduction will restart with a freshly sampled set of q-queries. The smoothness of the LDC guarantees that revealing any *single* query index reveals nothing about the ultimate desired index i. Note the inherent adaptivity of this approach.

We also establish information-theoretic separations between semi-PIR and PIR. These imply that some problems that we prove to be semi-PIR-hard are provably *not* PIR-hard in a strong sense, suggesting that semi-PIR captures the true complexity of some natural secure computation tasks rather than being an artifact of our proof techniques.

Our semi-PIR-hardness results apply to natural two-sided functions, whose output is revealed to both parties. A broad class of such examples is "optimal selection from a short-list", where a Receiver has a small list of candidate indices, and both parties learn the identity of the candidate with the maximum/minimum/most desired value. Situations of secure computation of such problems can be motivated by real-life scenarios in which the identity of the winning candidate (selected job applicant, purchased item, travel destination) is public information that cannot be hidden, yet one is interested in hiding the runner-ups or the choice criteria.

One such concrete problem is the Two-Sided Nearest Neighbor problem. Here a server holds a large database of points (x_i, y_i) in the Euclidean plane, say a list of restaurant locations, and a client holds a point (x, y), say representing its own location. The output of both parties is the point (x_i, y_i) which is closest to (x, y). As discussed above, the reason we consider here a two-sided output is that the selected restaurant can be publicly observed. And while this output may reveal a lot of partial information about the client's input, it is easy to imagine situations in which the client may wish to hide the exact location (x, y) from which the search has been conducted.

Two-sided versus one-sided functionalities. Unlike secure protocols realizing two-sided functionalities, secure protocols for one-sided functionalities must reveal no information about the output to one of the parties. This rules out iterative approaches in which partial information about the output is gradually revealed to both parties, allowing them to minimize the local computation by accessing only relevant portions of the input. However, as we show in this work, some natural two-sided functionalities exhibit an intermediate form of hardness captured by Semi-PIR. In such cases, both parties get the output, but one of the parties receives additional information only if some condition on the output is met (Table 1).

Local Compressibility. On the positive side, we identify a generic *local compressibility* property of combinatorial problems that directly permits efficient secure protocols for the problem, as well as any sufficiently "close" variant.

Loosely speaking, we say that a functionality $F : \{0,1\}^N \times \{0,1\}^N \to \{0,1\}^m \times \{0,1\}^m$ is locally compressible if there exists a preprocessing function $\mathsf{Pre} : \{0,1\}^N \to \{0,1\}^n$ for some $n \ll N$, for which it holds that

Table 1. Sample of our hardness results for combinatorial problems

Hardness	Combinatorial Problems
Easy	Two-Sided Locally Compressible Minimum Spanning Tree
	Two-Sided Locally Compressible High-Order Median Predicates
	Protocols from [AMP10,BS05,SV15]
Semi-PIR Hard	Two-Sided Single Source Single Destination Shortest Path
	Two-Sided Nearest Neighbor
	Two-Sided Closest Destination Problem
	Two-Sided Short-List Selection
PIR Hard	One-Sided Median
	One-Sided Approximate Set Cover
	One-Sided Convex Hull
	One-Sided Single Source Shortest Distances
	Two-sided Median Predicate

$F(X,Y) = F(\mathsf{Pre}(X), \mathsf{Pre}(Y))$. In such a case, an "easy" sublinear protocol for securely computing F can be achieved by first performing the local preprocessing, and then executing an arbitrary MPC for the circuit/program on the compressed inputs. This generality allows us to extend beyond the core functionality F itself, to provide an "easy" sublinear protocol for any composed function $G \circ F$ for which the circuit size of G is not too complex. This includes, for example, one-sided variants.

We demonstrate this local compressibility property in two example settings:

- **Minimum Spanning Tree.** The *MST* functionality accepts a set of weighted edges from each party (on n fixed vertices) and outputs the minimum spanning tree of the combined graph.
  A lightweight protocol for MST was given by [SV15] for the promise setting where all edge weights are distinct, as the corresponding MST promise problem falls within their "greedy-compatible" protocol framework. We observe that, within a similar promise setting, the MST of the combined graph is preserved when parties compute the MST of their local graphs first and then submit the resulting tree as their input to the MST functionality (i.e., $\mathsf{Pre}(X) = MST(X)$). Our approach thus yields "easy" protocols with sublinear communication for MST and related variants.
- **"High-Order" Median Predicates.** For any predicate function P that depends only on the highest-order bits of its input, we show that the median predicate functionality $P \circ Med$ is locally compressible. More specifically, consider the median problem for n inputs, and suppose P depends only on the $\ell \in o(\log n)$ most-significant bits of its input. Then a party's list of n input values can be compressed to a succinct $2^\ell \in o(n)$-size *count vector* corresponding to the number of occurrences of each length-ℓ prefix within the list. Since the high-order prefix of the median is equal to the median of

the corresponding high-order prefixes, this short count vector carries sufficient information to evaluate the desired functionality.

1.2 Organization of the Paper

Section 2 contains useful preliminaries. In Sect. 3 we present our formal notion of PIR-hardness, and PIR-hardness results for various combinatorial problem variants. Section 4 contains the definition and results pertaining to the notion of semi-PIR. Section 5 contains our positive local-compressibility results.

2 Preliminaries

Notation. We denote the security parameter by κ. We say that a function $\mu : \mathbb{N} \to \mathbb{N}$ is *negligible* if for every positive polynomial $p(\cdot)$ and all sufficiently large κ's it holds that $\mu(\kappa) < \frac{1}{p(\kappa)}$. We often use $[n]$ to denote the set $\{1, \ldots, n\}$. Moreover, we use $d \leftarrow \mathcal{D}$ to denote the process of sampling d from the distribution $\mathcal{D}$ or, if $\mathcal{D}$ is a set, a uniform choice from it. If $\mathcal{D}_1$ and $\mathcal{D}_2$ are two distributions, then we denote that they are statistically close by $\mathcal{D}_1 \approx_s \mathcal{D}_2$; we denote that they are computationally indistinguishable by $\mathcal{D}_1 \approx_c \mathcal{D}_2$; and we denote that they are identical by $\mathcal{D}_1 \equiv \mathcal{D}_2$.

Two-Party Computation. We assume familiarity with standard cryptographic primitives. For notational purposes, we recall here the basic working definitions. We refer to e.g. [Can01] for the formal definitions. A two-party protocol is cast by specifying a random process that maps pairs of inputs to pairs of outputs (one for each party). We refer to such a process as a functionality and denote it by $F : \{0,1\}^* \times \{0,1\}^* \to \{0,1\}^* \times \{0,1\}^*$ where $F = (F_1, F_2)$. That is, for every pair of inputs (x, y), the output-pair is a random variable $(F_1(x, y), F_2(x, y))$ ranging over pairs of strings. The first party (with input x) wishes to obtain $F_1(x, y)$ and the second party (with input y) wishes to obtain $F_2(x, y)$. The aim of a secure two-party protocol is to protect an honest party against dishonest behavior by the other party. In this paper, we consider semi-honest static adversaries which strengthens our impossibility results.

The security of a protocol is analyzed by comparing what an adversary can do in the protocol to what it can do in an ideal scenario that is secure by definition. This is formalized by considering an ideal computation involving an incorruptible trusted third party to whom the parties send their inputs. The trusted party computes the functionality on the inputs and returns to each party its respective output. Loosely speaking, a protocol is secure if any adversary interacting in the real protocol (where no trusted third party exists) can do no more harm than if it was involved in the above-described ideal computation.

Protocols in the Preprocessing or Correlated Randomness Model. We will also consider protocols for the preprocessing model. In the preprocessing model, the specification of a protocol also includes a joint distribution $P_{R_1 \cdots R_n}$

over $\mathcal{R}_1 \times \ldots \times \mathcal{R}_n$, where the $\mathcal{R}_i$'s are finite randomness domains. This distribution is used for sampling correlated random inputs $(r_1, \ldots, r_n) \leftarrow P_{R_1 \cdots R_n}$ received by the parties before the execution of the protocol. Therefore, the preprocessing is independent of the inputs. The actions of a party P_i in a given round may in this case depend on the private random input r_i received by P_i from the distribution $P_{R_1 \cdots R_n}$ and on its input x_i and the messages received in previous rounds. In addition, the action might depend on the statistical security parameter κ which is given as input to all parties along with x_i and r_i. Using the standard terminology of secure computation, the preprocessing model can be thought of as a hybrid model where the parties have one-time access to an ideal randomized functionality P (with no inputs) providing them with correlated, private random inputs r_i.

2.1 Private Information Retrieval

A (single-server) Private Information Retrieval (PIR) [CKGS98,KO97] protocol allows a client to retrieve a data item from a database held by a server while hiding which item it is after. More specifically, the database is modeled as an n-bit string z out of which the client retrieves the i-th bit z_i, while giving the server no information about the index i. The communication complexity of such a protocol is denoted by $c(n)$. A trivial PIR protocol would have the server sending the entire data string to the client (i.e. $c(n) = n$), thus satisfying the PIR privacy requirement in an information-theoretic way. We assume by default that any PIR protocol should be nontrivial in the sense that $c(n) < n$, and only consider computational security against semi-honest (passive) servers. We denote by $\mathsf{View}_S(z, i))$ the view of the PIR server in its interaction with the client on local inputs z, i and public input $n = |z|$, and by $\mathsf{Out}_C(z, i)$ the output of the client. Our definition treats the database size n as a public parameter that is also used as a security parameter.

Definition 1 (PIR). *Let (S, C) be an interactive protocol between a server S and a client C, where both S and C are PPT algorithms. We say that (S, C) is a* private information retrieval (PIR) *protocol if there exists a negligible function $\nu(n)$ such that:*

- **Correctness:** *For every $n \in \mathbb{N}$, $i \in [n]$, and $z = (z_1, \ldots, z_n) \in \{0, 1\}^n$,*

$$\Pr\left[\mathsf{Out}_C(z, i) = z_i\right] \geq 1 - \nu(n).$$

- **Security:** *For every non-uniform polynomial time distinguisher D, $n \in \mathbb{N}$, $i, j \in [n]$, and $z = (z_1, \ldots, z_n) \in \{0, 1\}^n$, it holds that $|p_i - p_j| \leq \nu(n)$, where*

$$p_i := \Pr\left[D(1^n, \mathsf{View}_S(z, i)) = 1\right],$$
$$p_j := \Pr\left[D(1^n, \mathsf{View}_S(z, j)) = 1\right].$$

- **Efficiency:** *The communication complexity $c(n)$ on a database $z \in \{0, 1\}^n$ is always required to be at most $n - 1$. We say that PIR protocol is* slightly sublinear *if $c(n) = O(n/\log^\gamma n)$ for every positive integer γ, and that it is* polylogarithmic *if $c(n) = O(\log^\gamma n)$ for some positive integer γ.*

We note that polylogarithmic single-server PIR protocols exist under (subexponential versions of) standard cryptographic assumptions [CMS99, Lip05, BV14]. On the other hand, PIR provably requires linear server computation in the database size [BIM04], and all known protocols make an intensive use of public key cryptography. Even in the fastest existing implementations of PIR [MBFK16], maximizing the speed of server (which is still at least an order of magnitude slower than a symmetric encryption of the entire database) has a high cost in communication.

Additional evidence for the hardness of PIR comes from the impossibility of realizing PIR information-theoretically in the OT-hybrid model or even using general correlated randomness [IKM+13]. This gives evidence against the possibility of using input-independent preprocessing or fast OT extension techniques [IKNP03] for amortizing the cost of PIR-based protocols, and should be contrasted with the fact that without the sublinear communication requirement, information-theoretic protocols exist in these models.

3   The PIR-Hardness Framework

We put forth a framework for separating "practical" sublinear computation protocols from "impractical" ones, by means of a notion of *PIR hardness*. PIR serves as an appealing benchmark metric for measuring protocol computation complexity in the sublinear communication regime: The functionality is natural and convenient to reduce to. And, since all known constructions make use of heavy public-key computations, this gives an indication that for any functionality which reduces to it, an analogous level of computation may be required. The high-level interpretation is thus that (given the current state of the art on PIR) saying that f is PIR-hard implies that evaluating f with a low communication complexity has a high computational cost. Even further, this computational cost cannot be amortized or moved to an input-independent preprocessing phase.

Definition 2 (PIR Hardness). *Let* $f : \{0,1\}^N \times \{0,1\}^N \to \{0,1\}^{m(N)} \times \{0,1\}^{m(N)}$ *be a two-party functionality.*

- *We say that f is $(n(N), \tau(N))$-PIR-hard if there is a single-server PIR protocol that makes $\tau(N)$ (expected) oracle calls to f on inputs of length N, where the PIR database size is $n(N)$ and, in addition to the oracle calls there is no additional communication.*
- *We say that f is non-interactively $n(N)$-PIR-hard if it is $(n(N), 1)$-PIR-hard, and that f is PIR-hard if it is non-interactively $n(N)$-PIR-hard for some $n(N) = \tilde{\Omega}(N) = N/polylog(N)$.*

Most (but not all) of the PIR hardness results obtained in this paper are of the simpler non-interactive type, namely the PIR protocol only applies a local mapping to the input of each party and then makes a single invocation of f with no additional interaction. The parameter $n(N)$ and $\tau(N)$ should be interpreted as a lower bound on the amount of expensive computation (which cannot be amortized or moved to a preprocessing phase) that is required for a sublinear-communication secure computation of f.

More concretely, we have the following easy corollary of PIR-hardness.

Claim. Suppose $f : \{0,1\}^N \times \{0,1\}^N \rightarrow \{0,1\}^{m(N)} \times \{0,1\}^{m(N)}$ is PIR-hard and has a protocol Π with $O(N^\beta)$ bits of communication for some $\beta < 1$. Then there is a nontrivial PIR protocol which on a database of size n makes a single invocation of Π on inputs of length $N = O(n)$ and uses no further interaction or assumptions.

The following remarks on our notion of PIR-hardness are in place:

Remark 1.

1. The above definition can be extended to allow extra sublinear communication beyond the f-oracle calls; however, our PIR-hardness results do not use this extension.
2. In the case of combinatorial problems involving graphs or other natural objects, the parameter N denotes the bit-length of a binary representation of the input for f. For example, in the case of a graph on ℓ nodes with polynomially bounded edge weights, we have $N = O(\ell^2 \log \ell)$. The polylogarithmic slackness in our default notion of PIR-hardness is meant to reduce the sensitivity of this notion to the way inputs are represented.

In the remainder of this section, we explore a general condition on functionalities which imply PIR hardness. We first consider functionalities f with *one-sided output*, i.e. where $f : \{0,1\}^N \times \{0,1\}^N \rightarrow \{\bot\} \times \{0,1\}^{m(N)}$ delivers output only to one of the two parties. We observe that in this setting, PIR hardness is tightly related to a combinatorial VC-dimension-style measure of complexity. We then extend this to demonstrate a sufficient condition for PIR-hardness of *two-sided predicate* functionalities.

3.1   VC-Dimension and Non-Interactive PIR-Hardness

In the case of one-sided output functionalities, where only one of the two parties receives output, the *privacy* property of PIR can be obtained immediately (namely, the server will play the role of the party who receives no output). PIR hardness of such a functionality then translates to a sufficient "combinatorial richness", capturing that the input-output behavior of the functionality is enough to encode the information of an entire database. We draw a connection between this property and a form of "efficient VC-dimension".

VC Dimension. We next define the Vapnik Chervonenkis (VC) dimension of a class of functions $\mathcal{F}$. The VC dimension gives a measure for the 'richness' of $\mathcal{F}$, which is useful in learning theory and computational complexity. We assume in the following that all functions in $\mathcal{F}$ are defined over the same input domain.

Definition 3 (VC-Dimension [VC71]). *Let $\mathcal{F}$ be a class of functions from some input domain D to $\{0,1\}$. We say that $\mathcal{F}$ shatters a point set $I \subset D$, if for every function $g : I \to \{0,1\}$, there is a function $f \in \mathcal{F}$ which agrees with g on I. The VC-dimension of $\mathcal{F}$, denoted by $VC(\mathcal{F})$, is the size of the largest point set I, that is shattered by $\mathcal{F}$.*

The VC-dimension can be extended to a class $\mathcal{F}$ of non-boolean functions from D to E. In this case, the set I is shattered if there exists a universal boolean (single-bit output) decoder $\gamma : E \to \{0,1\}$ such that I is shattered in the above sense by $\mathcal{F}' = \{\gamma(f(\cdot)) : f \in \mathcal{F}\}$.

For the generalization of VC dimension to functions with multi-bit outputs, a number of notions have been considered in the literature (e.g., [NAT89, BIKO12]). In this work, we handle multi-bit outputs applying a universal boolean decoder on the output of non-boolean functions, as was previously suggested in [BIKO12]. The work of [BIKO12] uses the relation between PIR and VC dimension to construct PIR protocols. We further develop this relation and use it to establish PIR-harndess.

Essentially, the VC-dimension of a multi-bit output function class is the maximum VC-dimension of the boolean function class $\gamma \circ \mathcal{F}$ over the choice of the boolean "decoder" function γ.

We observe that for a one-sided functionality $f : \{0,1\}^N \times \{0,1\}^N \to \{\perp\} \times \{0,1\}^{m(N)}$, non-interactive PIR hardness of f coincides directly with the following notion of *efficiently computable* VC-dimension of the induced function class $\mathcal{F} = \{f(x, \cdot)\}_{x \in \{0,1\}^N}$. Explicitly, a non-interactive construction of PIR of database size n from f corresponds directly to efficient procedures for: identifying a shattering set $I \subseteq \{0,1\}^N$ for $\mathcal{F}$ (dictating how the client maps his query index i to an input y to f), finding the appropriate function $f(x, \cdot)$ to yield the desired output string on the n inputs in I (dictating how the server maps his n-size database to an input x to f), and determining and evaluating the universal decoder γ (for converting the output of f to an output of the PIR query). Privacy of the resulting PIR scheme follows immediately, since the functionality does not output anything to the first party (server). Correctness of the PIR holds because this gives a mapping from $x \in \{0,1\}^N$ to a function $f(x, \cdot)$ and $i \in [N]$ to an input y for which $f(x, y) = x_i$.

Below is a proof of equivalence for non-interactive reductions for the case of boolean functionalities.

Theorem 2. *Let $f : \{0,1\}^N \times \{0,1\}^N \to \{\perp\} \times \{0,1\}$ be a one-sided functionality with inputs $x, y \in \{0,1\}^\kappa$ and a bit output. Let $\mathcal{F}_\kappa = \{f_\kappa(x, \cdot)\}$ for $x \in \{0,1\}^\kappa$. Then the set $\mathcal{S} = \{\mathcal{F}_\kappa\}_{\kappa \in \mathbb{N}}$ has efficiently computable VC-dimension h, where $h(\kappa) = \kappa$, if and only if f_κ is $(\kappa, 1)$-PIR-hard.*

Proof. If $VC(\mathcal{F}_\kappa) \geq h(\kappa)$ then for every κ both parties in the PIR protocol have access to shattered set I.[1] Then, the server given the database, which is an

---

[1] The shattered set must be exactly the same between the client and the server of the PIR protocol.

assignment of I, computes x for the function f_κ that satisfies the assignment. More specifically, if $\mathrm{VC}(\mathcal{F}_\kappa) \geq h(\kappa)$ then $\exists y = (y_1, \ldots, y_\kappa)$ such that for every assignment $(b_1, \ldots, b_\kappa)$ of y, $\exists\, x$ such that for every $i, f_\kappa(x, y_i) = b_i$. It is easy to see that both parties can run the PIR protocol on input x and y_i from the server and the client, respectively, such that only the client receives the i-th bit of the database b_i.

For the other direction, we need to show that given a PIR-hard function f_κ then $\mathrm{VC}(\mathcal{F}_\kappa) \geq h(\kappa)$. Based on the fact that the deterministic reduction of PIR on a κ-bit database is non-interactive and that the client has no information about the database the claim follows. In particular, since the client has no access to the database then for the i-th bit of the database the client will use the same y_i and the server will use the same x based on the database acting as the assignment. Therefore, $\mathrm{VC}(\mathcal{F}_\kappa) \geq h(\kappa)$ since PIR holds for all databases/assignments.

Two-Sided Predicates. The above additionally gives an approach for showing PIR-hardness of *two-sided predicate* functionalities, as we now describe.

Theorem 3. *Suppose the one-sided functionality* $f : \{0,1\}^N \times \{0,1\}^N \to \{\bot\} \times \{0,1\}$ *is* $(n(N), 1)$*-PIR-hard. Then the corresponding two-sided functionality* $f' : \{0,1\}^N \times \{0,1\}^N \to \{0,1\} \times \{0,1\}$ *that delivers the same output predicate* f *to both parties is* $(\lfloor n(N)/2 \rfloor, 1)$*-PIR-hard.*

Proof. By definition of $(n(N), 1)$-PIR-hardness, there exists an efficient non-interactive construction of single-server PIR on a $n(N)$-size database using a single execution of f. This corresponds to three efficient algorithms: (1) a mapping $C : [n] \to \{0,1\}^N$ taking the client's index $i \in [n]$ to some input $x \in \{0,1\}^N$ to submit to f, (2) a mapping $S : \{0,1\}^n \to \{0,1\}^N$ taking the server's database $z \in \{0,1\}^n$ to some input $y \in \{0,1\}^N$ to submit to f, and (3) a reconstruction procedure R (which may depend on state from the execution of C) translating the output bit of f to the queried value z_i.

Note that by the correctness of the existing PIR scheme for any database z (in particular, for a randomly chosen z), it must be that the output bit of f on inputs $(C(i), S(z))$ provides a full bit of entropy of information about the value of z_i. That is, the output bit of f must *be* either the value z_i or its negation, and the choice of which cannot be dependent on x (as this is unknown to the client).

We provide a construction of PIR on a $\lfloor n(N)/2 \rfloor$-size database using a single execution of f', corresponding to (C', S', R'). For notational simplicity, assume $\lfloor n/2 \rfloor = n/2$.

The transformation is as follows:

- C': The client encodes his input $i \in [n/2]$ as follows. First, sample a random bit $b \leftarrow \{0,1\}$. Then execute $C(i + b \cdot n/2)$.
- S': The server encodes his database $z \in \{0,1\}^{n/2}$ by executing $S(z||\bar{z})$; i.e., on the n-bit value formed by concatenating z with the bitwise negation of z.
- R': Given output w from the execution of f', output $b \oplus w$.

Correctness follows directly from the correctness of the underlying PIR (C, S, R). Security holds because the output bit w is distributed uniformly given the view of the server (i.e., given x).

A general version of Theorem 3 that applies to functionalities f with very short (sub-logarithmic length) outputs appears in the full version.

3.2   PIR-Hardness of Natural Combinatorial Problems

We demonstrate that in many cases even close variants of problems which admit practical sublinear protocols can be PIR-hard. In the following subsections, we consider variants of the Median, Convex Hull, Single-Source Shortest Path, and Approximate Set Cover problems.

Each of our reductions follows the approach and notation of the "efficient" VC-dimension connection described above, including the identification of shattered set I of the client's input space, and a universal decoder γ for converting the (possibly multi-bit) output of f to the output of the PIR. For each case, the corresponding mappings will indeed be efficiently computable, as required.

Revisiting the Median Protocol. For a subset $S \subset U$ of a totally ordered universe set U, the ρth-ranked element is the value $x \in S$ that is ranked ρ when the set S is sorted in increasing order. The median is the element with rank $\rho = \lceil |S|/2 \rceil$. Given two parties A and B with input sets $X_A, Y_B \subset U$, respectively, we consider the problem of privately computing the ρth-ranked element of $X_A \cup Y_B$. Aggarwal et al. [AMP10] described protocols for the *median* function with sublinear communication and computation overhead. Specifically, in the two-party case, let the size of U be polynomial in N (so that elements are described by polylog(N) bits), and let $|X_A|, |Y_B| = N$ be the total number of the input elements. Then, the protocol of Aggarwal et al. [AMP10] for securely evaluating the *median* entails a communication cost of $\tilde{\mathcal{O}}(\log N)$. We remark that the protocol of Aggarwal et al. [AMP10] finds the median on simplified input instances X_A and Y_B where $X_A \cap Y_B = \emptyset$ and $|X_A| = |Y_B|$.

The median two-party and multi-party protocols of [AMP10] are in the two-sided model, where both parties receive an output. Moreover, the security of their protocols relies on the fact that partial information is only leaked via the function output. We now show that secure protocols for the *one-sided* setting cannot enjoy such efficient sublinear-communication properties: namely, the one-sided median functionality is PIR-hard.

One-sided Median Functionality. In this one-sided model, given two parties A and B, only the first party A receives the output of the function while party B should not learn any information about the input of party A.

Definition 4 (One-sided MED functionality). *Let $N \in \mathbb{N}$. We define the two-party functionality* $\text{MED} : \{0,1\}^{\tilde{\mathcal{O}}(N)} \times \{0,1\}^{\tilde{\mathcal{O}}(N)} \to \{\perp\} \times \{0,1\}^{\tilde{\mathcal{O}}(\log N)}$ *by* $(X, Y) \mapsto (\perp, median(X \cup Y))$ *which on input two sets* $X, Y \subset \mathbb{Z}_{poly(N)}$, *from* *the sender and the receiver, respectively, outputs $\bot$ to the sender and the median of $X \cup Y$ to the receiver.*

Theorem 4. *The one-sided functionality* MED : $\{0,1\}^{\tilde{\mathcal{O}}(N)} \times \{0,1\}^{\tilde{\mathcal{O}}(N)} \rightarrow \{\bot\} \times \{0,1\}^{\tilde{\mathcal{O}}(\log N)}$ *is PIR-hard.*

Proof. We define a universal encoder γ that on input a bit-string outputs its Least-Significant Bit (LSB). We are going to find the point set I of size N, that is shattered by $\mathcal{F}'_{\text{MED}} = \{\gamma(\text{MED}(X,\cdot))\}_{X \in \mathbb{Z}_{poly(N)}}$.

Let (max, min) denote the maximum and the minimum element of $\mathbb{Z}_{poly(N)}$, respectively. Moreover, for each $i \in [N]$ let MIN_i (respectively, MAX_i) denote the multiset of size $|MIN_i| = i$ (resp. $|MAX_i| = i$) where each entry is equal to min (resp., max), respectively. Define $I = \{Y_1, \ldots, Y_N\}$ such that $Y_i = \{MIN_{N-i} \cup MAX_i\}$ for all $i \in [N]$. We will show that $\mathcal{F}'_{\text{MED}}$ shatters I. In particular, for each $g : I \rightarrow \{0,1\}$ we will show that $\exists X$ such that for every $Y_i \in I$, $\gamma(\text{MED}(X, Y_i)) = g(Y_i)$.

Let $g : I \rightarrow \{0,1\}$. Define $X = \{x_1, \ldots, x_N\}$ such that $x_i = (i)_2 \| 10 \cdots 0 \| g(Y_i) \in \mathbb{Z}_{poly(N)}$ where $(i)_2$ denotes the bit representation of i. More specifically, x_i is defined by concatenating a unique $\log N$-length prefix to each bit of $g(Y_i)$ to ensure that the resulting elements are sorted, and appending the binary representation of $N + 1$ (i.e., $\log N + 1$ bits) to ensure the existence of N distinct integers smaller than all the resulting values.[2]

It holds that $\forall i \in [N]$, $\gamma(\text{MED}(X, Y_i)) = g(Y_i)$ since $\text{MED}(X, Y_i) = x_i$ and $\gamma(\text{MED}(X, Y_i)) = LSB(x_i) = g(Y_i)$. That said, it follows that $VC(\mathcal{F}'_{\text{MED}}) \geq N$. Since all mappings are efficiently computable, it follows that MED is PIR-hard.

Revisiting the Convex Hull Protocol. In the convex hull algorithm, two parties securely compute the convex hull M of the union of their input sets of points G_A and G_B in an euclidean plane. Each element consists of two integers that represent the X and Y coordinates of the point. We are interested in cases where the convex hull has description size that is sublinear in the input size (as otherwise sublinear communication protocols are unachievable). We thus consider a promise problem variant of the functionality CH, defined as follows:

Definition 5 (One-sided CH functionality). *Let $N \in \mathbb{N}$. Define the two-party (promise problem) convex hull functionality* CH : $\{0,1\}^{\tilde{\mathcal{O}}(N)} \times \{0,1\}^{\tilde{\mathcal{O}}(N)} \rightarrow \{\bot\} \times \{0,1\}^{o(\log N)}$ *by* $\text{CH}(G_A, G_B) = (\bot, convexhull(G_A \cup G_B))$, *which on input two sets G_A, G_B of N points on the 2-dimensional euclidean plane, from party A and party B, respectively, outputs $\bot$ to party A and the convex hull of $G_A \cup G_B$ to party B.*

An efficient sublinear-communication protocol for the *two-sided* convex hull promise problem was given by [SV15] (as it fits into their "greedy compatible" framework), assuming slight additional promise restrictions on the inputs

---

[2] For the case where the set Y has to be distinct then $MIN_j = \{min, min + 1, \ldots, min_{j-1}\}$, $MAX_j = \{max, max + 1, \ldots, max_{j-1}\}$. Furthermore, in such a case $\forall I \in [N]$ compute $x_i = (i)_2 \| min \| 10 \cdots 0 \| g(Y_i) \in \mathbb{Z}_{2^\ell}$.

(namely, no two points have the same X or Y coordinate and no three-points are collinear). We prove that the one-sided convex hull problem is PIR-hard.

Theorem 5. *The one-sided functionality* $\mathrm{CH} : \{0,1\}^{\tilde{O}(N)} \times \{0,1\}^{\tilde{O}(N)} \to \{\bot\} \times \{0,1\}^{\tilde{O}(\log N)}$ *is PIR-hard.*

Proof. We define a universal encoder γ that on input a convex hull of four nodes identifies the longest edge and rotates it such that: (1) the longest edge is parallel to the X axes and (2) the shortest edge is above the longest edge. The encoder outputs 0 if the node of the shortest edge, which is closer to the longest edge, is the left one, otherwise output 1. We are going to find the point set I of size N, that is shattered by $\mathcal{F}'_{\mathrm{CH}} = \{\gamma(\mathrm{CH}(G,\cdot))\}_{G \subset S}$.

Let Cr be a circle with center the origin of the axes (with arbitrary radius) on the euclidean plane. Set $\psi = 2\pi/2N = \pi/N$. Let us define $I = \{Y_1, \ldots, Y_N\}$ for all $i \in [N]$. Consider the two points on the circle with angle $\phi_i = (2i) \cdot \psi$ and angle $\bar{\phi}_i = (2i+1) \cdot \psi$. Then, define by τ_1 and τ_2 the tangents of these two points, respectively. Tangents τ_1 and τ_2 intersect at point P_i. Consider the line e_i passing through the center of the circle and the point P_i. Denote the intersection points of the line e_i with the circle by Q_i, Q'_i such that point Q_i is closer to point P_i. Next, consider the tangent τ_3 of the point Q'_i and define by R_i, S_i the points created by the intersection of τ_1, τ_3 and τ_2, τ_3, respectively. The set Y_i includes points Q_i, R_i, S_i.

We will show that $\mathcal{F}'_{\mathrm{CH}}$ shatters I. In particular, for each $g : I \to \{0,1\}$ we will show that $\exists G$ such that for every $Y_i \in I$, $\gamma(\mathrm{CH}(G, Y_i)) = g(Y_i)$. Let $g : I \to \{0,1\}$ then for all $i \in \{0, \ldots, N-1\}$, assign each $g(Y_i)$ to the point T_i with angle $\phi_i = (2i + g(Y_i)) \cdot \psi$. Then, G consists for all points assigned to $g(Y_i)$.

It holds that $\forall i \in [N]$, $\gamma(\mathrm{CH}(G, Y_i)) = g(Y_i)$. In particular, the convex hull in each case contains the points (Q_i, R_i, S_i, T_i). By construction, the longest edge is drawn by nodes R_i, S_i and the shortest edge by nodes T_i, Q_i and point Q_i is closer to the longest edge. If $g(Y_i)$ is 0 then Q_i is closer to R_i and γ outputs 0. Thus, since each of the above mappings is efficiently computable, it follows that CH is PIR-hard.   □

Revisiting the Single-Source Shortest Distance Protocol. In the Single Source Shortest Distance (SSSD) protocol, two parties securely compute the shortest path distances from a source vertex s to all other vertices in a joint weighted graph. More specifically, let G_A and G_B be the two parties' respective weighted graphs. Assume that $G_A = (V_A, E_A, w_A)$ and $G_B = (V_B, E_B, w_B)$ are complete graphs on the same set of vertices. Let $w_A(e)$ and $w_B(e)$ represent the weight of edge e in G_A and G_B, respectively.[3] The goal is to output the list M which contains the shortest path distances from the source vertex s to all other vertices. If the input graphs (which may have quadratically many edges) are

---

[3] Note that we can also consider incomplete graphs and graphs that include disjoint edges by setting appropriate special values of $w(e)$ for the given edges e.

describable in $\tilde{\mathcal{O}}(N)$ bits, the output (which must have at most linearly many items) can be described by $\tilde{\mathcal{O}}(\sqrt{N})$ bits.

Definition 6 (One-sided SSSP functionality). *Define the two-party functionality* SSSP : $\{0,1\}^{\tilde{\mathcal{O}}(N)} \times \{0,1\}^{\tilde{\mathcal{O}}(N)} \to \{\perp\} \times \{0,1\}^{\tilde{\mathcal{O}}(\sqrt{N})}$ *by* SSSP(G_A, G_B) $= (\perp, shortestpaths(G_A, G_B))$ *which takes as input from A and B two complete, weighted graphs* G_A, G_B *respectively, on the same set of vertices. Then, it outputs* $\perp$ *to A and the list of shortest path distances from a source vertex s to all other vertices in the joint weighted directed graph to B.*

An efficient sublinear-communication protocol was given by [SV15] for the two-sided version of a related problem, of single-source all-destinations (SSAD), which outputs the list of shortest paths from s to each other node, as opposed to just the distance of these paths. (This follows from their "greedy compatible" framework, via Dijkstra's algorithm.)

We prove the one-sided SSSP problem is PIR-hard. As the information of one-sided SSSP can be directly inferred from the information of one-sided SSAD, this further implies PIR-hardness of the one-sided SSAD problem.

Theorem 6. *The one-sided functionality* SSSP : $\{0,1\}^{\tilde{\mathcal{O}}(N)} \times \{0,1\}^{\tilde{\mathcal{O}}(N)} \to$ $\{\perp\} \times \{0,1\}^{\tilde{\mathcal{O}}(\sqrt{N})}$ *is PIR-hard.*

Proof. We define a universal encoder γ that on input N integers and an index i outputs 0 if the ith integer is even, or 1 otherwise. We are going to define a set I of size $N(N-1)/2$, that is shattered by $\mathcal{F}'_{\text{SSSP}} = \{\gamma(\text{SSSP}(G, \cdot))\}_G$.

Let us define $I = \{Y_1, \ldots, Y_N\}$ for all $i \in [N]$. For each edge $i = (u,v)$ in the graph Y_i proceed as follows. The edge between the starting note s to u is set to the minimum weight i.e. $w_{Y_i}((s,u)) = 0$ and there is no weight assignment for (s,v). For every other edge $w \neq \{u,v\}$ connected to s, the weight on the edge (s,w) is assigned to N^2 i.e. $w_{Y_i}((s,w)) = N^2$.

We will show that $\mathcal{F}'_{\text{SSSP}}$ shatters I. In particular, for each $g : I \to \{0,1\}$ we will show that $\exists G$ such that for every $Y_i \in I$, $\gamma(\text{SSSP}(G, Y_i)) = g(Y_i)$. Let $g : I \to \{0,1\}$ then enumerate all the nodes from 1 up to N and for every edge $j \in \binom{N}{2}$ in the graph G assign each weight to $2N^2 + 2j + g(Y_i)$ (For the special case where the edge includes the starting point s there is no weight assignment).

It holds that $\forall i \in [N]$, $\gamma(\text{SSSP}(G, Y_i)) = g(Y_i)$. By construction the distance from the starting point s to v for $i = (u,v)$ is equal to $w_{Y_i}(u,v)$ which is equal to $2N^2 + 2j + g(Y_i)$. If $g(Y_i) = 0$ then w_{Y_i} is even. $\qquad\square$

Revisiting the Approximate Set Cover Protocol. Given a collection S of sets over a universe U, a set cover $C \subseteq S$ is a subcollection of the sets whose union is U. The set cover problem allows two parties A and B to securely find a minimum-cardinality set cover given S_A and S_B. While this problem is NP hard to solve exactly, it yields a natural greedy approximation algorithm. Namely, in each iteration, the algorithm takes the set of those remaining which contains the largest number of uncovered elements.

In what follows, the "Approximate Set Cover" functionality will refer to the output generated by running this greedy algorithm. As with previous problems, we will restrict our attention to a promise version of the problem, where the description size of the output set cover is sublinear in the input description size (as otherwise sublinear-communication protocols will not be possible).

Definition 7 (One-sided Approximate Set Cover SC functionality).
Let $N \in \mathbb{N}$. Given a universe U, we define the two-party functionality SC : $\{0,1\}^{\tilde{O}(N)} \times \{0,1\}^{\tilde{O}(N)} \to \{\bot\} \times \{0,1\}^{o(N)}$ *by* $SC(S_A, S_B) = (\bot, C)$ *which on input finite sets $S_A \subseteq U$ and $S_B \subseteq U$ from party A and party B, respectively, outputs the result $C \subseteq S_A \cup S_B$ of the greedy set cover algorithm to party B.*

An efficient sublinear-communication protocol for the *two-sided* greedy approximate set cover promise problem was given by [SV15], following their "greedy compatible" framework. We prove the corresponding one-sided problem is PIR-hard.

Theorem 7. *The one-sided functionality* SC : $\{0,1\}^{\tilde{O}(N)} \times \{0,1\}^{\tilde{O}(N)} \to \{\bot\} \times \{0,1\}^{o(N)}$ *is PIR-hard.*

Proof. We define a universal encoder γ that on input two sets outputs the minimum element that resides in both sets. We are going to define a set I of size $\Theta(N)$, that is shattered by $\mathcal{F}'_{SC} = \{\gamma(SC(S, \cdot))\}_S$.

Let $|U| = \ell + 2$ where $\binom{\ell}{\ell/2} \geq N$. In particular, let $U = \{0, 1, u_1, \dots, u_\ell\}$. Let $V = \{\{0,1\}^\ell\}^N$ be a vector with all bit-strings of length ℓ with hamming weight $1/2$ in lexicographical order. Denote by $V_{i,j}$ the bit of the j-th position of the i-th element of V. Define $I = \{Y_1, \dots, Y_N\}$ such that $Y_i = \{0,1\} \cup_{j \in \ell} \{u_j \mid V_{i,j} = 0\}$ for all $i \in [N]$.

We will show that $\mathcal{F}'_{SC}$ shatters I. In particular, for each $g : I \to \{0,1\}$ we will show that $\exists S = \{S_1, \dots, S_N\}$ such that for every $Y_i \in I$, $\gamma(SC(S, Y_i)) = g(Y_i)$. Let $g : I \to \{0,1\}$ then for all $i \in [N]$, set $S_i = \{g(Y_i)\} \cup_{j \in \ell} \{u_j \mid V_{i,j} = 1\}$.

It holds that $\forall i \in [N]$, $\gamma(SC(S, Y_i)) = g(Y_i)$. By construction the output collection consists of two sets, i.e., S_i and Y_i. If $g(Y_i) = 0$ then the common minimum element in both sets is 0.   □

4   Intermediate Hardness via Semi-PIR

There are natural two-sided functionalities that are provably not PIR-hard, but which instead imply the following notion of *semi-PIR*. Intuitively, semi-PIR is a relaxed version of PIR where the server is allowed to learn the output z_i and can furthermore learn the client's actual selection i only if $z_i = 0$. Note that a semi-PIR protocol with only two messages is necessarily a PIR protocol, but it is easy to convert any 2-message PIR protocol into (an artificial) 3-message semi-PIR protocol which is not a PIR protocol by having the client send i to the server if and only if $z_i = 0$.

The semi-PIR primitive is formally defined by making the following small change in the security requirement of PIR from Definition 1: instead of requiring indistinguishability between any $i, j \in [n]$, the requirement is only made for i, j such that $z_i = z_j = 1$.

One can roughly think of a semi-PIR protocol as a low-communication (passively) secure protocol for the functionality $\frac{1}{2}$PIR that maps (z, i) to (y, z_i), where $y = i$ if $z_i = 0$ and $y = \perp$ otherwise. Indeed, any semi-PIR protocol as above can be converted into a protocol for this functionality by having the client send y to the server in the end of the protocol.

4.1 Does Semi-PIR Imply PIR?

In this section we study the relation between semi-PIR and PIR. We show that a strong form of semi-PIR implies a weak form of PIR. Interestingly, this result is shown via an inherently *adaptive* reduction, which also exhibits some unusual tradeoffs between communication and computation. We then show that the semi-PIR functionality does not satisfy the default notion of PIR-hardness from Definition 2. In other words, one cannot construct a PIR protocol via a single non-interactive call to $\frac{1}{2}$PIR. While we leave open the possibility of constructing polylogarithmic PIR from polylogarithmic semi-PIR, we show that ruling out such a construction would imply a breakthrough in the achievable complexity of locally decodable codes.

Obtaining weak PIR from semi-PIR. We start by showing how to use a single invocation of semi-PIR to build a probabilistic PIR functionality that (on every selection i) leaks i to the server with probability $1/2$ (and lets the client know that leakage occurred), but otherwise reveals nothing to the server. We denote this probabilistic functionality by $\mathsf{Rand}\frac{1}{2}\mathsf{PIR}$.

Lemma 1. *There exists a protocol for* $\mathsf{Rand}\frac{1}{2}\mathsf{PIR}$ *that, on a database* $z \in \{0,1\}^n$, *uses a single invocation of* $\frac{1}{2}\mathsf{PIR}$ *on a database* $z' \in \{0,1\}^{2n}$ *and no additional interaction.*

Proof. The $\mathsf{Rand}\frac{1}{2}\mathsf{PIR}$ protocol proceeds as follows. The server maps z to $z' = (z, \bar{z})$. The client picks a random mask $r \in \{0,1\}$ and maps i to $i' = i + rn$. The parties then invoke the $\frac{1}{2}\mathsf{PIR}$ oracle on inputs (z', i'). The client's output in the $\mathsf{Rand}\frac{1}{2}\mathsf{PIR}$ protocol is $z'_{i'} \oplus r$, where $z'_{i'}$ is the output of the $\frac{1}{2}\mathsf{PIR}$. It is easy to check that the output is correct, and that the server learns nothing about i if $z'_{i'} = 0$, which happens with probability $1/2$ and is detectable by the client. $\square$

Given Lemma 1, it suffices to reduce PIR to $\mathsf{Rand}\frac{1}{2}\mathsf{PIR}$. Our reduction relies on the following strong form of locally decodable codes (LDCs), which can be viewed as 1-round multi-server PIR protocols with uniform queries of logarithmic size and a single answer bit. Using a general transformation of LDC to multi-server PIR from [KT00], such codes are implied by standard LDCs by allowing a small decoding error probability. For simplicity, we define here only the perfect notion which is satisfied by the best known LDC constructions.

Definition 8 (Perfect LDC). *We say that an encoding function $C : \{0,1\}^n \rightarrow \{0,1\}^N$ is a q-query perfect LDC, if there exists a probabilistic decoder algorithm $D(i)$ which probes q bits of the encoding such that the following properties hold:*

- **Correctness:** *For every $z \in \{0,1\}^n$ and $i \in [n]$, we have $\Pr[D^{C(z)}(i) = z_i] = 1$.*
- **Uniform queries:** *Letting $(i_1, \ldots, i_q) \in [N]^q$ be the sequence of indices read by $D(i)$, it holds that for every $j \in [q]$ the index i_j is uniformly distributed over $[N]$.*

Our construction of PIR from $\frac{1}{2}$PIR encodes the PIR database using a perfect LDC, and applies a "cautious" decoding strategy by repeatedly (and adaptively) using Rand$\frac{1}{2}$PIR to simulate the LDC decoder while ensuring that at most one query from each decoding attempt is leaked. This strategy yields the following theorem.

Theorem 8. *Let $n(N)$ and $q(N)$ be functions such that there is a $q(N)$-query perfect LDC $C : \{0,1\}^{n(N)} \rightarrow \{0,1\}^N$ in which both the encoder and the decoder can be implemented in time $poly(N)$. Then, there exists a protocol that, given a parameter N, implements in time $poly(N)$ PIR on a database $z \in \{0,1\}^{n(N)}$ by using an expected $O(q(N) \cdot 2^{q(N)})$ (adaptive) calls to $\frac{1}{2}$PIR on a database $z' \in \{0,1\}^N$ and no additional interaction.*

Proof. Let $q = q(N)$. The PIR protocol will make at most $q \cdot 2^q$ expected calls to Rand$\frac{1}{2}$PIR, which using Lemma 1 can be implemented using $q \cdot 2^{q+1}$ expected calls to $\frac{1}{2}$PIR. The protocol starts with the server encoding the PIR database $z \in \{0,1\}^n$ into a codeword $Z \in \{0,1\}^N$. The client and the server then repeatedly apply the following procedure until z_i is successfully recovered.

1. The client invokes the LDC decoder $D(i)$ to generate query indices $(i_1, \ldots, i_q)$.
2. For $j = 1, \ldots, q$ (sequentially), the client and the server invoke Rand$\frac{1}{2}$PIR with client input i_j and server input Z. The protocol restarts at Step 1 if i_j leaks (which occurs with probability $1/2$), otherwise it continues to the next j. If all indices Z_{i_j} have been successfully retrieved, the client invokes D to recover z_i.

Since the leakage events in different invocations of Rand$\frac{1}{2}$PIR are independent, the expected number of attempts until decoding is fully successful is 2^q, and so the expected number of Rand$\frac{1}{2}$PIR invocations is $q \cdot 2^q$. The (perfect) security of the protocol follows from the fact that in any invocation of D, at most a single index i_j is leaked. By the definition of perfect LDC, this index is uniformly distributed independently of i. □

Alternatively, one can implement a *worst-case* variant of the above reduction that runs σ copies in parallel, each with a constant failure probability. This results in a PIR to semi-PIR reduction that has $2^{-\Omega(\sigma)}$ error probability and makes $O(q(N) \cdot 2^{q(N)})$ rounds of calls to $\frac{1}{2}$PIR with a total number of $O(\sigma \cdot q(N) \cdot 2^{q(N)})$ of $\frac{1}{2}$PIR calls.

One can instantiate Theorem 8 by using known LDC constructions in several ways. In particular, using Reed-Muller LDCs with $q(N) = \Theta(\log N)$, one gets PIR with good communication complexity but super-polynomial computational complexity. To get slightly sublinear PIR with polynomial computational complexity, we rely on best constant-query LDC constructions from [Efr09].

Corollary 1 (polylogarithmic semi-PIR $\Rightarrow$ slightly sublinear PIR). *The existence of a polylogarithmic semi-PIR protocol implies the existence of a slightly sublinear PIR protocol. Moreover, if the semi-PIR protocol has constant round complexity then so does the PIR protocol.*

Proof. The LDC construction from [Efr09] is in fact a perfect LDC according to our definition, with the following parameters. For any positive integer α, there is a constant $q = q(\alpha)$, such that there is a q-query perfect LDC with $N(n) = \exp(\exp(\log^{1/\alpha} n))$, or $n(N) = \exp((\log\log N)^{\alpha}))$. Note that $n(N)$ is bigger than any polylogarithmic function in N. A slightly sublinear PIR is obtained by chopping a database of size N into blocks of size $n(N)$ and running the protocol guaranteed by Theorem 8 on each block. $\square$

We note that the existence of "dream LDC" with $q = O(1)$ queries and polynomial length $N(n)$ would imply a stronger reduction that constructs polylogarithmic PIR from polylogarithmic semi-PIR. Thus, ruling out such a reduction would imply ruling out such dream LDC, which would be considered a breakthrough in complexity theory.

Separating semi-PIR from PIR. On the other hand, we show that semi-PIR is *not* PIR hard. More broadly, we demonstrate limitations in the possibility of *non-adaptive* reductions from PIR to semi-PIR.

We begin by showing that with a single call to semi-PIR one cannot achieve secure PIR even with small non-trivial correctness.

Theorem 9. *There cannot exist any reduction from n-bit PIR to $\frac{1}{2}$PIR with correctness better than 0.6 which makes a single call to $\frac{1}{2}$PIR.*

Proof. Suppose towards a contradiction that there exists a reduction from PIR to $\frac{1}{2}$PIR via a single call with correctness 0.6. This corresponds to a (randomized) encoding E_{DB} from $x \in \{0,1\}^n$ to $\hat{x} \in \{0,1\}^{\hat{n}}$ and E_{index} from $i \in [n]$ to $j \in [\hat{n}]$, where the client learns $\hat{x}_j$ and the server learns j iff $\hat{x}_j = 1$ via $\frac{1}{2}$PIR. Since correctness is 0.6, there must exist $i \neq i' \in [n]$ for which the distributions $\{J \leftarrow E_{\mathsf{index}}(i)\}$ and $\{J' \leftarrow E_{\mathsf{index}}(i')\}$ are statistically far. By the privacy requirement, this means the resulting index j or j' cannot be revealed except with negligible probability. In turn, this implies $\hat{x}_j = 0$ except with negligible probability over $E_{\mathsf{DB}}, E_{\mathsf{index}}$. However, this implies that on a random database x the client has a negligible advantage in guessing x_i, yielding a contradiction. $\square$

We next build atop this result to further rule out the possibility of a reduction making *two* non-adaptive calls.

Theorem 10. *There cannot exist any reduction from PIR to $\frac{1}{2}$PIR which makes two parallel calls to $\frac{1}{2}$PIR.*

Proof. Consider any reduction achieving n-bit PIR, making 2 parallel calls to $\frac{1}{2}$PIR. This corresponds to a (randomized) encoding E_{DB} from $x \in \{0,1\}^n$ to $\hat{x} \in \{0,1\}^{\hat{n}}$ and E_{index} from $i \in [n]$ to $(i_1, i_2) \in [\hat{n}]^2$. By correctness, for every $i \in [n]$ there exists $i' \in [n]$ for which the distributions $\{(i_1, i_2) \leftarrow E_{\mathsf{index}}(i)\}$ and $\{(i'_1, i'_2) \leftarrow E_{\mathsf{index}}(i')\}$ are statistically far. Because of this, for each index i, it must be that the read values $(\hat{x}_{I_1}, \hat{x}_{i_2})$ take value $(1,1)$ with negligible probability over $E_{\mathsf{DB}}, E_{\mathsf{index}}$. Correctness of the final scheme implies that the values of $(\hat{x}_{I_1}, \hat{x}_{i_2})$ must have a full bit of entropy over a random database x; in particular, the value $(0,0)$ can occur with probability at most $1/2$. Then either $\hat{x}_{i_1}$ or $\hat{x}_{i_2}$ must equal 1 with probability at least $1/4$, without loss of generality say $\hat{x}_{i_1}$.

Consider, then, the following reduction which makes a single call to $\frac{1}{2}$PIR and achieves correctness $1/2 + 1/8 - \mathsf{negl}(n)$.

1. The server samples $\hat{x} \leftarrow E_{\mathsf{DB}}(x)$ and submits $\hat{x}$ to $\frac{1}{2}$PIR.
2. The client samples $(i_1, i_2) \leftarrow E_{\mathsf{index}}(i)$ and submits i_1 to $\frac{1}{2}$PIR.
3. The $\frac{1}{2}$PIR execution outputs $\hat{x}_{i_1}$ to the client and i_1 or $\perp$ to the server (depending on $\hat{x}_{i_1}$).
4. If $\hat{x}_{i_1} = 1$ then the client executes the decoding procedure for the original reduction on input $(1,0)$. Otherwise, he outputs a random bit.

Privacy of this construction follows from privacy of the original reduction. In the case that $\hat{x}_{i_1} = 1$, then with overwhelming probability we know that $\hat{x}_{i_2} = 0$, and thus the client computes the correct output. This means correctness of the overall scheme will hold with probability at least $1/4 + 3/4 \cdot 1/2 - \mathsf{negl}(n)$, contradicting Theorem 9. □

Because of the degradation in parameters, extending this separation to additional parallel queries will seem to require new ideas (e.g., for three queries ruling out $(1,1,1)$ gives a smaller boost in correctness when reducing to the two query case, which is insufficient to directly derive a contradiction). However, as a final note, we return to the $\mathsf{Rand}\frac{1}{2}$PIR functionality (used as an intermediate step in the earlier construction of PIR from $\frac{1}{2}$PIR), in which the input index is revealed with probability $1/2$. This setting yields a direct analysis, and we observe that even $O(\log n)$ parallel calls to $\mathsf{Rand}\frac{1}{2}$PIR cannot yield PIR.

Proposition 1. *There cannot exist any reduction from PIR to $\mathsf{Rand}\frac{1}{2}$PIR making $c \in O(\log n)$ parallel calls to $\mathsf{Rand}\frac{1}{2}$PIR with negligible correctness error.*

Proof. Consider any reduction achieving n-bit PIR, making $c \in O(\log n)$ parallel calls to $\mathsf{Rand}\frac{1}{2}$PIR. This corresponds to a (randomized) encoding E_{DB} from $x \in \{0,1\}^n$ to $\hat{x} \in \{0,1\}^{\hat{n}}$ and E_{index} from $i \in [n]$ to $(i_1, \ldots, i_c) \in [\hat{n}]^c$. By correctness, there exists $i \neq i' \in [n]$ for which the distributions $\{(i_1, \ldots, i_c) \leftarrow E_{\mathsf{index}}(i)\}$ and $\{(i'_1, \ldots, i'_c) \leftarrow E_{\mathsf{index}}(i')\}$ are statistically far. However, with noticeable probability $2^{-c} \in n^{-O(1)}$, all executions of $\mathsf{Rand}\frac{1}{2}$PIR will reveal the queried index, thus revealing the entire vector query $(i_1, \ldots, i_c)$, violating privacy of the scheme. □

4.2   Examples of Semi-PIR-Hard Problems

In this section we provide several natural examples for (two-sided) semi-PIR hard functionalities. The results of the previous section imply that any poly-logarithmic protocol for these functionalities would imply a slightly sublinear PIR.

Definition 9 (Two-sided Single Source Single Destination Shortest Path). *Let $N \in \mathbb{N}$. Define the two-party functionality* $\mathrm{SSSD}_{s,t} : \{0,1\}^{\widetilde{\mathcal{O}}(N^2)} \times \{0,1\}^{\widetilde{\mathcal{O}}(N^2)} \to \{0,1\}^{\log N} \times \{0,1\}^{\log N}$ *by* $\mathrm{SSSD}_{s,t}(G_A, G_B) = (shortestpath(G_A, G_B))$ *that expects as input from A and B two directed, complete, weighted graphs G_A, G_B respectively, on the same set of N vertices where each weight is in $\mathbb{N}$. The functionality outputs the shortest path from the source vertex s to the destination vertex t in the joint weighted directed graph to both A and B.*

Theorem 11. *Let $N \in \mathbb{N}$. The two-sided Single Source Single Destination shortest Path function* $\mathrm{SSSD}_{s,t} : \{0,1\}^{\widetilde{\mathcal{O}}(N^2)} \times \{0,1\}^{\widetilde{\mathcal{O}}(N^2)} \to \{0,1\}^{\log N} \times \{0,1\}^{\log N}$ *is semi-PIR hard.*

Proof. We construct a semi-PIR protocol Π_{SSSD}, by calling functionality SSSD. Let Z be the server's input set of size $\widetilde{\mathcal{O}}(N^2)$ where each element is a bit, and let i be the client's index. Moreover, let (G_A, G_B) be the two weighted input graphs provided to the SSSD functionality by the server and the client, respectively. Protocol $\Pi_{\mathrm{SSSD}}(Z, i)$ proceeds as follows:

Input Phase:
1. We split the nodes of G_A into two sets and the weight of each edge within each set is equal to infinity. Essentially, we form a complete bipartite graph with two extra vertices s, t. The source vertex s is connected to the vertices on the left side and a target vertex t connected to the vertices on the right side of the bipartite graph. We also consider an edge connecting s and t. The Server encodes the database Z on $\mathcal{O}(N^2)$ edges of the bipartite graph in G_A. In particular, the server assigns to edge j the weight $2Z_j$. The weight of the edge connecting s and t is set to 1.
2. The client sets up his graph G_B such that for the edge of interest $i = (u, v)$ the weight is set to $w_B(i) = 2$ and $w_B((s, u)) = 0$ and $w_B((v, t)) = 0$. The weights of all other edges are set to infinity.

Evaluation and Output Phase:
Invoke the two-sided SSSD functionality $\Pi_{\mathrm{SSSD}}(G_A, G_B)$ that outputs the shortest path. If the shortest path contains the edge connecting s and t then $Z_i = 1$, otherwise $Z_i = 0$.

If $Z_i = 1$ then the i'th edge weight is 2, and shortest path will consist of the single edge connecting s and t, hiding the identity of i. If $Z_i = 0$, the shortest path contains edge i revealing the index i to the Server.

Definition 10 (Two-sided Closest Destination Problem). *Let* $N \in \mathbb{N}$. *Define the two-party functionality* CDP $: \{0,1\}^{\widetilde{\mathcal{O}}(N)} \times \{0,1\}^{\widetilde{\mathcal{O}}(\log N)} \rightarrow \{0,1\}^{\log N}$ $\times \{0,1\}^{\log N}$ *by* CDP$(G,(s,t_1,t_2)) = (ClosestDest(G,(s,t_1,t_2)))$ *that expects as input a (sparse) graph G_A with size $\widetilde{\mathcal{O}}(N)$ description from party A and a source vertex s along with two target vertices t_1, t_2 with description size $\widetilde{\mathcal{O}}(\log N)$ from party B. Then, it outputs the identity of the closest destination from s to the one-out-of-two target vertices $dist(s, t_b) \leq dist(s, t_{1-b})$ to both A and B while t_{1-b} remains hidden.*

Theorem 12. *The two-sided Closest Destination Problem function* CDP $:$ $\{0,1\}^{\widetilde{\mathcal{O}}(N)} \times \{0,1\}^{\widetilde{\mathcal{O}}(\log N)} \rightarrow \{0,1\}^{\log N} \times \{0,1\}^{\log N}$ *is semi-PIR hard.*

Proof. We construct a semi-PIR protocol Π_{CDP}, calling functionality $f_{G_A,(s,t_1,t_2)_B}$. Let Z be the server's input set of size $\widetilde{\mathcal{O}}(N)$ where each element is a bit, and let i be the client's index. Moreover, let $G_A, (s,t_1,t_2)_B$ be the inputs to the CDP functionality by the server and the client, respectively. Π_{CDP} proceeds as follows:

Input Phase:
1. Without loss of generality the Server encodes the database Z on $\mathcal{O}(N)$ edges of a star graph G_A with $N+2$ vertices where the node s is connected to the other $N+1$ vertices. The server enumerates all these $N+1$ vertices from 1 up to $N+1$ and for $j \in [N]$ assigns the weight of the edge connecting s and j to $2Z_j$ and the edge connecting s to $N+1$ to 1.
2. The client chooses vertices s, i and $N+1$.

Evaluation and Output Phase:
Invoke the two-sided protocol Π_{CDP} that outputs a target destination. If the target is vertex i then $Z_i = 0$ and if the target is vertex $N+1$ then $Z_i = 1$. If $Z_i = 1$ then the output is independent of the index i and thus the identity of i is hidden.

Definition 11 (Two-sided Nearest Neighbor Problem). *Define the two-party functionality* NN $: \{0,1\}^{\widetilde{\mathcal{O}}(N)} \times \{0,1\}^{\widetilde{\mathcal{O}}(\log N)} \rightarrow \{0,1\}^{\mathcal{O}(\log N)} \times$ $\{0,1\}^{\mathcal{O}(\log N)}$ *by* NN(D,loc) *that expects as input a list D (of size N) of locations on the 2-dimensional euclidean plane from party A and a single location loc on the same plane from party B. Then, it outputs to both parties the location (x,y) in D that is nearest to location loc^A.*

Theorem 13. *The two-sided Nearest Neighbor function* NN $: \{0,1\}^{\widetilde{\mathcal{O}}(N)} \times$ $\{0,1\}^{\widetilde{\mathcal{O}}(\log N)} \rightarrow \{0,1\}^{\mathcal{O}(\log N)} \times \{0,1\}^{\mathcal{O}(\log N)}$ *is semi-PIR hard.*

Proof. We construct a semi-PIR protocol Π_{NN}, calling functionality NN. Let $Z \in \{0,1\}^N$ be the server's input database, and let i be the client's index. Moreover, let (D,loc) be the inputs to the NN functionality by the server and the client, respectively. Protocol Π_{NN} proceeds as follows:

Input Phase:

1. For $j \in [N]$, let $(a, b)_j$ be evenly spaced points on a circle with center c and radius r in the Euclidean plane. The Server generates his input D to NN with respect to these points in the following way. If $Z_j = 0$ then set the jth location $(x, y)_j = (a, b)_j$. If $Z_j = 1$ set the location $(x, y)_j$ arbitrary outside the circle. In addition, he includes the center point c.

2. The client outputs the location loc that intersects the line crossing from the centre c and location $(a, b)_i$ and the circle with center c and radius $r/2$.

Evaluation and Output Phase: Invoke the two-sided protocol Π_{NN} that outputs the nearest location to loc. If $Z_i = 0$ then the output is $(a, b)_i$. If $Z_i = 1$ then the output is the centre c which is independent of the index i. That said, in this case the identity of i is not leaked.

Definition 12 (Two-sided Short-List Selection). *Define the two-party functionality* SLS $: \{0, 1\}^{\tilde{\mathcal{O}}(N)} \times \{0, 1\}^{\tilde{\mathcal{O}}(\log N)} \to \{0, 1\}^{2 \log N} \times \{0, 1\}^{2 \log N}$ *by* SLS$(L, (\mathsf{idx}_0, \mathsf{idx}_1))$ *that expects as input a list L of size N and input domain $[N]$ from party A and two indices $(\mathsf{idx}_0, \mathsf{idx}_1)$ from party B. The output is idx_0 if $L_{\mathsf{idx}_0} < L_{\mathsf{idx}_1}$, idx_1 if $L_{\mathsf{idx}_0} > L_{\mathsf{idx}_1}$ or both $\mathsf{idx}_0, \mathsf{idx}_1$ if $L_{\mathsf{idx}_0} = L_{\mathsf{idx}_1}$.*

Theorem 14. *The two-sided Short-List Selection function* SLS $: \{0, 1\}^{\tilde{\mathcal{O}}(N)} \times \{0, 1\}^{\tilde{\mathcal{O}}(\log N)} \to \{0, 1\}^{2 \log N} \times \{0, 1\}^{2 \log N}$ *is semi-PIR hard.*

Proof. We construct a semi-PIR protocol Π_{SLS}, calling functionality SLS. Let Z be the server's input set of size N where each element is a bit, and let i be the client's index. Moreover, let $(L, \mathsf{idx}_0, \mathsf{idx}_1)$ be the inputs to the SLS functionality. Protocol Π_{SLS} proceeds as follows:

Input Phase:

1. The Server generates the list L of size $N + 1$ as follows. For $j \in [N]$, $L_j = Z_j$ and $L_{N+1} = 0$.

2. The client chooses indices i and $N + 1$.

Evaluation and Output Phase:

Invoke Π_{SLS} that outputs the index of the smallest entry or both indices in case of ties. If $Z_i = 0$ then the indices i and $N + 1$ are revealed. If $Z_i = 1$ only the $N + 1$ index is revealed which is independent of i.

Next, we observe that this problem is *not* PIR-hard by demonstrating it is implied by $\frac{1}{2}$PIR (which is separated from PIR in the above results).

Theorem 15. *If there exists a Semi-PIR protocol for a database of size $\mathcal{O}(N)$ that runs in k-rounds, then for every constant $c > 0$ there exists a protocol for the two-sided Short-List Selection function* SLS $: \{0, \ldots, c\}^{\tilde{\mathcal{O}}(N)} \times \{0, 1\}^{\tilde{\mathcal{O}}(\log N)} \to \{0, 1\}^{2 \log N} \times \{0, 1\}^{2 \log N}$ *that runs in $\mathcal{O}(c \cdot k)$ rounds.*

Proof. Let Π be the Semi-PIR protocol. Let L be the input list of f_{SLS}. Modify L such that each element is in its unary representation with $c+1$ bits. In particular, a number $n < c$ in the database is represented in unary by n ones. The rest of the $c + 1 - n$ most significant bits are set to 0. Construct the semi-PIR database D by storing all $N(c+1)$ bits of L in such a way that the element with index (idx, ℓ) is the ℓ-th bit of the element of L with index idx. Let $(\mathsf{idx}_0, \mathsf{idx}_1)$ be the input indices of party B to SLS. Then party A and party B run at most c sequential rounds, each one consisting of two parallel calls to Π. In the ℓ-th round, where $\ell \in [1, c]$, party B makes the following two parallel queries for every $b \in \{0, 1\}$.

$$\Pi\big(D, (\mathsf{idx}_b, \ell)\big) = \left\{ \begin{array}{ll} \big(\perp, ((\mathsf{idx}_b, \ell), D_{\mathsf{idx}_b, \ell})\big), & \text{if } D_{\mathsf{idx}_b, \ell} = 1 \\ \big((\mathsf{idx}_b, \ell), ((\mathsf{idx}_b, \ell), D_{\mathsf{idx}_b, \ell})\big), & \text{if } D_{\mathsf{idx}_b, \ell} = 0 \end{array} \right\}$$

If for some ℓ, b, $D_{\mathsf{idx}_b, \ell} = 0$ the protocol completes and there are no more adaptive calls. For the case where $D_{\mathsf{idx}_b, \ell} \neq D_{\mathsf{idx}_{1-b}, \ell}$ and $D_{\mathsf{idx}_b, \ell} = 0$ then $L_{\mathsf{idx}_{1-b}} > L_{\mathsf{idx}_b}$ and both parties receive idx_b. If $D_{\mathsf{idx}_b, \ell} = D_{\mathsf{idx}_{1-b}, \ell}$ then $L_{\mathsf{idx}_{1-b}} = L_{\mathsf{idx}_b}$ and both parties receive $(\mathsf{idx}_b, \mathsf{idx}_{1-b})$.

Combining Theorem 14 with Theorem 15, we obtain the following corollary:

Corollary 2 (Short-List Selection is not PIR-hard). *The two-sided Short-List Selection function* $\mathsf{SLS} : \{0,1\}^{\tilde{\mathcal{O}}(N)} \times \{0,1\}^{\tilde{\mathcal{O}}(\log N)} \rightarrow \{0,1\}^{2\log N} \times \{0,1\}^{2\log N}$ *is not PIR-hard.*

5   Low Communication Locally Compressible Problems

In this section, we show that it is actually possible to achieve semi-honest security for one-sided problems and beyond if the problem satisfies the following notion of input compressibility.

Definition 13 (Locally Compressible Inputs). *We say that a functionality* $F : \{0,1\}^N \times \{0,1\}^N \rightarrow \{0,1\}^m \times \{0,1\}^m$ *has* locally compressible inputs *if there exists a preprocessing function* $\mathsf{Pre} : \{0,1\}^N \rightarrow \{0,1\}^{N^\alpha}$ *with* $\alpha < 1$ *for which* $F(X, Y) = F(\mathsf{Pre}(X), \mathsf{Pre}(Y))$.

Local compressibility of the inputs can yield semi-honest secure non PIR-hard ("easy") protocols with reduced communication complexity by first executing the local preprocessing and then calling a generic two-party protocol on the preprocessed input data.

In the following section we show that two optimization problems that satisfy the above property admit low communication complexity and are *not* PIR-hard. The first problem is the minimum spanning tree and the second one is the median protocol for a certain predicate on the output specified in Sect. 5.2.

5.1   Revisiting the Minimum Spanning Tree Protocol

A Minimum Spanning Tree (MST) of an edge-weighted graph is a spanning tree whose weight is no larger than the weight of any other spanning tree. More formally, given a connected, undirected graph $G = (V, E)$, a *spanning tree* is an acyclic subset of edges $T \subseteq E$ that connects all the vertices together. Assuming that each edge e=(u,v) of G has a numeric weight or cost, $w(e)$, we define the cost of a spanning tree T to be the sum of edges in the spanning tree

$$w(T) = \sum_{(u,v) \in T} w(u, v).$$

MST is a spanning tree of minimum weight. Note that the MST may not in general be unique, but it is true that if all the edge weights are distinct, then the MST will be unique.

Definition 14 (MST functionality). *Let $N \in \mathbb{N}$. We define the two-party functionality $f_{\mathrm{MST}_N}(G_A, G_B) = (T, T)$ which on input two connected, undirected graphs $G_A = (V_A, E_A, w_A)$ and $G_B = (V_B, E_B, w_B)$ of size N with distinct edges where $V_A = V_B$ and $w_A(e), w_B(e)$ represent the weight of edge e in G_A and G_B, outputs a subset of edges $T \subseteq E_A \cup E_B$ that connect all the vertices together with the minimum weight $w(T) = \sum_{(e) \in T} w(e)$.*

An efficient sublinear-communication protocol for two-sided MST was given in [SV15].

Two-Sided Locally Compressible MST. In the sequel, we show that the MST protocol has locally compressible inputs and admits "easy" low communication secure protocols. Beyond the results of [SV15], this approach enables such protocols for secure computation of *functions of* the MST (whereas the [SV15] protocol only supports MST itself).

Theorem 16. *Let $n \in \mathbb{N}$, and let $\{0,1\}^\ell$ be the input domain of edge weights. Then for any function $g : \{0,1\}^{2\ell \cdot n} \to \{0,1\}^{n'}$ with circuit size $o(N)$, there exists a secure two-party computation protocol Π_{MST} for the functionality $g \circ f_{\mathrm{MST}_{n^2}} : \{0,1\}^{\ell \cdot n^2} \times \{0,1\}^{\ell \cdot n^2} \to \{0,1\}^{n'}$ which achieves statistical security in the preprocessing model, with communication complexity $\widetilde{\mathcal{O}}(n) \in o(N)$ (where $N = \ell \cdot n^2$).*

Proof. We proceed by constructing an MST protocol Π_{MST}, as per Definition 14, calling the preprocessing function $\mathsf{Pre} : \{0,1\}^{n^2} \to \{0,1\}^n$ as per Definition 13. Let (G_A, G_B) be the connected, undirected graphs provided to the Π_{MST} protocol by party A and party B, respectively.

Protocol $\Pi_{\mathrm{MST}}(G_A, G_B)$:

Input Phase:
The preprocessing function Pre on input a graph G outputs its MST, denoted
by $\mathrm{MST}(G)$. In this phase each party locally computes $\mathsf{Pre}(G_A)$ and $\mathsf{Pre}(G_B)$
to obtain $\mathrm{MST}(G_A)$ and $\mathrm{MST}(G_B)$, respectively.

Evaluation and Output Phase:
Given two graphs $G_1 = (V_1, E_1, w_1)$ and $G_2 = (V_2, E_2, w_2)$ we denote by
G_1 & G_2 the graph $G = (V, E, w)$ with $V = V_1, E = E_1 \cup E_2$ and for each
edge $e \in E$, $w(e) = min(w_1(e), w_2(e))$.

Let Π denote a generic two-party protocol in the is run in order to compute
and output $\mathrm{MST}((\mathrm{MST}(G_A)$ & $\mathrm{MST}(G_B))$ to both parties.

In order to prove correctness of the above protocol Π_{MST}, we need to
prove that the local compressibility does not alter the final output. More
specifically, we need to show that $\forall e \in \mathrm{MST}(G_A$ & $G_B)$ it is implied that
$e \in \mathrm{MST}((\mathrm{MST}(G_A)$ & $\mathrm{MST}(G_B))$.

Suppose for contradiction that there is an edge e in G_B that is in the
$\mathrm{MST}(G_A$ & $G_B)$ but not in $\mathrm{MST}(G_B)$. Consider the cut C of vertices (created by drawing a line that intersects the middle of the edge e), that contains
only the edge e of $\mathrm{MST}(G_A$ & $G_B)$ (it exists since by definition there no cycles
in the MST). It must be the case that e is the lightest edge of G_A & G_B in
this cut C, otherwise we can swap it out with a lighter edge and contradict the
minimality of $\mathrm{MST}(G_A$ & $G_B)$. A swap is defined by adding in e, forming a cycle
in the graph, therefore removing the other edge in this cut and cycle, which is
by assumption strictly heavier.

However, all edge weights in G_A & G_B are smaller or equal to the weights in
G_B, since we take the minimum weight at every edge. This means that e must
also be the lightest edge of G_B in this cut. But this contradicts minimality of
$\mathrm{MST}(G_B)$ since we could always swap some edge of $\mathrm{MST}(G_B)$ in this cut with
e to get a strictly cheaper MST. Finally, since without loss of generality we can
consider disjoint edges and connected graphs the edge e must also be included
in the final tree $\mathrm{MST}((\mathrm{MST}(G_A)$ & $\mathrm{MST}(G_B))$. This concludes the proof.

Security of the protocol Π_{MST} follows immediately from the security of the
Π protocol. Furthermore, it is clear that the communication complexity of the
$\Pi_{\mathrm{MST}}(G_A, G_B)$ in the RAM model is $\widetilde{\mathcal{O}}(n)$ since after the local compressibility
the input size to the generic two-party protocol Π is reduced to $\mathcal{O}(n)$, making
use of a generic statistically secure ORAM-based protocol.   □

5.2   Revisiting the Two-Sided Median Predicates Protocol

In the sequel, we focus on the case of predicates on the output of the median
protocol and in particular on the high-order bits of their input.

Theorem 17. *Let $N \in \mathbb{N}$ and let $\{0,1\}^\ell$ be the input domain. For any predicate*
$\mathcal{P} : \{0,1\}^\ell \to \{0,1\}$ *which depends only on the $o(\log N)$ most significant bits*

of the input, there exists a secure two-party computation protocol $\Pi_{\mathrm{pMED}_N^{\mathcal{P}\ell}}$ *for the functionality* $\mathrm{pMED}_N^{\mathcal{P}}$ *which achieves statistical security in the preprocessing model, with communication complexity* $o(N)$.

Proof. We proceed by constructing the $\Pi_{\mathrm{pMED}_N^{\mathcal{P}\ell}}$ protocol, calling the preprocessing function $\mathsf{Pre} : \{0,1\}^N \to \{0,1\}^{o(N)}$ as per Definition 13. Let $X, Y \subset (\{0,1\}^\ell)^N$ be two input sets from party A and party B, respectively, sorted in increasing order such that $|X \cup Y| = 2N$. Protocol $\Pi_{\mathrm{pMED}_N^{\mathcal{P}\ell}}(X, Y)$ proceeds as follows:

Input Phase:
The preprocessing function Pre on input a set S outputs a compressed output $\mathsf{Pre}(S)$ of $2^\ell \in o(N)$-size, denoted by ℓ'-size, *count vector* corresponding to the number of occurrences of each length-ℓ' prefix within the elements of the set. More specifically, since there are $2^{\ell'}$ different representations for the ℓ' most significant bits, party A computes a counter vector $\mathbf{c}^A = (c_1^A, \ldots c_{2^{\ell'}}^A)$ counting the appearance of each possible representation in the most significant bits of each element in the set X. Respectively, party B computes his counter vector $\mathbf{c}^B = (c_1^B, \ldots c_{2^{\ell'}}^B)$.

Evaluation and Output Phase:
Let Π denote a generic two-party protocol $\Pi(\mathbf{c}^A, \mathbf{c}^B)$ which on input the sets $\mathbf{c}^A, \mathbf{c}^B$, outputs to both parties the predicate result. For our purposes, protocol Π is computing the median of the prefixes as encoded by the counter vectors $\mathbf{c}^A, \mathbf{c}^B$.

Correctness of the above protocol $\Pi_{\mathrm{pMED}_N^{\mathcal{P}\ell}}$ follows from the correctness of the Π, which does output the correct output predicate guaranteed by the structure of $\mathbf{c}^A, \mathbf{c}^B$. More specifically, since the high-order prefix of the median is equal to the median of the corresponding high-order prefixes, this short count vector carries sufficient information to evaluate the desired output predicate. Security follows immediately from the security of the protocol Π. The communication complexity of the $\Pi_{\mathrm{pMED}_N^{\mathcal{P}\ell}}(X, Y)$ protocol is $o(N)$ since after the local compressibility the input size to the generic two-party protocol Π is set to $o(N)$. □

6   Concluding Remarks and Open Problems

Our work initiates an effort to design a rigorous complexity framework for identifying "hard" tasks, to which previous techniques for low-complexity sublinear MPC cannot possibly apply, making the first broad strokes of classifying natural problems as "hard" or "potentially easy." The framework we propose is not perfect, and indeed, problems that are "potentially easy" are not necessarily easy. This is also the case for the theory of NP-completeness, where some problems that are conjectured not to be NP-hard (such as integer factorization) are also conjectured to be not easy. However, again like NP-completeness, our framework does provide meaningful and useful separations between different flavors of natural problems that would otherwise look very similar. This can help understand and guide MPC solutions over big data.

There are many questions left to be studied. Whereas for one-sided functionalities, VC-dimension gives a good combinatorial characterization for PIR-hardness (restricted to deterministic, non-interactive reductions), the situation for two-sided functionalities is not as well understood unless the output is very short. Is there a natural analogue of VC-dimension that captures PIR-hardness and semi-PIR-hardness of two-sided functionalities? What about multi-party functionalities, or two-party functionalities that deliver different outputs to the two parties? What about extending our framework to the setting of security against malicious parties?

The relation between semi-PIR to PIR is also only partially understood. While we show that strong semi-PIR implies weak (but nontrivial) PIR, it is not clear that our reduction is the best possible. In particular, our reduction makes use of non-trivial machinery of locally decodable codes, it requires multiple rounds of calls to the semi-PIR oracle, and exhibits a tradeoff between communication and local computation. Are these nonstandard features inherent? For instance, can we rule out parallel reductions of this type, or prove that any reduction that makes few (sequential) calls to the semi-PIR oracle implies a locally decodable code with related parameters?

As discussed above, problems that escape our notions of hardness are not necessarily easy. It would be interesting to identify natural candidate problems of this kind and try to refine our hardness notions to capture them.

Finally, is there a useful hierarchy of hardness classes beyond PIR-hardness and Semi-PIR-hardness? For instance, one could try to capture different levels of "somewhat homomorphic encryption" that are more expensive to implement than PIR, say, corresponding to the circuit depth or algebraic degree.

Acknowledgements. The first author was supported by ISF grant 1861/16, AFOSR Award FA9550-17-1-0069, and ERC grants 307952, 742754. The second author was supported in part by ERC grant 742754, ISF grant 1709/14, NSF-BSF grant 2015782, and a grant from the Ministry of Science and Technology, Israel and Department of Science and Technology, Government of India. The third author was supported by NSF grants 1617676, 1526377 and 1618884, IBM under Agreement 4915013672 and the Packard Foundation under Grant 2015-63124.

References

[AMP10] Aggarwal, G., Mishra, N., Pinkas, B.: Secure computation of the median (and other elements of specified ranks). J. Cryptol. **23**(3), 373–401 (2010)

[BGW88] Ben-Or, M., Goldwasser, S., Wigderson, A.: Completeness theorems for non-cryptographic fault-tolerant distributed computation (extended abstract), pp. 1–10 (1988)

[BIKO12] Beimel, A., Ishai, Y., Kushilevitz, E., Orlov, I.: Share conversion and private information retrieval. In: Proceedings of the 27th Conference on Computational Complexity, CCC 2012, Porto, Portugal, 26–29 June 2012, pp. 258–268 (2012)

[BIM04] Beimel, A., Ishai, Y., Malkin, T.: Reducing the servers' computation in private information retrieval: PIR with preprocessing. J. Cryptol. **17**(2), 125–151 (2004)

[BIPW17] Boyle, E., Ishai, Y., Pass, R., Wootters, M.: Can we access a database both locally and privately? In: Kalai, Y., Reyzin, L. (eds.) TCC 2017, Part II. LNCS, vol. 10678, pp. 662–693. Springer, Cham (2017). https://doi.org/10.1007/978-3-319-70503-3_22

[BS05] Brickell, J., Shmatikov, V.: Privacy-preserving graph algorithms in the semi-honest model. In: Roy, B. (ed.) ASIACRYPT 2005. LNCS, vol. 3788, pp. 236–252. Springer, Heidelberg (2005). https://doi.org/10.1007/11593447_13

[BV14] Brakerski, Z., Vaikuntanathan, V.: Efficient fully homomorphic encryption from (standard) LWE. SIAM J. Comput. **43**(2), 831–871 (2014)

[Can01] Canetti, R.: Universally composable security: a new paradigm for cryptographic protocols. In: 42nd Annual Symposium on Foundations of Computer Science, FOCS 2001, Las Vegas, Nevada, USA, 14–17 October 2001, pp. 136–145 (2001)

[CCD88] Chaum, D., Crépeau, C., Damgård, I.: Multiparty unconditionally secure protocols (extended abstract). In: Proceedings of the 20th Annual ACM Symposium on Theory of Computing, Chicago, Illinois, USA, 2–4 May 1988, pp. 11–19 (1988)

[CHR17] Canetti, R., Holmgren, J., Richelson, S.: Towards doubly efficient private information retrieval. In: Kalai, Y., Reyzin, L. (eds.) TCC 2017, Part II. LNCS, vol. 10678, pp. 694–726. Springer, Cham (2017). https://doi.org/10.1007/978-3-319-70503-3_23

[CKGS98] Chor, B., Kushilevitz, E., Goldreich, O., Sudan, M.: Private information retrieval. J. ACM **45**(6), 965–981 (1998)

[CMS99] Cachin, C., Micali, S., Stadler, M.: Computationally private information retrieval with polylogarithmic communication. In: Stern, J. (ed.) EUROCRYPT 1999. LNCS, vol. 1592, pp. 402–414. Springer, Heidelberg (1999). https://doi.org/10.1007/3-540-48910-X_28

[Efr09] Efremenko, K.: 3-query locally decodable codes of subexponential length. In: Proceedings of the 41st Annual ACM Symposium on Theory of Computing, STOC 2009, Bethesda, MD, USA, 31 May–2 June 2009, pp. 39–44 (2009)

[Gen09] Gentry, C.: Fully homomorphic encryption using ideal lattices. In: Proceedings of the 41st Annual ACM Symposium on Theory of Computing, STOC 2009, Bethesda, MD, USA, 31 May–2 June 2009, pp. 169–178 (2009)

[GMW87] Goldreich, O., Micali, S., Wigderson, A.: How to play any mental game or a completeness theorem for protocols with honest majority. In: Proceedings of the 19th Annual ACM Symposium on Theory of Computing, New York, USA, pp. 218–229 (1987)

[IKM+13] Ishai, Y., Kushilevitz, E., Meldgaard, S., Orlandi, C., Paskin-Cherniavsky, A.: On the power of correlated randomness in secure computation. In: Sahai, A. (ed.) TCC 2013. LNCS, vol. 7785, pp. 600–620. Springer, Heidelberg (2013). https://doi.org/10.1007/978-3-642-36594-2_34

[IKNP03] Ishai, Y., Kilian, J., Nissim, K., Petrank, E.: Extending oblivious transfers efficiently. In: Boneh, D. (ed.) CRYPTO 2003. LNCS, vol. 2729, pp. 145–161. Springer, Heidelberg (2003). https://doi.org/10.1007/978-3-540-45146-4_9

[KO97] Kushilevitz, E., Ostrovsky, R.: Replication is NOT needed: SINGLE database, computationally-private information retrieval. In: 38th Annual Symposium on Foundations of Computer Science, FOCS 1997, Miami Beach, Florida, USA, 19–22 October 1997, pp. 364–373 (1997)

[KT00] Katz, J., Trevisan, L.: On the efficiency of local decoding procedures for error-correcting codes. In: Proceedings of the Thirty-Second Annual ACM Symposium on Theory of Computing, Portland, OR, USA, 21–23 May 2000, pp. 80–86 (2000)

[Lip05] Lipmaa, H.: An oblivious transfer protocol with log-squared communication. In: Zhou, J., Lopez, J., Deng, R.H., Bao, F. (eds.) ISC 2005. LNCS, vol. 3650, pp. 314–328. Springer, Heidelberg (2005). https://doi.org/10.1007/11556992_23

[MBFK16] Aguilar Melchor, C., Barrier, J., Fousse, L., Killijian, M.-O.: XPIR: private information retrieval for everyone. PoPETs 2016(2), 155–174 (2016)

[NAT89] Natarajan, B.K.: On learning sets and functions. Mach. Learn. 4, 67–97 (1989)

[SV15] Shelat, A., Venkitasubramaniam, M.: Secure computation from millionaire. In: Iwata, T., Cheon, J.H. (eds.) ASIACRYPT 2015, Part I. LNCS, vol. 9452, pp. 736–757. Springer, Heidelberg (2015). https://doi.org/10.1007/978-3-662-48797-6_30

[VC71] Vapnik, V.N., Chervonenkis, A.Y.: On the uniform convergence of relative frequencies of events to their probabilities. Theory Probab. Appl. 16(2), 264–280 (1971)

[Yao82] Yao, A.C.-C.: Protocols for secure computations (extended abstract). In: FOCS, pp. 160–164 (1982)

Garbling

Limits on the Power of Garbling Techniques for Public-Key Encryption

Sanjam Garg[1], Mohammad Hajiabadi[1,2], Mohammad Mahmoody[2(✉)], and Ameer Mohammed[2]

[1] University of California, Berkeley, Berkeley, USA
[2] University of Virginia, Charlottesville, USA
mohammad@virginia.edu

Abstract. Understanding whether public-key encryption can be based on one-way functions is a fundamental open problem in cryptography. The seminal work of Impagliazzo and Rudich [STOC'89] shows that black-box constructions of public-key encryption from one-way functions are impossible. However, this impossibility result leaves open the possibility of using non-black-box techniques for achieving this goal.

One of the most powerful classes of non-black-box techniques, which can be based on one-way functions (OWFs) alone, is Yao's garbled circuit technique [FOCS'86]. As for the non-black-box power of this technique, the recent work of Döttling and Garg [CRYPTO'17] shows that the use of garbling allows us to circumvent known black-box barriers in the context of identity-based encryption.

We prove that garbling of circuits that have OWF (or even random oracle) gates in them are insufficient for obtaining public-key encryption. Additionally, we show that this model also captures (non-interactive) zero-knowledge proofs for relations with OWF gates. This indicates that currently known OWF-based non-black-box techniques are perhaps insufficient for realizing public-key encryption.

1 Introduction

Public-key encryption (PKE) [15,33] is a fundamental primitive in cryptography and understanding what assumptions are sufficient for realizing it is a

S. Garg—Research supported in part from DARPA/ARL SAFEWARE Award W911NF15C0210, AFOSR Award FA9550-15-1-0274, AFOSR YIP Award, DARPA and SPAWAR under contract N66001-15-C-4065, a Hellman Award and research grants by the Okawa Foundation, Visa Inc., and Center for Long-Term Cybersecurity (CLTC, UC Berkeley). The views expressed are those of the author and do not reflect the official policy or position of the funding agencies.

M. Hajiabadi—Supported by NSF award CCF-1350939 and AFOSR Award FA9550-15-1-0274.

M. Mahmoody—Supported by NSF CAREER award CCF-1350939, a subcontract on AFOSR Award FA9550-15-1-0274, and University of Virginia's SEAS Research Innovation Award.

A. Mohammed—Supported by Kuwait University and the Kuwait Foundation for the Advancement of Science. Work done while at University of Virginia and visiting University of California, Berkeley.

© International Association for Cryptologic Research 2018
H. Shacham and A. Boldyreva (Eds.): CRYPTO 2018, LNCS 10993, pp. 335–364, 2018.
https://doi.org/10.1007/978-3-319-96878-0_12

foundational goal. Decades of research have provided us with numerous constructions of PKE from a variety of assumptions; see a recent survey by Barak [5]. However, all known constructions of PKE require computational assumptions that rely on rich structure and are stronger than what is necessary and sufficient for *private*-key cryptography, namely the mere existence of one-way functions (OWF). The seminal work of Impagliazzo and Rudich [23] provides some evidence that this gap between the assumption complexity of private-key and public-key encryption may be inherent. In particular, the work of [23] shows that there is no *black-box* construction of PKE from OWFs.[1]

When studying the *impossibility* of basing PKE on OWFs, focusing on a class of constructions (e.g., black-box constructions as in [23]) is indeed necessary. The reason is that to rule out "OWFs implying PKE" in a *logical* sense, we have to first prove the existence of OWFs unconditionally (thus, proving $\mathbf{P} \neq \mathbf{NP}$) and then rule out the existence of PKE altogether (thus breaking all assumptions under which PKE exists). That is why this line of separation results focuses on ruling out the possibility of using certain *techniques* or generic *proof methods* (here black-box techniques) as possible natural paths from OWFs to PKE.

Garbled circuits. Over the past few decades, garbling techniques [1,9,25, 35] (or randomized encodings [24] more generally) have been extensively used to build many cryptographic schemes. Roughly speaking, in a circuit garbling mechanism, a PPT encoder $\mathsf{Garb}(\mathsf{C})$ takes a circuit C as input, and outputs a *garbled circuit* $\widetilde{\mathsf{C}}$ and a set of *input labels* $\{\mathsf{label}_{i,b}\}_{i \in [m], b \in \{0,1\}}$ where m is the number of input wires of C. Using another algorithm $\mathsf{Eval}(\cdot)$, one can use the garbled circuit $\widetilde{\mathsf{C}}$ and input labels $\{\mathsf{label}_{i,x_i}\}_{i \in [m]}$ for an input $x = (x_1, \ldots, x_m)$, to compute $\mathsf{C}(x)$ without learning any other information. Note that if the original circuit C needs to run a cryptographic primitive f internally (e.g., a circuit C for a pseudorandom generator built from a OWF f), this use of garbling leads to a *non-black-box* construction. This is because the algorithm Garb needs to work with an actual circuit description of C, whose circuit description is in turn obtained by the circuit description of f, hence making non-black-box use of f.

Garbling, as a primitive, may itself be realized using one-way functions [25,35]. This puts forward the intriguing possibility of basing PKE solely on OWFs by making *black-box* use of garbling mechanisms over circuits that can run the one-way function. As stated above, such constructions will make non-black-box use of the underlying OWF (caused by garbling circuits that run the OWF internally) and hence the impossibility result of Impagliazzo and Rudich [23] has no bearing on such potential constructions. In fact, such non-black-box garbling techniques, combined with the Computational Diffie-Hellman assumption, have recently been used by Döttling and Garg [16] to circumvent black-box impossibility results [11,29] in the context of identity-based encryption (IBE). Thus, it is natural to ask:

Can non-black-box garbling techniques be used to realize PKE from OWFs?

---

[1] A (fully) black-box construction is one that treats the OWF as an oracle, and the security proof uses the OWF and the adversary both as oracles; see the surveys of [4,32] for formal definitions.

Our model. We study the above question in the model of Brakerski et al. [12] (see also follow up works [2,3,10]) which gives a general way of capturing non-black-box techniques via circuits with cryptographic gates (e.g., OWF gates). More formally, we will model the above-stated garbling-based non-black-box use of one-way functions as black-box use of garbling mechanisms that can take as input circuits C with one-way function (or even random oracle) gates planted in them. Such constructions are indeed *non-black-box* according to the taxonomy of [32] if viewed as standalone constructions solely based on the OWF itself. We stress that the allowed access to the garbling mechanism itself is black-box; the non-black-box feature arises from the fact that circuits with OWF gates may now be garbled.

A more sophisticated scenario is when the circuits being garbled have *garbling* gates, in addition to OWF gates, planted in them. We do not, however, consider such a recursive scenario and we leave it to future work. It is crucial to note that, to the best of our knowledge, all known constructions that make use of garbling schemes together with one-way functions (e.g., [8,17,26]) fall into our model, and thus, understanding the limitations of such techniques towards obtaining PKE is impactful.

1.1 Our Result

In this work, we show that black-box use of garbling mechanisms that allow circuits with OWF gates to be garbled is not sufficient for constructing PKE. More precisely, we prove the following.

Theorem 1 (Main result – informally stated). *There exists no black-box construction of public-key encryption schemes from any one-way function (or even a random oracle) f together with a garbling mechanism that can garble oracle-aided circuits with f-gates embedded in them.*

Comparison with prior work: impossibility from weaker garbling. The work of Asharov and Segev [2] showed that secret-key functional encryption with one-way function gates cannot be used (as a black-box) to obtain public-key encryption (or even key agreement). This result implies that for the special case of such weaker garbling schemes, called *non-decomposable* garbling, where the entire input is considered as a single unit (rather that as bit-by-bit input labels) is insufficient for realizing PKE.

On the other hand, throughout this work we use garbling to refer to a notion that supports bit-by-bit input labels, a notion that Bellare, Hoang and Rogaway [9] refer to as *projective garbling* (a.k.a. decomposable garbling). Under projective garbling, for a circuit C of input size m, one generates two garbled label $\{\mathsf{label}_{i,b}\}_{i\in[m],b\in\{0,1\}}$ for the ith input wire of the circuit. An important property enabled by this bit-by-bit garbling is the decomposability property: one can pick a garbled label for each input wire to form a garbled input for a long string. In contrast, under non-decomposable garbling, for each input X to the circuit, one independently generates a corresponding garbled input $\widetilde{X}$. As a result, different strings have independent garbled inputs.

We note that projective garbling is crucial for many applications of garbling. For example, even the most basic application of garbling in two party secure computation based on oblivious transfer uses the projective property. We refer the reader to [9, Fig. 3] for a detailed list of applications that require projective garbling. As a recent example, we note that the IBE construction of Döttling and Garg [16] (that circumvents a black-box impossibility result of Papakonstantinou et al. [29] using garbling) uses projective garbling crucially. Specifically, in [DG17] the encryptor provides a sequence of garbled circuits with no knowledge of what input each of those garbled circuits are later evaluated on by the decryptor. This *input-obliviousness* property is enabled by the encryptor sending all the bit-by-bit garbled labels in some encrypted form to the decryptor. Later, the decryptor can open exactly one garbled label for each input wire, hence obtaining a garbled input for the whole string. This input-obliviousness technique cannot be enabled using non-decomposable garbling. This is because a whole garbled input cannot be formed there by putting together smaller pieces. As a result, the party who generates a garbled circuit must be aware of the input on which this garbled circuit is to be evaluated on, in order for him to be able to provide the corresponding garbled input.

1.2   Extensions

Extension to key agreements. Our proof extends to rule out any black-box construction of *constant-round* key-agreement protocols from OWFs and garbling schemes for oracle-aided circuits. However, the proof of the separation for key-agreement beyond the case of two message protocols (which are equivalent to PKE) becomes much more involved. Therefore, for clarity of the presentation, and because the most interesting special case of constant-round key-agreement protocols happens to be PKE itself, in this presentation, we focus on the case of separation for PKE. See Section the full version for more details.

Resolving an open question of [12]. The work of [12] proved non-black-box limitations for one-way functions when used as part of zero knowledge (ZK) proofs for relations with one-way function gates. They showed that key-agreement protocols with *perfect* completeness cannot be realized in a 'black-box' way from oracles that provide a one-way function f together with ZK proofs of satisfiability for f-aided circuits. They left ruling out the possibility of protocols with imperfect (e.g., negligible) completeness as an open problem, as their techniques indeed crucially relied on the perfect completeness assumption. We demonstrate the power of our new techniques in this work by resolving the open problem of [12] along the way, for the case of PKE schemes (or even constant-round key-agreement schemes). In particular, in the full version of the paper, we observe that the oracles we use for proving oursuparations for the case of garbling, indeed imply the existence of NIZK proofs for satisfiability of circuits with OWF-gates. The extension of the result of [12] explained above then follows from the above observation.

1.3   Related Work and Future Directions

There are quite a few results that prove limitations for a *broad* class of non-black-box techniques [20,30,31], so long as the security reduction is black-box. In other words, these results are proved against basing certain primitives on any *falsifiable* assumption. However, when it comes to the case of non-black-box constructions of PKE from OWFs, no such general separations are known (and proving such results might in fact be impossible).

As described earlier, the works of [2,12] proved limitations of certain non-black-box constructions of PKE from OWFs. This is indeed the direction pursued in this work. The work of Dachman-Soled [14] takes yet another path, showing that certain non-black-box uses of one-way functions *in the security proof* are incapable of obtaining PKE from OWFs.

We note that we only consider a setting in which circuits with random oracle gates are garbled. We do not allow garbling of circuits which themselves include garbling gates. Such techniques are captured by the so called monolithic model of Garg, Mahmoody, and Mohammed [18,19]. We leave open the problem of ruling out such constructions.

Finally, as noted above, the extension of our results to the key-agreement setting (discussed in the full version) only cover the *constant-round* case. The reason is that, during the proof of our main result, we modify the protocol iteratively, once for each round, which increases the parameters of the protocol by a polynomial factor each time. We leave the extension to general polynomial-round protocols as an interesting future direction.

Organization. In Sect. 2 we give an overview of our approach and techniques. In Sect. 3 we give some definitions and basic lemmas. In Sect. 4 we will go over the proof steps of our main impossibility result. See the full version of the paper for full proofs of the main result and the extensions.

2   Technical Overview

For brevity, we refer to the primitive of a one-way function f and garbling circuits with f gates as GC-OWF. As usual in black-box separation results, we will prove our main theorem by providing an oracle O relative to which secure GC-OWF exists, but secure PKE does not.

2.1   Big Picture: Reducing the Problem to the Result of [23]

At a very high level, our approach is to reduce our problem to the result of [23]. Namely, we aim to show that one can always modify the PKE construction that is based on the GC-OWF oracle O into a new one that is almost as secure, but which no longer uses the garbling part of the oracle O. In other words, we modify the construction so that it becomes a construction from an OWF oracle alone. Our main result, then, follows from the impossibility result of [23] which rules out the possibility of getting PKE from one-way (or even random) functions. We call this process 'compiling out the garbling part' from the PKE construction.

As a technical remark, our transformation does not result in a normal black-box construction of PKE from OWFs, but rather results in an *inefficient* one which nonetheless makes a polynomial number of queries to the OWF oracle. The key point is that the *proof* of the work of Impagliazzo and Rudich [23] allows us to break any such (even inefficient, but still polynomial-query) constructions of PKE in the random oracle model using a polynomial number of queries during the attack. Our actual result follows by combining our compilation result with the result of [23], to get a polynomial query attack against the security of the original PKE. This will be sufficient for a black-box separation.

At a high level, our approach also bears similarities to recent impossibility results for indistinguishability obfuscation [18] as we also compile out the more powerful (and structured) parts of the oracle, ending up with a scheme that uses a much simpler oracle, for which an impossibility is known. However, there is a subtle distinction here. Unlike the results of [13,18,27,28], when we compile out the garbling-related queries from the PKE construction, we end up with an inefficient scheme that potentially runs in exponential time but nevertheless makes a polynomial number of queries. However, as mentioned above, this is fine for deriving our separation, because we can still rely on the fact that the result of [23] does something stronger and handles inefficient constructions as well.

2.2   Our Separating Idealized GC-OWF Oracle

In this subsection, we will first describe our oracle O that gives an intuitive way of obtaining GC-OWFs. The natural first version of this oracle is too strong as it also implies virtual black-box (VBB) obfuscation. We will then add a careful weakening subroutine to this oracle O to prevent it from implying obfuscation. In the next subsection we describe the ideas behind how to compile out the garbling-related subroutines of O from the PKE construction, while keeping the PKE construction "secure".

Our 1st oracle for GC-OWF. Our first version of the separating oracle $O = (f, L^f)$ will consist of a random oracle f (giving the OWF part) as well a garbling part $L^f = (\text{gc}, \text{eval}^f)$ with two subroutines. The encoding/garbling subroutine $\text{gc}(s, \text{C})$ is simply a random (injective) function that takes a seed s and a circuit C and maps them into a garbled circuit $\widetilde{\text{C}}$ as well as labels $\{\text{label}_{i,b}\}_{i \in [n], b \in \{0,1\}}$ for the input wires of C where n is the number of input wires in C.[2] The evaluation subroutine eval^f takes as input a garbled circuit $\widetilde{\text{C}}$

---

[2] In the main body, we will use two separate subroutines gc, gi for encoding circuits vs input labels, but for brevity here we combine them into one subroutine.

as well as a vector of input labels $\widetilde{X} = (\widetilde{x}_1 \cdots \widetilde{x}_n)$ and *only if* they were all encoded using the *same* seed s, eval^f returns the right output $\mathsf{C}^f(x_1, \ldots, x_n)$. Note that we include f in the representation of eval^f but not in that of gc; the reason is gc is simply a random oracle (independent of f), while eval^f needs to call f in order to compute $\mathsf{C}^f(x_1, \ldots, x_n)$.

Adding the weakening subroutine rev. It is easy to see that this first version of the oracle O as described above can realize a secure GC-OWF, but it can do much more than that! In fact this oracle implies even VBB obfuscation of circuits with f gates (which in turn *does* imply PKE [34]). We will, therefore, weaken the power of the oracle O later on by adding an extra subroutine to it (which we will call rev), which roughly speaking allows an attacker to break the garbling scheme if she has access to two labels for the same wire. We will describe this subroutine after it becomes clear how it will be useful for our main goal of compiling out the garbling aspect of O. Also note that, since we are defining our oracle *after* the (supposed) construction of PKE from GC-OWF is fixed, without loss of generality, the PKE construction from GC-OWF does *not* call the extra subroutine rev. This separation technique was also used before in the work of [21] and is reminiscent of the "two-oracle" approach of [22].

2.3 Compiling Out the Garbling Power of O from the Construction

Suppose $\mathcal{E}^O = (G^O, E^O, D^O)$ is a fully black-box construction of PKE using the oracle O described above. Our goal here is to 'reduce' our problem (of breaking $\mathcal{E}^O$ using a polynomial number of queries) to the result of [23] by compiling out the 'garbling power' of the oracle O from the scheme $\mathcal{E}^O$. But what subroutines do we have to compile out from $\mathcal{E}$? As it turns out, we do not have to eliminate both gc and eval^f subroutines; removing only eval^f queries will suffice.

Compiling out eval^f queries from PKE constructions $\mathcal{E}^O$. If we make sure that (the modified but "equally"-secure version of) $\mathcal{E}$ does not make any calls to the eval^f subroutine of the oracle O, it would be sufficient for our purposes, because the oracle $O' = (f, \mathsf{gc})$ is just a random oracle, and by the result of [23] we know that such oracle is not enough for getting PKE in a black-box way. Therefore, in what follows, our goal reduces to (solely) removing the eval^f queries from PKE constructions $\mathcal{E}^O$ in a way that we can argue the new construction is 'as secure' as the original one.

In order to make our proof more modular, we compile out eval^f queries from the different components (i.e., key-generation G, encryption E, and decryption D) of the construction $\mathcal{E}^O = (G^O, E^O, D^O)$ one at a time. First, we may easily see that G^O does not need to call eval^f at all. This is because G^O is the first algorithm to get executed in the system, and so G^O knows all the generated garbled circuits/labels. Therefore, G^O can, instead of calling eval^f queries, simply run $C^f(X)$ on its own by further calls to f.[3] Now, we proceed to compile out eval^f queries from the remaining two subroutines E and D in two steps. In each step, we assume that we are starting off with a construction that has no eval^f queries in some of its subroutines, and then we modify the construction to remove eval^f queries from the next subroutine.

- **Step 1:** Starting with $\mathcal{E} = (G^{f,\text{gc}}, E^O, D^O)$, we will compile $\mathcal{E}$ into a new scheme $\dot{\mathcal{E}} = (\dot{G}^{f,\text{gc}}, \dot{E}^{f,\text{gc}}, D^O)$, removing eval^f queries asked by E^O. We have to make sure $\dot{\mathcal{E}}$ is 'almost as secure' as the original scheme $\mathcal{E}$. This step is detailed in Sect. 4.1.
- **Step 2:** Given $\mathcal{E} = (G^{f,\text{gc}}, E^{f,\text{gc}}, D^O)$, we compile $\mathcal{E}$ into a new scheme $\ddot{\mathcal{E}} = (\ddot{G}^{f,\text{gc}}, \ddot{E}^{f,\text{gc}}, \ddot{D}^{f,\text{gc}})$, removing eval^f queries asked by D^O. Again, we have to make sure $\ddot{\mathcal{E}}$ is 'almost as secure' as the original one. This step is detailed in Sect. 4.2.

Once we accomplish both of the steps above, we will combine them into a single compiler that removes eval^f queries from $\mathcal{E}^O$, obtaining another PKE construction that is secure in the random oracle model (which we already know is impossible by the result of [23]).

Overview of Step 1. Let us start by looking at eval queries of the encryption algorithm $E^O(pk, b)$. Since the subroutine gc of oracle O is just a random mapping, for any eval query on inputs $(\widetilde{C}, \widetilde{X})$, denoted $qu = ((\widetilde{C}, \widetilde{X}) \xrightarrow[\text{eval}]{} ?)$, made by E^O and whose answer is not trivially $\bot$, we must have either of the following cases. Either (a) $\widetilde{C}$ was produced as a result of a gc query during the execution of E itself or (b) $\widetilde{C}$ was produced during the execution of G which has led to the generation of the public key pk. If case (a) holds, then E does not need to make that particular eval query at all. If case (b) holds, then in order to allow $\dot{E}^{f,\text{gc}}$ to simulate E^O without calling eval^f, the algorithm $\dot{E}^{f,\text{gc}}$ will resort to some 'hint list' H attached to pk by $\dot{G}^{f,\text{gc}}$. That is, a compiled public key $\dot{pk}$ produced by $\dot{G}^{f,\text{gc}}$ will now contain the original pk as well as a hint list H. Below, we further explain how the hint H is formed.

How $\dot{G}^{f,\text{gc}}$ forms the hint list H. A naive idea is to let H contain all the query/response pairs made by $G^{f,\text{gc}}$ to generate pk. This method hurts security. A better idea is to provide in H answers to individual eval queries like eval($\widetilde{C}, \widetilde{X}$) that are *likely* to be asked by $E^O(pk, b)$, and where $\widetilde{C}$ was generated by $G^{f,\text{gc}}$. That is, $\dot{G}^{f,\text{gc}}$ would run $E^O(pk, b)$ many times and would let H contain all encountered eval queries as well as their answers. Note that $\dot{G}^{f,\text{gc}}$ could simulate almost perfectly a random execution of $E^O(pk, b)$ without calling eval since $\dot{G}$ knows the randomness seeds of all the garbled circuits so far. However, this approach also fails! To see the difficulty, recall that a whole garbled input $\widetilde{X} =$

---

[3] More formally, because of the huge output space of gc, calling eval^f on a garbled circuit $\widetilde{C}$ that is produced on the fly is bound to be responded with $\bot$ with overwhelming probability as $\widetilde{C}$ will not be an encoding of any circuit.

$(\tilde{x}_1, \ldots, \tilde{x}_m)$ is made up of a sequence of garbled labels $\tilde{x}_i$, one for each input wire. Now imagine that for a garbled circuit $\tilde{\mathsf{C}}$ that was generated by $\dot{G}$, any new execution of $E^O(pk, b)$ calls the oracle eval on $\tilde{\mathsf{C}}$ and on a new garbled input $\tilde{X}$ that is formed by picking each of the garbled labels uniformly at random from the set of two labels for that corresponding input wire. labels. If E behaves this way, then no matter how many (polynomial) times we sample from $E^O(pk, b)$, we cannot hope to predict the garbled-input part of the eval query of the next execution of $E^O(pk, b)$. We refer to this as the *garbled-input unperdictability* problem, which stems from the decomposability nature of our garbling oracle. This is what makes our results different from those of [2], which dealt with non-decomposable garbling, for which such a complication is absent.

In short, we could only hope to predict the garbled circuit part of an eval query of $E^O(pk, b)$, and not necessarily the garbled-input part. To fix this garbled-input unpredictability problem, $\dot{G}^{f,\mathsf{gc}}$ will do the following trick: while sampling many executions of $E^O(pk, b)$, if $\dot{G}^{f,\mathsf{gc}}$ comes across two different eval queries $\mathsf{eval}(\tilde{\mathsf{C}}, \tilde{X}_1), \mathsf{eval}(\tilde{\mathsf{C}}, \tilde{X}_2)$ that are both answered with a value that is not $\bot$ (i.e., both are valid garbled circuits and inputs), then $\dot{G}^{f,\mathsf{gc}}$ releases the corresponding seed s and the *plain* circuit C of $\tilde{\mathsf{C}}$. That is, if $\mathsf{gc}(s, \mathsf{C}) = (\tilde{\mathsf{C}}, \cdots)$, then $\dot{G}^{f,\mathsf{gc}}$ puts the tuple $(s, \mathsf{C}, \tilde{\mathsf{C}})$ into the hint list H. If, however, during these sampling, $\tilde{\mathsf{C}}$ is evaluated upon *at most* one matching $\tilde{X}$, then $\dot{G}^{f,\mathsf{gc}}$ simply provides the answer to the query $\mathsf{eval}(\tilde{\mathsf{C}}, \tilde{X})$ in H.

Looking ahead, the algorithm $\dot{E}^{f,\mathsf{gc}}((pk, H), b)$, when facing an eval query $qu = \mathsf{eval}(\tilde{\mathsf{C}}, \tilde{X})$, will check whether qu is already answered in H, or whether the corresponding seed s and plain-circuit C of $\tilde{\mathsf{C}}$ could be retrieved from H. If so, $\dot{E}^{f,\mathsf{gc}}$ will reply to qu accordingly; otherwise, it will reply to qu with $\bot$.

Using the weakening subroutine rev to reduce the security of $\dot{\mathcal{E}}$ to $\mathcal{E}$. Note that $\dot{G}^{f,\mathsf{gc}}$ does not query any oracle subroutines beyond f and gc in order to form the hint list H attached to pk. This is because $\dot{G}^{f,\mathsf{gc}}$ has all the (otherwise-hidden) query-answer pairs used to produce pk, and thus for any encountered valid garbled circuit $\tilde{\mathsf{C}}$ during those sampled executions of E, $\dot{G}^{f,\mathsf{gc}}$ already knows the corresponding seed s and plain circuit C. Now we are left to show that this additional information H attached to pk does not degrade the security of the compiled scheme significantly. To this end, we will use the new weakening oracle intended to capture the natural use of garbling: the security of a garbled circuit $\tilde{\mathsf{C}}$ is guaranteed to hold so long as $\tilde{\mathsf{C}}$ is evaluated only on one garbled input. Capturing this, our new oracle rev takes as input a garbled circuit $\tilde{\mathsf{C}}$ and two garbled inputs $\tilde{X}_1$ and $\tilde{X}_2$, and if all of $\tilde{\mathsf{C}}$, $\tilde{X}_1$ and $\tilde{X}_2$ are encoded using the same seed s, then rev simply outputs (s, C), where $\mathsf{gc}(s, \mathsf{C}) = \tilde{\mathsf{C}}$. For security, we will show that any adversary against the semantic security of $\dot{\mathcal{E}}$ may be used in a black-box way, along with oracle access to $(f, \mathsf{gc}, \mathsf{eval}, \mathsf{rev})$, to mount an attack against the original scheme $\mathcal{E}$. This shows that the leakage caused by revealing H was also attainable in the original scheme (in which all parties including the attacker do have access to eval) if, in addition, access to the oracle rev — which reflects the intuitive way in which garbled circuits are supposed to be used — was also granted to the adversary. In our security proof we will crucially make use of the rev subroutine in order to construct tuples $(s, \mathsf{C}, \widetilde{\mathsf{C}})$ to store in the simulated hint list whenever $\widetilde{\mathsf{C}}$ should be evaluated on two different inputs. Tuples of the form $(\widetilde{\mathsf{C}}, \widetilde{X}, y)$ can in turn be simulated using oracle access to eval.

Overview of Step 2. The main idea is similar to Step (1): $\ddot{E}^{f,\mathsf{gc}}(pk, b)$ would first run $E^{f,\mathsf{gc}}(pk, b)$ to get the ciphertext c and then appropriately attach a hint H to c. The idea is that H should allow the eval-free algorithm $\ddot{D}^{f,\mathsf{gc}}((sk, \mathsf{H}), c)$ to simulate $D^{f,\mathsf{gc},\mathsf{eval}}(sk, c)$ well enough. Again, since we cannot simply copy the entire private view of $\ddot{E}^{f,\mathsf{gc}}(pk, b)$ into H (as that cannot be simulated by the security reduction and therefore would hurt security) we should instead ensure that w.h.p. all eval queries during the execution of $D^O(sk, c)$, whose garbled circuits were generated by $\ddot{E}^{f,\mathsf{gc}}(pk, b)$, can be answered using H. Let us call these eval queries $\ddot{E}$-*tied* queries. Unfortunately, when implementing this idea, we run into the following problem: $\ddot{E}^{f,\mathsf{gc}}(pk, b)$ cannot simply run $D^O(sk, c)$ to get a sense of eval queries because sk is private; this was absent in Step (1).

In order to resolve this new challenge, the algorithm $\ddot{E}^{f,\mathsf{gc}}(pk, b)$ needs to do some more offline work in order to get an idea of $\ddot{E}$-tied eval queries that come up during $D^O(sk, c)$. The main idea is that although the true secret key sk is unknown to $\ddot{E}^{f,\mathsf{gc}}(pk, b)$, in the eyes of $\ddot{E}^{f,\mathsf{gc}}(pk, b)$, the value of sk is equally likely to be any sk' that agrees with the entire view of $\ddot{E}^{f,\mathsf{gc}}(pk, b)$. Put differently, the probability that an $\ddot{E}$-tied garbled circuit comes up during $D^O(sk, c)$ is close to the probability that it comes up during the execution of $D^{O'}(sk', c)$, where O' is an offline oracle that agrees with all the private information of $\ddot{E}$, and also relative to which (pk, sk') is valid public-key/secret-key. As a result, such a fake sk' that is consistent with the view of $\ddot{E}^{f,\mathsf{gc}}(pk, b)$ will be used to learn the answers of the evaluation queries asked by $D^{O'}(sk', c)$[4].

Putting things together. Taken together, Steps (1) and (2) in conjunction with the result of Imagliazzo and Rudich [23] imply the following.

Lemma 2 (Informal). *The (claimed) semantic security of any candidate PKE construction $\mathcal{E}^{f,\mathsf{gc},\mathsf{eval}}$ can be broken by a poly-query adversary $\mathcal{A}^{f,\mathsf{gc},\mathsf{eval},\mathsf{rev}}$.*

Moreover, we can show that the oracle rev does not break the one-wayness or the garbling-security aspects of $(f, \mathsf{gc}, \mathsf{eval})$.

Lemma 3 (Informal). *The function f is one way against all poly-query adversaries with oracle access $(f, \mathsf{gc}, \mathsf{eval}, \mathsf{rev})$. Moreover, there exists a garbling scheme $L^{f,\mathsf{gc},\mathsf{eval}}$ for garbling circuits with f gates that remains secure against all poly-query adversaries $\mathcal{B}^{f,\mathsf{gc},\mathsf{eval},\mathsf{rev}}$.*

Now Lemmas 2 and 3 imply our main theorem, Theorem 1.

---

[4] The process of discovering such an sk' is what makes $\dot{E}$ an inefficient algorithm.

3 Preliminaries

We use κ for the security parameter. By PPT we mean a probabilistic polynomial time algorithm. By an *oracle* PPT/algorithm we mean a PPT that may make oracle calls. For any oracle algorithm A that has access to some oracle O, we denote a query qu asked by A to a subroutine T of O as $(qu \xrightarrow[T]{} ?)$. If the returned answer is β, then we denote the resulting query-answer pair as $(qu \xrightarrow[T]{} \beta)$. For a set S of query/answer pairs, we will use intuitive notation such as $(* \xrightarrow[T]{} \beta) \in$ S to mean that there exists a query qu such that $(qu \xrightarrow[T]{} \beta) \in$ S. We use $\|$ to concatenate strings and we use "," for attaching strings in a way they could be retrieved. Namely, one can uniquely identify x and y from (x, y). For example $(00\|11) = (0011)$, but $(0, 011) \neq (001, 1)$. For any given string x, we denote x_i to be the i'th string of x. For (family of) random variables $\{X_\kappa, Y_\kappa\}_\kappa$, by $X \stackrel{c}{\approx} Y$ we denote that they are computationally indistinguishable; namely, for any poly(κ)-time adversary A there is a negligible function negl(κ) such that $|\Pr[A(X_\kappa) = 1] - \Pr[A(Y_\kappa) = 1]| \leq$ negl(κ). When writing the probabilities, by putting an algorithm A in the subscript of the probability (e.g., $\Pr_A[\cdot]$) we emphasize that the probability is over A's randomness. For any given probability distribution $\mathbf{D}$, we denote $x \leftarrow \mathbf{D}$ as sampling from this distribution and obtaining a sample x from the support of $\mathbf{D}$. We may also use $x \in \mathbf{D}$ to mean that x is in the support of $\mathbf{D}$. For any two random variables X, Y, we denote $\Delta(X, Y)$ to be the statistical distance between the two random variables. Throughout the paper, whenever we write $f_1(\kappa) \leq f_2(\kappa)$ we mean that this inequality holds asymptotically; i.e., there exists κ_0 such that for all $\kappa \geq \kappa_0$, $f_1(\kappa) \leq f_2(\kappa)$.

3.1 Some Useful Lemmas

The following lemma shows that hitting the image of a sparse injective random function without having called the function on the corresponding preimage happens with negligible probability.

Lemma 4 (Hitting the image of random injective function). *Let A be an arbitrary polynomial-query algorithm with access to an oracle $O : \{0,1\}^\kappa \to \{0,1\}^{2\kappa}$ chosen uniformly at random from the set of all injective functions from $\{0,1\}^\kappa$ to $\{0,1\}^{2\kappa}$. We have*

$$\Pr[y \leftarrow A^O(1^\kappa) \mid \text{for some } x \colon y = O(x) \wedge (* \xrightarrow[O]{} y) \notin \mathsf{Q_A}] \leq 2^{-\kappa/2},$$

where the probability is taken over the random choice of O as well as A's random coins, and where $\mathsf{Q_A}$ is the set of all A's query-answer pairs.

We will also use the following standard information theoretic lemma frequently in the paper.

Lemma 5. *Let $X_1, \ldots, X_{t+1}$ be independent, Bernoulli random variables, where $\Pr[X_i = 1] = p$, for all $i \leq t + 1$. Then*

$$\Pr[X_1 = 0 \wedge \cdots \wedge X_t = 0 \wedge X_{t+1} = 1] \leq \frac{1}{t}.$$

3.2   Standard Primitives

The definition of a single-bit public key encryption scheme (G, E, D) with $(\frac{1}{2} + \delta)$-correctness is standard. For $\gamma = \gamma(\kappa)$ we say that an adversary $\mathcal{A}$ γ-breaks (G, E, D) if the advantage of the adversary in the standard semantic-security game is at least γ. See the full version for formal definitions.

Oracle aided circuits. A binary-output oracle-aided circuit C is a circuit with Boolean gates as well as oracle gates, and where the output of the circuit is a single bit. The input size, inpsize(C), is the number of input wires. The circuit size, denoted |C|, denotes the number of gates and input wires of the circuit. For a fixed function f we write C^f to denote the circuit C when the underlying oracle is fixed to f.

Definition 6 (Garbling schemes for oracle-aided circuits). *Fix a function f. A circuit garbling scheme for oracle-aided circuits relative to f (or with f gates) is a triple of algorithms* (Garb, Eval, Sim) *defined as follows:*

- Garb $(1^\kappa, C)$: *takes as input a security parameter κ, an oracle-aided circuit C and outputs a garbled circuit $\widetilde{C}$ with a set of labels* $\{\mathsf{label}_{i,b}\}_{i\in[m], b\in\{0,1\}}$, *where $m = $ inpsize(C).*
- Eval^f $\left(\widetilde{C}, \{\mathsf{label}_{i,b_i}\}_{i\in[m]}\right)$: *takes as input a garbled circuit $\widetilde{C}$ and a sequence of garbled input labels* $\{\mathsf{label}_{i,b_i}\}_{i\in[m]}$ *and outputs $y \in \{0,1\}^* \cup \{\bot\}$.*

  We define the following notions.

- **Correctness.** *For any oracle-aided circuit C and input $x \in \{0,1\}^m$, where $m = $ inpsize(C):*

$$\Pr\left[C^f(x) = \mathsf{Eval}^f\left(\widetilde{C}, \{\mathsf{label}_{i,x_i}\}_{i\in[m]}\right)\right] = 1$$

  where the probability is taken over Garb $(1^\kappa, C) \mapsto (\widetilde{C}, \{\mathsf{label}_{i,b}\}_{i\in[m], b\in\{0,1\}})$.
- **Security.** *For any polynomial $m = m(\kappa)$, any poly-size oracle circuit C with input size m, and any input $x \in \{0,1\}^m$:*

$$\left(\widetilde{C}, \{\mathsf{label}_{i,x_i}\}_{i\in[m]}\right) \stackrel{c}{\approx} \mathsf{Sim}\left(1^{|C|}, m, C^f(x)\right)$$

  where $(\widetilde{C}, \{\mathsf{label}_{i,b}\}_{i\in[m], b\in\{0,1\}}) \leftarrow$ Garb $(1^\kappa, C)$.

3.3   Black-Box Constructions

Now, we recall the standard notion of black-box constructions [4,23,32]. We do so in the context of building PKE from one-way functions and garbling.

Definition 7 (Black-box constructions of PKE from GC-OWF). *A fully black-box construction of a PKE scheme from a one-way function and a garbling scheme for circuits with one-way function gates (shortly, from GC-OWF) consists of a triple of PPT oracle algorithms (G, E, D) and a PPT oracle security-reduction $S = (S_1, S_2)$ such that for any function f and any correct garbling scheme $L = $ (Garb, Eval, Sim) relative to f, both the following hold:*

- **Correctness:** $\mathcal{E}^{f,L} = (G^{f,L}, E^{f,L}, D^{f,L})$ *is a* $(1 - \frac{1}{2^\kappa})$-*correct PKE scheme.*
  (See the remark after this definition.)
- **Security:** *For any adversary any A that breaks the semantic security of the PKE scheme* $\mathcal{E}^{f,L}$, *either*
  - $S_1^{f,L,A}$ *breaks the one-wayness of* f; *or*
  - $S_2^{f,L,A}$ *breaks the security of the scheme* $L = (\mathsf{Garb}, \mathsf{Eval}, \mathsf{Sim})$ *relative to* f. *That is, for some oracle-aided circuit* C *and input* x, $S_2^{f,L,A}$ *can distinguish between the tuple* $\left(\widetilde{\mathsf{C}}, \{\mathsf{label}_{i,x_i}\}_{i\in[m]}\right)$ *and* $\mathsf{Sim}\left(1^{|\mathsf{C}|}, 1^{|x|}, \mathsf{C}^f(x)\right)$, *where* $m = \mathsf{inpsize}(\mathsf{C})$ *and* $(\widetilde{\mathsf{C}}, \{\mathsf{label}_{i,b}\}_{i\in[m],b\in\{0,1\}}) \leftarrow \mathsf{Garb}\,(1^\kappa, \mathsf{C})$.

Remark about the correctness condition in Definition 7. In Definition 7, for correctness we require that the constructed PKE be $(1 - \frac{1}{2^\kappa})$ correct. This is without loss of generality since one may easily boost correctness using standard techniques; i.e., let the new public key be a tuple of public keys under the original scheme. Encrypt a given plaintext bit under each individual public key. For decryption, we decrypt all the ciphertexts and go with the majority bit. The semantic security of this expanded scheme reduces to that of the base scheme using a hybrid argument, which is a fully-black-box reduction.

Calling the base primitives on the same security parameter. For simplicity of exposition, for any given black-box construction $\mathcal{E}^{f,L}$ we assume that $\mathcal{E}^{f,L}$ on the security parameter 1^κ always calls f and L on the same security parameter 1^κ. There are standard techniques for doing away with this restriction, but those extensions will only complicate the proofs further. Looking ahead, when we define our oracles $(O', \mathsf{rev})_{\kappa,n}$ in Definition 10, which are parameterized over a security parameter κ and a circuit size $n = n(\kappa)$, the above restriction means that $\mathcal{E}^{O'}$ on the security parameter 1^κ always calls O' on parameters such as (κ, n_1), (κ, n_2), etc. That is, the value of κ will be the same across all queries, but each query may use a different value for n.

4   Separating Public-Key Encryption from OWF-Based Garbling

In this section, we state our main impossibility result and describe at a high-level the steps that we will take in order to prove our main theorem.

Theorem 8 (Main theorem). *There exists no fully black-box construction of a public-key encryption scheme from GC-OWFs; namely garbling schemes that garble circuits with one-way function gates in them (see Definition 7).*

Our theorem above follows from the following lemma.

Lemma 9. *There exists an oracle* $O = (f, \mathsf{gc}, \mathsf{gi}, \mathsf{eval}, \mathsf{rev})$ *for which the following holds (in what follows, let* $O' = (f, \mathsf{gc}, \mathsf{gi}, \mathsf{eval})$):

1. f is one-way relative to (O', rev). That is, f is one-way against all polynomial query (and even sub-exponential query) adversaries $\mathcal{A}^{O', \mathsf{rev}}$.

2. There exists a PPT GC-OWF construction $(\mathsf{Garb}^{O'}, \mathsf{Eval}^{O'}, \mathsf{Sim}^{O'})$ for f-aided circuits that is secure against any poly-query adversary $\mathcal{A}^{O', \mathsf{rev}}$.

3. For any PKE construction $\mathcal{E}^{O'}$ with access to the oracle O', there exists an attacker $\mathcal{A}^{O', \mathsf{rev}}$ that breaks the semantic security of $\mathcal{E}^{O'}$ using a polynomial number of queries.

Note that Lemma 9 immediately implies Theorem 8.

Roadmap: Proof of Lemma 9. As common in black-box impossibility results, we will show the existence of the oracles required by Lemma 9 by proving results with respect to oracles chosen randomly according to a distribution. We will describe our oracle distribution below and will then outline the main steps we will take in order to prove Lemma 9.

Definition 10 (The ideal model/oracle). Let $O = (f, \mathsf{gc}, \mathsf{gi}, \mathsf{eval}, \mathsf{rev})_{\kappa, n}$ be an ensemble of oracles parameterized by (κ, n), where κ denotes the security parameter and n denotes the size of a circuit which we want to garble. We describe the distribution $\mathbf{O}$ from which these oracles are sampled for fixed (κ, n).

- $f : \{0, 1\}^\kappa \to \{0, 1\}^\kappa$: a uniformly chosen random function.
- $\mathsf{gc}(s, F) : \{0, 1\}^\kappa \times \{0, 1\}^n \to \{0, 1\}^{2(\kappa + n)}$: an injective random function that, given a key $s \in \{0, 1\}^\kappa$ and a single-bit-output oracle-aided circuit F, outputs an encoding $\widetilde{F}$.
- $\mathsf{gi}(s, i, x_i) : \{0, 1\}^\kappa \times \{0, 1\}^{\log n} \times \{0, 1\} \to \{0, 1\}^{2(\kappa + \log n)}$: an injective random function that, given a key $s \in \{0, 1\}^\kappa$, an index $i \in \{0, 1\}^{\log n}$, an input-wire bit value $x_i \in \{0, 1\}$, outputs an encoding $\widetilde{x}_i$. As notation, for any $X = (x_1, \dots, x_n)$, we denote $\mathsf{gi}(s, X) := (\mathsf{gi}(s, i, x_i))_{i \in [n]} = \widetilde{X}$.
- $\mathsf{eval}(\widetilde{F}, \widetilde{X})$: given as input $\widetilde{F}$ and $\widetilde{X} = (\widetilde{x}_1, \dots, \widetilde{x}_m)$, if there is a string $s \in \{0, 1\}^\kappa$ and circuit F such that $\mathsf{gc}(s, F) = \widetilde{F}$, that $m = \mathsf{inpsize}(F)$ and that for every $i \in [m]$ there exists $x_i \in \{0, 1\}$ such that $\mathsf{gi}(s, i, x_i) = \widetilde{x}_i$, then it outputs $F^f(x_1 || \cdots || x_m)$. Otherwise, it outputs $\perp$.
- $\mathsf{rev}(\widetilde{F}, \widetilde{X}, \widetilde{X}')$: if there exists $s \in \{0, 1\}^\kappa$ and circuit F such that $\mathsf{gc}(s, F) = \widetilde{F}$ and that there exists $X, X' \in \{0, 1\}^{\mathsf{inpsize}(F)}$ such that $X \neq X'$, $\mathsf{gi}(s, X) = \widetilde{X}$ and $\mathsf{gi}(s, X') = \widetilde{X}'$, then it outputs (s, F). Otherwise, it outputs $\perp$.

Remark 11. The size of a garbled circuit outputted by the gc oracle is roughly twice the size of the corresponding input circuit. Current garbled circuits constructions are not capable of achieving such a short expansion factor. We are able to do this as we model the garbling mechanism as a totally random function. Nonetheless, working with such a short size expansion is without loss of generality, because a general black-box PKE construction out of GC-OWF should work with respect to any oracle that implements the GC-OWF securely. We should also mention that all our results hold (without having to make any changes) if the output of gc is bigger than the one specified in Definition 10.

First, we show that a random oracle $O = (f, \mathsf{gc}, \mathsf{gi}, \mathsf{eval}, \mathsf{rev})$ chosen according to the distribution $\mathbf{O}$ allows us to implement an ideal version of garbling for circuits with f gates. This is not surprising as O is indeed an idealized form of implementing this primitive.

Lemma 12 (Secure OWF and garbling exists relative to O). *Let $O = (f, \mathsf{gc}, \mathsf{gi}, \mathsf{eval}, \mathsf{rev})$ be as in Definition 10 and let $O' = (f, \mathsf{gc}, \mathsf{gi}, \mathsf{eval})$. Then, with probability (measure) one over the choice of O, the function f is one-way relative to O —i.e., f is one-way against any PPT oracle adversary with access to the oracle O. Moreover, there exists a PPT GC-OWF construction $(\mathsf{Garb}^{O'}, \mathsf{Eval}^{O'})$ for f-aided circuits which is secure relative to O with probability one over the choice of $O \leftarrow \mathbf{O}$.*

Proof. The fact that f is one-way relative to O with probability one over the choice of O is now standard (see [23]). Given any oracle $O = (O', \mathsf{eval})$, we now show how to construct a PPT garbling scheme $L^{O'} = (\mathsf{Garb}^{O'}, \mathsf{Eval}^{O'}, \mathsf{Sim}^{O'})$ for f-aided circuits. The algorithm $\mathsf{Garb}^{O'}$ on input $(1^\kappa, C)$ samples $s \leftarrow \{0,1\}^\kappa$, sets $m = \mathsf{inpsize}(C)$ and outputs the garbled circuit $\widetilde{C} = \mathsf{gc}(s, C)$ as well as a sequence of garbled inputs $(\widetilde{x}_{1,0}, \widetilde{x}_{1,1}, \ldots, \widetilde{x}_{m,0}, \widetilde{x}_{m,1})$, where for $i \in m$ and $b \in \{0,1\}$ we have $\widetilde{x}_{i,b} = \mathsf{gi}(s, i, b)$.

The algorithm $\mathsf{Eval}^{O'}(\widetilde{C}, \widetilde{x}_1 || \cdots || \widetilde{x}_m)$ simply outputs $\mathsf{eval}(\widetilde{C}, \widetilde{x}_1 || \cdots || \widetilde{x}_m)$. Correctness holds by definition of the oracle.

For security, we will define $\mathsf{Sim}^{O'}$ as follows: on input $(1^\kappa, n, m, y \in \{0,1\})$, where n denotes the size of the circuit, m denotes the number of input wires and y denotes the output value, we set C_0 to be a canonical circuit of size n and with m input wires that always outputs y. Sample $s \leftarrow \{0,1\}^\kappa$ and let $\widetilde{C} = \mathsf{gc}(s, C)$ and $\widetilde{X} = \mathsf{gi}(s, 0^m)$. Output $(\widetilde{C}, \widetilde{X})$. Simulation security follows from the random nature of the oracles. That is, for any polynomial-query distinguisher $\mathcal{A}^{O', \mathsf{rev}}$, for any n, m and any circuit C of size n and of input size m and any input $X \in \{0,1\}^m$, we have

$$\left| \Pr[\mathcal{A}^{O', \mathsf{rev}}(\widetilde{C}, \widetilde{X}) = 1] - \Pr[\mathcal{A}^{O', \mathsf{rev}}(\widetilde{C}', \widetilde{X}') = 1] \right| = \mathsf{negl}(\kappa), \qquad (1)$$

where $s \leftarrow \{0,1\}^\kappa$, $\widetilde{C} = \mathsf{gc}(s, C), \widetilde{X} = \mathsf{gi}(s, X), (\widetilde{C}', \widetilde{X}') \leftarrow \mathsf{Sim}^{O'}(1^\kappa, n, m, C^f(X))$. We omit the details of the proof of Eq. 1 as it can be obtained through a simple information theoretic argument.

We are left with proving Part 3 of Lemma 9. Proving this part is the main technical contribution of our paper, and is done via an oracle *reducibility* technique. In order to state this reducibility statement formally, we first need to define the notions of correctness and attack advantage in the *ideal model*.

Definition 13 (Correctness in the ideal model). *For a polynomial $p = p(\kappa)$ we say that a single-bit PKE scheme $\mathcal{E}^O = (G^O, E^O, D^O)$ is $\frac{1}{2} + \frac{1}{p}$ correct in the ideal model if for both $b \in \{0,1\}$:*

$$\Pr[D^O(sk, c) = b] \geq \frac{1}{2} + \frac{1}{p}, \qquad (2)$$

where the probability is over $O \leftarrow \mathbf{O}$, $(pk, sk) \leftarrow G^O(1^\kappa)$, $c \leftarrow E^O(pk, b)$.

Definition 14 (Ideal model attack advantage). *We say that an adversary $\mathcal{A}$ breaks the semantic security of a single-bit PKE (G^O, E^O, D^O) in the ideal model with probability γ (or with advantage γ) if $\Pr[A(pk, c) = b] \geq \gamma$, where the probability is taken over $O \leftarrow \mathbf{O}$, $(pk, sk) \leftarrow G^O(1^\kappa)$, $b \leftarrow \{0, 1\}$, $c \leftarrow E^O(pk, b)$ and over $\mathcal{A}$'s random coins.*

We are now ready to describe our oracle reducibility lemma.

Lemma 15 (Reducibility to the random oracle model). *Let $\mathcal{E}$ be a given PKE construction possibly making use of all the oracles $O' = (f, \mathsf{gc}, \mathsf{gi}, \mathsf{eval})$. There exists a compilation procedure and a polynomial-query security-reduction Red such that the compilation transforms $\mathcal{E}^{O'}$ into a new polynomial-query PKE construction $\ddot{\mathcal{E}}^{f, \mathsf{gc}, \mathsf{gi}}$, where $\ddot{\mathcal{E}}$ makes no eval queries and for which both the following hold:*

- *Correctness: If $\mathcal{E}^{O'}$ is $(1 - \frac{1}{2^\kappa})$ correct in the ideal model, the compiled scheme $\ddot{\mathcal{E}}^{f, \mathsf{gc}, \mathsf{gi}}$ has at least $(1 - \frac{1}{\kappa^7})$ correctness in the ideal model.*
- *Security reduction. For any constant c the following holds: if there exists an adversary $\mathcal{A}$ that breaks the semantic security of $\ddot{\mathcal{E}}^{f, \mathsf{gc}, \mathsf{gi}}$ in the ideal model with probability η, the algorithm $\mathsf{Red}^{O', \mathsf{rev}, \mathcal{A}}$ breaks the semantic security of $\mathcal{E}^{O'}$ in the ideal model with probability $\eta - \frac{1}{\kappa^c}$.*

Let us first show how to use Lemmas 12 and 15 to establish Lemma 9.

Completing proof of Lemma 9 and Theorem 8. Let $\mathcal{E} = (G, E, D)$ be a candidate PKE construction. We will show that with probability one over the choice of $(O', \mathsf{rev}) \leftarrow \mathbf{O}$, the PKE construction $\mathcal{E}^{O'}$ can be broken by a polynomial number of queries to (O', rev). Let us first show how to use this claim to complete the proof of Theorem 8, and we will then prove this claim. By Lemma 12, we know that with probability one over the choice of O we have (a) f is one-way relative to (O', rev) and (b) $(\mathsf{Garb}^{O'}, \mathsf{Eval}^{O'}, \mathsf{Sim}^{O'})$ is a secure GC-OWF construction for f-aided circuits against all polynomial-query adversaries with access to the oracles (O', rev). Thus, the foregoing claim coupled with Lemma 12 implies Lemma 9. In what follows we prove the foregoing claim.

By Definition 7 we know that $\mathcal{E}^{O'}$ has $(1 - \frac{1}{2^\kappa})$-correctness in the ideal model. Thus, by Lemma 15 there exists a compiled scheme $\ddot{\mathcal{E}}^{f, \mathsf{gc}, \mathsf{gi}}$ that has at least $(1 - \frac{1}{\kappa^7})$-correctness in the ideal model. Note that the oracles f, gc and gi are nothing but three independent random oracles. By the results of [7,23] there exists a polynomial query adversary $\mathcal{A}^{f, \mathsf{gc}, \mathsf{gi}}$ which breaks the semantic security of $\ddot{\mathcal{E}}^{f, \mathsf{gc}, \mathsf{gi}}$ in the ideal model with probability $(1 - \frac{1}{\kappa^6})$.[5] (See Definition 14 for the notion of "break in the ideal model".) Invoking Lemma 15 again and choosing the constant c appropriately, we will obtain a polynomial query adversary $\mathcal{B}^{O', \mathsf{rev}}$

---

[5] The results of [7,23] show how to break the semantic security of any key exchange (and hence PKE) construction in the random oracle model with a probability that is at most $\frac{1}{\kappa^{c'}}$ less than the correctness probability, for any arbitrary constant $c' > 0$.

which breaks the semantic security of $\mathcal{E}^{O'}$ in the ideal model with probability $(1 - \frac{1}{\kappa^5})$. That is,

$$\Pr_{O=(O',\text{rev}),pk,b,c} \left[\mathcal{B}^{O',\text{rev}}(pk,c) = b \right] \geq 1 - \frac{1}{\kappa^5}, \tag{3}$$

where $(pk, sk) \leftarrow G^{O'}(1^\kappa)$, $b \leftarrow \{0,1\}$ and $c \leftarrow E^{O'}(pk, b)$.

Using a simple averaging argument we have

$$\Pr_{O=(O',\text{rev})} \left[\Pr_{pk,b,c} \left[\mathcal{B}^{O',\text{rev}}(pk,c) = b \right] \geq 1 - \frac{1}{\kappa^3} \right] \geq 1 - \frac{1}{\kappa^2}. \tag{4}$$

Equation 4 implies that for at most $\frac{1}{\kappa^2}$ fraction of the oracles $O = (O', \text{rev})$, the adversary $\mathcal{B}^{O',\text{rev}}(pk,c)$, on security parameter κ, recovers b with probability less than $1 - \frac{1}{\kappa^3}$. Since $\sum_{i=1}^{\infty} \frac{1}{i^2}$ converges, by the Borel-Cantelli Lemma we have that for a measure-one fraction of oracles $O = (O', \text{rev}) \leftarrow \mathbf{O}$, the adversary $\mathcal{B}^{O',\text{rev}}$ breaks the semantic security of $\mathcal{E}^{O'}$. The proof is now complete. $\square$

Roadmap for the proof of Lemma 15. Finally, all that remains is proving Lemma 15 which shows that we can compile out eval queries from any PKE scheme without significantly hurting correctness or security. In the remainder of this paper, we show that such a compilation procedure exists. We obtain the compiled eval-free scheme $(\ddot{G}^{f,\text{gc},\text{gi}}, \ddot{E}^{f,\text{gc},\text{gi}}, \ddot{D}^{f,\text{gc},\text{gi}})$ in two steps. First, in Sect. 4.1, we show how to compile out eval queries from $E^{O'}$ only. In particular, we will prove the following lemma.

Lemma 16 (Compiling out eval from E). *Let δ be an arbitrary polynomial and parse $O = (O', \text{rev})$. There exists a compilation procedure that achieves the following for any constant c. Given any $(\frac{1}{2} + \delta)$-ideally-correct PKE scheme $\mathcal{E} = (G^{O'}, E^{O'}, D^{O'})$, the compiled PKE scheme $\dot{\mathcal{E}} = (\dot{G}^{f,\text{gc},\text{gi}}, \dot{E}^{f,\text{gc},\text{gi}}, D^{O'})$ is $(\frac{1}{2} + \delta - \frac{1}{\kappa^c})$-ideally-correct. Moreover, there exists a polynomial-query algorithm SecRed that satisfies the following: for any adversary $\mathcal{A}$ that breaks the semantic security of $\dot{\mathcal{E}}$ in the ideal model with advantage η, the adversary SecRed$^{\mathcal{A},O}$ breaks the semantic security of $\mathcal{E}$ in the ideal model with advantage at least $\eta - \frac{1}{\kappa^c}$.*

Then, in Sect. 4.2 we show how to compile out eval from $D^{O'}$, assuming neither of the algorithms G and E call eval. That is, we prove the following.

Lemma 17 (Compiling out eval from D). *Let δ be an arbitrary polynomial. There exists a compilation procedure that achieves the following for any constant c. Given any $(\frac{1}{2} + \delta)$-ideally-correct PKE scheme $\mathcal{E} = (G^{f,\text{gc},\text{gi}}, E^{f,\text{gc},\text{gi}}, D^{f,\text{gc},\text{gi},\text{eval}})$, the compiled PKE scheme $\ddot{\mathcal{E}} = (\ddot{G}^{f,\text{gc},\text{gi}}, \ddot{E}^{f,\text{gc},\text{gi}}, \ddot{D}^{f,\text{gc},\text{gi}})$ is $(\frac{1}{2} + \delta - \frac{1}{\kappa^c})$-ideally-correct. Moreover, there exists a polynomial-query algorithm SecRed that satisfies the following: for any adversary $\mathcal{A}$ that breaks the semantic security of $\ddot{\mathcal{E}}$ in the ideal model with advantage η, the adversary SecRed$^{\mathcal{A},O}$ breaks the semantic security of $\mathcal{E}$ in the ideal model with advantage at least $\eta - \frac{1}{\kappa^c}$.*

Th proof of Lemma 15 immediately follows from Lemmas 16 and 17.

4.1   Removing Garbling Evaluation Queries from Encryption

In this section, we will prove Lemma 16. Namely, we will show how to compile the PKE scheme $\mathcal{E} = (G^{f,\text{gc},\text{gi},\text{eval}}, E^{f,\text{gc},\text{gi},\text{eval}}, D^{f,\text{gc},\text{gi},\text{eval}})$ into a new PKE scheme $\dot{\mathcal{E}} = (\dot{G}^{f,\text{gc},\text{gi}}, \dot{E}^{f,\text{gc},\text{gi}}, D^{f,\text{gc},\text{gi},\text{eval}})$ with correctness and security comparable to the original scheme $\mathcal{E}$, but where $\dot{E}$ will not ask any eval queries. First, we may assume without loss of generality that G does not make queries to eval—it can predict the answer itself. Thus, we will focus on removing eval queries from E assuming that G does not make any eval queries.

Before describing the compilation process, we need to give some definitions.

Definition 18 (Valid outputs). *For any oracle* $O = (f, \text{gc}, \text{gi}, \text{eval}, \text{rev})$, *we say that* $\widetilde{F}$ *is a valid garbled circuit with respect to* O *if there exists* (s, F) *such that* $\text{gc}(s, F) = \widetilde{F}$. *Similarly, we say that* $\widetilde{X}$ *is a valid garbled input with respect to* O *if there exists* (s, X) *such that* $\text{gi}(s, X) = \widetilde{X}$.

We also define the notion of normal form with respect to oracle-aided algorithms. At a high-level, a normal form algorithm avoids asking any redundant queries if it already knows the answer to such queries.

Definition 19 (Normal form). *Let* A *be an oracle algorithm that accepts as input a query-answer set* $\mathsf{Q_S}$ *and let* $\mathsf{Q_A}$ *be the query-answer pairs that* A *has asked so far. We say that* A *is in* normal-form *if it satisfies these conditions:*

1. *A never asks duplicate queries.*
2. *Before it asks an* $((\widetilde{F}, \widetilde{X}) \xrightarrow{\text{eval}} ?)$ *query* qu, A *first checks if there exists a query-answer pair* $((s, F) \xrightarrow{\text{gc}} \widetilde{F})$ *in* $\mathsf{Q_A} \cup \mathsf{Q_S}$. *If that is the case then it would not issue* qu *to the oracle but would instead run* $F^f(X)$ *on its own where* X *can be obtained bit-by-bit by searching* $\text{gi}(s, i, x_i)$ *for every index position* $i \in n$ *and every bit* $x_i \in \{0, 1\}$.

Recall that our goal is to remove eval queries from E to obtain an eval-free algorithm $\dot{E}$. To make this transformation possible, the new algorithm $\dot{E}$ needs some help from its associated key generation algorithm $\dot{G}$ so as to make up for its lack of access to eval. This help is sent to $\dot{E}$ as part of a hint list H, attached to the public key, by the key generation algorithm $\dot{G}$. The following definition describes how $\dot{G}$ forms the hint list H based on its inside information Aux and based on information Q that G has collected about random executions of E.

Definition 20 (Building helper tuples). *We define a function* ConstHelp *that takes as input a query-answer set* Q *along with some query-answer set* Aux *and outputs a set* H *as follows:*

– *If there exists* $((\widetilde{F}, \widetilde{X}) \xrightarrow{\text{eval}} y \neq \bot) \in \mathsf{Q}$ *such that for no* $\widetilde{X}' \neq \widetilde{X}$ *do we have* $((\widetilde{F}, \widetilde{X}') \xrightarrow{\text{eval}} y' \neq \bot) \in \mathsf{Q}$, *then add* $((\widetilde{F}, \widetilde{X}) \xrightarrow{\text{eval}} y)$ *to* H.

- *If for two distinct $\widetilde{X}_1$ and $\widetilde{X}_2$ we have $((\widetilde{F}, \widetilde{X}_1) \xrightarrow[\text{eval}]{} y_1 \neq \bot) \in Q$ and $((\widetilde{F}, \widetilde{X}_2) \xrightarrow[\text{eval}]{} y_2 \neq \bot) \in Q$, then if for some (s, F) we have $((s, F) \xrightarrow[\text{gc}]{} \widetilde{F}) \in$ Aux, add $((s, F) \xrightarrow[\text{gc}]{} \widetilde{F})$ to H.*

Having a hint list H, we give the following definition that describes the idea of how the receiving algorithm $\dot{E}$ may use it to avoid making eval queries. In the following definition one may think of Q as a hint list.

Definition 21 (Emulating eval queries). *For $O = (f, \text{gc}, \text{gi}, \text{eval}, \text{rev})$, we define the function $\texttt{HandleEval}^{f,\text{gi}}$ to be a subroutine that takes as input a set Q of query-answer pairs to O and a query qu of the form $((\widetilde{F}, \widetilde{X}) \xrightarrow[\text{eval}]{} ?)$ then performs the following steps to answer qu:*

- *If there exists a tuple $((\widetilde{F}, \widetilde{X}) \xrightarrow[\text{eval}]{} y)$ in Q, then output y.*
- *If there exists $((s, F) \xrightarrow[\text{gc}]{} \widetilde{F}) \in Q$, then find X such that $\text{gi}(s, X) = \widetilde{X}$ and output $y = F^f(X)$.*
- *If neither of the above cases happen, then return $\bot$ as the answer to qu.*

We will also define the notion of a mixed oracle that uses O on non-eval queries but uses $\texttt{HandleEval}$ to answer eval queries without resorting to O. This oracle is constructed and used in the newly compiled algorithms when we want to avoid asking eval queries to O.

Definition 22 (Mixed oracle). *For an oracle $O = (f, \text{gc}, \text{gi}, \text{eval})$ and a set of query-answer pairs S, we denote $O[S]$ to be an Eval-mixed oracle that answers all f, gc, and gi queries by forwarding them to the real oracle O, but for any eval query qu it will emulate the answer by calling and returning $y = \texttt{HandleEval}^{f,\text{gi}}(S, qu)$.*

Compilation procedure. Let $\mathcal{E} = (G^{f,\text{gc},\text{gi}}, E^{f,\text{gc},\text{gi},\text{eval}}, D^{f,\text{gc},\text{gi},\text{eval}})$ be the give construction for which we want to remove eval queries from E. Without loss of generality, we assume that all the algorithms of $\mathcal{E}$ are in normal form (see Definition 19). For simplicity, we keep O as a superscript to all the algorithms of $\mathcal{E}$, but it is understood that the actual oracle access is of the form above.

We need the following definition as we will need to choose parameters in the compilation construction based on the query complexity of the construction.

Definition 23 (Parameter $q = q(\kappa)$: size-upperbound). *Throughout this section, fix $q = q(\kappa)$ to be an arbitrary polynomial that satisfies the following.*

1. *$q \geq \kappa$;*
2. *q is greater than the total number of queries that each of the algorithms (G^O, E^O, D^O) make on inputs corresponding to the security parameter 1^κ and on $O \leftarrow \mathbf{O}$; and*

3. *q is greater than the size of any query made by any of* (G^O, E^O, D^O) *on inputs corresponding to the security parameter* 1^κ *and on* $O \leftarrow \mathbf{O}$.

Construction 24 (compiled scheme $\dot{\mathcal{E}}$). *The compiled scheme* $(\dot{G}, \dot{E}, D)$ *is parameterized over a function* $t = t(\kappa)$, *which we will instantiate later.*

- $\dot{G}(1^\kappa)$: *Perform the following steps:*
    1. *Run* $(pk, sk) \leftarrow G^O(1^\kappa)$. *Add all query-answer pairs generated in this step to* OrigG.
    2. **Generating helper set** H *for* $\dot{E}$: *Set* LocalE $= \varnothing$.
        (a) *Do the following t times: Run* $E^{O[\mathsf{OrigG}]}(pk, 0)$ *and* $E^{O[\mathsf{OrigG}]}(pk, 1)$ *and keep adding all the resulting query-answer pairs to* LocalE.
        (b) *Set* H $:=$ ConstHelp(LocalE, OrigG $\cup$ LocalE).
    3. *Output* $\dot{pk} = (pk, \mathsf{H})$ *and* $\dot{sk} = sk$.
- $\dot{E}(\dot{pk}, b)$: *Parse* $\dot{pk} = (pk, \mathsf{H})$. *Run* $\dot{c} \leftarrow E^{O[\mathsf{H}]}(pk, b)$ *and add all the query-response pairs to* OrigE. *Return* $\dot{c}$.

Remark about $\dot{E}$. We note that, by the definition of $O[\mathsf{H}]$, all the eval queries of $E^{O[\mathsf{H}]}(pk, b)$ will be emulated using H. Thus, $\dot{E}$ will not issue any eval queries.

Query complexity of $\dot{\mathcal{E}}$. It is immediate to see that the query complexity of each of the compiled algorithms is polynomial in q and t, where q the query complexity of (G, E, D). Thus, we have the following lemma.

Lemma 25. *Let q be the size-upperbound of (G, E, D) as given in Definition 23. The query complexity of $\dot{\mathcal{E}} = (\dot{G}, \dot{E}, D)$ is at most* $q + (2q^2)t \leq 3tq^2$.

Correctness and security. We now give the correctness and security statements regarding the compiled scheme $\dot{\mathcal{E}} = (\dot{G}, \dot{E}, D)$ and prove them. By doing so, we complete the proof of Lemma 16.

Lemma 26 (Correctness of $\dot{\mathcal{E}}$). *Suppose the original scheme (G, E, D) is $(\frac{1}{2} + \delta)$ correct in the ideal model. The compiled scheme $(\dot{G}^O, \dot{E}^O, D^O)$ has at least $(\frac{1}{2} + \delta - \frac{2q}{t} - \mathrm{negl}(\kappa))$ correctness in the ideal model, where t is the number of iterations performed in $\dot{G}$.*

In particular, for any constant $c > 0$ by taking $t = q^{c+2}$, the compiled scheme $(\dot{G}^O, \dot{E}^O, D^O)$ has at least $(\frac{1}{2} + \delta - \frac{1}{\kappa^c})$ correctness.

Lemma 27 (Security of $\dot{\mathcal{E}}$). *There exists a polynomial-query algorithm* SecRed *that satisfies the following. For any adversary $\mathcal{A}$ that breaks the semantic security of $(\dot{G}^O, \dot{E}^O, \dot{D}^O)$ in the ideal model with probability at least γ, the algorithm* SecRed$^{\mathcal{A}, O}$ *breaks the semantic security of (G^O, E^O, D^O) with probability at least $\gamma - \frac{1}{2^{\kappa/4}} - \frac{1}{\kappa^c}$ for any constant $c > 0$.*

Proof of correctness for $\dot{\mathcal{E}}$. We first prove Lemma 26, which states that $\dot{\mathcal{E}} = (\dot{G}, \dot{E}, D)$ is still a correct PKE after removing the eval queries from E.

Parsing $\dot{pk} = (pk, \mathsf{H})$, recall that $\dot{E}^O(\dot{pk}, b)$ simply runs $E^{O[\mathsf{H}]}(pk, b)$. With this in mind, to prove Lemma 26, we give the following lemma, which shows that the distribution of outputs of $E^{O[\mathsf{H}]}(pk, b)$ and $E^O(pk, b)$ are close.

Lemma 28. *For $b \in \{0, 1\}$ we have*

$$\Pr_{O, r, pk, \mathsf{H}}[E^O(pk, b; r) \neq E^{O[\mathsf{H}]}(pk, b; r)] \leq \frac{2q}{t} + \frac{1}{2^{\kappa/3}}$$

where $O \leftarrow \mathbf{O}$, $((pk, \mathsf{H}), sk) \leftarrow \dot{G}^O(1^\kappa)$ and $r \leftarrow \{0, 1\}^*$.

We first show how to derive Lemma 26 from Lemma 28.

Proof (of Lemma 26). Parse $\dot{pk} = (pk, \mathsf{H})$. All the probabilities below are taken over the random choices of $\dot{pk}$, O and r. We have

$$\Pr[D^O(sk, \dot{E}^O(\dot{pk}, b; r)) \neq b] = \Pr[D^O(sk, E^{O[\mathsf{H}]}(pk, b; r)) \neq b]$$
$$\leq \Pr[D^O(sk, E^O(pk, b; r)) \neq b] + 2q/t + \mathrm{negl}(\kappa)$$
$$\leq \frac{1}{2} - \delta + 2q/t + \mathrm{negl}(\kappa)$$

where the first inequality follows from Lemma 28.   □

We now focus on proving Lemma 28. Fix $b \in \{0, 1\}$. For compactness, we define the following experiment that outputs some random variables that will be later used to define some events.

Experiment $\mathbf{Expr}(1^\kappa)$ for fixed $b \in \{0, 1\}$: Output the tuple of variables $\mathsf{Vars} = (pk, \mathsf{OrigG}, \mathsf{LocalE}, \mathsf{H}, r)$, where $pk, \mathsf{OrigG}, \mathsf{LocalE}$ and H are sampled as in $\dot{G}(1^\kappa)$ and $r \leftarrow \{0, 1\}^*$ is the randomness to $\dot{E}(pk, b)$.

We define the following bad events. Note that all these bad events as well as those that appear later are defined based on the output of Vars, and so we make this dependence implicit henceforth.

- Bad_1: The event that $E^O(pk, b; r)$ makes a query $qu = ((\widetilde{F}, \widetilde{X}) \xrightarrow[\text{eval}]{} ?)$, where $((*, *) \xrightarrow[\text{gc}]{} \widetilde{F}) \notin \mathsf{OrigG}$ and $\mathsf{eval}(\widetilde{F}, \widetilde{X}) \neq \bot$.

- Bad_2: The event that the execution of $E^O(pk, b; r)$ queries $qu = ((\widetilde{F}, \widetilde{X}) \xrightarrow[\text{eval}]{} ?)$ for which we have $((*, *) \xrightarrow[\text{gc}]{} \widetilde{F}) \in \mathsf{OrigG}$, $\mathsf{eval}(\widetilde{F}, \widetilde{X}) \neq \bot$ and $O[\mathsf{H}](qu) = \mathtt{HandleEval}(\mathsf{H}, qu) = \bot$.

Roadmap for the proof of Lemma 28. The proof of Lemma 28 now follows from the following lemmas.

Lemma 29. $\Pr_{O,\text{Vars}}[E^O(pk, b; r) \neq E^{O[\mathsf{H}]}(pk, b; r)] \leq \Pr[\mathsf{Bad}_1 \vee \mathsf{Bad}_2]$ *where* $O \leftarrow \mathbf{O}$ *and* $\text{Vars} = (pk, \text{OrigG}, \text{LocalE}, \mathsf{H}, r) \leftarrow \mathbf{Expr}(1^\kappa)$.

Lemma 30. $\Pr_{O,\text{Vars}}[\mathsf{Bad}_1] \leq \frac{1}{2^{\kappa/3}}$ *where* $O \leftarrow \mathbf{O}$ *and* $\text{Vars} \leftarrow \mathbf{Expr}(1^\kappa)$.

Lemma 31. $\Pr_{O,\text{Vars}}[\mathsf{Bad}_2 \wedge \overline{\mathsf{Bad}_1}] \leq \frac{2q}{t}$ *where* $O \leftarrow \mathbf{O}$ *and* $\text{Vars} \leftarrow \mathbf{Expr}(1^\kappa)$

The proof of Lemma 28 follows immediately from Lemmas 29, 30, and 31. We now prove all these lemmas below.

Proof (of Lemma 29). Let Bad be the event $E^O(pk, b; r) \neq E^{O[\mathsf{H}]}(pk, b; r)$. We show that whenever Bad holds, then either Bad_1 happens or Bad_2 happens, hence proving the lemma. Notice that the only difference between the executions of $E^O(pk, b; r)$ and $E^{O[\mathsf{H}]}(pk, b; r)$ is how eval queries are handled. Specifically, in $E^{O[\mathsf{H}]}(pk, b; r)$, the eval queries are simulated with respect to the set H whereas in $E^O(pk, b; r)$, the real oracle O is used to reply to these queries. All of f, gc, and gi queries will be handled identically in both experiments by forwarding them to O. Thus, we only need to consider what happens in either execution when a new query $qu = ((\widetilde{F}, \widetilde{X}) \xrightarrow[\text{eval}]{} ?)$ is asked.

Suppose Bad holds and let $qu = ((\widetilde{F}, \widetilde{X}) \xrightarrow[\text{eval}]{} ?)$ be the first eval query that will be answered differently between the two executions. That is, qu will be replied to with $\bot$ under $O[\mathsf{H}]$, but receives an answer $y \neq \bot$ from the real oracle O. We will now show that either Bad_1 or Bad_2 must hold. Consider two cases:

1. $((*, *) \xrightarrow[\text{gc}]{} \widetilde{F}) \notin \text{OrigG}$: In this case, the fact that $\text{eval}(\widetilde{F}, \widetilde{X}) \neq \bot$ implies that Bad_1 holds.
2. $((*, *) \xrightarrow[\text{gc}]{} \widetilde{F}) \in \text{OrigG}$: In this case the facts that $\text{eval}(\widetilde{F}, \widetilde{X}) \neq \bot$, that qu is a query during the execution of $E^O(pk, b; r)$, and that $O[\mathsf{H}](qu) = \bot$ imply that Bad_2 holds. □

Proof (of Lemma 30). The proof of this lemma follows by a simple reduction to Lemma 4. Letting $\alpha = \Pr[\mathsf{Bad}_1]$, we will show how to build an adversary $\mathcal{A}^{f,\text{gc},\text{gi}}(1^\kappa)$ in the sense of Lemma 4 that will win with probability $\alpha \cdot \frac{1}{\text{poly}(\kappa)}$.

Let i be the index of the first query qu during the execution of $E^O(pk, b; r)$ for which the event Bad_1 holds. Note that up to the query index i, the executions of $E^O(pk, b; r)$ and $E^{O[\text{OrigG}]}(pk, b; r)$ are identical. With this in mind, we build the adversary $\mathcal{A}^{f,\text{gc},\text{gi}}(1^\kappa)$ as follows.

The adversary $\mathcal{A}^{f,\text{gc},\text{gi}}(1^\kappa)$ samples $(pk, sk) \leftarrow G^{f,\text{gc},\text{gi}}(1^\kappa)$, forming the set of query/response pairs OrigG. Then $\mathcal{A}^{f,\text{gc},\text{gi}}$ guesses $i \leftarrow [q]$ and runs $E^{O[\text{OrigG}]}(pk, b; r)$ for a random r. Notice that $\mathcal{A}$ makes no queries to eval whatsoever, as it handles eval queries using OrigG. If the ith query of this execution is $((\widetilde{F}, *) \xrightarrow[\text{eval}]{} ?)$ for some $\widetilde{F}$, then $\mathcal{A}^{f,\text{gc},\text{gi}}$ returns $\widetilde{F}$; otherwise, $\mathcal{A}$ returns $\bot$.

$\mathcal{A}^{f,\text{gc},\text{gi}}(1^\kappa)$ wins with probability at least $\alpha \cdot \frac{1}{q}$. On the other hand, by Lemma 4 we know $\mathcal{A}$'s success probability is at most $\frac{1}{2^{\kappa/2}}$. Thus, we have $\alpha \leq \frac{1}{2^{\kappa/3}}$, and the proof is complete. □

Proof (of Lemma 31). We claim that whenever the event $\mathsf{Bad}_2 \wedge \overline{\mathsf{Bad}_1}$ holds then the event Miss, defined as follows, also holds. Miss is the event that during the execution of $E^{O[\mathsf{OrigG}]}(pk, b; r)$ there is a query $qu = ((\widetilde{F}, \widetilde{X}) \xrightarrow[\text{eval}]{} ?)$, such that

1. $((*, *) \xrightarrow[\text{gc}]{} \widetilde{F}) \in \mathsf{OrigG}$;
2. $\mathsf{eval}(\widetilde{F}, \widetilde{X}) \neq \bot$;
3. $((*, *) \xrightarrow[\text{gc}]{} \widetilde{F}) \notin \mathsf{H}$ and $((\widetilde{F}, \widetilde{X}) \xrightarrow[\text{eval}]{} *) \notin \mathsf{H}$.

The reason for the above claim is that if $\mathsf{Bad}_2 \wedge \overline{\mathsf{Bad}_1}$ holds, then $\overline{\mathsf{Bad}_1}$ must necessarily hold, and thus the two executions $E^O(pk, b; r)$ and $E^{O[\mathsf{OrigG}]}(pk, b; r)$ are identical. The rest follows by the definition of the event Bad_2. We will prove

$$\Pr[\mathsf{Miss}] \leq \frac{2q}{t}, \tag{5}$$

which yields the proof of this lemma. Thus, we focus on proving Eq. 5.

We break Miss into smaller events. We give some notation first. Let $i \in [n]$, $d \in \{0, 1\}$, F be circuit with input size n and let $\widetilde{F} = \mathsf{gc}(s, F)$, for some s. We say a garbled input $\widetilde{X} = (\widetilde{x}_1, \ldots, \widetilde{x}_n)$ is an (i, d)-*match* for $\widetilde{F}$ if $\widetilde{X}$ is a valid garbled input of $\widetilde{F}$ and the ith garbled bit of $\widetilde{X}$ corresponds to the bit d. Formally,

- for all $j \in [n]$ and $j \neq i$: $\widetilde{x}_j = \mathsf{gi}(s, j, 0)$ or $\widetilde{x}_j = \mathsf{gi}(s, j, 1)$;
- $\widetilde{x}_i = \mathsf{gi}(s, i, d)$.

We say that a set of query/response pairs U *contains an* (i, d)-*match for* $\widetilde{F}$ if there exists $((\widetilde{F}, \widetilde{X}) \xrightarrow[\text{eval}]{} *) \in \mathsf{U}$ such that $\widetilde{X}$ is an (i, d)-match for $\widetilde{F}$.

We also give the following notation. Recalling the way in which LocalE is constructed in $\dot{G}$ through t iterations, for $i \in [t]$ let LocalE_i be the set formed after the i-th iteration. Also, let OrigE^* be the set of all query/response pairs during the execution of $E^{O[\mathsf{OrigG}]}(pk, b; r)$.

We now define a series of events, $\mathsf{Miss}_{i,d}$, for $i \in [q]$ and $d \in \{0, 1\}$, and will show that if Miss holds then for some i and d the event $\mathsf{Miss}_{i,d}$ must hold.

Event $\mathsf{Miss}_{i,d}$ is the event that for some $\widetilde{F}$ that $((*, *) \xrightarrow[\text{gc}]{} \widetilde{F}) \in \mathsf{OrigG}$, both the following hold:

1. OrigE^* contains an (i, d)-match for $\widetilde{F}$;
2. none of the sets $\mathsf{LocalE}_1, \cdots, \mathsf{LocalE}_t$ do contain an (i, d)-match for $\widetilde{F}$.

We claim that if Miss holds then $\mathsf{Miss}_{i,d}$ must hold, for some $i \in [q]$ and $d \in \{0, 1\}$. Suppose the event Miss holds for the query $qu = ((\widetilde{F}, \widetilde{X}) \xrightarrow[\text{eval}]{} ?)$ (see above for the definition of Miss). We consider all possible cases:

- For no (i, d) does the set $\mathsf{LocalE} = \mathsf{LocalE}_1 \cup \cdots \cup \mathsf{LocalE}_t$ contain an (i, d)-match for $\widetilde{F}$. Since the set OrigE^* contains $((\widetilde{F}, \widetilde{X}) \xrightarrow[\text{eval}]{} *)$, there would be an (i, d)-match for $\widetilde{F}$ for all $i \in [q]$, so $\mathsf{Miss}_{i,d}$ holds for some d and all $i \in [q]$.

- There is one and only one garbled input $\widetilde{X}_1$ which is valid for $\widetilde{F}$ and for which we have $((\widetilde{F}, \widetilde{X}_1) \xrightarrow{\text{eval}} ?) \in \mathsf{LocalE}$. In this case, we must have $\widetilde{X}_1 \neq \widetilde{X}$, because otherwise we would have $((\widetilde{F}, \widetilde{X}) \xrightarrow{\text{eval}} *) \in \mathsf{H}$, a contradiction to the fact that Miss holds. Thus, for some (i, d) both the following must hold: (A) $\widetilde{X}$ is an (i, d)-match for $\widetilde{F}$ and (B) $\widetilde{X}_1$ is not an (i, d)-match for $\widetilde{F}$. Thus, for some i and d, the event $\mathsf{Miss}_{i,d}$ must hold.
- There are at least two different garbled inputs $\widetilde{X}_1$ and $\widetilde{X}_2$ which both are valid for $\widetilde{F}$ and which $((\widetilde{F}, \widetilde{X}_1) \xrightarrow{\text{eval}} ?) \in \mathsf{LocalE}$ and $((\widetilde{F}, \widetilde{X}_2) \xrightarrow{\text{eval}} ?) \in \mathsf{LocalE}$: This case cannot happen because otherwise we would have $(*, * \xrightarrow{\text{gc}} \widetilde{F}) \in \mathsf{H}$, a contradiction to the fact that Miss holds.

Having proved $\Pr[\mathsf{Miss}] \leq \sum_{i,d} \Pr[\mathsf{Miss}_{i,d}]$, we bound the probability of each individual $\mathsf{Miss}_{i,d}$. To bound the probability of the event $\mathsf{Miss}_{i,d}$, note that since all of $\mathsf{LocalE}_1, \ldots, \mathsf{LocalE}_t$ and OrigE^* are obtained via independent and identical processes, by Lemma 5 we have

$$\Pr[\mathsf{Miss}_{i,d}] \leq \frac{1}{t}.$$

Using a union bound, $\Pr[\mathsf{Miss}] \leq \frac{2q}{t}$, and Eq. 5 is now proved. This completes the proof. $\qquad\qquad\square$

Proof of security for $\dot{\mathcal{E}}$. We now give the proof of security.

Proof (of Lemma 27). To define the reduction algorithm SecRed we need to introduce the following procedure, overloading the definition of $\mathsf{ConstHelp}$ (Definition 20). In Definition 20 the procedure $\mathsf{ConstHelp}$ was given as input an auxiliary information set Aux which helps the procedure in finding answers to the eval queries provided in the given set Q. In the definition below, however, there is no auxiliary information set, but the procedure could use the oracle rev.

Definition 32. *Procedure* $\mathsf{ConstHelp}$:

- **Input:** *A set of query/answer pairs* Q.
- **Oracle:** $O = (f, \mathsf{gc}, \mathsf{gi}, \mathsf{eval}, \mathsf{rev})$.
- **Output:** *A "hint" set* H *formed as follows:*
  - *If there exists* $((\widetilde{F}, \widetilde{X}) \xrightarrow{\text{eval}} y \neq \bot) \in \mathsf{Q}$ *such that for no* $\widetilde{X}' \neq \widetilde{X}$ *do we have* $((\widetilde{F}, \widetilde{X}') \xrightarrow{\text{eval}} y' \neq \bot) \in \mathsf{Q}$, *then add* $((\widetilde{F}, \widetilde{X}) \xrightarrow{\text{eval}} y)$ *to* H.
  - *If for two distinct* $\widetilde{X}_1$ *and* $\widetilde{X}_2$ *we have* $((\widetilde{F}, \widetilde{X}_1) \xrightarrow{\text{eval}} y_1 \neq \bot) \in \mathsf{Q}$ *and* $((\widetilde{F}, \widetilde{X}_2) \xrightarrow{\text{eval}} y_2 \neq \bot) \in \mathsf{Q}$, *then add* $(((s, F)) \xrightarrow{\text{gc}} \widetilde{F})$ *to* H, *where* $(s, F) = \mathsf{rev}(\widetilde{F}, \widetilde{X}_1, \widetilde{X}_2)$.

We will now describe the attack oracle-aided algorithm SecRed against the semantic security of (G^O, E^O, D^O). The input to SecRed is pair of challenge (pk, c) sampled under $\mathcal{E}^O$. Moreover, SecRed has oracle access to O as well as an adversary against $\dot{\mathcal{E}}^O$.

Description of SecRed$^{\mathcal{A},O}(pk, c)$:

1. Initialize LocalE$^* = \varnothing$. For $i = [1, t]$, do the following: Run $E^O(pk, 0)$ and $E^O(pk, 1)$ and add all the resulting query-answer pairs to LocalE*.
2. Set H$^* \leftarrow$ ConstHelp^O(LocalE*).
3. Return $b' \leftarrow \mathcal{A}(pk, H^*, c)$.

We will now show that the following holds for both $b = 0$ and $b = 1$: The distribution Dist1 $= (pk, H^*, c)$ is statistically close to Dist2 $= (\dot{pk}, \dot{c})$, where $(pk, sk) \leftarrow G^O(1^\kappa)$, $c \leftarrow E^O(pk, b)$, and $(\dot{pk}, *) \leftarrow \dot{G}^O(1^\kappa)$ and $\dot{c} \leftarrow \dot{E}^O(\dot{pk}, b)$. Also, H* is sampled as in the execution of the security reduction SecRed$^{\mathcal{A},O}(pk, c)$. Let all the variables that appear below be sampled as in the above. First, it is easy to show that

$$\Delta((pk, \text{H}^*), \dot{pk}) \leq \text{poly}(\kappa) \times \frac{1}{2^{\kappa/2}} \leq \frac{1}{2^{\kappa/3}}.$$

Moreover, by Lemma 29 we have

$$\Delta(c, \dot{c}) \leq \frac{2q}{t} + \frac{1}{2^{\kappa/3}}. \tag{6}$$

Thus, SecRed$^{\mathcal{A},O}(pk, c)$ breaks the semantic security of (G^O, E^O, D^O) with probability at least $\gamma - \frac{2q}{t} - \frac{1}{2^{\kappa/4}}$.   □

4.2   Removing Garbling Evaluation Queries from Decryption

In this section, we will prove Lemma 17. Namely, we will present a procedure that compiles a PKE scheme $\mathcal{E} = (G^{f,\text{gc},\text{gi}}, E^{f,\text{gc},\text{gi}}, D^{f,\text{gc},\text{gi},\text{eval}})$ into a new PKE scheme $\ddot{\mathcal{E}} = (\ddot{G}^{f,\text{gc},\text{gi}}, \ddot{E}^{f,\text{gc},\text{gi}}, \ddot{D}^{f,\text{gc},\text{gi}})$ with correctness and security comparable to the original scheme $\mathcal{E}$, but where $\ddot{D}$ will not ask any eval queries.

Again, for simplicity we use the following convention where we keep the entire oracle O as a superscript to all the algorithms (G^O, E^O, D^O) as well as $(\ddot{G}^O, \ddot{E}^O, \ddot{D}^O)$ with the understanding that the actual oracle access is of the form given above. We also make the following assumption without loss of generality.

Assumption 33. *We assume that all the algorithms (G, E, D) are in normal form (Definition 19). Also, we assume w.l.o.g. that the secret key outputted by G contains all the query-response pairs made by G.*

Definition 34 (Query set). *For an oracle algorithm A^O, Query$(A^O(x; r))$ denotes the set of all queries asked during the execution of A^O on input x and randomness r. We write Query$(A^O(x))$ to indicate the random variable formed by returning Query$(A^O(x; r))$ for $r \leftarrow \{0, 1\}^*$.*

Definition 35 (Valid partial oracles). *We say that a partial oracle O_1 is valid if for some $O_2 \in \text{Supp}(\mathbf{O})$: $O_1 \subseteq O_2$.*

Definition 36 (Oracle consistency/sampling notation). *We say a partial oracle O_1 is consistent with a set of query/response pairs S if $O_1 \cup \mathsf{S}$ is valid.*

For a partial oracle O_1 and randomness r we say that (O_1, r) agrees with a public key pk if (1) $G^{O_1}(r) = (pk, *)$ and (2) all the queries in $\text{Query}(G^{O_1}(r))$ are defined in O_1. We say that (O_1, r) minimally agrees with pk if (1) (O_1, r) agrees with pk and (2) O_1 is defined only on the queries that occur during the execution and nothing more: namely, $O_1(qu)$ is defined iff $qu \in \text{Query}(G^{O_1}(r))$.

We let $\text{Partial}(pk, \mathsf{S})$ denote the set of all (O_1, r) where (1) (O_1, r) minimally agrees with pk and (2) O_1 agrees with S. We sometimes abuse notation and write $(O_1, sk) \leftarrow \text{Partial}(pk, \mathsf{S})$ to mean the following sampling: $(O_1, r) \leftarrow \text{Partial}(pk, \mathsf{S})$ and $(pk, sk) = G^{O_1}(r)$.

Definition 37 (Composed oracle). *Given a partial oracle O_p and full oracle O (an oracle that is defined on all points in its domain) we define $O_p \lozenge O$ to be the composed oracle that uses O_p to reply if the corresponding query is defined there, and uses O otherwise. Note that $O_p \lozenge O$ is not necessarily in $\text{Supp}(O)$.*

Compilation Procedure

Construction 38. *The scheme $\ddot{\mathcal{E}} = (\ddot{G}, \ddot{E}, \ddot{D})$ is parameterized over two functions $\varepsilon = \varepsilon(\kappa)$ and $t = t(\kappa)$, which we will instantiate later.*

- *$\ddot{G}(1^\kappa)$: Do the following steps:*
  1. *Set $\mathsf{OrigG} = \varnothing$. Run $(pk, sk) \leftarrow G^O(1^\kappa)$, and add all query-answer pairs that are encountered during this execution to OrigG.*
  2. *Set $\mathsf{LearnG} = \varnothing$. While there exists a query $qu \notin \mathsf{LearnG}$ such that*

$$\Pr_{O' \leftarrow \mathbf{O}}[qu \in \text{Query}(G^{O'}(1^\kappa)) \mid pk, \mathsf{LearnG}] \geq \varepsilon,$$

  then choose the lexicographically first such qu and add $(qu \xrightarrow{T} O'(qu))$ to LearnG. Note that $T \in \{f, \mathsf{gc}, \mathsf{gi}\}$.
  3. *Output $\ddot{pk} = (pk, \mathsf{LearnG})$ and $\ddot{sk} = sk$. (By Assumption 33, $\ddot{sk}$ contains OrigG.)*
- *$\ddot{E}(\ddot{pk}, b)$: Given $\ddot{pk} = (pk, \mathsf{LearnG})$ and $b \in \{0, 1\}$ do the following:*
  1. *Set $\mathsf{OrigE} = \varnothing$. Run $c \leftarrow E^O(pk, b)$ and add all the query-answer pairs that are observed during this execution to OrigE.*
  2. ***Generating helper set H for $\ddot{D}$:** Sample $t' \leftarrow [1, t]$. Set $\mathsf{S} = \mathsf{OrigE} \cup \mathsf{LearnG}$. For $i \in [1, t']$, do the following:*
     (a) ***Offline phase:** Sample $(\widehat{O}_i, \widehat{sk}_i) \leftarrow \text{Partial}(pk, S)$.*
     (b) ***Semi-online phase:** Run $D^{\widehat{O}_i \lozenge O[\mathsf{S}]}(\widehat{sk}_i, c)$ and add all query/response pairs made to the oracle O to the set S. Let $\widehat{\mathsf{OrigD}}_i$ be the set of all query-answer pairs made by this execution.*

After all iterations, set $\mathsf{H} := \mathtt{ConstHelp}(\widehat{\mathsf{OrigD}}, S)$, *where we define*
$\widehat{\mathsf{OrigD}} = \widehat{\mathsf{OrigD}}_1 \cup \cdots \cup \widehat{\mathsf{OrigD}}_{t'}$.

   3. Output $\ddot{c} = (c, \mathsf{H})$.
- $\ddot{D}(\ddot{pk}, \ddot{sk}, \ddot{c})$: *Given* $\ddot{pk} = (pk, \mathsf{LearnG})$, $\ddot{sk}$, *and* $\ddot{c} = (c, \mathsf{H})$, *output the value of*
$\tilde{b} \leftarrow D^{O[\mathsf{H} \cup \mathsf{LearnG}]}(\ddot{sk}, c)$.

Query Complexity of $\ddot{\mathcal{E}}$. The following lemma follows from the description of the compilation procedure of Construction 38.

Lemma 39. *Let* q *be as in Definition 23. Assuming* $\varepsilon = \frac{1}{\mathrm{poly}(\kappa)}$ *and* $t = \mathrm{poly}(\kappa)$, *all the algorithms of* $\ddot{\mathcal{E}}^O$ *make* $q^{O(1)}$ *queries. Concretely, the algorithm* $\ddot{E}$ *makes at most* $\nu := 4tq^2$ *queries.*

We note that by taking $\varepsilon = \frac{1}{\mathrm{poly}(\kappa)}$ the learning process of $\ddot{G}$ (i.e., for sampling LearnG) could be done by making a polynomial number of queries [6].

We give the correctness and security statements regarding the compiled scheme $\ddot{\mathcal{E}} = (\ddot{G}, \ddot{E}, \ddot{D})$. See the full version for the proofs.

Lemma 40 (Correctness of $\ddot{\mathcal{E}}$**).** *Suppose the original PKE scheme* (G, E, D) *is* $(\frac{1}{2} + \delta)$-*correct in the ideal model. The compiled PKE scheme* $(\ddot{G}, \ddot{E}, \ddot{D})$ *has at least* $(\frac{1}{2} + \delta - \eta)$ *correctness in the ideal model, where*

$$\eta = \frac{1}{2^{\kappa/5}} + \frac{2q}{t} + 3\varepsilon\nu t.$$

That is,

$$\Pr[\ddot{D}^O(sk, \ddot{c}) \neq b] \leq \tfrac{1}{2} - \delta + \eta \qquad (7)$$

where the probability is taken over $O \leftarrow \mathbf{O}$, $(\ddot{pk}, sk) \leftarrow \ddot{G}^O(1^\kappa)$, $b \leftarrow \{0,1\}$ *and* $\ddot{c} = (c, \mathsf{H}) \leftarrow \ddot{E}^O(\ddot{pk}, b)$. *Here* t *and* ε *are the underlying parameters of the compilation procedure, and* ν *is defined in Lemma 39. In particular, for any constant* $c > 0$ *by taking* $t = 2q^{c+2}$ *and* $\varepsilon = \frac{1}{q^{3c+8}}$, *the compiled scheme* $(\ddot{G}, \ddot{E}, \ddot{D})$ *has at least* $(\frac{1}{2} + \delta - 1/\kappa^c)$ *correctness in the ideal model.*

Lemma 41 (Security of $\ddot{\mathcal{E}}$**).** *Let* p *be an arbitrary polynomial which satisfies*

$$8tq^2\varepsilon + \frac{1}{2^{\kappa/2-1}} \leq \frac{1}{p}.$$

There exists a polynomial-query algorithm SecRed *that satisfies the following. For any adversary* $\mathcal{A}$ *that breaks the semantic security of* $(\ddot{G}^O, \ddot{E}^O, \ddot{D}^O)$ *in the ideal model* O *with probability at least* γ, *the adversary* $\mathsf{SecRed}^{\mathcal{A},O}$ *breaks the semantic security of* (G^O, E^O, D^O) *with probability at least* $\gamma - \beta$ *where*

$$\beta = t \cdot \left(\frac{1}{p-1} + 4tq^2\varepsilon + \frac{1}{q^{2c+4}} + \frac{1}{2^{\kappa/2-1}}\right).$$

In particular, for any constant c *by taking* $t = 2q^{c+2}$ *and* $\varepsilon = \frac{1}{q^{3c+8}}$ *we have the following: For any* $\mathcal{A}$ *breaking the semantic security of* $(\ddot{G}^O, \ddot{E}^O, \ddot{D}^O)$ *in the ideal model with probability at least* γ, *the (polynomial-query) adversary* $\mathsf{SecRed}^{\mathcal{A},O}$ *breaks the semantic security of* (G^O, E^O, D^O) *with probability at least* $\gamma - \frac{22}{\kappa^{2+c}}$.

References

1. Applebaum, B.: Garbled circuits as randomized encodings of functions: a primer. In: Lindell, Y. (ed.) Tutorials on the Foundations of Cryptography: Dedicated to Oded Goldreich. ISC, pp. 1–44. Springer, Cham (2017). https://doi.org/10.1007/978-3-319-57048-8_1. 336

2. Asharov, G., Segev, G.: Limits on the power of indistinguishability obfuscation and functional encryption. In: 2015 IEEE 56th Annual Symposium on Foundations of Computer Science (FOCS), pp. 191–209. IEEE (2015). 337, 339, 343

3. Asharov, G., Segev, G.: On constructing one-way permutations from indistinguishability obfuscation. In: Kushilevitz, E., Malkin, T. (eds.) TCC 2016, Part II. LNCS, vol. 9563, pp. 512–541. Springer, Heidelberg (2016). https://doi.org/10.1007/978-3-662-49099-0_19. 337

4. Baecher, P., Brzuska, C., Fischlin, M.: Notions of black-box reductions, revisited. In: Sako, K., Sarkar, P. (eds.) ASIACRYPT 2013. LNCS, vol. 8269, pp. 296–315. Springer, Heidelberg (2013). https://doi.org/10.1007/978-3-642-42033-7_16. 336, 346

5. Barak, B.: The complexity of public-key cryptography. In: Lindell, Y. (ed.) Tutorials on the Foundations of Cryptography, pp. 45–77. Springer, Cham (2017). https://doi.org/10.1007/978-3-319-57048-8_2. 336

6. Barak, B., Mahmoody-Ghidary, M.: Lower bounds on signatures from symmetric primitives. In: 48th Annual Symposium on Foundations of Computer Science, Providence, RI, USA, 20–23 October 2007, pp. 680–688. IEEE Computer Society Press (2007). 361

7. Barak, B., Mahmoody-Ghidary, M.: Merkle puzzles are optimal—an $O(n^2)$-query attack on any key exchange from a random oracle. In: Halevi, S. (ed.) CRYPTO 2009. LNCS, vol. 5677, pp. 374–390. Springer, Heidelberg (2009). https://doi.org/10.1007/978-3-642-03356-8_22. 350

8. Beaver, D.: Correlated pseudorandomness and the complexity of private computations. In: 28th Annual ACM Symposium on Theory of Computing, pp. 479–488, Philadelphia, PA, USA, 22–24 May 1996. ACM Press (1996). 337

9. Bellare, M., Hoang, V.T., Rogaway, P.: Foundations of garbled circuits. In: Yu, T., Danezis, G., Gligor, V.D. (eds.) ACM CCS 12: 19th Conference on Computer and Communications Security, Raleigh, NC, USA, 16–18 October 2012, pp. 784–796. ACM Press (2012). 336, 337, 338

10. Bitansky, N., Degwekar, A., Vaikuntanathan, V.: Structure vs. hardness through the obfuscation lens. In: Katz, J., Shacham, H. (eds.) CRYPTO 2017. LNCS, vol. 10401, pp. 696–723. Springer, Cham (2017). https://doi.org/10.1007/978-3-319-63688-7_23. 337

11. Boneh, D., Papakonstantinou, P.A., Rackoff, C., Vahlis, Y., Waters, B.: On the impossibility of basing identity based encryption on trapdoor permutations. In: 49th Annual Symposium on Foundations of Computer Science, Philadelphia, PA, USA, 25–28 October 2008, pp. 283–292. IEEE Computer Society Press (2008). 336

12. Brakerski, Z., Katz, J., Segev, G., Yerukhimovich, A.: Limits on the power of zero-knowledge proofs in cryptographic constructions. In: Ishai, Y. (ed.) TCC 2011. LNCS, vol. 6597, pp. 559–578. Springer, Heidelberg (2011). https://doi.org/10.1007/978-3-642-19571-6_34. 337, 338, 339

13. Canetti, R., Kalai, Y.T., Paneth, O.: On obfuscation with random oracles. In: Dodis, Y., Nielsen, J.B. (eds.) TCC 2015, Part II. LNCS, vol. 9015, pp. 456–467. Springer, Heidelberg (2015). https://doi.org/10.1007/978-3-662-46497-7_18. 340

14. Dachman-Soled, D.: Towards non-black-box separations of public key encryption and one way function. In: Hirt, M., Smith, A. (eds.) TCC 2016, Part II. LNCS, vol. 9986, pp. 169–191. Springer, Heidelberg (2016). https://doi.org/10.1007/978-3-662-53644-5_7. 339
15. Diffie, W., Hellman, M.E.: New directions in cryptography. IEEE Trans. Inf. Theory **22**(6), 644–654 (1976). 335
16. Döttling, N., Garg, S.: Identity-based encryption from the Diffie-Hellman assumption. In: Katz, J., Shacham, H. (eds.) CRYPTO 2017, Part I. LNCS, vol. 10401, pp. 537–569. Springer, Cham (2017). https://doi.org/10.1007/978-3-319-63688-7_18. 336, 338
17. Garg, S., Lu, S., Ostrovsky, R., Scafuro, A.: Garbled RAM from one-way functions. In: Servedio, R.A., Rubinfeld, R. (eds.) 47th Annual ACM Symposium on Theory of Computing, Portland, OR, USA, 14–17 June 2015, pp. 449–458. ACM Press (2015). 337
18. Garg, S., Mahmoody, M., Mohammed, A.: Lower bounds on obfuscation from all-or-nothing encryption primitives. In: Katz, J., Shacham, H. (eds.) CRYPTO 2017, Part I. LNCS, vol. 10401, pp. 661–695. Springer, Cham (2017). https://doi.org/10.1007/978-3-319-63688-7_22. 339, 340
19. Garg, S., Mahmoody, M., Mohammed, A.: When does functional encryption imply obfuscation? In: Kalai, Y., Reyzin, L. (eds.) TCC 2017, Part I. LNCS, vol. 10677, pp. 82–115. Springer, Cham (2017). https://doi.org/10.1007/978-3-319-70500-2_4. 339
20. Gentry, C., Wichs, D.: Separating succinct non-interactive arguments from all falsifiable assumptions. In: Fortnow, L., Vadhan, S.P. (eds.) 43rd Annual ACM Symposium on Theory of Computing, San Jose, CA, USA, 6–8 June 2011, pp. 99–108. ACM Press (2011). 339
21. Gertner, Y., Malkin, T., Reingold, O.: On the impossibility of basing trapdoor functions on trapdoor predicates. In: 42nd Annual Symposium on Foundations of Computer Science, Las Vegas, NV, USA, 14–17 October 2001, pp. 126–135. IEEE Computer Society Press (2001). 341
22. Hsiao, C.-Y., Reyzin, L.: Finding collisions on a public road, or do secure hash functions need secret coins? In: Franklin, M. (ed.) CRYPTO 2004. LNCS, vol. 3152, pp. 92–105. Springer, Heidelberg (2004). https://doi.org/10.1007/978-3-540-28628-8_6. 341
23. Impagliazzo, R., Rudich, S.: Limits on the provable consequences of one-way permutations. In: 21st Annual ACM Symposium on Theory of Computing, Seattle, WA, USA, 15–17 May 1989, pp. 44–61. ACM Press (1989). 336, 339, 340, 341, 342, 344, 346, 349, 350
24. Ishai, Y., Kushilevitz, E.: Randomizing polynomials: a new representation with applications to round-efficient secure computation. In: 41st Annual Symposium on Foundations of Computer Science, Redondo Beach, CA, USA, 12–14 November 2000, pp. 294–304. IEEE Computer Society Press (2000). 336
25. Lindell, Y., Pinkas, B.: A proof of security of Yao's protocol for two-party computation. J. Cryptol. **22**(2), 161–188 (2009). 336
26. Lu, S., Ostrovsky, R.: How to garble RAM programs? In: Johansson, T., Nguyen, P.Q. (eds.) EUROCRYPT 2013. LNCS, vol. 7881, pp. 719–734. Springer, Heidelberg (2013). https://doi.org/10.1007/978-3-642-38348-9_42. 337
27. Mahmoody, M., Mohammed, A., Nematihaji, S.: On the impossibility of virtual black-box obfuscation in idealized models. In: Kushilevitz, E., Malkin, T. (eds.) TCC 2016, Part I. LNCS, vol. 9562, pp. 18–48. Springer, Heidelberg (2016). https://doi.org/10.1007/978-3-662-49096-9_2. 340

28. Mahmoody, M., Mohammed, A., Nematihaji, S., Pass, R., Shelat, A.: Lower bounds on assumptions behind indistinguishability obfuscation. In: Kushilevitz, E., Malkin, T. (eds.) TCC 2016, Part I. LNCS, vol. 9562, pp. 49–66. Springer, Heidelberg (2016). https://doi.org/10.1007/978-3-662-49096-9_3. 340
29. Papakonstantinou, P.A., Rackoff, C.W., Vahlis, Y.: How powerful are the DDH hard groups? Cryptology ePrint Archive, Report 2012/653 (2012). http://eprint.iacr.org/2012/653. 336, 338
30. Pass, R.: Limits of provable security from standard assumptions. In: Fortnow, L., Vadhan, S.P. (eds.) 43rd Annual ACM Symposium on Theory of Computing, San Jose, CA, USA, 6–8 June 2011, pp. 109–118. ACM Press (2011). 339
31. Pass, R., Tseng, W.-L.D., Venkitasubramaniam, M.: Towards non-black-box lower bounds in cryptography. In: Ishai, Y. (ed.) TCC 2011. LNCS, vol. 6597, pp. 579–596. Springer, Heidelberg (2011). https://doi.org/10.1007/978-3-642-19571-6_35. 339
32. Reingold, O., Trevisan, L., Vadhan, S.: Notions of reducibility between cryptographic primitives. In: Naor, M. (ed.) TCC 2004. LNCS, vol. 2951, pp. 1–20. Springer, Heidelberg (2004). https://doi.org/10.1007/978-3-540-24638-1_1. 336, 337, 346
33. Rivest, R.L., Shamir, A., Adleman, L.M.: A method for obtaining digital signature and public-key cryptosystems. Commun. Assoc. Comput. Mach. 21(2), 120–126 (1978). 335
34. Sahai, A., Waters, B.: How to use indistinguishability obfuscation: deniable encryption, and more. In: Shmoys, D.B. (ed.) 46th Annual ACM Symposium on Theory of Computing, New York, NY, USA, 31 May–3 June 2014, pp. 475–484. ACM Press (2014). 341
35. Yao, A.C.-C.: How to generate and exchange secrets (extended abstract). In: 27th Annual Symposium on Foundations of Computer Science, Toronto, Ontario, Canada, 27–29 October 1986, pp. 162–167. IEEE Computer Society Press (1986). 336

Optimizing Authenticated Garbling
for Faster Secure Two-Party Computation

Jonathan Katz[1], Samuel Ranellucci[1,2], Mike Rosulek[3], and Xiao Wang[1(✉)]

[1] University of Maryland, College Park, USA
{jkatz,wangxiao}@cs.umd.edu,samuel@umd.edu
[2] George Mason University, Fairfax, USA
[3] Oregon State University, Corvallis, USA
rosulekm@eecs.oregonstate.edu

Abstract. Wang et al. (CCS 2017) recently proposed a protocol for malicious secure two-party computation that represents the state-of-the-art with regard to concrete efficiency in both the single-execution and amortized settings, with or without preprocessing. We show here several optimizations of their protocol that result in a significant improvement in the overall communication and running time. Specifically:

– We show how to make the "authenticated garbling" at the heart of their protocol compatible with the half-gate optimization of Zahur et al. (Eurocrypt 2015). We also show how to avoid sending an information-theoretic MAC for each garbled row. These two optimizations give up to a 2.6× improvement in communication, and make the communication of the online phase essentially equivalent to that of state-of-the-art *semi-honest* secure computation.
– We show various optimizations to their protocol for generating AND triples that, overall, result in a 1.5× improvement in the communication and a 2× improvement in the computation for that step.

1 Introduction

In recent years, we have witnessed amazing progress in secure two-party computation, in both the semi-honest and malicious settings. In the semi-honest case, there has been an orders-of-magnitude improvement in protocols based on Yao's garbled circuit [39] since the initial implementation by Malkhi et al. [26]. This has resulted from several important techniques, including oblivious-transfer extension [16], pipelining [13], hardware acceleration [6], free-XOR [19] and other improved garbling techniques [17,32], etc. Similarly, the concrete efficiency of secure two-party computation in the *malicious* case has also improved tremendously in both the single-execution [1,8,10,14,18,20–23,28,30,34–37,41] and amortized [15,24,25,31,33] settings. Whereas initial implementations in the malicious case could evaluate up to 1,000 gates at the rate of 1 gate/second [32], the current state-of-the-art protocol by Wang et al. [37] (the *WRK protocol*) can compute tens of millions of gates at a rate up to 700,000× faster. With this steady stream of improvements, it has become more and more difficult to

© International Association for Cryptologic Research 2018
H. Shacham and A. Boldyreva (Eds.): CRYPTO 2018, LNCS 10993, pp. 365–391, 2018.
https://doi.org/10.1007/978-3-319-96878-0_13

366   J. Katz et al.

Table 1. Communication complexity of different protocols (in MB) for evaluating an AES circuit. One-way communication refers to the maximum communication one party sends to the other; two-way communication refers to the sum of both parties' communication. The best prior number in each column is bolded for reference.

	One-way comm.		Two-way comm.	
	Dep. + online	Total	Dep. + online	Total
semi-honest	0.22	0.22	0.22	0.22
Single-execution setting				
[28]	**0.22**	15	**0.22**	15
[37]	0.57	3.43	0.57	6.29
[12]	3.39	**3.39**	3.39	**3.39**
This work, v. 1	0.33	2.24	0.33	4.15
This work, v. 2	0.22	2.67	0.22	5.12
Amortized setting (1024 executions)				
[33]	1.60	1.60	3.20	3.20
[28]	**0.22**	6.6	**0.22**	6.6
[37]	0.57	2.57	0.57	4.57
[18]	1.59	**1.59**	1.59	**1.59**
This work, v. 1	0.33	1.70	0.33	3.07
This work, v. 2	0.22	2.13	0.22	4.04

squeeze out additional performance gains; as an illustrative example, Zahur et al. [40] introduced a highly non-trivial optimization ("half-gates") just to reduce communication by 33%.

We show several improvements to the WRK protocol that, overall, improve its performance by 2–3×. Recall their protocol can be divided into three phases: a *function-independent phase* (Ind.) in which the parties know an upper bound on the number of gates in the circuit to be evaluated and the lengths of their inputs; a *function-dependent phase* (Dep.) in which the parties know the circuit, but not their inputs; and an *online phase* in which the parties evaluate the circuit on their respective inputs. Our results can be summarized as follows:

– We show how to make the "authenticated garbling" at the heart of the online phase of the WRK protocol compatible with the half-gate optimization of Zahur et al. We also show that it is possible to avoid sending an information-theoretic MAC for each garbled row. These two optimizations result in up to a 2.6× improvement in communication and, somewhat surprisingly, result in a protocol for malicious secure two-party computation in which the communication complexity of the online phase is essentially equivalent to that of state-of-the-art *semi-honest* secure computation.
– The function-dependent phase of the WRK protocol involves the computation of (shared) "AND triples" between the parties. We show various optimizations

of that step that result in a 1.5× improvement in the communication and a 2× improvement in the computation. Our optimizations also simplify the protocol significantly.

We can combine these improvements in various ways, and suggest in particular two instantiations of protocols with malicious security: one that minimizes the total communication across all phases, and one that trades off increased communication in the function-independent phase for reduced communication in the function-dependent phase. These protocols improve upon the state-of-the-art by a significant margin, as summarized in Table 1. For example, compared to the protocol of Nielsen et al. [28] we achieve the same communication across the function-dependent and online phases, but improve the total communication by more than 6×; compared to the prior work with the best total communication [12], we achieve a 1.5× improvement overall and, at the same time, push almost all communication to the function-independent preprocessing phase. (Our protocol also appears to be significantly better than that of Hazay et al. [12] in terms of computation. See Sect. 6 for a more detailed discussion.)

The multi-party case. It is natural to wonder whether we can extend our improved technique for authenticated garbling to the multi-party case, i.e., to improve upon [38]. Unfortunately, we have not yet been able to do so. In Sect. 7, we discuss some of the difficulties that arise.

1.1 Outline

In Sect. 2 we provide some background about the WRK protocol. We provide the high-level intuition behind our improvements in Sect. 3. In Sect. 4, we describe in detail our optimizations of the online phase of the WRK protocol, and in Sect. 5 we discuss our optimizations of the preprocessing phase. In Sect. 6, we compare our resulting protocols to prior work.

2 Background

We begin by describing some general background, followed by an in-depth review of the authenticated-garbling technique introduced in [37]. In the section that follows, we give a high-level overview of our optimizations and improvements.

We use κ and ρ to denote the computational and statistical security parameters, respectively. We sometimes use ":=" to denote assignment.

Information-theoretic MACs. As in prior work, we authenticate bits using a particular information-theoretic MAC. Let $\Delta_{\mathsf{B}} \in \{0,1\}^{\rho}$ be a value known to P_{B} that is chosen at the outset of the protocol. We say a bit b known to P_{A} is *authenticated to* P_{B} if P_{B} holds a key $\mathsf{K}[b]$ and P_{A} holds the corresponding tag $\mathsf{M}[b] = \mathsf{K}[b] \oplus b\Delta_{\mathsf{B}}$. We abstractly denote such a bit by $[b]_{\mathsf{A}}$; i.e., for some fixed Δ_{B}, when we say the parties hold $[b]_{\mathsf{A}}$ we mean that P_{A} holds $(b, \mathsf{M}[b])$ and P_{B} holds $\mathsf{K}[b]$ such that $\mathsf{M}[b] = \mathsf{K}[b] \oplus b\Delta_{\mathsf{B}}$. We analogously let $[b]_{\mathsf{B}}$ denote a bit b known to P_{B} and authenticated to P_{A}.

Functionality $\mathcal{F}_{\mathsf{abit}}$

Honest case:

1. Upon receiving init from both parties, choose uniform $\Delta_{\mathsf{A}}, \Delta_{\mathsf{B}} \in \{0,1\}^\rho$; send Δ_{A} to $\mathsf{P_A}$ and Δ_{B} to $\mathsf{P_B}$.
2. Upon receiving $(\mathsf{random}, \mathsf{A})$ from both parties, choose uniform $x \in \{0,1\}$ and $\mathsf{K}[x] \in \{0,1\}^\rho$, set $\mathsf{M}[x] := \mathsf{K}[x] \oplus x\Delta_{\mathsf{B}}$, and send $(x, \mathsf{M}[x])$ to $\mathsf{P_A}$ and $\mathsf{K}[x]$ to $\mathsf{P_B}$.
3. Upon receiving $(\mathsf{random}, \mathsf{B})$ from both parties, generate an authenticated bit for $\mathsf{P_B}$ in a manner symmetric to the above.

Corrupted parties: A corrupted party can specify the randomness used on its behalf by the functionality.

Global-key queries: A corrupted $\mathsf{P_A}$ (resp., $\mathsf{P_B}$) can, at any time, send Δ, and is told whether $\Delta = \Delta_{\mathsf{B}}$ (resp., $\Delta = \Delta_{\mathsf{A}}$).

Fig. 1. The authenticated-bits functionality.

A pair of authenticated bits $[b_1]_{\mathsf{A}}, [b_2]_{\mathsf{B}}$, each known to a different party, form an *authenticated share* of $b_1 \oplus b_2$. We denote this by $\langle b_1 \mid b_2 \rangle$, where the value in the left slot is known to $\mathsf{P_A}$, and the value in the right slot is known to $\mathsf{P_B}$. Both authenticated bits and authenticated shares are XOR-homomorphic.

Authenticated bits can be computed efficiently based on oblivious transfer [28,29]. We abstract away the particular protocol used to generate authenticated bits, and design our protocols in the $\mathcal{F}_{\mathsf{abit}}$-hybrid model (cf. Fig. 1) in which there is an ideal functionality that provides them.

Opening authenticated values. An authenticated bit $[b]_{\mathsf{A}}$ known to $\mathsf{P_A}$ can be opened by having $\mathsf{P_A}$ send b and $\mathsf{M}[b]$ to $\mathsf{P_B}$, who then verifies that $\mathsf{M}[b] = \mathsf{K}[b] \oplus b\Delta_{\mathsf{B}}$. As observed in prior work [9], it is possible to open n authenticated bits with less than n times the communication. Specifically, $\mathsf{P_A}$ can open $[b_1]_{\mathsf{A}}, \ldots, [b_n]_{\mathsf{A}}$ by sending $b_1, \ldots, b_n$ along with $h := H(\mathsf{M}[b_1], \ldots, \mathsf{M}[b_n])$, where H is a hash function modeled as a random oracle. $\mathsf{P_A}$ then simply checks whether $h = H(\mathsf{K}[b_1] \oplus b_1\Delta_{\mathsf{B}}, \ldots, \mathsf{K}[b_n] \oplus b_n\Delta_{\mathsf{B}})$.

We let $\mathsf{Open}([b_1]_{\mathsf{A}}, \ldots)$ denote the process of opening one or more authenticated bits in this way, and overload this notation so that $\mathsf{Open}(\langle b_1 \mid b_2 \rangle)$ denotes the process of having each party open its portion of an authenticated share.

Circuit-dependent preprocessing. We consider boolean circuits with gates represented as a tuple $(\alpha, \beta, \gamma, T)$, where α and β are (the indices of) the input wires of the gate, γ is the output wire of the gate, and $T \in \{\oplus, \wedge\}$ is the type of the gate. We use $\mathcal{W}$ to denote the output wires of all AND gates, $\mathcal{I}_1, \mathcal{I}_2$ to denote the input wires for each party, and $\mathcal{O}$ to denote the output wires.

Functionality $\mathcal{F}_{\mathsf{pre}}$

1. Choose uniform $\Delta_{\mathsf{A}}, \Delta_{\mathsf{B}} \in \{0, 1\}^\rho$. Send Δ_{A} to $\mathsf{P_A}$ and Δ_{B} to $\mathsf{P_B}$.
2. For each wire $w \in \mathcal{W} \cup \mathcal{I}$, generate a random authenticated share $\langle r_w \mid s_w \rangle$.
3. For each gate $\mathcal{G} = (\alpha, \beta, \gamma, T)$, in topological order:
   - If $T = \oplus$, generate a random authenticated share $\langle r_\gamma \mid s_\gamma \rangle$ for which $r_\gamma \oplus s_\gamma = r_\alpha \oplus s_\alpha \oplus r_\beta \oplus s_\beta$.
   - If $T = \wedge$, generate a random authenticated share $\langle r_\gamma^* \mid s_\gamma^* \rangle$ for which $r_\gamma^* \oplus s_\gamma^* = (r_\alpha \oplus s_\alpha) \wedge (r_\beta \oplus s_\beta)$.

Fig. 2. Preprocessing functionality for some fixed circuit.

Wang et al. [37] introduced an ideal functionality called $\mathcal{F}_{\mathsf{pre}}$ (cf. Fig. 2) that is used by the parties in a circuit-dependent, but input-*independent*, preprocessing phase. This functionality sets up information for the parties as follows:

1. For each wire w that is either an input wire of the circuit or an output wire of an AND gate, generate a random authenticated share $\langle r_w \mid s_w \rangle$. We refer to the value $\lambda_w \overset{\text{def}}{=} r_w \oplus s_w$ as the *mask* on wire w.
2. For the output wire γ of each XOR gate $(\alpha, \beta, \gamma, \oplus)$, generate a random authenticated share $\langle r_\gamma \mid s_\gamma \rangle$ whose value $r_\gamma \oplus s_\gamma$ is the XOR of the masks on the input wires α, β.
3. For each AND gate $(\alpha, \beta, \gamma, \wedge)$, generate a random authenticated share $\langle r_\gamma^* \mid s_\gamma^* \rangle$ such that

$$r_\gamma^* \oplus s_\gamma^* = (r_\alpha \oplus s_\alpha) \wedge (r_\beta \wedge s_\beta).$$

We refer to a triple of authenticated shares $(\langle r_\alpha \mid s_\alpha \rangle, \langle r_\beta \mid s_\beta \rangle, \langle r_\gamma^* \mid s_\gamma^* \rangle)$ for which $r_\gamma^* \oplus s_\gamma^* = (r_\alpha \oplus s_\alpha) \wedge (r_\beta \oplus s_\beta)$ as an *authenticated AND triple*. These are just (authenticated) Beaver triples [4] over the field $\mathbb{F}_2$.

Authenticated garbling. We now describe the idea behind the authenticated garbling technique from the WRK protocol. We assume the reader is familiar with basic concepts of garbled circuits, e.g., point-and-permute [5], free-XOR [19], etc.

Following the preprocessing phase described above, every wire w is associated with a secret mask λ_w, unknown to either party. If the actual value on that wire (when the circuit is evaluated on the parties' inputs) is z_w, then the *masked value* on that wire is defined to be $\hat{z}_w = z_w \oplus \lambda_w$. We focus on garbling a single AND gate $(\alpha, \beta, \gamma, \wedge)$. Assume $\mathsf{P_A}$ is the circuit garbler and $\mathsf{P_B}$ is the circuit evaluator. Say the garbled wire labels are $(\mathsf{L}_{\alpha,0}, \mathsf{L}_{\alpha,1})$ and $(\mathsf{L}_{\beta,0}, \mathsf{L}_{\beta,1})$ for wires α and β, respectively. Since we apply the free-XOR optimization, $\mathsf{P_A}$ also holds Δ such that $\mathsf{L}_{w,0} \oplus \mathsf{L}_{w,1} = \Delta$ for any wire w. The protocol inductively ensures that the evaluator $\mathsf{P_B}$ knows the wire labels $\mathsf{L}_{\alpha,\hat{z}_\alpha}, \mathsf{L}_{\alpha,\hat{z}_\beta}$ and masked values $\hat{z}_\alpha, \hat{z}_\beta$ for both input wires. Note that the correct masked value for the output wire is then

$$\hat{z}_\gamma = (\lambda_\alpha \oplus \hat{z}_\alpha) \wedge (\lambda_\beta \oplus \hat{z}_\beta) \oplus \lambda_\gamma,$$

and we need to ensure that $\mathsf{P_B}$ learns this value.

To achieve this, $\mathsf{P_A}$ generates a garbled gate consisting of 4 rows (one for each $u, v \in \{0, 1\}$)

$$G_{u,v} = H(\mathsf{L}_{\alpha,u}, \mathsf{L}_{\beta,v}) \oplus (r_{u,v}, \mathsf{M}[r_{u,v}], [\mathsf{L}_{\gamma,\hat{z}_{u,v}}]),$$

with bit $\hat{z}_{u,v}$ defined as

$$\hat{z}_{u,v} = (\lambda_\alpha \oplus u) \wedge (\lambda_\beta \oplus v) \oplus \lambda_\gamma.$$

Here, $[\mathsf{L}_{\gamma,\hat{z}_{u,v}}]$ is $\mathsf{P_A}$'s share of the garbled label; $r_{u,v}$ is $\mathsf{P_A}$'s share of the bit $\hat{z}_{u,v}$; and $\mathsf{P_B}$ holds the corresponding share $s_{u,v}$ such that $r_{u,v} \oplus s_{u,v} = \hat{z}_{u,v}$. The value $\mathsf{M}[r_{u,v}]$ is the MAC authenticating the underlying bit to $\mathsf{P_B}$. Also note that the definition of $\hat{z}_{u,v}$ indicates that when $u = \hat{z}_\alpha$ and $v = \hat{z}_\beta$ then $\hat{z}_{u,v} = \hat{z}_\gamma$.

Suppose the evaluator $\mathsf{P_B}$ holds $(u, \mathsf{L}_{\alpha,u})$ and $(v, \mathsf{L}_{\beta,v})$, where $u = \hat{z}_\alpha$ and $v = \hat{z}_\beta$. Then $\mathsf{P_B}$ can evaluate this AND gate by decrypting $G_{u,v}$ to obtain $r_{u,v}$ and $\mathsf{P_A}$'s share of $\mathsf{L}_{\gamma,\hat{z}_{u,v}}$. After verifying the MAC on $r_{u,v}$, party $\mathsf{P_B}$ can combine these values with its own shares to reconstruct the masked output value $\hat{z}_{u,v}$ (that is, $\hat{z}_\gamma$) and its corresponding label $\mathsf{L}_{\gamma,\hat{z}_{u,v}}$ (that is, $\mathsf{L}_{\gamma,\hat{z}_\gamma}$).

Assuming that the authenticated bits and shares of the labels can be computed securely, the above protocol is secure against malicious adversaries. In particular, even if $\mathsf{P_A}$ cheats and causes $\mathsf{P_B}$ to abort during evaluation, any such abort depends only on the *masked* values on the wires. Since the masks are random and unknown to either party, this means that any abort is input-independent. The MACs checked by $\mathsf{P_B}$ ensure correctness, namely that evaluation has resulted in the correct (masked) output-wire value.

From authenticated shares to shared labels. Another important optimization in the WRK protocol is to compute shares of labels efficiently using authenticated shares. Assume the parties hold an authenticated share $\langle r \mid s \rangle$ of some mask $\lambda = s \oplus r$. It is then easy to compute a share of $\lambda\Delta_\mathsf{A}$, since

$$\lambda\Delta_\mathsf{A} = (r \oplus s)\Delta_\mathsf{A} = \left(r\Delta_\mathsf{A} \oplus \mathsf{K}[s]\right) \oplus \left(\mathsf{M}[s]\right).$$

Since $\mathsf{P_A}$ has r, Δ_A, and $\mathsf{K}[s]$ while $\mathsf{P_B}$ has $\mathsf{M}[s]$, the two parties can locally compute shares of $\lambda\Delta_\mathsf{A}$ (namely, $[\lambda\Delta_\mathsf{A}]$) given only $\langle r \mid s \rangle$.

We can use this fact to compute shares of labels for a secret masked bit efficiently. Assuming the global authentication key (i.e., Δ_A) is also used as the free-XOR shift, then it holds that $\mathsf{L}_{\gamma,\hat{z}_{u,v}} = \mathsf{L}_{\gamma,0} \oplus \hat{z}_{u,v}\Delta_\mathsf{A}$. Therefore, the task of computing shares of labels reduces to the task of computing shares of $\hat{z}_{u,v}\Delta_\mathsf{A}$, since $\mathsf{L}_{\gamma,0}$ is known to $\mathsf{P_A}$.

Notice that

$$\hat{z}_{u,v}\Delta_\mathsf{A} = ((\lambda_\alpha \oplus u) \wedge (\lambda_\beta \oplus v) \oplus \lambda_\gamma)\,\Delta_\mathsf{A}$$
$$= \lambda_\alpha\lambda_\beta\Delta_\mathsf{A} \oplus u\lambda_\alpha\Delta_\mathsf{A} \oplus v\lambda_\beta\Delta_\mathsf{A} \oplus uv\Delta_\mathsf{A} \oplus \lambda_\gamma\Delta_\mathsf{A}.$$

If the parties hold an authenticated AND triple $(\langle r_\alpha \mid s_\alpha \rangle, \langle r_\beta \mid s_\beta \rangle, \langle r_\gamma^* \mid s_\gamma^* \rangle)$ and a random authenticated share $\langle r_\gamma \mid s_\gamma \rangle$ such that $\lambda_\alpha = r_\alpha \oplus s_\alpha$, $\lambda_\beta = r_\beta \oplus s_\beta$,

$\lambda_\alpha \wedge \lambda_\beta = r_\gamma^* \oplus s_\gamma^*$, and $\lambda_\gamma = r_\gamma \oplus s_\gamma$. The parties can then locally compute shares of $\lambda_\alpha \Delta_A$, $\lambda_\beta \Delta_A$, $\lambda_\gamma \Delta_A$, and $(\lambda_\alpha \wedge \lambda_\beta)\Delta_A$, and finally compute shares of $\hat{z}_{u,v}\Delta_A$ by linearly combining the above shares.

3   Overview of Our Optimizations

We separately discuss our optimizations for the authenticated garbling and the preprocessing phases. Details and proofs can be found in Sects. 4 and 5.

3.1   Improving Authenticated Garbling

As a high level, the key ideas behind authenticated garbling are that (1) it is possible to share garbled circuits such that neither party knows how rows in the garbled tables are permuted (since no party knows the masks on the wires); moreover, (2) information-theoretic MACs can be used to ensure correctness of the garbled tables. In the original protocol by Wang et al., these two aspects are tightly integrated: each garbled row includes an encryption of the corresponding MAC tag, so the evaluator only learns one such tag for each gate.

Here, we take a slightly different perspective on how authenticated garbling works. In particular, we (conceptually) divide the protocol into two parts:

- In the first part, the parties compute a shared garbled circuit, without any authentication, and let the evaluator reconstruct and evaluate that garbled circuit. We stress here that, even though there is no authentication, corrupting one or more garbled rows does not allow a selective-failure attack for the same reason as in the WRK protocol: any failure depends only on the *masked* wire values, but neither party knows those masks.
  This part is achieved by the encrypted wire labels alone, which have the form $H(\mathsf{L}_{\alpha,u}, \mathsf{L}_{\beta,v}) \oplus [\mathsf{L}_{\gamma,\hat{z}_{u,v}}]$. These require 4κ bits of communication per gate.
- In the second part, the evaluator holds masked wire values for every wire of the circuit. It then checks correctness of all these masked values. For example, it will ensure that for every AND gate, the underlying (real) values on the wires form an AND relationship. Such verification is needed for masked values that $\mathsf{P_B}$ obtains during the evaluation of the garbled circuit.
  The WRK protocol achieves this by encrypting *authenticated shares* of the form $H(\mathsf{L}_{\alpha,u}, \mathsf{L}_{\beta,v}) \oplus (r_{u,v}, \mathsf{M}[r_{u,v}])$ in each row of a garbled table. The evaluator decrypts one of the rows and checks the appropriate tag. These encrypted tags contribute 4ρ bits of communication per gate.

With this new way of viewing authenticated garbling, we can optimize each part independently. By doing so, we are able to reduce the communication of the first part to $2\kappa + 1$ bits per gate, and reduce the communication of the second part to 1 bit per gate. In the process, we also reduce the computation (in terms of hash evaluations) by about half. In the following, we discuss intuitively how these optimizations work.

Applying row-reduction techniques. In garbled circuits, *row reduction* refers to techniques that use fewer than four garbled rows per garbled gate [11,27,32, 40]. We review the simplest row-reduction technique here, describe the challenge of applying the technique to authenticated garbling, and then show how we overcome the challenge. This will serve as a warm-up to our final protocol that is compatible with the half-gate technique.

In classical garbling, a garbled AND gate can be written as (in our notation):

$$G_{0,0} = H(\mathsf{L}_{\alpha,0}, \mathsf{L}_{\beta,0}) \oplus \mathsf{L}_{\gamma,\hat{z}_{0,0}} = H(\mathsf{L}_{\alpha,0}, \mathsf{L}_{\beta,0}) \oplus \mathsf{L}_{\gamma,0} \oplus \hat{z}_{0,0}\Delta_\mathsf{A}$$
$$G_{0,1} = H(\mathsf{L}_{\alpha,0}, \mathsf{L}_{\beta,1}) \oplus \mathsf{L}_{\gamma,\hat{z}_{0,1}} = H(\mathsf{L}_{\alpha,0}, \mathsf{L}_{\beta,1}) \oplus \mathsf{L}_{\gamma,0} \oplus \hat{z}_{0,1}\Delta_\mathsf{A}$$
$$G_{1,0} = H(\mathsf{L}_{\alpha,1}, \mathsf{L}_{\beta,0}) \oplus \mathsf{L}_{\gamma,\hat{z}_{1,0}} = H(\mathsf{L}_{\alpha,1}, \mathsf{L}_{\beta,0}) \oplus \mathsf{L}_{\gamma,0} \oplus \hat{z}_{1,0}\Delta_\mathsf{A}$$
$$G_{1,1} = H(\mathsf{L}_{\alpha,1}, \mathsf{L}_{\beta,1}) \oplus \mathsf{L}_{\gamma,\hat{z}_{1,1}} = H(\mathsf{L}_{\alpha,1}, \mathsf{L}_{\beta,1}) \oplus \mathsf{L}_{\gamma,0} \oplus \hat{z}_{1,1}\Delta_\mathsf{A}.$$

The idea behind GRR3 row reduction [27] is to choose wire labels so $G_{0,0} = 0^\kappa$. That is, the garbler chooses

$$\mathsf{L}_{\gamma,0} := H(\mathsf{L}_{\alpha,0}, \mathsf{L}_{\beta,0}) \oplus \hat{z}_{0,0}\Delta_\mathsf{A}.$$

The garbler now needs to send only $(G_{0,1}, G_{1,0}, G_{1,1})$, reducing the communication from 4κ to 3κ bits. If the evaluator has input wires with masked values $(0,0)$, it can simply set $G_{0,0} = 0^\kappa$ and then proceed as before.

In authenticated garbling, the preprocessing results in shares of $\{\hat{z}_{u,v}\Delta_\mathsf{A}\}$. Hence, if $\mathsf{P_A}$ could compute $\mathsf{L}_{\gamma,0}$ then the parties could locally compute shares of the $\{G_{u,v}\}$ (since $\mathsf{P_A}$ knows all the $\mathsf{L}_{\alpha,u}, \mathsf{L}_{\beta,v}$ values and their hashes). $\mathsf{P_A}$ could then send its shares to $\mathsf{P_B}$ to allow $\mathsf{P_B}$ to recover the entire garbled gate. Unfortunately, $\mathsf{P_A}$ cannot compute $\mathsf{L}_{\gamma,0}$ because $\mathsf{P_A}$ does not know $\hat{z}_{0,0}$! Indeed, that value depends on the secret wire masks, unknown to either party.

Summarizing, row-reduction techniques in general compute one (or both) of the output-wire labels as a function of the input-wire labels **and** the secret masks, making them a challenge for authenticated garbling.

Our observation is that although $\mathsf{P_A}$ does not know $\hat{z}_{0,0}$, the garbling requires only $\hat{z}_{0,0}\Delta_\mathsf{A}$ for which the parties do have shares. Let S_A and S_B denote the parties' shares of this value, so that $S_A \oplus S_B = \hat{z}_{0,0}\Delta_\mathsf{A}$. Our main idea is for the parties to "shift" the entire garbling process by the value S_B, as follows:

1. $\mathsf{P_A}$ computes $\mathsf{L}_{\gamma,0} := H(\mathsf{L}_{\alpha,0}, \mathsf{L}_{\beta,0}) \oplus S_A$. Note this value differs from the standard garbling value by a shift of S_B. Intuitively, instead of choosing $\mathsf{L}_{\gamma,0}$ so that $G_{0,0} = 0^\kappa$, we set implicitly set $G_{0,0} = S_B$. Although $\mathsf{P_A}$ does not know S_B, it only matters that the evaluator $\mathsf{P_B}$ knows it.
2. Based on this value of $\mathsf{L}_{\gamma,0}$, the parties locally compute shares of the garbled gate $G_{0,1}, G_{1,0}, G_{1,1}$ defined above, and open them to $\mathsf{P_B}$.
3. When $\mathsf{P_B}$ evaluates the gate on input $\mathsf{L}_{\alpha,u}, \mathsf{L}_{\beta,v}$, if $(u,v) \neq (0,0)$ then evaluation is the same as usual. If $(u,v) = (0,0)$ then $\mathsf{P_B}$ sets $G_{0,0} = S_B$. This is equivalent to $\mathsf{P_B}$ doing the usual evaluation but shifting the result by S_B.

Using the half-gate technique. The state-of-the-art in semi-honest garbling is the half-gate construction of Zahur et al. [40]. It requires 2κ bits of communication per AND gate, while being compatible with free-XOR. We describe this idea, translated from the original work [40] to be written in terms of masks and masked wire values so as to match our notation.

The circuit garbler computes a garbled gate as:

$$G_0 := H(\mathsf{L}_{\alpha,0}) \oplus H(\mathsf{L}_{\alpha,1}) \oplus \lambda_\beta \Delta_\mathsf{A}$$
$$G_1 := H(\mathsf{L}_{\beta,0}) \oplus H(\mathsf{L}_{\beta,1}) \oplus \mathsf{L}_{\alpha,0} \oplus \lambda_\alpha \Delta_\mathsf{A},$$

and computes the 0-label for that gate's output wire as:

$$\mathsf{L}_{\gamma,0} := H(\mathsf{L}_{\alpha,0}) \oplus H(\mathsf{L}_{\beta,0}) \oplus (\lambda_\alpha \lambda_\beta \oplus \lambda_\gamma)\Delta_\mathsf{A}.$$

If the evaluator $\mathsf{P_B}$ holds masked values u, v and corresponding labels $\mathsf{L}_{\alpha,u}, \mathsf{L}_{\beta,v}$, it computes:

$$\mathsf{Eval}(u, v, \mathsf{L}_{\alpha,u}, \mathsf{L}_{\beta,v}) := H(\mathsf{L}_{\alpha,u}) \oplus H(\mathsf{L}_{\beta,v}) \oplus uG_0 \oplus v(G_1 \oplus \mathsf{L}_{\alpha,u}).$$

This results in the value

$$\begin{aligned}
\mathsf{Eval}(u, v, \mathsf{L}_{\alpha,u}, \mathsf{L}_{\beta,v}) &= H(\mathsf{L}_{\alpha,0}) \oplus H(\mathsf{L}_{\beta,0}) \oplus (uv \oplus v\lambda_\alpha \oplus u\lambda_\beta)\Delta_\mathsf{A} \\
&= H(\mathsf{L}_{\alpha,0}) \oplus H(\mathsf{L}_{\beta,0}) \oplus \Big((u \oplus \lambda_\alpha)(v \oplus \lambda_\beta) \oplus \lambda_\alpha \lambda_\beta\Big)\Delta_\mathsf{A} \\
&= H(\mathsf{L}_{\alpha,0}) \oplus H(\mathsf{L}_{\beta,0}) \oplus (\hat{z}_{u,v} \oplus \lambda_\alpha \lambda_\beta \oplus \lambda_\gamma)\Delta_\mathsf{A},
\end{aligned}$$

which is the correct output $\mathsf{L}_{\gamma,\hat{z}_{u,v}} = \mathsf{L}_{\gamma,0} \oplus \hat{z}_{u,v}\Delta_\mathsf{A}$.

As before, this garbling technique is problematic for authenticated garbling, because the garbler $\mathsf{P_A}$ cannot compute $\mathsf{L}_{\gamma,0}$ as specified. ($\mathsf{P_A}$ does not know the wire masks, so cannot compute the term $(\lambda_\alpha \lambda_\beta \oplus \lambda_\gamma)\Delta_\mathsf{A}$.)

However, the parties hold[1] shares of this value; say, $S_A \oplus S_B = (\lambda_\alpha \lambda_\beta \oplus \lambda_\gamma)\Delta_\mathsf{A}$. We can thus conceptually "shift" the entire garbling procedure by S_B to obtain the following interactive variant of half-gates:

1. $\mathsf{P_A}$ computes the output wire label as

$$\mathsf{L}_{\gamma,0} := H(\mathsf{L}_{\alpha,0}) \oplus H(\mathsf{L}_{\beta,0}) \oplus S_A,$$

   which is "shifted" by S_B from what the half-gates technique specifies.
2. The parties locally compute shares of G_0, G_1 as per the half-gates technique described above. These shares are opened to $\mathsf{P_B}$, so $\mathsf{P_B}$ learns (G_0, G_1).
3. To evaluate the gate on inputs $\mathsf{L}_{\alpha,u}, \mathsf{L}_{\beta,v}$, the evaluator $\mathsf{P_B}$ performs standard half-gates evaluation and then adds S_B as a correction value. This results in the correct output-wire label, since:

$$\begin{aligned}
\mathsf{Eval}(\mathsf{L}_{\alpha,u}, \mathsf{L}_{\beta,v}) \oplus S_B &= \mathsf{Eval}(\mathsf{L}_{\alpha,u}, \mathsf{L}_{\beta,v}) \oplus (\lambda_\alpha \lambda_\beta \oplus \lambda_\gamma)\Delta_\mathsf{A} \oplus S_A \\
&= H(\mathsf{L}_{\alpha,0}) \oplus H(\mathsf{L}_{\beta,0}) \oplus \hat{z}_{u,v}\Delta_\mathsf{A} \oplus S_A \\
&= \mathsf{L}_{\gamma,0} \oplus \hat{z}_{u,v}\Delta_\mathsf{A} \\
&= \mathsf{L}_{\gamma,\hat{z}_{u,v}}.
\end{aligned}$$

---

[1] Note that $(\lambda_\alpha \lambda_\beta \oplus \lambda_\gamma) = \hat{z}_{0,0}$, the same secret value as in the previous example.

Authentication almost for free. In the WRK scheme, suppose the actual values on the wires of an AND gate are $z_\alpha, z_\beta, z_\gamma$ with $z_\alpha \wedge z_\beta = z_\gamma$. During evaluation, P_B learn masked values $\hat{z}_\alpha = z_\alpha \oplus \lambda_\alpha$, $\hat{z}_\beta = z_\beta \oplus \lambda_\beta$, and $\hat{z}_\gamma = z_\gamma \oplus \lambda_\gamma$. For correctness it suffices to show that

$$z_\alpha \wedge z_\beta = z_\gamma \iff (\hat{z}_\alpha \oplus \lambda_\alpha) \wedge (\hat{z}_\beta \oplus \lambda_\beta) = (\hat{z}_\gamma \oplus \lambda_\gamma)$$
$$\iff \underbrace{(\hat{z}_\alpha \oplus \lambda_\alpha) \wedge (\hat{z}_\beta \oplus \lambda_\beta) \oplus \lambda_\gamma}_{\hat{z}_{\alpha,\beta}} = \hat{z}_\gamma.$$

Note the parties already have authenticated shares of $\lambda_\alpha, \lambda_\beta, \lambda_\gamma$, and $(\lambda_\alpha \wedge \lambda_\beta)$, so they can also derive authenticated shares of related values.

In the WRK scheme the garbler P_A prepares an authenticated share (MAC) of $\hat{z}_{\alpha,\beta}$ corresponding to each of the 4 possible values of $\hat{z}_\alpha, \hat{z}_\beta$. It encrypts this share so that it can only be opened using the corresponding wire labels. P_B can then decrypt and verify the relevant $\hat{z}_{\alpha,\beta}$ value (and take it to be the masked output value $\hat{z}_\gamma$).

Our approach is to apply a technique suggested for the SPDZ protocol [9]: evaluate the circuit without authentication and then perform batch authentication at the end. Thus, in our new protocol authentication works as follows:

1. P_B evaluates the circuit, obtaining (unauthenticated) masked values $\hat{z}_\alpha$ for every wire α.
2. P_B reveals the masked values of every wire (1 bit per wire). Revealing these to P_A does not affect privacy because the masks are hidden from both parties (except for certain input/output wires where one or both of the parties already know the underlying values).
3. P_A generates authenticated shares of only the relevant $\hat{z}_{\alpha,\beta}$ values and sends them. P_B verifies the authenticity of each share. This is equivalent to sending a MAC of P_A's shares. As described in Sect. 2, this can be done by sending only a hash of the MACs.

This technique for authentication adds an extra round, but it makes the authentication almost free in terms of communication. P_B sends 1 bit per wire and P_A sends only a single hash value to authenticate.

Details of the optimizations described above can be found in Sect. 4.

3.2   Improving the Preprocessing Phase

We also improve the efficiency of preprocessing in the WRK protocol significantly; specifically: (1) we design a new protocol for generating so-called leaky-AND triples. Compared to the best previous protocol by Wang et al., it reduces the number of hash calls by 2.5× and reduces communication by κ bits. (2) we propose a new function-dependent preprocessing protocol that can be computed much more efficiently. We remark that the second optimization is particularly suitable for RAM-model secure computation, where CPU circuits are fixed ahead of time.

To enable the above optimizations, we set $\mathsf{lsb}(\Delta_\mathsf{A}) := 1$ and $\mathsf{lsb}(\Delta_\mathsf{B}) := 0$, where $\mathsf{lsb}(x)$ denotes the least significant bit of x.

A new leaky-AND protocol. The output of a leaky-AND protocol is a random authenticated AND triple $(\langle r_\alpha \mid s_\alpha \rangle, \langle r_\beta \mid s_\beta \rangle, \langle r_\gamma^* \mid s_\gamma^* \rangle)$ with one caveat: the adversary can choose to guess the value of $r_\alpha \oplus s_\alpha$. A correct guess remains undetected while an incorrect guess will be caught. (See Fig. 4 for a formal definition.) The leaky-AND protocol by Wang et al. works in two steps. Two parties first run a protocol whose outputs are triples that are leaky without any correctness guarantee; then a checking procedure is run to ensure correctness. The leakage is later eliminated by bucketing. In our new protocol, we observe that these two steps can be computed at the same time, reducing the number of rounds as well as the amount of computation (i.e., H-evaluations). Moreover, computing and checking can be further improved by adopting ideas from the half-gate technique. Details are below.

Recall that in the half-gate approach, if a wire is associated with wire labels $(\mathsf{L}_0, \mathsf{L}_1 = \mathsf{L}_0 \oplus \Delta_\mathsf{A})$, the first row of the gate computed by the garbler has the form

$$G = H(\mathsf{L}_0) \oplus H(\mathsf{L}_1) \oplus C,$$

for some C. An evaluator holding (b, L_b) can evaluate it as

$$
\begin{aligned}
E &= bG \oplus H(\mathsf{L}_b) \\
&= b(H(\mathsf{L}_0) \oplus H(\mathsf{L}_1) \oplus C) \oplus H(\mathsf{L}_b) \\
&= b(H(\mathsf{L}_0) \oplus H(\mathsf{L}_1)) \oplus H(\mathsf{L}_b) \oplus bC \\
&= H(\mathsf{L}_0) \oplus bC.
\end{aligned}
\tag{1}
$$

Correctness ensures that $E \oplus H(\mathsf{L}_0) = bC$, which means that after the evaluation the two parties hold shares of bC. Note that when free-XOR is used with shift Δ_A, then a pair of garbled labels $(\mathsf{L}_0, \mathsf{L}_1)$ and the IT-MAC for a bit (i.e., $(\mathsf{K}[b], \mathsf{M}[b])$) have the same structure. Therefore the above can be reformulated and extended as follows:

$$
\begin{aligned}
G &= H(\mathsf{K}[b]) \oplus H(\mathsf{M}[b]) \oplus C_1 \\
E &= bG \oplus H(\mathsf{M}[b]) \oplus bC_2.
\end{aligned}
$$

Assuming the two parties have an authenticated bit $[b]_\mathsf{B}$, then $E \oplus H(\mathsf{K}[b]) = b(C_1 \oplus C_2)$. If we view C_1 and C_2 as shares of some value $C = C_1 \oplus C_2$, then this can be interpreted as a way to select on a shared value such that the selection bit b is known only to one party and at the same time the output (namely, $bC = H(\mathsf{K}[b]) \oplus E$) is still shared.

Now we are ready to present our protocol. We will start with a set of random authenticated bits $(\langle x_1 \mid x_2 \rangle, \langle y_1 \mid y_2 \rangle, \langle z_1 \mid r \rangle)$. We want the two parties to directly compute shares of

$$S = ((x_1 \oplus x_2) \wedge (y_1 \oplus y_2) \oplus z_1 \oplus r)(\Delta_A \oplus \Delta_B).$$

Assuming $\mathsf{lsb}(\Delta_A \oplus \Delta_B) = 1$, revealing $d = \mathsf{lsb}(S)$ allows the parties to "fix" these random authenticated shares to a valid triple (by computing $[z_2]_B = [r]_B \oplus d$). Once the parties hold shares of S (for example, P_A holds S_1 and P_B holds $S_2 = S \oplus S_1$), checking the correctness of d also becomes easy: d is valid if and only if $S_1 \oplus d\Delta_A$ from P_A equals to $S_2 \oplus d\Delta_B$ from P_B. A wrong d can pass the equality check only if the adversary guesses the other party's Δ value. Now the task is to compute shares of S, where S can be rewritten as

$$S = x_1(y_1 \oplus y_2)(\Delta_A \oplus \Delta_B) \oplus x_2(y_1 \oplus y_2)(\Delta_A \oplus \Delta_B) \oplus (z_1 \oplus r)(\Delta_A \oplus \Delta_B).$$

Here, we will focus on how to compute shares of

$$x_2(y_1\Delta_A \oplus y_1\Delta_B \oplus y_2\Delta_A \oplus y_2\Delta_B).$$

Now we apply the half-gate observation: P_A has $C_1 = y_1\Delta_A \oplus \mathsf{K}[y_2] \oplus \mathsf{M}[y_1]$ and P_B has $C_2 = y_2\Delta_B \oplus \mathsf{K}[y_1] \oplus \mathsf{M}[y_2]$, and we have

$$x_2(C_1 \oplus C_2) = x_2(y_1\Delta_A \oplus y_1\Delta_B \oplus y_2\Delta_A \oplus y_1\Delta_B).$$

Therefore, this value can be computed by P_A sending one ciphertext to P_B. Given the above observations, the final protocol can be derived in a straightforward way. Overall this new approach improves communication by $1.2\times$ and improves computation by $2\times$.

For details and a security proof corresponding to the above, see Sect. 5.1.

New function-dependent preprocessing. Here we show how to further improve the efficiency of function-dependent preprocessing. Recall that in the WRK protocol, each AND triple is derived from B leaky-AND triples, for $B \approx \frac{\rho}{\log C}$; these triples are then used to multiply authenticated masked values for each AND gate of the circuit. Our observation is that we can reduce the number of authenticated shares needed per gate from $3B+2$ to $3B-1$. This idea was initially used by Araki et al. [3] in the setting of honest-majority three-party computation. See Sect. 5.2 for details.

4   Technical Details: Improved Authenticated Garbling

Since we already discussed the main intuition of the protocol in the previous section, we will present our main protocol in the $\mathcal{F}_{\mathsf{pre}}$-hybrid model. Detailed protocol description is shown in Fig. 3. Each step in the protocol can be summarized as follows:

1. Parties generate circuit preprocessing information using $\mathcal{F}_{\mathsf{pre}}$.
2. P_A computes its own share of the garbled circuit and sends to P_B.
3-4. Parties process P_A and P_B's input and let P_B learn the corresponding masked input wire values and garbled labels.
5. P_B locally reconstructs the garbled circuit and evaluates it.

Protocol Π_{2pc}

Inputs: P_A holds $x \in \{0,1\}^{\mathcal{I}_1}$ and P_A holds $y \in \{0,1\}^{\mathcal{I}_2}$. Parties agree on a circuit for a function $f : \{0,1\}^{\mathcal{I}_1} \times \{0,1\}^{\mathcal{I}_2} \to \{0,1\}^{\mathcal{O}}$.

1. P_A and P_B call $\mathcal{F}_{pre}$, which sends Δ_A to P_A, Δ_B to P_B, and sends $\{\langle r_w \mid s_w \rangle\}_{w \in \mathcal{I} \cup \mathcal{W}}$, $\{\langle r_w^* \mid s_w^* \rangle\}_{w \in \mathcal{W}}$ to P_A and P_B. For each $w \in \mathcal{I}_1 \cup \mathcal{I}_2$, P_A also picks a uniform κ-bit string $L_{w,0}$.

2. Following the topological order of the circuit, for each gate $\mathcal{G} = (\alpha, \beta, \gamma, T)$,
   - If $T = \oplus$, P_A computes $L_{\gamma,0} := L_{\alpha,0} \oplus L_{\beta,0}$
   - If $T = \wedge$, P_A computes $L_{\alpha,1} := L_{\alpha,0} \oplus \Delta_A$, $L_{\beta,1} := L_{\beta,0} \oplus \Delta_A$, and
     $$G_{\gamma,0} := H(L_{\alpha,0}, \gamma) \oplus H(L_{\alpha,1}, \gamma) \oplus K[s_\beta] \oplus r_\beta \Delta_A$$
     $$G_{\gamma,1} := H(L_{\beta,0}, \gamma) \oplus H(L_{\beta,1}, \gamma) \oplus K[s_\alpha] \oplus r_\alpha \Delta_A \oplus L_{\alpha,0}$$
     $$L_{\gamma,0} := H(L_{\alpha,0}, \gamma) \oplus H(L_{\beta,0}, \gamma) \oplus K[s_\gamma] \oplus r_\gamma \Delta_A \oplus K[s_\gamma^*] \oplus r_\gamma^* \Delta_A$$
     $$b_\gamma := \mathsf{lsb}(L_{\gamma,0})$$
     P_A sends $G_{\gamma,0}, G_{\gamma,1}, b_\gamma$ to P_B.

3. For each $w \in \mathcal{I}_2$, two parties compute $r_w := \mathsf{Open}([r_w]_A)$. P_B then sends $y_w \oplus \lambda_w := y_w \oplus s_w \oplus r_w$ to P_A. Finally, P_A sends $L_{w, y_w \oplus \lambda_w}$ to P_B.

4. For each $w \in \mathcal{I}_1$, two parties compute $s_w := \mathsf{Open}([s_w]_B)$. P_A then sends $x_w \oplus \lambda_w := x_w \oplus s_w \oplus r_w$ and $L_{w, x_w \oplus \lambda_w}$ to P_B.

5. P_B evaluates the circuit in topological order. For each gate $\mathcal{G} = (\alpha, \beta, \gamma, T)$, P_B initially holds $(z_\alpha \oplus \lambda_\alpha, L_{\alpha, z_\alpha \oplus \lambda_\alpha})$ and $(z_\beta \oplus \lambda_\beta, L_{\beta, z_\beta \oplus \lambda_\beta})$, where z_α, z_β are the underlying values of the wires.
   (a) If $T = \oplus$, P_B computes $z_\gamma \oplus \lambda_\gamma := (z_\alpha \oplus \lambda_\alpha) \oplus (z_\beta \oplus \lambda_\beta)$ and $L_{\gamma, z_\gamma \oplus \lambda_\gamma} := L_{\alpha, z_\alpha \oplus \lambda_\alpha} \oplus L_{\beta, z_\beta \oplus \lambda_\beta}$.
   (b) If $T = \wedge$, P_B computes $G_0 := G_{\gamma,0} \oplus M[s_\beta]$, and $G_1 := G_{\gamma,1} \oplus M[s_\alpha]$. P_B evaluates the garbled table (G_0, G_1) to obtain the output label
     $$L_{\gamma, z_\gamma \oplus \lambda_\gamma} := H(L_{\alpha, z_\alpha \oplus \lambda_\alpha}, \gamma) \oplus H(L_{\beta, z_\beta \oplus \lambda_\beta}, \gamma) \oplus M[s_\gamma] \oplus M[s_\gamma^*]$$
     $$\oplus (z_\alpha \oplus \lambda_\alpha) G_0 \oplus (z_\beta \oplus \lambda_\beta)(G_1 \oplus L_{\alpha, z_\alpha \oplus \lambda_\alpha})$$
     and $z_\gamma \oplus \lambda_\gamma := b_\gamma \oplus \mathsf{lsb}(L_{\gamma, z_\gamma \oplus \lambda_\gamma})$

6. For each $w \in \mathcal{W}$, P_B sends $\hat{z}_w := z_w \oplus \lambda_w$ to P_A.

7. For each AND gates $(\alpha, \beta, \gamma, \wedge)$, both parties know $\hat{z}_\alpha = z_\alpha \oplus \lambda_\alpha$, $\hat{z}_\beta = z_\beta \oplus \lambda_\beta$, and $\hat{z}_\gamma = z_\gamma \oplus \lambda_\gamma$. Two parties compute authenticated share of bit c_γ defined as
   $$c_\gamma = (\hat{z}_\alpha \oplus \lambda_\alpha) \wedge (\hat{z}_\beta \oplus \lambda_\beta) \oplus (\hat{z}_\gamma \oplus \lambda_\gamma).$$
   Note that c_γ is a linear combination of $\lambda_\alpha, \lambda_\beta, \lambda_\gamma$ and $\lambda_\gamma^* = \lambda_\alpha \wedge \lambda_\beta$, therefore authenticated share of c_γ can be computed locally.

8. Two parties use Open to check that c_γ is 0 for all gates γ, and abort if any check fails.

9. For each $w \in \mathcal{O}$, two parties compute $r_w := \mathsf{Open}([r_w]_A)$. P_B computes $z_w := (\lambda_w \oplus z_w) \oplus r_w \oplus s_w$.

Fig. 3. The main protocol in the $\mathcal{F}_{pre}$ hybrid model

6–8. P_B sends all masked wire values (including all input, output, and internal wires) to P_A; two parties check the correctness of all masked wire values.

9. P_A reveals the masks of output wires to P_B, who can recover the output.

Note that steps 2 through 9 are performed in the online phase, with $2\kappa + 2$ bits of communication per AND gate, $\kappa + 1$ bits of communication per input bit, and 1 bit of communication per output bit.

4.1 Proof of Security

We start by stating our main theorem.

Theorem 1. *If H is modeled as a random oracle, the protocol in Fig. 3 securely computes f against malicious adversaries in the $\mathcal{F}_{\mathsf{pre}}$-hybrid model.*

Before proceeding to the formal proof, we first introduce two important lemmas. The first lemma addresses correctness of our distributed garbling scheme in the semi-honest case; the second lemma addresses correctness of the whole protocol when P_A is corrupted.

Lemma 1. *When both parties follow the protocol honestly then, after step 5, for each wire w in the circuit P_B holds $(z_w \oplus \lambda_w, \mathsf{L}_{w, z_w \oplus \lambda_w})$.*

Proof. We prove this by induction on the gates in the circuit.

Base case. It is easy to verify from step 3 and step 4 that the lemma holds for input wires.

Induction step. XOR-gates are trivial and so focus on an AND gate $(\alpha, \beta, \gamma, \wedge)$. First, the garbled tables are computed distributively, therefore we first write down the table after P_B merged its own share as follows. Note that we ignore the gate id (γ) for simplicity.

$$
\begin{aligned}
G_0 &= H(\mathsf{L}_{\alpha,0}) \oplus H(\mathsf{L}_{\alpha,1}) \oplus \mathsf{K}[s_\beta] \oplus r_\beta \Delta_A \oplus \mathsf{M}[s_\beta] \\
&= H(\mathsf{L}_{\alpha,0}) \oplus H(\mathsf{L}_{\alpha,1}) \oplus \lambda_\beta \Delta_A \\
G_1 &= H(\mathsf{L}_{\beta,0}) \oplus H(\mathsf{L}_{\beta,1}) \oplus \mathsf{K}[s_\alpha] \oplus r_\alpha \Delta_A \oplus \mathsf{M}[s_\alpha] \oplus \mathsf{L}_{\alpha,0} \\
&= H(\mathsf{L}_{\beta,0}) \oplus H(\mathsf{L}_{\beta,1}) \oplus \lambda_\alpha \Delta_A \oplus \mathsf{L}_{\alpha,0}.
\end{aligned}
$$

P_A locally computes the output garbled label for 0 values, namely $\mathsf{L}_{\gamma,0}$ as:

$$
\mathsf{L}_{\gamma,0} := H(\mathsf{L}_{\alpha,0}) \oplus H(\mathsf{L}_{\beta,0}) \oplus \mathsf{K}[s_\gamma] \oplus r_\gamma \Delta_A \oplus \mathsf{K}[s_\gamma^*] \oplus r_\gamma^* \Delta_A.
$$

P_B, who holds $(z_\alpha \oplus \lambda_\alpha, \mathsf{L}_{\alpha, z_\alpha \oplus \lambda_\alpha})$ and $(z_\beta \oplus \lambda_\beta, \mathsf{L}_{\beta, z_\beta \oplus \lambda_\beta})$ by the induction hypothesis, evaluates the circuit as follows:

$$
\begin{aligned}
\mathsf{L}_{\gamma, z_\gamma \oplus \lambda_\gamma} :=\ & H(\mathsf{L}_{\alpha, z_\alpha \oplus \lambda_\alpha}) \oplus H(\mathsf{L}_{\beta, z_\beta \oplus \lambda_\beta}) \oplus (z_\alpha \oplus \lambda_\alpha) G_0 \\
& \oplus (z_\beta \oplus \lambda_\beta)(G_1 \oplus \mathsf{L}_{\alpha, z_\alpha \oplus \lambda_\alpha}) \oplus \mathsf{M}[s_\gamma] \oplus \mathsf{M}[s_\gamma^*].
\end{aligned}
$$

Observe that

$$(z_\alpha \oplus \lambda_\alpha)G_0 \oplus H(\mathsf{L}_{\alpha, z_\alpha \oplus \lambda_\alpha})$$
$$= (z_\alpha \oplus \lambda_\alpha)\,(H(\mathsf{L}_{\alpha,0}) \oplus H(\mathsf{L}_{\alpha,1}) \oplus \lambda_\beta \varDelta_\mathsf{A}) \oplus H(\mathsf{L}_{\alpha, z_\alpha \oplus \lambda_\alpha})$$
$$= (z_\alpha \oplus \lambda_\alpha)\,(H(\mathsf{L}_{\alpha,0}) \oplus H(\mathsf{L}_{\alpha,1}) \oplus \lambda_\beta \varDelta_\mathsf{A}) \oplus (z_\alpha \oplus \lambda_\alpha)\,(H(\mathsf{L}_{\alpha,0}) \oplus H(\mathsf{L}_{\alpha,1})) \oplus H(\mathsf{L}_{\alpha,0})$$
$$= H(\mathsf{L}_{\alpha,0}) \oplus \lambda_\beta (z_\alpha \oplus \lambda_\alpha)\varDelta_\mathsf{A},$$

and

$$(z_\beta \oplus \lambda_\beta)(G_1 \oplus \mathsf{L}_{\alpha, z_\alpha \oplus \lambda_\alpha}) \oplus H(\mathsf{L}_{\beta, z_\beta \oplus \lambda_\beta})$$
$$= (z_\beta \oplus \lambda_\beta)\,(H(\mathsf{L}_{\beta,0}) \oplus H(\mathsf{L}_{\beta,1}) \oplus \lambda_\alpha \varDelta_\mathsf{A} \oplus (z_\alpha \oplus \lambda_\alpha)\varDelta_\mathsf{A}) \oplus H(\mathsf{L}_{\beta, z_\beta \oplus \lambda_\beta})$$
$$= (z_\beta \oplus \lambda_\beta)\,(H(\mathsf{L}_{\beta,0}) \oplus H(\mathsf{L}_{\beta,1}) \oplus z_\alpha \varDelta_\mathsf{A}) \oplus (z_\beta \oplus \lambda_\beta)\,(H(\mathsf{L}_{\beta,0}) \oplus H(\mathsf{L}_{\beta,1})) \oplus H(\mathsf{L}_{\beta,0})$$
$$= H(\mathsf{L}_{\beta,0}) \oplus (\lambda_\beta \oplus z_\beta)z_\alpha \varDelta_\mathsf{A}.$$

Therefore, we conclude that

$$\mathsf{L}_{\gamma,0} \oplus \mathsf{L}_{\gamma, z_\gamma \oplus \lambda_\gamma}$$
$$= H(\mathsf{L}_{\alpha,0}) \oplus H(\mathsf{L}_{\beta,0}) \oplus H(\mathsf{L}_{\alpha, z_\alpha \oplus \lambda_\alpha}) \oplus H(\mathsf{L}_{\beta, z_\beta \oplus \lambda_\beta}) \oplus (z_\alpha \oplus \lambda_\alpha)G_0$$
$$\oplus (z_\beta \oplus \lambda_\beta)(G_1 \oplus \mathsf{L}_{\alpha, z_\alpha \oplus \lambda_\alpha}) \oplus \lambda_\gamma \varDelta_\mathsf{A} \oplus (\lambda_\alpha \wedge \lambda_\beta)\varDelta_\mathsf{A}$$
$$= (\lambda_\alpha \oplus z_\alpha)\lambda_\beta \varDelta_\mathsf{A} \oplus (\lambda_\beta \oplus z_\beta)z_\alpha \varDelta_\mathsf{A} \oplus \lambda_\gamma \varDelta_\mathsf{A} \oplus (\lambda_\alpha \wedge \lambda_\beta)\varDelta_\mathsf{A}$$
$$= ((z_\alpha \wedge z_\beta) \oplus \lambda_\gamma)\varDelta_\mathsf{A} = (z_\gamma \oplus \lambda_\gamma)\varDelta_\mathsf{A}.$$

This means that, with respect to $\mathsf{P_A}$'s definition of $\mathsf{L}_{\gamma, z_\gamma \oplus \lambda_\gamma}$, $\mathsf{P_B}$'s label is always correct. The masked value is correct because the least-significant bit of $\varDelta_\mathsf{A}$ is 1; thus,

$$b_\gamma \oplus \mathsf{lsb}(\mathsf{L}_{\gamma, z_\gamma \oplus \lambda_\gamma}) = \mathsf{lsb}(\mathsf{L}_{\gamma,0}) \oplus \mathsf{lsb}(\mathsf{L}_{\gamma, z_\gamma \oplus \lambda_\gamma})$$
$$= \mathsf{lsb}(\mathsf{L}_{\gamma,0} \oplus \mathsf{L}_{\gamma, z_\gamma \oplus \lambda_\gamma})$$
$$= \mathsf{lsb}((z_\gamma \oplus \lambda_\gamma)\varDelta_\mathsf{A}) = z_\gamma \oplus \lambda_\gamma.$$

Lemma 2. *Let $x \overset{\mathrm{def}}{=} \hat{x}_w \oplus \lambda_w$ and $y \overset{\mathrm{def}}{=} \hat{y}_w \oplus \lambda_w$, where $\hat{x}_w$ is what $\mathsf{P_B}$ sends in step 3, $\hat{y}_w$ is what $\mathsf{P_A}$ sends in step 4, and λ_w is defined by $\mathcal{F}_{\mathsf{pre}}$. If $\mathsf{P_A}$ is malicious, then $\mathsf{P_B}$ either aborts or outputs $f(x, y)$.*

Proof. After step 5, $\mathsf{P_B}$ obtains a set of masked values $z_w \oplus \lambda_w$ for all wires w in the circuit. In the following, we will show that if these masked values are not correct, then $\mathsf{P_B}$ will abort with all but negligible probability.

Again we will prove by induction. Note that the lemma holds for all wires $w \in \mathcal{I}_1 \cup \mathcal{I}_2$, according to how x, y are defined, as well as for XOR-gates. In the following, we will focus on an AND gate $(\alpha, \beta, \gamma, \wedge)$. Now, according to induction hypothesis, we already know that $\mathsf{P_B}$ hold correct values of $(z_\alpha \oplus \lambda_\alpha, z_\beta \oplus \lambda_\beta)$.

Recall that the checking is done by computing

$$c = (\hat{z}_\alpha \oplus \lambda_\alpha) \wedge (\hat{z}_\beta \oplus \lambda_\beta) \oplus (\hat{z}_\gamma \oplus \lambda_\gamma).$$

The correctness of input masked values means that

$$c = z_\alpha \wedge z_\beta \oplus \hat{z}_\gamma \oplus \lambda_\gamma.$$

Since Open does not abort, $c = 0$, which means that $\hat{z}_\gamma = z_\alpha \wedge z_\beta \oplus \lambda_\gamma = z_\gamma \oplus \lambda_\gamma$. This means that the output masked wire value is also correct.

Given the above two lemmas, the proof of security of our main protocol is relatively easy. We provide all details below.

Proof. We consider separately a malicious P_A and P_B.

Malicious P_A. Let $\mathcal{A}$ be an adversary corrupting P_A. We construct a simulator $\mathcal{S}$ that runs $\mathcal{A}$ as a subroutine and plays the role of P_A in the ideal world involving an ideal functionality $\mathcal{F}$ evaluating f. $\mathcal{S}$ is defined as follows.

1. $\mathcal{S}$ plays the role of $\mathcal{F}_{pre}$ and records all values that $\mathcal{F}_{pre}$ sends to two parties.
2. $\mathcal{S}$ receives all values that $\mathcal{A}$ sends.
3. $\mathcal{S}$ acts as an honest P_B using input $y := 0$.
4. For each wire $w \in \mathcal{I}_1$, $\mathcal{S}$ receives $\hat{x}_w$ and computes $x_w := \hat{x}_w \oplus r_w \oplus s_w$, where r_w, s_w are the values used by $\mathcal{F}_{pre}$ in the previous steps.
6. $\mathcal{S}$ picks random bits for all $\hat{z}_w$ and send them to $\mathcal{A}$.
7–9. $\mathcal{S}$ acts as an honest P_B If an honest P_B would abort, $\mathcal{S}$ aborts; otherwise $\mathcal{S}$ computes the input x of $\mathcal{A}$. from the output of $\mathcal{F}_{pre}$ and the values $\mathcal{A}$ sent. $\mathcal{S}$ then sends x to $\mathcal{F}$.

We show that the joint distribution of the outputs of $\mathcal{A}$ and the honest P_B in the real world is indistinguishable from the joint distribution of the outputs of $\mathcal{S}$ and P_B in the ideal world. We prove this by considering a sequence of experiments, the first of which corresponds to the execution of our protocol and the last of which corresponds to execution in the ideal world, and showing that successive experiments are computationally indistinguishable.

Hybrid$_1$. This is the hybrid-world protocol, where we imagine $\mathcal{S}$ playing the role of an honest P_B using P_B's actual input y, while also playing the role of $\mathcal{F}_{pre}$.

Hybrid$_2$. Same as **Hybrid$_1$**, except that in step 6, for each wire $w \in \mathcal{I}_1$ the simulator $\mathcal{S}$ receives $\hat{x}_w$ and computes $x_w := \hat{x}_w \oplus r_w \oplus s_w$, where r_w, s_w are the values used by $\mathcal{F}_{pre}$. If an honest P_B would abort in any later step, $\mathcal{S}$ sends abort to $\mathcal{F}$; otherwise it sends $x = \{x_w\}_{w \in \mathcal{I}_1}$ to $\mathcal{F}$.

The distributions on the view of $\mathcal{A}$ in **Hybrid$_1$** and **Hybrid$_2$** are identical. The output P_B gets are the same due to Lemma 1 and Lemma 2.

Hybrid$_3$. Same as **Hybrid$_2$**, except that $\mathcal{S}$ uses $y' = 0$ in step 3 and ignore what $\mathcal{A}$ sends back. Then in step 6, $\mathcal{S}$ sends random bits instead of the value for $z_w \oplus \lambda_w$.

The distributions on the view of $\mathcal{A}$ in **Hybrid$_3$** and **Hybrid$_2$** are again identical (since the $\{s_w\}_{w \in \mathcal{I}_2}$ are uniform).

Note that **Hybrid$_3$** corresponds to the ideal-world execution described earlier. This completes the proof for a malicious P$_A$.

Malicious P$_B$. Let $\mathcal{A}$ be an adversary corrupting P$_B$. We construct a simulator $\mathcal{S}$ that runs $\mathcal{A}$ as a subroutine and plays the role of P$_B$ in the ideal world involving an ideal functionality $\mathcal{F}$ evaluating f. $\mathcal{S}$ is defined as follows.

1. $\mathcal{S}$ plays the role of $\mathcal{F}_{pre}$ and records all values sent to both parties.
2. $\mathcal{S}$ acts as an honest P$_A$ and send the shared garbled tables to P$_B$.
3. For each wire $w \in \mathcal{I}_2$, $\mathcal{S}$ receives $\hat{y}_w$ and computes $y_w := \hat{y}_w \oplus r_w \oplus s_w$, where r_w, s_w are the values used by $\mathcal{F}_{pre}$ in the previous steps.
4. $\mathcal{S}$ acts as an honest P$_A$ using input $x = 0$.
6–8. $\mathcal{S}$ acts as an honest P$_A$. If an honest P$_A$ would abort, $\mathcal{S}$ abort.
9. $\mathcal{S}$ sends y computed in step 3 to $\mathcal{F}$, which returns $z = f(x,y)$. $\mathcal{S}$ then computes $z' := f(0,y)$ and defines $r'_w = z_w \oplus z'_w \oplus r_w$ for each $w \in \mathcal{O}$. $\mathcal{S}$ then acts as an honest P$_A$ and opens values r'_w to $\mathcal{A}$. If an honest P$_A$ would abort, $\mathcal{S}$ $\mathcal{S}$ outputs whatever $\mathcal{A}$ outputs.

We now show that the distribution on the view of $\mathcal{A}$ in the real world is indistinguishable from the distribution on the view of $\mathcal{A}$ in the ideal world. (Note P$_A$ has no output.)

Hybrid$_1$. This is the hybrid-world protocol, where $\mathcal{S}$ acts as an honest P$_A$ using P$_A$'s actual input x, while playing the role of $\mathcal{F}_{pre}$.

Hybrid$_2$. Same as **Hybrid$_1$**, except that in step 3, $\mathcal{S}$ receives $\hat{y}_w$ and computes $y_w := \hat{y}_w \oplus r_w \oplus s_w$, where r_w, s_w are the values used by $\mathcal{F}_{pre}$. If an honest P$_A$ abort in any step, send abort to $\mathcal{F}$.

Hybrid$_3$. Same as **Hybrid$_2$**, except that in step 4, $\mathcal{S}$ acts as an honest P$_A$ with input $x = 0$. $\mathcal{S}$ sends x computed in step 3 to $\mathcal{F}$, which returns $z = f(x,y)$. $\mathcal{S}$ then computes $z' := f(0,y)$ and defines $r'_w = z_w \oplus z'_w \oplus r_w$ for each $w \in \mathcal{O}$. $\mathcal{S}$ then acts as an honest P$_A$ and opens values r'_w to $\mathcal{A}$. If an honest P$_A$ would abort, $\mathcal{S}$ $\mathcal{S}$ outputs whatever $\mathcal{A}$ outputs.

The distributions on the view of $\mathcal{A}$ in **Hybrid$_3$** and **Hybrid$_2$** are identical.

Note that **Hybrid$_3$** is identical to the ideal-world execution.

5   Technical Details: Improved Preprocessing

In this section, we provide details for our two optimizations of the preprocessing phase. The first optimization improves the efficiency to compute a leaky AND gate. Leaky AND gate is a key component towards a preprocessing with full security. This functionality ($\mathcal{F}_{Land}$) outputs triples with guaranteed correctness but the adversary can choose to guess the x value from the honest party: an incorrect guess will be caught immediately; while a correct guess remain undetected.

The second optimization focuses on how to combine leaky triples in a more efficient way. In particular, we observe that a recent optimization in the honest-majority secret sharing protocol by Araki et al. [3], can be applied to our setting too. As a result, we can roughly reduce the bucket size by one.

Functionality $\mathcal{F}_{\mathsf{Land}}$

Honest case:

1. Generate uniform $\langle x_1 \mid x_2 \rangle$, $\langle y_1 \mid y_2 \rangle$, $\langle z_1 \mid z_2 \rangle$ such that $z_1 \oplus z_2 = (x_1 \oplus x_2) \wedge (y_1 \oplus y_2)$, and send the respective shares to the two parties.
2. $\mathsf{P_A}$ can choose to send $(P_1, p_2, P_3) \in \{0,1\}^\kappa \times \{0,1\} \times \{0,1\}^\kappa$. The functionality checks

$$P_3 \oplus x_2 P_1 = (p_2 \oplus x_2 \mathsf{lsb}(P_1))\, \varDelta_\mathsf{B}.$$

   If the check fails, the functionality sends fail to both parties and abort. ($\mathsf{P_B}$ can do the same symmetrically.)

Corrupted parties: A corrupted party gets to specify the randomness used on its behalf by the functionality.

Fig. 4. Functionality $\mathcal{F}_{\mathsf{Land}}$ for computing a leaky AND triple.

5.1 Improved Leaky AND

Before giving the details, we point out a minor difference in the leaky-AND functionality ($\mathcal{F}_{\mathsf{Land}}$) as compared to [37]. As shown in Fig. 4, instead of letting $\mathcal{A}$ directly learn the value of x, the functionality allows $\mathcal{A}$ to send a query in a form of (P_1, p_2, P_3) and return if $P_3 \oplus x_2 P_1 = (p_2 \oplus x_2 \mathsf{lsb}(P_1))\varDelta_\mathsf{B}$. It can be seen that this special way is no more than a query on x and two queries on $\varDelta$, and the $\mathcal{A}$ cannot learn any information on y or z.

The main intuition of the protocol is already discussed in Sect. 3.2. We will proceed to present the protocol, in Fig. 5.

Theorem 2. *The protocol in Fig. 5 securely realizes $\mathcal{F}_{\mathsf{Land}}$ in the $(\mathcal{F}_{\mathsf{abit}}, \mathcal{F}_{\mathsf{eq}})$-hybrid model.*

Proof. As the first step, we will show that the protocol is correct if both parties are honest. We recall that

1. $G_1 := H(\mathsf{K}[x_2] \oplus \varDelta_\mathsf{A}) \oplus H(\mathsf{K}[x_2]) \oplus C_\mathsf{A}$
2. $G_2 := H(\mathsf{K}[x_1] \oplus \varDelta_\mathsf{B}) \oplus H(\mathsf{K}[x_1]) \oplus C_\mathsf{B}$
3. $C_\mathsf{A} := y_1 \varDelta_\mathsf{A} \oplus \mathsf{K}[y_2] \oplus \mathsf{M}[y_1]$
4. $C_\mathsf{B} := y_2 \varDelta_\mathsf{B} \oplus \mathsf{M}[y_2] \oplus \mathsf{K}[y_1]$

Note that

$$E_1 \oplus H(\mathsf{K}[x_2]) = x_2 G_1 \oplus H(\mathsf{M}[x_2]) \oplus x_2 C_\mathsf{B} \oplus H(\mathsf{K}[x_2]).$$

When $x_2 = 0$, we have

$$\begin{aligned} E_1 \oplus H(\mathsf{K}[x_2]) &= x_2 G_1 \oplus H(\mathsf{M}[x_2]) \oplus x_2 C_\mathsf{B} \oplus H(\mathsf{K}[x_2]) \\ &= H(\mathsf{M}[x_2]) \oplus H(\mathsf{K}[x_2]) \\ &= 0 = x_2(C_\mathsf{A} \oplus C_\mathsf{B}). \end{aligned}$$

Protocol Π_{Land}

Protocol:

1. P_A and P_B obtain random authenticated shares $(\langle x_1 \,|\, x_2 \rangle, \langle y_1 \,|\, y_2 \rangle, \langle z_1 \,|\, r \rangle)$.
   P_A locally computes $C_A := y_1 \Delta_A \oplus K[y_2] \oplus M[y_1]$, and
   P_B locally computes $C_B := y_2 \Delta_B \oplus M[y_2] \oplus K[y_1]$.
2. P_A sends $G_1 := H(K[x_2] \oplus \Delta_A) \oplus H(K[x_2]) \oplus C_A$ to P_B.
   P_B computes $E_1 := x_2 G_1 \oplus H(M[x_2]) \oplus x_2 C_B$.
3. P_B sends $G_2 := H(K[x_1] \oplus \Delta_B) \oplus H(K[x_1]) \oplus C_B$ to P_A.
   P_A computes $E_2 := x_1 G_2 \oplus H(M[x_1]) \oplus x_1 C_A$.
4. P_A computes $S_1 := H(K[x_2]) \oplus E_2 \oplus (z_1 \Delta_A \oplus K[r] \oplus M[z_1])$, P_B computes $S_2 := H(K[x_1]) \oplus E_1 \oplus (r \Delta_B \oplus M[r] \oplus K[z_1])$. P_A sends $\text{lsb}(S_1)$ to P_B; P_B sends $\text{lsb}(S_2)$ to P_A. Both parties computes $d := \text{lsb}(S_1) \oplus \text{lsb}(S_2)$.
5. P_A sends $L_1 := S_1 \oplus d \Delta_A$ to $\mathcal{F}_{\text{eq}}$, P_B sends $L_2 := S_2 \oplus d \Delta_B$ to $\mathcal{F}_{\text{eq}}$. If $\mathcal{F}_{\text{eq}}$ returns 0, parties abort, otherwise, they compute $[z_2]_B := [r]_B \oplus d$.

Fig. 5. Our improved leaky-AND protocol.

When $x_2 = 1$, we have

$$
\begin{aligned}
E_1 \oplus H(K[x_2]) &= x_2 G_1 \oplus H(M[x_2]) \oplus x_2 C_B \oplus H(K[x_2]) \\
&= x_2(G_1 \oplus C_B) \oplus H(M[x_2]) \oplus H(K[x_2]) \\
&= x_2(G_1 \oplus C_B) \oplus H(K[x_2] \oplus \Delta_A)) \oplus H(K[x_2]) \\
&= x_2(C_A \oplus C_B).
\end{aligned}
$$

Therefore,

$$
\begin{aligned}
E_1 \oplus H(K[x_2]) &= x_2(C_A \oplus C_B) \\
&= x_2(y_1 \Delta_A \oplus K[y_2] \oplus M[y_1] \oplus y_2 \Delta_B \oplus M[y_2] \oplus K[y_1])) \\
&= x_2(y_1 \Delta_A \oplus y_2 \Delta_A \oplus y_1 \Delta_B \oplus y_2 \Delta_B) \\
&= x_2(y_1 \oplus y_2)(\Delta_A \oplus \Delta_B).
\end{aligned}
$$

Similarly,

$$
E_2 \oplus H(K[x_1]) = x_1(y_1 \oplus y_2)(\Delta_A \oplus \Delta_B).
$$

Taking these two equations, we know that

$$
\begin{aligned}
S_1 \oplus S_2 &= (E_1 \oplus H(K[x_2])) \oplus (E_2 \oplus H(K[x_1])) \\
&\quad \oplus (z_1 \Delta_A \oplus K[r] \oplus M[z_1] \oplus r \Delta_B \oplus M[r] \oplus K[z_1]) \\
&= (x_1 \oplus x_2)(y_1 \oplus y_2)(\Delta_A \oplus \Delta_B) \\
&\quad \oplus (z_1 \Delta_A \oplus K[z_1] \oplus M[z_1] \oplus r \Delta_B \oplus K[r] \oplus M[r]) \\
&= (x_1 \oplus x_2)(y_1 \oplus y_2)(\Delta_A \oplus \Delta_B) \\
&\quad \oplus (z_1 \Delta_A \oplus z_1 \Delta_B \oplus r \Delta_B \oplus r \Delta_A)
\end{aligned}
$$

$$= (x_1 \oplus x_2)(y_1 \oplus y_2)(\Delta_A \oplus \Delta_B) \oplus (z_1 \oplus r)(\Delta_A \oplus \Delta_B)$$
$$= ((x_1 \oplus x_2) \wedge (y_1 \oplus y_2) \oplus z_1 \oplus r)(\Delta_A \oplus \Delta_B).$$

Since $\mathsf{lsb}(\Delta_A \oplus \Delta_B) = 1$, it holds that

$$d = \mathsf{lsb}(S_1 \oplus S_2) = (x_1 \oplus x_2) \wedge (y_1 \oplus y_2) \oplus z_1 \oplus r.$$

Therefore, $(x_1 \oplus x_2) \wedge (y_1 \oplus y_2) = d \oplus z_1 \oplus r = z_1 \oplus z_2$.

Now we will focus on the security of the protocol in the malicious setting. First note that the protocol is symmetric, therefore we only need to focus on the case of a malicious P_A. The local computation of both parties is deterministic, with all inputs sent from $\mathcal{F}_{abit}$. Therefore, all messages sent during the protocol can be anticipated (emulated) by $\mathcal{S}$ after $\mathcal{S}$ sending out the shares. This is not always possible if $\mathcal{A}$ uses local random coins or if $\mathcal{A}$ has private inputs. This fact significantly reduces the difficulty of the proof. Intuitively, $\mathcal{S}$ will be able to immediately catch $\mathcal{A}$ cheating by comparing what it sends with what it would have sent (which $\mathcal{S}$ knows by locally emulating). The majority of the work then is to extract $\mathcal{A}$'s attempt to perform a selective failure attack.

Define a simulator $\mathcal{S}$ as follows.

0a. $\mathcal{S}$ interacts with $\mathcal{F}_{\mathsf{Land}}$ and obtains P_A's share of $(\langle x_1 \,|\, x_2 \rangle, \langle y_1 \,|\, y_2 \rangle, \langle z_1 \,|\, z_2 \rangle)$. $\mathcal{S}$ also gets Δ_A from $\mathcal{F}_{abit}$. $\mathcal{S}$ randomly picks Δ_B and P_B's share of $(\langle x_1 \,|\, x_2 \rangle, \langle y_1 \,|\, y_2 \rangle, \langle z_1 \,|\, z_2 \rangle)$ in a way that makes it consistent with P_A's share. $\mathcal{S}$ now randomly picks d and computes $[r]_B := [z_2]_B \oplus d$.

0b. Using values $(\langle x_1 \,|\, x_2 \rangle, \langle y_1 \,|\, y_2 \rangle, \langle z_1 \,|\, r \rangle)$ from both parties, $\mathcal{S}$ locally emulates all messages sent by each party, namely (G_1, d_1, L_1) sent by an honest P_A and (G_2, d_2, L_2) sent by an honest P_B.

1. $\mathcal{S}$ plays the role of $\mathcal{F}_{abit}$ and sends out $(\langle x_1 \,|\, x_2 \rangle, \langle y_1 \,|\, y_2 \rangle, \langle z_1 \,|\, r \rangle)$ as defined above.

2. $\mathcal{S}$ acts as an honest P_B and receive G_1' sent by $\mathcal{A}$. $\mathcal{S}$ computes $P_1 = G_1' \oplus G_1$.

3. $\mathcal{S}$ randomly picks a G_2 and send it to $\mathcal{A}$.

4. $\mathcal{S}$ acts as an honest P_B and receives d_1'. $\mathcal{S}$ computes $p_2 := d_1' \oplus d_1$.

5. $\mathcal{S}$ plays the role of $\mathcal{F}_{eq}$ and obtain L_1'. $\mathcal{S}$ computes $P_3 = L_1' \oplus L_1$. $\mathcal{S}$ sends (P_1, p_2, P_3) to $\mathcal{F}_{\mathsf{Land}}$ as the selective failure attack query. If $\mathcal{F}_{\mathsf{Land}}$ abort, $\mathcal{S}$ plays the role of $\mathcal{F}_{eq}$ and aborts. If the value d in the protocol equals to r defined in step 0a, $\mathcal{F}_{eq}$ returns 0; otherwise $\mathcal{F}_{eq}$ returns 1.

6. $\mathcal{S}$ sends (P_1, p_2, P_3) to $\mathcal{F}_{\mathsf{Land}}$ as the selective failure query. If $\mathcal{F}_{\mathsf{Land}}$ returns fail, $\mathcal{S}$ sends 0 to $\mathcal{A}$ as the output of $\mathcal{F}_{eq}$.

Note that messages that $\mathcal{S}$ sends to $\mathcal{A}$ in the protocol are changed from (G_2, d_2, L_2) to $(G_2, d_2 \oplus x_2 \mathsf{lsb}(P_1), L_2 \oplus x_2 P_1 \oplus d' \Delta_B)$, where $d' = p_2 \oplus x_2 \cdot \mathsf{lsb}(P_1)$ and the equality checking in step 5 changed from comparing $L_1 = L_2$ to

$$L_1 \oplus P_3 = L_2 \oplus x_2 P_1 \oplus (p_2 \oplus x_2 \mathsf{lsb}(P_1)) \Delta_B,$$

that is

$$P_3 \oplus x_2 P_1 = (p_2 \oplus x_2 \mathsf{lsb}(P_1)) \Delta_B.$$

This is the same form as the selective failure query in $\mathcal{F}_{\mathsf{Land}}$.

Protocol Π_{pre}

Inputs: Two parties agree on a circuit for a function $f : \{0,1\}^{\mathcal{I}_1} \times \{0,1\}^{\mathcal{I}_2} \rightarrow \{0,1\}^{\mathcal{O}}$.

Protocol:

1. Two parties initialize $\mathcal{F}_{\mathsf{abit}}$, which sends $\varDelta_{\mathsf{A}}$ to P_{A} and $\varDelta_{\mathsf{B}}$ to P_{B}.
2. For each wire $w \in \mathcal{I}_1 \cup \mathcal{I}_2 \cup \mathcal{W}$, two parties obtain an authenticated share $\langle r_w \mid s_w \rangle$ from $\mathcal{F}_{\mathsf{abit}}$.
3. For each gate $\mathcal{G} = (\alpha, \beta, \gamma, \oplus)$, two parties compute $\langle r_\gamma \mid s_\gamma \rangle := \langle s_\alpha \mid r_\alpha \rangle \oplus \langle r_\beta \mid s_\beta \rangle$.
4. For each gate $\mathcal{G} = (\alpha, \beta, \gamma, \wedge)$, two parties have $(\langle r_\alpha \mid s_\alpha \rangle, \langle r_\beta \mid s_\beta \rangle)$, and run step 2 to step 5 in Π_{Land} to obtain $\langle r_\gamma^* \mid s_\gamma^* \rangle$, such that $r_\gamma^* \oplus s_\gamma^* = (r_\alpha \oplus s_\alpha) \wedge (r_\beta \oplus s_\beta)$
5. P_{A} and P_{B} call $\mathcal{F}_{\mathsf{Land}}$ to obtain $(B-1)|\mathcal{C}|$ number of leaky AND triples $(\langle x_1 \mid x_2 \rangle, \langle y_1 \mid y_2 \rangle, \langle z_1 \mid z_2 \rangle)$.
6. Two parties perform secure coin-flipping to determine a random permutation and permute the triples obtained in step 4. For each AND gate $\mathcal{G} = (\alpha, \beta, \gamma, \wedge)$ in the circuit, perform secure merging for $B - 1$ times.
   (a) Obtain the next triple in the permuted list, namely $(\langle x_1 \mid x_2 \rangle, \langle y_1 \mid y_2 \rangle, \langle z_1 \mid z_2 \rangle)$
   (b) Compute $\langle d_1 \mid d_2 \rangle := \langle y_1 \mid y_2 \rangle \oplus \langle r_\beta \mid s_\beta \rangle$, and $d := \mathsf{Open}(\langle d_1 \mid d_2 \rangle)$.
   (c) Update triple: $\langle r_\alpha \mid s_\alpha \rangle := \langle r_\alpha \mid s_\alpha \rangle \oplus \langle x_1 \mid x_2 \rangle$, $\langle r_\gamma^* \mid s_\gamma^* \rangle := \langle r_\gamma^* \mid s_\gamma^* \rangle \oplus \langle z_1 \mid z_2 \rangle \oplus d \langle x_1 \mid x_2 \rangle$.

Fig. 6. Protocol Π_{pre} instantiating $\mathcal{F}_{\mathsf{pre}}$ in the $(\mathcal{F}_{\mathsf{abit}}, \mathcal{F}_{\mathsf{Land}})$-hybrid model.

5.2 Improved Function-Dependent Preprocessing

In this section, we will focus on improving the preprocessing in the Leaky AND triple generation ($\mathcal{F}_{\mathsf{Land}}$) hybrid model. The main observation is that in the protocol of WRK, each wire is associated with a mask (in the authenticated share format). Then the AND of input masks are computed using one AND triple. This is a waste of randomness, since we also directly construct all triples in place for all wires. Note that the idea is similar to Araki et al. [3]. The detailed protocol is presented in Fig. 6.

Note that although the above optimization aims to reduce the overall cost of the protocol, but it turns out that even in this case, most of the computation and communication (including computation of all authenticated bits as well as all leaky-AND triples in step 5) can be still done in the function-independent phase. The function-dependent cost is increased by only κ bits per AND gate only. Therefore, here we have an option to trade-off between total communication and communication in the offline stage. By increasing the function-dependent cost by κ bits per gate, we reduce bucket size by 1. We believe both versions can be useful depending on the application, and the concrete cost of both versions of the protocol are presented in the performance section.

Table 2. Communication complexity of different protocols for evaluating AES, rounded to two significant figures. As in Table 1, one-way communication refers to the maximum communication one party sends to the other; two-way communication refers to the sum of both parties' communication. The best prior number in each column is bolded for reference.

	One-way Communication (Max)				Two-way Communication			
	Ind. (MB)	Dep. (MB)	Online (KB)	Total (MB)	Ind. (MB)	Dep. (MB)	Online (KB)	Total (MB)
	Single execution							
[28]	15	**0.22**	16	15	15	0.22	16	15
[37]	**2.9**	0.57	**4.9**	**3.4**	**5.7**	**0.57**	**6.0**	6.3
[12]	-	3.4	≥ 4.9	**3.4**	-	3.4	≥ 4.9	**3.4**
This work, v. 1	1.9	0.33	5.0	2.2	3.8	0.33	5.0	4.2
This work, v. 2	2.5	0.22	5.0	2.7	4.9	0.22	5.0	5.1
	Amortized cost over 1024 executions							
[33]	-	1.6	17	**1.6**	-	3.2	17	3.2
[28]	6.4	**0.22**	16	6.6	6.4	**0.22**	16	6.6
[18]	-	1.6	19	**1.6**	-	1.6	19	**1.6**
[37]	**2.0**	0.57	**4.9**	2.6	**4.0**	0.57	**6.0**	4.6
This work, v. 1	1.4	0.33	5.0	1.7	2.7	0.33	5.0	3.1
This work, v. 2	1.9	0.22	5.0	2.1	3.8	0.22	5.0	4.0

6   Performance

In this section, we discuss the concrete efficiency of our protocol. We consider two variants of our protocol that optimize the cost of different phases: The first version of our protocol is optimized to minimize the total communication; the second version is optimized to minimize the communication in the function-dependent phase. (The cost of the online phase is identical in both versions.)

6.1   Communication Complexity

Table 2 shows the communication complexity of recent two-party computation protocols in the malicious setting. Numbers for these protocols are obtained from the respective papers, while numbers for our protocol are calculated. We tabulate both one-way communication and total communication. If parties' data can be sent at the same time over a full-duplex network, then one-way communication is a better reflection of the running time. In general, for a circuit that requires a bucket size of B, we can obtain an estimation of the concrete communication cost: our first version has function dependent cost of 3κ per gate, and function independent cost of $(4B - 2)\kappa + (3B - 1)\rho$ per gate; our second version has a function dependent cost of 2κ per gate, and a function independent cost of $(4B + 2)\kappa + (3B + 2)\rho$ per gate.

We see that our protocol and the protocol by Nielsen et al. [28] are the only ones that, considering the function-dependent phase and the online phase, have cost similar to that of the state-of-the-art semi-honest garbled-circuit protocol. In other words, *the overhead induced by malicious security can be completely pushed to the preprocessing stage.* Compared to the protocol by Nielsen et al., we are able to reduce the communication in the preprocessing stage by 6× in the single-execution setting, and by 3.4× in the amortized setting. Our protocol also has the best total communication complexity in both settings, excepting the work of [18,33] which are 6% better but do not support function-independent preprocessing.

6.2   Computational Complexity

Since the WRK protocol represents the state-of-the-art as far as implementations are concerned, we compare the computational complexity of our protocol to theirs. We also include a comparison to the more recent protocol by Hazay et al. [12] (the *HIV protocol*), which has not yet been implemented.

Comparing to the WRK protocol. Our protocol follows the same high-level approach as the WRK protocol. Almost all H-evaluations in our protocol can be accelerated using fixed-key AES, as done in [6]. We tabulate the number of H-evaluations for both protocols in Table 3. Due to our improved $\mathcal{F}_{\mathsf{Land}}$, we are able to achieve a 2–2.5× improvement.

Table 3. Number of H-evaluations. We align the security parameters in both protocols and set $B = \rho/\log C + 1$ for a fair comparison.

	Ind.	Dep.	Online	Total
WRK	$10B$	8	2	$10B + 10$
This work, v. 1	$4B - 4$	8	2	$4B + 6$
This work, v. 2	$4B$	4	2	$4B + 6$

Comparing to the HIV protocol. As noted by the authors, the HIV protocol has polylogarithmic computational overhead compared to semi-honest garbled circuits. This is due to their use of the MPC-based zero-knowledge proof by Ames et al. [2]. On the other hand, in our protocol, the computation is linear in the circuit size. Furthermore, almost all cryptographic operations in our protocol can be accelerated using hardware AES instructions.

Taking an AES circuit as example, the ZK protocol by Ames et al. for a circuit of that size has a prover running time of around 70 ms and a verifier running time of around 30 ms. Therefore, even if we ignore the cost of computing and sending the garbled circuit, the oblivious transfers, and other operations, the end-to-end running time of the HIV protocol will still be at least 100 ms. On the other hand, the entire WRK protocol runs within 17 ms for the same circuit. As our protocol results in at least a 2× improvement, our protocol will be at least an order of magnitude faster than the HIV protocol.

7 Challenges in Extending to the Multi-Party Case

Wang et al. [38] have also shown how to extend their authenticated-garbling protocol to the multi-party case. In this section, we discuss the challenges involved in applying our new techniques to that setting. Note that Ben-Efraim [7] recently proposed new techniques for multi-party garbling, making it compatible with some of the half-gate optimizations. Despite being based on half-gates, they still require 4 garbled rows per AND gate, and thus their work still leaves open the question of reducing the communication complexity of the online phase in the multi-party case.

In the multi-party WRK protocol, there are $n - 1$ garbling parties and one evaluating party. For each wire, each garbler chooses their own set of wire labels (called "subkeys"). As in the 2-party case, the preprocessing defines some authenticated bits, and as a result all parties can locally compute additive shares of *any garbler's* subkey corresponding to *any authenticated value.*

In each gate, each garbler P_i generates standard Yao garbled gate consisting of 4 rows. Each row of P_i's gate is encrypted by only P_i's subkeys, and the payload of the row is P_i's shares of *all garblers'* subkeys. That way, the evaluator can decrypt the correct row of everyone's garbled gates, obtain everyone's shares of everyone's subkeys, and combine them to get everyone's appropriate subkey for the output wire.

Now suppose we modify things so each garbler generates a half-gates-style garbled gate instead of a standard Yao garbled gate. The half-gate uses garbler P_i's subkeys as its "keys" and encodes P_i's shares of all subkeys as its "payloads". Now the protocol may not be secure against an adversary corrupting the evaluator and a garbler. In particular, half-gates garbling defines $G_0 = H(\mathsf{L}_{\alpha,0}) \oplus H(\mathsf{L}_{\alpha,1}) \oplus \lambda_\beta \Delta$. When P_i is acting as garbler, these $\mathsf{L}_{\alpha,u}$ values correspond to P_i's subkeys. Now suppose P_i colludes with the evaluator. If the evaluator comes to learn G_0 (which is necessary to evaluate the gate in half of the cases), then the adversary can learn the secret mask λ_β since it is the only unknown term in G_0. Clearly revealing the secret wire mask breaks the privacy of the protocol. This is not a problem with Yao garbled gates, where each row can be written as $G_{u,v} = H(\mathsf{L}_{\alpha,u}, \mathsf{L}_{\beta,v}) \oplus$ [payload already known to garbler]. The secret masks do not appear in the garbled table, except indirectly through the payloads (subkey shares).

It is even unclear if row-reduction can be made possible. In the multi-party setting, the garbler has no control over the "payload" (i.e., output wire label) of the garbled gate when using row-reduction. Indeed, this is what makes it possible to reduce the size of a garbled gate. This is not a problem in the two-party case, where there is only one garbler who has control over all garbled gates and all wire labels. He generates a garbled table, and then computes his output wire label (subkey) as a function of the payload in the table. However, in the multi-party case, P_i generates a half-gate whose payloads include P_i's shares of P_j's subkeys! We would need P_j's choice of subkeys to depend on the payloads of P_i's garbling (for all i and j!). It is not clear how this can be done, and even if it were possible it would apparently require additional rounds proportional to the depth of the circuit.

Acknowledgments. This material is based on work supported by NSF awards #1111599, #1563722, #1564088, and #1617197. Portions of this work were also supported by DARPA and SPAWAR under contract N66001-15-C-4065. The U.S. Government is authorized to reproduce and distribute reprints for Governmental purposes not withstanding any copyright notation thereon. The views, opinions, and/or findings expressed are those of the authors and should not be interpreted as representing the official views or policies of the Department of Defense or the U.S. Government.

References

1. Afshar, A., Mohassel, P., Pinkas, B., Riva, B.: Non-interactive secure computation based on cut-and-choose. In: Nguyen, P.Q., Oswald, E. (eds.) EUROCRYPT 2014. LNCS, vol. 8441, pp. 387–404. Springer, Heidelberg (2014). https://doi.org/10.1007/978-3-642-55220-5_22

2. Ames, S., Hazay, C., Ishai, Y., Venkitasubramaniam, M.: Ligero: lightweight sublinear arguments without a trusted setup. In: ACM CCS 2017, pp. 2087–2104. ACM Press (2017)

3. Araki, T., Barak, A., Furukawa, J., Lichter, T., Lindell, Y., Nof, A., Ohara, K., Watzman, A., Weinstein, O.: Optimized honest-majority MPC for malicious adversaries - breaking the 1 billion-gate per second barrier. In: 2017 IEEE Symposium on Security and Privacy, San Jose, CA, USA, 22–26 May 2017, pp. 843–862. IEEE Computer Society Press (2017)

4. Beaver, D.: Efficient multiparty protocols using circuit randomization. In: Feigenbaum, J. (ed.) CRYPTO 1991. LNCS, vol. 576, pp. 420–432. Springer, Heidelberg (1992). https://doi.org/10.1007/3-540-46766-1_34

5. Beaver, D., Micali, S., Rogaway, P.: The round complexity of secure protocols. In: Proceedings of the Twenty-Second Annual ACM Symposium on Theory of Computing, pp. 503–513. ACM (1990)

6. Bellare, M., Hoang, V.T., Keelveedhi, S., Rogaway, P.: Efficient garbling from a fixed-key blockcipher. In: 2013 IEEE Symposium on Security and Privacy, Berkeley, CA, USA, 19–22 May 2013, pp. 478–492. IEEE Computer Society Press (2013)

7. Ben-Efraim, A.: On multiparty garbling of arithmetic circuits. Cryptology ePrint Archive, Report 2017/1186 (2017). https://eprint.iacr.org/2017/1186

8. Brandão, L.T.A.N.: Secure two-party computation with reusable bit-commitments, via a cut-and-choose with forge-and-lose technique. In: Sako, K., Sarkar, P. (eds.) ASIACRYPT 2013, Part II. LNCS, vol. 8270, pp. 441–463. Springer, Heidelberg (2013). https://doi.org/10.1007/978-3-642-42045-0_23

9. Damgård, I., Pastro, V., Smart, N.P., Zakarias, S.: Multiparty computation from somewhat homomorphic encryption. In: Safavi-Naini, R., Canetti, R. (eds.) CRYPTO 2012. LNCS, vol. 7417, pp. 643–662. Springer, Heidelberg (2012). https://doi.org/10.1007/978-3-642-32009-5_38

10. Frederiksen, T.K., Jakobsen, T.P., Nielsen, J.B., Nordholt, P.S., Orlandi, C.: MiniLEGO: efficient secure two-party computation from general assumptions. In: Johansson, T., Nguyen, P.Q. (eds.) EUROCRYPT 2013. LNCS, vol. 7881, pp. 537–556. Springer, Heidelberg (2013). https://doi.org/10.1007/978-3-642-38348-9_32

11. Gueron, S., Lindell, Y., Nof, A., Pinkas, B.: Fast garbling of circuits under standard assumptions. In: Ray, I., Li, N., Kruegel: C. (eds.) ACM CCS 2015, Denver, CO, USA, 12–16 October 2015, pp. 567–578. ACM Press (2015)

12. Hazay, C., Ishai, Y., Venkitasubramaniam, M.: Actively secure garbled circuits with constant communication overhead in the plain model. In: Kalai, Y., Reyzin, L. (eds.) TCC 2017, Part II. LNCS, vol. 10678, pp. 3–39. Springer, Cham (2017). https://doi.org/10.1007/978-3-319-70503-3_1
13. Huang, Y., Evans, D., Katz, J., Malka, L.: Faster secure two-party computation using garbled circuits. In: USENIX Security 2011 (2011)
14. Huang, Y., Katz, J., Evans, D.: Efficient secure two-party computation using symmetric cut-and-choose. In: Canetti, R., Garay, J.A. (eds.) CRYPTO 2013, Part II. LNCS, vol. 8043, pp. 18–35. Springer, Heidelberg (2013). https://doi.org/10.1007/978-3-642-40084-1_2
15. Huang, Y., Katz, J., Kolesnikov, V., Kumaresan, R., Malozemoff, A.J.: Amortizing garbled circuits. In: Garay, J.A., Gennaro, R. (eds.) CRYPTO 2014, Part II. LNCS, vol. 8617, pp. 458–475. Springer, Heidelberg (2014). https://doi.org/10.1007/978-3-662-44381-1_26
16. Ishai, Y., Kilian, J., Nissim, K., Petrank, E.: Extending oblivious transfers efficiently. In: Boneh, D. (ed.) CRYPTO 2003. LNCS, vol. 2729, pp. 145–161. Springer, Heidelberg (2003). https://doi.org/10.1007/978-3-540-45146-4_9
17. Kolesnikov, V., Mohassel, P., Rosulek, M.: FleXOR: flexible garbling for XOR gates that beats free-XOR. In: Garay, J.A., Gennaro, R. (eds.) CRYPTO 2014, Part II. LNCS, vol. 8617, pp. 440–457. Springer, Heidelberg (2014). https://doi.org/10.1007/978-3-662-44381-1_25
18. Kolesnikov, V., Nielsen, J.B., Rosulek, M., Trieu, N., Trifiletti, R.: DUPLO: unifying cut-and-choose for garbled circuits. In: ACM CCS 2017, pp. 3–20. ACM Press (2017)
19. Kolesnikov, V., Schneider, T.: Improved garbled circuit: free XOR gates and applications. In: Aceto, L., Damgård, I., Goldberg, L.A., Halldórsson, M.M., Ingólfsdóttir, A., Walukiewicz, I. (eds.) ICALP 2008, Part II. LNCS, vol. 5126, pp. 486–498. Springer, Heidelberg (2008). https://doi.org/10.1007/978-3-540-70583-3_40
20. Kreuter, B., Shelat, A., Shen, C.H.: Billion-gate secure computation with malicious adversaries. In: USENIX Security 2012 (2012)
21. Lindell, Y.: Fast cut-and-choose based protocols for malicious and covert adversaries. In: Canetti, R., Garay, J.A. (eds.) CRYPTO 2013, Part II. LNCS, vol. 8043, pp. 1–17. Springer, Heidelberg (2013). https://doi.org/10.1007/978-3-642-40084-1_1
22. Lindell, Y., Pinkas, B.: An efficient protocol for secure two-party computation in the presence of malicious adversaries. In: Naor, M. (ed.) EUROCRYPT 2007. LNCS, vol. 4515, pp. 52–78. Springer, Heidelberg (2007). https://doi.org/10.1007/978-3-540-72540-4_4
23. Lindell, Y., Pinkas, B.: Secure two-party computation via cut-and-choose oblivious transfer. In: Ishai, Y. (ed.) TCC 2011. LNCS, vol. 6597, pp. 329–346. Springer, Heidelberg (2011). https://doi.org/10.1007/978-3-642-19571-6_20
24. Lindell, Y., Riva, B.: Cut-and-choose Yao-based secure computation in the online/offline and batch settings. In: Garay, J.A., Gennaro, R. (eds.) CRYPTO 2014, Part II. LNCS, vol. 8617, pp. 476–494. Springer, Heidelberg (2014). https://doi.org/10.1007/978-3-662-44381-1_27
25. Lindell, Y., Riva, B.: Blazing fast 2PC in the offline/online setting with security for malicious adversaries. In: Ray, I., Li, N., Kruegel, C. (eds.) ACM CCS 2015, Denver, CO, USA, 12–16 October 2015, pp. 579–590. ACM Press (2015)
26. Malkhi, D., Nisan, N., Pinkas, B., Sella, Y.: Fairplay—a secure two-party computation system. In: USENIX Security 2004 (2004)

27. Naor, M., Pinkas, B., Sumner, R.: Privacy preserving auctions and mechanism design. In: 1st ACM Conference on Electronic Commerce (1999)
28. Nielsen, J., Schneider, T., Trifiletti, R.: Constant-round maliciously secure 2PC with function-independent preprocessing using LEGO. In: Network and Distributed System Security Symposium (NDSS) (2017)
29. Nielsen, J.B., Nordholt, P.S., Orlandi, C., Burra, S.S.: A new approach to practical active-secure two-party computation. In: Safavi-Naini, R., Canetti, R. (eds.) CRYPTO 2012. LNCS, vol. 7417, pp. 681–700. Springer, Heidelberg (2012). https://doi.org/10.1007/978-3-642-32009-5_40
30. Nielsen, J.B., Orlandi, C.: LEGO for two-party secure computation. In: Reingold, O. (ed.) TCC 2009. LNCS, vol. 5444, pp. 368–386. Springer, Heidelberg (2009). https://doi.org/10.1007/978-3-642-00457-5_22
31. Nielsen, J.B., Orlandi, C.: Cross and clean: amortized garbled circuits with constant overhead. In: Hirt, M., Smith, A.D. (eds.) TCC 2016, Part I. LNCS, vol. 9985, pp. 582–603. Springer, Heidelberg (2016). https://doi.org/10.1007/978-3-662-53641-4_22
32. Pinkas, B., Schneider, T., Smart, N.P., Williams, S.C.: Secure two-party computation is practical. In: Matsui, M. (ed.) ASIACRYPT 2009. LNCS, vol. 5912, pp. 250–267. Springer, Heidelberg (2009). https://doi.org/10.1007/978-3-642-10366-7_15
33. Rindal, P., Rosulek, M.: Faster malicious 2-party secure computation with online/offline dual execution. In: USENIX Security 2016 (2016)
34. Shelat, A., Shen, C.H.: Two-output secure computation with malicious adversaries. In: Paterson, K.G. (ed.) EUROCRYPT 2011. LNCS, vol. 6632, pp. 386–405. Springer, Heidelberg (2011). https://doi.org/10.1007/978-3-642-20465-4_22
35. Shelat, A., Shen, C.H.: Fast two-party secure computation with minimal assumptions. In: Sadeghi, A.R., Gligor, V.D., Yung, M. (eds.) ACM CCS 2013, Berlin, Germany, 4–8 November 2013, pp. 523–534. ACM Press (2013)
36. Wang, X., Malozemoff, A.J., Katz, J.: Faster secure two-party computation in the single-execution setting. In: Coron, J.-S., Nielsen, J.B. (eds.) EUROCRYPT 2017, Part II. LNCS, vol. 10212, pp. 399–424. Springer, Cham (2017). https://doi.org/10.1007/978-3-319-56617-7_14
37. Wang, X., Ranellucci, S., Katz, J.: Authenticated garbling and efficient maliciously secure two-party computation. In: ACM CCS 2017, pp. 21–37. ACM Press (2017)
38. Wang, X., Ranellucci, S., Katz, J.: Global-scale secure multiparty computation. In: ACM CCS 2017, pp. 39–56. ACM Press (2017)
39. Yao, A.C.C.: How to generate and exchange secrets (extended abstract). In: 27th FOCS, Toronto, Ontario, Canada, 27–29 October 1986, pp. 162–167. IEEE Computer Society Press (1986)
40. Zahur, S., Rosulek, M., Evans, D.: Two halves make a whole - Reducing Data Transfer in Garbled Circuits Using Half Gates. In: Oswald, E., Fischlin, M. (eds.) EUROCRYPT 2015, Part II. LNCS, vol. 9057, pp. 220–250. Springer, Heidelberg (2015). https://doi.org/10.1007/978-3-662-46803-6_8
41. Zhu, R., Huang, Y.: JIMU: faster LEGO-based secure computation using additive homomorphic hashes. In: Takagi, T., Peyrin, T. (eds.) ASIACRYPT 2017, Part II. LNCS, vol. 10625, pp. 529–572. Springer, Cham (2017). https://doi.org/10.1007/978-3-319-70697-9_19

Information-Theoretic MPC

Amortized Complexity of Information-Theoretically Secure MPC Revisited

Ignacio Cascudo[1(✉)], Ronald Cramer[2,3], Chaoping Xing[4], and Chen Yuan[2]

[1] Aalborg University, Aalborg, Denmark
ignacio@math.aau.dk
[2] CWI Amsterdam, Amsterdam, The Netherlands
{cramer,Chen.Yuan}@cwi.nl
[3] Leiden University, Leiden, The Netherlands
cramer@math.leidenuniv.nl
[4] Nanyang Technological University, Singapore, Singapore
xingcp@ntu.edu.sg

Abstract. A fundamental and widely-applied paradigm due to Franklin and Yung (STOC 1992) on Shamir-secret-sharing based general n-player MPC shows how one may trade the *adversary threshold* t against *amortized* communication complexity, by using a so-called packed version of Shamir's scheme. For e.g. the BGW-protocol (with active security), this trade-off means that if $t + 2k - 2 < n/3$, then k *parallel* evaluations of the *same* arithmetic circuit on different inputs can be performed at the overall cost corresponding to a *single* BGW-execution.

In this paper we propose a novel paradigm for amortized MPC that offers a *different* trade-off, namely with the size of the field of the circuit which is securely computed, instead of the adversary threshold. Thus, unlike the Franklin-Yung paradigm, this leaves the adversary threshold *unchanged*. Therefore, for instance, this paradigm may yield constructions enjoying the maximal adversary threshold $\lfloor (n-1)/3 \rfloor$ in the BGW-model (secure channels, perfect security, active adversary, synchronous communication).

Our idea is to compile an MPC for a circuit over an extension field to a parallel MPC of the same circuit but with inputs defined over its *base field* and with the *same* adversary threshold. Key technical handles are our notion of *reverse multiplication-friendly embeddings* (RMFE) and our proof, by algebraic-geometric means, that these are *constant-rate*, as well as efficient auxiliary protocols for creating "subspace-randomness" with good amortized complexity. In the BGW-model, we show that the latter can be constructed by combining our tensored-up linear secret sharing with protocols based on hyper-invertible matrices á la Beerliova-Hirt (or variations thereof). Along the way, we suggest alternatives for hyper-invertible matrices with the same functionality but which can be defined over a large enough constant size field, which we believe is of independent interest.

As a demonstration of the merits of the novel paradigm, we show that, in the BGW-model and with an optimal adversary threshold $\lfloor (n-1)/3 \rfloor$,

© International Association for Cryptologic Research 2018
H. Shacham and A. Boldyreva (Eds.): CRYPTO 2018, LNCS 10993, pp. 395–426, 2018.
https://doi.org/10.1007/978-3-319-96878-0_14

it is possible to securely compute a *binary circuit* with amortized complexity $O(n)$ of *bits per gate per instance*. Known results would give $n \log n$ bits instead. By combining our result with the Franklin-Yung paradigm, and assuming a *sub-optimal adversary* (i.e., an arbitrarily small $\epsilon > 0$ fraction below $1/3$), this is improved to $O(1)$ bits instead of $O(n)$.

1   Introduction

A fundamental and widely-applied paradigm due to Franklin and Yung [FY92] on Shamir-secret-sharing based general n-player MPC shows how one may trade the *adversary threshold* t against *amortized* communication complexity, by using a so-called packed version of Shamir's scheme. For e.g. the BGW-protocol [BGW88] (with active security), this trade-off means that if $t+2k-2 < n/3$, then k *parallel* evaluations of the *same* arithmetic circuit on different inputs can be performed at the overall cost corresponding to a *single* BGW-execution. In this paper we propose a novel paradigm for amortized MPC that offers a *different* trade-off, namely with the size of the field of the circuit which is securely computed, instead of the adversary threshold. In particular, unlike the Franklin-Yung paradigm, this leaves the adversary threshold *unchanged*.

We apply our paradigm in the *BGW-model*: secure channels, perfect security (privacy and correctness), active adversary, synchronous communication. Our aim is to achieve MPC that is efficient (in the amortized sense as discussed above), tolerates an adversary satisfying the maximal threshold (or close) and that evaluates *binary circuits*.

We motivate the latter choice as follows. Besides the fact that this is natural in applications to begin with, we note that many of the protocols in this model (such as [BGW88]) represent the target function to be computed as an arithmetic circuit over some finite field, and then process this circuit by distributed computing of each gate, using secret sharing. However, many of those protocols require that the size of the underlying finite field is larger the number of parties n (because of their use of Shamir's secret sharing scheme [Sha79] of some variant thereof). This means that applying those protocols to functions which are already naturally represented as a binary circuit requires to lift this circuit to a large enough extension field, wherewith the communication complexity incurs a multiplicative overhead of $\log n$. It is in these cases that our paradigm pays off by twisting this overhead into a vehicle for parallel evaluations.

Concretely, we get:

Theorem 1. *In the BGW-model, there is an efficient MPC protocol for n parties secure against the maximal number of active corruptions $\lfloor (n-1)/3 \rfloor$ that computes $\Omega(\log n)$ evaluations of a single binary circuit in parallel with an amortized communication complexity (per instance) of $O(n)$ bits per gate.*

The best known previous result of this kind and in this model is obtained from the MPC protocol by Beerliová-Trubíniová and Hirt [BH08] which communicates $O(n)$ *field elements* per gate, but requires the field size over which the arithmetic circuit is defined to be at least $2n$, and hence the computation of a binary circuit with that protocol requires $O(n \log n)$ bits of communication per gate. Our result will be proved by applying our paradigm to Beerliova-Hirt. Note that the Franklin-Yung paradigm does not apply here as we achieve security against a maximal adversary. *Combining* our result with the Franklin-Yung paradigm, however, we get the following:

Theorem 2. *In the BGW-model, for every $\epsilon > 0$, there is an efficient MPC protocol for n parties secure against a submaximal number of active corruptions $t < (1 - \epsilon)n/3$ that computes $\Omega(n \log n)$ evaluations of a single binary circuit in parallel with an amortized communication complexity (per instance) of $O(1)$ bits per gate.*

We note that, as opposed to Theorem 1, this theorem may plausibly and alternatively be argued without recourse to our novel paradigm: indeed, we could deploy the asymptotically good arithmetic secret sharing schemes from [CCCX09] (over a suitably large constant extension of $\mathbb{F}_2$ so as to get the desired adversary rate), combined with an overhaul of the complete (say) Beerliova-Hirt protocols so as to make them work over these schemes instead of Shamir's. In addition, this would use some of our present techniques to overcome the arising issue that the protocol tricks from Beerliova-Hirt involving hyper-invertible matrices (or an alternative that we discuss later on) are defined over an extension of the field of definition of the arithmetic secret sharing. Our present approach, however, in fact uses Beerliova-Hirt essentially as a *black-box*. Moreover, our approach covers *both* theorems with the same method.

As noted in [IKOS09], a complexity of $O(1)$ bits per gate can also be obtained in the non-amortized setting as long as the number of parties which provide inputs is constant, by combining the protocol from [DI06] with the aforementioned arithmetic secret sharing schemes from [CCCX09]. This would be secure against an active adversary corrupting $t = \Omega(n)$ parties, where the constant is in principle small, but this can be brought up to $t < (1 - \epsilon)n/3$, for any ϵ, by using Bracha's committees technique [Bra85] as described in [DIK10]. If we remove the assumption on the number of input-providing parties, then the best result in the non-amortized setting for suboptimal adversaries is given by [DIK10], where the communication complexity per gate is $O(\text{polylog}(n))$ bits. It is an interesting question to determine if the communication complexities obtained in Theorems 1 and 2 are optimal in this model[1].

We now give a brief preview of our paradigm and its technical challenges. First we introduce the notion of reverse multiplication friendly embeddings, which provide a way to embed the ring $\mathbb{F}_2^k$ into a field $\mathbb{F}_{2^m}$ so that coordinatewise products "map" to multiplications in the extension field in a certain manner that we will explain. Furthermore we will need to construct RMFEs with $m = O(k)$, so that the degree of the extension field does not explode.

---

[1] Some results on lower bounds for communication complexity of gate-by-gate protocols for arithmetic circuits like ours were obtained by [DNPR16] but they do not seem to be enough to claim such optimality.

Second, using such a map as a stepping stone, we construct a compiler that transforms a secure secret-sharing based computation protocol that evaluates an arithmetic circuit over a $\mathbb{F}_{2^m}$ (for example the one by [BH08]) into a secure protocol for the same number of parties and adversary that allows for parallel evaluation of $k = \Theta(m)$ of a related boolean circuit. Several obstacles appear when constructing this compiler, as a consequence of moving back and forth between the algebraic structures and we need to solve these issues with the introduction of several subprotocols. At the same time we need to ensure that these subprotocols do not require too much communication, so that this does not offset the gains from our embedding strategy.

It turns out that these subprotocols rely on a crucial step: we need to find a way to construct sharings of random elements in prescribed $\mathbb{F}_2$-subspaces of $(\mathbb{F}_{2^m})^v$. Our third contribution is a communication-efficient protocol to accomplish that. This in turn consists of the following ideas: first, we introduce a definition of generalized linear secret sharing scheme (GLSSS), where the secret and shares belong to vector spaces over the same field; then, we cast the strategy of random generation of sharings based on hyper-invertible matrices, introduced in [BH08] (and used in [DIK10]), in this language of GLSSS. This is still not good enough for our purposes, since our GLSSS is only linear over $\mathbb{F}_2$, while the hyper-invertible matrix is defined over $\mathbb{F}_{2^m}$. So the last idea we need consists in tensoring-up our scheme, that suitably transforms our $\mathbb{F}_2$-GLSSS into a $\mathbb{F}_{2^m}$-GLSSS. Along the way, we suggest alternatives for hyper-invertible matrices with the same functionality but which can be defined over a large enough constant size field, see Remarks 3 and 4. Although there is no overall advantage to our work (quantitatively), it does mean that, in the subprotocols where it is used, "amortization kicks in faster". Moreover, we believe it is of independent interest.

1.1   Main Ideas

We describe the general idea and the technical challenges we encounter more in detail.

Reverse multiplication friendly embeddings. As mentioned above, we want to embed several instances of the computation of a binary circuit into a single computation of an arithmetic circuit over an extension field. Ideally, we would wish that the ring $\mathbb{F}_2^k$, where the sum and product are defined coordinatewise, was isomorphic (as an $\mathbb{F}_2$-algebra) to the field $\mathbb{F}_{2^k}$, with the usual finite field sum and product; or said in a different way, that there would be a map $\eta : \mathbb{F}_2^k \to \mathbb{F}_{2^k}$ satisfying both $\eta(\mathbf{x} + \mathbf{y}) = \eta(\mathbf{x}) + \eta(\mathbf{y})$ and $\eta(\mathbf{x} * \mathbf{y}) = \eta(\mathbf{x}) \cdot \eta(\mathbf{y})$ for all $\mathbf{x}, \mathbf{y} \in \mathbb{F}_2^k$ (where $*$ denotes the coordinatewise product and $\cdot$ denotes the field product). If such a η existed, then embedding k evaluations of a boolean circuit into an evaluation of a circuit over $\mathbb{F}_{2^k}$ would be trivial: just define the arithmetic circuit C' to be the same as the boolean circuit, but substituting the sum and multiplication gates in $\mathbb{F}_2$ by gates performing the same operations in $\mathbb{F}_{2^k}$; then apply η to the vectors of boolean inputs, evaluate C' and map the result back to

$\mathbb{F}_2$ with η^{-1}. Furthermore, computing C securely would not be a problem, since the parties holding the inputs would secret-share them after applying η, and the result of the secure computation of C' would be opened prior to applying η^{-1}.

Unfortunately such an η does not exist: while $\mathbb{F}_2^k$ and $\mathbb{F}_{2^k}$ are isomorphic as $\mathbb{F}_2$-vector spaces, and hence the additive homomorphic condition can be satisfied, the multiplicative structures of $\mathbb{F}_2^k$ and $\mathbb{F}_{2^k}$ are however different for every $k \geq 2$ (e.g. the former structure has zero divisors, while the latter does not).

We then need to find some alternative weaker notion that allows us to travel back and forth between these two algebraic structures in a manner that it is still amenable to the secure computation protocols we want to adapt. In order to do this we introduce the notion of reverse multiplicative friendly embedding[2].

Definition 1. *Let q be a power of a prime and $\mathbb{F}_q$ a field of q elements, let $k, n \geq 1$ be integers. A pair (ϕ, ψ) is called an $(k, m)_q$-reverse multiplication friendly embedding (RMFE for short) if $\phi : \mathbb{F}_q^k \to \mathbb{F}_{q^m}$ and $\psi : \mathbb{F}_{q^m} \to \mathbb{F}_q^k$ are two $\mathbb{F}_q$-linear maps satisfying*

$$\mathbf{x} * \mathbf{y} = \psi(\phi(\mathbf{x}) \cdot \phi(\mathbf{y}))$$

for all $\mathbf{x}, \mathbf{y} \in \mathbb{F}_q^k$.

While this notion has not been explicitly defined[3] in the literature to the best of our knowledge, a construction for RMFEs was introduced in another work on secure multiparty computation, more precisely on the problem of correlation extraction [BMN17], where it is used to embed a number of instances of oblivious linear evaluation over a small field into one instance of OLE over the extension field. They obtain, for every prime power q and every integer $\ell \geq 1$, a $(2^\ell, 3^\ell)_q$. This implies that we can take $m = O(k^{\log 3/\log 2}) = O(k^{1.58\cdots})$. This construction is unfortunately not enough for our purposes, so in this paper, we show the existence of RMFEs with constant rate.

Theorem 3. *For every finite prime power q, there exists a family of $(k, m)_q$-RMFE where $m = \Theta(k)$.*

We show this result using techniques from algebraic geometry. We emphasize that this is the only point where algebraic geometry is used in our protocol.

As an aside, we also show, by some elementary results on polynomial interpolation, quite practical RMFE's for moderate values of m and with a reasonable rate m/k, indicating our main results may also have some practical value.

---

[2] The term "reverse" refers to the fact that multiplicative friendly embeddings where defined in [CCCX09]. The notions are similar but with the roles of the ring $\mathbb{F}_q^k$ and the field $\mathbb{F}_{q^m}$ swapped. Multiplicative friendly embeddings have been studied more extensively than their reverse counterpart, as they are a special case of bilinear multiplication algorithms [CC88]. They are also special cases of arithmetic codices [CCX12] (see also [CDN15]).

[3] Our original motivation for considering this notion (unpublished work, 2014) was to improve our result on arithmetic secret sharing [CCX11] from CRYPTO 2011.

The compiler. Given a $(k, m)_2$-RMFE (ϕ, ψ), we construct an information-theoretically secure protocol compiler that transforms a secure secret-sharing based computation protocol that evaluates an arithmetic circuit over a large enough finite field $\mathbb{F}_{2^m}$ into a secure protocol for the same number of parties and adversary that allows for the simultaneous evaluation of k instances of a related boolean circuit. The compiler introduces an overhead in the communication complexity of $O(nk)$ bits per multiplication gate of the circuit (hence $O(n)$ bits per multiplication). Our compiler requires that the MPC protocol for the arithmetic circuit over the extension field satisfies a number of properties, which are quite common and are fulfilled by most Shamir-secret-sharing based protocols in this model.

The first step of the compiler is to encode the input vectors as elements in $\mathbb{F}_{2^m}$ with the map ϕ. Now one could think that we proceed by evaluating the arithmetic circuit C' over $\mathbb{F}_{2^m}$ on the encoded inputs, and then decode the result with ψ. Unfortunately this idea does not work, for several reasons; even setting aside security considerations, note that this does not ensure correct computation: it is not even true that $\psi(\phi(\mathbf{x})) = \mathbf{x}$ for all $\mathbf{x}$, hence it does not compute the identity circuit correctly. Moreover $\mathbf{x}_1 * \mathbf{x}_2 * \ldots * \mathbf{x}_\ell = \psi(\phi(\mathbf{x}_1) \cdot \phi(\mathbf{x}_2) \cdot \ldots \cdot \phi(\mathbf{x}_\ell))$ does not necessarily hold when $\ell > 2$ either.

The way to correctly compute the result is instead as follows: encode the input vectors with ϕ and evaluate the arithmetic circuit C' over $\mathbb{F}_{2^m}$ on the encoded inputs, but with the additional step that, every time a multiplication gate its processed, we apply the composition $\phi \circ \psi$ to the output of the gate. We also need to slightly adjust the gates corresponding to a NOT gate in C: in C' these gates add the vector $\phi(1, 1, \ldots, 1)$; moreover if we have random gates in C (gates that produce a random bit), we will need to create random elements $\phi(r) \in \mathbb{F}_{2^m}$; we explain later how we can do this. By doing these transformations, we have the following invariant: at each wire of the $\mathbb{F}_{2^m}$-circuit C', the corresponding value is $\phi(\mathbf{w})$, where $\mathbf{w} = (w_1, \ldots, w_k)$ is the vector containing, for $i = 1, \ldots, k$, the bit w_i that would sit in the corresponding wire of the boolean circuit C on its i-th evaluation. Indeed note that

$$(\phi \circ \psi)(\phi(\mathbf{w}) \cdot \phi(\mathbf{w}')) = \phi(\psi(\phi(\mathbf{w}) \cdot \phi(\mathbf{w}'))) = \phi(\mathbf{w} * \mathbf{w}')$$

so multiplying two encoded vectors and then applying $\phi \circ \psi$ yields an encoding of the coordinatewise product. The rest of the gates obviously preserve this invariant.

At the last step, we decode the output by applying the inverse ϕ^{-1} of ϕ (which we insist, does not coincide with ψ). It is easy to derive from the definition of RMFE that ϕ is injective and hence ϕ^{-1} indeed exists.

Additional auxiliary protocols. However, several roadblocks are introduced when we want to transform a secret-sharing based secure computation protocol π' for C' into a secure computation protocol π that computes (k instances of) C: first, we want each of the input-holding parties to secret share a value $\phi(\mathbf{x}_i)$ with the secret sharing scheme used in π', but given that as we will see the image of

ϕ is not the full $\mathbb{F}_{2^m}$, the question is: how do we ensure that a party has not shared a value outside Im ϕ instead? Note such problem will not be detected by π', as this protocol will "accept" any sharing of a value in $\mathbb{F}_{2^m}$ as long as the sharing itself is correct, so we will need to add some type of zero-knowledge proof that ensures that the shared value is in Im ϕ. The second problem is that we need some protocol that transforms a sharing of an element a into a sharing of $(\phi \circ \psi)(a)$. The composition of ϕ and ψ is a $\mathbb{F}_2$-linear map, but this does not mean that it is $\mathbb{F}_{2^m}$-linear, and therefore it is not necessarily true that the parties can locally compute sharings for the output of this function by using the $\mathbb{F}_{2^m}$-linearity of the scheme[4].

We will show that we can reduce these two issues to the following problem: construct a secure multiparty protocol that outputs a sharing of a random element in a prescribed $\mathbb{F}_2$-subspace of $(\mathbb{F}_{2^m})^v$. That is, given an $\mathbb{F}_2$-vector space $V \subseteq \mathbb{F}_{2^m}^v$ the protocol should output $([r_1], \ldots, [r_v])$ where $(r_1, \ldots, r_v)$ is uniformly random in V, and $[x]$ denotes a sharing of x with the secret sharing scheme used in the protocol for C'.

GLSSS, tensoring-up and hyper-invertible matrices. In order to generate sharings of random elements of the given subspaces, we want to use a technique introduced in [BH08], based on so-called hyper-invertible matrices. In a nutshell this technique consist in the following. Suppose we want to create sharings of one of more random elements satisfying certain relation (we specify below what kind of relations are allowed). Then each party generates a sharing of random elements of their choice satisfying the said relation. Next the hyper-invertible matrix is applied (locally by each party) to the vector containing these n sharings. This creates n new sharings, which will obbey the same relation, if the parties have been honest. Next, some of these are opened to different parties, who check that relation indeed holds. The properties of the matrix will guarantee that if all honest parties declare themselves happy with this process, then the unopened sharings are guaranteed to satisfy the same relation.

However, the type of relations that are preserved by the hyper-invertible matrices are K-linear relations, where K is a field over which the matrix is defined. Unfortunately, the construction of hyper-invertible matrices from [BH08] is based on interpolation techniques, and it requires a field K which contains at least $2n$ elements. This clearly does not fit well with the $\mathbb{F}_2$-linear relations we are dealing with. Applying an $\mathbb{F}_{2^m}$-hyper-invertible matrix to a vector of elements in V will not necessarily output a vector of elements in V.

In order to solve this, we introduce several ideas. First, we formalize the idea of "sharing secrets which are bound by K-linear relations" by the notion of K-generalized linear secret sharing scheme (GLSSS), where the space of secrets

---

[4] To explain this further: we can think of the elements in $\mathbb{F}_{2^m}$ as polynomials in $\mathbb{F}_2[X]$ modulo a degree-m irreducible polynomial. Now consider for example the map that sends $u = a_0 + a_1 X + \cdots + a_{m-1} X^{m-1}$ to $F(u) := a_0$. This is a $\mathbb{F}_2$-linear map (it satisfies $F(u + v) = F(u) + F(v)$), but not $\mathbb{F}_{2^m}$-linear (otherwise there would exist $\lambda \in \mathbb{F}_{2^m}$ such that $F(u) = \lambda \cdot u$, and it is easy to see that this can not happen).

and the spaces of shares are (possibly different) K-vector spaces and the secret is determined by qualified sets of shares by means of a K-linear map. Our definition has the additional advantage that we do not need to worry about how encoding of the secret is done. We also show that the hyper-invertible-matrix technique fits naturally with the notion of GLSSS, since by means of this notion we can state one lemma that captures different instances of this technique in the literature [BH08, DIK10]. However, this is still not enough for our purposes, because this lemma only works if the hyper-invertible matrix is defined over the field K.

Then we introduce the concept of interleaved secret sharing scheme. Given a GLSSS Σ, the m-fold interleaved GLSSS $\Sigma^{\times m}$ is simply the n-player scheme naturally corresponding to m independent Σ-sharings of m secrets. The reason this is useful for our problem comes from the following observation, based on arguments from multilinear algebra (which we call tensoring-up lemma): if we start by a $\mathbb{F}_2$-GLSSS Σ with space of secrets V, there is a natural way in which we can see the interleaved GLSSS $\Sigma^{\times m}$ as a $\mathbb{F}_{2^m}$-GLSSS. Moreover, even though the space of secrets of the new scheme will be V^m, we can crucially access the individual Σ-sharings of each secret in V, since these are just the components of the sharing from $\Sigma^{\times m}$. This means that the hyper-invertible matrix methodology can be applied to $\Sigma^{\times m}$, where each party will bundle together m sharings of random elements in V as a sharing of a random element in V^m, apply the matrix to the resulting sharings using the $\mathbb{F}_2^m$-linear structure given by the tensoring up lemma, and "unzip" the result again into sharings of elements in V.

Putting things together. We will show that if we are using Shamir's scheme (or any secret sharing scheme where the size of the shares is the same as that of the secret) then our subprotocols require the communication of $O(n)$ field elements per gate. Because of our results on reverse multiplication friendly embeddings, this field will have 2^m elements where $m = \Theta(k)$.

On the other hand, to compute securely the arithmetic circuit over the extension field, we can use the protocol in [BH08], which also has communication complexity of $O(n)$ field elements per gate. Altogether we communicate $O(nk)$ bits per gate of the circuit to compute securely k evaluations of the circuit, an amortized cost of $O(n)$ bits per gate. In order for this amortization to work, we need that 2^m is at least n, hence we need to compute at least $k = \Omega(\log n)$ evaluations.

Remark 1 (On the Passive Case). One may possibly be inclined to believe that the case of *passive security* admits a much simpler solution that only involves the RFME's on top of standard protocols. Indeed, after the secure computation of the product of two ϕ-encodings, the secure computation of its $(\phi \circ \psi)$-image is *linear*. However, this is an $\mathbb{F}_2$-linear map defined on the extension field, *not* a linear combination of elements in this extension field. Therefore, the usual "secure computation of linear maps is for free" rule does not hold here. In particular, we still need the same auxiliary protocols with good amortized complexity (including the tensoring-up) as in the active case. That said, the hyper-invertible matrices can be replaced by standard privacy amplification based on error correcting codes over large enough extension fields so as to be able to handle $t < n/2$ with t maximal or arbitrarily close.

2   Abstract GLSSS and Hyper-Invertible Matrices

In our protocol application, we will have a linear secret sharing scheme Σ defined over some "small" finite field K that we wish to deploy in secure computations involving very useful randomization protocols for secret-sharings (i.e., based on hyper-invertible matrices) that, unfortunately, require their field of definition to be a *larger* extension field L. Below we explain how to treat a number of Σ-sharings as a single sharing according to a scheme Σ' that is defined over the extension field L and that has the same privacy and reconstruction properties as Σ. The rate does not change either. Furthermore, "constituent" Σ-sharings remain readily "accessible" from Σ'-sharings for our use in our protocol. Finally, L-linear operations on Σ'-sharings are easily emulated in terms of K-linear operations on constituent Σ-sharings. The way to achieve this is by exploiting basic properties of the tensor product from abstract multilinear algebra.

Another technical aspect of our protocol application is that our Σ is not explicitly constructed as a K-linear scheme in the most standard way (e.g., from a given K-linear error correcting code $C \subset K^n$ with convenient properties) but rather implicitly. Namely, Σ will turn out to be a K-*linear "subscheme"* of a standard L-linear scheme. Concretely, Σ will correspond to several, independent instances of Shamir's scheme over L, where the secrets satisfy K-linear relations. The resulting scheme is, by all means, K-linear. But instead of complicating matters by "forcing" this into a standard formulation where the secret (and the shares) are typically encoded "systematically," we will capture it formally by giving an equivalent but "coordinate-free" version of the usual definition of (general) linear secret sharing.

2.1   An Abstract Definition of GLSSS

For nonempty sets U and $\mathcal{I}$, we let $U^{\mathcal{I}}$ denote the *indexed Cartesian product* $\Pi_{i \in \mathcal{I}} U$. For a nonempty subset $A \subset \mathcal{I}$, the natural *projection* π_A maps a tuple $u = (u_i)_{i \in \mathcal{I}} \in U^{\mathcal{I}}$ to the tuple $(u_i)_{i \in A} \in U^A$. Let K be a field.

Definition 2. *(Abstract K-GLSSS)* A general K-linear secret sharing scheme Σ consists of the following data.

- *A player set $\mathcal{I} = \{1, \ldots, n\}$.*
- *A finite-dimensional K-vectorspace Z, the* secret-space, *and a finite-dimensional K-vectorspace U, the* share-space.
- *A K-linear subspace $C \subset U^{\mathcal{I}}$, where the latter is considered a K-vector space in the usual way (i.e., direct sum).*
- *A surjective K-linear map $\Phi : C \longrightarrow Z$, its* defining map.

Definition 3. *Suppose $A \subset \mathcal{I}$ is nonempty. Then A is a* privacy set *if the K-linear map*

$$(\Phi, \pi_A) : C \longrightarrow Z \times \pi_A(C), \quad x \mapsto (\Phi(x), \pi_A(x))$$

is surjective. Finally, A is a reconstruction set *if, for all $x \in C$, it holds that*

$$\pi_A(x) = 0 \Rightarrow \Phi(x) = 0.$$

Remark 2. The following observations follow directly from the definition.

- (Privacy) Suppose K is finite. If we fix an arbitrary secret $z \in Z$ and select $x \in C$ uniformly random such that $\Phi(x) = z$, then for each privacy set A, it holds that the distribution of the joint shares $\pi_A(x) \in U^A$ for A does not depend on the secret z.
  We denote such a random sharing of a secret z as $[z]$, as usual.
- (Reconstruction) For each reconstruction set A, there are K-linear *reconstruction maps* $\{\rho_i : U \to Z\}_{i \in A}$, depending on A, such that for all $x \in C$, it holds that

$$\sum_{i \in A} \rho_i(x_i) = \Phi(x).$$

By definition, $\mathcal{I}$ is a reconstruction set.

2.2 Randomization Based on Hyper-Invertible Matrices

Hyper-invertible matrices were introduced in [BH08]. Hyper-invertible matrices provide a way for several parties to jointly generate sharings of uniformly random secrets satisfying certain relations. In [BH08], this relation consists in the fact that the uniformly random element is shared with two different sharings (Shamir sharings of different thresholds). However, different uses have been found in other protocols: in [DIK10], the parties generate sharings of uniformly random elements together with a permutation of their coordinates.

In this section, we show that those two applications of hyper-invertible matrices can both be captured under a common framework through the notion of generalized linear secret sharing schemes. Moreover, this framework will also encompass other two applications of this strategy in our protocol.

Definition 4 ([BH08]). *A matrix $M \in K^{\ell \times \ell'}$ is* hyper-invertible *over K if every s-by-s submatrix of M is invertible in K, for every $1 \leq s \leq \min\{\ell, \ell'\}$.*

As in [BH08], we are only interested in this work in square hyper-invertible matrices, even though the results can be generalized easily to non-square ones. If K has at least 2ℓ elements, there is the following construction of an ℓ by ℓ hyper-invertible matrix.

Lemma 1 ([BH08]). *Let K be a finite field with $|K| \geq 2\ell$. Fix $\alpha_1, \ldots, \alpha_\ell, \beta_1, \ldots, \beta_\ell$ distinct elements in K. Let $M = (m_{i,j})$ where $m_{i,j} = \prod_{k \neq j} \frac{\beta_i - \alpha_k}{\alpha_j - \alpha_k}$. Then M is a ℓ by ℓ hyper-invertible matrix over K.*

The interest of square hyper-invertible matrices for secure multiparty computation protocols arises from the property that any combination of ℓ inputs/outputs of the K-linear map induced by M are uniquely determined by, and can be written as a linear function of, the other ℓ inputs/outputs. More formally we have the following.

Lemma 2 ([BH08]). *Let $M \in K^{\ell \times \ell}$ be a square hyper-invertible matrix over K. Consider two subsets $A, B \subseteq \{1, \ldots, \ell\}$ such that $|A| + |B| = \ell$. Then there is a linear map $f_{A,B} : K^\ell \to K^\ell$ such that for every $\mathbf{x} \in K^\ell$, we have $f_{A,B}(\{x_i\}_{i \in A}, \{y_i\}_{i \in B}) = (\{x_i\}_{i \notin A}, \{y_i\}_{i \notin B})$, where $\mathbf{y} = M\mathbf{x}$.*

An important observation is that, given an K-vector space Z, we can define the action of M on vectors from Z^ℓ. Moreover, it is trivial to see that the property above still holds.

Proposition 1. *Let $M \in K^{\ell \times \ell}$ be a square hyper-invertible matrix over K and let Z be a K-linear vector space. Consider two subsets $A, B \subseteq \{1, \ldots, \ell\}$ such that $|A| + |B| = \ell$. Then there is a linear map $f_{A,B} : Z^\ell \to Z^\ell$ such that for every $\mathbf{x} \in Z^\ell$, we have $f_{A,B}(\{x_i\}_{i \in A}, \{y_i\}_{i \in B}) = (\{x_i\}_{i \notin A}, \{y_i\}_{i \notin B})$, where $\mathbf{y} = M\mathbf{x}$.*

Consider now a K-GLSSS Σ with secret space Z, share space U and player set $\mathcal{I}$. Denote a sharing of an element $z \in Z$ by $[z]$. The goal of the following protocol is to generate random sharings of a set of uniformly random elements from Z.

Protocol RandEl

Protocol for a set of parties $\mathcal{I} = \{1, \ldots, n'\}$. Let M be a $n' \times n'$-hyper-invertible matrix over K. Let T be an integer with $1 \leq T \leq n'$.
Output: Sharings $[r^1], \ldots, [r^T]$ of uniformly random elements in Z.

- For $i \in \mathcal{I}$, player i selects a uniformly random element $s^i \in Z$ and shares it among $\mathcal{I}$ with Σ.
- The players locally compute $([r^1], \ldots, [r^{|\mathcal{I}|}])^T = M \cdot ([s^1], \ldots, [s^{|\mathcal{I}|}])^T$ (i.e., each party applies M to the vector of shares and interprets the resulting vector as containing shares to unknown elements $s^1, \ldots, s^{|\mathcal{I}|}$).
- For $i = T+1, \ldots, |\mathcal{I}|$ open r^i to party i. Party i checks that this is indeed a correct sharing to an element from Z and otherwise declares itself unhappy.
- Output the remaining unopened sharings $[r^1], \ldots, [r^T]$.

Proposition 2. *Suppose the active adversary has corrupted at most $t' \leq (n' - 1)/3$ parties from $\mathcal{I}$. Furthermore suppose Σ has t-privacy with $t \geq t'$, u-reconstruction with $u \leq n' - t'$, and assume that $T \leq n' - 2t'$.*

In these conditions if all honest players are happy after the execution of RandEl, then $[r^1], \ldots, [r^T]$ are correct sharings of uniformly random elements

$r^1, \ldots, r^T \in Z$ and the adversary has no information about these values, other than the fact that they belong to Z.

Proof. The proof essentially follows the steps of [BH08, Lemma 5].

We first consider robustness. First of all, by u-reconstruction, the shares of the (at least) $n' - t'$ honest parties uniquely determine the secrets, so cheating by the adversary by changing the shares corresponding to the corrupted parties will either be detected or not change the computation of the opened r^i. Assume that all honest players remain happy. Then all sharings opened to honest parties are valid sharings of elements r^i belong to Z. These are at least $n' - T - t'$. On the other hand the $n' - t'$ input sharings $[s^i]$ inputted by the honest parties are correct sharings of elements in Z. Note these are at least n' input/output values. By the properties of the hyper-invertible matrix the rest of inputs/outputs are a K-linear function of these values; a bit more precisely, for every honest party, her shares of the remaining inputs/outputs are a K-linear function of her shares of the honest inputs and the outputs opened by honest parties, and since the shares of honest parties fix the secrets (and Z is a K-linear vector space), the unopened secrets must be in Z.

We now consider privacy. First, by t-privacy, the shares of the adversary provide no information about the inputs provided by and the outputs opened to honest parties. The adversary does know the inputs s^i provided by corrupt parties and the outputs r^i opened to corrupt parties. But these are at most $2t'$ values. By Proposition 1, it is easy to see that these are completely independent of any set of $n' - 2t'$ other inputs/outputs, in particular the T outputted values.

This protocol generalizes the "double-sharing generation" strategy from [BH08] as well as the "generation of sharings of a random vector and a permutation of its coordinates" strategy from [DIK10]. In the first case, we can define a GLSSS that has K as space of secrets, K^2 as space of shares and where sharing in that scheme consists on independently sharing the secret with two standard Shamir secret sharing schemes of degrees d and d'. As long as $t' \leq \min d, d' < n - t'$, the conditions of the proposition are satisfied. In the second case, the secret space would be the ℓ-dimensional K-vector space $\{(\mathbf{x}, \pi(\mathbf{x})) : \mathbf{x} \in K^\ell\}$ where π is some known permutation of the coordinates of $\mathbf{x}$ and sharing means to share each coordinate individually with a K-linear scheme, and the proposition holds as long as this secret sharing scheme satisfies the privacy and reconstruction properties there.

2.3   Extending the Field of Definition of a GLSSS

Later on, we will encounter situations where our secret space is a $\mathbb{F}_2$-linear space, but not a $\mathbb{F}_{2^k}$-linear space, and hence the corresponding GLSSS is only linear over $\mathbb{F}_2$. This does not fit well with the fact that the hyper-invertible matrix will be defined over $\mathbb{F}_{2^m}$. Therefore we detail an strategy to extend the field of definition of a GLSSS.

To achieve the desired extension of the field of definition of a GLSSS, it is convenient to define *interleaved* GLSSS. Informally, the m-fold interleaved

GLSSS $\Sigma^{\times m}$ is the n-player scheme naturally corresponding to m Σ-sharings. In other words, with the notation of Sect. 2.1, a sharing in this scheme can be seen as an $m \times n$ matrix whose m rows each represent an element of C and whose n columns each represent an element of the K-vector space U^m, the share-space. We denote the K-linear subspace of $\times_{i \in I} U^m$ collecting these matrices as $C^{\times m}$. The defining K-linear map $\Phi^{\times m}$ is just the "row-wise" application of Φ and the secret-space is the K-vector space Z^m. Note that the privacy sets as well as the reconstruction sets coincide with those of Σ.

A Tensoring-Up Lemma. Let L be an extension field of K of degree m. We will explain later on how $\Sigma^{\times m}$ is in fact an L-linear GLSSS in a natural and convenient way, compatible with its K-linearity as already defined. To this end we need a brief intermezzo derived from basic multilinear algebra, specifically a special case of base change in tensor products.[5] We will give an explicit lemma that defers the use of tensor products and their relevant properties to the proof.

Definition 5. *For our purposes, a K-algebra is a ring having the field K as a subring. Suppose R, S are K-algebras. A K-algebra morphism $R \longrightarrow S$ is a ring morphism that fixes K, i.e., it is, in particular, a K-vector space morphism.*

Lemma 3. *Let L be an extension field of degree r over K and let V be a K-vector space. Then the following hold:*

1. *Let $K^{m,m}$ denote the matrix algebra over K consisting of all m x m matrices with entries in K, with the usual addition and (K-scalar-) multiplication. Then there is a (non-unique) injective K-algebra morphism*

$$\Phi : L \longrightarrow K^{m,m}; \quad \lambda \mapsto \Phi^{(\lambda)}.$$

In particular, the image of L is a field isomorphic to it.

2. *Each such Φ induces an L-vector space structure on V^m by defining L-scalar multiplication, for each $\lambda \in L$, as*

$$\lambda \cdot : V^m \longrightarrow V^m; \quad w \mapsto \Phi^{(\lambda)}(w),$$

where the action of $K^{m,m}$ on V^m is the natural one, i.e., multiplication of an m x m matrix with an m-(column)vector.

If $\lambda \in K$, it restricts to

$$\lambda \cdot : V^m \longrightarrow V^m; \quad w \mapsto \lambda \cdot w,$$

since Φ fixes K. Hence, this structure is compatible with the standard K-vector space structure on V^m, i.e., as a direct sum over V,

---

[5] For a treatment of abstract tensor-products aimed at a cryptographic audience, we refer to Ch. 10.9 (pp. 229–235) in [CDN15].

Proof. As to the first claim for each $\lambda \in L$, the multiplication-by-λ map on L is a K-vector space endomorphism on L (i.e., a morphism from L to L). The map Φ that sends $\lambda \in L$ to this associated morphism is clearly a K-algebra morphism from L to the K-algebra End of K-vector space endomorphisms of L. Note that Φ is injective as its domain is a field; the only possibility for its kernel is the trivial ideal (0) of L. Since L is a vector space of dimension r over K, it is clear that, once a basis is fixed, End may be given as $K^{m,m}$.

As to the second claim, an R-module M consists of an abelian group M, a ring R (with 1), together with a ring morphism mapping R to the ring of group endomorphisms of M. It is called a vector space if R is a field. In the present case, M is the direct sum V^m and R is L. By construction, (L, V^m, Φ) satisfies this condition. Finally, note that Φ maps $\lambda \in K$ to $\lambda \cdot I$, where I is the identity matrix.

As the lemma reflects one of the basic merits of tensor products, we verify it below in such terms and in more generality. By the very definition of tensor product, if M is an R-module (M and R take the role of V and K, resp.), where R is a commutative ring with 1, and if S is an extension ring (S takes the role of L), then the tensor product $S \otimes_R M$ is an R-module. By base change, we may, in fact, naturally view $S \otimes_R M$ as an S-module, compatible with the R-module structure already mentioned. Namely, for each $s \in S$, for each $r \in R$ and for each $m \in M$, define $s \cdot (r \otimes m) = (sr \otimes m)$ and extend this linearly to all of $S \otimes_R M$. If, in addition, S is free of rank r over R, then, as an R-module, the tensor product $S \otimes_R M$ is isomorphic to M^m. Since S-multiplication by a constant as defined above is, in particular, an endomorphism of R-modules, it is clear that such a map can be represented by an element of $K^{m,m}$.

Linearity over the Extension Field of the Interleaved Scheme. With the tensoring-up lemma in hand, we now explain how the m-fold interleaved GLSSS $\Sigma^{\times m}$ is L-linear, compatible with the K-linearity already pointed out. It is convenient, once again, to think of the elements of $C^{\times m}$ as matrices where each row is an element of C and where each column i collects the corresponding m shares for player i.

Take an arbitrary such matrix representing an element in $C^{\times m}$ and take an arbitrary $\lambda \in L$. Write the matrix map $\Phi^{(\lambda)}$ as $(\Phi_1, \ldots, \Phi_m)$ such that the image of $w \in U^m$ equals $(\Phi_1(w), \ldots, \Phi_m(w))$.

Then we simply replace each "column of shares" $w = (w_1, \ldots, w_m) \in U^m$ by the column $\lambda \cdot w = (\Phi_1(w), ..., \Phi_m(w))$. Since, the Φ_i are K-linear, it is immediate that application of the K-linear map $\Phi^{\times m}$ commutes with λ-multiplication. Thus, $\Psi^{\times m}$ is L-linear, compatible with its earlier mention K-linearity. Note that, for given Σ, this extension depends implicitly on the choice of a K-basis of L.

In summary:

Proposition 3. *Let L be a degree-m extension field of K and let Σ be a K-GLSSS. Then the m-fold interleaved K-GLSSS $\Sigma^{\times m}$ is naturally viewed as an L-GLSSS, compatible with its K-linearity.*

Use in Protocols. We work with this proposition as follows. Suppose we have a sharing in $\Sigma^{\times m}$, i.e., m Σ-sharings

$$([z_1], \ldots, [z_m]), \text{ with } z_1, \ldots, z_m \in Z.$$

If $\lambda \in L$, then

$$\lambda \cdot ([z_1], \ldots, [z_m]) = (\Phi_1([z_1], \ldots, [z_m]), \ldots, \Phi_r([z_1], \ldots, [z_m])),$$

which is λ times the given $\Sigma^{\times m}$-sharing, which, is, again, a $\Sigma^{\times m}$-sharing.

2.4   Alternatives for Hyper-Invertible Matrices: Same Functionality, but Constant-Size Field

We make two remarks about alternative approaches.

Remark 3. From known constructions, hyper-invertible matrices and their utility in MPC protocols appear tightly connected with polynomial evaluation codes (MDS codes) and therefore may seem to require a field of definition that grows as a linear function of n. However, we note that we found that there is an alternative coding-theoretic construction with essentially the same utility as that of hyper-invertible matrices but that allows constant-size finite fields, where the size should be large enough so as to make adversary rate $1/3$ possible. In a nutshell, the argument goes as follows: given a linear code C of length $2n'$, minimum distance d and minimum distance of its dual $d^\perp$ it is not difficult to see that every coordinate can be written as a linear function of any set of $2n' - d + 1$ coordinates, and that any set of $d^\perp - 1$ coordinates of a random codeword are uniformly distributed. Suppose in addition the code has dimension n', wlog assume its in systematic form and its generator matrix is $G = (I_{n'}|M)$. If we can take $d, d^\perp \geq (2/3 + \epsilon)n' \geq 2t' + \epsilon n'$, then Proposition 2 still holds for $T \leq \epsilon n'$. Taking random linear codes over $\mathbb{F}_{64}$ of rate $1/2$ should suffice for this purpose according to the Gilbert-Varshamov bound (and the secret sharing scheme should then be tensored-up to this field). Although there is no overall advantage to our work (quantitatively), it does mean that, in the subprotocols where it is used, "amortization kicks in faster."

Remark 4. Instead of tensoring-up the secret sharing scheme, we may have taken hyper-invertible matrices (or the alternative above) and have re-worked them to be defined over the base field, using the same technique as in tensoring-up (i.e., viewing the extension field as a matrix algebra over the base field) and making substitutions accordingly. This leads to a "block-wise" version of hyper-invertibility which is sufficient for our purposes. However, we feel that the present approach we took is more natural and leads to cleaner protocols.

3   The Protocol

In this section we detail our protocol π, that securely evaluates k instances of a binary circuit C in parallel by using a secure computation protocol π' that computes securely essentially the same arithmetic circuit defined over a extension field $\mathbb{F}_{2^m}$.

3.1   Framework

We consider a network of n parties who communicate via pairwise secure channels. Up to t of these parties are corrupted by an active adversary, where will require that $t < n/3$.

Let C be a boolean circuit consisting of input gates; computation gates which will be (unbounded fan-in) addition (XOR) gates, fan-in 2 multiplication (AND) gates and NOT gates (which we can think of as addition with the constant 1); random gates that output a uniformly random bit and an output gate. We assume that there is a single output gate for simplicity of notation only, as the generalization of our results to the case where there are more output gates is straightforward. Let c_I, c_R, c_M be the number of input, random and multiplication gates respectively.

Given a $(k, m)_2$-RMFE (ϕ, ψ), we define the following arithmetic circuit C_ϕ over the extension field $\mathbb{F}_{2^m}$: We replace the XOR and AND gates in C by gates implementing addition and multiplication in $\mathbb{F}_{2^m}$ and we replace the NOT gates by addition with the field element $\phi((1, 1, \ldots, 1)) \in \mathbb{F}_{2^m}$ (which may not coincide with the element 1 in $\mathbb{F}_{2^m}$). For consistency we replace the boolean random gates by gates which create random elements in $\mathbb{F}_{2^m}$; this does not really have too much importance, since we will entirely replace the computation of this gate by a subprotocol.

Preprocessing and player elimination. Our protocol π will have a pre-processing phase, which is independent of the inputs, and a computation phase.

In addition, π will use the player elimination framework. Player elimination, introduced in [HMP00], is a technique by which the computation (or part of it) is first divided in segments and in each segment, if at least one party has deviated from the protocol, a set of two parties is identified out of which at least one is a corrupt. The protocol then proceeds by eliminating these two parties and recompute the segment. This protocol works exactly as described in [BH08], so we refer the reader to that work for its detailed description. At every step of the protocol, we will denote by n' the number of active (not eliminated) parties, and by t', the number of active corrupted parties. Note that the invariant $t < n' - 2t'$ always holds.

It is important to mention that, as it occurs in other protocols such as [BH08], we will use player elimination in the preprocessing phase only.

Conditions on π'. We now describe the conditions that π' needs to satisfy so that we can apply our compiler and construct π. First of all, π' will be a secret-sharing based protocol and we need to make some assumptions on the underlying secret sharing scheme.

Given a secret sharing scheme with player set $\mathcal{I}$, by puncturing the scheme at a subset $A \subseteq \mathcal{I}$ we mean that we consider the secret sharing scheme where we remove the set A of parties (so the new player set is $\mathcal{I} \setminus A$ and sharing happens in the same way as in the original scheme, except that the shares that would correspond to the subset A are erased).

Definition 6. *We say that a secret sharing scheme is t-robust if there exists a polynomial-time algorithm that, when given as input all shares in a sharing $[x]$, among which at most t are erroneous, outputs x.*

We say that a secret sharing scheme on n parties is elimination-compatible t-robust, if for every $0 \leq u \leq t$, and any set of $2u$ parties, puncturing the scheme at those $2u$ parties results in a scheme on the other $n' = n - 2u$ parties which is t'-robust, where $t' = t - u$.

Remark 5. Note that a degree-t Shamir's secret sharing scheme for n parties is elimination-compatible t-robust as long as $3t + 1 \leq n$. Indeed, after player elimination, the set of possible sharings forms a Reed-Solomon code of length n' and dimension $t + 1$, and therefore minimum distance $n' - t$. There exist well known efficient algorithms that can correct any $e < (n' - t)/2$ errors. But the number of errors that can be introduced by the adversary is at most t', and as we noted above $t < n' - 2t'$ which implies $t' < (n - t)/2$.

We assume that the secure multiparty computation protocol π' to compute C_ϕ has the following features:

Assumptions on π'

- π' may have a preprocessing phase, which is independent of the inputs, and a computation phase. We allow the protocol to use player elimination in the preprocessing phase.
- π' is secure against an active adversary corrupting t parties.
- π' is a secret-sharing based secure multiparty computation protocol π' which uses a $\mathbb{F}_{2^m}$-secret sharing scheme with t-privacy and which is elimination-compatible t-robust (the sharing of an element x is denoted $[x]$)
- In the computation phase every input and intermediate computed value remain secret shared among the parties with this secret sharing scheme; the protocol creates these sharings as follows: at every addition gate, parties locally compute a sharing of the output of the gate from the sharings of the inputs using the linearity of the scheme; the same holds for addition and multiplication by known constants; multiplication gates are processed by a subprotocol Mult that on input $[a], [b]$ produces $[ab]$.

3.2 Result

In the rest of the section we prove the following theorem.

Theorem 4. *Assume there exists a $(k, m)_2$-reverse multiplication friendly embedding (ϕ, ψ), where $2^m \geq 2n$ and let π' be a secure multiparty computation protocol for the arithmetic circuit C_ϕ over $\mathbb{F}_{2^m}$ satisfying the assumptions above. Then there exists a multiparty computation protocol secure against* *an active adversary who corrupts at most t parties and which allows to compute k instances of the circuit C with communication complexity* $cc(\pi) = cc(\pi') + (c_I + c_M + c_R) \cdot O(n)$ *elements of* $\mathbb{F}_{2^m}$.

3.3   The General Structure of π

The general idea of the construction has been explained in the introduction: in the first step, for $i = 1, \ldots, n$, the i-th party, who has an input $\mathbf{x}_i = (x_i^{(1)}, \ldots, x_i^{(k)}) \in \mathbb{F}_2^k$, creates a sharing of $[\phi(\mathbf{x}_i)] \in \mathbb{F}_{2^m}$. A subprotocol CorrInput will ensure that this sharing is well constructed (in particular, it hides an element from Im ϕ). Then the parties execute π' on inputs $[\phi(\mathbf{x}_1)], \ldots, [\phi(\mathbf{x}_n)]$, but every time that there is a multiplication gate, the output of that gate, say $[a]$, will be re-encoded by applying a sub-protocol ReEncode that creates $[\phi(\psi(a))]$ from $[a]$. Therefore, at the end of the computation with π', the parties obtain $\phi(\mathbf{y})$, for $\mathbf{y} = (y^{(1)}, \ldots, y^{(k)})$. Here each $y^{(j)}$ is the output of the evaluation of C on $(x_1^{(j)}, \ldots, x_N^{(j)})$. Every party can now apply ϕ^{-1} to recover $\mathbf{y}$.

An additional detail is that for random gates we need to create sharings of uniformly random elements in Im ϕ. As we will see, this is exactly the main step in CorrInput too.

We explain the subprotocols in the following lines.

3.4   Auxiliary Protocols

We will now describe the subprotocols needed in π. We recall that since we use player elimination, at a given point of the protocol there will be n' active parties out of which t' are corrupted, where $n' = n - 2u$ and $t' = t - u$ for some $0 \le u \le t$, and that $t < n' - 2t'$ always holds. So we describe our protocols taking that into account (for the sake of notation the active parties are indexed by $1, \ldots, n'$). In particular, it will be understood that the secret sharing scheme at a given point of the protocol is the original secret sharing scheme punctured on $2u$ parties.

We start with the public reconstruction protocol ReconsPubl. One possibility could of course be simply to have every party send their share to each other, after which every party clearly can reconstruct, since the scheme is t'-robust. However, this incurs in a communication complexity of $\Theta(n^2)$ elements of the field. The following idea comes originally from [DN07] and allows to amortize the reconstruction, so that the communication complexity is still $\Theta(n^2)$ but $\Omega(n)$ sharings are simultaneously reconstructed.

Protocol ReconsPubl (from [DN07])

Input: $[a_1], [a_2], \ldots, [a_{n'-2t'}]$.

Output: All parties obtain $a_1, a_2, \ldots, a_{n'-2t'}$.

Fix $\beta_1, \ldots, \beta_{n'} \in \mathbb{F}_{2^m}$ pairwise distinct.

- Call $u_j := \sum_{i=1}^{n'-2t'} a_i \beta_j^i$. For all j, parties locally compute $[u_j] = \sum_{i=1}^{n'-2t'} [a_i] \beta_j^i$.

- For all i, all parties send their shares of u_i to P_i.
- For all i, P_i applies the robust reconstruction algorithm of the secret sharing scheme to obtain u_i.
- For all i, j, P_i sends u_i to P_j.
- For all j, P_j applies an standard error decoding algorithm for Reed-Solomon codes to recover $a_1, \ldots, a_{n'-2t'}$ from the values $\widetilde{u}_1, \ldots, \widetilde{u}_{n'}$ received in the previous step (using that $\widetilde{u}_i \neq u_i$ for at most t' values).

Remark 6. `ReconsPubl` allows to perfectly reconstruct $n' - 2t' = \Omega(n)$ sharings by communicating $2n'(n'-1) = O(n^2)$ elements of the field in total, an amortized cost of $O(n)$ elements of the field per reconstructed sharing.

As it has been mentioned before, both the subprotocols `CorrInput` and `ReEncode` need to use sharings of uniformly random elements in certain $\mathbb{F}_2$-subspaces. These will be generated in the preprocessing phase with the help of hyper-invertible matrices, by using the techniques introduced in Sect. 2.

We describe more explicitly how this works. Let $V \subseteq \mathbb{F}_{2^m}^v$ be a $\mathbb{F}_2$-subspace (in our protocols we will only encounter the cases $v = 1$, $v = 2$, but here we treat the problem more generally). The protocol `RandElSub`(V) generates sharings of uniformly random elements in V. Here, by a sharing of an element $u = (u_1, \ldots, u_v) \in V$ we refer to the generalized linear secret sharing scheme that consists in that each coordinate u_j is shared with the secret sharing scheme used in the protocol. This is an $\mathbb{F}_2$-generalized linear secret sharing scheme where the secret space is V and the share spaces are $\mathbb{F}_{2^m}^v$. We will call this secret sharing scheme Σ, but by abuse of notation we write $[u] = ([u_1], \ldots, [u_v])$.

We need a $n' \times n'$-hyper-invertible matrix M over some finite field. Since $|\mathbb{F}_{2^m}| \geq 2n'$ by assumption, we know how to construct such matrices over $\mathbb{F}_{2^m}$, by Lemma 1. However, the GLSSS described above is only linear over $\mathbb{F}_2$, so in order to apply the hyper-invertible matrix we need to tensor-up this GLSSS to a $\mathbb{F}_{2^m}$-linear one, using the techniques from Sect. 2. Recall that this will create the interleaved $\mathbb{F}_{2^m}$-GLSSS $\Sigma^{\times m}$ where the secrets are in V^m and each share is in $(\mathbb{F}_{2^m})^m$, to which we apply the hyper-invertible matrix technique. However, note that the interleaved scheme still allows to access easily the sharings of the individual elements in V. Namely the scheme has secrets $\mathbf{u} = (u_1, \cdots, u_m)$ where $u_j = (u_{j,1}, \ldots, u_{j,v}) \in V$ for $j = 1, \ldots, m$ and the sharing of $\mathbf{u}$ consists of independent sharings of all $u_{j,\ell} \in \mathbb{F}_{2^m}$ with the scheme used by π'. We abuse once more notation and denote $[\mathbf{u}] := ([u_1], \cdots, [u_m])$ where in turn $[u_i] = ([u_{i,1}], \ldots, [u_{i,v}])$.

The protocol is as follows.

Protocol RandElSub(V)

Parameter: Let T be an integer with $1 \leq T \leq n' - 2t'$.
Output: Sharings $[r_{j,\ell}^i]$ of elements $r_{j,\ell}^i \in \mathbb{F}_{2^m}$, $i = 1, \ldots, T$, $j = 1, \ldots, m$, $\ell = 1, \ldots, v$, where $r_j^i = (r_{j,1}^i, \ldots, r_{j,v}^i)$ are uniformly random elements from V.

Let $M \in \mathbb{F}_{2^m}^{n' \times n'}$ be a hyper-invertible matrix.

- For $i = 1, \ldots, n'$, P_i selects m uniformly random elements $s_1^i, \cdots, s_m^i \in V$ and creates a sharing $[\mathbf{s}^i] := ([s_1^i], \cdots, [s_m^i])$ with the interleaved secret sharing scheme $\Sigma^{\times m}$, where in turn $[s_j^i] := ([s_{j,1}^i], [s_{j,2}^i], ..., [s_{j,v}^i])$.
- Players locally compute $([\mathbf{r}^1], \ldots, [\mathbf{r}^{n'}]) = M([\mathbf{s}^1], \ldots, [\mathbf{s}^{n'}])$. Note that the entries of M are in $\mathbb{F}_{2^m}$ and that M acts on $[\mathbf{s}^i]$ as explained in Section 2, using the fact that $\Sigma^{\times m}$ is a $\mathbb{F}_{2^m} - GLSSS$.
- For $i = T + 1, \ldots, n'$, every party P_j sends its share of $[\mathbf{r}^i]$ to P_i. Note that $[\mathbf{r}^i]$ can always be parsed as $([r_1^i], \cdots, [r_m^i])$, so what P_j sends is her shares of m values shared with Σ. P_i verifies that the values received indeed are valid sharings of values $(r_1^i, \cdots, r_m^i)$, and that $r_1^i, \ldots, r_m^i \in V$. If any check fails, P_i gets unhappy.
- The remaining T sharings $[\mathbf{r}^1], \ldots, [\mathbf{r}^T]$ are outputted. Note that $[\mathbf{r}^i] = ([r_1^i], [r_2^i], \ldots, [r_m^i])$ where $r_j^i = (r_{j,1}^i, r_{j,2}^i, \ldots, r_{j,v}^i) \in V$ and $r_j^i = ([r_{j,1}^i], [r_{j,2}^i], \ldots, [r_{j,v}^i])$

Proposition 4. *If all honest players are happy after the execution of RandElSub, then $[\mathbf{r}^1], \ldots, [\mathbf{r}^T]$ are $\Sigma^{\times m}$-sharings of uniformly random vectors $\mathbf{r}^1, \ldots, \mathbf{r}^T \in V^m$, i.e., RandElSub produces Σ-sharings of mT uniformly random values $r_j^i \in V$, $j = 1, \ldots, m, i = 1, \ldots, T$, about which the adversary learns no information (other than the fact that they are elements from V). The total communication complexity of RandElSub is $(2n' - T)(n' - 1)mv$ field elements, which if $T = n' - 2t' = \Theta(n)$ yields an amortized cost of $O(nv)$ field elements per sharing of an element in V.*

The proof of this result consists in noticing that this protocol is RandEl from Sect. 2 applied to the $\mathbb{F}_{2^m}$-GLSSS $\Sigma^{\times m}$.

We will handle the case where parties declare themselves unhappy by means of the player elimination technique. For the moment we assume that enough sharings of random elements in the appropriate subspaces have been generated.

Next, we describe the protocol CorrInput which takes as input $[a]$ (where $a \in \mathbb{F}_{2^m}$) and whose goal is verifying that $a \in \text{Im } \phi$. For this the parties take a sharing $[r]$ of a uniformly random element $r \in \text{Im } \phi$, that have been generated by the protocol RandElSub(Im ϕ). Then they can use it to locally compute $[a + r]$, open this sharing and verify that $a + r \in \text{Im } \phi$, and since $\text{Im } \phi$ is a $\mathbb{F}_2$-vector subspace, $r, a + r \in \text{Im } \phi$ imply that $a \in \text{Im } \phi$. Moreover, since r is uniformly random in $\text{Im } \phi$, the opened value $a + r$ gives no additional information on a.

Protocol CorrInput

Input: $[a]$.
Output: Accept if $a \in \text{Im } \phi$. Reject otherwise.

- Take the next unused sharing $[r]$ produced by $\text{RandElSub}(\text{Im } \phi)$.
- Compute $[a + r] = [a] + [r]$ locally.
- Use ReconsPubl to open $[a + r]$. Let b be the opened value.
- Accept if $b \in \text{Im } \phi$. Reject otherwise.

It is quite straightforward that this protocol is secure. Note that all honest parties will receive the same output, because ReconsPubl will output the same value to all of them. Moreover, notice that if $[a]$ is a correct sharing, then ReconsPubl will succeed reconstructing $a + r$ even if malicious parties communicate false shares because ReconsPubl is robust.

Finally, we consider the protocol ReEncode, whose goal is to construct $[\phi(\psi(a))]$ from $[a]$, where $a \in \mathbb{F}_{2^m}$. We remark first that the composition $\phi \circ \psi : \mathbb{F}_{2^m} \to \mathbb{F}_{2^m}$ is an $\mathbb{F}_2$-linear map, but not an $\mathbb{F}_{2^m}$-linear map. Therefore we cannot use the $\mathbb{F}_{2^m}$-linearity of the secret sharing scheme to have parties locally compute $[\phi(\psi(a))]$ given $[a]$. Instead, we use a randomization technique, as in the case of CorrInput. Define the set

$$W = \{(x, \phi(\psi(x))) : x \in \mathbb{F}_{2^m}\} \subseteq (\mathbb{F}_{2^m})^2.$$

This is an $\mathbb{F}_2$-subspace of $(\mathbb{F}_{2^m})^2$. The parties will have called RandElSub on W in the preprocessing phase in order to create (at least) c_M sharings of random elements in W. They take a unused such sharing $[\mathbf{r}] = ([r], [\phi(\psi(r))])$. Then they can use it to locally compute $[a + r]$, open this value and then compute $[\phi(\psi(a))] = [\phi(\psi(a + r))] - [\phi(\psi(r))]$, where $[\phi(\psi(a + r))]$ is some default sharing of the public element $\phi(\psi(a + r))$, which can be computed from the opened information $a + r$. Note that this opened value $a + r$ gives no information about a, since r is uniform in $\mathbb{F}_{2^m}$.

Protocol ReEncode

Input: $[a]$.
Output: $[\phi(\psi(a))]$.
Let $W := \{(x, \phi(\psi(x))) : x \in \mathbb{F}_{2^m}\} \subseteq (\mathbb{F}_{2^m})^2$.

- Take the next unused sharing $[\mathbf{r}]$ produced by $\text{RandElSub}(W)$. Parse $[\mathbf{r}]$ as $([r], [s])$, where $s = \phi(\psi(r))$.

- Compute $[a + r] = [a] + [r]$ locally.
- Use ReconsPubl to open $[a + r]$. Let m be the opened value.
- Compute $[w] = \phi(\psi(m)) - [s]$.
- Output $[w]$.

3.5 Final Protocol

We describe our final protocol. The preprocessing phase will generate sharings of at least $c_I + c_R$ uniformly random values in Im ϕ, and at least c_M uniformly random values in W. In order to incorporate player elimination, we split the computation of these values in $(c_I + c_M + c_R)/t$ segments. After the computation of each segment, if some party is unhappy, then all values generated in that segment are discarded and player elimination is used to identify a set of two parties containing one malicious party. These two parties are eliminated and the computation of the segment is restarted with the updated values for n' and t' and all parties resetting their status to happy.

Protocol π

Inputs: $\mathbf{x}_i = (x_i^{(1)}, ..., x_i^{(k)}) \in \mathbb{F}_2^k$, $i = 1, \dots, N$, where each $\mathbf{x}_i$ is known to some party.

Output: All parties learn $\mathbf{y} = (y^{(1)}, \dots, y^{(k)})$ where $y^{(j)}$ is the evaluation of circuit C on input $(x_1^{(j)}, ..., x_N^{(j)})$.

(Input-independent) preprocessing phase:

- Generation of random elements in $\mathbb{F}_2$-subspaces. The following computation is splitted in $(c_I + c_M + c_R)/t$ segments. After each segment, if some party is unhappy, discard that computation, execute player elimination and restart the segment with the new set of parties.
  - The parties run RandElSub(Im ϕ) enough number of times to create sharings of at least $c_I + c_R$ random elements in Im ϕ.
  - The parties run RandElSub(W) enough number of times to create sharings of at least c_M random elements in Im ϕ.
- The parties execute the preprocessing phase of π', if there is any.

Computation phase:

- For $i = 1, \dots, N$, the party holding input $\mathbf{x}_i$ computes $\phi(\mathbf{x}_i)$ execute the subprotocol from π' to create $[\phi(\mathbf{x}_i)]$. The parties execute CorrInput, using the next unused sharing produced by RandElSub(Im ϕ).
- Parties execute the rest of the computation phase of π' on inputs $([\phi(\mathbf{x}_1)], \dots, [\phi(\mathbf{x}_N)])$ with the following changes:

> At every multiplication gate of C', after the parties execute Mult on inputs $([a], [b])$ and obtain $[ab]$, they apply subprotocol ReEncode to $[ab]$ and produce $[\phi(\psi(ab))]$. Each time ReEncode is called the next unused sharing produced by RandElSub(W) is used.
> At every random gate of C' the computation of the gate by π' is ignored and instead the next unused sharing $[\phi(r)]$ produced by RandElSub(Im ϕ) is used.
> – Let z be the output of π' in the execution of the protocol. The output of π is $\mathbf{y} = \phi^{-1}(z)$.

We consider the communication complexity of π. It executes one instance of π', one instance of CorrInput per input gate and one instance of ReEncode per multiplication gate of the circuit. In turn, both CorrInput and ReEncode execute the public reconstruction protocol of the secret sharing scheme and both subprotocols require one fresh sharing of a random element produced by RandElSub (invoked on $V = \mathrm{Im}\ \phi$ in the case of CorrInput and on $V = W$ in the case of ReEncode). Note that we can use RandElSub to create these sharings of random elements in batches of size $n \log n$ with a communication complexity $O(n^2 \log n)$, which gives an amortized complexity of $O(n)$ field elements per output sharing.

Therefore the communication complexity of the protocol π is $cc(\pi) = cc(\pi') + (c_I + c_M + c_R) \cdot O(n)$ field elements.

4 Reverse Multiplicative Friendly Embeddings

In this section, we show, by algebraic geometric means, effective $(k, m)_q$-RMFE's with $m = O(k)$ for every finite field $\mathbb{F}_q$. The hidden constant is actually quite small.

But first show that if the size of the base field q is larger than $k - 1$ we can construct a $(k, 2k - 1)_q$-RMFE's based on some elementary results on polynomial interpolation. Chaining these together by concatenation, we then show quite practical RMFE's for moderate values of m and reasonable rate m/k. This indicates that our main results may also have some practical value.

Lemma 4. *For all* $1 \leq k \leq q + 1$, *there exists a* $(k, 2k - 1)_q$-*RMFE.*

Proof. Let $\mathbb{F}_q[X]_{\leq m}$ denote the set of polynomials in $\mathbb{F}_q[X]$ of degree at most m and let ∞_{m+1} be a formal symbol such that $f(\infty_{m+1})$ is the coefficient of X^m in $f \in \mathbb{F}_q[X]_{\leq m}$. Let $x_1, \ldots, x_k$ be pairwise distinct elements in $\mathbb{F}_q \cup \{\infty_k\}$ and let $\alpha \in \mathbb{F}_{q^{2k-1}}$ be such that $\mathbb{F}_{q^{2k-1}} = \mathbb{F}_q(\alpha)$.

By [CDN15, Theorems 11.13, 11.96] the maps

$$\mathcal{E}_1 : \mathbb{F}_q[X]_{\leq k-1} \to \mathbb{F}_q^k; \quad f \mapsto (f(x_1), f(x_2), \ldots, f(x_k))$$

and

$$\mathcal{E}_2 : \mathbb{F}_q[X]_{\leq 2k-2} \rightarrow \mathbb{F}_{q^{2k-1}}; \quad f \mapsto f(\alpha)$$

are isomorphisms of $\mathbb{F}_q$-vector spaces.

Define also

$$\mathcal{E}'_1 : \mathbb{F}_q[X]_{\leq 2k-2} \rightarrow \mathbb{F}_q^k; \quad f \mapsto (f(x'_1), f(x'_2), \ldots, f(x'_k))$$

where $x'_i := x_i$ if $x_i \in \mathbb{F}_q$, and $x'_i := \infty_{2k-1}$ if $x_i = \infty_k$.

Now we define $\phi = \mathcal{E}_2 \circ \mathcal{E}_1^{-1}$ and $\psi = \mathcal{E}'_1 \circ \mathcal{E}_2^{-1}$ (where in the case of ϕ the composition makes sense because $\mathbb{F}_q[X]_{\leq k-1} \subseteq \mathbb{F}_q[X]_{\leq 2k-2}$). Then using that $fg(\alpha) = f(\alpha)g(\alpha)$ and $fg(x'_i) = f(x_i)g(x_i)$ for all $f, g \in \mathbb{F}_q[X]_{\leq k-1}$, it is immediate that (ϕ, ψ) is a $(k, 2k-1)_q$-RMFE.

Next, we show how to concatenate RMFEs over different finite fields.

Lemma 5. *Assume that* (ϕ_1, ψ_1) *is an* $(k_1, m_1)_{q^{m_2}}$-RMFE *and* (ϕ_2, ψ_2) *is an* $(k_2, m_2)_q$-RMFE. *Then*

$$\phi : \mathbb{F}_q^{k_1 k_2} \rightarrow \mathbb{F}_{q^{m_1 m_2}},$$

$$(\mathbf{x}_1, \ldots, \mathbf{x}_{k_1}) \mapsto (\phi_2(\mathbf{x}_1), \ldots, \phi_2(\mathbf{x}_{n_1})) \in \mathbb{F}_{q^{m_2}}^{k_1} \mapsto \phi_1(\phi_2(\mathbf{x}_1), \ldots, \phi_2(\mathbf{x}_{k_1}))$$

and

$$\psi : \mathbb{F}_{q^{m_1 m_2}} \rightarrow \mathbb{F}_q^{k_1 k_2},$$

$$\alpha \mapsto \psi_1(\alpha) = (\mathbf{u}_1, \ldots, \mathbf{u}_{k_1}) \in \mathbb{F}_{q^{m_2}}^{k_1} \mapsto (\psi_2(\mathbf{u}_1), \ldots, \psi_2(\mathbf{u}_{k_1}))$$

give an $(k_1 k_2, m_1 m_2)_q$-RMFE.

Proof. It is clear that both ϕ and ψ are $\mathbb{F}_q$-linear. For any $\mathbf{x}, \mathbf{y} \in \mathbb{F}_q^{k_1 k_2}$, we have

$$\begin{aligned}
\psi(\phi(\mathbf{x}) \cdot \phi(\mathbf{y})) &= \psi_2 \circ \psi_1(\phi_1(\phi_2(\mathbf{x}_1), \ldots, \phi_2(\mathbf{x}_{k_1})) \cdot \phi_1(\phi_2(\mathbf{y}_1), \ldots, \phi_2(\mathbf{y}_{k_1}))) \\
&= \psi_2((\phi_2(\mathbf{x}_1), \ldots, \phi_2(\mathbf{x}_{k_1})) * (\phi_2(\mathbf{y}_1), \ldots, \phi_2(\mathbf{y}_{k_1}))) \\
&= (\psi_2(\phi_2(\mathbf{x}_1) \cdot \phi_2(\mathbf{y}_1)), \ldots, \psi_2(\phi_2(\mathbf{x}_{k_1}) \cdot \phi_2(\mathbf{y}_{k_1}))) \\
&= (\mathbf{x}_1 * \mathbf{y}_1, \ldots, \mathbf{x}_{k_1} * \mathbf{y}_{k_1}) = \mathbf{x} * \mathbf{y}
\end{aligned}$$

This completes the proof.

Remark 7 ("On practical parameters"). As a consequence of applying the above two results we have the following embeddings of $\mathbb{F}_2^k$ into extensions of degree up to 325.

1. For all $r \leq 9$, there exists a $(2r, 6r-3)_2$-RMFE (obtained by concatenation of $(2, 3)_2$ and $(r, 2r-1)_8$-RMFEs, both promised by Lemma 4).
2. For all $r \leq 33$, there exists a $(3r, 10r-5)_2$-RMFE (obtained by concatenation of $(3, 5)_2$ and $(r, 2r-1)_{32}$-RMFEs, both promised by Lemma 4).

We now move to the asymptotic results, for which we need the methods from the theory of algebraic function fields. We will not give a detailed explanation of this area here, and refer the reader to the book by Stichtenoth [Sti09]. However, we sum up the facts that we need, ignoring some technical details.

A function field $F/\mathbb{F}_q$ is an algebraic extension of the rational function field $\mathbb{F}_q(x)$, that contains all fractions of polynomials in $\mathbb{F}_q[x]$. Associated to a function field, there is a non-negative integer $\mathfrak{g}$ called the genus, and an infinite set of "places" P, each having a degree $\deg P \in \mathbb{N}$. The number of places of a given degree is finite. The places of degree 1 are called rational places. Given a function $f \in F$ and a place P, two things can happen: either f has a pole in P, or f can be evaluated in P and the evaluation $f(P)$ can be seen as an element of the field $\mathbb{F}_{q^{\deg P}}$. If f and g do not have a pole in P then the evaluations satisfy the rules $\lambda(f(P)) = (\lambda f)(P)$ (for every $\lambda \in \mathbb{F}_q$), $f(P) + g(P) = (f + g)(P)$ and $f(P) \cdot g(P) = (f \cdot g)(P)$. Note that if P is a rational place (and f does not have a pole in P) then $f(P) \in \mathbb{F}_q$. The functions in F always have the same zeros and poles up to multiplicity (called order). An important fact of the theory of algebraic function fields is as follows: call $N_1(F)$ the number of rational places of F. Then over every finite field $\mathbb{F}_q$, there exists an infinite family of function fields $\{F_n\}$ such that their genus $\mathfrak{g}_n$ grow with n and $\lim N_1(F_n)/\mathfrak{g}_n = c_q$ with $c_q \in \mathbb{R}$, $c_q > 0$. The largest constant c_q satisfying the property above is called Ihara's constant $A(q)$ of $\mathbb{F}_q$. It is known that $0 < A(q) \leq \sqrt{q} - 1$ for every finite field $\mathbb{F}_q$. Moreover, $A(q) = \sqrt{q} - 1$ for q square and that for a prime p and any integer $a \geq 1$, $A(p^{2a+1}) \geq \frac{2(p^{a+1}-1)}{p+1+\epsilon}$ where $\epsilon = \frac{p-1}{p^a-1}$. These two results are constructive, since explicit families of function fields attaining these values are known, given in the first case by [GS95, GS96] and in the second case by [BBGS15].

A divisor G is a formal sum of places, $G = \sum c_P P$, such that $c_P \in \mathbb{Z}$ and $c_P = 0$ except for a finite number of P. We call this set of places where $c_P \neq 0$ the support of G, denoted $\mathrm{supp}(G)$. The degree of G is $\deg G := \sum c_P \deg P \in \mathbb{Z}$.

The Riemann-Roch space $\mathcal{L}(G)$ is the set of all functions in F with certain prescribed poles and zeros depending on G (together with the zero function). More precisely if $G = \sum c_P P$, every function $f \in \mathcal{L}(G)$ must have a zero of order at least $|c_P|$ in the places P with $c_P < 0$, and f can have a pole of order at most c_P in the places with $c_P > 0$. The space $\mathcal{L}(G)$ is a vector space over $\mathbb{F}_q$. Its dimension is governed by certain laws (given by the so-called Riemann-Roch theorem). A weaker version of that theorem called Riemann's theorem states that if $\deg G \geq 2\mathfrak{g} - 1$ then $\dim \mathcal{L}(G) = \deg(G) - \mathfrak{g} + 1$. On the other hand, if $\deg G < 0$, then $\dim \mathcal{L}(G) = 0$.

Given $f, g \in \mathcal{L}(G)$ its product $f \cdot g$ is in the space $\mathcal{L}(2G)$.

The following is a generalization of Lemma 4.

Lemma 6. *Let $F/\mathbb{F}_q$ be a function field of genus $\mathfrak{g}$ with k distinct rational places $P_1, P_2, \ldots, P_k$. Let G be a divisor of F such that $\mathrm{supp}(G) \cap \{P_1, \ldots, P_k\} = \emptyset$ and $\dim_{\mathbb{F}_q} \mathcal{L}(G) - \dim_{\mathbb{F}_q} \mathcal{L}(G - \sum_{i=1}^{k} P_i) = k$. If there is a place R of degree m with $m > 2 \deg(G)$, then there exists an $(k, m)_q$-RMFE.*

Proof. Consider the map

$$\pi : \mathcal{L}(G) \to \mathbb{F}_q^k; \quad f \mapsto (f(P_1), \ldots, f(P_k)).$$

Then the kernel of π is $\mathcal{L}(G - \sum_{i=1}^{k} P_i)$. Since $\dim_{\mathbb{F}_q} \text{Im}(\pi) = \dim_{\mathbb{F}_q} \mathcal{L}(G) - \dim_{\mathbb{F}_q} \mathcal{L}(G - \sum_{i=1}^{k} P_i) = k$, π is surjective. Choose a subspace W of $\mathcal{L}(G)$ of dimension k such that π induces an isomorphism between W and $\mathbb{F}_q^k$.

We write by $\mathbf{c}_f$ the vector $(f(P_1), \ldots, f(P_k))$, and by $f(R)$ the evaluation of f in the higher degree place R, for a function $f \in \mathcal{L}(2G)$. We now define

$$\phi: \pi(V) = \mathbb{F}_q^k \to \mathbb{F}_{q^m}; \quad \mathbf{c}_f \mapsto f(R) \in \mathbb{F}_{q^m}.$$

Note that the above $f \in W$ is uniquely determined by $\mathbf{c}_f$. Moreover ϕ is $\mathbb{F}_q$-linear and injective since $\deg(R) > \deg(G)$.

Define

$$\tau: \mathcal{L}(2G) \to \mathbb{F}_{q^m}; \quad f \mapsto f(R) \in \mathbb{F}_{q^m}.$$

Then τ is $\mathbb{F}_q$-linear and injective since $m = \deg(R) > \deg(2G)$.

Define the map

$$\psi': \text{Im}(\tau) \subseteq \mathbb{F}_{q^m} \to \mathbb{F}_q^k; \quad f(R) \mapsto (f(P_1), \ldots, f(P_k)) \in \mathbb{F}_q^k.$$

Note that the above $f \in \mathcal{L}(2G)$ is uniquely determined by $f(R)$. ψ is $\mathbb{F}_q$-linear and surjective (but not injective). We extend ψ' from $\text{Im}(\tau)$ to all of $\mathbb{F}_{q^m}$ linearly and call the resulting map ψ. We obtain thus the pair (ϕ, ψ).

For any $\mathbf{c}_f, \mathbf{c}_g \in \mathbb{F}_q^k$ we have

$$\psi(\phi(\mathbf{c}_f) \cdot \phi(\mathbf{c}_g)) = \psi(f(R) \cdot g(R)) = \psi((f \cdot g)(R)) = \mathbf{c}_{fg} = \mathbf{c}_f * \mathbf{c}_g,$$

where $f, g \in W$ are uniquely determined from $\mathbf{c}_f, \mathbf{c}_g$ as explained above. Note that $(fg)(R)$ belongs to $\text{Im}(\tau)$ since $fg \in \mathcal{L}(2G)$. We conclude that (ϕ, ψ) defined above is an $(k, m)_q$-RMFE.

Corollary 1. *Let $F/\mathbb{F}_q$ be a function field of genus $\mathfrak{g}$ with k distinct rational places and a place of degree $m \geq 2k + 4\mathfrak{g} - 1$. Then there exists an $(k, m)_q$-RMFE.*

Proof. We take G a divisor of degree $k + 2\mathfrak{g} - 1$ whose support is disjoint with the promised set of k rational places. Then, since both $\deg G \geq 2\mathfrak{g} - 1$ and $\deg(G - \sum_{i=1}^{k} P_i) \geq 2\mathfrak{g} - 1$ we can apply the Riemann Theorem to conclude that $\dim_{\mathbb{F}_q} \mathcal{L}(G) - \dim_{\mathbb{F}_q} \mathcal{L}(G - \sum_{i=1}^{k} P_i) = \deg(G) - \mathfrak{g} + 1 - (\deg(G) - \mathfrak{g} + 1 - k) = k$. We are then in the conditions of Lemma 6. $\square$

Proposition 5 ([Sti09], **Theorem 5.2.10 (c)**). *For every function field $F/\mathbb{F}_q$, and all $m \in \mathbb{N}$ with $2\mathfrak{g} + 1 \leq q^{(m-1)/2}(\sqrt{q} - 1)$, there exists a place in F of degree m. In particular this holds for every $m \geq 4\mathfrak{g} + 3$, regardless of q.*

This implies that the condition about the existence of the high degree place in Corollary 1 is in fact always satisfied as soon as $k \geq 2$, since any $m \geq 2k + 4\mathfrak{g} - 1$ satisfies the inequality in the proposition above.

Now we can show the main theorem of this section

Theorem 5. *There exists a family of $(k, m)_q$-RMFE with $k \to \infty$ and $m = O(k)$. More concretely*

$$\frac{m}{k} \to 2 + \frac{4}{A(q)}.$$

Proof. Take a family $\{F_\ell\}$ of function fields over $\mathbb{F}_q$ of growing genus $\mathfrak{g}_\ell \to \infty$ with $N_1(F_\ell)/\mathfrak{g}_\ell \to A(q)$. Since $N_1(F_\ell)$ is the number of distinct rational places of F_ℓ, we can take $k = N_1(F_\ell)$. Moreover we take $m = 2k + 4\mathfrak{g}_\ell - 1$. These parameters satisfy all conditions in Corollary 1 and therefore the construction above yields a $(k, m)_q$-RMFE.

For $q = 2$ a direct application of this result, together with the bound $A(2) \geq 97/376$ from [XY07] yields a family of $(k, m)_2$-RMFEs with

$$\frac{m}{k} \to 2 + \frac{4}{A(2)} \leq 2 + \frac{4 \times 376}{97} \approx 15.51.$$

4.1   An Explicit Construction over $\mathbb{F}_2$

The result above for $q = 2$ is not explicit, since the bound for $A(2)$ was attained by a non-explicit of function fields. In this section we will show an explicit construction of a family of RMFEs over $\mathbb{F}_2$ with a constant asymptotic ratio. This example also shows that, fortunately, as was the case for practical values of k, the expansion expressed by the asymptotic ratio m/k can be quite small.

Proposition 6. *There exists a constructive family of $(k, m)_{32}$-RMFE with $k \to \infty$ and $\frac{m}{k} \to 62/21$.*

Proof. This comes from applying Theorem 5, that implies the existence of a family of $(k, m)_{32}$-RMFE with $k \to \infty$ and $\frac{m}{k} \to 2 + \frac{4}{A(32)}$.

Now we use that for every prime p and every $a \geq 1$, we have $A(p^{2a+1}) \geq \frac{2(p^{a+1}-1)}{p+1+\epsilon}$ (where $\epsilon = \frac{p-1}{p^a-1}$) and that this is achieved for the explicit construction in [BBGS15]. In particular $p = 2$, $a = 2$ gives $A(32) \geq 21/5$. This means $2 + \frac{4}{A(32)} \leq 62/21$ and concludes the proof.

Corollary 2. *There exists a constructive family of $(k, m)_2$-RMFE with $k \to \infty$ and*

$$\frac{m}{k} \to 4.92...$$

Proof. Applying the concatenation in Lemma 5 to the $(3, 5)_2$-RMFE (from Lemma 4) and the family of $(k_1, m_1)_{32}$-RMFE with $\frac{m_1}{k_1} \to 62/21$ provides a family of $(3k_1, 5m_1)_2$-RMFE. Note that $\frac{5m_1}{3k_1} \to 5/3 \times 62/21 = 4.92...$

5 Proof of Theorem 1

The last step towards proving Theorem 1 is how to instantiate the protocol π' that securely computes the arithmetic circuit over $\mathbb{F}_{2^m}$. We use the protocol by Beerliová-Trubíniová and Hirt [BH08].

Theorem 6 ([BH08]). *There is a protocol π' which computes an arithmetic circuit over a field $\mathbb{F}_{2^m}$, where $|\mathbb{F}_{2^m}| > 2n$, with a communication complexity of $O((c_I + c_M + c_R) \cdot n + D_M \cdot n^2 + n^3)$ field elements, where D_M is the multiplicative depth of the circuit.*

The protocol π' satisfies all conditions in Sect. 3. In particular it is a secret-sharing based protocol where the secret sharing scheme used is degree t-Shamir's secret sharing scheme over $\mathbb{F}_{2^m}$.

Proof (of Theorem 2). We use Theorem 4 with a (ϕ, ψ) from the family of (k, m)-RMFEs with $m = \Theta(k)$ constructed in Sect. 4 and the protocol π' from Theorem 6. The total communication complexity is $O((c_I + c_M + c_R) \cdot n + D_M \cdot n^2 + n^3)$ elements of $\mathbb{F}_{2^m}$, and therefore $O(nm)$ bits per gate of the circuit. Note that this allows to compute $k = \Theta(m)$ evaluations of the circuit and therefore the amortized complexity is $O(n)$ bits per gate.

We point out one optimization that is possible when we combine our compiler with [BH08]. Indeed the input phase in [BH08] consists in selecting a sharing $[r]$ of a random element in $\mathbb{F}_{2^m}$ which has been generated in their preprocessing phase and opening this privately to the party P_i holding the input $a_i \in \mathbb{F}_{2^m}$, who broadcasts the difference of the random element and a_i so that the rest of the parties update their shares. If we use our compiler as described, in the next step P_i would prove $a_i \in \text{Im } \phi$. Rather than executing these two phases, we can merge these two processes in one step: instead of using $[r]$ for a uniformly random $r \in \mathbb{F}_{2^m}$, we can have parties take $[r']$ for a uniformly random $r' \in \text{Im } \phi$, generated in our preprocessing phase by RandElSub(Im ϕ), then open this to P_i, and have P_i broadcast the difference of $r' - a_i$. The other parties can now verify that $r' - a_i \in \text{Im } \phi$ and if so, update their shares accordingly.

6 Proof of Theorem 2

We combine our amortization technique with the packed secret sharing paradigm to further decrease the communication complexity in the case where the adversary is suboptimal.

The result is based on the observation that one can replace Shamir's secret sharing scheme by packed Shamir's secret sharing in the protocol from [BH08].

Theorem 7. *There is a multiparty computation protocol for n parties that evaluates $\ell = \Theta(n)$ instances of an arithmetic circuit over $\mathbb{F}_q$ (where $q \geq 2n$) with c_I input, c_R random and c_M multiplication gates, by communicating $O(c_I n + c_R n + c_M n + D_M n^2 + n^3)$ field elements, where D_M denotes the multiplicative depth of the circuit. The protocol is secure against an active adversary corrupting $t < (n - 2\ell + 2)/3$ players.*

In order to sketch an argument for this result, we briefly describe how [BH08] works. This protocol has a preprocessing phase and a computation phase. In the computation phase, all inputs and intermediate values are shared among the network of parties using Shamir's secret sharing of degree t, denoted by $[\cdot]_t$. In order to process multiplication gates, the protocol uses the well known randomization technique due to Beaver [Bea91], which relies on auxiliary shared triplets $([a]_t, [b]_t, [c]_t)$, where a, b are random field elements and $c = ab$; these have been computed in the preprocessing phase.

The preprocessing phase uses player elimination and its goal is to generate the aforementioned triplets as well as "individual" sharings of random elements that are used in input and random gates. The crucial step in order to obtain these triplets is to be able to generate "double sharings" of random elements, more specifically one needs to generate pairs $[r]_t, [r]_{t'}$ and $[r]_t, [r]_{2t'}$ (where t' as always is the updated corruption tolerance after player elimination). This is done by means of hyper-invertible matrices in a way we have already sketched in Sect. 2. Here an important point underlying the protocol is that the product of two degree-t polynomials is a degree-$2t$ polynomial. A small detail is that at some points of the computation the parties need to generate, from a publicly known value x, the 0-degree sharing $[x]_0$. This is simply that each party defines as share the value x.

Finally, the other important point to notice regards reconstruction of secrets: throughout the protocol two reconstruction protocols are used for the secret sharing scheme: `ReconsPriv` reconstructs the secret privately towards a party, and consists on all other parties sending their shares to her. The protocol `ReconsPubl`, which we have already detailed in this paper, reconstructs a batch of secrets publicly, with amortized communication. Given a sharing $[\cdot]_d$, the secret can be reconstructed (with either protocol) t'-robustly if $d < n' - 2t'$ and t'-detectably (meaning that either the correct secret is reconstructed or the party detects the sharing is erroneous) if $d < n' - t'$. Hence t and t'-degree sharings can be robustly reconstructed and $2t'$-degree sharings can be detectably reconstructed. This is enough for the purposes of [BH08].

We describe how this would be adapted so that packed Shamir secret sharing is used instead. We recall how packed Shamir secret sharing for n parties and with secrets in $\mathbb{F}_q^\ell$ (where $\ell < n$), is defined; by assumption $\mathbb{F}_q$ has at least $n + \ell < 2n$ elements. Fix $\omega_1, \ldots, \omega_\ell, \alpha_1, \ldots, \alpha_n$ pairwise distinct points in $\mathbb{F}_q$. Then, for a degree $d \geq \ell - 1$, degree d-packed Shamir secret sharing works as follows: given $\mathbf{s} = (s_1, \ldots, s_\ell) \in \mathbb{F}_q^\ell$, a polynomial $f \in \mathbb{F}_q[X]$ is chosen uniformly at random among all polynomials of degree $\leq d$ with $f(\omega_j) = s_j$ for $j = 1, \ldots, \ell$. Then $[\mathbf{s}]_d^\ell$ is the vector $(f(\alpha_1), \ldots, f(\alpha_n))$ where $f(\alpha_i)$ is sent to the i-th player. This packed scheme has $(d-\ell+1)$-privacy: any set of $d-\ell+1$ shares gives no information about the secret. On the other hand, it has exactly the same reconstruction properties (even in the presence of errors) as degree d-standard Shamir. In particular, it has $d+1$-reconstruction ($d+1$ honest shares determine the secret), it is t-robust as long as $d < n - 2t$ and it has t-detectable reconstruction as long as $d < n - t$.

We can turn [BH08] into a protocol that computes ℓ parallel evaluations of an arithmetic circuit over $\mathbb{F}_q$ with $O(1)$ field elements communicated per gate by doing the following modifications. The standard Shamir sharings $[\cdot]_t$, $[\cdot]_{t'}$, $[\cdot]_0$ and $[\cdot]_{2t'}$ in [BH08] are substituted by packed Shamir sharings $[\cdot]_{t+\ell-1}^\ell$, $[\cdot]_{t'+\ell-1}^\ell$, $[\cdot]_{\ell-1}^\ell$ and $[\cdot]_{2t'+2\ell-2}^\ell$, respectively. Multiplication of secrets in $\mathbb{F}_q$ becomes now componentwise multiplication in $\mathbb{F}_q^\ell$.

One can then verify that all properties we need are still preserved: first, we have that $2t' + 2\ell - 2 = 2(t' + \ell - 1)$, which is needed in the shared triplets generation; moreover, the main scheme is now $[\cdot]_{t+\ell-1}^\ell$, which is still t-private; furthermore, under the assumption that $t < (n - 2\ell + 2)/3$, we have $2t' + 2\ell - 2 < n' - t'$ and $t' + \ell - 1 \leq t + \ell - 1 < n' - 2t'$, so $[\cdot]_{t+\ell-1}^\ell$, $[\cdot]_{t'+\ell-1}^\ell$ are have t'-robust reconstruction and $[\cdot]_{2t'+2\ell-2}^\ell$ has t'-detectable reconstruction; finally $[\cdot]_{\ell-1}^\ell$ is a degenerate secret sharing scheme that takes the unique polynomial of degree $\ell - 1$ that interpolates the secret and generates the corresponding shares, i.e., every party can compute her share given the secret, so it plays exactly the role which is needed from $[\cdot]_0$ in the original protocol. The double sharing generation via hyper-invertible matrices still works, because it can be still captured with our notion of GLSSS. Indeed we will have a $\mathbb{F}_q$-GLSSS where the secret is now in $\mathbb{F}_q^\ell$ but each of the shares in $\mathbb{F}_q^2$ and consists of a share with $[\cdot]_d^\ell$, and another with $[\cdot]_d'^\ell$ (the protocol will need to invoke this with $d = t + \ell - 1$, $d' = t' + \ell - 1$ and with $d = t + \ell - 1$, $d' = 2t' + 2\ell - 2$).

This establishes Theorem 7 given that the communication complexity of this modified protocol is the same as that of [BH08], but it computes ℓ evaluations of the arithmetic circuit under the weaker assumption that $t < (n - 2\ell + 2)/3$.

Now we show Theorem 2.

Proof (of Theorem 2). We describe a secure multiparty computation protocol for n parties with perfect security against an adversary corrupting $t < (n - 2\ell + 2)/3$ parties that computes simultaneously $k\ell$ evaluations of the binary circuit C with communication $O(k\ell)$ bits per gate of the circuit, and hence $O(1)$ bits per gate per instance. We recover the theorem by taking $\ell = \epsilon n/2$.

We briefly describe how to modify our compiler from Sect. 3 so that it works with packed Shamir secret sharing.

Take (ϕ, ψ) from a family of $(k, m)_2$-RMFE with $m = \Theta(k)$ and such that $2^m > 2n$. Define $\Phi : \mathbb{F}_2^{k\ell} \rightarrow (\mathbb{F}_{2^m})^\ell$ and $\Psi : (\mathbb{F}_{2^m})^\ell \rightarrow \mathbb{F}_2^{k\ell}$ that respectively consist in applying ϕ to each block of k coordinates of the input and ψ to each coordinate of the input. Parties now encode their vectors of inputs with Φ and provide these to the protocol π' (for example the vers and they need to prove that their inputs are in Φ. In order to do this the parties need to apply RandElSub to $\text{Im}\,\Phi = (\text{Im}\,\phi)^\ell$ in the preprocessing phase. At multiplication gates, the parties need to compute $[\Phi(\Psi(\mathbf{a}))]$ from $[\mathbf{a}]$ which can be done in similar fashion as in Sect. 3 but using random sharings generated by applying RandElSub to the $\mathbb{F}_2$-subspace $\mathcal{W} = \{(\mathbf{x}, \Phi(\Psi(\mathbf{x}))) : \mathbf{x} \in \mathbb{F}_{2^m}^\ell\}$ in the preprocessing phase. We also need to use that ReconsPubl is t'-robust as explained above.

Because the secret sharing scheme is now the packed version of Shamir's, we attain the same complexity as in our protocol, but now we are computing

$k\ell = \Theta(kn)$ evaluations of the circuit. The amortized complexity per gate per instance of the compiler is therefore $O(1)$ bits. Using this in combination with the packed version of [BH08] described above as protocol π' proves the theorem.

Acknowledgements. The work of Ronald Cramer and Chen Yuan was supported in part by ERC Advanced Grant No. 74079 (ALGSTRONGCRYPTO). Part of Chen Yuan's work was performed while he was employed at NTU in Singapore. The authors thank Martin Hirt, Ivan Damgård, Yuval Ishai, and Jesper Buus Nielsen for helpful discussions and the anonymous reviewers for their valuable comments.

References

[BBGS15] Bassa, A., Beelen, P., Garcia, A., Stichtenoth, H.: Towers of function fields over non-prime finite fields. Moscow Math. J. **15**(1), 1–29 (2015)

[Bea91] Beaver, D.: Efficient multiparty protocols using circuit randomization. In: Feigenbaum, J. (ed.) CRYPTO 1991. LNCS, vol. 576, pp. 420–432. Springer, Heidelberg (1992). https://doi.org/10.1007/3-540-46766-1_34

[BGW88] Ben-Or, M., Goldwasser, S., Wigderson, A.: Completeness theorems for non-cryptographic fault-tolerant distributed computation (extended abstract). In: Proceedings of the 20th Annual ACM Symposium on Theory of Computing, Chicago, Illinois, USA, 2–4 May 1988, pp. 1–10 (1988)

[BH08] Beerliová-Trubíniová, Z., Hirt, M.: Perfectly-secure MPC with linear communication complexity. In: Canetti, R. (ed.) TCC 2008. LNCS, vol. 4948, pp. 213–230. Springer, Heidelberg (2008). https://doi.org/10.1007/978-3-540-78524-8_13

[BMN17] Block, A.R., Maji, H.K., Nguyen, H.H.: Secure computation based on leaky correlations: high resilience setting. In: Katz, J., Shacham, H. (eds.) CRYPTO 2017, Part II. LNCS, vol. 10402, pp. 3–32. Springer, Cham (2017). https://doi.org/10.1007/978-3-319-63715-0_1

[Bra85] Bracha, G.: An o(log n) expected rounds randomized byzantine generals protocol. In: Proceedings of the 17th Annual ACM Symposium on Theory of Computing, Providence, Rhode Island, USA, 6–8 May 1985, pp. 316–326 (1985)

[CC88] Chudnovsky, D., Chudnovsky, G.: Algebraic complexities and algebraic curves over finite fields. J. Complex. **4**, 285–316 (1988)

[CCCX09] Cascudo, I., Chen, H., Cramer, R., Xing, C.: Asymptotically good ideal linear secret sharing with strong multiplication over *Any* fixed finite field. In: Halevi, S. (ed.) CRYPTO 2009. LNCS, vol. 5677, pp. 466–486. Springer, Heidelberg (2009). https://doi.org/10.1007/978-3-642-03356-8_28

[CCX11] Cascudo, I., Cramer, R., Xing, C.: The torsion-limit for algebraic function fields and its application to arithmetic secret sharing. In: Rogaway, P. (ed.) CRYPTO 2011. LNCS, vol. 6841, pp. 685–705. Springer, Heidelberg (2011). https://doi.org/10.1007/978-3-642-22792-9_39

[CCX12] Cascudo, I., Cramer, R., Xing, C.: The arithmetic codex. In: 2012 IEEE Information Theory Workshop, Lausanne, Switzerland, 3–7 September 2012, pp. 75–79 (2012)

[CDN15] Cramer, R., Damgård, I., Nielsen, J.B.: Secure Multiparty Computation and Secret Sharing. Cambridge University Press, Cambridge (2015)

[DI06] Damgård, I., Ishai, Y.: Scalable secure multiparty computation. In: Dwork, C. (ed.) CRYPTO 2006. LNCS, vol. 4117, pp. 501–520. Springer, Heidelberg (2006). https://doi.org/10.1007/11818175_30

[DIK10] Damgård, I., Ishai, Y., Krøigaard, M.: Perfectly secure multiparty computation and the computational overhead of cryptography. In: Gilbert, H. (ed.) EUROCRYPT 2010. LNCS, vol. 6110, pp. 445–465. Springer, Heidelberg (2010). https://doi.org/10.1007/978-3-642-13190-5_23

[DN07] Damgård, I., Nielsen, J.B.: Scalable and unconditionally secure multiparty computation. In: Menezes, A. (ed.) CRYPTO 2007. LNCS, vol. 4622, pp. 572–590. Springer, Heidelberg (2007). https://doi.org/10.1007/978-3-540-74143-5_32

[DNPR16] Damgård, I., Nielsen, J.B., Polychroniadou, A., Raskin, M.: On the communication required for unconditionally secure multiplication. In: Robshaw, M., Katz, J. (eds.) CRYPTO 2016, Part II. LNCS, vol. 9815, pp. 459–488. Springer, Heidelberg (2016). https://doi.org/10.1007/978-3-662-53008-5_16

[FY92] Franklin, M.K., Yung, M.: Communication complexity of secure computation (extended abstract). In: Proceedings of the 24th Annual ACM Symposium on Theory of Computing, Victoria, British Columbia, Canada, 4–6 May 1992, pp. 699–710 (1992)

[GS95] García, A., Stichtenoth, H.: A tower of Artin-Schreier extensions of function fields attaining the Drinfeld-Vlăduţ bound. Invent. Math. 121(1), 211–222 (1995)

[GS96] Garcia, A., Stichtenoth, H.: On the asymptotic behaviour of some towers of function fields over finite fields. J. Number Theory 61(2), 248–273 (1996)

[HMP00] Hirt, M., Maurer, U.M., Przydatek, B.: Efficient secure multi-party computation. In: Okamoto, T. (ed.) ASIACRYPT 2000. LNCS, vol. 1976, pp. 143–161. Springer, Heidelberg (2000). https://doi.org/10.1007/3-540-44448-3_12

[IKOS09] Ishai, Y., Kushilevitz, E., Ostrovsky, R., Sahai, A.: Zero-knowledge proofs from secure multiparty computation. SIAM J. Comput. 39(3), 1121–1152 (2009)

[Sha79] Shamir, A.: How to share a secret. Commun. ACM 22(11), 612–613 (1979)

[Sti09] Stichtenoth, H.: Algebraic Function Fields and Codes. Graduate Texts in Mathematics, vol. 254, 2nd edn. Springer, Berlin (2009). https://doi.org/10.1007/978-3-540-76878-4

[XY07] Xing, C., Yeo, S.L.: Algebraic curves with many points over the binary field. J. Algebra 311(2), 775–780 (2007)

Private Circuits: A Modular Approach

Prabhanjan Ananth[1(✉)], Yuval Ishai[2], and Amit Sahai[3]

[1] CSAIL, MIT, Cambridge, USA
prabhanjan@csail.mit.edu
[2] Technion, Haifa, Israel
yuvali@cs.technion.ac.il
[3] UCLA, Los Angeles, USA
sahai@cs.ucla.edu

Abstract. We consider the problem of protecting general computations against constant-rate random leakage. That is, the computation is performed by a randomized boolean circuit that maps a randomly encoded input to a randomly encoded output, such that even if the value of every wire is independently leaked with some constant probability $p > 0$, the leakage reveals essentially nothing about the input.

In this work we provide a conceptually simple, modular approach for solving the above problem, providing a simpler and self-contained alternative to previous constructions of Ajtai (STOC 2011) and Andrychowicz et al. (Eurocrypt 2016). We also obtain several extensions and generalizations of this result. In particular, we show that for every leakage probability $p < 1$, there is a finite basis $\mathbb{B}$ such that leakage-resilient computation with leakage probability p can be realized using circuits over the basis $\mathbb{B}$. We obtain similar positive results for the stronger notion of *leakage tolerance*, where the input is not encoded, but the leakage from the entire computation can be simulated given random p'-leakage of input values alone, for any $p < p' < 1$. Finally, we complement this by a negative result, showing that for every basis $\mathbb{B}$ there is some leakage probability $p < 1$ such that for any $p' < 1$, leakage tolerance as above *cannot* be achieved in general.

We show that our modular approach is also useful for protecting computations against *worst case* leakage. In this model, we require that leakage of any $\mathbf{t}$ (adversarially chosen) wires reveal nothing about the input. By combining our construction with a previous derandomization technique of Ishai et al. (ICALP 2013), we show that security in this setting can be achieved with $O(\mathbf{t}^{1+\varepsilon})$ random bits, for every constant $\varepsilon > 0$. This (near-optimal) bound significantly improves upon previous constructions that required more than $\mathbf{t}^3$ random bits.

1 Introduction

Ishai, Sahai, and Wagner [ISW03] introduced the fundamental notion of a leakage-resilient circuit compiler, which in its simplest form is defined as follows. The compiler consists of a triple of algorithms (Compile, Encode, Decode). Given any circuit C, the compiled version of the circuit $\hat{C} = \mathsf{Compile}(C)$ takes a

© International Association for Cryptologic Research 2018
H. Shacham and A. Boldyreva (Eds.): CRYPTO 2018, LNCS 10993, pp. 427–455, 2018.
https://doi.org/10.1007/978-3-319-96878-0_15

randomly encoded input $\hat{x} = \mathsf{Encode}(x)$ and (using additional fresh randomness) produces an encoded output $\hat{y}$ such that $C(x) = \mathsf{Decode}(\hat{y})$. Furthermore, suppose each wire in the compiled circuit $\hat{C}$ leaks its value[1] with some probability $p > 0$, independently for each wire. Then, informally speaking, we require that the leaked wire values reveal essentially nothing about the input x to the circuit.

The above notion of resilience to random leakage can be seen as a natural cryptographic analogue of the classical notion of fault-tolerant computation due to von Neumann [vN56] and Pippenger [Pip85], where every gate in a circuit can *fail* with some constant probability. In addition to being of theoretical interest, the random leakage model is motivated by the fact that resilience to a notion of "noisy leakage", which captures many instances of real-life side channel attacks, can be reduced to resilience to random leakage [DDF14]. The random leakage model is also motivated by its application to "oblivious zero-knowledge PCPs", where every proof symbol is queried independently with probability p, which in turn are useful for constructing zero-knowledge proofs that only involve unidirectional communication over noisy channels [GIK+15].

We turn to discuss the state of the art on constructing leakage-resilient circuit compilers with respect to leakage probability p. The original work of [ISW03] only achieved security for values of p that vanish both with the circuit size and the level of security. Ajtai [Ajt11] achieved the first leakage-resilient circuit compiler that tolerated some (unspecified) constant probability of leakage p. However, to say the least, Ajtai's result is quite intricate and poorly understood. A more recent work of Andrychowicz, Dziembowski, and Faust [ADF16] obtained a simpler derivation of Ajtai's result. However, their construction is still quite involved and relies on heavy tools such as expander graphs (also used in Ajtai's construction) and algebraic geometric codes. The present work is motivated by the following, informally stated, question:

Is there a "simple" method of building leakage-resilient circuit compilers that can tolerate some constant *probability of leakage $p > 0$?*

1.1 Our Contribution

Our main contribution is an affirmative answer to the above question. We present a conceptually simple, modular approach for solving the above problem, providing a simpler and self-contained alternative to the constructions from [Ajt11, ADF16]. In particular, our construction avoids the use of explicit constant-degree expanders or algebraic geometric codes.

Roughly speaking, our construction uses a recursive amplification technique that starts with a constant-size gadget, which only achieves a weak level of security, and amplifies security by a careful composition of the gadget with itself. The existence of the finite gadget, in turn, follows readily from results on information-theoretic secure multiparty computation (MPC), such as the initial feasibility

---

[1] The original model of [ISW03] considers the worst-case notion of **t**-private circuits, where the leakage consists of an adversarially chosen set of **t** wires. We will discuss this alternative model later.

results from [BOGW88, CCD88]. We refer the reader to Sect. 1.2 for a more detailed overview of our technique.

We then extend the above result and generalize it in several directions, and also present some negative results. Concretely, we obtain the following results regarding constant-rate random leakage:

- For every leakage probability $p < 1$, there is a finite basis $\mathbb{B}$ such that leakage-resilient computation with leakage probability p can be realized using circuits over the basis $\mathbb{B}$.
- We obtain a similar positive result for the stronger[2] notion of *leakage tolerance*, where the input is not encoded, but the leakage from the entire computation can be simulated given random p'-leakage of input values alone, for any $p < p' < 1$.
- Finally, we complement this by a negative result, showing that for every basis $\mathbb{B}$ there is some leakage probability $p = p_{\mathbb{B}} < 1$ such that for any $p' < 1$, leakage tolerance as above *cannot* be achieved in general, where $p_{\mathbb{B}}$ tends to 1 as $\mathbb{B}$ grows. The negative result is based on impossibility results for information-theoretic MPC without an honest majority [CK91].

Our work leaves open two natural open questions. First, in the case of binary circuits, there is a huge gap between the tiny leakage probability guaranteed by the analysis of our construction (roughly $p = 2^{-14}$) and the best one could hope for. This is the case even in the stronger model of leakage tolerance, where our negative result only rules out constructions that tolerate $p > 0.8$ leakage probability.

A second question is the possibility of tolerating higher leakage probability (arbitrarily close to 1) for the weaker notion of *leakage-resilient* circuits with input encoder. A partial explanation for the difficulty of this question is the possibility of using the input encoder to generate correlated randomness that enables information-theoretic MPC with no honest majority.[3]

Private Circuits with Near-Optimal Randomness. As an unexpected application of our technique, we show that the modular approach is also useful for protecting computations in the more standard model of *worst case leakage*. Indeed, we show that essentially the same construction that is secure in the random probing model is also secure in the worst case leakage model with threshold t. Using this observation and a certain "randomness locality" feature of our construction, and building on robust local pseudo-random generators [IKL+13], we

---

[2] Note that leakage-tolerance can be easily used to achieve leakage-resilience by letting the encoder apply to the input a secret sharing scheme that tolerates a p'-fraction of leakage, where the compiler is applied to an augmented circuit that starts by reconstructing the input from its shares.

[3] Indeed, the technique of Beaver [Bea91] can be used to obtain resilience to an arbitrary leakage probability $p < 1$, but at the cost of allowing the output of the input encoder to be bigger than the circuit size. In contrast, our definition of leakage-resilient circuit compiler requires the output of the input encoder to be a fixed polynomial in the input length, independently of the size of the circuit.

obtain leakage tolerant circuit compilers with leakage parameter t that use only $O(t^{1+\varepsilon})$ random bits, for any constant $\varepsilon > 0$. We show that this bound is nearly tight by observing that at least t random bits are required to protect computations against worst case leakage. Our upper bound on the randomness complexity is a major improvement over the best previous upper bound of $O(t^{3+\varepsilon})$ from [IKL+13].

We present our results formally in Sect. 3.3.

1.2 Technical Overview

In this section, we give a high level overview of the composition-based approach that we utilize to get our main result. We use the composition-based approach to achieve constructions of leakage-resilient and leakage tolerant circuit compilers in both the worst-case probing and random probing settings. For the most part of the current discussion, we focus on achieving leakage resilient circuit compilers in the random probing setting.

In the composition-based approach, we start with a leakage-resilient circuit compiler CC_0 secure against $\mathbf{p}$-random probing attacks and has constant simulation error ε. By $\mathbf{p}$-random probing attacks, we mean that every wire in the compiled circuit is leaked with probability $\mathbf{p}$. We refer to this leakage-resilient circuit compiler as a base gadget. The goal is to recursively compose this base gadget to obtain a leakage-resilient circuit compiler also secure against $\mathbf{p}$-random probing attacks but the failure probability is negligible (in the size of the circuit being compiled).

First Attempt. A naive approach to compose is as follows: to compile a circuit C, compute $CC_0.\mathsf{Compile}(\cdots CC_0.\mathsf{Compile}(C)\cdots)$. In the kth step, $CC_0.\mathsf{Compile}$ is executed for k levels of recursion. Its easy to see that leakage on the resulting compiled circuit cannot be simulated only if it holds that the simulation of $CC_0.\mathsf{Compile}$ fails for every level of recursion. That is, the failure probability of the resulting circuit compiler is ε^k for k levels of recursion. If we set k to be the size of C then we obtain negligible simulation error, as desired. However, as the simulation error reduces with every recursion step, the size of the compiled circuit increases with every recursion step. Even if the compiled circuit in the base gadget had constant overhead, the size of the compiled circuit obtained after k steps grows exponential in k. This means that we need to devise a composition mechanism where the error probability degrades much faster than the size growth of the compiled circuit.

Our Approach: In a Nutshell. Our idea is to cleverly compose n gadgets, each with simulation error ε, in such a way that the composed gadget fails only if at least t of the gadgets fail, for some parameters t, n with $t < n$. Our composition mechanism ensures that the size of the composed gadget incurs a constant blowup whereas the simulation error degrades exponentially in $\frac{1}{\varepsilon}$.

To realize such a composition mechanism, we employ techniques from Cohen et al. [CDI+13]. Cohen et al. showed how to employ player emulation strategy [HM00] to achieve a conceptually simpler construction of secure MPC in the honest majority setting. While the goal of Cohen et al. is seemingly unrelated to the problem we are trying to solve, we show that the player emulation strategy employed by their work can be adapted to our context.

We first recall their approach. They showed how to transform a threshold formula, composed solely of threshold gates, into a secure MPC protocol. In more detail, they start with a T-out-N threshold formula composed of t-out-n threshold gates. They then show how to transform a secure MPC protocol for n parties tolerating t corruptions into a MPC protocol for N parties tolerating at most T corruptions (also written as T-out-N secure MPC). At a high level, their transformation proceeds as follows: they replace the topmost t-out-n threshold gate with a T-out-N secure MPC. That is, every input wire of the topmost gate corresponds to a party in the secure MPC protocol. Every party in this MPC is emulated by a T-out-N secure MPC. In other words, for every gate input to the topmost gate, the corresponding player is replaced with a t-out-n secure MPC. For instance, if the topmost gate had exactly N gates as its children then the resulting MPC has n^2 number of parties and can tolerate at most t^2 number of corruptions. This process can be continued as long as the secure MPC protocol still satisfies polynomial efficiency.

Armed with their methodology, we show how to construct a leakage-resilient circuit compiler. We start with a t-out-n secure MPC protocol Π in the passive security model. The functionality associated with this protocol takes as input n shares of two bits (a, b) and outputs n shares of $\text{NAND}(a, b)$[4]. This secure MPC protocol will be our base gadget for NAND with respect to some constant probability of wire leakage and constant simulation error. We then compose this base gadget as follows: in the kth level of recursion, we start with Π and *emulate* the computation of every gate in Π with an inner gadget computed from $(k-1)^{th}$ level of recursion. Why is this secure? the hope is that the resulting gadget can be simulated by simulating all the inner gadgets. Unfortunately, this doesn't work since some of the inner gadgets can fail. However, we can map the inner gadgets that fail to corrupting the corresponding parties in Π. And thus, as long as at most t inner gadgets fail, we can invoke the simulator of Π to simulate the composed gadget. We can show that the probability that at most t inner gadgets fail degrades exponentially in $\frac{1}{\varepsilon_{k-1}}$, where ε_{k-1} is the simulation error of the inner gadget. On the other hand, the size of the composed gadget grows only by a constant factor. Expanding this out, we can conclude that after k steps the size grows exponential in k whereas the simulation error degrades *doubly* exponential in k. Substituting k to be logarithmic in the size of C, we attain the desired result. While the current discussion focusses on the analysis for the random probing setting, similar (and a much simpler) analysis can also be done for the worst-case probing setting. Specifically, we can show that after k levels of recursion, the circuit compiler is secure against worst case probing attacks with leakage parameter t^k.

---

[4] We consider NAND gates because they are universal gates. In fact we can substitute NAND with any other universal basis.

Security Issues. Recall that the simulation of the composed gadget requires simulating all the inner gadgets. Since the inner gadgets are connected to each other, we need to ensure that these different simulations are consistent with each other. To give an example, suppose there are two inner gadgets connected by a wire w. The simulators for these two different inner gadgets could assign conflicting values to w. At its core, we handle this problem by keeping a budget of wires "in reserve", and define a notion of composable simulation that can make use of this flexibility to resolve conflicts between simulators for components that share wires. For example, if two simulators S_1 and S_2 "want to disagree" about a wire w, we will break the tie by allowing simulator S_1 to decide the value in wire w, and asking the other simulator S_2 to use one of the reserve wires to make up for the fact that S_2 did not get its wish for the value of wire w. This is possible because of the flexibility inherent in the secret sharing schemes underlying the MPC protocols of the base gadget. Similar notions of composable leakage-resilient circuit compliers were considered in [BBD+16, BBP+16, BBP+17].

From NAND *to arbitrary circuits.* So far the above approach shows how to design a gadget for NAND tolerating constant wire leakage probability and with negligible simulation error. The fact that we design gadgets just for NAND gates is crucially used to argue that the size of the composed gadget blows up only by a constant factor in each step. We show how to use this gadget to design a gadget for any circuit over NAND basis: to compile C, we replace every gate in C with a gadget for NAND. We then show how to stitch these different gadgets together to obtain a gadget for C.

Final Template. We now lay out our template. We first define a special case of leakage-resilient circuit compilers, called *composable* circuit compilers. This notion will incorporate the composition simulation mechanism mentioned earlier.

- The first step is to design a composable circuit compiler for NAND tolerating constant wire leakage probability and has constant simulation error.
- We then apply our composition approach to obtain a composable circuit compiler for NAND tolerating constant wire leakage probability and has negligible simulation error.
- Finally, we show how to bootstrap a composable circuit compiler for NAND to obtain a composable circuit compiler for any circuit. The resulting compiler still tolerates constant wire leakage probability and has negligible simulation error.

A leakage tolerant circuit compiler can be constructed by additionally designing a leakage resilient input encoder.

Randomness Complexity. As discussed above, an unexpected feature of our construction is that it allows us to obtain leakage tolerant circuit compilers in the worst case probing setting with near-optimal randomness complexity. This application relies on the fact that after k levels of recursion, the compiled circuit has *randomness locality* of $O(k)$. (The randomness locality of a circuit compiler is said to be d if the value assigned to every wire during the evaluation of a compiled circuit depends on the inputs and at most d randomness gates.) In particular, we can construct a compiler with randomness locality $O(\log(\mathbf{t}))$ that is secure against $\mathbf{t}$-worst case probing attacks. This can be argued by observing that the initial compiled circuit has constant randomness locality and in every recursion step, the randomness locality increases by a constant. Combining this with a result from [IKL+13], we obtain a circuit compiler secure in the worst case probing model with threshold $\mathbf{t}$ and randomness complexity $\mathbf{t}^{1+\varepsilon}$. This improves upon the bound of $\mathbf{t}^{3+\varepsilon}$ in [IKL+13].

Organization. We first present the necessary preliminaries in Sect. 2. We then define the notion of circuit compilers in Sect. 3. We define leakage resilience and leakage tolerance in the same section. The notion of composable circuit compilers, that will be a building block for both leakage tolerant and leakage resilient circuit compilers, is presented in Sect. 4.1. We present the starting step (base case) in the composition step in Sect. 4.2. The composition step itself is presented in Sect. 4.3. The result of the composition step doesn't quite meet our efficiency requirements and so we present the exponential-to-polynomial transformation in Sect. 4.4. Finally, we combine all these steps to present the main construction of a composable circuit compiler in Sect. 4.5.

Armed with a construction of composable circuit compiler, we present a construction of leakage *tolerant* circuit compilers in Sect. 5. We also present negative results that upper bounds the leakage rate in the random probing model in the same section. We show that the construction of leakage tolerant circuit compiler can be transformed to have small randomness complexity. This is shown in Sect. 7. In the same section, we show a lower bound on randomness complexity of leakage tolerant circuit compilers.

We show implication of composable circuit compilers to leakage *resilient* circuit compilers in Sect. 6.

2   Preliminaries

We use the abbreviation PPT for probabilistic polynomial time. Some notational conventions are presented below.

- Suppose A is a probabilistic algorithm. We use the notation $y \leftarrow A(x)$ to denote that the output of an execution of A on input x is y.
- Suppose $\mathcal{D}$ is a probability distribution with support $\mathcal{V}$. We denote the sampling algorithm associated with $\mathcal{D}$ to be Sampler. We denote by $x \xleftarrow{\$} \mathsf{Sampler}$ if the output of an execution of Sampler is x. For every $x \in \mathcal{V}$, Sampler outputs x with probability p_x, as specified by $\mathcal{D}$. Unless specified otherwise, we only consider efficiently sampleable distributions. We also consider parameterized distributions of the form $\mathcal{D} = \{\mathcal{D}_{aux}\}$. In this case, there is a sampling algorithm Sampler defined for all these distributions. Sampler takes as input aux and outputs an element in the support of $\mathcal{D}_{aux}$.

– Consider two probability distributions $\mathcal{D}_0$ and $\mathcal{D}_1$ with discrete support $\mathcal{V}$ and let their associated sampling algorithms be $\mathsf{Sampler}_1$ and $\mathsf{Sampler}_2$. We denote $\mathcal{D}_0 \approx_{s,\varepsilon} \mathcal{D}_1$ if the distributions $\mathcal{D}_0$ and $\mathcal{D}_1$ are ε-statistically close. That is, $\sum_{v \in \mathcal{V}} |\Pr[v \leftarrow \mathsf{Sampler}_1] - \Pr[v \leftarrow \mathsf{Sampler}_2]| \leq 2\varepsilon$.

Circuits. A deterministic boolean circuit C is a directed acyclic graph whose vertices are boolean gates and whose edges are wires. The boolean gates belong to a basis $\mathbb{B}$. An example of a basis is $\mathbb{B} = \{\mathbf{AND}, \mathbf{OR}, \mathbf{NOT}\}$. We will assume without loss of generality that every gate has fan-in (the number of input wires) at most 2 and fan-out[5] (the number of output wires) at most 2. A randomized circuit is a circuit augmented with random-bit gates. A random-bit gate, denoted by $\mathbf{RAND}$, is a gate with fan-in 0 that produces a random bit and sends it along its output wire; the bit is selected uniformly and independently of everything else afresh for each invocation of the circuit. We also consider basis consisting of functions (possibly randomized) on finite domains (as opposed to just boolean gates). The size of a circuit is defined to be the number of gates in the circuit.

2.1   Information Theoretic Secure MPC

We now provide the necessary background of secure multiparty computation. In this work, we focus on information theoretic security. We first present the syntax and then the security definitions.

Syntax. We define a secure multiparty computation protocol Π for n parties $P_1, \ldots, P_n$ associated with an n-party functionality $F : \{0,1\}^{\ell_1} \times \cdots \times \{0,1\}^{\ell_n} \times \{0,1\}^{\ell_r} \rightarrow \{0,1\}^{\ell_{y_1}} \times \cdots \times \{0,1\}^{\ell_{y_n}}$. We denote ℓ_i to be the length of the ith party's input, ℓ_{y_i} to be the length of the ith party's output and ℓ_r is the length of the randomness input to F. In any given execution of the protocol, the ith party receives as input $x_i \in \{0,1\}^{\ell_i}$ and all the parties jointly compute the functionality $F(x_1, \ldots, x_n; r)$, where $r \in \{0,1\}^{\ell_r}$ is sampled uniformly at random. In the end, party P_i outputs y_i, where $(y_1, \ldots, y_n) = F(x_1, \ldots, x_n; r)$.

We defined such n-party functionalities that additionally receive the randomness as input to be *randomized functionalities*. In this work we only consider randomized n-party functionalities and henceforth, the input randomness will be implicit in the description of the functionality.

Semi-honest Adversaries. We consider the adversarial model where the adversaries follow the instructions of the protocol. That is, they receive their inputs from the environment, behave as prescribed by the protocol and finally output their view of the protocol. Such type of adversaries are referred to as semi-honest adversaries.

We define semi-honest security below. Denote $\mathsf{Real}^{\Pi}_{F,S}(x_1, \ldots, x_n)$ to be the joint distribution over the outputs of all the parties along with the views of the parties indexed by the set S.

---

[5] If a circuit has arbitrary fan-out, then this can be transformed into another circuit of fan-out 2 with a loss of logarithmic factor in the depth.

Definition 1 (Semi-Honest Security). *Consider a n-party functionality F as defined above. Fix a set of inputs $(x_1, \ldots, x_n)$, where $x_i \in \{0,1\}^{\ell_i}$ and let r_i be the randomness of the ith party. Let Π be a n-party protocol implementing F. We say that Π satisfies ε-statistical security against semi-honest adversaries if for every subset of parties S, there exists a PPT simulator Sim such that:*

$$\left\{ \left(\{y_i\}_{i \notin S}, \mathsf{Sim}\left(\{y_i\}_{i \in S}, \{x_i\}_{i \in S} \right) \right) \right\} \approx_{s,\varepsilon} \left\{ \mathsf{Real}_{F,S}^{\Pi}(x_1, \ldots, x_n) \right\},$$

*where y_i is the ith output of $F(x_1, \ldots, x_n)$. If the above two distributions are identical, then we say that Π satisfies **perfect security against semi-honest adversaries**.*

Starting with the work of [BOGW88, CCD88], several constructions construct semi-honest secure multi-party computation protocol in the information-theoretic setting assuming that a majority of the parties are honest.

We consider the notion of randomness locality of a secure MPC protocol.

Definition 2 (Randomness Locality). *A semi-honest secure multiparty computation protocol for a functionality F is said to have randomness locality d if every value computed in the protocol is determined by the inputs of all parties and at most d random bits (either as input to the functionality or to the parties).*

3 Circuit Compilers

We define the notion of circuit compilers. This notion allows for transforming an input x, a circuit C (See Sect. 2 for a definition of circuits) into an encoded input $\widehat{x}$ and a randomized circuit $\widehat{C}$ such that evaluation of $\widehat{C}$ on $\widehat{x}$ yields an encoding $\widehat{C(x)}$. The decode algorithm then decodes $\widehat{C(x)}$ to yield $C(x)$.

Definition 3 (Circuit Compilers). *A circuit compiler CC defined for a class of circuits $\mathcal{C}$ comprises of the following algorithms (Compile, Encode, Decode) defined below:*

- **Circuit Compilation,** Compile(C): *It is a deterministic algorithm that takes as input circuit C and outputs a randomized circuit $\widehat{C}$.*
- **Input Encoding,** Encode(x): *This is a probabilistic algorithm that takes as input x and outputs an encoded input $\widehat{x}$.*
- **Output Decoding,** Decode$(\widehat{y})$: *This is a deterministic algorithm that takes as input an encoding $\widehat{y}$ and outputs the plain text string y.*

The algorithms defined above satisfies the following properties:

- **Correctness of Evaluation**: *For every circuit $C \in \mathcal{C}$ of input length ℓ, every $x \in \{0,1\}^\ell$, it always holds that $y = C(x)$, where:*
  - $\widehat{C} \leftarrow$ Compile(C).
  - $\widehat{x} \leftarrow$ Encode(x).
  - $\widehat{y} \leftarrow \widehat{C}(\widehat{x})$.

- $y \leftarrow \mathsf{Decode}(\widehat{y})$.
- **Efficiency**: *Consider a parameter $k \in \mathbb{N}$. We require that the running time of* $\mathsf{Compile}(C)$ *to be* $\mathrm{poly}(k, |C|)$, *the running time of* $\mathsf{Encode}(x)$ *to be* $\mathrm{poly}(k, |x|)$ *and the running time of* $\mathsf{Decode}(\widehat{C(x)})$ *to be* $\mathrm{poly}(k, |C(x)|)$. *We emphasize that the encoding complexity only grow poly-logarithmically in terms of the size of C. Typically, k will be set to* $\mathrm{poly}(\log(|C|))$.

Few remarks are in order.

Remark 1. The standard basis we consider in this work is $\{\mathbf{AND}, \mathbf{XOR}\}$. Unless otherwise specified, all the circuits considered in this work will be defined over the standard basis. Also unless otherwise specified, the compiled circuit is over the same basis as the original circuit.

Remark 2. Later, we also consider circuit compilers with relaxed efficiency guarantees, where we allow for the running time of the algorithms to be exponential in the parameter k.

Additional Properties. We are interested in circuit compilers that have (i) low randomness locality: every value in the execution of the compiled circuit depends only on few random bits and, (ii) low randomness complexity: only a small amount of randomness should be used in the evaluation of the compiled circuit.
     We capture these two properties formally below.

Definition 4 (Randomness Locality). *Consider a circuit compiler* CC *defined for a class of circuits $\mathcal{C}$ comprising of the following algorithms* ($\mathsf{Compile}$, $\mathsf{Encode}, \mathsf{Decode}$). CC *has d-randomness locality if for every circuit $C \in \mathcal{C}$, input x, the value of every wire in the computation of $\widehat{C}$ on $\widehat{x}$ is determined by at most d random-bit gates in $\widehat{C}$ and $\widehat{x}$, where (i) $\widehat{C} \leftarrow \mathsf{Compile}(C)$ and, (ii) $\widehat{x} \leftarrow \mathsf{Encode}(x)$.*

Definition 5 (Randomness Complexity). *Consider a circuit compiler* CC *defined for a class of circuits $\mathcal{C}$ comprising of the following algorithms* ($\mathsf{Compile}$, $\mathsf{Encode}, \mathsf{Decode}$). CC *has randomness complexity r if the number of random-bit gates in the compiled circuit is at most r.*

Non-Boolean Basis. In this work, we also consider a setting where the compiled circuit is defined over a basis that is different from the basis of the original circuit (before compilation). We define this formally below.

Definition 6. *Consider two collections of finite functions $\mathbb{B}'$ and $\mathbb{B}$. A circuit compiler* $\mathsf{CC} = (\mathsf{Compile}, \mathsf{Encode}, \mathsf{Decode})$ *is defined over $\mathbb{B}'$ (written CC over $\mathbb{B}'$) for a class of circuits $\mathcal{C}$ over $\mathbb{B}$ if it holds that for every $C \in \mathcal{C}$ over basis $\mathbb{B}$, the compiled circuit $\widehat{C}$, generated as $\widehat{C} \leftarrow \mathsf{Compile}(C)$, is defined over basis $\mathbb{B}'$.*

We next define the security guarantees associated with circuit compilers.

3.1   Leakage Resilience

We adopt the definition of leakage resilient circuit compilers from [GIM+16].

Definition 7. *A circuit compiler* CC = (Compile, Encode, Decode) *for a class of circuits $\mathcal{C}$ is said to be ε-leakage resilient against a class of randomized leakage functions $\mathcal{L}$ if the following holds:*
    There exists a PPT simulator Sim *such that for every circuit $C : \{0,1\}^{\ell} \to \{0,1\}$ and $C \in \mathcal{C}$, input $x \in \{0,1\}^{\ell}$, leakage function $L_{comp} \in \mathcal{L}$, the distribution $L_{comp}(\widehat{C}, \widehat{x})$ is ε-statistically close to* Sim (C)*, where $\widehat{C} \leftarrow$ Compile(C) and $\widehat{x} \leftarrow$ Encode(x).*

Informally, the above definition states that the leakage L_{comp} on the computation of the compiled circuit $\widehat{C}$ on encoded input $\widehat{x}$ reveals no information about the input x.

Remark 3. While the above notion considers leakage only on a single computation, this notion already implies the stronger multi-leakage setting where there are multiple encoded inputs and a leakage function is computed on every computation of $\widehat{C}$. This follows from a standard hybrid argument[6].

p-*Random Probing Attacks* [ISW03, Ajt11, ADF16]. In this work, we are interested in the following probabilistic leakage function: every wire in the computation of the compiled circuit $\widehat{C}$ on the encoded input $\widehat{x}$ is leaked independently with probability **p**.
    More formally, denote the leakage function $\mathcal{L}_{\mathbf{p}} = \{L_{comp}\}$, where the probabilistic function L_{comp} is defined below.

$L_{comp}\left(\widehat{C}, \widehat{x}\right)$: construct the set of leaked values $\mathcal{S}_{\mathsf{leak}}^{C}$ as follows. For every wire w (input wires included) in $\widehat{C}$ and value v_w assigned to w during the computation of $\widehat{C}$ on $\widehat{x}$, include (w, v_w) with probability **p** in $\mathcal{S}_{\mathsf{leak}}^{C}$. Also, include (w', v_w) in $\mathcal{S}_{\mathsf{leak}}^{C}$, if w' and w are two output wires of the same gate. Output $\mathcal{S}_{\mathsf{leak}}^{C}$.
We define leakage resilient circuit compilers with respect to the leakage function defined above.

Definition 8 (Leakage Resilience Against Random Probing Attacks).
A circuit compiler CC = (Compile, Encode, Decode) *for a family of circuits $\mathcal{C}$ is said to be $(\mathbf{p}, \varepsilon)$-leakage resilient against random probing attacks if* CC *is ε-leakage resilient against $\mathcal{L}_{\mathbf{p}}$. Moreover, we define the leakage rate of* CC *to be* **p**.

t-*Probing (Worst Case Probing) Attacks.* We also consider **t**-probing attacks, where the adversary is allowed to observe any t wires in the computation of the compiled circuit. We define the class of leakage functions $\mathcal{L}_{\mathbf{t}} = \{L_{comp}^{S}\}_{|S| \leq t}$, where L_{comp}^{S} is defined below.

---

[6] Here we use the fact that the circuit compilation algorithm is deterministic.

$L_{comp}^{S}\left(\widehat{C}, \widehat{x}\right)$: construct the set of leaked values $\mathcal{S}_{\mathsf{leak}}^{C}$ as follows. For every wire $w \in S$ and v_w assigned to w during the computation of $\widehat{C}$ on $\widehat{x}$, include (w, v_w) in $\mathcal{S}_{\mathsf{leak}}^{C}$. Also, include (w', v_w) in $\mathcal{S}_{\mathsf{leak}}^{C}$, if w' and w are two output wires of the same gate. Output $\mathcal{S}_{\mathsf{leak}}^{C}$.

Definition 9 (Leakage Resilience Against Worst Case Probing Attacks). *A circuit compiler* CC = (Compile, Encode, Decode) *for a family of circuits $\mathcal{C}$ is said to be leakage resilient against* t*-probing attacks if* CC *is leakage resilient against* $\mathcal{L}_t$. *Moreover, we define the leakage parameter of* CC *to be* t.

3.2   Leakage Tolerance

Another notion we study is leakage tolerant circuit compilers. In this notion, unlike leakage resilient circuit compilers, Encode is an identity function. Consequently, we need to formalize the security definition so that the leakage on the computation of $\widehat{C}$ on x can be simulated with bounded leakage on the input x.

Definition 10. *A circuit compiler* CC = (Compile, Encode, Decode) *for a class of circuits $\mathcal{C}$ is said to be ε-leakage tolerant against a class of leakage functions $\mathcal{L}$ if the following two conditions hold:*

- Encode *is an identity function.*
- *There exists a simulator* Sim *such that for every circuit $C : \{0,1\}^\ell \to \{0,1\}$ and $C \in \mathcal{C}$, input $x \in \{0,1\}^\ell$, leakage function $L = (L_{comp}, L_{inp}) \in \mathcal{L}$, the distribution $L_{comp}(\widehat{C}, \widehat{x})$ is ε-statistically close to* Sim $(C, L_{inp}(x))$, *where $\widehat{C} \leftarrow$ Compile(C) and $\widehat{x} \leftarrow$ Encode(x).*

Henceforth, we omit Encode *algorithm and denote a leakage tolerant circuit compiler to consist of* (Compile, Decode).

(p, p′)-*Random Probing Attacks.* As before, we are interested in the following probabilistic leakage function: every wire in the computation of the compiled circuit $\widehat{C}$ on the encoded input $\widehat{x}$ is leaked independently with probability **p**.

More formally, denote the leakage function $\mathcal{L}_{\mathbf{p},\mathbf{p}'} = \{(L_{comp}, L_{inp})\}$, where the probabilistic functions L_{comp} is as defined in Sect. 3.1 and L_{inp} is defined below.

$L_{inp}(x)$: construct the set of leaked values $\mathcal{S}_{\mathsf{leak}}^{I}$ as follows. For every input wire w carrying the ith bit of x, include (w, x_i) in $\mathcal{S}_{\mathsf{leak}}^{I}$ with probability **p′**. If (w, x_i) is included, also include (w', x_i) in $\mathcal{S}_{\mathsf{leak}}^{I}$, where w' is the other input wire carrying x_i. Output $\mathcal{S}_{\mathsf{leak}}^{I}$.

We define leakage tolerance against random probing attacks below.

Definition 11 (Leakage Tolerance Against Random Probing Attacks). *A circuit compiler* CC = (Compile, Decode) *for a family of circuits $\mathcal{C}$ is said to be* (**p, p′**, ε)*-leakage tolerant against random probing attacks if* CC *is ε-leakage tolerant against $\mathcal{L}_{\mathbf{p},\mathbf{p}'}$. Moreover, we define the leakage rate of* CC *to be* **p**.

t-*Probing (Worst Case Probing) Attacks.* As before, we are interested in the class of leakage functions where the adversary is allowed to query a **t**-sized subset of wire values in the circuit. We consider the class of leakage functions $\mathcal{L}_{\mathbf{t}} = \{(L_{comp}^{S}, L_{inp}^{S'})\}_{|S'| \leq \mathbf{t}}$, where L_{comp}^{S} is as defined in Sect. 3.1 and $L_{inp}^{S'}$ is defined below.

$L_{inp}^{S'}\left(\widehat{C}, \widehat{x}\right)$: construct the set of leaked values $\mathcal{S}_{\mathsf{leak}}^{I}$ as follows. include (w, x_i) in $\mathcal{S}_{\mathsf{leak}}^{I}$ if and only if $w \in S'$ and wire w carries the ith bit of x. If w' also carries the ith bit of x, include (w', x_i) in $\mathcal{S}_{\mathsf{leak}}^{I}$. Output the set $\mathcal{S}_{\mathsf{leak}}^{I}$.

Definition 12 (Leakage Tolerance Against Worst Case Probing Attacks). *A circuit compiler* CC = (Compile, Encode, Decode) *for a family of circuits* $\mathcal{C}$ *is said to be leakage tolerant against* **t**-*probing attacks if* CC *is leakage tolerant against* $\mathcal{L}_{\mathbf{t}}$. *Moreover, we define the leakage parameter of* CC *to be* **t**.

3.3 Our Results

We state our results below.

WORST CASE PROBING:

Randomness Complexity. We prove positive and negative results on the randomness complexity of leakage tolerant circuit compilers. We prove this is in the worst case probing regime. The proofs for both the theorems can be found in Sect. 7.

Theorem 1 (Randomness Complexity: Positive Result). *There is a leakage tolerant circuit compiler such that given a circuit of size s and worst-case leakage bound* **t**, *the compiler outputs a circuit of size $s \cdot \mathrm{poly}(\mathbf{t})$ which is perfectly secure against* **t** *(worst-case) probing attacks and uses only* $\mathbf{t}^{1+\varepsilon}$ *random bits.*

Theorem 2 (Randomness Complexity: Negative Result). *The number of random bits used in any leakage tolerant circuit compiler secure against* **t**-*probing attacks is at least* **t**.

En route to proving the above positive result, we prove that there is a construction of leakage tolerant circuit compiler that has randomness locality $\log(\mathbf{t})$. This is shown in Sect. 5.2.

Lemma 1 (Randomness Locality). *There is a leakage tolerant circuit compiler secure against* **t**-*probing attacks satisfying* $O(\log(\mathbf{t}))$-*randomness locality.*

RANDOM PROBING:

Leakage Tolerance: Positive Results. We show the following results in Sect. 3.2.

Theorem 3 (Boolean Basis). *There exist constants* $0 < \mathbf{p} < \mathbf{p}' < 1$ *such that there is a* $(\mathbf{p}, \mathbf{p}', \epsilon)$-*leakage tolerant circuit compiler, where* ϵ *is negligible in the circuit size.*

Theorem 4 (Finite Basis). *For any* $0 < \mathbf{p}' < \mathbf{p} < 1$ *there is a basis* $\mathbb{B}$ *over which there is a* $(\mathbf{p}, \mathbf{p}', \epsilon)$-*leakage tolerant circuit compiler, where* ϵ *is negligible in the circuit size.*

Leakage Tolerance: Negative Result. The following theorem upper bounds the rate of a leakage tolerant circuit compiler in the random probing model. We prove this theorem in the full version.

Theorem 5. *For any basis $\mathbb{B}$ there is $0 < \mathbf{p} < 1$, such that for any $0 < \mathbf{p'} < 1$, there is no $(\mathbf{p}, \mathbf{p'}, 0.1)$-leakage tolerant circuit compiler over $\mathbb{B}$.*

Leakage Resilience: Positive Results. We demonstrate a construction of leakage resilient circuit compiler over boolean basis. Both the theorems below are shown in Sect. 6.

Theorem 6 (Boolean Basis). *There is a constant $0 < \mathbf{p} < 1$ such that there is a $(\mathbf{p}, \epsilon)$-leakage resilient circuit compiler and ϵ is negligible in the circuit size.*

We prove a result about finite basis in the full version.

Theorem 7 (Finite Basis). *For any $0 < \mathbf{p} < 1$ there is a basis $\mathbb{B}$ over which there is a $(\mathbf{p}, \epsilon)$-leakage resilient circuit compiler, where ϵ is negligible in the circuit size.*

4   Composition Theorem: Intermediate Step

We present a composition theorem, a key step in our constructions of leakage tolerant and leakage resilient circuit compilers. We identify a type of circuit compilers satisfying some properties, that we call *composable circuit compilers*. This notion will be associated with 'composition-friendly' properties.

Before we formally define the properties, we motivate the use of composable circuit compilers.

- In our composition theorem, we need to 'attach' different composable circuit compiler gadgets. For instance, the output wires of composable compiler CC_1 will be the input wires of another compiler CC_2. In order to ensure correctness, we need to make sure that the output encoding of CC_1 is the same as the input encoding of CC_2. We guarantee this by introducing XOR encoding property that states that the input encoding and output encoding are additive secret shares.
- While the above bullet resolves the issue of correctness, this raises some security concerns. In particular, when we simulate CC_1 and CC_2 separately, conflicting values could be assigned to the wires that join CC_1 and CC_2. These issues have been studied in the prior works, mainly in the context of worst case leakage [BBD+16, BBP+16, BBP+17]. And largely, this was not formally studied for the random probing setting. We formulate the following simulation definition to handle this issue in the probabilistic setting: the simulator $Sim = (Sim_1, Sim_2)$ (termed as partial simulator) will work in two main steps:
  - In the first step, the simulator first determines the wires to be leaked. Then, Sim_1 determines a 'shadow' of input and output wires that additionally need to be simulated.

- In the second step, the values for the input and output wires selected in the above step is assigned values. Then Sim_2 is executed to assign the internal wire values.

At a high level Sim works as follows: first $\mathsf{CC}_1.\mathsf{Sim}_1$ and $\mathsf{CC}_2.\mathsf{Sim}_1$ is executed to obtain the shadow of input and output wires that need to be simulated. At this point, we take the union of the output wires of CC_1 and input wires of CC_1 that need to be simulated. Then, we assign the values to all the wires. Once this is done, we independently execute $\mathsf{CC}_1.\mathsf{Sim}_2$ and $\mathsf{CC}_2.\mathsf{Sim}_2$ to obtain the simulated wire values in both CC_1 and CC_2, as desired.

4.1   Composable Circuit Compilers

The syntax of composable circuit compilers is the same as that of circuit compilers (Definition 3). In addition, it is required to satisfy the properties stated next.

XOR Encoding Property. We start with XOR encoding property. This property states that the input encoding (resp., output encoding) is an additive secret sharing of the inputs (resp., outputs).

Definition 13 (N-XOR Encoding). *A circuit compiler* (Compile, Encode, Decode) *for a family of circuits $\mathcal{C}$ is said to have N-**XOR encoding property** if the following always holds: for every circuit $C \in \mathcal{C}, x \in \{0,1\}^\ell$,*

- $\mathsf{Encode}(x)$ *computes XOR secret sharing of x_i for every $i \in [\ell]$, where x_i is the ith input bit of x. It then outputs the concatenation of the XOR secret shares of all the bits of x.*
  It outputs $\widehat{x} = (\widehat{x}^1, \ldots, \widehat{x}^\ell) \in \{0,1\}^{\ell N}$, where $x_i = \oplus_{j=1}^N \widehat{x}_j^i$. That is, x_i is a XOR secret sharing of $\{\widehat{x}_j^i\}_{j\in[N]}$.
- *Let $\widehat{x} \leftarrow \mathsf{Encode}(x)$ and $\widehat{C} \leftarrow \mathsf{Compile}(C)$. Upon evaluation, denote the output encoding to be $\widehat{y} \leftarrow \widehat{C}(\widehat{x})$. Suppose $C(x) = y \in \{0,1\}^{\ell'}$ and $\widehat{y} = (\widehat{y}^1, \ldots, \widehat{y}^{\ell'}) \in \{0,1\}^{\ell' N}$. We require that $\{\widehat{y}_j^i\}$ is a XOR secret sharing of y_i, i.e., $y_i = \oplus_{j=1}^N \widehat{y}_i^j$.*

When N is clear from the context, we drop it from the notation.

Composable Security (Random Probing Setting). Next, we define the composable security property. We first deal with the random probing setting. There are two parts associated with this security property.

- **Partial simulation:** This states that, conditioned on the simulator not aborting, the leakage of all the wires in the compiled circuit can be perfectly simulated by the leakage of a fraction of values assigned to the input and output wires alone.
- **Simulation with Abort:** We require that the simulator aborts with small probability.

Before stating the formal definition of composable security, we first set up some notation. We formalize the leakage function L_{comp} defined in the previous section in terms of the following sampler algorithm, $\mathsf{RPDistr}_{\mathbf{p}}^{w}(\cdot, \cdot)$[7].

SAMPLER $\mathsf{RPDistr}_{\mathbf{p}}^{w}(\widehat{C}, \widehat{x})$: Denote the set of wires in $\widehat{C}$ as $\mathcal{W}$. Consider the computation of $\widehat{C}$ on input encoding $\widehat{x}$. For every wire $w \in \mathcal{W}$, denote $\mathbf{val}(w)$ to be the value assigned to w during the evaluation of $\widehat{C}$ on $\widehat{x}$.

We construct the set $\mathcal{S}_{\mathsf{leak}}$ as follows: initially $\mathcal{S}_{\mathsf{leak}}$ is assigned to be $\{\}$. For every $w \in \mathcal{W}$, with probability $\mathbf{p}$, include $(w, \mathbf{val}(w))$ in $\mathcal{S}_{\mathsf{leak}}$ (i.e., with probability $(1 - \mathbf{p})$, the pair $(w, \mathbf{val}(w))$ is not included). Output $\mathcal{S}_{\mathsf{leak}}$.

We define the notion of partial simulator below.

Definition 14 (Partial Simulator: Random Probing). *A partial simulator* Sim *defined by a deterministic polynomial time algorithm* Sim_1 *and probabilistic polynomial time algorithm* Sim_2 *executes as follows: On input a circuit* $\widehat{C}$,

- *Denote* $\mathcal{W}$ *to be the set of wires in* $\widehat{C}$. *Construct a set* $\mathcal{W}_{lk}$ *as follows: include every wire* $w \in \mathcal{W}$ *in the set* $\mathcal{W}_{lk}$ *with probability* $\mathbf{p}$.
- $\mathsf{Sim}_1(\widehat{C}, \mathcal{W}_{lk})$ *outputs* $(\mathcal{W}^{inp}, \mathcal{W}^{out}, I)$. $\mathcal{W}^{inp}$ *is a subset of input wires,* $\mathcal{W}^{out}$ *is a subset of output wires and* I *denotes a set of indices.*
- *For every wire* $w \in \mathcal{W}^{inp}$, *include* $(w, v_w) \in S^{inp}$ *such that* v_w *is a bit sampled uniformly at random. Similarly, construct the set* S^{out}.
- $\mathsf{Sim}_2 \left(\widehat{C}, \mathcal{W}_{lk}, \mathcal{W}^{inp}, S^{inp}, \mathcal{W}^{out}, S^{out}, I \right)$ *outputs* $\mathcal{S}_{lk}$.

Finally, Sim *outputs* $\mathcal{S}_{lk}$.

We now define the notion of composable security in the random probing model.

Definition 15 (Composable Security: Random Probing). *A circuit compiler* CC = (Compile, Encode, Decode) *for* $\mathcal{C}$, *consisting of circuits of input length* ℓ, *is said to be* $(\mathbf{p}, \varepsilon)$-**composable secure** *against random probing attacks if there exists a probabilistic polynomial time partial simulator* Sim = $(\mathsf{Sim}_1, \mathsf{Sim}_2)$ *such that the following holds:*

- $\mathbf{p}$-**Partial Simulation:** *for every circuit* $C \in \mathcal{C}$, *input* $x \in \{0, 1\}^{\ell}$,

$$\left\{ \mathsf{RPDistr}_{\mathbf{p}}^{w} \left(\widehat{C}, \widehat{x} \right) \right\} \equiv \left\{ \mathsf{Sim}(\widehat{C}) \big|_{L \leftarrow \mathsf{Sim}(\widehat{C}) \wedge L \neq \perp} \right\},$$

where, $\widehat{C} \leftarrow$ Compile(C) *and* $\widehat{x} \leftarrow$ Encode(x). *That is, conditioned on the simulator not aborting, its output distribution is identical to* $\mathsf{RPDistr}_{\mathbf{p}}^{w}(\widehat{C}, \widehat{x})$.
- ε-**Simulation with Abort:** *For every* $C \in \mathcal{C}$, $\mathsf{Sim}(\widehat{C})$ *aborts with probability* ε.

---

[7] The superscript w is used to signify leakage of wire values.

Composable Security (Worst Case Probing). We define the composable security in the worst case probing setting. This will be defined along the same lines as in the random probing setting.

Intuitively, we want to capture the following guarantee: simulation of a subset of wires in the circuit can be carried out given a subset of input wire values and a subset of output wire values. We formalize this in terms of partial simulator below.

Definition 16 (Partial Simulator: Worst Case Probing). *A partial simulator* Sim, *associated with a parameter* t, *defined by a deterministic polynomial time algorithm* Sim_1 *and probabilistic polynomial time algorithm* Sim_2 *executes as follows: On input a circuit* $\widehat{C}$ *and a set of wires* $\mathcal{W}_{lk}$ *of size at most* t,

- $Sim_1(\widehat{C}, \mathcal{W}_{lk})$ *outputs* $(\mathcal{W}^{inp}, \mathcal{W}^{out})$. *The sets* $\mathcal{W}^{inp}$ *and* $\mathcal{W}^{out}$ *(of size at most* t) *respectively denote the subset of input and output wires whose values are necessary to simulate the values of the wires in* $\mathcal{W}_{lk}$.
- *For every wire* $w \in \mathcal{W}^{inp}$, *include* $(w, v_w) \in S^{inp}$ *such that* v_w *is a bit sampled uniformly at random. Similarly, construct the set* S^{out}.
- $Sim_2\left(\widehat{C}, \mathcal{W}_{lk}, \mathcal{W}^{inp}, S^{inp}, \mathcal{W}^{out}, S^{out}\right)$ *outputs* S_{lk}.

Finally, Sim *outputs* S_{lk}.

We now define the notion of composable security in the context of worst case probing. Before that, we formalize the leakage function L_{comp} defined in the previous section in terms of the following algorithm $WCDistr_S^w$, parameterized by a t-sized set S.

SAMPLER $WCDistr_S^w(\widehat{C}, \widehat{x})$: On input circuit $\widehat{C}$, input encoding $\widehat{x}$, construct the set S_{leak} as follows: For every wire $w \in \widehat{C}$, let v_w be the value assigned to the wire w during the execution of $\widehat{C}$ on $\widehat{x}$. Include (w, v_w) in S_{leak} for every $w \in S$. Output S_{leak}.

Definition 17 (Composable Security: Worst Case Probing). *A circuit compiler* CC = (Compile, Encode, Decode) *for a class of circuits* $\mathcal{C}$ *is said to be* **t-composable secure** *against* t-*probing attacks if there exists a probabilistic polynomial time partial simulator* Sim = (Sim_1, Sim_2), *associated with a parameter* t, *such that the following holds:*

- **t-Partial Simulation:** *for every circuit* $C \in \mathcal{C}$, *input* $x \in \{0,1\}^\ell$,

$$\left\{ WCDistr_{\mathcal{W}_{lk}}^w \left(\widehat{C}, \widehat{x}\right) \right\} \equiv \left\{ Sim(\widehat{C}, \mathcal{W}_{lk}) \right\},$$

where $\widehat{C} \leftarrow$ Compile(C), $\widehat{x} \leftarrow$ Encode(x) *and* $\mathcal{W}_{lk}$ *is any subset of wires in* $\widehat{C}$ *of size at most* t.

Main Definition. We now give definitions of composable circuit compilers for the random probing and the worst case probing models.

Definition 18 (Composable Circuit Compilers: Random Probing). *A circuit compiler* CC = (Compile, Encode, Decode) *is said to be a* $(\mathbf{p}, \varepsilon)$-*secure composable circuit compiler in the random probing model if* CC *satisfies:*

- *XOR encoding property.*
- $(\mathbf{p}, \varepsilon)$-*composable security.*

We refer to CC *as a secure composable circuit compiler and in particular, omit* $(\mathbf{p}, \varepsilon)$ *if this is clear from the context.*

Definition 19 (Composable Circuit Compilers: Worst Case Probing). *A circuit compiler* CC = (Compile, Encode, Decode) *is said to be a* t-*secure composable circuit compiler in the worst case probing model if* CC *satisfies:*

- *XOR encoding property.*
- t-*composable security.*

We refer to CC *as a secure composable circuit compiler and in particular, omit* t *if this is clear from the context.*

L-efficient Composable CC. En route to constructing composable circuit compiler, we construct an intermediate composable circuit compiler that produces exponentially sized compiled circuits. We define the following notion to capture this step.

Definition 20 (L-efficient Composable CC). *A circuit compiler* CC = (Compile, Encode, Decode) *is an L-efficient composable circuit compiler for a class of circuits* $\mathcal{C}$ *if for every* $C \in \mathcal{C}$, *we have* $|\widehat{C}| \leq L(|C|)$, *where* $\widehat{C} \leftarrow$ Compile(C).

In particular, CC *is a composable circuit compiler if* L *is a polynomial.*

4.2 Base Case: Constant Simulation Error

We construct a composable circuit compiler CC = (Compile, Encode, Decode) for a class of circuits $\mathcal{C}$. Let Π be a perfectly semi-honest secure n-party computation protocol for an n-party randomized[8] functionality $F = F[C]$ (defined in Fig. 1) tolerating t number of corruptions.

---

[8] Recall that a randomized n-party functionality is one that in addition to taking n inputs, also takes as input randomness.

---

n-party functionality, $F[C]$

Input: $(\widehat{x}_1^1 || \cdots || \widehat{x}_1^\ell \; ; \; \cdots \; ; \; \widehat{x}_n^1 || \cdots || \widehat{x}_n^\ell)$, where ℓ is the input length of C.

- It then computes $x_i = \oplus_{j=1}^n \widehat{x}_j^i$ for every $i \in [\ell]$. Denote x to be a bit string, where the ith bit of x is x_i.
- It then computes $C(x)$ to obtain y. Let y_i be the ith output bit of y. Let the length of y be ℓ_y.
- Sample bits $\widehat{y}_j^i$ uniformly at random such that $y_i = \oplus_{j=1}^n \widehat{y}_j^i$ for every $i \in [\ell_y]$. Set $\widehat{\mathbf{y}}^i = (\widehat{y}_1^i, \ldots, \widehat{y}_n^i)$, for every $i \in [n]$. Output $(\widehat{\mathbf{y}}^1, \ldots, \widehat{\mathbf{y}}^{\ell_y})$.

Fig. 1. Functionality $F[C]$, parameterized by a circuit C.

We describe the scheme below.

Circuit Compilation, Compile(C): This algorithm takes as input circuit $C \in \mathcal{C}$. We associate a boolean circuit Ckt_Π with Π such that the following holds:

- Protocol Π on input $(\widehat{\mathbf{x}}^1; \ldots; \widehat{\mathbf{x}}^n)$, where $\widehat{\mathbf{x}}^i$ is i^{th} party's input, outputs $(\widehat{\mathbf{y}}^1; \ldots; \widehat{\mathbf{y}}^n)$ if and only if Ckt_Π on input $\widehat{\mathbf{x}}^1 || \cdots || \widehat{\mathbf{x}}^n$ outputs $(\widehat{\mathbf{y}}^1; \ldots; \widehat{\mathbf{y}}^n)$.
- Furthermore, the gates of Ckt_Π can be partitioned into n sub-circuits such that the ith sub-circuit implements the ith party in Π. Denote the ith sub-circuit to be Ckt_i. Also, denote the number of gates in Ckt_Π to be $\mathsf{N_g}$.
- The wires between the sub-circuits are analogous to the communication channels between the corresponding parties.

Output $\widehat{C} = \mathsf{Ckt}_\Pi$.

Input encoding, Encode(x): On input $x \in \{0,1\}^\ell$, it outputs the encoding $\widehat{x} = ((x_1^1, \ldots, x_\ell^1), \ldots, (x_1^n, \ldots, x_\ell^n))$, where $x_i = \oplus_{j=1}^n x_i^j$.

Output decoding, Decode($\widehat{y}$): It takes as input encoding $\widehat{y} = ((y_1^1, \ldots, y_{\ell'}^1), \ldots, (y_1^n, \ldots, y_{\ell'}^n))$. It then outputs y, where the ith bit of y is $y_i = \oplus_{j=1}^n y_i^j$.

We prove the following two propositions in the full version.

Proposition 1 (Worst Case Probing). *Let Π be a perfectly semi-honest secure n-party computation protocol for n-party functionality F (defined in Fig. 1) tolerating t corruptions and having randomness locality d. Then, CC is a t-secure composable circuit compiler secure against t-probing attacks. Moreover, the randomness locality of CC is d.*

Proposition 2 (Random Probing). *Let Π be a perfectly semi-honest secure n-party computation protocol for n-party functionality F (defined in Fig. 1)* *tolerating t corruptions and having randomness locality d. Then there is a constant $\mathbf{p} > 0$ such that CC is a $(\mathbf{p}, \varepsilon_0)$-secure composable circuit compiler, where $\varepsilon_0 = e^{-\frac{(1+t)^2}{12N_g} \cdot \frac{1}{\mathbf{p}}}$. Moreover, the randomness locality of CC is d.*

4.3 Composition Step

We present the main composition step in this section. It allows for transforming a composable circuit compiler CC_K satisfying $(\mathbf{p}, \varepsilon_K)$-composable security in the random probing setting (resp., $\mathbf{t}_K$-composable security in the worst case) into CC_{K+1} satisfying $(\mathbf{p}, \varepsilon_{K+1})$-composable security (resp., $t \cdot \mathbf{t}_K$-composable security in the worst case), where ε_{K+1} is (exponentially) smaller than ε_K. In terms of efficiency, the efficiency of CC_{K+1} degrades by a constant factor. The main tool we use to prove the composition theorem is a perfectly secure MPC protocol that tolerates at most t corruptions.

We first present the transformation of CC_K into CC_{K+1}. Let $\mathsf{CC}_K = (\mathsf{Compile}_K, \mathsf{Encode}_K, \mathsf{Decode}_K)$ be a composable circuit compiler. We now build CC_{K+1} as follows:

Circuit Compilation, $\mathsf{CC}_{K+1}.\mathsf{Compile}(C)$: It takes as input a circuit C and outputs a compiled circuit $\widehat{C}$. There are two steps involved in the construction of $\widehat{C}$. In Step I, we first consider a MPC protocol Π^9 for a randomized functionality F and using this we construct a circuit Ckt_Π. In Step II, we convert Ckt_Π into another circuit Ckt_Π^*. In this step, we make use of the compiler CC_K. The output of this algorithm is $\widehat{C} = \mathsf{Ckt}_\Pi^*$.

STEP I: CONSTRUCTING Ckt_Π. Consider a n-party functionality $F = F[C]$; see Fig. 1.

Let Π denote a n-party information theoretically secure computation protocol for F. Construct Ckt_Π as done in Sect. 4.2.

STEP II: TRANSFORMING Ckt_Π INTO Ckt_Π^*. Replace every gate in Ckt_Π with the CC_K gadgets and then show how to "stitch" all these gadgets together.

- Replacing Gate by CC_K gadget: For every gate G in the circuit Ckt_Π, we execute the compiler $\mathsf{CC}_K.\mathsf{Compile}(G)$ to obtain $\widehat{G}$.
- "Stitching" Gadgets: We created CC_K gadgets for every gate in the circuit. Now we show how to connect these gadgets with each other.

Let G_k be a gate in Ckt_Π. Let G_k' and G_k'' be two gates such that the output wires from these two gates are inputs to G_k. Let $\widehat{G_k} \leftarrow \mathsf{CC}_K.\mathsf{Compile}(G_k)$, $\widehat{G_k'} \leftarrow \mathsf{CC}_K.\mathsf{Compile}(G_k')$ and $\widehat{G_k''} \leftarrow \mathsf{CC}_K.\mathsf{Compile}(G_k'')$. We connect the output of $\widehat{G_k'}$ and $\widehat{G_k''}$ with the input of $\widehat{G_k}$. That is, the output encodings of $\widehat{G_k'}$ and $\widehat{G_k''}$ form

---

[9] The parties in this protocol are equipped with randomness gates.

the input encoding to $\widehat{G_k}$. Here, we use the fact that the output encoding and the input encoding are computed using the same secret sharing scheme, and in particular we use the XOR secret sharing scheme.

We perform the above operation for every gate in Ckt_Π.

We denote the result of applying Step I and II to Ckt_Π to be the circuit Ckt_Π^*. Furthermore, we denote Ckt_i^* to be the circuit obtained by applying Steps I and II to sub-circuits Ckt_i. Note that Ckt_i^* is a sub-circuit of Ckt_Π. Moreover, Ckt_i^* takes as input XOR secret sharing of the ith party's input and outputs XOR secret sharing of the ith party's output.

Output $\widehat{C} = \mathsf{Ckt}_\Pi^*$.

Input Encoding, $\mathsf{CC}_{K+1}.\mathsf{Encode}(x)$: On input x, compute $(x_{1,1}, \ldots, x_{\ell,1})$, $\ldots, (x_{1,n}, \ldots, x_{\ell,n}))$, where $x_i = \oplus_{j=1}^{n} x_{i,j}$. Compute $\widehat{x_{i,j}} \leftarrow \mathsf{CC}_K.\mathsf{Encode}(x_{i,j})$, for every $i \in [\ell]$ and $j \in [n]$. Output $\left(\{\widehat{x_{i,j}}\}_{i \in [\ell], j \in [n]} \right)$.

Output Encoding, $\mathsf{CC}_{K+1}.\mathsf{Decode}(\widehat{y})$: On input $\left(\{\widehat{y_{i,j}}\}_{i \in [\ell'], j \in [n]} \right)$, first compute $\mathsf{CC}_K.\mathsf{Decode}(\widehat{y_{i,j}})$ to obtain $y_{i,j}$, for every $i \in [\ell'], j \in [n]$. It computes y, where the the ith bit of the output is computed as $y_i = \oplus_{j=1}^{n} \widehat{y_j^i}$. Output $y = y_1 || \cdots || y_n$.

We prove the following two propositions in the full version.

Proposition 3 (Worst Case Probing). *Suppose CC_K is $\mathbf{t}_K$-composable secure against $\mathbf{t}_K$-probing attacks and Π is perfectly secure tolerating t number of corruptions. Then, CC_{K+1} is $t \cdot \mathbf{t}_K$-composable secure against $\mathbf{t}$-probing attacks. If CC_K has randomness locality d_K and Π has randomness locality d then CC_{K+1} has randomness locality $2d + d_K$.*

Proposition 4 (Random Probing). *Let CC_K satisfy $(\mathbf{p}, \varepsilon_K)$-composable security property. Then, CC_{K+1} satisfies $(\mathbf{p}, \varepsilon_{K+1})$-composable security property, where $\varepsilon_{K+1} = e^{-\frac{(1+t)^2}{12\mathsf{N_g}} \cdot \frac{1}{\varepsilon_K}}$. If CC_K has randomness locality d_K and Π has randomness locality d then CC_{K+1} has randomness locality $2d + d_K$.*

4.4 Stitching Transformation: Exp to Poly Efficiency

Consider a L_{exp}-efficient composable circuit compiler $\mathsf{CC}_{\mathsf{exp}}$ for a basis of gates $\mathbb{B}$, where L_{exp} is a exponential function. We construct a L_{poly}-efficient composable circuit compiler $\mathsf{CC}_{\mathsf{poly}}$ for a class of all circuits $\mathcal{C}$ over the basis $\mathbb{B}$, where L_{poly} is a polynomial.

We describe the construction below.

Circuit compilation, $\mathsf{CC}_{\mathsf{poly}}.\mathsf{Compile}(C)$: It takes as input circuit $C \in \mathcal{C}$. For every gate G in C, it computes $\widehat{G} \leftarrow \mathsf{CC}_{\mathsf{exp}}.\mathsf{Compile}(G)$ to obtain the gadget $\widehat{G}$.

Once it computes all the gadgets, it then 'stitches' all the gadgets together. The stitching operation is performed as follows: let G_k be a gate in C. Let G'_k and G''_k be two gates such that the output wires from these two gates are inputs to G_k. We connect the output of $\widehat{G'_k}$ and $\widehat{G''_k}$ with the input of $\widehat{G_k}$. That is, the output encodings of $\widehat{G'_k}$ and $\widehat{G''_k}$ form the input encoding to $\widehat{G_k}$. Here, we use the fact that the output encoding and the input encoding are computed using the same secret sharing scheme, i.e., the XOR secret sharing scheme. Denote the resulting circuit obtained after stitching all the gadgets together to be $\widehat{C}$. Output $\widehat{C}$.

Input Encoding, $\mathsf{CC}_{\mathsf{poly}}.\mathsf{Encode}(x)$: It takes as input x and then computes the XOR secret sharing of every bit of x. Output the concatenation of the XOR secret shares of all the bits of x, denoted by $\widehat{x}$.

Output Decoding, $\mathsf{CC}_{\mathsf{poly}}.\mathsf{Decode}(\widehat{y})$: On input $\widehat{y}$, parse it as $((\widehat{y}_1^1,\ldots,\widehat{y}_n^1),\ldots,(\widehat{y}_1^{\ell'},\ldots,\widehat{y}_n^{\ell'}))$. Reconstruct the ith bit of the output as $y_i = \oplus_{j=1}^n \widehat{y}_j^i$. Output $y = y_1||\cdots||y_n$.

We prove the following two propositions in the full version.

Proposition 5 (Worst Case Probing). *Suppose $\mathsf{CC}_{\mathsf{exp}}$ satisfies t-composable security. Then $\mathsf{CC}_{\mathsf{poly}}$ satisfies t-composable security. If $\mathsf{CC}_{\mathsf{exp}}$ has randomness locality d then $\mathsf{CC}_{\mathsf{poly}}$ has randomness locality d.*

Proposition 6 (Random Probing). *Let $\mathsf{CC}_{\mathsf{exp}}$ satisfies $(\mathbf{p}, \varepsilon_{\mathsf{exp}})$-composable security. $\mathsf{CC}_{\mathsf{poly}}$, associated with circuits of size s, satisfies $(\mathbf{p}, s \cdot \varepsilon_{\mathsf{exp}})$-composable security. If $\mathsf{CC}_{\mathsf{exp}}$ has randomness locality d then $\mathsf{CC}_{\mathsf{poly}}$ has randomness locality d.*

4.5   Main Construction: Formal Description

We now combine all the components we developed in the previous sections to obtain a construction of composable circuit compiler. In particular, the main construction consists of the following main steps:

- Start with a secure MPC protocol Π for a constant number of parties.
- Apply the base case compiler to obtain a composable circuit compiler, which has constant simulation error in the case of random probing model and tolerates constant threshold in the case of worst case probing model.
- Recursively apply the composition step on the base compiler obtain from the above bullet. The resulting compiler, after sufficiently many iterations, satisfies negligible error in the random probing setting and satisfies a large threshold in the case of worst case probing model.
- The disadvantage with the compiler resulting from the previous step is that the size of the compiled circuit could be exponentially larger than the original circuit. To improve the efficiency from exponential to polynomial, we apply the exponential-to-polynomial transformation.

Proof: Worst Case Probing

We sketch the construction in Fig. 2.

Construction of CC_{main}

- **Circuit compilation, CC_{main}.Compile(C):** On input a circuit C, it executes the following steps:

    - It transforms Π into a composable circuit compiler CC_{base} satisfying **t**-composable security, where $\mathbf{t} = t$ and L_1-efficiency.

    - Set $CC_1 = CC_{base}$ with $\mathbf{t}_0 = t$. Repeat the following process for $i = 1, \ldots, K$: Using the composition theorem, satisfying $\mathbf{t}_i$-composable security, it transforms CC_i into a composable circuit compiler CC_{i+1} satisfying $\mathbf{t}_{i+1}$-composable security. Moreover, $\mathbf{t}_K = t^K$.

    - It transforms CC_K into a composable circuit compiler CC^* satisfying $f \cdot L_1^K(k)$-efficiency and t^K-composable security property, where f is a linear function.

    - It finally executes $CC^*(C)$ to obtain the compiled circuit $\widehat{C}$.

    - Output $\widehat{C}$.

- **Input encoding, CC_{main}.Encode(x):** It computes the XOR secret sharing of every bit of x. Output the concatenation of the XOR secret shares of all the bits of x, denoted by $\widehat{x}$.

- **Output encoding, CC_{main}.Decode($\widehat{y}$):** It reconstructs the XOR secret sharing of every bit of y. Output y.

Fig. 2. Construction of CC_{main}

Proposition 7. *Let $K \in \mathbb{N}$. Consider a MPC protocol Π for a n-party functionality F (Fig. 1) and tolerating at most t with randomness locality d. Then, CC_{main} is a t^K-composable secure composable circuit compiler secure against worst case probing attacks for all circuits satisfying $(L_1(k))^K \cdot f$-efficiency, where:*

- *$L_1(k)$ is a constant and f is a linear function,*
- *c is a constant,*
- *Moreover, the randomness locality of CC_{main} is $O(K)$.*

Instantiation. By instantiating the tools in the above proposition, we get the following proposition.

Proposition 8. *Consider a parameter* $\mathbf{t} > 0$. *There is a composable circuit compiler satisfying* $\mathbf{t}$*-composable security against worst case probing attacks satisfying randomness locality* $O(\log(\mathbf{t}))$.

Proof. Suppose we have a MPC protocol Π for the n-party functionality F (Fig. 1) tolerating at most t corruptions, for some constant n (for instance, [BOGW88,CCD88]). We then obtain a circuit compiler CC_{main}, which is t^K-composable secure and satisfy $c^K \cdot f$-efficiency, where c is a constant and f is a linear function. Setting $K = \lceil \frac{\log(\mathbf{t})}{\log(t)} \rceil$, we have that CC_{main} is $\mathbf{t}$-composable secure and satisfying polynomial efficiency, as desired. Moreover, the randomness locality of CC_{main} is $O(K) = O(\log(\mathbf{t}))$. This completes the proof.

We present the constructions in the worst case and random probing models below. The proofs are deferred to the full version.

Proof: Random Probing. We now present a construction (Fig. 3) of composable circuit compiler for a class of circuits $\mathcal{C}$ over basis $\mathbb{B}$ starting from a MPC protocol Π for the n-party functionality F that can tolerate t semi-honest adversaries. We denote this construction by CC_{main}.

Proposition 9. *Let* $K \in \mathbb{N}$. *Consider a MPC protocol* Π *for a* n-party *functionality* F *and tolerating at most* t *corruptions with randomness locality* d *satisfying the property that* $e^{\frac{12N_g}{(1+t)^2}} \geq \left(\frac{12N_g}{(1+t)^2} \right)^4$, *where* N_g *is the number of gates in the implementation of* Π.

Then, CC_{main} *is a* $(\mathbf{p}, c^{c^K})$*-secure composable circuit compiler for all circuits satisfying* $(L_1(k))^K \cdot f$*-efficiency, where:*

- $\mathbf{p} = \frac{(1+t)^2}{48N_g \ln\left(\frac{12N_g}{(1+t)^2} \right)}$
- $L_1(k)$ *is a constant and* f *is a linear function,*
- c *is a constant,*
- N_g *is the number of gates in the circuit* Ckt_Π

Moreover, the randomness complexity of CC_{main} *is* $O(K)$.

Instantiation. We use a specific instantiation of the MPC protocol in the above proposition to get the following result.

Proposition 10. *There is a construction of a composable circuit compiler for* $\mathcal{C}$ *satisfying* $(\mathbf{p}, \mathsf{negl})$*-composable security, where* $\mathbf{p} = 6.5 \times 10^{-5}$.

5   Leakage Tolerant Circuit Compilers

In this section, we present a construction of leakage tolerant circuit compiler with constant leakage rate. Later, we present a negative result on the leakage rate of a leakage tolerant circuit compiler.

Construction of CC_{main}

- **Circuit compilation, $\mathsf{CC}_{main}.\mathsf{Compile}(C)$:** On input a circuit C, it executes the following steps:

  - It transforms Π into a composable circuit compiler $\mathsf{CC}_{\mathsf{base}}$ satisfying $(\mathbf{p}, \varepsilon_1)$-composable security, where $\varepsilon_1 = e^{-\frac{(1+t)^2}{12\mathbf{N_g}} \cdot \frac{1}{\mathbf{p}}}$ and L_1-efficiency.

  - Set $\mathsf{CC}_1 = \mathsf{CC}_{\mathsf{base}}$. Repeat the following process for $i = 1, \ldots, K$: Using the composition step, it transforms CC_i into a composable circuit compiler CC_{i+1} satisfying $(\mathbf{p}, \varepsilon_{i+1})$-security.

  - Using the exponential-to-polynomial transformation, it transforms CC_K into a composable circuit compiler CC^* satisfying $f \cdot L_1^K(k)$-efficiency and $(\mathbf{p}, s \cdot \varepsilon_K)$-composable security property, where f is a linear function.

  - It finally executes $\mathsf{CC}^*(C)$ to obtain the compiled circuit $\widehat{C}$.

  - Output $\widehat{C}$.

- **Input encoding, $\mathsf{CC}_{main}.\mathsf{Encode}(x)$:** It computes the XOR secret sharing of every bit of x. Output the concatenation of the XOR secret shares of all the bits of x, denoted by $\widehat{x}$.

- **Output encoding, $\mathsf{CC}_{main}.\mathsf{Decode}(\widehat{y})$:** It reconstructs the XOR secret sharing of every bit of y. Output y.

Fig. 3. Construction of CC_{main}

5.1 Construction: Random Probing

We prove the following proposition.

Proposition 11. *Let CC_{comp} be a composable compiler for a class of circuits $\mathcal{C}$ satisfying $(\mathbf{p}, \varepsilon)$-composable security. Then, CC_{LT} is a $(\mathbf{p}, \mathbf{p}', \varepsilon')$-leakage tolerant circuit compiler for $\mathcal{C}$ secure against random probing attacks, where $\mathbf{p}' = (1 + \eta)^2 \left(1 - (1 - \mathbf{p})^6\right)$ and $\varepsilon' = \varepsilon + \frac{1}{e^{c \cdot n}}$, for arbitrarily small constant $\eta > 0$.*

To prove the above theorem, we start with a composable secure circuit compiler and then attach a leakage tolerant circuit that computes the additive shares of input. In particular, we need to prove that the leakage of values in the sharing circuit can be simulated with leakage on the input bits.

Combining with Proposition 10 obtain the following proposition.

Proposition 12. *Consider a basis $\mathbb{B}$. There is a construction of $(\mathbf{p}, \mathbf{p}', \mathsf{negl})$-leakage tolerant circuit compiler against random probing attacks for all circuits over $\mathbb{B}$ of size s, where $\mathbf{p} = 6.5 \times 10^{-5}$ and $\mathbf{p}' = 3.9 \times 10^{-4}$.*

Non-Boolean Basis. We show how to achieve a leakage tolerant compiler with leakage rate arbitrarily close to 1 with the compiled circuit defined over a non-boolean basis. The starting point is a composable circuit compiler where the compiled circuit with leakage rate arbitrarily close to 1 and over a large basis.

Proposition 13. *Let $\delta > 0$. Consider a basis $\mathbb{B}'$ consisting of all randomized functions mapping n bits to n bits. Suppose there is a construction of a composable circuit compiler $\mathsf{CC_{NB}}$ over $\mathbb{B}'$ for $\mathcal{C}$ over $\mathbb{B}$ satisfying $(\mathbf{p}, \varepsilon)$-composable security. Then there is a construction of $(\mathbf{p}, \mathbf{p}', \varepsilon')$-secure leakage tolerant circuit compiler over $\mathbb{B}'$ for $\mathcal{C}$ over $\mathbb{B}$, where $\mathbf{p}' = 1 - ((1 - \mathbf{p})^2) \cdot (1 - \mathbf{p}^n)^2)$ and $\varepsilon' = \varepsilon + \frac{1}{e^{c \cdot n}}$, for some constant c.*

5.2   Construction: Worst Case Probing

We present the construction of a leakage tolerant circuit compiler in the worst case probing model.

Proposition 14. *For any basis $\mathbb{B}$ and any $\mathbf{t} > 0$, there is a construction of leakage tolerant circuit compiler secure against $\mathbf{t}$-probing attacks. Moreover, this compiler has randomness locality $O(\log(\mathbf{t}))$.*

Proof. From Proposition 8, there is a construction of $\mathbf{t}$-secure composable circuit compiler CC_{comp}. We construct a leakage tolerant circuit compiler CC_{LT} as follows:

- Compile(C): On input C, it does the following:
  - Compute CC_{comp}.Compile(C) to obtain the compiled circuit $\mathsf{CC}_{comp}.\widehat{C}$.
  - Constructs a circuit $\widehat{C}$ that takes as input x,
    * Computes N shares of every bit of x, where N is determined the input length of $\mathsf{CC}_{comp}.\widehat{C}$. In particular, for every i, it computes shares of x_i as follows: $(x_i \oplus r_1, r_1 \oplus r_2, \ldots, r_{N-2} \oplus r_{N-1}, r_{N-1})$, where r_i is sampled freshly at random. For every i^{th} bit, since there are two input wires carrying x_i, we perform the sharing process twice.
    * Compute $\mathsf{CC}_{comp}.\widehat{C}$ on the shares of x as computed in the bullet above.
- Decode($\widehat{y}$): It parses $\widehat{y}$ as $(\widehat{y}^1, \ldots, \widehat{y}^\ell)$ and reconstructs the shares in $\widehat{y}^i$ to obtain the value y_i.

We claim that CC_{comp} is a $\mathbf{t}$-secure leakage tolerant circuit compiler. The correctness and efficiency properties of CC_{comp} follow from the respective properties of CC_{LT}. To argue security, we first note that any $\mathbf{t}$ wires of leakage in the sharing circuit can be simulated with $\mathbf{t}$ input and output wires of leakage of the sharing circuit (this follows from the fact that every wire in the sharing circuit is either an input or an output wire). The $\mathbf{t}$-composable security of CC_{comp} then implies the security of CC_{LT}.

Next, we show that CC_{comp} has randomness locality $O(\log(\mathbf{t}))$. We first note that the sharing circuit has constant randomness locality. This combined with the fact that CC has $O(\log(\mathbf{t}))$ randomness locality proves the result.

6   Leakage Resilient Circuit Compilers

In this section, we give upper bounds for leakage resilient circuit compilers. Note that any structural circuit compiler for circuit class $\mathcal{C}$ is also a leakage resilient circuit compiler for $\mathcal{C}$. Using this fact, we state the following theorem.

Theorem 8. *There is a construction of* $(\mathbf{p}, \exp(-s))$-*leakage resilient circuit compiler for all circuits over* $\mathbb{B}$ *of size* s, *secure against random probing attacks, where* $\mathbf{p} = 6.5 \times 10^{-5}$.

The proof of the above theorem follows from Proposition 10.

7   Randomness Complexity

We present a construction of leakage tolerant circuit compiler with near optimal randomness complexity. To show this, we use two lemmas from [IKL+13]. We first state a lemma about the existence of explicit robust r-wise PRGs. We refer the reader to [IKL+13] for the definition of strong $(\mathbf{t}, q)$ robust r-wise PRGs.

Lemma 2 ([IKL+13]). *For any* $\eta > 0$, *there exists* $\delta, c > 0$, *such that for any* $m \leq \exp n^{\delta}$, *there is an explicit d-strong* $(n^{1-\eta}, 21)$-*robust r-wise independent PRG* $G : \{0,1\}^n \rightarrow \{0,1\}^m$ *for* $r = n^{1-\eta}$ *and* $d \leq \log^c(m)$.

The following theorem[10] states that any $\mathbf{t}$-leakage tolerant circuit compiler establishes the connection between randomness locality and randomness complexity.

Lemma 3 ([IKL+13]). *Consider a* $q \cdot \mathbf{t}$-*leakage tolerant circuit compiler. Suppose the compiled circuit uses* m *random bits and makes an d-local use of its randomness. Let* $G : \{0,1\}^n \rightarrow \{0,1\}^m$ *be a strong* $(\mathbf{t}, q)$-*robust r-wise PRG with* $r \geq \mathbf{t} \cdot \max(d, q)$. *Then there is a leakage tolerant circuit compiler secure against* $\mathbf{t}$-*probing attacks which uses* n *random bits.*

Recall that the leakage tolerant compiler in Theorem 14 has randomness locality $O(\log(\mathbf{t}))$. This fact along with the above two lemmas yields the following theorem.

Theorem 9. *For any* $\mathbf{t} > 0$, *there is a construction of leakage tolerant circuit compiler secure against* $\mathbf{t}$-*probing attacks using* $\mathbf{t}^{1+\varepsilon} \cdot \mathrm{polylog}(|C|)$ *random bits.*

Acknowledgements. We thank Jean-Sébastien Coron, Stefan Dziembowski, and Sebastian Faust for helpful discussions. The second author was supported in part by ERC grant 742754, ISF grant 1709/14, NSF-BSF grant 2015782, and a grant from the Ministry of Science and Technology, Israel and Department of Science and Technology, Government of India. The third author's research is supported in part from a DARPA/ARL SAFEWARE award, NSF Frontier Award 1413955, and NSF grant 1619348, BSF grant 2012378, a Xerox Faculty Research Award, a Google Faculty

---

[10] They phrase this in the language of private circuits and so we rephrase their theorem in our language.

Research Award, an equipment grant from Intel, and an Okawa Foundation Research Grant. This material is based upon work supported by the Defense Advanced Research Projects Agency through the ARL under Contract W911NF-15-C-0205. The views expressed are those of the authors and do not reflect the official policy or position of the Department of Defense, the National Science Foundation, or the U.S. Government.

References

[ADF16] Andrychowicz, M., Dziembowski, S., Faust, S.: Circuit compilers with $O(1/\log(n))$ leakage rate. In: Fischlin, M., Coron, J.-S. (eds.) EURO-CRYPT 2016. LNCS, vol. 9666, pp. 586–615. Springer, Heidelberg (2016). https://doi.org/10.1007/978-3-662-49896-5_21

[Ajt11] Ajtai, M.: Secure computation with information leaking to an adversary. In: Proceedings of the Forty-Third Annual ACM Symposium on Theory of Computing, pp. 715–724. ACM (2011)

[BBD+16] Barthe, G., Belaïd, S., Dupressoir, F., Fouque, P.-A., Grégoire, B., Strub, P.-Y., Zucchini, R.: Strong non-interference and type-directed higher-order masking. In: Proceedings of the 2016 ACM SIGSAC Conference on Computer and Communications Security, pp. 116–129. ACM (2016)

[BBP+16] Belaïd, S., Benhamouda, F., Passelègue, A., Prouff, E., Thillard, A., Vergnaud, D.: Randomness complexity of private circuits for multiplication. In: Fischlin, M., Coron, J.-S. (eds.) EUROCRYPT 2016. LNCS, vol. 9666, pp. 616–648. Springer, Heidelberg (2016). https://doi.org/10.1007/978-3-662-49896-5_22

[BBP+17] Belaïd, S., Benhamouda, F., Passelègue, A., Prouff, E., Thillard, A., Vergnaud, D.: Private multiplication over finite fields. In: Katz, J., Shacham, H. (eds.) CRYPTO 2017. LNCS, vol. 10403, pp. 397–426. Springer, Cham (2017). https://doi.org/10.1007/978-3-319-63697-9_14

[Bea91] Beaver, D.: Efficient multiparty protocols using circuit randomization. In: Feigenbaum, J. (ed.) CRYPTO 1991. LNCS, vol. 576, pp. 420–432. Springer, Heidelberg (1992). https://doi.org/10.1007/3-540-46766-1_34

[BOGW88] Ben-Or, M., Goldwasser, S., Wigderson, A.: Completeness theorems for non-cryptographic fault-tolerant distributed computation. In: Proceedings of the Twentieth Annual ACM Symposium on Theory of Computing, pp. 1–10. ACM (1988)

[CCD88] Chaum, D., Crépeau, C., Damgård, I.: Multiparty unconditionally secure protocols. In: Proceedings of the Twentieth Annual ACM Symposium on Theory of Computing, pp. 11–19. ACM (1988)

[CDI+13] Cohen, G., et al.: Efficient multiparty protocols via log-depth threshold formulae. In: Canetti, R., Garay, J.A. (eds.) CRYPTO 2013. LNCS, vol. 8043, pp. 185–202. Springer, Heidelberg (2013). https://doi.org/10.1007/978-3-642-40084-1_11

[CK91] Chor, B., Kushilevitz, E.: A zero-one law for boolean privacy. SIAM J. Discret. Math. 4(1), 36–47 (1991)

[DDF14] Duc, A., Dziembowski, S., Faust, S.: Unifying leakage models: from probing attacks to noisy leakage. In: Nguyen, P.Q., Oswald, E. (eds.) EURO-CRYPT 2014. LNCS, vol. 8441, pp. 423–440. Springer, Heidelberg (2014). https://doi.org/10.1007/978-3-642-55220-5_24

[GIK+15] Garg, S., Ishai, Y., Kushilevitz, E., Ostrovsky, R., Sahai, A.: Cryptography with one-way communication. In: Gennaro, R., Robshaw, M. (eds.) CRYPTO 2015, Part II. LNCS, vol. 9216, pp. 191–208. Springer, Heidelberg (2015). https://doi.org/10.1007/978-3-662-48000-7_10

[GIM+16] Goyal, V., Ishai, Y., Maji, H.K, Sahai, A., Sherstov, A.A.: Bounded-communication leakage resilience via parity-resilient circuits. In: 2016 IEEE 57th Annual Symposium on Foundations of Computer Science (FOCS), pp. 1–10. IEEE (2016)

[HM00] Hirt, M., Maurer, U.: Player simulation and general adversary structures in perfect multiparty computation. J. Cryptol. **13**(1), 31–60 (2000)

[IKL+13] Ishai, Y., et al.: Robust pseudorandom generators. In: Fomin, F.V., Freivalds, R., Kwiatkowska, M., Peleg, D. (eds.) ICALP 2013. LNCS, vol. 7965, pp. 576–588. Springer, Heidelberg (2013). https://doi.org/10.1007/978-3-642-39206-1_49

[ISW03] Ishai, Y., Sahai, A., Wagner, D.: Private circuits: securing hardware against probing attacks. In: Boneh, D. (ed.) CRYPTO 2003. LNCS, vol. 2729, pp. 463–481. Springer, Heidelberg (2003). https://doi.org/10.1007/978-3-540-45146-4_27

[Pip85] Pippenger, N.: On networks of noisy gates. In: FOCS, pp. 30–38 (1985)

[vN56] von Neumann, J.: Probabilistic logics and synthesis of reliable organisms from unreliable components. Autom. Stud. **34**, 43–98 (1956)

Various Topics

A New Public-Key Cryptosystem via Mersenne Numbers

Divesh Aggarwal[1]($\boxtimes$), Antoine Joux[2], Anupam Prakash[3,4],
and Miklos Santha[4,5]

[1] School of Computing and Centre for Quantum Technologies,
National University of Singapore, Singapore, Singapore
divesh.aggarwal@gmail.com
[2] Chaire de Cryptologie de la Fondation SU, Sorbonne Université,
Institut de Mathématiques de Jussieu-Paris Rive Gauche, Inria,
CNRS, Univ Paris Diderot, Paris, France
[3] School of Physical and Mathematical Sciences,
Nanyang Technological University, Singapore, Singapore
[4] Centre for Quantum Technologies,
National University of Singapore, Singapore, Singapore
[5] IRIF, Université Paris Diderot, CNRS, Paris, France

Abstract. In this work, we propose a new public-key cryptosystem whose security is based on the computational intractability of the following problem: Given a Mersenne number $p = 2^n - 1$, where n is a prime, a positive integer h, and two n-bit integers T, R, decide whether their exist n-bit integers F, G each of Hamming weight less than h such that $T = F \cdot R + G$ modulo p.

1 Introduction

1.1 Motivation

Since the seminal work of Diffie and Hellman [DH76] which presented the fundamentals of public-key cryptography, one of the most important goal of cryptographers has been to construct secure and practically efficient public-key cryptosystems. Rivest, Shamir, and Adleman [RSA78] came up with the first practical public-key cryptosystem based on the hardness of factoring integers, and it remains the most popular scheme till date.

Shor [Sho97] gave a quantum algorithm that solves the abelian hidden subgroup problem and as a result solves both discrete logarithms and factoring. Back in 1994, this was not considered a real threat to the practical cryptographic schemes since quantum computers were far from being a reality. However, given the recent advances in quantum computing, there is serious effort in both the industry and the scientific community to make information security systems resistant to quantum computing. In fact, the National Institute of Standards and Technology (NIST) is now beginning to prepare for the transition into quantum-resistant cryptography and has announced a project where they are accepting submissions for quantum-resistant public-key cryptographic algorithms [NIS17].

© International Association for Cryptologic Research 2018
H. Shacham and A. Boldyreva (Eds.): CRYPTO 2018, LNCS 10993, pp. 459–482, 2018.
https://doi.org/10.1007/978-3-319-96878-0_16

In the recent years, some presumably quantum-safe public-key cryptosystems have been proposed in the literature. Perhaps the most promising among these are those based on the hardness of lattice problems like Learning with Errors (LWE) based cryptosystems [Reg09], Ring-LWE based cryptosystems [LPR10] and NTRU [HPS98]. While these cryptosystems have so far resisted any classical or quantum attacks, it cannot be excluded that such attacks are possible in the future. In fact, there have been some, albeit unsuccessful, attempts at a quantum algorithm solving the LWE problem [ES16]. In particular, there is no unifying complexity-theoretic assumption (like NP-hardness) that relates the difficulty of breaking all these cryptosystems. Thus, it is desirable to come up with promising new proposals for public-key cryptosystems.

It is worthwhile to note that even though the concept of public-key cryptography was introduced four decades ago, the number of existing public-key cryptographic schemes whose hardness does not depend on the hardness of factoring or finding short vectors in lattices is not very large [KLC+00, McE78, LvTMW09, GWO+13, NS97]. This is not an exhaustive list but it illustrates the various approaches that have been tried. The rarity of proposals for potentially quantum safe public key cryptosystems further motivates the problem of constructing such cryptosystems.

1.2   Our Cryptosystem

Our cryptosystem is based on arithmetic modulo so called Mersenne numbers, i.e., numbers of the form $p = 2^n - 1$, where n is a prime. These numbers have an extremely useful property: For any number x modulo p, and $y = 2^z$, where z is a positive integer, $x \cdot y$ is a cyclic shift of x by z positions and thus the Hamming weight of x is unchanged under multiplication by powers of 2. Our encryption scheme is based on the simple observation that, given a uniformly random n-bit string R, when we consider $T = F \cdot R + G \pmod{p}$, where the binary representation of F and G modulo p has low Hamming weight, then T looks pseudorandom, i.e., it is hard to obtain any non-trivial information about F, G from R, T.

The public-key is chosen to be the pair (R, T), and the secret key is the string F. The encryption scheme also requires an efficient error correcting code with encoding function $\mathcal{E} : \{0,1\}^k \rightarrow \{0,1\}^n$ and decoding function $\mathcal{D} : \{0,1\}^n \rightarrow \{0,1\}^k$. In order to encrypt a message $m \in \{0,1\}^k$, the encryption algorithm chooses three random numbers A, B_1, B_2 of low Hamming weight modulo p and then outputs

$$C := (C_1, C_2),$$

where $C_1 = A \cdot R + B_1$, and $C_2 = (A \cdot T + B_2) \oplus \mathcal{E}(m)$ where $\oplus$ denotes the bitwise XOR operation. Given the private key, one can compute

$$C_2^* := C_1 \cdot F = (A \cdot T + B_2) - A \cdot G - B_2 + B_1 \cdot F.$$

Since A, B_1, B_2, F, G have low Hamming weight, the Hamming distance between $A \cdot T + B_2$ and C_2^* is expected to be low, and so we get that $\mathcal{D}(C_2 \oplus C_2^*)$ is equal to to m with high probability. For more details on our scheme and the underlying security assumption, we refer the reader to Sects. 4 and 5.

1.3 Related Work

The Mersenne cryptosystem can be seen as belonging to a family that started with the Ntru cryptosystem and as been instantiated in many ways [HPS98, Reg09, LPR10, MTSB13]. The common idea behind all these cryptosystems is to work with elements in a ring which are hidden by adding some small noise. This notion of smallest needs to be somewhat preserved under the arithmetic operation. At the same time, it should be somewhat unnatural and not fully compatible with the ring structure in order to lead to hard problems.

Our goal in designing the Mersenne cryptosystem was to find a very simple instantiation of this paradigm based on the least complicated ring we could find. This led us to consider numbers modulo a prime together with the Hamming weight to measure smallest. In this context, it is natural to restrict ourselves to Mersenne primes, since reduction modulo such a prime cannot increase Hamming weights. Moreover, our cryptosystem relies on a conceptually simpler ring of numbers modulo a prime and its description only requires very elementary mathematics.

Our first proposal using this structure [AJPS17] only allowed us to encrypt a single bit at a time. The security parameters in [AJPS17] were based on the assumption that there is no attack on the cryptosystem that runs faster than the trivial attack that runs in time $\binom{n}{h}$. Subsequent works showed that this assumption was incorrect. In particular, [BCGN17] showed a non-trivial guess-and-determine attack based on a low-dimension lattice reduction subroutine that runs in time $(2 + \varepsilon + o(1))^{2h}$ for some small constant ε, and [dBDJdW17] gave a meet-in-the-middle attack that runs in time $O\left(\binom{n-1}{h-1}^{1/2}\right)$ on classical computers and $O\left(\binom{n-1}{h-1}^{1/3}\right)$ on quantum machines. While these attacks could be circumvented by choosing the parameter h to be as large as the security parameter, this would make our cryptosystem inefficient.

Fortunately, in the present proposal, we are able to overcome this difficulty. We describe a variant that allows us to encrypt many bits at a time. This allows in turn to choose much larger parameters which resist the attacks in [BCGN17, dBDJdW17], even in their Groverized quantum form while still maintaining the efficiency of our cryptosystem. Such quantum attacks would have complexity larger than 2^h where h is the Hamming weight we allow for low Hamming weight numbers. This explains our choice of h to be equal to the desired quantum security level.

Since it is well-known that a cryptosystem of this type can be easily vulnerable to chosen-ciphertext attack, it is extremely important to bind them together with a CCA-secure wrapper. We chose to present the system as a key encapsulation mechanism because this makes the design of the CCA wrapper very simple.

1.4   Organization of the Paper

In Sect. 2 we introduce some preliminaries about Mersenne primes and security definitions. In Sect. 3 we provide a semantically secure basic bit by bit encryption scheme. In Sect. 4 we give a semantically secure blockwise encryption scheme. In Sect. 5, we prove the semantic security for the scheme presented in Sect. 4. In Sect. 6 we discuss the known cryptanalytic attacks against this scheme. In Sect. 7 we give the final key encapsulation scheme secure against chosen ciphertext attacks in the random oracle model. In Sects. 8 and 9 we provide an instantiation for the error correcting codes used in our encryption/key encapsulation schemes.

2   Preliminaries

Notations. For any distinguisher D that outputs a bit $b \in \{0,1\}$, the distinguishing advantage to distinguish between two random variables X and Y is defined as:
$$\Delta^D(X \; ; \; Y) := |\Pr[D(X) = 1] - \Pr[D(Y) = 1]|.$$

The following lemma is well known and easy to see.

Lemma 1. *Given a probabilistic polynomial time computable function f on two random variables X and Y, if there is a probabilistic polynomial time distinguisher D that distinguishes between $f(X)$ and $f(Y)$ with advantage δ, then there is a probabilistic polynomial time distinguisher D' that distinguishes between X, and Y with advantage δ.*

2.1   Mersenne Numbers and Mersenne Primes

Let n be a positive integer, and let $p = 2^n - 1$. When n is a prime, p is called a Mersenne number, and if $2^n - 1$ is itself a prime number, then it is called a Mersenne prime. Note that if n is a composite number of the form $n = k\ell$, then $2^k - 1$ and $2^\ell - 1$ divide p, and hence p is not a prime. The smallest Mersenne primes are
$$2^2 - 1, 2^3 - 1, 2^5 - 1, 2^7 - 1, 2^{13} - 1, 2^{17} - 1, \ldots$$

We denote by $\mathbb{Z}_p$ the ring of integers modulo p. We index binary strings from right to left, that is for $x \in \{0,1\}^n$ we write x as $x_n \ldots x_1$. The Hamming weight of an n-bit string y is the total number of 1's in y and is denoted by $\mathsf{Ham}(y)$. Let $\mathsf{seq} : \mathbb{Z}_p \to \{0,1\}^n$ be the map which to $x \in \mathbb{Z}_p$ associates the binary string $\text{seq}(x)$ representing x. The map $\text{int} : \{0,1\}^n \to \mathbb{Z}_p$ sends a string y into the integer represented by y modulo p. Clearly seq and int are inverse functions between $\mathbb{Z}_p$ and $\{0,1\}^n \setminus \{1^n\}$, and $\text{int}(1^n) = 0$. We use this bijection between $\mathbb{Z}_p$ and $\{0,1\}^n \setminus \{1^n\}$ to define *addition* and *multiplication* over $\{0,1\}^n$ in the natural way: for $y, y' \in \{0,1\}^n$, let $y + y' = \text{seq}(\text{int}(y) + \text{int}(y'))$, and let $y \cdot y' = \text{seq}(\text{int}(y) \cdot \text{int}(y'))$. It is easy to see that both operations remain associative and commutative, and the distributivity of the multiplication over the addition also holds. We also set $(-1) \cdot y = -y = \text{seq}(-(\text{int}(y))$. Observe that addition is invariant by rotation, that is if $\text{rot}_k(y)$ denotes the circular rotation of y by k positions to the left, then $\text{rot}_k(y + y') = \text{rot}_k(y) + \text{rot}_k(y')$.

Lemma 2. *Let $p = 2^n - 1$. For all $A, B \in \{0,1\}^n$, we have*

1. $\text{Ham}(A + B) \leq \text{Ham}(A) + \text{Ham}(B)$.
2. $\text{Ham}(A \cdot B) \leq \text{Ham}(A) \cdot \text{Ham}(B)$.
3. *If $A \neq 0^n$ then $\text{Ham}(-A) = n - \text{Ham}(A)$.*

Proof.

1. If $A = 1^n$ the result is obviously true. When $A \neq 1^n$, we prove the result by induction on the Hamming weight of B. If $B = 0^n$ the statement is obviously true.

   For the induction step we first prove the claim when $\text{Ham}(B) = 1$. Let i be the index on which B takes the value 1. Since addition is invariant by rotation, we may assume that $i = 1$ and thus $B = 0^{n-1}1$. A can be written as $C01^j$ for some $0 \leq j \leq n - 1$, and $A + B = C10^j$. Thus $\text{Ham}(A + B) = \text{Ham}(A) - j + 1 \leq \text{Ham}(A) + 1$.
   Let $\text{Ham}(B) = k > 1$. Then we can decompose B as $B_1 + B_2$, where $\text{Ham}(B_1) = k - 1$ and $\text{Ham}(B_2) = 1$. By the previous claim and the induction hypothesis we get:

   $$\text{Ham}(A+B) = \text{Ham}((A+B_1)+B_2) \leq \text{Ham}(A+B_1)+1 \leq \text{Ham}(A)+(k-1)+1,$$

   and the result follows.
2. If $B = 0^n$ the statement is obviously true. Otherwise, for some $k \geq 1$, we can decompose B as $B_1 + \cdots + B_k$, where each B_i has Hamming weight 1, for $1 \leq i \leq k$. Let j_i be the index of the position where B_i takes the value 1. Then $A \cdot B_i = \text{rot}_{j_i-1}(A)$. Thus $\text{Ham}(A \cdot B_i) = \text{Ham}(A)$, and by distributivity we get $A \cdot B = A \cdot B_1 + \cdots + A \cdot B_k$. The result then follows from part (1).
3. If $A \neq 0^n$ then $-A$ is the binary string obtained from A by replacing 0's by 1's and 1's by 0's.

$\square$

2.2 Security Definitions

Public-Key Encryption. A public key encryption scheme comprises three algorithms: the key generation algorithm KeyGen, the encryption algorithm Enc, and the decryption algorithm Dec. The KeyGen algorithm outputs a public-key pk, and a secret key sk. The encryption algorithm Enc takes as input a message m, and pk, and outputs a ciphertext C. The decryption algorithm takes as input a ciphertext C and sk, and outputs a message m' or a special symbol $\perp$ indicating rejection. We say that the encryption scheme is $1 - \delta$ correct if for all m, $\Pr[\mathsf{Dec}(\mathsf{sk}, \mathsf{Enc}(\mathsf{pk}, m)) = m] \geq 1 - \delta$, where the probability is over the randomness of pk, sk and the encryption algorithm.

We denote the security parameter by λ. All other parameters including key lengths and ciphertext size are polynomial functions of λ.

Definition 1. *The public-key encryption scheme*

$$PKE = (\mathsf{KeyGen}, \mathsf{Enc}, \mathsf{Dec})$$

is said to be semantically secure *if for any probabilistic polynomial time distinguisher and any pair of messages m_0, m_1 of equal length, given the public key pk, the advantage for distinguishing $C_0 = \mathsf{Enc}(\mathsf{pk}, m_0)$ and $C_1 = \mathsf{Enc}(\mathsf{pk}, m_1)$ is at most $\frac{\mathrm{poly}(|C_i|)}{2^\lambda}$ for some polynomial poly.*

Definition 2. *The public-key encryption scheme*

$$PKE = (\mathsf{KeyGen}, \mathsf{Enc}, \mathsf{Dec})$$

is said to be secure under chosen ciphertext attacks *if for any probabilistic polynomial time distinguisher that is given access to an oracle that decrypts any given ciphertext, the following holds: For any pair of messages m_0, m_1 of equal length, given the public key pk, the advantage for distinguishing $C_0 = \mathsf{Enc}(\mathsf{pk}, m_0)$ and $C_1 = \mathsf{Enc}(\mathsf{pk}, m_1)$ is at most $\frac{\mathrm{poly}(|C_i|)}{2^\lambda}$ for some polynomial poly under the assumption that the distinguisher does not query the oracle with C_0 or C_1.*

Key Encapsulation Mechanism. A key-encapsulation mechanism (KEM) comprises three algorithms: the key generation algorithm KeyGen, the encapsulation algorithm Encaps, and the decapsulation algorithm Decaps, and a key space $\mathcal{K}$. The KeyGen algorithm outputs a public-key pk, and a secret key sk. The encapsulation algorithm Encaps takes as input a public key pk to produce a ciphertext C and a key $K \in \mathcal{K}$. The decapsulation algorithm Decaps takes as input a ciphertext C and sk, and outputs a key K' or a special symbol $\perp$ indicating rejection. We say that the KEM is $(1 - \delta)$-correct if

$$\Pr[\mathsf{Decaps}(\mathsf{sk}, C) = K \; : \; (C, K) \leftarrow \mathsf{Encaps}(\mathsf{pk})] \geq 1 - \delta,$$

where the probability is over the randomness of pk, sk and the encapsulation algorithm.

We denote the security parameter by λ. All other parameters including key lengths and ciphertext size are polynomial functions of λ.

Definition 3. The key-encapsulation mechanism

$$KEM = (\mathsf{KeyGen}, \mathsf{Encaps}, \mathsf{Decaps})$$

is said to be *semantically secure* if for any probabilistic polynomial time distinguisher, given the public key pk, the advantage for distinguishing (C, K_0) and (C, K_1), where $(C, K_0) \leftarrow \mathsf{Encaps}(\mathsf{pk})$ and K_1 is uniform and independent of C is at most $\frac{\mathrm{poly}(|C|,|K_0|)}{2^\lambda}$ for some polynomial poly.

Definition 4. The key-encapsulation mechanism

$$KEM = (\mathsf{KeyGen}, \mathsf{Encaps}, \mathsf{Decaps})$$

is said to be *secure under chosen ciphertext attacks* if for any probabilistic polynomial time distinguisher that is given access to the decapsulation oracle and the public key pk, the advantage for distinguishing (C, K_0) and (C, K_1), where $(C, K_0) \leftarrow \mathsf{Encaps}(\mathsf{pk})$ and K_1 is uniform and independent of C is at most $\frac{\mathrm{poly}(|C|,|K_0|)}{2^\lambda}$ for some polynomial poly under the assumption that the distinguisher does not query the oracle with C.

2.3 Security Assumptions

The semantic security of our encryption scheme is based on the following assumption.

Definition 5. The *Mersenne Low Hamming Combination Assumption* states that given an n-bit Mersenne prime $p = 2^n - 1$, and an integer h, the advantage of any probabilistic polynomial time adversary running in time $\mathrm{poly}(n)$ in attempting to distinguish between

$$\left(\begin{bmatrix} R_1 \\ R_2 \end{bmatrix} , \begin{bmatrix} R_1 \\ R_2 \end{bmatrix} \cdot A + \begin{bmatrix} B_1 \\ B_2 \end{bmatrix} \right) \text{ and } \left(\begin{bmatrix} R_1 \\ R_2 \end{bmatrix} , \begin{bmatrix} R_3 \\ R_4 \end{bmatrix} \right)$$

is at most $\frac{\mathrm{poly}(n)}{2^\lambda}$, where R_1, R_2, R_3, R_4 are independent and uniformly random n-bit strings, and A, B_1, B_2, are independently chosen n-bit strings each having Hamming weight h.

We note that the assumption has some striking similarity to the learning with errors assumption by Regev [Reg09], where A corresponds to the secret, and B_1, B_2 correspond to the small error. The Mersenne Low Hamming Combination Assumption, in particular implies that one cannot obtain any useful information about A, B from the pair $(R_1, A \cdot R_1 + B)$. Notice that if the pair $(R_1, A \cdot R_1 + B)$ is assumed to be pseudorandom, then so is the pair $(R_1, A \cdot (-R_1) + B)$, and so one cannot obtain any useful information about A, B from the pair $(R_1, -A \cdot R_1 + B)$. The Mersenne Low Hamming Ratio Assumption is a homogeneous version of this assumption in the sense that we state that no useful information about A, B can be obtained from $(R_1, -A \cdot R_1 + B)$ given that $-A \cdot R_1 + B = 0$. It is required for the semantic security of the bit-by-bit encryption scheme that we describe in the next section, and was introduced in a previous version of this paper.

Definition 6. *The* Mersenne Low Hamming Ratio Assumption *states that given an n-bit Mersenne prime $p = 2^n - 1$, and an integer h, the advantage of any probabilistic polynomial time adversary running in time poly(n) in attempting to distinguish between* $\mathrm{seq}(\frac{\mathrm{int}(A)}{\mathrm{int}(B)})$ *and R is at most $\frac{\mathrm{poly}(n)}{2^\lambda}$, where R is a uniformly random n-bit string, and A, B, are independently chosen n-bit strings each having Hamming weight h.*

3 Basic Bit-by-Bit Encryption

In the following, we describe a basic encryption scheme to encrypt a single bit $b \in \{0, 1\}$.

Key Generation.

- Given the security parameter λ, choose a Mersenne prime $p = 2^n - 1$ and an integer h such that $\binom{n}{h} \geq 2^\lambda$ and $4h^2 < n$.
- Choose F, G to be two independent n-bit strings chosen uniformly at random from all n-bit strings of Hamming weight h.
- Set $\mathsf{pk} := H = \mathrm{seq}(\frac{\mathrm{int}(F)}{\mathrm{int}(G)})$, and $\mathsf{sk} := G$.

Encryption. The encryption algorithm chooses two independent strings A, B uniformly at random from all strings with Hamming weight h. A bit b is encrypted as
$$C = \mathsf{Enc}(\mathsf{pk}, b) := (-1)^b (A \cdot H + B).$$

Decryption. The decryption algorithm computes $d = \mathsf{Ham}(C \cdot G)$. If $d \leq 2h^2$, then output 0; if $d \geq n - 2h^2$, then output 1. Else output $\bot$.

For the correctness of the decryption note that $C \cdot G = (-1)^b \cdot (A \cdot F + B \cdot G)$ which, by Lemma 2, has Hamming weight at most $2h^2$ if $b = 0$, and at least $n - 2h^2$ if $b = 1$.

The basic bit-by-bit encryption scheme can be viewed as a simple proposal for a cryptosystem based on arithmetic modulo the Mersenne primes, however it is not efficient with respect to ciphertext size. Since this is not our final proposed encryption scheme, we do not analyze its security although it will easily follow from the Mersenne Low Hamming Ratio and the Mersenne Low Hamming Combination Assumption and appeared in a previous version of this paper. In the next section, we describe a scheme for encrypting longer message blocks.

4 Our Main Semantically Secure Public-Key Cryptosystem

It is reasonable to choose the message block length to be the same as the security parameter in practice. For this reason, we describe below a scheme for encrypting a message block $m \in \{0, 1\}^\lambda$.

Key Generation.

- Given the security parameter λ, choose a Mersenne prime $p = 2^n - 1$ such that $h = \lambda$ and $n > 10h^2$.
- Let F, G to be two independent n-bit strings chosen uniformly at random from all n-bit strings of Hamming weight h. Let R be a uniformly random n-bit string.
- Set $\mathsf{pk} := (R, F \cdot R + G) := (R, T)$, and $\mathsf{sk} := F$.

Encryption. The encryption algorithm chooses three strings A, B_1, B_2 independently and uniformly at random from all strings with Hamming weight h. Let $(\mathcal{E}, \mathcal{D})$ be the encoding and decoding algorithms of an error correcting code that we choose later. The message $m \in \{0,1\}^\lambda$ is encrypted as,

$$\mathsf{Enc}(\mathsf{pk}, m) := (C_1, C_2) := (A \cdot R + B_1, (A \cdot T + B_2) \oplus \mathcal{E}(m)).$$

Here $\mathcal{E} : \{0,1\}^\lambda \to \{0,1\}^n$ is a suitably chosen error correcting code and $\oplus$ denotes the bitwise XOR operation.

Decryption. The decryption algorithm computes $\mathcal{D}((F \cdot C_1) \oplus C_2)$.

In order to say that the scheme is $(1 - \delta)$-correct, we need to choose the error correcting code such that $\Pr_{(C_1,C_2) \leftarrow \mathsf{Enc}(\mathsf{pk},m)}[\mathcal{D}((F \cdot C_1) \oplus C_2) = m] \geq 1 - \delta$, where the probability is over the randomness of the encryption algorithm and the choice of pk, sk. For concrete instantiations of error correcting codes that satisfy this for a small enough δ, see Sect. 8.

5   Semantic Security of the Cryptosystem

In this section, we prove the semantic security of the PKE scheme in Sect. 4.

Theorem 1. *The encrpytion scheme* (Enc, Dec) *described in Sect. 4 is semantically secure under the Mersenne Low Hamming Combination Assumption.*

Proof. In the following, let $A, B_1, B_2, F, G, R, R', R'', R'''$ be independently chosen such that A, B_1, B_2, F, G are chosen uniformly from all strings of Hamming weight h, and R, R', R'', R''' are uniformly random strings. Let $T = F \cdot R + G$. By the Mersenne Low Hamming Combination Assumption, for any probabilistic polynomial time distinguisher D running in time poly(n),

$$\Delta^D(R, T \; ; \; R, R') \leq \frac{\mathrm{poly}(n)}{2^\lambda}.$$

Now, from Lemma 1, we have that for any probabilistic polynomial time distinguisher D' running in time poly(n),

$$\Delta^{D'}(R, T, A \cdot R + B_1, A \cdot T + B_2; R, R', A \cdot R + B_1, A \cdot R' + B_2) \leq \frac{\mathrm{poly}(n)}{2^\lambda}.$$

Again, by the Mersenne Low Hamming Combination Assumption, we have that

$$\Delta^{D'}(R, R', A \cdot R + B_1, A \cdot R' + B_2 \; ; \; R, R', R'', R''') \leq \frac{\text{poly}(n)}{2^\lambda}.$$

Using the triangle inequality, we get that

$$\Delta^{D'}(R, T, A \cdot R + B_1, A \cdot T + B_2 \; ; \; R, R', R'', R''') \leq \frac{2\,\text{poly}(n)}{2^\lambda}.$$

This implies that for any message m,

$$\Delta^{D'}(R, T, A \cdot R + B_1, A \cdot T + B_2 \oplus \mathcal{E}(m) \; ; \; R, R', R'', R''') \leq \frac{2\,\text{poly}(n)}{2^\lambda},$$

since R, R', R'', R''' and $R, R', R'', R''' \oplus \mathcal{E}(m)$ are identically distributed. This implies the required semantic security. □

6   Analysis of Our Security Assumption

6.1   Attempts at Cryptanalysis

In this section, we mention the known approaches to break our security assumption and thereby mention the conjectured security guarantee for our scheme. For cryptanalysis, it is often more convenient to talk about search problems. We introduce the following search problem whose solution would imply an attack on our cryptosystem.

Definition 7 (Mersenne Low Hamming Combination Search Problem). *For an n-bit Mersenne number $p = 2^n - 1$ and an integer h, given tuple $(R, FR + G \pmod{p})$ where R is a uniformly random n-bit string and F, G have Hamming weight h, find F, G.*

For the remainder of the paper, we call this problem $\mathcal{P}$. It is easy to see that if one can efficiently solve the problem $\mathcal{P}$, then one can break the assumption in Definition 5, and hence the security of our cryptosystem. It is therefore important to study the hardness of this problem.

Hamming Distance Distribution. Let R be a uniformly random n bit string and $Y = FR + G$ where F, G are chosen uniformly at random from n bit strings with Hamming weight h. A basic test for the assumption that Y is pseudorandom given R is to check that the distribution of $\mathsf{Ham}(R, R')$ is close to the distribution of $\mathsf{Ham}(R, T)$ where T is a uniformly random n bit string.

If R is a fixed string and X is a uniformly random n bit string, the random variable $f_R(X) = \frac{\mathsf{Ham}(X, R) - n/2}{\sqrt{n/4}}$ is approximated by the standard normal random variable $N(0, 1)$. We generated R at random and then obtained samples $Y = $

$FR + G$ where F, G are uniformly distributed over strings of Hamming weight $\sqrt{n}$. A quantile-quantile plot of $f_R(Y_i)$ against samples from $N(0, 1)$ is close to a straight line and does not show significant deviations from normality.

One could also perform more advanced statistical tests, such as the NIST suite [RSN+01] to verify the pseudorandomness of Y given R. However, in the context of cryptographic schemes, such tests only serve as sanity checks and it is preferable to focus on dedicated cryptanalysis.

Weak key attack. Following the appearance of a preliminary version of this paper, [BCGN17] found a weak key attack on the Mersenne Low Hamming Ratio search problem where given $H = \text{seq}(\frac{\text{int}(F)}{\text{int}(G)}) \mod P$ with F, G having low Hamming weight, the goal is to find F and G.

The weak key attack of [BCGN17] is based on rational reconstruction. If all the bits of F and G are in the right half of the bits, then both F and G are smaller than $\sqrt{P}$ and they can easily be recovered using a continued fraction expansion of H/P. The weak key attack also extends to the Mersenne Low Hamming Combination search problem, we choose parameters such that the success probability for this attack is negligible.

Generalization using LLL. The authors of the above weak attack also proposed in [BCGN17] a generalization based on guessing a decomposition of F and G into windows of bits such that in any window all the '1's are on the right. Using such a decompostion and replacing the use of continued fraction by LLL in relatively small dimension they can recover F and G from any compatible window decomposition.

A careful analysis of this method and its cost is presented in [dBDJdW17] and concludes that its running time is $2^{(2+\epsilon)h}$ for some small constant h. For simplicity, we assume that the cost of this attack is 2^{2h} on a classical computer.

Even if this attack was developed for the homogeneous Mersenne Low Hamming Ratio assumption, it is likely that it generalizes to the Mersenne Low Hamming Combination Assumption. We thus assume that it is the case. To the best of our knowledge, this is the most efficient known attack on our security assumption and the security parameters proposed in Sect. 8 have been revised to withstand it.

Quantum Speedup via Grover's Algorithm. With access to a quantum computer, one could use Grover's algorithm [Gro96] to obtain a quadratic speedup over the above attack.

Note that the attack performs a lattice reduction step for each guess of window decomposition and concludes that they are correct if the lattice reduction step succeeds. The Groverized version of the algorithm would prepare a superposition over possible guesses of window decompositions, use a unitary operator that performs lattice reduction to mark the good guesses for window decomposition and then amplify the success probability using Grover's search. This would certainly need very sophisticated universal quantum computers and it may well be infeasible for near term quantum devices. However, in view of this potential quantum attack and potential cryptananalytic improvements, we take this attack into account. With this constraint, our cryptosystem can only be secure if we make sure that h is at least equal to the desired security level. For simplicity, we just set $h = \lambda$ and assume that the best possible attack on the Mersenne Low Hamming Combination problem has complexity at least 2^h to derive security estimates in Sect. 8.

Meet in the middle attack. A recent work [dBDJdW17] gave a non-trivial meet-in-the-middle attack that makes use of locality-sensitive hash functions. Its complexity is $O\left(\binom{n-1}{h-1}^{1/2}\right)$ on classical computers and $O\left(\binom{n-1}{h-1}^{1/3}\right)$ on quantum machines.

For our choice of parameters, this is much bigger than 2^h and thus doesn't affect the security level.

Attacking the system if n is not a prime. We mention here that it seems quite important to choose $2^n - 1$ to be a prime for our cryptosystem. There is at least a partial attack when n is not prime. Indeed if n_0 divides n, then $q = 2^{n_0} - 1$ divides $p = 2^n - 1$, and also F, G have Hamming weight at most h modulo q. Thus, given $Y = FR + G \mod q$, one can try to guess the secret key G modulo q, which can be done in $\sqrt{\binom{n_0}{h}}$ time using a quantum algorithm. This also reveals F modulo q and we can likely use it to guess F, G modulo p much faster than the attacks that work in the prime case.

6.2 Active Attacks

Active attacks and/or decryption errors attacks are powerful tools that can be used to attack our bit-by-bit encryption. We recall that the basic idea of such attacks is to ask for the decryption of incorrectly formed ciphertext and use the answers to recover information about the key.

For example, incorrect ciphertexts can be obtained by picking a random bitstring, by modifying a valid one or encrypting in a non conformant way. Here, we review the attack in the context of a single bit, but it is important to note that the encryption of many bits remain vulnerable to such attacks, even if plaintext redundancy in the style of OAEP paddings [Sho02] is added. We show in Sect. 7 how to withstand such attacks using appropriate checks of ciphertext validity.

For simplicity, assume that we have access to a decryption oracle. Forming pseudo ciphertexts of the form $A^*H + B^*$ with A^* and B^* with low but not conformant Hamming weights can leak information about the private key. In particular, one might incrementally add '1' bits into B^* (or A^*) until decryption transitions from 0 to $\perp$. We did not concretely write down a full working attack along this line, but it is clear that our encryption scheme would be vulnerable to such attacks.

7   Mersenne Key Encapsulation Mechanism

Since we have seen in Sect. 6.2 that the semantically secure cryptosystem described in Sect. 4 cannot offer resistance to chosen-ciphertext attack, we need to integrate it into a more complex scheme with this ability. A first approach would be to use an existing generic transformation for this purpose. However, this is not a simple matter, indeed, systems such as OAEP or REACT [OP01] perform checks at the plaintext level and thus cannot protect against the attack strategy of Sect. 6.2. The Naor-Yung paradigm [NY90, CHK10] would be more suitable but the introduction of dual-encryption and non-interactive proofs is too costly for our purpose.

In this section, we specify a full cryptosystem that achieves this level of resistance using a transformation specifically designed for our encryption scheme. We present our cryptosystem as a key encapsulation mechanism. It can be turned into an public key encryption scheme using a standard transformation.

Let $\mathsf{Enc}, \mathsf{Dec}$ be the encryption and decryption algorithms as defined in Sect. 4. In addition to this, our transformation uses a random oracle $\mathcal{H}$ that takes as input λ-bit strings, and outputs a uniformly random string that is long enough to compute a λ-bit string, and three n-bit strings, each chosen uniformly over all strings of Hamming weight h, and such that all four strings are independent. Let $\mathcal{H}_0(k), \mathcal{H}_1(k), \mathcal{H}_2(k), \mathcal{H}_3(k)$ be the four such outputs obtained from the random oracle on input k. As usual, every output is randomly selected whenever a fresh query is asked.

Key Generation. The key generation is identical to the semantically secure cryptosystem and produces $\mathsf{pk} := R, T := F \cdot R + G$, and $\mathsf{sk} := F$ where R is a uniformly random n-bit string, and F, G are chosen uniformly at random from n-bit strings of Hamming weight h.

Key Encapsulation. Given the public key $\mathsf{pk} = (R, T)$, the algorithm Encaps proceeds as follows:

1. Pick a uniformly random λ-bit string K.
2. Let $S = \mathcal{H}_0(K)$.
3. Let $A = \mathcal{H}_1(K)$, $B_1 = \mathcal{H}_2(K)$, and $B_2 = \mathcal{H}_3(K)$.
4. Let $C = (C_1, C_2)$, where $C_1 = A \cdot R + B_1$, and $C_2 = \mathcal{E}(K) \oplus (A \cdot T + B_2)$.
5. Output C, S.

Decapsulation. Given a ciphertext $C = (C_1, C_2)$, and $\mathsf{sk} = F$, the decapsulation algorithm Decaps algorithm proceeds as follows:

1. Compute $K' = \mathcal{D}((F \cdot C_1) \oplus C_2)$.
2. Let $A' = \mathcal{H}_1(K')$, $B_1' = \mathcal{H}_2(K')$, and $B_2 = \mathcal{H}_3(K')$.
3. Let $C' = (C_1', C_2')$, where $C_1' = A' \cdot R + B_1'$, and $C_2' = \mathcal{E}(K') \oplus (A' \cdot T + B_2')$.

4. If $C = C'$, output $\mathcal{H}_0(K')$, else output $\perp$.

A proof of the CCA security of our transformation is nearly identical to that of [HHK17]. We include the proof below for completeness.

Theorem 2. *Assume that $\mathcal{H}$ is a random oracle and that the scheme from Sect. 4 is semantically secure. Then the above mentioned key encapsulation mechanism is secure against chosen-ciphertext attacks.*

Proof. We need to show that chosen-ciphertext queries are not helping the adversary, i.e. that they can be simulated without significantly degrading the adversary's advantage. Once this is done, the semantic security suffices to conclude our result.

For this, we consider the behavior of the decapsulation oracle when receiving a ciphertext $C^\star = (C_1^\star, C_2^\star)$. We want to conclude, that unless the ciphertext was produced by a procedure functionally equivalent to the encapsulation specification, the decapsulation oracle outputs $\perp$ with overwhelming probability.

The decapsulation oracle, on input $C^\star = (C_1^\star, C_2^\star)$ computes $K^\star = \mathcal{D}((F \cdot C_1^\star) \oplus C_2^\star)$, and then calls the encapsulation algorithm with input $\widetilde{K}$ to obtain $\widetilde{C} = (\widetilde{C}_1, \widetilde{C}_2)$. If $\widetilde{C} = C^\star$, then the oracle outputs $\mathcal{H}_0(\widetilde{K})$, and the oracle outputs $\perp$ otherwise.

If the random oracle was previously queried with the seed $\widetilde{K}$ by the adversary, then since the encapsulation procedure is a deterministic function of $\widetilde{K}$, the output of the decapsulation oracle could be efficiently simulated by the adversary. On the other hand, if the random oracle was never queried with the key $\widetilde{K}$, then we have that $\widetilde{C}_1 = A \cdot R + B_1$, where $A = \mathcal{H}_1(\widetilde{K})$ and $B_1 = \mathcal{H}_2(\widetilde{K})$. Since $\mathcal{H}_1, \mathcal{H}_2$ are random oracles, A, B are assumed to be independent of everything else, and hence the probability that the decapsulation oracle does not output $\perp$ is at most

$$\Pr[A \cdot R + B_1 = C_1^\star] = \Pr[B_1 = C_1^\star - A \cdot R] \leq \frac{1}{\binom{n}{h}}.$$

$\square$

8   Instantiating Error Correcting Code in Our Scheme

In this section, we give a concrete choice of parameters, instantiate error correcting codes in our scheme, and analyze the probability of decryption error.

We will set the security parameter to $\lambda = 256$. This is the one of the most acceptable choices in the cryptographic community given the current computational powers.

As we discussed in Sect. 6, the best known efficient attack on our cryptosystem succeeds runs in time $O(2^{2h})$. We assume, somewhat conservatively, that even with future advancements in cryptanalysis of our scheme, the running time cannot be improved beyond $O(2^h)$. Under this assumption, we set h to be the security parameter λ. Thus, $\lambda = h = 256$. Also, in order to prevent against unforeseen attacks that exploit the factorization of p, we choose $p = 2^n - 1$ to be a Mersenne prime.

8.1 Instantiation Based on Deterministic Error-Correction Codes

We will need the following result. We prove this in Sect. 9.

Theorem 3. *Let U be a random variable having uniform distribution on strings of length n. For every n-bit string x of Hamming weight Δ and for every $\varepsilon > 0$,*

$$\Pr[\mathsf{Ham}(U, U + x) \geq 2(1 + \varepsilon)\Delta] \leq 2^{-2\Delta(\varepsilon - \ln(1+\varepsilon))}.$$

We now bound the Hamming distance between $F \cdot (A \cdot R + B_1)$ and $A \cdot (F \cdot R + G) + B_2$. Using Theorem 3, and Lemma 2, we get that for any $\varepsilon \in (0, 1)$,

$$\Pr[\mathsf{Ham}(F \cdot (A \cdot R + B_1), F \cdot (A \cdot R)) \geq 2h^2(1 + \varepsilon)] \leq 2^{-2h^2(\varepsilon - \ln(1+\varepsilon))},$$

and

$$\Pr[\mathsf{Ham}(A \cdot (F \cdot R + G) + B_2, F \cdot (A \cdot R)) \geq 2(h^2 + h)(1 + \varepsilon)] \leq 2^{-(2h^2+h)(\varepsilon - \ln(1+\varepsilon))}.$$

Using union bound, and triangle inequality, we get that

$$\Pr[\mathsf{Ham}(F \cdot (A \cdot R + B_1), A \cdot (F \cdot R + G) + B_2) \geq (4h^2 + 2h)(1 + \varepsilon)]$$

is at most

$$2^{-(2h^2-1)(\varepsilon - \ln(1+\varepsilon))} \leq 2^{-(2h^2-1)(\varepsilon^2/2 - \varepsilon^3/3)} \leq 2^{-(2h^2-1)(\varepsilon^2/6)},$$

where the second to last inequality follows from the Taylor series expansion of $\ln(1 + \varepsilon)$.

This our scheme is $1 - \delta$-correct if the error correction code $(\mathcal{E}, \mathcal{D})$ corrects up to $(4h^2 + 2h)(1 + \varepsilon)$ errors where ε is chosen such that $2^{-(2h^2-1)(\varepsilon^2/6)} < \delta$.

This implies that by choosing an appropriate error-correction code, we get that for any $\delta > 0$, and for $n = ch^2$, for a large enough constant c, our scheme is $1 - \delta$-correct. In particular, we can instantiate our scheme with $n \geq 2^\ell - 1 = 2^{21} - 1 = 32h^2 - 1$, and using Dual-BCH Codes [MS77], we can encode a message of length $k = 256$, such that the parameter where $t = \lceil k/\ell \rceil = 13$, and the scheme corrects up to at least

$$\frac{n}{4} - \frac{(t-1) \cdot 2^{\ell/2}}{2} - 1 \geq 8h^2 - 40h$$

errors. Thus, choosing $\varepsilon = \frac{8h^2 - 40h}{4h^2 + 2h} - 1$ gives an instantiation of our scheme with decryption error as low as $2^{-h^2/4}$.

Notice that the bound on the Hamming weight of $F \cdot B_1$, $A \cdot G$, and also the bound on the Hamming distance in Theorem 3 is not tight, and perhaps it will be difficult to prove much tighter bounds. Moreover, the error distribution is randomized, and exploiting this fact could perhaps lead to better error correction as we discuss in the next section.

8.2   Instantiation Based on Repetition Codes

In the previous section we considered dual-BCH codes which correct a certain fraction of errors no matter how these errors are distributed. On the other hand we observe that for our particular application, the error is "quite" random, and even though this distribution is difficult to mathematically analyze, it is reasonable to conjecture that the error pattern is somewhat similar to the model where each bit is flipped with probability $q < \frac{(4h^2 + 2h)(1+\varepsilon)}{n}$. As we stated in the previous section, the bounds on the Hamming weight of $F \cdot B_1$ and $A \cdot G$, and also the bound in Theorem 3 are not tight, which means q will likely be sufficiently smaller than $\frac{4h^2 + 2h}{n}$.

Thus if we choose $n > 10h^2$, and we encode each bit b of the message $m \in \{0,1\}^k$ using a repetition code of length ρ (where $k \cdot \rho < n$) as $bb \cdots b \in \{0^\rho, 1^\rho\}$, then we expect the number of bits flipped to be smaller than $\rho/2$ with very high probability. Thus, we could decrypt correctly by looking at blocks of length ρ, and decode 1 if the number of 1s in this block of length ρ is more than $\rho/2$, and is 0, otherwise.

We analyzed the error probability when we choose $n = 756839$, $k = 256$, and $\rho = 2048$. At the present time, we are unable to provide a tight rigorous analysis of the decryption error probability. In order to give a satisfactory bound, we would need either to enlarge the parameters (as discussed in the previous section) of the scheme again or to replace the very simple repetition encoding that we are using by a more complex one. One very simple option would be to combine the repetition encoding with a random permutation of the bits of C_2 which are used to mask the encoded value at encryption time. This random permutation could be built from C_1 using the XOF provided by NIST. However, this would make the cryptosystem too slow and add an extra layer of complexity that is really undesirable.

Thus, we propose a heuristic analysis of the decryption error probability. This analysis is based on the distribution of the Hamming weights that are encountered in the decryption blocks corresponding to a single bit. Since, with our choice of parameters every bit is encoded into $\rho = 2048$ bits, we want to see how often a bit might cross the Hamming weight 1024 boundary. It is easy to equip the code and count the Hamming weights encountered during decryption. We performed experiments involving 10000 of each key generation, encapsulation and decapsulation in order to collect the distribution shown in Fig. 1. We see that the distribution looks like a superposition of two Gaussian distributions one corresponding to encryptions of a 0 and one to encryptions of a 1. Our heuristic assumption is that the probability of decryption failure is very close to the one corresponding to these Gaussian distributions. More precisely, we fitted a Gaussian G_0 corresponding to zeroes by searching for best fitting values of p and σ in:

$$G_0(x) = \frac{1}{2\sigma\sqrt{2\pi}}e^{-\frac{(x-p)^2}{2\sigma^2}}.$$

Note the extra $1/2$ compared to a usual normal distribution. This is due to the fact that half of the encrypted bits are zeroes and half are ones. By symmetry, the Gaussian distribution corresponding to ones is simply $G_1(x) = G_0(\rho - x)$. We found that taking $p = 499.6$ and $\sigma = 28.64$ yields the very good approximation shown on Fig. 2 where the two Gaussian are superposed with the measured data.

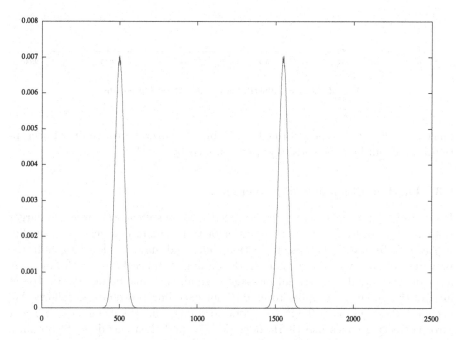

Fig. 1. Density distribution of Hamming Weights during decryption

As a consequence, the probability that a single bit crosses the 1024 boundary is approximated by:

$$0.5\operatorname{erfc}(\frac{1024-p}{\sigma\sqrt{2}}) < 2^{-247}.$$

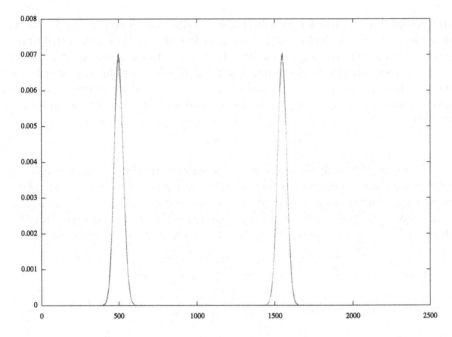

Fig. 2. Density distribution with fitted Gaussians

Since the encrypted value is formed of 256-bits, the overall probability of decryption failure can be heuristically upper bounded by 2^{-239}.

8.3   Further Efficiency Improvements

If we need to use repetition codes, we need n to be sufficiently large, say larger than $10h^2$, in order for the decryption error to be small. If we choose a smaller n (say $n \approx 4h^2$), and then use repetition codes and majority decoding as in the previous section, we expect that the Hamming distance between the message after decoding and the original message is small (but maybe non-zero). To get around this issue, we propose to modify our encoding procedure as follows. We encode a message $m \in \{0,1\}^k$ to a codeword $c_1 \in \{0,1\}^{n_1}$ using some efficient error correcting codes like BCH codes [MS77], and then encode c_1 to obtain a codeword $c_2 \in \{0,1\}^{n_2}$ with $n_2 \leq n$, using repetition codes, then the errors remaining after majority decoding can be corrected by decoding the modified BCH code. Notice that the choice of a smaller n does not alter the security of the scheme, since the security of the scheme depends on the parameter h.

Again we cannot rigorously analyse the decryption error probability, but we can obtain a heuristic analysis similar to the one in the previous section. Concretely, we obtain the following parameters.

Encoding $k = 256$ *bits, with* $n = 216091$. We can choose the next smaller Mersenne prime $p = 2^{216091} - 1$. In this case, we use a BCH code that encodes k-bit messages to $n_1 = 2^9 - 1 = 511$ bit messages. Each of these n_1 bits is encoded using a repetition code which repeats each bit 422 times. We again performed an experiment with 10000 key generation, encapsulation, and decapsulation and observed that the distribution of the Hamming weight for each 422 bits looks like a superposition of two Gaussian distributions one corresponding to encryptions of a 0 and one to encryptions of a 1. In particular, the Gaussian-like distribution corresponding to encryption of 1 has mean $\mu = 234.65$, and variance $\sigma^2 = 132.47$, and hence the probability of decoding a bit incorrectly under the heuristic assumption is

$$0.5\,\mathrm{erfc}(\frac{234.65 - 211}{\sigma\sqrt{2}}) < 0.02.$$

The BCH code corrects up to $\lfloor \frac{511-256}{9} \rfloor = 28$ errors [MS77]. Thus, assuming that the Hamming weight of each block of 422 bits is distributed independently, the probability that there is a decapsulation error is at most

$$\sum_{i=29}^{511} \binom{511}{i} \cdot 0.02^i \cdot 0.98^{511-i},$$

which can be estimated to be at most 2^{-25}.

Encoding $k = 256$ *bits, with* $n = 86243$, *and* $h = 128$. We cannot choose $n = 86243$ if $h = 256$, since n must be significantly larger than h^2 for the scheme to work. However, if we are willing to relax the security requirement to 128-bit security, then we can choose a much smaller Mersenne prime, and the scheme is extremely efficient. In particular, we can choose the Mersenne prime $p = 2^{86243} - 1$. Again, the BCH code encodes k-bit messages to $n_1 = 2^9 - 1 = 511$ bit messages. Each of these n_1 bits is encoded using a repetition code which repeats each bit 168 times. We again performed experiment with 10000 key generation, encapsulation, and decapsulation and again observed that the distribution of the Hamming weight for each 168 bits looks like a superposition of two Gaussian distributions one corresponding to encryptions of a 0 and one to encryptions of a 1. The mean and variance of this distribution are $\mu = 104.55$, and $\sigma^2 = 68.91$, respectively, and hence the probability of decoding a bit incorrectly under the heuristic assumption is

$$0.5\,\mathrm{erfc}(\frac{234.65 - 211}{\sigma\sqrt{2}}) < 0.005.$$

Thus, assuming that the Hamming weight of each block of 168 bits is distributed independently, the probability that there is a decapsulation error is at most

$$\sum_{i=29}^{511} \binom{511}{i} \cdot 0.005^i \cdot 0.995^{511-i},$$

which can be estimated to be at most 2^{-60}.

9   Proof of Theorem 3

Let x be an arbitrary n-bit string of Hamming weight Δ, for some positive integer Δ. We can decompose x as $x_1 + \ldots x_\Delta$ where for all $1 \leq i \leq \Delta$, the string x_i has Hamming distance 1 whose single 1 bit is in position j_i, and $j_1 < \ldots < j_\Delta$. Let $U = U^0$ be the random variable which takes an n-bit binary string with uniform distribution. For $1 \leq i \leq \Delta$, we define the random variables $U^i = U^{i-1} + x_i$ and $Y_i = \mathsf{Ham}(U, U^i) - \mathsf{Ham}(U, U^{i-1})$. The main result in this section is an upper bound the tail of the random variable measuring the Hamming distance of U and $U + x$, that is U and U^Δ.

Theorem 4. *Let U be a random variable having uniform distribution on strings of length n. For every n-bit string x of Hamming weight Δ and for every $\varepsilon > 0$,*

$$\Pr[\mathsf{Ham}(U, U + x) \geq 2(1 + \varepsilon)\Delta] \leq 2^{-2\Delta(\varepsilon - \ln(1+\varepsilon))}.$$

Observe from the Taylor series $\ln(1 + \varepsilon) = \varepsilon - \frac{\varepsilon^2}{2} + \frac{\varepsilon^3}{3} - \ldots$ that, for small ε, we can well approximate the right hand side of the above inequality by $2^{-\Delta \varepsilon^2}$.

Proof. The string U^h is constructed from U in Δ steps, where in every step we add a new string of Hamming weight 1 to the string obtained in the previous steps. Our first lemma bounds the tail of the random variable measuring the increase in the Hamming distance in one step.

Lemma 3. *For every n-bit string x of Hamming weight Δ, and for all integers $s, y_1, \ldots, y_{\Delta-1}$, we have*

$$\Pr[Y_\Delta \geq s | Y_1 = y_1, \ldots, Y_{\Delta-1} = y_{\Delta-1}] \leq \min\{1, 2^{-(s-1)}\}.$$

Proof. Observe that for $s \leq 1$ the statement is trivial, therefore we only consider $s \geq 2$. For every $r \geq s$, and for every Δ, we well determine $\Pr[Y_\Delta = r | Y_1 = y_1, \ldots, Y_{\Delta-1} | Y_1 = y_1, \ldots, Y_{\Delta-1} = y_{\Delta-1}]$, and then we will sum up these values. Let $Z^{\Delta-1}$ denote the event $Y_1 = y_1, \ldots, Y_{\Delta-1} = y_{\Delta-1}$.

Since addition, Hamming distance and the uniform distribution are invariant under rotation, we can suppose without generality that $x_1 = 1$. Under the condition that U and x don't have a 1 in the same position, $Y_\Delta = 1$ with probability 1, and the statement follows. Therefore we can work under the condition that U and x have a common 1, and we can suppose, again without loss of generality, that $U_1 = 1$.

First we consider the case $\Delta = 1$. For $2 \leq r \leq n$, the random variable Y_1 is r when $U_r = 0$ and $U_{r-1} = \ldots = U_2 = 1$. Thus $\Pr[Y_1 = r] = 2^{-r+1}$.

We suppose now that $\Delta \geq 2$. We say that $i \leq \Delta$ is a wrap-around step if $U_n^{i-1} = 1$ and $U_n^i = 0$. Observe that in that case U_1^{i-1} and U_1^i are different. We define the random variable t_Δ as $n + 1$ if i is a wrap-around step for some

$1 \leq i \leq \Delta - 1$. Otherwise, let t_Δ be the smallest integer such that all but the first (from right) t_Δ bits are the same in U and $U^{\Delta-1}$. Since the single 1 bit in $x_{\Delta-1}$ is in position $j_{\Delta-1}$, it follows from the definition that $t_\Delta \geq j_{\Delta-1}$, and that $U^{\Delta-1}_{t_\Delta-1} = \ldots = U^{\Delta-1}_{j_{\Delta-1}} = 0$. In addition, if $t_\Delta \leq n$ then $U^{\Delta-1}_{t_\Delta} = 1$.

Case 1: $j_\Delta \leq t_\Delta - 1$. Since $j_\Delta > j_{\Delta-1}$, we have then $U^{\Delta-1}_{j_\Delta} = 0$. Therefore $U^{\Delta-1}$ and U^Δ differ only in one position, implying Y_Δ is never more than 1.

Case 2: $j_\Delta \geq t_\Delta$. Then $t_\Delta \leq n$ and therefore none of the previous steps was a wrap-around step. Thus $U^{\Delta-1}_1 = \ldots = U^1_1 = U_{t_\Delta} = 0$ and $U^{\Delta-1}_{t_\Delta} = 1$.

Step Δ is a wrap-around step when $U^{\Delta-1}_{j_\Delta} = \ldots = U^{\Delta-1}_n = 1$. When $j_\Delta > t_\Delta$, this is equivalent to $U_{j_\Delta} = \ldots = U_n = 1$, and happens with probability $2^{-(n-j_\Delta+1)}$. In that case $U^{\Delta-1}$ and U^Δ differ in positions $1, j_\Delta \ldots, n$. Among these positions U^Δ and U differ at $j_\Delta, \ldots, n$ but $U^\Delta_1 = U_1 = 1$, and therefore $Y_\Delta = n - j_\Delta$. When $j_\Delta = t_\Delta$, this is equivalent to $U_{j_\Delta+1} = \ldots = U_n = 1$, and happens with probability $2^{-(n-j_\Delta)}$. In that case $U^{\Delta-1}$ and U^Δ differ in positions $1, j_\Delta \ldots, n$. Among these positions U^Δ and U differ at $j_\Delta + 1, \ldots, n$ but $U^\Delta_1 = U_1 = 1$, and therefore $Y_\Delta = n - j_\Delta - 1$.

Step Δ is not a wrap-around step when $U^{\Delta-1}_\ell = 0$ and $U^{\Delta-1}_{\ell-1} = \ldots = U^{\Delta-1}_{j_\Delta} = 1$, for some $j_\Delta \leq \ell \leq n$. This never happens when $j_\Delta = n$. When $t_\Delta < j_\Delta < n$, this is equivalent to $U_\ell = 0$ and $U^{\Delta-1}_{j_\Delta} = \ldots = U^{\Delta-1}_{\ell-1} = 1$, which happens with probability $2^{-(\ell-j_\Delta+1)}$. In that case $U^{\Delta-1}$ and U^Δ differ in positions $j_\Delta, \ldots, \ell$, where $U^{\Delta-1}$ coincides with U and U^Δ differs from U, implying $Y_\Delta = \ell - j_\Delta + 1$. When $j_\Delta = t_\Delta$, ℓ must be at least $j_\Delta + 1$, and the condition is equivalent to $U_\ell = 0$ and $U^{\Delta-1}_{j_\Delta+1} = \ldots = U^{\Delta-1}_{\ell-1} = 1$, which happens with probability $2^{-(\ell-j_\Delta)}$. In that case $U^{\Delta-1}$ and U^Δ differ in positions $j_\Delta+1, \ldots, \ell$, where $U^{\Delta-1}$ coincides with U and U^Δ differs from U, implying $Y_\Delta = \ell - j_\Delta$.

All together, $\Pr[Y_\Delta = r | Z^{\Delta-1}] \neq 0$ in the following cases. When $j_\Delta > t_\Delta$, we have $\Pr[Y_\Delta = r | Z^{\Delta-1}] = 2^{-r}$ for $r \in \{2, \ldots n - j_\Delta - 1, n - j_\Delta + 1\}$ and $\Pr[Y_\Delta = n - j_\Delta] = 2^{-(n-j_\Delta+1)} + 2^{-(n-j_\Delta)}$. When $j_\Delta = t_\Delta$, we have $\Pr[Y_\Delta = r | Z^{\Delta-1}] = 2^{-r}$ for $r \in \{2, \ldots n-j_\Delta-2, n-j_\Delta\}$ and $\Pr[Y_\Delta = n-j_\Delta-1 | Z^{\Delta-1}] = 2^{-(n-j_\Delta)} + 2^{-(n-j_\Delta-1)}$. The statement follows by summing up these probabilities for $r \geq s$.   □

Observe that $\mathsf{Ham}(U, U + x) = Y_1 + \ldots + Y_\Delta$. In order to bound the tail of $Y_1 + \ldots + Y_\Delta$, we introduce the independent random variables $X_1, \ldots X_\Delta$, where X_i is a geometric random variable with success probability $\frac{1}{2}$, for each $1 \leq i \leq \Delta$. This means that by definition, for every positive integer r, we have $\Pr[X_i = r] = 2^{-r}$. The definition immediately implies that for every integer s, we also have $\Pr[X_i \geq s] = \min\{1, 2^{-(s-1)}\}$. Our next lemma states that the tail of $Y_1 + \ldots + Y_\Delta$ can be upper bounded by the tail of $X_1 + \ldots + X_\Delta$.

Lemma 4. *For every n-bit string x of Hamming weight Δ, for every integer s,*

$$\Pr[Y_1 + \ldots + Y_\Delta \geq s] \leq \Pr[X_1 + \ldots + X_\Delta \geq s].$$

Proof. We prove it by induction on Δ. When $\Delta = 1$, from Lemma 3 we have

$$\Pr[Y_1 \geq s] \leq \min\{1, 2^{-(s-1)}\} = \Pr[X_1 \geq s].$$

When $\Delta \geq 2$, we have the following series of (in)equalities:

$$\Pr[\sum_{i=1}^{\Delta} Y_i \geq s] \leq \sum_{y_1,\ldots,y_{\Delta-1}} \Pr[Y_1 = y_1, \ldots, Y_{\Delta-1} = y_{\Delta-1}] \Pr[X_\Delta \geq s - \sum_{i=1}^{\Delta-1} y_i]$$

$$= \Pr[\sum_{i=1}^{\Delta-1} Y_i + X_\Delta \geq s]$$

$$= \sum_{y} \Pr[\sum_{i=1}^{\Delta-1} Y_i \geq y] \Pr[X_\Delta = s - y]$$

$$\leq \sum_{y} \Pr[\sum_{i=1}^{\Delta-1} X_i \geq y] \Pr[X_\Delta = s - y]$$

$$= \Pr[\sum_{i=1}^{\Delta} X_i \geq s].$$

The first inequality follows from Lemma 3 and the second inequality from the inductive hypothesis. For the third equality we have used that X_Δ is independent from the random variables Y_i. □

Our final lemma is a special case of Theorem 2.3 in the artice [Jan17] on tail bounds for sums of geometric and exponential variables.

Lemma 5 *[Jan17]. Let $X_1, \ldots X_\Delta$ be independent geometric random variables with success probability $\frac{1}{2}$, and let $\varepsilon > 0$. Then*

$$\Pr[\sum_{i=1}^{\Delta} X_i \geq 2(1+\varepsilon)\Delta] \leq 2^{-2\Delta(\varepsilon - \ln(1+\varepsilon))}.$$

Putting together Lemmas 4 and 5, we immediately obtain our bound on the Hamming distance of U and U_Δ, which concludes the proof. □

10   Conclusion

In this paper, we propose a simple new public-key encryption scheme. As with other public-key cryptosystems, the security of our cryptosystem relies on unproven assumptions mentioned in Definition 5. In Sect. 6.1, we summarized the known cryptanalytic attacks against this scheme. The proposed cryptosystem is based on a relatively new assumption, and it will require more cryptanalytic effort before one can be reasonably confident about the security assumption.

Acknowledgments. This research was partially funded by the Singapore Ministry of Education and the National Research Foundation, also through the Tier 3 Grant "Random numbers from quantum processes", MOE2012-T3-1-009. This work has been supported in part by the European Union's H2020 Programme under grant agreement number ERC-669891 and the French ANR Blanc program under contract ANR-12-BS02-005 (RDAM project). The second author is grateful to CQT where the work has started during his visit.

References

[AJPS17] Aggarwal, D., Joux, A., Prakash, A., Santha, M.: A new public-key cryptosystem via mersenne numbers. Cryptology ePrint Archive, Report 2017/481, version:20170530.072202 (2017)

[BCGN17] Beunardeau, M., Connolly, A., Géraud, R., Naccache, D.: On the hardness of the Mersenne Low Hamming Ratio assumption. Technical report, Cryptology ePrint Archive, 2017/522 (2017)

[CHK10] Cramer, R., Hofheinz, D., Kiltz, E.: A twist on the Naor-Yung paradigm and its application to efficient CCA-secure encryption from hard search problems. In: Micciancio, D. (ed.) TCC 2010. LNCS, vol. 5978, pp. 146–164. Springer, Heidelberg (2010). https://doi.org/10.1007/978-3-642-11799-2_10

[dBDJdW17] de Boer, K., Ducas, L., Jeffery, S., de Wolf, R.: Attacks on the AJPS mersenne-based cryptosystem. Technical report, Cryptology ePrint Archive, Report 2017/1171 (2017). https://eprint.iacr.org/2017/1171

[DH76] Diffie, W., Hellman, M.: New directions in cryptography. IEEE Trans. Inf. Theory **22**(6), 644–654 (1976)

[ES16] Eldar, L., Shor, P.W.: An efficient quantum algorithm for a variant of the closest lattice-vector problem. arXiv preprint arXiv:1611.06999 (2016)

[Gro96] Grover, L.K.: A fast quantum mechanical algorithm for database search. In: Proceedings of the Twenty-Eighth Annual ACM Symposium on Theory of Computing, pp. 212–219 (1996)

[GWO+13] Lize, G., Wang, L., Ota, K., Dong, M., Cao, Z., Yang, Y.: New public key cryptosystems based on non-abelian factorization problems. Secur. Commun. Netw. **6**(7), 912–922 (2013)

[HHK17] Hofheinz, D., Hövelmanns, K., Kiltz, E.: A modular analysis of the Fujisaki-Okamoto transformation. In: Kalai, Y., Reyzin, L. (eds.) TCC 2017. LNCS, vol. 10677, pp. 341–371. Springer, Cham (2017). https://doi.org/10.1007/978-3-319-70500-2_12

[HPS98] Hoffstein, J., Pipher, J., Silverman, J.H.: NTRU: a ring-based public key cryptosystem. In: Buhler, J.P. (ed.) ANTS 1998. LNCS, vol. 1423, pp. 267–288. Springer, Heidelberg (1998). https://doi.org/10.1007/BFb0054868

[Jan17] Janson, S.: Tail bounds for sums of geometric and exponential variables. arXiv preprint arXiv:1709.08157 (2017)

[KLC+00] Ko, K.H., Lee, S.J., Cheon, J.H., Han, J.W., Kang, J., Park, C.: New public-key cryptosystem using braid groups. In: Bellare, M. (ed.) CRYPTO 2000. LNCS, vol. 1880, pp. 166–183. Springer, Heidelberg (2000). https://doi.org/10.1007/3-540-44598-6_10

[LPR10]   Lyubashevsky, V., Peikert, C., Regev, O.: On ideal lattices and learning with errors over rings. In: Gilbert, H. (ed.) EUROCRYPT 2010. LNCS, vol. 6110, pp. 1–23. Springer, Heidelberg (2010). https://doi.org/10.1007/978-3-642-13190-5_1

[LvTMW09] Lempken, W., van Tran, T., Magliveras, S.S., Wei, W.: A public key cryptosystem based on non-abelian finite groups. J. Cryptol. **22**(2), 62–74 (2009)

[McE78]   McEliece, R.J.: A public-key cryptosystem based on algebraic coding theory. Coding Thv **4244**, 114–116 (1978)

[MS77]   MacWilliams, F.J., Sloane, N.J.A.: The Theory of Error-Correcting Codes. Elsevier, New York (1977)

[MTSB13]   Misoczki, R., Tillich, J.-P., Sendrier, N., Barreto, P.S.: MDPC-McEliece: new McEliece variants from moderate density parity-check codes. In: 2013 IEEE International Symposium on Information Theory Proceedings (ISIT), pp. 2069–2073. IEEE (2013)

[NIS17]   NIST. Post quantum crypto project (2017). http://csrc.nist.gov/groups/ST/post-quantum-crypto/. Accessed 19 May 2017

[NS97]   Naccache, D., Stern, J.: A new public-key cryptosystem. In: Fumy, W. (ed.) EUROCRYPT 1997. LNCS, vol. 1233, pp. 27–36. Springer, Heidelberg (1997). https://doi.org/10.1007/3-540-69053-0_3

[NY90]   Naor, M., Yung, M.: Public-key cryptosystems provably secure against chosen ciphertext attacks. In: Proceedings of the Twenty-Second Annual ACM Symposium on Theory of Computing, STOC 1990, pp. 427–437. ACM, New York (1990)

[OP01]   Okamoto, T., Pointcheval, D.: REACT: rapid enhanced-security asymmetric cryptosystem transform. In: Naccache, D. (ed.) CT-RSA 2001. LNCS, vol. 2020, pp. 159–174. Springer, Heidelberg (2000). https://doi.org/10.1007/3-540-45353-9_13

[Reg09]   Regev, O.: On lattices, learning with errors, random linear codes, and cryptography. J. ACM **56**(6), 34, 40 (2009)

[RSA78]   Rivest, R.L., Shamir, A., Adleman, L.: A method for obtaining digital signatures and public-key cryptosystems. Commun. ACM **21**(2), 120–126 (1978)

[RSN+01]   Rukhin, A., Soto, J., Nechvatal, J., Smid, M., Barker, E.: A statistical test suite for random and pseudorandom number generators for cryptographic applications. Technical report, DTIC Document (2001)

[Sho97]   Shor, P.W.: Polynomial-time algorithms for prime factorization and discrete logarithms on a quantum computer. SIAM J. Comput. **26**(5), 1484–1509 (1997)

[Sho02]   Shoup, V.: OAEP reconsidered. J. Cryptol. **15**(4), 223–249 (2002)

Fast Homomorphic Evaluation of Deep Discretized Neural Networks

Florian Bourse[1], Michele Minelli[2,3(✉)],
Matthias Minihold[4], and Pascal Paillier[5]

[1] Orange Labs, Applied Crypto Group, Cesson-Sévigné, France
[2] DIENS, École normale supérieure, CNRS, PSL Research University, Paris, France
michele.minelli@ens.fr
[3] Inria, Paris, France
[4] Horst Görtz Institut für IT-Security, Ruhr-Universität Bochum, Bochum, Germany
[5] CryptoExperts, Paris, France

Abstract. The rise of machine learning as a service multiplies scenarios where one faces a privacy dilemma: either sensitive user data must be revealed to the entity that evaluates the cognitive model (e.g., in the Cloud), or the model itself must be revealed to the user so that the evaluation can take place locally. Fully Homomorphic Encryption (FHE) offers an elegant way to reconcile these conflicting interests in the Cloud-based scenario and also preserve non-interactivity. However, due to the inefficiency of existing FHE schemes, most applications prefer to use Somewhat Homomorphic Encryption (SHE), where the complexity of the computation to be performed has to be known in advance, and the efficiency of the scheme depends on this global complexity.

In this paper, we present a new framework for homomorphic evaluation of neural networks, that we call FHE–DiNN, whose complexity is strictly linear in the depth of the network and whose parameters can be set beforehand. To obtain this scale-invariance property, we rely heavily on the bootstrapping procedure. We refine the recent FHE construction by Chillotti *et al.* (ASIACRYPT 2016) in order to increase the message space and apply the sign function (that we use to activate the neurons in the network) during the bootstrapping. We derive some empirical results, using TFHE library as a starting point, and classify encrypted images from the MNIST dataset with more than 96% accuracy in less than 1.7 s.

Finally, as a side contribution, we analyze and introduce some variations to the bootstrapping technique of Chillotti *et al.* that offer an improvement in efficiency at the cost of increasing the storage requirements.

Keywords: Fully homomorphic encryption · Neural networks
Bootstrapping · MNIST

H. Shacham and A. Boldyreva (Eds.): CRYPTO 2018, LNCS 10993, pp. 483–512, 2018.
https://doi.org/10.1007/978-3-319-96878-0_17

1 Introduction

Fully Homomorphic Encryption (FHE). An FHE scheme provides a way to encrypt data while supporting computations through the encryption envelope. Given an encryption of a plaintext x, one can compute an encryption of $f(x)$ for any computable function f. This operation does not require intermediate decryption or knowledge of the decryption key and therefore can be performed based on public information only. Applications of FHE are numerous but one particular use of interest is the privacy-preserving delegation of computations to a remote service. The first construction of FHE dates back to 2009 and is due to Gentry [Gen09]. A number of improvements have followed [vDGHV10, SS10, SV10, BV11a, BV11b, BGV12, GHS12, GSW13, BV14], leading to a biodiversity of techniques, features and complexity assumptions.

All known FHE schemes are obtained by first building a leveled Somewhat Homomorphic Encryption (SHE) scheme, which can evaluate circuits of a-priori bounded depth (usually, only the multiplicative depth is considered, because the noise growth introduced by additions is negligible compared to that introduced by multiplications). In order to obtain unbounded computation capabilities on encrypted values, an FHE scheme can be built from an SHE scheme with a technique called *bootstrapping*, which intuitively means using the homomorphic properties of the scheme to decrypt and then re-encrypt, refreshing the ciphertext to enable further computation. However, this process is very costly. Hence, there have been numerous works on trying to obtain more efficient bootstrappings [AP13, AP14, DM15, CGGI16b, CGGI17], and on trying to minimize the number of bootstrappings required for evaluating a circuit [LP13, PV16, BLMZ17]. Another approach is to simply avoid bootstrapping altogether and use an SHE scheme, adjusting the parameters to be able to carry out the desired computation.

In practice, there are now two main freely available libraries for fully homomorphic encryption. The first one, HElib [HS14, HS15], which implements the BGV scheme [BGV12], is the most widely used in applications. It allows for packing of ciphertexts and SIMD computations, amortizing the cost for certain tasks. It is able to perform additions and multiplications in an efficient way, but the bootstrapping operation is significantly slow. In practice, it is often used as a somewhat homomorphic scheme. The second one, TFHE [CGGI16a], features a very efficient bootstrapping operation but, as a downside, this has to be applied after every gate computation. This library is more efficient than HElib when used for realizing an FHE. However, for simple tasks requiring small computational depth, HElib used as an SHE will perform better. Moreover, TFHE is currently not capable of amortizing large SIMD computations as well as HElib does.

The quest for privacy-preserving machine learning. Machine Learning As a Service (MLAS) is becoming popular because of its versatility. These applications typically have high computation and data-storage requirements, which make them less suitable as client-side technologies. Moreover, since the process of training a cognitive model is time and resource-consuming, the trained prediction algorithm is often considered critical intellectual property by its owner, who is typically not willing to share its technology or proprietary tools, resulting in that machine learning algorithms are most conveniently cloud-based.

However, this setting raises new issues concerning the privacy of the uploaded input data. Users want to send their encrypted data to a cloud service that offers privacy-preserving predictions, and fulfills this task using its powerful yet undisclosed, state-of-the-art predictive models. In this paper, we put forward a new and versatile FHE framework that makes it efficient for the cloud to operate a neural network dedicated to some specific machine learning task. The network, previously trained on plaintext dataset, does not have access to the input data in the clear, but is only given user-provided encrypted inputs and returns encrypted predictions.

Obviously, encrypting the user's data ensures its confidentiality, since the private key under which the data is encrypted is assumed never to leave the owner's controlled domain. In this setting, only the legitimate owner of the secret key can decrypt the result returned by the delegated computation that has been homomorphically performed in the cloud. The cloud service only learns superficial information, but can still charge the user for using the service.

Neural networks (NNs) are often built from medical, financial or otherwise sensitive data. They are usually trained to solve a *classification* problem: all possible observations are categorized into classes and, given a training dataset of observation/class pairs, the network should be able to assign the correct class to new observations. Such framework can be easily applied to problems like establishing a diagnosis from medical observations.

In this work we do not consider the problem of privacy-preserving datamining, intended as training a neural network over encrypted data, which can be addressed, e.g., with the approach of [AS00]. Instead, we assume that the neural network is trained with data in the clear and we focus on the evaluation part.

Another potential concern for the service provider is that users might be sending malicious requests in order to either learn what is considered a company secret (the neural network itself), or specific sensitive information encoded in the weights (which could be a breach into the privacy of the training dataset). In this latter case, a statistical database can be used in the training phase, as is discussed in the differential privacy literature [Dwo06].

Prior works. Cryptonets [DGBL+16] was the first initiative to address the challenge of achieving blind, non-interactive classification. The main idea consists in applying a leveled SHE scheme such as BGV [BGV12] to the network inputs and propagating the signals across the network homomorphically, thereby consuming levels of homomorphic evaluation whenever non-linearities are met.

In NNs, non-linearities come from activation functions which are usually picked from a small set of non-linear functions of reference (logistic sigmoid, hyperbolic tangent, . . .) chosen for their mathematical convenience. To optimally accommodate the underlying SHE scheme, Cryptonets replace their standard activation by the (depth 1) square function, which only consumes one level but does not resemble the typical sigmoidal shape. A number of subsequent works have followed the same approach and improved it, typically by adopting higher degree polynomials as activation functions for more training stability [ZYC16], or by renormalizing weighted sums prior to applying the approximate function, so that its degree can be kept as low as possible [CdWM+17]. Practical experiments have shown that training can accommodate approximated activations and generate NNs with very good accuracy.

However, this approach suffers from an inherent limitation: the homomorphic computation, local to a single neuron, depends on the total number of levels required to implement the network, which is itself roughly proportional to the number of its activated layers. Therefore, the overall performance of the homomorphic classification heavily depends on the total multiplicative depth of the circuit and rapidly becomes prohibitive as the number of layers increases. This approach does not scale well and is not adapted to deep learning, where neural networks can contain tens, hundreds or sometimes thousands of layers [HZRS15, ZK16].

Finally, we note that other approaches based on multiparty computation (MPC) have been proposed, e.g., [BPTG15, MZ17, MRSV17], but they require interactivity between the party that holds the data and the party that performs the blind classification. Even though practical performances of MPC-based solutions have been impressive compared to FHE-based solutions, they incur other issues like network latency and high bandwidth usage. Because of these downsides, FHE-based solutions seem more scalable for real-life applications. In this work, we focus on a non-interactive, blind evaluation, and we rely on FHE.

Our contributions. We adopt a scale-invariant approach to the problem. In our framework, called FHE–DiNN, each neuron's output is refreshed through bootstrapping, resulting in that arbitrarily deep networks can be homomorphically evaluated. Of course, the entire homomorphic evaluation of the network will take time proportional to the number of its neurons or, if parallelism is involved, to the number of its layers. Evaluating one neuron is now essentially independent of the dimensions of the network: it just relies on system-wide parameters.

In FHE–DiNN, unlike in standard neural networks, the weights and biases, as well as the domain and range of the activation function cannot be real-valued and must be discretized. We call such networks *Discretized Neural Networks* or DiNNs. This particular form of neural networks is somehow inspired by a more restrictive one, referred to in the literature as *Binarized Neural Networks* (BNNs) [CB16] where signals and weights are restricted to the set $\{-1, 1\}$ instead of $\mathbb{Z}$

as in the case of DiNNs (so BNNs are a special case of DiNNs). Interestingly, it has been empirically observed by [CB16] that BNNs can achieve accuracies close to the ones obtained with state-of-the-art classical NNs, at the price of an overhead in the total network size, which is largely compensated by the obtained performance gains. For the sake of scale-invariance, we decided to choose as activation function the sign, so the signal which is propagated has values in $\{-1, 1\}$, and cannot grow out of control. So the evaluation of DiNNs boils down to repeatedly computing the sign of a weighted sum of ± 1 inputs.

In order to perform this classification on encrypted data, we adapt the recent construction by Chillotti *et al.*, known as TFHE [CGGI16b] to support sign and weighted sum as the two basic operations of the scheme, the sign being computed during a bootstrapping procedure in order to refresh the ciphertext.

As a side contribution, we also present a few techniques to optimize the usage of TFHE in applications: how to reduce the required bandwidth, how to reduce the overall noises in the ciphertexts, and a slightly faster alternative to the bootstrapping procedure that also produces ciphertexts with less noise, at the expense of a bigger bootstrapping key.

Finally, we conducted experiments on the MNIST dataset [LBBH98]. We used the library keras [C+15] to train two simple neural networks with one hidden layer containing 30 (respectively, 100) neurons and we converted them into DiNNs by simply discretizing the weights and using the sign as activation function. Of course, this introduced a loss in accuracy, and although much better accuracies could certainly be obtained through various optimizations or by directly training a DiNN (rather than converting a canonical neural network), this was not the goal of this work. Our aim was conducting experiments to measure the accuracy of the homomorphic classification and comparing it to that in the clear. We found that, for a security level of 80 bits, our implementation takes about 0.49 s (respectively, 1.65 s) seconds per classification (with no underlying parallelism whatsoever) and achieves 93.71% (respectively, 96.35%) accuracy when evaluated homomorphically.

Comparison with cryptonets [DGBL+16]. In Cryptonets, propagated signals are reals properly encoded into compatible plaintexts and a single encrypted input (i.e., an image pixel) takes $2 \cdot 382 \cdot 8192$ bits ($=766$ kB). Therefore, an entire image takes $28 \cdot 28 \cdot 766$ kB ≈ 586 MB. However, with the same storage requirements, Cryptonets can batch 8192 images together, so that the amortized size of an encrypted image is reduced to 73.3 kB. In the case of FHE–DiNN, we are able to exploit the batching technique *on a single image*, resulting in that each encrypted image takes ≈ 8.2 kB. In the case of Cryptonets, the complete homomorphic evaluation of the network takes 570 s, whereas in our case it takes 0.49 s (or 1.6 s in the case of a slightly larger network). However, it should be noted that (a) the networks that we use for our experiments are considerably smaller than that used in Cryptonets, so we also compare the time-per-neuron and, in this case, our solution is faster by roughly a factor 36; moreover (b) once again Cryptonets support image batching, so 8192 images can be classified in 570 s, resulting in only 0.07 s per image. Cryptonets' ability to batch images together can be useful in some applications where the same user wants to classify a large number of samples together. In the simplest case where the user only wants a single image to be classified, this feature does not help.

Regarding classification accuracy, the NN used by Cryptonets achieves 98.95% of correctly classified samples, when evaluated on the MNIST dataset. In our case, a loss of accuracy occurs due to the preliminary simplification of the MNIST images, and especially because of the discretization of the network. We stress however that our prime goal was not accuracy but to achieve a qualitatively better homomorphic evaluation at the neuron level.

Finally, we also achieve scale-invariance, meaning that we can keep on computing over the encrypted outputs of our network, whereas Cryptonets are bounded by the initial choice of parameters. In Table 1 we present a detailed comparison with Cryptonets.

Table 1. Comparison with Cryptonets and its amortized version (denoted by Cryptonets*). FHE–DiNN30 and FHE–DiNN100 refer to neural networks with one hidden layer composed of 30 and 100 neurons, respectively.

	Neurons	Size of ct.	Accuracy	Time enc	Time eval	Time dec
Cryptonets	945	586 MB	98.95%	122 s	570 s	5 s
Cryptonets*	945	73.3 kB	98.95%	0.015 s	0.07 s	0.0006 s
FHE–DiNN30	30	≈8.2 kB	93.71%	0.000168 s	0.49 s	0.0000106 s
FHE–DiNN100	100	≈8.2 kB	96.35%	0.000168 s	1.65 s	0.0000106 s

Outline of the paper. The paper is organized as follows: in Sect. 2 we define our notation and we introduce notions about fully homomorphic encryption and artificial neural networks; in Sect. 3 we present our Discretized Neural Networks and show a simple technique to build these models; in Sect. 4 we explain how to homomorphically evaluate a DiNN and present our main result; in Sect. 5 we present some technical refinements that allow us to improve the efficiency of the evaluation and that can be useful also for other FHE-based solutions; finally, in Sect. 6 we give experimental results on data in the clear and on encrypted inputs, draw some conclusions and identify several open problems.

2   Preliminaries

In this section we clarify our notation and recall some definitions and constructions that are going to be useful in the rest of the paper.

2.1 Notation

We denote the real numbers by $\mathbb{R}$, the integers by $\mathbb{Z}$ and use $\mathbb{T}$ to indicate $\mathbb{R}/\mathbb{Z}$, i.e., the torus of real numbers modulo 1. We use $\mathbb{B}$ to denote the set $\{0,1\}$, and we use $\mathcal{R}[X]$ for polynomials in the variable X with coefficients in $\mathcal{R}$, for any ring $\mathcal{R}$. We use $\mathbb{R}_N[X]$ to denote $\mathbb{R}[X]/\left(X^N+1\right)$ and $\mathbb{Z}_N[X]$ to denote $\mathbb{Z}[X]/\left(X^N+1\right)$ and we write their quotient as $\mathbb{T}_N[X] = \mathbb{R}_N[X]/\mathbb{Z}_N[X]$, i.e., the ring of polynomials in X quotiented by $\left(X^N+1\right)$, with real coefficients modulo 1. Vectors are denoted by lower-case bold letters, and we use $\|\cdot\|_1$ and $\|\cdot\|_2$ to denote the L_1 and the L_2 norm of a vector, respectively. Given a vector $\mathbf{a}$, we denote its i-th entry by a_i. We use $\langle \mathbf{a}, \mathbf{b} \rangle$ to denote the inner product between vectors $\mathbf{a}$ and $\mathbf{b}$.

Given a set A, we write $a \xleftarrow{\$} A$ to indicate that a is sampled uniformly at random from A. If $\mathcal{D}$ is a probability distribution, we will write $d \leftarrow \mathcal{D}$ to denote that d is sampled according to $\mathcal{D}$.

2.2 Fully Homomorphic Encryption over the Torus

Learning with errors. The Learning with Errors (LWE) problem was introduced by Regev in [Reg05]. Let n be a positive integer and χ be a probability distribution over $\mathbb{R}$ for the noise. For any vector $\mathbf{s} \in \{0,1\}^n$, we define the LWE distribution $\mathsf{lwe}_{n,\mathbf{s},\chi}$ as $(\mathbf{a}, b)$, where $\mathbf{a} \xleftarrow{\$} \mathbb{T}^n$ and $b = \langle \mathbf{s}, \mathbf{a} \rangle + e \in \mathbb{T}$, with $e \leftarrow \chi$.

Then the LWE assumption states that, for $\mathbf{s} \xleftarrow{\$} \{0,1\}^n$, it is hard to distinguish between $(\mathbf{a}, b)$ and $(\mathbf{u}, v)$, for $(\mathbf{a}, b) \leftarrow \mathsf{lwe}_{n,\mathbf{s},\chi}$ and $(\mathbf{u}, v) \xleftarrow{\$} \mathbb{T}^{n+1}$.

Sub-Gaussians. Let $\sigma > 0$ be a real Gaussian parameter. We define the Gaussian function with parameter σ as $\rho_\sigma(x) = \exp\left(-\pi |x|^2 / \sigma^2\right)$ for any $x \in \mathbb{R}$. Then we say that a distribution $\mathcal{D}$ is sub-Gaussian with parameter σ if there exists $M > 0$ such that for all $x \in \mathbb{R}$,

$$\mathcal{D}(x) \le M \cdot \rho_\sigma(x).$$

Lemma 2.1 (Pythagorean additivity of sub-Gaussians). *Let $\mathcal{D}_1$ and $\mathcal{D}_2$ be sub-Gaussian distributions with parameters σ_1 and σ_2, respectively. Then $\mathcal{D}^+$, obtained by sampling $\mathcal{D}_1$ and $\mathcal{D}_2$ and summing the results, is a sub-Gaussian with parameter $\sqrt{\sigma_1^2 + \sigma_2^2}$.*

LWE-based private-key encryption scheme. We recall the Regev encryption scheme from [Reg05]. Let $\mu \in \{0,1\}$ be a message and λ the security parameter; we encrypt and decrypt as follows:

Setup (λ): for a security parameter λ, fix $n = n(\lambda)$ and return $\mathbf{s} \xleftarrow{\$} \{0,1\}^n$
Enc $(\mathbf{s}, \mu)$: return $(\mathbf{a}, b)$, with $\mathbf{a} \xleftarrow{\$} \mathbb{T}^n$ and $b = \langle \mathbf{s}, \mathbf{a} \rangle + e + \frac{\mu}{2}$, where $e \leftarrow \chi$
Dec $(\mathbf{s}, (\mathbf{a}, b))$: return $\lfloor 2(b - \langle \mathbf{s}, \mathbf{a} \rangle) \rceil$

We usually refer to e as the *noise* of the ciphertext, and say that a ciphertext is a valid encryption of μ if it decrypts to μ with overwhelming probability.

We now give some notions on the formulation of FHE over the torus and the bootstrapping procedure. The following part is based on [CGGI16b].

TLWE. TLWE is a generalization of LWE and Ring-LWE [LPR10]. Let $k \geq 1$ be an integer, N be a power of 2 and χ be an error distribution over $\mathbb{R}_N[X]$. A TLWE secret key $\bar{s} \in \mathbb{B}_N[X]^k$ is a vector of k polynomials over $\mathbb{Z}_N[X]$ with binary coefficients. Given a message encoded as a polynomial $\mu \in \mathbb{T}_N[X]$, a fresh TLWE encryption of μ under the key $\bar{s}$ is a sample $(\mathbf{a}, b) \in \mathbb{T}_N[X]^k \times \mathbb{T}_N[X]$, with $\mathbf{a} \xleftarrow{\$} \mathbb{T}_N[X]^k$ and $b = \bar{s} \cdot \mathbf{a} + \mu + e$, where $e \leftarrow \chi$.

From a TLWE encryption $\bar{c}$ of a polynomial $\mu \in \mathbb{T}_N[X]$ under a TLWE key $\bar{s}$ we can extract a LWE encryption $c' = \mathsf{Extract}(\bar{c})$ of the constant term of μ under an extracted key $\mathbf{s}' = \mathsf{ExtractKey}(\bar{s})$. For the details of the algorithms Extract and ExtractKey, we refer the reader to [CGGI16b, Definition 4.1].

TGSW. TGSW is a generalized version of the GSW FHE scheme [GSW13]. The key concept here is that TGSW can be seen as the matrix equivalent of TLWE, just like GSW can be seen as the matrix equivalent of LWE. More details can be found in [CGGI16b].

As in previous works, our average-case noise analysis relies on the following heuristic. This assumption matches empirical results [DM15, CGGI16b]. Note that the worst-case bounds do not require this heuristic.

Assumption 1. *We assume that all the error coefficients of TLWE or TGSW samples of the linear combinations we consider are independent and concentrated. In particular, we assume that they are sub-Gaussian where σ is the square-root of their variance.*

Overview of the bootstrapping procedure. The core idea for the efficiency of the new bootstrapping procedure is the so-called external product $\boxdot$, that performs the following mapping

$$\boxdot : \mathsf{TGSW} \times \mathsf{TLWE} \rightarrow \mathsf{TLWE}.$$

Roughly speaking, the external product of a TGSW encryption of a polynomial $\mu_1 \in \mathbb{T}_N[X]$ and a TLWE encryption of a polynomial $\mu_2 \in \mathbb{T}_N[X]$ is a TLWE encryption of $(\mu_1 \cdot \mu_2) \in \mathbb{T}_N[X]$.

Now the bootstrapping procedure of an n-LWE sample (here, n denotes the dimension) consists of the 3 following functions:

BlindRotate: $\text{TGSW}^n \times \text{TLWE} \times n\text{-LWE} \to \text{TLWE}$
On input TGSW encryptions of $(s_i)_{i \in [n]}$, a (possibly noiseless) TLWE encryption of testVector and an n-LWE sample $(\mathbf{a}, b)$, computes a TLWE encryption of $X^\phi \cdot$ testVector, where $\phi = b - \langle \mathbf{s}, \mathbf{a} \rangle$;

Extract: $\text{TLWE} \to N\text{-LWE}$
On input a TLWE encryption of polynomial $\mu \in \mathbb{T}_N[X]$, computes an N-LWE encryption of the constant term $\mu(0)$;

KeySwitch: $n\text{-LWE}^N \times N\text{-LWE} \to n\text{-LWE}$
On input n-LWE encryptions of $(s_i')_{i \in [N]}$, and an N-LWE sample $(\mathbf{a}, b)$ computes an n-LWE encryption of $b - \langle \mathbf{s}', \mathbf{a} \rangle$.

Then we can define a function $\text{Bootstrap}\,(\cdot, \cdot, \cdot)$ that takes as input a bootstrapping key bk, a keyswitching key ksk, and a ciphertext and outputs a new ciphertext. Roughly speaking,

$$\text{Bootstrap} = \text{KeySwitch} \circ \text{Extract} \circ \text{BlindRotate}.$$

We note that BlindRotate works on LWE samples with values in $[2N]$ instead of $\mathbb{T}$, thus the first step is to map $\mathbb{T}$ to $[2N]$ by multiplying and rounding.

When studying the noise distribution during this operation, and to measure the impact of our changes on this procedure, we note that there are actually two different relevant noises: the *overhead noise* which is added to the input ciphertext before its virtual decryption and the *output noise*, which is the one in the final output ciphertext.

2.3 Artificial Neural Networks

An artificial neural network is a computing system inspired by biological brains. Here, we consider a neural network (NN) that is composed of a population of artificial neurons arranged in layers. Each neuron of a dense layer accepts n_I real-valued inputs $\mathbf{x} = (x_1, \dots, x_{n_I})$ and performs the following two computations:

1. It computes a real value $y = \sum_{i=1}^{n_I} w_i x_i + \beta$, which is a weighted sum of the inputs with real values called *weights*: w_i is the weight associated to the input x_i, and β, also real-valued, is referred to as the *bias* of the neuron.
2. It applies a non-linear function f, the *activation* function, and returns $f(y)$.

The neuron's output can be written as $f(\langle \mathbf{w}, \mathbf{x} \rangle) = f(\sum_{i=0}^{n_I} w_i x_i)$ if one extends the inputs and the neuron's weights vector by setting $\mathbf{w} = (\beta, w_1, \dots, w_{n_I})$ and $\mathbf{x} = (1, x_1, \dots, x_{n_I})$. The neurons of a neural network are organized in successive layers, which are categorized according to their activation function. Neurons of one layer are connected to the neurons of the next layer by paths that are associated to weights. An input layer composed of the network's inputs as well as an output layer made of the network's output values are also added to the network. Internal layers are called *hidden*, since they are not directly accessible from the external world.

NNs are usually composed of layers of various types: fully connected (every neuron of the layer takes all incoming signals as inputs), convolutional (it applies a convolution to its input), pooling, and so forth. Neural networks could in principle be recurrent systems, as opposed to the purely feed-forward ones, where each neuron is only evaluated once. The *universal approximation theorem* (see, e.g., [Hor91, Cyb89]) states that a neural network with a single hidden layer that contains a finite amount of neurons, can approximate any continuous function. Despite this, the number of neurons in that layer can grow exponentially. Instead, a deep neural network has several layers of non-linearities, which allow to extract increasingly complex features of the input and can lead to a better ability to generalize, especially in the case of more complex tasks.

The FHE–DiNN framework presented in this work is able to evaluate NNs of arbitrary depth, comprising possibly many hidden layers.

2.4   The MNIST Dataset

The MNIST database (Modified National Institute of Standards and Technology database) is a dataset of images representing digits handwritten by more than 500 different writers, and is commonly used as a benchmark for machine learning systems [LBBH98]. The MNIST database contains 60 000 training images and 10 000 testing images. The format of the images is 28×28 and the value of each pixel represents a level of gray. Moreover, each image is labeled with the digit it depicts.

A typical neural network for the MNIST dataset has $28 \cdot 28 = 784$ input nodes (one per pixel), an arbitrary number of hidden layers with an arbitrary number of neurons per layer, and finally 10 output nodes (one per possible digit). The output values can be interpreted as "scores" given by the NN: the classification is then given by the digit that achieves the highest score.

Over the years, the MNIST dataset has been a typical benchmark for classifiers, and many approaches have been applied: linear classifiers, principal component analysis, support vector machines, neural networks, convolutional neural networks, etc. For a more complete review on these approaches, we refer the reader to, e.g., [LBBH98]. Neural networks are known to perform well on this dataset. For example, [LBBH98] proposes different architectures for neural networks and obtains more than 97% of correct classifications. More recent works even surpassed 99% of accuracy [CMS12]. For a nice overview on the results obtained on this dataset and on the techniques that were used, we refer the reader to [LCB98].

3   Discretized Neural Networks (DiNN)

In this section we formally define DiNNs and we explain how they differ from a traditional neural network and how to simply convert a NN into a DiNN.

3.1   Definition of a Discretized Neural Network

First of all, we recall that state-of-the-art fully homomorphic encryption schemes cannot support operations over real messages. Traditional neural networks have real-valued weights, and this incompatibility motivates investigating alternative architectures.

Definition 3.1. *A Discretized Neural Network (DiNN) is a feed-forward artificial neural network whose inputs are integer values in $\{-I, \ldots, I\}$ and whose weights are integer values in $\{-W, \ldots, W\}$, for some $I, W \in \mathbb{N}$. For every neuron of the network, the activation function maps the inner product between the incoming inputs vector and the corresponding weights to integer values in $\{-I, \ldots, I\}$.*

In particular, for this paper we chose $\{-1, 1\}$ as the input space and $\mathsf{sign}\,(\cdot)$ as the activation function for the hidden layers:

$$\mathsf{sign}\,(x) = \begin{cases} -1, & x < 0, \\ +1, & x \geq 0. \end{cases} \tag{3.1}$$

These choices are inspired by the fact that we designed the model with the idea of performing homomorphic evaluations over encrypted input. As a consequence, we wanted the message space to be as small as possible, which, in turn, would allow us to increase the efficiency of the overall evaluation.

We also note that using an activation function whose output is in the same range as the network's input allows us to maintain the same semantics across different layers. In our case, what enters a neuron is always a weighted sum of values in $\{-1, 1\}$. In order to make the evaluation of the network compatible with FHE schemes, discretizing the input space is not sufficient: we also need to have discrete values for the weights of the network[1].

3.2   Simple Conversion from a Traditional Neural Network to a DiNN

In this subsection we show a very simple method to convert an already-trained canonical neural network (i.e., with real weights) into a DiNN. This method is not guaranteed to be the best way to obtain such a conversion; it indeed intro-

---

[1] As all the computations are done over the torus (i.e., modulo 1), scaling a ciphertext by any integer factor preserves the relations that make the decryption correct. However, this does not hold for non-integer factors.

duces a visible loss in the classification accuracy and would probably be best used as a first step in the conversion procedure. However, we remind the reader that this work is aimed at the homomorphic evaluation of a network, thus we decided not to put too much effort in the construction of a sophisticated cleartext model. This procedure allows us to obtain a network which respects our constraints and that can be evaluated over encrypted inputs, so it is sufficient for our purposes.

It turns out that the only thing that we need to do is discretizing the weights and biases of the network. To this purpose, we define the function

$$\mathsf{processWeight}\,(w, \tau) = \tau \cdot \left\lceil \frac{w}{\tau} \right\rceil \qquad (3.2)$$

where $\tau \in \mathbb{N}$ is a parameter that controls the precision of the discretization. In the following, we implicitly take all the weights as discretized after being processed through the formula in Eq. 3.2. After fixing a value τ, the network obtained by applying $\mathsf{processWeight}\,(\cdot, \tau)$ to all the weights and biases is a DiNN. The parameter τ has to be chosen carefully, since it defines the message space that our encryption scheme must support. Thus, we want the bound on $\langle \mathbf{w}, \mathbf{x} \rangle$ to be small for all neurons, where $\mathbf{w}$ and $\mathbf{x}$ are the discretized weights and the inputs associated to the neuron, respectively. In Fig. 1, we show the evaluation of a single neuron: we first compute $\langle \mathbf{w}, \mathbf{x} \rangle$, which we refer to as a *multisum*, and then apply the sign function to the result.

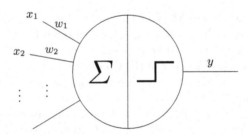

Fig. 1. Evaluation of a single neuron. The output value is $y = \mathsf{sign}\,(\langle \mathbf{w}, \mathbf{x} \rangle)$, where w_i are the discretized weights associated to the incoming wires and x_i are the corresponding input values.

4   Homomorphic Evaluation of a DiNN

We now give a high level description of our procedure to homomorphically evaluate a DiNN, called FHE–DiNN. We basically need two ingredients: we need to be able to compute the multisum between the encrypted inputs and the weights and we need to homomorphically extract the sign of the result. In order to maintain the scalability of our scheme across the layers of a given DiNN, we perform a bootstrapping operation for every neuron in hidden layers. This ensures that the ciphertext encrypting the sign of the result after applying one layer of the DiNN can be used for further computations without an initially fixed limit on the number of layers that the network can contain. Hence we can choose parameters that are independent of the number of layers and evaluate arbitrarily deep neural networks.

4.1   Evaluating the Multisum

In our framework, the weights of the network are available in clear, so we can evaluate the multisum just by using homomorphic additions. The only things that need our attention are the message space of our encryption scheme, which has to be large enough to accommodate for all possible values of the multisums, and the noise level that might grow too much and lead to incorrect results.

Extending the message space. In order for our FHE scheme to be able to correctly evaluate the multisum, we need all the possible values of the multisum to be inside our message space. To this end, we extend our LWE encryption scheme as follows. This idea was already used in previous works such as [PW08, KTX08, ABDP15, ALS16].

Construction 1 (Extended LWE-based private-key encryption scheme). *Let B be a positive integer and let $m \in [-B, B]$ be a message. Then we split the torus into $2B + 1$ slices, one for each possible message, and we encrypt and decrypt as follows:*

Setup (λ): *for a security parameter λ, fix $n = n(\lambda)$, $\sigma = \sigma(\lambda)$; return $\mathbf{s} \xleftarrow{\$} \mathbb{T}^n$*
Enc $(\mathbf{s}, m)$: *return $(\mathbf{a}, b)$, with $\mathbf{a} \xleftarrow{\$} \mathbb{T}^n$ and $b = \langle \mathbf{s}, \mathbf{a} \rangle + e + \frac{m}{2B+1}$, where $e \leftarrow \chi_\sigma$*
Dec $(\mathbf{s}, (\mathbf{a}, b))$: *return $\lfloor (b - \langle \mathbf{s}, \mathbf{a} \rangle) \cdot (2B + 1) \rceil$*

An input message is mapped to the center of its corresponding torus slice by scaling it by $1/(2B + 1)$ during encryption, and decoded by scaling it by $2B + 1$ during decryption.

Correctness of homomorphically evaluating the multisum. Note that ciphertexts can be homomorphically added and scaled by a known integer constant: for any two messages $m_1, m_2 \in [-B, B]$, any secret key $\mathbf{s}$, any $c_1 = (\mathbf{a}_1, b_1) \leftarrow \mathsf{Enc}(\mathbf{s}, m_1)$, $c_2 = (\mathbf{a}_2, b_2) \leftarrow \mathsf{Enc}(\mathbf{s}, m_2)$, and constant $w \in \mathbb{Z}$, we have that

$$\mathsf{Dec}(\mathbf{s}, c_1 + w \cdot c_2) = \mathsf{Dec}(\mathbf{s}, (\mathbf{a}_1 + w \cdot \mathbf{a}_2, b_1 + w \cdot b_2)) = m_1 + w \cdot m_2$$

as long as (1) $m_1 + w \cdot m_2 \in [-B, B]$, and (2) the noise did not grow too much.
The first condition is easily met by choosing $B \geq \|\mathbf{w}\|_1$ for all weight vectors $\mathbf{w}$ in the network (e.g., we can take the max).

Fixing the noise. Increasing the message space has an impact on the choice of parameters. Evaluating the multisum with a given weight vector $\mathbf{w}$ means that, if the standard deviation of the initial noise is σ, then the standard deviation of the output noise can be as high as $\|\mathbf{w}\|_2 \cdot \sigma$ (see Lemma 2.1), which in turn means that our initial standard deviation must be smaller than the one in [CGGI16b] by a factor $\max_{\mathbf{w}} \|\mathbf{w}\|_2$. Moreover, for correctness to hold, we need the noise to remain smaller than half a slice of the torus. As we are splitting the torus into $2B + 1$ slices rather than 2, we need to further decrease the noise by a factor B. Special attention must be paid to security: taking a smaller noise might in fact compromise the security of the scheme. In order to mitigate this problem, we can increase the dimension of the LWE problem n, but this in turn induces more noise overhead in the bootstrapping procedure due to rounding errors.

4.2   Homomorphic Computation of the Sign Function

We take advantage of the flexibility of the bootstrapping technique introduced by Chillotti *et al.* [CGGI16b] in order to perform the sign extraction and the bootstrapping at the same time. Concretely, in the call to BlindRotate, we change the value of testVector to

$$\frac{-1}{2B+1} \sum_{i=0}^{N-1} X^i.$$

Then, if the value of the phase $b - \langle \mathbf{s}, \mathbf{a} \rangle$ is between 1 and N (positive), the output will be an encryption of 1, otherwise if it is between $N + 1$ and $2N$ (negative), the output will be an encryption of -1.

In order to give more intuition, we present an illustration of the bootstrapping technique in Fig. 2. The first step of the bootstrapping basically consists in mapping the torus $\mathbb{T}$ to an object that we will refer to as the wheel. This wheel is split into $2N$ "ticks" that are associated to the possible values that are encrypted in the bootstrapped ciphertext. The bootstrapping procedure then consists in choosing a value for each tick, rotating the wheel by $b - \langle \mathbf{s}, \mathbf{a} \rangle$ ticks counter-clockwise, and picking the value of the rightmost tick. We note that the values on the wheel are encoded in the testVector variable, which contains values for the ticks on the top part of the wheel. The bottom values are then fixed by the anticyclic property of $\mathbb{T}_N[X]$ (the value at tick $N + i$ is minus the value at tick i).

From now on, we say that a bootstrapping is *correct* if, given a valid encryption of a message μ, its output is a valid encryption of $\mathsf{sign}\,(\mu)$ with overwhelming probability.

4.3   Scale-Invariance

If the parameters are set correctly then, by using the two operations described above, we can homomorphically evaluate neural networks of any depth. In particular, the choice of parameters is independent of the depth of the neural network.

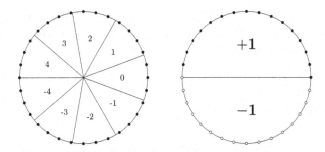

Fig. 2. On the left, we show the first step of the bootstrapping, which consists in mapping the torus (the continuous circle) to the wheel (the $2N$ ticks on it) by rounding to the closest tick. Each slice corresponds to one of the possible results of the multisum operation. On the right we show the final result of the bootstrapping: each tick of the top part of the wheel is mapped to its sign which is $+1$ and each tick of the bottom part to -1. This can roughly be seen as embedding the wheel back to the torus.

This result cannot be achieved with previous techniques relying on somewhat homomorphic evaluations of the network. In fact, they have to choose parameters that accommodate for the whole computation, whereas our method only requires the parameters to accommodate for the evaluation of a single neuron. The rest of the computation follows by induction. More precisely, our choice of parameters only depends on bounds on the norms ($\|\cdot\|_1$ and $\|\cdot\|_2$) of the input weights of a neuron. In the following, we denote these bounds by M_1 and M_2, respectively.

We say that the homomorphic evaluation of the neural network is *correct* if the decryptions of its output scores are equal to the scores given by its evaluation in the clear with overwhelming probability. Then, the scale-invariance is formally defined by the following theorem:

Theorem 4.1 (Scale-invariance of our homomorphic evaluation). *For any DiNN of any depth, any correctly generated bootstrapping key* bk *and keyswitching key* ksk, *and any ciphertext c, let σ be a Gaussian parameter such that the noise of* Bootstrap (bk, ksk, c) *is sub-Gaussian with parameter σ. Then, if the bootstrapping is correct on input ciphertexts with sub-Gaussian noise of parameter $\frac{\sigma}{M_2}$ and message space larger than $2M_1 + 1$, the result of the homomorphic evaluation of the DiNN is correct.*

Proof. The proof is a simple induction on the structure of the neural network. First, the correctness of the evaluation of the first layer is implied by the choice of parameters for the encryption[2].

---

[2] If it is not, we can bootstrap all input ciphertexts in order to ensure this holds.

If the evaluation is correct for all neurons of the ℓ-th layer, then the correctness for all neurons of the $(\ell+1)$-th layer follows from the two observations made in the previous subsections:

- The result of the homomorphic evaluation of the multisum is a valid encryption of the multisum;
- The result of the bootstrapping is a valid encryption of the sign of the multisum.

The first fact is implied by the choice of the message space, since the multisum value is contained in $[-M_1, M_1]$. The second one comes directly from the correctness of the bootstrapping, because the homomorphic computation of the multisum on ciphertexts with sub-Gaussian noise of parameter σ yields a ciphertext with sub-Gaussian noise of parameter at most σM_2 (cf. Lemma 2.1).

Then, the correctness of the encryption scheme ensures that the final ciphertexts are valid encryptions of the scores.   □

5   Refinements of TFHE

In this section, we present several improvements that helped us achieving better efficiency for the actual FHE–DiNN implementation. These various techniques can without any doubt be applied in other FHE-based applications.

5.1   Reducing Bandwidth Usage

One of the drawbacks of our evaluation process is that encrypting individual values for each input neuron yields a very large ciphertext, which is inconvenient from a user perspective, as a high bandwidth requirement is the direct consequence. In order to mitigate this issue, we "pack" multiple values into one ciphertext. We use the standard technique of encrypting a polynomial (using the TLWE scheme instead of LWE) whose coefficients correspond to the different values we want to encrypt:

$$ct = \text{TLWE.Encrypt}\left(\sum_i x_i X^i\right),$$

where the x_i's represent the values of the input neurons to be encrypted[3]. This packing technique is what made Ring-LWE an attractive variant to the standard LWE problem, as was already presented in [LPR10], and is widely used in FHE applications to amortize the cost of operations [HS14, HS15].

---

[3] If the number of input neurons is bigger than the maximal degree of the polynomials N, we can pack the ciphertext by groups of N, compute partial multisums with our technique, and aggregate them afterwards.

Then, we observe that for each neuron in the first hidden layer, we can compute the multisum with coefficients w_i by scaling the input TLWE ciphertext by a factor

$$\sum_i w_i X^{-i}.$$

Indeed, it is easy to verify that the constant term of $\left(\sum_i x_i X^i\right) \cdot \left(\sum_i w_i X^{-i}\right)$ is $\sum_i w_i x_i$, and we can obtain an LWE encryption of this value by invoking Extract.

Remark 1. We note that this computation is actually equivalent to doing the multisum directly on LWE ciphertexts, so the resulting noise growth of this approach is *exactly* the same as before. We end up saving bandwidth usage (by a factor up to N, the degree of the polynomials) basically for free. Furthermore, as the weights of the neural network never change, we can precompute and store the FFT representation of the polynomials $\sum w_i X^{-i}$, thus saving time during the online classification.

In a nutshell, we reduce the size of the ciphertexts for N elements from N LWE ciphertexts to 1 TLWE ciphertext. In terms of numbers of elements in $\mathbb{T}$, the cost dropped from $N(n+1)$ to $N(k+1)$.

We remark that the resulting ciphertext is an LWE ciphertext in dimension N, and not the original n, thus requiring key-switching to become a legitimate ciphertext. However, this is not a problem thanks to the trick presented in the following subsection.

5.2 Moving KeySwitch Around

The main goal of key-switching here is to reduce the LWE dimension. The benefits in memory usage and efficiency of this reduction are extremely important, since the size of the bootstrapping key, the final noise level, and the number of external products (the most costly operation) all depend linearly on this parameter. However, we noticed that reducing this dimension in the beginning of the bootstrapping procedure instead of the end gave much better results, hence the new bootstrapping function:

$$\text{Bootstrap} = \text{Extract} \circ \text{BlindRotate} \circ \text{KeySwitch}.$$

The intuition is that, with this technique, the noise produced by KeySwitch will not be multiplied by $\|\mathbf{w}\|_2$ when performing the computation of the multisum, but will only be added at the end. Basically, we moved the noise of the output ciphertext produced by KeySwitch to an overhead noise.

Doing this, we reverse the usage of the two underlying LWE schemes: everything is now done on high dimensional N-LWE, whereas the low dimensional n-LWE scheme is only used during the bootstrapping operation. Since the noise in the key-switching key is not used for any computation anymore, we can allow it to be bigger, thus reducing the dimension we need for the same security to hold and, in turn, gaining in time per bootstrapping.

The only downside is that working with higher dimensional N-LWE samples means slightly more memory usage for the server, bigger output ciphertext[4], and slightly slower addition of ciphertexts. However, as this operation is instantaneous when compared to other operations such as bootstrapping, this is not an issue.

5.3 Dynamically Changing the Message Space

In Sect. 4, we showed how to evaluate the whole neural network by induction, using a message space of $2B + 1$ slices, where B is a bound on the values of the multisums across the whole evaluation. However, in order to be able to reduce the probability of errors along the way, we are able to use different message spaces for each layer of the DiNN, and adapt the number of slots to the values given by the local computations, depending on the values of the weights $\mathbf{w}$. In order to do so, we change the value of testVector to

$$\frac{-1}{2B_\ell + 1} \sum_{i=0}^{N-1} X^i,$$

where B_ℓ is now indexed by the current layer ℓ, and is a bound on the values of the multisums for the next layer $\ell + 1$. The point of this manoeuvre is that if the number of slots is smaller, the slices are bigger, and the noise would have to be bigger in order to change the plaintext message. This trick might seem superfluous, because it decreases a probability that is already negligible. However sometimes, in practical scenarios, the correctness of the scheme is relaxed, and this trick allows us to obtain results closer to the expected values without costing any extra computation or storage.

5.4 Alternative BlindRotate Implementations

Following the technique of [ZYL+17], we try to gain efficiency in the bootstrapping by reducing the number of external products that we have to compute. In order to do so, they slightly unfold the loop computing $X^{\langle \mathbf{s},\mathbf{a}\rangle}$ in the BlindRotate algorithm. They group the terms of the sum two by two, using the following formula for each of the new terms:

$$X^{as+a's'} = ss'X^{a+a'} + s(1 - s')X^a + (1 - s)s'X^{a'} + (1 - s)(1 - s').$$

In order to compute this new function, they change the bootstrapping key to contain encryptions of the values $ss', s(1 - s'), (1 - s)s'$, and $(1 - s)(1 - s')$,

---

[4] This can be circumvented by applying one last round of KeySwitch at the end of the protocol, if needed.

Algorithm 1. Alternative BlindRotate algorithm.
***
Input: an n-LWE ciphertext $(\mathbf{a}, b)$ with coefficients in $\mathbb{Z}_{2N}$, a (possibly noiseless) TLWE encryption $\mathbf{C}$ of testVector, the bootstrapping key bk such that for all i in $[n/2]$, $\mathsf{bk}_{3i}, \mathsf{bk}_{3i+1}$, and bk_{3i+2} are respectively TGSW encryptions of $s_{2i}s_{2i+1}, s_{2i}(1 - s_{2i+1})$, and $s_{2i+1}(1 - s_{2i})$
Output: a TLWE encryption of $X^{b-\langle \mathbf{s}, \mathbf{a}\rangle} \cdot$ testVector
1: $ACC \leftarrow X^b \cdot \mathbf{C}$
2: **for** $i = 1 \ldots n/2$ **do**
3:    $ACC \leftarrow ((X^{a_{2i}+a_{2i+1}} - 1)\mathsf{bk}_{3i} + (X^{a_{2i}} - 1)\mathsf{bk}_{3i+1} + (X^{a_{2i+1}} - 1)\mathsf{bk}_{3i+2}) \boxdot ACC$
4: **end for**
5: **return** ACC
***

thus expanding the size of the bootstrapping key by a factor 2. Using this idea, they cut the number of iterations of the loop by half, thus computing only half the amount of external products, which is the most costly operation of the bootstrapping. However, by doing so, they introduce the computation of 4 scalings of TGSW ciphertexts (which are matrices) by constant polynomials, and 3 TGSW additions, when TFHE's BlindRotate only needed 1 scaling of a TLWE ciphertext, and 1 TLWE addition. Another benefit is that the homomorphic computation of $\langle \mathbf{s}, \mathbf{a}\rangle$ induces rounding errors on only $n/2$ terms instead of n. The noise of the output ciphertext is also different. On the bright side, the technique of [ZYL+17] reduces the noise induced by the precision errors during the gadget decomposition by a factor 2. On the other hand, it increases the noise coming from the bootstrapping key by a factor 2.

In this work, we suggest to use another formula in order to compute each term of the slightly unfolded sum. Observing that $ss' + s(1 - s') + (1 - s)s' + (1 - s)(1 - s') = 1$, we can save 1 element in the bootstrapping key:

$$X^{as+a's'} = ss'(X^{a+a'} - 1) + s(1 - s')(X^a - 1) + (1 - s)s'(X^{a'} - 1) + 1.$$

The resulting BlindRotate algorithm is described in Algorithm 1. Having a 1 in the decomposition is a valuable advantage, because it means that we can move it out of the external product and instead add the previous value of the accumulator to the result. Thus, efficiency-wise, we halved the number of external products at the cost of only 3 scalings of TGSW ciphertexts by constant polynomials, 2 TGSW additions, and 1 TLWE addition. We note that while multiplying naively by a monomial might be faster than multiplying by a degree 2 polynomial, the implementation pre-computes and stores the FFT representation of the bootstrapping keys in order to speed up polynomial multiplication. Thus, multiplying by a polynomial of any degree has the same cost. The size of the bootstrapping key is now $3/2$ times larger than the size of the one in TFHE, which is a compromise between the two previous methods. As in [ZYL+17], the noise induced by precision errors and roundings is halved compared to TFHE. On the other hand, now we increase the noise coming from the bootstrapping

Table 2. Comparison of the three alternative BlindRotate algorithms. n denotes the LWE dimension after keyswitching; δ refers to the noise introduced by rounding the LWE samples into $[2N]$ before we can BlindRotate; N is the degree of the polynomials in the TLWE scheme; k is the dimension of the TLWE ciphertexts; ε is the precision $(1/2\beta)^\ell/2$ of the gadget matrix (tensor product between the identity Id_{k+1} and the powers of $1/2\beta$ arranged as ℓ-dimensional vector $(1/2\beta, \ldots, (1/2\beta)^\ell)$); σ_{bk} is the standard deviation of the noise of the TGSW encryptions in the bootstrapping key, and $\mathcal{A}_{bk}$ is a bound on this noise. These values were derived using the theorems for noise analysis in [CGGI17]

		TFHE	ZYLZD17	FHE–DiNN
Efficiency	External products	n	$n/2$	$n/2$
	Scaled TGSW add.	0	4	3
	Scaled TLWE add.	1	0	1
	Noise overhead	δ	$\delta/2$	$\delta/2$
Out noise (average)	roundings	$n(1+kN)\varepsilon^2$	$\frac{n}{2}(1+kN)\varepsilon^2$	$\frac{n}{2}(1+kN)\varepsilon^2$
	from BK	$n(k+1)\ell N\beta^2\sigma_{bk}^2$	$2n(k+1)\ell N\beta^2\sigma_{bk}^2$	$3n(k+1)\ell N\beta^2\sigma_{bk}^2$
Out noise (worst)	roundings	$n(1+kN)\varepsilon$	$\frac{n}{2}(1+kN)\varepsilon$	$\frac{n}{2}(1+kN)\varepsilon$
	from BK	$n(k+1)\ell N\beta\mathcal{A}_{bk}$	$2n(k+1)\ell N\beta\mathcal{A}_{bk}$	$3n(k+1)\ell N\beta\mathcal{A}_{bk}$
Storage	TGSW in the BK	n	$2n$	$3n/2$

key by a factor 3 instead. However, we note that it is possible to reduce this noise without impacting efficiency by reducing the noise in the bootstrapping key, trading off security (depending on what the bottleneck for security of the scheme is, this could come for free), whereas in order to reduce the noise induced by the precision errors, efficiency will be impacted. We recapitulate these numbers on Table 2.

We note that this idea could be generalized to unfoldings consisting of more than two terms, yielding more possible trade-offs, but we did not explore further because of the dissuasive exponential growth in the number of operands in the general formula.

6   Experimental Results and Conclusions

We implemented the proposed approach to test its accuracy and efficiency. This section is divided into two main parts: the first one describes the training of the neural network over data in the clear and the second one details the results obtained when evaluating the network over encrypted inputs.

6.1   Pre-processing the MNIST Database

In order to respect the constraint of having inputs in $\{-1, 1\}$, we binarized all the images with a threshold value equal to 128: any pixel whose value is smaller than the threshold is mapped to -1; the others are mapped to $+1$. This actually reduces the amount of information available, as each 8-bit grayscale value is clamped to a single bit, and one could wonder if this could impact the accuracy of the classification. Although this is possible, a quick visual inspection of the result shows that the digits depicted in the images are still clearly recognizable.

6.2   Building a DiNN from Data in the Clear

In order to train the neural network, we first chose its topology, i.e., the number of hidden layers and neurons per hidden layer. We experimented with several values, always keeping in mind that a smaller number of neurons per layer is preferable: having more neurons means that the value of the multisum will be potentially higher, thus requiring a larger message space in the homomorphic evaluation, which in turn forces to choose bigger parameters for the scheme. After some tries, we decided to show the feasibility of our approach through the homomorphic evaluation of two neural networks. Both have 784 neurons in the input layer (one per pixel), a single hidden layer, and an output layer composed of 10 neurons (one per class). The difference between the two models is the size of the hidden layer: the first network has 30 neurons, while the second has 100.

In order to build a DiNN, we use the simple approach described in Subsect. 3.2: we (1) train a traditional neural network (i.e., with real weights and biases), and then we (2) discretize all the values by applying the function in Eq. 3.2. For step (1) we take advantage of the library keras [C+15] with Tensorflow [AAB+15], which offers a simple and highly customizable framework for defining, training and evaluating even complex models of neural networks. Through a farly simple Python script and in little time, we are able to define and train our models as desired. Given its similarity with (a scaled and shifted version of) the sign function, as an activation function we used the version of hard_sigmoid defined in Tensorflow. The reason behind this choice is that we know we will substitute this activation function with the true $\text{sign}(x)$. Thus, using a function which is already similar to it helps reducing the errors introduced by this switch.

Once we obtain the trained model, we proceed to choose a value $\tau \in \mathbb{N}$ and discretize the weights and the biases of the network, as per Eq. 3.2, thus finally obtaining a DiNN that we can later evaluate over encrypted inputs. The choice of τ is an important part of the process: on one hand, picking a very small value will give little resolution to the network[5], potentially degrading the accuracy largely; on the other hand, picking a very large value will minimize the loss in accuracy but increase the message space that we will need to support for homomorphic evaluation, thus forcing us to choose larger parameters and making the overall evaluation less efficient. Also, note that it is possible to choose different values of the parameter τ for different layers of the network. Although there might be better choices, we did not invest too much efforts in optimizing the cleartext model and simply chose the value $\tau = 10$ for both layers of each model. Finally, we switched all the activation functions from hard_sigmoid $(\cdot)$ to sign $(\cdot)$. In order to assess the results of the training and how the accuracy varies because of these changes, in Table 3 we report the accuracies obtained on the MNIST test set. Note that these values are referred to the evaluation *over cleartext inputs*.

---

[5] This means that the number of values that the weights will be able to take will be fairly limited.

Table 3. Accuracy obtained when evaluating the models in the clear on the MNIST test set. The first value refers to the evaluation of the model as output by the training; the second refers to the model where all the values for weights and biases have been discretized; the third refers to the same model, but with sign $(\cdot)$ as the activation function for all the neurons in the hidden layer.

	Original NN	DiNN + hard_sigmoid	DiNN + sign
30 neurons	94.76%	93.76% (-1%)	93.55% (-1.21%)
100 neurons	96.75%	96.62% (-0.13%)	96.43% (-0.32%)

6.3   Classifying Encrypted Inputs

Implementing the homomorphic evaluation of the neural network over encrypted input was more than a mere coding exercise, but allowed us to discover several interesting properties of our DiNNs.

The starting point was the TFHE library by Chillotti *et al.*, which is freely available on GitHub [CGGI16a] and which was used to efficiently perform the bootstrapping operation. The library takes advantage of FFT processors for fast polynomial multiplication and, although not parallelized, achieves excellent timing results. We extended the code to apply this fast bootstrapping procedure to our use case.

Parameters. We now present our setting of the parameters, following the notation of [CGGI16b], to which we refer the reader for extra details. In Table 4 we highlight the main security parameters regarding our ciphertexts, together with an estimate of the security level that this setting achieves. Other additional parameters, related to the various operations we need to perform, are the following:

Table 4. The security parameters we use for the different kinds of ciphertexts. The estimated security has been extracted from the plot in [CGGI16b] and later verified with the estimator from Albrecht *et al.* [APS15].

Ciphertext	Dimension	α	Estimated security
input	1024	2^{-30}	>150 bits
keyswitching key	450	2^{-17}	>80 bits
bootstrapping key	1024	2^{-36}	>100 bits

- Degree of the polynomials in the ring: $N = 1024$;
- Dimension of the TLWE problem: $k = 1$;
- Basis for the decomposition of TGSW ciphertexts: $Bg = 1024$;
- Length of the decomposition of TGSW ciphertexts: $\ell = 3$;
- Basis for the decomposition during key switching: 8;
- Length of the decomposition during key switching: $t = 5$;

With this choice of parameters, we achieve a minimum security level of 80 bits and a single bootstrapping operation takes roughly 15 ms on a single core of an Intel Core i7-4720HQ CPU @ 2.60 GHz. Also, we note that by exploiting the packing technique presented in Subsect. 5.1, we save a factor 172 in the size of the input ciphertext: instead of having $784 \cdot (450 + 1)$ torus elements (corresponding to a 450-LWE ciphertext for each of the 784 pixels in an image), we now have only $2 \cdot 1024$ torus elements (corresponding to the two polynomials that form a TLWE sample).

Finally, we calculated the maximum value of the norms of the weight vectors associated to each neuron, both for the first and the second layer. These values, which can be computed at setup time (since the weights are available in the clear), define the theoretical bounds on the message space that our scheme should be able to support. In practice, we evaluated the actual values of the multisums on the training set, and took a message space slightly larger[6] than what we computed. We note that with this method, it is possible that some input could make the multisum go out of bounds, but this was not observed when evaluating the network on the test set. Moreover, this allows us to take a considerably smaller message space in some cases, and thus reduce the probability of errors. In Table 5 we report the theoretical message space we would need to support and the message space we actually used for our implementation.

In order to pinpoint our noise parameters, we also calculated the maximum L_2-norms of the weight vectors in each layer: for the network with 30 hidden neurons, we have $\max_{\mathbf{w}} \|\mathbf{w}\|_2 \approx 119$ for the first layer and ≈ 85 for the second layer; for the network with 100 hidden neurons, we have $\max_{\mathbf{w}} \|\mathbf{w}\|_2 \approx 69$ for the first layer and ≈ 60 for the second layer.

Table 5. Message space: theoretically required values and how we set them in our experiments with FHE–DiNN.

	FHE–DiNN30			FHE–DiNN100		
	$\max_{\mathbf{w}} \|\mathbf{w}\|_1$	theor.	exp.	$\max_{\mathbf{w}} \|\mathbf{w}\|_1$	theor.	exp.
1st layer	2338	4676	2500	1372	2744	1800
2nd layer	399	798	800	488	976	1000

Evaluation. Our homomorphic evaluation follows the outline presented in Fig. 3 in order to classify an encrypted image,

1. Encrypt the image as a TLWE ciphertext;
2. Multiply the TLWE ciphertext by the polynomial which encodes the weights associated to the hidden layer. This operation takes advantage of FFT for speeding up the calculations;
3. From each of the so-computed ciphertexts, extract a 1024-LWE ciphertext, which encrypts the constant term of the result;

---

[6] As we do not achieve perfect correctness with our parameters, the message can be shifted. This fact has to be taken into account when choosing the number of slots.

4. Perform a key switching in order to move from a 1024-LWE ciphertext to a 450-LWE one;
5. Bootstrap to decrease the noise level. By setting the testVector, this operation also applies the sign function and changes the message space of our encryption scheme for free.
6. Perform the multisum of the resulting ciphertext and the weights leading to the output layer, through the technique showed in Subsect. 4.1.[7]
7. Return the 10 ciphertexts corresponding to the 10 scores assigned by the neural network. These ciphertext can be decrypted and the argmax can be computed to obtain the classification given by the network.

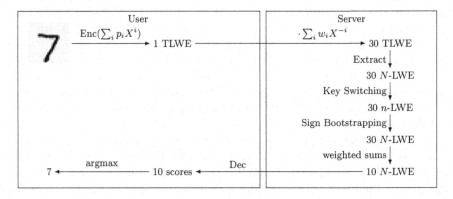

Fig. 3. Refined homomorphic evaluation of a 784:30:10 neural network with activation function sign. The whole image (784 pixels) is packed into 1 TLWE ciphertext to minimize bandwidth usage. After evaluation, the user recovers 10 ciphertexts corresponding to the scores assigned by the network to each digit.

In Table 6 we present the complete results of our experiments, both when using the original BlindRotate algorithm from [CGGI16b] (denoted by or) and when using the modified algorithm presented in Subsect. 5.4 (denoted by un, unfolded).

The homomorphic evaluation of the network on the entire test set was compared to its classification in the clear and we observed the following facts:

Observation 1. *The accuracy achieved when classifying encrypted images is close to that obtained when classifying images in the clear.*

In the case of the network with 30 hidden neurons, we obtain a classification accuracy of 93.55% in the clear (cf. Table 3) and of 93.71% homomorphically. In the case of the network with 100 hidden neurons, we have 96.43% accuracy

---

[7] Note that we do not apply any activation function to the output neurons: we are only interested in being able to retrieve the scores and sorting them to recover the classification given by the network.

in the clear and 96.35% on encrypted inputs. These gaps are explained by the following observations.

Observation 2. *During the evaluation, some signs are flipped during the bootstrapping but this does not significantly harm the accuracy of the network.*

We use aggressive internal parameters (e.g., N and, in general, all the parameters that control the precision) for the homomorphic evaluation, knowing that this could sometimes lead the bootstrapping procedure to return an incorrect result when extracting the sign of a message. In fact, we conjectured that the neural network would be resilient to perturbations and experimental results proved that this is indeed the case: when running our experiment over the full test set, we noticed that the number of wrong bootstrappings is 3383 (respectively, 9088) but this did not change the outcome of the classification in more than 196 (respectively, 105) cases (cf. Table 6).

Table 6. Results of homomorphic evaluation of two DiNNs on the full test set. The second column gives the number of disagreements (images classified differently) between the evaluation in the clear and the homomorphic one; the numbers in parentheses give the disagreements in favor of the cleartext evaluation and those in favor of the homomorphic evaluation, respectively. The third column gives the number of wrong bootstrapping, i.e., when the sign is flipped. The fourth value gives the number of disagreements in which at least one bootstrapping was wrong. Finally, the last column gives the time required to classify a single image.

	Accur.	Disag.	Wrong BS	Disag. (wrong BS)	Time
30 or	93.71%	273 (105–121)	3383/300000	196/273	0.515 s
30 un	93.46%	270 (119–110)	2912/300000	164/270	0.491 s
100 or	96.26%	127 (61–44)	9088/1000000	105/127	1.679 s
100 un	96.35%	150 (66–58)	7452/1000000	99/150	1.64 s

Observation 3. *The classification of an encrypted image might disagree with the classification of the same image in the clear but this does not significantly worsen the overall accuracy.*

This is a property that we expected during the implementation phase and our intuition to explain this fact is the following: the network is assigning 10 scores to each image, one per digit, and when two scores are close (i.e., the network is hesitating between two classes), it can happen that the classification in the clear is correct and the one over the encrypted image is wrong. But the opposite can also be true, thus leading to classifying correctly an encrypted sample that was misclassified in the clear. We experimentally verified that disagreements between the evaluations do not automatically imply that the homomorphic classification is worse than the one in the clear: out of 273 (respectively, 127) disagreements, the classification in the clear was correct 105 (respectively, 61) times, against 121 (respectively, 44) times in favor of the homomorphic one[8] (cf. Table 6).

---

[8] In the remaining cases, the classifications were different but they were both wrong.

Observation 4. *Using the modified version of the BlindRotate algorithm presented in Subsect. 5.4 decreases the number of wrong bootstrappings.*

Before stating some open problems, we conclude with the following note: using a bigger neural network generally leads to a better classification accuracy, at the cost of performing more calculations and, above all, more bootstrapping operations. However, the evaluation time will always grow linearly with the number of neurons. Although it is true that evaluating a bigger network is computationally more expensive, we stress that the bootstrapping operations are independent of each other and can thus be performed in parallel. Ideally, parallelizing the execution across a number of cores equal to the number of neurons in a layer (30 or 100 in our work) would result in that the evaluation of the layer would take roughly the time of a bootstrapping (i.e., around 15 ms).

Future directions and open problems. This work opens a number of possibilities and, thus, raises several interesting open problems. The first one is about the construction of our DiNNs. In this work, we did not pay too much attention to this step and, as a consequence, we considerably worsened the accuracy when moving from a canonical neural network to a DiNN. In order to improve the classification given by these discretized networks, it would be interesting to *train* a DiNN, rather than simply discretizing an already-trained model. Using discrete values and the sign function for the activation makes some calculations (e.g., some derivatives) impossible. Techniques to overcome these limitations have already been proposed in the literature (e.g., [CB16]) and they can be applied to our DiNNs as well. Also, another potentially interesting approach would be mixing these two ways of constructing a DiNN, for example by first discretizing a given model and then training the resulting network to refine it. Another natural question is whether we can batch several bootstrappings together, in order to improve the overall efficiency of the evaluation. Moreover, the speed of the evaluation would benefit from taking advantage of multi-core processing units, like GPUs.

Most interestingly, our FHE–DiNN framework is flexible and can be adapted to more generic cognitive architectures: we leave this as an interesting open problem. In particular, excellent results have been obtained by using Convolutional Neural Networks (see e.g., [LBBH98]), and we believe that trying to apply FHE–DiNN to these models would be an interesting line of research. Achieving this goal would require extending the current capabilities of FHE. For example, we would need to be able to homomorphically evaluate the max function, which is required to construct the widely-used max pooling layers. To the best of our knowledge, a technique for an efficient homomorphic evaluation of the max function is currently not known. Finally, the methodology presented in this work is by no means limited to image recognition, but can be applied to other machine learning problems as well.

Acknowledgments. Florian Bourse was supported by the European Research Council under the European Community's Seventh Framework Programme (FP7/2007-2013 Grant Agreement no. 339563 – CryptoCloud), and by the French ANR Project ANR-16-CE39-0014 PERSOCLOUD. Part of this work was done while the author was employed by CNRS and visiting CryptoExperts.

Michele Minelli and Matthias Minihold were supported by European Union's Horizon 2020 research and innovation programme under grant agreement No H2020-MSCA-ITN-2014-643161 ECRYPT-NET. This work was done while the authors were visiting CryptoExperts. The authors would like to thank CRYPTO's anonymous reviewers for providing useful suggestions and helping improve the paper.

References

[AAB+15] Abadi, M., Agarwal, A., Barham, P., Brevdo, E., Chen, Z., Citro, C., Corrado, G.S., Davis, A., Dean, J., Devin, M., Ghemawat, S., Goodfellow, I., Harp, A., Irving, G., Isard, M., Jia, Y., Jozefowicz, R., Kaiser, L., Kudlur, M., Levenberg, J., Mané, D., Monga, R., Moore, S., Murray, D., Olah, C., Schuster, M., Shlens, J., Steiner, B., Sutskever, I., Talwar, K., Tucker, P., Vanhoucke, V., Vasudevan, V., Viégas, F., Vinyals, O., Warden, P., Wattenberg, M., Wicke, M., Yu, Y., Zheng, X.: TensorFlow: large-scale machine learning on heterogeneous systems (2015). Software: tensorflow.org

[ABDP15] Abdalla, M., Bourse, F., De Caro, A., Pointcheval, D.: Simple functional encryption schemes for inner products. In: Katz, J. (ed.) PKC 2015. LNCS, vol. 9020, pp. 733–751. Springer, Heidelberg (2015). https://doi.org/10.1007/978-3-662-46447-2_33

[ALS16] Agrawal, S., Libert, B., Stehlé, D.: Fully secure functional encryption for inner products, from standard assumptions. In: Robshaw, M., Katz, J. (eds.) CRYPTO 2016, Part III. LNCS, vol. 9816, pp. 333–362. Springer, Heidelberg (2016). https://doi.org/10.1007/978-3-662-53015-3_12

[AP13] Alperin-Sheriff, J., Peikert, C.: Practical bootstrapping in quasilinear time. In: Canetti, R., Garay, J.A. (eds.) CRYPTO 2013, Part I. LNCS, vol. 8042, pp. 1–20. Springer, Heidelberg (2013). https://doi.org/10.1007/978-3-642-40041-4_1

[AP14] Alperin-Sheriff, J., Peikert, C.: Faster bootstrapping with polynomial error. In: Garay, J.A., Gennaro, R. (eds.) CRYPTO 2014, Part I. LNCS, vol. 8616, pp. 297–314. Springer, Heidelberg (2014). https://doi.org/10.1007/978-3-662-44371-2_17

[APS15] Albrecht, M.R., Player, R., Scott, S.: On the concrete hardness of learning with errors. Cryptology ePrint Archive, Report 2015/046 (2015). http://eprint.iacr.org/2015/046

[AS00] Agrawal, R., Srikant, R.: Privacy-preserving data mining. SIGMOD Rec. **29**(2), 439–450 (2000)

[BGV12] Brakerski, Z., Gentry, C., Vaikuntanathan, V.: (Leveled) fully homomorphic encryption without bootstrapping. In: ITCS 2012, pp. 309–325. ACM, January 2012

[BLMZ17] Benhamouda, F., Lepoint, T., Mathieu, C., Zhou, H.: Optimization of bootstrapping in circuits. In: Proceedings of the Twenty-Eighth Annual

ACM-SIAM Symposium on Discrete Algorithms, SODA 2017, Philadelphia, PA, USA, pp. 2423–2433. Society for Industrial and Applied Mathematics (2017)

[BPTG15] Bost, R., Popa, R.A., Tu, S., Goldwasser, S.: Machine learning classification over encrypted data. In: NDSS 2015. The Internet Society, February 2015

[BV11a] Brakerski, Z., Vaikuntanathan, V.: Efficient fully homomorphic encryption from (standard) LWE. In: 52nd FOCS, pp. 97–106. IEEE Computer Society Press, October 2011

[BV11b] Brakerski, Z., Vaikuntanathan, V.: Fully homomorphic encryption from ring-LWE and security for key dependent messages. In: Rogaway, P. (ed.) CRYPTO 2011. LNCS, vol. 6841, pp. 505–524. Springer, Heidelberg (2011). https://doi.org/10.1007/978-3-642-22792-9_29

[BV14] Brakerski, Z., Vaikuntanathan, V.: Lattice-based FHE as secure as PKE. In: ITCS 2014, pp. 1–12. ACM, January 2014

[C+15] Chollet, F., et al.: Keras (2015). https://github.com/keras-team/keras

[CB16] Courbariaux, M., Bengio, Y.: Binarynet: training deep neural networks with weights and activations constrained to +1 or −1. CoRR, abs/1602.02830 (2016)

[CdWM+17] Chabanne, H., de Wargny, A., Milgram, J., Morel, C., Prouff, E.: Privacy-preserving classification on deep neural network. IACR Cryptology ePrint Archive 2017:35 (2017)

[CGGI16a] Chillotti, I., Gama, N., Georgieva, M., Izabachène, M.: TFHE: Fast Fully Homomorphic Encryption Library over the Torus (2016). https://github.com/tfhe/tfhe

[CGGI16b] Chillotti, I., Gama, N., Georgieva, M., Izabachène, M.: Faster fully homomorphic encryption: bootstrapping in less than 0.1 seconds. In: Cheon, J.H., Takagi, T. (eds.) ASIACRYPT 2016, Part I. LNCS, vol. 10031, pp. 3–33. Springer, Heidelberg (2016). https://doi.org/10.1007/978-3-662-53887-6_1

[CGGI17] Chillotti, I., Gama, N., Georgieva, M., Izabachène, M.: Faster packed homomorphic operations and efficient circuit bootstrapping for TFHE. In: Takagi, T., Peyrin, T. (eds.) ASIACRYPT 2017, Part I. LNCS, vol. 10624, pp. 377–408. Springer, Cham (2017). https://doi.org/10.1007/978-3-319-70694-8_14

[CMS12] Cireşan, D., Meier, U., Schmidhuber, J.: Multi-column deep neural networks for image classification. ArXiv e-prints, February 2012

[Cyb89] Cybenko, G.: Approximation by superpositions of a sigmoidal function. Math. Control Sig. Syst. **2**(4), 303–314 (1989)

[DGBL+16] Dowlin, N., Gilad-Bachrach, R., Laine, K., Lauter, K., Naehrig, M., Wernsing, J.: CryptoNets: applying neural networks to encrypted data with high throughput and accuracy. Technical report, February 2016

[DM15] Ducas, L., Micciancio, D.: FHEW: bootstrapping homomorphic encryption in less than a second. In: Oswald, E., Fischlin, M. (eds.) EUROCRYPT 2015, Part I. LNCS, vol. 9056, pp. 617–640. Springer, Heidelberg (2015). https://doi.org/10.1007/978-3-662-46800-5_24

[Dwo06] Dwork, C.: Differential privacy (invited paper). In: Bugliesi, M., Preneel, B., Sassone, V., Wegener, I. (eds.) ICALP 2006, Part II. LNCS, vol. 4052, pp. 1–12. Springer, Heidelberg (2006). https://doi.org/10.1007/11787006_1

[Gen09] Gentry, C.: A fully homomorphic encryption scheme. Ph.D. thesis, Stanford University (2009). crypto.stanford.edu/craig

[GHS12] Gentry, C., Halevi, S., Smart, N.P.: Homomorphic evaluation of the AES circuit. In: Safavi-Naini, R., Canetti, R. (eds.) CRYPTO 2012. LNCS, vol. 7417, pp. 850–867. Springer, Heidelberg (2012). https://doi.org/10.1007/978-3-642-32009-5_49

[GSW13] Gentry, C., Sahai, A., Waters, B.: Homomorphic encryption from learning with errors: conceptually-simpler, asymptotically-faster, attribute-based. In: Canetti, R., Garay, J.A. (eds.) CRYPTO 2013, Part I. LNCS, vol. 8042, pp. 75–92. Springer, Heidelberg (2013). https://doi.org/10.1007/978-3-642-40041-4_5

[Hor91] Hornik, K.: Approximation capabilities of multilayer feedforward networks. Neural Netw. 4(2), 251–257 (1991)

[HS14] Halevi, S., Shoup, V.: Algorithms in HElib. In: Garay, J.A., Gennaro, R. (eds.) CRYPTO 2014, Part I. LNCS, vol. 8616, pp. 554–571. Springer, Heidelberg (2014). https://doi.org/10.1007/978-3-662-44371-2_31

[HS15] Halevi, S., Shoup, V.: Bootstrapping for HElib. In: Oswald, E., Fischlin, M. (eds.) EUROCRYPT 2015, Part I. LNCS, vol. 9056, pp. 641–670. Springer, Heidelberg (2015). https://doi.org/10.1007/978-3-662-46800-5_25

[HZRS15] He, K., Zhang, X., Ren, S., Sun, J.: Deep residual learning for image recognition. CoRR, abs/1512.03385 (2015)

[KTX08] Kawachi, A., Tanaka, K., Xagawa, K.: Concurrently secure identification schemes based on the worst-case hardness of lattice problems. In: Pieprzyk, J. (ed.) ASIACRYPT 2008. LNCS, vol. 5350, pp. 372–389. Springer, Heidelberg (2008). https://doi.org/10.1007/978-3-540-89255-7_23

[LBBH98] LeCun, Y., Bottou, L., Bengio, Y., Haffner, P.: Gradient-based learning applied to document recognition. Proc. IEEE 86(11), 2278–2324 (1998)

[LCB98] LeCun, Y., Cortes, C., Burges, C.: The MNIST database of handwritten digits (1998). http://yann.lecun.com/exdb/mnist/

[LP13] Lepoint, T., Paillier, P.: On the minimal number of bootstrappings in homomorphic circuits. In: Adams, A.A., Brenner, M., Smith, M. (eds.) FC 2013 Workshops. LNCS, vol. 7862, pp. 189–200. Springer, Heidelberg (2013). https://doi.org/10.1007/978-3-642-41320-9_13

[LPR10] Lyubashevsky, V., Peikert, C., Regev, O.: On ideal lattices and learning with errors over rings. In: Gilbert, H. (ed.) EUROCRYPT 2010. LNCS, vol. 6110, pp. 1–23. Springer, Heidelberg (2010). https://doi.org/10.1007/978-3-642-13190-5_1

[MRSV17] Makri, E., Rotaru, D., Smart, N.P., Vercauteren, F.: PICS: private image classification with SVM. Cryptology ePrint Archive, Report 2017/1190 (2017). https://eprint.iacr.org/2017/1190

[MZ17] Mohassel, P., Zhang, Y.: SecureML: a system for scalable privacy-preserving machine learning. In: 2017 IEEE Symposium on Security and Privacy, pp. 19–38. IEEE Computer Society Press, May 2017

[PV16] Paindavoine, M., Vialla, B.: Minimizing the number of bootstrappings in fully homomorphic encryption. In: Dunkelman, O., Keliher, L. (eds.) SAC 2015. LNCS, vol. 9566, pp. 25–43. Springer, Cham (2016). https://doi.org/10.1007/978-3-319-31301-6_2

[PW08] Peikert, C., Waters, B.: Lossy trapdoor functions and their applications. In 40th ACM STOC, pp. 187–196. ACM Press, May 2008

512   F. Bourse et al.

[Reg05] Regev, O.: On lattices, learning with errors, random linear codes, and cryptography. In: 37th ACM STOC, pp. 84–93. ACM Press, May 2005

[SS10] Stehlé, D., Steinfeld, R.: Faster fully homomorphic encryption. In: Abe, M. (ed.) ASIACRYPT 2010. LNCS, vol. 6477, pp. 377–394. Springer, Heidelberg (2010). https://doi.org/10.1007/978-3-642-17373-8_22

[SV10] Smart, N.P., Vercauteren, F.: Fully homomorphic encryption with relatively small key and ciphertext sizes. In: Nguyen, P.Q., Pointcheval, D. (eds.) PKC 2010. LNCS, vol. 6056, pp. 420–443. Springer, Heidelberg (2010). https://doi.org/10.1007/978-3-642-13013-7_25

[vDGHV10] van Dijk, M., Gentry, C., Halevi, S., Vaikuntanathan, V.: Fully homomorphic encryption over the integers. In: Gilbert, H. (ed.) EUROCRYPT 2010. LNCS, vol. 6110, pp. 24–43. Springer, Heidelberg (2010). https://doi.org/10.1007/978-3-642-13190-5_2

[ZK16] Zagoruyko, S., Komodakis, N.: Wide residual networks. CoRR, abs/1605.07146 (2016)

[ZYC16] Zhang, Q., Yang, L.T., Chen, Z.: Privacy preserving deep computation model on cloud for big data feature learning. IEEE Trans. Comput. 65(5), 1351–1362 (2016)

[ZYL+17] Zhou, T., Yang, X., Liu, L., Zhang, W., Ding, Y.: Faster bootstrapping with multiple addends. Cryptology ePrint Archive, Report 2017/735 (2017). http://eprint.iacr.org/2017/735

Oblivious Transfer

Adaptive Garbled RAM from Laconic Oblivious Transfer

Sanjam Garg[1], Rafail Ostrovsky[2], and Akshayaram Srinivasan[1(✉)]

[1] University of California, Berkeley, Berkeley, USA
{sanjamg,akshayaram}@berkeley.edu
[2] UCLA, Los Angeles, USA
rafail@cs.ucla.edu

Abstract. We give a construction of an adaptive garbled RAM scheme. In the adaptive setting, a client first garbles a "large" persistent database which is stored on a server. Next, the client can provide garbling of multiple adaptively and adversarially chosen RAM programs that execute and modify the stored database arbitrarily. The garbled database and the garbled program should reveal nothing more than the running time and the output of the computation. Furthermore, the sizes of the garbled database and the garbled program grow only linearly in the size of the database and the running time of the executed program respectively (up to poly logarithmic factors). The security of our construction is based on the assumption that laconic oblivious transfer (Cho et al., CRYPTO 2017) exists. Previously, such adaptive garbled RAM constructions were only known using indistinguishability obfuscation or in random oracle model. As an additional application, we note that this work yields the first constant round secure computation protocol for persistent RAM programs in the malicious setting from standard assumptions. Prior works did not support persistence in the malicious setting.

1 Introduction

Over the years, garbling methods [Yao86, LP09, AIK04, BHR12b, App17] have been extremely influential and have engendered an enormous number of applications in cryptography. Informally, garbling a function f and an input x, yields

S. Garg—Research supported in part from DARPA/ARL SAFEWARE Award W911 NF15C0210, AFOSR Award FA9550-15-1-0274, AFOSR YIP Award, DARPA and SPAWAR under contract N66001-15-C-4065, a Hellman Award and research grants by the Okawa Foundation, Visa Inc., and Center for Long-Term Cybersecurity (CLTC, UC Berkeley). The views expressed are those of the author and do not reflect the official policy or position of the funding agencies.

R. Ostrovsky—Research supported in part by NSF grant 1619348, DARPA SPAWAR contract N66001-15-1C-4065, US-Israel BSF grant 2012366, OKAWA Foundation Research Award, IBM Faculty Research Award, Xerox Faculty Research Award, B. John Garrick Foundation Award, Teradata Research Award, and Lockheed-Martin Corporation Research Award. The views expressed are those of the authors and do not reflect position of the Department of Defense or the U.S. Government.

© International Association for Cryptologic Research 2018
H. Shacham and A. Boldyreva (Eds.): CRYPTO 2018, LNCS 10993, pp. 515–544, 2018.
https://doi.org/10.1007/978-3-319-96878-0_18

the function encoding $\widehat{f}$ and the input encoding $\widehat{x}$. Given $\widehat{f}$ and $\widehat{x}$, there exists an efficient decoding algorithm that recovers $f(x)$. The security property requires that $\widehat{f}$ and $\widehat{x}$ do not reveal anything about f or x except $f(x)$. By now, it is well established that realizing garbling schemes [BHR12b, App17] is an important cryptographic goal.

One shortcoming of standard garbling techniques has been that the size of the function encoding grows linearly in the size of the circuit computing the function and thus leads to large communication costs. Several methods have been devised to overcome this constraint.

- Lu and Ostrovsky [LO13] addressed the question of garbling RAM program execution on a persistent garbled database. Here, the efficiency requirement is that the size of the function encoding grows only with the running time of the RAM program. This work has lead to fruitful line of research [GHL+14, GLOS15, GLO15, LO17] that reduces the communication cost to grow linearly with running times of the programs executed, rather that the corresponding circuit sizes. A key benefit of this approach is that it has led to constructions based on one-way functions.
- Goldwasser, Kalai, Popa, Vaikuntanathan, and Zeldovich [GKP+13] addressed the question of reducing the communication cost by reusing the encodings. Specifically, they provided a construction of reusable garbled circuits based on standard assumptions (namely learning-with-errors). However, their construction needs input encoding to grow with the depth of the circuit being garbled.
- Finally, starting with Gentry, Halevi, Raykova, and Wichs [GHRW14], a collection of works [CHJV15, BGL+15, KLW15, CH16, CCHR16, ACC+16] have attempted to obtain garbling schemes where the size of the function encoding only grows with its description size and is otherwise independent of its running time on various inputs. However, these constructions are proven secure only assuming indistinguishability obfuscation [BGI+01, GGH+13].

A recurring theme in all the above research efforts has been the issue of *adaptivity*: Can the adversary adaptively choose the input after seeing the function encoding?

This task is trivial if one reveals both the function encoding and the input encoding together after the input is specified. However, this task becomes highly non-trivial if we require the size of the input encoding to only grow with the size of the input and independent of the complexity of computing f. The first solution to this problem was provided by Bellare, Hoang and Rogaway [BHR12a] for the case of circuits in the random oracle model [BR93]. Subsequently, several adaptive circuit garbling schemes have been obtained in the standard model from (i) one-way functions [HJO+16, JW16, JKK+17],[1] or (ii) using laconic OT [GS18a] which relies on public-key assumptions [CDG+17, DG17, DGHM18, BLSV18].

However, constructing adaptively secure schemes for more communication constrained settings has proved much harder. In this paper, we focus on the

---

[1] A drawback of these works is that the size of the input encoding grows with the width/depth of the circuit computing f.

case of RAM programs. More specifically, adaptively secure garbled RAM is known only using random oracles (e.g. [LO13, GLOS15]) or under very strong assumptions such as indistinguishability obfuscation [CCHR16, ACC+16]. In this work, we ask:

Can we realize adaptively secure garbled RAM from standard assumptions?

Further motivating the above question, is the tightly related application of constructing *constant round* secure RAM computation over a persistent database in the malicious setting. More specifically, as shown by Beaver, Micali and Rogaway [BMR90] garbling techniques can be used to realize constant round secure computation [Yao82, GMW87] constructions. Similarly, above-mentioned garbling schemes for RAM programs also yield constant round, communication efficient secure computation solutions [HY16, Mia16, GGMP16, KY18]. However, preserving persistence of RAM programs in the malicious setting requires the underlying garbling techniques to provide adaptive security.[2]

1.1   Our Results

In this work, we obtain a construction of adaptively secure garbled RAM based on the assumption that laconic oblivious transfer [CDG+17] exists. Laconic oblivious transfer can be based on a variety of public-key assumptions such as (i) Computation Diffie-Hellman Assumption [DG17], (ii) Factoring Assumption [DG17], or (iii) Learning-With-Errors Assumption [BLSV18, DGHM18]. In our construction, the size of the garbled database and the garbled program grow only linearly in the size of the database and the running time of the executed program respectively (up to poly logarithmic factors). The main result in our paper is:

Theorem 1 (Informal). *Assuming either the Computational Diffie-Hellman assumption or the Factoring assumption or the Learning-with-Errors assumption, there exists a construction of adaptive garbled RAM scheme where the time required to garble a database, a program and an input grows linearly (upto poly logarithmic factors) with the size of the database, running time of the program and length of the input respectively.*[3]

Additionally, plugging our adaptively secure garbled RAM scheme into a malicious secure constant round secure computation protocol yields a maliciously secure constant round secure RAM computation protocol [IKO+11, ORS15, BL18, GS18b] for a persistent database. Again, this construction is based on the assumption that laconic OT exists and the underlying assumptions needed for the constant round protocol.

---

[2] We note that adaptive security is not essential for obtaining protocols with round complexity that grows with the running time of the executed programs [OS97, GKK+12, WHC+14].

[3] As in the case of adaptively secure garbled circuits, the size of the input encoding must also grow with the output length of the program. Here, we implicitly assume that the input and the outputs have the same length.

2   Our Techniques

In this section, we outline the main challenges and the techniques used in our construction of adaptive garbled RAM.

Starting Point. In a recent result, Garg and Srinivasan [GS18a] gave a construction of adaptively secure garbled circuit transfer where the size of the input encoding grows only with the input and the output length. The main idea behind their construction is a technique to "linearize" a garbled circuit. Informally, a garbled circuit is said to be linearized if the simulation of particular garbled gate depends only on simulating one other gate (or in other words, the simulation dependency graph is a line). In order to linearize a garbled circuit, their work transforms a circuit into a sequence of CPU step circuits that can make read and write accesses at *fixed* locations in an external memory. The individual step circuits are garbled using a (plain) garbling scheme and the access to the memory is mediated using a laconic OT.[4] The use of laconic OT enables the above mentioned garbling scheme to have "linear" structure wherein the simulation of a particular CPU step depends only on simulating the previous step circuit.

A Generalization. Though the approach of Garg and Srinivasan shares some similarities with a garbling a RAM program (like garbling a sequence of CPU step circuits), there are some crucial differences.

1. The first difference is that unlike a circuit, the locations that are accessed by a RAM program are dynamically chosen depending on the program's input.
2. The second difference is that the locations that are accessed might leak information about the program and the input and a garbled RAM scheme must protect against such leakages.

The first step we take in constructing an adaptive garbled RAM scheme is to generalize the above approach of Garg and Srinivasan [GS18a] to construct an adaptively secure garbled RAM scheme with weaker security guarantees. The security that we achieve is that of unprotected memory access [GHL+14]. Informally, a garbled RAM scheme is said to have unprotected memory access if both the contents of the database and the memory locations that are accessed are revealed in the clear. This generalization is given in Sect. 4.

In the non-adaptive setting, there are standard transformations (outlined in [GHL+14]) from a garbled RAM with unprotected memory access to a standard garbled RAM scheme where both the memory contents and the access patterns are hidden. This transformation involves the additional use of an ORAM scheme. Somewhat surprisingly, these transformations fail in the adaptive setting! The details follow.

---

[4] A laconic OT scheme allows to compress a large database/memory to a small digest. The digest in some sense binds the entire database. In particular, given the digest there exists efficient algorithms that can read/update particular memory locations. The time taken by these algorithms grow only logarithmically with the size of the database.

Challenges. To understand the main challenges, let us briefly explain how the security proof goes through in the work of Garg and Srinivasan [GS18a]. In a typical construction of a garbled RAM program, using a sequence of garbled circuits, one would expect that the simulation of garbled circuits would be done from the first CPU step to the last CPU step. However, in [GS18a] proof, the simulation is done in a rather unusual manner, from the last CPU step to the first CPU step. Of course, it is not possible to simulate the last CPU step directly. Thus, the process of simulating the last CPU step itself involves a sequence of hybrids that simulate and "un-simulate" the garbling of the previous CPU steps. Extending this approach so that the memory contents and the access patterns are both hidden faces the following two main challenges.

- **Challenge 1:** In the Garg and Srinivasan construction [GS18a], memory contents were encrypted using one-time pads. Since the locations that each CPU step (for a circuit) reads from and write to are fixed, the one-time pad corresponding to that location could be hardwired to those CPU steps. On the other hand, in the case of RAM programs the locations being accessed are dynamically chosen and thus it is not possible to hard-wire the entire one-time pad into each CPU step as this would blow up the size of these CPU steps.
  It is instructive to note that encrypting the memory using an encryption scheme and decrypting the read memory contents does not suffice. See more on this in preliminary attempt below.
- **Challenge 2:** In the non-adaptive setting, it is easy to amplify unprotected memory access security to the setting where memory accesses are hidden using an oblivious RAM scheme [Gol87,Ost90,GO96]. However, in the adaptive setting this transformation turns out to be tricky. In a bit more detail, the Garg and Srinivasan [GS18a] approach of simulating CPU step circuits from the last to the first ends up in conflict with the security of the ORAM scheme where the simulation is typically done from the first to the last CPU steps. We note here that the techniques of Canetti et al. [CCHR16] and Ananth et al. [ACC+16], though useful, do not apply directly to our setting. In particular, in the Canetti et al. [CCHR16] and Ananth et al. [ACC+16] constructions, CPU steps where obfuscated using an indistinguishability obfuscation scheme. Thus, in their scheme the obfuscation for any individual CPU step could be changed independently. For example, the PRF key used in any CPU step could be punctured independent of the other CPU steps. On the other hand, in our construction, inspite of each CPU step being garbled separately, its input labels are hardwired in the previous garbled circuit. Therefore, a change in hardwired secret value (like a puncturing a key) in a CPU step needs an intricate sequence of hybrids for making this change. For instance, in the case of the example above, it is not possible to puncture the PRF key hardwired in a particular CPU step in one simple hybrid step. Instead any change in this CPU step must change the CPU step before it and so on. In summary, in our case, any such change would involve a new and intricate hybrid argument.

2.1   Solving Challenge 1

In this subsection, we describe our techniques to solve challenge 1.

Preliminary Attempt. A very natural approach to encrypting external memory would be to use a pseudorandom function to encrypt memory content in each location. More precisely, a data value d in location L is encrypted using the key $\mathsf{PRF}_K(L)$ where K is the PRF key. The key K for this pseudorandom function is hardwired in each CPU step so that it first decrypts the ciphertext that is read from the memory and uses the underlying data for further processing. This approach to solving Challenge 1 was in fact used in the works of Canetti et al. [CCHR16] and Ananth et al. [ACC+16] (and several other prior works) in a similar context. However, in order to use the security of this PRF, we must first remove the hardwired key from each of the CPU steps. This is easily achieved if we rely on indistinguishability obfuscation. Indeed, a single hybrid change is sufficient to have the punctured key to be hardwired in each of the CPU steps. However, in our setting this does not work! In particular, we need to puncture the PRF key in each of the CPU step circuits by simulating them individually and the delicate dependencies involved in garbling each CPU step blows up the size of the garbled input to grow with the running time of the program.[5] Due to the same reason, the approaches of encrypting the memory by maintaining a tree of secret keys [GLOS15, GLO15] do not work.

Our New Idea: A Careful Timed Encryption Mechanism. From the above attempts, the following aspect of secure garbled RAM arise. Prior approaches for garbling RAM programs use PRF keys that in some sense "decrease in power"[6] as hybrids steps involve sequential simulation of the CPU steps starting with the first CPU step and ending in the last CPU step. However, in the approach of [GS18a], the hybrids do a backward pass, from the last CPU step circuit to the first CPU step circuit. Therefore, we need a mechanism wherein the hardwired key for encryption in some sense "strengthens" along the first to the last CPU step.

Location vs. Time. In almost all garbled RAM constructions, the data stored at a particular location is encrypted using a location dependent key (e.g. [GLOS15]). This was not a problem when the keys are being weakened across CPU steps. However, in our case we need the key to be strengthened in power across CPU steps. Thus, we need a special purpose encryption scheme where the keys are derived based on time rather than the locations. Towards this goal, we construct

---

[5] For the readers who are familiar with [GS18a], the number of CPU steps that have to be maintained in the input dependent simulation for puncturing the PRF key grows with the number of CPU steps that last wrote to this location and this could be as large as the running time of the program.

[6] The tree-based approaches of storing the secret keys use the mechanism wherein the hardwired secret keys decrease in power in subsequent CPU steps. In particular, the secret key corresponding to the root can decrypt all the locations, the secret keys corresponding to its children can only decrypt a part of the database and so on.

a special purpose encryption scheme called as a *timed encryption* scheme. Let us explain this in more detail.

Timed Encryption. A timed encryption scheme is just like any (plain) symmetric key encryption except that every message is encrypted with respect to a timestamp. Additionally, there is a special key constrain algorithm that constrains a key to only decrypt ciphertexts that are encrypted within a specific timestamp. The security requirement is that the constrained key does not help in distinguishing ciphertexts of two messages that are encrypted with respect to some future timestamp. We additionally require the encryption using a key constrained with respect to a timestamp time to have the same distribution as an encryption using an unconstrained key as long as the timestamp to which we are encrypting is less than or equal to time. For efficiency, we require that the size of the constrained key to grow only with the length of the binary representation of the timestamp.

Solving Challenge 1. Timed encryption provides a natural approach to solving challenge 1. In every CPU step, we hardwire a time constrained key that allows that CPU step to decrypt all the memory updates done by the prior CPU steps. The last CPU step in some sense has the most powerful key hardwired, i.e., it can decrypt all the updates made by all the prior CPU steps and the first CPU step has the least powerful key hardwired. Thus, the hardwired secret key strengthens from the first CPU step to the last CPU step. In the security proof, a backward pass of simulating the last CPU step to the first CPU step conforms well with the semantics and security properties of a timed encryption scheme. This is because we remove the most powerful keys first and the rest of the hardwired secret keys in the previous CPU steps do not help in distinguishing between encryptions of the actual value that is written and some junk value. We believe that the notion timed encryption might have other applications and be of independent interest.

Constructing Timed Encryption. We give a construction of a timed encryption scheme from any one-way function. Towards this goal, we introduce a notion called as *range constrained PRF*. A range constrained PRF is a special constrained PRF [BW13] where the PRF key can be constrained to evaluate input points that fall within a particular range. The ranges that we will be interested in are of the form $[0, x]$. That is, the constrained key can be used to evaluate the PRF on any $y \in [0, x]$. For efficiency, we require that the size of the constrained key to only grow with the binary representation of x. Given such a PRF, we can construct a timed encryption scheme as follows. The key generation samples a range constrained PRF key. The encryption of a message m with respect to a timestamp time proceeds by evaluating the PRF on time to derive sk and then using sk as a key for symmetric encryption scheme to encrypt the message m. The time constraining algorithm just constrains the PRF key with respect to the range $[0, \text{time}]$. Thus, the goal of constructing a timed encryption scheme reduces to the goal of constructing a range constrained PRF. In this work, we give a construction of range constrained PRF by adding a range constrain algorithm to the tree-based PRF scheme of Goldreich, Goldwasser and Micali [GGM86].

2.2   Solving Challenge 2

Challenge 1 involves protecting the contents of the memory whereas challenge 2 involves protecting the access pattern. As mentioned before, in the non-adaptive setting, this problem is easily solved using an oblivious RAM scheme. However, in our setting we need an oblivious RAM scheme with some special properties.

The works of Canetti et al. [CCHR16] and Ananth et al. [ACC+16] define a property of an ORAM scheme as *strong localized randomness property* and then use this property to hide their access patterns. Informally, an ORAM scheme is said to have a strong localized randomness property if the locations of the random tape accessed by an oblivious program in simulating each memory access are disjoint. Further, the number of locations touched for simulating each memory access must be poly logarithmic in the size of the database. These works further proved that the Chung-Pass ORAM scheme [CP13] satisfies the strong localized randomness property. Unfortunately, this strong localized randomness property alone is not sufficient for our purposes. Let us give the details.

To understand why the strong localized randomness property alone is not sufficient, we first recall the details of the Chung-Pass ORAM (henceforth, denoted as CP ORAM) scheme. The CP ORAM is a tree-based ORAM scheme where the leaves of this tree are associated with the actual memory. A position map associates each data block in the memory with a random leaf node. Accessing a memory location involves first reading the position map to get the address of the leaf where this data block resides. Then, the path from the root to this particular leaf is traversed and the content of the this data block is read. It is guaranteed that the data block is located somewhere along the path from the root to leaf node. The read data block is then placed in the root and the position map is updated so that another random leaf node is associated with this data block. To balance the memory, an additional flush is performed but for the sake of this introduction we ignore this step. The CP ORAM scheme has strong localized randomness as the randomness used in each memory accesses involves choosing a random leaf to update the position map. Let us now explain why this property alone is not sufficient for our purpose.

Recall that in the security proof of [GS18a], the CPU steps are simulated from the last step to the first. A simulation of a CPU step involves changing the bit written by the step to some junk value and the changing the location accessed to a random location. We can change the bit to be written to a junk value using the security of the timed encryption scheme, however changing the location accessed to random is problematic. Note that the location that is being accessed in the CP ORAM is a random root to leaf path. However, the address of this leaf is stored in the memory via the position map. Therefore, to simulate a particular CPU step, we must first change the contents of the position map. This change must be performed in those CPU steps that last updated this memory location. Unfortunately, timed encryption is not useful in this setting as we can use its security only after removing all the secret keys that are hardwired in the future time steps. However, in our case, the CPU steps that last updated this particular location might be so far into the past that removing all the intermediate encryption keys might blow up the cost of the input encoding to be as large as the program running time.

To solve this issue, we modify the Chung-Pass ORAM to additionally have the CPU steps to encrypt the data block that is written using a puncturable PRF. Unlike the previous approaches of encrypting the data block with respect to the location, we encrypt it with respect to the time step that modifies the location. This helps in circumventing the above problem as we can first puncture the PRF key (which in turn involves a careful set of hybrids) and use its security to change the position map to contain an encryption of the junk value instead of the actual address of the leaf node.[7] Once this change is done, the locations that the concerned CPU step is accessing is a random root to leaf path.

3 Preliminaries

Let λ denote the security parameter. A function $\mu(\cdot) : \mathbb{N} \to \mathbb{R}^+$ is said to be negligible if for any polynomial $\mathsf{poly}(\cdot)$ there exists $\lambda_0 \in \mathbb{N}$ such that for all $\lambda > \lambda_0$ we have $\mu(\lambda) < \frac{1}{\mathsf{poly}(\lambda)}$. For a probabilistic algorithm A, we denote $A(x; r)$ to be the output of A on input x with the content of the random tape being r. When r is omitted, $A(x)$ denotes a distribution. For a finite set S, we denote $x \leftarrow S$ as the process of sampling x uniformly from the set S. We will use PPT to denote Probabilistic Polynomial Time. We denote $[a]$ to be the set $\{1, \ldots, a\}$ and $[a, b]$ to be the set $\{a, a+1, \ldots, b\}$ for $a \leq b$ and $a, b \in \mathbb{Z}$. For a binary string $x \in \{0, 1\}^n$, we will denote the ith bit of x by x_i. We assume without loss of generality that the length of the random tape used by all cryptographic algorithms is λ. We will use $\mathsf{negl}(\cdot)$ to denote an unspecified negligible function and $\mathsf{poly}(\cdot)$ to denote an unspecified polynomial function.

We assume reader's familiarity with the notions of a puncturable PRF and selectively secure garbled circuits and omit the formal definitions here for the lack of space.

3.1 Updatable Laconic Oblivious Transfer

In this subsection, we recall the definition of updatable laconic oblivious transfer from [CDG+17].

We give the formal definition below from [CDG+17]. We generalize their definition to work for blocks of data instead of bits. More precisely, the reads and the updates happen at the block-level rather than at the bit-level.

Definition 1 ([CDG+17]). *An updatable laconic oblivious transfer consists of the following algorithms:*

---

[7] Unlike in the location based encryption scheme, it is sufficient to change the encryption only in the CPU steps that last modified this location.

- crs ← crsGen($1^\lambda, 1^N$): *It takes as input the security parameter 1^λ (encoded in unary) and a block size N and outputs a common reference string* crs.
- (d, $\widehat{D}$) ← Hash(crs, D): *It takes as input the common reference string* crs *and database $D \in \{\{0,1\}^N\}^*$ as input and outputs a digest* d *and a state $\widehat{D}$. We assume that the state $\widehat{D}$ also includes the database D.*
- e ← Send(crs, d, L, $\{m_{i,0}, m_{i,1}\}_{i \in [N]}$): *It takes as input the common reference string* crs, *a digest* d, *and a location $L \in \mathbb{N}$ and set of messages $m_{i,0}, m_{i,1} \in \{0,1\}^{p(\lambda)}$ for every $i \in [N]$ and outputs a ciphertext* e.
- ($m_1, \ldots, m_N$) ← Receive$^{\widehat{D}}$(crs, e, L): *This is a RAM algorithm with random read access to $\widehat{D}$. It takes as input a common reference string* crs, *a ciphertext* e, *and a location $L \in \mathbb{N}$ and outputs a set of messages $m_1, \ldots, m_N$.*
- e_w ← SendWrite(crs, d, L, $\{b_i\}_{i \in [N]}$, $\{m_{j,0}, m_{j,1}\}_{j=1}^{|d|}$): *It takes as input the common reference string* crs, *a digest* d, *and a location $L \in \mathbb{N}$, bits $b_i \in \{0,1\}$ for each $i \in [N]$ to be written, and $|d|$ pairs of messages $\{m_{j,0}, m_{j,1}\}_{j=1}^{|d|}$, where each $m_{j,c}$ is of length $p(\lambda)$ and outputs a ciphertext e_w.*
- $\{m_j\}_{j=1}^{|d|}$ ← ReceiveWrite$^{\widehat{D}}$(crs, L, $\{b_i\}_{i \in [N]}$, e_w): *This is a RAM algorithm with random read/write access to $\widehat{D}$. It takes as input the common reference string* crs, *a location L, a set of bits $b_1, \ldots, b_N \in \{0,1\}$ and a ciphertext e_w. It updates the state $\widehat{D}$ (such that $D[L] = b_1 \ldots b_N$) and outputs messages $\{m_j\}_{j=1}^{|d|}$.*

We require an updatable laconic oblivious transfer to satisfy the following properties.

Correctness: *We require that for any database D of size at most $M = \mathsf{poly}(\lambda)$, any memory location $L \in [M]$, any set of messages $(m_{i,0}, m_{i,1}) \in \{0,1\}^{p(\lambda)}$ for each $i \in [N]$ where $p(\cdot)$ is a polynomial that*

$$\Pr\left[\forall i \in [N], \ m_i = m_{i,D[L,i]} \ \middle| \ \begin{array}{ll} \mathsf{crs} & \leftarrow \mathsf{crsGen}(1^\lambda) \\ (\mathsf{d}, \widehat{D}) & \leftarrow \mathsf{Hash}(\mathsf{crs}, D) \\ e & \leftarrow \mathsf{Send}(\mathsf{crs}, \mathsf{d}, L, \{m_{i,0}, m_{i,1}\}_{i \in [N]}) \\ (m_1, \ldots, m_N) & \leftarrow \mathsf{Receive}^{\widehat{D}}(\mathsf{crs}, e, L) \end{array}\right] = 1,$$

where $D[L, i]$ denotes the ith bit in the Lth block of D.

Correctness of Writes: *Let database D be of size at most $M = \mathsf{poly}(\lambda)$ and let $L \in [M]$ be any two memory locations. Let D^* be a database that is identical to D except that $D^*[L, i] = b_i$ for all $i \in [N]$ some sequence of $\{b_j\} \in \{0,1\}$. For any sequence of messages $\{m_{j,0}, m_{j,1}\}_{j \in [\lambda]} \in \{0,1\}^{p(\lambda)}$ we require that*

$$\Pr\left[\begin{array}{l} m'_j = m_{j,\mathsf{d}^*_j} \\ \forall j \in [|\mathsf{d}|] \end{array} \ \middle| \ \begin{array}{ll} \mathsf{crs} & \leftarrow \mathsf{crsGen}(1^\lambda, 1^N) \\ (\mathsf{d}, \widehat{D}) & \leftarrow \mathsf{Hash}(\mathsf{crs}, D) \\ (\mathsf{d}^*, \widehat{D}^*) & \leftarrow \mathsf{Hash}(\mathsf{crs}, D^*) \\ e_w & \leftarrow \mathsf{SendWrite}(\mathsf{crs}, \mathsf{d}, L, \{b_i\}_{i \in [N]}, \{m_{j,0}, m_{j,1}\}_{j=1}^{|\mathsf{d}|}) \\ \{m'_j\}_{j=1}^{|\mathsf{d}|} & \leftarrow \mathsf{ReceiveWrite}^{\widehat{D}}(\mathsf{crs}, L, \{b_i\}_{i \in [N]}, e_w) \end{array}\right] = 1,$$

Sender Privacy: *There exists a PPT simulator* $\mathsf{Sim}_{\ell\mathsf{OT}}$ *such that the for any non-uniform PPT adversary* $\mathcal{A} = (\mathcal{A}_1, \mathcal{A}_2)$ *there exists a negligible function* $\mathsf{negl}(\cdot)$ *s.t.,*

$$\left| \Pr[\mathsf{Expt}^{\mathsf{real}}(1^\lambda, \mathcal{A}) = 1] - \Pr[\mathsf{Expt}^{\mathsf{ideal}}(1^\lambda, \mathcal{A}) = 1] \right| \leq \mathsf{negl}(\lambda)$$

where $\mathsf{Expt}^{\mathsf{real}}$ *and* $\mathsf{Expt}^{\mathsf{ideal}}$ *are described in Fig. 1.*

$\mathsf{Expt}^{\mathsf{real}}[1^\lambda, \mathcal{A}]$	$\mathsf{Expt}^{\mathsf{ideal}}[1^\lambda, \mathcal{A}]$
1. $\mathsf{crs} \leftarrow \mathsf{crsGen}(1^\lambda, 1^N)$.	1. $\mathsf{crs} \leftarrow \mathsf{crsGen}(1^\lambda)$.
2. $(D, L, \{m_{i,0}, m_{i,1}\}_{i \in [N]}, \mathsf{st}) \leftarrow$ $\mathcal{A}_1(\mathsf{crs})$.	2. $(D, L, \{m_{i,0}, m_{i,1}\}_{i \in [N]}, \mathsf{st}) \leftarrow$ $\mathcal{A}_1(\mathsf{crs})$.
3. $(\mathsf{d}, \widehat{D}) \leftarrow \mathsf{Hash}(\mathsf{crs}, D)$.	3. $(\mathsf{d}, \widehat{D}) \leftarrow \mathsf{Hash}(\mathsf{crs}, D)$.
4. Output	4. Output
$\mathcal{A}_2(\mathsf{st}, \mathsf{Send}(\mathsf{crs}, \mathsf{d}, L, \{m_{i,0}, m_{i,1}\}_{i \in [N]}))$.	$\mathcal{A}_2(\mathsf{st}, \mathsf{Sim}_{\ell\mathsf{OT}}(\mathsf{crs}, D, L, \{m_{i, D[L,i]}\}_{i \in [N]}))$.

Fig. 1. Sender privacy security game

Sender Privacy for Writes: *There exists a PPT simulator* $\mathsf{Sim}_{\ell\mathsf{OTW}}$ *such that the for any non-uniform PPT adversary* $\mathcal{A} = (\mathcal{A}_1, \mathcal{A}_2)$ *there exists a negligible function* $\mathsf{negl}(\cdot)$ *s.t.,*

$$\left| \Pr[\mathsf{WriSenPrivExpt}^{\mathsf{real}}(1^\lambda, \mathcal{A}) = 1] - \Pr[\mathsf{WriSenPrivExpt}^{\mathsf{ideal}}(1^\lambda, \mathcal{A}) = 1] \right| \leq \mathsf{negl}(\lambda)$$

where $\mathsf{WriSenPrivExpt}^{\mathsf{real}}$ *and* $\mathsf{WriSenPrivExpt}^{\mathsf{ideal}}$ *are described in Fig. 2.*

$\mathsf{WriSenPrivExpt}^{\mathsf{real}}[1^\lambda, \mathcal{A}]$	$\mathsf{WriSenPrivExpt}^{\mathsf{ideal}}[1^\lambda, \mathcal{A}]$
1. $\mathsf{crs} \leftarrow \mathsf{crsGen}(1^\lambda, 1^N)$.	1. $\mathsf{crs} \leftarrow \mathsf{crsGen}(1^\lambda, 1^N)$.
2. $(D, L, \{b_i\}_{i \in [N]}, \{m_{j,0}, m_{j,1}\}_{j \in [\lambda]}, \mathsf{st})$ $\leftarrow \mathcal{A}_1(\mathsf{crs})$.	2. $(D, L, \{b_i\}_{i \in [N]}, \{m_{j,0}, m_{j,1}\}_{j \in [\lambda]}, \mathsf{st})$ $\leftarrow \mathcal{A}_1(\mathsf{crs})$.
3. $(\mathsf{d}, \widehat{D}) \leftarrow \mathsf{Hash}(\mathsf{crs}, D)$.	3. $(\mathsf{d}, \widehat{D}) \leftarrow \mathsf{Hash}(\mathsf{crs}, D)$.
	4. $(\mathsf{d}^*, \widehat{D}^*) \leftarrow \mathsf{Hash}(\mathsf{crs}, D^*)$ where D^* be a database that is identical to D except that $D^*[L, i] = b_i$ for each $i \in [N]$.
4. $e_w \leftarrow \mathsf{SendWrite}(\mathsf{crs}, \mathsf{d}, L, \{b_i\}_{i \in [N]},$ $\{m_{j,0}, m_{j,1}\}_{j=1}^{|\mathsf{d}|})$	5. $e_w \leftarrow \mathsf{Sim}_{\ell\mathsf{OTW}}(\mathsf{crs}, D, L, \{b_i\}_{i \in [N]},$ $\{m_{j, \mathsf{d}_j^*}\}_{j \in [\lambda]})$
5. Output $\mathcal{A}_2(\mathsf{st}, e_w)$.	6. Output $\mathcal{A}_2(\mathsf{st}, e_w)$.

Fig. 2. Sender privacy for writes security game

Efficiency: *The algorithm Hash runs in time* $|D|\mathsf{poly}(\log|D|, \lambda)$. *The algorithms* Send, SendWrite, Receive, ReceiveWrite *run in time* $N \cdot \mathsf{poly}(\log|D|, \lambda)$.

Theorem 2 ([CDG+17,DG17,BLSV18,DGHM18]).   *Assuming either the Computational Diffie-Hellman assumption or the Factoring assumption or the Learning with Errors assumption, there exists a construction of updatable laconic oblivious transfer.*

Remark 1. We note that the security requirements given in Definition 1 is stronger than the one in [CDG+17] as we require the crs to be generated before the adversary provides the database D and the location L. However, the construction in [CDG+17] already satisfies this definition since in the proof, we can guess the location by incurring a $1/|D|$ loss in the security reduction.

3.2   Somewhere Equivocal Encryption

We now recall the definition of Somewhere Equivocal Encryption from the work of [HJO+16]. Informally, a somewhere equivocal encryption allows to create a simulated ciphertext encrypting a message m with certain positions of the message being "fixed" and the other positions having a "hole". The simulator can later fill these "holes" with arbitrary message values by deriving a suitable decryption key. The main efficiency requirement is that the size of the decryption key grows only with the number of "holes" and is otherwise independent of the message size. We give the formal definition below.

Definition 2 ([HJO+16]). *A somewhere equivocal encryption scheme with block-length s, message length n (in blocks) and equivocation parameter t (all polynomials in the security parameter) is a tuple of probabilistic polynomial algorithms $\Pi = (\mathsf{KeyGen}, \mathsf{Enc}, \mathsf{Dec}, \mathsf{SimEnc}, \mathsf{SimKey})$ such that:*

- key $\leftarrow$ KeyGen(1^λ): *It is a PPT algorithm that takes as input the security parameter (encoded in unary) and outputs a key* key.
- $\bar{c} \leftarrow$ Enc(key, $m_1 \ldots m_n$): *It is a PPT algorithm that takes as input a key* key *and a vector of messages $\overline{m} = m_1 \ldots m_n$ with each $m_i \in \{0,1\}^s$ and outputs a ciphertext $\bar{c}$.*
- $\overline{m} \leftarrow$ Dec(key, $\bar{c}$): *It is a deterministic algorithm that takes as input a key* key *and a ciphertext $\bar{c}$ and outputs a vector of messages $\overline{m} = m_1 \ldots m_n$.*
- (st, $\bar{c}$) $\leftarrow$ SimEnc(($m_i)_{i \notin I}, I$): *It is a PPT algorithm that takes as input a set of indices $I \subseteq [n]$ and a vector of messages $(m_i)_{i \notin I}$ and outputs a ciphertext $\bar{c}$ and a state* st.
- key$'$ $\leftarrow$ SimKey(st, $(m_i)_{i \in I}$): *It is a PPT algorithm that takes as input the state information* st *and a vector of messages $(m_i)_{i \in I}$ and outputs a key* key$'$.

and satisfies the following properties:

Correctness. *For every* key $\leftarrow$ KeyGen(1^λ), *for every $\overline{m} \in \{0,1\}^{s \times n}$ it holds that:*

$$\mathsf{Dec}(\mathsf{key}, \mathsf{Enc}(\mathsf{key}, \overline{m})) = \overline{m}$$

Simulation with No Holes. *We require that the distribution of $(\bar{c}, \text{key})$ computed via $(\text{st}, \bar{c}) \leftarrow \text{SimEnc}(\overline{m}, \emptyset)$ and $\text{key} \leftarrow \text{SimKey}(\text{st}, \emptyset)$ to be identical to $\text{key} \leftarrow \text{KeyGen}(1^\lambda)$ and $\bar{c} \leftarrow \text{Enc}(\text{key}, m_1 \dots m_n)$. In other words, simulation when there are no holes (i.e., $I = \emptyset$) is identical to honest key generation and encryption.*

Security. *For any PPT adversary $\mathcal{A}$, there exists a negligible function $\nu = \nu(\lambda)$ such that:*

$$\left| \Pr[\text{Exp}_{\mathcal{A},\Pi}^{\text{simenc}}(1^\lambda, 0) = 1] - \Pr[\text{Exp}_{\mathcal{A},\Pi}^{\text{simenc}}(1^\lambda, 1) = 1] \right| \leq \nu(\lambda)$$

where the experiment $\text{Exp}_{\mathcal{A},\Pi}^{\text{simenc}}$ is defined as follows:

Experiment $\text{Exp}_{\mathcal{A},\Pi}^{\text{simenc}}$

1. *The adversary $\mathcal{A}$ on input 1^λ outputs a set $I \subseteq [n]$ s.t. $|I| < t$, a vector $(m_i)_{i \notin I}$, and a challenge $j \in [n] \setminus I$. Let $I' = I \cup \{j\}$.*
2. *   – If $b = 0$, compute $\bar{c}$ as follows: $(\text{st}, \bar{c}) \leftarrow \text{SimEnc}((m_i)_{i \notin I}, I)$.*
   *   – If $b = 1$, compute $\bar{c}$ as follows: $(\text{st}, \bar{c}) \leftarrow \text{SimEnc}((m_i)_{i \notin I'}, I')$.*
3. *Send $\bar{c}$ to the adversary $\mathcal{A}$.*
4. *The adversary $\mathcal{A}$ outputs the set of remaining messages $(m_i)_{i \in I}$.*
   *   – If $b = 0$, compute key as follows: $\text{key} \leftarrow \text{SimKey}(\text{st}, (m_i)_{i \in I})$.*
   *   – If $b = 1$, compute key as follows: $\text{key} \leftarrow \text{SimKey}(\text{st}, (m_i)_{i \in I'})$*
5. *Send key to the adversary.*
6. *$\mathcal{A}$ outputs b' which is the output of the experiment.*

Theorem 3 ([HJO+16]). *Assuming the existence of one-way functions, there exists a somewhere equivocal encryption scheme for any polynomial message-length n, black-length s and equivocation parameter t, having key size $t \cdot s \cdot \text{poly}(\lambda)$ and ciphertext of size $n \cdot s \cdot \text{poly}(\lambda)$ bits.*

3.3   Random Access Machine (RAM) Model of Computation

We start by describing the Random Access Machine (RAM) model of computation in Sect. 3.3. Most of this subsection is taken verbatim from [CDG+17].

Notation for the RAM Model of Computation. The RAM model consists of a CPU and a memory storage of M blocks where each block has length N. The CPU executes a program that can access the memory by using read/write operations. In particular, for a program P with memory of size M, we denote the initial contents of the memory data by $D \in \{\{0,1\}^N\}^M$. Additionally, the program gets a "short" input $x \in \{0,1\}^n$, which we alternatively think of as the initial state of the program. We use $|P|$ to denote the running time of program P. We use the notation $P^D(x)$ to denote the execution of program P with initial memory contents D and input x. The program P can read from and write to various locations in memory D throughout its execution.[8]

---

[8] In general, the distinction between what to include in the program P, the memory data D and the short input x can be somewhat arbitrary. However as motivated by our applications we will typically be interested in a setting where the data D is large while the size of the program $|P|$ and input length x is small.

We will also consider the case where several different programs are executed sequentially and the memory persists between executions. We denote this process as $(y_1, \ldots, y_\ell) = (P_1(x_1), \ldots, P_\ell(x_\ell))^D$ to indicate that first $P_1^D(x_1)$ is executed, resulting in some memory contents D_1 and output y_1, then $P_2^{D_1}(x_2)$ is executed resulting in some memory contents D_2 and output y_2 etc. As an example, imagine that D is a huge database and the programs P_i are database queries that can read and possibly write to the database and are parameterized by some values x_i.

CPU-Step Circuit. Consider an execution of a RAM program which involves at most T CPU steps. We represent a RAM program P via T small *CPU-Step Circuits* each of which executes one CPU step. In this work we will denote one CPU step by:[9]

$$C_{\mathsf{CPU}}^P(\mathsf{state}, \mathsf{rData}) = (\mathsf{state}', \mathsf{R/W}, L, \mathsf{wData})$$

This circuit takes as input the current CPU state state and $\mathsf{rData} \in \{0,1\}^N$. Looking ahead the data rData will be read from the memory location that was requested by the previous CPU step. The circuit outputs an updated state state', a read or write $\mathsf{R/W}$, the next location to read/write from $L \in [M]$, and data wData to write into that location ($\mathsf{wData} = \bot$ when reading). The sequence of locations accessed during the execution of the program collectively form what is known as the *access pattern*, namely $\mathsf{MemAccess} = \{(\mathsf{R/W}^\tau, L^\tau) : \tau = 1, \ldots, T\}$. We assume that the CPU state state contains information about the location that the previous CPU step requested to read from. In particular, $\mathsf{lastLocation}(\mathsf{state})$ outputs the location that the previous CPU step requested to read and it is $\bot$ if the previous CPU step was a write.

Note that in the description above without loss of generality we have made some simplifying assumptions. We assume that each CPU-step circuit always reads from or writes to some location in memory. This is easy to implement via a dummy read and write step. Moreover, we assume that the instructions of the program itself are hardwired into the CPU-step circuits.

Representing RAM computation by CPU-Step Circuits. The computation $P^D(x)$ starts with the initial state set as $\mathsf{state}_1 = x$. In each step $\tau \in \{1, \ldots T\}$, the computation proceeds as follows: If $\tau = 1$ or $\mathsf{R/W}^{\tau-1} = \mathsf{write}$, then $\mathsf{rData}^\tau := \bot$; otherwise $\mathsf{rData}^\tau := D[L^{\tau-1}]$. Next it executes the CPU-Step Circuit $C_{\mathsf{CPU}}^{P,\tau}(\mathsf{state}^\tau, \mathsf{rData}^\tau) = (\mathsf{state}^{\tau+1}, \mathsf{R/W}^\tau, L^\tau, \mathsf{wData}^\tau)$. If $\mathsf{R/W}^\tau = \mathsf{write}$, then set $D[L^\tau] = \mathsf{wData}^\tau$. Finally, when $\tau = T$, then $\mathsf{state}^{\tau+1}$ is the output of the program.

3.4   Oblivious RAM

In this subsection, we recall the definition of oblivious RAM [Gol87, Ost90, GO96].

---

[9] In the definition below, we model each C_{CPU} as a deterministic circuit. Later, we extend the definition to allow each C_{CPU} to have access to random coins.

Definition 3 (Oblivious RAM). *An* Oblivious RAM *scheme consists of two procedures* (OProg, OData) *with the following syntax:*

- $P^* \leftarrow$ OProg($1^\lambda, 1^{\log M}, 1^T, P$): *Given a security parameter λ, a memory size M, a program P that runs in time T,* OProg *outputs an probabilistic oblivious program P^* that can access D^* as RAM. A probabilistic RAM program is modeled exactly as a deterministic program except that each step circuit additionally take random coins as input.*
- $D^* \leftarrow$ OData($1^\lambda, D$): *Given the security parameter λ, the contents of the database $D \in \{\{0,1\}^N\}^M$, outputs the oblivious database D^*. For convenience, we assume that* OData *works by compiling a program P that writes D to the memory using* OProg *to obtain P^*. It then evaluates the program P^* by using uniform random tape and outputs the contents of the memory as D^*.*

Efficiency. *We require that the run-time of* OData *should be $M \cdot N \cdot$ poly($\log(MN)$) $\cdot$ poly(λ), and the run-time of* OProg *should be $T \cdot$ poly(λ) $\cdot$ poly($\log(MN)$). Finally, the oblivious program P^* itself should run in time $T' = T \cdot$ poly(λ) $\cdot$ poly($\log(MN)$). Both the new memory size $M' = |D^*|$ and the running time T' should be efficiently computable from $M, N, T,$ and λ.*

Correctness. *Let $P_1, \ldots, P_\ell$ be programs running in polynomial times $t_1, \ldots, t_\ell$ on memory D of size M. Let $x_1, \ldots, x_\ell$ be the inputs and λ be a security parameter. Then we require that:*

$$\Pr[(P_1^*(x_1), \ldots, P_\ell^*(x_\ell))^{D^*} = (P_1(x_1), \ldots, P_\ell(x_\ell))^D] = 1$$

where $D^* \leftarrow$ OData($1^\lambda, D$), $P_i^* \leftarrow$ OProg($1^\lambda, 1^{\log M}, 1^T, P_i$) and $(P_1^*(x_1), \ldots, P_\ell^*(x_\ell))^{D^*}$ indicates running the ORAM programs on D^* sequentially using an uniform random tape.

Security. *For security, we require that there exists a PPT simulator* Sim *such that for any sequence of programs $P_1, \ldots, P_\ell$ (running in time $t_1, \ldots, t_\ell$ respectively), initial memory data $D \in \{\{0,1\}^N\}^M$, and inputs $x_1, \ldots, x_\ell$ we have that:*

$$\mathsf{MemAccess} \overset{s}{\approx} \mathsf{Sim}(1^\lambda, \{1^{t_i}\}_{i=1}^\ell)$$

where $(y_1, \ldots, y_\ell) = (P_1(x_1), \ldots, P_\ell(x_\ell))^D$, $D^* \leftarrow$ OData($1^\lambda, 1^N, D$), $P_i^* \leftarrow$ OProg($1^\lambda, 1^{\log M}, 1^T, P_i$) and MemAccess *corresponds to the access pattern of the CPU-step circuits during the sequential execution of the oblivious programs $(P_1^*(x_1), \ldots, P_\ell^*(x_\ell))^{D^*}$ using an uniform random tape.*

3.4.1 Strong Localized Randomness For our construction of adaptively secure garbled RAM, we need an additional property called as strong localized randomness property [CCHR16] from an ORAM scheme. We need a slightly stronger formalization than the one given in [CCHR16] (refer to footnote 10).

Strong Localized Randomness. Let $D \in \{\{0,1\}^N\}^M$ be any database and (P, x) be any program/input pair. Let $D^* \leftarrow$ OData($1^\lambda, 1^N, D$) and $P^* \leftarrow$

$\mathsf{OProg}(1^\lambda, 1^{\log M}, 1^T, P)$. Further, let the step circuits of P^* be indicated by $\{C_{\mathsf{CPU}}^{P^*, \tau}\}_{\tau \in [T']}$. Let R be the contents of the random tape used in the execution of P^*.

Definition 4 ([CCHR16]). *We say that an ORAM scheme has strong localized randomness property if there there exists a sequence of efficiently computable values $\tau_1 < \tau_2 < \ldots < \tau_m$ where $\tau_1 = 1$, $\tau_m = T'$ and $\tau_t - \tau_{t-1} \leq \mathsf{poly}(\log MN)$ for all $t \in [2, m]$ such that:*

1. *For every $j \in [m-1]$ there exists an interval I_j (efficiently computable from j) of size $\mathsf{poly}(\log MN, \lambda)$ s.t. for any $\tau \in [\tau_j, \tau_{j+1})$, the random tape accessed by $C_{\mathsf{CPU}}^{P^*, \tau}$ is given by R_{I_j} (here, R_{I_j} denotes the random tape restricted to the interval I_j).*
2. *For every $j, j' \in [m-1]$ and $j \neq j'$, $I_j \cap I_{j'} = \emptyset$.*
3. *Further, for every $j \in [m]$, there exists an $k < j$ such that given $R_{\backslash \{I_k \cup I_j\}}$ (where $R_{\backslash \{I_k \cup I_j\}}$ denotes the content of the random tape except in positions $I_j \cup I_k$) and the output of step circuits $C_{\mathsf{CPU}}^{P^*, \tau}$ for $\tau \in [\tau_k, \tau_{k+1})$, the memory access made by step circuits $C_{\mathsf{CPU}}^{P^*, \tau}$ for $\tau \in [\tau_j, \tau_{j+1})$ is computationally indistinguishable to random. This k is efficiently computable given the program P and the input x.*[10]

We argue in the full version of our paper that the Chung-Pass ORAM scheme [CP13] where the contents of the database are encrypted using a special encryption scheme satisfies the above definition of strong localized randomness. We now give details on this special encryption scheme. The key generation samples a puncturable PRF key $K \leftarrow \mathsf{PP.KeyGen}(1^\lambda)$. If the τth step-circuit has to write a value wData to a location L, it first samples $r \leftarrow \{0,1\}^\lambda$ and computes $c = (\tau \| r, \mathsf{PP.Eval}(K, \tau \| r) \oplus \mathsf{wData})$. It writes c to location L. The decryption algorithm uses K to first compute $\mathsf{PP.Eval}(K, \tau \| r)$ and uses it compute wData.

Remark 2. For the syntax of the ORAM scheme to be consistent with this special encryption scheme, we will use a puncturable PRF to generate the random tape of P^*. This key will also be used implicitly used to derive the key for this special encryption scheme.

3.5   Adaptive Garbled RAM

We now give the definition of adaptive garbled RAM.

Definition 5. *An adaptive garbled RAM scheme GRAM consists of the following PPT algorithms satisfying the correctness, efficiency and security properties (Fig. 3).*

---

[10] Here, we require that the memory access to be indistinguishable to random even given the outputs of the step circuits $C_{\mathsf{CPU}}^{P^*, \tau}$ for $\tau \in [\tau_k, \tau_{k+1})$. This is where we differ from the definition of [CCHR16].

- GRAM.Memory($1^\lambda, D$): *It is a PPT algorithm that takes the security parameter 1^λ and a database $D \in \{0,1\}^M$ as input and outputs a garbled database $\widetilde{D}$ and a secret key SK.*
- GRAM.Program(SK, i, P): *It is a PPT algorithm that takes as input a secret key SK, a sequence number i, and a program P as input (represented as a sequence of CPU steps) and outputs a garbled program $\widetilde{P}$.*
- GRAM.Input(SK, i, x): *It is a PPT algorithm that takes as input a secret key SK, a sequence number i and a string x as input and outputs the garbled input $\widetilde{x}$.*
- GRAM.Eval$^{\widetilde{D}}$(st, $\widetilde{P}, \widetilde{x}$): *It is a RAM program with random read write access to $\widetilde{D}$. It takes the state information st, garbled program $\widetilde{P}$ and the garbled input $\widetilde{x}$ as input and outputs a string y and updated database $\widetilde{D}'$.*

Correctness. *We say that a garbled RAM GRAM is correct if for every database D, $t = \mathsf{poly}(\lambda)$ and every sequence of program and input pair $\{(P_1, x_1), \ldots, (P_t, x_t)\}$ we have that*

$$\Pr[\mathsf{Expt}_{\mathsf{correctness}}(1^\lambda, \mathsf{UGRAM}) = 1] \leq \mathsf{negl}(\lambda)$$

where $\mathsf{Expt}_{\mathsf{correctness}}$ is defined in Fig. 5.

Adaptive Security. *We say that GRAM satisfies adaptive security if there exists (stateful) simulators (SimD, SimP, SimIn) such that for all t that is polynomial in the security parameter λ and for all polynomial time (stateful) adversaries $\mathcal{A}$, we have that*

$$\left| \Pr[\mathsf{Expt}_{\mathsf{real}}(1^\lambda, \mathsf{GRAM}, \mathcal{A}) = 1] - \Pr[\mathsf{Expt}_{\mathsf{ideal}}(1^\lambda, \mathsf{Sim}, \mathcal{A}) = 1] \right| \leq \mathsf{negl}$$

where $\mathsf{Expt}_{\mathsf{real}}, \mathsf{Expt}_{\mathsf{ideal}}$ are defined in Fig. 4.

Efficiency. *We require the following efficiency properties from a UGRAM scheme.*

- $(\widetilde{D}, SK) \leftarrow$ GRAM.Memory($1^\lambda, D$).
- Set $D_1 := D$, $\widetilde{D}_1 := \widetilde{D}$ and st $= \bot$.
- **for** every i from 1 to t
  - $\widetilde{P}_i \leftarrow$ GRAM.Program(SK, i, P_i) .
  - $\widetilde{x}_i \leftarrow$ GRAM.Input(SK, i, x_i).
  - Compute $(y_i, D_{i+1}) := P_i^{D_i}(x_i)$ and $(\widetilde{y}_i, \widetilde{D}_{i+1}, \mathsf{st}) :=$ UGRAM.Eval$^{\widetilde{D}_i}(i, \mathsf{st}, \widetilde{P}_i, \widetilde{x}_i)$.
- Output 1 if there exists an $i \in [t]$ such that $\widetilde{y}_i \neq y_i$.

Fig. 3. Correctness experiment for GRAM

$$
\begin{array}{ll}
\mathsf{Expt}_{\mathsf{real}}[1^\lambda, \mathsf{GRAM}, \mathcal{A}] & \mathsf{Expt}_{\mathsf{ideal}}[1^\lambda, \mathsf{Sim}, \mathcal{A}]
\end{array}
$$

- $D \leftarrow \mathcal{A}(1^\lambda)$ where $D \in \{0,1\}^M$
- $(\widetilde{D}, SK) \leftarrow \mathsf{GRAM.Memory}(1^\lambda, D)$.
- **for** every i from 1 to t
  - $P_i \leftarrow \mathcal{A}(\widetilde{D}, \{(\widetilde{P}_1, \widetilde{x}_1), \ldots, (\widetilde{P}_{i-1}, \widetilde{x}_{i-1})\})$.
  - $\widetilde{P}_i \leftarrow \mathsf{GRAM.Program}(SK, i, P_i)$.
  - $x_i \leftarrow \mathcal{A}(\widetilde{D}, \{(\widetilde{P}_1, \widetilde{x}_1), \ldots, (\widetilde{P}_{i-1}, \widetilde{x}_{i-1})\}, \widetilde{P}_i)$.
  - $\widetilde{x}_i \leftarrow \mathsf{GRAM.Input}(SK, i, x_i)$.
- Output $\mathcal{A}(\{(\widetilde{P}_1, \widetilde{x}_1), \ldots, (\widetilde{P}_t, \widetilde{x}_t)\})$.

- $D \leftarrow \mathcal{A}(1^\lambda)$ where $D \in \{0,1\}^M$.
- $(\widetilde{D}, \mathsf{st}) \leftarrow \mathsf{SimD}(1^\lambda, 1^M)$.
- **for** every i from 1 to t
  - $P_i \leftarrow \mathcal{A}(\widetilde{D}, \{(\widetilde{P}_1, \widetilde{x}_1), \ldots, (\widetilde{P}_{i-1}, \widetilde{x}_{i-1})\})$.
  - $(\widetilde{P}_i, \mathsf{st}) \leftarrow \mathsf{SimP}(1^{|P_i|}, \mathsf{st})$.
  - $x_i \leftarrow \mathcal{A}(\widetilde{D}, \{(\widetilde{P}_1, \widetilde{x}_1), \ldots, (\widetilde{P}_{i-1}, \widetilde{x}_{i-1})\}, \widetilde{P}_i)$.
  - $(y_i, D_{i+1}) := P_i^{D_i}(x_i)$ where $D_1 := D$.
  - $\widetilde{x}_i \leftarrow \mathsf{SimIn}(\mathsf{st}, y_i)$.
- Output $\mathcal{A}(\{(\widetilde{P}_1, \widetilde{x}_1), \ldots, (\widetilde{P}_t, \widetilde{x}_t)\})$.

Fig. 4. Adaptive security experiment for GRAM

- *The running time of* GRAM.Memory *should be bounded by* $M \cdot \mathsf{poly}(\log M) \cdot \mathsf{poly}(\lambda)$.
- *The running time of* GRAM.Program *should be bounded by* $T \cdot \mathsf{poly}(\log M) \cdot \mathsf{poly}(\lambda)$ *where* T *is the number of CPU steps in the description of the program* P.
- *The running time of* GRAM.Input *should be bounded by* $|x| \cdot \mathsf{poly}(\log M, \log T) \cdot \mathsf{poly}(\lambda)$.
- *The running time of* GRAM.Eval *should be bounded by* $T \cdot \mathsf{poly}(\log M) \cdot \mathsf{poly}(\lambda)$ *where* T *is the number of CPU steps in the description of the program* P.

4   Adaptive Garbled RAM with Unprotected Memory Access

Towards our goal of constructing an adaptive garbled RAM, we first construct an intermediate primitive with weaker security guarantees. We call this primitive as *adaptive garbled RAM with unprotected memory access*. Informally, a garbled RAM scheme has unprotected memory access if both the contents of the database and the access to the database are revealed in the clear to the adversary. We differ from the security definition given in [GHL+14] in three aspects. Firstly, we give an indistinguishability style definition for security whereas [GHL+14] give a simulation style definition. The indistinguishability based definition makes it easier to get full-fledged adaptive security later. Secondly and most importantly, we allow the adversary to adaptively choose the inputs based on the garbled program. Thirdly, we also require the garbled RAM scheme to satisfy a special property called as *equivocability*. Informally, equivocability requires that the real garbling of a program P is indistinguishable to a simulated garbling where the simulator is not provided with the description of the step circuits for a certain number of time steps (this number is given by the equivocation parameter). Later, when the input is specified, the simulator is given the output of these step circuits and must come-up with an appropriate garbled input.

We now give the formal definition of this primitive.

Definition 6. *An adaptive garbled RAM scheme with unprotected memory access UGRAM consists of the following PPT algorithms satisfying the correctness, efficiency and security properties.*

- UGRAM.Memory$(1^\lambda, 1^n, D)$: *It is a PPT algorithm that takes the security parameter 1^λ, an equivocation parameter n and a database $D \in \{\{0,1\}^N\}^M$ as input and outputs a garbled database $\widetilde{D}$ and a secret key SK.*
- UGRAM.Program(SK, i, P): *It is a PPT algorithm that takes as input a secret key SK, a sequence number i, and a program P as input (represented as a sequence of CPU steps) and outputs a garbled program $\widetilde{P}$.*
- UGRAM.Input(SK, i, x): *It is a PPT algorithm that takes as input a secret key SK, a sequence number i and a string x as input and outputs the garbled input $\widetilde{x}$.*
- UGRAM.Eval$^{\widetilde{D}}(\mathsf{st}, \widetilde{P}, \widetilde{x})$: *It is a RAM program with random read write access to $\widetilde{D}$. It takes the state information st, garbled program $\widetilde{P}$ and the garbled input $\widetilde{x}$ as input and outputs a string y and updated database $\widetilde{D}'$.*

Correctness. *We say that a garbled RAM UGRAM is correct if for every database D, $t = \mathsf{poly}(\lambda)$ and every sequence of program and input pair $\{(P_1, x_1), \ldots, (P_t, x_t)\}$ we have that*

$$\Pr[\mathsf{Expt}_{\mathsf{correctness}}(1^\lambda, \mathsf{UGRAM}) = 1] \leq \mathsf{negl}(\lambda)$$

where $\mathsf{Expt}_{\mathsf{correctness}}$ is defined in Fig. 5.

- $(\widetilde{D}, SK) \leftarrow$ UGRAM.Memory$(1^\lambda, 1^n, D)$.
- Set $D_1 := D$, $\widetilde{D}_1 := \widetilde{D}$ and $\mathsf{st} = \perp$.
- **for** every i from 1 to t
    - $\widetilde{P}_i \leftarrow$ UGRAM.Program(SK, i, P_i) .
    - $\widetilde{x}_i \leftarrow$ UGRAM.Input(SK, i, x_i).
    - Compute (y_i, D_{i+1}) := $P_i^{D_i}(x_i)$ and $(\widetilde{y}_i, \widetilde{D}_{i+1}, \mathsf{st})$:= UGRAM.Eval$^{\widetilde{D}_i}(i, \mathsf{st}, \widetilde{P}_i, \widetilde{x}_i)$.
- Output 1 if there exists an $i \in [t]$ such that $\widetilde{y}_i \neq y_i$.

Fig. 5. Correctness experiment for UGRAM

Security. *We require the following two properties to hold.*

- **Equivocability.** *There exists a simulator* Sim *such that for any non-uniform PPT stateful adversary* $\mathcal{A}$ *and* $t = \text{poly}(\lambda)$ *we require that:*

$$\left|\Pr[\text{Expt}_{\text{equiv}}(1^\lambda, \mathcal{A}, 0) = 1] - \Pr[\text{Expt}_{\text{equiv}}(1^\lambda, \mathcal{A}, 1) = 1]\right| \leq \text{negl}(\lambda)$$

where $\text{Expt}_{\text{equiv}}(1^\lambda, \mathcal{A}, b)$ *is described in Fig. 6.*
- **Adaptive Security.** *For any non-uniform PPT stateful adversary* $\mathcal{A}$ *and* $t = \text{poly}(\lambda)$ *we require that:*

$$\left|\Pr[\text{Expt}_{\text{UGRAM}}(1^\lambda, \mathcal{A}, 0) = 1] - \Pr[\text{Expt}_{\text{UGRAM}}(1^\lambda, \mathcal{A}, 1) = 1]\right| \leq \text{negl}(\lambda)$$

where $\text{Expt}_{\text{UGRAM}}(1^\lambda, \mathcal{A}, b)$ *is described in Fig. 7.*

---

1. $D \leftarrow \mathcal{A}(1^\lambda, 1^n)$.
2. $\widetilde{D}$ is computed as follows:
   (a) If $b = 0$: $(\widetilde{D}, SK) \leftarrow \text{UGRAM.Memory}(1^\lambda, 1^n, D)$.
   (b) If $b = 1$: $\widetilde{D} \leftarrow \text{Sim}(1^\lambda, 1^n, D)$.
3. **for** each i from t:
   (a) $(P_i, I) \leftarrow \mathcal{A}(\widetilde{D}, \{\widetilde{P}_j, \widetilde{x}_j\}_{j \in [i-1]})$ where $I \subset [\|P_i\|]$ and $|I| \leq n$.
   (b) $\widetilde{P}_i$ is computed as follows:
       i. If $b = 0$: $\widetilde{P}_i \leftarrow \text{UGRAM.Program}(SK, i, P_i)$.
       ii. If $b = 1$: $\widetilde{P}_i \leftarrow \text{Sim}(\{C_{\text{CPU}}^{P_i,t}\}_{t \notin I})$
   (c) $x_i \leftarrow \mathcal{A}(\{\widetilde{P}_j, \widetilde{x}_j\}_{j \in [i-1]}, \widetilde{P}_i)$.
   (d) $\widetilde{x}_i$ is computed as follows:
       i. If $b = 0$: $\widetilde{x}_i \leftarrow \text{UGRAM.Input}(SK, i, x_i)$
       ii. If $b = 1$: $\widetilde{x}_i \leftarrow \text{Sim}(x_i, \{y_t\}_{t \in I})$ where y_t is the o/p of $C_{\text{CPU}}^{P_i,t}$ when P_i is executed with x_i.
4. $b' \leftarrow \mathcal{A}(\{\widetilde{P}_j, \widetilde{x}_j\}_{j \in [t]})$.
5. Output b'.

Fig. 6. $\text{Expt}_{\text{equiv}}(1^\lambda, \mathcal{A}, b)$

Efficiency. *We require the following efficiency properties from a* UGRAM *scheme.*

- *The running time of* UGRAM.Memory *should be bounded by* $MN \cdot \text{poly}(\log MN) \cdot \text{poly}(\lambda)$.
- *The running time of* UGRAM.Program *should be bounded by* $T \cdot \text{poly}(\log MN) \cdot \text{poly}(\lambda)$ *where* T *is the number of CPU steps in the description of the program* P.
- *The running time of* UGRAM.Input *should be bounded by* $n \cdot |x| \cdot \text{poly}(\log MN, \log T) \cdot \text{poly}(\lambda)$.

---

1. $D \leftarrow \mathcal{A}(1^\lambda, 1^n)$.
2. $(\widetilde{D}, SK) \leftarrow \mathsf{UGRAM.Memory}(1^\lambda, 1^n, D)$.
3. **for** x each i from t:
   (a) $(P_{i,0}, P_{i,1}) \leftarrow \mathcal{A}(\widetilde{D}, \{\widetilde{P}_j, \widetilde{x}_j\}_{j \in [i-1]})$.
   (b) $\widetilde{P}_i$ is computed as follows:
       i. If $b = 0 : \widetilde{P}_i \leftarrow \mathsf{UGRAM.Program}(SK, i, P_{i,0})$.
       ii. If $b = 1 : \widetilde{P}_i \leftarrow \mathsf{UGRAM.Program}(SK, i, P_{i,1})$
   (c) $x_i \leftarrow \mathcal{A}(\{\widetilde{P}_j, \widetilde{x}_j\}_{j \in [i-1]}, \widetilde{P}_i)$.
   (d) $\widetilde{x}_i \leftarrow \mathsf{UGRAM.Input}(i, SK, x_i)$
4. $b' \leftarrow \mathcal{A}(\{\widetilde{P}_j, \widetilde{x}_j\}_{j \in [t]})$.
5. Output b' if the output of each step circuit in $P_{i,0}^D(x_i)$ is same as $P_{i,1}^D(x_i)$ for every $i \in [t]$.

Fig. 7. $\mathsf{Expt}_{\mathsf{UGRAM}}(1^\lambda, \mathcal{A}, b)$

- *The running time of* UGRAM.Eval *should be bounded by* $T \cdot \mathsf{poly}(\log MN, \log T) \cdot \mathsf{poly}(\lambda)$ *where T is the number of CPU steps in the description of the program P.*

4.1 Construction

In this subsection, we give a construction of adaptive garbled RAM with unprotected memory access from updatable laconic oblivious transfer, somewhere equivocal encryption and garbling scheme for circuits with selective security using the techniques developed in the construction of adaptive garbled circuits [GS18a]. Our main theorem is:

Theorem 4. *Assuming the existence of updatable laconic oblivious transfer, somewhere equivocal encryption, a pseudorandom function and garbling scheme for circuits with selective security, there exists a construction of adaptive garbled RAM with unprotected memory access.*

Construction. We give the formal description of the construction in Fig. 8. We use a somewhere equivocal encryption with block length set to $|\widetilde{\mathsf{SC}}_\tau|$ where $\widetilde{\mathsf{SC}}_\tau$ denotes the garbled version of the step circuit SC described in Fig. 9, the message length to be T (which is the running time of the program P) and the equivocation parameter to be $t + \log T$ where t is the actual equivocation parameter for the UGRAM scheme.

Correctness. The correctness of the above construction follows from a simple inductive argument that for each step $\tau \in [\|P\|]$, the state and the database are updated correctly at the end of the execution of $\widetilde{\mathsf{SC}}_\tau$. The base case is $\tau = 0$. In order to prove the inductive step for a step τ, observe that if the step τ outputs a read then labels recovered in Step 4.(c).(ii) of SS-EvalCkt correspond to data

UGRAM.Memory$(1^\lambda, 1^t, D)$: On input a database $D \in \{\{0,1\}^N\}^M$ do:
1. Sample crs $\leftarrow$ crsGen$(1^\lambda, 1^N)$ and $K \leftarrow$ PRFKeyGen(1^λ) defining PRF$_K$: $\{0,1\}^{2\lambda+1} \rightarrow \{0,1\}^\lambda$.
2. For each $k \in [\lambda]$ and $b \in \{0,1\}$, compute $\mathsf{lab}^1_{k,b} := \mathsf{PRF}_K(1\|k\|b)$.
3. Compute $(d, \widehat{D}) = \mathsf{Hash}(\mathsf{crs}, D)$.
4. Output $\widehat{D}, \{\mathsf{lab}^1_{k,d_k}\}_{k\in[\lambda]}$ as the garbled memory and (K, crs) as the secret key.

UGRAM.Program(SK, i, P): On input $SK = (K, \mathsf{crs})$, sequence number i, and a program P (with T step-circuits) do:
1. For each step $\tau \in [2, T]$, $k \in [\lambda + n + N]$ and $b \in \{0,1\}$,
   (a) Sample $\mathsf{lab}^\tau_{k,b} \leftarrow \{0,1\}^\lambda$.
   (b) Set $\mathsf{lab}^1_{k,b} := \mathsf{PRF}_K(i\|k\|b)$ and $\mathsf{lab}^{T+1}_{k,b} := \mathsf{PRF}_K((i+1)\|k\|b)$.
   We use $\{\mathsf{lab}^\tau_{k,b}\}$ to denote $\{\mathsf{lab}^\tau_{k,b}\}_{k\in[\lambda+n+N],b\in\{0,1\}}$.
2. **for** each τ from T down to 1 **do**:
   (a) Compute $\widetilde{\mathsf{SC}}_\tau \leftarrow \mathsf{GarbleCkt}\left(1^\lambda, \mathsf{SC}[\mathsf{crs}, \tau, \{\mathsf{lab}^{\tau+1}_{k,b}\}], \{\mathsf{lab}^\tau_{k,b}\}\right)$ where the step-circuit SC is described in Figure 9.
3. Compute key = KeyGen$(1^\lambda; \mathsf{PRF}_K(i\|0^\lambda\|0))$
4. Compute $c \leftarrow \mathsf{Enc}(\mathsf{key}, \{\widetilde{\mathsf{SC}}_\tau\}_{\tau\in[T]})$ and output $\widetilde{P} := c$.

UGRAM.Input(SK, i, x) : On input the secret key $SK = (K, \mathsf{crs})$, sequence number i and a string $x \in \{0,1\}^n$ do:
1. For each $k \in [\lambda + n + N]$ and $b \in \{0,1\}$, compute $\mathsf{lab}^1_{k,b} := \mathsf{PRF}_K(i\|k\|b)$.
2. Compute key = KeyGen$(1^\lambda; \mathsf{PRF}_K(i\|0^\lambda\|0))$.
3. Output $\widetilde{x} := \left(\mathsf{key}, \{\mathsf{lab}^1_{k,x_k}\}_{k\in[\lambda+1,\lambda+n]}, \{\mathsf{lab}^1_{k,0}\}_{k\in[n+\lambda+1,n+\lambda+N]}\right)$.

UGRAM.Eval$^{\widetilde{D}}(i, \mathsf{st}, \widetilde{P}, \widetilde{x})$: On input i, state st, the garbled program $\widetilde{P}$, and garbled input $\widetilde{x}$ do:
1. Parse $\widetilde{x}$ as $\left(\mathsf{key}, \{\mathsf{lab}_k\}_{k\in[\lambda+1,n+\lambda+N]}\right)$ and $\widetilde{P}$ as c.
2. If $i = 1$, obtain $\{\mathsf{lab}_k\}_{k\in[\lambda]}$ from garbled memory; else, parse st as $\{\mathsf{lab}_k\}_{k\in[\lambda]}$.
3. Compute $\{\widetilde{\mathsf{SC}}_\tau\}_{\tau\in[T]} := \mathsf{Dec}(\mathsf{key}, c)$ and set $\overline{\mathsf{lab}} := \{\mathsf{lab}_k\}_{k\in[n+\lambda+N]}$.
4. **for** each τ from 1 to T **do**:
   (a) Compute $(R/W, L, A, \{\mathsf{lab}_k\}_{k\in[\lambda+1,n+\lambda]}, B) := \mathsf{EvalCkt}(\widetilde{\mathsf{SC}}_\tau, \overline{\mathsf{lab}})$.
   (b) If $R/W = \mathsf{write}$,
      i. Parse A as (e_w, wData) and B as $\{\mathsf{lab}_k\}_{k\in[\lambda+1,n+\lambda+N]}$.
      ii. $\{\mathsf{lab}_k\}_{k\in[\lambda]} \leftarrow \mathsf{ReceiveWrite}^{\widehat{D}}(\mathsf{crs}, L, \mathsf{wData}, e_w)$
   (c) **else**,
      i. Parse A as $\{\mathsf{lab}_k\}_{k\in[n+\lambda]}$ and B as e.
      ii. $\{\mathsf{lab}_k\}_{k\in[n+\lambda+1,n+\lambda+N]} \leftarrow \mathsf{Receive}^{\widehat{D}}(\mathsf{crs}, L, e)$
   (d) Set $\overline{\mathsf{lab}} := \{\mathsf{lab}_k\}_{k\in[n+\lambda+N]}$.
5. Parse $\overline{\mathsf{lab}}$ as $\{\mathsf{lab}_k\}_{k\in[n+\lambda+N]}$. Output $\{\mathsf{lab}_k\}_{k\in[\lambda+1,n+\lambda]}$ and st $:= \{\mathsf{lab}_k\}_{k\in[\lambda]}$.

Fig. 8. Adaptive garbled RAM with unprotected memory access

Step Circuit SC

Input: A digest d, state state and a block rData.
Hardcoded: The common reference string crs, the step number τ and a set of labels $\{\mathsf{lab}_{k,b}\}$.

1. Compute $(\mathsf{state}', \mathsf{R/W}, L, \mathsf{wData}) := C_{\mathsf{CPU}}^{P,\tau}(\mathsf{state}, \mathsf{rData})$.
2. If $\tau = T$, reset $\mathsf{lab}_{k,b} = b$ for all $k \in [\lambda+1, \lambda+n]$ and $b \in \{0,1\}$.
3. **if** R/W = write **do:**
   (a) Compute $e_w \leftarrow \mathsf{SendWrite}(\mathsf{crs}, \mathsf{d}, L, \mathsf{wData}, \{\mathsf{lab}_{k,b}\}_{k \in [\lambda], b \in \{0,1\}})$.
   (b) Output $(\mathsf{R/W}, L, e_w, \mathsf{wData}, \{\mathsf{lab}_{k,\mathsf{state}'_{k-\lambda}}\}_{k \in [\lambda+1, \lambda+n]}, \{\mathsf{lab}_{k,0}\}_{k \in [n+\lambda+1, n+\lambda+N]})$.
4. **else,**
   (a) Compute $e \leftarrow \mathsf{Send}(\mathsf{crs}, \mathsf{d}, L, \{\mathsf{lab}_{k,b}\}_{k \in [n+\lambda+1, n+\lambda+N], b \in \{0,1\}})$.
   (b) $(\mathsf{R/W}, L, \{\mathsf{lab}_{k,\mathsf{d}_k}\}_{k \in [\lambda]}, \{\mathsf{lab}_{k,\mathsf{state}'_{k-\lambda}}\}_{k \in [\lambda+1, \lambda+n]}, e)$.

Fig. 9. Description of the step circuit

block in the location requested. Otherwise, the labels recovered in Step 4(b).(ii) of SS-EvalCkt corresponds to the updated value of the digest with the corresponding block written to the database.

Efficiency. The efficiency of our construction directly follows from the efficiency of updatable laconic oblivious transfer and the parameters set for the somewhere equivocal encryption. In particular, the running time of UGRAM.Memory is $D \cdot \mathsf{poly}(\lambda)$, UGRAM.Program is $T \cdot \mathsf{poly}(\log MN, \lambda)$ and that of UGRAM.Input is $n|x| \cdot \mathsf{poly}(\log M, \log T, \lambda)$. The running time of UGRAM.Eval is $T \cdot \mathsf{poly}(\log M, \log T, \lambda)$.

Security. We prove the security of this construction in the full version of our paper.

5   Timed Encryption

In this section, we give the definition and construction of a timed encryption scheme. We will use a timed encryption scheme in the construction of adaptive garbled RAM in the next section.

A timed encryption scheme is a symmetric key encryption scheme with some special properties. In this encryption scheme, every message is encrypted with respect to a timestamp time. Additionally, there is a special algorithm called as constrain that takes an encryption key K and a timestamp time' as input and outputs a time constrained key $K[\mathsf{time}']$. A time constrained key $K[\mathsf{time}']$ can be used to decrypt any ciphertext that is encrypted with respect to timestamp time < time'. For security, we require that knowledge of a time constrained key does not help an adversary to distinguish between encryptions of two messages that are encrypted with respect to some future timestamp.

Definition 7. *A timed encryption scheme is a tuple of algorithms* (TE.KeyGen, TE.Enc, TE.Dec, TE.Constrain) *with the following syntax.*

- TE.KeyGen(1^λ): *It is a randomized algorithm that takes the security parameter* 1^λ *and outputs a key* K.
- TE.Constrain(K, time): *It is a deterministic algorithm that takes a key* K *and a timestamp* time $\in [0, 2^\lambda - 1]$ *and outputs a time-constrained key* $K[\text{time}]$.
- TE.Enc(K, time, m): *It is a randomized algorithm that takes a key* K, *a times-tamp* time *and a message* m *as input and outputs a ciphertext* c *or* $\perp$.
- TE.Dec(K, c): *It is a deterministic algorithm that takes a key* K *and a cipher-text* c *as input and outputs a message* m.

We require a timed encryption scheme to follow the following properties.

Correctness. *We require that for all messages* m *and for all timestamps* $\text{time}_1 \leq \text{time}_2$:

$$\Pr[\text{TE.Dec}(K[\text{time}_2], c) = m] = 1$$

where $K \leftarrow \text{TE.KeyGen}(1^\lambda)$, $K[\text{time}_2] := \text{TE.Constrain}(K, \text{time}_2)$ *and* $c \leftarrow \text{TE.Enc}(K, \text{time}_1, m)$.

Encrypting with Constrained Key. *For any message* m *and timestamps* $\text{time}_1 \leq \text{time}_2$, *we require that:*

$$\{\text{TE.Enc}(K, \text{time}_1, m)\} \approx \{\text{TE.Enc}(K[\text{time}_2], \text{time}_1, m)\}$$

where $K \leftarrow \text{TE.KeyGen}(1^\lambda)$, $K[\text{time}_2] := \text{TE.Constrain}(K, \text{time}_2)$ *and* $\approx$ *denotes that the two distributions are identical.*

Security. *For any two messages* m_0, m_1 *and timestamps* $(\text{time}, \{\text{time}_i\}_{i \in [t]})$ *where* $\text{time}_i < \text{time}$ *for all* $i \in [t]$, *we require that:*

$$\{\{K[\text{time}_i]\}_{i \in [t]}, \text{TE.Enc}(K, \text{time}, m_0)\} \overset{c}{\approx} \{\{K[\text{time}_i]\}_{i \in [t]}, \text{TE.Enc}(K, \text{time}, m_1)\}$$

where $K \leftarrow \text{TE.KeyGen}(1^\lambda)$ *and* $K[\text{time}_i] := \text{TE.Constrain}(K, \text{time}_i)$ *for every* $i \in [t]$.

We prove the following theorem in the full version of our paper.

Theorem 5. *Assuming the existence of one-way functions, there exists a con-struction of timed encryption.*

6   Construction of Adaptive Garbled RAM

In this section, we give a construction of adaptive garbled RAM. We make use of the following primitives.

- A timed encryption scheme (TE.KeyGen, TE.Enc, TE.Dec, TE.Constrain). Let N be the output length of TE.Enc when encrypting single bit messages.

GRAM.Memory($1^\lambda, D$): On input the database $D \in \{0,1\}^M$:
    1. Sample $K \leftarrow$ TE.KeyGen(1^λ) and $S \leftarrow$ PRFKeyGen(1^λ) defining PRF_S :
       $\{0,1\}^\lambda \to \{0,1\}^n$ (where n is the input length of each program).
    2. Initialize an empty array $\widehat{D}$ of M blocks with block length N.
    3. **for** each i from 1 to M **do:**
       (a) Set $\widehat{D}[i] \leftarrow$ TE.Enc($K, 0^\lambda, D[i]$).
    4. $D^* \leftarrow$ OData($1^\lambda, 1^N, \widehat{D}$).
    5. $(\widetilde{D}, SK) \leftarrow$ UGRAM.Memory($1^\lambda, 1^t, D^*$) where $t = \mathsf{poly}(\log MN)$.
    6. Output $\widetilde{D}$ as the garbled memory and (K, S, SK) as the secret key.
GRAM.Program(SK', i, P): On input $SK' = (K, S, SK)$, sequence number i, and
    a program P:
    1. Sample $K' \leftarrow$ PP.KeyGen(1^λ)
    2. $P^* \leftarrow$ OProg($1^\lambda, 1^{\log M}, 1^T, P$) where P^* runs in time T'.
    3. For each $\tau \in [T']$, compute $K[(i\|\tau)] \leftarrow$ TE.Constrain($K, (i\|\tau)$) where $(i\|\tau)$
       is expressed as a λ-bit string.
    4. Compute $r := \mathsf{PRF}_S(i)$.
    5. Let $\tau_1, \ldots, \tau_m$ be the sequence of values guaranteed by strong localized
       randomness.
    6. **for** each $\tau \in [T']$ **do:**
       (a) Let $j \in [m-1]$ be such that $\tau \in [\tau_j, \tau_{j+1})$.
       (b) Let $C_{\mathsf{CPU}}^\tau := \mathsf{SC}_\tau[i, \tau, K[(i\|\tau)], I_j, K', r']$ where $r' = r$ if $\tau = T'$, else
          $r' = \bot$. The step circuit SC is described in Figure 11
    7. Construct a RAM program P' with step-circuits given by $\{C_{\mathsf{CPU}}^\tau\}$.
    8. $\widetilde{P} \leftarrow$ UGRAM.Program(SK, i, P').
    9. Output $\widetilde{P}$.
GRAM.Input(SK', i, P): On input $SK' = (K, S, SK)$, i and x:
    1. Compute $r = \mathsf{PRF}_S(i)$
    2. Compute $\widehat{x} \leftarrow$ UGRAM.Input(SK, i, x)
    3. Output $\widetilde{x} = (\widehat{x}, r)$.
GRAM.Eval$^{\widetilde{D}}(i, \mathsf{st}, \widetilde{P}, \widetilde{x})$: On input state st, the garbled program $\widetilde{P}$, and garbled
    input $\widetilde{x}$:
    1. Compute $(y, \mathsf{st}') \leftarrow$ UGRAM.Eval$^{\widetilde{D}}(\mathsf{st}, \widetilde{P}, \widehat{x})$ and update st to st'. Output
       $y \oplus r$.

Fig. 10. Construction of adaptive GRAM

- A puncturable pseudorandom function (PP.KeyGen, PP.Eval, PP.Punc).
- An oblivious RAM scheme (OData, OProg) with strong localized randomness.
- An adaptive garbled RAM scheme UGRAM with unprotected memory access.

The formal description of our construction appears in Fig. 10.

Correctness. We give an informal argument for correctness. The only difference
between UGRAM and the construction we give in Fig. 10 is that we encrypt the
database using a timed encryption scheme and encode it using a ORAM scheme.
To argue the correctness of our construction, it is sufficient to argue that each

---

Step Circuit SC_τ

Input: A ciphertext c_{CPU} and a data block $X \in \{0,1\}^N$.
Hardcoded: The sequence number i, step number τ, the constrained key $K[(i\|\tau)]$, the interval I_j, the key K' and a string r'.

1. Compute $rData := TE.Dec(K[(i\|\tau)], X)$ and $state = TE.Dec(K[(i\|\tau)], c_{CPU})$.
2. Compute $R_{I_j} = PP.Eval(K', I_j)$.
3. Compute $(R/W, L, state', wData) := C_{CPU}^{P^*, \tau}(state, rData, R_{I_j})$.
4. **if** $\tau = T'$, then output $c'_{CPU} = state' \oplus r'$; else $c'_{CPU} = TE.Enc(K[(i\|\tau)], state')$.
5. **else if** $R/W = $ write **do:**
   (a) Compute $X' \leftarrow TE.Enc(K[i\|\tau], (i, \tau), wData)$.
   (b) Output $(c'_{CPU}, R/W, L, X')$.
6. **else if** $R/W = $ read, output $(c'_{CPU}, R/W, L, \perp)$.

Fig. 11. Description of the step circuit

step circuit SC faithfully emulates the corresponding step circuit of P^*. Let $SC^{i,\tau}$ be the step circuit that corresponds to the τth step of the ith program P_i. We observe that any point in time the Lth location of the database $\widehat{D}$ is an encryption of the actual data bit with respect to timestamp $time := (i'\|\tau')$ where $SC^{i',\tau'}$ last wrote at the Lth location. It now follows from this invariant and the correctness of the timed encryption scheme that the hardwired constrained key $K[i\|\tau]$ in $SC^{i,\tau}$ can be used to decrypt the read block X as the step that last modified this block has a timestamp that is less than $(i\|\tau)$.

Efficiency. We note that setting the equivocation parameter $n = poly(\log MN)$, we obtain that the running time of $GRAM.Input$ is $|x| \cdot poly(\lambda, \log MN)$. The rest of the efficiency criterion follow directly from the efficiency of adaptive garbled RAM with unprotected memory access.

Security. We give the proof of security in the full version of our paper.

References

[ACC+16] Ananth, P., Chen, Y.-C., Chung, K.-M., Lin, H., Lin, W.-K.: Delegating RAM computations with adaptive soundness and privacy. In: Hirt, M., Smith, A.D. (eds.) TCC 2016-B, Part II. LNCS, vol. 9986, pp. 3–30. Springer, Heidelberg (2016). https://doi.org/10.1007/978-3-662-53644-5_1

[AIK04] Applebaum, B., Ishai, Y., Kushilevitz, E.: Cryptography in NC^0. In: 45th FOCS, pp. 166–175. IEEE Computer Society Press, October 2004

[App17] Applebaum, B.: Garbled circuits as randomized encodings of functions: a primer. IACR Cryptology ePrint Archive, 2017:385 (2017)

[BGI+01] Barak, B., et al.: On the (im)possibility of obfuscating programs. In: Kilian, J. (ed.) CRYPTO 2001. LNCS, vol. 2139, pp. 1–18. Springer, Heidelberg (2001). https://doi.org/10.1007/3-540-44647-8_1

[BGL+15] Bitansky, N., Garg, S., Lin, H., Pass, R., Telang, S.: Succinct randomized encodings and their applications. In: Servedio, R.A., Rubinfeld, R. (eds.) 47th ACM STOC, pp. 439–448. ACM Press, June 2015

[BHR12a] Bellare, M., Hoang, V.T., Rogaway, P.: Adaptively secure garbling with applications to one-time programs and secure outsourcing. In: Wang, X., Sako, K. (eds.) ASIACRYPT 2012. LNCS, vol. 7658, pp. 134–153. Springer, Heidelberg (2012). https://doi.org/10.1007/978-3-642-34961-4_10

[BHR12b] Bellare, M., Hoang, V.T., Rogaway, P.: Foundations of garbled circuits. In: Yu, T., Danezis, G., Gligor, V.D. (eds.) ACM CCS 2012, pp. 784–796. ACM Press, October 2012

[BL18] Benhamouda, F., Lin, H.: k-round multiparty computation from k-round oblivious transfer via garbled interactive circuits. In: Nielsen, J.B., Rijmen, V. (eds.) EUROCRYPT 2018, Part II. LNCS, vol. 10821, pp. 500–532. Springer, Cham (2018). https://doi.org/10.1007/978-3-319-78375-8_17

[BLSV18] Brakerski, Z., Lombardi, A., Segev, G., Vaikuntanathan, V.: Anonymous IBE, leakage resilience and circular security from new assumptions. In: Nielsen, J.B., Rijmen, V. (eds.) EUROCRYPT 2018. LNCS, vol. 10820, pp. 535–564. Springer, Cham (2018). https://doi.org/10.1007/978-3-319-78381-9_20. https://eprint.iacr.org/2017/967

[BMR90] Beaver, D., Micali, S., Rogaway, P.: The round complexity of secure protocols (extended abstract). In: 22nd ACM STOC, pp. 503–513. ACM Press, May 1990

[BR93] Bellare, M., Rogaway, P.: Random oracles are practical: a paradigm for designing efficient protocols. In: Ashby, V. (ed.) ACM CCS 1993, pp. 62–73. ACM Press, November 1993

[BW13] Boneh, D., Waters, B.: Constrained pseudorandom functions and their applications. In: Sako, K., Sarkar, P. (eds.) ASIACRYPT 2013, Part II. LNCS, vol. 8270, pp. 280–300. Springer, Heidelberg (2013). https://doi.org/10.1007/978-3-642-42045-0_15

[CCHR16] Canetti, R., Chen, Y., Holmgren, J., Raykova, M.: Adaptive succinct garbled RAM or: how to delegate your database. In: Hirt, M., Smith, A. (eds.) TCC 2016-B, Part II. LNCS, vol. 9986, pp. 61–90. Springer, Heidelberg (2016). https://doi.org/10.1007/978-3-662-53644-5_3

[CDG+17] Cho, C., Döttling, N., Garg, S., Gupta, D., Miao, P., Polychroniadou, A.: Laconic oblivious transfer and its applications. In: Katz, J., Shacham, H. (eds.) CRYPTO 2017. LNCS, vol. 10402, pp. 33–65. Springer, Cham (2017). https://doi.org/10.1007/978-3-319-63715-0_2

[CH16] Canetti, R., Holmgren, J.: Fully succinct garbled RAM. In: Sudan, M. (ed.) ITCS 2016, pp. 169–178. ACM, January 2016

[CHJV15] Canetti, R., Holmgren, J., Jain, A., Vaikuntanathan, V.: Succinct garbling and indistinguishability obfuscation for RAM programs. In: Servedio, R.A., Rubinfeld, R. (eds.) 47th ACM STOC, pp. 429–437. ACM Press, June 2015

[CP13] Chung, K.-M., Pass, R.: A simple ORAM. Cryptology ePrint Archive, Report 2013/243 (2013). https://eprint.iacr.org/2013/243

[DG17] Döttling, N., Garg, S.: Identity-based encryption from the Diffie-Hellman assumption. In: Katz, J., Shacham, H. (eds.) CRYPTO 2017. LNCS, vol. 10401, pp. 537–569. Springer, Cham (2017). https://doi.org/10.1007/978-3-319-63688-7_18

[DGHM18] Döttling, N., Garg, S., Hajiabadi, M., Masny, D.: New constructions of identity-based and key-dependent message secure encryption schemes. In: Abdalla, M., Dahab, R. (eds.) PKC 2018. LNCS, vol. 10769, pp. 3–31. Springer, Cham (2018). https://doi.org/10.1007/978-3-319-76578-5_1. https://eprint.iacr.org/2017/978

[GGH+13] Garg, S., Gentry, C., Halevi, S., Raykova, M., Sahai, A., Waters, B.: Candidate indistinguishability obfuscation and functional encryption for all circuits. In: 54th FOCS, pp. 40–49. IEEE Computer Society Press, October 2013

[GGM86] Goldreich, O., Goldwasser, S., Micali, S.: How to construct random functions. J. ACM **33**(4), 792–807 (1986)

[GGMP16] Garg, S., Gupta, D., Miao, P., Pandey, O.: Secure multiparty RAM computation in constant rounds. In: Hirt, M., Smith, A. (eds.) TCC 2016-B, Part I. LNCS, vol. 9985, pp. 491–520. Springer, Heidelberg (2016). https://doi.org/10.1007/978-3-662-53641-4_19

[GHL+14] Gentry, C., Halevi, S., Lu, S., Ostrovsky, R., Raykova, M., Wichs, D.: Garbled RAM revisited. In: Nguyen, P.Q., Oswald, E. (eds.) EUROCRYPT 2014. LNCS, vol. 8441, pp. 405–422. Springer, Heidelberg (2014). https://doi.org/10.1007/978-3-642-55220-5_23

[GHRW14] Gentry, C., Halevi, S., Raykova, M., Wichs, D.: Outsourcing private RAM computation. In: 55th FOCS, pp. 404–413. IEEE Computer Society Press, October 2014

[GKK+12] Gordon, S.D., Katz, J., Kolesnikov, V., Krell, F., Malkin, T., Raykova, M., Vahlis, Y.: Secure two-party computation in sublinear (amortized) time. In: Yu, T., Danezis, G., Gligor, V.D. (eds.) ACM CCS 2012, pp. 513–524. ACM Press, October 2012

[GKP+13] Goldwasser, S., Kalai, Y.T., Popa, R.A., Vaikuntanathan, V., Zeldovich, N.: Reusable garbled circuits and succinct functional encryption. In: Boneh, D., Roughgarden, T., Feigenbaum, J. (eds.) 45th ACM STOC, pp. 555–564. ACM Press, June 2013

[GLO15] Garg, S., Lu, S., Ostrovsky, R.: Black-box garbled RAM. In: Guruswami, V. (ed.) 56th FOCS, pp. 210–229. IEEE Computer Society Press, October 2015

[GLOS15] Garg, S., Lu, S., Ostrovsky, R., Scafuro, A.: Garbled RAM from one-way functions. In: Servedio, R.A., Rubinfeld, R. (eds.) 47th ACM STOC, pp. 449–458. ACM Press, June 2015

[GMW87] Goldreich, O., Micali, S., Wigderson, A.: How to play any mental game or a completeness theorem for protocols with honest majority. In: Aho, A. (ed.) 19th ACM STOC, pp. 218–229. ACM Press, May 1987

[GO96] Goldreich, O., Ostrovsky, R.: Software protection and simulation on oblivious rams. J. ACM **43**(3), 431–473 (1996)

[Gol87] Goldreich, O.: Towards a theory of software protection and simulation by oblivious RAMs. In: Aho, A. (ed.) 19th ACM STOC, pp. 182–194. ACM Press, May 1987

[GS18a] Garg, S., Srinivasan, A.: Adaptively secure garbling with near optimal online complexity. In: Nielsen, J.B., Rijmen, V. (eds.) EUROCRYPT 2018. LNCS, vol. 10821, pp. 535–565. Springer, Cham (2018). https://doi.org/10.1007/978-3-319-78375-8_18

[GS18b] Garg, S., Srinivasan, A.: Two-round multiparty secure computation from minimal assumptions. In: Nielsen, J.B., Rijmen, V. (eds.) EUROCRYPT 2018, Part II. LNCS, vol. 10821, pp. 468–499. Springer, Cham (2018). https://doi.org/10.1007/978-3-319-78375-8_16

[HJO+16] Hemenway, B., Jafargholi, Z., Ostrovsky, R., Scafuro, A., Wichs, D.: Adaptively secure garbled circuits from one-way functions. In: Robshaw, M., Katz, J. (eds.) CRYPTO 2016, Part III. LNCS, vol. 9816, pp. 149–178. Springer, Heidelberg (2016). https://doi.org/10.1007/978-3-662-53015-3_6

[HY16] Hazay, C., Yanai, A.: Constant-round maliciously secure two-party computation in the RAM Model. In: Hirt, M., Smith, A. (eds.) TCC 2016-B, Part I. LNCS, vol. 9985, pp. 521–553. Springer, Heidelberg (2016). https://doi.org/10.1007/978-3-662-53641-4_20

[IKO+11] Ishai, Y., Kushilevitz, E., Ostrovsky, R., Prabhakaran, M., Sahai, A.: Efficient non-interactive secure computation. In: Paterson, K.G. (ed.) EUROCRYPT 2011. LNCS, vol. 6632, pp. 406–425. Springer, Heidelberg (2011). https://doi.org/10.1007/978-3-642-20465-4_23

[JKK+17] Jafargholi, Z., Kamath, C., Klein, K., Komargodski, I., Pietrzak, K., Wichs, D.: Be adaptive, avoid overcommitting. In: Katz, J., Shacham, H. (eds.) CRYPTO 2017. LNCS, vol. 10401, pp. 133–163. Springer, Cham (2017). https://doi.org/10.1007/978-3-319-63688-7_5

[JW16] Jafargholi, Z., Wichs, D.: Adaptive security of yao's garbled circuits. In: Hirt, M., Smith, A. (eds.) TCC 2016-B, Part I. LNCS, vol. 9985, pp. 433–458. Springer, Heidelberg (2016). https://doi.org/10.1007/978-3-662-53641-4_17

[KLW15] Koppula, V., Lewko, A.B., Waters, B.: Indistinguishability obfuscation for turing machines with unbounded memory. In: Servedio, R.A., Rubinfeld, R. (eds.) 47th ACM STOC, pp. 419–428. ACM Press, June 2015

[KY18] Keller, M., Yanai, A.: Efficient maliciously secure multiparty computation for RAM. In: Nielsen, J.B., Rijmen, V. (eds.) EUROCRYPT 2018. LNCS, vol. 10822, pp. 91–124. Springer, Cham (2018). https://doi.org/10.1007/978-3-319-78372-7_4. https://eprint.iacr.org/2017/981

[LO13] Lu, S., Ostrovsky, R.: How to garble RAM programs? In: Johansson, T., Nguyen, P.Q. (eds.) EUROCRYPT 2013. LNCS, vol. 7881, pp. 719–734. Springer, Heidelberg (2013). https://doi.org/10.1007/978-3-642-38348-9_42

[LO17] Lu, S., Ostrovsky, R.: Black-box parallel garbled RAM. In: Katz, J., Shacham, H. (eds.) CRYPTO 2017. LNCS, vol. 10402, pp. 66–92. Springer, Cham (2017). https://doi.org/10.1007/978-3-319-63715-0_3

[LP09] Lindell, Y., Pinkas, B.: A proof of security of Yao's protocol for two-party computation. J. Cryptol. 22(2), 161–188 (2009)

[Mia16] Miao, P.: Cut-and-choose for garbled RAM. Cryptology ePrint Archive, Report 2016/907 (2016). http://eprint.iacr.org/2016/907

[ORS15] Ostrovsky, R., Richelson, S., Scafuro, A.: Round-optimal black-box two-party computation. In: Gennaro, R., Robshaw, M. (eds.) CRYPTO 2015, Part II. LNCS, vol. 9216, pp. 339–358. Springer, Heidelberg (2015). https://doi.org/10.1007/978-3-662-48000-7_17

[OS97] Ostrovsky, R., Shoup, V.: Private information storage (extended abstract). In: 29th ACM STOC, pp. 294–303. ACM Press, May 1997

[Ost90] Ostrovsky, R.: Efficient computation on oblivious RAMs. In: 22nd ACM STOC, pp. 514–523. ACM Press, May 1990

[WHC+14] Wang, X.S., Huang, Y., Chan, T.H.H., Shelat, A., Shi, E.: SCORAM: oblivious RAM for secure computation. In: Ahn, G.-J., Yung, M., Li, N. (eds.) ACM CCS 2014, pp. 191–202. ACM Press, November 2014

[Yao82] Yao, A.C.-C.: Protocols for secure computations (extended abstract). In: 23rd FOCS, pp. 160–164. IEEE Computer Society Press, November 1982

[Yao86] Yao, A.C.-C.: How to generate and exchange secrets (extended abstract). In: 27th FOCS, pp. 162–167. IEEE Computer Society Press, October 1986

On the Round Complexity of OT Extension

Sanjam Garg[1]([✉]), Mohammad Mahmoody[2], Daniel Masny[1], and Izaak Meckler[1]

[1] University of California, Berkeley, Berkeley, USA
sanjamg@berkeley.edu
[2] University of Virginia, Charlottesville, USA

Abstract. We show that any OT extension protocol based on one-way functions (or more generally any symmetric-key primitive) either requires an additional round compared to the base OTs or must make a non-black-box use of one-way functions. This result also holds in the semi-honest setting or in the case of certain setup models such as the common random string model. This implies that OT extension in any secure computation protocol must come at the price of an additional round of communication or the non-black-box use of symmetric key primitives. Moreover, we observe that our result is tight in the sense that positive results can indeed be obtained using non-black-box techniques or at the cost of one additional round of communication.

1 Introduction

Multiparty secure computation (MPC) [Yao82, GMW87] allows mutually distrustful parties to compute a joint function on their inputs, from which the parties learn their corresponding outputs but nothing more. Oblivious transfer (OT) [Rab81, EGL85, BCR87, Kil88, IPS08] is the fundamental building block for two and multiparty secure computation.

An OT protocol is a two-party protocol between a sender with inputs x_0, x_1 and a receiver with input bit b. An OT protocol allows the receiver to only learn x_b while b remains hidden from the sender. OT is a very powerful tool and is

S. Garg—Research supported in part from DARPA/ARL SAFEWARE Award W911NF15C0210, AFOSR Award FA9550-15-1-0274, AFOSR YIP Award, DARPA and SPAWAR under contract N66001-15-C-4065, a Hellman Award and research grants by the Okawa Foundation, Visa Inc., and Center for Long-Term Cybersecurity (CLTC, UC Berkeley). The views expressed are those of the author and do not reflect the official policy or position of the funding agencies.

M. Mahmoody—Supported by NSF CAREER award CCF-1350939, a subcontract on AFOSR Award FA9550-15-1-0274, and University of Virginia's SEAS Research Innovation Award.

D. Masny—Supported by the Center for Long-Term Cybersecurity (CLTC, UC Berkeley).

H. Shacham and A. Boldyreva (Eds.): CRYPTO 2018, LNCS 10993, pp. 545–574, 2018.
https://doi.org/10.1007/978-3-319-96878-0_19

sufficient to realize any secure computation functionality [Kil88,IPS08]. Nevertheless, all known constructions of OT have the drawback of being significantly less efficient than "symmetric-key primitives" like block ciphers or hash functions. This comparatively low efficiency seems to be unavoidable as black-box constructions of OT from one-way functions are known to be impossible [IR89]. Overcoming this difficulty, one promising approach is to use OT *extension*. OT extension allows a sender and a receiver to extend a relatively small number of base OTs to a much larger number of OTs using only symmetric-key primitives (e.g., one-way functions, pseudorandom generators, collision-resistant hash functions, etc.), which are indeed much cheaper.

Beaver first proposed the idea of such an OT extension protocol [Bea96]. Beaver's protocol solely relied on a security parameter number of base OTs and, perhaps surprisingly, only on a pseudorandom generator (PRG). This insight – that a small number of inefficient base OTs could be efficiently extended to a large number of OTs – has been a crucial step in overcoming the efficiency limitation of OT in particular and multiparty computation in general. Beaver's construction, however, made an expensive *non-black-box* use of the underlying PRG leading to inefficient protocols.

In an influential work, Ishai, Kilian, Nissim and Pentrank [IKNP03] obtained an OT extension (referred to as IKNP) which made only *black-box* use of the underlying cryptographic primitive, which could be realized using a random oracle. This yielded a significantly more efficient protocol in comparison to Beaver's protocol. They also observed that the random oracle in their construction can be relaxed to the notion of a correlation robust hash function. Follow up works on OT extension achieve security against stronger adversaries [NNOB12,ALSZ15] or reduce communication and computation costs [KK13].

The practical impact of the OT extension protocols has been enormous. OT extension can be used to improve the computational efficiency of virtually any implementation of secure MPC. In particular, the standard recipe for realizing efficient secure computation protocols is as follows. We start with the OT-hybrid model where everyone has access to an ideal OT functionality called OT-hybrid. Then instantiate an OT extension using the OT-hybrid, which implies that only black-box access to the OTs is used. An efficient secure computation protocol is then realized using OT extension to minimize the number of public-key operations. Use of OT extension yields remarkable efficiency gains for many implemented protocols (e.g. see [ALSZ13]).

In addition to the computational efficiency, round complexity is another parameter of concern in the construction of efficient secure computation protocols. Significant research effort has been made toward realizing round efficient OT [NP01,AIR01,HK12,PVW08] and round efficient two-party [KO04, ORS15] and multiparty [BMR90,AJL+12,GGHR14,MW16,GMPP16,GS17a, BL17,GS17b] secure computation protocols. Several of these protocols are also black-box in the use of the underlying cryptographic primitives. However, all these works only yield protocols (with a given round complexity) using a large number public-key operations. Ideally, we would like to construct OT extension protocols that can be used to reduce the number of public-key operations needed in these protocols while preserving the round complexity and the black-box nature of the underlying protocol. This brings us to our following main question:

> *Can we realize a round-preserving OT extension protocol which*
> *makes only black-box use of "symmetric-key" cryptographic primitives?*

The random oracle model (ROM) accurately captures the black-box use of such symmetric-key primitives, as it directly provides us with ideally strong hash functions as well as block-ciphers or even ideal ciphers [CPS08, HKT11]. Therefore, in order to answer the above question, we study the possibility of OT extension protocols in the ROM that preserve the round complexity.[1]

1.1   Our Results

We provide a negative answer to the above main question. In other words, we show that any OT extension protocol based on so called symmetric-key primitives, requires either an additional round compared to the base OTs or must make a non-black-box use of symmetric-key primitives. We capture black-box use of one-way functions, or even correlation-robust hash functions, as well as common random string setup[2] by proving our impossibility result under the idealized notion of these primitives which is provided by a random oracle. In particular, we prove the following theorem.

Theorem 1 (Impossibility of round-preserving OT extension in ROM– *Informally Stated*). *Suppose a sender S and a receiver $\mathcal{R}$ want to perform m OTs in r rounds using a random oracle, and they both have access to n, r-round OTs (i.e. the receiver obtains its outputs at the end of round r) where $n < m$. Then, if S and $\mathcal{R}$ can ask polynomially many more queries to the random oracle, one of them could always break security of the m OTs.*

Theorem 1 holds even for an extension from n *string* OTs to $m = n + 1$ *bit* OTs, and even for the setting of semi-honest security. It also gives an alternative, and arguably simpler, proof to Beaver's impossibility result that information-theoretically secure OT extension does not exist in the plain model [Bea96]. We sketch the main ideas in Sect. 1.2 and provide the details in Sect. 3.

Additionally, we observe that our results are tight in two different ways. First, the IKNP protocol [IKNP03] realizes black-box OT extension using one additional round. Second, our result is also tight with respect to the black-box use of the symmetric-key primitives captured by random oracles. Beaver's original protocol provided OT extension in which the receiver has no control over which input he receives (it will be chosen at random). This notion of OT is often referred to as "random" OT. The known generic way of going from "random"

---

[1] The only symmetric-key primitive not directly implied by a random oracle is one-way permutations. However, most negative results in the random oracle model, including our work, extend to one-way permutations using standard techniques [IR89].

[2] Note that a random oracle also provides a common random string for free.

OTs to "chosen" OTs will add another round [EGL85]. We observe (see the full version [GMMM17]) that Beaver's original non-black-box OT extension protocol [Bea96], which only relies on a PRG, can be modified to provide round-preserving "chosen-input" OTs, but this result will require *non-black-box* use of the PRG.

We remark that our results have implications in several other settings, for example, in the plain model under malicious security. In this setting, an OT protocol takes at least 4 rounds [KO04, ORS15]. Therefore, our results imply that in this setting, black-box OT extension protocols must be at least five rounds while a non-black-box construction with four rounds can be realized. Another example is the correlated setup model [FKN94, IKM+13, BGI+14] where our results imply that there is no non-interactive OT extension even in the presence of a random oracle. Interestingly, this setting behaves very differently from a setting of shared randomness, where the amount of shared randomness can be easily increased by using the random oracle as a PRG. On the contrary, in case of a single communication round, the IKNP protocol [IKNP03] can be used to increase the amount of correlated randomness in this setting.

Finally, we note that our impossibility result of Theorem 1 also holds for the case of random *permutation* oracle model. The proof of Theorem 1 directly extends to this setting using the standard trick introduced in [IR89]. Namely, the attacker can always ask all the oracle queries of input lengths at most $c \cdot \log \kappa$ for sufficiently large constant c, in which case, the probability of the honest parties, the simulator, or the attacker (of the random oracle model) itself getting a collision while accessing the random oracle on input of length $> c \log \kappa$ is sufficiently small. However, without collisions, (length preserving) random oracles and random permutation oracles are the same.

1.2   Technical Overview

In this section, we explain the key ideas behind the proof of our main impossibility result of Theorem 1. For a formal treatment, see Sect. 3. In a nutshell, we first present an entropy-based information-theoretic argument for the plain model, where there are no oracles involved beyond the hybrid OTs. We then extend our attack to the random oracle model, by making use of the 'dependency learner' of [IR89, BMG09, HOZ16, BM17], which is a algorithms that allows us to 'approximate' certain plain-model arguments also in the random oracle model. As we will see, the combination of these two steps will make crucial use of the *round-preserving* property of the (presumed) OT-extension construction.

To explain the core ideas behind our proofs, it is instructive to even define a 2-round-preserving OT extension protocol, to see how the definition accurately models the concept of round-preserving OT extension, and because we are particularly interested in ruling out black-box 2-round OT extension protocols. Below, we first describe the notation and the simplifying assumptions for this special case (of 2-round extensions), before going over the ideas behind the proof.

Notation and the simplified setting. Here we define some basic notations and also state some simplifying assumptions, some of which are without loss of generality when we focus on 2-round-preserving OT extensions, and the rest are relaxed when proving the formal attack in Sect. 3. Here we focus on the case of extending n instances of OT, into m instances for some $m \gg n$. (This is without loss of generality as even "one-more" OT extension, i.e., $m = n + 1$, can be used to get polynomially many more OTs – e.g., see [LZ13].)

Inputs and outputs: Let $[m] := \{1, \ldots, m\}$. Suppose $\vec{b} = (b_1, \ldots, b_m) \in \{0,1\}^m$ are the choice bits of the receiver $\mathcal{R}$ and $x = (x_i^0, x_i^1)_{i \in [m]} \in \{0,1\}^{2m}$ are the pairs of bits that the sender holds as its input. (Our main negative result holds even if the hybrid $\mathbb{OT}_n$ provides string OTs, but in this simplified exposition, we work with bit OTs.) The receiver $\mathcal{R}$ wishes to get output $(x_i^{b_i})_{i \in [m]}$.

The oracle and OT hybrid: The two parties have access to a random oracle $\mathbf{H}$ as well as n instances of a OT-hybrid functionality for *bit* inputs which we denote with $\mathbb{OT}_n$. When using $\mathbb{OT}_n$, $\mathcal{R}$ and $\mathcal{S}$ will *not* reverse their roles such that $\mathcal{R}$ always receives the output from $\mathbb{OT}_n$. This is without loss of generality for a round preserving OT extension for the following reason. First note that the last message of the constructed OT should be from the sender to the receiver, as otherwise it could be dropped. Moreover, we use a hybrid $\mathbb{OT}_n$ that *requires* (here) two rounds to deliver its output. Therefore, if both the used hybrid OTs and constructed OTs have the same (here two) rounds, the last messages of the hybrid and the constructed OTs should both go from the sender to the receiver. Thus, we model the 2-round-preserving OT extension in the ROM as follows.

1. $\mathcal{R}$ sends a single message t_1 to $\mathcal{S}$, and it also chooses and submits the input $\vec{c} = (c_1, \ldots, c_n)$ to the hybrid $\mathbb{OT}_n$.
2. $\mathcal{S}$ sends a single message t_2 to $\mathcal{R}$, and it also chooses and submits inputs $(y_i^0, y_i^1)_{i \in [n]}$ to the hybrid $\mathbb{OT}_n$.
3. $\mathcal{R}$ also receives $\gamma = (y_i^{c_i})_{i \in [n]}$ from $\mathbb{OT}_n$.
4. $\mathcal{R}$ outputs what is supposed to be $(x_i^{b_i})_{i \in [m]}$.

We assume in this simplified exhibition that the protocol has *perfect* completeness, namely the receiver obtains the correct answer with probability one.

An information theoretic attack for the no-oracle setting. Our starting point is an inefficient (information theoretic) attack on OT extension when there are no oracles involved. The fact that OT extension protocols, regardless of their round complexity, can *not* be information theoretically secure was already shown by Beaver [Bea96], and the work of Lindell and Zarosim [LZ13] improved that result to derive one-way functions from OT extensions. As we will see, our information theoretic attack has the main feature that in the *round-preserving* OT extension setting, it can be adapted to the random oracle model by also using tools and ideas from [IR89,BMG09,HOZ16] where new challenges arise.

Now we describe an attack for the sender and an attack for the receiver in the case that they pick their inputs $\vec{b}, x$ uniformly at random, and will show that at least one of these attacks will succeed with non-negligible probability. Ruling out the possibility of secure OT for the random-inputs case is stronger and it rules out the general (selected-input) case as well. Also note that when we refer to *attacking* parties here, what we formally mean, is a semi-honest execution of the protocol, followed by a distinguisher (as part of the attack) who is able to use the view of the honest execution to make distinguishing predictions that shall not be possible in case of semi-honest security. For simplicity, we combine these two steps and simply refer to them as the attacker.

- **Attacking sender $\widehat{\mathcal{S}}$.** Since $\widehat{\mathcal{S}}$ gets no output, in a secure protocol the random input $\vec{b} \in \{0,1\}^m$ of the receiver shall remain indistinguishable from a uniform $\mathbf{U}_m$ in eyes of the receiver who knows the transcript $T = (t_1, t_2)$. (See Lemma 19 for a formalization.) Therefore, a natural attacking strategy for the sender $\widehat{\mathcal{S}}$ is to look at the transcript T at the end, and based on that information, try to distinguish the true $\vec{b}$ (in case it is revealed to him) from a random uniform string $\mathbf{U}_m$ of length m.[3] Thus, if the distribution of $(\vec{b}, T)$, is ε-far from $(\mathbf{U}_m, T)$ for *non-negligible* ε, the protocol is not secure, because given the transcript T an efficient sender can distinguish $\vec{b}$ from $\mathbf{U}_m$.

- **Attacking receiver $\widehat{\mathcal{R}}$.** After running the protocol honestly to get the actual output for the honestly chosen input $\vec{b}$, the cheating receiver $\widehat{\mathcal{R}}$ tries to also find *another* input $\vec{b}' \neq \vec{b}$ together with its corresponding correct output $\{x_i^{b_i'}\}_{i \in [m]}$. If $\widehat{\mathcal{R}}$ could indeed do so, it would be a successful attack since in at least one of the locations $i \in [m]$, the receiver will read both of (x_i^0, x_i^1), though that shall not be possible for semi-honest secure protocols. (See Lemma 20 for a formalization.) By the perfect completeness of the protocol,[4] all $\widehat{\mathcal{R}}$ needs to do is to find another fake view $V'_{\mathcal{R}}$ for the receiver such that: (1) $V'_{\mathcal{R}}$ contains $\vec{b}' \neq \vec{b}$ as its input, and that (2) $V'_{\mathcal{R}}$ is consistent with the transcript T, the input c given to $\mathbb{OT}_n$, as well as the output γ obtained from it. Finding such $V'_{\mathcal{R}}$ efficiently, violates sender's security.

One of $\widehat{\mathcal{S}}, \widehat{\mathcal{R}}$ succeeds: an entropy-based argument. If the attacking sender $\widehat{\mathcal{S}}$ described above does not succeed with a non-negligible advantage, it means that $(\vec{b}, T)$, as a random variable, is statistically close to $(\mathbf{U}_m, T)$, which in turn implies that (with high probability over T) conditioned on the transcript T, the receiver's input $\vec{b}$ has close to (full) m bits of entropy.[5] (See Lemma 14 for a formalization of this argument.) Therefore, if the malicious receiver $\widehat{\mathcal{R}}$, after finishing the honest execution encoded in the view $\mathbf{V}_{\mathcal{R}}$, "re-samples" a fake view $V'_{\mathcal{R}}$ from the distribution of its view conditioned on T, denoted $(\mathbf{V}_{\mathcal{R}} \mid T)$, then it will get a different $\vec{b}' \neq \vec{b}$, as part of $V'_{\mathcal{R}}$ with some noticeable probability. (See Lemma 15 for a formalization of this argument.) However, as described above, the attacking receiver $\widehat{\mathcal{R}}$ also needs to condition its sampled view

---

[3] Technically, the true input $\vec{b}$ or independent random input $\mathbf{U}_n$ are not given to the sender in an actual execution of the protocol, but for a secure protocol, these two shall remain indistinguishable even if revealed (see Lemma 19).

[4] Our formal proof of Sect. 3 does not assume perfect completeness.

[5] This is why we choose to work with Shannon entropy, as we want distributions close to $\mathbf{U}_m$ to have almost full entropy; this does not hold e.g., for min-entropy.

$V'_{\mathcal{R}} \leftarrow (\mathbf{V}'_{\mathcal{R}} \mid T, \vec{c}, \gamma)$ on its input c given to $\mathbb{OT}_n$ and the output γ obtained from it to get a *correct* output $\{x_i^{b'_i}\}_{i \in [m]}$ for the new fake input $\vec{b}' \neq \vec{b}$. It can be shown that if $m > |\vec{c}| + |\gamma| = 2n$, then there is still enough entropy left in the sampled $\vec{b}'$, even after further conditioning on $\vec{c}, \gamma$ (and transcript T). Therefore, if $m \gg 2n$, then at least one of the attacks succeeds with non-negligible probability.

Polynomial-query attacks in the random oracle model. The above information theoretic argument for the no-oracle case no longer works when we move to the ROM for the following simple reason. A fresh fake sample $V'_{\mathcal{R}}$ for the receiver's view that is consistent with the transcript T and OT-hybrid inputs c and output γ might be *inconsistent* with oracle query-answer pairs that already exist in sender's view, because the fake view $V'_{\mathcal{R}}$ might make up some answers to some oracle queries that are also asked by the sender but received a *different* answer from the actual oracle. Therefore, we will not have any guarantee that the faked sampled view of the sender leads to *correct* outputs for the new fake input $\vec{b}'$. In fact, because we already know that OT extension in the random oracle model *is* possible [IKNP03], the above issue is inherent when we try to extend our attack to the ROM. However, we have not yet used the fact that we are aiming at attacks that succeed for *round-preserving* OT extensions. Below, we first describe a natural (yet insufficient) idea for extending our information-theoretic attack to the ROM, and then will extend this idea further by also relying on the round-preserving aspect of the construction.

1st try: using "dependency learner" of [IR89,BMG09,HOZ16,BM17]**.** As described above, when we move to the oracle setting, the random oracle $\mathbf{H}$ creates further correlation between the views of $\mathcal{S}$ and $\mathcal{R}$ beyond what the transcript (or $\mathbb{OT}_n$) does. One natural idea for removing the correlation made by a random oracle between the views of two parties is to use the so-called 'dependency learner' Eve algorithm of [BMG09,BMG09,HOZ16,BM17] (see Theorem 17). The Eve algorithm is a deterministic algorithm such that for any inputless, two-party, protocol $\mathcal{A}, \mathcal{B}$ in the ROM, given the public transcript T of the interaction between $\mathcal{A}, \mathcal{B}$, Eve asks polynomially-many oracle queries from the random oracle $\mathbf{H}$ in a way that conditioned on the view of Eve (that includes T and its oracle query-answer pairs $P_{\mathcal{E}}$) the views of $\mathcal{A}, \mathcal{B}$ become close to *independent* random variables.[6] The magic of the algorithm Eve is that, because both parties can run it at the end, the parties can pretend that $P_{\mathcal{E}}$ is also part of the transcript, and thus we get an augmented transcript $V_{\mathcal{E}} = (T, P_{\mathcal{E}})$ that includes (almost) all of the correlation between the views of the two parties.

The above simple idea fails, however, because of the additional involvement of $\mathbb{OT}_n$ in the protocol, which creates further correlation between the views of the parties. Consequently, this seemingly simple issue prevents us from being able to run the Eve algorithm to (almost) eliminate the correlation between $\mathcal{S}, \mathcal{R}$ views, as the Eve algorithm only applies to inputless protocols in the ROM that have *no other* source of communication other than the transcript.

---

[6] In the plain model, the views of two interacting parties *are* independent given the transcript, and this enables the information theoretic attack against OT extension.

2nd try: using the dependency learner over a *shortened* protocol. Recall that we are dealing with round-preserving OT extensions, and have not used this property yet! One consequence of this assumption is that we can now assume that the OT-hybrid output γ is sent to the receiver *after* the last message t_2 is sent. Now, if we *stop* the execution of $\mathcal{R}$ right after t_2 is sent and call this modified variant of $\mathcal{R}$ the algorithm $\mathcal{R}_1$, even though the input $\vec{c}$ is submitted to $\mathbb{OT}_n$ by $\mathcal{R}_1$, no output is received by $\mathcal{R}_1$ from $\mathbb{OT}_n$ yet, therefore we would not have any correlated randomness distributed between the parties through $\mathbb{OT}_n$ hybrid. Therefore, our new modified two party protocol $\mathcal{S}, \mathcal{R}_1$ would be a simple inputless protocol in the ROM over which we can execute the dependency learner Eve over its transcript $T = (t_1, t_2)$. Indeed, if we run Eve with respect to $\mathcal{S}, \mathcal{R}_1$, Eve will gather enough information about the oracle encoded in its oracle query-answer set $(P_\mathcal{E})$ so that the views of $\mathcal{S}$ and $\mathcal{R}_1$, conditioned on Eve's view $(T, P_\mathcal{E})$, would be *close* to a product distribution. Therefore, we can hope to again use an approximate version of our information theoretic argument in the no-oracle setting by interpreting $T' = (T, P_\mathcal{E})$ as the new 'transcript'.

Finishing receiver's execution. The above argument (of applying the dependency learner Eve over a shortened variant $\mathcal{R}_1$ of $\mathcal{R}$) still does not lead to an actual attack, because we need to *finish* the execution of $\mathcal{R}_1$, which is only a *partial* execution of the receiver, to obtain the actual output corresponding to the fake input $\vec{b}'$. Only then, we can call $\widehat{\mathcal{R}}$ a successful attack. With this goal in mind, let us call $\mathcal{R}_2$ to be the rest of the execution of the cheating receiver which starts right after finishing the first part $\mathcal{R}_1$. Namely, $\mathcal{R}_2$ takes over the computation of $\mathcal{R}_1$ to finish it, and the first thing it receives is the output γ of $\mathbb{OT}_n$. However, to obtain the actual output, there might be further necessary oracle calls to the random oracle $\mathbf{H}$. Since $\widehat{\mathcal{R}}$ is interested to know the output $\vec{b}'$ planted in the fake view $V'_\mathcal{R}$, the execution of $\mathcal{R}_2$ using $V'_\mathcal{R}$ needs to pretend that $V'_\mathcal{R}$ is the *actual view* of the receiver, which in turn implies pretending that the original honest view $V_\mathcal{R}$ does not exist.

Leveraging on the lack of intersection queries. Interestingly, it turns out that another crucial property of the dependency learner algorithm Eve (i.e. Part 2 of Theorem 17) allows us to get a consistent execution of $\mathcal{R}_2$ using the fake view $V'_\mathcal{R}$ while pretending that the original honest (non-fake) execution of the receiver (encoded in view $V_\mathcal{R}$) does not exist. Namely, Eve's algorithm guarantees that, with high probability over $T' = (T, P_\mathcal{E})$, there will be no 'intersection queries' between the set of queries asked by the honest sender and the original (i.e., honest) partial execution of the receiver that obtains the first output (of input $\vec{b}$) for the attacker $\widehat{\mathcal{R}}$. In a nutshell, what we do to finish the execution of $\mathcal{R}_2$ is to answer with fresh random strings, any query q that is *not* learned by Eve but *is* in the view of the original honest receiver's execution. In Sect. 3 we show that a careful case analysis proves this strategy to lead to a correct continuation of the fake view $V'_\mathcal{R}$ obtaining the right output for the fake input $\vec{b}'$.

Organization. In Sect. 2 we describe the basic notation, main definitions, and some useful lemmas. In Sect. 3 we formalize and prove our main impossibility result of Theorem 1. In the full version [GMMM17] we observe that Beaver's non-black-box round-preserving OT extension [Bea96] could be "chosen input".

2 Preliminaries

Logarithms in this work are taken base 2. For a bit b, we denote bit $1 - b$ by $\bar{b}$. We use PPT to denote a probabilistic, polynomial-time Turing machine.

Notation on random variables. All the distributions and random variables in this paper are *finite*. We use **bold font** to denote random variables. We usually use the same non-bold letter for samples form the random variables, so by $Q \leftarrow \mathbf{Q}$ we indicate that Q is sampled from the distribution of the random variable $\mathbf{Q}$. By $(\mathbf{X}, \mathbf{Y})$ we denote a *joint* distribution over random variables $\mathbf{X}$ and $\mathbf{Y}$. By $\mathbf{X} \equiv \mathbf{Y}$ we denote that $\mathbf{X}$ and $\mathbf{Y}$ are identically distributed. For jointly distributed $(\mathbf{X}, \mathbf{Y}, \mathbf{Z})$, when random variable $\mathbf{Z}$ is clear from the context, by $((\mathbf{X}, \mathbf{Y}) \mid Z)$ we denote the distribution of $(\mathbf{X}, \mathbf{Y})$ conditioned on $\mathbf{Z} = Z$. By $(\mathbf{X} \times \mathbf{Y})$ we denote a product distribution in which $\mathbf{X}$ and $\mathbf{Y}$ are sampled *independently* from their marginal distributions. For jointly distributed $(\mathbf{X}, \mathbf{Y}, \mathbf{Z})$ and any $Z \leftarrow \mathbf{Z}$, we denote $((\mathbf{X}|Z) \times (\mathbf{Y}|Z))$ by $(\mathbf{X} \times \mathbf{Y})|Z$. For a finite set S, by $x \leftarrow$ S we denote that x is sampled from S uniformly at random. By $\mathrm{Supp}(\mathbf{X})$ we denote the *support set* of the random variable $\mathbf{X}$, defined as $\mathrm{Supp}(\mathbf{X}) := \{x \mid \Pr[\mathbf{X} = x] > 0\}$. $\mathbf{U}_n$ is the uniform distribution over $\{0, 1\}^n$.

Notation on events. An event B is simply a set, so for any random variable $\mathbf{X}$, the probability $\Pr[\mathbf{X} \in \mathsf{B}] := \Pr[\mathbf{X} \in \mathsf{B} \cap \mathrm{Supp}(\mathbf{X})]$ is well defined. More formally, we assume $\mathsf{B} \subseteq \mathsf{U}$ is a subset of the 'universe' set U where $\mathrm{Supp}(\mathbf{X}) \subseteq \mathsf{U}$ for any 'relevant' random variable $\mathbf{X}$ (in our analyses). In particular, we could refer to the same event B across different random variables. For any particular sample $X \leftarrow \mathbf{X}$, we say that the event B *holds over* X iff $X \in \mathsf{B}$.[7] For an event B by $\bar{\mathsf{B}}$ we denote to the complement (with respect to the underlying universe U). Therefore, $\Pr[\mathbf{X} \in \bar{\mathsf{B}}] = 1 - \Pr[\mathbf{X} \in \mathsf{B}]$ is always well defined. By $\Pr_{\mathcal{D}}[\mathsf{B}]$ or $\Pr[\mathsf{B}; \mathcal{D}]$ we mean the probability of B for sampling process described by $\mathcal{D}$.

2.1 Lemmas About Statistical Distance and Mutual Dependency

Definition 2 ((Conditional) statistical distance). *By* $\mathrm{SD}(\mathbf{X}, \mathbf{Y})$ *we denote the statistical distance between random variables* $\mathbf{X}, \mathbf{Y}$ *defined as*

$$\mathrm{SD}(\mathbf{X}, \mathbf{Y}) = \max_{\mathsf{B}} \Pr[\mathbf{X} \in \mathsf{B}] - \Pr[\mathbf{Y} \in \mathsf{B}] = \frac{1}{2} \cdot \sum_{Z} |\Pr[\mathbf{X} = Z] - \Pr[\mathbf{Y} = Z]|.$$

We call $\mathbf{X}$ *and* $\mathbf{Y}$ ε-*close, denoted by* $\mathbf{X} \approx_\varepsilon \mathbf{Y}$, *if* $\mathrm{SD}(\mathbf{X}, \mathbf{Y}) \leq \varepsilon$.

---

[7] In this terminology, B is seen as a *property* that holds for all $X \in \mathsf{B}$, but not for the rest. In fact, we define our events B using properties over objects in the universe.

For an event A, we let $\mathsf{SD}_A(\mathbf{X},\mathbf{Y}) = \mathsf{SD}((\mathbf{X} \mid A),(\mathbf{Y} \mid A))$, denote the conditional *statistical distance of* $\mathbf{X},\mathbf{Y}$, *and for correlated random variable* $\mathbf{Z}$, *by* $\mathsf{SD}_Z(\mathbf{X},\mathbf{Y})$ *we denote* $\mathsf{SD}((\mathbf{X} \mid \mathbf{Z} = Z),(\mathbf{Y} \mid \mathbf{Z} = Z))$, *and we also let*

$$\mathsf{SD}_{\mathbf{Z}}(\mathbf{X},\mathbf{Y}) = \mathop{\mathbb{E}}_{Z \leftarrow \mathbf{Z}} \mathsf{SD}_Z(\mathbf{X},\mathbf{Y}).$$

In the following lemma, is a well-known[8] fact stating that statistical distance is the maximum advantage of distinguishing two distributions.

Lemma 3. *Let* D *be any potentially randomized (distinguishing) algorithm. Then:* $\Pr[D(\mathbf{X}) = 1] - \Pr[D(\mathbf{Y}) = 1] \leq \mathsf{SD}(\mathbf{X},\mathbf{Y})$ *and the equality can be achieved by any 'canonical' distinguisher such that:* $C(Z) = 1$ *if* $\Pr[\mathbf{X} = Z] > \Pr[\mathbf{Y} = Z]$, *and* $C(Z) = 0$ *if* $\Pr[\mathbf{X} = Z] < \Pr[\mathbf{Y} = Z]$.

The following well-known lemma[9] states that statistically close distributions could be sampled jointly while they are equal with high probability.

Lemma 4 (Coupling vs. statistical distance). $\mathsf{SD}(\mathbf{X},\mathbf{Y}) \leq \varepsilon$ *iff there is a way to* jointly *sample* $(\mathbf{X},\mathbf{Y})$ *such that* $\Pr[\mathbf{X} = \mathbf{Y}] \geq 1 - \varepsilon$.

The following lemma says that if $\mathbf{X} \equiv \mathbf{X}'$ in two pairs of jointly distributed random variables $(\mathbf{X},\mathbf{Y}),(\mathbf{X}',\mathbf{Y}')$, then the statistical distance of the two pairs could be written as a linear combination of conditional probabilities.

Proposition 5. $\mathsf{SD}((\mathbf{X},\mathbf{Y}),(\mathbf{X},\mathbf{Y}')) = \mathbb{E}_{X \leftarrow \mathbf{x}} \mathsf{SD}((\mathbf{Y} \mid X),(\mathbf{Y}' \mid X))$. *Moreover, if* $\mathsf{SD}((\mathbf{X},\mathbf{Y}),(\mathbf{X},\mathbf{Y}')) = \varepsilon$, *any canonical distinguisher* D *of the following form* ε-*distinguishes* $(\mathbf{X},\mathbf{Y})$ *from* $(\mathbf{X},\mathbf{Y}')$:

- If $\Pr[\mathbf{Y} = Y \mid X] > \Pr[\mathbf{Y}' = Y \mid X]$, then $D(X,Y) = 1$.
- If $\Pr[\mathbf{Y} = Y \mid X] < \Pr[\mathbf{Y}' = Y \mid X]$, then $D(X,Y) = 0$.
- If $\Pr[\mathbf{Y} = Y \mid X] = \Pr[\mathbf{Y}' = Y \mid X]$, then $D(X,Y) \in \{0,1\}$ *arbitrarily.*

Proof. We prove both parts using Lemma 3. By Lemma 3, $\mathsf{SD}((\mathbf{X},\mathbf{Y}),(\mathbf{X},\mathbf{Y}'))$ equals the maximum advantage by which a distinguisher D can distinguish the two distributions $(\mathbf{X},\mathbf{Y}),(\mathbf{X},\mathbf{Y}')$. Now, such D is always given a sample $X \leftarrow \mathbf{X}$ from $\mathbf{X} \equiv \mathbf{X}'$ first, conditioned on which it has to maximize $\Pr[D(\mathbf{Y} \mid X) = 1] - \Pr[D(\mathbf{Y}' \mid X) = 1]$. However, for each X, the maximum of $\Pr[D(\mathbf{Y} \mid X) = 1] - \Pr[D(\mathbf{Y}' \mid X) = 1]$ is again described by Lemma 3 to be equal to $\mathsf{SD}((\mathbf{Y} \mid X),(\mathbf{Y}' \mid X))$. Furthermore, the canonical distinguisher described above works due to the canonical distinguisher of Lemma 3 □

The following definition from [BM17] is a measure of correlation between jointly distributed pairs of random variables.

---

[8] For example, see Exercise 8.61 from [Sho09].
[9] For example, see Lemma 3.6 of [Ald83] for a proof.

Definition 6 ((Conditional) mutual dependency [BM17]). *For a joint distribution* $(\mathbf{X}, \mathbf{Y})$, *we define their* mutual-dependency *as* $\mathsf{MutDep}(\mathbf{X}, \mathbf{Y}) = \mathsf{SD}((\mathbf{X}, \mathbf{Y}), (\mathbf{X} \times \mathbf{Y}))$. *For correlated* $(\mathbf{X}, \mathbf{Y}, \mathbf{Z})$, *and for* $Z \leftarrow \mathbf{Z}$, *we define*

$$\mathsf{MutDep}_Z(\mathbf{X}, \mathbf{Y}) = \mathsf{SD}_Z((\mathbf{X}, \mathbf{Y}), (\mathbf{X} \times \mathbf{Y})) = \mathsf{SD}(((\mathbf{X}, \mathbf{Y})|Z), (\mathbf{X}|Z \times \mathbf{Y}|Z))$$

to be the mutual dependency of $\mathbf{X}, \mathbf{Y}$ *conditioned on the given* Z, *and we let*

$$\mathsf{MutDep}_{\mathbf{Z}}(\mathbf{X}, \mathbf{Y}) = \underset{Z \leftarrow \mathbf{Z}}{\mathbb{E}} \, \mathsf{MutDep}_Z(\mathbf{X}, \mathbf{Y}).$$

The following proposition follows from Proposition 5 and Definition 6.

Proposition 7. *It holds that* $\mathsf{MutDep}(\mathbf{X}, \mathbf{Y}) = \mathbb{E}_{X \leftarrow \mathbf{x}} \, \mathsf{SD}((\mathbf{Y} \mid X), \mathbf{Y})$.

Lemma 8. *For a joint distribution* $(\mathbf{X}, \mathbf{Y})$, *the statistical distance between the following distributions is at most* $2 \cdot \mathsf{MutDep}(\mathbf{X}, \mathbf{Y})$. *(Note how* Y, Y' *are flipped.)*

1. *Sample* $(X, Y) \leftarrow (\mathbf{X}, \mathbf{Y})$, *independently sample* $Y' \leftarrow \mathbf{Y}$, *output* (X, Y, Y').
2. *Sample* $(X, Y) \leftarrow (\mathbf{X}, \mathbf{Y})$, *independently sample* $Y' \leftarrow \mathbf{Y}$, *output* (X, Y', Y).

Proof. The following hybrid distribution is $\mathsf{MutDep}(\mathbf{X}, \mathbf{Y})$-far from either of the distributions in Lemma 8. Sample $X \leftarrow \mathbf{X}, Y_1, Y_2 \leftarrow \mathbf{Y}$ all independently and output (X, Y_1, Y_2). Therefore, the claim follows from the triangle inequality. □

Lemma 9. *Let* $\mathbf{X} = (\mathbf{A}, \mathbf{B}, \mathbf{C})$ *be correlated random variables. Let another joint distribution* $\mathbf{X}' = (\mathbf{A}', \mathbf{B}', \mathbf{C}')$ *be defined as follows.*

– *Sample* $A' \leftarrow \mathbf{A}$, *then* $C' \leftarrow (\mathbf{C} \mid \mathbf{A} = A')$, *then* $B' \leftarrow (\mathbf{B} \mid \mathbf{C} = C')$, *and output the sample* $X' = (A', B', C')$.

Then $\mathsf{SD}(\mathbf{X}, \mathbf{X}') = \mathsf{MutDep}_{\mathbf{C}}(\mathbf{A}, \mathbf{B})$. *Furthermore, if* $\mathbf{C} = f(\mathbf{B})$ *is a function of only* $\mathbf{B}$ *(in the joint distribution* $\mathbf{X}$) *then* $\mathsf{SD}(\mathbf{X}, \mathbf{X}') \leq 2 \cdot \mathsf{MutDep}(\mathbf{A}, \mathbf{B})$.

Remark 10. Before proving Lemma 9, note that the second conclusion would be false if $\mathbf{C}$ could also depend on $\mathbf{A}$. For example, consider the case where $\mathbf{A}, \mathbf{B}, \mathbf{C}$ are all random bits conditioned on $\mathbf{A} \oplus \mathbf{B} \oplus \mathbf{C} = 0$. In that case, without conditioning on $\mathbf{C}$, $\mathsf{MutDep}(\mathbf{A}, \mathbf{B}) = 0$ as $\mathbf{A}, \mathbf{B}$ are independent. However, given any specific bit $C \leftarrow \mathbf{C}$, the distributions of $\mathbf{A}, \mathbf{B}$ would be correlated, and their conditional mutual-dependency would be $1/2$, so $\mathsf{MutDep}_{\mathbf{C}}(\mathbf{A}, \mathbf{B}) = 1/2$.

Proof (of Lemma 9). First, we show $\mathsf{SD}(\mathbf{X}, \mathbf{X}') = \mathsf{MutDep}_{\mathbf{C}}(\mathbf{A}, \mathbf{B})$. Note that $\mathbf{C} \equiv \mathbf{C}'$, so we can apply Proposition 7. For a given $C \leftarrow \mathbf{C} \equiv \mathbf{C}'$, for $(\mathbf{X} \mid C)$ we will sample $(\mathbf{A}, \mathbf{B})$ jointly, while in $(\mathbf{X}' \mid \mathbf{C}' = C)$ we will sample from $(\mathbf{A} \mid C) \equiv (\mathbf{A}' \mid \mathbf{C}' = C)$ and $(\mathbf{B} \mid C) = (\mathbf{B}' \mid \mathbf{C}' = C)$ independently from their marginal distributions. Now, we show that $\mathsf{SD}(\mathbf{X}, \mathbf{X}') \leq 2 \cdot \mathsf{MutDep}(\mathbf{X}, \mathbf{Y})$, if we further know that $\mathbf{C}$ is only a function of $\mathbf{B}$. Consider a third joint distribution $\mathbf{X}'' = (\mathbf{A}'', \mathbf{B}'', \mathbf{C}'') \equiv (\mathbf{A} \times (\mathbf{B}, \mathbf{C}))$; namely, $(\mathbf{B}'', \mathbf{C}'') \equiv (\mathbf{B}, \mathbf{C})$, and $\mathbf{A}''$ is sampled from the marginal distribution of $\mathbf{A}$. Firstly, note that for every $A \leftarrow \mathbf{A}, B \leftarrow \mathbf{B}$, it holds that $(\mathbf{C}'' \mid \mathbf{A}'' = A, \mathbf{B}'' = B) \equiv (\mathbf{C} \mid \mathbf{B} = B) \equiv (\mathbf{C} \mid \mathbf{A} = A, \mathbf{B} = B)$, because $\mathbf{A}''$ is independently sampled from $(\mathbf{B}'', \mathbf{C}'')$, and that $\mathbf{C} = f(\mathbf{B})$ is only a function of $\mathbf{B}$. Therefore, because the conditional distribution of $\mathbf{C} \equiv \mathbf{C}''$ is the same given $(\mathbf{A}'' = A, \mathbf{B}'' = B)$ or $(\mathbf{A} = A, \mathbf{B} = B)$, by Lemma 12,

$$\mathsf{SD}(\mathbf{X}, \mathbf{X}'') = \mathsf{SD}((\mathbf{A}, \mathbf{B}), (\mathbf{A}'', \mathbf{B}'')) = \mathsf{MutDep}(\mathbf{A}, \mathbf{B}). \tag{1}$$

Secondly, for all $A \leftarrow \mathbf{A}, C \leftarrow \mathbf{C}$, it holds that $(\mathbf{B}'' \mid \mathbf{A}'' = A, \mathbf{C}'' = C) \equiv (\mathbf{B} \mid \mathbf{C} = C) \equiv (\mathbf{B}' \mid \mathbf{A}' = A, \mathbf{C}' = C)$, so by Lemma 12, it holds that

$$\mathsf{SD}(\mathbf{X}', \mathbf{X}'') = \mathsf{MutDep}(\mathbf{A}, \mathbf{C}) \leq \mathsf{SD}(\mathbf{X}, \mathbf{X}'') = \mathsf{MutDep}(\mathbf{A}, \mathbf{B}). \tag{2}$$

Therefore, by the triangle inequality and Eqs. (1) and (2), it holds that $\mathsf{SD}(\mathbf{X}, \mathbf{X}') \leq \mathsf{SD}(\mathbf{X}, \mathbf{X}'') + \mathsf{SD}(\mathbf{X}', \mathbf{X}'') \leq 2 \cdot \mathsf{MutDep}(\mathbf{A}, \mathbf{B})$. □

Variations of the following lemma are used in previous works.[10] It states an intuitive way to bound the statistical distance of sequences of random variables in systems where there exist some low-probability 'bad' events, and conditioned on those bad events not happening the two systems proceed statistically closely. Here we only need this specific variant for random systems with two blocks.

Lemma 11 (Bounding statistical distance of pairs). *Let* $\mathbf{X} = (\mathbf{X}_1, \mathbf{X}_2)$ *and* $\mathbf{X}' = (\mathbf{X}'_1, \mathbf{X}'_2)$ *be two jointly distributed pairs of random variables where* $\mathsf{SD}(\mathbf{X}_1, \mathbf{X}'_1) \leq \alpha$. *Let* B *be an event (i.e. an arbitrary set) such that for every* $X_1 \in \mathsf{Supp}(\mathbf{X}_1) \cap \mathsf{Supp}(\mathbf{X}'_1) \setminus$ B *it holds that* $\mathsf{SD}((\mathbf{X}_2 \mid \mathbf{X}_1 = X_1), (\mathbf{X}'_2 \mid \mathbf{X}'_1 = X_1)) \leq \beta$. *Then, it holds that*

$$\mathsf{SD}(\mathbf{X}, \mathbf{X}') \leq \alpha + \beta + \Pr[\mathbf{X}_1 \in \mathsf{B}].$$

Proof. Using two direct applications of Lemma 4, we show how to sample $(\mathbf{X}, \mathbf{X}')$ *jointly* in a way that $\Pr[\mathbf{X} = \mathbf{X}'] \geq 1 - (\alpha + \beta + \rho)$ where $\Pr[\mathbf{X}_1 \in \mathsf{B}] = \rho$. Then Lemma 11 follows (again by an application of Lemma 4).

Firstly, by Lemma 4 we can sample $(\mathbf{X}_1, \mathbf{X}'_1)$ jointly, while $\Pr[\mathbf{X}_1 = \mathbf{X}'_1] \geq 1 - \alpha$. Now, we *expand* the joint sampling of $(\mathbf{X}_1, \mathbf{X}'_1)$ to a full joint sampling of $(\mathbf{X}, \mathbf{X}') \equiv (\mathbf{X}_1, \mathbf{X}_2, \mathbf{X}'_1, \mathbf{X}'_2)$ as follows. We first sample $(X_1, X'_1) \leftarrow (\mathbf{X}_1, \mathbf{X}'_1)$ from their joint distribution. Then, for each sampled (X_1, X'_1), we sample the distributions $(\mathbf{X}_2, \mathbf{X}'_2 \mid X_1, X'_1)$ also *jointly* such that $\Pr[\mathbf{X}_2 = \mathbf{X}'_2 \mid X_1, X'_1] = 1 - \mathsf{SD}((\mathbf{X}_2 \mid X_1), (\mathbf{X}'_2 \mid X'_1))$. We can indeed do such joint sampling, again by applying Lemma 4, but this time we apply that lemma to the conditional distributions $(\mathbf{X}_2 \mid X_1, X'_1) \equiv (\mathbf{X}_2 \mid X_1)$ and $(\mathbf{X}'_2 \mid X_1, X'_1) \equiv (\mathbf{X}'_2 \mid X'_1)$.

Now, we lower bound $\Pr[\mathbf{X}_1 = \mathbf{X}'_1 \wedge \mathbf{X}_2 = \mathbf{X}'_2]$ when we sample all the blocks through the joint distribution $(\mathbf{X}_1, \mathbf{X}_2, \mathbf{X}'_1, \mathbf{X}'_2)$ defined above. First, we know that $\Pr[\mathbf{X}_1 = \mathbf{X}'_1] \geq 1 - \alpha$ and $\Pr[\mathbf{X}_1 \notin \mathsf{B}] \geq 1 - \rho$, therefore $\Pr[\mathbf{X}_1 = \mathbf{X}'_1 \notin \mathsf{B}] \geq 1 - \alpha - \rho$. Moreover, for any such $X_1 \in \mathsf{Supp}(\mathbf{X}_1) \cap \mathsf{Supp}(\mathbf{X}'_1) \setminus \mathsf{B}$, we have

$$\Pr[\mathbf{X}_2 = \mathbf{X}'_2 \mid \mathbf{X}_1 = \mathbf{X}'_1 = X_1] \geq 1 - \mathsf{SD}((\mathbf{X}_2 \mid X_1), (\mathbf{X}'_2 \mid \mathbf{X}'_1 = X_1)) \geq 1 - \beta.$$

Therefore, the lemma follows by a union bound. □

---

[10] For example see Lemma 2.2 of [GKLM12].

The following useful lemma could be derived as a special case of Lemma 11 above by letting $B = \mathrm{Supp}(\mathbf{X}_1) \cup \mathrm{Supp}(\mathbf{X}_1')$ and $\beta = 0$.

Lemma 12. *If* $(\mathbf{X}, \mathbf{Y}), (\mathbf{X}', \mathbf{Y}')$ *are joint distributions and* $(\mathbf{Y} \mid X) \equiv (\mathbf{Y}' \mid \mathbf{X}' = X)$ *for all* $X \in \mathrm{Supp}(\mathbf{X}) \cap \mathrm{Supp}(\mathbf{X}')$, *then* $\mathsf{SD}((\mathbf{X}, \mathbf{Y}), (\mathbf{X}', \mathbf{Y}'')) = \mathsf{SD}(\mathbf{X}, \mathbf{X}')$.

2.2 Lemmas About Shannon Entropy

Definition 13 ((Conditional) Shannon entropy). *For a random variable* $\mathbf{X}$, *its Shannon entropy is defined as* $\mathrm{H}(\mathbf{X}) = \mathbb{E}_{X \leftarrow \mathbf{X}} \log(1/\Pr[\mathbf{X} = X])$. *The conditional (Shannon) entropy is defined as* $\mathrm{H}(\mathbf{X} \mid \mathbf{Y}) = \mathbb{E}_{Y \leftarrow \mathbf{Y}} \mathrm{H}(\mathbf{X} \mid Y)$. *The binary (Shannon) entropy function* $\mathrm{H}(\varepsilon) = -p \log p - (1 - p) \log(1 - p)$ *is equal to the entropy of a Bernoulli process with probability* ε.[11]

Jensen's inequality implies that we always have $\mathrm{H}(\mathbf{X}) \geq \mathrm{H}(\mathbf{X} \mid \mathbf{Y}) \geq \mathbf{0}$.

Lemma 14 (Lower bounding entropy using statistical distance). *Suppose* $\mathsf{SD}(\mathbf{X}, \mathbf{U}_n) \leq \varepsilon$. *Then* $\mathrm{H}(\mathbf{X}) \geq (1 - \varepsilon) \cdot n - \mathrm{H}(\varepsilon)$.

Proof. Since $\mathsf{SD}(\mathbf{X}, \mathbf{U}_n) \leq \varepsilon$, using Lemma 4 we can sample $(\mathbf{X}, \mathbf{U}_n)$ jointly such that $\Pr[\mathbf{X} \neq \mathbf{U}_n] \leq \varepsilon$. In this case, we have

$$n = \mathrm{H}(\mathbf{U}_n) \leq \mathrm{H}(\mathbf{X}_n, \mathbf{U}_n) = \mathrm{H}(\mathbf{X}) + \mathrm{H}(\mathbf{U}_n \mid \mathbf{X}) \leq \mathrm{H}(\mathbf{X}) + \mathrm{H}(\varepsilon) + \varepsilon \cdot \log(2^n - 1)$$

where the last inequality follows from Fano's lemma [Fan68]. Therefore, we get $\mathrm{H}(\mathbf{X}) \geq (1 - \varepsilon) \cdot n - \mathrm{H}(\varepsilon)$. □

Lemma 15 (Upper-bounding collision probability using (conditional) Shannon entropy). *Suppose* $\mathrm{Supp}(\mathbf{X}) \subseteq \{0, 1\}^n$.

1. *If* $\mathrm{H}(\mathbf{X}) \geq 2/3$, *then it holds that*

$$\Pr_{X_1, X_2 \leftarrow \mathbf{X}} [X_1 \neq X_2] \geq \frac{1}{10n}.$$

2. *If* $\mathrm{H}(\mathbf{X} \mid \mathbf{Y}) \geq 5/6$ *for a jointly distributed* $(\mathbf{X}, \mathbf{Y})$, *then it holds that*

$$\Pr_{Y \leftarrow \mathbf{Y}, X_1, X_2 \leftarrow (\mathbf{X}|Y)} [X_1 \neq X_2] \geq \frac{1}{60 \cdot n^2}.$$

Proof. First, we prove Part 1. In the following let $\varepsilon = 1/(10n) \leq 1/10$. Our first goal is to show that $\Pr_{X_1, X_2 \leftarrow \mathbf{X}}[X_1 \neq X_2] \geq \varepsilon$. There are two cases to consider:

---

[11] The notation is well defined: If the input ε is a real number, by $\mathrm{H}(\varepsilon)$ we mean the binary entropy, and otherwise we mean the entropy of a random variable.

1. Case (1): Suppose first that there is some $A \subseteq \mathrm{Supp}(\mathbf{X})$ with $\varepsilon \leq p_A = \Pr_{X \leftarrow \mathbf{X}}[X \in A] \leq 1 - \varepsilon$. Then, letting $B = \mathrm{Supp}(\mathbf{X}) \setminus A$, we also have $\varepsilon \leq \Pr_{X \leftarrow \mathbf{X}}[X \in B] \leq 1 - \varepsilon$. Since A and B are disjoint, we have

$$\Pr_{X_1, X_2 \leftarrow \mathbf{X}}[X_1 \neq X_2] \geq \Pr_{X_1, X_2 \leftarrow \mathbf{X}}[X_1 \in A, X_2 \in B \text{ or } X_1 \in B, X_2 \in A]$$
$$= 2 \cdot p_A \cdot (1 - p_A) \geq 2 \cdot \varepsilon \cdot (1 - \varepsilon) = 2 \cdot \varepsilon - 2 \cdot \varepsilon^2 \geq \varepsilon.$$

The last inequality follows from $\varepsilon \leq 1/10$, which implies $\varepsilon \geq 2\varepsilon^2$.

2. If we are not in Case (1) above, then for every $A \subseteq \mathrm{Supp}(\mathbf{X})$, $\Pr_{X \leftarrow \mathbf{X}}[X \in A] < \varepsilon$ or $\Pr_{X \leftarrow \mathbf{X}}[X \in A] > 1 - \varepsilon$. In particular, for every $X \in \mathrm{Supp}(\mathbf{X})$, we have $\Pr[\mathbf{X} = X] < \varepsilon$ or $\Pr[\mathbf{X} = X] > 1 - \varepsilon$. Now there are two cases:

   (a) For all $X \in \mathrm{Supp}(\mathbf{X})$, $\Pr[\mathbf{X} = X] < \varepsilon$. In this case, because $\varepsilon < 1/10$, we can build a set $A \subseteq \mathrm{Supp}(\mathbf{X})$ that implies being in Case (1). Namely, let $A_0, A_1, \ldots, A_m$ be a sequence of sets where $A_i = \{1, \ldots, i\} \subseteq [m] = \mathrm{Supp}(\mathbf{X})$. Suppose i is the smallest number for which $\Pr[\mathbf{X} \in A_i] \geq \varepsilon$, which means $\Pr[\mathbf{X} \in A_{i-1}] < \varepsilon$. In this case we have:

   $$\Pr[\mathbf{X} \in A_i] \leq \Pr[\mathbf{X} \in A_{i-1}] + \Pr[\mathbf{X} = i] < 2\varepsilon < 1 - \varepsilon$$

   where the last inequality follows from $\varepsilon < 1/10$.

   (b) There is some $X \in \mathrm{Supp}(\mathbf{X})$ where $\Pr[\mathbf{X} = x] > 1 - \varepsilon$. Now suppose we sample $\mathbf{X}$ jointly with a Boolean $\mathbf{B}$ where $\mathbf{B} = 0$ iff $\mathbf{X} = X$. So, we get:

   $$2/3 \leq \mathrm{H}(\mathbf{X}) \leq \mathrm{H}(\mathbf{B}) + \mathrm{H}(\mathbf{X} \mid \mathbf{B})$$
   $$= \mathrm{H}(\mathbf{B}) + \Pr[\mathbf{B} = 0] \cdot \mathrm{H}(\mathbf{X} \mid \mathbf{B} = 0) + \Pr[\mathbf{B} = 1] \cdot \mathrm{H}(\mathbf{X} \mid \mathbf{B} = 1)$$
   $$< \mathrm{H}(\varepsilon) + \Pr[\mathbf{B} = 0] \cdot 0 + \varepsilon \cdot n$$
   $$\leq \mathrm{H}(1/10) + 1/10$$
   $$< 1/2 + 1/10 \quad (\text{because } H(1/10) < 1/2)$$

   which is a contradiction.

Now we prove Part 2. Because we have $\mathrm{H}(\mathbf{X} \mid \mathbf{Y}) \geq 5/6$, and because $\mathrm{H}(\mathbf{X} \mid Y) \leq n$ for any $Y \leftarrow \mathbf{Y}$, by an averaging argument it holds that $\Pr_{Y \leftarrow \mathbf{Y}}[\mathrm{H}(\mathbf{X} \mid Y) > 2/3] \geq 1/(6n)$. That is because otherwise, $\mathrm{H}(\mathbf{X} \mid \mathbf{Y})$ would be at most $(2/3) \cdot (1 - 1/(6n)) + n \cdot (1/(6n)) < 5/6$. Therefore, with probability at least $1/(6n)$ we get $Y \leftarrow \mathbf{Y}$ for which we have

$$\Pr[X_1 \neq X_2; Y \leftarrow \mathbf{Y}, X_1, X_2 \leftarrow (\mathbf{X} \mid Y)] \geq 1/(10n).$$

The claim then follows by using the chain rule. $\qquad \square$

2.3 Lemmas About the Random Oracle Model

Definition 16 (Random Oracles). *A random oracle $\mathbf{H}(\cdot)$ is a randomized function such that for all $x \in \{0, 1\}^*$, $\mathbf{H}(x)$ is independently mapped to a random string of the same length $|x|$.*

Even though the above definition is for infinite random oracles, in this work we are only interested and only use *finite* random oracles, as there is always an upper bound (based on the security parameter) on the maximum length of the queries asked by a polynomial time algorithm.

Notation on oracle-aided algorithms. For any view $V_{\mathcal{A}}$ of a party $\mathcal{A}$ with access to some oracle O, by $\mathcal{Q}(V_{\mathcal{A}})$ we refer to the set of queries to O in the view $V_{\mathcal{A}}$, and by $\mathcal{P}(V_{\mathcal{A}})$ we denote the set of oracle query-answer pairs in $V_{\mathcal{A}}$. So, $\mathcal{Q}(\cdot), \mathcal{P}(\cdot)$ are operators that extract the queries or query-answer pairs. When it is clear from the context, we might simply use $Q_{\mathcal{A}} = \mathcal{Q}(V_{\mathcal{A}})$ and by $P_{\mathcal{A}} = \mathcal{P}(V_{\mathcal{A}})$. When $\mathcal{A}$ is an interactive algorithm, if $\mathcal{A}$ has no inputs and uses randomness $r_{\mathcal{A}}$, and if T is the transcript of the interaction, then $V_{\mathcal{A}} = (r_{\mathcal{A}}, T, P_{\mathcal{A}})$.

Variants of the following lemma were implicit in [IR89,BMG09] and stated in [DLMM11]. See the works of [HOZ16,BM17] for formal proofs.

Theorem 17 (Dependency learner [IR89,BMG09,HOZ16,BM17]**).** *Let* $(\mathcal{A}, \mathcal{B})$ *be an interactive protocol between Alice and Bob in which they might use private randomness (but no inputs otherwise) and they each ask at most m queries to a random oracle* $\mathbf{H}$. *Then, there is a deterministic eavesdropping algorithm Eve (whose algorithm might depend on Alice and Bob and) who gets as input $\delta \in [0,1]$ and the transcript T of the messages exchanged between Alice and Bob, asks at most* $\mathrm{poly}(m/\delta)$ *queries to the random oracle* $\mathbf{H}$, *and we have:*

1. *The average of the statistical distance between $(\mathbf{V}_{\mathcal{A}}, \mathbf{V}_{\mathcal{B}})$ and $(\mathbf{V}_{\mathcal{A}} \times \mathbf{V}_{\mathcal{B}})$ conditioned on $\mathbf{V}_{\mathcal{E}}$ is at most δ. Namely,*

$$\mathsf{MutDep}_{\mathbf{V}_{\mathcal{E}}}(\mathbf{V}_{\mathcal{A}}, \mathbf{V}_{\mathcal{B}}) = \mathop{\mathbb{E}}_{V_{\mathcal{E}} \leftarrow \mathbf{V}_{\mathcal{E}}} \mathsf{MutDep}((\mathbf{V}_{\mathcal{A}} \mid V_{\mathcal{E}}), (\mathbf{V}_{\mathcal{B}} \mid V_{\mathcal{E}})) \leq \delta.$$

2. *The probability that Alice and Bob have an 'intersection query' outside of the queries asked by Eve to the random oracle is bounded as follows:*

$$\Pr[\mathcal{Q}(\mathbf{V}_{\mathcal{A}}) \cap \mathcal{Q}(\mathbf{V}_{\mathcal{B}}) \not\subseteq \mathcal{Q}(\mathbf{V}_{\mathcal{E}})] \leq \delta.$$

The two parts of Theorem 17 could be derived from each other, but doing that is not trivial and involves asking *more* oracle queries from the oracle. We will use both of the properties in our formal proof of our main result in Sect. 3.

Notation for indistinguishability in the ROM. For families of random variables $\{\mathbf{X}_{\kappa}\}, \{\mathbf{Y}_{\kappa}\}$ by $\mathbf{X}_{\kappa} \equiv_c \mathbf{Y}_{\kappa}$ we mean that $\{\mathbf{X}_{\kappa}\}, \{\mathbf{Y}_{\kappa}\}$ are indistinguishable against nonuniform PPT algorithms. When we are in the random oracle model, we use the same notation $\mathbf{X}_{\kappa} \equiv_c \mathbf{Y}_{\kappa}$ when the distinguishers are $\mathrm{poly}(\kappa)$-query algorithms. Namely, for any $\mathrm{poly}(\kappa)$-query oracle-aided algorithm D there is a negligible function ε, such that $\Pr[D(\mathbf{X}_{\kappa}) = 1] - \Pr[D(\mathbf{Y}_{\kappa}) = 1] \leq \varepsilon(\kappa)$, where the probabilities are over the inputs $\mathbf{X}_{\kappa}, \mathbf{Y}_{\kappa}$ and the randomness of D and the oracle $\mathbf{H}$. When κ is clear from the context, we write $\mathbf{X} \equiv_c \mathbf{Y}$ for simplicity.

2.4   OT and its Multi-Input Variant k-OT in the ROM

In this subsection, we recall the notions of OT and its multi-input version on k inputs, denoted k-OT. We will also prove basic lemmas that allows us to prove the existence of *attacks* against semi-honest security of k-OT.

We start by defining (multi-) oblivious transfer (OT) formally.

Definition 18 (k-OT). *A k-parallel 1-out-of-2 oblivious transfer (OT). functionality (k-OT) is a two-party functionality between a sender $\mathcal{S}$ and a receiver $\mathcal{R}$ as follows. The sender has input $\{x_i^0, x_i^1\}_{i \in [k]}$ which are arbitrary strings, and the receiver has the input $\vec{b} \in \{0,1\}^k$. The sender receives no output at the end, while the receiver receives $\{x_i^{b_i}\}_{i \in [k]}$.*

Semi-honest security of k-OT. We use standard definition of simulation-based security, see e.g. [Lin16]. In particular, for any semi-honest secure OT protocol between $\mathcal{S}$ and $\mathcal{R}$, there are two PPT simulator $\mathsf{Sim}_\mathcal{S}, \mathsf{Sim}_\mathcal{R}$ such that for any input $\vec{b}$ of $\mathcal{R}$ and any input $x = \{x_i^0, x_i^1\}_{i \in [k]}$ for $\mathcal{S}$, it holds that:

$$\mathsf{Sim}_\mathcal{S}(x) \equiv_c \mathbf{V}_\mathcal{S}(x, \vec{b}) \quad \text{and} \quad \mathsf{Sim}_\mathcal{R}(\vec{b}, \{x_i^{b_i}\}_{i \in [k]}) \equiv_c \mathbf{V}_\mathcal{R}(x, \vec{b}).$$

Plain model security vs. the ROM security. In the plain model all the parties (including the simulator and the adversary and the distinguishers) are PPT algorithms. In the random random oracle model we, the honest parties and the simulators are oracle-aided PPTs, while the *distinguishers* are poly(κ)-query (computationally unbounded) algorithms accessing the random oracle $\mathbf{H}$.[12] Recall that by the notation defined at the end of Sect. 2 we can use the same notation $\equiv_c$ for indistinguishably against poly(κ) distinguishers in the ROM.

Sufficient conditions for breaking the semi-honest security of k-OT. We now state and prove two simple lemmas showing that the attacks that we construct in Sect. 3 are indeed attacks according to the standard definition of simulation-based security, see e.g. [Lin16]. The following lemma, states the intuitive fact that in any OT protocol, the input of the sender should remain indistinguishable from a random string, if the receiver chooses its input randomly.

Lemma 19. *Let $(\mathcal{S}, \mathcal{R})$ be a semi-honest secure m-OT protocol in the plain model (resp., in the ROM) in which the receiver's inputs are chosen uniformly at random from $\{0,1\}^m$, and in which $\mathcal{S}, \mathcal{R}$ are PPTs (resp., oracle-aided PPTs). Fix any input x for the sender. Let $\vec{b} \equiv \mathbf{U}_m$ be the uniformly random inputs of the receiver and $\mathbf{V}_\mathcal{S}(x, \vec{b})$ the random variable denoting the view of the sender (for inputs $x, \vec{b}$ being used by the sender and the receiver). Then we have*

$$(\mathbf{V}_\mathcal{S}(x, \vec{b}), \vec{b}) \equiv_c (\mathbf{V}_\mathcal{S}(x, \vec{b}) \times \mathbf{U}_m).$$

(Recall that in the ROM, the distinguisher is an poly(κ)-query, computationally unbounded, algorithm for security parameter κ.)

---

[12] Note that this definition is for a setting where the random oracle is the sole source of hardness. E.g., this is how the security of the protocol in [IKNP03] could be proved.

Proof. We prove the lemma for the computational setting in the plain model where the distinguishers are PPT algorithms. The same proof holds for the random oracle model in which the distinguishers are $\mathrm{poly}(\kappa)$-query oracle-aided algorithms accessing a random oracle $\mathbf{H}$, where κ is the security parameter.

By the security definition of OT, there is a PPT simulator $\mathsf{Sim}_\mathcal{S}$ such that for any input $\vec{b}$ of $\mathcal{R}$ it simulates the view of $\mathcal{S}$:

$$\mathsf{Sim}_\mathcal{S}(x) \equiv_c \mathbf{V}_\mathcal{S}(x, \vec{b}).$$

Hence, by averaging over $\vec{b} \leftarrow \vec{\mathbf{b}}$, we have $(\mathbf{V}_\mathcal{S}(x, \vec{\mathbf{b}}), \vec{\mathbf{b}}) \equiv_c (\mathsf{Sim}_\mathcal{S}(x), \vec{\mathbf{b}})$ for uniform $\vec{\mathbf{b}}$. (In other words, if the latter two were distinguishable, one could distinguish $\mathsf{Sim}_\mathcal{S}(x)$ from $\mathbf{V}_\mathcal{S}(x, \vec{b})$ by the same advantage for some $\vec{b}$.) Since $\mathsf{Sim}_\mathcal{S}(x)$ is independent of the receiver's input $\vec{b}$, we conclude

$$(\mathbf{V}_\mathcal{S}(x, \vec{\mathbf{b}}), \vec{\mathbf{b}}) \equiv_c (\mathsf{Sim}_\mathcal{S}(x), \vec{\mathbf{b}}) \equiv \mathsf{Sim}_\mathcal{S}(x) \times \mathbf{U}_m \equiv_c \mathbf{V}_\mathcal{S}(x, \vec{\mathbf{b}}) \times \mathbf{U}_m. \qquad \square$$

Lemma 20. *Let $(\mathcal{S}, \mathcal{R})$ be a semi-honest secure m-OT protocol in which the sender's inputs are chosen uniformly at random and in which $\mathcal{S}, \mathcal{R}$ are PPTs (resp., oracle-aided PPTs). Fix any input vector $\vec{b}$ for the receiver. Let $\mathbf{x} = \{\mathbf{x}_i^0, \mathbf{x}_i^1\}_{i \in [m]}$ be uniformly random inputs for the sender, and let $\mathbf{V}_\mathcal{R}(\mathbf{x}, \vec{b})$ be the random variable denoting the view of the receiver (when the inputs $\mathbf{x}, \vec{b}$ are used by the two parties). Then, it holds that*

$$(\mathbf{V}_\mathcal{R}(\mathbf{x}, \vec{b}), \{\mathbf{x}_i^{\bar{b}_i}\}_{i \in [m]}) \equiv_c (\mathbf{V}_\mathcal{R}(\mathbf{x}, \vec{b}) \times \{\mathbf{x}_i'\}_{i \in [m]})$$

where $\mathbf{x}_i'$'s are independent uniformly random strings of the appropriate length.

Proof. As in the proof of Lemma 19, we only prove the lemma for the computational setting in the plain model where the distinguishers are PPT algorithms. The same proof holds for random oracle model in which the distinguishers are $\mathrm{poly}(\kappa)$-query algorithms in the random oracle model, for security parameter κ.

By the security definition of m-OT, there is a PPT simulator $\mathsf{Sim}_\mathcal{R}$ such that for any input $\{x_i^0, x_i^1\}_{i \in [m]}$ of $\mathcal{S}$ it simulates the view of $\mathcal{R}$:

$$\mathsf{Sim}_\mathcal{R}(\vec{b}, \{x_i^{b_i}\}_{i \in [m]}) \equiv_c \mathbf{V}_\mathcal{R}(x, \vec{b}).$$

Hence $(\mathbf{V}_\mathcal{R}(\mathbf{x}, \vec{b}), \{\mathbf{x}_i^{\bar{b}_i}\}_{i \in [m]}) \equiv_c (\mathsf{Sim}_\mathcal{R}(\vec{b}, \{\mathbf{x}_i^{b_i}\}_{i \in [m]}), \{\mathbf{x}_i^{\bar{b}_i}\}_{i \in [m]})$ holds for uniform $\mathbf{x}$. Since $\mathsf{Sim}_\mathcal{R}(\vec{b}, \{\mathbf{x}_i^{b_i}\}_{i \in [m]})$ is independent of $\{\mathbf{x}_i^{\bar{b}_i}\}_{i \in [m]}$ (i.e., the sender's input that is not learned by receiver), similarly to Lemma 19, we conclude that

$$(\mathbf{V}_\mathcal{R}(\mathbf{x}, \vec{b}), \{\mathbf{x}_i^{\bar{b}_i}\}_{i \in [m]}) \equiv_c \mathbf{V}_\mathcal{R}(\mathbf{x}, \vec{b}) \times \{\mathbf{x}_i'\}_{i \in [m]}. \qquad \square$$

Remark 21. In Sect. 3, we will use Lemma 19 for getting a (computationally unbounded) $\mathrm{poly}(\kappa)$-query attacking sender. Namely, instead of directly breaking the semi-honest security definition of k-OT, our attacking semi-honest sender (or more accurately, the distinguisher), will pursue a different goal based on

Lemma 19. Namely, based on his own view, the attacking sender will try to distinguish the receiver's actual input $\vec{b}$ from an independently uniform string.

Similarly, we will use Lemma 20 to get a (computationally unbounded) poly(κ)-query attacking receiver. Namely, our attacking semi-honest receiver (i.e., the distinguisher) will find another input $\vec{b}' \neq \vec{b}$ and read sender's inputs according to $\vec{b}'$. Doing so would be a successful attack by Lemma 20.

3   Impossibility of Round-preserving OT Extension in the Random Oracle Model

In this section we formally state and prove our main impossibility result, Theorem 1. We start by formalizing the model for round-preserving OT extension.

OT extension is the task of using a limited number of "base OTs" to generate an increased number of OTs. The weakest possible form of OT extension is using n base OTs to construct $n + 1$ OTs, but doing so is also sufficient as this can be repeated to get further extension (e.g., see [LZ13]). In our definition of OT extension, we model base OTs with an OT-hybrid functionality. This functionality can be seen as a trusted third party that receives the inputs of sender and receiver over a perfectly secure channel and sends to the receiver the output of the base OTs. The presence of an OT-hybrid functionality is often referred to as the OT-hybrid model [IKNP03].

Here, we are particularly interested in the notion of a *round-preserving* OT extension protocol. Intuitively, this is an OT extension which uses the same number of rounds as the base OTs that implement the OT-hybrid functionality. Given an r-round-preserving OT extension protocol E from n-OTs to $(n + 1)$-OTs, one may then instantiate $\mathbb{OT}_n$ with a concrete r-round OT to obtain $(n+1)$-OT that also works in r rounds.

The following definition formalizes the hybrid model using which we model OT extension protocols that preserve the round complexity of the base OTs. We first describe the model, and then we will discuss some subtle aspects of it.

Definition 22 (Round-preserving OT extension). *A round-preserving OT extension protocol is a 2-party protocol with the following form.*

1. *$\mathcal{S}$ has input $\{x_i^0, x_i^1\}_{i \in [n+1]}$ and $\mathcal{R}$ has input $\vec{b} = (b_1, \ldots, b_{n+1})$.*
2. *Both of $\mathcal{S}, \mathcal{R}$ can query the random oracle $\mathbf{H}$ at any time.*
3. *$\mathcal{R}$ and $\mathcal{S}$ exchange $r = \mathrm{poly}(\kappa)$ number of messages $t_1, \ldots, t_r$.*
4. *By the time $\mathcal{S}$ sends the final message t_r to $\mathcal{R}$, $\mathcal{S}$ has submitted its inputs $\{y_i^0, y_i^1\}_{i \in [n]}$ and $\mathcal{R}$ has submitted its input $\vec{c} = (c_1, \ldots, c_n)$ to $\mathbb{OT}_n$.*
5. *Right after $\mathcal{S}$ sends the final message to $\mathcal{R}$, $\mathcal{R}$ receives $\{y_i^{c_i}\}_{i \in [n]}$ from $\mathbb{OT}_n$.*
6. *$\mathcal{R}$ outputs, perhaps after more queries to $\mathbf{H}$, what is supposed to be $\{x_i^{b_i}\}_{i \in [n+1]}$.*

The completeness and semi-honest security of OT extension is defined based on the semi-honest security of k-OT (Definition 18) for $k = n + 1$.

When to submit inputs to hybrid $\mathbb{OT}_n$. We emphasize that the output from the OT-hybrid functionality is received only after the final message has been sent. This is the case because the OT-hybrid functionality in an r-round OT extension protocol is implemented using an r-round base OT protocol, which produces its output after receiving the final message. In this definition, the parties choose their inputs for $\mathbb{OT}_n$ at some points before the last message. Note that, "naturally" the inputs to a r-round OT functionality should be submitted at the beginning, but allowing the parties to choose their inputs to $\mathbb{OT}_n$ more flexibly only makes our impossibility result stronger.

In Definition 22, messages exchanged in an extension protocol are not allowed to depend on the intermediate messages of the base OT protocol. This is justified since these messages are simulatable. Moreover, without loss of generality, we assume that $\mathbb{OT}_n$ is never used in the "opposite" direction (with the sender acting as the receiver and the receiver as the sender), because then there would be not enough rounds for the output of $\mathbb{OT}_n$ affecting any message sent to the receiver, who is the only party with an output. Indeed, not surprisingly, the known protocols [WW06] for switching the sender/receiver roles of the OT require additional rounds. This role-switching is used in the OT extension of the IKNP protocol [IKNP03], which also requires one more round. In fact, our impossibility result shows that the result of [IKNP03] is round-optimal (though it is not round-preserving) among all black-box protocols for OT extension using symmetric-key primitives.

Based on Definition 22 above, we can now state Theorem 1 formally.

Theorem 23. *Let $(\mathcal{S}, \mathcal{R})$ be a round-preserving OT extension protocol (according to Definition 22) with security parameter κ using random oracle $\mathbf{H}$ as follows.*

1. *The $n \leq \text{poly}(\kappa)$ OTs modeled by $\mathbb{OT}_n$ are allowed to be string OTs.*
2. *$(\mathcal{S}, \mathcal{R})$ implement bit $(n + 1)$-OT with $\lambda = \text{negl}(\kappa)$ completeness error.*
3. *Either of $(\mathcal{S}, \mathcal{R})$ ask at most $m = \text{poly}(\kappa)$ queries to the random oracle $\mathbf{H}$.*

Then the constructed $(n + 1)$-OT cannot be (even semi-honest) secure for both of $\mathcal{S}$ or $\mathcal{R}$ against adversaries who can ask $\text{poly}(m \cdot n) \leq \text{poly}(\kappa)$ queries to $\mathbf{H}$.

In particular, either of $\mathcal{S}$ or $\mathcal{R}$ can execute the protocol honestly, then ask $\text{poly}(\kappa)$ more queries, and then break the (semi-honest) security of the constructed bit $(n + 1)$-OT by advantage $\frac{1}{\text{poly}(n)} \geq \frac{1}{\text{poly}(\kappa)}$ according to either of the attacks described in Lemma 19 or Lemma 20.

The above theorem proves that for any round-preserving OT extension protocol, there is always a $\text{poly}(\kappa)$-query attack by one of the parties that succeeds in breaking the semi-honest security of the protocol with non-negligible advantage $1/\text{poly}(\kappa)$. In fact, we show how to break the security of such protocols even when the main inputs (but not those of the hybrid $\mathbb{OT}_n$) are chosen at random.

3.1   Proving Theorem 23

In the rest of this section, we prove Theorem 23 above.

Notation. First we clarify our notation used.

- $\vec{b} = (b_1, \ldots, b_{n+1}) \in \{0,1\}^{n+1}$ is $\mathcal{R}$'s own input, and it submits $\vec{c} = (c_1, \ldots, c_n) \in \{0,1\}^n$ as its input to $\mathbb{OT}_n$ during the execution of the protocol.
- $(x_i^0, x_i^1)_{i \in [n+1]}$ is $\mathcal{S}$'s input, and it submits $\{y_i^0, y_i^1\}_{i \in [n]}$ as its input to $\mathbb{OT}_n$.
- For $r \in \mathbb{N}$, $T = (t_1, \ldots, t_r)$ is the transcript of the protocol.
- γ is the output of $\mathbb{OT}_n$ that $\mathcal{R}$ receives after t_r is sent to $\mathcal{R}$.
- $V_{\mathcal{S}}$ and $V_{\mathcal{R}}$ denote, in order, the views of $\mathcal{S}$ and $\mathcal{R}$, where $V_{\mathcal{R}}$ only includes the receiver's view *before receiving* γ from $\mathbb{OT}_n$.

We will show that by asking $\text{poly}(\kappa)$ queries after executing the protocol honestly: either the sender can distinguish the receiver's *uniformly random* input from an actual independent random string, which is an attack by Lemma 19, or the receiver can read both of sender's inputs for an index i with non-negligible probability[13]), which is an attack by Lemma 20.

We first define each party's attack and then will prove that one of them will succeed with non-negligible probability. Both attacks will make heavy use of the 'dependency learning' attack of Theorem 17. We will use that lemma for some sufficiently small parameter δ that will be chosen when we analyze the attacks.

Construction 24 (Sender's attack $\widehat{\mathcal{S}}$). *Here $\widehat{\mathcal{S}}$ tries to distinguish between an independently sampled random string from $\{0,1\}^{n+1}$ and the actual input $\vec{b}$ (chosen at random and then) used by the receiver, based on the transcript T of the (honestly executed protocol) and its knowledge about the random oracle $\mathbf{H}$.*

1. *$\widehat{\mathcal{S}}$ chooses its own input $x = (x_i^0, x_i^1)_{i \in [n+1]}$ uniformly at random.*
2. *After the last message t_r is sent, $\widehat{\mathcal{S}}$ runs the Eve algorithm of Theorem 17 over the full transcript $T = (t_1, \ldots, t_r)$ for sufficiently small δ (to be chosen later) over the following modified version $(\mathcal{S}, \mathcal{R}_1)$ of the original protocol, to learn a set of oracle query-answer pairs $P_{\mathcal{E}}$.*
   - *$\mathcal{S}$ and $\mathcal{R}$ choose their inputs uniformly at random.*
   - *$\mathcal{R}_1$ stops right after the last message is sent (right before γ is delivered). Note that even though $\mathcal{S}, \mathcal{R}_1$ submit some inputs to $\mathbb{OT}_n$, because no outputs are received by $\mathcal{R}_1$ and because all inputs are chosen at random, this is a randomized "inputless" protocol between $\mathcal{S}, \mathcal{R}_1$ for which we can indeed run the attacker Eve of Theorem 17.*
3. *$\widehat{\mathcal{S}}$ then considers the distribution $(\mathbf{V}_{\mathcal{R}} \mid \mathbf{V}_{\mathcal{E}} = V_{\mathcal{E}})$ conditioned on the obtained Eve view $V_{\mathcal{E}} = (T, P_{\mathcal{E}})$, where T is the transcript and $P_{\mathcal{E}}$ are the oracle query-answer pairs learned by Eve.[14] Then, given an input from $\{0,1\}^{n+1}$, $\widehat{\mathcal{S}}$ tries*

---

[13] One can always guess a bit with probability $1/2$, however, if the receiver specifies explicitly that she has found both inputs of the sender correctly with non-negligible probability, this is a violation of security and cannot be simulated efficiently in the ideal world. Our attacking receiver will indeed specify when she succeeds.

[14] More formally, the distinguishing task is done by the distinguisher, and thus $\widehat{\mathcal{S}}$ tries to obtain a view that is not simulatable. However, for simplicity of the exposition, we combine the semi-honest attacker and the distinguisher.

to use the maximum-likelihood method to distinguish receiver's input $\vec{b}$ from a random string. Namely, given a string β, $\widehat{S}$ outputs 1 if $\Pr[\vec{\mathbf{b}} = \beta \mid V_{\mathcal{E}}] > 2^{-(n+1)}$, where $\vec{\mathbf{b}}$ is the random variable denoting the receiver's input $\vec{b}$, and it outputs 0 otherwise. In other words, $\widehat{S}$, outputs 1 if the given β, from the eyes of Eve, is more likely to be the actual receiver's input $\vec{b}$ than being sampled from $\mathbf{U}_{n+1}$ independently.

An interesting thing about the above attack is that here the sender somehow chooses to 'forget' about its own view and only considers Eve's view (which still includes the transcript), but doing this is always possible since Eve's view is part of the attacking sender's view.

Construction 25 (Receiver's attack $\widehat{\mathcal{R}}$). $\widehat{\mathcal{R}}$ *follows the protocol honestly, denoted by the honest execution $\mathcal{R}$, but its goal is to obtain also another output not corresponding to its original input $\vec{b}$. (Doing this would establish an attack by Lemma 20.) In order to get to this goal, in addition to executing $\mathcal{R}$ honestly to obtain the 'default' output $(x_i^{b_i})_{i\in[n+1]}$ with respect to $\vec{b}$, the cheating receiver $\widehat{\mathcal{R}}$ also runs the following algorithm, denoted by $\mathcal{R}'$, that tries to find the output with respect to some other input $\vec{b}' \neq \vec{b}$. $\mathcal{R}'$ will try to pick $\vec{b}' \neq \vec{b}$ in a way that it remains consistent with the transcript T as well as the received OT-hybrid output γ (by enforcing the consistency with the OT-hybrid input $\vec{c}$), so that the obtained output is correct with respect to $\vec{b}'$. Formally, the algorithm $\mathcal{R}'$ is equal to $\mathcal{R}$ until the last message t_r is sent from $\mathcal{S}$ (i.e., we refer to this partial execution as $\mathcal{R}_1$), but then $\mathcal{R}'$ (as part of the attack $\widehat{\mathcal{R}}$) diverges from $\mathcal{R}$'s execution as follows.*

1. *After the last message t_r is sent by the sender $\mathcal{S}$, the cheating receiver $\widehat{\mathcal{R}}$ runs the Eve algorithm of Theorem 17 over the same input-less protocol $(\mathcal{S}, \mathcal{R}_1)$ used by $\widehat{S}$ in Construction 24 (where inputs are chosen at random and the protocol ends when t_r is sent) to obtain Eve's view $V_{\mathcal{E}} = (T, P_{\mathcal{E}})$ for the same δ used by $\widehat{S}$ in Construction 24.*

2. *$\widehat{\mathcal{R}}$ then samples from the distribution $V'_{\mathcal{R}} \leftarrow (\mathbf{V}_{\mathcal{R}} \mid \mathbf{V}_{\mathcal{E}} = V_{\mathcal{E}}, \vec{\mathbf{c}} = \vec{c})$ where $\mathbf{V}_{\mathcal{R}}$ denotes the random variable encoding the view of the inputless protocol $(\mathcal{S}, \mathcal{R}_1)$ over which the Eve algorithm is executed. Now, $\widehat{\mathcal{R}}$ interprets $V'_{\mathcal{R}}$ as the (partial) execution of $\mathcal{R}'$ till t_r is sent (i.e., only reflecting the $\mathcal{R}_1$ part), and it continues executing $\mathcal{R}'$ to a full execution of the receiver as follows.*

3. *Upon receiving γ from $\mathbb{OT}_n$, $\mathcal{R}'$ continues the protocol (as the receiver) using the partial view $V'_{\mathcal{R}}$ and γ as follows. Note that in order to finish the execution, all we have to do is to describe how each oracle query q made by the (remaining execution of) $\mathcal{R}'$ is answered. Let $\mathcal{L}$ be an empty set and then update it inductively, whenever a new query q is asked by $\mathcal{R}'$, as follows.*
   (a) *If $q \in \mathcal{Q}(V'_{\mathcal{R}})$, then use the corresponding answer specified in $V'_{\mathcal{R}}$.*
   (b) *Otherwise, if $(q, a) \in \mathcal{L}$ for some a, use a as answer to q.*
   (c) *Otherwise, if $q \in \mathcal{Q}(V_{\mathcal{R}}) \setminus (\mathcal{Q}(V_{\mathcal{E}}) \cup \mathcal{Q}(V'_{\mathcal{R}}))$,[15] pick a random answer a for query q, and also add (q, a) to $\mathcal{L}$ for the future.*

---

[15] To have $q \in \mathcal{Q}(V_{\mathcal{R}}) \setminus (\mathcal{Q}(V_{\mathcal{E}}) \cup \mathcal{Q}(V'_{\mathcal{R}}))$ means that q is not asked by Eve and it is not in the fake receiver's view $V'_{\mathcal{R}}$ (for partial execution $\mathcal{R}_1$), but q is in the *honest* original execution of $\mathcal{R}_1$.

(d) *Otherwise, ask q from the real random oracle* **H**.

When the emulation of $\mathcal{R}'$ is completed, output whatever is obtained as the output of $\mathcal{R}'$ corresponding to the input $\vec{b}'$ described in $V'_\mathcal{R}$.

Now we show that at least one of the attacks $\widehat{S}, \widehat{\mathcal{R}}$ above succeeds.

Claim 1. *Either the attacking sender $\widehat{S}$ of Construction 24 will distinguish $\vec{b}$ from $\mathbf{U}_{n+1}$ with advantage at least $\Omega(1/n)$, or the attacking receiver $\widehat{\mathcal{R}}$ of Construction 25 can obtain correct outputs corresponding to its random $\vec{b}$ as well as some $\vec{b}' \neq \vec{b}$ with probability at least $\Omega(1/n^2) - O(\lambda + \delta)$, where λ is the completeness error of the protocol and δ is the selected Eve parameter.*

Proving Theorem 23 using Claim 1. Because $\lambda = \text{negl}(\kappa) < o(1/n^2)$, by choosing $\delta = o(1/n^2)$ in Claim 1, either the attacking sender of Construction 24 will break the security by Lemma 19, or the attacking receiver of Construction 25 succeeds in breaking the security with advantage $\Omega(1/n^2)$ (by asking $\text{poly}(\kappa)$ oracle queries) by Lemma 20. In the following, we will prove Claim 1.

3.2   Proof of Claim 1

In this subsection, we will prove Claim 1. Let $\varepsilon = 1/(1000n + 1000)$.

When $\widehat{S}$ succeeds. If it holds that $\text{SD}_{\mathbf{V}_\varepsilon}(\vec{b}, \mathbf{U}_{n+1}) \geq \varepsilon$, then because the attacking $\widehat{S}$ of Construction 24 is indeed using the canonical distinguisher of Proposition 5 (i.e., the maximum likelihood predicate), by Lemma 3 and Proposition 5, $\widehat{S}$ will be able to ε-distinguish the true randomly chosen input $\vec{b}$ of the receiver $\mathcal{R}$ from a uniform string $\mathbf{U}_{n+1}$ by advantage at least ε. Therefore, by Lemma 19, $\widehat{\mathcal{R}}$ succeeds in breaking the security with non-negligible advantage ε.

So, in what follows we assume that $\widehat{S}$ does not succeed, and based on this we show that $\widehat{\mathcal{R}}$ does indeed succeed in its attack.

When $\widehat{\mathcal{R}}$ succeeds. In what follows we always assume

$$\mathbb{E}_{V_\mathcal{E} \leftarrow \mathbf{V}_\mathcal{E}} \text{SD}((\vec{b} \mid V_\mathcal{E}), \mathbf{U}_{n+1}) = \text{SD}_{\mathbf{V}_\varepsilon}(\vec{b}, \mathbf{U}_{n+1}) < \varepsilon = \frac{1}{1000n + 1000} \quad (3)$$

and we will show, using Inequality (3) and Lemma 20, that the receiver's attacker $\widehat{\mathcal{R}}$ will succeed with the non-negligible probability. First note that by just continuing the protocol honestly, the receiver will indeed find the right output for its sampled $\vec{b}$ with probability at least $1 - \lambda$ where λ is the completeness error. So all we have to prove is that with probability $\Omega(1/n^2) - O(\delta) - \lambda$, it will simultaneously hold that (1) $\vec{b}' \neq \vec{b}$ and (2) the receiver $\mathcal{R}'$ gets the output corresponding to $\vec{b}'$ (and sender's actual input x). To prove this, it will suffice to prove the following two statements:

- $\Pr[\vec{b}' \neq \vec{b}] \geq \Omega(1/n^2)$ where $\vec{b}$ and $\vec{b}'$ are the random variables denoting the original and the fake inputs of $\widehat{\mathcal{R}}$.
- The receiver will get the right answer for $\vec{b}'$ with probability $1 - O(\delta) - \lambda$.

Then, by a union bound, we can conclude that the $\widehat{\mathcal{R}}$ will indeed manage to launch a successful attack with probability $\Omega(1/n^2) - O(\delta + \lambda)$. In the following we will formalize and prove the above two claims in forms of Claims 2 and 3.

Claim 2. *If Inequality (3) holds, then* $\Pr[\vec{b}' \neq \vec{b}] \geq \Omega(1/n^2)$ *where the probability is over the randomness of the sender $\mathcal{S}$, cheating receiver $\widehat{\mathcal{R}}$, and $\mathbf{H}$.*

Proof. By sampling the components of the system 'in reverse', we can imagine that first $(T, P_{\mathcal{E}}) = V_{\mathcal{E}} \leftarrow \mathbf{V}_{\mathcal{E}}$ is sampled from its corresponding marginal distribution, then $\vec{c} \leftarrow (\vec{\mathbf{c}} \mid V_{\mathcal{E}})$ is sampled, then $(V_{\mathcal{S}}, V_{\mathcal{R}}) \leftarrow ((\mathbf{V}_{\mathcal{S}}, \mathbf{V}_{\mathcal{R}}) \mid V_{\mathcal{E}}, \vec{c})$, and finally $V_{\mathcal{R}}' \leftarrow (\mathbf{V}_{\mathcal{R}} \mid V_{\mathcal{E}}, \vec{c})$ are sampled, each conditioned on previously sampled components of the system. We will rely on this order of sampling in our arguments below. However, we can ignore the sampling of $V_{\mathcal{S}}$, when we want to compare the components $V_{\mathcal{R}}, V_{\mathcal{R}}'$ and the relation between $\vec{b}, \vec{b}'$. Thus, we can think of $V_{\mathcal{R}}, V_{\mathcal{R}}'$ as two independent samples from the same distribution $(\mathbf{V}_{\mathcal{R}} \mid V_{\mathcal{E}}, \vec{c})$. Consequently, $\vec{b}, \vec{b}'$ are also two independent samples from $(\vec{\mathbf{b}} \mid V_{\mathcal{E}}, \vec{c})$.

By Inequality (3) and an averaging argument over the sampled $V_{\mathcal{E}} \leftarrow \mathbf{V}_{\mathcal{E}}$, with probability at least $1 - 1/10$ over the choice of $V_{\mathcal{E}} \leftarrow \mathbf{V}_{\mathcal{E}}$, it holds that $\mathrm{SD}_{V_{\mathcal{E}}}(\vec{\mathbf{b}}, \mathbf{U}_{n+1}) < \varepsilon' = \frac{1}{100n+100}$. We call such $V_{\mathcal{E}}$ a 'good' sample. For any good $V_{\mathcal{E}}$, using Lemma 14 it holds that $\mathrm{H}(\vec{\mathbf{b}} \mid V_{\mathcal{E}}) \geq (1 - \varepsilon') \cdot (n+1) - \mathrm{H}(\varepsilon')$, and since the length of $\vec{c}$ is n, by further conditioning on random variable $\vec{\mathbf{c}}$ we have:

$$\mathrm{H}(\vec{\mathbf{b}} \mid V_{\mathcal{E}}, \vec{c}) \geq (1 - \varepsilon') \cdot (n+1) - n - \mathrm{H}(\varepsilon') = 1 - \varepsilon' \cdot (n+1) - \mathrm{H}(\varepsilon') \geq 9/10$$

where the last inequality follows from $\varepsilon' \leq 1/200$, and $\mathrm{H}(1/200) < 1/20$. Therefore, by Lemma 15 (using $\mathbf{X} = \vec{\mathbf{b}}, \mathbf{Y} = (V_{\mathcal{E}}, \vec{c})$) we conclude that the event $\vec{b} \neq \vec{b}'$ happens with probability at least $\Omega(1/n^2)$. Finally, since $V_{\mathcal{E}}$ is a good sample with probability $\Omega(1)$, we can still conclude that $\vec{b} \neq \vec{b}'$ happens with probability at least $\Omega(1/n^2)$, finishing the proof of Claim 2. $\square$

Claim 3. *If Inequality (3) holds, then with probability $1 - \lambda - O(\delta)$ (over the randomness of the honest sender $\mathcal{S}$, the cheating receiver $\widehat{\mathcal{R}}$, and the oracle $\mathbf{H}$) the cheating receiver $\mathcal{R}'$ obtains the correct answer for $\vec{b}'$ (i.e., $x_1^{b_1'}, \ldots, x_{n+1}^{b_{n+1}'}$).*

Proof. We want to argue that the full sampled view of the fake receiver $\mathcal{R}'$ (including $V_{\mathcal{E}}'$ followed by the computation as described in the fake execution $\mathcal{R}'$ as part of $\widehat{\mathcal{R}}$) will be statistically close to an actual honest execution of the protocol (i.e., a full execution of $\mathcal{R}$ over random input). For this goal, we define and compare the outcomes of the following experiments. For clarity, and because we use the same names for random variables in different experiments, we might use $\langle \mathbf{X} \rangle_Z$ to emphasize that we are referring to $\mathbf{X}$ in the experiment Z.

Outputs of experiments. The output of the experiments below are vectors with six components. Therefore, the order of the elements in these vectors is very important, and e.g., if we change their order, that changes the actual output.

- Real **experiment.** This experiment outputs $\langle V_{\mathcal{E}}, \vec{c}, V_{\mathcal{S}}, V_{\mathcal{R}}, V_{\mathcal{R}}', P' \rangle_{\mathsf{Real}}$ where $V_{\mathcal{E}}$ is Eve's view, $V_{\mathcal{S}}$ is sender's view, $V_{\mathcal{R}}$ is receiver's honestly generated view (till last message is sent), $V_{\mathcal{R}}'$ is the sampled fake view of $\mathcal{R}'$ only till last message is sent ($V_{\mathcal{R}}, V_{\mathcal{R}}'$ are both part of the view of $\widehat{\mathcal{R}}$), and P' is the set of query-answer pairs that $\mathcal{R}'$ generates after γ (i.e., the message coming from $\mathbb{OT}_n$ after the last message is sent) is sent (some of which are answered using real oracle $\mathbf{H}$ and the rest are emulated using random coin tosses).

- Ideal **experiment.** In this experiment, we also sample a fake receiver's view $V_{\mathcal{R}}'$ the same as in the Real experiment, but then there is no real attack happening and we use the real oracle $\mathbf{H}$ to obtain the query-answer pairs P to finish the computation of $\mathcal{R}$ (which is the original honest execution) using the honest partial view $V_{\mathcal{R}}$. At the end we output $\langle V_{\mathcal{E}}, \vec{c}, V_{\mathcal{S}}, V_{\mathcal{R}}', V_{\mathcal{R}}, P \rangle_{\mathsf{Ideal}}$. Other the change from P' to P, note the crucial that we are switching the locations of the real and fake receiver views $V_{\mathcal{R}}, V_{\mathcal{R}}'$ in the output vector.

Remark 26 (Why not containing γ explicitly in outputs of experiments?). Note that even though γ is not included explicitly in the output of the experiment, it is implicitly there, because γ is a deterministic function of $V_{\mathcal{S}}$ and $\vec{c}$. In particular, because both $V_{\mathcal{R}}, V_{\mathcal{R}}'$ are consistent with $\vec{c}$, they can both lead to correct answers for sender inputs $\vec{b}, \vec{b}'$. In addition, if we *did* include γ in the outputs of the experiments, it would *not* change their statistical distance.

Remark 27 (Why outputting $V_{\mathcal{R}}, V_{\mathcal{R}}'$ both?). Note that our final goal is to show that the fake view $V_{\mathcal{R}}'$ in the Real experiment 'behaves closely' to the actual honest view $V_{\mathcal{R}}$ in the Ideal experiment. So, one might wonder why we include both in the analysis of the experiments. The reason is that the honest and fake views $V_{\mathcal{R}}, V_{\mathcal{R}}'$ in the Real experiment are *not* independent of each other, so if we want to continue the execution of $V_{\mathcal{R}}'$ in the Real experiment to finish the view of $\mathcal{R}'$ (to get the output corresponding to the fake input $\vec{b}'$) we need to be aware of the oracle queries whose answers are already fixed as part of the view of $V_{\mathcal{R}}$. The reason is that we have to answer (some of them) intentionally at random, because corresponding queries in the Ideal experiment are being asked *for the first time.* In order to answer such queries the same way that they are answered in the Ideal experiment, we need to keep track of them in both experiments and avoid some 'bad' events that prevent us from answering from the right distribution.

To prove Claim 3, **it is enough to prove** $O(\delta)$**-closeness of experiments.** If we show that the outputs of the two experiments are $O(\delta)$ (statistically) close, then by the completeness error in the ideal word, which is at most λ, we could conclude that the completeness error in the real world over the randomness of $\langle \mathbf{V}_{\mathcal{E}}, \mathbf{V}_{\mathcal{S}}, \mathbf{V}_{\mathcal{R}}, \mathbf{P} \rangle_{\mathsf{Ideal}}$ is at most $\lambda + O(\delta)$, where the completeness now means that the fake view of the attacking receiver is obtaining the right answer!

To prove that the two experiments' outputs are $O(\delta)$ close, we do the following:

1. We first prove that $\langle \mathbf{V}_{\mathcal{E}}, \vec{c}, \mathbf{V}_{\mathcal{S}}, \mathbf{V}_{\mathcal{R}}, \mathbf{V}_{\mathcal{R}}' \rangle_{\mathsf{Real}} \approx_{O(\delta)} \langle \mathbf{V}_{\mathcal{E}}, \vec{c}, \mathbf{V}_{\mathcal{S}}, \mathbf{V}_{\mathcal{R}}', \mathbf{V}_{\mathcal{R}} \rangle_{\mathsf{Ideal}}$.
2. Then we show that $\Pr[\langle \mathbf{V}_{\mathcal{E}}, \vec{c}, \mathbf{V}_{\mathcal{S}}, \mathbf{V}_{\mathcal{R}}, \mathbf{V}_{\mathcal{R}}' \rangle_{\mathsf{Real}} \in \mathsf{B}] \le \delta$ for some 'bad' event B. (Recall that an event in this work is simply a set, and the same set can be used as an event for different random variables, as long as their samples are

inside a universe where B is also defined.) Intuitively, the bad event captures the event fact that an 'intersection' query exists between the views of the sender and the receiver that is missed by Eve. Indeed, we could also bound the probability of the same event B in the Ideal experiment, however we simply bound it in Real and that turns out to be enough.

3. Finally, we show that as long as the event B does not happen over the sampled $\alpha = \langle V_{\mathcal{E}}, \vec{c}, V_{\mathcal{S}}, V_{\mathcal{R}}, V'_{\mathcal{R}} \rangle_{\mathsf{Real}} \leftarrow \langle \mathbf{V}_{\mathcal{E}}, \vec{c}, \mathbf{V}_{\mathcal{S}}, \mathbf{V}_{\mathcal{R}}, \mathbf{V}'_{\mathcal{R}} \rangle_{\mathsf{Real}}$ (i.e., $\alpha \notin$ B) and if the sampled prefixes of the outputs are equal $\alpha = \langle V_{\mathcal{E}}, \vec{c}, V_{\mathcal{S}}, V'_{\mathcal{R}}, V_{\mathcal{R}} \rangle_{\mathsf{Ideal}} = \langle V_{\mathcal{E}}, \vec{c}, V_{\mathcal{S}}, V_{\mathcal{R}}, V'_{\mathcal{R}} \rangle_{\mathsf{Real}}$, then the corresponding distributions

$$(\langle \mathbf{P} \rangle_{\mathsf{Ideal}} \mid \langle V_{\mathcal{E}}, \vec{c}, V_{\mathcal{S}}, V'_{\mathcal{R}}, V_{\mathcal{R}} \rangle_{\mathsf{Ideal}}) \equiv (\langle \mathbf{P}' \rangle_{\mathsf{Real}} \mid \langle V_{\mathcal{E}}, \vec{c}, V_{\mathcal{S}}, V_{\mathcal{R}}, V'_{\mathcal{R}} \rangle_{\mathsf{Real}})$$

will be identically distributed.

If we prove the above 3 claims, the $O(\delta)$ closeness of the experiments' outputs will follow from Lemma 11, which will finish the proof of Claim 3. To apply Lemma 11, we let $\mathbf{X}_1 = \langle \mathbf{V}_{\mathcal{E}}, \vec{c}, \mathbf{V}_{\mathcal{S}}, \mathbf{V}_{\mathcal{R}}, \mathbf{V}'_{\mathcal{R}} \rangle_{\mathsf{Real}}, \mathbf{X}_2 = \langle \mathbf{P}' \rangle_{\mathsf{Real}}, \mathbf{X}'_1 = \langle \mathbf{V}_{\mathcal{E}}, \vec{c}, \mathbf{V}_{\mathcal{S}}, \mathbf{V}'_{\mathcal{R}}, \mathbf{V}_{\mathcal{R}} \rangle_{\mathsf{Ideal}}, \mathbf{X}'_2 = \langle \mathbf{P} \rangle_{\mathsf{Ideal}}$. We will prove the above 3 items through Claims 4, 5 and 6 below.

Claim 4. $\langle \mathbf{V}_{\mathcal{E}}, \vec{c}, \mathbf{V}_{\mathcal{S}}, \mathbf{V}_{\mathcal{R}}, \mathbf{V}'_{\mathcal{R}} \rangle_{\mathsf{Real}} \approx_{O(\delta)} \langle \mathbf{V}_{\mathcal{E}}, \vec{c}, \mathbf{V}_{\mathcal{S}}, \mathbf{V}'_{\mathcal{R}}, \mathbf{V}_{\mathcal{R}} \rangle_{\mathsf{Ideal}}$.

Proof. By Part 1 of Theorem 17 it holds that in the real world:

$$\mathop{\mathbb{E}}_{(V_{\mathcal{E}}) \leftarrow (\mathbf{V}_{\mathcal{E}}, \vec{c})} \mathsf{MutDep}((\mathbf{V}_{\mathcal{S}}, \mathbf{V}_{\mathcal{R}})_{\mathsf{Real}} \mid V_{\mathcal{E}}) \leq \delta.$$

By averaging over $V_{\mathcal{E}} \leftarrow \mathbf{V}_{\mathcal{E}}$ and then using Lemma 9 (and letting $\mathbf{C} := \vec{c}, \mathbf{B} := \mathbf{V}_{\mathcal{R}}, \mathbf{A} := \mathbf{V}_{\mathcal{S}}$) and noting that $\vec{c}$ is only a function of $V_{\mathcal{R}}$, it holds that

$$\mathop{\mathbb{E}}_{(V_{\mathcal{E}}, \vec{c}) \leftarrow (\mathbf{V}_{\mathcal{E}}, \vec{c})} \mathsf{MutDep}((\mathbf{V}_{\mathcal{S}}, \mathbf{V}_{\mathcal{R}})_{\mathsf{Real}} \mid V_{\mathcal{E}}, \vec{c}) \leq 2\delta.$$

For a fixed $(V_{\mathcal{E}}, \vec{c}) \leftarrow (\mathbf{V}_{\mathcal{E}}, \vec{c})$, we can use Lemma 8 (by letting $\mathbf{X} \equiv (\mathbf{V}_{\mathcal{S}} \mid V_{\mathcal{E}}, \vec{c})$ and $\mathbf{Y} \equiv (\mathbf{V}_{\mathcal{R}} \mid V_{\mathcal{E}}, \vec{c}))$ and then average over $(V_{\mathcal{E}}, \vec{c}) \leftarrow (\mathbf{V}_{\mathcal{E}}, \vec{c})$ to conclude

$$\mathop{\mathbb{E}}_{(V_{\mathcal{E}}, \vec{c}) \leftarrow (\mathbf{V}_{\mathcal{E}}, \vec{c})} \mathsf{SD}(((\mathbf{V}_{\mathcal{S}}, \mathbf{V}_{\mathcal{R}}, \mathbf{V}'_{\mathcal{R}})_{\mathsf{Real}} \mid V_{\mathcal{E}}, \vec{c}), ((\mathbf{V}_{\mathcal{S}}, \mathbf{V}'_{\mathcal{R}}, \mathbf{V}_{\mathcal{R}})_{\mathsf{Ideal}} \mid V_{\mathcal{E}}, \vec{c})) \leq 4\delta.$$

Finally, by Proposition 5, the left side of the above inequality is the same as $\mathsf{SD}(\langle \mathbf{V}_{\mathcal{E}}, \vec{c}, \mathbf{V}_{\mathcal{S}}, \mathbf{V}_{\mathcal{R}}, \mathbf{V}'_{\mathcal{R}} \rangle_{\mathsf{Real}}, \langle \mathbf{V}_{\mathcal{E}}, \vec{c}, \mathbf{V}_{\mathcal{S}}, \mathbf{V}'_{\mathcal{R}}, \mathbf{V}_{\mathcal{R}} \rangle_{\mathsf{Ideal}})$, finishing the proof. □

In the definition below, roughly speaking, the 'bad' event B contains possible outputs of the experiments for which some intersection queries exist between the views of the sender $\mathcal{S}$ and the receiver $\mathcal{R}$ that are missed by the Eve algorithm.

Definition 28 (The bad event B). *Let* U *be a 'universe' containing all possible outputs of the two experiments (and maybe more elements) defined as follows:*

$$\{\langle z_1, \ldots z_5 \rangle \mid z_1 \in \mathsf{Supp}(\mathbf{V}_{\mathcal{E}}), z_2 \in \mathsf{Supp}(\vec{c}), z_3 \in \mathsf{Supp}(\mathbf{V}_{\mathcal{S}}), z_4, z_5 \in \mathsf{Supp}(\mathbf{V}_{\mathcal{R}})\}.$$

Let the 'bad' event $\mathsf{B} \subseteq \mathsf{U}$ *be the set that:*

$$\mathsf{B} = \{\alpha = \langle z_1, z_2, z_3, z_4, z_5 \rangle \mid \alpha \in \mathsf{U}, \mathcal{Q}(z_4) \cap \mathcal{Q}(z_3) \not\subseteq \mathcal{Q}(z_1)\}$$

Namely, if we interpret z_1, z_3, z_4 *as views of oracle-aided algorithms and extract their queries, it holds that* $\mathcal{Q}(z_4) \cap \mathcal{Q}(z_3) \not\subseteq \mathcal{Q}(z_1)$.

The following claim implies that with high probability, a sample from the output of the Real experiment does not fall into B. (In other words, the property by which B is defined, does not hold over the sampled output).

Claim 5. $\Pr[\langle \mathbf{V}_{\mathcal{E}}, \vec{c}, \mathbf{V}_{\mathcal{S}}, \mathbf{V}_{\mathcal{R}}, \mathbf{V}'_{\mathcal{R}} \rangle_{\mathsf{Real}} \in \mathsf{B}] \leq \delta$.

Proof. The claim directly follows from the second property of Eve's algorithm (i.e., Part 2 in Theorem 17). Namely, a sample

$$\alpha = \langle z_1, z_2, z_3, z_4, z_5 \rangle \leftarrow \langle \mathbf{V}_{\mathcal{E}}, \vec{c}, \mathbf{V}_{\mathcal{S}}, \mathbf{V}_{\mathcal{R}}, \mathbf{V}'_{\mathcal{R}} \rangle_{\mathsf{Real}}$$

will have components corresponding to $z_1 = V_{\mathcal{E}}, z_3 = V_{\mathcal{S}}, z_4 = V_{\mathcal{R}}$, and so by Part 2 of Theorem 17 we know that with probability at least $1 - \delta$ it holds that $\mathcal{Q}(V_{\mathcal{S}}) \cap \mathcal{Q}(V_{\mathcal{R}}) \subseteq \mathcal{Q}(V_{\mathcal{E}})$. Therefore, $\alpha \in \mathsf{B}$ would happen in Real experiment with probability at most δ. □

Remark 29 (Other possible choices for defining bad event B *and stating Claim 5).* One can also define an alternative version B′ of the bad event B based on the modified condition $\mathcal{Q}(z_5) \cap \mathcal{Q}(z_3) \not\subseteq \mathcal{Q}(z_1)$ (i.e., using z_5 instead of z_4), and one can also choose either of Real or Ideal experiments for bounding the probability of the bad event (B or B′) by $O(\delta)$. This gives rise to four possible ways of defining the bad event and bounding it in an experiment. We note that all four cases above (i.e., both variations of the bad event B or B′ in both of the Real and the Ideal) experiments can be proved to happen with probability at most $O(\delta)$. Furthermore, all of these four possible choices could be used (together with Lemma 11) for bounding the statistical distance of the output of experiments Real and Ideal by $O(\delta)$. In fact, once we show that statistical distance of the output of experiments Real and Ideal is $O(\delta)$, we can go back and derive all four combinations (of choosing the bad event from B or B′ and stating Claim 5 in either of Real or Ideal experiments) to be true. Thus, basically all of these four choices are "equivalent" up to constant factors in the bound we get in Claim 5. Nonetheless, among these four choices, we found the choice of the bad event B according to Definition 28 and stating Claim 5 in the Real experiment to be the simplest choice to prove (using Theorem 17) and use for proving Claim 3 (by bounding the statistical distance of the outputs of experiments using Lemma 11).

Claim 6. *If samples* $\alpha = \langle V_{\mathcal{E}}, \vec{c}, V_{\mathcal{S}}, V'_{\mathcal{R}}, V_{\mathcal{R}} \rangle_{\mathsf{Ideal}} = \langle V_{\mathcal{E}}, \vec{c}, V_{\mathcal{S}}, V_{\mathcal{R}}, V'_{\mathcal{R}} \rangle_{\mathsf{Real}}$ *are equal, and if event* B *does not happen over the sample* α *(i.e.,* $\alpha \notin \mathsf{B}$*), then*

$$(\langle \mathbf{P} \rangle_{\mathsf{Ideal}} \mid \langle V_{\mathcal{E}}, \vec{c}, V_{\mathcal{S}}, V'_{\mathcal{R}}, V_{\mathcal{R}} \rangle_{\mathsf{Ideal}}) \equiv (\langle \mathbf{P}' \rangle_{\mathsf{Real}} \mid \langle V_{\mathcal{E}}, \vec{c}, V_{\mathcal{S}}, V_{\mathcal{R}}, V'_{\mathcal{R}} \rangle_{\mathsf{Real}}).$$

Proof. We show that conditioned on the same sample α being the prefix of the outputs of the two experiments, the random process that generates the last components $\langle \mathbf{P'} \rangle_{\mathsf{Real}}$ and $\langle \mathbf{P} \rangle_{\mathsf{Ideal}}$ are identically distributed in the two experiments.

After sampling $\alpha = \langle V_{\mathcal{E}}, \vec{c}, V_{\mathcal{S}}, V'_{\mathcal{R}}, V_{\mathcal{R}} \rangle_{\mathsf{Ideal}}$, every new query q will be answered as follows in Ideal: If q is already in $\mathcal{Q}(V_{\mathcal{E}}) \cup \mathcal{Q}(V_{\mathcal{S}}) \cup \mathcal{Q}(V_{\mathcal{R}})$ then the answer is already fixed and that answer will be used, otherwise q will be answered at random (by the random oracle $\mathbf{H}$). Since we are assuming $\langle V_{\mathcal{E}}, \vec{c}, V_{\mathcal{S}}, V'_{\mathcal{R}}, V_{\mathcal{R}} \rangle_{\mathsf{Ideal}} = \langle V_{\mathcal{E}}, \vec{c}, V_{\mathcal{S}}, V_{\mathcal{R}}, V'_{\mathcal{R}} \rangle_{\mathsf{Real}}$, we would like to prove that in the Real experiment, q is answered similarly. Indeed, we will prove that in the Real experiment, if q is already in $\langle \mathcal{Q}(V_{\mathcal{E}}) \cup \mathcal{Q}(V_{\mathcal{S}}) \cup \mathcal{Q}(V'_{\mathcal{R}}) \rangle_{\mathsf{Real}}$ then the fixed answer will be used, and otherwise q will be answered at random. We make the following case study in the Real experiment based on the algorithm of $\widehat{\mathcal{R}}$ from Construction 25. (In the second case below we make a crucial use of the fact that the event B has not happened over the current sample $\langle V_{\mathcal{E}}, \vec{c}, V_{\mathcal{S}}, V'_{\mathcal{R}}, V_{\mathcal{R}} \rangle_{\mathsf{Ideal}} = \langle V_{\mathcal{E}}, \vec{c}, V_{\mathcal{S}}, V_{\mathcal{R}}, V'_{\mathcal{R}} \rangle_{\mathsf{Real}}$.)

1. If $q \in \langle \mathcal{Q}(V'_{\mathcal{R}}) \rangle_{\mathsf{Real}}$, then $\widehat{\mathcal{R}}$ uses the answer stated in $V'_{\mathcal{R}}$. Otherwise:
2. if $q \in \langle \mathcal{Q}(V_{\mathcal{R}}) \setminus (\mathcal{Q}(V_{\mathcal{E}}) \cup \mathcal{Q}(V'_{\mathcal{R}})) \rangle_{\mathsf{Real}}$, $\widehat{\mathcal{R}}$ answers q at random (and keeps its answer in a list $\mathcal{L}$ to reuse in case of being asked again). In the ideal world, this query q would be part of the *fake* view $\langle V'_{\mathcal{R}} \rangle_{\mathsf{Ideal}}$ (recall the fake and real views are switched across the Real vs. Ideal experiments) which is *ignored* in the Ideal world when we generate $\langle P \rangle_{\mathsf{Ideal}}$, and so we have two cases:
    (a) If q is already in $\langle \mathcal{Q}(V_{\mathcal{S}}) \rangle_{\mathsf{Real}}$, it means that $\alpha \in \mathsf{B}$ for $\alpha = \langle V_{\mathcal{E}}, \vec{c}, V_{\mathcal{S}}, V'_{\mathcal{R}}, V_{\mathcal{R}} \rangle_{\mathsf{Ideal}} = \langle V_{\mathcal{E}}, \vec{c}, V_{\mathcal{S}}, V_{\mathcal{R}}, V'_{\mathcal{R}} \rangle_{\mathsf{Real}}$ which is not true.
    (b) Otherwise, $q \notin \langle \mathcal{Q}(V_{\mathcal{S}}) \rangle_{\mathsf{Real}} = \langle \mathcal{Q}(V'_{\mathcal{S}}) \rangle_{\mathsf{Ideal}}$, which means that q is a new query in the ideal world, and so it is answered at random, just like how it is answered in the real world by the attacker $\widehat{\mathcal{R}}$.
3. If above cases do not happen, but q is still part of $\langle \mathcal{Q}(V_{\mathcal{E}}) \cup \mathcal{Q}(V_{\mathcal{S}}) \rangle_{\mathsf{Real}}$, $\widehat{\mathcal{R}}$ would forward this query to be asked from the actual random oracle $\mathbf{H}$ which would also get the correct answer (i.e., the same answer stated in $V_{\mathcal{E}}$ or $V_{\mathcal{S}}$).

Therefore, in all cases q will be answered from the same distribution across the Real and Ideal experiments. This shows that the process of generating the last component of the output of these experiments is identically distributed. $\quad\square$

This finishes the proof of Claim 3. $\quad\square$

References

[AIR01] Aiello, W., Ishai, Y., Reingold, O.: Priced oblivious transfer: how to sell digital goods. In: Pfitzmann, B. (ed.) EUROCRYPT 2001. LNCS, vol. 2045, pp. 119–135. Springer, Heidelberg (2001). https://doi.org/10.1007/3-540-44987-6_8

[AJL+12] Asharov, G., Jain, A., López-Alt, A., Tromer, E., Vaikuntanathan, V., Wichs, D.: Multiparty computation with low communication, computation and interaction via threshold FHE. In: Pointcheval, D., Johansson, T. (eds.) EUROCRYPT 2012. LNCS, vol. 7237, pp. 483–501. Springer, Heidelberg (2012). https://doi.org/10.1007/978-3-642-29011-4_29

[Ald83]   Aldous, D.: Random walks on finite groups and rapidly mixing Markov chains. In: Azéma, J., Yor, M. (eds.) Séminaire de Probabilités XVII 1981/82. LNM, vol. 986, pp. 243–297. Springer, Heidelberg (1983). https://doi.org/10.1007/BFb0068322

[ALSZ13] Asharov, G., Lindell, Y., Schneider, T., Zohner, M.: More efficient oblivious transfer and extensions for faster secure computation. In: Sadeghi, A.-R., Gligor, V.D., Yung, M. (eds.) ACM CCS 2013, Berlin, Germany, 4–8 November 2013, pp. 535–548. ACM Press (2013)

[ALSZ15] Asharov, G., Lindell, Y., Schneider, T., Zohner, M.: More efficient oblivious transfer extensions with security for malicious adversaries. In: Oswald, E., Fischlin, M. (eds.) EUROCRYPT 2015, Part I. LNCS, vol. 9056, pp. 673–701. Springer, Heidelberg (2015). https://doi.org/10.1007/978-3-662-46800-5_26

[BCR87]   Brassard, G., Crepeau, C., Robert, J.-M.: All-or-nothing disclosure of secrets. In: Odlyzko, A.M. (ed.) CRYPTO 1986. LNCS, vol. 263, pp. 234–238. Springer, Heidelberg (1987). https://doi.org/10.1007/3-540-47721-7_17

[Bea96]   Beaver, D.: Correlated pseudorandomness and the complexity of private computations. In: 28th ACM STOC, Philadephia, PA, USA, 22–24 May 1996, pp. 479–488. ACM Press (1996)

[BGI+14] Beimel, A., Gabizon, A., Ishai, Y., Kushilevitz, E., Meldgaard, S., Paskin-Cherniavsky, A.: Non-interactive secure multiparty computation. In: Garay, J.A., Gennaro, R. (eds.) CRYPTO 2014, Part II. LNCS, vol. 8617, pp. 387–404. Springer, Heidelberg (2014). https://doi.org/10.1007/978-3-662-44381-1_22

[BL17]   Benhamouda, F., Lin, H.: k-round multiparty computation from k-round oblivious transfer via garbled interactive circuits. Cryptology ePrint Archive, Report 2017/1125 (2017). EUROCRYPT 2018

[BM17]   Barak, B., Mahmoody, M.: Merkle's key agreement protocol is optimal: an $O(n^2)$ attack on any key agreement from random oracles. J. Cryptol. **30**(3), 699–734 (2017)

[BMG09]   Barak, B., Mahmoody-Ghidary, M.: Merkle puzzles are optimal — an $O(n^2)$-query attack on any key exchange from a random oracle. In: Halevi, S. (ed.) CRYPTO 2009. LNCS, vol. 5677, pp. 374–390. Springer, Heidelberg (2009). https://doi.org/10.1007/978-3-642-03356-8_22

[BMR90]   Beaver, D., Micali, S., Rogaway, P.: The round complexity of secure protocols (extended abstract). In: 22nd ACM STOC, Baltimore, MD, USA, 14–16 May 1990, pp. 503–513. ACM Press (1990)

[CPS08]   Coron, J.-S., Patarin, J., Seurin, Y.: The random oracle model and the ideal cipher model are equivalent. In: Wagner, D. (ed.) CRYPTO 2008. LNCS, vol. 5157, pp. 1–20. Springer, Heidelberg (2008). https://doi.org/10.1007/978-3-540-85174-5_1

[DLMM11] Dachman-Soled, D., Lindell, Y., Mahmoody, M., Malkin, T.: On the black-box complexity of optimally-fair coin tossing. In: Ishai, Y. (ed.) TCC 2011. LNCS, vol. 6597, pp. 450–467. Springer, Heidelberg (2011). https://doi.org/10.1007/978-3-642-19571-6_27

[EGL85]   Even, S., Goldreich, O., Lempel, A.: A randomized protocol for signing contracts. Commun. ACM **28**(6), 637–647 (1985)

[Fan68]   Fano, R.M.: Transmission of Information. A Ststistical Theory of Communications. MIT Press, Cambridge (1968)

[FKN94]   Feige, U., Kilian, J., Naor, M.: A minimal model for secure computation (extended abstract). In: 26th ACM STOC, Montréal, Québec, Canada, 23–25 May 1994, pp. 554–563. ACM Press (1994)

[GGHR14] Garg, S., Gentry, C., Halevi, S., Raykova, M.: Two-round secure MPC from indistinguishability obfuscation. In: Lindell, Y. (ed.) TCC 2014. LNCS, vol. 8349, pp. 74–94. Springer, Heidelberg (2014). https://doi.org/10.1007/978-3-642-54242-8_4

[GKLM12] Goyal, V., Kumar, V., Lokam, S., Mahmoody, M.: On black-box reductions between predicate encryption schemes. In: Cramer, R. (ed.) TCC 2012. LNCS, vol. 7194, pp. 440–457. Springer, Heidelberg (2012). https://doi.org/10.1007/978-3-642-28914-9_25

[GMMM17] Garg, S., Mahmoody, M., Masny, D., Meckler, I.: On the round complexity of OT extension. Cryptology ePrint Archive, Report 2017/1187 (2017). https://eprint.iacr.org/2017/1187

[GMPP16] Garg, S., Mukherjee, P., Pandey, O., Polychroniadou, A.: The exact round complexity of secure computation. In: Fischlin, M., Coron, J.-S. (eds.) EUROCRYPT 2016, Part II. LNCS, vol. 9666, pp. 448–476. Springer, Heidelberg (2016). https://doi.org/10.1007/978-3-662-49896-5_16

[GMW87]   Goldreich, O., Micali, S., Wigderson, A.: How to play any mental game or a completeness theorem for protocols with honest majority. In: Aho, A. (ed.) 19th ACM STOC, New York City, NY, USA, 25–27 May 1987, pp. 218–229. ACM Press (1987)

[GS17a]   Garg, S., Srinivasan, A.: Garbled protocols and two-round MPC from bilinear maps. In: 58th FOCS, pp. 588–599. IEEE Computer Society Press (2017)

[GS17b]   Garg, S., Srinivasan, A.: Two-round multiparty secure computation from minimal assumptions. Cryptology ePrint Archive, Report 2017/1156 (2017). EUROCRYPT 2018

[HK12]   Halevi, S., Kalai, Y.T.: Smooth projective hashing and two-message oblivious transfer. J. Cryptol. 25(1), 158–193 (2012)

[HKT11]   Holenstein, T., Künzler, R., Tessaro, S.: The equivalence of the random oracle model and the ideal cipher model, revisited. In: Fortnow, L., Vadhan, S.P. (eds.) 43rd ACM STOC, San Jose, CA, USA, 6–8 June 2011, pp. 89–98. ACM Press (2011)

[HOZ16]   Haitner, I., Omri, E., Zarosim, H.: Limits on the usefulness of random oracles. J. Cryptol. 29(2), 283–335 (2016)

[IKM+13] Ishai, Y., Kushilevitz, E., Meldgaard, S., Orlandi, C., Paskin-Cherniavsky, A.: On the power of correlated randomness in secure computation. In: Sahai, A. (ed.) TCC 2013. LNCS, vol. 7785, pp. 600–620. Springer, Heidelberg (2013). https://doi.org/10.1007/978-3-642-36594-2_34

[IKNP03] Ishai, Y., Kilian, J., Nissim, K., Petrank, E.: Extending oblivious transfers efficiently. In: Boneh, D. (ed.) CRYPTO 2003. LNCS, vol. 2729, pp. 145–161. Springer, Heidelberg (2003). https://doi.org/10.1007/978-3-540-45146-4_9

[IPS08]   Ishai, Y., Prabhakaran, M., Sahai, A.: Founding cryptography on oblivious transfer – efficiently. In: Wagner, D. (ed.) CRYPTO 2008. LNCS, vol. 5157, pp. 572–591. Springer, Heidelberg (2008). https://doi.org/10.1007/978-3-540-85174-5_32

[IR89]   Impagliazzo, R., Rudich, S.: Limits on the provable consequences of one-way permutations. In: Proceedings of the 21st Annual ACM Symposium on Theory of Computing (STOC), pp. 44–61. ACM Press (1989)

[Kil88] Kilian, J.: Founding cryptography on oblivious transfer. In: Proceedings of the 20th Annual ACM Symposium on Theory of Computing (STOC), pp. 20–31 (1988)

[KK13] Kolesnikov, V., Kumaresan, R.: Improved OT extension for transferring short secrets. In: Canetti, R., Garay, J.A. (eds.) CRYPTO 2013, Part II. LNCS, vol. 8043, pp. 54–70. Springer, Heidelberg (2013). https://doi.org/10.1007/978-3-642-40084-1_4

[KO04] Katz, J., Ostrovsky, R.: Round-optimal secure two-party computation. In: Franklin, M. (ed.) CRYPTO 2004. LNCS, vol. 3152, pp. 335–354. Springer, Heidelberg (2004). https://doi.org/10.1007/978-3-540-28628-8_21

[Lin16] Lindell, Y.: How to simulate it - a tutorial on the simulation proof technique. Cryptology ePrint Archive, Report 2016/046 (2016)

[LZ13] Lindell, Y., Zarosim, H.: On the feasibility of extending oblivious transfer. In: Sahai, A. (ed.) TCC 2013. LNCS, vol. 7785, pp. 519–538. Springer, Heidelberg (2013). https://doi.org/10.1007/978-3-642-36594-2_29

[MW16] Mukherjee, P., Wichs, D.: Two round multiparty computation via multi-key FHE. In: Fischlin, M., Coron, J.-S. (eds.) EUROCRYPT 2016, Part II. LNCS, vol. 9666, pp. 735–763. Springer, Heidelberg (2016). https://doi.org/10.1007/978-3-662-49896-5_26

[NNOB12] Nielsen, J.B., Nordholt, P.S., Orlandi, C., Burra, S.S.: A new approach to practical active-secure two-party computation. In: Safavi-Naini, R., Canetti, R. (eds.) CRYPTO 2012. LNCS, vol. 7417, pp. 681–700. Springer, Heidelberg (2012). https://doi.org/10.1007/978-3-642-32009-5_40

[NP01] Naor, M., Pinkas, B.: Efficient oblivious transfer protocols. In: Kosaraju, S.R. (ed.) 12th SODA, Washington, DC, USA, 7–9 January 2001, pp. 448–457. ACM-SIAM (2001)

[ORS15] Ostrovsky, R., Richelson, S., Scafuro, A.: Round-optimal black-box two-party computation. In: Gennaro, R., Robshaw, M. (eds.) CRYPTO 2015, Part II. LNCS, vol. 9216, pp. 339–358. Springer, Heidelberg (2015). https://doi.org/10.1007/978-3-662-48000-7_17

[PVW08] Peikert, C., Vaikuntanathan, V., Waters, B.: A framework for efficient and composable oblivious transfer. In: Wagner, D. (ed.) CRYPTO 2008. LNCS, vol. 5157, pp. 554–571. Springer, Heidelberg (2008). https://doi.org/10.1007/978-3-540-85174-5_31

[Rab81] Rabin, M.: How to exchange secrets by oblivious transfer. Technical report TR-81, Harvard Aiken Computation Laboratory (1981)

[Sho09] Shoup, V.: A Computational Introduction to Number Theory and Algebra. Cambridge University Press, Cambridge (2009)

[WW06] Wolf, S., Wullschleger, J.: Oblivious transfer is symmetric. In: Vaudenay, S. (ed.) EUROCRYPT 2006. LNCS, vol. 4004, pp. 222–232. Springer, Heidelberg (2006). https://doi.org/10.1007/11761679_14

[Yao82] Yao, A.C.-C.: Protocols for secure computations (extended abstract). In: 23rd FOCS, Chicago, Illinois, 3–5 November 1982, pp. 160–164. IEEE Computer Society Press (1982)

Non-malleable Codes

Non-Malleable Codes for Partial
Functions with Manipulation Detection

Aggelos Kiayias[1], Feng-Hao Liu[2], and Yiannis Tselekounis[1(✉)]

[1] University of Edinburgh, Edinburgh, UK
akiayias@inf.ed.ac.uk, ytselekounis@ed.ac.uk
[2] Florida Atlantic University, Boca Raton, USA
fenghao.liu@fau.edu

Abstract. Non-malleable codes were introduced by Dziembowski, Pietrzak and Wichs (ICS '10) and its main application is the protection of cryptographic devices against tampering attacks on memory. In this work, we initiate a comprehensive study on non-malleable codes for the class of partial functions, that read/write on an arbitrary subset of codeword bits with specific cardinality. Our constructions are efficient in terms of information rate, while allowing the attacker to access asymptotically almost the entire codeword. In addition, they satisfy a notion which is stronger than non-malleability, that we call non-malleability with manipulation detection, guaranteeing that any modified codeword decodes to either the original message or to ⊥. Finally, our primitive implies All-Or-Nothing Transforms (AONTs) and as a result our constructions yield efficient AONTs under standard assumptions (only one-way functions), which, to the best of our knowledge, was an open question until now. In addition to this, we present a number of additional applications of our primitive in tamper resilience.

1 Introduction

Non-malleable codes (NMC) were introduced by Dziembowski, Pietrzak and Wichs [27] as a relaxation of error correction and error detection codes, aiming to provide strong privacy but relaxed correctness. Informally, non-malleability guarantees that any modified codeword decodes either to the original message or to a completely unrelated one, with overwhelming probability. The definition of non-malleability is simulation-based, stating that for any tampering function f, there exists a simulator that simulates the tampering effect by only accessing f, i.e., without making any assumptions on the distribution of the encoded message.

The main application of non-malleable codes that motivated the seminal work by Dziembowski et al. [27] is the protection of cryptographic implementations

A. Kiayias—Research partly supported by the H2020 project FENTEC (# 780108).
F.-H. Liu—Research supported by the NSF Award #CNS-1657040.
Y. Tselekounis—Research partly supported by the H2020 project PANORAMIX (# 653497).

H. Shacham and A. Boldyreva (Eds.): CRYPTO 2018, LNCS 10993, pp. 577–607, 2018.
https://doi.org/10.1007/978-3-319-96878-0_20

from *active physical attacks* against memory, known as *tampering attacks*. In this setting, the adversary modifies the memory of the cryptographic device, receives the output of the computation, and tries to extract sensitive information related to the private memory. Security against such types of attacks can be achieved by encoding the private memory of the device using non-malleable codes. Besides that, various applications of non-malleable codes have been proposed in subsequent works, such as CCA secure encryption schemes [20] and non-malleable commitments [4].

Due to their important applications, constructing non-malleable codes has received a lot of attention over recent years. As non-malleability against general functions is impossible [27], various subclasses of tampering functions have been considered, such as split-state functions [1–3,26,27,36,37], bit-wise tampering and permutations [4,5,27], bounded-size function classes [32], bounded depth/fan-in circuits [6], space-bounded tampering [29], and others (cf. Sect. 1.4). One characteristic shared by those function classes is that they allow *full access* to the codeword, while imposing structural or computational restrictions to the way the function computes over the input. In this work we initiate a comprehensive study on non-malleability for functions that receive *partial access* over the codeword, which is an important yet overlooked class, as we elaborate below.

The class of partial functions. The class of *partial functions* contains all functions that read/write on an arbitrary subset of codeword bits with specific cardinality. Concretely, let c be a codeword with length ν. For $\alpha \in [0, 1)$, the function class $\mathcal{F}^{\alpha\nu}$ (or $\mathcal{F}^{\alpha}$ for brevity) consists of all functions that operate over any subset of bits of c with cardinality at most $\alpha\nu$, while leaving the remaining bits intact. The work of Cheraghchi and Guruswami [18] explicitly defines this class and uses a subclass (the one containing functions that always touch the first $\alpha\nu$ bits of the codeword) in a negative way, namely as the tool for deriving capacity lower bounds for *information-theoretic* non-malleable codes against split-state functions. Partial functions were also studied implicitly by Faust et al. [32], while aiming for non-malleability against bounded-size circuits.[1]

Even though capacity lower bounds for partial functions have been derived (cf. [18]), our understanding about *explicit* constructions is still limited. Existential results can be derived by the probabilistic method, as shown in prior works [18,27][2], but they do not yield explicit constructions. On the other hand, the capacity bounds do not apply to the computational setting, which could potentially allow more practical solutions. We believe that this is a direction that needs to be explored, as besides the theoretical interest, partial functions is

---

[1] Specifically, in [32], the authors consider a model where a common reference string (CRS) is available, with length roughly logarithmic in the size of the tampering function class; as a consequence, the tampering function is allowed to read/write the whole codeword while having only partial information over the CRS.

[2] Informally, prior works [18,27] showed existence of non-malleable codes for classes of certain bounded cardinalities. The results cover the class of partial functions.

a natural model that complies with existing attacks that require partial access to the registers of the cryptographic implementation [8, 10–12, 44].[3]

Besides the importance of partial functions in the active setting, i.e., when the function is allowed to partially *read/write* the codeword, the passive analogue of the class, i.e., when the function is only given *read access* over the codeword, matches the model considered by All-Or-Nothing Transforms (AONTs), which is a notion originally introduced by Rivest [41], providing security guarantees similar to those of leakage resilience: reading an arbitrary subset (up to some bounded cardinality) of locations of the codeword does not reveal the underlying message. As non-malleable codes provide privacy, non-malleability for partial functions is the active analogue of (and in fact implies) AONTs, that find numerous applications [13, 14, 40, 41, 43].

Plausibility. At a first glance one might think that partial functions better comply with the framework of error-correction/detection codes (ECC/EDC), as they do not touch the whole codeword. However, if we allow the adversary to access asymptotically almost the entire codeword, it is conceivable it can use this generous *access rate*, i.e., the fraction of the codeword that can be accessed (see below), to create correlated encodings, thus we believe solving non-malleability in this setting is a natural question. Additionally, non-malleability provides simulation based security, which is not considered by ECC/EDC.

We illustrate the separation between the notions using the following example. Consider the set of partial functions that operate either on the right or on the left half of the codeword (the function chooses if it is going to be left or right), and the trivial encoding scheme that on input message s outputs (s, s). The decoder, on input (s, s'), checks if $s = s'$, in which case it outputs s, otherwise it outputs $\perp$. This scheme is clearly an EDC against the aforementioned function class,[4] as the output of the decoder is in $\{s, \perp\}$, with probability 1; however, it is malleable since the tampering function can create encodings whose validity depends on the message. On the other hand, an ECC would provide a trivial solution in this setting, however it requires restriction of the adversarial access fraction to $1/2$ (of the codeword); by accessing more than this fraction, the attacker can possibly create invalid encodings depending on the message, as general ECCs do not provide privacy. Thus, the ECC/EDC setting is inapt when aiming for simulation based security in the presence of attackers that access almost the entire codeword. Later in this section, we provide an extensive discussion on challenges of non-malleability for partial functions.

Besides the plausibility and the lack of a comprehensive study, partial functions can potentially allow stronger primitives, as constant functions are excluded from the class. This is similar to the path followed by Jafargholi and Wichs [34], aiming to achieve *tamper detection* (cf. Sect. 1.4) against a class of

---

[3] The attacks by [8, 11, 12] require the modification of a single (random) memory bit, while in [10] a single error per each round of the computation suffices. In [44], the attack requires a single faulty byte.

[4] It is not an ECC as the decoder does not know which side has been modified by the tampering function.

functions that implicitly excludes constant functions and the identity function. In this work we prove that this intuition holds, by showing that partial functions allow a stronger primitive that we define as *non-malleability* with *manipulation detection* (MD-NMC), which in addition to simulation based security, it also guarantees that any tampered codeword will either decode to the original message or to $\perp$. Again, and as in the case of ECC/EDC, we stress out that manipulation/tamper-detection codes do not imply MD-NMC, as they do not provide simulation based security (cf. Sect. 1.4).[5]

Given the above, we believe that partial functions is an interesting and well-motivated model. The goal of this work is to answer the following (informally stated) question:

Is it possible to construct efficient (high information rate) non-malleable codes for partial functions, while allowing the attacker to access almost the entire codeword?

We answer the above question in the affirmative. Before presenting our results (cf. Sect. 1.1) and the high level ideas behind our techniques (cf. Sect. 1.2), we identify the several challenges that are involved in tackling the problem.

Challenges. We first define some useful notions used throughout the paper.

- *Information rate*: the ratio of message to codeword length, as the message length goes to infinity.
- *Access rate*: the fraction of the number of bits that the attacker is allowed to access over the total codeword length, as the message length goes to infinity.

The access rate measures the effectiveness of a non-malleable code in the partial function setting and reflects the level of adversarial access to the codeword. In this work, we aim at constructing non-malleable codes for partial functions with high *information rate* and high *access rate*, i.e., both rates should approach 1 simultaneously. Before discussing the challenges posed by this requirement, we first review some known impossibility results. First, non-malleability for partial functions with concrete access rate 1 is impossible, as the function can fully decode the codeword and then re-encode a related message [27]. Second, information-theoretic non-malleable codes with constant information rate (e.g., 0.5) are not possible against partial functions with constant access rate [18][6], and consequently, solutions in the information-theoretic settings such as ECC and Robust Secret Sharing (RSS) do not solve our problem. Based on these facts, in order to achieve our goal, the only path is to explore the computational setting, aiming for access rate at most $1 - \epsilon$, for some $\epsilon > 0$.

At a first glance one might think that non-malleability for partial functions is easier to achieve, compared to other function classes, as partial functions

---

[5] Clearly, MD-NMC imply manipulation/error-detection codes.

[6] Informally, in [18] (Theorem 5.3) the authors showed that any information-theoretic non-malleable code with a constant access rate and a constant information rate must have a constant distinguishing probability.

cannot touch the whole codeword. Having that in mind, it would be tempting to conclude that existing designs/techniques with minor modifications are sufficient to achieve our goal. However, we will show that this intuition is misleading, by pointing out why prior approaches fail to provide security against partial functions with high access rate.

The current state of the art in the computational setting considers tools such as (Authenticated) Encryption [1,22,24,28,36,37], *non-interactive zero-knowledge* (NIZK) proofs [22,28,30,37], and ℓ-more extractable collision resistant hashes (ECRH) [36], where others use KEM/DEM techniques [1,24]. Those constructions share a common structure, incorporating a short secret key sk (or a short encoding of it), as well as a long ciphertext, e, and a proof π (or a hash value). Now, consider the partial function f that gets full access to the secret key sk and a constant number of bits of the ciphertext e, partially decrypts e and modifies the codeword depending on those bits. Then, it is not hard to see that non-malleability falls apart as the security of the encryption no longer holds. The attack requires access rate only $O((|sk|)/(|sk| + |e| + |\pi|))$, for [22,28,37] and $O(\text{poly}(k)/|s|)$ for [1,24,36]. A similar attack applies to [30], which is in the continual setting.

One possible route to tackle the above challenges, is to use an encoding scheme over the ciphertext, such that partial access over it does not reveal the underlying message.[7] The guarantees that we need from such a primitive resemble the properties of AONTs, however this primitive does not provide security against active, i.e., tampering, attacks. Another approach would be to use Reconstructable Probabilistic Encodings [6], which provide error-correcting guarantees, yet still it is unknown whether we can achieve information rate 1 for such a primitive. In addition, the techniques and tools for protecting the secret key can be used to achieve optimal information rate as they are independent of the underlying message, yet at the same time, they become the weakest point against partial functions with high access rate. Thus, the question is how to overcome the above challenges, allowing access to almost the entire codeword.

In this paper we solve the challenges presented above based on the following observation: in existing solutions the structure of the codeword is fixed and known to the attacker, and independently of the primitives that we use, the only way to resolve the above issues is by hiding the structure via randomization. This requires a structure recovering mechanism that can either be implemented by an "external" source, or otherwise the structure needs to be reflected in the codeword in some way that the attacker cannot exploit. In the present work we implement this mechanism in both ways, by first proposing a construction in the *common reference string* (CRS) model, and then we show how to remove the CRS using slightly bigger alphabets. Refer to Sect. 1.2 for a technical overview.

---

[7] In the presence of NIZKs we can have attacks with low access rate that read sk, e, and constant number of bits from the proof.

1.1   Our Results

We initiate the study of *non-malleable codes* with *manipulation-detection* (MD-NMC), and we present the first (to our knowledge) construction for this type of codes. We focus on achieving simultaneously high *information rate* and high *access rate*, in the partial functions setting, which by the results of [18], it can be achieved only in the computational setting.

Our contribution is threefold. First, we construct an information rate 1 non-malleable code in the CRS model, with access rate $1 - 1/\Omega(\log k)$, where k denotes the security parameter. Our construction combines Authenticated Encryption together with an inner code that protects the key of the encryption scheme (cf. Sect. 1.2). The result is informally summarized in the following theorem.

Theorem 1.1 (Informal). *Assuming one-way functions, there exists an explicit computationally secure MD-NMC in the CRS model, with information rate 1 and access rate $1 - 1/\Omega(\log k)$, where k is the security parameter.*

Our scheme, in order to achieve security with error $2^{-\Omega(k)}$, produces codewords of length $|s| + O(k^2 \log k)$, where $|s|$ denotes the length of the message, and uses a CRS of length $O(k^2 \log k \log(|s| + k))$. We note that our construction does not require the CRS to be fully tamper-proof and we refer the reader to Sect. 1.2 for a discussion on the topic.

In our second result we show how to remove the CRS by slightly increasing the size of the alphabet. Our result is a computationally secure MD-NMC in the standard model, achieving information and access rate $1 - 1/\Omega(\log k)$. Our construction is proven secure by a reduction to the security of the scheme presented in Theorem 1.1. Below, we informally state our result.

Theorem 1.2 (Informal). *Assuming one-way functions, there exists an explicit, computationally secure MD-NMC in the standard model, with alphabet length $O(\log k)$, information rate $1 - 1/\Omega(\log k)$ and access rate $1 - 1/\Omega(\log k)$, where k is the security parameter.*

Our scheme produces codewords of length $|s|(1 + 1/O(\log k)) + O(k^2 \log^2 k)$.

In Sect. 1.2, we consider security against continuous attacks. We show how to achieve a weaker notion of continuous security, while avoiding the use of a self-destruct mechanism, which was originally achieved by [28]. Our notion is weaker than full continuous security [30], since the codewords need to be updated. Nevertheless, our update operation is deterministic and avoids the full re-encoding process [27,37]; it uses only shuffling and refreshing operations, i.e., we avoid cryptographic computations such as group operations and NIZKs. We call such an update mechanism a "light update." Informally, we prove the following result.

Theorem 1.3 (Informal). *One-way functions imply continuous non-malleable codes with deterministic light updates and without self-destruct, in the standard model, with alphabet length $O(\log k)$, information rate $1 - 1/\Omega(\log k)$ and access rate $1 - 1/\Omega(\log k)$, where k is the security parameter.*

As we have already stated, non-malleable codes against partial functions imply AONTs [41]. The first AONT was presented by Boyko [13] in the random oracle model, and then Canetti et al. [14] consider AONTs with public/private parts as well as a secret-only part, which is the full notion. Canetti et al. [14] provide efficient constructions for both settings, yet the fully secure AONT (called "secret-only" in that paper) is based on non-standard assumptions.[8]

Assuming one-way functions, our results yield efficient, fully secure AONTs, in the standard model. This resolves, the open question left in [14], where the problem of constructing AONT under standard assumptions was posed. Our result is presented in the following theorem.

Theorem 1.4 (Informal). *Assuming one-way functions, there exists an explicit secret-only AONT in the standard model, with information rate 1 and access rate $1 - 1/\Omega(\log k)$, where k is the security parameter.*

The above theorem is derived by the Informal Theorem 1.1 yielding an AONT whose output consists of both the CRS and the codeword produced by the NMC scheme in the CRS model. A similar theorem can be derived with respect to the Informal Theorem 1.2. Finally, and in connection to AONTs that provide leakage resilience, our results imply leakage-resilient codes [37] for partial functions.

In the full version of the paper we provide concrete instantiations of our constructions, using textbook instantiations [35] for the underlying authenticated encryption scheme. For completeness, we also provide information theoretic variants of our constructions that maintain high access rate and thus necessarily sacrifice information rate.

1.2 Technical Overview

On the manipulation detection property. In the present work we exploit the fact that the class of partial functions does not include constant functions and we achieve a notion that is stronger than non-malleability, which we call *non-malleability* with *manipulation detection*. We formalize this notion as a strengthening of non-malleability and we show that our constructions achieve this stronger notion. Informally, manipulation detection ensures that any tampered codeword will either decode to the original message or to ⊥.

A MD-NMC in the CRS model. For the exposition of our ideas, we start with a naive scheme (which does not work), and then show how we resolve all the challenges. Let $(\mathsf{KGen}, \mathsf{E}, \mathsf{D})$ be a (symmetric) authenticated encryption scheme and consider the following encoding scheme: to encode a message s, the encoder computes $(sk\|e)$, where $e \leftarrow \mathsf{E}_{sk}(s)$ is the ciphertext and $sk \leftarrow \mathsf{KGen}(1^k)$, is the secret key. We observe that the scheme is secure if the tampering function can only read/write on the ciphertext, e, assuming the authenticity property

---

[8] In [43] the authors present a deterministic AONT construction that provides weaker security.

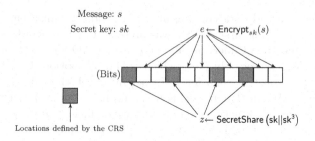

Fig. 1. Description of the scheme in the CRS model.

of the encryption scheme, however, restricting access to sk, which is short, is unnatural and makes the problem trivial. On the other hand, even partial access to sk, compromises the authenticity property of the scheme, and even if there is no explicit attack against the non-malleability property, there is no hope for proving security based on the properties of $(\mathsf{KGen}, \mathsf{E}, \mathsf{D})$, in black-box way.

A solution to the above problems would be to protect the secret key using an inner encoding, yet the amount of tampering is now restricted by the capabilities of the inner scheme, as the attacker knows the exact locations of the *"sensitive"* codeword bits, i.e., the non-ciphertext bits. In our construction, we manage to protect the secret key while avoiding the bottleneck on the access rate by designing an inner encoding scheme that provides limited security guarantees when used standalone, still when it is used in conjunction with a *shuffling technique* that permutes the inner encoding and ciphertext bit locations, it guarantees that any attack against the secret key will create an invalid encoding with overwhelming probability, even when allowing access to *almost the entire* codeword.

Our scheme is depicted in Fig. 1 and works as follows: on input message s, the encoder (i) encrypts the message by computing $sk \leftarrow \mathsf{KGen}(1^k)$ and $e \leftarrow \mathsf{E}_{sk}(s)$, (ii) computes an m-out-of-m secret sharing z of $(sk\|sk^3)$ (interpreting both sk and sk^3 as elements in some finite field),[9] and outputs a random shuffling of $(z\|e)$, denoted as $P_{\Sigma}(z\|e)$, according to the common reference string Σ. Decoding proceeds as follows: on input c, the decoder (i) inverts the shuffling operation by computing $(z\|e) \leftarrow P_{\Sigma}^{-1}(c)$, (ii) reconstructs $(sk\|sk')$, and (iii) if $sk^3 = sk'$, outputs $\mathsf{D}_{sk}(e)$, otherwise, it outputs $\bot$.

In Sect. 3 we present the intuition behind our construction and a formal security analysis. Our instantiation yields a rate 1 computationally secure MD-NMC in the CRS model, with access rate $1 - 1/\Omega(\log k)$ and codewords of length $|s| + O(k^2 \log k)$, under mild assumptions (e.g., one way functions).

On the CRS. In our work, the tampering function, and consequently the codeword locations that the function is given access to, are fixed before sampling the

---

[9] In general, any polynomial of small degree, e.g., sk^c, would suffice, depending on the choice of the underlying finite field. Using sk^3 suffices when working over fields of characteristic 2. We could also use sk^2 over fields of characteristic 3.

CRS and this is critical for achieving security. However, proving security in this setting is non-trivial. In addition, the tampering function receives full access to the CRS when tampering with the codeword. This is in contrast to the work by Faust et al. [32] in the information-theoretic setting, where the (internal) tampering function receives partial information over the CRS.

In addition, our results tolerate adaptive selection of the codeword locations, with respect to the CRS, in the following way: each time the attacker requests access to a location, he also learns if it corresponds to a bit of z or e, together with the index of that bit in the original string. In this way, the CRS is gradually disclosed to the adversary while picking codeword locations.

Finally, our CRS sustains a substantial amount of tampering that depends on the codeword locations chosen by the attacker: an attacker that gets access to a sensitive codeword bit is allowed to modify the part of the CRS that defines the location of that bit in the codeword. The attacker is allowed to modify all but $O(k \log(|s| + k))$ bits of the CRS, that is of length $O(k^2 \log k \log(|s| + k))$. To our knowledge, this is the first construction that tolerates, even partial modification of the CRS. In contrast, existing constructions in the CRS model are either using NIZKs [22,28,30,37], or they are based on the *knowledge of exponent assumption* [36], thus tampering access to the CRS might compromise security.

Removing the CRS. A first approach would be to store the CRS inside the codeword together with $P_\Sigma(z\|e)$, and give to the attacker read/write access to it. However, the tampering function, besides getting direct (partial) access to the encoding of sk, it also gets indirect access to it by (partially) controlling the CRS. Then, it can modify the CRS in way such that, during decoding, ciphertext locations of its choice will be treated as bits of the inner encoding, z, increasing the tampering rate against z significantly. This makes the task of protecting sk hard, if not impossible (unless we restrict the access rate significantly).

To handle this challenge, we embed the structure recovering mechanism inside the codeword and we emulate the CRS effect by increasing the size of the alphabet, giving rise to a block-wise structure.[10] Notice that, non-malleable codes with large alphabet size (i.e., poly$(k) + |s|$ bits) might be easy to construct, as we can embed in each codeword block the verification key of a signature scheme together with a secret share of the message, as well as a signature over the share. In this way, partial access over the codeword does not compromise the security of the signature scheme while the message remains private, and the simulation is straightforward. This approach however, comes with a large overhead, decreasing the information rate and access rate of the scheme significantly. In general, and similar to error correcting codes, we prefer smaller alphabet sizes – the larger the size is, the more coarse access structure is required, i.e., in order to access individual bits we need to access the blocks that contain them. In this work, we aim at minimizing this restriction by using small alphabets, as we describe below.

---

[10] Bigger alphabets have been also considered in the context of error-correcting codes, in which the codeword consists of symbols.

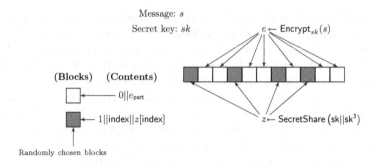

Fig. 2. Description of the scheme in the standard model.

Our approach on the problem is the following. We increase the alphabet size to $O(\log k)$ bits, and we consider two types of blocks: (i) *sensitive blocks*, in which we store the inner encoding, z, of the secret key, sk, and (ii) *non-sensitive* blocks, in which we store the ciphertext, e, that is fragmented into blocks of size $O(\log k)$. The first bit of each block indicates whether it is a sensitive block, i.e., we set it to 1 for sensitive blocks and to 0, otherwise. Our encoder works as follows: on input message s, it computes z, e, as in the previous scheme and then uses rejection sampling to sample the indices, $\rho_1, \ldots, \rho_{|z|}$, for the sensitive blocks. Then, for every $i \in \{1, \ldots, |z|\}$, ρ_i is a sensitive block, with contents $(1\|i\|z[i])$, while the remaining blocks keep ciphertext pieces of size $O(\log k)$. Decoding proceeds as follows: on input codeword $C = (C_1, \ldots, C_{\mathsf{bn}})$, for each $i \in [\mathsf{bn}]$, if C_i is a non-sensitive block, its data will be part of e, otherwise, the last bit of C_i will be part of z, as it is dictated by the index stored in C_i. If the number of sensitive blocks is not the expected, the decoder outputs $\perp$, otherwise, z, e, have been fully recovered and decoding proceeds as in the previous scheme. Our scheme is depicted in Fig. 2.

The security of our construction is based on the fact that, due to our shuffling technique, the position mapping will not be completely overwritten by the attacker, and as we prove in Sect. 4, this suffices for protecting the inner encoding over sk. We prove security of the current scheme (cf. Theorem 4.4) by a reduction to the security of the scheme in the CRS model. Our instantiation yields a rate $1 - 1/\Omega(\log k)$ MD-NMC in the standard model, with access rate $1 - 1/\Omega(\log k)$ and codewords of length $|s|(1 + 1/O(\log k)) + O(k^2 \log^2 k)$, assuming one-way functions.

It is worth pointing out that the idea of permuting blocks containing sensitive and non-sensitive data was also considered by [42] in the context of list-decodable codes, however the similarity is only in the fact that a permutation is being used at some point in the encoding process, and our objective, construction and proof are different.

Continuously non-malleable codes with light updates. We observe that the codewords of the block-wise scheme can be updated efficiently, using shuffling and refreshing operations. Based on this observation, we prove that our code is secure against continuous attacks, for a notion of security that is weaker than the original one [30], as we need to update our codeword. However, our update mechanism is using cheap operations, avoiding the full decoding and re-encoding of the message, which is the standard way to achieve continuous security [27,37]. In addition, our solution avoids the usage of a self-destruction mechanism that produces ⊥ in all subsequent rounds after the first round in which the attacker creates an invalid codeword, which was originally achieved by [28], and makes an important step towards practicality.

The update mechanism works as follows: in each round, it randomly shuffles the blocks and refreshes the randomness of the inner encoding of sk. The idea here is that, due to the continual shuffling and refreshing of the inner encoding scheme, in each round the attacker learns nothing about the secret key, and every attempt to modify the inner encoding, results to an invalid key, with overwhelming probability. Our update mechanism can be made deterministic if we further encode a seed of a PRG together with the secret key, which is similar to the technique presented in [37].

Our results are presented in Sect. 5 (cf. Theorem 5.3), and the rates for the current scheme match those of the one-time secure, block-wise code.

1.3   Applications

Security against passive attackers - AONTs. Regarding the passive setting, our model and constructions find useful application in all settings where AONTs are useful (cf. [13,14,40,41]), e.g., for increasing the security of encryption without increasing the key-size, for improving the efficiency of block ciphers and constructing remotely keyed encryption [13,41], and also for constructing computationally secure secret sharing [40]. Other uses of AONTs are related to optimal asymmetric encryption padding [13].

Security against memory tampering - (Binary alphabets, Logarithmic length CRS). As with every NMC, the most notable application of the proposed model and constructions is when aiming for protecting cryptographic devices against memory tampering. Using our CRS based construction we can protect a large tamperable memory with a small (logarithmic in the message length) tamperproof memory, that holds the CRS.

The construction is as follows. Consider any device performing cryptographic operations, e.g., a smart card, whose memory is initialized when the card is being issued. Each card is initialized with an independent CRS, which is stored in a tamper-proof memory, while the codeword is stored in a tamperable memory. Due to the independency of the CRS values, it is plausible to assume that the adversary is not given access to the CRS prior to tampering with the card; the full CRS is given to the tampering function while it tampers with the codeword during computation. This idea is along the lines of the *only computation leaks information* model [38], where data can only be leaked during computation, i.e., the attacker learns the CRS when the devices performs computations that depend on it. We note that in this work we allow the tampering function to read the full CRS, in contrast to [32], in which the tampering function receives partial information over it (our CRS can also be tampered, cf. the above discussion). In subsequent rounds the CRS and the codeword are being updated by the device, which is the standard way to achieve security in multiple rounds while using a one-time NMC [27].

Security against memory tampering - (Logarithmic length alphabets, no CRS). In modern architectures data is stored and transmitted in chunks, thus our block-wise encoding scheme can provide tamper-resilience in all these settings. For instance, consider the case of arithmetic circuits, having memory consisting of consecutive blocks storing integers. Considering adversaries that access the memory of such circuits in a block-wise manner, is a plausible scenario. In terms of modeling, this is similar to tamper-resilience for arithmetic circuits [33], in which the attacker, instead of accessing individual circuit wires carrying bits, it accesses wires carrying integers. The case is similar for RAM computation where the CPU operates over 32 or 64 bit registers (securing RAM programs using NMC was also considered by [22–24,31]). We note that the memory segments in which the codeword blocks are stored do not have to be physically separated, as partial functions output values that depend on the whole input in which they receive access to. This is in contrast to the split-state setting in which the tampering function tampers with each state independently, and thus the states need to be physically separated.

Security against adversarial channels. In Wiretap Channels [9,39,45] the goal is to communicate data privately against eavesdroppers, under the assumption that the channel between the sender and the adversary is "noisier" than the channel between the sender and the receiver. The model that we propose and our block-wise construction can be applied in this setting to provide privacy against a wiretap adversary under the assumption that due to the gap of noise there is a small (of rate $o(1)$) fraction of symbols that are delivered intact to the receiver and dropped from the transmission to the adversary. This enables private, key-less communication between the parties, guaranteeing that the receiver will either receive the original message, or $\perp$. In this way, the communication will be non-malleable in the sense that the receiver cannot be lead to output $\perp$ depending on any property of the plaintext. Our model allows the noise in the receiver side to depend on the transmission to the wiretap adversary, that tampers with a large (of rate $1 - o(1)$) fraction of symbols, leading to an "active" variant of the wiretap model.

1.4   Related Work

Manipulation detection has been considered independently of the notion of non-malleability, in the seminal paper by Cramer et al. [21], who introduced the notion of *algebraic manipulation detection* (AMD) codes, providing security against additive attacks over the codeword. A similar notion was considered by Jafargholi and Wichs [34], called *tamper detection*, aiming to detect malicious modifications over the codeword, independently of how those affect the output of the decoder. Tamper detection ensures that the application of any (admissible) function to the codeword leads to an invalid decoding.

Non-malleable codes for other function classes have been extensively studied, such as constant split-state functions [17,25], block-wise tampering [15,19], while the work of [2] develops beautiful connections among various function classes. In addition, other variants of non-malleable codes have been proposed, such as continuous non-malleable codes [30], augmented non-malleable codes [1], locally decodable/updatable non-malleable codes [16,22–24,31], and non-malleable codes with split-state refresh [28]. In [7] the authors consider AC0 circuits, bounded-depth decision trees and streaming, space-bounded adversaries. Leakage resilience was also considered as an additional feature, e.g., by [16,24,28,37].

2 Preliminaries

In this section we present basic definitions and notation that will be used throughout the paper.

Definition 2.1 (Notation). *Let t, i, j, be non-negative integers. Then, $[t]$ is the set $\{1, \ldots, t\}$. For bit-strings x, y, $x\|y$, is the concatenation of x, y, $|x|$ denotes the length of x, for $i \in [|x|]$, $x[i]$ is the i-th bit of x, $\|_{j=1}^{t} x_j := x_1\| \ldots \|x_t$, and for $i \leq j$, $x[i : j] = x[i]\| \ldots \|x[j]$. For a set I, $|I|$, $\mathcal{P}(I)$, are the cardinality and power set of I, respectively, and for $I \subseteq [|x|]$, $x_{|I}$ is the projection of the bits of x with respect to I. For a string variable c and value v, $c \leftarrow v$ denotes the assignment of v to c, and $c[I] \leftarrow v$, denotes an assignment such that $c_{|I}$ equals v. For a distribution D over a set $\mathcal{X}$, $x \leftarrow D$, denotes sampling an element $x \in \mathcal{X}$, according to D, $x \leftarrow \mathcal{X}$ denotes sampling a uniform element x from $\mathcal{X}$, $U_{\mathcal{X}}$ denotes the uniform distribution over $\mathcal{X}$ and $x_1, \ldots, x_t \xleftarrow{\text{rs}} \mathcal{X}$ denotes sampling a uniform subset of $\mathcal{X}$ with t distinct elements, using rejection sampling. The statistical distance between two random variables X, Y, is denoted by $\Delta(X, Y)$, "$\approx$" and "$\approx_c$", denote statistical and computational indistinguishability, respectively, and $\mathsf{negl}(k)$ denotes an unspecified, negligible function, in k.*

Below, we define coding schemes, based on the definitions of [27,37].

Definition 2.2 (Coding scheme [27]). *A (κ, ν)-coding scheme, $\kappa, \nu \in \mathbb{N}$, is a pair of algorithms $(\mathsf{Enc}, \mathsf{Dec})$ such that: $\mathsf{Enc} : \{0,1\}^{\kappa} \to \{0,1\}^{\nu}$ is an encoding algorithm, $\mathsf{Dec} : \{0,1\}^{\nu} \to \{0,1\}^{\kappa} \cup \{\bot\}$ is a decoding algorithm, and for every $s \in \{0,1\}^{\kappa}$, $\Pr[\mathsf{Dec}(\mathsf{Enc}(s)) = s] = 1$, where the probability runs over the randomness used by $(\mathsf{Enc}, \mathsf{Dec})$.*

We can easily generalize the above definition for larger alphabets, i.e., by considering $\mathsf{Enc} : \{0,1\}^{\kappa} \to \Gamma^{\nu}$ and $\mathsf{Dec} : \Gamma^{\nu} \to \{0,1\}^{\kappa} \cup \{\bot\}$, for some alphabet Γ.

Definition 2.3 (Coding scheme in the Common Reference String (CRS) Model [37]). *A (κ, ν)-coding scheme in the CRS model, $\kappa, \nu \in \mathbb{N}$,* *is a triple of algorithms* (Init, Enc, Dec) *such that:* Init *is a randomized algorithm which receives* 1^k, *where k denotes the security parameter, and produces a common reference string* $\Sigma \in \{0,1\}^{\text{poly}(k)}$, *and* (Enc($1^k, \Sigma, \cdot$), Dec($1^k, \Sigma, \cdot$)) *is a* (κ, ν)*-coding scheme,* $\kappa, \nu = \text{poly}(k)$.

For brevity, 1^k will be omitted from the inputs of Enc and Dec.

Below we define *non-malleable codes* with *manipulation detection*, which is a stronger notion than the one presented in [27], in the sense that the tampered codeword will always decode to the original message or to $\bot$. Our definition is with respect to alphabets, as in Sect. 4 we consider alphabets of size $O(\log k)$.

Definition 2.4 (Non-Malleability with Manipulation Detection (MD-NMC)). *Let Γ be an alphabet, let* (Init, Enc, Dec) *be a (κ, ν)-coding scheme in the common reference string model, and $\mathcal{F}$ be a family of functions $f : \Gamma^\nu \to \Gamma^\nu$. For any $f \in \mathcal{F}$ and $s \in \{0,1\}^\kappa$, define the tampering experiment*

$$\mathsf{Tamper}_s^f := \left\{ \begin{array}{c} \Sigma \leftarrow \mathsf{Init}(1^k), c \leftarrow \mathsf{Enc}(\Sigma, s), \tilde{c} \leftarrow f_\Sigma(c), \tilde{s} \leftarrow \mathsf{Dec}(\Sigma, \tilde{c}) \\ Output: \tilde{s}. \end{array} \right\}$$

which is a random variable over the randomness of Enc, Dec *and* Init. *The coding scheme* (Init, Enc, Dec) *is non-malleable with manipulation detection with respect to the function family $\mathcal{F}$, if for all, sufficiently large k and for all $f \in \mathcal{F}$, there exists a distribution $D_{(\Sigma, f)}$ over $\{0,1\}^\kappa \cup \{\bot, \mathsf{same}^*\}$, such that for all $s \in \{0,1\}^\kappa$, we have:*

$$\left\{ \mathsf{Tamper}_s^f \right\}_{k \in \mathbb{N}} \approx \left\{ \begin{array}{c} \tilde{s} \leftarrow D_{(\Sigma, f)} \\ Output\ s\ if\ \tilde{s} = \mathsf{same}^*,\ and\ \bot\ otherwise \end{array} \right\}_{k \in \mathbb{N}}$$

where $\Sigma \leftarrow \mathsf{Init}(1^k)$ and $D_{(\Sigma, f)}$ is efficiently samplable given access to f, Σ. Here, "$\approx$" may refer to statistical, or computational, indistinguishability.

In the above definition, f is parameterized by Σ to differentiate tamper-proof input, i.e., Σ, from tamperable input, i.e., c.

Below we define the tampering function class that will be used throughout the paper.

Definition 2.5 (The class of partial functions $\mathcal{F}_\Gamma^{\alpha\nu}$ (or $\mathcal{F}^\alpha$)). *Let Γ be an alphabet, $\alpha \in [0, 1)$ and $\nu \in \mathbb{N}$. Any $f \in \mathcal{F}_\Gamma^{\alpha\nu}$, $f : \Gamma^\nu \to \Gamma^\nu$, is indexed by a set $I \subseteq [\nu], |I| \leq \alpha\nu$, and a function $f' : \Gamma^{\alpha\nu} \to \Gamma^{\alpha\nu}$, such that for any $x \in \Gamma^\nu$, $(f(x))_{|_I} = f'(x_{|_I})$ and $(f(x))_{|_{I^c}} = x_{|_{I^c}}$, where $I^c := [\nu] \backslash I$.*

For simplicity, in the rest of the text we will use the notation $f(x)$ and $f(x_{|_I})$ (instead of $f'(x_{|_I})$). Also, the length of the codeword, ν, according to Γ, will be omitted from the notation and whenever Γ is omitted we assume that $\Gamma = \{0, 1\}$. In Sect. 3, we consider $\Gamma = \{0, 1\}$, while in Sect. 4, $\Gamma = \{0, 1\}^{O(\log k)}$, i.e., the tampering function operates over blocks of size $O(\log k)$. When considering the CRS model, the functions are parameterized by the common reference string.

The following lemma is useful for proving security throughout the paper.

Lemma 2.6. *Let* (Enc, Dec) *be a* (κ, ν)-*coding scheme and* $\mathcal{F}$ *be a family of functions. For every* $f \in \mathcal{F}$ *and* $s \in \{0,1\}^{\kappa}$, *define the tampering experiment*

$$\mathsf{Tamper}_s^f := \left\{ \begin{array}{c} c \leftarrow \mathsf{Enc}(s), \tilde{c} \leftarrow f(c), \tilde{s} \leftarrow \mathsf{Dec}(\tilde{c}) \\ \textit{Output } \mathsf{same}^* \textit{ if } \tilde{s} = s, \textit{ and } \tilde{s} \textit{ otherwise.} \end{array} \right\}$$

which is a random variable over the randomness of Enc *and* Dec. (Enc, Dec) *is an* MD-NMC *with respect to* $\mathcal{F}$, *if for any* $f \in \mathcal{F}$ *and all sufficiently large* k: (i) *for any pair of messages* $s_0, s_1 \in \{0,1\}^{\kappa}$, $\left\{ \mathsf{Tamper}_{s_0}^f \right\}_{k \in \mathbb{N}} \approx \left\{ \mathsf{Tamper}_{s_1}^f \right\}_{k \in \mathbb{N}}$, *and* (ii) *for any* s, $\Pr \left[\mathsf{Tamper}_s^f \notin \{\bot, s\} \right] \leq \mathsf{negl}(k)$. *Here,* "$\approx$" *may refer to statistical, or computational, indistinguishability.*

The proof of the above lemma is provided in the full version of the paper. For coding schemes in the CRS model the above lemma is similar, and Tamper_s^f internally samples $\Sigma \leftarrow \mathsf{Init}(1^k)$.

3   An MD-NMC for Partial Functions, in the CRS Model

In this section we consider $\Gamma = \{0,1\}$ and we construct a rate 1 MD-NMC for $\mathcal{F}^{\alpha}$, with access rate $\alpha = 1 - 1/\Omega(\log k)$. Our construction is defined below and depicted in Fig. 1.

Construction 3.1. *Let* $k, m \in \mathbb{N}$, *let* (KGen, E, D) *be a symmetric encryption scheme,* (SS$_m$, Rec$_m$) *be an* m-*out-of-*m *secret sharing scheme, and let* $l \leftarrow 2m|sk|$, *where* sk *follows* KGen(1^k). *We define an encoding scheme* (Init, Enc, Dec), *that outputs* $\nu = l + |e|$ *bits,* $e \leftarrow \mathsf{E}_{sk}(s)$, *as follows:*

- Init(1^k): *Sample* $r_1, \ldots, r_l \xleftarrow{\text{rs}} \{0,1\}^{\log(\nu)}$, *and output* $\Sigma = (r_1, \ldots, r_l)$.
- Enc$(\Sigma, \cdot)$: *for input message* s, *sample* $sk \leftarrow$ KGen(1^k), $e \leftarrow \mathsf{E}_{sk}(s)$.
  - **(Secret share)** *Sample* $z \leftarrow \mathsf{SS}_m(sk\|sk^3)$, *where* $z = \|_{i=1}^{2|sk|} z_i$, $z \in \{0,1\}^{2m|sk|}$, *and for* $i \in [|sk|]$, z_i *(resp.* $z_{|sk|+i}$*) is an* m-*out-of-*m *secret sharing of* $sk[i]$ *(resp.* $sk^3[i]$*).*
  - **(Shuffle)** *Compute* $c \leftarrow P_{\Sigma}(z\|e)$ *as follows:*
    1. **(Sensitive bits):** *Set* $c \leftarrow 0^{\nu}$. *For* $i \in [l]$, $c[r_i] \leftarrow z[i]$.
    2. **(Ciphertext bits):** *Set* $i \leftarrow 1$. *For* $j \in [l + |e|]$, *if* $j \notin \{r_p \mid p \in [l]\}$, $c[j] \leftarrow e[i]$, $i{++}$.

  Output c.
- Dec$(\Sigma, \cdot)$: *on input* c, *compute* $(z\|e) \leftarrow P_{\Sigma}^{-1}(c)$, $(sk\|sk') \leftarrow \mathsf{Rec}_m(z)$, *and if* $sk^3 = sk'$, *output* $\mathsf{D}_{sk}(e)$, *otherwise output* $\bot$.

The set of indices of z_i *in the codeword will be denoted by* Z_i.

In the above we consider all values as elements over $\mathbf{GF}(2^{\mathrm{poly}(k)})$.

Our construction combines authenticated encryption with an inner encoding that works as follows. It interprets sk as an element in the finite field $\mathbf{GF}(2^{|sk|})$ and computes sk^3 as a field element. Then, for each bit of $(sk\|sk^3)$, it computes an m-out-of-m secret sharing of the bit, for some parameter m (we note that elements in $\mathbf{GF}(2^{|sk|})$ can be interpreted as bit strings). Then, by combining the inner encoding with the shuffling technique, we get a encoding scheme whose security follows from the observations that we briefly present below:

– For any tampering function which does not have access to all m shares of a single bit of $(sk\|sk^3)$, the tampering effect on the secret key can be expressed essentially as a linear shift, i.e., as $((sk + \delta)\|(sk^3 + \eta))$ for some $(\delta, \eta) \in \mathbf{GF}(2^{|sk|}) \times \mathbf{GF}(2^{|sk|})$, independent of sk.
– By permuting the locations of the inner encoding and the ciphertext bits, we have that with overwhelming probability any tampering function who reads/writes on a $(1 - o(1))$ fraction of codeword bits, will not learn any single bit of $(sk\|sk^3)$.
– With overwhelming probability over the randomness of sk and CRS, for non-zero η and δ, $(sk + \delta)^3 \neq sk^3 + \eta$, and this property enables us to design a consistency check mechanism whose output is simulatable, without accessing sk.
– The security of the final encoding scheme follows by composing the security of the inner encoding scheme with the authenticity property of the encryption scheme.

Below we present the formal security proof of the above intuitions.

Theorem 3.2. *Let k, $m \in \mathbb{N}$ and $\alpha \in [0, 1)$. Assuming $(\mathsf{SS}_m, \mathsf{Rec}_m)$ is an m-out-of-m secret sharing scheme and $(\mathsf{KGen}, \mathsf{E}, \mathsf{D})$ is 1-IND-CPA[11] secure, authenticated encryption scheme, the code of Construction 3.1 is a MD-NMC against $\mathcal{F}^\alpha$, for any α, m, such that $(1 - \alpha)m = \omega(\log(k))$.*

Proof. Let I be the set of indices chosen by the attacker and $I^c = [\nu]\backslash I$, where $\nu = 2m|sk| + |e|$. The tampered components of the codeword will be denoted using the character "~" on top of the original symbol, i.e., we have $\tilde{c} \leftarrow f(c)$, the tampered secret key sk (resp. sk^3) that we get after executing $\mathsf{Rec}_m(\tilde{z})$ will be denoted by $\widetilde{sk}$ (resp. $\widetilde{sk}'$). Also the tampered ciphertext will be $\tilde{e}$. We prove the needed using a series of hybrid experiments that are depicted in Fig. 3. Below, we describe the hybrids.

– $\mathsf{Exp}_0^{\Sigma, f, s}$: We prove security of our code using Lemma 2.6, i.e., by showing that (i) for any s_0, s_1, $\mathsf{Tamper}_{s_0}^f \approx \mathsf{Tamper}_{s_1}^f$, and (ii) for any s, $\Pr\left[\mathsf{Tamper}_s^f \notin \{\bot, s\}\right] \leq \mathsf{negl}(k)$, where Tamper_s^f is defined in Lemma 2.6. For any f, s, $\Sigma \leftarrow \mathsf{Init}(1^k)$, the first experiment, $\mathsf{Exp}_0^{\Sigma, f, s}$, matches the experiment Tamper_s^f in the CRS model, i.e., Σ is sampled inside Tamper_s^f.

---

[11] This is an abbreviations for indistinguishability under chosen plaintext attack, for a single pre-challenge query to the encryption oracle.

$\mathsf{Exp}_0^{\Sigma,f,s}:$
$c \leftarrow \mathsf{Enc}(\Sigma, s), \tilde{c} \leftarrow 0^{\nu}$
$\tilde{c}[I] \leftarrow f_{\Sigma}(c_{|_I}), \tilde{c}[I^c] \leftarrow c_{|_{I^c}}$
$\tilde{s} \leftarrow \mathsf{Dec}(\tilde{c})$

Output same* if $\tilde{s} = s$ and $\tilde{s}$ otherwise.

$\mathsf{Exp}_1^{\Sigma,f,s}:$
$c \leftarrow \mathsf{Enc}(\Sigma, s), \tilde{c} \leftarrow 0^{\nu}$
$\tilde{c}[I] \leftarrow f_{\Sigma}(c_{|_I}), \tilde{c}[I^c] \leftarrow c_{|_{I^c}}$
If $\exists i : |(I \cap Z_i)| = m$:
   $\tilde{s} \leftarrow \perp$
Else:
   $\tilde{s} \leftarrow \mathsf{Dec}(\tilde{c})$

Output same* if $\tilde{s} = s$ and $\tilde{s}$ otherwise.

$\mathsf{Exp}_2^{\Sigma,f,s}:$
$sk \leftarrow \mathsf{KGen}(1^k), e \leftarrow \mathsf{E}_{sk}(s)$
$z^* \leftarrow \bar{\mathsf{SS}}_m^f(\Sigma, sk), c \leftarrow P_{\Sigma}(z^*\|e)$
$\tilde{c} \leftarrow 0^{\nu}, \tilde{c}[I] \leftarrow f_{\Sigma}(c_{|_I}), \tilde{c}[I^c] \leftarrow c_{|_{I^c}}$

If $\exists i : |(I \cap Z_i)| = m$:
   $\tilde{s} \leftarrow \perp$
Else:
   If $\exists i : \bigoplus_{j \in (I \cap Z_i)} c[j] \neq \bigoplus_{j \in (I \cap Z_i)} \tilde{c}[j]$:
      $\tilde{s} \leftarrow \perp$
   Else:
      $\tilde{s} \leftarrow \mathsf{D}_{sk}(\tilde{e})$

Output same* if $\tilde{s} = s$ and $\tilde{s}$ otherwise.

$\mathsf{Exp}_3^{\Sigma,f,s}:$
$sk \leftarrow \mathsf{KGen}(1^k), e \leftarrow \mathsf{E}_{sk}(s)$
$z^* \leftarrow \bar{\mathsf{SS}}_m^f(\Sigma, sk), c \leftarrow P_{\Sigma}(z^*\|e)$

$\tilde{c} \leftarrow 0^{\nu}, \tilde{c}[I] \leftarrow f_{\Sigma}(c_{|_I})$

If $\exists i : |(I \cap Z_i)| = m$:
   $\tilde{s} \leftarrow \perp$
Else:
   If $\exists i : \bigoplus_{j \in (I \cap Z_i)} c[j] \neq \bigoplus_{i \in (I \cap Z_i)} \tilde{c}[j]$:
      $\tilde{s} \leftarrow \perp$
   Else: $\tilde{s} \leftarrow \perp$
      If $\tilde{e} = e$:
         $\tilde{s} \leftarrow$ same*

Output $\tilde{s}$.

Fig. 3. The hybrid experiments for the proof of Theorem 3.2.

- $\mathsf{Exp}_1^{\Sigma,f,s}$: In the second experiment we define Z_i, $i \in [2|sk|]$, to be the set of codeword indices in which the secret sharing z_i is stored, $|Z_i| = m$. The main difference from the previous experiment is that the current one outputs $\perp$, if there exists a bit of sk or sk^3 for which the tampering function reads all the shares of it, while accessing at most $\alpha\nu$ bits of the codeword. Intuitively, and as we prove in Claim 3.3, by permuting the location indices of $z\|e$, this event happens with probability negligible in k, and the attacker does not learn any bit of sk and sk^3, even if he is given access to $(1-o(1))\nu$ bits of the codeword.
- $\mathsf{Exp}_2^{\Sigma,f,s}$: By the previous hybrid we have that for all $i \in [2|sk|]$, the tampering function will not access all bits of z_i, with overwhelming probability. In the third experiment we unfold the encoding procedure, and in addition, we substitute the secret sharing procedure SS_m with $\bar{\mathsf{SS}}_m^f$ that computes shares z_i^* that reveal no information about $sk\|sk^3$; for each i, $\bar{\mathsf{SS}}_m^f$ simply "drops" the bit of z_i with the largest index that is not being accessed by f. We formally define $\bar{\mathsf{SS}}_m^f$ below.

$\tilde{\mathsf{SS}}_m^f(\Sigma, sk)$:

1. Sample $(z_1, \ldots, z_{2|sk|}) \leftarrow \mathsf{SS}_m(sk||sk^3)$ and set $z_i^* \leftarrow z_i$, $i \in [2|sk|]$.
2. For $i \in [2|sk|]$, let $l_i := \max_d \{d \in [m] \wedge \mathsf{Ind}(z_i[d]) \notin I)\}$, where Ind returns the index of $z_i[d]$ in c, i.e., l_i is the largest index in $[m]$ such that $z_i[l_i]$ is not accessed by f.
3. (**Output**): For all i set $z_i^*[l_i] = *$, and output $z^* := ||_{i=1}^{2|sk|} z_i^*$.

In $\mathsf{Exp}_1^{\Sigma, f, s}$, $z = ||_{i=1}^{2|sk|} z_i$, and each z_i is an m-out-of-m secret sharing for a bit of sk or sk^3. From Claim 3.3, we have that for all i, $|I \cap Z_i| < m$ with overwhelming probability, and we can observe that the current experiment is identical to the previous one up to the point of computing $f(c_{|_I})$, as $c_{|_I}$ and $f(c_{|_I})$ depend only on z^*, that carries no information about sk and sk^3.

Another difference between the two experiments is in the external "Else" branch: $\mathsf{Exp}_1^{\Sigma, f, s}$ makes a call on the decoder while $\mathsf{Exp}_2^{\Sigma, f, s}$, before calling $\mathsf{D}_{sk}(\tilde{e})$, checks if the tampering function has modified the shares in a way such that the reconstruction procedure $((\tilde{sk}, \tilde{sk}') \leftarrow \mathsf{Rec}_m(\tilde{z}))$ will give $\tilde{sk} \neq sk$ or $\tilde{sk}' \neq sk'$. This check is done by the statement "If $\exists i : \bigoplus_{j \in (I \cap Z_i)} c[j] \neq \bigoplus_{j \in (I \cap Z_i)} \tilde{c}[j]$", without touching sk or sk^3.[12] In case modification is detected the current experiments outputs $\bot$. The intuition is that an attacker that partially modifies the shares of sk and sk^3, creates shares of $\tilde{sk}$ and $\tilde{sk}'$, such that $\tilde{sk}^3 = \tilde{sk}'$, with negligible probability in k. We prove this by a reduction to the 1-IND-CPA security of the encryption scheme: any valid modification over the inner encoding of the secret key gives us method to compute the original secret key sk, with non-negligible probability. The ideas are presented formally in Claim 3.4.

- $\mathsf{Exp}_3^{\Sigma, f, s}$: The difference between the current experiment and the previous one is that instead of calling the decryption $\mathsf{D}_{sk}(\tilde{e})$, we first check if the attacker has modified the ciphertext, in which case the current experiment outputs $\bot$, otherwise it outputs same*. By the previous hybrid, we reach this newly introduced "Else" branch of $\mathsf{Exp}_3^{\Sigma, f, s}$, only if the tampering function didn't modify the secret key. Thus, the indistinguishability between the two experiments follows from the authenticity property of the encryption scheme in the presence of z^*: given that $\tilde{sk} = sk$ and $\tilde{sk}' = sk'$, we have that if the attacker modifies the ciphertext, then with overwhelming probability $\mathsf{D}_{sk}(\tilde{e}) = \bot$, otherwise, $\mathsf{D}_{sk}(\tilde{e}) = s$, and the current experiment correctly outputs same* or $\bot$ (cf. Claim 3.5).

- Finally, we prove that for any $f \in \mathcal{F}^\alpha$, and message s, $\mathsf{Exp}_3^{\Sigma, f, s}$ is indistinguishable from $\mathsf{Exp}_3^{\Sigma, f, 0}$, where 0 denotes the zero-message. This follows by the semantic security of the encryption scheme, and gives us the indistinguishability property of Lemma 2.6. The manipulation detection property is derived by the indistinguishability between the hybrids and the fact that the output of $\mathsf{Exp}_3^{\Sigma, f, s}$ is in the set $\{\text{same}^*, \bot\}$.

---

[12] Recall that our operations are over $\mathbf{GF}(2^{\mathrm{poly}(k)})$.

In what follows, we prove indistinguishability between the hybrids using a series of claims.

Claim 3.3. *For k, $m \in \mathbb{N}$, assume $(1-\alpha)m = \omega(\log(k))$. Then, for any $f \in \mathcal{F}^\alpha$ and any message s, we have $\mathsf{Exp}_0^{\Sigma,f,s} \approx \mathsf{Exp}_1^{\Sigma,f,s}$, where the probability runs over the randomness used by* Init, Enc.

Proof. The difference between the two experiments is that $\mathsf{Exp}_1^{\Sigma,f,s}$ outputs $\perp$ when the attacker learns all shares of some bit of sk or sk^3, otherwise it produces output as $\mathsf{Exp}_0^{\Sigma,f,s}$ does. Let E the event "$\exists i : |(I \cap Z_i)| = m$". Clearly, $\mathsf{Exp}_0^{\Sigma,f,s} = \mathsf{Exp}_1^{\Sigma,f,s}$ conditioned on $\neg E$, thus the statistical distance between the two experiments is bounded by $\Pr[E]$. In the following we show that $\Pr[E] \leq \mathsf{negl}(k)$. We define by E_i the event in which f learns the entire z_i. Assuming the attacker reads n bits of the codeword, we have that for all $i \in [2|sk|]$,

$$\Pr_\Sigma[E_i] = \Pr_\Sigma[\; |I \cap Z_i| = m \;] = \prod_{j=0}^{m-1} \frac{n-j}{\nu-j} \leq \left(\frac{n}{\nu}\right)^m.$$

We have $n = \alpha\nu$ and assuming $\alpha = 1 - \epsilon$ for $\epsilon \in (0,1]$, we have $\Pr[E_i] \leq (1-\epsilon)^m \leq 1/e^{m\epsilon}$ and $\Pr[E] = \Pr_\Sigma\left[\bigcup_{i=1}^{2|sk|} E_i\right] \leq \frac{2|sk|}{e^{m\epsilon}}$, which is negligible when $(1-\alpha)m = \omega(\log(k))$, and the proof of the claim is complete. ∎

Claim 3.4. *Assuming* $(\mathsf{KGen}, \mathsf{E}, \mathsf{D})$ *is 1-IND-CPA secure, for any $f \in \mathcal{F}^\alpha$ and any message s, $\mathsf{Exp}_1^{\Sigma,f,s} \approx \mathsf{Exp}_2^{\Sigma,f,s}$, where the probability runs over the randomness used by* Init, Enc.

Proof. In $\mathsf{Exp}_2^{\Sigma,f,s}$ we unfold the encoding procedure, however instead of calling SS_m, we make a call to $\bar{\mathsf{SS}}_m^f$. As we have already stated above, this modification does not induce any difference between the output of $\mathsf{Exp}_2^{\Sigma,f,s}$ and $\mathsf{Exp}_1^{\Sigma,f,s}$, with overwhelming probability, as z^* is indistinguishable from z in the eyes of f. Another difference between the two experiments is in the external "Else" branch: $\mathsf{Exp}_1^{\Sigma,f,s}$ makes a call on the decoder while $\mathsf{Exp}_2^{\Sigma,f,s}$, before calling $\mathsf{D}_{sk}(\tilde{e})$, checks if the tampering function has modified the shares in a way such that the reconstruction procedure will give $\tilde{sk} \neq sk$ or $\tilde{sk}' \neq sk'$. This check is done by the statement "If $\exists i : \bigoplus_{j\in(I\cap Z_i)} c[j] \neq \bigoplus_{j\in(I\cap Z_i)} \tilde{c}[j]$", without touching sk or sk^3 (cf. Claim 3.3).[13] We define the events E, E' as follows

$$E : \mathsf{Dec}(\tilde{c}) \neq \perp, E' : \exists i : \bigoplus_{j\in(I\cap Z_i)} c[j] \neq \bigoplus_{j\in(I\cap Z_i)} \tilde{c}[j].$$

Clearly, conditioned on $\neg E'$ the two experiments are identical, since we have $\tilde{sk} = sk$ and $\tilde{sk}' = sk'$, and the decoding process will output $\mathsf{D}_{sk}(\tilde{e})$ in both experiments. Thus, the statistical distance is bounded by $\Pr[E']$. Now, conditioned on $E' \wedge \neg E$, both experiments output $\perp$. Thus, we need to bound

---

[13] Recall that our operations are over $\mathbf{GF}(2^{\mathrm{poly}(k)})$.

$\Pr[E \wedge E']$. Assuming $\Pr[E \wedge E'] > p$, for $p = 1/\mathrm{poly}(k)$, we define an attacker $\mathcal{A}$ that simulates $\mathrm{Exp}_2^{\Sigma, f, s}$, and uses f, s to break the 1-IND-CPA security of $(\mathsf{KGen}, \mathsf{E}, \mathsf{D})$ in the presence of z^*, with probability at least $1/2 + p''/2$, for $p'' = 1/\mathrm{poly}(k)$.

First we prove that any 1-IND-CPA secure encryption scheme, remains secure even if the attacker receives $z^* \leftarrow \bar{\mathsf{SS}}_m^f(\Sigma, sk)$, as z^* consists of $m - 1$ shares of each bit of sk and sk^3, i.e., for the entropy of sk we have $\mathbf{H}(sk|z^*) = \mathbf{H}(sk)$. Towards contradiction, assume there exists $\mathcal{A}$ that breaks the 1-IND-CPA security of $(\mathsf{KGen}, \mathsf{E}, \mathsf{D})$ in the presence of z^*, i.e., there exist s, s_0, s_1 such that $\mathcal{A}$ distinguishes between $(z^*, \mathsf{E}_{sk}(s), \mathsf{E}_{sk}(s_0))$ and $(z^*, \mathsf{E}_{sk}(s), \mathsf{E}_{sk}(s_1))$, with non-negligible probability p. We define an attacker $\mathcal{A}'$ that breaks the 1-IND-CPA security of $(\mathsf{KGen}, \mathsf{E}, \mathsf{D})$ as follows: $\mathcal{A}'$, given $(\mathsf{E}_{sk}(s), \mathsf{E}_{sk}(s_b))$, for some $b \in \{0, 1\}$, samples $\hat{sk} \leftarrow \mathsf{KGen}(1^k)$, $\hat{z}^* \leftarrow \bar{\mathsf{SS}}_m^f(\Sigma, \hat{sk})$ and outputs $b' \leftarrow \mathcal{A}(z^*, \mathsf{E}_{sk}(s), \mathsf{E}_{sk}(s_b))$. Since $(z^*, \mathsf{E}_{sk}(s), \mathsf{E}_{sk}(s_b)) \approx (\hat{z}^*, \mathsf{E}_{sk}(s), \mathsf{E}_{sk}(s_b))$ the advantage of $\mathcal{A}'$ in breaking the 1-IND-CPA security of the scheme is the advantage of $\mathcal{A}$ in breaking the 1-IND-CPA security of the scheme in the presence of z^*, which by assumption is non-negligible, and this completes the current proof. We note that the proof idea presented in the current paragraph also applies for proving that other properties that will be used in the rest of the proof, such as semantic security and authenticity, of the encryption scheme, are retained in the presence of z^*.

Now we prove our claim. Assuming $\Pr[E \wedge E'] > p$, for $p = 1/\mathrm{poly}(k)$, we define an attacker $\mathcal{A}$ that breaks the 1-IND-CPA security of $(\mathsf{KGen}, \mathsf{E}, \mathsf{D})$ in the presence of z^*, with non-negligible probability. $\mathcal{A}$ receives the encryption of s, which corresponds to the oracle query right before receiving the challenge ciphertext, the challenge ciphertext $e \leftarrow \mathsf{E}_{sk}(s_b)$, for uniform $b \in \{0, 1\}$ and uniform messages s_0, s_1, as well as z^*. $\mathcal{A}$ is defined below.

$$\mathcal{A}\left(z^* \leftarrow \bar{\mathsf{SS}}_m^f(\Sigma, sk), e' \leftarrow \mathsf{E}_{sk}(s), e \leftarrow \mathsf{E}_{sk}(s_b)\right):$$

1. **(Define the shares that will be accessed by f):** For $i \in [2|sk|]$, define $w_i := (z_i^*)_{|[m] \setminus \{l_i\}}$ and for $i \in [m - 1]$ define $C_i = \|_{j=1}^{|sk|} w_j[i]$, $D_i = \|_{j=|sk|+1}^{2|sk|} w_j[i]$.

2. **(Apply f)** Set $c \leftarrow P_{\Sigma}(z^* \| e)$, compute $\tilde{c}[I] \leftarrow f_{\Sigma}(c_{|I})$ and let $\tilde{C}_i, \tilde{D}_i, i \in [m]$, be the tampered shares resulting after the application of f to $c_{|I}$.

3. **(Guessing the secret key)** Let $U = \sum_{i=1}^{m-1} C_i$, $V = \sum_{i=1}^{m-1} D_i$, i.e., U, V denote the sum of the shares that are being accessed by the attacker (maybe partially), and $\tilde{U} = \sum_{i=1}^{m-1} \tilde{C}_i$, $\tilde{V} = \sum_{i=1}^{m-1} \tilde{D}_i$, are the corresponding tampered values after applying f on U, V. Define

$$p(X) := (U - \tilde{U})X^2 + (U^2 - \tilde{U}^2)X + (U^3 - \tilde{U}^3 - V + \tilde{V}),$$

and compute the set of roots of $p(X)$, denoted as $\mathcal{X}$, which are at most two. Then set

$$\hat{\mathcal{SK}} := \{x + U | x \in \mathcal{X}\}. \tag{1}$$

4. **(Output)** Execute the following steps,
   (a) For $\hat{sk} \in \hat{\mathcal{SK}}$, compute $s' \leftarrow \mathsf{D}_{\hat{sk}}(e')$, and if $s' = s$, compute $s'' \leftarrow \mathsf{D}_{\hat{sk}}(e)$. Output b' such that $s_{b'} = s''$.
   (b) Otherwise, output $b' \leftarrow \{0, 1\}$.

In the first step $\mathcal{A}$ removes the dummy symbol "$*$" and computes the shares that will be partially accessed by f, denoted as C_i for sk and as D_i for sk^3.[3] In the second step, it defines the tampered shares, $\tilde{C}_i$, $\tilde{D}_i$. Conditioned on E', it is not hard to see that $\mathcal{A}$ simulates perfectly $\mathsf{Exp}_2^{\Sigma, f, s}$. In particular, it simulates perfectly the input to f as it receives $e \leftarrow \mathsf{E}_{sk}(s)$ and all but $2|sk|$ of the actual bit-shares of sk, sk^3. Part of those shares will be accessed by f. Since for all i, $|I \cap Z_i| < m$, the attacker is not accessing any single bit of sk, sk^3. Let C_m, D_m, be the shares (not provided by the encryption oracle) that completely define sk and sk^3, respectively. By the definition of the encoding scheme and the fact that sk, $sk^3 \in \mathbf{GF}(2^{\mathrm{poly}(k)})$, we have $\sum_{i=1}^{m} C_i = sk$, $\sum_{i=1}^{m} D_i = sk^3$, and

$$(U + C_m)^3 = V + D_m. \tag{2}$$

In order for the decoder to output a non-bottom value, the shares created by the attacker must decode to $\tilde{sk}$, $\tilde{sk}'$, such that $\tilde{sk}^3 = \tilde{sk}'$, or in other words, if

$$\left(\tilde{U} + C_m\right)^3 = \tilde{V} + D_m. \tag{3}$$

From 2 and 3 we receive

$$(U - \tilde{U})C_m^2 + (U^2 - \tilde{U}^2)C_m + (U^3 - \tilde{U}^3) = V - \tilde{V}. \tag{4}$$

Clearly, $\Pr[E \wedge E' \wedge (U = \tilde{U})] = 0$. Thus, assuming $\Pr[E \wedge E'] > p$, for $p > 1/\mathrm{poly}(k)$, we receive

$$
\begin{aligned}
p < \Pr\left[E \wedge E' \wedge (U \neq \tilde{U})\right] &\leq \Pr\left[\mathsf{Dec}(\tilde{c}) \neq \perp \wedge E' \wedge U \neq \tilde{U}\right] \\
&\leq \Pr\left[\tilde{sk}^3 = \tilde{sk}' \wedge E' \wedge (U \neq \tilde{U})\right] \\
&\overset{(4,1)}{=} \Pr\left[C_m \in \mathcal{X}\right] \overset{(1)}{\leq} \Pr\left[sk \in \hat{\mathcal{SK}}\right],
\end{aligned} \tag{5}
$$

and $\mathcal{A}$ manages to recover C_m, and thus sk, with non-negligible probability $p' \geq p$. Let W be the event of breaking 1-IND-CPA security. Then,

$$
\begin{aligned}
\Pr[W] &= \Pr[W|sk \in \hat{\mathcal{SK}}] \cdot \Pr[sk \in \hat{\mathcal{SK}}] \\
&\quad + \Pr[W|sk \notin \hat{\mathcal{SK}}] \cdot \Pr[sk \notin \hat{\mathcal{SK}}] \\
&\overset{(5)}{=} p' + \frac{1}{2}(1 - p') = \frac{1}{2} + \frac{p'}{2},
\end{aligned} \tag{6}
$$

and the attacker breaks the IND-CPA security of $(\mathsf{KGen}, \mathsf{E}, \mathsf{D})$. Thus, we have $\Pr[E \wedge E'] < \mathsf{negl}(k)$, and both experiments output $\perp$ with overwhelming probability. ∎

Claim 3.5. *Assuming the authenticity property of* $(\mathsf{KGen}, \mathsf{E}, \mathsf{D})$, *for any* $f \in \mathcal{F}^\alpha$ *and any message* s, $\mathsf{Exp}_2^{\Sigma, f, s} \approx \mathsf{Exp}_3^{\Sigma, f, s}$, *where the probability runs over the randomness used by* Init, KGen *and* E.

Proof. Before proving the claim, recall that the authenticity property of the encryption scheme is preserved under the presence of z^* (cf. Claim 3.4). Let E be the event $\tilde{sk} = sk \wedge \tilde{sk}' = sk^3$ and E' be the event $\tilde{e} \neq e$. Conditioned on $\neg E$, the two experiments are identical, as they both output $\bot$. Also, conditioned on $E \wedge \neg E'$, both experiments output same^*. Thus, the statistical distance between the two experiments is bounded by $\Pr[E \wedge E']$. Let B be the event $\mathsf{D}_{sk}(\tilde{e}) \neq \bot$. Conditioned on $E \wedge E' \wedge \neg B$ both experiments output $\bot$. Thus, we need to bound $\Pr[E \wedge E' \wedge B]$.

Assuming there exist s, f, for which $\Pr[E \wedge E' \wedge B] > p$, where $p = 1/\mathrm{poly}(k)$, we define an attacker $\mathcal{A} = (\mathcal{A}_1, \mathcal{A}_2)$ that simulates $\mathsf{Exp}_3^{\Sigma, f, s}$ and breaks the authenticity property of the encryption scheme in the presence of z^*, with non-negligible probability. $\mathcal{A}$ is defined as follows: sample $(s, st) \leftarrow \mathcal{A}_1(1^k)$, and then, on input (z^*, e, st), where $e \leftarrow \mathsf{E}_{sk}(s)$, $\mathcal{A}_2$, samples $\Sigma \leftarrow \mathsf{Init}(1^k)$, sets $\tilde{c} \leftarrow 0^\nu$, $c \leftarrow P_\Sigma(z^*\|e)$, computes $\tilde{c}[I] \leftarrow f(c_{|_I})$, $\tilde{c}[I^c] \leftarrow c_{|_{I^c}}$, $(\tilde{z}^*\|\tilde{e}) \leftarrow P_\Sigma^{-1}(\tilde{c})$, and outputs $\tilde{e}$. Assuming $\Pr[E \wedge E' \wedge B] > p$, we have that $\mathsf{D}_{sk}(\tilde{e}) \neq \bot$ and $\tilde{e} \neq e$, with non-negligible probability and the authenticity property of $(\mathsf{KGen}, \mathsf{E}, \mathsf{D})$ breaks. ∎

Claim 3.6. *Assuming* $(\mathsf{KGen}, \mathsf{E}, \mathsf{D})$ *is semantically secure, for any* $f \in \mathcal{F}^\alpha$ *and any message* s, $\mathsf{Exp}_3^{\Sigma, f, s} \approx \mathsf{Exp}_3^{\Sigma, f, 0}$, *where the probability runs over the randomness used by* Init, KGen, E. "$\approx$" *may refer to statistical or computational indistinguishability, and* 0 *is the zero-message.*

Proof. Recall that $(\mathsf{KGen}, \mathsf{E}, \mathsf{D})$ is semantically secure even in the presence of $z^* \leftarrow \bar{\mathsf{SS}}_m^f(\Sigma, sk)$ (cf. Claim 3.4), and towards contradiction, assume there exist $f \in \mathcal{F}^\alpha$, message s, and PPT distinguisher D such that

$$\left| \Pr\left[\mathsf{D}\left(\Sigma, \mathsf{Exp}_3^{\Sigma, f, s}\right) = 1 \right] - \Pr\left[\mathsf{D}\left(\Sigma, \mathsf{Exp}_3^{\Sigma, f, 0}\right) \right] = 1 \right| > p,$$

for $p = 1/\mathrm{poly}(k)$. We are going to define an attacker $\mathcal{A}$ that breaks the semantic security of $(\mathsf{KGen}, \mathsf{E}, \mathsf{D})$ in the presence of z^*, using $s_0 := s$, $s_1 := 0$. $\mathcal{A}$, given z^*, e, executes Program.

> Program(z^*, e) :
> $c \leftarrow P_\Sigma(z^*\|e), \tilde{c} \leftarrow 0^\nu, \tilde{c}[I] \leftarrow f(c_{|_I})$
> If $\exists i : |(I \cap Z_i)| = m$: $\tilde{s} \leftarrow \bot$
> Else:
> > If $\exists i : \bigoplus_{j \in (I \cap Z_i)} c[j] \neq \bigoplus_{j \in (I \cap Z_i)} \tilde{c}[j]$: $\tilde{s} \leftarrow \bot$
> > Else: $\tilde{s} \leftarrow \bot$ and If $\tilde{e} = e$: $\tilde{s} \leftarrow \mathsf{same}^*$
> Output $\tilde{s}$.

It is not hard to see that $\mathcal{A}$ simulates $\mathsf{Exp}_3^{\Sigma, f, s_b}$, thus the advantage of $\mathcal{A}$ against the semantic security of $(\mathsf{KGen}, \mathsf{E}, \mathsf{D})$ is the same with the advantage of D in distinguishing between $\mathsf{Exp}_3^{\Sigma, f, s_0}$, $\mathsf{Exp}_3^{\Sigma, f, s_1}$, which by assumption is non-negligible. We have reached a contradiction and the proof of the claim is complete. ∎

From the above claims we have that for any $f \in \mathcal{F}^\alpha$ and any s, $\mathsf{Exp}_0^{\Sigma, f, s} \approx \mathsf{Exp}_3^{\Sigma, f, 0}$, thus for any $f \in \mathcal{F}^\alpha$ and any s_0, s_1, $\mathsf{Exp}_0^{\Sigma, f, s_0} \approx \mathsf{Exp}_0^{\Sigma, f, s_1}$. Also, by the indistinguishability between $\mathsf{Exp}_0^{\Sigma, f, s}$ and $\mathsf{Exp}_3^{\Sigma, f, 0}$, the second property of Lemma 2.6 has been proven as the output of $\mathsf{Exp}_3^{\Sigma, f, 0}$ is in $\{s, \bot\}$, with overwhelming probability, and non-malleability with manipulation detection of our code follows by Lemma 2.6, since $\mathsf{Exp}_0^{\Sigma, f, s}$ is identical to Tamper_s^f of Lemma 2.6.   ∎

4   Removing the CRS

In this section we increase the alphabet size to $O(\log(k))$ and we provide a computationally secure, rate 1 encoding scheme in the standard model, tolerating modification of $(1 - o(1))\nu$ blocks, where ν is the total number of blocks in the codeword. Our construction is defined below and the intuition behind it has already been presented in the Introduction (cf. Sect. 1, Fig. 2). In the following, the projection operation will be also used with respect to bigger alphabets, enabling the projection of blocks.

Construction 4.1. *Let k, $m \in \mathbb{N}$, let $(\mathsf{KGen}, \mathsf{E}, \mathsf{D})$ be a symmetric encryption scheme and $(\mathsf{SS}_m, \mathsf{Rec}_m)$ be an m-out-of-m secret sharing scheme. We define an encoding scheme $(\mathsf{Enc}^*, \mathsf{Dec}^*)$, as follows:*

- $\mathsf{Enc}^*(1^k, \cdot)$: *for input message s, sample $sk \leftarrow \mathsf{KGen}(1^k)$, $e \leftarrow \mathsf{E}_{sk}(s)$.*
  - **(Secret share)** *Sample $z \leftarrow \mathsf{SS}_m(sk \| sk^3)$, where $z = \|_{i=1}^{2|sk|} z_i$, $z \in \{0,1\}^{2m|sk|}$, and for $i \in [|sk|]$, z_i (resp. $z_{|sk|+i}$) is an m-out-of-m secret sharing of $sk[i]$ (resp. $sk^3[i]$).*
  - **(Construct blocks & permute)** *Set $l \leftarrow 2m|sk|$, $\mathsf{bs} \leftarrow \log l + 2$, $d \leftarrow |e|/\mathsf{bs}$, $\mathsf{bn} \leftarrow l + d$, sample $\rho := (\rho_1, \ldots, \rho_l) \xleftarrow{\mathsf{rs}} \{0,1\}^{\log(\mathsf{bn})}$ and compute $C \leftarrow \Pi_\rho(z \| e)$ as follows:*
    1. *Set $t \leftarrow 1$, $C_i \leftarrow 0^{\mathsf{bs}}$, $i \in [\mathsf{bn}]$.*
    2. **(Sensitive blocks)** *For $i \in [l]$, set $C_{r_i} \leftarrow (1 \| i \| z[i])$.*
    3. **(Ciphertext blocks)** *For $i \in [\mathsf{bn}]$, if $i \neq r_j$, $j \in [l]$, $C_i \leftarrow (0 \| e[t : t + (\mathsf{bs} - 1)])$, $t \leftarrow t + (\mathsf{bs} - 1)$.[14]*
  Output $C := (C_1 \| \ldots \| C_{\mathsf{bn}})$.
- $\mathsf{Dec}^*(1^k, \cdot)$: *on input C, parse it as $(C_1 \| \ldots \| C_{\mathsf{bn}})$, set $t \leftarrow 1$, $l \leftarrow 2m|sk|$, $z \leftarrow 0^l$, $e \leftarrow 0$, $\mathcal{L} = \emptyset$ and compute $(z \| e) \leftarrow \Pi^{-1}(C)$ as follows:*
  - *For $i \in [\mathsf{bn}]$,*
    * **(Sensitive block)** *If $C_i[1] = 1$, set $j \leftarrow C_i[2 : \mathsf{bs} - 1]$, $z[j] \leftarrow C_i[\mathsf{bs}]$, $\mathcal{L} \leftarrow \mathcal{L} \cup \{j\}$.*
    * **(Ciphertext block)** *Otherwise, set $e[t : t + \mathsf{bs} - 1] = C_i[2 : \mathsf{bs}]$, $t \leftarrow t + \mathsf{bs} - 1$.*
  - *If $|\mathcal{L}| \neq l$, output $\bot$, otherwise output $(z \| e)$.*

---

[14] Here we assume that $\mathsf{bs} - 1$, divides the length of the ciphertext e. We can always achieve this property by padding the message s with zeros, if necessary.

If $\Pi^{-1}(C) = \bot$, output $\bot$, otherwise, compute $(sk||sk') \leftarrow \mathsf{Rec}_m(z)$, and if $sk^3 = sk'$, output $\mathsf{D}_{sk}(e)$, otherwise output $\bot$.

The set of indices of the blocks in which z_i is stored will be denoted by Z_i.

We prove security for the above construction by a reduction to the security of Construction 3.1. We note that that our reduction is non-black box with respect to the coding scheme in which security is reduced to; a generic reduction, i.e., non-malleable reduction [2], from the standard model to the CRS model is an interesting open problem and thus out of the scope of this work.

In the following, we consider $\Gamma = \{0,1\}^{O(\log(k))}$. The straightforward way to prove that $(\mathsf{Enc}^*, \mathsf{Dec}^*)$ is secure against $\mathcal{F}_\Gamma^\alpha$ by a reduction to the security of the bit-wise code of Sect. 3, would be as follows: for any $\alpha \in \{0,1\}$, $f \in \mathcal{F}_\Gamma^\alpha$ and any message s, we have to define α', $g \in \mathcal{F}^{\alpha'}$, such that the output of the tampered execution with respect to $(\mathsf{Enc}^*, \mathsf{Dec}^*)$, f, s, is indistinguishable from the tampered execution with respect to $(\mathsf{Init}, \mathsf{Enc}, \mathsf{Dec})$, g, s, and g is an admissible function for $(\mathsf{Init}, \mathsf{Enc}, \mathsf{Dec})$. However, this approach might be tricky as it requires the establishment of a relation between α and α' such that the sensitive blocks that f will receive access to, will be simulated using the sensitive bits accessed by g. Our approach is cleaner: for the needs of the current proof we leverage the power of Construction 3.1, by allowing the attacker to choose adaptively the codeword locations, as long as it does not request to read all shares of the secret key. Then, for every block that is accessed by the block-wise attacker f, the bit-wise attacker g requests access to the locations of the bit-wise code that enable him to fully simulate the input to g. We formally present our ideas in the following sections. In Sect. 4.1 we introduce the function class $\mathcal{F}_{\mathsf{ad}}$ that considers adaptive adversaries with respect to the CRS and we prove security of Construction 3.1 in Corollary 4.3 against a subclass of $\mathcal{F}_{\mathsf{ad}}$, and then, we reduce the security of the block-wise code $(\mathsf{Enc}^*, \mathsf{Dec}^*)$ against $\mathcal{F}_\Gamma^\alpha$ to the security of Construction 3.1 against $\mathcal{F}_{\mathsf{ad}}$ (cf. Sect. 4.2).

4.1   Security Against Adaptive Adversaries

In the current section we prove that Construction 3.1 is secure against the class of functions that request access to the codeword adaptively, i.e., depending on the CRS, as long as they access a bounded number of sensitive bits. Below, we formally define the function class $\mathcal{F}_{\mathsf{ad}}$, in which the tampering function picks up the codeword locations depending on the CRS, and we consider $\Gamma = \{0,1\}$.

Definition 4.2 (The function class $\mathcal{F}_{\mathsf{ad}}^\nu$). *Let $(\mathsf{Init}, \mathsf{Enc}, \mathsf{Dec})$ be an (κ, ν)-coding scheme and let Σ be the range of $\mathsf{Init}(1^k)$. For any $g = (g_1, g_2) \in \mathcal{F}_{\mathsf{ad}}^\nu$, we have $g_1 : \Sigma \to \mathcal{P}([\nu])$, $g_2^\Sigma : \{0,1\}^{|\mathsf{range}(g_1)|} \to \{0,1\}^{|\mathsf{range}(g_1)|} \cup \{\bot\}$, and for any $c \in \{0,1\}^\nu$, $g^\Sigma(c) = g_2\left(c|_{g_1(\Sigma)}\right)$. For brevity, the function class will be denoted as $\mathcal{F}_{\mathsf{ad}}$.*

Construction 3.1 remains secure against functions that receive full access to the ciphertext, as well as they request to read all but one shares for each bit of

sk and sk^3. The result is formally presented in the following corollary and its proof, which is along the lines of the proof of Theorem 3.2, is given in the full version of the paper.

Corollary 4.3. *Let $k, m \in \mathbb{N}$. Assuming $(\mathsf{SS}_m, \mathsf{Rec}_m)$ is an m-out-of-m secret sharing scheme and $(\mathsf{KGen}, \mathsf{E}, \mathsf{D})$ is 1-IND-CPA secure authenticated encryption scheme, the code of Construction 4.1 is a MD-NMC against any $g = (g_1, g_2) \in \mathcal{F}_{\mathsf{ad}}$, assuming that for all $i \in [2|sk|]$, $(Z_i \cap g_1(\Sigma)) < m$, where $sk \leftarrow \mathsf{KGen}(1^k)$ and $\Sigma \leftarrow \mathsf{Init}(1^k)$.*

4.2   MD-NM Security of the Block-Wise Code

In the current section we prove security of Construction 4.1 against $\mathcal{F}_\Gamma^\alpha$, for $\Gamma = \{0,1\}^{O(\log(k))}$.

Theorem 4.4. *Let $k, m \in \mathbb{N}$, $\Gamma = \{0,1\}^{O(\log(k))}$ and $\alpha \in [0,1)$. Assuming $(\mathsf{SS}_m, \mathsf{Rec}_m)$ is an m-out-of-m secret sharing scheme and $(\mathsf{KGen}, \mathsf{E}, \mathsf{D})$ is a 1-IND-CPA secure authenticated encryption scheme, the code of Construction 4.1 is a MD-NMC against $\mathcal{F}_\Gamma^\alpha$, for any α, m, such that $(1 - \alpha)m = \omega(\log(k))$.*

$g_1(\Sigma = (r_1, \ldots, r_l))$:
- **(Simulate block shuffling)**:
  Sample $\rho := (\rho_1, \ldots, \rho_l) \xleftarrow{\text{rs}} \{0,1\}^{\log(\mathsf{bn})}$
- **(Construct I)**: Set $I = \emptyset$,
  * **(Add ciphertext locations to I)**:
    For $j \in [|e| + l]$, if $j \notin \{r_i | i \in [l]\}$, $I \leftarrow (I \cup j)$.
  * **(Add sensitive bit locations to I according to I_b)**:
    For $j \in [\mathsf{bn}]$, if $j \in I_b$ and $\exists i \in [l]$ such that $j = \rho_i$, $I \leftarrow (I \cup r_i)$.
- **Output**: Output I.

Fig. 4. The function g_1 that appears in the hybrid experiments of Fig. 7.

$g_2^\Sigma(c_{|_I})$:
  $t \leftarrow 1$, $C_i^* \leftarrow 0^{\mathsf{bs}}$, $i \in [\mathsf{bn}]$.
- **(Reconstruct I)**: Compute $I \leftarrow g_1(\Sigma)$.
- **(Simulate ciphertext blocks)**:
  For $i \in [\mathsf{bn}]$, if $i \neq \rho_j$, $j \in [l]$, $C_i^* \leftarrow (0||e[t : t + (\mathsf{bs} - 1)])$, $t \leftarrow t + (\mathsf{bs} - 1)$.
- **(Simulate sensitive blocks)**:
    * For $i \in [|I|]$, if $\exists j \in [l]$, such that $\mathsf{Ind}(c_{|_I}[i]) = r_j$, set $C_{\rho_j}^* \leftarrow \left(1||j||c_{|_I}[i]\right)$.
    * Set $C^* := (C_1^* || \ldots || C_{\mathsf{bn}}^*)$ and $\tilde{C}^* := C^*$.
- **(Apply f)**: compute $\tilde{C}^*[I_b] \leftarrow f(C_{|_{I_b}}^*)$.
- **(Output)**: Output $\tilde{C}_{|_{I_b}}^*$.

Fig. 5. The function g_2 that appears in the hybrid experiments of Fig. 7.

Proof. Following Lemma 2.6, we prove that for any $f \in \mathcal{F}_P^\alpha$, and any pair of messages s_0, s_1, $\mathsf{Tamper}_{s_0}^f \approx \mathsf{Tamper}_{s_1}^f$, and for any s, $\Pr\left[\mathsf{Tamper}_s^f \notin \{\bot, s\}\right] \leq \mathsf{negl}(k)$, where Tamper denotes the experiment defined in Lemma 2.6 with respect to the encoding scheme of Construction 4.1, $(\mathsf{Enc}^*, \mathsf{Dec}^*)$. Our proof is given by a series of hybrids depicted in Fig. 7. We reduce the security $(\mathsf{Enc}^*, \mathsf{Dec}^*)$, to the security of Construction 3.1, $(\mathsf{Init}, \mathsf{Enc}, \mathsf{Dec})$, against $\mathcal{F}_{\mathsf{ad}}$ (cf. Corollary 4.3). The idea is to move from the tampered execution with respect to $(\mathsf{Enc}^*, \mathsf{Dec}^*)$, f, to a tampered execution with respect to $(\mathsf{Init}, \mathsf{Enc}, \mathsf{Dec})$, g, such that the two executions are indistinguishable and $(\mathsf{Init}, \mathsf{Enc}, \mathsf{Dec})$ is secure against g.

Let I_b be the set of indices of the blocks that f chooses to tamper with, where $|I_b| \leq \alpha\nu$, and let $l \leftarrow 2m|sk|$, $\mathsf{bs} \leftarrow \log l + 2$, $\mathsf{bn} \leftarrow l + |e|/\mathsf{bs}$. Below we describe the hybrids of Fig. 7.

- $\mathsf{Exp}_0^{f,s}$: The current experiment is the experiment Tamper_s^f, of Lemma 2.6, with respect to $(\mathsf{Enc}^*, \mathsf{Dec}^*)$, f, s.
- $\mathsf{Exp}_1^{(g_1,g_2),s}$: The main difference between $\mathsf{Exp}_0^{f,s}$ and $\mathsf{Exp}_1^{(g_1,g_2),s}$, is that in the latter one, we introduce the tampering function (g_1, g_2), that operates over codewords of $(\mathsf{Init}, \mathsf{Enc}, \mathsf{Dec})$ and we modify the encoding steps so that the experiment creates codewords of the bit-wise code $(\mathsf{Init}, \mathsf{Enc}, \mathsf{Dec})$. (g_1, g_2) simulates partially the block-wise codeword C, while given partial access to the bit-wise codeword $c \leftarrow \mathsf{Enc}(s)$. As we prove in the full version, it simulates perfectly the tampering effect of f against $C \leftarrow \mathsf{Enc}^*(s)$.
  g_1 operates as follows (cf. Fig. 4): it simulates perfectly the randomness for the permutation of the block-wise code, denoted as ρ, and constructs a set of indices I, such that g_2 will receive access to, and tamper with, $c_{|_I}$. The set I is constructed with respect to the set of blocks I_b, that f chooses to read, as well as Σ, that reveals the original bit positions, i.e., the ones before permuting $(z||e)$. g_2 receives $c_{|_I}$, reconstructs I, simulates partially the blocks of the block-wise codeword, C, and applies f on the simulated codeword. The code of g_2 is given in Fig. 5. In the full version we show that g_2, given $c_{|_I}$, simulates perfectly $C_{|_{I_b}}$, which implies that $g_2^\Sigma(c_{|_I}) = f(C_{|_{I_b}})$, and the two executions are identical.
- $\mathsf{Exp}_2^{(g_1,g_3),s}$: In the current experiment, we substitute the function g_2 with g_3, and Dec^* with Dec, respectively. By inspecting the code of g_2 and g_3 (cf. Figs. 5 and 6, respectively), we observe that latter function executes the code of the former, plus the "Check labels and simulate $\tilde{c}[I]$" step. Thus the two experiments are identical up to the point of computing $f(C_{|_{I_b}})$. The main idea here is that we want the current execution to be with respect to $(\mathsf{Init}, \mathsf{Enc}, \mathsf{Dec})$ against (g_1, g_3). Thus, we substitute Dec^* with Dec, and we expand the function g_2 with some extra instructions/checks that are missing from Dec. We name the resulting function as g_3 and we prove that the two executions are identical.
- Finally, we prove that for any f and any s, $\mathsf{Exp}_2^{(g_1,g_3),s} \approx \mathsf{Exp}_2^{(g_1,g_3),0}$ and $\Pr\left[\mathsf{Exp}_2^{(g_1,g_3),s} \notin \{\bot, s\}\right] \leq \mathsf{negl}(k)$. We do so by proving that (g_1, g_3) is

admissible for $(\mathsf{Init}, \mathsf{Enc}, \mathsf{Dec},)$, i.e., $(g_1, g_3) \in \mathcal{F}_{\mathsf{ad}}$, and g_3 will not request to access more that $m - 1$ shares for each bit of sk, sk^3 (cf. Corollary 4.3). This implies security according to Lemma 2.6.

$g_3^{\Sigma}(c_{|_I})$:

$\quad t \leftarrow 1$, $C_i^* \leftarrow 0^{\mathsf{bs}}$, $i \in [\mathsf{bn}]$.
- **(Reconstruct I)**: Compute $I \leftarrow g_1(\Sigma)$.
- **(Simulate ciphertext blocks)**:
  For $i \in [\mathsf{bn}]$, if $i \neq \rho_j$, $j \in [l]$, $C_i^* \leftarrow (0||e[t : t + (\mathsf{bs} - 1)])$, $t \leftarrow t + (\mathsf{bs} - 1)$.
- **(Simulate sensitive blocks)**:
  * For $i \in [|I|]$, if $\exists j \in [l]$, such that $\mathsf{Ind}(c_{|_I}[i]) = r_j$, set $C_{\rho_j}^* \leftarrow \left(1||j||c_{|_I}[i]\right)$.
  * Set $C^* := (C_1^*||\ldots||C_{\mathsf{bn}}^*)$ and $\tilde{C}^* := C^*$.
- **(Apply f)**: compute $\tilde{C}^*[I_b] \leftarrow f(C_{|_{I_b}}^*)$.
- **(Check labels and simulate $\tilde{c}[I]$)**: If $\Pi^{-1}(\tilde{C}^*) = \bot$, set $d \leftarrow 1$, otherwise set $(\tilde{z}^*||\tilde{e}) \leftarrow \Pi^{-1}(\tilde{C}^*)$, $\tilde{c}^* \leftarrow P_{\Sigma}(\tilde{z}^*||\tilde{e})$.
- **(Output)**: If $d = 1$ output $\bot$, otherwise output $\tilde{c}_{|_I}^*$.

Fig. 6. The function g_3 that appears in the hybrid experiments of Fig. 7.

$\mathsf{Exp}_0^{f,s}$:
$sk \leftarrow \mathsf{KGen}(1^k)$, $e \leftarrow \mathsf{E}_{sk}(s)$
$z \leftarrow \mathsf{SS}_m(sk||sk^3)$

$\rho := (\rho_1, \ldots, \rho_l) \xleftarrow{\mathsf{n}} \{0,1\}^{\log(\mathsf{bn})}$
$C \leftarrow \Pi_\rho(z||e)$, $\tilde{C} \leftarrow C$
$\tilde{C}[I_b] \leftarrow f(C_{|_{I_b}})$

$\tilde{s} \leftarrow \mathsf{Dec}^*(\tilde{C})$

Output **same*** if $\tilde{s} = s$ and $\tilde{s}$ otherwise.

$\mathsf{Exp}_1^{(g_1,g_2),s}$:
$sk \leftarrow \mathsf{KGen}(1^k)$, $e \leftarrow \mathsf{E}_{sk}(s)$
$z \leftarrow \mathsf{SS}_m(sk||sk^3)$

$\Sigma \leftarrow \mathsf{Init}^*(1^k)$, $c \leftarrow P_{\Sigma}(z||e)$
$I \leftarrow g_1(\Sigma)$
$C \leftarrow \Pi_\rho(z||e)$, $\tilde{C} \leftarrow C$
$\tilde{C}[I_b] \leftarrow g_2^{\Sigma}(c_{|_I})$

$\tilde{s} \leftarrow \mathsf{Dec}^*(\tilde{C})$

Output **same*** if $\tilde{s} = s$ and $\tilde{s}$ otherwise.

$\mathsf{Exp}_2^{(g_1,g_3),s}$:
$\Sigma \leftarrow \mathsf{Init}^*(1^k)$
$sk \leftarrow \mathsf{KGen}(1^k)$, $e \leftarrow \mathsf{E}_{sk}(s)$
$z \leftarrow \mathsf{SS}_m(sk||sk^3)$

$c \leftarrow P_{\Sigma}(z||e)$, $\tilde{c} \leftarrow c$
$I \leftarrow g_1(\Sigma)$
$\tilde{c}[I] \leftarrow g_3^{\Sigma}(c_{|_I})$

$\tilde{s} \leftarrow \mathsf{Dec}(\Sigma, \tilde{c})$

Output **same*** if $\tilde{s} = s$ and $\tilde{s}$ otherwise.

Fig. 7. The hybrid experiments for the proof of Theorem 4.4.

The indistinguishability between the hybrids is given in the full version of the paper. ∎

5 Continuous MD-NMC with Light Updates

In this section we enhance the block-wise scheme of Sect. 4 with an update mechanism, that uses only shuffling and refreshing operations. The resulting code is secure against continuous attacks, for a notion of security that is weaker than the original one [30], as we need to update our codeword. Below we define the update mechanism, which is denoted as Update^*.

Construction 5.1. *Let k, $m \in \mathbb{N}$, $(\mathsf{KGen}, \mathsf{E}, \mathsf{D})$, $(\mathsf{SS}_m, \mathsf{Rec}_m)$ be as in Construction 4.1. We define the update procedure, Update^*, for the encoding scheme of Construction 4.1, as follows:*

- Update$^*(1^k, \cdot)$: *on input C, parse it as $(C_1||\ldots||C_{bn})$, set $l \leftarrow 2m|sk|$, $\hat{\mathcal{L}} = \emptyset$, and set $\hat{C} := (\hat{C}_1||\ldots||\hat{C}_{bn})$ to 0.*
  - **(Secret share $0^{2|sk|}$)**: *Sample $z \leftarrow SS_m\left(0^{2|sk|}\right)$, where $z = ||_{i=1}^{2|sk|} z_i$, $z \in \{0,1\}^{2m|sk|}$, and for $i \in [2|sk|]$, z_i is an m-out-of-m secret sharing of the 0 bit.*
  - **(Shuffle & Refresh)**: *Sample $\rho := (\rho_1, \ldots, \rho_l) \overset{\text{rs}}{\leftarrow} \{0,1\}^{\log(bn)}$. For $i \in [bn]$,*
    - **(Sensitive block)** *If $C_i[1] = 1$,*
      - **(Shuffle)**: *Set $j \leftarrow C_i[2 : bs - 1]$, $\hat{C}_{\rho_j} \leftarrow C_i$.*
      - **(Refresh)**: *Set $\hat{C}_{\rho_j}[bs] \leftarrow \hat{C}_{\rho_j}[bs] \oplus z[j]$.*
    - **(Ciphertext block)**
      If $C_i[1] = 0$, set $j \leftarrow \min_n \left\{ n \in [bn] \big| n \notin \hat{\mathcal{L}}, n \neq \rho_i, i \in [l] \right\}$, and $\hat{C}_j \leftarrow C_i$, $\hat{\mathcal{L}} \leftarrow \hat{\mathcal{L}} \cup \{j\}$.

Output $\hat{C}$.

The following definition of security is along the lines of the one given in [30], adapted to the notion of non-malleability with manipulation detection. Also, after each invocation the codewords are updated, where in our case the update mechanism is only using shuffling and refreshing operations. In addition, there is no need for self-destruct after detecting an invalid codeword [28].

Definition 5.2 (Continuously MD-NMC with light updates). *Let $CS = (Enc, Dec)$ be an encoding scheme, $\mathcal{F}$ be a functions class and $k, q \in \mathbb{N}$. Then, CS is a q-continuously non-malleable (q-CNM) code, if for every, sufficiently large $k \in \mathbb{N}$, any pair of messages $s_0, s_1 \in \{0,1\}^{\text{poly}(k)}$, and any PPT algorithm $\mathcal{A}$, $\left\{ \text{Tamper}_{s_0}^{\mathcal{A}}(k) \right\}_{k \in \mathbb{N}} \approx \left\{ \text{Tamper}_{s_1}^{\mathcal{A}}(k) \right\}_{k \in \mathbb{N}}$, where,*

$$
\begin{aligned}
&\text{Tamper}_s^{\mathcal{A}}(k): \\
&\quad C \leftarrow Enc(s), \tilde{s} \leftarrow 0 \\
&\quad \text{For } \tau \in [q]: \\
&\quad\quad f \leftarrow \mathcal{A}(\tilde{s}), \tilde{C} \leftarrow f(C), \tilde{s} \leftarrow Dec(\tilde{C}) \\
&\quad\quad \text{If } \tilde{s} = s: \tilde{s} \leftarrow \text{same}^* \\
&\quad\quad C \leftarrow \text{Update}^*(1^k, C) \\
&\quad out \leftarrow \mathcal{A}(\tilde{s}) \\
&\quad \textbf{Return}: out
\end{aligned}
$$

and for each round the output of the decoder is not in $\{s, \perp\}$ with negligible probability in k, over the randomness of $\text{Tamper}_s^{\mathcal{A}}$.

In the full version of the paper we prove the following statement.

Theorem 5.3. *Let $q, k, m, \in \mathbb{N}$, $\Gamma = \{0,1\}^{O(\log(k))}$ and $\alpha \in [0,1)$. Assuming (SS_m, Rec_m) is an m-out-of-m secret sharing scheme and $(KGen, E, D)$ is a 1-IND-CPA, authenticated encryption scheme, the scheme of Construction 5.1 is a continuously MD-NMC with light updates, against $\mathcal{F}_\Gamma^\alpha$, for any α, m, such that $(1 - \alpha)m = \omega(\log(k))$.*

In the above theorem, q can be polynomial (resp. exponential) in k, assuming the underlying encryption scheme is computationally (resp. unconditionally) secure.

References

1. Aggarwal, D., Agrawal, S., Gupta, D., Maji, H.K., Pandey, O., Prabhakaran, M.: Optimal computational split-state non-malleable codes. In: Kushilevitz, E., Malkin, T. (eds.) TCC 2016. LNCS, vol. 9563, pp. 393–417. Springer, Heidelberg (2016). https://doi.org/10.1007/978-3-662-49099-0_15
2. Aggarwal, D., Dodis, Y., Kazana, T., Obremski, M.: Non-malleable reductions and applications. In: STOC, pp. 459–468 (2015)
3. Aggarwal, D., Dodis, Y., Lovett, S.: Non-malleable codes from additive combinatorics. In: STOC, pp. 774–783 (2014)
4. Agrawal, S., Gupta, D., Maji, H.K., Pandey, O., Prabhakaran, M.: Explicit non-malleable codes against bit-wise tampering and permutations. In: Gennaro, R., Robshaw, M. (eds.) CRYPTO 2015. LNCS, vol. 9215, pp. 538–557. Springer, Heidelberg (2015). https://doi.org/10.1007/978-3-662-47989-6_26
5. Agrawal, S., Gupta, D., Maji, H.K., Pandey, O., Prabhakaran, M.: A rate-optimizing compiler for non-malleable codes against bit-wise tampering and permutations. In: Dodis, Y., Nielsen, J.B. (eds.) TCC 2015. LNCS, vol. 9014, pp. 375–397. Springer, Heidelberg (2015). https://doi.org/10.1007/978-3-662-46494-6_16
6. Ball, M., Dachman-Soled, D., Kulkarni, M., Malkin, T.: Non-malleable codes for bounded depth, bounded fan-in circuits. In: Fischlin, M., Coron, J.-S. (eds.) EUROCRYPT 2016. LNCS, vol. 9666, pp. 881–908. Springer, Heidelberg (2016). https://doi.org/10.1007/978-3-662-49896-5_31
7. Ball, M., Dachman-Soled, D., Kulkarni, M., Malkin, T.: Non-malleable codes from average-case hardness: AC^0, decision trees, and streaming space-bounded tampering. Cryptology ePrint Archive, Report 2017/1061 (2017)
8. Bao, F., Deng, R.H., Han, Y., Jeng, A., Narasimhalu, A.D., Ngair, T.: Breaking public key cryptosystems on tamper resistant devices in the presence of transient faults. In: Christianson, B., Crispo, B., Lomas, M., Roe, M. (eds.) Security Protocols 1997. LNCS, vol. 1361, pp. 115–124. Springer, Heidelberg (1998). https://doi.org/10.1007/BFb0028164
9. Bellare, M., Tessaro, S., Vardy, A.: Semantic security for the wiretap channel. In: Safavi-Naini, R., Canetti, R. (eds.) CRYPTO 2012. LNCS, vol. 7417, pp. 294–311. Springer, Heidelberg (2012). https://doi.org/10.1007/978-3-642-32009-5_18
10. Biham, E., Shamir, A.: Differential fault analysis of secret key cryptosystems. In: Kaliski, B.S. (ed.) CRYPTO 1997. LNCS, vol. 1294, pp. 513–525. Springer, Heidelberg (1997). https://doi.org/10.1007/BFb0052259
11. Boneh, D., DeMillo, R.A., Lipton, R.J.: On the importance of checking cryptographic protocols for faults. In: Fumy, W. (ed.) EUROCRYPT 1997. LNCS, vol. 1233, pp. 37–51. Springer, Heidelberg (1997). https://doi.org/10.1007/3-540-69053-0_4
12. Boneh, D., DeMillo, R.A., Lipton, R.J.: On the importance of eliminating errors in cryptographic computations. J. Cryptol. 14(2), 101–119 (2001)
13. Boyko, V.: On the security properties of OAEP as an all-or-nothing transform. In: Wiener, M. (ed.) CRYPTO 1999. LNCS, vol. 1666, pp. 503–518. Springer, Heidelberg (1999). https://doi.org/10.1007/3-540-48405-1_32

14. Canetti, R., Dodis, Y., Halevi, S., Kushilevitz, E., Sahai, A.: Exposure-resilient functions and all-or-nothing transforms. In: Preneel, B. (ed.) EUROCRYPT 2000. LNCS, vol. 1807, pp. 453–469. Springer, Heidelberg (2000). https://doi.org/10.1007/3-540-45539-6_33

15. Chandran, N., Goyal, V., Mukherjee, P., Pandey, O., Upadhyay, J.: Block-wise non-malleable codes. IACR Cryptology ePrint Archive, p. 129 (2015)

16. Chandran, N., Kanukurthi, B., Raghuraman, S.: Information-theoretic local non-malleable codes and their applications. In: Kushilevitz, E., Malkin, T. (eds.) TCC 2016-A. LNCS, vol. 9563, pp. 367–392. Springer, Heidelberg (2016). https://doi.org/10.1007/978-3-662-49099-0_14

17. Chattopadhyay, E., Zuckerman, D.: Non-malleable codes against constant split-state tampering. In: FOCS, pp. 306–315 (2014)

18. Cheraghchi, M., Guruswami, V.: Capacity of non-malleable codes. In: ITCS 2014 (2014)

19. Choi, S.G., Kiayias, A., Malkin, T.: BiTR: built-in tamper resilience. In: Lee, D.H., Wang, X. (eds.) ASIACRYPT 2011. LNCS, vol. 7073, pp. 740–758. Springer, Heidelberg (2011). https://doi.org/10.1007/978-3-642-25385-0_40

20. Coretti, S., Maurer, U., Tackmann, B., Venturi, D.: From single-bit to multi-bit public-key encryption via non-malleable codes. In: Dodis, Y., Nielsen, J.B. (eds.) TCC 2015. LNCS, vol. 9014, pp. 532–560. Springer, Heidelberg (2015). https://doi.org/10.1007/978-3-662-46494-6_22

21. Cramer, R., Dodis, Y., Fehr, S., Padró, C., Wichs, D.: Detection of algebraic manipulation with applications to robust secret sharing and fuzzy extractors. In: Smart, N. (ed.) EUROCRYPT 2008. LNCS, vol. 4965, pp. 471–488. Springer, Heidelberg (2008). https://doi.org/10.1007/978-3-540-78967-3_27

22. Dachman-Soled, D., Kulkarni, M., Shahverdi, A.: Locally decodable and updatable non-malleable codes in the bounded retrieval model. Cryptology ePrint Archive, Report 2017/303 (2017). http://eprint.iacr.org/2017/303

23. Dachman-Soled, D., Kulkarni, M., Shahverdi, A.: Tight upper and lower bounds for leakage-resilient, locally decodable and updatable non-malleable codes. In: Fehr, S. (ed.) PKC 2017. LNCS, vol. 10174, pp. 310–332. Springer, Heidelberg (2017). https://doi.org/10.1007/978-3-662-54365-8_13

24. Dachman-Soled, D., Liu, F.-H., Shi, E., Zhou, H.-S.: Locally decodable and updatable non-malleable codes and their applications. In: Dodis, Y., Nielsen, J.B. (eds.) TCC 2015. LNCS, vol. 9014, pp. 427–450. Springer, Heidelberg (2015). https://doi.org/10.1007/978-3-662-46494-6_18

25. Döttling, N., Nielsen, J.B., Obremski, M.: Information theoretic continuously non-malleable codes in the constant split-state model. Cryptology ePrint Archive, Report 2017/357 (2017). http://eprint.iacr.org/2017/357

26. Dziembowski, S., Kazana, T., Obremski, M.: Non-malleable codes from two-source extractors. In: Canetti, R., Garay, J.A. (eds.) CRYPTO 2013. LNCS, vol. 8043, pp. 239–257. Springer, Heidelberg (2013). https://doi.org/10.1007/978-3-642-40084-1_14

27. Dziembowski, S., Pietrzak, K., Wichs, D.: Non-malleable codes. In: ICS (2010)

28. Faonio, A., Nielsen, J.B.: Non-malleable codes with split-state refresh. In: Fehr, S. (ed.) PKC 2017. LNCS, vol. 10174, pp. 279–309. Springer, Heidelberg (2017). https://doi.org/10.1007/978-3-662-54365-8_12

29. Faust, S., Hostáková, K., Mukherjee, P., Venturi, D.: Non-malleable codes for space-bounded tampering. In: Katz, J., Shacham, H. (eds.) CRYPTO 2017. LNCS, vol. 10402, pp. 95–126. Springer, Cham (2017). https://doi.org/10.1007/978-3-319-63715-0_4

30. Faust, S., Mukherjee, P., Nielsen, J.B., Venturi, D.: Continuous non-malleable codes. In: Lindell, Y. (ed.) TCC 2014. LNCS, vol. 8349, pp. 465–488. Springer, Heidelberg (2014). https://doi.org/10.1007/978-3-642-54242-8_20

31. Faust, S., Mukherjee, P., Nielsen, J.B., Venturi, D.: A tamper and leakage resilient von neumann architecture. In: Katz, J. (ed.) PKC 2015. LNCS, vol. 9020, pp. 579–603. Springer, Heidelberg (2015). https://doi.org/10.1007/978-3-662-46447-2_26

32. Faust, S., Mukherjee, P., Venturi, D., Wichs, D.: Efficient non-malleable codes and key-derivation for poly-size tampering circuits. In: Nguyen, P.Q., Oswald, E. (eds.) EUROCRYPT 2014. LNCS, vol. 8441, pp. 111–128. Springer, Heidelberg (2014). https://doi.org/10.1007/978-3-642-55220-5_7

33. Genkin, D., Ishai, Y., Prabhakaran, M.M., Sahai, A., Tromer, E.: Circuits resilient to additive attacks with applications to secure computation. In: STOC 2014, pp. 495–504 (2014)

34. Jafargholi, Z., Wichs, D.: Tamper detection and continuous non-malleable codes. In: Dodis, Y., Nielsen, J.B. (eds.) TCC 2015. LNCS, vol. 9014, pp. 451–480. Springer, Heidelberg (2015). https://doi.org/10.1007/978-3-662-46494-6_19

35. Katz, J., Lindell, Y.: Introduction to Modern Cryptography (2007)

36. Kiayias, A., Liu, F.-H., Tselekounis, Y.: Practical non-malleable codes from l-more extractable hash functions. In: Proceedings of the 2016 ACM SIGSAC Conference on Computer and Communications Security, CCS 2016, pp. 1317–1328. ACM, New York (2016)

37. Liu, F.-H., Lysyanskaya, A.: Tamper and leakage resilience in the split-state model. In: Safavi-Naini, R., Canetti, R. (eds.) CRYPTO 2012. LNCS, vol. 7417, pp. 517–532. Springer, Heidelberg (2012). https://doi.org/10.1007/978-3-642-32009-5_30

38. Micali, S., Reyzin, L.: Physically observable cryptography. In: Naor, M. (ed.) TCC 2004. LNCS, vol. 2951, pp. 278–296. Springer, Heidelberg (2004). https://doi.org/10.1007/978-3-540-24638-1_16

39. Ozarow, L.H., Wyner, A.D.: Wire-tap channel II. AT&T Bell Lab. Tech. J. **63**(10), 2135–2157 (1984)

40. Resch, J.K., Plank, J.S.: AONT-RS: blending security and performance in dispersed storage systems. In: FAST 2011 (2011)

41. Rivest, R.L.: All-or-nothing encryption and the package transform. In: Biham, E. (ed.) FSE 1997. LNCS, vol. 1267, pp. 210–218. Springer, Heidelberg (1997). https://doi.org/10.1007/BFb0052348

42. Shaltiel, R., Silbak, J.: Explicit list-decodable codes with optimal rate for computationally bounded channels. In: APPROX/RANDOM 2016 (2016)

43. Stinson, D.R.: Something about all or nothing (transforms). Des. Codes Crypt. **22**(2), 133–138 (2001)

44. Tunstall, M., Mukhopadhyay, D., Ali, S.: Differential fault analysis of the advanced encryption standard using a single fault. In: Ardagna, C.A., Zhou, J. (eds.) WISTP 2011. LNCS, vol. 6633, pp. 224–233. Springer, Heidelberg (2011). https://doi.org/10.1007/978-3-642-21040-2_15

45. Wyner, A.D.: The wire-tap channel. Bell Syst. Tech. J. **54**(8), 1355–1387 (1975)

Continuously Non-Malleable Codes in the Split-State Model from Minimal Assumptions

Rafail Ostrovsky[1], Giuseppe Persiano[2], Daniele Venturi[3(✉)], and Ivan Visconti[4]

[1] Computer Science Department, University of California Los Angeles, Los Angeles, USA
[2] DISA-MIS, University of Salerno, Fisciano, Italy
[3] Computer Science Department, Sapienza University of Rome, Rome, Italy
venturi@di.uniroma1.it
[4] DIEM, University of Salerno, Fisciano, Italy

Abstract. At ICS 2010, Dziembowski, Pietrzak and Wichs introduced the notion of *non-malleable codes*, a weaker form of error-correcting codes guaranteeing that the decoding of a tampered codeword either corresponds to the original message or to an unrelated value. The last few years established non-malleable codes as one of the recently invented cryptographic primitives with the highest impact and potential, with very challenging open problems and applications.

In this work, we focus on so-called *continuously* non-malleable codes in the split-state model, as proposed by Faust *et al.* (TCC 2014), where a codeword is made of two shares and an adaptive adversary makes a polynomial number of attempts in order to tamper the target codeword, where each attempt is allowed to modify the two shares independently (yet arbitrarily). Achieving continuous non-malleability in the split-state model has been so far very hard. Indeed, the only known constructions require strong setup assumptions (i.e., the existence of a common reference string) and strong complexity-theoretic assumptions (i.e., the existence of non-interactive zero-knowledge proofs and collision-resistant hash functions).

As our main result, we construct a continuously non-malleable code in the split-state model without setup assumptions, requiring only one-to-one one-way functions (i.e., essentially optimal computational assumptions). Our result introduces several new ideas that make progress

Research supported in part by "GNCS - INdAM", FARB 300392FRB15VISCO, NSF grant 1619348, DARPA SafeWare subcontract to Galois Inc., DARPA SPAWAR contract N66001-15-1C-4065, US-Israel BSF grant 2012366, OKAWA Foundation Research Award, IBM Faculty Research Award, Xerox Faculty Research Award, B. John Garrick Foundation Award, Teradata Research Award, and Lockheed-Martin Corporation Research Award. The views expressed are those of the authors and do not reflect position of the Department of Defense or the U.S. Government. Work partially done while the second and fourth authors were visiting UCLA.

H. Shacham and A. Boldyreva (Eds.): CRYPTO 2018, LNCS 10993, pp. 608–639, 2018.
https://doi.org/10.1007/978-3-319-96878-0_21

towards understanding continuous non-malleability, and shows interesting connections with protocol-design and proof-approach techniques used
in other contexts (e.g., look-ahead simulation in zero-knowledge proofs,
non-malleable commitments, and leakage resilience).

Keywords: Continuously non-malleable codes · Split-state model
Minimal assumptions

1 Introduction

Dziembowski, Pietrzak and Wichs introduced the notion of a non-malleable code
(NMC) in [27]. Their new notion generated tremendous interest in recent years
both for the challenging theoretical questions raised by such codes, and for their
interesting applications in cryptography. An NMC is a key-less procedure that
allows to encode a message m in such a way that, upon input the encoding of
m, it is not possible (or it is hard in the computational case) to produce an
encoding of a value related to m.[1]

Obviously an NMC requires some restrictions on the view of the adversary.
Indeed, as the encoding/decoding are key-less procedures, an adversary could
always decode a codeword, change the underlying message to a related value, and
encode the result. For this reason, non-malleability is typically parameterized by
the set of allowed modifications Φ that can be applied by the adversary to a target
encoding, and previous work on NMCs focused on constructing non-malleable
codes for restricted (yet meaningful) classes Φ.

The split-state model. One of the most natural and investigated models is to
assume that a codeword c consists of two shares $c = (c_0, c_1)$, and that each
tampering attempt $\phi = (\phi_0, \phi_1) \in \Phi$ is characterized by two arbitrary functions
that can be applied to each share independently. Note that the two tampering
functions cannot run the decoding procedure, because both shares are needed in
order to decode a codeword, whereas each of the functions ϕ_0, ϕ_1 can access only
one share. This setting is often called the *split-state model* and is the focus of this
paper; we often use the terminology *split-state code* to denote a code in the split-
state model. We refer the reader to Sect. 1.5 for an overview of known constructions of non-malleable codes for different classes Φ. Previous work showed how
to construct split-state non-malleable codes, both for the information-theoretic
setting [3,4,6,7,13,18,25,27,37] and the computational setting [2,22,30,38].

1.1 Continuous Non-Malleability

The original notion of NMCs provides a security guarantee only against adversaries that try to tamper the codeword once. The more general case of *continuous*

---

[1] In this paper, we will only focus on *efficient* NMCs where both the encoding and
decoding procedures run in polynomial time.

non-malleability was introduced by Faust *et al.* [30], with the goal of guaranteeing non-malleability even after multiple (adaptively chosen) tampering attempts;
that is, the adversary is allowed to choose the tampering functions to apply in the
next round based on the answers obtained in the previous rounds. As pointed out
also in [30], continuously non-malleable codes (CNMCs) are arguably the most
natural generalization of standard NMCs, and allow to significantly strengthen
their applications [19, 20, 31].

Different flavors of non-malleability. The work of [27] considered a default and a
strong flavor of non-malleability. In both cases, the adversary is allowed to see the
decoding $\tilde{m}$ of the modified codeword $\tilde{c} = \phi(c)$. However, the default notion only
guarantees non-malleability as long as the decoded message is different from the
original message, i.e. it might be possible for the attacker to create an encoding
$\tilde{c} \neq c$ such that $\tilde{c}$ still decodes to the original message m. In contrast, this is not
allowed in the case of *strong* non-malleability which guarantees that whenever
$\tilde{c} \neq c$ the decoded value $\tilde{m}$ will be unrelated to m. An even stronger flavor,
known as *super* non-malleability [30, 32, 36], ensures that $\tilde{c}$ is independent of c
whenever $\tilde{c} \neq c$ is a valid codeword. This is modeled by allowing the adversary
to actually see $\tilde{c}$ (as long as $\tilde{c} \neq c$ and $\tilde{c}$ is valid).

Clearly, the above flavors of non-malleability can also be considered in the
continuous setting. In this paper, we focus only on the "default flavor" of continuous non-malleability. This in contrast to previous work on continuously non-
malleable codes (except [19, 20, 28]), which instead by default considered continuous super non-malleability. While the notion we consider is strictly weaker
than continuous strong or super non-malleability, to the best of our knowledge,
it is sufficient for all known applications of continuously non-malleable codes, in
particular [19, 20, 28, 30, 31].

Depending on the tampering functions being applied always to the initial
encoding c, or to the result of the previous tampering attempt, one can also
have notions called *non-persistent* and *persistent* tampering. In this paper we
focus on the setting of non-persistent tampering, which is the strongest[2] flavor
of continuous non-malleability (and also the variant most useful for applications).

Self destruction. Unfortunately, even for very simple classes Φ, continuous non-
malleability as hinted above is actually impossible to achieve. Indeed, a simple
attack—proposed for the first time by Gennaro *et al.* [33] in the context of
"Algorithmic Tamper-Proof Security"—allows to completely recover a target
encoding by simply trying to guess each of its bits individually: the output
of the decoding corresponding to each tampering attempt will yield either the
original message or the special symbol $\perp$ (denoting an invalid codeword), thus
revealing the entire codeword (and thus the underlying message) in a bit-wise
fashion. Remarkably, such an attack can be performed by looking at each bit of
the encoding independently, which is a special case of split-state tampering.

---

[2] In fact, note that a persistent continuous attack specified as a sequence of (deterministic) tampering functions $\phi, \phi', \phi'', \cdots$, can always be emulated by a non-persistent
continuous attack specified as $\phi(\cdot), \phi'(\phi(\cdot)), \phi''(\phi'(\phi(\cdot))), \cdots$.

The standard way out is to relax continuous non-malleability, therefore circumventing the above impossibility result, assuming a special "self-destruct" feature: After the first invalid encoding is processed, the system "blows-up" and stops processing further queries.[3]

Message uniqueness. It is not hard to show that any code achieving continuous non-malleability in the split-state model must satisfy a property called *message uniqueness.* Informally, message uniqueness means that if we fix the left share c_0 of an encoding, it should be hard to come up with two distinct right shares $c_1, \bar{c}_1$ such that both (c_0, c_1) and $(c_0, \bar{c}_1)$ are valid codewords decoding to two *distinct* messages, say m and $\bar{m}$ respectively.[4] (An analogous guarantee must hold in case we fix the right share.)

To see why uniqueness is needed, assume it is possible to efficiently find two encodings (c_0, c_1) and $(c_0, \bar{c}_1)$ violating message uniqueness, and let (c_0^*, c_1^*) be the target encoding that we want to maul via a split-state attack. Then, in a continuous attack, we can simply consider the tampering functions $(\phi_0^{(i)}, \phi_1^{(i)})$ that always fix the left share to c_0 (regardless of c_0^*) and, depending on the i-th bit of c_1^* either overwrite c_1^* with c_1 or with $\bar{c}_1$. The sequence of decoded messages produced by such an attack allows an adversary to recover c_1^* without the risk of incurring a self-destruct. After c_1^* is available, an additional tampering query easily allows to encode a related value.[5]

The state of the art: trusted setup and strong computational assumptions. The attack based on uniqueness implies that information-theoretic continuous non-malleability in the split-state model is impossible. This is because message uniqueness in the information-theoretic setting means that each share of a split-state encoding must completely determine the message. So, an unbounded tampering function accessing a single share of the codeword could just recover the underlying message by simply brute forcing all possible values for the missing share, and running the decoding algorithm until a valid message is found. Afterwards, it can complete the attack by setting (along with the other tampering function that hardwires correlated randomness and performs the same steps) an encoding of a related message.

The only known constructions of a CNMC in the split-state model (therefore also achieving message uniqueness) are the codes of [28,30], but unfortunately

---

[3] In practice self-destruct could be implemented using a single (untamperable) bit of public state, or by having the device overwrite its own memory in case of an invalid encoding.

[4] Since [30] by default considered continuous super non-malleability, they require an even stronger form of uniqueness called *codeword uniqueness,* which intuitively says that it should be hard to find $(c_0, c_1, \bar{c}_1)$ such that both (c_0, c_1) and $(c_0, \bar{c}_1)$ are valid, and $c_1 \neq \bar{c}_1$, even if the two codewords encode the same message. This flavor of uniqueness is not needed in this paper.

[5] Message uniqueness is, instead, not necessary for the simpler case of continuous non-malleability against *persistent* tampering. Split-state codes achieving such a weaker security guarantee were recently constructed unconditionally in [7].

these constructions rely on both trusted setup and strong computational assumptions. Indeed such codes require: (i) a "common reference string", i.e., the existence of a honestly generated string (with some given distribution) that is assumed to be untamperable, and that is available to the encoding and decoding functions, and to the adversary; (ii) the existence of non-interactive zero-knowledge proofs and either collision-resistant hash functions [30] or public-key encryption resilient to continual leakage [24,28] (which we only know how to obtain under concrete number-theoretic assumptions over bi-linear groups).

The open problem. Unfortunately, in practical situations, trusted setup is very difficult to come by (and also expensive to implement). Moreover, one should always try to get the best possible security, limiting or avoiding the trust on other parties, on some setup, and on strong computational assumptions. This leads to the following major open question:

Q1: *Can we construct a split-state CNMC under minimal complexity-theoretic assumptions in the plain model (i.e., without trusted setup)?*

Towards the above main question, one might also be interested in the following natural question:

Q2: *Is any split-state code satisfying both message uniqueness and one-time non-malleability also continuously non-malleable?*

1.2   Our Contribution

In this paper we give definitive answers to the above questions. Our main contribution is a positive answer to question **Q1**, therefore providing the first construction of a split-state CNMC (for non-persistent tampering) without assuming any trusted third party, or strong computational assumption. Indeed, we will show that the sole existence of one-to-one one-way functions already suffices for our purpose.

Theorem 1 (Informal). *If one-to-one one-way functions exists, there is a construction of a split-state code that satisfies continuous non-malleability in the* plain model.

In addition, we also give a negative answer to question **Q2**. In particular, we show that there exist (albeit contrived) split-state codes that are one-time non-malleable and satisfy message uniqueness, but can be broken by a simple continuous attack.

Theorem 2 (Informal). *If one-to-one one-way function exists, then there is a construction of a split-state code that satisfies both (perfect) message uniqueness and one-time non-malleability in the* plain model, *but that is insecure for two tampering queries.*

We notice that the computational assumption that we use is essentially optimal. In fact, each of the two shares of an encoding of a split-state non-malleable code satisfying message uniqueness implicitly defines a non-interactive commitment, and, as shown in [39], there is no black-box construction of a non-interactive commitment scheme from general one-way functions.

1.3 Positive Result

Our positive result introduces new ideas that make progress towards understanding continuous non-malleability, and shows interesting connections with protocol-design and proof-approach techniques used in other contexts (e.g., look-ahead simulation in zero-knowledge proofs [46], non-malleable commitments [43], and leakage resilience [26]). We highlight some of the challenges below.

Hardness of constructing one-time NMCs with message uniqueness. Before describing our encoding scheme, let us give some intuition why the problem of obtaining both non-malleability and message uniqueness in the plain model might be hard to tackle (even using non-standard assumptions). Let $c = (c_0, c_1)$ be a split-state codeword. Since we want to achieve message uniqueness, the left share must completely determine the encoded message; an analogous property must also hold for the right share. We can thus interpret each of the two shares produced by the encoding as a non-interactive perfectly[6] binding commitment. On the other hand, $c = (c_0, c_1)$ must also be non-malleable.

Now, consider the following natural candidate inspired by the recent construction of [12]. We let $c_0 = (\gamma_0, r_1)$ and $c_1 = (\gamma_1, r_0)$, where γ_0 and γ_1 are perfectly binding non-interactive non-malleable commitments of a message m, using randomness r_0 and r_1 (respectively). In the plain model, such commitments can be based on *adaptive* one-way functions [41], and, as shown by Pass [42], they cannot be constructed under falsifiable assumptions (in a black-box sense).

Although, at least intuitively, the above scheme should satisfy both properties of non-malleability and message uniqueness, we now argue that this might be very hard to prove. Recall that the experiment defining one-time non-malleability in the split-state model proceeds as follows: First the adversary chooses two messages m_0, m_1, and then it is allowed to specify a single pair of tampering functions $\phi = (\phi_0, \phi_1)$ that is applied to an encoding $c = (c_0, c_1)$ of m_b, for hidden bit b that the adversary needs to guess, upon which the attacker receives the decoded value corresponding to the tampered codeword. (Unless such value equals one of m_0, m_1, in which case the adversary obtains a special output same*.) Consider the following pair of split-state functions $\phi = (\phi_0, \phi_1)$.

– Function ϕ_0, by looking at γ_0 recovers some of the bits of r_0; function ϕ_1 acts similarly, i.e. it recovers some of the bits of r_1 by looking at γ_1.[7]

---

[6] It is easy to see that, in the plain model, computational uniqueness implies *perfect* uniqueness.

[7] The assumption that the tampering functions can recover some bits of the randomness, is justified by the fact that we do not know of any (even non-adaptive) one-way function that hides all the bits of its input.

- As a consequence, ϕ_0 and ϕ_1 have some shared randomness ρ (coming from the coins of the commitments). Let us now be generous, and further assume that functions ϕ_0, ϕ_1 can recover the encoded value m_b by looking at γ_0 and γ_1 (respectively).[8] Clearly, any reduction basing one-time non-malleability of the above scheme on the assumption that the commitments are non-malleable must in particular work for such a strong split-state attack.
- Finally, the functions (ϕ_0, ϕ_1) use the shared randomness ρ to coordinate as follows: with probability $1/2$ (with common coins derived from ρ) they replace (c_0, c_1) with an encoding $(\tilde{c}_0, \tilde{c}_1)$ of a value $\tilde{m}$ related to m_b (computed using common randomness derived from ρ), and with probability $1/2$ (again with common coins derived from ρ) they replace (c_0, c_1) with uncorrelated encodings (therefore decoding to $\perp$) of a value $\tilde{m}$ related to m_{1-b}. The decoded message will be related to m_b with probability $1/2$, and to m_{1-b} with probability $1/2$.

The above attack is clearly successful. Consider now the reduction that, given a target commitment γ, samples a random string r, runs $(\tilde{\gamma}, \tilde{r}) = \phi_0(\gamma, r)$, and returns $\tilde{\gamma}$ as mauled commitment. The advantage of such a reduction is zero, as both ϕ_0 and ϕ_1 return either a commitment to a message related to m_b (with probability $1/2$) or a commitment to a message related to m_{1-b} (still with probability $1/2$). It is, thus, not clear how such functions could help in breaking the non-malleability of the commitment scheme.

Message uniqueness and one-time non-malleability via commitments and leakage resilience. Our first idea is to circumvent the problem that the adversary might be able to coordinate ϕ_0 and ϕ_1 using common randomness coming from the commitment, by hiding such randomness. To this end, we make use of a (non-unique) primitive: an auxiliary split-state non-malleable code which encodes the message m concatenated with the randomness r used to compute the commitment. The reason why we can count on this non-unique tool is that in the security proof we can have a first hybrid experiment where we disconnect the randomness r of the commitment from the input of the auxiliary non-malleable code. Next, non-malleability follows by a reduction to the hiding property of the commitment scheme.

Remarkably, our proof works even if the underlying commitment is *malleable*; hence, we can instantiate our construction based on standard cryptographic assumptions, such as the existence of one-to-one one-way functions (which imply standard perfectly binding and computationally hiding non-interactive commitments [34]). Intuitively, the reason is that mauling the commitment does not help, since the message is also input to the auxiliary non-malleable code. The above trick is inspired by a beautiful idea of Pass and Rosen [44,45]. Indeed, they constructed non-malleable commitments by composing regular (i.e., potentially malleable) commitment schemes and non-malleable zero-knowledge arguments of knowledge. One can see our technique as one more (though completely different)

---

[8] Note that both functions recover the same value m_b, because the commitments are perfectly binding.

application of the Pass-Rosen trick. We stress that despite the common spirit of their and our technique, our construction has to deal with several difficulties that go much beyond the simple use of the above trick.

In order to reduce a successful attack to our code to the security of the inner auxiliary NMC, we need to use the tampering functions (ϕ_0, ϕ_1) chosen by the adversary to define the tampering functions (ϕ_0', ϕ_1') against the underlying code. This requires two adjustments: (i) the input to the functions (ϕ_0', ϕ_1') must be enriched adding a commitment; (ii) the output of the functions (ϕ_0, ϕ_1) must be shrunk removing the commitment. While the former adjustment is pretty straightforward (indeed it can be accomplished by just hardwiring a commitment and a description of ϕ_0, ϕ_1 in the description of ϕ_0', ϕ_1'), the latter is more complicated since we can't simply remove the commitment. In fact, the commitments produced by the tampering functions could play an important role for the success of the adversary! This issue will be resolved by additionally assuming that the inner NMC be a leakage-resilient NMC,[9] which allows us to obtain (via a leakage query) the modified commitment as generated by the tampering functions (ϕ_0, ϕ_1) chosen by the adversary. As we show, this leakage can be used by the distinguisher of the inner auxiliary NMC to simulate consistently the view of the distinguisher attacking the full code, thus reaching a contradiction.

The tough continuous case: we are short on leakage queries! The above technique consisting of using a leakage query to adjust the output of the distinguisher can be applied because the leaked information (i.e., a commitment) is small compared to the size of the codewords, and such a small leakage is tolerated by known constructions.

Consider now a continuous attack, where the adversary picks several tampering functions adaptively. A naive adaptation of the above trick would clearly result in too much leakage, since there is no a-priori fixed bound on the number of tampering queries made by the adversary, and each query requires to leak the corresponding modified commitment. Hence, the proof approach discussed so far fails in the case of a continuous attack.

We overcome this obstacle using two additional ideas: (i) A new proof strategy based on optimistic answers and rewinding simulation exploiting look-ahead threads, and (ii) a special leakage-resilient NMC with unconditional security.

Optimistic answers and simulation through look-ahead threads. Our proof strategy borrows the rewinding simulation used in zero-knowledge proofs, and combines it with optimistic answers in order to save on the overall amount of leakage queries. Recall that the main reason to use a leakage query is to obtain the modified commitments that are part of the tampered codewords produced by the tampering functions chosen by the adversary. Note that, once the commitments are leaked, they can also be decommitted via brute force search, since the goal

---

[9] Roughly, this means that the code remains non-malleable even given some bounded, independent (yet arbitrary), leakage on the two shares of a target encoding. See Sect. 3 for a precise definition. Suitable codes were recently constructed in [6].

is now to break *unconditional security* of the underlying leakage-resilient NMC, and therefore the reduction is allowed to run in exponential time.

In order to save on leakage queries, we simulate the answers to the adversary's tampering queries by using an optimistic approach, essentially returning the value that had more chances to be encoded by the tampering functions. Such a value can be computed through brute-force search, by applying the tampering functions to all possible encodings and returning the decoded message that appears more often. This sequence of "simulated" answers can be seen as a look-ahead thread [46], where the reduction tries to understand the correct answers to be played in the main thread of the interaction with the adversary. Indeed, when the adversary stops, the reduction will run a special leakage query in order to learn the first point j of the simulation where the optimistic answer was wrong, and the commitment γ that should have been considered instead. This information implies that all answers up to the j-th query were correct, therefore the reduction can complete the current lookahead thread, return to the main thread, simulate the answer to the first j queries as before, and decommit γ through brute-force search in order to answer the $(j + 1)$-th query. Next, the reduction starts another look-ahead thread, and so on, until all queries have been answered correctly. Through an induction argument, we will show that the reduction can successfully break the underlying one-time NMC by carefully adjusting the last pair of tampering functions chosen by the adversary.

The tricky bit is the following: How do we bound the number of look-ahead threads? Indeed, there is a leakage query for each look-ahead thread, and therefore without bounding the number of threads we can not contradict security of the underlying leakage-resilient NMC.

The small mutual information of [6]. The number of leakage queries is proportional to the number of look-ahead threads, and thus to the number of errors done by the reduction when giving optimistic answers. Hence, it is crucial to study the consequences of a wrong optimistic answer.

Whenever an optimistic answer is wrong, we have that the two tampering functions sent by the adversary modify the target codeword yielding a value that is not the most likely outcome. Intuitively, this means that for each look-ahead thread the adversary risks as decoding the special value $\perp$ (leading to self-destruct) with probability at least $1/2$. In fact, notice that if one tampering function sets a value that is not the most likely one, then with probability at least $1/2$ the other tampering function will set a different value, and therefore the decoding will return $\perp$. Clearly, if the adversary is risking $\perp$ with probability at least $1/2$, the number of such look-ahead threads is at most poly-logarithmic.

While the above argument is intuitively appealing, the difficulty is that the two tampering functions could coordinate their outputs using some correlated information encoded in their inputs. In such a case, they could produce two valid shares that encode a message which is different from the most likely outcome, and still the probability of self-destruct is less than $1/2$. We circumvent this complication by assuming two additional properties of the underlying one-time NMC, namely that the mutual information between the two shares of an encoding is not too high, and further that codewords are uniform over a subset of all possible encodings. Hence, we argue that any such tampering query (yielding a message that is different from the optimistic answer and incurring in a probability of self-destruct less than $1/2$) will cost one bit of correlated information, and thus, after a small number of such queries, the mutual information becomes zero and the probability of $\perp$ for each additional query is at least $1/2$. The latter allows our reduction to succeed.

Finally, we show that the code of [6] satisfies all the properties we need, and moreover, by carefully selecting the parameters, it tolerates enough leakage in order to apply our reduction.

1.4 Negative Result

Since in the split-state model continuous non-malleability implies message uniqueness, a natural question is whether the two properties are actually equivalent. We show that they are not equivalent in a very strong sense, indeed in Sect. 5 we describe a code that is one-time non-malleable and satisfies message uniqueness, but that is already insecure for 2 tampering queries. The scheme makes black-box use of any one-time non-malleable code in the split-state model additionally satisfying message uniqueness (such as our scheme from Sect. 4). The idea is to encode both the message m and some random pad κ using the underlying non-malleable code. Let us write (c_0^1, c_1^1) and (c_0^2, c_1^2) for the corresponding encodings. The obtained codeword has $c_0^* := (c_0^1, c_0^2, \delta)$ as left share, and $c_1^* := (c_1^1, c_1^2, \delta)$ as right share, where $\delta = m \oplus \kappa$ is a one-time pad encryption of the message m using pad κ. The decoding simply decodes the first component of each share (i.e., the pair (c_0^1, c_1^1)) using the decoding procedure of the underlying non-malleable code (completely ignoring all other elements).

On the one hand, one can show that the modified scheme inherits both message uniqueness and one-time non-malleability from the underlying auxiliary code. Intuitively, this is because successfully mauling a codeword (c_0^*, c_1^*) still requires to maul (c_0^1, c_1^1), which is hard by the one-time non-malleability of the underlying NMC. (We refer the reader to Sect. 5 for a detailed proof sketch.) On the other hand, using a first tampering query, a split-state adversary can swap c_0^1 with c_0^2 on the left, and c_1^1 with c_1^2 on the right, thus obtaining the random pad κ in the clear as a response. Once the pad is known, the second tampering query can hard-wire the value κ, recover the message $m = \delta \oplus \kappa$ in the clear (both from the left and the right share), and finally encode a related value.

1.5 Additional Related Work

Non-malleable codes. Only a few constructions of continuously non-malleable codes are known (besides the already mentioned constructions of [28,30]). In particular, continuous non-malleability is known to be achievable in the information-theoretic setting, for the simpler cases of bit-wise independent tampering [19,20] (where each bit of the codeword is tampered independently), and constant-state tampering [5]. Jafargholi and Wichs [36] obtain different flavors of continuous non-malleability for the case of tampering functions with high min-entropy or few fixed points. Aggarwal *et al.* [7] show that split-state continuous non-malleability is achievable in the information-theoretic setting, when tampering is persistent. Finally, Chattopadhyay *et al.* [14] construct *one-many non-malleable codes* that are secure with respect to an adversary that can specify many tampering functions to be applied to the one target codeword; the adversary succeeds if at least one of the tampering functions produces a valid encoding of a related message. Importantly, this notion does not rely on the self-destruct mechanism, but the total number of tampering attempts must be a-priori bounded.

Several other constructions of (one-time) non-malleable codes exist in the literature, achieving security for a plethora of tampering models, including: bit-wise independent tampering and permutations [8,9,18], circuits of polynomial size [17,27,32], constant-state tampering [16], block-wise tampering [12], space-bounded algorithms [11,29], and bounded-depth circuits [10,15].

Applications. The typical application of non-malleable codes is the protection of cryptographic algorithms from tampering attacks against the memory [27,30,38]. Non-malleable codes were also used to protect arbitrary computations (and not only storage) against tampering [13,22,31].

A recent line of work shows interesting connections between the notions of non-malleable codes and non-malleable commitments. In particular, [12] proves that block-wise non-malleable codes (for two blocks) are equivalent to non-interactive non-malleable commitments (w.r.t. opening). Recently Goyal *et al.* [35] showed how to construct 3-round non-malleable commitments from standard assumptions when the adversary plays left and right sessions in parallel. Their scheme crucially relies on the power of split-state non-malleable codes.

Non-malleable codes can also be used to tackle the question of domain extension for non-malleable public-key encryption [19,20,40] and non-malleable commitments [8].

2   Overview of Techniques

2.1   Description of Our Code

Our code $\Pi = (\mathsf{Enc}, \mathsf{Dec})$ is formally depicted in Fig. 1 on page 21, and it is based on a non-interactive commitment scheme with message space $\mathcal{M} := \{0,1\}^k$, randomness space $\mathcal{R} := \{0,1\}^\rho$ and commitment space $\Gamma \subseteq \{0,1\}^\ell$, and on an auxiliary split-state code $\Pi' = (\mathsf{Enc}', \mathsf{Dec}')$ mapping bitstrings of length $(k + \rho)$ into bitstrings of length $2n'$; the length of a codeword will be $2n = 2n' + 2\ell$. We denote by Commit the commitment function. (We refer the reader to Sect. 3 for the standard definitions of continuously non-malleable codes and non-interactive commitments.)

Intuitively, the encoding algorithm Enc constructs a commitment $\gamma \in \{0,1\}^\ell$ of the message $m \in \{0,1\}^k$ using randomness $r \in \{0,1\}^\rho$. Then it encodes the string $m\|r$ via Enc', obtaining (c_0', c_1'). Finally, it outputs the split-state encoding $((\gamma, c_0'), (\gamma, c_1'))$, of length $2n = 2n' + 2\ell$. The decoding algorithm first checks that the commitment γ on the left and right shares is equal, in which case it decodes (c'_0, c'_1) obtaining a value $m||r$, and outputs m if and only if (m, r) is a valid opening of the commitment.

For the security proof, we need the commitment scheme to be computationally hiding, and the underlying code Π' to be a split-state non-malleable code with unconditional security (under a single tampering query), and that additionally Π' satisfies leakage resilience, and two additional properties on the distribution of the codewords. The first property, which we call *codewords uniformity*, intuitively says that the two shares of an encoding under Π' are uniform over the set of all possible shares when considered in isolation, whereas their joint distribution is uniform over a smaller subset of the codewords space. The second property, which we call *conditional independence*, intuitively says that the mutual information between the left and right share is bounded. We show how to instantiate our construction in Sect. 4.

2.2 Proof Intuition

We next give an overview of the proof of non-malleability. Note that we do not make any assumption on the malleability of the commitment scheme. Let us write $\mathbf{T}(b, q)$ for the random variable corresponding to the tampering experiment defining continuous non-malleability of the above defined encoding scheme Π, with hidden bit b, and where the adversary asks q tampering queries. In this experiment, the adversary can adaptively choose up to q split-state tampering queries that are applied to a target encoding $c = ((\gamma, c'_0), (\gamma, c'_1))$ of message m_b; after each tampering query, the adversary learns the outcome corresponding to decoding the modified codeword. Importantly, both $\mathbf{T}(0, q)$ and $\mathbf{T}(1, q)$ are additionally parameterized by messages m_0, m_1, and moreover the output of the experiments is defined to be same^* in case the tampered codeword decodes to either of m_0, m_1; furthermore, in case the answer to a tampering query is equal to $\perp$ (i.e., the modified codeword is invalid), all future queries are answered with $\perp$ (i.e., the experiment self-destructs).

Our goal is to show $\mathbf{T}(0, q) \approx_c \mathbf{T}(1, q)$, for all polynomials $q(\lambda)$. The main idea is to consider a hybrid experiment $\mathbf{H}(b, q)$ where we decouple the randomness used to define the commitment in the target codeword from the input of the inner encoding scheme Π'. Namely, in experiment $\mathbf{H}(b, q)$ the target codeword has the form $c := ((\gamma, c'_0), (\gamma, c'_1))$ where γ is a commitment to m_b using randomness r (as before), and (c'_0, c'_1) is an encoding of a random uncorrelated value $s' \leftarrow_\$ \{0,1\}^{k+\rho}$ (instead of the string $m_b||r$). We then argue that $\mathbf{T}(0, q) \approx_s \mathbf{H}(0, q) \approx_c \mathbf{H}(1, q) \approx_s \mathbf{T}(1, q)$, as outlined in the following subsections.

2.3 First Step

We start by showing that $\mathbf{T}(b, q) \approx_s \mathbf{H}(b, q)$, for all $b \in \{0, 1\}$ and for all $q \in \mathrm{poly}(\lambda)$, down to the non-malleability of the underlying encoding scheme Π'. This part of the proof is completely information-theoretic, and moreover it relies on the two additional properties of codewords uniformity and conditional independence discussed above. Fix $b = 0$ (the proof for the other case is analogous). We use induction on the number of tampering queries $q(\lambda)$, as explained below.

Induction Basis. The base case of the induction requires to show that $\mathbf{T}(0, 1) \approx_s \mathbf{H}(0, 1)$. We consider a reduction having access to a target encoding $c' = (c'_0, c'_1)$ that is either an encoding of $s'_0 := m_0 || r$ or an encoding of a random string $s'_1 := s'$. Note that, since the reduction knows both m_0 and r, it can perfectly simulate the distribution of the target codeword $c = ((\gamma, c'_0), (\gamma, c'_1))$ for experiments $\mathbf{T}(0, 1)$ and $\mathbf{H}(0, 1)$ inside the tampering oracle; this is done by computing offline $\gamma = \mathsf{Commit}(m_0; r)$, and by hard-wiring this value into the tampering function.

Thus, the reduction can perfectly simulate the *input* for a tampering query as it would be done in $\mathbf{T}(0, 1)$ and $\mathbf{H}(0, 1)$. The difficulty, however, is that the reduction only gets to see the decoding of the value $\tilde{s}$ corresponding to the tampered codeword $\tilde{c}' = (\tilde{c}'_0, \tilde{c}'_1)$, which is not directly the same as the output of the experiment in $\mathbf{T}(0, 1)$ and $\mathbf{H}(0, 1)$. For instance, in case $\tilde{s} \notin \{\mathtt{same}^*, \bot, m_0 || \tilde{r}, m_1 || \tilde{r}\}$, for any $\tilde{r} \in \{0, 1\}^\rho$, the reduction knows that $\tilde{c}'$ is a valid encoding of some string $\tilde{s} := \tilde{m} || \tilde{r} \in \{0, 1\}^{k+\rho}$, but the output of experiment $\mathbf{T}(0, 1)$ and $\mathbf{H}(0, 1)$ is either equal to $\tilde{m}$ or $\bot$ depending on whether $\tilde{m}$ and $\tilde{r}$ are consistent with the modified commitment $\tilde{\gamma}$.

In order to overcome this obstacle, we exploit the leakage resilience property of Π'; in particular, we let the reduction leak the value $\tilde{\gamma}$ (as defined above). Our analysis shows that this is all one needs in order to complete the simulation in a perfect manner (with all but a negligible probability).

Inductive Step. Next, we assume that $\mathbf{T}(0, i) \approx_s \mathbf{H}(0, i)$ for some $i \in [q - 1]$, and we show that this implies $\mathbf{T}(0, i + 1) \approx_s \mathbf{H}(0, i + 1)$. This is achieved once again via a reduction to the underlying one-time non-malleable code. Notice that, as before, the reduction can perfectly simulate the distribution of the target codeword $c = ((\gamma, c'_0), (\gamma, c'_1))$ for experiments $\mathbf{T}(0, i + 1)$ and $\mathbf{H}(0, i + 1)$ inside the tampering oracle. However two new problems arise. First, in the experiments $\mathbf{T}(0, i + 1)$ and $\mathbf{H}(0, i + 1)$ the adversary can ask up to $i + 1$ tampering queries, whereas the reduction can play only one query, and so it needs to simulate the answer to all other i tampering queries on its own (and in a consistent manner). Second, even if the reduction were able to answer all other queries, it is a priori unclear how to choose which of the $i + 1$ tampering functions the reduction should use in order to break one-time non-malleability of the code Π'.

The solution to the second problem comes immediately from the induction hypothesis. In fact, we know that, with overwhelming probability, the adversary cannot be successful after just i queries, as this would contradict our assumption that $\mathbf{T}(0, i) \approx_s \mathbf{H}(0, i)$. Using this observation, our strategy will be to simulate the answer to the first i tampering queries in a consistent manner, and later rely on the $(i + 1)$-th query in order to violate security of the code Π'.

The solution to the first problem, instead, is more complicated. Essentially our reduction plays the following strategy:

1. At setup, compute all possible encodings $\hat{c} := (\hat{c}_0, \hat{c}_1)$ of the challenge messages $s'_0 = m_0 \| r$ and $s'_1 = s'$, and store $\hat{c}$ in an initially empty array $\hat{\mathcal{S}}^{(1)} := \hat{\mathcal{S}}_0^{(1)} \times \hat{\mathcal{S}}_1^{(1)}$, where $\hat{\mathcal{S}}_0^{(1)}$ and $\hat{\mathcal{S}}_1^{(1)}$ are the sub-arrays containing, respectively, all the left shares $\hat{c}_0$ and all the right shares $\hat{c}_1$.

2. Upon input a tampering query $(\phi_0^{(j)}, \phi_1^{(j)})$ from the adversary, for any $j \leq i$, answer as follows:
   - For all codewords $\hat{c} = (\hat{c}_0, \hat{c}_1) \in \hat{\mathcal{S}}_0^{(j)} \times \hat{\mathcal{S}}_1^{(j)}$, decode the corresponding tampered codeword $(\phi_0^{(j)}(\gamma, \hat{c}_0)), \phi_1^{(j)}(\gamma, \hat{c}_1))$.
   - Let m^* be the most likely outcome, and answer the query with $\tilde{m}^{(j)} = m^*$.
   - Define $\hat{\mathcal{S}}_0^{(j+1)}, \hat{\mathcal{S}}_1^{(j+1)}$ to be the sub-arrays of $\hat{\mathcal{S}}_0^{(j)}, \hat{\mathcal{S}}_1^{(j)}$ containing all possible codewords which are compatible with the answer to the j-th query being m^*.

3. Make sure all answers $\tilde{m}^{(1)}, \ldots, \tilde{m}^{(i)}$ are correct whenever the corresponding codewords produced by the tampering functions are valid. This is achieved as follows:
   - Define a leakage query that hardwires all answers $(\tilde{m}^{(1)}, \ldots, \tilde{m}^{(i)})$, as well as the tampering queries $(\phi_0^{(1)}, \ldots, \phi_0^{(i)})$ and the arrays $\hat{\mathcal{S}}_0^{(1)}, \ldots, \hat{\mathcal{S}}_0^{(i)}$, and returns the first index j (if any) such that the target left share (γ, c'_0) is contained in the j-th array, but is not contained in the $(j + 1)$-th array. (An analogous check is performed on the target right share (γ, c'_1), using $\phi_1^{(1)}, \ldots, \phi_1^{(i)}$ and $\hat{\mathcal{S}}_1^{(1)}, \ldots, \hat{\mathcal{S}}_1^{(i)}$.) In case such an index is found, the leakage query additionally returns the correct answer $\hat{m}$.[10]
   - Rewind the adversary to step 2, at the iteration where it asked the j-th query, and modify the answer using the leaked value. Additionally, update the arrays $\hat{\mathcal{S}}_0^{(j+1)}, \hat{\mathcal{S}}_1^{(j+1)}$ consistently[11] with the answer of the j-th tampering query being $\hat{m}$, and go back to step 2 continuing from the $(j + 1)$-th tampering query.

4. Upon input the final tampering query $(\phi_0^{(i)}, \phi_1^{(i)})$ from the adversary, use this query to define the tampering query (ϕ'_0, ϕ'_1) to be applied to the target encoding; this is done in exactly the same way as discussed above for the base case of the induction.

In order to conclude the proof, we need to show two things. First, we need to argue that the total number of rewinds performed by the reduction is somewhat limited, so that the reduction does not exceed the total leakage bound supported by the underlying non-malleable code. Second, we need to ensure that the simulation performed by the reduction generates a distribution that is

---

[10] This is achieved by leaking the commitment $\tilde{\gamma}$ corresponding to the tampering query $(\phi_0^{(j)}, \phi_1^{(j)})$, and by having the reduction find the corresponding (unique) message via brute force.

[11] The new $\hat{\mathcal{S}}_0^{(j+1)}, \hat{\mathcal{S}}_1^{(j+1)}$ are obtained from $\hat{\mathcal{S}}_0^{(j)}, \hat{\mathcal{S}}_1^{(j)}$ by removing the encodings that are not compatible with the answer of the j-th tampering query being $\hat{m}$.

indistinguishable from what the adversary would expect in a real execution of experiments $\mathbf{T}(0, i+1)$ and $\mathbf{H}(0, i+1)$. We deal with these issues as follows.

Challenge #1: Bounding the Leakage. Let $(\phi_0^{(j)}, \phi_1^{(j)})$ be a tampering query provoking one of the rewinds. Denote by $\tilde{c}_0 := (\tilde{\gamma}, \tilde{c}_0') = \phi_0^{(j)}(\gamma, c_0')$ the corresponding modified left share. By message uniqueness, which for our code easily follows by the perfect binding property of the commitment, $\tilde{c}_0$ is a valid left share of at most one message $\tilde{m} \in \{0,1\}^k$. A counting argument shows that the probability associated to the output of the decoding being $\tilde{m}$ is $\tilde{p} \leq 1/2$. Hence, intuitively, we would like to argue that since $\tilde{m}$ is not the most likely outcome, there is a probability of at least $1/2$ that the modified right share $\tilde{c}_1 := (\tilde{\gamma}, \tilde{c}_1') = \phi_1^{(j)}(\gamma, c_1')$ will correspond to a message different from $\tilde{m}$, and thus every such query yields a self-destruct with probability at least $1/2$.

Unfortunately, it is unclear how to complete the above argument using any one-time unconditionally secure non-malleable code. In fact, the left and right shares of the inner encoding $c' = (c_0', c_1')$ are correlated, and a tampering query could exploit such correlation in order to generate an output which is not the most likely outcome, and yet the probability of self-destruct is smaller than $1/2$. We solve this problem by relying on the two additional properties of codewords uniformity and conditional independence. In particular, by a careful information-theoretic argument, we can show that codewords uniformity implies that every tampering query evading the above argument decreases the mutual information between the left and right share of c' by at least one bit. By conditional independence, the maximum number of such queries is bounded, after which the mutual information between c_0' and c_1' is zero, and any further tampering query causing a rewind will incur a probability of self-destruct of at least $1/2$.

Challenge #2: Arguing indistinguishability. As for indistinguishability, note that the corrected answers $\tilde{m}^{(1)}, \ldots, \tilde{m}^{(i)}$ might still be inconsistent, due to the fact that the tampered inner codeword $(\tilde{c}_0', \tilde{c}_1')$ decodes to $\perp$ for some of the queries. Indeed, such an invalid codeword can not be detected using the above leakage queries since they allow only to read the commitments computed by the tampering function in the two shares. The adversary might notice this inconsistency, and could for instance instruct the distinguisher to flip its output in order to make the reduction fail.

We circumvent this obstacle as follows. First off, let us assume w.l.o.g. that the distinguisher satisfies the following invariant: it outputs 0 (resp. 1) whenever it believes the target codeword is an encoding of m_0 (resp. m_1). Hence, we let the reduction ask an additional leakage query, leaking a single bit, that hard-wires a description of the distinguisher and of the final tampering query $(\phi_0^{(i+1)}, \phi_1^{(i+1)})$, together with all the answers $\tilde{m}^{(1)}, \ldots, \tilde{m}^{(i)}$ to the first i queries, the commitment γ, and the final arrays $\hat{\mathcal{S}}_0^{(i+1)}, \hat{\mathcal{S}}_1^{(i+1)}$. The goal of the leakage query is to allow the reduction to check that the output of the distinguisher on the simulated view satisfies the above invariant. This is achieved as follows. For each

$\hat{c}_1 \in \hat{\mathcal{S}}_1^{(i+1)}$,[12] let $\hat{b} \in \{0, 1\}$ be such that $(c_0', \hat{c}_1')$ is an encoding of $m_{\hat{b}}$; the leakage query computes the answer $\tilde{m}^*$ to the $(i + 1)$-th tampering query by applying $(\phi_0^{(i+1)}, \phi_1^{(i+1)})$ to $((\gamma, c_0'), (\gamma, \hat{c}_1'))$, and then it returns $\hat{\delta} = 1$ iff the output of the distinguisher upon $(\tilde{m}^{(1)}, \ldots, \tilde{m}^{(i)}, \tilde{m}^*)$ is more often equal to $\hat{b}$.

Finally, in case $\hat{\delta} = 0$, the reduction returns a random guess, whereas if $\hat{\delta} = 1$, it uses the output of the distinguisher on the simulated view, the intuition being that the outcome of the distinguisher is used only if no inconsistency was introduced during the simulation of each tampering query. The proof shows that this allows us to keep the non-negligible advantage of the distinguisher, thus contradicting one-time unconditional non-malleability of the code Π'.

2.4 Second Step

In a second step we show that $\mathbf{H}(0, q) \approx_c \mathbf{H}(1, q)$, down to the hiding property of the commitment scheme. This step is significantly easier, because in both experiments $\mathbf{H}(0, q)$ and $\mathbf{H}(1, q)$ the input of the commitment and of the inner encoding algorithm are completely independent. Hence, in the reduction we can embed a target commitment γ (which is either a commitment to m_0 or a commitment to m_1) and complete the codeword by sampling a fresh encoding (c_0', c_1') of a random value $s' \in \{0, 1\}^{k+\rho}$. This way, we can easily turn a distinguisher between the two hybrids into an adversary breaking the hiding property of the commitment scheme.

Note that in this case the reduction can perfectly simulate the view of the distinguisher, as it has a perfectly distributed target codeword (either w.r.t. $\mathbf{H}(0, q)$ or w.r.t. $\mathbf{H}(1, q)$) "in its hands".

3 Preliminaries

3.1 Notation

For a string x, we denote its length by $|x|$; if $\mathcal{X}$ is a set, $|\mathcal{X}|$ represents the number of elements in $\mathcal{X}$. When x is chosen randomly in $\mathcal{X}$, we write $x \leftarrow_\$ \mathcal{X}$. When A is a randomized algorithm, we write $y \leftarrow_\$ \mathsf{A}(x)$ to denote a run of A on input x and output y; in this case, the value y is a random variable and $\mathsf{A}(x; r)$ denotes a run of A on input x and randomness r. An algorithm A is *probabilistic polynomial-time* (PPT) if A is randomized and for any input $x, r \in \{0, 1\}^*$ the computation of $\mathsf{A}(x; r)$ terminates in at most poly$(|x|)$ steps.

We denote with $\lambda \in \mathbb{N}$ the security parameter. A function $\nu : \mathbb{N} \to [0, 1]$ is negligible in the security parameter (or simply negligible) if it vanishes faster than the inverse of any polynomial in λ, i.e. $\nu(\lambda) = \lambda^{-\omega(1)}$. We sometimes write negl(λ) (resp., poly(λ)) to denote all negligible functions (resp., polynomial functions) in the security parameter. All algorithms are implicitly assumed to take the security parameter as input.

---

[12] Without loss of generality we describe the leakage function as a leakage query on the left share.

For a random variable $\mathbf{X}$, we write $\mathbb{P}[\mathbf{X} = x]$ for the probability that $\mathbf{X}$ takes on a particular value $x \in \mathcal{X}$ (with $\mathcal{X}$ the set where $\mathbf{X}$ is defined). The statistical distance between two random variables $\mathbf{X}$ and $\mathbf{X}'$ defined over the same set $\mathcal{X}$ is defined as $\Delta(\mathbf{X}; \mathbf{X}') = \frac{1}{2}\sum_{x \in \mathcal{X}} |\mathbb{P}[\mathbf{X} = x] - \mathbb{P}[\mathbf{X}' = x]|$. The mutual information between $\mathbf{X}$ and $\mathbf{Y}$ is a measure of their mutual dependence, and it is defined as $\mathbb{I}(\mathbf{X}; \mathbf{Y}) = \mathbb{H}(\mathbf{X}) - \mathbb{H}(\mathbf{X}|\mathbf{Y})$, where $\mathbb{H}(\cdot)$ denotes the Shannon's entropy.

Given two ensembles of random variables $\mathbf{X} = \{\mathbf{X}_\lambda\}_{\lambda \in \mathbb{N}}$ and $\mathbf{Y} = \{\mathbf{Y}_\lambda\}_{\lambda \in \mathbb{N}}$, we write $\mathbf{X} \equiv \mathbf{Y}$ to denote that the two ensembles are identically distributed, $\mathbf{X} \approx_s \mathbf{Y}$ to denote that the two ensembles are statistically close (i.e., $\Delta(\mathbf{X}_\lambda; \mathbf{Y}_\lambda) \in \mathsf{negl}(\lambda)$), and $\mathbf{X} \approx_c \mathbf{Y}$ to denote that the two ensembles are computationally indistinguishable (i.e., $|\mathbb{P}[\mathsf{D}(\mathbf{X}_\lambda) = 1] - \mathbb{P}[\mathsf{D}(\mathbf{Y}_\lambda) = 1]| \in \mathsf{negl}(\lambda)$ for all PPT distinguishers D).

3.2   Non-Malleable Codes

Introduced by Dziembowski, Pietrzak, and Wichs [27], non-malleable codes allow to encode a message in such a way that the decoding of a tampered codeword (according to a restricted class of modifications) either yields the original message or an unrelated value.

Definition 1 (Encoding scheme). *A (k, n)-code $\Pi = (\mathsf{Enc}, \mathsf{Dec})$ consists of a pair of algorithms specified as follows: (i) The (randomized) encoding algorithm Enc takes as input a string $s \in \{0,1\}^k$ and returns a codeword $c \in \{0,1\}^n$; (ii) The (deterministic) decoding algorithm Dec takes as input a codeword $c \in \{0,1\}^n$ and outputs a value in $\{0,1\}^k \cup \{\bot\}$, where $\bot$ denotes an invalid codeword. A codeword $c \in \{0,1\}^n$ such that $\mathsf{Dec}(c) \neq \bot$ is called a valid codeword.*

The code Π satisfies correctness if, for all $s \in \{0,1\}^k$, we have that $\mathsf{Dec}(\mathsf{Enc}(s)) = s$ with overwhelming probability over the randomness of the encoding algorithm.

Standard non-malleability, as defined in [27], allows an adversary to maul a target encoding only once. Continuous non-malleability [30] extends the basic non-malleability requirement by allowing the adversary to tamper multiple times, where tampering might either be non-persistent (i.e., the adversary always mauls the *same* target encoding) or persistent (i.e., the current tampering function is applied to the encoding resulting from the previous mauling attempt). Throughout this paper, we always assume that tampering is non-persistent (which is the more challenging scenario).

Split-state model. Below we recall the definition of continuous non-malleability in the so-called *split-state model*. Here a codeword $c \in \{0,1\}^{2n}$ consists of two *shares* $c_0 \in \{0,1\}^n$ and $c_1 \in \{0,1\}^n$.[13] We call such codes *split-state $(k, 2n)$-codes*. In the split-state model, the tampering functions $\phi : \{0,1\}^{2n} \to \{0,1\}^{2n}$ can be

---

[13] More generally, the encoding might not be symmetric in which case $c_0 \in \{0,1\}^{n_0}$ and $c_1 \in \{0,1\}^{n_1}$, for arbitrary values $n_0, n_1 \in \mathbb{N}$ such that $n_0 + n_1 = n$; while this generalization is immediate, it is not needed in this paper.

described as pairs $\phi := (\phi_0, \phi_1)$ of functions with $\phi_0, \phi_1 : \{0,1\}^n \to \{0,1\}^n$. Tampering function ϕ, when applied to codeword c modifies it into $\tilde{c} := \phi(c)$ defined as

$$\tilde{c} := (\phi_0(c_0), \phi_1(c_1)).$$

To define the notion of continuous non-malleability, we introduce experiment $\mathbf{Tamper}_{s_0,s_1}^{\Pi,\mathsf{A}}$ that is parameterized by a split-state code Π, by a PPT adversary A, and by two messages s_0 and s_1, and takes as inputs the security parameter λ, a bit b, and a value $q \in \mathbb{N}$. In this experiment the adversary has access to two leakage oracles $\mathcal{O}^\ell$, and one tampering oracle $\mathcal{O}_{\mathrm{cnm}}$.

Definition 2 (Leakage oracle). *A leakage oracle $\mathcal{O}^\ell$ is a stateful oracle that maintains a counter ct that is initially set to 0. When $\mathcal{O}^\ell$ is invoked for a string c and a leakage function ψ, value $\psi(c)$ is computed, its length is added to ct and if $\mathsf{ct} \le \ell$ then $\psi(c)$ is returned; otherwise, $\bot$ is returned.*

Definition 3 (Tampering oracle). *A tampering oracle $\mathcal{O}_{\mathrm{cnm}}^{s_0,s_1}$ is a stateful oracle (implicitly) parameterized by a split-state code $\Pi = (\mathsf{Enc}, \mathsf{Dec})$ and two strings s_0 and s_1, with state st initialized to $\mathsf{st} = \mathtt{Active}$. The oracle takes as input a codeword $c = (c_0, c_1)$ and a split-state tampering function $\phi = (\phi_0, \phi_1)$ and its output is defined as follows.*

> *Oracle $\mathcal{O}_{\mathrm{cnm}}^{s_0,s_1}((c_0, c_1), (\phi_0, \phi_1))$:*
> ___
> *If state $= \mathtt{SelfDestruct}$, return $\bot$*
> *Let $(\tilde{c}_0, \tilde{c}_1) := (\phi_0(c_0), \phi_1(c_1))$*
> *If $\tilde{s} = \mathsf{Dec}(\tilde{c}_0, \tilde{c}_1) \in \{s_0, s_1\}$, return $\mathtt{same}^*$*
> *If $\mathsf{Dec}(\tilde{c}_0, \tilde{c}_1) = \bot$, set state $= \mathtt{SelfDestruct}$ and return $\bot$*
> *Else, return $\tilde{s}$*

Definition 4 (Continuous non-malleability). *Let $\Pi = (\mathsf{Enc}, \mathsf{Dec})$ be a split-state $(k, 2n)$-code. We say that Π is ℓ-leakage-resilient q-time non-malleable in the split-state model if for all $s_0, s_1 \in \{0,1\}^k$ and for all PPT adversaries A asking at most q tampering queries, we have that*

$$\left\{\mathbf{Tamper}_{s_0,s_1}^{\Pi,\mathsf{A}}(\lambda, 0, q)\right\}_{\lambda \in \mathbb{N}} \approx_c \left\{\mathbf{Tamper}_{s_0,s_1}^{\Pi,\mathsf{A}}(\lambda, 1, q)\right\}_{\lambda \in \mathbb{N}}, \tag{1}$$

where, for $b \in \{0,1\}$,

$$\mathbf{Tamper}_{s_0,s_1}^{\Pi,\mathsf{A}}(\lambda, b, q) := \left\{ \begin{array}{c} (c_0, c_1) \leftarrow_\$ \mathsf{Enc}(s_b); \\ out_\mathsf{A} \leftarrow \mathsf{A}^{\mathcal{O}^\ell(c_0,\cdot), \mathcal{O}^\ell(c_1,\cdot), \mathcal{O}_{\mathrm{cnm}}^{s_0,s_1}((c_0,c_1),(\cdot,\cdot))}(1^\lambda) \end{array} \right\}.$$

Without loss of generality, we can assume that the value out_A consists of the adversary's view. In case Eq. (1) only holds for $q = 1$, we write that the encoding scheme is *one-time* non-malleable, whereas if Eq. (1) holds for an arbitrary polynomial $q(\cdot)$, we say that encoding scheme is *continuously* non-malleable; it is also worth noting that for $q = 0$ (i.e., no tampering allowed) the above notion collapses to the definition of leakage-resilient codes [23], which have turned useful in several constructions o non-malleable codes [30,32]. Also note that we can cast information-theoretic security by simply requiring that Eq. (1) holds for the statistical distance, for all possibly unbounded distinguishers, where now also the tampering functions ϕ specified by the adversary, as well as the leakage functions ψ, need not be polynomial-time computable.

As explained in the introduction, our definition of continuous non-malleability is strictly weaker than the one originally considered in [30] (and afterwards also in [32,36]), in that the adversary at the end of each tampering query only obtains the decoding of the tampered codeword (unless this happens to be equal to one of s_0, s_1), and not the tampered codeword itself (as long as $\tilde{c} \neq c$ is valid). We further observe that (continuous) non-malleability can also be stated through the existence of an efficient simulator, however the two formulations are equivalent for messages of super-polynomial size [27]. (This fact was proven for the case of one-time non-malleability, but it holds more generally for the case of continuous non-malleability and when considering leakage.)

Message uniqueness. As shown in [30] continuous non-malleability in the split-state model is impossible to achieve in the information-theoretic setting.[14] In the computational setting, in order to be continuously non-malleable, a split-state code must[15] satisfy a special property called *message uniqueness.* Informally, message uniqueness says that it should be hard to fix one part of an encoding, say $c_0 \in \{0,1\}^n$, and compute two *distinct* other parts $c_1, \bar{c}_1 \in \{0,1\}^n$ such that both (c_0, c_1) and $(c_0, \bar{c}_1)$ are *valid* encodings of two *different* messages.

Definition 5 (Message uniqueness). *Let $\Pi = (\mathsf{Enc}, \mathsf{Dec})$ be a split-state code. We say that Π satisfies* perfect message uniqueness *if, for all $\beta \in \{0,1\}$, there do not exist values $(c_\beta, c_{1-\beta}, \bar{c}_{1-\beta})$ such that $c_{1-\beta} \neq \bar{c}_{1-\beta}$ and, at the same time,*

$$\bot \neq \mathsf{Dec}(c_\beta, c_{1-\beta}) \neq \mathsf{Dec}(c_\beta, \bar{c}_{1-\beta}) \neq \bot.$$

Remark 1 (On perfect uniqueness). One could define a computational or statistical variant of the uniqueness property, where tuples of values violating message uniqueness exist but are hard to find. We note, however, that in the plain model assuming perfect message uniqueness is w.l.o.g. In fact, if a tuple $(c_\beta, c_{1-\beta}, \bar{c}_{1-\beta})$ violating message uniqueness exists (i.e., uniqueness is not perfect), we can always consider the specific PPT adversary that has such a tuple hard-wired in its code (and that thus contradicts computational and statistical message uniqueness).

Remark 2 (On message versus codeword uniqueness). An even stronger flavor of uniqueness, not needed in this paper and known as codeword uniqueness, requires that, for all $\beta \in \{0,1\}$, there do not exist values $(c_\beta, c_{1-\beta}, \bar{c}_{1-\beta})$ such that $c_{1-\beta} \neq \bar{c}_{1-\beta}$ and, at the same time, $\mathsf{Dec}(c_\beta, c_{1-\beta}) \neq \bot$ and $\mathsf{Dec}(c_\beta, \bar{c}_{1-\beta}) \neq \bot$,

---

[14] Information-theoretic security is, instead, possible in other settings, such as bit-wise independent tampering [19,20], constant-state tampering [5], and split-state *persistent* tampering [7].

[15] Otherwise a generic attack is possible; see Sect. 1 for an informal description.

but eventually $\mathsf{Dec}(c_\beta, c_{1-\beta}) = \mathsf{Dec}(c_\beta, \bar{c}_{1-\beta})$. Codeword uniqueness is a strictly stronger property than message uniqueness, and is known to be necessary for achieving the stronger flavor of continuous super non-malleability [30].

3.3 Non-Interactive Commitments

A non-interactive commitment scheme is a randomized efficient algorithm Commit taking as input a message $m \in \mathcal{M}$ and random coins $r \in \mathcal{R}$, and outputting a commitment $\gamma \in \Gamma$. A decommitment of γ consists simply of revealing m and r. The sets $\mathcal{M}$, $\mathcal{R}$ and Γ are called (respectively) the message space, the randomness space, and the commitment space. A commitment scheme satisfies two properties called *hiding* and *binding*. We recall such properties below.

The binding property says that it is hard to open a given commitment $\gamma \in \Gamma$ in two different ways. Exactly as for the case of uniqueness, the assumption of perfect binding is w.l.o.g. in the plain model.

Definition 6 (Binding). *We say that a non-interactive commitment* Commit *is perfectly binding if there do not exist pairs* $(m_0, r_0), (m_1, r_1)$ *such that* $m_0 \neq m_1$ *and, at the same time,* $\mathsf{Commit}(m_0; r_0) = \mathsf{Commit}(m_1; r_1)$.

The hiding property says that for any pair of messages m_0, m_1 it is hard to tell whether a given commitment γ is for m_0 or for m_1.

Definition 7 (Hiding). *We say that a non-interactive commitment* Commit *is computationally hiding if for all messages* $m_0, m_1 \in \mathcal{M}$ *the following holds:*

$$\left\{ \gamma : \ \gamma \leftarrow_\$ \mathsf{Commit}(1^\lambda, m_0) \right\}_{\lambda \in \mathbb{N}} \approx_c \left\{ \gamma : \ \gamma \leftarrow_\$ \mathsf{Commit}(1^\lambda, m_1) \right\}_{\lambda \in \mathbb{N}}.$$

4 Code Construction

In this section we present a construction of a split-state code that achieves continuous non-malleability. The scheme is in the plain model, and can be based on any (possibly malleable) non-interactive commitment scheme (cf. Sect. 3.3), and on an information-theoretic one-time non-malleable and leakage-resilient split-state code (cf. Sect. 3.2) satisfying a few additional properties (see below).

Note that the first assumption is necessary, meaning that a continuously non-malleable code in the split-state model implies a non-interactive commitment scheme. In fact, recall that any continuously non-malleable code must satisfy message uniqueness. Given a non-malleable split-state code $\Pi = (\mathsf{Enc}, \mathsf{Dec})$ with message uniqueness, consider the non-interactive commitment scheme where, in order to commit to message m, the committer computes a split-state encoding (c_0, c_1) of m using algorithm Enc. The left part c_0 constitutes the commitment, and the right part c_1 is the decommitment. The receiver verifies that c_1 is the correct opening of c_0 as m, by running Dec on input (c_0, c_1) and verifying that the output is indeed m. Binding follows by the fact that Π satisfies message uniqueness, and hiding follows by the non-malleability of Π.

4.1   Additional Properties

For our construction, we will rely on a split-state code meeting two non-standard requirements that we formally define below. The first property intuitively says that, for any message, the encoder outputs codewords that are uniformly random over some subset of all possible codewords.

Definition 8 (Codewords uniformity). *Let $\Pi = (\mathsf{Enc}, \mathsf{Dec})$ be a split-state $(k, 2n)$-code, and denote by $\mathbf{C} = (\mathbf{C}_0, \mathbf{C}_1)$ the random variable corresponding to the output of the encoding algorithm upon input some value $s \in \{0,1\}^k$. We say that Π satisfies* codewords uniformity *if, for all values $s \in \{0,1\}^k$, we have that each of $\mathbf{C}_0$ and $\mathbf{C}_1$ in isolation is uniform, respectively, over subsets $\mathcal{C}_0 \subseteq \{0,1\}^n$ and $\mathcal{C}_1 \subseteq \{0,1\}^n$, whereas $(\mathbf{C}_0, \mathbf{C}_1)$ is uniformly distributed over some subset $\overline{\mathcal{C}} := \overline{\mathcal{C}}_0 \times \overline{\mathcal{C}}_1 \subset \mathcal{C}_0 \times \mathcal{C}_1$.*

Let Commit be a non-interactive commitment scheme with message space $\mathcal{M} := \{0,1\}^k$, randomness space $\mathcal{R} := \{0,1\}^\rho$, and commitment space $\Gamma \subseteq \{0,1\}^\ell$. Let $\Pi' = (\mathsf{Enc}', \mathsf{Dec}')$ be a split-state $(k + \rho, 2n')$-code. Define the following split-state $(k, 2n)$-code, where $n := n' + \ell$.

Encoding: Upon input a value $m \in \{0,1\}^k$, sample random coins $r \leftarrow_\$ \{0,1\}^\rho$ and compute $\gamma := \mathsf{Commit}(m; r)$ and $(c_0', c_1') \leftarrow_\$ \mathsf{Enc}(m\|r)$. Return the codeword $c = (c_0, c_1) := ((\gamma, c_0'), (\gamma, c_1'))$.

Decoding: Upon input a codeword $c \in \{0,1\}^{2n}$, parse $c := (c_0, c_1) := ((\gamma_0, c_0'), (\gamma_1, c_1'))$. Hence, proceed as follows:
  (a) If $\gamma_0 \neq \gamma_1$, return $\bot$; else, let $\gamma = \gamma_0 = \gamma_1$.
  (b) Run $s = \mathsf{Dec}'(c_0', c_1')$; if $s = \bot$ return $\bot$.
  (c) Parse $s := m\|r$; if $\gamma = \mathsf{Commit}(m; r)$ return m, else return $\bot$.

Fig. 1. Description of our code.

The second property captures the fact that, for any message, the distribution of the left and right share of a codeword have limited dependence (in terms of their mutual information).

Definition 9 (Conditional independence). *Let $\Pi = (\mathsf{Enc}, \mathsf{Dec})$ be a split-state $(k, 2n)$-code, and denote by $\mathbf{C} = (\mathbf{C}_0, \mathbf{C}_1)$ the random variable corresponding to the output of the encoding algorithm upon input some value $s \in \{0,1\}^k$. We say that Π satisfies α-conditional independence if, for all values $s \in \{0,1\}^k$, we have that $\mathbb{I}(\mathbf{C}_0; \mathbf{C}_1) \leq \alpha$.*

4.2   Theorem Statement

Consider the split-state $(k, 2n)$-code $\Pi = (\mathsf{Enc}, \mathsf{Dec})$ depicted in Fig. 1, based on a non-interactive commitment scheme Commit with message space $\mathcal{M} := \{0,1\}^k$, randomness space $\mathcal{R} := \{0,1\}^\rho$ and commitment space $\Gamma \subseteq \{0,1\}^\ell$, and on an auxiliary split-state $(k+\rho, 2n')$-code $\Pi' = (\mathsf{Enc}', \mathsf{Dec}')$. The properties we require from each building block are directly stated in Theorem 3 below.

Intuitively, the encoding algorithm Enc constructs a commitment $\gamma \in \{0,1\}^\ell$ of the message $m \in \{0,1\}^k$ using randomness $r \in \{0,1\}^\rho$. Then it encodes the string $m\|r$ via Enc', obtaining (c_0', c_1'). Finally, it outputs the split-state encoding $((\gamma, c_0'), (\gamma, c_1'))$.

Theorem 3 (Theorem 1, restated). *Assume that* Commit *is a non-interactive perfectly binding and computationally hiding commitment scheme, with message space* $\mathcal{M} := \{0,1\}^k$, *randomness space* $\mathcal{R} := \{0,1\}^\rho$ *and commitment space* $\Gamma \subseteq \{0,1\}^\ell$. *Let* Π' *be a split-state* $(k + \rho, 2n')$-code *that is unconditionally* ℓ'-*leakage-resilient one-time non-malleable, for* $\ell' = (2\ell + O(\log \lambda)) \cdot (\alpha + O(\log \lambda))$, *and that additionally satisfies the properties of* codewords uniformity *and* α-conditional independence. *Then, the encoding scheme* Π *described in Fig. 1 is a split-state* $(k, 2(n' + \ell))$-code *satisfying continuous non-malleability.*

Instantiating the scheme. For the commitment scheme we can rely on the standard construction based on one-to-one one-way functions [34]. If the message m is k-bit long, the resulting commitment will have $\ell \in O(k^2)$ bits.

For the underlying non-malleable code we use a scheme constructed in [6], which we briefly recall below. Let $\mathbb{F}$ be a finite field. The encoder first encodes the underlying message $m' \in \{0,1\}^k$ using an auxiliary one-time split-state non-malleable code with unconditional security, obtaining shares $(c_0'', c_1'') \in [N] \times [N]$, where $[N]$ is a sparse subset of $\mathbb{F}$ with size $N \ll |\mathbb{F}|$. Hence, each share c_0'', c_1'' is processed using a slight variant of the inner-product extractor, i.e. c_0'' (resp. c_1'') is encoded via two additional shares $(c_{0,0}'', c_{0,1}'') \in \mathbb{F}^{2t}$ (resp. $(c_{1,0}'', c_{1,1}'') \in \mathbb{F}^{2t}$) such that $\xi(\langle c_{0,0}'', c_{0,1}'' \rangle) = c_0''$ (resp. $\xi(\langle c_{1,0}'', c_{1,1}'' \rangle) = c_1''$), where $\xi : \mathbb{F} \to [N]$ is an arbitrary bijection. The final encoding is then defined to be $c' = (c_0', c_1') = ((c_{0,0}'', c_{1,0}''), (c_{0,1}'', c_{1,1}'')) \in \mathbb{F}^{2t} \times \mathbb{F}^{2t}$.

By plugging in the above construction the split-state non-malleable code of [1,4], which has $\log N \in O(k^7)$, and choosing statistical error $\varepsilon := 2^{-k^2}$, we obtain a leakage-resilient one-time split-state non-malleable code with unconditional security and with leakage parameter $\ell' \approx k^{14}/12$ (cf. [6, Corollary 4.2]). It is important to note that the definition of leakage-resilient non-malleability considered in [6] is simulation based, and not indistinguishability based as our Definition 4. However, the former implies the latter: This was originally proven in [27] without considering leakage, but the same statement holds true, with basically the same proof, for the case of leakage, as long as the indistinguishability between the real and simulated experiment holds for the joint distribution of the leakage and the decoding of the tampered codeword. The latter requirement is fulfilled by the construction in [6], as the outer layer of their encoding is a split-state leakage-resilient code [23].

The above code is also easily seen to satisfy codewords uniformity (as c_0' and c_1' are uniform over the entire space of valid codewords when taken in isolation, whereas (c_0', c_1') is jointly uniform over a subset of the space of all valid codewords), and α-conditional independence, for $\alpha \in O(k^7)$ (as both $(c''_{0,0}, c''_{0,1})$ and $(c''_{1,0}, c''_{1,1})$ are uniform subject to their inner product being, respectively, c''_0 and c''_1, and moreover $|c''_0|, |c''_1| \in O(k^7)$). Hence, the leakage bound in Theorem 3 is satisfied too, as the required leakage is roughly $4k^9 + 2k^7 + 2k^2 \ll k^{14}/12$ (neglecting constant and logarithmic terms).

4.3   Security Analysis

For simplicity, let us define $\mathbf{T}_{m_0,m_1}(\lambda, b, q) \equiv \mathbf{Tamper}^{\Pi,A}_{m_0,m_1}(\lambda, b, q)$. We need to show that for all messages $m_0, m_1 \in \{0,1\}^k$ and for all PPT adversaries A asking $q(\lambda) \in \mathrm{poly}(\lambda)$ tampering queries, there exists a negligible function $\nu : \mathbb{N} \to [0,1]$ such that for all PPT distinguishers D:

$$|\mathbb{P}\left[D(\mathbf{T}_{m_0,m_1}(\lambda, 0, q)) = 1\right] - \mathbb{P}\left[D(\mathbf{T}_{m_0,m_1}(\lambda, 1, q)) = 1\right]| \leq \nu(\lambda).$$

Message Uniqueness. We start by showing that our code meets perfect message uniqueness.

Lemma 1. *The code of Fig. 1 satisfies perfect message uniqueness.*

Proof. Since our code is symmetric, it suffices to prove message uniqueness for the case $\beta = 0$. Assume there exist values $(c_0, c_1, \bar{c}_1)$ such that both (c_0, c_1) and $(c_0, \bar{c}_1)$ are *valid* codewords satisfying

$$\mathsf{Dec}(c_0, c_1) = m_0 \neq m_1 = \mathsf{Dec}(c_0, \bar{c}_1).$$

Write $c_0 = (\gamma, c'_0)$, $c_1 = (\gamma, c'_1)$, and $\bar{c}_1 = (\bar{\gamma}, \bar{c}'_1)$. By the fact that $(c_0, \bar{c}_1)$ is valid, it follows that $\bar{\gamma} = \gamma$. Let $s_0 := m_0 || r_0 = \mathsf{Dec}'(c'_0, c'_1)$ and $s_1 := m_1 || r_1 = \mathsf{Dec}'(c'_0, \bar{c}'_1)$ be obtained, respectively, by decoding (c'_0, c'_1) and $(c'_0, \bar{c}'_1)$. Note that both s_0 and s_1 are different from $\perp$, as (c_0, c_1) and $(c_0, \bar{c}_1)$ are valid codewords. We conclude that

$$\mathsf{Commit}(m_0; r_0) = \gamma = \mathsf{Commit}(m_1; r_1)$$

with $m_0 \neq m_1$, which contradicts the fact that Commit is perfectly binding.

First Hybrid Step. Consider the hybrid experiment $\mathbf{H}_{m_0,m_1}(\lambda, b, q)$ that is identical to $\mathbf{T}_{m_0,m_1}(\lambda, b, q)$, except that we let the auxiliary code $(\mathsf{Enc}', \mathsf{Dec}')$ encode a random string $s' \in \{0,1\}^{k+\rho}$ (instead of the string $m||r$). The experiment is described formally in Fig. 2.

We will now prove that, as long as the number of tampering queries is polynomial, the above hybrid experiment is statistically close to the original experiment. The proof is by induction on the number of tampering queries $q(\lambda) \in \mathrm{poly}(\lambda)$. The lemma below constitutes the induction basis.

Hybrid $\mathbf{H}_{m_0,m_1}(\lambda,b,q)$:

The experiment is parameterized by messages $m_0, m_1 \in \{0,1\}^k$, security parameter $\lambda \in \mathbb{N}$, a secret bit $b \in \{0,1\}$, and the number of tampering queries $q(\lambda) \in \text{poly}(\lambda)$. It proceeds as follows:

- It first computes $\gamma := \mathsf{Commit}(m_b; r)$, for random coins $r \leftarrow_{\$} \{0,1\}^\rho$, and then it sets $(c_0', c_1') \leftarrow_{\$} \mathsf{Enc}'(s')$ for random $s' \leftarrow_{\$} \{0,1\}^{k+\rho}$.
- The target encoding is defined to be $(c_0, c_1) := ((\gamma, c_0'), (\gamma, c_1'))$.
- Upon input the i-th tampering query $(\phi_0^{(i)}, \phi_1^{(i)})$, let $(\tilde{c}_0, \tilde{c}_1) = (\phi_0^{(i)}(c_0), \phi_1^{(i)}(c_1))$ be such that $\tilde{c}_0 := (\tilde{\gamma}_0, \tilde{c}_0')$ and $\tilde{c}_1 := (\tilde{\gamma}_1, \tilde{c}_1')$. Thus:
  (a) If $\tilde{\gamma}_0 \neq \tilde{\gamma}_1$, return $\perp$; else let $\tilde{\gamma} = \tilde{\gamma}_0 = \tilde{\gamma}_1$ and run $\tilde{s} = \mathsf{Dec}'(\tilde{c}_0', \tilde{c}_1')$.
  (b) If $\tilde{s} = \perp$, return $\perp$.
  (c) If $\tilde{s} = s'$ return $\mathtt{same}^*$ in case $\tilde{\gamma} = \gamma$, and $\perp$ otherwise.
  (d) Else, parse $\tilde{s} := \tilde{m} \| \tilde{r}$. If $\tilde{\gamma} \neq \mathsf{Commit}(\tilde{m}; \tilde{r})$, return $\perp$; otherwise, return $\mathtt{same}^*$ if $\tilde{m} \in \{m_0, m_1\}$, and else return $\tilde{m}$.

Fig. 2. Hybrid experiment in the proof of Theorem 3.

Lemma 2. *For all messages $m_0, m_1 \in \{0,1\}^k$, for all values $b \in \{0,1\}$, and for all unbounded adversaries* A, *we have that*

$$\{\mathbf{T}_{m_0,m_1}(\lambda,b,1)\}_{\lambda \in \mathbb{N}} \approx_s \{\mathbf{H}_{m_0,m_1}(\lambda,b,1)\}_{\lambda \in \mathbb{N}}.$$

Proof. We show the proof for the case $b = 0$, the proof for the other case being analogous. Assume that there exist a pair of messages $m_0, m_1 \in \{0,1\}^k$, an unbounded adversary A, an unbounded distinguisher D, and a polynomial $p(\cdot)$ such that, for infinitely many values of $\lambda \in \mathbb{N}$, we have

$$|\mathbb{P}[\mathsf{D}(\mathbf{T}_{m_0,m_1}(\lambda,0,1)) = 1] - \mathbb{P}[\mathsf{D}(\mathbf{H}_{m_0,m_1}(\lambda,0,1)) = 1]| \geq 1/p(\lambda).$$

Note that the probabilities in the above equation are taken over the random coin tosses of (A, D), over the choice of $r \leftarrow_{\$} \{0,1\}^\rho$ and $s' \leftarrow_{\$} \{0,1\}^{k+\rho}$, and over the randomness of algorithm Enc'. By an averaging argument, this means that there must exist at least two values $r \in \{0,1\}^\rho$ and $s' \in \{0,1\}^{k+\rho}$ such that the above equation holds when we fix these particular values of r and s'. We build an unbounded adversary A' and an unbounded distinguisher D' such that

$$\left|\mathbb{P}\left[\mathsf{D}'(\mathbf{T}'_{s_0',s_1'}(\lambda,0,1)) = 1\right] - \mathbb{P}\left[\mathsf{D}'(\mathbf{T}'_{s_0',s_1'}(\lambda,1,1)) = 1\right]\right| \geq 1/p(\lambda) - \nu(\lambda),$$

where $s_0' := m_0 \| r$ and $s_1' := s'$, $\nu(\lambda) \in \text{negl}(\lambda)$ is a negligible function, and where we wrote $\mathbf{T}'_{s_0',s_1'}(\lambda,b,1)$ as a shorthand for $\mathbf{Tamper}_{s_0',s_1'}^{\Pi',\mathsf{A}'}(\lambda,b,1)$. This will contradict the one-time unconditional non-malleability of $(\mathsf{Enc}', \mathsf{Dec}')$, and thus will conclude the proof of the lemma.

Let $c' := (c_0', c_1')$ be the target encoding in the tampering experiment relative to $(\mathsf{Enc}', \mathsf{Dec}')$. Here, c' is either an encoding of s_0' or an encoding of s_1'. Adversary A', on input $(1^\lambda, m_0, s_0', s_1')$, proceeds as follows:

- Parse $s_0' := m_0 || r$ and compute $\gamma := \mathsf{Commit}(m_0; r)$.
- Run $\mathsf{A}(1^\lambda)$, obtaining a pair of polynomial-time computable functions (ϕ_0, ϕ_1), where $\phi_0, \phi_1 : \{0,1\}^{n'+\ell} \to \{0,1\}^{n'+\ell}$.
- Define the polynomial-time computable leakage function ψ_0' (resp. ψ_1') that hardwires γ and ϕ_0 (resp. ϕ_1), and, upon input c_0' (resp. c_1') returns the value $\tilde{\gamma}_0$ (resp. $\tilde{\gamma}_1$) defined by $\phi_0(\gamma, c_0') := (\tilde{\gamma}_0, \tilde{c}_0')$ (resp. $\phi_1(\gamma, c_1') := (\tilde{\gamma}_1, \tilde{c}_1')$).
- Forward ψ_0' to $\mathcal{O}^\ell(c_0')$ and ψ_1' to $\mathcal{O}^\ell(c_1')$, obtaining values $\tilde{\gamma}_0, \tilde{\gamma}_1$.
- Define the polynomial-time computable tampering function ϕ_0' (resp. ϕ_1') that hardwires γ and ϕ_0 (resp. ϕ_1), and, upon input c_0' (resp. c_1'), returns the value $\tilde{c}_0'$ (resp. $\tilde{c}_1'$) defined by $\phi_0'(\gamma, c_0') := (\tilde{\gamma}_0, \tilde{c}_0')$ (resp. $\phi_1'(\gamma, c_1') := (\tilde{\gamma}_1, \tilde{c}_1')$).
- Forward (ϕ_0, ϕ_1) to $\mathcal{O}_{\mathsf{cnm}}^{s_0', s_1'}$, obtaining a value $\tilde{s} \in \{0,1\}^k \cup \{\bot, \mathsf{same}^*\}$.

Notice that attacker A' asks a single (split-state) leakage query yielding exactly 2ℓ bits, and a single (split-state) tampering query, as required. Distinguisher D', upon input $(1^\lambda, m_0, m_1, r, s')$, and upon receiving a pair $(\tilde{\gamma}_0, \tilde{\gamma}_1)$ in response of A's leakage query, and a value $\tilde{s} \in \{0,1\}^{k+\rho} \cup \{\mathsf{same}^*, \bot\}$ in response of A's tampering query, proceeds as follows.

- If $\tilde{s} = \bot$, return $\mathsf{D}(\bot)$.
- If $\tilde{s} = \mathsf{same}^*$:
  - In case $\tilde{\gamma}_0 = \tilde{\gamma}_1 = \mathsf{Commit}(m_0; r)$, return $\mathsf{D}(\mathsf{same}^*)$.
  - Else return $\mathsf{D}(\bot)$.
- If $\tilde{s} \notin \{\mathsf{same}^*, \bot\}$:
  - Parse $\tilde{s} := \tilde{m} || \tilde{r}$;
  - In case $\tilde{\gamma}_0 \neq \mathsf{Commit}(\tilde{m}; \tilde{r})$ or $\tilde{\gamma}_1 \neq \mathsf{Commit}(\tilde{m}; \tilde{r})$, return $\mathsf{D}(\bot)$;
  - In case $\tilde{m} \in \{m_0, m_1\}$ return $\mathsf{D}(\mathsf{same}^*)$;
  - Else return $\mathsf{D}(\tilde{m})$.

For the analysis, we next prove that the simulation performed by $(\mathsf{A}', \mathsf{D}')$ is perfect with overwhelming probability. First, depending on the target encoding (c_0', c_1') being either an encoding of s_0' or an encoding of s_1', the view of A's tampering functions is identical to the distribution of the target codeword in either experiment $\mathbf{T}_{m_0, m_1}(\lambda, 0, 1)$ or $\mathbf{H}_{m_0, m_1}(\lambda, 0, 1)$, with our fixed choice of r and s'. Second, the view of D is simulated correctly, with all but a negligible probability. Indeed:

- If $(\tilde{c}_0', \tilde{c}_1')$ yields $\bot$, both $\mathbf{T}_{m_0, m_1}(\lambda, 0, 1)$ and $\mathbf{H}_{m_0, m_1}(\lambda, 0, 1)$ would return $\bot$, which is perfectly emulated by the reduction.
- If $(\tilde{c}_0', \tilde{c}_1')$ yields same^*, it means that the inner codeword decodes to either $s_0' = m_0 || r$ or to $s_1' = s' := m' || r'$. Without loss of generality, assume further that the commitments in the tampered share satisfy $\tilde{\gamma}_0 = \tilde{\gamma}_1 := \tilde{\gamma}$. (In fact, if this is not the case, both experiments return $\bot$, which is once again perfectly emulated by the reduction.) There are 4 possible cases: either both experiments output s_0', or both experiments output s_1', or one experiment outputs s_0' while the other outputs s_1'. However, since the view in the real experiment is independent of the value m', we can condition on the event that the real experiment does not output s'. Thus, there are only two cases to consider:

(i) Experiment $\mathbf{T}_{m_0,m_1}(\lambda, 0, 1)$ and $\mathbf{H}_{m_0,m_1}(\lambda, 0, 1)$ both return $s'_0 = m_0 || r$.
(ii) Experiment $\mathbf{T}_{m_0,m_1}(\lambda, 0, 1)$ returns $s'_0 = m_0 || r$, but $\mathbf{H}_{m_0,m_1}(\lambda, 0, 1)$ returns $s'_1 = m' || r'$.
In both cases, the output of experiments $\mathbf{T}_{m_0,m_1}(\lambda, 0, 1)$ and $\mathbf{H}_{m_0,m_1}(\lambda, 0, 1)$ is equal to same* if $\tilde{\gamma} = \gamma$, and else both experiments return $\perp$. This is exactly what the reduction does. So, depending on the target codeword being either an encoding of s'_0 or an encoding of s'_1, the reduction simulates, except with negligible probability 2^{-k}, the outcome of either experiment $\mathbf{T}_{m_0,m_1}(\lambda, 0, 1)$ or $\mathbf{H}_{m_0,m_1}(\lambda, 0, 1)$.

- If $(\tilde{c}'_0, \tilde{c}'_1)$ yields some value $\tilde{s} = \tilde{m} || \tilde{r} \notin \{$same*$, \perp\}$, it means in particular that $\tilde{s} \notin \{s'_0, s'_1\}$. In such a case both experiments $\mathbf{T}_{m_0,m_1}(\lambda, 0, 1)$ and $\mathbf{H}_{m_0,m_1}(\lambda, 0, 1)$ would return $\perp$ in case the modified commitments $\tilde{\gamma}_0, \tilde{\gamma}_1$ do not match the opening $\tilde{m}, \tilde{r}$. Otherwise, it means that the modified codeword produced by A leads to a valid encoding of some message $\tilde{m} \in \{0, 1\}^k$. Hence, the output of both experiments would either be same* or $\tilde{m}$ (depending on $\tilde{m}$ being equal to one of the two messages m_0, m_1 or not).

To summarize, depending on the target encoding (c'_0, c'_1) being either an encoding of $s'_0 := m_0 || r$ or an encoding of $s'_1 := s'$, the view of (A, D) is identical, except with negligible probability, to the view in either experiment $\mathbf{T}_{m_0,m_1}(\lambda, 0, 1)$ or $\mathbf{H}_{m_0,m_1}(\lambda, 0, 1)$, for our fixed choice of r and s'. Thus, the advantage of $(\mathsf{A}', \mathsf{D}')$ is negligibly close to that of (A, D). This concludes the proof of the lemma.

The next lemma constitutes the inductive step. The proof appears in the full version.

Lemma 3. *Assume that for all messages $m_0, m_1 \in \{0, 1\}^k$, for all $b \in \{0, 1\}$, and for all unbounded adversaries A, it holds that*

$$\{\mathbf{T}_{m_0,m_1}(\lambda, b, i)\}_{\lambda \in \mathbb{N}} \approx_s \{\mathbf{H}_{m_0,m_1}(\lambda, b, i)\}_{\lambda \in \mathbb{N}},$$

where $i \in [q - 1]$ and $q \in \mathrm{poly}(\lambda)$. Then, for all messages $m_0, m_1 \in \{0, 1\}^k$, for all $b \in \{0, 1\}$, and for all unbounded adversaries A, we have that

$$\{\mathbf{T}_{m_0,m_1}(\lambda, b, i+1)\}_{\lambda \in \mathbb{N}} \approx_s \{\mathbf{H}_{m_0,m_1}(\lambda, b, i+1)\}_{\lambda \in \mathbb{N}},$$

By combining Lemma 2 and Lemma 3, we have shown that the hybrid experiment of Fig. 2 is statistically indistinguishable from the original tampering experiment:

Lemma 4. *For all messages $m_0, m_1 \in \{0, 1\}^k$, for all values $b \in \{0, 1\}$, for all $q(\lambda) \in \mathrm{poly}(\lambda)$, and for all unbounded adversaries A, we have that*

$$\{\mathbf{T}_{m_0,m_1}(\lambda, b, q)\}_{\lambda \in \mathbb{N}} \approx_s \{\mathbf{H}_{m_0,m_1}(\lambda, b, q)\}_{\lambda \in \mathbb{N}}.$$

Second Hybrid Step. Finally, we show that the view in experiment $\mathbf{H}_{m_0,m_1}(\lambda, b, q)$ is (computationally) independent of the hidden bit $b \in \{0, 1\}$. The proof appears in the full version.

Lemma 5. *For all messages* $m_0, m_1 \in \{0,1\}^k$, *for all* $q(\lambda) \in \text{poly}(\lambda)$, *and for all PPT adversaries* A, *we have that*

$$\{\mathbf{H}_{m_0,m_1}(\lambda, 0, q)\}_{\lambda \in \mathbb{N}} \approx_c \{\mathbf{H}_{m_0,m_1}(\lambda, 1, q)\}_{\lambda \in \mathbb{N}}.$$

Putting it Together. By combining Lemma 4 and Lemma 5, we obtain that for all $m_0, m_1 \in \{0,1\}^k$, for all $q(\lambda) \in \text{poly}(\lambda)$, and for all PPT adversaries A:

$$\{\mathbf{T}_{m_0,m_1}(\lambda, 0, q)\}_{\lambda \in \mathbb{N}} \approx_s \{\mathbf{H}_{m_0,m_1}(\lambda, 0, q)\}_{\lambda \in \mathbb{N}}$$
$$\approx_c \{\mathbf{H}_{m_0,m_1}(\lambda, 1, q)\}_{\lambda \in \mathbb{N}} \approx_s \{\mathbf{T}_{m_0,m_1}(\lambda, 1, q)\}_{\lambda \in \mathbb{N}},$$

which concludes the proof of the theorem.

5 Uniqueness $\not\Rightarrow$ Continuous Non-Malleability

As mentioned earlier, the property of message uniqueness is necessary for constructing continuously non-malleable codes in the split-state model. It is a natural question whether message uniqueness is also sufficient, namely any split-state code that satisfies message uniqueness and one-time non-malleability is also continuously non-malleable.

Here, we give a negative answer to the above question, by exhibiting a contrived split-state code that satisfies both message uniqueness and one-time non-malleability, but can be broken with a simple continuous attack. The constructed code makes black-box use of any split-state code satisfying both (perfect) message uniqueness and computational one-time non-malleability (as, e.g., our encoding scheme from Sect. 4). Our counter-example is "tight", in the sense that the attack breaking continuous non-malleability requires only two tampering queries.

The code. Consider the following split-state $(k, 4n+2k)$-code $\Pi^* = (\text{Enc}^*, \text{Dec}^*)$, based on an auxiliary split-state (k, n)-code $\Pi = (\text{Enc}, \text{Dec})$. The properties we require from each building block are directly stated in Theorem 4 below.

Encoding: Upon input a value $m \in \{0,1\}^k$, sample a random string $\kappa \leftarrow_s \{0,1\}^k$, compute $\delta := m \oplus \kappa$, and return the codeword

$$c^* = (c_0^*, c_1^*) := ((c_0^1, c_0^2, \delta), (c_1^1, c_1^2, \delta)), \tag{2}$$

where $(c_0^1, c_1^1) \leftarrow_s \text{Enc}(m)$ and $(c_0^2, c_1^2) \leftarrow_s \text{Enc}(\kappa)$.

Decoding: Upon input a codeword $c^* \in \{0,1\}^{4n+2k}$, parse $c^* := (c_0^*, c_1^*)$ as defined in Eq. (2) and return the same as $\text{Dec}(c_0^1, c_1^1)$.

Note that the decoding process simply decodes the first encoding (c_0^1, c_1^1) contained in c^*, completely ignoring the rest of the codeword.

Theorem 4 (Theorem 2, restated). *Assume that Π = (Enc, Dec) is a split-state (k, n)-code satisfying (perfect) message uniqueness and (computational) one-time non-malleability. Then the encoding scheme Π^* = (Enc*, Dec*) described above is a split-state $(k, 4n+2k)$-code meeting the following conditions:*

(i) Π^* satisfies (perfect) message uniqueness;
(ii) Π^* satisfies (computational) one-time non-malleability;
(iii) Π^* is not 2-non-malleable.

Proof overview. Before coming to the proof, let us discuss some intuition. The proof of Theorem 4 can be found in the full version. Here, we give the main intuition. The proof of property (i) follows almost directly by message uniqueness of Π. As for the proof of property (iii), it is sufficient to consider the tampering function that simply swaps c_0^1 with c_0^2 and c_1^1 with c_1^2. Note that the decoded message corresponding to such a query is equal to the value κ; hence, we can hard-wire κ in the second tampering query which allows to unmask the message computing $m = \delta \oplus \kappa$ and thus encode a related value.

To prove property (ii) we consider two hybrid experiments $\mathbf{H}_1^*$ and $\mathbf{H}_2^*$, and show that $\mathbf{T}_0^* \approx_c \mathbf{H}_1^* \approx_c \mathbf{H}_2^* \approx_c \mathbf{T}_1^*$ where $\mathbf{T}_b^*$ denotes the random variable corresponding to the non-malleability experiment with Π^* using hidden bit $b \in \{0,1\}$. Here, the difference between $\mathbf{T}_0^*$ and $\mathbf{H}_1^*$ is that in the latter we replace the codeword (c_0^1, c_1^1) with an encoding of m_1 (instead of m_0); in $\mathbf{H}_2^*$, instead, we change the distribution of δ to $\delta := m_1 \oplus \kappa$ and additionally we now let (c_0^2, c_1^2) be an encoding of $\kappa' := \kappa \oplus m_0 \oplus m_1$. To argue the indistinguishability of the hybrids, we then proceed as follows:

- In a first step we show that $\mathbf{T}_0^* \approx_c \mathbf{H}_1^*$, down to the non-malleability of the underlying encoding scheme Π. The reduction has access to a target codeword $c^1 = (c_0^1, c_1^1)$ that is either an encoding of m_0 or an encoding of m_1, and, given m_0, it can perfectly simulate the distribution of a target codeword for either experiment $\mathbf{T}_0^*$ or $\mathbf{H}_1^*$ inside the tampering oracle $\mathcal{O}_{\mathrm{cnm}}^{m_0, m_1}(c^1, \cdot)$. To do so, the reduction can sample offline a random κ, define $\delta = m_0 \oplus \kappa$, and set $c^2 := (c_0^2, c_1^2)$ to be an encoding of κ.
  Notice that the reduction gets to see the output of the decoding corresponding to the modified pair $(\tilde{c}_0^1, \tilde{c}_1^1)$, which is a perfect simulation for the output of either experiment $\mathbf{T}_0^*$ or $\mathbf{H}_1^*$.
- In a second step we show that $\mathbf{H}_1^* \equiv \mathbf{H}_2^*$; this is because if κ is random so is κ', and moreover $\kappa' \oplus m_0 = m_1 \oplus \kappa$; thus the two distributions are identical.
- In a third step we show that $\mathbf{H}_2^* \approx_c \mathbf{T}_1^*$, down to the non-malleability of the underlying encoding scheme Π. The reduction has access to a target codeword $c^2 = (c_0^2, c_1^2)$ that is either an encoding of $\kappa' := \kappa \oplus m_0 \oplus m_1$ or an encoding of κ, and, as before, it can perfectly simulate the distribution of the target codeword in either experiment $\mathbf{H}_2^*$ or $\mathbf{T}_1^*$ inside the tampering oracle $\mathcal{O}_{\mathrm{cnm}}^{\kappa, \kappa'}(c^2, \cdot)$. After computing the codeword $((\tilde{c}_0^1, \tilde{c}_0^2, \tilde{\delta}_0), (\tilde{c}_1^1, \tilde{c}_1^2, \tilde{\delta}_1))$, the tampering function defined by the reduction swaps $\tilde{c}_0^1$ with $\tilde{c}_0^2$ and $\tilde{c}_1^1$ with $\tilde{c}_1^2$; this way it obtains the decoding of $(\tilde{c}_0^1, \tilde{c}_1^1)$, which is what one needs in order to simulate the output of the two experiments.

An additional difficulty is that the experiment in which the reduction runs is parameterized by messages (κ, κ'), whereas the emulated experiments $\mathbf{H}_2^*$ or $\mathbf{T}_1^*$ are parameterized by (m_0, m_1). This means, for instance, that if the reduction obtains same* it cannot directly conclude that the simulated output should also be same*, but it needs to carefully adjust the received output in order to make the simulation go through.

6 Conclusion and Open Problems

We have shown a construction of a split-state continuously non-malleable code in the plain model. Our construction can be instantiated under the assumption that one-to-one one-way functions exist. Additionally, we have clarified that message uniqueness, albeit being necessary for obtaining continuous non-malleability in the split-state model, is not sufficient for constructing such codes.

Interesting open questions related to our work are, for instance, whether continuous non-malleability can be achieved, under minimal assumptions, together with additional properties, such as strong non-malleability [27], super-non-malleability [32], augmented non-malleability [2], and locality [13,21,22], or whether the rate of our code construction can be improved.

References

1. Aggarwal, D.: Affine-evasive sets modulo a prime. Inf. Process. Lett. **115**(2), 382–385 (2015)
2. Aggarwal, D., Agrawal, S., Gupta, D., Maji, H.K., Pandey, O., Prabhakaran, M.: Optimal computational split-state non-malleable codes. In: Kushilevitz, E., Malkin, T. (eds.) TCC 2016. LNCS, vol. 9563, pp. 393–417. Springer, Heidelberg (2016). https://doi.org/10.1007/978-3-662-49099-0_15
3. Aggarwal, D., Dodis, Y., Kazana, T., Obremski, M.: Non-malleable reductions and applications. In: ACM STOC, pp. 459–468 (2015)
4. Aggarwal, D., Dodis, Y., Lovett, S.: Non-malleable codes from additive combinatorics. In: ACM STOC, pp. 774–783 (2014)
5. Aggarwal, D., Dottling, N., Nielsen, J.B., Obremski, M., Purwanto, E.: Continuous non-malleable codes in the 8-split-state model. Cryptology ePrint Archive, Report 2017/357 (2017). https://eprint.iacr.org/2017/357
6. Aggarwal, D., Dziembowski, S., Kazana, T., Obremski, M.: Leakage-resilient non-malleable codes. In: Dodis, Y., Nielsen, J.B. (eds.) TCC 2015. LNCS, vol. 9014, pp. 398–426. Springer, Heidelberg (2015). https://doi.org/10.1007/978-3-662-46494-6_17
7. Aggarwal, D., Kazana, T., Obremski, M.: Inception makes non-malleable codes stronger. In: Kalai, Y., Reyzin, L. (eds.) TCC 2017. LNCS, vol. 10678, pp. 319–343. Springer, Cham (2017). https://doi.org/10.1007/978-3-319-70503-3_10
8. Agrawal, S., Gupta, D., Maji, H.K., Pandey, O., Prabhakaran, M.: Explicit non-malleable codes against bit-wise tampering and permutations. In: Gennaro, R., Robshaw, M. (eds.) CRYPTO 2015. LNCS, vol. 9215, pp. 538–557. Springer, Heidelberg (2015). https://doi.org/10.1007/978-3-662-47989-6_26

9. Agrawal, S., Gupta, D., Maji, H.K., Pandey, O., Prabhakaran, M.: A rate-optimizing compiler for non-malleable codes against bit-wise tampering and permutations. In: Dodis, Y., Nielsen, J.B. (eds.) TCC 2015. LNCS, vol. 9014, pp. 375–397. Springer, Heidelberg (2015). https://doi.org/10.1007/978-3-662-46494-6_16

10. Ball, M., Dachman-Soled, D., Kulkarni, M., Malkin, T.: Non-malleable codes for bounded depth, bounded fan-in circuits. In: Fischlin, M., Coron, J.-S. (eds.) EUROCRYPT 2016. LNCS, vol. 9666, pp. 881–908. Springer, Heidelberg (2016). https://doi.org/10.1007/978-3-662-49896-5_31

11. Ball, M., Dachman-Soled, D., Kulkarni, M., Malkin, T.: Non-malleable codes from average-case hardness: AC^0, decision trees, and streaming space-bounded tampering. In: Nielsen, J.B., Rijmen, V. (eds.) EUROCRYPT 2018. LNCS, vol. 10822, pp. 618–650. Springer, Cham (2018). https://doi.org/10.1007/978-3-319-78372-7_20

12. Chandran, N., Goyal, V., Mukherjee, P., Pandey, O., Upadhyay, J.: Block-wise non-malleable codes. In: ICALP, pp. 31:1–31:14 (2016)

13. Chandran, N., Kanukurthi, B., Raghuraman, S.: Information-theoretic local non-malleable codes and their applications. In: Kushilevitz, E., Malkin, T. (eds.) TCC 2016. LNCS, vol. 9563, pp. 367–392. Springer, Heidelberg (2016). https://doi.org/10.1007/978-3-662-49099-0_14

14. Chattopadhyay, E., Goyal, V., Li, X.: Non-malleable extractors and codes, with their many tampered extensions. In: ACM STOC, pp. 285–298 (2016)

15. Chattopadhyay, E., Li, X.: Non-malleable codes and extractors for small-depth circuits, and affine functions. In: ACM STOC, pp. 1171–1184 (2017)

16. Chattopadhyay, E., Zuckerman, D.: Non-malleable codes against constant split-state tampering. In: IEEE FOCS, pp. 306–315 (2014)

17. Cheraghchi, M., Guruswami, V.: Capacity of non-malleable codes. In: Innovations in Theoretical Computer Science, pp. 155–168 (2014)

18. Cheraghchi, M., Guruswami, V.: Non-malleable coding against bit-wise and split-state tampering. In: Lindell, Y. (ed.) TCC 2014. LNCS, vol. 8349, pp. 440–464. Springer, Heidelberg (2014). https://doi.org/10.1007/978-3-642-54242-8_19

19. Coretti, S., Dodis, Y., Tackmann, B., Venturi, D.: Non-malleable encryption: simpler, shorter, stronger. In: Kushilevitz, E., Malkin, T. (eds.) TCC 2016. LNCS, vol. 9562, pp. 306–335. Springer, Heidelberg (2016). https://doi.org/10.1007/978-3-662-49096-9_13

20. Coretti, S., Maurer, U., Tackmann, B., Venturi, D.: From single-bit to multi-bit public-key encryption via non-malleable codes. In: Dodis, Y., Nielsen, J.B. (eds.) TCC 2015. LNCS, vol. 9014, pp. 532–560. Springer, Heidelberg (2015). https://doi.org/10.1007/978-3-662-46494-6_22

21. Dachman-Soled, D., Kulkarni, M., Shahverdi, A.: Tight upper and lower bounds for leakage-resilient, locally decodable and updatable non-malleable codes. In: Fehr, S. (ed.) PKC 2017. LNCS, vol. 10174, pp. 310–332. Springer, Heidelberg (2017). https://doi.org/10.1007/978-3-662-54365-8_13

22. Dachman-Soled, D., Liu, F.-H., Shi, E., Zhou, H.-S.: Locally decodable and updatable non-malleable codes and their applications. In: Dodis, Y., Nielsen, J.B. (eds.) TCC 2015. LNCS, vol. 9014, pp. 427–450. Springer, Heidelberg (2015). https://doi.org/10.1007/978-3-662-46494-6_18

23. Davì, F., Dziembowski, S., Venturi, D.: Leakage-resilient storage. In: Garay, J.A., De Prisco, R. (eds.) SCN 2010. LNCS, vol. 6280, pp. 121–137. Springer, Heidelberg (2010). https://doi.org/10.1007/978-3-642-15317-4_9

24. Dodis, Y., Lewko, A.B., Waters, B., Wichs, D.: Storing secrets on continually leaky devices. In: IEEE FOCS, pp. 688–697 (2011)

25. Dziembowski, S., Kazana, T., Obremski, M.: Non-malleable codes from two-source extractors. In: Canetti, R., Garay, J.A. (eds.) CRYPTO 2013. LNCS, vol. 8043, pp. 239–257. Springer, Heidelberg (2013). https://doi.org/10.1007/978-3-642-40084-1_14

26. Dziembowski, S., Pietrzak, K.: Leakage-resilient cryptography. In: IEEE FOCS, pp. 293–302 (2008)

27. Dziembowski, S., Pietrzak, K., Wichs, D.: Non-malleable codes. In: Innovations in Computer Science, pp. 434–452 (2010)

28. Faonio, A., Nielsen, J.B., Simkin, M., Venturi, D.: Continuously non-malleable codes with split-state refresh. In: Preneel, B., Vercauteren, F. (eds.) ACNS 2018. LNCS, vol. 10892, pp. 1–19. Springer, Cham (2018). https://doi.org/10.1007/978-3-319-93387-0_7

29. Faust, S., Hostáková, K., Mukherjee, P., Venturi, D.: Non-malleable codes for space-bounded tampering. In: Katz, J., Shacham, H. (eds.) CRYPTO 2017. LNCS, vol. 10402, pp. 95–126. Springer, Cham (2017). https://doi.org/10.1007/978-3-319-63715-0_4

30. Faust, S., Mukherjee, P., Nielsen, J.B., Venturi, D.: Continuous non-malleable codes. In: Lindell, Y. (ed.) TCC 2014. LNCS, vol. 8349, pp. 465–488. Springer, Heidelberg (2014). https://doi.org/10.1007/978-3-642-54242-8_20

31. Faust, S., Mukherjee, P., Nielsen, J.B., Venturi, D.: A tamper and leakage resilient von neumann architecture. In: Katz, J. (ed.) PKC 2015. LNCS, vol. 9020, pp. 579–603. Springer, Heidelberg (2015). https://doi.org/10.1007/978-3-662-46447-2_26

32. Faust, S., Mukherjee, P., Venturi, D., Wichs, D.: Efficient non-malleable codes and key-derivation for poly-size tampering circuits. In: Nguyen, P.Q., Oswald, E. (eds.) EUROCRYPT 2014. LNCS, vol. 8441, pp. 111–128. Springer, Heidelberg (2014). https://doi.org/10.1007/978-3-642-55220-5_7

33. Gennaro, R., Lysyanskaya, A., Malkin, T., Micali, S., Rabin, T.: Algorithmic tamper-proof (ATP) security: theoretical foundations for security against hardware tampering. In: Naor, M. (ed.) TCC 2004. LNCS, vol. 2951, pp. 258–277. Springer, Heidelberg (2004). https://doi.org/10.1007/978-3-540-24638-1_15

34. Goldreich, O., Micali, S., Wigderson, A.: Proofs that yield nothing but their validity for all languages in NP have zero-knowledge proof systems. J. ACM **38**(3), 691–729 (1991)

35. Goyal, V., Pandey, O., Richelson, S.: Textbook non-malleable commitments. In: ACM STOC, pp. 1128–1141 (2016)

36. Jafargholi, Z., Wichs, D.: Tamper detection and continuous non-malleable codes. In: Dodis, Y., Nielsen, J.B. (eds.) TCC 2015. LNCS, vol. 9014, pp. 451–480. Springer, Heidelberg (2015). https://doi.org/10.1007/978-3-662-46494-6_19

37. Li, X.: Improved non-malleable extractors, non-malleable codes and independent source extractors. In: ACM STOC, pp. 1144–1156 (2017)

38. Liu, F.-H., Lysyanskaya, A.: Tamper and leakage resilience in the split-state model. In: Safavi-Naini, R., Canetti, R. (eds.) CRYPTO 2012. LNCS, vol. 7417, pp. 517–532. Springer, Heidelberg (2012). https://doi.org/10.1007/978-3-642-32009-5_30

39. Mahmoody, M., Pass, R.: The curious case of non-interactive commitments – on the power of black-box vs. non-black-box use of primitives. In: Safavi-Naini, R., Canetti, R. (eds.) CRYPTO 2012. LNCS, vol. 7417, pp. 701–718. Springer, Heidelberg (2012). https://doi.org/10.1007/978-3-642-32009-5_41

40. Matsuda, T., Hanaoka, G.: An asymptotically optimal method for converting bit encryption to multi-bit encryption. In: Iwata, T., Cheon, J.H. (eds.) ASIACRYPT 2015. LNCS, vol. 9452, pp. 415–442. Springer, Heidelberg (2015). https://doi.org/10.1007/978-3-662-48797-6_18

41. Pandey, O., Pass, R., Vaikuntanathan, V.: Adaptive one-way functions and applications. In: Wagner, D. (ed.) CRYPTO 2008. LNCS, vol. 5157, pp. 57–74. Springer, Heidelberg (2008). https://doi.org/10.1007/978-3-540-85174-5_4
42. Pass, R.: Unprovable security of perfect NIZK and non-interactive non-malleable commitments. In: Sahai, A. (ed.) TCC 2013. LNCS, vol. 7785, pp. 334–354. Springer, Heidelberg (2013). https://doi.org/10.1007/978-3-642-36594-2_19
43. Pass, R., Rosen, A.: Concurrent non-malleable commitments. In: IEEE FOCS, pp. 563–572 (2005)
44. Pass, R., Rosen, A.: Concurrent nonmalleable commitments. SIAM J. Comput. **37**(6), 1891–1925 (2008)
45. Pass, R., Rosen, A.: New and improved constructions of nonmalleable cryptographic protocols. SIAM J. Comput. **38**(2), 702–752 (2008)
46. Richardson, R., Kilian, J.: On the concurrent composition of zero-knowledge proofs. In: Stern, J. (ed.) EUROCRYPT 1999. LNCS, vol. 1592, pp. 415–431. Springer, Heidelberg (1999). https://doi.org/10.1007/3-540-48910-X_29

Zero Knowledge

Non-Interactive Zero-Knowledge Proofs for Composite Statements

Shashank Agrawal[1(✉)], Chaya Ganesh[2], and Payman Mohassel[1]

[1] Visa Research, Palo Alto, USA
{shaagraw,pmohasse}@visa.com
[2] Aarhus University, Aarhus, Denmark
ganesh@cs.au.dk

Abstract. The two most common ways to design non-interactive zero-knowledge (NIZK) proofs are based on Sigma protocols and QAP-based SNARKs. The former is highly efficient for proving algebraic statements while the latter is superior for arithmetic representations.

Motivated by applications such as privacy-preserving credentials and privacy-preserving audits in cryptocurrencies, we study the design of NIZKs for composite statements that compose algebraic and arithmetic statements in arbitrary ways. Specifically, we provide a framework for proving statements that consist of ANDs, ORs and function compositions of a mix of algebraic and arithmetic components. This allows us to explore the full spectrum of trade-offs between proof size, prover cost, and CRS size/generation cost. This leads to proofs for statements of the form: knowledge of x such that $SHA(g^x) = y$ for some public y where the prover's work is 500 times fewer exponentiations compared to a QAP-based SNARK at the cost of increasing the proof size to 2404 group and field elements. In application to anonymous credentials, our techniques result in 8 times fewer exponentiations for the prover at the cost of increasing the proof size to 298 elements.

1 Introduction

Zero-knowledge proofs provide the ability to convince a verifier that a statement is true without revealing the secrets involved. Since their conception in the mid 1980s, zero-knowledge proofs have emerged as a fundamental object in modern cryptography, with connections to the theory of computation [7,36,41,61]. Zero-knowledge proofs (ZKPs) have found numerous applications as a building block in other cryptographic constructions such as identification schemes [32], group signature schemes [19], public-key encryption [55], anonymous credentials [17], voting [23], and secure multi-party computation [42]. Most recently, ZKPs have been used as a core component in digital cryptocurrencies such as ZCash and Monero to make the transactions private and anonymous [8,56].

Zero-knowledge proofs exist for all languages in NP [41], but not all such constructions are efficiently implementable. Indeed, a large body of work has

---

C. Ganesh—Work done as an intern at Visa Research.

© International Association for Cryptologic Research 2018
H. Shacham and A. Boldyreva (Eds.): CRYPTO 2018, LNCS 10993, pp. 643–673, 2018.
https://doi.org/10.1007/978-3-319-96878-0_22

been devoted to the design and implementation of efficient ZKPs for a variety of statements. In case of Non-Interactive Zero-Knowledge (NIZK) proofs, which is the focus of this paper, the most practical approaches are based on (i) Sigma protocols (with the Fiat-Shamir transform), (ii) zk-SNARKs and (iii) "MPC-in-the-head" techniques, each with their own efficiency properties, advantages and shortcomings. While the MPC-in-the-head technique [48] has led to (Boolean) circuit-friendly NIZKs [6,20,40], this line of work produces large proofs. In this paper we focus on Sigma protocols and zk-SNARKs, and elaborate on these next.

Sigma Protocols. Many of the statements we prove in cryptographic constructions are efficiently representable as algebraic functions over some group $\mathcal{G}$, such as an elliptic-curve group where the discrete-logarithm problem is hard. For example, Alice may want to convince Bob that she knows an x such that $g^x = y$ for publicly known values $g, y \in \mathcal{G}$ (knowledge of discrete log), or she may like to show that x lies between two public integers a and b (range proof).

Sigma protocol-based ZKPs are extremely efficient for such statements. They yield short proof sizes, require a constant number of public-key operations, and do not impose trusted common reference string (CRS) generation [26,38,45,46,59,60]. Moreover, they can be made non-interactive, i.e. only a single message from prover to verifier, using the efficient Fiat-Shamir transformation [34].

While Sigma protocols are efficient for algebraic statements, they are significantly slower when it comes to non-algebraic ones. Consider a cryptographic hash function or a block cipher represented by a Boolean or arithmetic circuit C, and suppose Alice wants to show that she knows an input x such that $C(x) = y$ for some public y. Alice can treat each gate of C as an algebraic function and provide a proof that the input and output wires of each gate satisfy the associated algebraic relation, to show that she indeed knows x, but this would be prohibitively expensive. In particular, both the proving/verification time and the proof size would grow linearly with the size of circuit which in case of hash functions and block-ciphers can be *tens of thousands* of exponentiations and group elements.

zk-SNARKS. There has been a series of works on constructing zero-knowledge *Succinct* Non-interactive ARguments of Knowledge (zk-SNARKs) [9,10,12,39,44,51,52,57]. Starting with the construction of Kilian [50] based on probabilistically checkable proofs (PCPs), made non-interactive by Micali [53], there has been further works [11,29,43] that construct succinct arguments by removing interaction in Kilian's PCP-based protocol. Despite these advances, PCPs remain concretely expensive and current implementations along this line are not yet efficient. A more effective approach for proving statements about functions represented as Boolean or arithmetic circuits is based on Quadratic Arithmetic Programs (QAPs) [39] and throughout the paper, we will be concerned with QAP-based zk-SNARK proofs. Such proofs are very short and have fast verification time. More precisely, the proofs have constant size and can be verified in time that is linear in the length of the input x, rather than the length of the circuit C. Thus, zk-SNARKs are better suited for proving statements about hash functions or block ciphers than (non-interactive) Sigma protocols.

In principle, zk-SNARKs could also be used to prove algebraic statements, such as knowledge of discrete-log in a cyclic group by representing the exponentiation circuit as a QAP. The circuit for computing a single exponentiation is in the order of *thousands or millions* of gates depending on the group size. In zk-SNARKs based on QAP, the prover cost is linear in the size of circuit and an honestly generated common reference string (CRS) is needed, whose size also grows proportional to the circuit size. This makes them extremely inefficient for algebraic statements. In contrast, Sigma protocols can be used to prove knowledge of discrete-log with a constant number of exponentiations.

Another disadvantage of zk-SNARKs is that the CRS is generated with respect to a particular circuit C and, in the most efficient instantiations, needs to be regenerated when proving a new statement represented with a different circuit C'. This is not desirable since in current applications such as ZCash, where CRS is generated using an expensive secure multi-party computation (MPC) protocol in order to guarantee soundness of the proof system [4]. In contrast, Sigma protocols have constant-size untrusted CRSs that can be used to prove arbitrary statements and can be generated inexpensively (without an MPC).

1.1 Composite Statements and Applications

Composite statements that include multiple algebraic and arithmetic components appear in various applications. We discuss three important cases here.

Proof of Solvency. Consider privacy-preserving proofs of solvency for Bitcoin exchanges [27,62]. Here an exchange wants to prove to its customers that it has enough reserves to cover its liabilities, or, in simple words, that it is solvent. A proof of reserves in the Bitcoin network amounts to showing that the exchange has control over certain Bitcoin addresses. A Bitcoin address is a 160-bit hash of the public portion of a public/private ECDSA keypair [2], where the public portion is derived from the private key by doing an exponentiation operation on the secp256k1 curve [1][1]. Thus the exchange wants to show that it knows the private keys corresponding to some hashed public keys available on the blockchain. Furthermore, the proof should not reveal the public keys themselves otherwise an adversary would be able to track the movement of exchange's funds.

In particular, the exchange wants to show that it knows a secret x such that $H(g^x) = y$ where H is a hash function such as SHA-256. The statement has both algebraic (g^x) and Boolean (hash function H) parts. One can express the composite function (exponentiate then hash) as a purely algebraic or Boolean function and then use a Sigma protocol or zk-SNARK respectively, but, in the former case, the proof size and verification time will be quite large, while in the

---

[1] Most cryptocurrencies generate public/private keys and define an address in a similar manner. Apart from Bitcoin and its fork Bitcoin Cash, Ethereum is another prominent example.

latter, the proof generation time will increase substantially and a much larger CRS is needed. Ideally, one would like to use a Sigma protocol for the algebraic part and a zk-SNARK for the Boolean part, and then combine the two proofs so that no extra information about x is revealed (beyond the fact that $H(g^x) = y$).

Thus any proof of solvency for a Bitcoin exchange must deal with a zero-knowledge proof that combines both Boolean and algebraic statements. Existing proposals for proofs of solvency get around this problem by assuming (incorrectly) that public keys themselves are available on the blockchain so that Sigma protocols alone suffice [27]. As we will see later, our efficient techniques allow designing NIZKs for proving knowledge of x given $H(g^x)$ that require roughly 500 times fewer exponentiations for the prover compared to proving the same statement using a QAP-based SNARK.

Privacy-Preserving Credentials. Digital certificates (X.509) are commonly used to identify entities over the Internet. They include a message m that may contain various identifying information about a user or a machine, and a digital signature (by a certificate authority) on the message attesting to its authenticity. The signature can then be verified by anyone who holds the public verification key. Typically, certificates reveal the message m and hence the identity of their owner. Anonymous credentials [22] provide the same authentication guarantees without revealing the identifying message, and are widely studied due to their strong privacy guarantees. A main ingredient for making digital certificates anonymous is a ZKP of knowledge of a message m and a signature σ, where σ is a valid signature on message m with respect to the verification key vk. The ZKP ensures that we do not leak any information about m beyond the knowledge of a valid signature. A large body of work has studied anonymous credentials, but only a handful of techniques can turn commonly used X.509 certificates into anonymous credentials. The main challenge is that the ZKP statement being proven is a hybrid statement containing both algebraic (RSA or elliptic-curve operations) and Boolean functions (hashing), since the message is hashed before being algebraically signed. The work of Delignat-Lavaud et al. [30] constructs a proof for such a hybrid statement using only zk-SNARKs which, as discussed earlier, is inefficient for the algebraic component, while the work of Chase et al. [21] design such ZKP proofs in the interactive setting where the prover and verifier exchange multiple messages. Efficient NIZK for composite statements based on both zk-SNARKs and Sigma protocols would yield more efficient anonymous credential systems. Using our techniques for RSA signature results in prover's work that is about 8 times fewer group exponentiations compared to Cinderella [30].

zk-SNARKs with composable CRSs. Anonymous decentralized digital crypto-currencies such as ZCash use zk-SNARKs to prove a massive statement containing many different smaller components. For example, at a high level, one of the statement being proven in ZCash is of the form: I have knowledge of x_i's such that $H(x_1 \| H(x_2 \| \ldots H(x_n))) = y$ for a large value of n. The CRS generated for proving this statement is extremely large (about a gigabyte for ZCash [3]) and cannot be reused to prove any other statement. A better alternative is to generate a much smaller CRS for proving a statement of the form: I have knowledge of x, y such that $H(x||H(y))$, combined with a technique for composing many such proofs. More generally, one can envision a general system with CRSs for small size statements $C_1, \ldots, C_n$ that enables NIZKs for arbitrary composition of these statements without having to generate new CRSs for each new composition. This yields a trade-off between proof size and the CRS size (and its reusability).

1.2 Contributions

Motivated by the above applications, we study the design of NIZKs for composite statements that compose algebraic and arithmetic statements in arbitrary ways. Specifically, we provide new protocols for statements that consist of ANDs, ORs and function compositions of a mix of algebraic and arithmetic components. In doing so, our goal is to maintain the invariant that algebraic components are proven using Sigma protocols, and arithmetic statements using QAP-based zk-SNARKs. This allows us to explore the full spectrum of trade-offs between proof size (verification cost), prover cost, and CRS size (and cost of generation) for composite statements.

More precisely, we propose new NIZKs for proof of knowledge of x, x_1, x_2, y_1, y_2 such that

- $f_1(x_1, f_2(x_2)) = z$,
- $f_1(x, y_1) = z_1$ AND $f_2(x, y_2) = z_2$,
- $f_1(x, y_1) = z_1$ OR $f_2(x, y_2) = z_2$,

for public values z, z_1, z_2, and where f_1 and f_2 can be either algebraic or arithmetic. Given our NIZKs for these compositions, it is easy to handle arbitrary composite statements. This is the first work that directly addresses the question of non-interactive proofs for composite statements and how disparate techniques can be used to prove them in zero-knowledge efficiently. We note that in this paper we primarily focus on elliptic curves as our algebraic group, as they are the most efficient for instantiating both zk-SNARKs and Sigma protocols.

2 Preliminaries

Notation. Throughout the paper, we use κ to denote the security parameter or level. A function is negligible if for all large enough values of the input, it is smaller than the inverse of any polynomial. We use negl to denote a negligible function. We write $\mathcal{X}_\kappa \equiv \mathcal{Y}_\kappa$ to mean that distributions $\mathcal{X}_\kappa$ and $\mathcal{Y}_\kappa$ are identical. We use $[1, n]$ to represent the set of numbers $\{1, 2, \ldots, n\}$. If Alg is a randomized algorithm, we use $y \leftarrow \text{Alg}(x)$ to denote that y is the output of Alg on x. We write $x \xleftarrow{R} \mathcal{X}$ to mean sampling a value x uniformly from the set $\mathcal{X}$.

We denote an interactive protocol between two parties A and B by $\langle A, B \rangle$. $\langle A(x), B(y) \rangle (z)$ denotes a protocol where A has input x, B has input y and z is a common input. Also, view$_A$ denotes the "view" of A in an interaction with B, which consists of the input to A, its random coins, and the messages sent by B (view$_B$ is defined in a similar manner).

Bilinear groups. Let GroupGen be an asymmetric pairing group generator that on input 1^κ, outputs description of three cyclic groups $\mathbb{G}, \widetilde{\mathbb{G}}, \mathbb{G}_T$ of prime order $p = \Theta(2^\kappa)$ equipped with a non-degenerate efficiently computable bilinear map $e : \mathbb{G} \times \widetilde{\mathbb{G}} \to \mathbb{G}_T$, and generators g and $\tilde{g}$ for $\mathbb{G}$ and $\widetilde{\mathbb{G}}$ respectively. The discrete logarithm assumption is said to hold in $\mathbb{G}$ relative to GroupGen if for all PPT algorithms $\mathcal{A}$, $\Pr[x \leftarrow \mathcal{A}(\mathbb{G}, p, g, h) \mid (\mathbb{G}, \widetilde{\mathbb{G}}, \mathbb{G}_T) \leftarrow \mathsf{GroupGen}; x \xleftarrow{R} \mathbb{Z}_p; h := g^x]$ is $\mathsf{negl}(\kappa)$.

In this paper, we primarily consider elliptic curves as our algebraic group. Let E be an elliptic curve defined over a field $\mathbb{F}_t$. The set of points on the curve form a group under the point addition operation, and we denote the group by $E(\mathbb{F}_t)$. For an element $P \in E(\mathbb{F}_t)$ of prime order p, P_x and P_y represent the x and y co-ordinates of the point P respectively. In some constructions, we use additive notation and write $Q = \alpha P$ for a scalar $\alpha \in \mathbb{F}_p$. The discrete logarithm assumption is believed to hold in well chosen elliptic curve groups where group elements are represented with $O(\kappa)$ bits. In our constructions, we use asymmetric bilinear groups where $\mathbb{G} \neq \widetilde{\mathbb{G}}$, and discrete logarithm is hard in $\mathbb{G}$. We also rely on q-type assumptions similar to Parno et al. [57] (but in asymmetric groups).

Zero-knowledge Proofs. Let R be an efficiently computable binary relation which consists of pairs of the form (s, w) where s is a statement and w is a witness. Let $\mathcal{L}$ be the language associated with R, i.e., $\mathcal{L} = \{s \mid \exists w \text{ s.t. } R(s, w) = 1\}$.

A zero-knowledge proof for $\mathcal{L}$ lets a prover P convince a verifier V that $s \in \mathcal{L}$ for a common input s without revealing w. A proof of knowledge captures not only the truth of a statement $s \in \mathcal{L}$, but also that the prover "possesses" a witness w to this fact. We are concerned with non-interactive proofs in this paper where P sends only one message to V, and V decides whether to accept or not based on its input, the message, and any public parameters. We define them formally below.

2.1 Non-interactive Zero-knowledge Proofs

Non-interactive zero-knowledge (NIZK) proofs are usually studied in the common reference string (CRS) model, wherein a string of a special structure is generated in a setup phase, and made available to everyone to prove/verify statements.

Definition 2.1 (Non-interactive Zero-knowledge Argument [13,33]). *A NIZK argument for an NP relation R consists of a triple of polynomial time algorithms* (Setup, Prove, Verify) *defined as follows.*

- Setup(1^κ) *takes a security parameter κ and outputs a CRS Σ.*
- Prove(Σ, s, w) *takes as input the CRS Σ, a statement s, and a witness w, and outputs an argument π.*
- Verify(Σ, s, π) *takes as input the CRS Σ, a statement s, and a proof π, and outputs either 1 accepting the argument or 0 rejecting it.*

The algorithms above should satisfy the following properties.

1. *Completeness. For all* $\kappa \in \mathbb{N}$, $(s,w) \in R$,

$$\Pr\left(\mathsf{Verify}(\varSigma, s, \pi) = 1 : \begin{array}{l} \varSigma \leftarrow \mathsf{Setup}(1^\kappa) \\ \pi \leftarrow \mathsf{Prove}(\varSigma, s, w) \end{array}\right) = 1.$$

2. *Computational soundness. For all PPT adversaries* $\mathcal{A}$, *the following probability is negligible in* κ:

$$\Pr\left(\begin{array}{l} \mathsf{Verify}(\varSigma, \tilde{s}, \tilde{\pi}) = 1 \\ \wedge \; \tilde{s} \notin L \end{array} : \begin{array}{l} \varSigma \leftarrow \mathsf{Setup}(1^\kappa) \\ (\tilde{s}, \tilde{\pi}) \leftarrow \mathcal{A}(1^\kappa, \varSigma) \end{array}\right).$$

3. *Zero-knowledge. There exists a PPT simulator* $(\mathcal{S}_1, \mathcal{S}_2)$ *such that* $\mathcal{S}_1$ *outputs a simulated CRS* $\varSigma$ *and trapdoor* τ; $\mathcal{S}_2$ *takes as input* $\varSigma$, *a statement* s *and* τ, *and outputs a simulated proof* π; *and, for all PPT adversaries* $(\mathcal{A}_1, \mathcal{A}_2)$, *the following probability is negligible in* κ:

$$\left| \Pr\left(\begin{array}{l} (s,w) \in R \; \wedge \\ \mathcal{A}_2(\pi, \mathsf{st}) = 1 \end{array} : \begin{array}{l} \varSigma \leftarrow \mathsf{Setup}(1^\kappa) \\ (s, w, \mathsf{st}) \leftarrow \mathcal{A}_1(1^\kappa, \varSigma) \\ \pi \leftarrow \mathsf{Prove}(\varSigma, s, w) \end{array}\right) \right.$$

$$\left. - \Pr\left(\begin{array}{l} (s,w) \in R \; \wedge \\ \mathcal{A}_2(\pi, \mathsf{st}) = 1 \end{array} : \begin{array}{l} (\varSigma, \tau) \leftarrow \mathcal{S}_1(1^\kappa) \\ (s, w, \mathsf{st}) \leftarrow \mathcal{A}_1(1^\kappa, \varSigma) \\ \pi \leftarrow \mathcal{S}_2(\varSigma, \tau, s) \end{array}\right) \right|.$$

Definition 2.2 (Non-interactive Zero-knowledge Argument of Knowledge). *A NIZK argument of knowledge for a relation* R *is a NIZK argument for* R *with the following additional extractability property:*

- *Extraction. For any PPT adversary* $\mathcal{A}$, *random string* $r \xleftarrow{R} \{0,1\}^*$, *there exists a PPT algorithm* Ext *such that the following probability is negligible in* κ:

$$\Pr\left(\begin{array}{l} \mathsf{Verify}(\varSigma, \tilde{s}, \tilde{\pi}) = 1 \\ \wedge \; R(\tilde{s}, w') = 0 \end{array} : \begin{array}{l} \varSigma \leftarrow \mathsf{Setup}(1^\kappa) \\ (\tilde{s}, \tilde{\pi}) \leftarrow \mathcal{A}(1^\kappa, \varSigma; r) \\ w' = \mathsf{Ext}(\varSigma, \tilde{s}, \tilde{\pi}; r) \end{array}\right).$$

Definition 2.3 (zero-knowledge Succinct Non-interactive ARgument of Knowledge (zk-SNARK)). *A zk-SNARK for a relation* R *is a non-interactive zero-knowledge argument of knowledge for* R *with the following additional property:*

- *Succinctness. For any* s *and* w, *the length of the proof* π *is given by* $|\pi| = \mathsf{poly}(\kappa) \cdot \mathsf{polylog}(|s| + |w|)$.

2.2 Sigma Protocols

Sigma protocols are two-party interactive protocols of a specific structure. Let P (the prover) and V (the verifier) be two parties with common input s and a private input w for P. In a Sigma protocol, P sends a message a, V replies with a random κ-bit string r, P then sends a message e, and V decides to accept or reject based on the transcript (a, r, e). If V accepts (outputs 1), then the transcript is called accepting.

Definition 2.4 (Sigma protocol [28]). *An interactive protocol between a prover P and a verifier V is a Σ protocol for a relation R if the following properties are satisfied:*

1. *It is a three move public coin protocol.*
2. *Completeness: If P and V follow the protocol then $\Pr[\langle P(w), V \rangle (s) = 1] = 1$ whenever $(s, w) \in R$.*
3. *Special soundness: There exists a polynomial time algorithm called the extractor which when given s and two transcripts (a, r, e) and (a, r', e') that are accepting for s, with $r \neq r'$, outputs w' such that $(s, w') \in R$.*
4. *Special honest verifier zero knowledge: There exists a polynomial time simulator which on input s and a random r outputs a transcript (a, r, e) with the same probability distribution as that generated by an honest interaction between P and V on (common) input s.*

Fiat-Shamir transform. A Σ protocol can be efficiently compiled into a non-interactive zero-knowledge proof of knowledge (in the random oracle model) through the Fiat-Shamir transform [34]. Not only the transformation removes interaction from the protocol, but also makes it zero-knowledge against malicious verifiers. At a high level, the transform works by having the prover compute the verifier's message by applying an appropriate hash function, modeled as a random oracle in the security proof, to the prover's first message to obtain a random challenge.

OR composition of Σ-protocols. In Cramer et al. [26], the authors devise an OR composition technique for Sigma protocols. Essentially, a prover can efficiently show $((x_0 \in \mathcal{L}) \vee (x_1 \in \mathcal{L}))$ without revealing which x_i is in the language. More generally, the OR transform can handle two different relations R_0 and R_1.

Theorem 2.5 (OR-composition [26]). *If Π_0 is a Σ-protocol for R_0 and Π_1 a Σ-protocol for R_1, then there is a Σ-protocol Π_{OR} for the relation R_{OR} given by $\{((x_0, x_1), w) : ((x_0, w) \in R_0) \vee ((x_1, w) \in R_1)\}$.*

Pedersen commitment. Throughout the paper, we use algebraic commitment schemes that allow proving linear relationships among committed values. The Pedersen commitment scheme [58] is one such example which gives unconditional hiding and computational binding properties based on the hardness of computing discrete logarithm in a group $\mathcal{G}$, say of order q. Given two random generators $g, h \in \mathcal{G}$ such that $\log_g h$ is unknown, a value $x \in \mathbb{Z}_q$ is committed to by choosing r randomly from $\mathbb{Z}_q$, and computing $g^x h^r$. We write $\mathsf{Com}_q(x)$ to denote a Pedersen commitment to x in a group of order q.

Sigma protocols are known in literature to prove knowledge of a committed value, equality of two committed values, and so on, and these protocols can be combined in natural ways. In particular, linear relationships between Pedersen commitments can be shown through existing techniques [18,19,37,60]. For example, one could show that $y = ax + b$ for some public values a and b, given $\mathsf{Com}_q(x)$ and $\mathsf{Com}_q(y)$.

We use $\mathsf{PK}\{(x, y, \ldots) : statements \text{ about } x, y, \ldots\}$ to denote a proof of knowledge of $x, y, \ldots$ that satisfies *statements* [19]. Other values in *statements* are public.

2.3   SNARK Construction from QAP

The work of Gennaro et al. [39] showed how to encode computations as quadratic programs. They show how to convert any Boolean circuit into a Quadratic Span Program (QSP) and any arithmetic circuit into a Quadratic Arithmetic Program (QAP). In this work, we will only use the latter definition. Even though QSPs are designed for Boolean circuits, arithmetic *split gates* defined in Parno et al. [57] translate an arithmetic wire into binary output wires, and Boolean functions may be computed using arithmetic gates. Parno et al. also note that such an arithmetic embedding results in a smaller QAP compared to the QSP of the original Boolean circuit. In the rest of the paper, we assume that Boolean functions are computed by a QAP defined over an arithmetic field, and hence will only be concerned with QAP.

Definition 2.6 (Quadratic Arithmetic Program [39]). *A quadratic arithmetic program (QAP) Q over a field $\mathbb{F}$ consists of three sets of polynomials $V = \{v_k(x) : k \in \{0, \ldots, m\}\}, W = \{w_k(x) : k \in \{0, \ldots, m\}\}, Y = \{y_k(x) : k \in \{0, \ldots, m\}\}$ and a target polynomial $t(x)$, all in $\mathbb{F}[X]$.*

Let $f : \mathbb{F}^n \to \mathbb{F}^{n'}$ be a function with input variables labeled $1, \ldots, n$ and output variables labeled $m - n' + 1, \ldots, m$. A QAP Q is said to compute f if the following holds: $a_1, \ldots, a_n, a_{m-n'+1}, \ldots, a_m \in \mathbb{F}^{n+n'}$ is a valid assignment to the input and output variables of f (i.e., $f(a_1, \ldots, a_n) = (a_{m-n'+1}, \ldots, a_m)$) iff there exist $(a_{n+1}, \cdots, a_{m-n'}) \in \mathbb{F}^{m-n-n'}$ such that $t(x)$ divides $p(x)$, where

$$p(x) = \left(v_0(x) + \sum_{k=1}^{m} a_k v_k(x) \right) \cdot \left(w_0(x) + \sum_{k=1}^{m} a_k w_k(x) \right) - \left(y_0(x) + \sum_{k=1}^{m} a_k y_k(x) \right).$$

The size of the QAP Q is m, and degree is $deg(t(x))$.

The polynomials $v_k(x), w_k(x), y_k(x)$ have degree at most $deg(t(x)) - 1$, since they can be reduced modulo $t(x)$ without affecting the divisibility check.

3   NIZK on Committed IO for Algebraic Statements

In this section, we design Sigma protocols for knowledge of inputs and outputs of algebraic statements where the inputs and outputs are committed to. In other words, we enable proof of knowledge of x_i given commitments $\mathsf{Com}(x_i)$ to inputs and a commitment $\mathsf{Com}(\Pi g_i^{P_i(x_i)})$ to the output of an algebraic function where g_is are public generators in an elliptic curve group and P_is are public single-variable polynomials. An important ingredient in this is a proof of knowledge of double discrete log which we elaborate on next.

3.1   Proof of Knowledge of Double Discrete Logarithm

Our goal is to prove the equality of a committed value and the discrete logarithm of another committed value. When the commitments are in elliptic curve groups, the known techniques for double discrete logarithm proofs will not work [19, 54]. This is because a group element cannot be naturally interpreted as a field element, as can be done in integer groups. Towards this end, we first describe a protocol to prove that the sum of two elliptic curve points that are committed to, is another public point on the curve.

In this section, we consider the family of curves E given by

$$y^2 = x^3 + ax + b, \tag{1}$$

where $a, b \in \mathbb{F}_t$, but the techniques we describe below would extend to other curve families like Edwards [31]. The curve sec256k1 used by Bitcoin has the form of Eq. 1 with $a = 0, b = 7$.

The point addition relation is defined by the point addition equation specific to the curve family. Let $P = (x_1, y_1), Q = (x_2, y_2), P, Q \in E(\mathbb{F}_t)$ for the family E above. For distinct $P, Q, P \neq -Q$, $(x_3, y_3) = P + Q$ is given by

$$x_3 = \left(\frac{y_2 - y_1}{x_2 - x_1} \right)^2 - x_1 - x_2, \tag{2}$$

$$y_3 = \frac{y_2 - y_1}{x_2 - x_1}(x_1 - x_3) - y_1. \tag{3}$$

We use $\mathsf{addFormula}(P, Q)$ to denote (x_3, y_3) computed in this way. When $P = Q$, the operation is doubling of the point P, denoted by $\mathsf{doubleFormula}(P)$. In this case, (x_3, y_3) is given by

$$x_3 = \left(\frac{3x_1^2 + a}{2y_1} \right)^2 - 2x_1, \tag{4}$$

$$y_3 = \frac{3x_1^2 + a}{2y_1}(x_1 - x_3) - y_1. \tag{5}$$

We could prove the above relations for committed x_1, x_2, y_1, y_2 using known Sigma protocol techniques. But since the point addition computation is over $\mathbb{F}_t$, the commitments to the coordinates have to be in a group of order t, which is not necessarily the same as p, the order of the group $E(\mathbb{F}_t)$. The Complex Multiplication (CM) method could be used to find elliptic curve groups of a specific order. However, it is quite inefficient for large orders and would make our protocols impractical. We avoid the CM method by proposing a protocol that does not need to find a group of a given order.

We rewrite the point addition formula (Eqs. 2 and 3) as

$$x_3 x_2^2 + x_3 x_1^2 + x_1^3 + x_2^3 + 2y_1 y_2 = y_2^2 + y_1^2 + x_1^2 x_2 + x_1 x_2^2 + 2x_1 x_2 x_3, \quad (6)$$

$$x_2 y_3 + x_3 y_2 + x_2 y_1 = x_1 y_2 + x_3 y_1 + x_1 y_3. \quad (7)$$

Let L_x and R_x denote the left-hand side and right-hand side respectively of Eq. 6, and L_y and R_y of Eq. 7. That is:

$$L_x(x_1, y_1, x_2, y_2) = x_3 x_2^2 + x_3 x_1^2 + x_1^3 + x_2^3 + 2y_1 y_2,$$
$$R_x(x_1, y_1, x_2, y_2) = y_2^2 + y_1^2 + x_1^2 x_2 + x_1 x_2^2 + 2x_1 x_2 x_3,$$
$$L_y(x_1, y_1, x_2, y_2) = x_2 y_3 + x_3 y_2 + x_2 y_1,$$
$$R_y(x_1, y_1, x_2, y_2) = x_1 y_2 + x_3 y_1 + x_1 y_3.$$

We use Sigma protocols to prove that L_x, R_x, L_y and R_y satisfy the above relations using committed intermediate values. To do so, in addition to linear relationships, our protocol needs to prove that a committed value is the product of two committed values: given $C_1 = \mathsf{Com}(a) = g^a h^{r_1}, C_2 = \mathsf{Com}(b) = g^b h^{r_2}, C_3 = \mathsf{Com}(c) = g^c h^{r_3}$, prove $c = ab$. This can be done by proving knowledge of b such that the discrete logarithm of C_4 with respect to C_1 is equal to the committed value in C_2, and the equality of committed values in C_4 and C_3, where $C_4 = C_1^b$. The prover computes and sends $C_4 = C_1^b$ with the following proof: $\mathsf{PK}\{(a, b, c, b', c', r_1, r_2, r_3, r_4) : C_1 = g^a h^{r_1} \wedge C_2 = g^b h^{r_2} \wedge C_3 = g^c h^{r_3} \wedge C_4 = C_1^{b'} \wedge C_4 = g^{c'} h^{r_4} \wedge b' = b \wedge c' = c\}$. In general, Sigma protocols for polynomial relationships among committed values were given by Camenisch and Michels [18].

Let G_2 be an elliptic-curve group of order q such that $q > 2t^3$, and P', Q' be points in G_2. We commit to the coordinates and the intermediate values necessary for the proof in G_2, and since the largest intermediate value in Eqs. 6 and 7 is cubic, the choice of q ensures there is no wrap around when the computation is modulo q. Since all computation on committed values will now be modulo q, and the addition equations are to be computed modulo t, we use division with remainder. We prove equality of L_x and R_x modulo q, divide them by t taking away multiples of t, and prove that the remainders are equal. When used together with appropriate range proofs to prove that the remainder does not exceed the divisor, and that the committed coordinates are in the desired range, we get equality modulo t. (There are several known techniques to build range proofs [14,16], that is, to prove that $x \in [0, S]$ for a public S and committed x, including the recent, very efficient technique called Bulletproof [15].)

The protocol addition given in Fig. 1 proves that the addition formula holds for committed points P, Q and their sum T. We show that addition is secure in the full version. The protocol's cost is dominated by the range proofs in steps 4, 5, 6 and the proof for polynomial relationships in steps 2 and 3. addition roughly has a proof size of $75 + \log \log t$ elements, and prover's work $60 + \log t$ exponentiations.

Let $\mathsf{C}_P = \mathsf{Com}_q(P) = (\mathsf{Com}_q(P_x), \mathsf{Com}_q(P_y))$ denote a commitment to a point $P = (P_x, P_y)$.

Given $T = (T_x, T_y)$,
$C_1 = \mathsf{Com}_q(P_x), C_2 = \mathsf{Com}_q(P_y), C_3 = \mathsf{Com}_q(Q_x), C_4 = \mathsf{Com}_q(Q_y)$, prove that
$T = P + Q$, where $P = (P_x, P_y), Q = (Q_x, Q_y), T \in E(\mathbb{F}_t)$ and $q > 2t^3$.

1. Let $L_x(P_x, P_y, Q_x, Q_y) = k_1 t + r_1, R_x(P_x, P_y, Q_x, Q_y) = k'_1 t + r'_1, L_y(P_x, P_y, Q_x, Q_y) = k_2 t + r_2, R_y(P_x, P_y, Q_x, Q_y) = k'_2 t + r'_2$, for
$k_1, k'_1, k_2, k'_2 < \frac{q}{t}$ and $r_1, r'_1, r_2, r'_2 < t$.
Compute and send commitments $C_4 = \mathsf{Com}_q(L_x), C_5 = \mathsf{Com}_q(R_x), C_6 = \mathsf{Com}_q(L_y), C_7 = \mathsf{Com}_q(R_y), C_8 = \mathsf{Com}_q(k_1), C_9 = \mathsf{Com}_q(r_1), C_{10} = \mathsf{Com}_q(k'_1), C_{11} = \mathsf{Com}_q(r'_1), C_{12} = \mathsf{Com}_q(k_2), C_{13} = \mathsf{Com}_q(r_2), C_{14} = \mathsf{Com}_q(k'_2), C_{15} = \mathsf{Com}_q(r'_2)$.
2. Prove that $(P_x, P_y), (Q_x, Q_y)$ and (T_x, T_y) satisfy the addition equation for the x-coordinate.
   π_1 : $\mathsf{PK}\{(P_x, P_y, Q_x, Q_y, L_x, R_x) : C_1 = \mathsf{Com}_q(P_x) \wedge C_2 = \mathsf{Com}_q(P_y) \wedge C_3 = \mathsf{Com}_q(Q_x) \wedge C_4 = \mathsf{Com}_q(Q_y) \wedge C_4 = \mathsf{Com}_q(L_x) \wedge C_5 = \mathsf{Com}_q(R_x) \wedge L_x = T_x Q_x^2 + T_x P_x^2 + P_x^3 + P_y^3 + 2P_y Q_y \wedge R_x = Q_y^2 + P_y^2 + P_x^2 Q_x + P_x Q_x^2 + 2P_x Q_x T_x\}$
3. Prove that $(P_x, P_y), (Q_x, Q_y)$ and (T_x, T_y) satisfy the addition equation for the y-coordinate.
   π_2 : $\mathsf{PK}\{(P_x, P_y, Q_x, Q_y, L_y, R_y) : C_1 = \mathsf{Com}_q(P_x) \wedge C_2 = \mathsf{Com}_q(P_y) \wedge C_3 = \mathsf{Com}_q(Q_x) \wedge C_4 = \mathsf{Com}_q(Q_y) \wedge C_6 = \mathsf{Com}_q(L_y) \wedge C_7 = \mathsf{Com}_q(R_y) \wedge L_y = Q_x T_y + T_x Q_y + Q_x P_y \wedge R_y = P_x Q_y + T_x P_y + P_x T_y\}$
4. Prove that the coordinates are in the correct range.
   π_3 : $\mathsf{PK}\{(P_x, P_y, Q_x, Q_y) : C_1 = \mathsf{Com}_q(P_x) \wedge C_2 = \mathsf{Com}_q(P_y) \wedge C_3 = \mathsf{Com}_q(Q_x) \wedge C_4 = \mathsf{Com}_q(Q_y) \wedge Q_x < t \wedge Q_y < t \wedge P_x < t \wedge P_y < t\}$
5. Prove that L_x and R_x are equal modulo t, by dividing each side by t, showing correct range for the quotients and the remainders, and proving the remainders are equal.
   π_4 : $\mathsf{PK}\{(L_x, R_x, k_1, k'_1, r_1, r'_1) : C_4 = \mathsf{Com}_q(L_x) \wedge C_5 = \mathsf{Com}_q(R_x) \wedge C_8 = \mathsf{Com}_q(k_1) \wedge C_9 = \mathsf{Com}_q(r_1) \wedge C_{10} = \mathsf{Com}_q(k'_1) \wedge C_{11} = \mathsf{Com}_q(r'_1) \wedge L_x = k_1 t + r_1 \wedge R_x = k'_1 t + r'_1 \wedge r_1 < t \wedge r'_1 < t \wedge k_1 < \frac{q}{t} \wedge k'_1 < \frac{q}{t} \wedge r_1 - r'_1 = 0\}$
6. Prove that L_y and R_y are equal modulo t, by dividing each side by t, showing correct range for the quotients and the remainders, and proving the remainders are equal.
   π_5 : $\mathsf{PK}\{(L_y, R_y, k_2, k'_2, r_2, r'_2) : C_6 = \mathsf{Com}_q(L_y) \wedge C_7 = \mathsf{Com}_q(R_y) \wedge C_{12} = \mathsf{Com}_q(k_2) \wedge C_{13} = \mathsf{Com}_q(r_2) \wedge C_{14} = \mathsf{Com}_q(k'_2) \wedge C_{15} = \mathsf{Com}_q(r'_2) \wedge L_y = k_2 t + r_2 \wedge R_y = k'_2 t + r'_2 \wedge r_2 < t \wedge r'_2 < t \wedge k_2 < \frac{q}{t} \wedge k'_2 < \frac{q}{t} \wedge r_2 - r'_2 = 0\}$

Fig. 1. addition : $\mathsf{PK}\{(P = (P_x, P_y), Q = (Q_x, Q_y)) : T = (T_x, T_y) =$ addFormula$(P, Q) \wedge C_1 = \mathsf{Com}_q(P_x) \wedge C_2 = \mathsf{Com}_q(P_y) \wedge C_3 = \mathsf{Com}_q(Q_x) \wedge C_4 = \mathsf{Com}_q(Q_y)\}$

Theorem 3.1 *Let $E(\mathbb{F}_t)$ be an elliptic curve given by Eq. 1, $T \in E$ and $q > 2t^3$. Then, addition in Fig. 1 is a Σ-protocol for the relation $R = \{((T, C_P, C_Q), (P, Q)) : C_P = \mathsf{Com}_q(P) \wedge C_Q = \mathsf{Com}_q(Q) \wedge T =$ addFormula$(P, Q) \wedge P, Q \in E\}$.*

Using techniques similar to the above protocol addition, we obtain a protocol double to prove that doubling formula holds, i.e. $T = \mathsf{doubleFormula}(P)$. Now, we can handle all cases of point addition through the following statement:

$$(P \neq Q \wedge P \neq -Q \wedge T = \mathsf{addFormula}(P,Q)) \vee$$
$$(P = Q \wedge T = \mathsf{doubleFormula}(P)) \vee (P = -Q \wedge T = 0).$$

This statement can be proved using OR composition of Sigma protocols: protocol addition for the first part of the OR statement, protocol double for the second, and simple Sigma protocols for the last component. We denote the proof of point addition of two committed points by pointAddition.

$$\mathsf{pointAddition} : \mathsf{PK}\{(P,Q) : \mathsf{C}_P = \mathsf{Com}_q(P) \wedge \mathsf{C}_Q = \mathsf{Com}_q(Q) \wedge P, Q \in E \wedge$$
$$((P \neq Q \wedge P \neq -Q \wedge T = \mathsf{addFormula}(P,Q)) \vee$$
$$(P = Q \wedge T = \mathsf{doubleFormula}(P)) \vee (P = -Q \wedge T = 0))\}$$

For curves with a complete formula like Edwards, a point addition proof will not have different cases based on the relationship between P and Q.

Theorem 3.2. *Let $E(\mathbb{F}_t)$ be an elliptic curve given by Eq. 1, $T \in E$ and $q > 2t^3$. Then, pointAddition is a Σ-protocol for the relation $R = \{((T, \mathsf{C}_P, \mathsf{C}_Q), (P, Q)) : \mathsf{C}_P = \mathsf{Com}_q(P) \wedge \mathsf{C}_Q = \mathsf{Com}_q(Q) \wedge T = P + Q \wedge P, Q \in E\}$.*

We note that the protocol addition may be modified to prove point addition for a committed point T in the following way. The proofs π_1 and π_2 are on committed coordinates (T_x, T_y), and the range proof π_3 also includes proving the range of coordinates of T. We denote the point addition proof $\mathsf{PK}\{(P, Q, T) : \mathsf{C}_P = \mathsf{Com}_q(P) \wedge \mathsf{C}_Q = \mathsf{Com}_q(Q) \wedge \mathsf{C}_T = \mathsf{Com}_q(T) \wedge T = P + Q \wedge P, Q, T \in E\}$ on all committed inputs by comPointAddition.

We now construct a protocol to prove the equality of a committed value and the discrete logarithm of another committed value using the point addition proof. The double discrete logarithm proof is given in Fig. 2. (See the full version for a proof of security.) While the prover's work is dominated by the protocol pointAddition, we note that the range proofs for each challenge bit may be batched [15]. For soundness 2^{-60}, the protocol ddlog incurs proof size of about $2370 + \log\log t$ elements and prover's work of $1800 + 30\log t$ exponentiations.

Theorem 3.3. *Let $E(\mathbb{F}_t)$ be an elliptic curve given by Eq. 1, and $P \in E$ be an element of prime order p. Then, ddlog is a Σ-protocol for the relation $R = \{(P, \mathsf{C}, \mathsf{C}_h, (\lambda, h)) : \mathsf{C} = \mathsf{Com}(\lambda) \wedge \mathsf{C}_h = \mathsf{Com}(h) \wedge h = \lambda P, 0 < \lambda < p\}$ with soundness $1/2$.*

3.2 Sigma Protocols on Committed Outputs

In this section, we construct Sigma protocols for committed output. First, we note a simpler construction when the output is a single bit. (This simpler variant is used in our OR compositions.) In particular, given an algebraic commitment to

Given $\mathsf{C}_1 = \mathsf{Com}_p(\lambda), \mathsf{C}_2 = \mathsf{Com}_q(x), \mathsf{C}_3 = \mathsf{Com}_q(y)$, for $q > 2t^3$, prove that $(x, y) = \lambda P$, where $P \in E$ is an element of prime order p, $0 < \lambda < p$, P', Q', points in G_2 of order q.

1. The prover computes the following values: $a_1 = \mathsf{Com}_p(\alpha) = \alpha P + \beta_1 Q, a_2 = \mathsf{Com}_q(\gamma_1) = \gamma_1 P' + \beta_2 Q', a_3 = \mathsf{Com}_q(\gamma_2) = \gamma_2 P' + \beta_3 Q'$ where $\alpha \in \mathbb{F}_p$ is chosen at random, and $(\gamma_1, \gamma_2) = \alpha P$.
   and sends a_1, a_2, a_3 to the verifier.
2. The verifier chooses a random challenge bit c and sends it to the prover.
3. For challenge c,
   - If $c = 0$, compute $z_1 = \alpha, z_2 = \beta_1, z_3 = \beta_2, z_4 = \beta_3$. Send the tuple (z_1, z_2, z_3, z_4)
   - If $c = 1$, compute $z_1 = \alpha - \lambda$. Let $T = z_1 P = (t_1, t_2)$. The prover uses pointAddition (Figure 1) to prove that $T = (\gamma_1, \gamma_2) - (x, y)$.
   $\pi : \mathsf{PK}\{(x, y, \gamma_1, \gamma_2) : T = (\gamma_1, \gamma_2) - (x, y)\}$. Send (z_1, π)
4. Verification: Compute $(t_1, t_2) = z_1 P$. If $c = 0$, check if $a_1 = z_1 P + z_2 Q, a_2 = t_1 P' + z_3 Q', a_3 = t_2 P' + z_4 Q'$. If $c = 1$, verify proof π.

Fig. 2. ddlog : $\mathsf{PK}\{(\lambda, x, y, r, r_1, r_2) : \mathsf{Com}_p(\lambda) = \lambda P + rQ \wedge \mathsf{Com}_q(x) = xP' + r_1 Q' \wedge \mathsf{Com}_q(y) = yP' + r_2 Q' \wedge (x, y) = \lambda P\}$

private input x, public y and an efficient Sigma protocol to prove that $f(x, y) = 1$, we show how to construct an efficient Sigma protocol to prove $f(x, y) = b$, for a committed bit b. Let $f : \mathbb{Z}_q^{n+m} \to \{0, 1\}$, and let C be a commitment to the input x. Let f_{com} be the relation, $f_{\mathsf{com}} = \{(y, (x, b)) : ((x, y) \in \mathcal{L}_f \wedge b = 1) \vee (b = 0)\}$. The Sigma protocol for the relation f_{com} is given by the proof $\mathsf{PK}\{(b, x) : f(x, y) = b \wedge D_b = g^b h^{r_1} \wedge C = g^x h^r\}$. Let $\mathcal{G}$ be a group of order q, g a generator of $\mathcal{G}$, and h a random element of $\mathcal{G}$ such that the discrete logarithm of h with respect to g is unknown to the prover. Let Π be a Σ-protocol for the relation f. The Σ-protocol for f_{com} is shown in Fig. 3.

Given $y, C = \mathsf{Com}(x), D_b = \mathsf{Com}(b)$, prove that $f(x, y) = b$.

- The prover uses the protocol Π for f, Σ-protocol for proving knowledge of committed values, and the OR-transform to prove the following statement:

$$\mathsf{PK}\{(b, x) : \big(f(x, y) = 1 \wedge b = 1 \wedge D_b = g^b h^{r_1} \wedge C = g^x h^r\big)$$
$$\vee \big(b = 0 \wedge D_b = g^b h^{r_1} \wedge C = g^x h^r\big)\}$$

Fig. 3. comBitSigma : $\mathsf{PK}\{(b, x) : f(x, y) = b \wedge D_b = g^b h^{r_1} \wedge C = g^x h^r\}$

Theorem 3.4. *If Π is a Σ-protocol for f, then* comBitSigma *is a Σ-protocol for f_{com}.*

To generalize the above to the case where output is a group element and not a single bit, we need one more building block.

Proof of Point Addition and Discrete Log on Committed Points. Suppose we want to prove that a committed point is the sum of two group elements. But the challenge is that the input group elements are secret and are committed to, hence the prover also needs to prove knowledge of discrete logarithms of the input points with respect to a public base. Specifically, our goal is to design a protocol to prove knowledge of discrete logarithms of two committed points such that their sum is another committed point which we do using comPointAddition. Let E be an elliptic curve defined over $\mathbb{F}_t$, and let $P \in E$ be an element of prime order p. Let $q > 2t^3$ be a prime. The protocol comSum : $\mathsf{PK}\{(\gamma, \alpha, \beta, x_1, x_2) : \gamma = \alpha + \beta \wedge \alpha = x_1 P \wedge \beta = x_2 P\}$ for $0 < x_1, x_2 < p$ is shown in Fig. 4.

- The prover computes commitments $c_1 = \mathsf{Com}_p(x_1), c_2 = \mathsf{Com}_p(x_2), c_3 = \mathsf{Com}_q(\alpha), c_4 = \mathsf{Com}_q(\beta), c_5 = \mathsf{Com}_q(\gamma)$
- The prover uses ddlog to give the following proof.
  $\mathsf{PK}\{(x_1, \alpha) : \alpha = x_1 P \wedge c_3 = \mathsf{Com}_q(\alpha) \wedge c_1 = \mathsf{Com}_p(x_1)\}$
- The prover uses ddlog to give the following proof.
  $\mathsf{PK}\{(x_2, \beta) : \beta = x_2 P \wedge c_4 = \mathsf{Com}_q(\beta) \wedge c_2 = \mathsf{Com}_p(x_2)\}$
- The prover uses comPointAddition to give the following proof, given the commitments $c_3 = (\mathsf{Com}_q(\alpha_x), \mathsf{Com}_q(\alpha_y)), c_4 = (\mathsf{Com}_q(\beta_x), \mathsf{Com}_q(\beta_y)), c_5 = (\mathsf{Com}_q(\gamma_x), \mathsf{Com}_q(\gamma_y))$ and the point addition formula for the elliptic curve that defines the group (Equations 6,7).
  $\mathsf{PK}\{(\gamma, \alpha, \beta) : \gamma = \alpha + \beta \wedge c_3 = \mathsf{Com}_q(\alpha) \wedge c_4 = \mathsf{Com}_q(\beta) \wedge c_5 = \mathsf{Com}_q(\gamma)\}$

Fig. 4. comSum : $\mathsf{PK}\{(\gamma, \alpha, \beta, x_1, x_2) : \gamma = \alpha + \beta \wedge \alpha = x_1 P \wedge \beta = x_2 P\}$

When Committed Output is a Group Element. In the following discussion, similar to before, for a group element $\alpha = (\alpha_x, \alpha_y)$, where α_x, α_y are the two coordinates of the elliptic curve point, the commitment to the point is performed by committing to its two coordinates in the proper group, i.e. $\mathsf{Com}(\alpha) = (\mathsf{Com}(\alpha_x), \mathsf{Com}(\alpha_y))$.

We observe that given the above-mentioned building blocks i.e. ddlog and comSum, we can construct Sigma protocol on a committed output group element for algebraic statements of the form $f(x_1, \ldots, x_n) = \Pi g_i^{P_i(x_i)}$. We sketch the ideas at a high-level for some simple functions. Let $f : \mathbb{Z}_p^n \to \mathcal{G}$, where $\mathcal{G}$ is a group $E(\mathbb{F}_t)$ of order p. When $f(x) = g^x$, then this reduces to the ddlog proof. For $f(x_1, x_2) = g_1^{x_1} g_2^{x_2}$, it suffices to commit to $g_1^{x_1}$ and $g_2^{x_2}$ separately and call the comSum proof. To consider higher degree polynomials in the exponent let us consider $f(x) = g^{x^2}$. To construct a proof $\mathsf{PK}\{(x,y) : g^{x^2} = y \wedge C_1 = \mathsf{Com}(x) \wedge C_2 = \mathsf{Com}(y)\}$, the prover computes the commitments $C_1 = \mathsf{Com}_p(x)$, $C_2 = \mathsf{Com}_p(x^2)$ and $C_3 = \mathsf{Com}_q(k) = (\mathsf{Com}_q(k_x), \mathsf{Com}_q(k_y))$, where $k = g^{x^2} = (k_x, k_y)$, for the choice of q as discussed in Sect. 3.1. Now, the prover gives the following proofs. $\mathsf{PK}\{(x_2, k) : k = g^{x_2} \wedge C_2 = \mathsf{Com}_p(x_2) \wedge C_3 = \mathsf{Com}_q(k)\}$ using ddlog, and a Sigma protocol for $\mathsf{PK}\{(x_1, x_2) : x_2 = x_1^2 \wedge C_1 = \mathsf{Com}_p(x_1) \wedge C_2 = \mathsf{Com}_p(x_2)\}$. Given the above building blocks, it is easy to see that we can extend the techniques to devise proofs comSigma for $f(x_1, \ldots x_n) = \Pi g_i^{P_i(x_i)}$.

4   NIZK on Committed IO for Non-Algebraic Statements

In this section we instantiate the following two building blocks which are critical for our NIZKs for composite statements.

- *zk-SNARK on committed input.* Given an algebraic commitment $C = g^x h^r$, and a circuit f, a zk-SNARK proof that $f(x, z) = b$.
- *zk-SNARK on committed input and output.* Given algebraic commitments $C_1 = g^x h^r, C_2 = g^b h^r$, and a circuit f, a zk-SNARK proof that $f(x, z) = b$.

We first give a brief high-level description of our central ideas. Our starting point is a SNARK where the proof consists of multi-exponentiation that resembles a Pedersen commitment. We identify what part of the proof allows commitments to a private input (witness) and private output (for hiding intermediate values of a larger computation) by suitably separating the input/output wires so there are corresponding distinct proof elements in the SNARK. We then commit to the private input and output of the SNARK proof independently using Pedersen commitment, and show equality of the committed values and the values in the multi-exponentiation proof element. While this observation has been used in prior works in verifiable computation [24,35], it has been in different contexts and for different purposes. We briefly discuss how our ideas relate to two such ideas.

In [24], the authors present a verifiable computation scheme called Geppetto where the prover can share state across proofs. They generalize QAPs to create MultiQAPs which allow one to commit to data, and use it in many proofs. But crucially, all the proofs are for statements still represented as circuits while we also utilize the commitment to switch to sigma protocol proofs.

In [35], certain proof elements of a SNARK act as "accumulated" value of inputs in the context of large data size. The multi-exponentiations computed by the verifier in [35] act as a hash on data and different computations may be performed (verifiably) on it. The verifier computes the hash, and the proof verification involves checking the proof is consistent with the hash along with checks that the computation was performed correctly on the data using only the hash that was computed. On the other hand, in our setting, the multi-exponentiation is part of the proof, and computed by the prover, whose consistency across proofs must be shown. Additionally, these proofs could be different sigma protocols proving a variety of algebraic relations among some subset of the input used in the SNARK. Though our idea of exploiting a proof element with a certain structure is similar to the above works, we use it towards a different end.

For concreteness, we describe our protocol using the verifiable computation protocol Pinocchio [57] as a starting point. But our techniques carry over to other SNARK constructions as well. The key property we need from a SNARK construction is that the proof contains a multi-exponentiation of the input/output. Given this, we separate the circuit wires and obtain in a non-blackbox way, commitments as part of the SNARK proof.

Before giving the description of the above building blocks, we introduce an important ingredient: a protocol for proving equality of the discrete logarithms $(a_1, \ldots, a_n)$ in $y = \prod_{i=1}^{n} G_i^{a_i}$ and individual algebraic commitments to them. Using the standard notation, we denote the protocol by $\mathsf{PK}\{(a_1, \ldots, a_n, r_1, \ldots, r_n) : y = \prod_{i=1}^{n} G_i^{a_i} \wedge C_1 = g^{a_1} h^{r_1} \wedge \cdots \wedge C_n = g^{a_n} h^{r_n}\}$. We include the steps of the protocol in the full version.

4.1 zk-SNARK on Committed Inputs

Recall that at a high level, each polynomial of the quadratic program (Definition 2.6), say, $v_k(x) \in \mathbb{F}[x]$ is mapped to an element in a bilinear group, $g^{v_k(s)}$, where s is a secret value chosen during CRS generation. Given these group elements and the values a_i on the circuit wires which are the coefficients of the quadratic program, the prover can compute "in the exponent" to obtain $g^{v(s)}$, where $v(s) = \sum a_i v_k(s)$. The verifier uses the bilinear map to verify that the divisibility check of the QAP holds. We assume the computations are over large fields, that is, the QAP is defined over $\mathbb{F}_p$ for a large p. The size of the field is exponential in the security parameter. We omit p in all further descriptions of the field.

Let $f : \mathbb{F}^N \to \mathbb{F}^{n'}$ be a function with input/output values from $\mathbb{F}$, computed by an arithmetic circuit C with input wires labeled $1, \ldots, N$, output wires labeled $m - n' + 1, \ldots, m$. Let $\mathcal{Q}$ be a QAP of size m and degree d corresponding to C. We separate the circuit wires I into private input, public input, intermediate values, and output wires. Let $I_{com} \subseteq \{1, \ldots, N\}$ be the set of indices corresponding to the private inputs $a_1, \ldots, a_n$, I_{pub} the indices for the public input wires, and I_{out} the indices for the public output. Then let $I_{mid} = \{1, \ldots, m\} \setminus (I_{pub} \cup I_{com} \cup I_{out})$ be the indices of the intermediate wires. This way there are separate CRS elements corresponding to the private input and public input allowing the prover to compute corresponding proof elements. The divisibility check can still proceed, and we include additional span checks for the new proof elements. Now, we bind the multi-exponentiation corresponding to the private input in the proof to the value committed to in a Pedersen commitment using the protocol comEq. Let $C_i = g^{a_i} h^{r_i}$ be a Pedersen commitment to the ith input a_i. The construction comInSnark : $\mathsf{PK}\{(a_1, \ldots, a_n, r_1, \ldots, r_n) : f(a_1, \ldots a_n, z_1, \ldots, z_{N-n}) = (b_1, \ldots, b_{n'}) \wedge C_1 = g^{a_1} h^{r_1} \wedge \cdots \wedge C_n = g^{a_n} h^{r_n}\}$ is given in Fig. 5.

Given commitments to private inputs $C_i = g^{a_i} h^{r_i}$ for $i \in [n]$, public inputs $z_1, \ldots, z_{N-n}$, and public outputs $b_1, \ldots, b_{n'}$.

1. CRS generation: Run $\mathsf{GroupGen}(1^\kappa)$ to get $(p, \mathbb{G}, \widetilde{\mathbb{G}}, \mathbb{G}_T, g, \tilde{g}, e)$. Choose $r_v, r_w, \alpha_v, \alpha_w, \alpha_y, s, \beta, \gamma \xleftarrow{R} \mathbb{F}$. Set $r_y = r_v r_w, g_v = g^{r_v}, g_w = g^{r_w}, \tilde{g}_w = \tilde{g}^{r_w}, g_y = g^{r_y}$.
   Set the CRS to be:

$$\mathsf{crs} = \Big(\{g_v^{v_k(s)}\}_{k \in I_{com}}, \{g_v^{v_k(s)}\}_{k \in I_{mid}}, \{\tilde{g}_w^{w_k(s)}\}_{k \in I_{com}},$$
$$\{\tilde{g}_w^{w_k(s)}\}_{k \in I_{mid}}, \{g_y^{y_k(s)}\}_{k \in I_{com}}, \{g_y^{y_k(s)}\}_{k \in I_{mid}}, \{g_v^{\alpha_v v_k(s)}\}_{k \in I_{com}},$$
$$\{g_v^{\alpha_v v_k(s)}\}_{k \in I_{mid}}, \{\tilde{g}_w^{\alpha_w w_k(s)}\}_{k \in I_{com}}, \{\tilde{g}_w^{\alpha_w w_k(s)}\}_{k \in I_{mid}},$$
$$\{g_y^{\alpha_y y_k(s)}\}_{k \in I_{com}}, \{g_y^{\alpha_y y_k(s)}\}_{k \in I_{mid}}, \{g^{s^i}\}_{i \in [d]}, \{\tilde{g}^{s^i}\}_{i \in [d]},$$
$$\{g^{\alpha_v s^i}\}_{i \in [d]}, \{\tilde{g}^{\alpha_v s^i}\}_{i \in [d]}, \{g^{\alpha_w s^i}\}_{i \in [d]}, \{\tilde{g}^{\alpha_w s^i}\}_{i \in [d]}, \{g^{\alpha_y s^i}\}_{i \in [d]},$$
$$\{\tilde{g}^{\alpha_y s^i}\}_{i \in [d]}, \{g_v^{\beta v_k(s)} g_w^{\beta w_k(s)} g_y^{\beta y_k(s)}\}_{k \in I_{com}}, \{g_v^{\beta v_k(s)} g_w^{\beta w_k(s)} g_y^{\beta y_k(s)}\}_{k \in I_{mid}} \Big)$$

   Set the short verification CRS to be:

$$\mathsf{shortcrs} = \Big(g, \tilde{g}, \tilde{g}^{\alpha_v}, g^{\alpha_w}, \tilde{g}^{\alpha_y}, \tilde{g}^\gamma, g^{\beta\gamma}, \tilde{g}^{\beta\gamma}, g^{t(s)},$$
$$\{g_v^{v_k(s)}\}_{k \in I_{com}}, \{g_v^{v_k(s)}\}_{k \in I_{pub} \cup I_{out}}, \{\tilde{g}_w^{w_k(s)}\}_{k \in I_{pub} \cup I_{out}}, \{g_y^{y_k(s)}\}_{k \in I_{pub} \cup I_{out}} \Big)$$

2. Prove: On input $z_1, \ldots, z_{N-n}$, witness $a_1, \ldots, a_n$, and crs, the prover evaluates the QAP to obtain $\{a_i\}_{i \in [m]}$. (Equivalently, evaluates the circuit to obtain the values on the circuit wires). The prover solves for the quotient polynomial h such that $p(x) = h(x)t(x)$. Let $v_{com}(x) = \sum_{k \in I_{com}} a_k v_k(x)$, $v_{mid}(x) = \sum_{k \in I_{mid}} a_k v_k(x)$ and similarly define $w_{com}(x), w_{mid}(x), y_{com}(x)$ and $y_{mid}(x)$.
   – The prover computes the proof π:

$$\Big(g_v^{v_{com}(s)}, g_v^{v_{mid}(s)}, \tilde{g}_w^{w_{com}(s)}, \tilde{g}_w^{w_{mid}(s)}, g_y^{y_{com}(s)}, g_y^{y_{mid}(s)}, \tilde{g}^{h(s)},$$
$$\tilde{g}_v^{\alpha_v v_{com}(s)}, \tilde{g}_v^{\alpha_v v_{mid}(s)}, g_w^{\alpha_w w_{com}(s)}, g_w^{\alpha_w w_{mid}(s)}, \tilde{g}_y^{\alpha_y y_{com}(s)}, \tilde{g}_y^{\alpha_y y_{mid}(s)}$$
$$g_v^{\beta v_{com}(s)} g_w^{\beta w_{com}(s)} g_y^{\beta y_{com}(s)}, g_v^{\beta v_{mid}(s)} g_w^{\beta w_{mid}(s)} g_y^{\beta y_{mid}(s)} \Big)$$

   – Prove input consistency with commitment: The prover uses the Sigma protocol comEq to compute π_{in}: $\mathrm{PK}\{(a_1, \ldots, a_n, r_1, \ldots, r_n) : y = \prod_{i=1}^{n} G_i^{a_i} \wedge C_1 = g^{a_1} h^{r_1} \wedge \cdots \wedge C_n = g^{a_n} h^{r_n}\}$, for $G_i = g_v^{v_i(s)}, i \in I_{com}$, and $y = g_v^{v_{com}(s)}$.

3. Verify:
   – On input $\mathsf{shortcrs}$, z, and proofs π, π_{in} parse π as

$$\pi = \Big(g^{V_{com}}, g^{V_{mid}}, \tilde{g}^{W_{com}}, \tilde{g}^{W_{mid}}, g^{Y_{com}}, g^{Y_{mid}}, \tilde{g}^H,$$
$$\tilde{g}^{V'_{com}}, \tilde{g}^{V'_{mid}}, g^{W'_{com}}, g^{W'_{mid}}, \tilde{g}^{Y'_{com}}, \tilde{g}^{Y'_{mid}}, g^{Z_{com}}, g^{Z_{mid}} \Big)$$

Fig. 5. $\mathsf{comInSnark}$: $\mathrm{PK}\{(a_1, \ldots, a_n, r_1, \ldots, r_n) : f(a_1, \ldots a_n, z_1, \ldots, z_{N-n}) = (b_1, \ldots, b_{n'}) \wedge C_1 = g^{a_1} h^{r_1} \wedge \ldots \wedge C_n = g^{a_n} h^{r_n}\}$

– Divisibility check. Compute $g_v^{v_{io}(s)} = \prod_{k \in I_{pub} \cup I_{out}} (g_v^{v_k(s)})^{a_k}$. Similarly, compute $\tilde{g}_w^{w_{io}(s)}$ and $g_y^{y_{io}(s)}$. Verify that

$$e\left(g_v^{v_0(s)} g_v^{v_{io}(s)} g^{V_{com}} g^{V_{mid}}, \tilde{g}_w^{w_0(s)} \tilde{g}_w^{w_{io}(s)} \tilde{g}^{W_{com}} \tilde{g}^{W_{mid}}\right)$$

$$= e\left(g^{t(s)}, \tilde{g}^H\right) \cdot e\left(g_y^{y_0(s)} g_y^{y_{io}(s)} g^{Y_{com}} g^{Y_{mid}}, \tilde{g}\right).$$

– Verify that the linear combinations are in correct spans.
$e\left(g^{V_{com}}, \tilde{g}^{\alpha_v}\right) = e\left(g, \tilde{g}^{V'_{com}}\right)$, $e\left(g^{V_{mid}}, \tilde{g}^{\alpha_v}\right) = e\left(g, \tilde{g}^{V'_{mid}}\right)$,
$e\left(g^{W'_{com}}, \tilde{g}\right) = e\left(g^{\alpha_w}, \tilde{g}^{W_{com}}\right)$, $e\left(g^{W'_{mid}}, \tilde{g}\right) = e\left(g^{\alpha_w}, \tilde{g}^{W_{mid}}\right)$,
$e\left(g^{Y_{com}}, \tilde{g}^{\alpha_y}\right) = e\left(g, \tilde{g}^{Y'_{com}}\right)$, $e\left(g^{Y_{mid}}, \tilde{g}^{\alpha_y}\right) = e\left(g, \tilde{g}^{Y'_{mid}}\right)$.

– Verify same coefficients in all linear combinations.
   (a) $e\left(g^{Z_{com}}, \tilde{g}^\gamma\right) = e\left(g^{V_{com}} g^{Y_{com}}, \tilde{g}^{\beta\gamma}\right) \cdot e\left(g^{\beta\gamma}, \tilde{g}^{W_{com}}\right)$
   (b) $e\left(g^{Z_{mid}}, \tilde{g}^\gamma\right) = e\left(g^{V_{mid}} g^{Y_{mid}}, \tilde{g}^{\beta\gamma}\right) \cdot e\left(g^{\beta\gamma}, \tilde{g}^{W_{mid}}\right)$

– Verify input consistency with commitment: Set $G_i = g_v^{v_i(s)}$, $i \in I_{com}$, and $y = g^{V_{com}}$. Verify the proof π_{in}.

Fig. 5. (*continued*)

Zero-knowledge. We make our construction zero-knowledge, and obtain zkcomInSnark, by randomizing the elements in the proof π such that the checks verify and the proof is statistically indistinguishable from random group elements. Specifically, the prover chooses random $\delta_v, \delta_w, \delta_y \leftarrow \mathbb{F}$, and adds $\delta_v t(s)$ in the exponent to $v_{com}(s)$, $v_{mid}(s)$; $\delta_w t(s)$ to $w_{com}(s)$, $w_{mid}(s)$; and $\delta_y t(s)$ to $y_{com}(s)$, $y_{mid}(s)$. It is easy to see that the modified value of $p(x)$ remains divisible by $t(x)$. The following terms are added to crs: $g_v^{t(s)}$, $\tilde{g}_w^{t(s)}$, $g_y^{t(s)}$, $g_v^{\alpha_v t(s)}$, $g_w^{\alpha_w t(s)}$, $g_y^{\alpha_y t(s)}$, $g_v^{\beta t(s)}$, $g_w^{\beta t(s)}$, $g_y^{\beta t(s)}$ ($g_v^{t(s)}$ is also added to shortcrs). Prover can now compute the new values in π from crs, and they are verified in the same manner as before. The proof π_{in} now proves a slightly different statement: $\mathrm{PK}\{(a_1, \ldots, a_n, \delta, r_1, \ldots, r_n) : y = H^\delta \prod_{i=1}^n G_i^{a_i} \wedge C_1 = g^{a_1} h^{r_1} \wedge \ldots \wedge C_n = g^{a_n} h^{r_n}\}$. To verify it, the verifier uses $g_v^{t(s)}$ from shortcrs.

Theorem 4.1. *If q-PDH, 2q-SDH and d-PKE assumptions hold for* GroupGen *for $q \geq 4d + 4$, then zk-*comInSnark *instantiated with a QAP of degree d is secure under Definition 2.2.*

A proof of Theorem 4.1 can be found in the full version. Similarly, by separating the circuit wires into private input, public input, intermediate values and private output, we obtain zk-SNARK on committed input and output. We state the theorem below.

Theorem 4.2. *If q-PDH, 2q-SDH and d-PKE assumptions hold for* GroupGen *for $q \geq 4d + 4$, and discrete logarithm assumption holds in $\mathbb{G}$, then zk-*comIOSnark *instantiated with a QAP of degree d is secure under Definition 2.2.*

5   Constructions for Compound Statements

In this section we use the building blocks we constructed in Sects. 4 and 3, to devise proofs for compound statements. In the following, we distinguish between functions that have an efficient algebraic representation versus functions that are efficiently represented as an arithmetic circuit over a field. Of course, any algebraic function can be written as a circuit over some field. But certain functions, modular exponentiation for instance, have a large circuit size and hence it is more desirable to not use a circuit in computing them. Therefore, when we say *algebraic* or *arithmetic* for functions below, we really mean the efficient representation of the function for computation. We say a function f is arithmetic if an arithmetic circuit is used to compute f, and say f is algebraic if it is represented algebraically. In this section, we show how to prove compound statements involving function compositions, OR, and AND. In our compositions, the SNARK used for the circuit could use a group whose order does not match with the group of the sigma protocol for the algebraic part. We construct a building block Eq to prove equality of committed values in different groups, given in the full version, which we use in our compositions.

5.1   Function Composition

We assume that the commitments we use in the following are in groups of correct order for the computation, so as to focus on the ideas for the composition. Wlog., our compositions hold even when the scalar field of the elliptic curve group, the field the curve is defined over and the field of the arithmetic circuit are all different, since we can prove equality of committed values in different groups using the protocol Eq. We present the interactive variant for ease of presentation but note that all our constructions can be made non-interactive by running all the proofs in parallel and invoking the standard Fiat-Shamir transform (see Sect. 2.1). The constructions below also easily generalize to functions that have more input/output elements than shown, i.e. we can obtain constructions for statements of the form $\mathsf{PK}\{(x_1,\ldots,x_n,y_1,\ldots,y_m) : f_1(x_1,\ldots,x_n, f_2(y_1,\ldots,y_m)) = z\}$ where f_1, f_2 may each be arithmetic or algebraic. We give constructions composition by elaborating on the four possible compositions next:

1. f_1 and f_2 are functions represented as arithmetic circuits. Let $f_1 : \mathbb{F}_p^2 \to \mathbb{F}_p$, and $f_2 : \mathbb{F}_p \to \mathbb{F}_p$, and we want to prove knowledge of secrets x_1, x_2 such that $f_1(x_1, f_2(x_2)) = z$ for a public z. An example is proof of knowledge of x_1 and x_2 such that $H(x_1 \| H(x_2)) = z$ where H is a collision resistant hash function such as SHA256. Such a composition can help reduce the size of CRS by composing the same or a few SNARK systems multiple times to obtain more complex statements without an increase in CRS size.

> – The prover commits to x_1, x_2 and $x_3 = f_2(x_2)$ by computing $c_1 = \mathsf{Com}_p(x_1), c_2 = \mathsf{Com}_p(x_2), c_3 = \mathsf{Com}_p(x_3)$. The prover sends c_1, c_2, c_3 to the verifier.

- The prover uses zk-comIOSnark to give a proof that $f_2(x_2) = x_3$, given c_2 and c_3. $\mathsf{PK}\{(x_2, x_3, r_2, r_3) : f_2(x_2) = x_3 \wedge c_2 = \mathsf{Com}_p(x_2) \wedge c_3 = \mathsf{Com}_p(x_3)\}$.
- The prover uses zk-comInSnark to give a proof that $f_1(x_1, x_3) = z$ given c_1, c_3 and z. $\mathsf{PK}\{(x_1, x_3, r_1, r_3) : f_1(x_1, x_3) = z \wedge c_1 = \mathsf{Com}_p(x_1) \wedge c_3 = \mathsf{Com}_p(x_3)\}$.

2. f_1 is an arithmetic circuit and f_2 is algebraic. Let $f_1 : \mathbb{F}_p^3 \to \mathbb{F}_p, f_2 : \mathbb{Z}_q \to \mathcal{G}$ and $T : \mathcal{G} \to \mathbb{F}_p^2$. In this proof, we assume the algebraic function is over an elliptic curve group and assume the natural transformation for mapping an elliptic curve point to a tuple of field elements, i.e. its coordinates. Let $\mathcal{G}$ be an elliptic curve group of prime order q, and let $T(k) = (k_x, k_y)$ for $k \in \mathcal{G}$, where (k_x, k_y) are the coordinates of the elliptic curve point. The following is a protocol for $\mathsf{PK}\{(x_1, x_2) : f_1(x_1, T(f_2(x_2))) = z\}$. An example is proving knowledge of x such that $H(g^x) = z$.

- The prover commits to x_1, x_2 and $k = f_2(x_2)$ by computing $c_1 = \mathsf{Com}_p(x_1), c_2 = \mathsf{Com}_q(x_2), c_3 = \mathsf{Com}_p(k) = (\mathsf{Com}_p(k_x), \mathsf{Com}_p(k_y))$, and sends c_1, c_2, c_3 to the verifier.
- The prover uses the protocols ddlog and the sigma protocol on committed group element comSigma to give the following proof: $\mathsf{PK}\{(x_2, k, r_2, r_3) : f_2(x_2) = k \wedge c_2 = \mathsf{Com}_q(x_2) \wedge c_3 = \mathsf{Com}_p(k)\}$.
- The prover uses zk-comInSnark to prove $f_1(x_1, T(k)) = z$ given c_1, c_3, c_4. $\mathsf{PK}\{(x_1, k, r_1, r_3) : f_1(x_1, T(k)) = z \wedge c_1 = \mathsf{Com}_p(x_1) \wedge c_3 = \mathsf{Com}_p(k)\}$.

3. f_1 is algebraic, and f_2 is an arithmetic circuit. Let $f_1 : \mathbb{Z}_q^2 \to \mathcal{G}, f_2 : \mathbb{F}_p \to \mathbb{F}_p$. Let Π be a Σ-protocol for f_1. The following is a protocol for $\mathsf{PK}\{(x_1, x_2) : f_1(x_1, f_2(x_2)) = z\}$. An example is proving knowledge of x such that $g^{H(x)} = z$ where H is a hash function. This composition commonly appears when proving knowledge of a digitally signed message.

- The prover commits to $x_1, x_2, x_3 = f_2(x_2)$ by computing $c_1 = \mathsf{Com}_q(x_1), c_2 = \mathsf{Com}_p(x_2), c_3 = \mathsf{Com}_q(x_3), c_3' = \mathsf{Com}_p(x_3)$. c_3 is committed to twice, in groups of order p and q. The prover sends c_1, c_2, c_3, c_3' to the verifier.
- The prover uses zk-comIOSnark to give a proof that $f_2(x_2) = x_3$, given c_2 and c_3'. $\mathsf{PK}\{(x_2, x_3', r_2, r_3') : f_2(x_2) = x_3' \wedge c_2 = \mathsf{Com}_p(x_2) \wedge c_3' = \mathsf{Com}_p(x_3')\}$.
- The prover uses the sigma protocol Π to give the following proof. $\mathsf{PK}\{(x_1, x_3, r_1, r_3) : f_1(x_1, x_3) = z \wedge c_1 = \mathsf{Com}_q(x_1) \wedge c_3 = \mathsf{Com}_q(x_3)\}$.
- The prover uses the protocol Eq to prove that c_3' and c_3 are commitments to the same value. $\mathsf{PK}\{(x_3, x_3', r_3, r_3') : x_3 \equiv x_3' \pmod{q} \wedge c_3 = \mathsf{Com}_q(x_3) \wedge c_3' = \mathsf{Com}_p(x_3')\}$

4. f_1 and f_2 are algebraic. Let $f_1 : \mathbb{Z}_p^3 \to \mathcal{G}_1, f_2 : \mathbb{Z}_q \to \mathcal{G}_2$, where $\mathcal{G}_1$ and $\mathcal{G}_2$ are elliptic curve groups of prime order p and q respectively. Let $T(k) = (k_x, k_y)$ for $k \in \mathcal{G}_2$, where (k_x, k_y) are the coordinates of the elliptic curve

point. Let Π_1 be a Σ-protocol for f_1. Let $x_1 \in \mathbb{Z}_p, x_2 \in \mathbb{Z}_q$. An example is proving knowledge of x such that $g_1^{T(g_2^x)}$ for generators g_1 and g_2 for two different groups and a valid transformation T for mapping from one group to another. These statements often occur in anonymous credential constructions or proving statements about accumulators but the only previous constructions are for RSA groups.

- The prover commits to x_1, x_2 and $k = f_2(x_2)$ by computing $c_1 = \mathsf{Com}_p(x_1), c_2 = \mathsf{Com}_q(x_2), c_3 = \mathsf{Com}_p(k) = (\mathsf{Com}_p(k_x), \mathsf{Com}_p(k_y))$, and sends c_1, c_2, c_3 to the verifier.
- The prover uses the protocols ddlog and the sigma protocol on committed group element comSigma for f_2 to give the following proof: $\mathsf{PK}\{(x_2, k, r_2, r_3) : f_2(x_2) = k \wedge c_2 = \mathsf{Com}_q(x_2) \wedge c_3 = \mathsf{Com}_p(k)\}$.
- The prover uses the sigma protocol Π_1 to give the following proof. $\mathsf{PK}\{(x_1, k, r_1, r_3) : f_1(x_1, T(k)) = z \wedge c_1 = \mathsf{Com}_p(x_1) \wedge c_3 = \mathsf{Com}_p(k)\}$.

Theorem 5.1 (Function Composition). *The constructions* composition *are non-interactive zero-knowledge arguments* $\mathsf{PK}\{(x_1, \ldots, x_n, y_1, \ldots, y_m) : f_1(x_1, \ldots, x_n, f_2(y_1, \ldots, y_m)) = z\}$, *as per Definition 2.2, for any* $f_1, f_2 \in \{\text{algebraic}, \text{arithmetic}\}$ *assuming the security of* zk-comInSnark, zk-comIOSnark, ddlog, Eq.

5.2   OR Composition

Consider the OR composition where a prover wants to show that $f_1(x_1, x_2) = 1$ or $f_2(x_1, x_3) = 1$ *but* without revealing which one is true. We give constructions compoundOR : $\mathsf{PK}\{(x_1, x_2, x_3) : f_1(x_1, x_2) \vee f_2(x_1, x_3) = 1\}$, where the f_is could have either an arithmetic or algebraic representation, and could have shared secret inputs.

1. f_1 and f_2 are functions represented as arithmetic circuits. Let $f_1 : \mathbb{F}_p^2 \to \{0, 1\}$, and $f_2 : \mathbb{F}_q^2 \to \{0, 1\}$, $q < p$. An example is composing proofs for two SNARK systems that work over different elliptic curve groups.

- The prover commits to the inputs by computing, $c_1 = \mathsf{Com}_p(x_1), c_1' = \mathsf{Com}_q(x_1), c_2 = \mathsf{Com}_p(x_2), c_3 = \mathsf{Com}_q(x_3)$, and to the output bits $b_1 = f_1(x_1, x_2), b_2 = f_1(x_1, x_3)$, $c_4 = \mathsf{Com}_p(b_1), c_5 = \mathsf{Com}_q(b_2), c_5' = \mathsf{Com}_p(b_2)$. x_1 and b_2 are committed to in both groups of order p and q.
- The prover uses zk-comIOSnark to give proofs. $\mathsf{PK}\{(x_1, x_2, b_1, r_1, r_2, r_4) : f_1(x_1, x_2) = b_1 \wedge c_1 = \mathsf{Com}_p(x_1) \wedge c_2 = \mathsf{Com}_p(x_2) \wedge c_4 = \mathsf{Com}_p(b_1)\}$. $\mathsf{PK}\{(x_1', x_3, b_2, r_1', r_3, r_5) : f_2(x_1', x_3) = b_2 \wedge c_1' = \mathsf{Com}_q(x_1') \wedge c_3 = \mathsf{Com}_q(x_3) \wedge c_5 = \mathsf{Com}_q(b_2)\}$.

- The prover uses the protocol Eq to prove that c_1' and c_1 are commitments to the same value.
  $\mathsf{PK}\{(x_1, x_1' r_1, r_1') : x_1 \equiv x_1' \pmod{q} \wedge c_1 = \mathsf{Com}_p(x_1) \wedge c_1' = \mathsf{Com}_q(x_1)\}$
- The prover uses the protocol Eq to prove that c_5' and c_5 are commitments to the same value.
  $\mathsf{PK}\{(b_2, b_2', r_5, r_5') : b_2 \equiv b_2' \pmod{q} \wedge c_5 = \mathsf{Com}_q(b_2) \wedge c_5' = \mathsf{Com}_p(b_2')\}$
- The prover uses the Sigma protocol OR-transform to give the following proof.
  $\mathsf{PK}\{(b_1, b_2, r_4, r_5) : (b_1 = 1 \wedge c_4 = \mathsf{Com}_p(b_1)) \vee (b_2 = 1 \wedge c_5' = \mathsf{Com}_p(b_2))\}$

2. One of them is an arithmetic circuit and the other is an algebraic relation. Wlog., f_1 is represented as an arithmetic circuit and f_2 is an algebraic statement. Let $f_1 : \mathbb{F}_p^2 \to \{0,1\}, f_2 : \mathbb{Z}_q^2 \to \{0,1\}, q < p$. Let Π be a Σ-protocol for f_2. An example is proving knowledge of x such that $H(x) = y$ OR $g^x = z$.

- The prover commits to the inputs, $c_1 = \mathsf{Com}_q(x_1), c_1' = \mathsf{Com}_p(x_1), c_2 = \mathsf{Com}_p(x_2), c_3 = \mathsf{Com}_q(x_3)$. The prover computes the outputs $b_1 = f_1(x_1, x_2), b_2 = f_1(x_1, x_3)$ and commits to them by computing $c_4 = \mathsf{Com}_p(b_1), c_5 = \mathsf{Com}_q(b_2), c_5' = \mathsf{Com}_p(b_2)$.
- The prover uses comIOSnark to give the following proof.
  $\mathsf{PK}\{(x_1', x_2, b_1, r_1', r_2, r_4) : f_1(x_1', x_2) = b \wedge c_1' = \mathsf{Com}_p(x_1) \wedge c_2 = \mathsf{Com}_p(x_2) \wedge c_4 = \mathsf{Com}_p(b_1)\}.$
- The prover uses the protocol Π and protocol comBitSigma (Fig. 3) to prove the following.
  $\mathsf{PK}\{(x_1, x_3, b_2, r_1, r_3, r_5) : f_2(x_1, x_3) = b_2 \wedge c_1 = \mathsf{Com}_q(x_1) \wedge c_3 = \mathsf{Com}_q(x_3) \wedge c_5 = \mathsf{Com}_q(b_2)\}$
- The prover uses the protocol Eq to prove that c_1' and c_1 are commitments to the same value.
  $\mathsf{PK}\{(x_1, x_1' r_1, r_1') : x_1 \equiv x_1' \pmod{q} \wedge c_1 = \mathsf{Com}_q(x_1) \wedge c_1' = \mathsf{Com}_p(x_1)\}$
- The prover uses the protocol Eq to prove that c_5' and c_5 are commitments to the same value.
  $\mathsf{PK}\{(b_2, b_2', r_5, r_5') : b_2 \equiv b_2' \pmod{q} \wedge c_5 = \mathsf{Com}_q(b_2) \wedge c_5' = \mathsf{Com}_p(b_2')\}$
- The prover uses the Sigma protocol OR-transform to prove the following.
  $\mathsf{PK}\{(b_1, b_2, r_4, r_5) : (b_1 = 1 \wedge c_4 = \mathsf{Com}_q(b_1)) \vee (b_2 = 1 \wedge c_5 = \mathsf{Com}_q(b_2))\}.$

Let f_{OR} be the relation given by $f_{\mathsf{OR}} = \{((f_1, f_2), (x_1, x_2, x_3)) : ((x_1, x_2) \in R_{f_1}) \vee ((x_1, x_3) \in R_{f_2})\}$.

Theorem 5.2 (OR Composition). *The constructions* compoundOR *are non-interactive zero-knowledge arguments* $\mathsf{PK}\{(x_1, x_2, x_3) : f_1(x_1, x_2) \vee f_2(x_1, x_3) = 1\}$, *as per Definition 2.2, for the relation* f_{OR}, *for any* $f_1, f_2 \in \{\mathsf{algebraic}, \mathsf{arithmetic}\}$, *assuming the security of* zk-comInSnark, *zk-*comIOSnark, comBitSigma, Eq.

5.3 AND Composition

Techniques shown in Sect. 5.2 extend for proofs of the form, $\mathsf{PK}\{(x_1, x_2, x_3) : f_1(x_1, x_2) \wedge f_2(x_1, x_3) = 1\}$ for all combinations of f_1 and f_2 being arithmetic and algebraic. In particular, to prove the AND of multiple statements, we use our building blocks comInSnark for the arithmetic part, Σ-protocol for the algebraic part, and Eq to switch between groups.

6   Applications

6.1   Privacy-preserving Audits of Bitcoin Exchanges

In this section, we show how to use our constructions for proving composite statements in zero-knowledge to build a privacy-preserving proof of solvency for Bitcoin exchanges. A proof of solvency demonstrates that an exchange controls sufficient reserves to settle each customer's account. If the exchange loses a large amount of money in an attack, it would not be able to provide such a proof. Thus customers will find out about the attack very soon and take necessary actions.

A proof of solvency consists of three components:

- A *proof of liabilities* that allows customers to verify that their accounts are included in the total.
- A *proof of assets* which shows that the exchange has a certain amount of reserves.
- A proof that the reserves cover the liabilities to an acceptable degree.

Let g, h be fixed public generators of a group G of order q. For a Bitcoin public key y, $x \in \mathbb{Z}_q$ is the corresponding secret key such that $y = g^x$. In the proof of assets below, for a group element $k = (k_x, k_y)$, we write $\mathsf{Com}(k)$ to mean a commitment to the coordinates of k, i.e. $\mathsf{Com}(k) = (\mathsf{Com}(k_x), \mathsf{Com}(k_y))$. The Bitcoin address corresponding to a key y is given by $\mathsf{h} = H(y)$, where H hashes y to a more compact representation. We denote the balance associated with an address h by $\mathsf{bal}(\mathsf{h})$.

Proof of assets. We give the proof of assets in Fig. 6, which allows an exchange to generate a commitment to its total assets along with a zero-knowledge proof that the exchange knows the private keys for a set of Bitcoin addresses whose total value is equal to the committed value. The exchange creates a set of hashes $\mathcal{PK}$ to serve as an anonymity set: $\mathcal{PK} = \{\mathsf{h}_1, \cdots, \mathsf{h}_n\}$ from the public data available on the blockchain. Let $x_1, \cdots, x_n$ be the corresponding secret keys, so that $\mathsf{h}_i = H(g^{x_i})$, s_i indicates whether the exchange knows the ith secret key. The total assets can now be expressed as $\mathsf{Assets} = \sum_{i=1}^{n} s_i \cdot \mathsf{bal}(\mathsf{h}_i)$. The public data available on the blockchain is $\mathsf{h}_i = H(y_i), p_i = g^{\mathsf{bal}(\mathsf{h}_i)}$ for all $i \in [1, n]$.

Zero-knowledge and soundness of the proof of assets follow from properties of our constructions for compound statements (Theorems 5.1 and 5.2) and properties of the Sigma protocols used. Proofs of liabilities and solvency have been moved to the full version because they are very similar to Provisions. We compare the trade-off between proof size and prover's work in our approach versus Provisions and a full SNARK solution in Table 1 in Appendix A.

- The exchange computes the commitments. For $i \in [1, n]$, commit to x_i by publishing $\alpha_i = \mathsf{Com}_q(x_i) = g^{x_i} h^{r_i}$, and commit to y_i by publishing $\beta_i = \mathsf{Com}_q(y_i)$.
- The exchange commits to the balance in each address for the public keys he controls and to 0 otherwise, by publishing $u_i = \mathsf{Com}_q(s_i \cdot \mathsf{bal}(\mathsf{h}_i)) = g^{s_i \cdot \mathsf{bal}(\mathsf{h}_i)} h^{t_i}$, $s_i \in \{0, 1\}$, where $s_i = 1$ if the exchange knows x_i such that $y_i = g^{x_i}$.
- The exchange uses protocols ddlog, comIOSnark and the constructions for function composition and OR composition, composition and compoundOR respectively, to prove the following for each i,

$$\pi_i : \mathsf{PK}\{(x_i, y_i, s_i, r_i, a_i, b_i, t_i) : \big(\alpha_i = \mathsf{Com}_q(x_i) \wedge \beta_i = \mathsf{Com}_q(y_i) \wedge$$
$$u_i = \mathsf{Com}_q(s_i \cdot \mathsf{bal}(\mathsf{h}_i)) \wedge f_1(f_2(x_i), \mathsf{h}_i) = s_i \wedge s_i = 1\big) \vee \big(s_i = 0\big)\}$$

where $f_2(x) = g^x$ and $f_1(y, \mathsf{h}) = 1$ if $H(y) = \mathsf{h}$ and 0 otherwise.
- Compute and publish $Z_{Assets} = \prod_{i=1}^n u_i$.

Fig. 6. Proof of assets

6.2 Privacy-Preserving Credentials

Another application of our compositions for compound statements is in privacy-preserving verification of credentials. A credential system allows a user to obtain credentials from an organization or a Certificate Authority, and later prove to a verifier that she has been given appropriate credentials. Typically, the user's credentials will contain a set of attributes, and the verifier will require that the user prove that the attributes in his credential satisfy certain policy. Many different constructions have been proposed for anonymous credential systems built around sigma protocols. The signatures used, therefore, are specially designed so that a sigma protocol can be used to prove knowledge of the signature on a committed message. If we want to base anonymous credentials on standard signatures, like RSA signatures, we will need to prove a compound statement involving an algebraic relation (for the exponentiation), and a circuit-based statement (for the hash function). The recent work of [30] achieves privacy-preserving verification of X.509 certificates by using zk-SNARKs, and this involves representing the exponentiation in an RSA group as a circuit. Here, we use our composition constructions to build an efficient proof avoiding expensive circuit representation of algebraic statements.

Given a SHA hash digest of a message m, a candidate RSA signature σ, and an RSA modulus N, verification involves checking whether $\sigma^e \bmod n = h$, where $h = \mathsf{padding}(\mathsf{SHA}(m))$. The construction given in Fig. 7 achieves privacy-preserving verification for credentials based on RSA signatures. We compare the trade off between the proof size and prover's work in our approach versus other methods in Table 2 in Appendix A. Our compositions and similar techniques extend to yield efficient privacy-preserving verification for credentials based on existing infrastructure like standard RSA-PSS, RSA-PKCS etc.

- The prover commits to the message m, the digest h, and the signature σ by computing $c_1 = \mathsf{Com}_p(m)$, $c_2 = \mathsf{Com}_p(h)$, $c_3 = \mathsf{Com}_n(\sigma)$, $c_4 = \mathsf{Com}_n(h)$ for $p < n$.
- The prover uses zk-comIOSnark to give a proof that the hash digest is correct, given c_1 and c_2.
  $\mathsf{PK}\{(m, h, r_1, r_2) : \mathsf{padding}(\mathsf{SHA}(m)) = h \wedge c_1 = \mathsf{Com}_p(m) \wedge c_2 = \mathsf{Com}_p(h)\}$.
- The prover uses a sigma protocol to prove knowledge of e-th root of a committed value [19].
  $\mathsf{PK}\{(h, \sigma, r_2, r_3) : \sigma^e \bmod n = h \wedge c_2 = \mathsf{Com}_n(h) \wedge c_3 = \mathsf{Com}_n(\sigma)\}$.
- The prover uses the protocol Eq to prove that the commitments c_2 and c_4 are to the same value: $\mathsf{PK}\{(h, h', r_2, r_4) : c_2 = \mathsf{Com}_p(h) \wedge c_4 = \mathsf{Com}_n(h') \wedge h \equiv h' \bmod p\}$.

Fig. 7. RSA signature verification

References

1. Secp256k1. https://en.bitcoin.it/wiki/Secp256k1
2. Technical background of version 1 bitcoin addresses. https://en.bitcoin.it/wiki/Technical_background_of_version_1_Bitcoin_addresses
3. Zcash 1.0 "Sprout" Guide. https://github.com/zcash/zcash/wiki/1.0-User-Guide
4. Zcash Parameter Generation. https://z.cash/technology/paramgen.html
5. Aho, A. (ed.): 19th ACM STOC. ACM Press, May 1987
6. Ames, S., Hazay, C., Ishai, Y., Venkitasubramaniam, M.: Ligero: lightweight sublinear arguments without a trusted setup. In: Thuraisingham, B.M., Evans, D., Malkin, T., Xu, D. (eds.) ACM CCS 17, pp. 2087–2104. ACM Press, October/November 2017
7. Ben-Or, M., Goldreich, O., Goldwasser, S., Håstad, J., Kilian, J., Micali, S., Rogaway, P.: Everything provable is provable in zero-knowledge. In: Goldwasser, S. (ed.) CRYPTO 1988. LNCS, vol. 403, pp. 37–56. Springer, New York (1990). https://doi.org/10.1007/0-387-34799-2_4
8. Ben-Sasson, E., Chiesa, A., Garman, C., Green, M., Miers, I., Tromer, E., Virza, M.: Zerocash: decentralized anonymous payments from bitcoin. In: 2014 IEEE Symposium on Security and Privacy, pp. 459–474. IEEE Computer Society Press, May 2014
9. Ben-Sasson, E., Chiesa, A., Genkin, D., Tromer, E., Virza, M.: SNARKs for C: verifying program executions succinctly and in zero knowledge. In: Canetti, R., Garay, J.A. (eds.) CRYPTO 2013, Part II. LNCS, vol. 8043, pp. 90–108. Springer, Heidelberg (2013). https://doi.org/10.1007/978-3-642-40084-1_6
10. Ben-Sasson, E., Chiesa, A., Tromer, E., Virza, M.: Succinct non-interactive zero knowledge for a von Neumann architecture. In: 23rd USENIX Security Symposium (USENIX Security 14), pp. 781–796. USENIX Association, San Diego, CA (2014)

11. Bitansky, N., Canetti, R., Chiesa, A., Tromer, E.: From extractable collision resistance to succinct non-interactive arguments of knowledge, and back again. In: Goldwasser, S. (ed.) ITCS 2012, pp. 326–349. ACM, January 2012
12. Bitansky, N., Chiesa, A., Ishai, Y., Paneth, O., Ostrovsky, R.: Succinct noninteractive arguments via linear interactive proofs. In: Sahai, A. (ed.) TCC 2013. LNCS, vol. 7785, pp. 315–333. Springer, Heidelberg (2013). https://doi.org/10.1007/978-3-642-36594-2_18
13. Blum, M., Feldman, P., Micali, S.: Non-interactive zero-knowledge and its applications (extended abstract). In: 20th ACM STOC, pp. 103–112. ACM Press, May 1988
14. Boudot, F.: Efficient proofs that a committed number lies in an interval. In: Preneel, B. (ed.) EUROCRYPT 2000. LNCS, vol. 1807, pp. 431–444. Springer, Heidelberg (2000). https://doi.org/10.1007/3-540-45539-6_31
15. Bünz, B., Bootle, J., Boneh, D., Poelstra, A., Wuille, P., Maxwell, G.: Bulletproofs: efficient range proofs for confidential transactions. Cryptology ePrint Archive, Report 2017/1066 (2017). https://eprint.iacr.org/2017/1066
16. Camenisch, J., Chaabouni, R., Shelat, A.: Efficient protocols for set membership and range proofs. In: Pieprzyk, J. (ed.) ASIACRYPT 2008. LNCS, vol. 5350, pp. 234–252. Springer, Heidelberg (2008). https://doi.org/10.1007/978-3-540-89255-7_15
17. Camenisch, J., Lysyanskaya, A.: An efficient system for non-transferable anonymous credentials with optional anonymity revocation. In: Pfitzmann, B. (ed.) EUROCRYPT 2001. LNCS, vol. 2045, pp. 93–118. Springer, Heidelberg (2001). https://doi.org/10.1007/3-540-44987-6_7
18. Camenisch, J., Michels, M.: Proving in zero-knowledge that a number is the product of two safe primes. In: Stern, J. (ed.) EUROCRYPT 1999. LNCS, vol. 1592, pp. 107–122. Springer, Heidelberg (1999). https://doi.org/10.1007/3-540-48910-X_8
19. Camenisch, J., Stadler, M.: Efficient group signature schemes for large groups (extended abstract). In: Kaliski Jr. [49], pp. 410–424
20. Chase, M., Derler, D., Goldfeder, S., Orlandi, C., Ramacher, S., Rechberger, C., Slamanig, D., Zaverucha, G.: Post-quantum zero-knowledge and signatures from symmetric-key primitives. In: Proceedings of the 2017 ACM SIGSAC Conference on Computer and Communications Security, pp. 1825–1842. ACM (2017)
21. Chase, M., Ganesh, C., Mohassel, P.: Efficient zero-knowledge proof of algebraic and non-algebraic statements with applications to privacy preserving credentials. In: Robshaw, M., Katz, J. (eds.) CRYPTO 2016, Part III. LNCS, vol. 9816, pp. 499–530. Springer, Heidelberg (2016). https://doi.org/10.1007/978-3-662-53015-3_18
22. Chaum, D.: Blind signatures for untraceable payments. In: Chaum, D., Rivest, R.L., Sherman, A.T. (eds.) CRYPTO 1982, pp. 199–203. Plenum Press, New York (1982). https://doi.org/10.1007/978-1-4757-0602-4_18
23. Cohen, J.D., Fischer, M.J.: A robust and verifiable cryptographically secure election scheme (extended abstract). In: 26th FOCS, pp. 372–382. IEEE Computer Society Press, October 1985
24. Costello, C., Fournet, C., Howell, J., Kohlweiss, M., Kreuter, B., Naehrig, M., Parno, B., Zahur, S.: Geppetto: versatile verifiable computation. In: 2015 IEEE Symposium on Security and Privacy, pp. 253–270. IEEE Computer Society Press, May 2015
25. Cramer, R. (ed.): TCC 2012. LNCS, vol. 7194. Springer, Heidelberg (2012)

26. Cramer, R., Damgård, I., Schoenmakers, B.: Proofs of partial knowledge and simplified design of witness hiding protocols. In: Desmedt, Y.G. (ed.) CRYPTO 1994. LNCS, vol. 839, pp. 174–187. Springer, Heidelberg (1994). https://doi.org/10.1007/3-540-48658-5_19

27. Dagher, G.G., Bünz, B., Bonneau, J., Clark, J., Boneh, D.: Provisions: privacy-preserving proofs of solvency for bitcoin exchanges. In: Ray, I., Li, N., Kruegel, C. (eds.) ACM CCS 2015, pp. 720–731. ACM Press, October 2015

28. Dåmgard, I.: On Sigma Protocols. http://www.cs.au.dk/~ivan/Sigma.pdf

29. Damgård, I., Faust, S., Hazay, C.: Secure two-party computation with low communication. In: Cramer [25], pp. 54–74

30. Delignat-Lavaud, A., Fournet, C., Kohlweiss, M., Parno, B.: Cinderella: turning shabby X.509 certificates into elegant anonymous credentials with the magic of verifiable computation. In: 2016 IEEE Symposium on Security and Privacy, pp. 235–254. IEEE Computer Society Press, May 2016

31. Edwards, H.: A normal form for elliptic curves. Bull. Am. Math. Soc. 44(3), 393–422 (2007)

32. Feige, U., Fiat, A., Shamir, A.: Zero knowledge proofs of identity. In: Aho [5], pp. 210–217

33. Feige, U., Lapidot, D., Shamir, A.: Multiple non-interactive zero knowledge proofs based on a single random string (extended abstract). In: 31st FOCS, pp. 308–317. IEEE Computer Society Press, October 1990

34. Fiat, A., Shamir, A.: How to prove yourself: practical solutions to identification and signature problems. In: Odlyzko, A.M. (ed.) CRYPTO 1986. LNCS, vol. 263, pp. 186–194. Springer, Heidelberg (1987). https://doi.org/10.1007/3-540-47721-7_12

35. Fiore, D., Fournet, C., Ghosh, E., Kohlweiss, M., Ohrimenko, O., Parno, B.: Hash first, argue later: adaptive verifiable computations on outsourced data. In: Weippl, E.R., Katzenbeisser, S., Kruegel, C., Myers, A.C., Halevi, S. (eds.) ACM CCS 2016, pp. 1304–1316. ACM Press, October 2016

36. Fortnow, L.: The complexity of perfect zero-knowledge (extended abstract). In: Aho [5], pp. 204–209

37. Fujisaki, E., Okamoto, T.: Statistical zero knowledge protocols to prove modular polynomial relations. In: Kaliski Jr. [49], pp. 16–30

38. Garay, J.A., MacKenzie, P.D., Yang, K.: Strengthening zero-knowledge protocols using signatures. J. Cryptol. 19(2), 169–209 (2006)

39. Gennaro, R., Gentry, C., Parno, B., Raykova, M.: Quadratic span programs and succinct NIZKs without PCPs. In: Johansson, T., Nguyen, P.Q. (eds.) EUROCRYPT 2013. LNCS, vol. 7881, pp. 626–645. Springer, Heidelberg (2013). https://doi.org/10.1007/978-3-642-38348-9_37

40. Giacomelli, I., Madsen, J., Orlandi, C.: Zkboo: faster zero-knowledge for boolean circuits. In: 25th USENIX Security Symposium, USENIX Security 2016, Austin, TX, USA, 10–12 August 2016 (2016)

41. Goldreich, O., Micali, S., Wigderson, A.: Proofs that yield nothing but their validity and a methodology of cryptographic protocol design (extended abstract). In: 27th FOCS, pp. 174–187. IEEE Computer Society Press, October 1986

42. Goldreich, O., Micali, S., Wigderson, A.: How to play any mental game or a completeness theorem for protocols with honest majority. In: Aho [5], pp. 218–229

43. Goldwasser, S., Lin, H., Rubinstein, A.: Delegation of computation without rejection problem from designated verifier CS-Proofs. Cryptology ePrint Archive, Report 2011/456 (2011). http://eprint.iacr.org/2011/456

44. Groth, J.: Short pairing-based non-interactive zero-knowledge arguments. In: Abe, M. (ed.) ASIACRYPT 2010. LNCS, vol. 6477, pp. 321–340. Springer, Heidelberg (2010). https://doi.org/10.1007/978-3-642-17373-8_19
45. Groth, J., Kohlweiss, M.: One-out-of-many proofs: or how to leak a secret and spend a coin. In: Oswald, E., Fischlin, M. (eds.) EUROCRYPT 2015, Part II. LNCS, vol. 9057, pp. 253–280. Springer, Heidelberg (2015). https://doi.org/10.1007/978-3-662-46803-6_9
46. Guillou, L.C., Quisquater, J.-J.: A practical zero-knowledge protocol fitted to security microprocessor minimizing both transmission and memory. In: Barstow, D., et al. (eds.) EUROCRYPT 1988. LNCS, vol. 330, pp. 123–128. Springer, Heidelberg (1988). https://doi.org/10.1007/3-540-45961-8_11
47. 2013 IEEE Symposium on Security and Privacy. IEEE Computer Society Press, May 2013
48. Ishai, Y., Kushilevitz, E., Ostrovsky, R., Sahai, A.: Zero-knowledge from secure multiparty computation. In: Johnson, D.S., Feige, U. (eds.) 39th ACM STOC, pp. 21–30. ACM Press, June 2007
49. Kaliski Jr., B.S. (ed.): CRYPTO 1997. LNCS, vol. 1294. Springer, Heidelberg (1997)
50. Kilian, J.: A note on efficient zero-knowledge proofs and arguments. In: Proceedings of the Twenty-fourth Annual ACM Symposium on Theory of Computing, pp. 723–732. ACM (1992)
51. Lipmaa, H.: Progression-free sets and sublinear pairing-based non-interactive zero-knowledge arguments. In: Cramer [25], pp. 169–189
52. Lipmaa, H.: Succinct non-interactive zero knowledge arguments from span programs and linear error-correcting codes. In: Sako, K., Sarkar, P. (eds.) ASIACRYPT 2013, Part I. LNCS, vol. 8269, pp. 41–60. Springer, Heidelberg (2013). https://doi.org/10.1007/978-3-642-42033-7_3
53. Micali, S.: Computationally sound proofs. SIAM J. Comput. **30**(4), 1253–1298 (2000)
54. Miers, I., Garman, C., Green, M., Rubin, A.D.: Zerocoin: anonymous distributed E-cash from Bitcoin. In: IEEE S&P 2013 [47], pp. 397–411
55. Naor, M., Yung, M.: Public-key cryptosystems provably secure against chosen ciphertext attacks. In: 22nd ACM STOC, pp. 427–437. ACM Press, May 1990
56. Noether, S., Mackenzie, A., Team, M.C.: Ring confidential transactions. https://lab.getmonero.org/pubs/MRL-0005.pdf
57. Parno, B., Howell, J., Gentry, C., Raykova, M.: Pinocchio: nearly practical verifiable computation. In: IEEE S&P 2013 [47], pp. 238–252
58. Pedersen, T.P.: Non-interactive and information-theoretic secure verifiable secret sharing. In: Feigenbaum, J. (ed.) CRYPTO 1991. LNCS, vol. 576, pp. 129–140. Springer, Heidelberg (1992). https://doi.org/10.1007/3-540-46766-1_9
59. Pointcheval, D., Stern, J.: Security proofs for signature schemes. In: Maurer, U. (ed.) EUROCRYPT 1996. LNCS, vol. 1070, pp. 387–398. Springer, Heidelberg (1996). https://doi.org/10.1007/3-540-68339-9_33
60. Schnorr, C.P.: Efficient signature generation by smart cards. J. Cryptol. **4**(3), 161–174 (1991)
61. Vadhan, S.P.: A study of statistical zero-knowledge proofs. Ph.D. thesis, Massachusetts Institute of Technology (1999)
62. Wilcox, Z.: Proving bitcoin reserves. https://iwilcox.me.uk/2014/proving-bitcoin-reserves

A   Efficiency

We briefly discuss the estimated cost of some of the building blocks. The ddlog proof is dominated by the cost of the range proofs in steps 4, 5, 6 of pointAddition protocol in Fig. 1. In a recent work [15], it was shown how to prove that a committed value is in a range using only a number of field elements that is logarithmic in the bit length of the range. Using these proofs to instantiate all the necessary range proofs in protocol pointAddition, the prover's work is $30 \log t + 1800$ group exponentiations, the verifier's work is $10 \log t$ exponentiations, and the proof size is $2370 + \log \log t$ elements where the proof is for a curve defined over $\mathbb{F}_t$. The cost of comInSnark is the cost of the comEq in addition to the cost incurred by separating the wires in the underlying SNARK construction. The proof size of comInSnark is 15 group elements, and 2 field elements for every committed value (input/output). In the case of our following applications, the proof size is 17 elements. The prover's work is the number of exponentiations for computing the SNARK proof and an additional 2 exponentiations for the comEq proof. The verifier's work is 2 exponentiations and 21 pairings. Similarly, comIOSnark has proof size 26 elements, the prover's work, in addition to the exponentiations for the SNARK proof is 4 exponentiations and the verifier's work is 4 exponentiations and 30 pairings.

Proof of solvency. In Table 1, we compare the proof size and prover's work of Provisions with our protocol and a solution that uses zk-SNARK for the entire statement. The proof size and prover's work are dominated by the range proofs; the numbers below give only the dominating terms ignoring small constants and are assuming that the range proofs are realized using Bulletproofs.

Table 1. Comparison of prover work and proof sizes for proof of solvency using different methods. n is the size of the anonymity set, c is the number of customer accounts, m is $\lceil \log \mathsf{Max} \rceil = 51$, p is the bit length of the modulus for exponentiation (size of the field over which the curve is defined). For $n = 500,000$ and $c = 2$ million, the proof size and prover's work in Provisions is $5 * 10^6$ and $4 * 10^7$ respectively. For the same parameters, our approach gives proof size of 10^9 and prover's work 10^{10}, while also achieving the additional pay-to-hash functionality. A fully zk-SNARK solution requires prover's work roughly 10^{13}. (Exp. stands for exponentiations.)

zk technique	Functionality	Proof size (in elements)	Prover
Provisions	pay-to-pub	$10n + \log m + \log c$	$5n + 4mc$ exp.
SNARK	pay-to-pub, pay-to-hash	7	$(\lvert H \rvert + p^3)n + c$ exp.
Our composition techniques	pay-to-pub, pay-to-hash	$2396n + \log p + \log n$	$(\lvert H \rvert + 30p + 1800)n + c$ exp.

Privacy preserving credentials. In Table 2, we compare the proof size and prover's work in privacy-preserving credentials for Cinderella, the interactive protocol of [21], and our composition.

Table 2. Comparison of prover work and proof sizes for credential verification using different methods. p is the order of the group in which commitments are computed, $|m|$ is the bit length of the message. For $e = 65537, \log p = 256, |H| = 23785$, we note an 87% decrease in prover's work compared to Cinderella at the cost of increasing the proof size to 298 from 7 group elements. (Exp. stands for exponentiations.)

zk technique	Feature	Proof size	Prover
Cinderella	non-interactive	7	$|H|$+ additional 164,826 equations for RSA (as optimized in Cinderella)
GC + Sigma [21]	interactive	$|H|$	$|m| + |h|$ exp. + $|H|$ symmetric-key operations
Our composition techniques	non-interactive	$42 + \log p$	$|H| + \log p + 16$ exp.

From Laconic Zero-Knowledge
to Public-Key Cryptography
Extended Abstract

Itay Berman[1], Akshay Degwekar[1], Ron D. Rothblum[1,2(✉)],
and Prashant Nalini Vasudevan[1]

[1] MIT, Cambridge, USA
{itayberm,akshayd,prashvas,ronr}@mit.edu
[2] Northeastern University, Boston, USA

Abstract. Since its inception, public-key encryption (PKE) has been
one of the main cornerstones of cryptography. A central goal in
cryptographic research is to understand the foundations of public-
key encryption and in particular, base its existence on a natural and
generic complexity-theoretic assumption. An intriguing candidate for
such an assumption is the existence of a cryptographically hard language
$\mathcal{L} \in \mathsf{NP} \cap \mathsf{SZK}$.

In this work we prove that public-key encryption can be based on
the foregoing assumption, as long as the (honest) prover in the zero-
knowledge protocol is *efficient* and *laconic*. That is, messages that the
prover sends should be efficiently computable (given the NP witness) and
short (i.e., of sufficiently sub-logarithmic length). Actually, our result
is stronger and only requires the protocol to be zero-knowledge for
an *honest-verifier* and sound against computationally bounded cheating provers.

Languages in NP with such laconic zero-knowledge protocols are
known from a variety of computational assumptions (e.g., Quadratic
Residuocity, Decisional Diffie-Hellman, Learning with Errors, etc.).
Thus, our main result can also be viewed as giving a unifying framework for constructing PKE which, in particular, captures many of the
assumptions that were already known to yield PKE.

We also show several extensions of our result. First, that a certain
weakening of our assumption on laconic zero-knowledge is actually *equivalent* to PKE, thereby giving a complexity-theoretic characterization
of PKE. Second, a mild strengthening of our assumption also yields a
(2-message) oblivious transfer protocol.

1   Introduction

Underlying symmetric key encryption is a centuries-old idea: *shared secrets
enable secure communication*. This idea takes many forms: the Caeser cipher,
the unconditionally secure one-time pads, fast heuristic constructions like AES,

Full version available at: https://eccc.weizmann.ac.il/report/2017/172.

H. Shacham and A. Boldyreva (Eds.): CRYPTO 2018, LNCS 10993, pp. 674–697, 2018.
https://doi.org/10.1007/978-3-319-96878-0_23

and a multitude of candidates based on the hardness of a variety of problems. The discovery of *public-key* encryption, by Diffie and Hellman [DH76] and Rivest, Shamir and Adleman [RSA78], was revolutionary as it gave us the ability to communicate securely without any shared secrets. Needless to say, this capability is one of the cornerstones of secure communication in today's online world.

As is typically the case in cryptography, we are currently very far from establishing the security of public-key cryptography unconditionally. Rather, to establish security, we rely on certain computational intractability assumptions. Despite four decades of extensive research, we currently only know constructions of public-key encryption from a handful of assumptions, most notably assumptions related to the hardness of factoring, finding discrete logarithms and computational problems related to lattices (as well as a few more exotic assumptions).

One of the central open problems in cryptography is to place public-key encryption on firmer complexity-theoretic grounding, ideally by constructing public-key encryption from the minimal assumption that one-way functions exist. Such a result seems well beyond current techniques, and by the celebrated result of Impagliazzo and Rudich [IR89] requires a non-blackbox approach. Given that, a basic question that we would like to resolve is the following:

From what general complexity-theoretic assumptions can we construct public-key cryptography?

Our motivation for asking this question is twofold. First, we seek to understand: Why is it the case that so few assumptions give us public-key encryption? What kind of "structured hardness" is required? Secondly, we hope that this understanding can guide the search for new concrete problems that yield public-key encryption.

1.1 Our Results

Our main result is a construction of a public-key encryption scheme from a general complexity-theoretic assumption: namely, the existence of a cryptographically hard language $\mathcal{L} \in$ NP that has a *laconic* (honest-verifier) statistical zero-knowledge argument-system. We first discuss the notions mentioned above, and then proceed to state the main result more precisely (yet still informally).

By a **cryptographically hard language** we mean an NP language that is average-case hard to decide with a solved instance generator. Namely, that there are two distributions Y and N, over YES and NO instances of the language respectively, such that (1) Y and N are computationally indistinguishable; and (2) there exists an efficient solved instance generator for the YES distribution.[1] A proof-system is *laconic* [GH98, GVW02] if the number of bits sent *from the prover*

---

[1] Loosely speaking, a solved-instance generator for the YES distribution Y of an average-case hard language $\mathcal{L} \in$ NP is an algorithm that generates samples $(x, w) \in \mathcal{R}_{\mathcal{L}}$ (where $\mathcal{R}_{\mathcal{L}}$ is the NP relation) and where x is distributed according to Y.

to the verifier is very small.[2] An **argument-system** is similar to an interactive proof, except that soundness is only required to hold against *computationally bounded* (i.e., polynomial time) cheating provers. Honest verifier zero-knowledge means that the *honest* verifier learns no more in the interaction than the fact that $x \in \mathcal{L}$ (i.e., the verifier can simulate the honest interaction by itself). Thus, our main result can be stated as follows:

Theorem 1.1 (Informally Stated, see Theorem 2.6). *Assume that there exists a cryptographically hard language* $\mathcal{L} \in$ NP *with an r-round statistical honest-verifier zero-knowledge argument-system, with constant soundness, that satisfies the following two requirements:*

- **Efficient Prover:** *The strategy of the honest prover can be implemented in polynomial-time, given the* NP *witness.*[3]
- **Laconic Prover:** *The prover sends at most q bits in each of the r rounds, such that* $r^2 \cdot q^3 = O(\log n)$, *where n is the input length.*

Then, there exists a public-key encryption (PKE) scheme.

We emphasize that requiring only *honest-verifier* zero-knowledge (as opposed to full-fledged zero-knowledge) and *computational soundness* (i.e., an argument-system) weakens our assumption, and therefore only strengthens our main result. We also comment that we can handle provers that are less laconic (i.e., send longer messages) by assuming that the language $\mathcal{L}$ is sub-exponentially hard. Lastly, we remark the assumption in Theorem 1.1 may be viewed as a generalization of the notion of *hash proof systems* [CS02].[4] We discuss this point in more detail in Sect. 1.2.

1.1.1   Instantiations

Many concrete assumptions (which are already known to yield public-key encryption schemes) imply the conditions of Theorem 1.1. First, number-theoretic assumptions such as Quadratic Residuosity (QR) and Decisional Diffie-Hellman (DDH) can be shown to imply the existence of a cryptographically hard NP language with a laconic and efficient SZK argument-system and therefore satisfy the conditions of Theorem 1.1 (these and the other implications mentioned below are proven in the full version of this paper).

---

[2] Laconic proof-systems with constant soundness and very short communication (e.g., just a single bit) are indeed known. As a matter of fact, many of the known hard problems that are known to yield public-key encryption schemes have such laconic SZK proof-systems (see Sect. 1.1.1).

[3] In the context of *argument-systems* (in contrast to general interactive proofs), the assumption that the honest prover is efficient goes without saying. Nevertheless, we wish to emphasize this point here.

[4] As a matter of fact, hash proof systems can be viewed as a special case of our assumption in which the (honest) prover is *deterministic* or, equivalently, sends only a single bit. In contrast, we handle arbitrary *randomized* provers (that are sufficiently laconic) and indeed most of the technical difficulty arises from handling this more general setting. See additional details in Sect. 1.2.

We can also capture assumptions related to lattices and random linear codes by slightly relaxing the conditions of Theorem 1.1. Specifically, Theorem 1.1 holds even if we relax the completeness, soundness and zero-knowledge conditions of the argument-system to hold only for *most* (but not necessarily all) of the instances (chosen from the average-case hard distribution). We call arguments with these weaker properties *average-case* SZK arguments.

It is not hard to see that *lossy encryption* [PVW08, BHY09] yields such an *average-case* laconic and efficient zero-knowledge argument-system. Recall that a PKE scheme is *lossy* if its public-keys are indistinguishable from so-called "lossy keys" such that a ciphertext generated using such a lossy key does not contain information about the underlying plaintext. Consider the following proof-system for the language consisting of all valid public-keys: given an allegedly valid public-key, the verifier sends to the prover an encryption of a random bit b and expects to get in response the value b. It is not hard to see that this protocol is a laconic and efficient average-case SZK argument-system.

Several concrete assumptions yield cryptographically hard languages with *average-case* laconic and efficient SZK arguments (whether via lossy encryption or directly). Most notably, Learning With Errors (LWE) [Reg05], Learning Parity with Noise (LPN) with small errors [Ale03] and most of the assumptions used by Applebaum *et al.* [ABW10] to construct PKE, all imply the existence of such languages.

Thus, Theorem 1.1 gives a common framework for constructing public-key encryption based on a variety of different intractability assumptions (all of which were already known to yield public-key encryption via a variety of somewhat ad hoc techniques), see also Fig. 1.

One notable hardness assumption that we do not know to imply our assumption (even the average-case variant) is *integer factorization* (and the related RSA assumption). We consider a further weakening of our assumption that captures also the factoring and RSA assumptions. As a matter of fact, we show that this further relaxed assumption is actually *equivalent* to the existence of a public-key encryption scheme. We discuss this in more detail in Sect. 1.1.3.

1.1.2   Perspective — From SZK-Hardness to Public-Key Encryption

As noted above, one of the central goals in cryptography is to base public-key encryption on a general notion of structured hardness. A natural candidate for such structure is the class SZK of statistical zero-knowledge proofs, since many of the assumptions that are known to yield public-key encryption have SZK proof-systems. Indeed, it is enticing to believe that the following dream version of Theorem 1.1 holds:

Open Problem 1.1. *Assume that there exists a cryptographically-hard language $\mathcal{L} \in$ NP $\cap$ SZK. Then, there exists a public-key encryption scheme.*

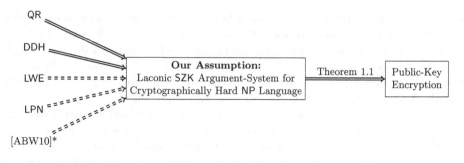

Fig. 1. Instantiations of our assumption. Dashed arrows means that we only obtain average-case completeness, soundness and zero-knowledge. The (*) sign means that most, but not all, assumptions from [ABW10] imply our assumption.

(Here by SZK we refer to the class of languages having statistical zero-knowledge *proof-systems* rather than argument-systems as in Theorem 1.1. Assuming this additional structure only makes the statement of Open Problem 1.1 weaker and therefore easier to prove.)

Solving Open Problem 1.1 would be an outstanding breakthrough in cryptography. For instance, it would allow us to base public-key cryptography on the intractability of the discrete logarithm (DLOG) problem,[5] since (a decision problem equivalent to) DLOG has a perfect zero-knowledge proof-system[6] [GK93], or under the plausible quasi-polynomial average-case[7] hardness of the graph isomorphism problem (via the perfect zero-knowledge protocol of [GMW87]).

We view Theorem 1.1 as an initial step toward solving Open Problem 1.1. At first glance, it seems that Theorem 1.1 must be strengthened in *two* ways in order to solve Open Problem 1.1. Namely, we need to get rid of the requirements that the (honest) prover is (1) efficient and (2) laconic. However, it turns out that it suffices to remove only *one* of these restrictions, no matter which one, in order to solve Open Problem 1.1. We discuss this next.

Handling Inefficient Provers. Sahai and Vadhan [SV03] showed a problem, called *statistical distance*, which is both (1) complete for SZK, and (2) has an extremely laconic honest-verifier statistical zero-knowledge proof in which the prover only sends a *single* bit (with constant soundness error). The immediate implication

---

[5] Public-key schemes based on assumptions related to discrete log such as the decisional (or even computational) Diffie Hellman assumption are known to exist. Nevertheless, basing public-key encryption solely on the hardness of discrete log has been open since the original work of Diffie and Hellman [DH76].

[6] That proof-system is actually laconic but it is unclear how to implement the prover efficiently.

[7] Graph isomorphism is in fact known to be solvable in polynomial-time for many natural distributions, and the recent breakthrough result of Babai [Bab16] gives a quasi-polynomial worst-case algorithm. Nevertheless, it is still conceivable that Graph Isomorphism is average-case quasi-polynomially hard (for some efficiently samplable distribution).

is that any SZK protocol can be compressed to one in which the prover sends only a single bit.

Unfortunately, the foregoing transformation does not seem to maintain the computational efficiency of the prover. Thus, removing the requirement that the prover is efficient from Theorem 1.1 (while maintaining the laconism requirement) would solve Open Problem 1.1.

Handling Non-Laconic Provers. Suppose that we managed to remove the laconism requirement from Theorem 1.1 and only required the prover to be efficient. It turns out that the latter would actually imply an even stronger result than that stated in Open Problem 1.1. Specifically, assuming only the existence of one-way functions, Haitner *et al.* [HNO+09] construct (non-laconic) statistical zero-knowledge *arguments* for any NP language, with an efficient prover. Thus, removing the laconism requirement from Theorem 1.1 would yield public-key encryption based merely on the existence of one-way functions.

In fact, even a weaker result would solve Open Problem 1.1. Suppose we could remove the laconism requirement from Theorem 1.1 while insisting that the proof-system has *statistical soundness* (rather than computational). Such a result would solve Open Problem 1.1 since Nguyen and Vadhan [NV06] showed that every language in NP ∩ SZK has an SZK protocol in which the prover is efficient (given the NP witness).

To summarize, removing the laconism requirement from Theorem 1.1, while still considering an argument-system, would yield public-key encryption from one-way functions (via [HNO+09]). On the other hand, removing the laconism requirement while insisting on statistical soundness would solve Open Problem 1.1 (via [NV06]). (Note that neither the [NV06] nor [HNO+09] proof-systems are laconic, so they too cannot be used directly together with Theorem 1.1 to solve Open Problem 1.1.)

1.1.3   Extensions

We also explore the effect of strengthening and weakening our assumption. A natural strengthening gives us oblivious transfer, and as mentioned above, a certain weakening yields a complete complexity-theoretic characterization of public-key encryption.

A Complexity-Theoretic Characterization. The assumption from which we construct public-key encryption (see Theorem 1.1) requires some underlying hard *decision* problem. In many cryptographic settings, however, it seems more natural to consider hardness of *search* problems (e.g., integer factorization). Thus, we wish to explore the setting of laconic SZK arguments when only assuming the hardness of computing a witness for an instance sampled from a solved instance generator. Namely, an NP relation for which it is hard, given a random instance, to find a corresponding witness.

We introduce a notion of (computationally sound) proof-systems for such NP search problems, which we call *arguments of weak knowledge* (AoWK). Loosely speaking, this argument-system convinces the verifier that the prover with which it is interacting has at least some partial knowledge of some witness. Or in other words, no efficient cheating prover can convince the verifier to accept given *only* the input. We further say that an AoWK is *zero-knowledge* if the verifier learns nothing beyond the fact that the prover has the witness.

We show that Theorem 1.1 still holds under the weaker assumption that there is an efficient and laconic SZK-AoWK (with respect to some hard solved instance generator). Namely, the latter assumption implies the existence of PKE. Furthermore, we also show that the same assumption is also *implied* by any PKE scheme, thus establishing an equivalence between the two notions which also yields a certain complexity-theoretic characterization of public-key encryption.

Oblivious Transfer. *Oblivious Transfer* (OT) is a fundamental cryptographic primitive, which is complete for the construction of general secure multiparty computation (MPC) protocols [GMW87, Kil88]. We show that by making a slightly stronger assumption, Theorem 1.1 can extended to yield a (two-message) semi-honest OT protocol.

For our OT protocol, in addition to the conditions of Theorem 1.1, we need to further assume that there is a way to sample instances x such that it is hard to tell whether $x \in \mathcal{L}$ or $x \notin \mathcal{L}$ *even given the coins of the sampling algorithm.*[8] We refer to this property as *enhanced cryptographic hardness* in analogy to the notion of *enhanced* trapdoor permutations.

1.2 Related Works

Cryptography and Hardness of SZK. Ostrovsky [Ost91] showed that the existence of a language in SZK with average-case hardness implies the existence of one-way functions. Our result can be interpreted as an extension of Ostrovsky's result: By assuming additional structure on the underlying SZK protocol, we construct a public-key encryption scheme. In fact, some of the ideas underlying our construction are inspired by Ostrovsky's one-way function.

Average-case SZK hardness also implies constant-round statistically hiding commitments [OV08], a primitive not implied by one-way functions in a black-box way [HHRS15]. Assuming the existence of an average-case hard language in a *subclass* of SZK (i.e., of languages having perfect randomized encodings), Applebaum and Raykov [AR16] construct Collision Resistant Hash functions.

In the other direction, some cryptographic primitives like homomorphic encryption [BL13], lossy encryption, witness encryption and indistinguishability obfuscators [KMN+14, PPS15], and PIR (computational private information

---

[8] In particular, the sampling algorithm that tosses a coin $b \in \{0, 1\}$ and outputs $x \in \mathcal{L}$ if $b = 0$ and $x \notin \mathcal{L}$ otherwise does not satisfy the requirement (since the value of b reveals whether $x \in \mathcal{L}$).

retrieval) [LV16] imply the existence of average-case hard problems in SZK.[9] We also mention that many other primitives, such as one-way functions, public-key encryption and oblivious transfer do not imply the existence of average-case hard problems in SZK (under black-box reductions) [BDV16].

Hash Proof-Systems. Hash Proof-Systems, introduced by Cramer and Shoup [CS02], are a cryptographic primitive which, in a nutshell, can be described as a cryptographically hard language in NP with a one-round SZK protocol in which the honest prover is efficient given the NP witness and *deterministic* (and without loss of generality sends only a single bit). This is precisely what we assume for our main result except that we can handle *randomized* provers that send more bits of information (and the protocol can be multi-round). This special case of deterministic provers is significantly simpler to handle (and will serve as a warmup when describing our techniques). Our main technical contribution is handling arbitrary *randomized* provers.

Public-key encryption schemes have been shown to imply the existence of certain *weak* hash proof-systems [HLWW16]. Hash proof-systems were also shown in [GOVW12] to yield *resettable* statistical zero-knowledge proof-systems.

Laconic Provers. A study of interactive proofs in which the prover is laconic (i.e., transmits few bits to the verifier) was initiated by Goldreich and Håstad [GH98] and was further explored by Goldreich, Vadhan and Wigderson [GVW02]. These focus in these works is on general interactive proofs (that are not necessarily zero-knowledge) and their main results are that laconic interactive proofs are much weaker than general (i.e., non-laconic) interactive proofs.

1.3   Techniques

To illustrate the techniques used, we sketch the proof of a slightly simplified version of Theorem 1.1. Specifically, we construct a PKE given a cryptographically hard language $\mathcal{L}$ with a *single-round* efficient-prover and laconic SZK argument-system (we shall briefly mention the effect of more rounds where it is most relevant). For simplicity, we also assume that the SZK protocol has *perfect* completeness and zero-knowledge. In the actual construction, given in the technical sections, we handle constant completeness error, negligible simulation error, and more rounds of interaction. Lastly, since we find the presentation more appealing, rather than presenting a public-key scheme, we construct a *single-round key-agreement* protocol.[10] Any such protocol can be easily transformed into a public-key encryption scheme.

---

[9] On a somewhat related note, we mention that combining [BL13] with our result gives a construction of public-key encryption from symmetric-key additively homomorphic encryption. This was already shown in [Rot11] via a direct construction.

[10] Loosely speaking, a key agreement protocol allows Alice and Bob to agree on a common key that is *unpredictable* to an external observer that has wire tapped their communication lines.

Let $\mathcal{L} \in$ NP be a cryptographically hard language with an SZK argument-system with prover $\mathbf{P}$, verifier $\mathbf{V}$ and simulator $\mathbf{S}$. We assume that the argument-system has perfect completeness, no simulation error and soundness error s, for some $s > 0$. Let $\mathsf{Y}_{\mathcal{L}}$ be a solved-instance generator for $\mathcal{L}$ producing samples of the form (x, w), where $x \in \mathcal{L}$ and w is a valid witness for x. The fact that $\mathcal{L}$ is cryptographically hard means that there exists a sampler $\mathsf{N}_{\mathcal{L}}$ that generates NO instances for $\mathcal{L}$ that are computationally indistinguishable from the YES instances generated by $\mathsf{Y}_{\mathcal{L}}$.

Deterministic Prover. As a warmup, we assume first that the honest prover in the SZK argument-system is *deterministic*. As will be shown below, this case is significantly easier to handle than the general case, but it is a useful step toward our eventual protocol.

We construct a key-agreement protocol between Alice and Bob as follows. First Alice generates a solved instance-witness pair $(x, w) \leftarrow \mathsf{Y}_{\mathcal{L}}$. Alice then sends x across to Bob. Bob runs the simulator $\mathbf{S}(x)$ to generate a transcript (a', b', r'), where a' corresponds to the verifier's message, b' corresponds to the prover's message and r' correspond to the simulated random string for the verifier.[11] Bob sends the first message a' across to Alice. Bob then outputs the simulated second message b'. Alice uses the witness w to generate the prover's response b (i.e., the prover $\mathbf{P}$'s actual response given the message a' from the verifier) and outputs b. The protocol is also depicted in Fig. 2.

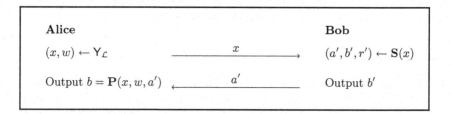

Fig. 2. Key agreement from deterministic provers

To argue that Fig. 2 constitutes a key-agreement protocol, we need to show that Alice and Bob output the same value, and that no efficient eavesdropper Eve (who only sees their messages) can predict this output with good probability.

That they agree on the same value follows from the fact that the prover is deterministic and the simulation is perfect. More specifically, since the simulation is perfect, the distribution of the simulated verifier's message a' is the same as that of the actual verifier's message; and now since the prover is deterministic, given (x, w, a'), the prover's response b, which is also Alice's output, is fixed.

---

[11] Throughout this paper, we use the convention that primed symbols are for objects associated with a simulated (rather than real) execution of the protocol.

Since the simulation is perfect and $x \in \mathcal{L}$, if the simulator outputs (a', b', r'), then b', which is Bob's output, is necessarily equal to b.

Next, we show that any eavesdropper Eve who is able to guess Bob's output in the protocol can be used to break the cryptographic hardness of $\mathcal{L}$. Suppose Eve is able to guess Bob's output in the protocol with probability p. This means that given only x and a', where (a', b', r') is produced by the simulator $\mathbf{S}(x)$, Eve is able to find the message b':

$$\Pr_{\substack{(x,\cdot)\leftarrow Y_{\mathcal{L}} \\ (a',b',r')\leftarrow \mathbf{S}(x)}} [b' = b'' \text{ where } b'' \leftarrow \text{Eve}(x, a')] = p.$$

As the SZK argument has perfect completeness, and the simulation is also perfect, the transcripts produced by the simulator (on YES instances) are always accepted by the verifier. As Eve is able to produce the same prover messages as the simulator, her messages will also be accepted by the verifier. Namely,

$$\Pr_{\substack{(x,\cdot)\leftarrow Y_{\mathcal{L}} \\ (a',b',r')\leftarrow \mathbf{S}(x)}} [\mathbf{V}(x, a', b''; r') = 1 \text{ where } b'' \leftarrow \text{Eve}(x, a')] \geq p.$$

Again using the fact that the simulation is perfect, we can replace the simulated message a' and simulated coin tosses r' with a verifier message a and coins r generated by a real execution of the protocol:

$$\Pr_{\substack{(x,\cdot)\leftarrow Y_{\mathcal{L}} \\ a\leftarrow \mathbf{V}(x;r)}} [\mathbf{V}(x, a, b''; r) = 1 \text{ where } b'' \leftarrow \text{Eve}(x, a)] \geq p.$$

Recall that $N_{\mathcal{L}}$ samples no-instances that are computationally indistinguishable from the YES instances generated by $Y_{\mathcal{L}}$. If x had been a NO instance sampled using $N_{\mathcal{L}}$, then the (computational) soundness of the SZK argument implies that the verifier would reject with probability $1 - s$:

$$\Pr_{\substack{x\leftarrow N_{\mathcal{L}} \\ a\leftarrow \mathbf{V}(x;r)}} [\mathbf{V}(x, a, b''; r) = 1 \text{ where } b'' \leftarrow \text{Eve}(x, a)] < s,$$

where s is the soundness error. If p is larger than s by a non-negligible amount, then we have a distinguisher, contradicting the cryptographic hardness of $\mathcal{L}$. So, no efficient eavesdropper can recover the agreed output value with probability noticeably more than s, the soundness error of the SZK argument.

Notice that so far we have only guaranteed that the probability of success of the eavesdropper is s, which may be as large as a constant (rather than negligible).[12] Nevertheless, using standard amplification techniques (specifically those of Holenstein and Renner [HR05]) we can compile the latter to a full-fledged key-agreement protocol.

---

[12] This error can be made negligible by parallel repetition [BIN97] (recall that parallel repetition preserves *honest-verifier* zero-knowledge). Doing so however makes the prover's messages longer. While this is not an issue when dealing with deterministic provers, it will prove to be problematic in the general case of a randomized prover.

Randomized Prover. So far we have handled deterministic provers. But what happens if the prover were *randomized*? Agreement is now in jeopardy as the prover's message b is no longer completely determined by the instance x and the verifier's message a. Specifically, after Alice receives the simulated verifier message a' from Bob, she still does not know the value of b' that Bob obtained from the simulator – if she ran $\mathbf{P}(x, w, a')$, she could get one of several possible b's, any of which could be the correct b'. Roughly speaking, Alice only has access to the *distribution* from which b' was sampled (but not to the specific value that was sampled).

Eve, however, has even less to work with than Alice; we can show, by an approach similar to (but more complex than) the one we used to show that no polynomial-time eavesdropper can guess b' in the deterministic prover case, that no polynomial-time algorithm can sample from any distribution that is close to the true distribution of b' for most x's and a''s.

We make use of this asymmetry between Alice and Eve in the knowledge of the *distribution* of b' (given x and a) to perform key agreement. We do so by going through an intermediate useful technical abstraction, which we call a *Trapdoor Pseudoentropy Generator*, that captures this asymmetry. We first construct such a generator, and then show how to use any such generator to do key agreement.

Trapdoor Pseudoentropy Generator. A distribution is said to possess *pseudoentropy* [HILL99] if it is computationally indistinguishable from another distribution that has higher entropy[13]. We will later claim that in the protocol in Fig. 2 (when used with a randomized prover), the distribution of b' has some pseudoentropy for the eavesdropper who sees only x and a'. In contrast, Alice, who knows the witness w, can *sample* from the distribution that b' was drawn from. This set of properties is what is captured by our notion of a trapdoor pseudoentropy generator.

A trapdoor pseudoentropy generator consists of three algorithms. The key generation algorithm KeyGen outputs a public and secret key pair $(\mathsf{pk}, \mathsf{sk})$. The *encoding*, given a public key pk, outputs a pair of strings (u, v), where we call u the public message and v the private message.[14] The *decoding* algorithm Dec, given as input the corresponding secret key and the public message u, outputs a value v'. These algorithms are required to satisfy the following properties (simplified here for convenience):

- **Correctness:** The distributions of v and v' are identical, given pk, sk, and u.
- **Pseudoentropy:** The distribution of v has some pseudoentropy given pk and u.

---

[13] By default, the measure of entropy employed is that of Shannon entropy. The Shannon entropy of a variable X given Y is defined as: $\mathrm{H}(X|Y) = \mathrm{E}_y \left[-\sum_x \Pr[X = x|y] \cdot \log(\Pr[X = x|y]) \right]$.

[14] We refer to this procedure as an encoding algorithm because we think of the public message as an encoding of the private message.

Correctness here only means that the secret key can be used to sample from the distribution of the private message v corresponding to the given public message u. This captures the weaker notion of agreement observed in the protocol earlier when Alice had sampling access to the distribution of Bob's output.

The pseudoentropy requirement says that without knowledge of the secret key, the private message v seems to have more entropy – it looks "more random" than it actually is. This is meant to capture the asymmetry of knowledge between Alice and Eve mentioned earlier.

Constructing a Trapdoor Pseudoentropy Generator. Our construction of a trapdoor pseudoentropy generator is described in Fig. 3. It is an adaptation of the earlier key exchange protocol for deterministic provers (from Fig. 2). The public key is an instance x in the language $\mathcal{L}$ and the corresponding secret key is a witness w for x – these are sampled using the solved-instance generator. To encode with public key x, the simulator from the SZK argument for $\mathcal{L}$ is run on x and the simulated verifier message a' is set to be the public message, while the simulated prover message b' is the private message. To decode given x, w and a', the actual prover is run with this instance, witness and verifier message, and the response it generates is output.

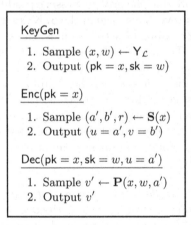

Fig. 3. Trapdoor pseudoentropy generator

Now we argue that this is a valid pseudoentropy generator. Since we will need to be somewhat precise, for the rest of this section, we introduce the jointly-distributed random variables X, A and B, where X represents the instance (sampled from $Y_{\mathcal{L}}$), A represents the verifier's message (with respect to X), and B represents the prover's response (with respect to X and A). Note that since the simulation in the SZK argument is perfect, A and B represent the distributions of the messages output by the simulator as well.

The correctness of our construction follows from the perfect zero knowledge of the underlying SZK argument – the private message v produced by Enc here is the simulated prover's message b', while the output of Dec is the actual prover's response b with the same instance and verifier's message. Both of these have the same distribution, which corresponds to that of B conditioned on $X = x$ and $A = a'$.

In order to satisfy the pseudoentropy condition, the variable B needs to have some pseudoentropy given X and A. What we know, as mentioned earlier, is that B is *unpredictable* given X and A – that no polynomial-time algorithm, given x and a', can sample from a distribution close to that of the corresponding prover's message b. Towards this end, we will use a result of Vadhan and Zheng [VZ12], who give a tight equivalence between unpredictability and pseudoentropy. Applied to our case, their results say what we want – that the variable B has additional pseudoentropy $\log(1/s)$ given X and A, where s is the soundness error from the SZK argument. More precisely, there exists a variable C such that:

$$(X, A, B) \approx_c (X, A, C) \quad \text{and} \quad \mathrm{H}(C|X, A) > \mathrm{H}(B|X, A) + \log(1/s), \qquad (1)$$

where the above expressions refer to *Shannon entropy*. The result of Vadhan and Zheng applies only when the variable B has a polynomial-sized domain, which holds since the proof-system is *laconic* (this is the first out of several places in which we use the laconism of the proof-system). The above shows that the construction in Fig. 3 is indeed a trapdoor pseudoentropy generator. Finally, and this will be crucial ahead, note that the private message produced by Enc is short (i.e., the same length as the prover's message in the SZK argument we started with).

In the case of an SZK protocol with r rounds, the above construction would be modified as follows. The encoder Enc samples a transcript from $\mathbf{S}(x)$, picks $i \in [r]$ at random, sets the public message u to be all the messages in the transcript upto the verifier's message in the ith round, and the private message v to be the prover's message in the ith of the transcript. The decoder Dec samples v' by running the prover on the partial transcript u to get the actual prover's response in the ith round.[15] Zero knowledge ensures that v' and v are distributed identically, and unpredictability arguments similar to the ones above tell us that v' has pseudoentropy at least $\log(1/s)/r$.

From Laconic Trapdoor Pseudoentropy Generator to Key Agreement. Next, given a trapdoor pseudoentropy generator, such as the one in Fig. 3, we show how to construct a single-round key agreement protocol. We start with a pseudoentropy generator in which the public key is pk, the private key is sk, the public message is u, the private message is v, and the output of Dec is v'. The random variables corresponding to these are the same symbols in upper case. v and v' come from the distribution $V_{\mathsf{pk},u}$ (V conditioned on $PK = \mathsf{pk}$ and $U = u$), and V has additional pseudo-Shannon-entropy η given PK and U, where η can be thought of as a constant (η was $\log(1/s)$ in the foregoing construction).

In the key agreement protocol, first Alice samples a key pair (pk, sk) for the pseudoentropy generator and sends the public key pk to Bob. Bob runs $(u, v) \leftarrow \mathsf{Enc}(\mathsf{pk})$, keeps the private message v and sends the public message u

---

[15] For simplicity, assume that the prover is stateless so it can be run on a partial transcript. In the actual proof we handle stateful provers as well.

to Alice. We would like for Alice and Bob to agree on the string v. In order for this to be possible, Bob needs to send more information to Alice so as to specify the specific v that was sampled from $V_{\mathsf{pk},u}$. A natural idea is for Bob to send, along with the message u, a hash $h(v)$ of v, where h is a sampled from a pairwise independent hash function family $\mathcal{H}$.

Alice, on receiving the hash function h and the hash value $h(v)$, uses rejection sampling to find v. She can sample freely from the distribution $V_{\mathsf{pk},u}$ by running $\mathsf{Dec}(\mathsf{sk}, u)$ because she knows the secret key sk of the pseudoentropy generator and the public message u. She keeps drawing samples v' from $V_{\mathsf{pk},u}$, until she finds one that hashes to $h(v)$. Note that this brute force search is only feasible if the number of strings in the support of V is small, which is the case if the number of bits in v is small – considering the big picture, this is one of the reasons we want the prover from the SZK argument to be laconic.

The main question now is how to set the length of the hash function. On the one hand, having a long hash helps agreement, as more information is revealed to Alice about v. On the other hand, security demands a short hash that does not leak "too much" information about v.

For agreement, roughly speaking, if the hash length were more than the *max-entropy*[16] of V given PK and U, which we denote by $\mathrm{H}_{\max}(V|PK, U)$, then the set of possible prover responses is being hashed to a set of comparable size, so with good probability, the hash value $h(v)$ will have a unique pre-image, which Alice can identify.

For security we would like to argue, using the Leftover Hash Lemma, that to any eavesdropper $h(v)$ looks uniformly random given (pk, u, h). This would be true if the hash length were less than the *min-entropy*[17] of V given PK and U, which we denote by $\mathrm{H}_{\min}(V|PK, U)$. Unfortunately, both of the above conditions cannot hold simultaneously because the min-entropy is upper-bounded by the max-entropy.

At this point we use the fact that Eve is computationally bounded. Hence, a *computational* analogue of high min-entropy, which we will call *pseudo-min-entropy*, would suffice for security. Concretely, consider a random variable C such that (PK, U, C) is computationally indistinguishable from (PK, U, V). Furthermore, suppose that the min-entropy of C given PK and U is considerably larger than the hash length. We can then use the Leftover Hash Lemma to argue that $h(V)$ looks uniform to efficient eavesdroppers:

$$(PK, U, h, h(V)) \approx_c (PK, U, h, h(C)) \approx_s (PK, U, h, R)$$

where R is the uniform distribution over the range of h.

The benefit of this observation is that, since C is only required to be computationally close and not statistically close to V, the min-entropy of C given

---

[16] The max entropy corresponds to the logarithm of the support size. The *conditional* max entropy of a random variable X given Y is defined as: $\mathrm{H}_{\max}(X|Y) = \max_y \log(|Supp(X|Y = y)|)$.

[17] The min-entropy of a variable X given Y is defined as: $\mathrm{H}_{\min}(X|Y) = -\log(\max_{x,y} \Pr[X = x|Y = y])$.

PK and U could be much larger than that of V given PK and U. And if we can find a C such that $\mathrm{H_{min}}(C|PK,U)$ is sufficiently larger than $\mathrm{H_{max}}(V|PK,U)$, then we will indeed be able to choose a hash length that is both large enough for agreement and small enough for security.

Also notice that for the agreement to work, it is not necessary for the hash length to be larger than the max-entropy of V (given PK and U) itself – instead, if there was another variable D such that (PK,U,D) is *statistically* close to (PK,U,V), and also Alice is somehow able to sample from D given $PK = \mathsf{pk}$ and $U = u$, then it is sufficient for the hash to be longer than $\mathrm{H_{max}}(D|PK,U)$. Given such a variable, Bob will operate as he did earlier, but Alice can assume that he is actually sampling from $D_{\mathsf{pk},u}$ instead of $V_{\mathsf{pk},u}$, and since these two distributions are close most of the time, the probability of Alice's subsequent computation going wrong is small. This helps us because now we might be able to find such a D that has lower max-entropy given PK and U than V, and then $\mathrm{H_{min}}(C|PK,U)$ would only have to be larger than this.

Following these observations, we set ourselves the following objective: find variables C and D such that:

$$(PK,U,D) \approx_s (PK,U,V) \approx_c (PK,U,C)$$

$$\text{and} \tag{2}$$

$$\mathrm{H_{max}}(D|PK,U) < \mathrm{H_{min}}(C|PK,U)$$

What we do know about V is that it has some pseudo-Shannon-entropy given PK and U. That is, there is a variable C such that:

$$(PK,U,V) \approx_c (PK,U,C) \quad \text{and} \quad \mathrm{H}(C|PK,U) > \mathrm{H}(V|PK,U) + \eta \tag{3}$$

The rest of our construction deals with using this pseudo-Shannon-entropy to achieve the objectives above. This we do using a technique from Information Theory dating back to Shannon [Sha48] which is often referred to in the cryptography literature as *flattening of distributions*, which we describe next. We note that this technique has found use in cryptography before [HILL99, GV99, SV03].

Flattening and Typical Sets. The central idea here is that repetition essentially reduces the general case to the case where the distribution is "almost flat". Namely, if we start with a distribution that has Shannon entropy ξ and repeat it k times, then the new distribution is close to being uniform on a set whose size is roughly $2^{k\xi}$. This set is called the *typical set*; it consists of all elements whose probability is close to $2^{-k\xi}$.

In our case, consider the distribution (PK^k, U^k, V^k), which is the k-fold product repetition of (PK,U,V). Roughly speaking, we define the *typical set* of V^k conditioned on any $(\mathbf{pk}, \mathbf{u})$ in the support[18] of (PK^k, U^k) as follows[19]:

---

[18] The support of (PK^k, U^k) consists of vectors with k elements. We represent vectors by bold symbols, e.g., $\mathbf{v}$.

[19] The actual definition quantifies how different from 2^{-kH} the probability is allowed to be.

$$\mathcal{T}_{V^k|\mathbf{pk},\mathbf{u}} = \Big\{ \mathbf{v} : \ \Pr\big[V^k = \mathbf{v} \big| (PK^k, U^k) = (\mathbf{pk}, \mathbf{u})\big] \approx 2^{-kH(V|PK,U)} \Big\}$$

Considering the typical set is useful for several reasons. On the one hand, the typical set is quite small (roughly $2^{kH(V|PK,U)}$) in size, which means that any distribution supported within it has somewhat low max-entropy. On the other hand, there is an upper bound on the probability of any element that occurs in it, which could be useful in lower bounding min-entropy, which is what we want to do.

The most important property of the typical set it that it contains most of the probability mass of the conditional repeated distribution. That is, for most $(\mathbf{pk}, \mathbf{u}, \mathbf{v})$ sampled from (PK^k, U^k, V^k), it holds that $\mathbf{v}$ lies in the typical set conditioned on $(\mathbf{pk}, \mathbf{u})$; quantitatively, Holenstein and Renner [HR11] show the following:

$$\Pr_{(\mathbf{pk},\mathbf{u},\mathbf{v})\leftarrow(PK^k,U^k,V^k)} \big[\mathbf{v} \notin \mathcal{T}_{V^k|\mathbf{pk},\mathbf{u}}\big] < 2^{-\Omega(k/q^2)} \tag{4}$$

where q is the number of bits in each sample from V. Recall that in our earlier construction of the trapdoor pseudoentropy generator, this corresponds to the length of the prover's message in the SZK argument we started with. We want the above quantity to be quite small, which requires that $k \gg q^2$. This is one of the considerations in our ultimate choice of parameters, and is another reason we want the prover's messages to not be too long.

Back to PKE *Construction.* We shall use the above facts to now show that V^k has pseudo-min-entropy given PK^k and U^k. Let C be the random variable from the expression (3) above that we used to show that V has pseudo-Shannon-entropy. After repetition, we have that:

$$(PK^k, U^k, V^k) \approx_c (PK^k, U^k, C^k) \quad \text{and}$$
$$H(C^k|PK^k, U^k) = k \cdot H(C|PK, U) > k \cdot (H(V|PK, U) + \eta).$$

Next, consider the variable C' that is obtained by restricting, for each $\mathbf{pk}$ and $\mathbf{u}$, the variable C^k to its typical set conditioned on $(\mathbf{pk}, \mathbf{u})$. By applying the bound of Holenstein and Renner (4) with an appropriate choice of k, we infer that:

$$(PK^k, U^k, C^k) \approx_s (PK^k, U^k, C').$$

Further, the upper bound on the probabilities of elements in the typical set tells us that C' has high min-entropy[20] given PK^k and U^k:

---

[20] $H_{\min}(C'|PK^k, U^k)$ could actually be slightly less than the approximate lower bound presented here because there is some slack allowed in the definition of the typical set – it can contain elements whose probabilities are slightly larger than $2^{-kH(C|PK,U)}$. We need to pick this slack carefully – if it is too large, C' loses its min-entropy, and if it is too small the typical set also becomes too small and the bound in (4), which actually depends on this slack, becomes meaningless. This is another constraint on our choice of parameters.

$$H_{\min}(C'|PK^k, U^k) \approx H(C^k|PK^k, U^k) \geq k \cdot (H(V|PK, U) + \eta).$$

Putting the above few expressions together tells us that V^k has some pseudo-min-entropy given PK^k and U^k, which is in fact somewhat more than its Shannon entropy:

$$(PK^k, U^k, V^k) \approx_c (PK^k, U^k, C')$$

and

$$H_{\min}(C'|PK^k, U^k) \gtrsim H(V^k|PK^k, U^k) + k \cdot \eta. \tag{5}$$

This satisfies our objective of getting a variable – V^k here – that has high pseudo-min-entropy (given PK^k and U^k). Our goal is now to find another variable that is statistically close to V^k given PK^k and U^k, and also has small max-entropy given PK^k and U^k. We do this using the same approach as above. Consider the variable V' that is constructed from V^k in the same way C' was from C^k – for each $(\mathbf{pk}, \mathbf{u})$, restrict V^k to its typical set conditioned on $(\mathbf{pk}, \mathbf{u})$. Again, bound (4) tells us that the new distribution is close to the old one. And also, because of the upper bound on the size of the typical set, we have an upper bound on the max-entropy[21] of V' given PK^k and U^k.

$$(PK^k, U^k, V^k) \approx_s (PK^k, U^k, V')$$

and

$$H_{\max}(V'|PK^k, U^k) \lesssim H(V^k|PK^k, U^k). \tag{6}$$

Putting together expressions (5) and (6), we find that the relationship we want between these entropies of C' and V' is indeed satisfied:

$$H_{\min}(C'|PK^k, U^k) \gtrsim H_{\max}(V'|PK^k, U^k) + k \cdot \eta.$$

To summarize, we manage to meet the conditions of expression (2) with respect to (PK^k, U^k, V^k) (instead of (PK, U, V)) with C' taking the role of C and V' taking the role of D. We can now finally fix the length of our hash – call it ℓ – to be between $H_{\max}(V'|PK^k, U^k)$ and $H_{\min}(C'|PK^k, U^k)$, which can be done by setting it to a value between $H(V^k|PK^k, U^k)$ and $H(V^k|PK^k, U^k) + k\eta$ for an appropriate k, and emulate the earlier protocol. We will be able to use the Leftover Hash Lemma as desired to argue security and use the low max-entropy of V' to argue agreement.

The final key agreement protocol from a trapdoor pseudoentropy generator is presented in Fig. 4.

---

[21] The same caveats as in Footnote 20 regarding the min-entropy of C' apply here as well.

Alice		Bob
$\{(\mathsf{pk}_i, \mathsf{sk}_i) \leftarrow \mathsf{KeyGen}\}_{i \in [k]}$	$\xrightarrow{\ \mathbf{pk} = (\mathsf{pk}_1, \mathsf{pk}_2 \ldots \mathsf{pk}_k)\ }$	$\{(u_i, v_i) \leftarrow \mathsf{Enc}(\mathsf{pk}_i)\}_{i \in [k]}$
Use the samplers $\{\mathsf{Dec}(\mathsf{pk}_i, \mathsf{sk}_i, u_i)\}$ to recover		
the distribution of V^k conditioned on $(\mathbf{pk}, \mathbf{u})$.		
Find $\mathbf{v}'$ such that:	$\xleftarrow{\ \mathbf{u}, h, h(\mathbf{v})\ }$	$h \leftarrow \mathcal{H}_\ell$
1. $\mathbf{v}'$ is in the typical set of this distribution		
2. $h(\mathbf{v}') = h(\mathbf{v})$		
Output $\mathbf{v}'$		Output $\mathbf{v}$

Fig. 4. Key agreement from trapdoor pseudoentropy generator

How Laconic? To examine how long the prover's message can be, lets recall the restrictions of our construction. First, we need both parties to be efficient. While Bob is clearly efficient, Alice performs an exhaustive search over the domain of possible prover messages. The size of this domain is $2^{q \cdot k}$ because the parties repeat the underlying protocol k times and the length of each prover's message is q bits. For Alice to be efficient, this domain has to be polynomial-sized, requiring that $q \cdot k = O(\log n)$, where n is the input length. Second, we need that the concentration bound for the typical set (Eq. (4)) to be meaningful; that is, we need k/q^2 to be at least a constant. Together, these imply that q^3 needs to be $O(\log n)$. Lastly, this setting of parameters also suffices for the [VZ12] result that we used in Eq. (1).

2   The Assumption and Main Theorem

In this section, we specify our assumption on the existence of laconic zero-knowledge proof-systems, and state our main theorem regarding its implication for public-key encryption. Due to space limitations, the formal descriptions of our constructions and the proof of our theorem are deferred to the full version of this paper.

We first introduce some necessary definitions and notations. Throughout this section, we use $\mathcal{L}$ to denote an NP language with witness relation $\mathcal{R}_\mathcal{L}$. We use $\mathsf{Y}_\mathcal{L}$ and $\mathsf{N}_\mathcal{L}$ to denote probabilistic polynomial-time algorithms that are to be seen as sampling algorithms for YES and NO instances of $\mathcal{L}$. More specifically, the sampler $\mathsf{Y}_\mathcal{L}(1^\lambda)$ outputs samples of the form (x, w) such that with all but negligible probability (in λ), it holds that $(x, w) \in \mathcal{R}_\mathcal{L}$. We call $\mathsf{Y}_\mathcal{L}$ a solved instance generator. On the other hand, $\mathsf{N}_\mathcal{L}(1^\lambda)$ outputs samples x such that with all but negligible probability, $x \notin \mathcal{L}$. We shall not rely on the fact that the NO sampler $\mathsf{N}_\mathcal{L}$ is an *efficient* algorithm. Still we find it easier to present it as such for symmetry with $\mathsf{Y}_\mathcal{L}$ (which must be efficient).

We shall be concerned with properties of the tuple $(\mathcal{L}, \mathsf{Y}_\mathcal{L}, \mathsf{N}_\mathcal{L})$ – the language $\mathcal{L}$ equipped with (efficiently sampleable) distributions over its YES and NO instances (where YES instances come with corresponding witnesses). Since the choice of YES and NO distributions is always clear from the context, we often simply refer to the above tuple as the language (although we actually mean the language $\mathcal{L}$ with these specific distributions over its instances). We start by defining what we mean when we say that such a language is *cryptographically hard*.

Definition 2.1 (Cryptographic Hardness). *Let* $t = t(\lambda) \in \mathbb{N}$ *and* $\varepsilon = \varepsilon(\lambda) \in [0, 1]$. *The language* $(\mathcal{L}, \mathsf{Y}_{\mathcal{L}}, \mathsf{N}_{\mathcal{L}})$ *is* (t, ε)-*cryptographically hard if* $\mathsf{Y}_{\mathcal{L}}$ *is a solved instance generator, and for every probabilistic algorithm* A *that on input* $(1^{\lambda}, x)$ *runs in time* $t(\lambda)$ *and for all sufficiently large* $\lambda \in \mathbb{N}$ *it holds that:*

$$\left| \Pr_{(x, \cdot) \leftarrow \mathsf{Y}_{\mathcal{L}}(1^{\lambda})} \left[\mathsf{A}(1^{\lambda}, x) = 1 \right] - \Pr_{x \leftarrow \mathsf{N}_{\mathcal{L}}(1^{\lambda})} \left[\mathsf{A}(1^{\lambda}, x) = 1 \right] \right| \leq \varepsilon(\lambda),$$

where the above probabilities are also over the random coins of A. *We say that* $(\mathcal{L}, \mathsf{Y}_{\mathcal{L}}, \mathsf{N}_{\mathcal{L}})$ *is* cryptographically hard *if it is* $(\lambda^c, 1/\lambda^c)$-*hard for every constant* $c > 0$.

Being *cryptographically hard* is a stronger requirement than the usual notion of *average-case hardness* (the latter means that it is hard to distinguish a random YES instance from a random NO instance). Specifically, cryptographic hardness requires both (1) average-case hardness *and* (2) the existence of a solved instance generator (wrt the average-case hard distribution). In particular, the existence of a cryptographically hard language is equivalent to the existence of one-way functions.[22] As noted above, when we say that the language $\mathcal{L}$ is cryptographically hard we are actually implicitly referring to the sampling algorithms $\mathsf{Y}_{\mathcal{L}}$ and $\mathsf{N}_{\mathcal{L}}$.

Next we define honest-verifier statistical zero-knowledge (SZK) arguments, which are similar to statistical honest-verifier zero-knowledge proofs but the soundness condition is only required to hold against malicious provers that run in polynomial-time. We remark that since we will be using the existence of SZK arguments to construct other objects, both the relaxations that we employ (namely requiring only *computational* soundness and *honest verifier* zero knowledge) only strengthen our results.

Below, we use $(\mathbf{P}, \mathbf{V})(1^{\lambda}, x)$ to refer to the transcript of an execution of an interactive protocol with prover $\mathbf{P}$ and verifier $\mathbf{V}$ on input $(1^{\lambda}, x)$. We also use $(\mathbf{P}(w), \mathbf{V})(1^{\lambda}, x)$ to denote a similar execution where the prover is additionally given a witness w as an auxiliary input. In both cases, we sometimes also use the

---

[22] That YES instances are indistinguishable from NO instances implies that it is hard to compute a witness for a YES instance. Given this, a function that takes coins for $\mathsf{Y}_{\mathcal{L}}$ and outputs the instance (but not the witness) generated by $\mathsf{Y}_{\mathcal{L}}$ is one-way (c.f., [Gol08, Proposition 7.2]). For the other direction, assuming that one-way functions exist implies the existence of a linear-stretch pseudorandom generators (PRG) G [HILL99]. The language that is cryptographically hard contains those strings that are in the range of G. The solved instance generator samples a random string r and outputs $G(r)$ as the input and r as the witness. The corresponding NO distribution is that of a random string in the range of the PRG.

same notation to refer to the result (i.e., verifier's output) of such an execution – the appropriate interpretation will be clear from context.

Definition 2.2 (SZK Arguments). *Let* $c = c(\lambda) \in [0,1]$ *and* $s = s(\lambda) \in [0,1]$. *An interactive protocol* $(\mathbf{P}, \mathbf{V})$ *is an* Honest Verifier SZK Argument *with* completeness error c *and* soundness error s *for a language* $\mathcal{L} \in$ NP, *with witness relation* $\mathcal{R}_{\mathcal{L}}$, *if the following properties hold:*

- **Efficiency:** *Both* $\mathbf{P}$ *and* $\mathbf{V}$ *are probabilistic polynomial-time algorithms.*
- **Completeness:** *For any* $(x, w) \in \mathcal{R}_{\mathcal{L}}$, *and all large enough* λ:

$$\Pr\big[(\mathbf{P}(w), \mathbf{V})(1^{\lambda}, x) \ accepts\big] \geq 1 - c(\lambda),$$

  where the parameter c *is called the* completeness error.
- **Soundness:** *For any probabilistic polynomial-time cheating prover* $\mathbf{P}^*$, *any* $x \notin \mathcal{L}$, *and large enough* λ:

$$\Pr\big[(\mathbf{P}^*, \mathbf{V})(1^{\lambda}, x) \ accepts\big] \leq s(\lambda),$$

  where the parameter s *is called the* soundness error.
- **Honest Verifier Statistical Zero Knowledge:** *There is a probabilistic polynomial-time algorithm* $\mathbf{S}$ *(called the simulator) that when given any* $x \in \mathcal{L}$ *simulates the transcript of the interactive proof on input* x. *That is, for any* $(x, w) \in \mathcal{R}_{\mathcal{L}}$ *and for all sufficiently large* λ:

$$\mathrm{SD}\big((\mathbf{P}(w), \mathbf{V})(1^{\lambda}, x), \mathbf{S}(1^{\lambda}, x)\big) \leq \mathsf{negl}(\lambda).$$

Note that our definition only deals with NP languages and requires that the prover is efficient. Typically, when defining an SZK *proof* (rather than argument) this is not done, and the honest prover is allowed to be computationally unbounded. However, this is the natural choice since we focus on *argument* systems (where the soundness requirement is only against malicious provers that are also efficient).

Remark 2.3 (Restricted-view Simulation). For our main result, it suffices that the simulator only simulates the transcript of the interactive proof and not the random-coins of the verifier. The standard definition of simulation is stronger – it also requires that the simulator output random-coins for the verifier that are consistent with the transcript. Ostrovsky [Ost91] called the weaker notion *restricted-view* simulation, and showed that average-case hard languages with honest-verifier SZK proofs with restricted-view simulation (without efficient provers) imply the existence of one-way functions.

We will be dealing with SZK arguments that have additional properties captured by the next definition. Recall that a *round* in an interactive proof is a pair of messages, the first one (possibly empty) from $\mathbf{V}$ to $\mathbf{P}$, and the next the other way.

Definition 2.4 (Laconism). *Let $q = q(\lambda) \in \mathbb{N}$ and $r = r(\lambda) \in \mathbb{N}$. An interactive protocol $(\mathbf{P}, \mathbf{V})$ is said to be r-round and q-laconic if it has at most $r(\lambda)$ rounds, and each message from $\mathbf{P}$ to $\mathbf{V}$ is at most $q(\lambda)$ bits long when run on any input $(1^\lambda, x)$, for large enough λ.*

We can now state our main assumption as follows.

Assumption 2.5. *There exists a cryptographically hard language $(\mathcal{L}, \mathsf{Y}_{\mathcal{L}}, \mathsf{N}_{\mathcal{L}})$ for which there is an r-round and q-laconic honest-verifier SZK argument with completeness error c and soundness error s such that:*

- *There is a constant $\beta > 0$ such that $1 - c(\lambda) > s(\lambda) + \beta$, for large enough $\lambda \in \mathbb{N}$.*
- *q and r are such that $r^2 \cdot q^3 = O(\log(\lambda))$.*

Our main result is given in the next theorem.

Theorem 2.6. (PKE from Laconic SZK). *If Assumption 2.5 holds, then there exists a public-key encryption scheme.*

The construction of our public-key encryption scheme from Assumption 2.5, and the proof of Theorem 2.6, are presented in the full version of this paper. There, in addition, we consider two relaxations of Assumption 2.5, each of which still suffices for our construction. We also present a comparison of our assumptions to concrete assumptions that have been used in the past to construct public-key encryption.

Acknowledgments. We thank Vinod Vaikuntanathan for his encouragement and for helpful discussions. We thank the anonymous reviewers for very useful comments and in particular for suggesting the abstraction of trapdoor pseudoentropy generator.

Research supported in part by NSF Grants CNS-1413920 and CNS-1350619, and by the Defense Advanced Research Projects Agency (DARPA) and the U.S. Army Research Office under contracts W911NF-15-C-0226 and W911NF-15-C-0236. The third author was also supported by the SIMONS Investigator award agreement dated 6-5-12 and the Cybersecurity and Privacy Institute at Northeastern University.

References

[ABW10] Applebaum, B., Barak, B., Wigderson, A.: Public-key cryptography from different assumptions. In: Proceedings of the 42nd ACM Symposium on Theory of Computing, STOC 2010, Cambridge, Massachusetts, USA, 5–8 June 2010, pp. 171–180 (2010)

[Ale03] Alekhnovich, M.: More on average case vs approximation complexity. In: Proceedings of the 44th Symposium on Foundations of Computer Science (FOCS 2003), Cambridge, MA, USA, 11–14 October 2003, pp. 298–307. IEEE Computer Society (2003)

[AR16] Applebaum, B., Raykov, P.: On the relationship between statistical zero-knowledge and statistical randomized encodings. In: Robshaw, M., Katz, J. (eds.) CRYPTO 2016. LNCS, vol. 9816, pp. 449–477. Springer, Heidelberg (2016). https://doi.org/10.1007/978-3-662-53015-3_16

[Bab16] Babai, L.: Graph isomorphism in quasipolynomial time [extended abstract]. In Proceedings of the 48th Annual ACM SIGACT Symposium on Theory of Computing, STOC 2016, Cambridge, MA, USA, 18–21 June 2016, pp. 684–697 (2016)

[BDV16] Bitansky, N., Degwekar, A., Vaikuntanathan, V.: Structure vs hardness through the obfuscation lens. IACR Cryptology ePrint Archive 2016:574 (2016)

[BHY09] Bellare, M., Hofheinz, D., Yilek, S.: Possibility and impossibility results for encryption and commitment secure under selective opening. In: Joux, A. (ed.) EUROCRYPT 2009. LNCS, vol. 5479, pp. 1–35. Springer, Heidelberg (2009). https://doi.org/10.1007/978-3-642-01001-9_1

[BIN97] Bellare, M., Impagliazzo, R., Naor, M.: Does parallel repetition lower the error in computationally sound protocols? In: 38th Annual Symposium on Foundations of Computer Science, FOCS 1997, Miami Beach, Florida, USA, 19–22 October 1997, pp. 374–383 (1997)

[BL13] Bogdanov, A., Lee, C.H.: Limits of Provable Security for Homomorphic Encryption. In: Canetti, R., Garay, J.A. (eds.) CRYPTO 2013, Part I. LNCS, vol. 8042, pp. 111–128. Springer, Heidelberg (2013). https://doi.org/10.1007/978-3-642-40041-4_7

[CS02] Cramer, R., Shoup, V.: Universal hash proofs and a paradigm for adaptive chosen ciphertext secure public-key encryption. In: Knudsen, L.R. (ed.) EUROCRYPT 2002. LNCS, vol. 2332, pp. 45–64. Springer, Heidelberg (2002). https://doi.org/10.1007/3-540-46035-7_4

[DH76] Diffie, W., Hellman, M.E.: New directions in cryptography. IEEE Trans. Inf. Theory 22(6), 644–654 (1976)

[GH98] Goldreich, O., Håstad, J.: On the complexity of interactive proofs with bounded communication. Inf. Process. Lett. 67(4), 205–214 (1998)

[GK93] Goldreich, O., Kushilevitz, E.: A perfect zero-knowledge proof system for a problem equivalent to the discrete logarithm. J. Cryptol. 6(2), 97–116 (1993)

[GMW87] Goldreich, O., Micali, S., Wigderson, A.: How to play any mental game or a completeness theorem for protocols with honest majority. In: Proceedings of the 19th Annual ACM Symposium on Theory of Computing, New York, New York, USA, pp. 218–229 (1987)

[Gol08] Goldreich, O.: Computational Complexity - A Conceptual Perspective. Cambridge University Press, Cambridge (2008)

[GOVW12] Garg, S., Ostrovsky, R., Visconti, I., Wadia, A.: Resettable statistical zero knowledge. In: Cramer, R. (ed.) TCC 2012. LNCS, vol. 7194, pp. 494–511. Springer, Heidelberg (2012). https://doi.org/10.1007/978-3-642-28914-9_28

[GV99] Goldreich, O., Vadhan, S.P.: Comparing entropies in statistical zero knowledge with applications to the structure of SZK. In: Proceedings of the 14th Annual IEEE Conference on Computational Complexity, Atlanta, Georgia, USA, 4–6 May 1999, p. 54 (1999)

[GVW02] Goldreich, O., Vadhan, S., Wigderson, A.: On interactive proofs with a laconic prover. Comput. Complex. 11(1–2), 1–53 (2002)

[HHRS15] Haitner, I., Hoch, J.J., Reingold, O., Segev, G.: Finding collisions in interactive protocols–tight lower bounds on the round and communication complexities of statistically hiding commitments. SIAM J. Comput. 44(1), 193–242 (2015)

[HILL99] Håstad, J., Impagliazzo, R., Levin, L.A., Luby, M.: A pseudorandom generator from any one-way function. SIAM J. Comput. **28**(4), 1364–1396 (1999)

[HLWW16] Hazay, C., López-Alt, A., Wee, H., Wichs, D.: Leakage-resilient cryptography from minimal assumptions. J. Cryptol. **29**(3), 514–551 (2016)

[HNO+09] Haitner, I., Nguyen, M.-H., Ong, S.H., Reingold, O., Vadhan, S.P.: Statistically hiding commitments and statistical zero-knowledge arguments from any one-way function. SIAM J. Comput. **39**(3), 1153–1218 (2009)

[HR05] Holenstein, T., Renner, R.: One-way secret-key agreement and applications to circuit polarization and immunization of public-key encryption. In: Shoup, V. (ed.) CRYPTO 2005. LNCS, vol. 3621, pp. 478–493. Springer, Heidelberg (2005). https://doi.org/10.1007/11535218_29

[HR11] Holenstein, T., Renner, R.: On the randomness of independent experiments. IEEE Trans. Inf. Theory **57**(4), 1865–1871 (2011)

[IR89] Impagliazzo, R., Rudich, S.: Limits on the provable consequences of one-way permutations. In: Proceedings of the Twenty-First Annual ACM Symposium on Theory of Computing, pp. 44–61. ACM (1989)

[Kil88] Kilian, J.: Founding crytpography on oblivious transfer. In: Proceedings of the Twentieth Annual ACM Symposium on Theory of Computing, pp. 20–31. ACM (1988)

[KMN+14] Komargodski, I., Moran, T., Naor, M., Pass, R., Rosen, A., Yogev, E.: One-way functions and (im)perfect obfuscation. In: 55th IEEE Annual Symposium on Foundations of Computer Science, FOCS 2014, Philadelphia, PA, USA, 18–21 October 2014, pp. 374–383. IEEE Computer Society (2014)

[LV16] Liu, T., Vaikuntanathan, V.: On basing private information retrieval on NP-hardness. In: Kushilevitz, E., Malkin, T. (eds.) TCC 2016, Part I. LNCS, vol. 9562, pp. 372–386. Springer, Heidelberg (2016). https://doi.org/10.1007/978-3-662-49096-9_16

[NV06] Nguyen, M.-H., Vadhan, S.P.: Zero knowledge with efficient provers. In: Proceedings of the 38th Annual ACM Symposium on Theory of Computing, Seattle, WA, USA, 21–23 May 2006, pp. 287–295 (2006)

[Ost91] Ostrovsky, R.: One-way functions, hard on average problems, and statistical zero-knowledge proofs. In: Proceedings of the Sixth Annual Structure in Complexity Theory Conference, Chicago, Illinois, USA, 30 June - 3 July 1991, pp. 133–138 (1991)

[OV08] Ong, S.J., Vadhan, S.: An equivalence between zero knowledge and commitments. In: Canetti, R. (ed.) TCC 2008. LNCS, vol. 4948, pp. 482–500. Springer, Heidelberg (2008). https://doi.org/10.1007/978-3-540-78524-8_27

[PPS15] Pandey, O., Prabhakaran, M., Sahai, A.: Obfuscation-based non-black-box simulation and four message concurrent zero knowledge for NP. In: Dodis, Y., Nielsen, J.B. (eds.) TCC 2015, Part II. LNCS, vol. 9015, pp. 638–667. Springer, Heidelberg (2015). https://doi.org/10.1007/978-3-662-46497-7_25

[PVW08] Peikert, C., Vaikuntanathan, V., Waters, B.: A framework for efficient and composable oblivious transfer. In: Wagner, D. (ed.) CRYPTO 2008. LNCS, vol. 5157, pp. 554–571. Springer, Heidelberg (2008). https://doi.org/10.1007/978-3-540-85174-5_31

[Reg05] Regev, O.: On lattices, learning with errors, random linear codes, and cryptography. In: Gabow, H.N., Fagin, R. (eds.) Proceedings of the 37th Annual ACM Symposium on Theory of Computing, Baltimore, MD, USA, 22–24 May 2005, pp. 84–93. ACM (2005)

[Rot11] Rothblum, R.: Homomorphic encryption: from private-key to public-key. In: Ishai, Y. (ed.) TCC 2011. LNCS, vol. 6597, pp. 219–234. Springer, Heidelberg (2011). https://doi.org/10.1007/978-3-642-19571-6_14

[RSA78] Rivest, R.L., Shamir, A., Adleman, L.M.: A method for obtaining digital signatures and public-key cryptosystems. Commun. ACM **21**(2), 120–126 (1978)

[Sha48] Shannon, C.E.: A mathematical theory of communication. Bell Syst. Tech. J. **27**(3), 379–423 (1948)

[SV03] Sahai, A., Vadhan, S.: A complete problem for statistical zero knowledge. J. ACM (JACM) **50**(2), 196–249 (2003)

[VZ12] Vadhan, S., Zheng, C.J.: Characterizing pseudoentropy and simplifying pseudorandom generator constructions. In: Proceedings of the Forty-Fourth Annual ACM Symposium on Theory of Computing, pp. 817–836. ACM (2012)

Updatable and Universal Common Reference Strings with Applications to zk-SNARKs

Jens Groth[1], Markulf Kohlweiss[2,3], Mary Maller[1,2(✉)], Sarah Meiklejohn[1], and Ian Miers[2,4]

[1] University College London, London, UK
{j.groth,mary.maller.15,s.meiklejohn}@ucl.ac.uk
[2] Microsoft Research Cambridge, Cambridge, UK
[3] University of Edinburgh, Edinburgh, UK
markulf.kohlweiss@ed.ac.uk
[4] Cornell Tech, New York, USA
imiers@cs.jhu.edu

Abstract. By design, existing (pre-processing) zk-SNARKs embed a secret trapdoor in a relation-dependent common reference strings (CRS). The trapdoor is exploited by a (hypothetical) simulator to prove the scheme is zero knowledge, and the secret-dependent structure facilitates a linear-size CRS and linear-time prover computation. If known by a real party, however, the trapdoor can be used to subvert the security of the system. The structured CRS that makes zk-SNARKs practical also makes deploying zk-SNARKS problematic, as it is difficult to argue why the trapdoor would not be available to the entity responsible for generating the CRS. Moreover, for pre-processing zk-SNARKs a new trusted CRS needs to be computed every time the relation is changed.

In this paper, we address both issues by proposing a model where a number of users can update a universal CRS. The updatable CRS model guarantees security if at least one of the users updating the CRS is honest. We provide both a negative result, by showing that zk-SNARKs with private secret-dependent polynomials in the CRS cannot be updatable, and a positive result by constructing a zk-SNARK based on a CRS consisting only of secret-dependent monomials. The CRS is of quadratic size, is updatable, and is universal in the sense that it can be specialized into one or more relation-dependent CRS of linear size with linear-time prover computation.

J. Groth—The research leading to these results has received funding from the European Research Council under the European Union's Seventh Framework Programme (FP/2007-2013)/ERC Grant Agreement no. 307937.
This work was done in part while Mary Maller was an intern at Microsoft Research Cambridge, and she is funded by Microsoft Research Cambridge.
S. Meiklejohn—Supported in part by EPSRC Grant EP/N028104/1.
This work was done in part while Ian Miers was visiting Microsoft Research Cambridge.

H. Shacham and A. Boldyreva (Eds.): CRYPTO 2018, LNCS 10993, pp. 698–728, 2018.
https://doi.org/10.1007/978-3-319-96878-0_24

1   Introduction

Since their introduction three decades ago, zero-knowledge proofs have been constructed in a variety of different models. Arguably the simplest setting is the Uniform Random String (URS) model, introduced by Blum, Feldman, and Micali [BFM88] and used heavily since [FLS99, Dam92, SP92, KP98, SCP00, GO14, Gro10a, GGI+15]. In the URS model both the prover and verifier have access to a string sampled uniformly at random and it enables the prover to send a single non-interactive zero-knowledge (NIZK) proof that convinces the verifier. This model is limited, however, so many newer NIZK proof systems are instead in the Common Reference String (CRS) model [CF01, Dam00, FF00, GOS12, GS12]. Here, the reference string must have some structure based on *secret* random coins (e.g., be of the form $G^s, G^{s^2}, G^{s^3}, \ldots$) and the secret (e.g., the value s) must be discarded after generation. This makes CRS generation an inherently trusted process.

Until recently, little consideration had been given to how to generate common reference strings in practice, and it was simply assumed that a trusted party could be found. The introduction of zk-SNARKs (zero-knowledge Succinct Non-interactive ARguments of Knowledge) in the CRS model [Gro10b], however, and subsequent academic and commercial usage has brought this issue front and center. In particular, zk-SNARKs are of considerable interest for cryptocurrencies given their usage in both Zcash [BCG+14], which relies on them in order to preserve privacy, and Ethereum, which recently integrated support for them [Buc17]. In these decentralized settings in which real monetary value is at stake, finding a party who can be widely accepted as trusted is nearly impossible.

Ben-Sasson et al. [BCG+15] and subsequently Bowe et al. [BGG17] examined the use of multi-party computation to generate a CRS, where only one out of n parties needs to be honest but the participants must be selected in advance. In concurrent work, Bowe et al. [BGM17] propose a protocol that avoids the pre-selection requirement and as a result scales to more participants. Both protocols, however, result in a CRS for a fixed circuit with a fixed set of participants. This raises issues about who the participants are and how they were selected, which are compounded by the fact that upgrades for increased performance or functionality require a new circuit and thus a new invocation of the protocol. This offers both renewed opportunities for adversarial subversion and loss of faith in the integrity of the parameters. Despite multi-party CRS generation, CRS setup (and particularly the cost it imposes on upgrading protocols), is thus a major obstacle to the practical deployment and usage of zk-SNARKs.

Motivated by this issue of trusted setup, several works have recently examined alternatives to CRS-based pre-processing SNARKS in the URS and random oracle model, despite the associated performance disadvantages. Current proposed alternatives [BSBHR18, WTas+17, AHIV17, BCG+17, BCC+16, BBB+18], either have proofs that even for modest circuit sizes, range into the hundreds of kilobytes or have verification times that are linear in the size of the circuit and make verification of large statements impractical for many applications. In contrast, (Quadratic Arithmetic Program) QAP-based zk-SNARKs offer quasi-constant-size proofs and verification times in the tens of milliseconds. Thus, modulo the barrier of having a trusted CRS setup, they are ideally suited to applications such as blockchains where space and bandwidth are highly constrained and proofs are expected to be verified many times in a performance-critical process.

Our contributions. To provide a middle ground between the fully trusted and fully subverted CRS models, we introduce and explore a new setup model for NIZK proofs: the *updatable CRS model.* In the updatable CRS model, any user can at any point choose to update the common reference string, provided that they also prove they have done the update correctly. If the proof of correctness verifies, then the new CRS resulting from the update can be considered trustworthy (i.e., uncorrupted) as long as either the old CRS or the updater was honest. If multiple users participate in this process, then it is possible to get a sequence of updates by different people over a period of time. If any one update is honest at any point in the sequence, then the scheme is sound.

We introduce our model for updatable zero-knowledge proofs in Sect. 3, where we also relate it to the classical CRS model (which we can think of as weaker) and the models for subversion-resistant proofs [BFS16, ABLZ17] (which we can think of as stronger).

Since Bellare et al. showed that it was impossible to achieve both subversion soundness and even standard zero-knowledge, it follows that it is also impossible to achieve subversion soundness and updatable zero-knowledge. With this in mind, we next explore the space of NIZK proofs that achieve subversion zero-knowledge (and thus updatable zero-knowledge) and updatable soundness.

We first observe that the original pairing-based zk-SNARK construction due to Groth [Gro10b] can be made updatably sound. His construction, however, has a quadratic-sized reference string, resulting in quadratic prover complexity. Our positive result in Sect. 5 provides a construction of an updatable QAP-based zk-SNARK that uses a quadratic-sized universal CRS, but allows for the derivation of linear-sized relation-dependent CRSs (and thus linear prover complexity). Because our universal CRS consists solely of monomials, our construction gets around our negative result in Sect. 6, which demonstrates that it is impossible to achieve updatable soundness for any pairing-based NIZK proof that relies on embedding non-monomials in the reference string (e.g., uses terms G^{s^2+s}). In particular, this shows that QAP-based zk-SNARKs such as Pinocchio [PHGR13] do not satisfy updatable soundness.

Applications. Updatable common reference strings are a natural model for parameter generation in a cryptocurrency, or other blockchain-based settings. Informally, in a blockchain, blocks of data are agreed upon by peers in a global network according to some consensus protocol, with different blocks of data being contributed by different users.

If each block (or one out of every n blocks) contains an update to the CRS performed by the creator of the block, then assuming the blockchain as a whole is correct, the CRS is sound. Indeed, we achieve a stronger property than the blockchain itself: assuming one single block was honestly generated, then the CRS is sound even if all other blocks are generated by dishonest parties.

While updatable security thus seems to be a natural fit for blockchain-based settings, there would be considerable work involved in making the construction presented in this paper truly practical. As our construction is compatible with several techniques designed to achieve efficiency (e.g., pruning of the blockchain) and does not require replication of the entire sequence of updated CRSs in order to perform verification, we believe this is a promising avenue for future research.

Knowledge assumptions. Our approach to proving that the updates are carried out correctly is to prove the existence of a correct CRS update under a knowledge extractor assumption. Knowledge assumptions define conditions under which extractors can retrieve the internal 'knowledge' of the adversary, in this case secret randomness used to update the CRS correctly. While less reassuring than standard model assumptions, the security of zk-SNARKs typically rely on knowledge assumptions anyway (and must be based on non-falsifiable assumptions [GW11]), and our construction is proven updatably sound under the same assumptions as those that are used to prove standard soundness. We assume that an adversary does not subvert our scheme by hiding a trapdoor in the groups. Choosing such elliptic curve groups is a contentious affair [BCC+14] and outside the scope of this paper, but one option for guaranteeing the adversary does not implant a trapdoor is to use a deterministic group generation algorithm.

Updatable CRS vs. URS model. The updatable CRS model is closer to the URS model than the CRS model, but it is important to acknowledge the differences. In the URS model, given a valid proof and a URS, a verifier only needs to be convinced that the URS was sampled at random (e.g. via a hash function in the random oracle model). An updatable CRS, in contrast, allows a skeptical verifier to trust proofs made with respect to a CRS that they themselves updated (or contributed to via a previous update). This is a weaker property than the URS model, as it cannot help with proofs formed before this update. On the other hand, updatable CRS schemes inherit the efficiency and expressiveness of the CRS model, without fully inheriting its reliance on a trusted setup.

2   Related Work

In addition to the works referenced in the introduction, we compare here with the research most closely related to our own.

In terms of acknowledging the potential for an adversary to compromise the CRS, Bellare, Fuchsbauer and Scafuro [BFS16] ask what security can be maintained for NIZK proofs when the CRS is subverted. They formalise the different notions of subversion resistance and then investigate their possibility. Using similar techniques to Goldreich et al. [GOP94], they show that soundness in this setting cannot be achieved at the same time as (standard) zero-knowledge. Building on the notions of Bellare et al., two recent papers [ABLZ17, Fuc17] discuss how to

Table 1. Comparison for pairing-based zk-SNARKs for boolean and arithmetic circuit satisfiability with ℓ-element known circuit inputs, m wires, and n gates, of which $n_\times$ are multiplication gates. $\mathbb{G}$ means group elements in either source group, Ex means group exponentiations, $M_\mathbb{G}$ means group multiplications, and P means pairings.

Scheme	Universal CRS	Circuit CRS	Size	Prover comp	Verifier comp
[Gro10b] ($\mathbb{F}_2$)	$O(n^2)$ $\mathbb{G}$	—	42 $\mathbb{G}$	$O(n^2)$ Ex	$36P + nM_\mathbb{G}$
[PHGR13] ($\mathbb{F}_q$)	—	$O(n_\times + m - \ell)$ $\mathbb{G}$	8 $\mathbb{G}$	$O(n_\times + m - \ell)$ Ex	$12P + \ell$ Ex
[Gro16] ($\mathbb{F}_q$)	—	$O(n_\times + m)$ $\mathbb{G}$	3 $\mathbb{G}$	$O(n_\times + m - \ell)$ Ex	$3P + \ell$ Ex
This work ($\mathbb{F}_q$)	$O(n_\times^2)$ $\mathbb{G}$	$O(n_\times + m - \ell)$ $\mathbb{G}$	3 $\mathbb{G}$	$O(n_\times + m - \ell)$ Ex	$5P + \ell$ Ex

achieve subversion zero-knowledge for zk-SNARKs. None of these schemes, however, can avoid the impossibility result and they do not simultaneously preserve soundness and zero-knowledge under subversion.

The multi-string model by Groth and Ostrovsky [GO14] addresses the problem of subversion by designing protocols that require only the majority of the parties contributing multiple reference strings to be honest. Their construction gives statistically sound proofs but they are of linear size in both the number of reference strings and the size of the instance.

In terms of zk-SNARKs, some of the most efficient constructions in the literature [Lip13, PHGR13, BCTV14, DFGK14, Gro16, GM17] use the quadratic span program (QSP) or quadratic arithmetic program (QAP) approach of Gennaro et al. [GGPR13]. The issue with this approach when it comes to updatability is that it requires embedding arbitrary polynomials in the exponents of group elements in the common reference string. However, we show in Sect. 6 that if it is possible to update these polynomial embeddings, then it is possible to compute all the constituent monomials in the polynomials. Uncovering the underlying monomials, however, would completely break those zk-SNARKs, so QSP-based and QAP-based updatable zk-SNARKs require a fundamentally new technique.

Two early zk-SNARKs by Groth [Gro10b] and Lipmaa [Lip12] do, however, use only monomials. The main drawback of [Gro10b] is that it has a quadratic-sized CRS and quadratic prover computation, but it has a CRS that consists solely of monomials, and thus is updatable. Lipmaa still has quadratic prover computation, however he suggested the use of progression-free sets to construct NIZK arguments with a CRS consisting of $n^{(1+o(1))}$ group elements. It uses progression-free sets to give an elegant product argument and a permutation argument, which are then combined to give a circuit satisfiability argument.

We give a performance comparison of pairing-based zk-SNARKs in Table 1, comparing the relative size of the CRS and the proof, and the computation required for the prover and verifier. We compare Groth's original zk-SNARK, two representative QAP-based zk-SNARKs, and our updatable and specializable QAP-based zk-SNARK. As can be seen, our efficiency is comparable to the QAP-based schemes, but our universal reference string is not restricted to proving a pre-specified circuit. For the QAP-based SNARKs one could use Valiant's universal circuit construction [Val76, LMS16] to achieve universality but this would introduce a $\log n$ multiplicative overhead. We pose as an interesting open question whether updatable zk-SNARKs with a shorter universal CRS exist.

In concurrent work, Bowe et al. [BGM17] propose a two-phase protocol for the generation of a zk-SNARK reference string that is player-replaceable [GHM+17]. Like our protocol, the first phase of their protocol also computes monomials with parties operating in a similar one-shot fashion. However, there are several differences. First, their protocol does so under the stronger assumption of a random oracle, whereas we prove the security of our updatable zk-SNARK directly under the same assumptions as a trusted setup zk-SNARK. More significantly, to create a full CRS which does not have quadratic prover time, Bowe et al. require a second phase. As one party in each phase must be honest and the second phase depends on the first, the final CRS is not updatable. There is no way to increase the number of parties in the first phase after the second phase has started and, restarting the first phase means discarding the participants in the second phase. As a result, the protocol is still a multi-party computation to produce a fixed CRS with a fixed set of participants, albeit with the set of participants fixed midway through the protocol instead of at the start. In contrast, we produce a CRS with linear overhead from a quadratic-sized universal updatable CRS via an untrusted *specialization* process. Thus our CRS can be continuously updated without discarding past participation.

3   Defining Updatable and Universal CRS Schemes

In this section, we begin by presenting some notation and revisiting the basic definitions of non-interactive zero-knowledge proofs in the common reference string model, in which the reference string must be run by a trusted third party. We then present our new definitions for an updatable CRS scheme, which relaxes the CRS model by allowing the adversary to either fully generate the reference string itself, or at least contribute to its computation as one of the parties performing updates. In this our model is related to subversion-resistant proofs [BFS16], which we also present and compare to our own model.

3.1   Notation

If x is a binary string then $|x|$ denotes its bit length. If S is a finite set then $|S|$ denotes its size and $x \xleftarrow{\$} S$ denotes sampling a member uniformly from S and assigning it to x. We use $\lambda \in \mathbb{N}$ to denote the security parameter and 1^λ to denote its unary representation. We use ε to denote the empty string.

Algorithms are randomized unless explicitly noted otherwise. "PPT" stands for "probabilistic polynomial time" and "DPT" stands for "deterministic polynomial time." We use $y \leftarrow A(x; r)$ to denote running algorithm A on inputs x and random coins r and assigning its output to y. We write $y \xleftarrow{\$} A(x)$ or $y \xleftarrow{r} A(x)$ (when we want to refer to r later on) to denote $y \leftarrow A(x; r)$ for r sampled uniformly at random. $\mathcal{A}.\mathsf{rt}(\lambda)$, and sample $r \xleftarrow{\$} \{0, 1\}^{\mathcal{A}.\mathsf{rl}(\lambda)}$.

We use code-based games in security definitions and proofs [BR06]. A game $\mathsf{Sec}_{\mathcal{A}}(\lambda)$, played with respect to a security notion Sec and adversary $\mathcal{A}$, has a MAIN procedure whose output is the output of the game. The notation $\Pr[\mathsf{Sec}_{\mathcal{A}}(\lambda)]$ is used to denote the probability that this output is 1.

3.2 NIZK Proofs in the CRS Model

Let Setup be a setup algorithm that takes as input a security parameter 1^{λ} and outputs a common reference string $\mathtt{crs}$ sampled from some distribution $\mathcal{D}$. Let R be a polynomial time decidable relation with triples $(\mathtt{crs}, \phi, w)$. We say w is a witness to the instance ϕ being in the relation defined by $\mathtt{crs}$ when $(\mathtt{crs}, \phi, w) \in R$.

Non-interactive zero-knowledge (NIZK) proofs and arguments in the CRS model are comprised of three algorithms $(\mathsf{Setup}, \mathsf{Prove}, \mathsf{Verify})$, and satisfy completeness, zero-knowledge, and (knowledge) soundness. Perfect completeness requires that for all reference strings output by setup $\mathtt{crs} \xleftarrow{\$} \mathsf{Setup}(1^{\lambda})$, whenever $(\mathtt{crs}, \phi, w) \in R$ we have that $\mathsf{Verify}(\mathtt{crs}, \phi, \mathsf{Prove}(\mathtt{crs}, \phi, w)) = 1$. Soundness requires that an adversary cannot output a proof that verifies with respect to an instance not in the language, and knowledge soundness goes a step further and for any prover producing a valid proof there is an extractor $\mathcal{X}$ that can extract a valid witness. Finally, zero knowledge requires that there exists a pair $(\mathsf{SimSetup}, \mathsf{SimProve})$ such that an adversary cannot tell if it is given an honest CRS and honest proofs, or a simulated CRS and simulated proofs (in which the simulator does not have access to the witness, but does have a simulation trapdoor). We present these notions more formally below.

3.3 Updating Common Reference Strings

In our definitions we relax the CRS model by allowing the adversary to either fully generate the reference string itself, or at least contribute to its computation as one of the parties performing updates. Informally, we can think of this as having the adversary interact with the Setup algorithm. More formally, we can define an updatable CRS scheme that consists of PPT algorithms Setup, Update and a DPT algorithm $\mathsf{VerifyCRS}$ that behave as follows:

- $(\mathtt{crs}, \rho) \xleftarrow{\$} \mathsf{Setup}(1^{\lambda})$ takes as input the security parameter and returns a common reference string and a proof of correctness.
- $(\mathtt{crs}', \rho') \xleftarrow{\$} \mathsf{Update}(1^{\lambda}, \mathtt{crs}, (\rho_i)_{i=1}^{n})$ takes as input the security parameter, a common reference string, and a list of update proofs for the common reference string. It outputs an updated common reference string and a proof of the correctness of the update.
- $b \leftarrow \mathsf{VerifyCRS}(1^{\lambda}, \mathtt{crs}, (\rho_i)_{i=1}^{n})$ takes as input the security parameter, a common reference string, and a list of proofs. It outputs a bit indicating acceptance, $b = 1$, or rejection $b = 0$.

Definition 1. *An updatable CRS scheme is perfectly correct if*

- *for all* $(\mathsf{crs}, \rho) \xleftarrow{\$} \mathsf{Setup}(1^\lambda)$ *we have* $\mathsf{VerifyCRS}(1^\lambda, \mathsf{crs}, \rho) = 1$;
- *for all* $(\lambda, \mathsf{crs}, (\rho_i)_{i=1}^n)$ *such that* $\mathsf{VerifyCRS}(1^\lambda, \mathsf{crs}, (\rho)_{i=1}^n) = 1$ *we have for* $(\mathsf{crs}', \rho_{n+1}) \xleftarrow{\$} \mathsf{Update}(1^\lambda, \mathsf{crs}, (\rho_i)_{i=1}^n)$ *that* $\mathsf{VerifyCRS}(1^\lambda, \mathsf{crs}', (\rho)_{i=1}^{n+1}) = 1$.

Please observe that a standard trusted setup is a special case of an updatable setup with $\rho = \varepsilon$ as the update proof where the verification algorithm accepts anything. For a subversion-resistant setup the proof ρ can be considered as extra elements included in the CRS solely to make the CRS verifiable.

3.4 Security Properties

We recall the notions of zero-knowledge, soundness, and knowledge soundness associated with NIZK proof systems. In addition to considering the standard setting with a trusted reference string, we also capture the subversion-resistant setting, in which the adversary generates the reference string [BFS16, ABLZ17, Fuc17], and introduce our new updatable reference string setting.

For each security property, the game in the left column of Fig. 1 resembles the usual security game for zero-knowledge, soundness, and knowledge soundness. The difference is in the creation of the CRS crs, which is initially set to $\perp$. We then model the process of generating the CRS as an interaction between the adversary and a setup oracle $\mathcal{O}_s$, at the end of which the oracle sets this value crs and returns it to the adversary.

In principle, this process of creating the CRS can look like anything: it could be trusted, or even a more general MPC protocol. For the sake of this paper, however, we focus on three types of setup: (1) a trusted setup (T) where the setup generator ignores the adversary when generating crs; (2) a subvertible setup (S) where the setup generator gets crs from the adversary and uses it after checking that it is well formed; and (3) a model in between that we call an updatable setup (U). In this new model, an adversary can adaptively generate sequences of CRSs and arbitrarily interleave its own malicious updates into them. The only constraints on the final CRS are that it is well formed and that at least one honest participant has contributed to it by providing an update.

In the definition of zero-knowledge, we require the existence of a PPT simulator consisting of algorithms (SimSetup, SimUpdate, SimProve) that share state with each other. The idea is that it can be used to simulate the generation of common reference strings and simulate proofs without knowing the corresponding witnesses.

Definition 2. *Let* $\mathrm{P} = (\mathsf{Setup}, \mathsf{Update}, \mathsf{VerifyCRS}, \mathsf{Prove}, \mathsf{Verify})$ *be a non-interactive argument for the relation* R. *Then the argument is* X-*secure, for* $\mathsf{X} \in \{\mathsf{T}, \mathsf{U}, \mathsf{S}\}$, *if it satisfies each of the following:*

- P *is complete, if for all PPT algorithms* $\mathcal{A}$ *the advantage* $|1 - \Pr[\mathsf{COMP}_{\mathcal{A}}(\lambda)]|$ *is negligible in* λ.

$$\frac{\text{MAIN } \mathsf{COMP}_{\mathcal{A}}(\lambda)}{}$$

MAIN $\mathsf{COMP}_{\mathcal{A}}(\lambda)$

$(\mathrm{crs}, (\rho_i)_{i=1}^n, \phi, w) \leftarrow \mathcal{A}(1^\lambda)$
$b \leftarrow \mathsf{VerifyCRS}(1^\lambda, \mathrm{crs}, (\rho_i)_{i=1}^n)$
if $b = 0$ or $(\mathrm{crs}, \phi, w) \notin R$ return 1
$\pi \xleftarrow{\$} \mathsf{Prove}(\mathrm{crs}, \phi, w)$
return $\mathsf{Verify}(\mathrm{crs}, \phi, \pi)$

MAIN $\mathsf{X\text{-}ZK}_{\mathcal{A}, \mathsf{Sim}_{\mathcal{A}}}(\lambda)$

$b \xleftarrow{\$} \{0, 1\}$
if $b = 0$
$\quad \mathsf{Setup} \leftarrow \mathsf{SimSetup}$
$\quad \mathsf{Update} \leftarrow \mathsf{SimUpdate}$
$\mathrm{crs} \leftarrow \bot; Q \leftarrow \emptyset$
$\mathrm{state} \xleftarrow{r} \mathcal{A}^{\mathsf{X}\text{-}\mathcal{O}_\mathsf{s}}(1^\lambda)$
$b' \xleftarrow{\$} \mathcal{A}^{\mathcal{O}_\mathsf{pf}}(\mathrm{state})$
return 1 if $b' = b$ else return 0

$\underline{\mathcal{O}_{\mathsf{pf}}(\phi, w)}$
if $(\mathrm{crs}, \phi, w) \notin R$ return $\bot$
if $b = 0$ return $\mathsf{SimProve}_{\mathcal{A}}(\mathrm{crs}, r, \phi)$
else return $\mathsf{Prove}(\mathrm{crs}, \phi, w)$

MAIN $\mathsf{X\text{-}SND}_{\mathcal{A}}(\lambda)$

$\mathrm{crs} \leftarrow \bot; Q \leftarrow \emptyset$
$(\phi, \pi) \xleftarrow{\$} \mathcal{A}^{\mathsf{X}\text{-}\mathcal{O}_\mathsf{s}}(1^\lambda)$
return $\mathsf{Verify}(\mathrm{crs}, \phi, \pi) \wedge \phi \notin L_R$

MAIN $\mathsf{X\text{-}KSND}_{\mathcal{A}, \mathcal{X}_{\mathcal{A}}}(\lambda)$

$\mathrm{crs} \leftarrow \bot, Q \leftarrow \emptyset$
$(\phi, \pi) \xleftarrow{r} \mathcal{A}^{\mathsf{X}\text{-}\mathcal{O}_\mathsf{s}}(1^\lambda)$
$w \xleftarrow{\$} \mathcal{X}_{\mathcal{A}}(\mathrm{crs}, r)$
return $\mathsf{Verify}(\mathrm{crs}, \phi, \pi) \wedge (\phi, w) \notin R$

$\underline{\mathsf{T}\text{-}\mathcal{O}_\mathsf{s}(x)}$
if $\mathrm{crs} \neq \bot$ return $\bot$
$(\mathrm{crs}, \rho) \xleftarrow{\$} \mathsf{Setup}(1^\lambda)$
return (crs, ρ)

$\underline{\mathsf{U}\text{-}\mathcal{O}_\mathsf{s}(\mathrm{intent}, \mathrm{crs}_n, (\rho_i)_{i=1}^n)}$
if $\mathrm{crs} \neq \bot$ return $\bot$
if $\mathrm{intent} = \mathrm{setup}$
$\quad (\mathrm{crs}_1, \rho_1) \xleftarrow{\$} \mathsf{Setup}(1^\lambda)$
$\quad Q \leftarrow \{\rho_1\}$
$\quad$ return (crs_1, ρ_1)
if $\mathrm{intent} = \mathrm{update}$
$\quad b \leftarrow \mathsf{VerifyCRS}(1^\lambda, \mathrm{crs}_n, (\rho_i)_{i=1}^n) = 0$
$\quad$ if $b = 0$ return $\bot$
$\quad (\mathrm{crs}', \rho') \xleftarrow{\$} \mathsf{Update}(1^\lambda, \mathrm{crs}_n, (\rho_i)_{i=1}^n)$
$\quad Q \leftarrow Q \cup \{\rho'\}$
$\quad$ return (crs', ρ')
if $\mathrm{intent} = \mathrm{final}$
$\quad b \leftarrow \mathsf{VerifyCRS}(1^\lambda, \mathrm{crs}_n, (\rho_i)_{i=1}^n)$
$\quad$ if $b = 0$ or $Q \cap \{\rho_i\}_i = \emptyset$ return $\bot$
$\quad$ set $\mathrm{crs} \leftarrow \mathrm{crs}_n$ and return crs
else return $\bot$

$\underline{\mathsf{S}\text{-}\mathcal{O}_\mathsf{s}(\mathrm{crs}_n, (\rho_i)_{i=1}^n)}$
if $\mathrm{crs} \neq \bot$ return $\bot$
$b \leftarrow \mathsf{VerifyCRS}(1^\lambda, \mathrm{crs}_n, (\rho_i)_{i=1}^n) = 0$
if $b = 0$ return $\bot$
set $\mathrm{crs} \leftarrow \mathrm{crs}_n$ and return crs

Fig. 1. The left games define completeness, zero-knowledge (X-ZK), soundness (X-SND), and knowledge soundness (X-KSND). The right oracles define the notions $\mathsf{X} \in \{\mathsf{T}, \mathsf{U}, \mathsf{S}\}$; i.e., trusted, updatable, and subvertible CRS setups. A complete game is constructed by using an oracle from the right side in the game on the left side.

- P *is* X-*zero-knowledge, if for all PPT algorithms* $\mathcal{A}$ *there exists a simulator* $\mathsf{Sim}_{\mathcal{A}} = (\mathsf{SimSetup}, \mathsf{SimUpdate}, \mathsf{SimProve}_{\mathcal{A}})$ *where the advantage* $|2\Pr[\text{X-ZK}_{\mathcal{A},\mathsf{Sim}_{\mathcal{A}}}(1^{\lambda}) = 1] - 1|$ *is negligible in* λ.
- P *is* X-*sound if for all PPT algorithms* $\mathcal{A}$ *the probability* $\Pr[\text{X-SND}_{\mathcal{A}}(1^{\lambda}) = 1]$ *is negligible in* λ.
- P *is* X-*knowledge-sound if for all PPT algorithms* $\mathcal{A}$ *there exists a PPT extractor* $\mathcal{X}_{\mathcal{A}}$ *such that the probability* $|\Pr[\text{X-KSND}_{\mathcal{A},\mathcal{X}_{\mathcal{A}}}(1^{\lambda})]|$ *is negligible in* λ.

Moreover, if a definition holds with respect to an adversary with unbounded computation, we say it holds statistically, and if the advantage is exactly 0, we say it holds perfectly.

One of the main benefits of our model is its flexibility. For example, a slightly weaker but still trusted setup could be defined that would allow the adversary to pick some parameters (e.g., the number of gates in an arithmetic circuit or a specific finite field) and then run the setup on those. In addition to different types of setup assumptions, it also would be easy to incorporate additional security notions into this framework, such as simulation soundness.

Our definition of subversion-resistant security is adapted from that of Abdolmaleki et al. [ABLZ17], and our definition of update security is itself adapted from this definition. We stress that this new notion of setup security is necessary: while we prove that our construction in Sect. 5 satisfies subversion zero-knowledge, this is known to be mutually exclusive with subversion soundness [BFS16], so update security provides the middle ground in which we can obtain positive results. In terms of relating these notions, it is fairly straightforward that updatable security implies trusted security, and that subversion-resistant security implies updatable security (for all security notions).

The proofs for the following lemmas are included in the full version of the paper [GKM+18].

Lemma 1. *A proof system that satisfies a security notion with updatable setup also satisfies the security notion with trusted setup.*

Lemma 2. *A proof system that satisfies a security notion with subvertible setup also satisfies the security notion with updatable setup.*

3.5 Specializing Common Reference Strings

Consider a CRS for a universal relation that can be used to prove any arithmetic circuit is satisfiable. Instances of the relation specify both wiring and inputs freely. For a specific arithmetic circuit it is desirable to use the large CRS to derive a smaller circuit-specific CRS for a relation with fixed wiring but flexible inputs, as this might lead to more efficient prover and verifier algorithms. This can be seen as a form of pre-computation on the large CRS to get better efficiency, but there are conceptual advantages in giving the notion a name so in the following we formalize the idea of *specializing a universal CRS*.

Let Φ be a DPT decidable set of relations, with each relation $R_\phi \in \Phi$ being itself DPT decidable. The universal relation R for Φ defines a language with instances $\phi = (R_\phi, u)$ such that $((R_\phi, u), w) \in R$ if and only if $R_\phi \in \Phi$ and $(u, w) \in R_\phi$. We say that a setup generates specializable universal reference strings crs for R if there exists a DPT algorithm $\mathsf{crs}_{R_\phi} \leftarrow \mathsf{Derive}^\star(\mathsf{crs}, R_\phi)$ and algorithms Prove and Verify can be defined in terms of algorithms $\pi \leftarrow \mathsf{Prove}^\star(\mathsf{crs}_{R_\phi}, u, w)$ and $b \leftarrow \mathsf{Verify}^\star(\mathsf{crs}_{R_\phi}, u, \pi)$ as follows:

- $\mathsf{Prove}(\mathsf{crs}, \phi, w)$ parses $\phi = (R_\phi, u)$, asserts $R_\phi \in \Phi$, derives $\mathsf{crs}_{R_\phi} \leftarrow \mathsf{Derive}^\star(\mathsf{crs}, R_\phi)$, and returns the proof generated by $\mathsf{Prove}^\star(\mathsf{crs}_{R_\phi}, u, w)$.
- $\mathsf{Verify}(\mathsf{crs}, \phi, \pi)$ first parses $\phi = (R_\phi, u)$, checks $R_\phi \in \Phi$, derives $\mathsf{crs}_{R_\phi} \leftarrow \mathsf{Derive}^\star(\mathsf{crs}, R_\phi)$, and returns $\mathsf{Verify}^\star(\mathsf{crs}_{R_\phi}, u, \pi)$.

Existing zk-SNARKs for boolean and arithmetic circuit verification have different degrees of universality. Groth [Gro10b] is universal and works for any boolean circuit, i.e., the wiring of the circuit can be specified in the instance, while subsequent SNARKs such as [GGPR13] and descendants have reference strings that are for circuits with fixed wiring.

Schemes with specializable CRS derivation aim to achieve the generality of the former and the performance of the latter. As the Derive algorithm operates only on public information, it can be executed by protocol participants whenever necessary. This has two advantages. First, one can transform any attack against a prover and verifier employing a specialized CRS into an attack on the universal CRS and we thus do not need any special security notions. Second, it makes it easier to design efficient updatable schemes as being able to update the universal CRS that does not yet have a relation-dependent structure and publicly derive an efficient circuit-specific CRS after the update. We will exploit this in the second half of the paper, where we present an updatable zk-SNARK that avoids our own impossibility result in Sect. 6. We will employ a quadratic-size CRS that is universal for all QAPs, but then specialize it to obtain a linear-size CRS and linear-time prover computation.

4   Background

Let $\mathcal{G}(1^\lambda)$ be a DPT[1] bilinear group generator that given the security parameter 1^λ produces bilinear group parameters $bp = (p, \mathbb{G}_1, \mathbb{G}_2, \mathbb{G}_T, e, G, H)$. $\mathbb{G}_1, \mathbb{G}_2, \mathbb{G}_T$ are groups of prime order p with generators $G \in \mathbb{G}_1$, $H \in \mathbb{G}_2$ and $e : \mathbb{G}_1 \times \mathbb{G}_2 \to \mathbb{G}_T$ is a non-degenerative bilinear map, which means $e(G^a, H^b) = e(G, H)^{ab}$ and $e(G, H)$ generates $\mathbb{G}_T$.

---

[1] Often the cryptographic literature allows for probabilistic bilinear group generation, but for our purpose it is useful to have *deterministic* parameter generation that cannot be influenced by the adversary.

4.1   Knowledge and Computational Assumptions

The knowledge-of-exponent assumption (KEA) introduced by Damgård [Dam91] says that given $G, \hat{G} = G^{\alpha}$ it is infeasible to create $C, \hat{C}$ such that $\hat{C} = C^{\alpha}$ without knowing an exponent c such that $C = G^{c}$ and $\hat{C} = \hat{G}^{c}$. Bellare and Palacio [BP04] extended this to the KEA3 assumption, which says that given $G, G^{\alpha}, G^{s}, G^{\alpha s}$ it is infeasible to create C, C^{α} without knowing c_0, c_1 such that $C = G^{c_0}(G^{s})^{c_1}$. This assumption has been used also in symmetric bilinear groups by Abe and Fehr [AF07], who called it the extended knowledge-of-exponent assumption.

The *bilinear knowledge of exponent assumption* (B-KEA), which Abdolmaleki et al. [ABLZ17] refer to as the BDH-KE assumption, generalizes further to asymmetric groups. It states that it is infeasible to compute $C, \hat{C}$ such that $e(C, \hat{G}) = e(G, \hat{C})$ without knowing s such that $(C, \hat{C}) = (G^{s}, \hat{G}^{s})$. It corresponds to the special case of $q = 0$ of the *q-power knowledge of exponent* (q-PKE) assumption in asymmetric bilinear groups introduced by Groth [Gro10b].

We introduce the *q-monomial knowledge assumption*, as a generalization of q-PKE to multi-variate monomials. We note that our construction in Sect. 5 could be made uni-variate by employing higher powers which would allow the use of the ungeneralised q-PKE assumption.

Assumption 1 (The $q(\lambda)$-Monomial Knowledge Assumption $(q(\lambda)$-MK)). *Let $\boldsymbol{a} = \{a_i(\boldsymbol{X})\}_{i=1}^{n_a}$ and $\boldsymbol{b} = \{a_i(\boldsymbol{X})\}_{i=1}^{n_b}$ be sets of n-variate monomials with the degree, the number of monomials n_a, n_b, and the number of variables n all bounded by $q(\lambda)$. Let $\mathcal{A}$ be an adversary and $\mathcal{X}_{\mathcal{A}}$ be an extractor. Define the advantage $\mathsf{Adv}_{\mathcal{G},q(\lambda),a,b,\mathcal{A},\mathcal{X}_{\mathcal{A}}}^{\mathsf{MK}}(\lambda) = \Pr[\mathsf{MK}_{\mathcal{G},q(\lambda),a,b,\mathcal{A},\mathcal{X}_{\mathcal{A}}}(\lambda)]$ where $\mathsf{MK}_{\mathcal{G},q(\lambda),a,b,\mathcal{A},\mathcal{X}_{\mathcal{A}}}$ is defined as*

$$\frac{\text{MAIN } \mathsf{MK}_{\mathcal{G},q(\lambda),a,b,\mathcal{A},\mathcal{X}_{\mathcal{A}}}(\lambda)}{bp = (p, \mathbb{G}_1, \mathbb{G}_2, \mathbb{G}_T, e, G, H) \leftarrow \mathcal{G}(1^{\lambda})}$$

$$\boldsymbol{x} \xleftarrow{\$} \mathbb{F}_p^s$$
$$(G^a, H^b) \xleftarrow{r} \mathcal{A}(bp, \{G^{a_i(\mathsf{x})}\}_{i=1}^{n_1}, \{H^{b_i(\mathsf{x})}\}_{i=1}^{n_2})$$
$$(c_0, c_1, \ldots, c_{n_b}) \leftarrow \mathcal{X}_{\mathcal{A}}(bp, \{G^{a_i(\mathsf{x})}\}_{i=1}^{n_1}, \{H^{b_i(\mathsf{x})}\}_{i=1}^{n_2}; r)$$
$$\text{return } a = b \text{ and } b \neq c_0 + \textstyle\sum_i c_i \cdot b_i(\boldsymbol{x})$$

The MK assumption holds relative to $\mathcal{G}$ if for all PPT adversaries $\mathcal{A}$ there exists a PPT extractor $\mathcal{X}_{\mathcal{A}}$ such that $\mathsf{Adv}_{\mathcal{G},q(\lambda),a,b,\mathcal{A},\mathcal{X}_{\mathcal{A}}}^{\mathsf{MK}}(\lambda)$ is negligible in λ.

The following multi-variate computational assumption is closely related to the uni-variate *q-bilinear gap* assumption of Ghadafi and Groth [GG17] and is implied by the *computational polynomial assumption* of Groth and Maller [GM17].

Assumption 2 (The $q(\lambda)$-Monomial Computational Assumption $(q(\lambda)$-MC)). *Let $\boldsymbol{a} = \{a_i(\boldsymbol{X})\}_{i=1}^{n_a}$ and $\boldsymbol{b} = \{a_i(\boldsymbol{X})\}_{i=1}^{n_b}$ be sets of n variate monomials with the degree, the number of monomials n_a, n_b, and the number of variables n all bounded by $q(\lambda)$. Let $\mathcal{A}$ be a PPT algorithm, and define the advantage $\mathsf{Adv}_{\mathcal{G},q(\lambda),a,b,\mathcal{A}}^{\mathsf{MC}}(\lambda) = \Pr[\mathsf{MC}_{\mathcal{G},q(\lambda),a,b,\mathcal{A}}(\lambda)]$ where $\mathsf{MC}_{\mathcal{G},q(\lambda),a,b,\mathcal{A}}$ is defined as*

$$\underline{\text{MAIN } \mathsf{MC}_{\mathcal{G},q(\lambda),a,b,\mathcal{A}}(\lambda)}$$
$$bp = (p, \mathbb{G}_1, \mathbb{G}_2, \mathbb{G}_T, e, G, H) \leftarrow \mathcal{G}(1^\lambda)$$
$$\boldsymbol{x} \leftarrow \mathbb{F}_p^n$$
$$(A, a(X)) \leftarrow \mathcal{A}(bp, \{G^{a_i(\boldsymbol{x})}\}_{i=1}^{n_1}, \{H^{b_i(\boldsymbol{x})}\}_{i=1}^{n_2})$$
$$\text{return } 1 \text{ if } A = G^{a(\boldsymbol{x})} \text{ and } a(X) \notin \mathsf{span}\{1, a_1(X), \dots, a_{n_1}(X)\}$$
$$\text{else return } 0$$

The MC assumption holds relative to $\mathcal{G}$ if for all PPT adversaries $\mathcal{A}$ we have $\mathsf{Adv}^{\mathsf{MC}}_{\mathcal{G},q(\lambda),a,b,\mathcal{A}}(\lambda)$ is negligible in λ.

4.2   A QAP-Based zk-SNARK Recipe

Here we describe a generalised approach for using Quadratic Arithmetic Programs (QAPs) to construct a SNARK scheme for arithmetic circuit satisfiability. A similar approach can be used with Quadratic Span Programs (QSPs). In both cases, zero-knowledge is obtained by ensuring that all of the commitments are randomised. We show in Sect. 6 that the recipe is unlikely to directly lead to updatable zk-SNARKs. However, by modifying the recipe in Sect. 5 we are able to construct updatable zk-SNARKs.

Arithmetic Circuits: Arithmetic circuits are a means to describe computations that consist solely of field additions and multiplications. We will now describe an arithmetic circuit over a field $\mathbb{F}$ with n multiplication gates and m wires. Such a circuit consists of gates connected together by wires. The gates specify an operation (either addition or multiplication) and the wires contain values in $\mathbb{F}$. Each gate has a left input wire and a right input wire leading into it, and an output wire leading from it. The circuit can have split wires i.e. the same wire leads into multiple gates. The circuit is satisfied if for every gate, the operation applied to the input wires is equal to the output wire.

Any NP relation can be described with a family of arithmetic circuits that decide which statement and witness pairs are included. In a relation described by an arithmetic circuit, an instance is defined by a value assignment to ℓ fixed input wires. The witness is the values of the remaining $m - \ell$ wires such that the arithmetic circuit is satisfied.

Fix the circuit: We label the n gates with unique distinct values $r_1, \dots, r_n \in \mathbb{F}$. We will convert the arithmetic circuit into equations over polynomials, and these values will serve as points on which formal polynomials representing the circuit will be evaluated.

Describe all m wires using three sets of m polynomials with degree at most $n - 1$. These polynomials determine for which gates each wire behaves as a left input wire, a right input wire, and an output wire. They also determine whether the wires have been split, and whether there are any additions before a wire is fed into a multiplication gate. The three sets of polynomials are: $U = \{u_i(X)\}_{i=0}^m$ describes the left input wires; $V = \{v_i(X)\}_{i=0}^m$ describes the right input wires;

and $W = \{w_i(X)\}_{i=0}^m$ describes the output wires. We will throughout the paper fix $u_0(X) = v_0(X) = w_0(X) = 1$. The polynomials are designed such that they are equal to 1 at each of the values of the multiplication gates which they lead into/ out of and 0 at all other gate values.

Commit to wire values: Suppose there are m wires with values $(a_1, \ldots, a_m)$ and that the witness wires run from $\{\ell+1, \ldots, m\}$. The common reference string includes the values

$$\{G^{u_i(x)}, G^{v_i(x)}, G^{w_i(x)}\}_{i=\ell+1}^m$$

for some x chosen at random. The commitment to the left input, right, and output wires will include the values

$$C_L = G^{\sum_{i=\ell+1}^m a_i u_i(x)}, \ C_R = G^{\sum_{i=\ell+1}^m a_i v_i(x)}, \ C_O = G^{\sum_{i=\ell+1}^m a_i w_i(x)}.$$

Prove that repeated wires are consistent: If a wire is split into two left inputs, there is no need to do anything because of the design of the wire polynomials. However, it is necessary to check that split wires that split into at least one left input wire and at least one right input wire are consistent. This is done by including terms in the common reference string of the form

$$\left\{G^{\alpha_u u_i(x) + \alpha_v v_i(x)}\right\}_{i=\ell+1}^m$$

for some unknown α_u, α_v, and then requiring the prover to provide an element Y such that $\alpha_u C_L + \alpha_v C_R = Y$. For some schemes $\alpha_0 = \alpha_1$.

Prove that output wires are consistent with input wires: This can be done together with proving consistency of repeated wires. The common reference string includes terms of the form

$$\left\{G^{\alpha_u u_i(x) + \alpha_v v_i(x) + \alpha_w w_i(x)}\right\}_{i=\ell+1}^m$$

for some unknown $\alpha_u, \alpha_v, \alpha_w$. The prover is required to provide an element Y such that $\alpha_u C_L + \alpha_v C_R + \alpha_w C_O = Y$.

Prove the commitments are well formed: There are values in the common reference string that should not be included in the commitments generated by the prover, such as the $\{a_i u_i(x)\}_{i=1}^\ell$ values related to the instance. This can be checked using the same approach as descried above for the consistency proof.

Prove that gates are evaluated correctly: Determine a quadratic polynomial equation that checks that the gates are evaluated correctly. There is a unique degree n polynomial $t(X)$ which is equal to 0 at each of the gate values $(r_1, \ldots, r_n)$. Suppose that $a_1, \ldots, a_m$ are the wire values. Then

$$\left(\sum_{i=0}^{m} a_i u_i(X)\right) \cdot \left(\sum_{i=0}^{m} a_i v_i(X)\right) - \sum_{i=0}^{m} a_i w_i(X)$$

is equal to 0 when evaluated at the gate values if and only if the multiplication gates are evaluated correctly. This polynomial expressions shares its zeros with $t(X)$, which means that $t(X)$ divides it. Hence the prover is required to show that at the unknown point x,

$$\left(G^{\sum_{i=0}^{\ell} a_i u_i(x)} C_L\right) \otimes \left(G^{\sum_{i=0}^{\ell} a_i v_i(x)} C_R\right) = G^{t(x)+\sum_{i=0}^{\ell} a_i w_i(x)} C_O$$

for $\otimes$ a function that finds the product of the values inside the two encodings.

5 An Updatable QAP-Based zk-SNARK

In this section we give a construction for an updatable QAP-based zk-SNARK that makes use of a universal reference string. We then prove it satisfies subversion zero knowledge and updatable knowledge soundness under the knowledge-of-exponent assumptions introduced in Sect. 4.

We let the security parameter 1^λ (deterministically) determine parameters (d, m, ℓ, bp), where $bp = (p, \mathbb{G}_1, \mathbb{G}_2, \mathbb{G}_T, e, G, H)$, with $\mathbb{G}_1, \mathbb{G}_2, \mathbb{G}_T$ groups of prime order p with generators $G \in \mathbb{G}_1$, $H \in \mathbb{G}_2$ and $e : \mathbb{G}_1 \times \mathbb{G}_2 \to \mathbb{G}_T$ a non-degenerative bilinear map. Here d is the degree of the QAP, m is number of input variables, out of which ℓ are part of the instance formed of public field elements to a QAP.

Recall from Sect. 4.2, a QAP for the given parameters is defined by polynomials $\{u_i(x), v_i(x), w_i(x)\}_{i=0}^{m}$ of degree less than d, and $t(x)$ of degree d. The QAP defines a relation R_{QAP} containing pairs of instances and witnesses $(a_1, \ldots, a_\ell)$ and $(a_{\ell+1}, \ldots, a_m)$ such that, with $a_0 = 1$,

$$\left(u_0(x) + \sum_{i=1}^{m} a_i u_i(x)\right) \cdot \left(v_0(x) + \sum_{i=1}^{m} a_i v_i(x)\right) \equiv w_0(x) + \sum_{i=1}^{m} a_i w_i(x) \mod t(x).$$

The sequence of parameters indexed by the security parameter define a universal relation R consisting of all pairs of QAPs and instances as described above that have a matching witness. In the notation from Sect. 3.5 let Φ be all possible QAPs for the parameters, then the universal relation R for Φ contains instances $\phi = (R_{\text{QAP}}, u = (a_1, \ldots, a_\ell))$, with matching witnesses $w = (a_{\ell+1}, \ldots, a_m)$.

5.1 Reworking the QAP Recipe

Our final scheme is formally given in Figs. 2 and 3. In this section we describe some of the technical ideas behind it. Due to our impossibility result in Sect. 6, many of the usual tricks behind the QAP-based approach are not available to us, which means we need something new. To obtain this we first switch to a multi-variate scheme, where the proof elements need to satisfy equations in the indeterminates X, Y, Z. We can then prove the well-formedness of our proof elements using a subspace argument for our chosen sums of witness QAP polynomials. Once we have that the proof elements are well formed, we show that the exponents of two of them multiply to get an exponent in the third proof element such that (1) the sum of all the terms where Y has given power j is equal to the QAP expression in the X indeterminate, and (2) the value Y^j is not given in the universal CRS. For our final scheme, we use $j = 7$.

Fix the circuit: The circuit need only be fixed upon running the CRS derivation algorithm. At this point, the circuit is described as a QAP like that described in Sect. 4; i.e., for $a_0 = 1$, the field elements $(a_1, \ldots, a_m) \in R_{\text{QAP}}$ if and only if

$$\left(\sum_{i=0}^{m} a_i u_i(X) \right) \cdot \left(\sum_{i=0}^{m} a_i v_i(X) \right) = \sum_{i=0}^{m} a_i w_i(X) + q(X)t(X)$$

for some degree $(d-2)$ polynomial $q(X)$.

Prove the commitments are well formed: In our scheme an honest prover outputs group elements (A, B, C) such that

$$\log(A) = \log(B) = q(x)y + \sum_{i=0}^{m} a_i(w_i(x)y^2 + u_i(x)y^3 + v_i(x)y^4) - y^5 - t(x)y^6.$$

Ensuring that $\log(A) = \log(B)$ can be achieved with a pairing equation of the form $e(A, H) = e(G, B)$. Thus we need to show only that A is of the correct form.

Usually, as described in Sect. 4, this is done by encoding only certain polynomials in the CRS and forcing computation to use linear combinations of elements in the CRS. Since we cannot do this and allow updates, we instead construct a new subspace argument. First we subtract out the known elements in the instance using a group element S which the verifier computes in order to obtain a new group element with the exponent

$$q(x)y + \sum_{i=\ell+1}^{m} a_i(w_i(x)y^2 + u_i(x)y^3 + v_i(x)y^4).$$

Set M be the $(m + d - \ell) \times 4d$ matrix that contains the coefficients of $\{(w_i(x)y^2 + u_i(x)y^3 + v_i(x)y^4)\}_{i=\ell+1}^{m}, \{x^i y\}_{i=0}^{d-1}$ with respect to monomials $\{x^i y^j\}_{(i,j)=(0,1)}^{(d-1,4)}$. We denote these coefficients by $m_l(x, y) = \sum_{i,j} M_{l,(i,j)} \cdot x^i y^j$, e.g., $m_1(x, y) = (w_{\ell+1}(x)y^2 + u_{\ell+1}(x)y^3 + v_{\ell+1}(x)y^4)$. Then we set the corresponding null-matrix be N such that $MN = 0$. We address the rows of N by the corresponding monomial degrees in M. The columns of this matrix defines polynomials $n_k(x, y) = \sum_{i,j} N_{(i,j),k} \cdot x^{d-i}y^{4-j}$, such that in the convolution of $m_l(x, y) \cdot n_k(x, y)$ the $(d, 4)$ degree terms disappear. If we introduce the variable z, and set $\hat{N} = H^{\sum_k n_k(x,y)z^k}$, then the pairing $e(AS, \hat{N})$ yields a target group element with 0 coefficients for all $x^d y^4 z^k$ terms exactly when A is chosen from the right subspace. Thus, given a CRS that does not contain any $x^d y^4 z^k$ terms for $k > 1$, and a verification equation that checks that, $(\log A + \log S) \cdot \log(\hat{N}) = \log C_1$ the prover can only compute the component C_1 if A is correctly formed.

Prove that the QAP is satisfied: Assuming that A and B are of the correct form, we have that $\log(A) \cdot \log(B)$ is equal to

$$
\left(q(x)y + \sum_{i=0}^{m} a_i(w_i(x)y^2 + u_i(x)y^3 + v_i(x)y^4) - y^5 - t(x)y^6 \right)^2 .
$$

which, for terms involving y^7, yields

$$
t(x)q(x) - \sum_{i=0}^{m} a_i w_i(x) + \left(\sum_{i=0}^{m} a_i u_i(X) \right) \cdot \left(\sum_{i=0}^{m} a_i v_i(X) \right) .
$$

The terms in other powers of y can be considered as computable garbage and are cancelled out in other proof components. The equation above is satisfied for some polynomial $q(X)$ if and only if the QAP is satisfied. Thus, given a CRS that does not contain any y^7 terms, and a verification equation that checks that, $\log A \cdot \log B = \log C_2$ we ensure that the proof element C_2 is computable if and only if the QAP is satisfied.

Remark 1. It is always possible to make everything univariate in x by choosing y, z as suitable powers of x, but we find it conceptually easier and more readable to give them different names.

Derivation of a Linear Common Reference String: Astute readers may note that these techniques require the CRS to have quadratic set of monominals in order to compute the null matrix. We resolve this by providing an untrusted derive function which can be seen as a form of precomputation in order to find the linear common reference string for a fixed relation. Using the linear common reference string, our prover also makes a linear number of group exponentiations in the circuit size.

5.2 Updatability of the Universal Common Reference String

In this section we describe the universal common reference string and how to update it. We then prove that for any adversary that computes a valid common reference string, either through setup or through updates, we can extract the randomness it used. In Sect. 5.3, we show that – for our construction – proving security for an adversary that makes one update to a freshly generated CRS is equivalent to proving the full version of updatable security, in which an adversary makes all but one update in the sequence.

$\underline{\mathsf{Setup}(1^\lambda)}$

$x, y, z \xleftarrow{\$} \mathbb{F}_p^*; \quad \rho \leftarrow (G^x, G^y, G^z, G^x, G^y, G^z, H^x, H^y, H^z)$

$$\mathsf{crs} \leftarrow \left(\begin{array}{c} G,\ G^x,\ G^z,\ \{G^{x^i y^j}\}_{i=0,j=1,j\neq 7}^{2d,12},\ \{G^{x^i y^j z^k}\}_{i=0,j=1,k=1,(i,j)\neq(d,4)}^{2d,6,3d}, \\ \{G^{x^i y^j z^{6d}}\}_{i=0,j=1}^{d,4}\ H,\ H^x, \{H^{x^i y^j}\}_{i=0,j=1}^{d,6}, \{H^{x^i y^j z^k}\}_{i=0,j=0,k=1}^{d,2,3d},\ H^{z^{6d}} \end{array} \right)$$

$\underline{\mathsf{Update}(1^\lambda, \mathsf{crs}, \{\rho_i\}_{i=1}^n)}$

$$\text{parse} \left(\begin{array}{c} G,\ G_{1,0,0},\ G_{0,0,1},\ \{G_{i,j,0}\}_{i=0,j=1,j\neq 7}^{2d,12}, \\ \{G_{i,j,k}\}_{i=0,j=1,k=1,(i,j)\neq(d,4)}^{2d,6,3d},\ \{G_{i,j,6d}\}_{i=0,j=1}^{d,4} \\ H,\ H_{1,0,0}, \{H_{i,j,0}\}_{i=0,j=1}^{d,6}, \{H_{i,j,k}\}_{i=0,j=0,k=1}^{d,2,3d},\ H_{0,0,6d} \end{array} \right) \leftarrow \mathsf{crs}$$

$\alpha, \beta, \gamma \xleftarrow{\$} \mathbb{F}_p^*$

$$\mathsf{crs}' \leftarrow \left(\begin{array}{c} G,\ G_{1,0,0}^\alpha,\ G_{0,0,1}^\gamma,\ \{G_{i,j,0}^{\alpha^i \beta^j}\}_{i=0,j=1,j\neq 7}^{2d,12},\ \{G_{i,j,k}^{\alpha^i \beta^j \gamma^k}\}_{i=0,j=1,k=1,(i,j)\neq(d,4)}^{2d,6,3d}, \\ \{G_{i,j,6d}^{\alpha^i \beta^j \gamma^{6d}}\}_{i=0,j=1}^{d,4},\ H,\ H_{1,0,0}^\alpha, \{H_{i,j,0}^{\alpha^i \beta^j}\}_{i=0,j=1}^{d,6}, \{H_{i,j,k}^{\alpha^i \beta^j \gamma^k}\}_{i=0,j=0,k=1}^{d,2,3d}, \\ H_{0,0,6d}^{\gamma^{6d}} \end{array} \right)$$

$\rho \leftarrow (G_{1,0,0}^\alpha, G_{0,1,0}^\beta, G_{0,0,1}^\gamma, G^\alpha, G^\beta, G^\gamma, H^\alpha, H^\beta, H^\gamma)$

$\underline{\mathsf{VerifyCRS}(1^\lambda, \mathsf{crs}, \{\rho_i\}_{i=1}^n)}$

$$\text{parse} \left(\begin{array}{c} G,\ G_{1,0,0},\ G_{0,0,1},\ \{G_{i,j,0}\}_{i=0,j=1,j\neq 7}^{2d,12}, \\ \{G_{i,j,k}\}_{i=0,j=1,k=1,(i,j)\neq(d,4)}^{2d,6,3d},\ \{G_{i,j,6d}\}_{i=0,j=1}^{d,4}\ H, \\ H_{1,0,0}, \{H_{i,j,0}\}_{i=0,j=1}^{d,6}, \{H_{i,j,k}\}_{i=0,j=0,k=1}^{d,2,3d},\ H_{0,0,6d} \end{array} \right) \leftarrow \mathsf{crs}$$

parse $\{(A_i, B_i, C_i, \bar{A}_i, \bar{B}_i, \bar{C}_i, \hat{A}_i, \hat{B}_i, \hat{C}_i)\}_{i=1}^n \leftarrow \{\rho\}_{i=1}^n$

assert the proofs are correct:

$\quad A_1 = \bar{A}_1,\ B_1 = \bar{B}_1,\ C_1 = \bar{C}_1$

$\quad$ for $2 \leq i \leq n$: $\quad e(A_i, H) = e(A_{i-1}, \hat{A}_i)$

$\qquad\qquad\qquad\qquad \wedge\ e(B_i, H) = e(B_{i-1}, \hat{B}_i)\ \wedge\ e(C_i, H) = e(C_{i-1}, \hat{C}_i)$

$\quad e(\bar{A}_n, H) = e(G, \hat{A}_n)\ \wedge\ e(\bar{B}_n, H) = e(G, \hat{B}_n)\ \wedge\ e(\bar{C}_n, H) = e(G, \hat{C}_n)$

$\quad A_n = G_{1,0,0} \neq 1\ \wedge\ B_n = G_{0,1,0} \neq 1\ \wedge\ C_n = G_{0,0,1} \neq 1$

assert the exponents supposed to be y^j are correct:

$\quad$ for $1 \leq j \leq 6$: $e(G_{0,j,0}, H) = e(G, H_{0,j,0})$

$\quad$ for $1 \leq j \leq 5$: $e(G, H_{0,j+1,0}) = e(G_{0,1,0}, H_{0,j,0})$

$\quad$ for $8 \leq j \leq 12$: $e(G_{0,j,0}, H) = e(G_{0,6,0}, H_{0,j-6,0})$

assert the exponents supposed to be $x^i y^j$ are correct:

$\quad e(G_{1,0,0}, H) = e(G, H_{1,0,0})$

$\quad$ for $1 \leq i \leq d, 1 \leq j \leq 6, 8 \leq j \leq 12$: $e(G_{i,j,0}, H) = e(G_{i-1,j,0}, H_{1,0,0})$

$\quad$ for $1 \leq i \leq d, 1 \leq j \leq 6$: $e(G_{i,j,0}, H) = e(G, H_{i,j,0})$

assert the exponents supposed to be $x^i y^j z^k$ are correct:

$\quad e(G_{0,0,1}, H) = e(G, H_{0,0,1})$

$\quad$ for $1 \leq k \leq 3d$: $\quad e(G_{0,1,k}, H) = e(G_{0,1,0}, H_{0,0,k})$

$\quad$ for $0 \leq i \leq d, j = 0, 1, 2, k = 1 \leq k \leq 3d$: $\quad e(G_{i,j,0}, H_{0,0,k}) = e(G, H_{i,j,k})$

$\quad$ for $0 \leq i \leq d, 1 \leq j \leq 6, 1 \leq k \leq 3d, (i,j) \neq (d,4)$:

$\qquad e(G_{i,j,k}, H) = e(G_{i,j,0}, H_{0,0,k})$

$\quad$ for $d+1 \leq i \leq 2d, 1 \leq j \leq 6, 1 \leq k \leq 3d$: $e(G_{i,j,k}, H) = e(G_{i-d,0,k}, H_{d,j,0})$

$\quad e(G_{0,1,3d}, H_{0,0,3d}) = e(G_{0,1,0}, H_{0,0,6d})$

$\quad$ for $0 \leq i \leq d, 1 \leq j \leq 4$: $e(G_{i,j,0}, H_{0,0,6d}) = e(G_{i,j,6d}, H)$

Fig. 2. The setup process, along with the algorithms to create updates, and verify the setups and updates.

The universal CRS contains base G exponents $\{x^i y^j z^k\}_{(i,j,k)\in S_1}$ where

$$S_1 = \begin{pmatrix} \{(1,0,0),(0,1,0),(0,0,1)\} \\ \cup\{(i,j,0) : i \in [0,2d], j \in [1,12], j \neq 7\} \\ \cup\{(i,j,k) : i \in [0,2d], j \in [1,6], k \in [1,3d], (i,j) \neq (d,4)\} \\ \cup\{(i,j,6d) : i \in [0,d], j \in [1,4]\} \end{pmatrix}$$

and base H exponents $\{x^i y^j z^k\}_{(i,j,k)\in S_2}$ where

$$S_2 = \begin{pmatrix} \{(1,0,0),(0,1,0),(0,0,1),(0,0,6d)\} \\ \cup\{(i,j,0) : i \in [0,d], j \in [1,6]\} \\ \cup\{(i,j,k) : i \in [0,d], j \in [0,2], k \in [1,3d]\} \end{pmatrix}.$$

We begin with two lemmas about completeness, proofs of which can be found in the full version of the paper.

Lemma 3 (Correctness of the CRS generation). *The scheme is perfectly correct in the sense that*

$$\Pr[(\mathsf{crs}, \rho) \leftarrow \mathsf{Setup}(1^\lambda) : \mathsf{VerifyCRS}(1^\lambda, \mathsf{crs}, \rho) = 1] = 1;$$

$$\Pr\left[\begin{array}{l} (\mathsf{crs}', \rho_{n+1}) \leftarrow \mathsf{Update}(1^\lambda, \mathsf{crs}, \{\rho_i\}_{i=1}^n) : \\ \mathsf{VerifyCRS}(1^\lambda, \mathsf{crs}, \{\rho_i\}_{i=1}^n) = 1 \wedge \mathsf{VerifyCRS}(1^\lambda, \mathsf{crs}', \{\rho_i\}_{i=1}^{n+1}) \neq 1 \end{array} \right] = 1.$$

We now give two lemmas used to prove the full security of our construction and the update security of each component. These lemmas prove that even a dishonest updater needs to know their contribution to the trapdoor. Again, proofs can be found in the full version of the paper.

Lemma 4 (Trapdoor extraction for subvertible CRSs). *Suppose that there exists a PPT adversary $\mathcal{A}$ that outputs a crs, ρ such that $\mathsf{VerifyCRS}(1^\lambda, \mathsf{crs}, \rho) = 1$ with non-negligible probability. Then, by the 0-MK assumption (equivalent to the B-KEA assumption) there exists a PPT extractor $\mathcal{X}$ that, given the random tape of $\mathcal{A}$ as input, outputs (x, y, z) such that $(\mathsf{crs}, \rho) = \mathsf{Setup}(1^\lambda; (x, y, z))$.*

This lemma proves that even when given an honestly generated CRS as input, updaters need to know their contribution to the trapdoor. In this way security against the updater is linked to an honest CRS.

Lemma 5 (Trapdoor extraction for updatable CRSs). *Suppose that there exists a PPT adversary $\mathcal{A}$ such that given $(\mathsf{crs}, \rho_1) \xleftarrow{\$} \mathsf{Setup}(1^\lambda)$, $\mathcal{A}$ queries $\mathsf{U}\text{-}\mathcal{O}_s$ on $(\mathsf{final}, \mathsf{crs}', \{\rho_1, \rho_2\})$ where $\mathsf{VerifyCRS}(R, \mathsf{crs}', \{\rho_1, \rho_2\}) = 1$ with non-negligible probability. Then, with $\boldsymbol{a} = \{X^i Y^j Z^k : (i,j,k) \in S_1\}$ and $\boldsymbol{b} = \{X^i Y^j Z^k : (i,j,k) \in S_2\}$, the q-MK and the q-MC assumptions imply that there exists a PPT extractor $\mathcal{X}$ that, given the randomness of $\mathcal{A}$ as input, outputs (α, β, γ) such that $\bar{A}_2 = G^\alpha$, $\bar{B}_2 = G^\beta$, and $\bar{C}_2 = G^\gamma$.*

5.3   Single Adversarial Updates Imply Updatable Security

The following lemma relates updatable security to a model in which the adversary can make only a single update after an honest setup. This is because it is much cleaner to prove the security of our construction in this latter model (as we do in Theorem 4), but we would still like to capture the generality of the former.

We already know from Lemma 4 that it is possible to extract the adversary's contribution to the trapdoor when the adversary generates the CRS itself, and from Lemma 5 that it is possible to extract it when the adversary updates an honest CRS. To collapse chains of honest updates into an honest setup it is convenient that the trapdoor contributions of Setup and Update commute in our scheme. As the trapdoor in our scheme consists of all the randomness used by these algorithms, we will from now on refer to chains of honest updates and (single) honest setups interchangeably.

Trapdoor contributions cannot just be commuted but also combined; that is, for τ, τ' and τ'', $\mathsf{Update}'(1^\lambda, \mathsf{Update}'(1^\lambda, \mathsf{Setup}'(1^\lambda; \tau); \tau'); \tau'') = \mathsf{Setup}'(1^\lambda; \tau \otimes \tau' \otimes \tau'') = \mathsf{Update}'(1^\lambda, \mathsf{Update}'(1^\lambda, \mathsf{Setup}'(1^\lambda; \tau''); r'); r)$. Moreover, in our construction the proof ρ depends only on the relation and the randomness of the update algorithm. In particular it is independent of the reference string being updated. This enables the following simulation: Given the trapdoor $\tilde{\tau} = (x, y, z)$ of $\mathtt{crs}$, and the elements $(G_{1,0,0}, G_{0,1,0}, G_{0,0,1}, H_{1,0,0}, H_{0,1,0}, H_{0,0,1})$ of $\mathtt{crs}'$ we can simulate a proof $\rho_2 = (A_2, B_2, C_2, \bar{A}_2, \bar{B}_2, \bar{C}_2, \hat{A}_2, \hat{B}_2, \hat{C}_2)$ of $\mathtt{crs}'$ being an update of $\mathtt{crs}$ using $A_2 \leftarrow G_{1,0,0}$, $B_2 \leftarrow G_{0,1,0}$, $C_2 \leftarrow G_{0,0,1}$, $\bar{A}_2 \leftarrow G_{1,0,0}^{x^{-1}}$, $\bar{B}_2 \leftarrow G_{0,1,0}^{y^{-1}}$, $\bar{C}_2 \leftarrow G_{0,0,1}^{z^{-1}}$, $\hat{A}_2 \leftarrow H_{1,0,0}^{x^{-1}}$, $\hat{B}_2 \leftarrow H_{0,1,0}^{y^{-1}}$, $\hat{C}_2 \leftarrow H_{0,0,1}^{z^{-1}}$. We refer to this as $\rho(\mathtt{crs}')^{\tau^{-1}}$ in our reduction.

These properties together allow us to prove the result. We here give a detailed proof for knowledge soundness, as this is the most involved notion. Moreover, given that knowledge soundness implies soundness and we prove subversion zero-knowledge directly, it is the only notion we need.

Lemma 6 (Single adversarial updates imply full updatable knowledge soundness). *If our construction is U-KSND secure for adversaries that can query on* (Setup, $\emptyset$) *only once and then on* (final, S) *for a set S such that $|S| \leq 2$, then under the assumptions of Lemma 4 and Lemma 5 it is (fully) U-KSND-secure.*

Proof. We need to show that when the advantage is negligible for all PPT adversaries $\mathcal{B}$ with knowledge extractors $\mathcal{X}_\mathcal{B}$ in the restricted game, then the advantage is negligible for all adversaries $\mathcal{A}$ with knowledge extractors $\mathcal{X}_\mathcal{A}$ in the unrestricted game.

In our representation we split $\mathcal{A}$ into two stages $\mathcal{A}_1$ and $\mathcal{A}_2$, where the first stage ends with the successful query with intent final (i.e., the query that sets $\mathtt{crs}$). Let $\mathcal{A}_1, \mathcal{A}_2$ be an adversary against the U-KSND game. Let $\mathcal{B}$ be the following adversary against the restricted U-KSND game.

$\underline{\mathcal{B}^{\mathsf{U}\text{-}\mathcal{O}_\mathsf{s}}(1^\lambda)}$

$(\mathsf{crs}_h, \rho_h) \xleftarrow{\$} \mathsf{U}\text{-}\mathcal{O}_\mathsf{s}(\mathsf{Setup}, \emptyset)$
$\mathsf{st} \xleftarrow{r} \mathcal{A}_1^{\mathcal{O}_\mathsf{s}^{\mathsf{sim}}}(1^\lambda)$
$\{\rho_i, \mathsf{crs}_i\}_{i=1}^n \leftarrow S_{\mathsf{final}}$
find largest ℓ such that $(\rho_\ell, \tau_\ell) \in Q_c$
for all $i \in [\ell+1, n]$
$\quad \tau_i \leftarrow \mathcal{X}_{\mathcal{D}_i}(1^\lambda, r\|t)$
$S \leftarrow \{(\mathsf{crs}_h, \rho_h), \mathsf{Update}(1^\lambda, \mathsf{crs}_h, \{\rho_h\}; \prod_{i=\ell}^n \tau_i)\}$
$\mathsf{crs} \xleftarrow{\$} \mathsf{U}\text{-}\mathcal{O}_\mathsf{s}(\mathsf{final}, S)$
return $\mathcal{A}_2(\mathsf{st})$

$\underline{\mathcal{O}_\mathsf{s}^{\mathsf{sim}}((\mathsf{intent}, S))}$

if $\mathsf{crs} \neq \bot$ return $\bot$
if $\mathsf{intent} = \mathsf{setup}$ // initialise a CRS sequence
$\quad (\mathsf{crs}', \rho') \xleftarrow{\tau} \mathsf{Update}(1^\lambda, \mathsf{crs}_h, \{\rho_h\})$
$\quad t \leftarrow t\|\tau; \; Q_c \leftarrow Q_c \cup \{(\rho', \tau)\}$
$\quad$ return (crs', ρ')
if $\mathsf{intent} = \mathsf{update}$ // update a sequence
$\quad \tilde{\tau} \leftarrow \mathcal{X}_{\mathcal{C}}(1^\lambda, r\|t)$
$\quad \mathsf{crs}' \xleftarrow{\tau} \mathsf{Update}(1^\lambda, \mathsf{crs}_h, \{\rho_h\})$
$\quad \rho' \leftarrow \rho(\mathsf{crs}_h)^{\tau/\tilde{\tau}}$
$\quad t \leftarrow t\|\tau; \; Q_c \leftarrow Q_c \cup \{(\rho', \tau)\}$
$\quad$ return (crs', ρ')
// $\mathsf{intent} = \mathsf{final}$ finalise sequence
$b \leftarrow \mathsf{VerifyCRS}(1^\lambda, S) \wedge$
$\qquad\qquad Q_c \cap \{(\rho_i, *)\}_i \neq \emptyset$
if b: $\mathsf{crs} \leftarrow \mathsf{crs}_n$
$\qquad S_{\mathsf{final}} \leftarrow S; \; \text{return } \mathsf{crs}_n$
return $\bot$

Our adversary $\mathcal{B}$ can query its own oracle $\mathsf{U}\text{-}\mathcal{O}_\mathsf{s}$ only once on the empty set, so it does this upfront to receive an honest reference string crs_h. It then picks randomness r and runs $\mathcal{A}$ in a simulated environment in which $\mathcal{B}$ itself answers oracle queries. We keep track of the randomness $\mathcal{B}$ uses in the simulation in t.

$\mathcal{B}$ embeds the honest reference string in every query with $\mathsf{intent} \neq \mathsf{final}$. For this we exploit the fact that CRSs in our scheme are fully re-randomizable. On setup queries (i.e., when $S = \emptyset$), we simply return a randomized crs_h.

On general update queries, $\mathcal{B}$ additionally needs to compute a valid update proof ρ. To do this, let $\mathcal{C}$ be the algorithm that, given crs_h, runs $\mathcal{A}$ and the simulated oracles up to the update query and returns crs_n. To extract the trapdoor for the set S, we use either the subversion trapdoor extractor $\mathcal{X}_{\mathcal{C}}$ for adversary $\mathcal{C}$ that is guaranteed to exist by Lemma 4 (if S does not contain randomized honest reference strings), or the update trapdoor extractor that is guaranteed to exist by Lemma 5 (if it does). This latter extractor provides the update trapdoor, with respect to crs_h, of the reference string crs_n provided by the adversary. While $\mathcal{A}$ can make use of values returned in prior queries, the randomness used by these queries is contained in t and thus also available to $\mathcal{X}_{\mathcal{C}}$.

Next, $\mathcal{A}$ finalizes n reference strings. Now, the goal of $\mathcal{B}$ is to return a single update of crs_h, so it needs to compress the entire sequence of updates $\{\rho_i\}_{i=\ell+1}^n$ into one. To extract the randomness that went into each individual update, $\mathcal{B}$ builds adversaries $\mathcal{D}_i$, $i \in [\ell+1, n]$, from $\mathcal{A}$ that return only (crs_i, ρ_i). By Lemma 5 there exist extractors $\mathcal{X}_{\mathcal{D}_i}$ that extract only the randomness that went into these individual updates; i.e., $\delta_i = (x_i, y_i, z_i)$ such that $\rho_{i-1}, \mathsf{crs}_i = \mathsf{Update}(1^\lambda, \mathsf{crs}_{i-1}; \delta_i)$. Using these extractors, $\mathcal{B}$ computes $(\mathsf{crs}'_h, \rho'_h) \leftarrow \mathsf{Update}(1^\lambda, \mathsf{crs}_h, \{\rho_h\}; \prod_{i=\ell+1}^n \delta_i)$, sets $S \leftarrow \{\mathsf{crs}'_h, \{\rho_h, \rho'_h\}\}$, and calls $\mathcal{O}_\mathsf{s}(\mathsf{final}, S)$ to finalize its own CRS. By construction, $\mathsf{crs}'_h = \mathsf{crs}_n$. In the rest of the game $\mathcal{B}$ behaves like $\mathcal{A}$.

We build extractor $\mathcal{X}_{\mathcal{A}}$ from the extractor $\mathcal{X}_{\mathcal{B}}$ which is guaranteed to exist. In our definitions, knowledge extractors share state with setup algorithms. Here the main implication of this is that the extractor has access to the challenger's randomness, and thus can re-execute the challenger to retrieve its internal state. $\mathcal{X}_\mathcal{A}(r, t \| \tau)$ runs $\mathcal{X}_\mathcal{B}(r \| t, \tau)$. Thus the construction of $\mathcal{X}_\mathcal{A}$ simply uses $\mathcal{X}_\mathcal{B}$ but shifts the randomness of the simulation into the randomness of the challenger. As the simulation is perfect, $\mathcal{A}$ will behave identically. Furthermore, $r \| t$ is a valid randomness string for $\mathcal{B}$ and $\mathcal{X}_\mathcal{B}$ receives input that is consistent with a restricted game with $\mathcal{B}$. From this point onward $\mathcal{B}$ behaves exactly like $\mathcal{A}_2$. As $\mathcal{B}$ has negligible success probability against $\mathcal{X}_\mathcal{B}$ in the restricted U-KSND$_{\mathcal{B}, \mathcal{X}_\mathcal{B}}(1^\lambda)$ game, $\mathcal{A}$ thus has negligible success probability against $\mathcal{X}_\mathcal{A}$ in the unrestricted U-KSND$_{\mathcal{A}, \mathcal{X}_\mathcal{A}}(1^\lambda)$ game. $\qquad\square$

5.4   The zk-SNARK Scheme

In this section we construct a zk-SNARK for QAP satisfiability given the universal common reference string in Sect. 5.2. First we derive a QAP specific CRS from the universal CRS with which we can construct efficient prove and verify algorithms.

Lemma 7. *The derive algorithm is computable in polynomial time and the proof system has perfect completeness if QAP is such that $t(x) \neq y^{-1}$.*

A proof of this lemma can be found in the full version of the paper [GKM+18].

Theorem 3. *The proof system has perfect subversion zero-knowledge if QAP is such that $t(x) \neq y^{-1}$.*

Proof. To prove subversion zero-knowledge, we need to both show the existence of an extractor $\mathcal{X}_\mathcal{A}$, and describe a SimProve algorithm that produces indistinguishable proofs when provided the extracted trapdoor (which it can compute given the randomness of both $\mathcal{A}$ and the honest algorithms). The simulator knows x, y, z and picks $r \leftarrow \mathbb{F}_p$ and sets $A = G^r, B = H^r$ and $C = G^{r^2 + (r + y^5 + t(x)y^6 - \sum_{i=0}^{\ell} a_i(w_i(x)y^2 + u_i(x)y^3 + v_i(x)y^4)) \cdot n(x,y,z)}$. The simulated proof has the same distribution as a real proof, since $y \neq 0$ and $t(x) \neq y^{-1}$ and thus the randomisation of A given in $r(y - t(x)y^2)$ makes A uniformly random. Given A the verification equations uniquely determine B, C. So both real and simulated proofs have uniformly random A and satisfy the equations. Consequently, subversion zero-knowledge follows from the extraction of the trapdoor, which can be extracted by Lemma 4. $\qquad\square$

Theorem 4. *The proof system has update knowledge soundness assuming the q-MK and the q-MC assumptions hold with $a = \{X^i Y^j Z^k : (i, j, k) \in S_1\}$ and $b = \{X^i Y^j Z^k : (i, j, k) \in S_2\}$.*

Proof. To prove this it suffices, by the results in Sect. 5.3, to prove security in the setting in which the adversary makes only one update to the CRS. Imagine we have a PPT adversary $\mathcal{A}^{\mathsf{U}\text{-}\mathcal{O}_s}$ that after querying U-$\mathcal{O}_s$ on (Setup, $\emptyset$) to get crs, then queries on (final, crs', $\{\rho, \rho'\})$), and outputs u, π that gets accepted;

Derive(crs, QAP)
_____
parse $(\ell, \{u_i(X), v_i(X), w_i(X)\}_{i=0}^m, t(X)) \leftarrow$ QAP
assert $G^{y-t(x)y^2} \neq 1$
let $s_i(X, Y) = w_i(X)Y^2 + u_i(X)Y^3 + v_i(X)Y^4$ for $i = 0, \dots, m$
let $s_{m+j}(X, Y) = t(X)Y^{j+1}$ for $j = 1, 2, 3$
compute polynomials $n_1(X, Y), \dots, n_{3d-m+\ell}(X, Y)$ such that
    for all $i = \{\ell+1, \dots, m+3\}, k \in \{1, \dots, 3d-m+\ell\}$ the product
    $s_i(X, Y) \cdot n_k(X, Y)$ has coefficient 0 for the term $X^d Y^4$
    for all $p(X, Y) \cdot Y^2 \notin \mathrm{span}\{s_i(X, Y)\}_{i=\ell+1}^{m+3}$ there exists $k \in \{1, \dots, 3d-m+\ell\}$
    such that the product $p(X, Y) \cdot Y^2 \cdot n_k(X, Y)$ has non-zero coefficient for the
    term $X^d Y^4$
let $n(X, Y, Z) = Z^{6d} + \sum_{k=1}^{3d-m+\ell} n_k(X, Y)Z^k$

$$\mathrm{crs_{QAP}} \leftarrow \begin{pmatrix} \mathrm{QAP},\, G,\, \{G^{x^i y^j}\}_{i=0, j=1, j\neq 7}^{2d, 12},\, G^{y-t(x)y^2}, \\ \{G^{w_i(x)y^2 + u_i(x)y^3 + v_i(x)y^4}\}_{i=0}^m, G^{y^5},\, G^{t(x)y^6},\, \{G^{x^i y \cdot n(x,y,z)}\}_{i=0}^d, \\ G^{(y-t(x)y^2) \cdot n(x,y,z)},\, \{G^{(w_i(x)y^2 + u_i(x)y^3 + v_i(x)y^4) \cdot n(x,y,z)}\}_{i=\ell+1}^m\, H, \\ \{H^{x^i y}\}_{i=0}^d,\, H^{y-t(x)y^2},\, \{H^{w_i(x)y^2 + u_i(x)y^3 + v_i(x)y^4}\}_{i=0}^m, H^{y^5}, \\ H^{t(x)y^6},\, H^{n(x,y,z)} \end{pmatrix}$$

Prove(crs$_{\mathrm{QAP}}$, u, w)
_____
assert $H^{y^5} \neq H^{t(x)y^6}$
set $a_0 = 1$ and parse $(a_1, \dots, a_\ell) \leftarrow u$ and $(a_{\ell+1}, \dots, a_m) \leftarrow w$
let $q(X) = \frac{\sum_{i=0}^m a_i u_i(X) \cdot \sum_{i=0}^m a_i v_i(X) - \sum_{i=0}^m a_i w_i(X)}{t(X)}$
pick $r \xleftarrow{\$} \mathbb{F}_p$ and compute $A \leftarrow G^{a(x,y)}, B \leftarrow H^{b(x,y)}, C \leftarrow G^{c(x,y,z)}$, where
$a(x, y) = b(x, y)$
    $= q(x)y + r(y - t(x)y^2) + \sum_{i=0}^m a_i(w_i(x)y^2 + u_i(x)y^3 + v_i(x)y^4) - y^5 - t(x)y^6,$
$c(x, y, z) =$
    $a(x, y) \cdot b(x, y) +$
    $\big(q(x) \cdot y + r \cdot (y - t(x)y^2) + \sum_{i=\ell+1}^m a_i(w_i(x)y^2 + u_i(x)y^3 + v_i(x)y^4)\big) \cdot n(x, y, z).$
return $\pi = (A, B, C)$

Verify(crs$_{\mathrm{QAP}}$, u, π)
_____
set $a_0 = 1$ and parse $(a_1, \dots, a_\ell) \leftarrow u$ and $(A, B, C) \leftarrow \pi$
assert $e(A, H) = e(G, B)$
assert $e(A, B) \cdot e(AG^{y^5 + t(x)y^6 - \sum_{i=0}^\ell a_i(w_i(x)y^2 + u_i(x)y^3 + v_i(x)y^4)}, H^{n(x,y,z)})$
    $= e(C, H)$

Fig. 3. An updatable and specializable zk-SNARK for QAP

i.e., such that VerifyCRS$(R, \mathrm{crs}', \{\rho, \rho'\}) = 1$, $\mathrm{crs_{QAP}} \leftarrow$ Derive$(\mathrm{crs}', \mathrm{QAP})$, and Verify$(\mathrm{crs_{QAP}}, u, \pi) = 1$. Set $a_0 = 1$ and parse the instance as $u = (a_1, \dots, a_\ell)$ and the proof as (A, B, C). By Lemma 5, because the updated CRS verifies, there exists an extractor $\mathcal{X}_A$ that outputs $\tau = (\alpha, \beta, \gamma)$ such that Update$(1^\lambda, \mathrm{crs}, \{\rho\}; \tau) = (\mathrm{crs}', \rho')$.

From the first verification equation we have $e(A, H) = e(G, B)$, which means there is an $a \in \mathbb{F}_p$ such that $A = G^a$ and $B = H^a$. From the q-MK assumption there exists a PPT extractor $\mathcal{X}_A$ for $\mathcal{A}$ that outputs field elements

$\{a_{i,j,k}\}_{(i,j,k)\in\{(0,0,0)\}\cup S_1}$ defining a formal polynomial $a(X,Y,Z)$ equal to

$$a_{0,0,0} + a_{1,0,0}X + \sum_{i=0,j=1}^{d,6} a_{i,j,0}X^iY^j + \sum_{i=0,j=0,k=1}^{2d,3,3d} a_{i,j,k}X^iY^jZ^k + a_{0,0,6d}Z^{6d}$$

such that $B = H^{a(x,y,z)}$.

Taking the adversary and extractor together, we can see them as a combined algorithm that outputs A, B, C and the formal polynomial $a(X,Y,Z)$ such that $A = G^{a(x,y,z)}$. By the q-MC assumption this has negligible probability of happening unless $a(X,Y,Z)$ is in the span of $\{0,0,0\} \cup S_1 \cap S_2$

$$\left\{1, X, Z, \{X^iY^j\}_{i=0,j=1,j\neq7}^{2d,12}, \{X^iY^jZ^k\}_{i=0,j=1,k=1,(i,j)\neq(d,4)}^{2d,6,3d}, \{X^iY^jZ^{6d}\}_{i=0,j=1}^{d,4}\right\}.$$

This means

$$a(X,Y,Z) = a_{0,0,0} + a_{1,0,0}X + \sum_{i=0,j=1}^{d,6} a_{i,j,0}X^iY^j + \sum_{i=0,j=1,k=1}^{d,3,3d} a_{i,j,k}X^iY^jZ^k.$$

From the second verification equation we get $C = G^{f(x,y,z)}$ where $f(x,y,z)$ is given by

$$a(x,y,z)^2 + \left(a(x,y,z) + \beta^5y^5 + t(\alpha x)\beta^6y^6\right.$$

$$\left. - \sum_{i=0}^{\ell} a_i(w_i(\alpha x)\beta^2y^2 + u_i(\alpha x)\beta^3y^3 + v_i(\alpha x)\beta^4y^4)\right) \cdot n(\alpha x, \beta y, \gamma z).$$

By the q-MC assumption this means

$$a(X,Y,Z)^2 + \left(a(X,Y,Z) + \beta^5Y^5 + t(\alpha X)\beta^6Y^6\right.$$

$$\left. - \sum_{i=0}^{\ell} a_i(w_i(\alpha X)\beta^2Y^2 + u_i(\alpha X)\beta^3Y^3 + v_i(\alpha X)\beta^4Y^4)\right) \cdot (\gamma^{6d}Z^{6d} + \sum_{k=1}^{3d-m+\ell} n_k(\alpha X, \beta Y)\gamma^kZ^k)$$

also belongs to the span of

$$\left\{1, X, Z, \{X^iY^j\}_{i=0,j=1,j\neq7}^{2d,12}, \{X^iY^jZ^k\}_{i=0,j=1,k=1,(i,j)\neq(d,4)}^{2d,6,3d}, \{X^iY^jZ^{6d}\}_{i=0,j=1}^{d,4}\right\}.$$

Set $a'_{i,j,k} = \frac{a_{i,j,0}}{\alpha^i\beta^j\gamma^k}$ and observe that

$$a(X,Y,Z) = \sum_{i,j,k} a_{i,j,k}X^iY^jZ^k = \sum_{i,j,k} a'_{i,j,k}(\alpha X)^i(\beta Y)^j(\gamma Z)^k = a'(\alpha X, \beta Y, \gamma Z).$$

W.l.o.g. we can then rename the variables $\alpha X, \beta Y, \gamma Z$ by X, Y, Z to get that

$$a'(X,Y,Z)^2 + \left(a'(X,Y,Z) + Y^5 + t(X)Y^6\right.$$

$$\left. - \sum_{i=0}^{\ell} a_i(w_i(X)Y^2 + u_i(X)Y^3 + v_i(X)Y^4)\right) \cdot (Z^{6d} + \sum_{k=1}^{3d-m+\ell} n_k(X,Y)Z^k)$$

The span has no monomials of the form $X^i Y^j Z^k$ for $k > 6d$. Looking at the sub-part $a'(X, Y, Z) Z^{6d}$ we deduce that $a'_{i,j,k} = 0$ for all $k \neq 0$, which means

$$a'(X, Y, Z) = a'_{0,0,0} + a_{1,0,0} X' + \sum_{i=0, j=1}^{d,6} a'_{i,j,0} X^i Y^j.$$

There is also no Z^{6d} or XZ^{6d} monimials in the span, so we get $a'_{0,0,0} = 0$ and $a'_{1,0,0} = 0$. We are now left with

$$a'(X, Y, Z) = \sum_{i=0, j=1}^{d,6} a'_{i,j,0} X^i Y^j.$$

Define $q(X), p(X, Y)$ such that

$$q(X) \cdot Y + p(X, Y) \cdot Y^2 = \sum_{i=0, j=1}^{d,6} a'_{i,j,0} X^i Y^j + Y^5 + t(X) Y^6$$
$$- \sum_{i=0}^{\ell} a_i (w_i(X) Y^2 + u_i(X) Y^3 + v_i(X) Y^4).$$

Looking at the remaining terms of the form $X^i Y^j Z^k$ we see that for $k = 0, \ldots, 3d - m + \ell$

$$\left(q(X) \cdot Y + p(X, Y) \cdot Y^2 \right) \cdot n_k(X, Y) \in \text{span}\{X^i Y^j\}_{i=0, j=1, (i,j) \neq (d,4)}^{2d,6}.$$

Since $n_k(X, Y)$ has at most degree 2 in Y this implies $p(X, Y) \cdot Y^2 \cdot n_k(X, Y)$ has coefficient 0 for the term $X^d Y^4$. Recall the $n_k(X, Y)$ polynomials had been constructed such that this is only possible if $p(X, Y) \cdot Y^2$ can be written as

$$\sum_{i=\ell+1}^{m} a_i (w_i(X) Y^2 + u_i(X) Y^3 + v_i(X) Y^4) + r_1 t(X) Y^2 + r_2 t(X) Y^3 + r_3 t(X) Y^4.$$

Finally, we look at terms of the form $X^i Y^7$. These do not exist in the span, so all the terms of that form in $a(X, Y, Z)^2$ should sum to zero. This implies

$$\left(\begin{array}{c} q(X) \cdot Y + \sum_{i=0}^{m} a_i (w_i(X) Y^2 + u_i(X) Y^3 + v_i(X) Y^4) \\ + r_1 t(X) Y^2 + r_2 t(X) Y^3 + r_3 t(X) Y^4 - Y^5 - t(X) Y^6 \end{array} \right)^2$$

should have no $x^i Y^7$ terms. This in turn implies

$$2 \left(\begin{array}{c} (r_3 \sum_{i=0}^{m} a_i u_i(X) + r_2 \sum_{i=0}^{m} a_i v_i(X) - r_1 - q(X)) \cdot t(X) \\ - \sum_{i=0}^{m} a_i w_i(X) + \sum_{i=0}^{m} a_i u_i(X) \cdot \sum_{i=0}^{m} a_i v_i(X) \end{array} \right) = 0$$

By definition of QAP we now have that $(a_{\ell+1}, \ldots, a_m)$ is a witness for the instance $(a_1, \ldots, a_\ell)$. $\qquad \square$

6 Updating a Reference String Reveals the Monomials

In this section we show a negative result; namely, that for any updatable pairing-based NIZK with polynomials encoded into the common reference string, it must also be allowed (which often it is not) for an adversary to know encodings of the monomials that make up the polynomials. The reason for this is that from the encodings of the polynomials, we can construct an adversary that uses the update algorithm in order to extract the monomials. After describing our monomial extractor, we give one example (for the sake of brevity) of how to use our monomial extractor to break a QAP-based zk-SNARK, namely Pinocchio [PHGR13]. Due to the similarity in the approaches, however, we believe that the same techniques could be used to show that most other QSP/QAP-based zk-SNARKs in the literature also cannot be made updatable. As our universal CRS does consist of monomials, we can avoid this impossibility result yet still achieve linear-size specialized CRSs for proving specific relations.

Due to space constraints, we present our monomial extractor in the full version of the paper, which shows that if a NIZK scheme has an update algorithm, it can be used to extract all monomials from the common reference string. Intuitively, the existence of this monomial extractor would break most pairing-based NIZK proofs using QAPs or QSPs. This is because these arguments typically depend on the instance polynomials and the witness polynomials being linearly independent from each other. Here we give an example by demonstrating how to break the knowledge soundness of Pinocchio [PHGR13].

Example 1 (We cannot update the common reference string for Pinocchio). Consider the zk-SNARK in Pinocchio [PHGR13]. The scheme runs over a QAP relation described by

$$R = \{(p, \mathbb{G}, \mathbb{G}_T, e), \{v_k(X), w_k(X), y_k(X)\}_{k=0}^m, t(X)\}$$

where $t(X)$ is a degree n polynomial, $u_k(X), v_k(X), w_k(X)$ are degree $n-1$ polynomials and $(p, \mathbb{G}, \mathbb{G}_T, e)$ is a bilinear group. The instance $(c_1, \ldots, c_\ell)$ is in the language if and only if there is a witness of the form $(c_{\ell+1}, \ldots, c_m)$ such that, where c_0 is set to 1,

$$\left(\sum_{i=0}^m c_k u_k(X)\right) \cdot \left(\sum_{i=0}^m c_k v_k(X)\right) = \sum_{i=0}^m c_k w_k(X) + h(X)t(X)$$

for $h(X)$ some degree $n-1$ polynomial.

Here we switch to symmetric pairings, as Pinocchio was originally described in the symmetric setting (i.e. where $\mathbb{G}_1 = \mathbb{G}_2$.

The common reference string is given by

$$\left(\begin{array}{l} G, G^{\alpha_w} G^\gamma, G^{\beta\gamma}, G^{r_u r_v t(s)}, \{G^{s^i}\}_{i=1}^n \left\{G^{r_u u_k(s)}, G^{r_v v_k(s)}, G^{r_u r_v w_k(s)}\right\}_{k=0}^m, \\ \left\{G^{r_u \alpha_u u_k(s)}, G^{r_v \alpha_v v_k(s)}, G^{r_u r_v \alpha_w w_k(s)}, G^{\beta(r_u u_k(s) + r_v v_k(s) + r_u r_v w_k(s))}\right\}_{k=\ell+1}^m \end{array}\right)$$

where $r_u, r_v, s, \alpha_u, \alpha_v, \alpha_w, \beta, \gamma$ are random field elements and $G \in \mathbb{G}$. Hence, for $\mathrm{Ec}(x) = G^x$, there exists a matrix $\hat{X}$ such that $\mathrm{crs} = \hat{X}\mathrm{Ec}(\boldsymbol{\tau})$ for

$$\boldsymbol{\tau} = \begin{pmatrix} \alpha_w, \gamma, \beta\gamma, \{r_u r_v s^i, s^i\}_{i=0}^n, \\ \{r_u s^i, r_v s^i, r_u \alpha_u s^i, r_v \alpha_v s^i, r_u r_v \alpha_w s^i, r_u \beta s^i, r_v \beta s^i, r_u r_v \beta s^i\}_{i=0}^{n-1} \end{pmatrix}. \quad (1)$$

Lemma 8. *For* $\mathrm{crs} = G^{\boldsymbol{\tau}}$ *where* $\boldsymbol{\tau}$ *is as in* (1)*, there exists an adversary that can find a verifying proof for any instance* $(c_1, \ldots, c_\ell) \in \mathbb{F}_p$.

Proof. The verifier in Pinocchio

$$0/1 \leftarrow \mathsf{Verify}(\mathrm{crs}; c_1, \ldots, c_\ell; A_1, A_2, A_3, B_1, B_2, B_3, H, Z)$$

returns 1 if and only the following equations are satisfied

$$e(G^{r_u \sum_{k=0}^{\ell} c_k u_k(s)} A_1, G^{r_v \sum_{k=0}^{\ell} c_k v_k(s)} A_2) = e(G^{r_u r_v t(s)}, H)e(G^{r_u r_v \sum_{k=0}^{\ell} c_k w_k(s)} A_3, G)$$

$$e(B_1, G) = e(A_1, G^{\alpha_u})$$
$$e(B_2, G) = e(A_2, G^{\alpha_v})$$
$$e(B_3, G) = e(A_1, G^{\alpha_w})$$
$$e(Z, G^{\gamma}) = e(A_1 A_2 A_3, G^{\beta\gamma}).$$

Suppose the adversary sets the degree $n-1$ polynomials $\nu(X), \omega(X), \xi(X)$ as

$$\nu(X) \leftarrow \sum_{k=0}^{\ell} c_k v_k(X)$$
$$\omega(X) \leftarrow \sum_{k=0}^{\ell} c_k w_k X$$
$$\xi(X) \leftarrow \sum_{k=0}^{\ell} c_k y_k(X)$$

It then sets the components H, A_1, A_2, A_3 by

$$H = G, \ A_1 = G^{r_u s} G^{-r_u \nu(s)}, \ A_2 = G^{r_v s^{n-1}} G^{-r_v \omega(s^i)},$$

$$A_3 = G^{-r_u r_v(t(s) - s^n) - r_u r_v \xi(s)}$$

Direct verification shows that A_1, A_2, A_3 satisfy the first verification equation. Note that $\boldsymbol{\tau}$ does not include the value $\alpha_w r_u r_v s^n$, so the final coefficient of $t(s)$ cannot be included in A_3, else the algorithm could not satisfy the fifth verification equation. Instead we include $r_u s$ in A_1 and r_v in A_2, so that the LHS of the first verification equation returns the sole component not cancelled on the RHS: $e(G, G)^{r_u r_v s^n}$.

To satisfy verification equations 2–4 the algorithm sets

$$B_1 = G^{\alpha_u r_u s} G^{-\alpha_u r_u \nu(s)}, \ B_2 = G^{\alpha_v r_v s^{n-1}} G^{-\alpha_v r_v \omega(s)},$$

$$B_3 = G^{-\alpha_w r_u r_v(t(s) - s^n) - \alpha_w r_u r_v \xi(s)}$$

and to satisfy the fifth and final verification equation the algorithm sets

$$Z = G^{\beta r_u s} G^{\beta r_v s^{n-1}} G^{-\beta r_u \nu(s)} G^{-\beta r_v \omega(s)} G^{-\beta r_u r_v(t(s) - s^n) - \beta r_u r_v \xi(s)}.$$

We then have that $\mathsf{Verify}(\mathrm{crs}; c_1, \ldots, c_\ell; A_1, A_2, A_3, B_1, B_2, B_3, H, Z) = 1$. $\square$

Theorem 5. *If there exists an update algorithm for Pinocchio, then either the relation is easy or the scheme is not knowledge-sound.*

Proof. Suppose that $\mathsf{crs} \leftarrow \mathsf{Setup}(1^\lambda)$; i.e., $\mathsf{crs} = \hat{X}G^\tau$ for τ as in Eq. 1. Suppose that $(c_1, \dots, c_\ell) \in \mathbb{F}_p$.

The polynomials $u_k(X), v_k(X), w_k(X)$ are Lagrange polynomials, meaning that each and every one of the components τ are used in the crs. This means that the RREF of $\hat{X}$, which we shall call $\hat{R}$, is such that for $1 \leq i \leq \text{length}(\hat{R})$, there exists some j such that $\hat{R}[i][j] \neq 0$. Hence by running MonoExtract, an adversary $\mathcal{A}$ can calculate G^τ. By Lemma 8, the adversary $\mathcal{A}$ can continue, and calculate a verifying proof for $(c_1, \dots, c_\ell)$. Hence either there is a PPT extractor that can output a valid witness for any instance (meaning the language is easy), or there is no extractor and $\mathcal{A}$ breaks knowledge-soundness. $\square$

References

[ABLZ17] Abdolmaleki, B., Baghery, K., Lipmaa, H., Zając, M.: A subversion-resistant SNARK. In: Takagi, T., Peyrin, T. (eds.) ASIACRYPT 2017. LNCS, vol. 10626, pp. 3–33. Springer, Cham (2017). https://doi.org/10.1007/978-3-319-70700-6_1

[AF07] Abe, M., Fehr, S.: Perfect NIZK with adaptive soundness. In: TCC (2007)

[AHIV17] Ames, S., Hazay, C., Ishai, Y., Venkitasubramaniam, M.: Ligero: lightweight sublinear arguments without a trusted setup. In: Proceedings of ACM CCS (2017)

[BBB+18] Bünz, B., Bootle, J., Boneh, D., Poelstra, A., Maxwell, G.: Bulletproofs: short proofs for confidential transactions and more. In: Proceedings of the IEEE Symposium on Security & Privacy (2018)

[BCC+14] Bernstein, D.J., Chou, T., Chuengsatiansup, C., Hülsing, A., Lange, T., Niederhagen, R., van Vredendaal, C.: How to manipulate curve standards: a white paper for the black hat. Cryptology ePrint Archive, Report 2014/571 (2014). http://eprint.iacr.org/2014/571

[BCC+16] Bootle, J., Cerulli, A., Chaidos, P., Groth, J., Petit, C.: Efficient zero-knowledge arguments for arithmetic circuits in the discrete log setting. In: Fischlin, M., Coron, J.-S. (eds.) EUROCRYPT 2016. LNCS, vol. 9666, pp. 327–357. Springer, Heidelberg (2016). https://doi.org/10.1007/978-3-662-49896-5_12

[BCG+14] Ben-Sasson, E., Chiesa, A., Garman, C., Green, M., Miers, I., Tromer, E., Virza, M.: Zerocash: decentralized anonymous payments from Bitcoin. In: Proceedings of the IEEE Symposium on Security & Privacy (2014)

[BCG+15] Ben-Sasson, E., Chiesa, A., Green, M., Tromer, E., Virza, M.: Secure sampling of public parameters for succinct zero knowledge proofs. In: Proceedings of the IEEE Symposium on Security & Privacy (2015)

[BCG+17] Bootle, J., Cerulli, A., Ghadafi, E., Groth, J., Hajiabadi, M., Jakobsen, S.K.: Linear-time zero-knowledge proofs for arithmetic circuit satisfiability. In: Takagi, T., Peyrin, T. (eds.) ASIACRYPT 2017. LNCS, vol. 10626, pp. 336–365. Springer, Cham (2017). https://doi.org/10.1007/978-3-319-70700-6_12

[BCTV14] Ben-Sasson, E., Chiesa, A., Tromer, E., Virza, M.: Scalable zero knowledge via cycles of elliptic curves. In: Garay, J.A., Gennaro, R. (eds.) CRYPTO 2014. LNCS, vol. 8617, pp. 276–294. Springer, Heidelberg (2014). https://doi.org/10.1007/978-3-662-44381-1_16

[BFM88] Blum, M., Feldman, P., Micali, S.: Non-interactive zero-knowledge and its applications (extended abstract). In: STOC, pp. 103–112 (1988)

[BFS16] Bellare, M., Fuchsbauer, G., Scafuro, A.: NIZKs with an untrusted CRS: security in the face of parameter subversion. In: Cheon, J.H., Takagi, T. (eds.) ASIACRYPT 2016. LNCS, vol. 10032, pp. 777–804. Springer, Heidelberg (2016). https://doi.org/10.1007/978-3-662-53890-6_26

[BGG17] Bowe, S., Gabizon, A., Green, M.: A multi-party protocol for constructing the public parameters of the Pinocchio zk-SNARK. Cryptology ePrint Archive, Report 2017/602 (2017)

[BGM17] Bowe, S., Gabizon, A., Miers, I.: Scalable multi-party computation for zk-SNARK parameters in the random beacon model. Cryptology ePrint Archive, Report 2017/1050 (2017). https://eprint.iacr.org/2017/1050

[BP04] Bellare, M., Palacio, A.: Towards plaintext-aware public-key encryption without random oracles. In: Lee, P.J. (ed.) ASIACRYPT 2004. LNCS, vol. 3329, pp. 48–62. Springer, Heidelberg (2004). https://doi.org/10.1007/978-3-540-30539-2_4

[BR06] Bellare, M., Rogaway, P.: The security of triple encryption and a framework for code-based game-playing proofs. In: Vaudenay, S. (ed.) EUROCRYPT 2006. LNCS, vol. 4004, pp. 409–426. Springer, Heidelberg (2006). https://doi.org/10.1007/11761679_25

[BSBHR18] Ben-Sasson, E., Bentov, I., Horesh, Y., Riabzev, M.: Scalable, transparent, and post-quantum secure computational integrity. Cryptology ePrint Archive, Report 2018/046 (2018). https://eprint.iacr.org/2018/046

[Buc17] Buck, J.: Ethereum upgrade Byzantium is live, verifies first ZK-Snark proof. https://cointelegraph.com/news/ethereum-upgrade-byzantium-is-live-verifies-first-zk-snark-proof. Accessed Sept 2017

[CF01] Canetti, R., Fischlin, M.: Universally composable commitments. Cryptology ePrint Archive, Report 2001/055 (2001). http://eprint.iacr.org/2001/055

[Dam91] Damgård, I.: Towards practical public key systems secure against chosen ciphertext attacks. In: Feigenbaum, J. (ed.) CRYPTO 1991. LNCS, vol. 576, pp. 445–456. Springer, Heidelberg (1992). https://doi.org/10.1007/3-540-46766-1_36

[Dam92] Damgård, I.: Non-interactive circuit based proofs and non-interactive perfect zero-knowledge with preprocessing. In: Rueppel, R.A. (ed.) EUROCRYPT 1992. LNCS, vol. 658, pp. 341–355. Springer, Heidelberg (1993). https://doi.org/10.1007/3-540-47555-9_28

[Dam00] Damgård, I.: Efficient concurrent zero-knowledge in the auxiliary string model. In: Preneel, B. (ed.) EUROCRYPT 2000. LNCS, vol. 1807, pp. 418–430. Springer, Heidelberg (2000). https://doi.org/10.1007/3-540-45539-6_30

[DFGK14] Danezis, G., Fournet, C., Groth, J., Kohlweiss, M.: Square span programs with applications to succinct NIZK arguments. In: Sarkar, P., Iwata, T. (eds.) ASIACRYPT 2014. LNCS, vol. 8873, pp. 532–550. Springer, Heidelberg (2014). https://doi.org/10.1007/978-3-662-45611-8_28

[FF00]   Fischlin, M., Fischlin, R.: Efficient non-malleable commitment schemes. In: Bellare, M. (ed.) CRYPTO 2000. LNCS, vol. 1880, pp. 413–431. Springer, Heidelberg (2000). https://doi.org/10.1007/3-540-44598-6_26

[FLS99]   Feige, U., Lapidot, D., Shamir, A.: Multiple noninteractive zero knowledge proofs under general assumptions. SIAM J. Comput. **29**(1), 1–28 (1999)

[Fuc17]   Fuchsbauer, G.: Subversion-zero-knowledge SNARKs. Cryptology ePrint Archive, Report 2017/587 (2017)

[GG17]   Ghadafi, E., Groth, J.: Towards a classification of non-interactive computational assumptions in cyclic groups. In: Takagi, T., Peyrin, T. (eds.) ASIACRYPT 2017, Part II. LNCS, vol. 10625, pp. 66–96. Springer, Cham (2017). https://doi.org/10.1007/978-3-319-70697-9_3

[GGI+15]   Gentry, C., Groth, J., Ishai, Y., Peikert, C., Sahai, A., Smith, A.D.: Using fully homomorphic hybrid encryption to minimize non-interative zero-knowledge proofs. J. Cryptol. **28**(4), 820–843 (2015)

[GGPR13]   Gennaro, R., Gentry, C., Parno, B., Raykova, M.: Quadratic span programs and succinct NIZKs without PCPs. In: Johansson, T., Nguyen, P.Q. (eds.) EUROCRYPT 2013. LNCS, vol. 7881, pp. 626–645. Springer, Heidelberg (2013). https://doi.org/10.1007/978-3-642-38348-9_37

[GHM+17]   Gilad, Y., Hemo, R., Micali, S., Vlachos, G., Zeldovich, N.: Algorand: scaling Byzantine agreements for cryptocurrencies. In: SOSP (2017)

[GKM+18]   Groth, J., Kohlweiss, M., Maller, M., Meiklejohn, S., Miers, I.: Updatable and universal common reference strings with applications to zk-SNARKS. Cryptology ePrint Archive, Report 2018/280 (2018). https://eprint.iacr.org/2018/280

[GM17]   Groth, J., Maller, M.: Snarky signatures: minimal signatures of knowledge from simulation-extractable SNARKs. In: Katz, J., Shacham, H. (eds.) CRYPTO 2017. LNCS, vol. 10402, pp. 581–612. Springer, Cham (2017). https://doi.org/10.1007/978-3-319-63715-0_20

[GO14]   Groth, J., Ostrovsky, R.: Cryptography in the multi-string model. J. Cryptol. **27**(3), 506–543 (2014)

[GOP94]   Goldreich, O., Ostrovsky, R., Petrank, E.: Computational complexity and knowledge complexity. In: Electronic Colloquium on Computational Complexity (ECCC), vol. 1, no. 7 (1994)

[GOS12]   Groth, J., Ostrovsky, R., Sahai, A.: New techniques for noninteractive zero-knowledge. J. ACM **59**(3), 11:1–11:35 (2012)

[Gro10a]   Groth, J.: Short non-interactive zero-knowledge proofs. In: Abe, M. (ed.) ASIACRYPT 2010. LNCS, vol. 6477, pp. 341–358. Springer, Heidelberg (2010). https://doi.org/10.1007/978-3-642-17373-8_20

[Gro10b]   Groth, J.: Short pairing-based non-interactive zero-knowledge arguments. In: Abe, M. (ed.) ASIACRYPT 2010. LNCS, vol. 6477, pp. 321–340. Springer, Heidelberg (2010). https://doi.org/10.1007/978-3-642-17373-8_19

[Gro16]   Groth, J.: On the size of pairing-based non-interactive arguments. In: Fischlin, M., Coron, J.-S. (eds.) EUROCRYPT 2016. LNCS, vol. 9666, pp. 305–326. Springer, Heidelberg (2016). https://doi.org/10.1007/978-3-662-49896-5_11

[GS12]   Groth, J., Sahai, A.: Efficient noninteractive proof systems for bilinear groups. SIAM J. Comput. **41**(5), 1193–1232 (2012)

[GW11]   Gentry, C., Wichs, D.: Separating succinct non-interactive arguments from all falsifiable assumptions. In: STOC, pp. 99–108 (2011)

[KP98] Kilian, J., Petrank, E.: An efficient noninteractive zero-knowledge proof system for NP with general assumptions. J. Cryptol. **11**(1), 1–27 (1998)

[Lip12] Lipmaa, H.: Progression-free sets and sublinear pairing-based noninteractive zero-knowledge arguments. In: TCC, pp. 169–189 (2012)

[Lip13] Lipmaa, H.: Succinct non-interactive zero knowledge arguments from span programs and linear error-correcting codes. In: Sako, K., Sarkar, P. (eds.) ASIACRYPT 2013, Part I. LNCS, vol. 8269, pp. 41–60. Springer, Heidelberg (2013). https://doi.org/10.1007/978-3-642-42033-7_3

[LMS16] Lipmaa, H., Mohassel, P., Sadeghian, S.S.: Valiant's universal circuit: improvements, implementation, and applications. IACR Cryptology ePrint Archive 2016:17 (2016)

[PHGR13] Parno, B., Howell, J., Gentry, C., Raykova, M.: Pinocchio: nearly practical verifiable computation. In: Proceedings of the IEEE Symposium on Security & Privacy (2013)

[SCP00] De Santis, A., Di Crescenzo, G., Persiano, G.: Necessary and sufficient assumptions for non-iterative zero-knowledge proofs of knowledge for all NP relations. In: 27th International Colloquium on Automata, Languages and Programming (ICALP), pp. 451–462 (2000)

[SP92] De Santis, A., Persiano, G.: Zero-knowledge proofs of knowledge without interaction (extended abstract). In: 33rd Annual Symposium on Foundations of Computer Science, pp. 427–436 (1992)

[Val76] Valiant, L.G.: Universal circuits (preliminary report). In: Proceedings of the 8th Annual ACM Symposium on Theory of Computing, pp. 196–203 (1976)

[WTas+17] Wahby, R.S., Tzialla, I., Shelat, A., Thaler, J., Walfish, M.: Doubly-efficient zk-SNARKs without trusted setup. Cryptology ePrint Archive, Report 2017/1132 (2017). https://eprint.iacr.org/2017/1132

Obfuscation

A Simple Obfuscation Scheme
for Pattern-Matching with Wildcards

Allison Bishop[1,2($\boxtimes$)], Lucas Kowalczyk[2], Tal Malkin[2], Valerio Pastro[2,3],
Mariana Raykova[3], and Kevin Shi[2]

[1] IEX, New York, USA
[2] Columbia University, New York, USA
{allison,luke,tal,valerio,kshi}@cs.columbia.edu
[3] Yale University, New Haven, USA
mariana.raykova@yale.edu

Abstract. We give a simple and efficient method for obfuscating pattern matching with wildcards. In other words, we construct a way to check an input against a secret pattern, which is described in terms of prescribed values interspersed with unconstrained "wildcard" slots. As long as the support of the pattern is sufficiently sparse and the pattern itself is chosen from an appropriate distribution, we prove that a polynomial-time adversary cannot find a matching input, except with negligible probability. We rely upon the generic group heuristic (in a regular group, with no multilinearity). Previous work [9,10,32] provided less efficient constructions based on multilinear maps or LWE.

1 Introduction

The discipline of cryptography is fundamentally about the separation of seemingly intertwined information and abilities: how do we separate the ability the compute a function from the ability to invert a function? How do we separate the ability to encrypt from the ability to decrypt? How do we separate partial knowledge of a key through a side-channel attack from the ability to compromise a cryptographic scheme? The study of cryptographic obfuscation is born from the question: how do we separate the ability to run code from the ability to read code? Since the seminal work of [7] that placed this question firmly on a rigorous theoretical foundation, it has been clear that this kind of separation would be powerful, both inside and outside the typical reach of the discipline of cryptography.

If we can hide secrets inside functioning software, we can protect cryptographic keys, and many of cryptography's disparate and hard won achievements follow as a consequence. We can also protect intellectual property, and the inner workings of critical code like software patches, which in their unprotected form might leak information that could be used to attack remaining vulnerable

---

V. Pastro—Work done while the author was a postdoc at Columbia University and Yale University.

© International Association for Cryptologic Research 2018
H. Shacham and A. Boldyreva (Eds.): CRYPTO 2018, LNCS 10993, pp. 731–752, 2018.
https://doi.org/10.1007/978-3-319-96878-0_25

machines. But as with any cryptographic primitive, the suitability of program obfuscation for any particular task depends on three main axes by which we must evaluate proposed constructions: (1) efficiency, (2) the underlying computational and architectural assumptions, and (3) the derived security guarantees.

Two possibilities for (3), defined in [7], are the notion of virtual black box obfuscation (VBB) and the notion of indistinguishability obfuscation (IO). Virtual black box obfuscation is a very powerful and intuitive notion, which requires that anything that can be done by an attacker in possession of the obfuscated code can also be done by a simulator who can only run the software in a "black box", with no access to intermediary values or other properties of the computation between input ingestion and output production. This notion would be suitable for virtually[1] all possible applications of obfuscation, but it is shown in [7] that it is impossible to achieve for general functionalities. The notion of IO requires something weaker, merely that an attacker in possession of two different obfuscations of the *same* functionality cannot tell them apart. In other words, we only enforce indistinguishability for program descriptions that may differ internally but whose external input/output behavior is *identical*.

At the time of its introduction by [7], IO was neither shown to be impossible, nor shown to be particularly useful. Progress instead was made for VBB obfuscation of very basic functionalities, such as point obfuscation [27,31] and hyperplane membership [12], which lie below the reach of the impossibility result for VBB. But following the unprecedented construction of cryptographic multilinear maps in [17], two breakthroughs occurred in quick succession. A first candidate construction for indistinguishability obfuscation of general functions was proposed in [18], and the flexible technique of "punctured programming" was developed for deriving meaningful cryptographic results from the IO security guarantee [30].

Since then, the cryptographic research community has been riding out wave of positive and negative results: increasingly powerful constructions employing idealized models on multilinear maps or new, complex assumptions [2,4–6,18–20,23–26,33], attacks on the underlying multilinear maps [3,13–16,28], and a steady stream of works deriving applications and consequences from various forms of obfuscation(e.g., [1,21,22], and many more).

Our work is focused on the goal of obfuscating a modest but well-motivated functionality, one that does not require the use of multilinear maps, and hence does not inherit the risks of their still volatile security assumptions or the inefficiency that currently comes with using such a general-purpose tool. We consider the problem of *pattern matching with wildcards:* suppose there is an input binary string S of length n, and a pattern specification P also of length n, where for each bit P either dictates a particular bit value, or has a wildcard $*$, indicating that either value is allowed. For example, with $n = 5$, a pattern P would look like: $00 * 11$, and there would be two "matching" input strings S in this case, 00011 and 00111. The function we will obfuscate is the final "yes" or "no" outcome: for each P, we define the associated function $f_P(S)$ that outputs 1 when S matches P and outputs 0 otherwise.

---

[1] Pun intended.

This kind of functionality might appear, for instance, in a context like software patching. If a pattern P represents a problematic type of user input, say, that needs to be filtered out, we can obfuscate this function f_P to reject bad inputs without unnecessarily revealing P in full and helping attackers learn how to design such bad inputs. If the input length n is reasonably long and the number of matches to the pattern is not too dense in the space of inputs, we can hope that an attacker who queries a polynomial number of input strings will never manage to find a "bad" input that matches the pattern. We find these situations (where the adversary does not have enough information to identify the function being obfuscated) to be the most compelling subset of the standard VBB obfuscation security guarantee (as opposed to the subset involving simulating an adversary that already knows the function being obfuscated). Accordingly, we demonstrate that our construction satisfies a distributional security notion from [9,10,32]: if the pattern P is chosen from a suitable random distribution (and the number of wildcards $w \le 0.75n$), then a PPT attacker will not be able to distinguish our obfuscation of f_P from an obfuscation of a function that always outputs 0.

Our construction uses only the basic tools of group operations and polynomial interpolation, and so is quite efficient. Our security analysis will be in the generic group model, for a regular cyclic group, with no multilinearity required. It remains an interesting open problem to obtain a security analysis in the standard model, using standard assumptions like DDH, for instance. [29] showed that the easier problem of bounded Hamming distance decoding is at least as hard as the DDH problem. While the result is not applicable to the obfuscation construction, the intermediary problem of finding nontrivial representations of the identity element first described by [11] is potentially applicable.

The functionality of pattern matching with wildcards has been previously obfuscated in [9,10]. These constructions rely on multiplicative encoding schemes that enable multiplication of the encoded values and also zero-testing, i.e. checking whether an encoded value is zero. Unlike multilinear maps, these encoding schemes do not need to have additive properties. This functionality has been realized either through the use of general multilinear maps [9] or through lattice-based encodings relying on a new instance dependent assumption called entropic LWE [10]. A recent work by Wichs and Zirdelis [32] provides an obfuscation construction for a more general high entropy class, called compute-and-compare functions, from LWE. This class includes our pattern matching with wildcards. We view our construction as a simple and highly efficient alternative to such an LWE-based construction, and this is in line with the long tradition of analogous functionalities being achieved in the discrete-logarithm and LWE regimes.

To keep our scheme as intuitive and as efficient as possible, we start from additive basics. Let's first consider a pattern P with no wildcards. In this case, our function f_P is just a point function, since there is only one input string that matches the fully prescriptive pattern. Here we can work over Z_p and choose uniformly random values $a_1, \ldots, a_{n-1} \in Z_p$ and set $a_n = -(a_1 + \cdots + a_{n-1})$. We can choose additional random values $r_1, \ldots, r_n \in Z_p$. Now our obfuscated program can be comprised of $2n$ elements of Z_p, which we will label as $x_{i,b}$ where

$i \in [n]$ and $b \in [0,1]$. For each input bit position i, if the pattern value P is b, we set $x_{i,b} := a_i$ and $x_{i,1-b} = r_i$. To evaluate the obfuscated program on an input string, the evaluator simply selects the value corresponding to each input bit, and takes the sum modulo p. If it is 0, the output is 1. Otherwise the output is 0. Given these $2n$ values, if an attacker wants to find the pattern P, they are essentially trying to solve the subset sum problem (this is a slight variant since we have this kind of pair structure on the elements, but still the security intuition is the same).

Now if we want to introduce wildcards, it is clear we cannot simply give out a_i for both values for input bit i, since this will be noticed. The next thing we might try is to choose a random polynomial F of degree n over Z_p whose constant term is 0. Now we can set $x_{i,b} = F(2i + b)$ for positions that match the pattern, including both values of b in a wildcard position i. Our desired functionality can now be evaluated through polynomial interpolation. However, we quickly start to run into attacks based on list-decoding or regular decoding of Reed-Solomon codes, which can enable an attacker to recover the polynomial F once there are enough valid evaluations due to the wild cards.

A key observation at this point is that these decoding-style attacks rely upon non-linear functions of the given values, while the honest evaluation of the intended program needs only linear operations. This allows us to place the values $x_{i,b}$ in the exponent of a group $G = \langle g \rangle$ where discrete-log is difficult, and give out $g^{x_{i,b}}$ instead. This stops the decoding attacks without preventing honest evaluations. In the generic group model, the attacker is essentially limited to linear functions of the given exponents, so we can indeed formalize this intuition and obtain a security proof.

The hardness of noisy polynomial interpolation in the exponent was previously analyzed by [29], who gave a generic group argument concerning the problem of interpolating a polynomial with a slightly different error distribution. Our work follows a similar idea, but the specific wildcard structure we employ for our application creates some subtle differences, so we give a full argument here for completeness. We also provide a more rigorous exposition of the generic group proof argument.

It is an interesting problem to prove security for such a scheme without resorting to a generic group analysis. It seems that we should need a computational assumption like subset sum to assert that even though the group operations allow a discovery of the hidden structure, it is too sparse inside a combinatorially large space of possible input evaluations to be efficiently found. It also seems that we should need a computational assumption like DDH to explain exactly how the group blocks non-linear attacks. However, assumptions like DDH allow us to hide structure that is already non-linear, but requires us to preserve any structure that is linear, since linear structure on any small number of group elements can be discovered by brute force by an attacker. We could try to formulate some new assumption that is a strengthening of the subset sum assumption to the kind of intertwined linear structures that arise from polynomial evaluation, but this doesn't yet seem to yield insight beyond asserting security of the scheme itself.

We would ideally like to see a hybrid argument that combined simple subset-sum like steps with simple DDH-like steps, but designing such a reduction remains an intriguing challenge. Given that LWE-based approaches in the standard model are known, this represents a new test case on the boundary of the analogies we know between DDH-hard groups and the LWE setting. We expect that further study of this disconnect in proof technology between the LWE setting and the DDH setting may yield general insights into the inherent relationships (or lack thereof) between these different mathematical underpinnings.

2 Preliminaries

2.1 The Generic Group Model

We will prove the security of our construction against *generic adversaries*, which interact with group elements via the generic group model as defined in [8]. In this model, an adversary can only interact with the group via oracle calls to its group operation and zero test functionality. Group elements are represented by "handles", which are uniformly random strings long enough that the small probability of collision between handles representing different group elements can be ignored. A generic group operation oracle takes as input two group handles and returns a new handle representing the group element that is the result of the group operation on the two inputs (and is consistent with all handles previously used). Note that such an oracle can be efficiently simulated using a lookup table.

We use $\mathcal{G}$ to denote such a generic group operation oracle that answers adversary calls. $\mathcal{A}^{\mathcal{G}}$ will denote an adversary given access to this oracle and $\mathcal{O}^{\mathcal{G}}$ will denote the set of handles generated by $\mathcal{G}$ corresponding to the group elements in the construction $\mathcal{O}$.

2.2 Distributional Virtual Black-Box Obfuscation in the Generic Group Model

We will use a definition of *distributional virtual black-box (VBB) obfuscation in the generic group model* which is essentially the definition of [9], except using the generic group model instead of the random graded encoding model:

Definition 1 (Distributional VBB Obfuscator). *Let $\mathcal{C} = \{C_n\}_{n \in \mathbb{N}}$ be a family of polynomial-size circuits, where C_n is a set of boolean circuits operating on inputs of length n, and let $\mathcal{O}$ be a ppt algorithm which takes as input an input length $n \in \mathbb{N}$ and a circuit $C \in \mathcal{C}$ and outputs a boolean circuit $\mathcal{O}(C)$ (not necessarily in $\mathcal{C}$). Let $\mathcal{D} = \{\mathcal{D}_n\}_{n \in \mathbb{N}}$ be an ensemble of distribution families $\mathcal{D}_n$ where each $D \in \mathcal{D}_n$ is a distribution over C_n.*

$\mathcal{O}$ is a distributional VBB obfuscator for the distribution class $\mathcal{D}$ over the circuit family $\mathcal{C}$ if it has the following properties:

1. Functionality-Preserving: For every $n \in \mathbb{N}$, $C \in C_n$, and $\boldsymbol{x} \in \{0,1\}^n$, with all but $negl(n)$ probability over the coins of $\mathcal{O}$:

$$(\mathcal{O}(C, 1^n)(\boldsymbol{x}) = C(\boldsymbol{x})$$

2. *Polynomial Slowdown: For every* $n \in \mathbb{N}$ *and* $C \in \mathcal{C}_n$, *the evaluation of* $\mathcal{O}(C, 1^n)$ *can be performed in time* $poly(|C|, n)$.
3. *Distributional Virtual Black-Box in Generic Group Model: For every polynomial (in* n*) time generic adversary* $\mathcal{A}$, *there exists a polynomial time simulator* $\mathcal{S}$, *such that for every* $n \in \mathbb{N}$, *every distribution* $D \in \mathcal{D}_n$ *(a distribution over* $\mathcal{C}_n$, *and every predicate* $P : \mathcal{C}_n \rightarrow \{0, 1\}$:

$$\left| \Pr_{C \leftarrow \mathcal{D}_n, \mathcal{G}, \mathcal{O}^\mathcal{G}, \mathcal{A}} [\mathcal{A}^\mathcal{G}(\mathcal{O}^\mathcal{G}(C, 1^n)) = P(C)] - \Pr_{C \leftarrow \mathcal{D}_n, \mathcal{S}} [\mathcal{S}^C(1^{|C|}, 1^n) = P(C)] \right| = negl(n)$$

Remark 1. As in [9], we remark that a stronger notion of functionality-preserving exists in the literature, where the obfuscated program must agree with $C(\boldsymbol{x})$ on all inputs $\boldsymbol{x}$ *simultaneously*. We use the relaxed requirement that for every input (individually), the obfuscated circuit is correct except for negligible probability. We also note that our construction can be modified to achieve the stronger property by using a group of sufficiently large size (2^{2n}) and the union bound over each of the 2^n inputs.

2.3 Schwartz-Zippel Lemma

A key step in our hybrid proof of security relies on the Schwartz-Zippel Lemma, which we will reproduce here:

Lemma 1. *Let* $\mathbb{Z}_p$ *be a finite field of size* p *and let* $P \in \mathbb{Z}_p[x_1, \ldots, x_n]$ *be a non-zero polynomial of degree* $\leq d$. *Let* $r_1, \ldots, r_n$ *be selected at random independently and uniformly from* $\mathbb{Z}_p$. *Then:* $\Pr[P(r_1, \ldots, r_n) = 0] \leq \frac{d}{p}$.

3   Obfuscating Pattern Matching with Wildcards

The class of functions for pattern matching with wildcards is parametrized by $(n, \boldsymbol{y}, \mathcal{W})$, where $\mathcal{W} \subset [n]$ is an index set and $f_{\boldsymbol{y}} : \{0, 1\}^{n-|\mathcal{W}|} \longrightarrow \{0, 1\}$ is a point function over $n - |\mathcal{W}|$ input variables that outputs 1 on the single input $\boldsymbol{y} \in \{0, 1\}^{n-|\mathcal{W}|}$. The function $\Pi_{\mathcal{W}^c} : \{0, 1\}^n \longrightarrow \{0, 1\}^{n-|\mathcal{W}|}$ projects a boolean vector of length n onto only the entries not in the index set $\mathcal{W}$. $f_{\boldsymbol{y}, \mathcal{W}}$, the function for pattern $\boldsymbol{y}$ with wildcard slots $\mathcal{W}$, is defined to be $f_{\boldsymbol{y}, \mathcal{W}}(x) := f_{\boldsymbol{y}}(\Pi_{\mathcal{W}^c}(x))$. Our obfuscation scheme for the class of functions for pattern matching with wildcards is as follows:

Setup(n): sample $a_1, \cdots, a_{n-1} \sim \mathbb{Z}_p$ uniformly at random and construct the fixed polynomial $F(x) := a_1 x + a_2 x^2 + \cdots + a_{n-1} x^{n-1}$. Let G be a group with generator g of prime order $p > 2^n$.

Construction($n, \boldsymbol{y}, \mathcal{W}$): the obfuscator outputs $2n$ elements arranged in a $2 \times n$ table of n columns corresponding to the n input variables with two entries each corresponding to the two possible boolean values of each input. For each slot h_{ij} where $(i, j) \in \{0, 1\}^n \times \{0, 1\}$, if either $i \in \mathcal{W}$ or $y_i = j$, then the obfuscator releases the element $h_{ij} = g^{F(2i+j)}$. Otherwise, the obfuscator releases h_{ij} as a uniformly random element of G.

Evaluation($\boldsymbol{x}$): to evaluate $f_{\boldsymbol{y},\mathcal{W}}(\boldsymbol{x})$, for each $i = 1, \cdots, n$, compute:

$$C_i := \prod_{j \neq i} \frac{-2j - x_j}{2i - x_i - x_j + 2j}$$

choose the elements h_{ix_i}, and compute:

$$T := \prod_{i=0}^{n-1} (h_{ix_i})^{C_i}$$

Output 1 if $T = g^0$ and 0 otherwise.

Functionality-Preserving: The fact that this obfuscation scheme is functionality-preserving follows from the fact that, if $\boldsymbol{x}$ is an accepting input of f ($f(\boldsymbol{x}) = 1$), then the chosen handles form n proper evaluations of the polynomial $F(x)$ on distinct elements. Further, the C_i scalars used in evaluation are Lagrange coefficients, making the evaluation a polynomial interpolation that returns $F(0) = 0$ in this case, causing $T = g^0$ and the evaluation to output 1 (with probability 1).

$$\prod_{i=0}^{n-1} (h_{ix_i})^{C_i} = \prod_{i=0}^{n-1} g^{C_i F(2i + x_i)}$$
$$= g^{\sum_{i=0}^{n-1} C_i F(2i + x_i)}$$
$$= g^{F(0)}$$
$$= g^0$$

On the other hand, if even one input bit was not accepting (so $f(\boldsymbol{x}) = 0$), then at least one of the h_{ix_i}'s used in interpolation would be a uniformly random group element (not $g^{F(2i+j)}$). Thus, the evaluation product would be a product that includes a uniformly random group element raised to some power, which would result in $T = g^0$ with negligible probability $\frac{1}{p}$.

Polynomial Slowdown: Given a the set of $2n$ group elements, assuming group operations can be performed in $poly(n)$ time, the computation of C_i and T described in the **Evaluation** procedure can be performed in polynomial time.

Distributional Virtual Black-Box: We give a proof of our construction's distributional VBB security in the generic group model in Sect. 4 in Theorem 1.

4   Distributional VBB Security in the Generic Group Model

This section will prove Theorem 1, which establishes the distributional virtual black box security of our construction in the generic group model over the class of uniform distributions for point functions with wildcards. Our framework for reasoning in the generic group setting draws from [8].

In a generic group proof, there are many closely related but technically distinct kinds of objects that are often conflated. There are the underlying group elements, which can be associated with their exponents in $\mathbb{Z}_p$ relative to the common base. There are the handles that the group oracle associates to these elements. There are formal polynomials which may track known or unknown relationships between group elements. There are subsets of handles which the adversary has previously seen, and other handles whose distribution remains independent of the adversary's view so far. In order to make our proof as rigorous and precise as possible, we will keep explicit track of all of these various objects, and the maps between them.

We define an equivalent security game where an adversary calls two oracles simultaneously, one of whose behavior is already completely known. The purpose of incorporating a known oracle into the security game is to rigorously define when the unknown oracle deviates from expected behavior, and thus, when the adversary has distinguishing power. Given that a low probability failure event does not occur, any algorithm's behavior when interacting with either of these oracles should be identical. The actual calculation of the probability of such a failure event is conceptually simple and done by many previous works for different noise distributions. On the other hand, in order to properly describe the notion of "identical behavior" we introduce some basic technical machinery from category theory.

We establish some notation before proceeding. Let bold letters denote symbolic variables and non-bold letters denote the sampled random values for the corresponding variable. Let $f \in \mathbb{Z}_p[\mathbf{a}_1, \cdots, \mathbf{a}_n, \mathbf{x}]$ be a fixed polynomial of degree $n-1$ in $\mathbf{x}$ which is linear in each $\mathbf{a}_i$ individually. Let $\mathcal{H}_S$ and $\mathcal{H}_M$ be two identical copies of the same space of strings corresponding to handles in the generic group model.

Since our proof takes place in the generic group model, and our obfuscated program consists of a set of group elements, we will use the notation $\mathcal{G}_S, \mathcal{G}_M, \mathcal{G}_E$ to denote three different ways that an adversary can be supplied with handles representing an obfuscated program and how requests to the generic group operation oracle are answered. $\mathcal{G}_S$ will implement faithful interaction with the true construction in the generic group model. $\mathcal{G}_M$ implements a hybrid setting that we will show is indistinguishable from $\mathcal{G}_S$ to the adversary. Finally, $\mathcal{G}_E$ implements a setting that can be simulated without knowledge of the function drawn from the distribution (and is indistinguishable from $\mathcal{G}_M$).

The high level structure of our proof is pretty typical for a generic group argument. The group oracle $\mathcal{G}_M$ will behave similarly to $\mathcal{G}_S$, but instead of sampling random exponents according to the proscribed polynomial structure, it will work with formal polynomials representing this structure, hence ignoring any spurious relationship arises from a particular choice at the sampling stage. Arguing that $\mathcal{G}_S$ and $\mathcal{G}_M$ are indistinguishable is where we use the Schwartz-Zippel Lemma. An adversary will only receive a different distribution of handles if it manages to find a spurious relationship while interacting with $\mathcal{G}_S$, which must mean that the sampling happened to choose a root of a non-trivial, low degree formal polynomial. The Schwartz-Zippel Lemma allows us to conclude that this will occur with only negligible probability over the sampling employed by $\mathcal{G}_S$.

To argue that $\mathcal{G}_M$ and $\mathcal{G}_E$ are indistinguishable, we will need to argue that the adversary cannot (except with negligible probability), detect the remaining formal polynomial structure in $\mathcal{G}_M$, since doing so requires referencing many correctly structured elements and avoiding the random elements completely. As long as the wildcards are not too dense, this is an intractable combinatorial problem for the adversary.

Definition 2 ($\mathcal{G}_S$: Oracle *Start*).
First, sample the following uniformly at random:

- $\mathcal{W} = \{i_1, \cdots, i_w\} \subset [n]$
- $y_i \in \{0,1\}$ *for each* $i \notin \mathcal{W}$
- $a_1, \cdots, a_n \in \mathbb{Z}_p$
- *Random embedding* $\Phi_S : G \hookrightarrow \mathcal{H}_S$

For the initial set of handles representing the 2n group elements in the obfuscation of $f_{y,\mathcal{W}}$*, for each entry* $(i,j) \in [n] \times \{0,1\}$*:*

- *If* $i \in \mathcal{W}$ *or* $y_i = j$ *(i.e. the input bit is part of an accepting string), output*
  $\Phi_S \left(g^{F(a_1, \cdots, a_n, 2i+j)} \right)$
- *Otherwise sample a uniformly random exponent* ρ_{ij} *and output* $\Phi_S(g^{\rho_{ij}})$

Given a group operation query on (h_1, h_2)*:*

- *Find* $g_1 = \Phi_S^{-1}(h_1)$ *and* $g_2 = \Phi_S^{-1}(h_2)$*. If either does not exist, ignore the query.*
- *Return* $\Phi_S(g_1 \cdot g_2)$

Note that $\mathcal{G}_S$ faithfully instantiates our construction described in Sect. 3 in the generic group model. We will now describe an alternative oracle implementation that uses symbolic variables instead of group elements to produce the generic group functionality:

Definition 3 ($\mathcal{G}_M$: Oracle *Middle*).
First, sample the following uniformly at random:

- $\mathcal{W} = \{i_1, \cdots, i_w\} \subset [n]$
- $y_i \in \{0,1\}$ *for each* $i \notin \mathcal{W}$
- *Random embedding* $\Phi_M : \mathbb{Z}_p[a_1, \cdots, a_n, b_1, \cdots, b_{n-w}] \hookrightarrow \mathcal{H}_M$.

Let $\sigma : \{0,1\}^n \times \{0,1\} \rightarrow [n-w]$ *be an arbitrary ordering of the* $(n-w)$ *coordinate pairs* (i,j) *where* $i \notin \mathcal{W}$ *and* $j \neq y_i$*, and which is not defined on the other coordinate pairs.*

For the initial set of handles representing the 2n group elements in the obfuscation of $f_{y,\mathcal{W}}$*, for each entry* $(i,j) \in [n] \times \{0,1\}$*:*

- If $i \in \mathcal{W}$ or $y_i = j$ (i.e. the input bit is part of an accepting string), output
  $\Phi_M(F(\mathbf{a}_1, \cdots, \mathbf{a}_n, 2i + j))$
- Otherwise output the label $\Phi_M(\mathbf{b}_{\sigma(ij)})$

Given a group operation query on (h_1, h_2):

- Find $p_1 = \Phi_M^{-1}(h_1)$ and $p_2 = \Phi_M^{-1}(h_2)$. If either does not exist, ignore the query.
- Return $\Phi_M(p_1 + p_2)$

The two oracles are related by the existence of the following *evaluation map in the exponent*:

$$\phi : \mathbb{Z}[\mathbf{a}_1, \cdots, \mathbf{a}_n, \mathbf{b}_1, \cdots, \mathbf{b}_{n-w}] \longrightarrow G$$
$$F(\mathbf{a}_1, \cdots, \mathbf{a}_n, \mathbf{a}_n, \mathbf{b}_1, \cdots, \mathbf{b}_{n-w}) \longmapsto g^{F(a_1, \cdots, a_n, b_1, \cdots, b_{n-w})}$$

where $b_k = \rho_{\sigma^{-1}(k)}$ are the values of the random exponents sampled by Oracle S for the non-accepting slots. Only the existence of this evaluation map is necessary for the proof, so its dependence on unknown random values is not an issue.

In particular ϕ is a surjective group homomorphism of $(\mathbb{Z}_p[\mathbf{a}_1, \cdots, \mathbf{a}_n, \mathbf{b}_1, \cdots, \mathbf{b}_{n-w}], +)$ into $(G, \times)$, since it is a composition of an evaluation map with an exponential map, which are both surjective group homomorphisms.

The idea behind defining such an evaluation map is to define the failure event as a substructure of a larger structure which may then be used to formalize when the behavior is identical. In particular, we will see that the failure event corresponds to the kernel of this evaluation map that we just defined.

Simultaneous Oracle Game. Rather than proving that the difference in any adversary's output probabilities when interacting with $(\mathcal{G}_S$ vs. $\mathcal{G}_M)$ or $(\mathcal{G}_M$ vs. $\mathcal{G}_E)$ is small directly, we will define another security game and exhibit a reduction to the desired statements. In this new security game, the adversary simultaneously queries two oracles for operations on group elements: one oracle $\mathcal{G}_M$ is known and serves as a convenience for formalizing the generic group oracle, and the second $\mathcal{G}_*$ is the unknown that the adversary wishes to identify. We define the game with oracles $(\mathcal{G}_S, \mathcal{G}_M)$ below and note that the game and reduction for oracles $(\mathcal{G}_M, \mathcal{G}_E)$ is symmetric.

Definition 4 (Simultaneous Oracle Game). *An adversary is given access to a pair of oracles $(\mathcal{G}_M, \mathcal{G}_*)$, where $\mathcal{G}_*$ is $\mathcal{G}_M$ with probability $1/2$ and $\mathcal{G}_S$ with probability $1/2$. In each round, the adversary asks the same query to both oracles. The adversary wins the game if he guesses correctly the identity of $\mathcal{G}_*$.*

To make precise the notion of an adversary playing both oracles simultaneously and asking the same queries, the adversary maintains two sets $\mathcal{H}_S^t$ and $\mathcal{H}_M^t$ which are the sets of handles returned by the oracles after t query rounds. The adversary then maintains a function $\Psi : \mathcal{H}_M^t \to \mathcal{H}_S^t$. Initially, the adversary sets $\Psi(h^b_{ij}) = h^a_{ij}$ for each initial slot location $(i,j) \in \{1,n\} \times \{0,1\}$, where h^a_{ij} is the handle corresponding to the slot (i,j) in oracles S and h^b_{ij} the handle in oracles M. After each query $h^m = \mathcal{G}_M(h^b_1, h^b_2)$ and $h^s = \mathcal{G}_S(\Psi(h^b_1), \Psi(h^b_2))$ the adversary updates the function with the definition $\Psi(h^s) = h^m$.

Lemma 2. *Suppose there exists an algorithm $\mathcal{A}$ such that*

$$\left| Pr[\mathcal{A}^{\mathcal{G}_M}(\mathcal{O}^{\mathcal{G}_M}) = 1] - Pr[\mathcal{A}^{\mathcal{G}_S}(\mathcal{O}^{\mathcal{G}_S}) = 1] \right| \geq \delta$$

Then an adversary can win the simultaneous oracle game with probability at least $\frac{1}{2} + \frac{\delta}{2}$ for any pair of oracles $(\mathcal{G}_M, \mathcal{G}_* = \mathcal{G}_M/\mathcal{G}_S)$.

Proof. Let $p = Pr[\mathcal{A}^{\mathcal{G}_M}(\mathcal{O}^{\mathcal{G}_M}) = 1]$ and $q = Pr[\mathcal{A}^{\mathcal{G}_S}(\mathcal{O}^{\mathcal{G}_S}) = 1]$. The adversary can estimate these parameters to within a bounded polynomial of the true parameter by simulating each oracle and $\mathcal{A}$'s behavior on each.

Without loss of generality, we can assume that $p \geq q$. Otherwise, we can define p, q to be the inverse quantities $Pr[\mathcal{A}^{\mathcal{G}_M}(\mathcal{O}^{\mathcal{G}_M}) = 0], Pr[\mathcal{A}^{\mathcal{G}_S}(\mathcal{O}^{\mathcal{G}_S}) = 0]$ respectively.

The adversary will guess $\mathcal{G}_* = \mathcal{G}_M$ if $\mathcal{A}^{\mathcal{G}_*}(\mathcal{O}^{\mathcal{G}_*}) = 1$ and $\mathcal{G}_* = \mathcal{G}_S$ if $\mathcal{A}^{\mathcal{G}_*}(\mathcal{O}^{\mathcal{G}_*}) = 0$. The probability of success is given by

$$\begin{aligned}
Pr[\mathcal{A}^{\mathcal{G}_*}(\mathcal{O}^{\mathcal{G}_*}) = \mathcal{G}_*] &= Pr[\mathcal{G}_* = \mathcal{G}_M] Pr[\mathcal{A}^{\mathcal{G}_M}(\mathcal{O}^{\mathcal{G}_M}) = 1] \\
&\quad + Pr[\mathcal{G}_* = \mathcal{G}_S] Pr[\mathcal{A}^{\mathcal{G}_S}(\mathcal{O}^{\mathcal{G}_S}) = 0] \\
&= \frac{1}{2} + \frac{1}{2}(p - q) \\
&\geq \frac{1}{2} + \frac{\delta}{2}
\end{aligned}$$

Indistinguishability between *Start* and *Middle*. The following gives a criteria for overall indistinguishability of the output handle distributions.

Definition 5. *The pair (h^s, h^m) of answers returned by $(\mathcal{G}_S, \mathcal{G}_M)$ after query number t is called identical if it satisfies one of the following:*

1. $h^s \notin \mathcal{H}^t_S$ and $h^m \notin \mathcal{H}^t_M$
2. *The oracles return handles $h^s \in \mathcal{H}_S, h^m \in \mathcal{H}_M$ respectively such that $\Psi(h^m) = h^s$*

Note that in case (1), h^s and h^m are both freshly sampled uniformly random strings and their distributions are equal.

Lemma 3. *In the simultaneous oracle game with $\mathcal{G}_* = \mathcal{G}_S$, suppose for every query (h^m_1, h^m_2) to oracle M and corresponding query $(\Psi(h^m_1), \Psi(h^m_2))$ to oracle S, the answers returned are identical. Then for any algorithm $\mathcal{A}$, we have*

$$Pr[\mathcal{A}^{\mathcal{G}_S}(\mathcal{O}^{\mathcal{G}_S}) = 1] = Pr[\mathcal{A}^{\mathcal{G}_M}(\mathcal{O}^{\mathcal{G}_M}) = 1]$$

742 A. Bishop et al.

Proof (Proof of Lemma 3). If we had swapped the oracles $\mathcal{G}_S$ and $\mathcal{G}_M$ and the adversary had used Ψ^{-1} instead of Ψ, the answer distributions would have been identical and $\mathcal{A}$ would have to produce the same output distribution.

Remark 2. Note that this argument does not depend on the particular implementations of $\mathcal{G}_S, \mathcal{G}_M$, and therefore the lemma also holds for the pair of oracles $\mathcal{G}_M, \mathcal{G}_E$ (to be defined later in Definition 6).

Thus it suffices to show that

Lemma 4. *Suppose an adversary makes an arbitrary sequence of queries and receives answers*

$$\{h_t^s = \mathcal{G}_S(\Psi(h_{t1}^m), \Psi(h_{t2}^m))\}_{t=1}^Q$$
$$\{h_t^m = \mathcal{G}_M(h_{t1}^m, h_{t2}^m)\}_{t=1}^Q$$

Then with overall probability at least $1 - \dfrac{(Q+2n)^2}{p}$, *for every* t, h_t^s *and* h_t^m *are identical as defined in Definition 5.*

Proof. Initially each set of $2n$ handles given by each oracle are uniformly random strings and hence indistinguishable. The proof is by induction under the following hypothesis:

Suppose the adversary has made t queries so far and has $\mathcal{H}_S^t, \mathcal{H}_M^t$ satisfying the following:

1. For each query made so far, the answer distributions have been identical.
2. For every $h^s \in \mathcal{H}_S^t$, there exists a unique $f \in \mathbb{Z}_p[\mathbf{a}_1, \cdots, \mathbf{a}_n]$ such that $\Phi_S \circ \phi(f) = \Phi_M^{-1}(f)$.

We can state this inductive hypothesis this in the following commutative diagram:

$$
\begin{array}{ccccc}
\mathbb{Z}_p[\mathbf{a}, \mathbf{b}] & \xrightarrow{\Phi_M, \cong} & \mathrm{Im}(\Phi_M) & \xleftarrow{i_M} & \mathcal{H}_M^t \\
\phi \downarrow & & {\scriptstyle \exists !} & & \downarrow \Psi, = \\
G & \xrightarrow{\Phi_S, \cong} & \mathrm{Im}(\Phi_S) & \xleftarrow{i_S} & \mathcal{H}_S^t
\end{array}
$$

Here $\mathrm{Im}(\Phi_M), \mathrm{Im}(\Phi_S)$ are the relevant handles in the handle spaces. Commutativity of the lower triangle under the unique lift means that for all $h^s \in \mathcal{H}_S^t, \exists! f \in \mathbb{Z}_p[\mathbf{x}]$ such that $i_S(h^s) = \Phi_S \circ \phi(f)$. Note that the upper triangle trivially commutes because the unique lift is defined by the composition $\Phi_M \circ i_M \circ \Psi^{-1}$. To ease the notation a little, we'll omit the inclusion maps from here on when it is obvious the handle is in $\mathcal{H}_*^t$.

Now assuming the inductive hypothesis, suppose the $(t+1)$th query is the group operation of $h_1, h_2 \in \mathcal{H}_M^t$ and $\Psi(h_1), \Psi(h_2) \in \mathcal{H}_S^t$. Oracle M will output the handle $h^m = \Phi_M\left(\Phi_M^{-1}(h_1) + \Phi_M^{-1}(h_2)\right) =: h_1 \cdot h_2$, and Oracle S will output the handle $h^s = \Phi_S\left(\Phi_S^{-1}(\Psi(h_1)) \times \Phi_S^{-1}(\Psi(h_2))\right) =: \Psi(h_1) \cdot \Psi(h_2)$. The $(\cdot)$ notation on handles is justified by the fact that $\mathrm{Im}(\Phi_M) \subset \mathcal{H}_M$ is trivially isomomorphic as a group to $\mathbb{Z}_p[\mathbf{a}_1, \cdots, \mathbf{a}_n]$, where its group operation is obtained by pulling back by Φ_M, and likewise for $\mathrm{Im}(\Phi_S) \subset \mathcal{H}_S$.

We have the following two cases:

1. $h^m \in \mathcal{H}_M^t$ (i.e. this handle was seen previously). Then

$$
\begin{aligned}
\Psi(h_1) \cdot \Psi(h_2) &= (\Phi_S \circ \phi \circ \Phi_M^{-1})(h_1) \cdot (\Phi_S \circ \phi \circ \Phi_M^{-1})(h_2) \\
&= (\Phi_S \circ \phi \circ \Phi_M^{-1})(h_1 \cdot h_2) \\
&= (\Phi_S \circ \phi \circ \Phi_M^{-1})(h^m) \\
&= \Psi(h^m)
\end{aligned}
$$

where we use commutativity of the diagram on each factor handle, the homomorphism property of the maps, the definition of oracle M's output, and commutativity of the diagram on the output handle (which we can do since the handle was previously defined).

Thus the handles in the output pair have the same distribution, and since no new handles are created, the inductive hypothesis trivially remains satisfied.

2. $h^m \notin \mathcal{H}_M^t$ (i.e. this is a new handle).
   (a) If $h^s \notin \mathcal{H}_S^t$ is also a new handle, then the unique lift simply extends to map h^s to $\Phi_M^{-1}(h^m)$, and both $\mathcal{H}_M^t$ and $\mathcal{H}_S^t$ are augmented by one element. The handles in the output pair are new and uniformly distributed, and the inductive hypothesis is satisfied.
   (b) If $h^s \in \mathcal{H}_S^t$, then by the inductive hypothesis, h^s lifts to some $f_s \in \mathbb{Z}_p[\mathbf{x}]$ which maps to some $\tilde{h}^b = \Psi^{-1}(h^s)$. However we also have $f_m = \Phi_M^{-1}(h^m) \neq f_s$, since $h^m \notin \mathcal{H}_M^t$. Thus both f_s and f_m are lifts of h^s which make the diagram commute, so after this query the inductive hypothesis is no longer satisfied for the next query.
   This event only happens if $f_s - f_m \in \ker \phi$ and $f_s - f_m$ is nontrivial. Thus the proof is complete as long as we show this event happens with low probability.

Now consider the following sequential variant of the game. The adversary plays the game using the real Oracle M and his own simulation of Oracle S obtained by outputting a uniformly random string when $\mathcal{G}_M$ does and using the Ψ map when $\mathcal{G}_M$ outputs an existing string. He then plays the exact same sequence to the real Oracle S and compares these answers to the ones produced by the real Oracle M. As long as the bad event does not occur, the sequence of queries asked in this sequential game is identical to the sequence of queries asked playing the real pair of oracles.

Note that the occurrence of the bad event is decided by the initial random sampling of $a_1, \cdots, a_n \in \mathbb{Z}_p$, and thus the bad event either occurs in both the sequential and parallel variants or in neither. So it suffices to just bound the probability of the bad event occurring at any time in the sequential game.

For each pair (f_s, f_m), $f_s - f_m$ is a degree-1 polynomial in n variables over $\mathbb{Z}_p$. Thus the bad event happens with probability at most $\frac{1}{p}$ by Lemma 1, the Schwartz-Zippel lemma. Thus by a union bound, after Q queries of either type, there are at most $(Q + 2n)^2$ pairs of symbolic polynomials, so with probability at most $\frac{(Q+2n)^2}{p}$ the two distributions of handles are distinguishable.

We remark that everything in the proof only relied on diagram arguments and did not care about the actual structure of the underlying objects, except for analyzing when $f_s - f_m \in \ker \phi$ occurred. Thus in the proceeding reductions between other oracles, all this automatically follows provided we can define an appropriate evaluation map ϕ, and we only need to analyze the kernel of the corresponding evaluation map.

Lemma 5. *For an adversary $\mathcal{A}$ in the generic group model which makes Q queries to the generic group oracle,*

$$\left| \Pr_{\substack{C \leftarrow \mathcal{D}_n, \\ \mathcal{G}_S, \mathcal{O}, \mathcal{A}}} [\mathcal{A}^{\mathcal{G}_S}(\mathcal{O}^{\mathcal{G}_S}(C, 1^n)) = P(C)] - \Pr_{\substack{C \leftarrow \mathcal{D}_n, \\ \mathcal{G}_M, \mathcal{O}, \mathcal{A}}} [\mathcal{A}^{\mathcal{G}_M}(\mathcal{O}^{\mathcal{G}_M}(C, 1^n)) = P(C)] \right| \leq \frac{(Q + 2n)^2}{2^n}$$

Proof. From Lemma 3 we have that:

$$\Pr[\mathcal{A}^{\mathcal{G}_S}(\mathcal{O}^{\mathcal{G}_S}) = 1] = \Pr[\mathcal{A}^{\mathcal{G}_M}(\mathcal{O}^{\mathcal{G}_M}) = 1]$$

as long as all queries to the generic group oracles are *identical* as defined in Definition 5.

Lemma 4 tells us that the probabilities of all queries not being identical during the simultaneous oracle game between $(\mathcal{G}_S, \mathcal{G}_M)$ is at most $\dfrac{(Q + 2n)^2}{p}$, where Q is the number of the adversary's queries to the generic group oracle and $p > 2^n$ is the order of the group.

Therefore, the difference $\Pr[\mathcal{A}^{\mathcal{G}_M}(\mathcal{O}^{\mathcal{G}_M}) = 1] - \Pr[\mathcal{A}^{\mathcal{G}_S}(\mathcal{O}^{\mathcal{G}_S}) = 1]$ is at most $\dfrac{(Q + 2n)^2}{2^n}$, and so an adversary's advantage in the simultaneous oracle game between $(\mathcal{G}_M, \mathcal{G}_S)$ and $(\mathcal{G}_M, \mathcal{G}_M)$ is:

$$\begin{aligned}
\Pr[\mathcal{A}^{\mathcal{G}_*}(\mathcal{O}^{\mathcal{G}_*}) = \mathcal{G}_*] &= \Pr[\mathcal{G}_* = \mathcal{G}_M] \Pr[\mathcal{A}^{\mathcal{G}_M}(\mathcal{O}^{\mathcal{G}_M}) = 1] \\
&\quad + \Pr[\mathcal{G}_* = \mathcal{G}_S] \Pr[\mathcal{A}^{\mathcal{G}_S}(\mathcal{O}^{\mathcal{G}_S}) = 0] \\
&= \frac{1}{2} + \frac{1}{2}(\Pr[\mathcal{A}^{\mathcal{G}_M}(\mathcal{O}^{\mathcal{G}_M}) = 1] - \Pr[\mathcal{A}^{\mathcal{G}_S}(\mathcal{O}^{\mathcal{G}_S}) = 1]) \\
&\leq \frac{1}{2} + \frac{(Q + 2n)^2}{2 \cdot 2^n}
\end{aligned}$$

This, plugged into the reduction from Lemma 2, tells us that for all adversaries:

$$\left| Pr[\mathcal{A}^{\mathcal{G}_M}(\mathcal{O}^{\mathcal{G}_M}) = 1] - Pr[\mathcal{A}^{\mathcal{G}_S}(\mathcal{O}^{\mathcal{G}_S}) = 1] \right| \leq \frac{(Q + 2n)^2}{2^n}$$

Game between *Middle* and *End*

Definition 6 ($\mathcal{G}_E$: Oracle *End*).
First, sample the following uniformly at random:

- *Random embedding $\Phi_E : \mathbb{Z}_p[\mathbf{c}_1, \cdots, \mathbf{c}_{2n}] \hookrightarrow \mathcal{H}_E$.*

For the initial set of handles representing the $2n$ group elements in the obfuscation of $f_{y,\mathcal{W}}$, for each entry $(i,j) \in [n] \times \{0,1\}$:

- *Output $\Phi_E(\mathbf{c}_{2i+j})$*

Given a group operation query on (h_1, h_2):

- *Find $p_1 = \Phi_E^{-1}(h_1)$ and $p_2 = \Phi_E^{-1}(h_2)$. If either does not exist, ignore the query.*
- *Return $\Phi_E(p_1 + p_2)$*

Oracle M and Oracle E are related by the following *evaluation map* which is defined on the generators of $\mathbb{Z}_p[\mathbf{c}_1, \cdots, \mathbf{c}_{2n}]$ and extended by linearity.

$$
\phi : \mathbb{Z}_p[\mathbf{c}_1, \cdots, \mathbf{c}_{2n}] \longrightarrow \mathbb{Z}_p[\mathbf{a}_1, \cdots, \mathbf{a}_n, \mathbf{b}_1, \cdots, \mathbf{b}_{n-w}]
$$
$$
\mathbf{c}_k \longmapsto \mathbf{b}_{\sigma(\lfloor k/2 \rfloor, k \bmod 2)} \text{ if } \sigma \text{ is defined here}
$$
$$
\mathbf{c}_k \longmapsto F(\mathbf{a}_1, \cdots, \mathbf{a}_n, k) \text{ otherwise}
$$

In other words the monomial c_k is mapped to the same symbolic polynomial that Oracle *Middle* assigned to the slot $(\lfloor k/2 \rfloor, k \bmod 2)$, which is either a symbolic variable $\mathbf{b}$ or a symbolic polynomial $F(\mathbf{a}_1, \cdots, \mathbf{a}_n, k)$. Since the $\mathbf{c}_k$'s generate the entire additive group $\mathbb{Z}_p[\mathbf{c}_1, \cdots, \mathbf{c}_{2n}]$, this extends to a group homomorphism of $(\mathbb{Z}_p[\mathbf{c}_1, \cdots, \mathbf{c}_{2n}], +)$ into $(\mathbb{Z}_p[\mathbf{a}_1, \cdots, \mathbf{a}_n, \mathbf{b}_1, \cdots, \mathbf{b}_{n-w}], +)$.

Lemma 6. *Suppose an adversary makes an arbitrary sequence of queries and receives answers*

$$
\{h_t^m = \mathcal{G}_S(\Psi(h_{t1}^e), \Psi(h_{t2}^e))\}_{t=1}^Q
$$
$$
\{h_t^e = \mathcal{G}_M(h_{t1}^e, h_{t2}^e)\}_{t=1}^Q
$$

If $w/n \leq 3/4$, then with overall probability at least $1 - \frac{2}{2^{0.0613n}}$ for every t, h_t^s and h_t^m are identical as defined in Definition 5.

The proof of this lemma starts with the same setup as the proof of 4. The adversary maintains a function $\Psi : \mathcal{H}_E \to \mathcal{H}_M$ and two sets of handles $\mathcal{H}_E^t, \mathcal{H}_M^t$.

Proof. Inductively, after t queries, assume the following commutative diagram is true:

$$
\begin{array}{ccccc}
\mathbb{Z}_p[\mathbf{c}_1, \cdots, \cdots, \mathbf{c}_{2n}] & \xrightarrow{\Phi_E, \cong} & \mathrm{Im}(\Phi_E) & \xleftarrow{i_E} & \mathcal{H}_E^t \\
\downarrow{\phi} & & \downarrow{\exists!} & & \downarrow{\Psi, =} \\
\mathbb{Z}_p[\mathbf{a}_1, \cdots, \mathbf{a}_n, \mathbf{b}_1, \cdots, \mathbf{b}_{n-w}] & \xrightarrow[\Phi_M, \cong]{} & \mathrm{Im}(\Phi_M) & \xleftarrow{i_M} & \mathcal{H}_M^t
\end{array}
$$

The same diagram chase from the proof of (4) tells us that the next pair of query answers (h^e, h^m) only fails to satisfy the inductive hypothesis if h^m lifts to $f_m \in \mathbb{Z}_p[\mathbf{c}]$ by the inductive hypothesis, but $f_m \neq \Phi_E^{-1}(h^e) =: f_e$, so $f_m - f_e \in \ker \phi$ and $f_m - f_e$ is nontrivial. Necessary but not sufficient conditions for $f_m - f_e$ to be in the kernel of ϕ are:

1. $f_m - f_e$ must have a zero coefficient in front of any $\mathbf{c}_k$ that is defined under the σ map, since each free variable $\mathbf{b}_j$ has a unique preimage.
2. $f_m - f_e$ must have at least $n - 1$ nonzero coefficients

As with the proof of (4), we analyze the sequential variant where the adversary plays a sequence of queries to $\mathcal{G}_E$ and then plays the exact same sequence of queries to $\mathcal{G}_M$. After Q queries the adversary has at most $Q + 2n$ symbolic polynomials in $\mathbb{Z}_p[\mathbf{c}]$. For each pair of polynomials f_m, f_e in this set, the variables $\mathbf{c}_k$ are mapped by the initial random sampling of the wildcard slots by Oracle M.

Now suppose the adversary fixes a polynomial containing $n - 1$ nonzero coefficients of the $\mathbf{c}_k$'s such that m columns in the original table of $2n$ entries have nonzero coefficients for both entries in the column. This means that the oracle must necessarily choose those m columns to be wildcard slots, since otherwise one of the two entries in the column will not be in the kernel of the ϕ map.

This means that the probability over the initialization of the oracle that these m columns are all chosen to be wildcard slots is $\frac{\binom{n-m}{w-m}}{\binom{n}{w}}$. The remaining $n - 1 - 2m$ columns each must either match the entry chosen by the adversary or be a wildcard slot. There are $(n - 1 - 2m) - (w - m) = (n - 1 - w) - m$ slots that cannot be wildcard slots and thus have at most probability $1/2$ each of matching the entry chosen by the adversary. Thus the probability that this polynomial is in the kernel of ϕ is

$$\frac{\binom{n-m}{w-m}}{\binom{n}{w}} \left(\frac{1}{2}\right)^{n-1-w-m} \tag{1}$$

An upper bound for this can be computed by maximizing the expression with respect to the adversary's choice of m. If we increment m by 1, the first factor is multiplied by $\frac{w-m}{n-m}$ while the second factor is multiplied by 2. Note that $\frac{w-m}{n-m}$ is monotonically decreasing in m; thus, this quantity is maximized when m is the largest possible integer such that $\frac{w-m}{n-m} > 1/2$ is still true. Note that when $w < n/2$, then the optimal choice is $m = 0$. Assuming $w > n/2$ and solving for this inequality we obtain that $m = 2w - n$. Now the problem also has a physical constraint that $m \leq n/2$ since the adversary can choose at most $n/2$ slots. Thus there are three parameter regimes based on α:

1. $\alpha \leq n/2$: the optimal choice is $m = 0$
2. $n/2 \leq \alpha \leq 3n/4$: the optimal choice is $m = 2w - n$
3. $n > 3n/4$: the optimal choice is $m = n/2$

In case 1, the probability is then clearly bounded by $(1/2)^{n-1-w}$.

In case 2, making the substitution $m = 2w - n$ and $w = \alpha n$ where $\alpha \in [0, 1)$ in the expression (1), we obtain

$$\frac{\binom{2(1-\alpha)n}{(1-\alpha)n}}{\binom{n}{\alpha n}} 2^{(3\alpha-2)n} = \frac{[2(1-\alpha)n]!}{[(1-\alpha)n]![(1-\alpha)n]!} \frac{[\alpha n]![(1-\alpha)n]!}{n!} 2^{(3\alpha-2)n}$$

$$= \frac{[2(1-\alpha)n]![\alpha n]!}{[(1-\alpha)n]!n!} 2^{(3\alpha-2)n}$$

Recall that for all integers k the following is true by Sterling's formula:

$$\sqrt{2\pi}\sqrt{k}\left(\frac{k}{e}\right)^k \leq k! \leq e\sqrt{k}\left(\frac{k}{e}\right)^k$$

We can absorb the factors of $\sqrt{2\pi}$ and e in front into a small constant term less than 2. Note that since each factorial is a constant multiple of n, then the $\sqrt{k}$ term also yields a constant term, so we only need to compute the $(k/e)^k$ terms. This gives

$$\frac{[2(1-\alpha)n/e]^{2(1-\alpha)n}[\alpha n/e]^{\alpha n}}{[(1-\alpha)n/e]^{(1-\alpha)n}[n/e]^n} 2^{(3\alpha-2)n} = \left(\frac{[2(1-\alpha)n/e]^{2(1-\alpha)}[\alpha n/e]^{\alpha}}{[(1-\alpha)n/e]^{(1-\alpha)}[n/e]^1} 2^{(3\alpha-2)}\right)^n$$

We just need to show that the base is a constant bounded away from 1. Collecting terms in this, we obtain

$$2^{2(1-\alpha)}[1-\alpha]^{2(1-\alpha)}[n/e]^{2(1-\alpha)}\alpha^\alpha[n/e]^\alpha[1-\alpha]^{-(1-\alpha)}[n/e]^{-(1-\alpha)}[n/e]^{-1}2^{3\alpha-2}$$

$$= [n/e]^{2(1-\alpha)n-(1-\alpha)+\alpha-1}[1-\alpha]^{2(1-\alpha)-(1-\alpha)}\alpha^\alpha 2^{2(1-\alpha)+3\alpha-2}$$

$$= (1-\alpha)^{1-\alpha}\alpha^\alpha 2^\alpha$$

Taking $\log_2$ we obtain $(1-\alpha)\log_2(1-\alpha)+\alpha\log_2\alpha+\alpha \leq -0.0613$ when $\alpha \leq 3/4$, so the probability of success is bounded by $\frac{2}{2^{0.0613n}}$.

Finally in case 3, substituting $m = n/2$ in the expression (1) gives

$$\frac{\binom{n/2}{(\alpha-1/2)n}}{\binom{n}{\alpha n}} 2^{(\alpha-1/2)n} = \frac{[n/2]!}{[(1-\alpha)n]![(\alpha-1/2)n]!} \frac{[\alpha n]![(1-\alpha)n]!}{n!} 2^{(\alpha-1/2)n}$$

$$= \frac{[n/2]![\alpha n]!}{n![(\alpha-1/2)n]!} 2^{(\alpha-1/2)n}$$

Applying the Sterling approximation, we obtain

$$\frac{[n/e]^{n/2}2^{-n/2}[\alpha n/e]^{\alpha n}}{[n/e]^n[(\alpha-1/2)n/e]^{(\alpha-1/2)n}}2^{(\alpha-1/2)n} = \left(\frac{[n/e]^{1/2}2^{-1/2}[\alpha n/e]^{\alpha}}{[n/e]^1[(\alpha-1/2)n/e]^{(\alpha-1/2)}}2^{(\alpha-1/2)}\right)^n$$

The base of the exponent is

$$[n/e]^{1/2+\alpha-1-(\alpha-1/2)}[\alpha]^{\alpha}[\alpha-1/2]^{(1/2-\alpha)}2^{\alpha-1} = \alpha^{\alpha}(\alpha-1/2)^{1/2-\alpha}2^{\alpha-1}$$

Again taking $\log_2$ we obtain the condition $(1/2-\alpha)\log_2(\alpha-1/2)+\alpha\log\alpha+\alpha-1 < 0$, which is satisfied when $\alpha < 0.774$. This does not give much of an improvement over the previous constraint of $\alpha \le 3/4$, so we state our final result just in that regime.

Apply a union bound of this probability over all $(Q+2n)^2$ pairs of symbolic polynomials to get the statement in the theorem.

Lemma 7. *For an adversary $\mathcal{A}$ in the generic group model which makes Q queries to the generic group oracle,*

$$\left|\Pr_{\substack{C\leftarrow\mathcal{D}_n,\\ \mathcal{G}_M,\mathcal{O},\mathcal{A}}}[\mathcal{A}^{\mathcal{G}_M}(\mathcal{O}^{\mathcal{G}_M}(C,1^n)) = P(C)] - \Pr_{\substack{C\leftarrow\mathcal{D}_n,\\ \mathcal{G}_E\mathcal{O},\mathcal{A}}}[\mathcal{A}^{\mathcal{G}_E}(\mathcal{O}^{\mathcal{G}_E}(C,1^n)) = P(C)]\right| \le \frac{1}{2^{0.0613n}}$$

Proof. Uses Lemmas 3 (recalling that the statement also holds for the pair $\mathcal{G}_M,\mathcal{G}_E$) and 6 plugged into the reduction from Lemma 2.

From Lemma 3 (recalling that the statement also holds for the pair $\mathcal{G}_M,\mathcal{G}_S$) we have that:

$$\Pr[\mathcal{A}^{\mathcal{G}_M}(\mathcal{O}^{\mathcal{G}_M}) = 1] = \Pr[\mathcal{A}^{\mathcal{G}_E}(\mathcal{O}^{\mathcal{G}_E}) = 1]$$

as long as all queries to the generic group oracles are *identical* as defined in Definition 5.

Lemma 4 tells us that the probabilities of all queries not being identical during the simultaneous oracle game between $(\mathcal{G}_M,\mathcal{G}_E)$ is at most $\frac{2}{2^{0.0613n}}$.

Therefore, the difference $\Pr[\mathcal{A}^{\mathcal{G}_E}(\mathcal{O}^{\mathcal{G}_E}) = 1] - \Pr[\mathcal{A}^{\mathcal{G}_M}(\mathcal{O}^{\mathcal{G}_M}) = 1]$ is at most $\frac{2}{2^{0.0613n}}$, and so an adversary's advantage in the simultaneous oracle game between $(\mathcal{G}_M,\mathcal{G}_E)$ and $(\mathcal{G}_M,\mathcal{G}_M)$ is:

$$\Pr[\mathcal{A}^{\mathcal{G}_*}(\mathcal{O}^{\mathcal{G}_*}) = \mathcal{G}_*] = \Pr[\mathcal{G}_* = \mathcal{G}_E]\Pr[\mathcal{A}^{\mathcal{G}_E}(\mathcal{O}^{\mathcal{G}_E}) = 1]$$
$$+ \Pr[\mathcal{G}_* = \mathcal{G}_M]\Pr[\mathcal{A}^{\mathcal{G}_M}(\mathcal{O}^{\mathcal{G}_M}) = 0]$$
$$= \frac{1}{2} + \frac{1}{2}(\Pr[\mathcal{A}^{\mathcal{G}_E}(\mathcal{O}^{\mathcal{G}_E}) = 1] - \Pr[\mathcal{A}^{\mathcal{G}_M}(\mathcal{O}^{\mathcal{G}_M}) = 1])$$
$$\le \frac{1}{2} + \frac{2}{2^{0.0613n}}$$

This, plugged into the reduction from Lemma 2, tells us that for all adversaries:

$$\left|\Pr[\mathcal{A}^{\mathcal{G}_M}(\mathcal{O}^{\mathcal{G}_M}) = 1] - \Pr[\mathcal{A}^{\mathcal{G}_E}(\mathcal{O}^{\mathcal{G}_E}) = 1]\right| \le \frac{1}{2^{0.0613n}}$$

Theorem 1. *The obfuscator for pattern matching with wildcards defined in Sect. 3 satisfies distributional VBB security for the ensemble of uniform distributions over $\{0,1\}^n$.*

Proof. For any adversary $\mathcal{A}$ in the Distributional VBB game (in the generic group model), consider the following Simulator $\mathcal{S}$ which simply runs $\mathcal{A}$ on input produced by and interacted with like in Oracle *End* and outputs the same. Note that none of the behavior in Oracle *End* is dependent on the actual function $f_{y,W}$ obfuscated. Therefore a simulator with no access to the function $f_{y,W}$ drawn from the distribution is able to simulate $\mathcal{A}$ as described.

$\mathcal{S}$ then perfectly simulates the behavior of $\mathcal{A}$ interacting with oracle $\mathcal{O}_E$:

$$\Pr_{C \leftarrow \mathcal{D}_n, \mathcal{S}}[\mathcal{S}^C(1^{|C|}, 1^n) = P(C)] = \Pr_{C \leftarrow \mathcal{D}_n, \mathcal{G}_E, \mathcal{O}, \mathcal{A}}[\mathcal{A}^{\mathcal{G}_E}(\mathcal{O}^{\mathcal{G}_E}(C, 1^n)) = P(C)]$$

From Lemma 7, we have that the difference in output probabilities between $\mathcal{A}^{\mathcal{G}_E}(\mathcal{O}^{\mathcal{G}_E})$ and $\mathcal{A}^{\mathcal{G}_M}(\mathcal{O}^{\mathcal{G}_M})$ in the distributional VBB game in the generic group model is at most $\frac{1}{2^{0.0613n}}$:

$$\left| \Pr_{\substack{C \leftarrow \mathcal{D}_n, \\ \mathcal{G}_E, \mathcal{O}, \mathcal{A}}}[\mathcal{A}^{\mathcal{G}_E}(\mathcal{O}^{\mathcal{G}_E}(C, 1^n)) = P(C)] - \Pr_{\substack{C \leftarrow \mathcal{D}_n, \\ \mathcal{G}_M, \mathcal{O}, \mathcal{A}}}[\mathcal{A}^{\mathcal{G}_M}(\mathcal{O}^{\mathcal{G}_M}(C, 1^n)) = P(C)] \right| \leq \frac{1}{2^{0.0613n}}$$

From Lemma 5, we have that the difference in output probabilities between $\mathcal{A}^{\mathcal{G}_M}(\mathcal{O}^{\mathcal{G}_M})$ and $\mathcal{A}^{\mathcal{G}_S}(\mathcal{O}^{\mathcal{G}_S})$ in the distributional VBB game in the generic group model is at most $\frac{(Q + 2n)^2}{2^n}$:

$$\left| \Pr_{\substack{C \leftarrow \mathcal{D}_n, \\ \mathcal{G}_M, \mathcal{O}, \mathcal{A}}}[\mathcal{A}^{\mathcal{G}_M}(\mathcal{O}^{\mathcal{G}_M}(C, 1^n)) = P(C)] - \Pr_{\substack{C \leftarrow \mathcal{D}_n, \\ \mathcal{G}_S, \mathcal{O}, \mathcal{A}}}[\mathcal{A}^{\mathcal{G}_S}(\mathcal{O}^{\mathcal{G}_S}(C, 1^n)) = P(C)] \right| \leq \frac{(Q + 2n)^2}{2^n}$$

Now, recall that $\mathcal{G}_S$ faithfully instantiates $\mathcal{O}$ in the generic group model. Therefore, using the triangle inquality we have:

$$\left| \Pr_{\substack{C \leftarrow \mathcal{D}_n, \\ \mathcal{G}, \mathcal{O}, \mathcal{A}}}[\mathcal{A}^{\mathcal{G}}(\mathcal{O}^{\mathcal{G}}(C, 1^n)) = P(C)] - \Pr_{\substack{C \leftarrow \mathcal{D}_n, \\ \mathcal{S}}}[\mathcal{S}^C(1^{|C|}, 1^n) = P(C)] \right| \leq \frac{(Q + 2n)^2}{2^n} + \frac{1}{2^{0.0613n}}$$

which is a negligible function of n since the number of an adversary's generic group queries Q is a polynomial function of n, and so $\mathcal{O}$ satisfies distributional VBB security in the generic group model.

Acknowledgments. The second, third, fourth, and fifth authors are supported in part by the Defense Advanced Research Project Agency (DARPA) and Army Research Office (ARO) under Contract W911NF-15-C-0236.

The second and third authors are supported in part by NSF grants CNS-1445424 and CCF1423306, and the Leona M. & Harry B. Helmsley Charitable Trust.

The fourth and fifth authors are supported in part by NSF grants CNS-1633282, 1562888, 1565208, and DARPA SafeWare W911NF-16-1-0389.

The first and second authors are supported in part by NSF grant CNS-1552932.

The second author is supported in part by an NSF Graduate Research Fellowship DGE-16-44869.

Any opinions, findings and conclusions or recommendations expressed are those of the authors and do not necessarily reflect the views of the Defense Advanced Research Projects Agency, Army Research Office, the National Science Foundation, or the U.S. Government.

The authors wish to thank Cong Zhang for discussions in preliminary stages of this work.

References

1. Ananth, P., Jain, A., Sahai, A.: Patchable indistinguishability obfuscation: $i\mathcal{O}$ for evolving software. In: Coron, J.-S., Nielsen, J.B. (eds.) EUROCRYPT 2017, Part III. LNCS, vol. 10212, pp. 127–155. Springer, Cham (2017). https://doi.org/10.1007/978-3-319-56617-7_5

2. Ananth, P., Sahai, A.: Projective arithmetic functional encryption and indistinguishability obfuscation from degree-5 multilinear maps. In: Coron, J.-S., Nielsen, J.B. (eds.) EUROCRYPT 2017, Part I. LNCS, vol. 10210, pp. 152–181. Springer, Cham (2017). https://doi.org/10.1007/978-3-319-56620-7_6

3. Apon, D., Döttling, N., Garg, S., Mukherjee, P.: Cryptanalysis of indistinguishability obfuscations of circuits over GGH13. In: 44th International Colloquium on Automata, Languages, and Programming, ICALP 2017, Warsaw, Poland, 10–14 July 2017, pp. 38:1–38:16 (2017)

4. Applebaum, B., Brakerski, Z.: Obfuscating circuits via composite-order graded encoding. In: Dodis, Y., Nielsen, J.B. (eds.) TCC 2015, Part II. LNCS, vol. 9015, pp. 528–556. Springer, Heidelberg (2015). https://doi.org/10.1007/978-3-662-46497-7_21

5. Badrinarayanan, S., Miles, E., Sahai, A., Zhandry, M.: Post-zeroizing obfuscation: new mathematical tools, and the case of evasive circuits. In: Fischlin, M., Coron, J.-S. (eds.) EUROCRYPT 2016, Part II. LNCS, vol. 9666, pp. 764–791. Springer, Heidelberg (2016). https://doi.org/10.1007/978-3-662-49896-5_27

6. Barak, B., Garg, S., Kalai, Y.T., Paneth, O., Sahai, A.: Protecting obfuscation against algebraic attacks. In: Nguyen, P.Q., Oswald, E. (eds.) EUROCRYPT 2014. LNCS, vol. 8441, pp. 221–238. Springer, Heidelberg (2014). https://doi.org/10.1007/978-3-642-55220-5_13

7. Barak, B., et al.: On the (im)possibility of obfuscating programs. In: Kilian, J. (ed.) CRYPTO 2001. LNCS, vol. 2139, pp. 1–18. Springer, Heidelberg (2001). https://doi.org/10.1007/3-540-44647-8_1

8. Boneh, D., Boyen, X., Goh, E.-J.: Hierarchical identity based encryption with constant size ciphertext. In: Cramer, R. (ed.) EUROCRYPT 2005. LNCS, vol. 3494, pp. 440–456. Springer, Heidelberg (2005). https://doi.org/10.1007/11426639_26

9. Brakerski, Z., Rothblum, G.N.: Obfuscating conjunctions. In: Canetti, R., Garay, J.A. (eds.) CRYPTO 2013, Part II. LNCS, vol. 8043, pp. 416–434. Springer, Heidelberg (2013). https://doi.org/10.1007/978-3-642-40084-1_24

10. Brakerski, Z., Vaikuntanathan, V., Wee, H., Wichs, D.: Obfuscating conjunctions under entropic ring LWE. In: Proceedings of the 2016 ACM Conference on Innovations in Theoretical Computer Science, Cambridge, MA, USA, 14–16 January 2016, pp. 147–156 (2016)

11. Brands, S.: Untraceable off-line cash in wallet with observers. In: Stinson, D.R. (ed.) CRYPTO 1993. LNCS, vol. 773, pp. 302–318. Springer, Heidelberg (1994). https://doi.org/10.1007/3-540-48329-2_26

12. Canetti, R., Rothblum, G.N., Varia, M.: Obfuscation of hyperplane membership. In: Micciancio, D. (ed.) TCC 2010. LNCS, vol. 5978, pp. 72–89. Springer, Heidelberg (2010). https://doi.org/10.1007/978-3-642-11799-2_5
13. Chen, Y., Gentry, C., Halevi, S.: Cryptanalyses of candidate branching program obfuscators. In: Coron, J.-S., Nielsen, J.B. (eds.) EUROCRYPT 2017, Part III. LNCS, vol. 10212, pp. 278–307. Springer, Cham (2017). https://doi.org/10.1007/978-3-319-56617-7_10
14. Cheon, J.H., Han, K., Lee, C., Ryu, H., Stehlé, D.: Cryptanalysis of the multilinear map over the integers. In: Oswald, E., Fischlin, M. (eds.) EUROCRYPT 2015, Part I. LNCS, vol. 9056, pp. 3–12. Springer, Heidelberg (2015). https://doi.org/10.1007/978-3-662-46800-5_1
15. Coron, J.-S., et al.: Zeroizing without low-level zeroes: new MMAP attacks and their limitations. In: Gennaro, R., Robshaw, M. (eds.) CRYPTO 2015, Part I. LNCS, vol. 9215, pp. 247–266. Springer, Heidelberg (2015). https://doi.org/10.1007/978-3-662-47989-6_12
16. Coron, J.-S., Lee, M.S., Lepoint, T., Tibouchi, M.: Zeroizing attacks on indistinguishability obfuscation over CLT13. In: Fehr, S. (ed.) PKC 2017, Part I. LNCS, vol. 10174, pp. 41–58. Springer, Heidelberg (2017). https://doi.org/10.1007/978-3-662-54365-8_3
17. Garg, S., Gentry, C., Halevi, S.: Candidate multilinear maps from ideal lattices. In: Johansson, T., Nguyen, P.Q. (eds.) EUROCRYPT 2013. LNCS, vol. 7881, pp. 1–17. Springer, Heidelberg (2013). https://doi.org/10.1007/978-3-642-38348-9_1
18. Garg, S., Gentry, C., Halevi, S., Raykova, M., Sahai, A., Waters, B.: Candidate indistinguishability obfuscation and functional encryption for all circuits. In: FOCS (2013)
19. Garg, S., Miles, E., Mukherjee, P., Sahai, A., Srinivasan, A., Zhandry, M.: Secure obfuscation in a weak multilinear map model. In: Hirt, M., Smith, A. (eds.) TCC 2016. LNCS, vol. 9986, pp. 241–268. Springer, Heidelberg (2016). https://doi.org/10.1007/978-3-662-53644-5_10
20. Gentry, C., Lewko, A.B., Sahai, A., Waters, B.: Indistinguishability obfuscation from the multilinear subgroup elimination assumption. In: IEEE 56th Annual Symposium on Foundations of Computer Science, FOCS 2015, Berkeley, CA, USA, 17–20 October 2015, pp. 151–170 (2015)
21. Goyal, R., Koppula, V., Waters, B.: Lockable obfuscation. In: 58th IEEE Annual Symposium on Foundations of Computer Science, FOCS 2017, Berkeley, CA, USA, 15–17 October 2017, pp. 612–621 (2017)
22. Hofheinz, D., Jager, T., Khurana, D., Sahai, A., Waters, B., Zhandry, M.: How to generate and use universal samplers. In: Cheon, J.H., Takagi, T. (eds.) ASIACRYPT 2016, Part II. LNCS, vol. 10032, pp. 715–744. Springer, Heidelberg (2016). https://doi.org/10.1007/978-3-662-53890-6_24
23. Lin, H.: Indistinguishability obfuscation from constant-degree graded encoding schemes. In: Fischlin, M., Coron, J.-S. (eds.) EUROCRYPT 2016, Part I. LNCS, vol. 9665, pp. 28–57. Springer, Heidelberg (2016). https://doi.org/10.1007/978-3-662-49890-3_2
24. Lin, H.: Indistinguishability obfuscation from SXDH on 5-linear maps and locality-5 PRGs. In: Katz, J., Shacham, H. (eds.) CRYPTO 2017, Part I. LNCS, vol. 10401, pp. 599–629. Springer, Cham (2017). https://doi.org/10.1007/978-3-319-63688-7_20

25. Lin, H., Tessaro, S.: Indistinguishability obfuscation from trilinear maps and block-wise local PRGs. In: Katz, J., Shacham, H. (eds.) CRYPTO 2017, Part I. LNCS, vol. 10401, pp. 630–660. Springer, Cham (2017). https://doi.org/10.1007/978-3-319-63688-7_21

26. Lin, H., Vaikuntanathan, V.: Indistinguishability obfuscation from DDH-like assumptions on constant-degree graded encodings. In: IEEE 57th Annual Symposium on Foundations of Computer Science, FOCS 2016, Hyatt Regency, New Brunswick, New Jersey, USA, 9–11 October 2016, pp. 11–20 (2016)

27. Lynn, B., Prabhakaran, M., Sahai, A.: Positive results and techniques for obfuscation. In: Cachin, C., Camenisch, J.L. (eds.) EUROCRYPT 2004. LNCS, vol. 3027, pp. 20–39. Springer, Heidelberg (2004). https://doi.org/10.1007/978-3-540-24676-3_2

28. Miles, E., Sahai, A., Zhandry, M.: Annihilation attacks for multilinear maps: cryptanalysis of indistinguishability obfuscation over GGH13. In: Robshaw, M., Katz, J. (eds.) CRYPTO 2016, Part II. LNCS, vol. 9815, pp. 629–658. Springer, Heidelberg (2016). https://doi.org/10.1007/978-3-662-53008-5_22

29. Peikert, C.: On error correction in the exponent. In: Halevi, S., Rabin, T. (eds.) TCC 2006. LNCS, vol. 3876, pp. 167–183. Springer, Heidelberg (2006). https://doi.org/10.1007/11681878_9

30. Sahai, A., Waters, B.: How to use indistinguishability obfuscation: deniable encryption, and more. In: STOC (2014)

31. Wee, H.: On obfuscating point functions. In: Proceedings of the 37th Annual ACM Symposium on Theory of Computing, Baltimore, MD, USA, 22–24 May 2005, pp. 523–532 (2005)

32. Wichs, D., Zirdelis, G.: Obfuscating compute-and-compare programs under LWE. IACR Cryptology ePrint Archive 2017:276 (2017)

33. Zimmerman, J.: How to obfuscate programs directly. In: Oswald, E., Fischlin, M. (eds.) EUROCRYPT 2015, Part II. LNCS, vol. 9057, pp. 439–467. Springer, Heidelberg (2015). https://doi.org/10.1007/978-3-662-46803-6_15

On the Complexity of Compressing Obfuscation

Gilad Asharov[1], Naomi Ephraim[2(✉)], Ilan Komargodski[1], and Rafael Pass[1]

[1] Cornell Tech, New York, NY 10044, USA
{asharov,komargodski}@cornell.edu, rafael@cs.cornell.edu
[2] Cornell University, Ithaca, NY 14853, USA
nephraim@cs.cornell.edu

Abstract. Indistinguishability obfuscation has become one of the most exciting cryptographic primitives due to its far reaching applications in cryptography and other fields. However, to date, obtaining a plausibly secure construction has been an illusive task, thus motivating the study of seemingly weaker primitives that imply it, with the possibility that they will be easier to construct.

In this work, we provide a systematic study of compressing obfuscation, one of the most natural and simple to describe primitives that is known to imply indistinguishability obfuscation when combined with other standard assumptions. A compressing obfuscator is roughly an indistinguishability obfuscator that outputs just a slightly compressed encoding of the truth table. This generalizes notions introduced by Lin et al. (PKC 2016) and Bitansky et al. (TCC 2016) by allowing for a broader regime of parameters.

We view compressing obfuscation as an independent cryptographic primitive and show various positive and negative results concerning its power and plausibility of existence, demonstrating significant differences from full-fledged indistinguishability obfuscation.

First, we show that as a cryptographic building block, compressing obfuscation is weak. In particular, when combined with one-way functions, it cannot be used (in a black-box way) to achieve public-key encryption, even under (sub-)exponential security assumptions. This is in sharp contrast to indistinguishability obfuscation, which together with one-way functions implies almost all cryptographic primitives.

Second, we show that to construct compressing obfuscation with perfect correctness, one only needs to assume its existence with a very weak correctness guarantee and polynomial hardness. Namely, we show a correctness amplification transformation with optimal parameters that

G. Asharov—Supported by a Junior Fellow award from the Simons Foundation.
N. Ephraim—Supported by an AFOSR grant FA9550-15-1-0262.
I. Komargodski—Supported in part by a Packard Foundation Fellowship and by an AFOSR grant FA9550-15-1-0262.
R. Pass—Supported in part by NSF Award CNS-1561209, NSF Award CNS-1217821, NSF Award CNS-1704788, AFOSR Award FA9550-15-1-0262, a Microsoft Faculty Fellowship, and a Google Faculty Research Award.

H. Shacham and A. Boldyreva (Eds.): CRYPTO 2018, LNCS 10993, pp. 753–783, 2018.
https://doi.org/10.1007/978-3-319-96878-0_26

relies only on polynomial hardness assumptions. This implies a universal construction assuming only polynomially secure compressing obfuscation with approximate correctness. In the context of indistinguishability obfuscation, we know how to achieve such a result only under sub-exponential security assumptions together with derandomization assumptions.

Lastly, we characterize the existence of compressing obfuscation with *statistical* security. We show that in some range of parameters and for some classes of circuits such an obfuscator *exists*, whereas it is unlikely to exist with better parameters or for larger classes of circuits. These positive and negative results reveal a deep connection between compressing obfuscation and various concepts in complexity theory and learning theory.

1   Introduction

Program obfuscation is an intriguing and powerful concept in modern cryptography. A program obfuscator is a compiler that "scrambles" programs into ones that are hard to reverse engineer, while preserving their functionality. The predominant notion that captures the above concept is *indistinguishability obfuscation*, introduced in the seminal work of Barak et al. [14], which has inspired a vibrant area of research in recent years. Informally, indistinguishability obfuscation (iO) guarantees that the obfuscations of two functionally equivalent circuits of the same size are computationally indistinguishable.

There are two main reasons why iO has become such a central primitive— its potential to exist and its power. As opposed to stronger notions of obfuscation that are known not to exist for all circuits (such as *virtual black-box* obfuscation [14]), general purpose iO might be realizable, and in fact, since the work of Garg et al. [38] many candidate constructions of iO have emerged [5,8,13,27,38,42,44,68,73]. As for its power, iO serves as a hub for an impressive number of cryptographic primitives, ranging from classical concepts such as one-way functions [53], public-key encryption [70], trapdoor permutations [19], ZAPs and non-interactive witness-indistinguishable proofs [18], to ones that are still far beyond the reach of any other assumption, such as deniable encryption [70], fully-secure multi-input functional encryption [45], and many others.

Despite immense efforts to construct iO from concrete assumptions, all currently known candidate constructions have been shown to be vulnerable to attacks [7,12,23,32,33,43,62,66].[1] Another line of work shows how to construct iO from some seemingly "simpler" or "weaker" generic cryptographic primitives (together with more standard assumptions). These include primitives such as low-degree multilinear maps [4,55,56,59], compact functional encryption schemes [3,20], compact randomized encodings [58], and variants of

---

[1] Some of the attacks apply directly to the candidate construction while some only apply to the underlying graded encoding scheme [34,35,42]. See Ananth et al. [1, Appendix A] for an overview.

exponentially-efficient indistinguishability obfuscation [17,57], all of which have no known instantiations from standard assumptions.

The difficulty of constructing iO motivates the study of such seemingly weaker cryptographic primitives, with the hope that such a study could elucidate the foundations of iO. In this paper, we focus on the primitive which is arguably the simplest to define and the closest in its nature to iO: indistinguishability obfuscation with nontrivial compression, or in short, *compressing obfuscation*.

Compressing obfuscation. For functions $t(s, n)$ and $\ell(s, n)$, we say that an obfuscator $\mathcal{O}$ is (t, ℓ)-compressing if, when given a circuit C of size s on n inputs, the obfuscator $\mathcal{O}(C)$ runs in time $t(s, n)$ and has output length $\ell(s, n)$. In the case of iO, both t and ℓ are polynomial in s and n, but in general, we allow them to be super-polynomial, or even (sub-)exponential. This definition generalizes existing relaxations of iO (such as XiO and SXiO which we discuss below) and allows us to characterize the extent to which efficiency impacts the existence, applications, and limitations of obfuscation. Throughout this work, we mostly focus on the following two settings of parameters, which intuitively, are relaxed versions of iO that only allow obfuscating circuits with logarithmic input size:

- **XiO.** The first (and weaker) setting of parameters is that of *exponentially-efficient iO* (XiO), introduced by Lin et al. [57]. XiO allows the running time of the obfuscator to be as large as the truth table of the circuit to be obfuscated, but requires the size of the obfuscated circuit to be slightly smaller than its truth table. More formally, for a function c (which denotes the compression of XiO), we say that c-XiO is a (t, ℓ)-compressing obfuscator with $t(s, n) = \mathsf{poly}(2^n, s)$ and $\ell(s, n) = c(n) \cdot \mathsf{poly}(s)$. When there exists a constant $\epsilon > 0$ such that $c(n) = 2^{n(1-\epsilon)}$, we denote c-XiO simply by XiO. Lin et al. [57] showed that XiO for all circuits and Learning With Errors (LWE), both with sub-exponential security, imply iO.

- **SXiO.** The second (and stronger) setting of parameters is that of *strong XiO* (SXiO), introduced by Bitansky et al. [17]. SXiO requires that the time to obfuscate a circuit is slightly smaller than the truth table of the circuit. More formally, for a function c, we say that c-SXiO is a (t, ℓ)-compressing obfuscator with $t(s, n) = \ell(s, n) = c(n) \cdot \mathsf{poly}(s)$. Similar to the above case, when there exists some constant $\epsilon > 0$ such that $c(n) = 2^{n(1-\epsilon)}$, we denote this simply by SXiO. Bitansky et al. [17] showed that SXiO and any public-key encryption, both with sub-exponential security, imply iO.

These two settings of parameters have seemingly minor differences, but nevertheless, are not known to be equivalent. Moreover, as mentioned above, their known implications illustrate the richness of the world of compressing obfuscation, and indicate that efficiency is a fundamental property of obfuscation. Since the regime of parameters for compressing obfuscation is somewhat non-standard (especially, the distinction between time and output length in XiO), it has not received adequate attention, and as a result we know very little about it.

In this work, we provide a systematic study of compressing obfuscation as an independent cryptographic primitive, and thus characterize the extent to which efficiency plays a role in obfuscation.

1.1 Our Results

Our results span a wide range of topics concerning compressing obfuscation, including limitations of its power, existence in an information-theoretic setting, constructions for limited classes of functions, and correctness amplification.

XiO vs. PKE. We start by exploring the power of XiO as an independent cryptographic primitive. One the one hand, we know that when combined with LWE it implies full-fledged iO (which in turn implies almost all cryptographic primitives). On the other hand, as opposed to iO [53], we do not even know whether XiO by itself[2] implies one-way functions — the most basic cryptographic primitive.

One of the original applications of obfuscation, which was proposed by Diffie and Hellman back in 1976 [36], is to transform private-key encryption into public-key encryption. When combined with one-way functions, iO can be used to perform such a transformation, as shown by [38,70]. This raises the same question regarding XiO: Can it bridge the gap between the world of private-key cryptography and that of public-key cryptography? We provide evidence that it cannot, and thus show a concrete lower bound on its potential power.

Theorem 1.1 (informal). *There is no fully black-box construction of a perfectly correct key-agreement protocol from one-way functions and perfectly correct $2^{(1-\epsilon)n}$-XiO for any constant $\epsilon > 0$, even with sub-exponential security.*

The result is obtained by following the black-box framework of [9,10,15], where they consider obfuscation for *oracle-aided* circuits. This captures exactly the flavor of constructions which give public-key encryption from one-way functions and iO [70]. We make various modifications to this framework to capture the notion of XiO for oracle-aided circuits.

Previously, by combining [9,17], the above result follows for the case of $2^{(1-\epsilon)n}$-XiO where $0 < \epsilon \leq 1/2$ (i.e., the obfuscator has only somewhat *weak* compression).[3] In contrast, our separation works even when given access to an obfuscator with very strong compression (i.e. any constant $\epsilon > 0$) and even if the obfuscator satisfies perfect correctness.

The frameworks that this result is based on are rooted in the ideas of Impagliazzo and Rudich [51], who show a separation between one-way permutations and key-agreement. Their result holds both for the case of key-agreement with perfect or imperfect completeness. Nevertheless, we note that our separation does not hold for imperfect key-agreement, and we leave the extension to future work.

---

[2] Assuming any average- or worst-case hardness assumption. This is necessary as XiO exists unconditionally if P = NP.

[3] Indeed, [9] showed a separation of perfect key-agreement from imperfect private-key FE, and [17] showed a black-box construction of $2^{n/2}$-XiO from private-key FE.

Statistical security. Our result that it is unlikely that key-agreement can be constructed from XiO and one-way functions can be viewed as "good news", as it hints that XiO is a somewhat "weak" primitive, and therefore it might be possible to base its existence on well-studied assumptions. In fact, it might even be possible that compressing obfuscation exists unconditionally (even if $P \neq NP$). Toward this end, we show almost matching upper and lower bounds for the existence of compressing obfuscation with statistical security, both for the case of perfect correctness and that of approximate correctness. Our results show tight connections between compressing obfuscation and various concepts in complexity theory and learning and thus we view this as one of the central takeaways of this work.

For the case of approximate correctness, we show a 2^{n^ϵ}-SXiO for $\epsilon > 0$ for small classes of circuits (such as AC^0). On the other hand, we show that such an obfuscator cannot exist for larger classes of circuits that contain a (puncturable) PRF, unless $\overline{SAT} \in AM[2^{n^\epsilon}]$, where $\overline{SAT}$ is the problem of deciding whether a formula is unsatisfiable and $AM[t(n)]$ is the class of all languages on instances of size n that have an AM protocol in which the running time of the verifier and the message sizes are at most $t(n)$.

Theorem 1.2 (informal). *There exists a statistically secure and approximately correct 2^{n^ϵ}-SXiO for AC^0 and $\epsilon > 0$. On the contrary, unless $\overline{SAT} \in AM[2^{n^\epsilon}]$, there is no such obfuscator for any class that contains a (puncturable) PRF.*

This result naturally leads to the question of whether we can get a similar statement for the case of perfect correctness. We are unable to get such a result for SXiO, but we do get it for XiO, albeit with worse compression.[4]

Theorem 1.3 (informal). *There exists a $2^{n(1-\epsilon)}$-XiO for $\epsilon \in 1/\mathsf{poly}\log(n)$ with statistical security and perfect correctness for AC^0.*

Ruling out statistically secure XiO with any compression is left as an open problem. We do show that unless $\overline{SAT} \in AM[2^{c(1-\epsilon)n}]$ for a universal constant $c \in \mathbb{N}$, there is no statistically secure and perfectly correct $2^{n(1-\epsilon)}$-SXiO for AC^0 (see Theorem 5.2). It is known, by the recent result of Williams [72], that $\overline{SAT} \in AM[\tilde{O}(2^{n/2})]$. However, it might be that for larger values of ϵ (such as $\epsilon = 1 - (0.1/c)$ or even $\epsilon = 1 - o(1)$) it holds that $\overline{SAT} \notin AM[2^{c(1-\epsilon)n}]$.

The positive results are based on classical (PAC) learning algorithms [60,71] and the circuit compression algorithm of [31]. Both negative results above rely on and (carefully) extend analogous arguments from the iO literature [24,47,53]. Goldwasser and Rothblum [47] showed that statistical iO with perfect correctness cannot exist unless $NP \subseteq SZK$. Brakerski, Brzuska, and Fleischhacker [24] extend the result to handle statistical iO with *approximate* correctness by showing that (assuming additionally one-way functions) unless $coNP \subseteq AM$, it cannot exist.

---

[4] The obfuscator we get is weak due to two reasons. First, the class for which we obtain XiO does not contain (puncturable) PRFs and thus is not sufficient for known transformations to iO. Second, the compression we achieve is not enough for cryptographic applications.

Correctness amplification. Our results above suggest that approximate correctness might be easier to achieve than perfect correctness, in an information theoretic setting. Is this the case also in the computational setting? To address this question, we show a transformation from approximately correct XiO to perfectly correct XiO, assuming the original XiO applies to a large enough class of circuits. This transformation achieves optimal parameters and only incurs polynomial security loss, indicating that correctness is not the bottleneck in constructing XiO from standard assumptions.

Theorem 1.4 (informal). *If there exists an XiO scheme for all polynomial size circuits which is correct with probability $(1/2 + 1/\mathsf{poly})$ over the the inputs and the obfuscation, then there exists a perfectly correct XiO scheme, assuming polynomially-secure LWE and NIZKs.*

Prior to this result, there were no correctness amplification procedures for XiO which required only polynomial security or achieved optimal parameters. Correctness amplifications for related primitives, such as those of [2,21] for iO, do not apply to XiO, since they involve a random self-reducibility step which inherently requires running the obfuscator on polynomial-size inputs. The transformation of Bitansky et al. [16] shows how to transform an XiO which is correct with probability 0.99 over the inputs and the obfuscation to a weak notion of functional encryption. This notion of functional encryption was known to imply a relaxed notion of XiO, namely, XiO with preprocessing [57]. Our transformation works for a much weaker notion of correctness (as opposed to .99) and results in full-fledged, perfectly correct XiO (as opposed to XiO with preprocessing).

Technically, our regime of parameters introduces many difficulties which require us to tailor a construction that is based on a delicate combination of various types of error-correcting codes together with cryptographic primitives (inspired by [65]).

While we show this transformation for the case of XiO, our result extends naturally to the case of SXiO. In particular, we can obtain perfectly correct XiO from the transformation, or SXiO which is correct on all but a negligible fraction of obfuscations.

Universal construction. Using our correctness amplification procedure, we obtain a universal construction of an XiO (resp. SXiO), assuming only the mere existence of XiO (resp. SXiO) with *polynomial* security and only (very weak) approximate correctness. For XiO, the resulting universal construction satisfies perfect correctness. Note that in the context of iO, perfect correctness is known to be achievable using only derandomization assumptions [22]. Our result is obtained by adapting the robust combiner of Ananth et al. [1] to the setting of XiO (resp. SXiO) and then using our correctness amplification transformation.

1.2   Related Work

Universal construction and robust combiners. It was shown in [48] that, in general, a robust combiner implies the existence of a universal construction. A robust combiner for a cryptographic primitive takes several candidate constructions of the primitive and outputs one construction that is as good as any of the input constructions (see also [49,50]). A combiner for encryption appears already in [11], and perhaps the most known universal construction is that of one-way functions, due to [54].

Combiners for obfuscation were given in [1,2,37]. The work of [1] shows a robust combiner for indistinguishability obfuscation with sub-exponential security loss, and assuming either LWE or DDH. The work of [2] removes the sub-exponential assumption, but does not go all the way to iO—it shows a transforming combiner from candidates for indistinguishability obfuscation of which one of them is polynomially secure to a secure functional encryption scheme.

Existence of iO. Mahmoody et al. [63] showed that iO cannot be based on random oracles or on constant degree multilinear maps (in a black-box way). Garg et al. [40] showed that iO cannot be constructed from any type of encryption that has an "all-or-nothing" type of security (as in PKE or Witness Encryption). Lastly, Garg et al. [41] studied the minimal compactness needed from a functional encryption scheme to imply iO, and giving matching constructions, following [3,20].

Limitations on the power of iO were studied by Asharov and Segev [9,10] and by Bitansky, Degwekar and Vaikuntanathan [15]. So far, we know that iO and one-way functions do not imply collision-resistant hash functions [9], domain-invariant one-way permutations [10], and hardness in NP ∩ coNP [15]. Also, iO and one-way permutations do not imply hardness in SZK [15].

Relaxations of iO. In addition to (S)XiO, another relaxation of iO is *decomposable obfuscation* (dO), which was recently introduced by Liu and Zhandry [61]. Decomposable obfuscation relaxes the security requirement of iO by requiring that obfuscations of circuits which satisfy a new notion of functional equivalence are indistinguishable. In particular, it is efficient to verify if two circuits satisfy their notion of functional equivalence, unlike traditional functional equivalence. This is similar to the case of XiO, because it is applied on circuits with only logarithmic input size for polynomial time applications. In [61], they question whether iO with efficiently verifiable functional equivalence implies public-key encryption. In fact, they have to assume the existence of public-key encryption for all the applications of dO that they show which imply public-key encryption. As mentioned above, we show a separation from XiO and OWFs to public key encryption. Therefore, our result serves as further evidence to the hypothesis that (non) efficiently checkable functional equivalence is one of the key factors which distinguishes iO from notions like XiO and dO.

Compressing primitives. Recently, compressing witness encryption (WE) was studied by Brakerski et al. [25]. Witness encryption, introduced by Garg et al. [39], allows encrypting a message relative to a statement $x \in L$ for a language $L \in \mathsf{NP}$ such that anyone holding a witness to the statement can decrypt the message, but if $x \notin L$, then it is computationally hard to decrypt. A compressing WE is such that the encryption time (and thus size) is less than the time it takes to solve the NP instance. Brakerski et al. showed that such a WE scheme can be constructed under "standard" assumptions (such as LWE or bilinear maps with sub-exponential security). This is in sharp contrast to SXiO (or even XiO).

Paper organization. We proceed with a technical overview of our results. We refer the reader to the full version of the paper for important preliminaries and definitions. In Sect. 3 we show our correctness amplification transformation, and in Sect. 4 we prove our impossibility result on constructing key-agreement from XiO and OWFs. In Sect. 5 we present our positive and negative results regarding statistically secure compressing obfuscation. Most of the technical material is omitted and appears in the full version.

2   Technical Overview

In this section we provide a high level overview of our results. We start with the correctness amplification (and its application to universal constructions) in Sect. 2.1. We proceed with the limitations on the power of XiO in Sect. 2.2, and conclude with our constructions and impossibilities of statistically secure XiO in Sect. 2.3.

2.1   Correctness Amplification

Our correctness amplification for XiO is a transformation from an approximately correct XiO scheme to an XiO scheme that is perfectly correct. Here, by approximately correct, we mean an XiO scheme which is correct with probability $(1/2 + 1/\mathsf{poly})$ over the inputs and the obfuscation, and by perfectly correct, we mean an XiO scheme which is correct on all inputs and all obfuscations with probability 1. The starting point for our correctness amplification is the transformation of Bitansky et al. [16], which transforms an XiO scheme which is correct with probability .99 over the obfuscation and the inputs to a functional encryption (FE) scheme which is correct on all inputs (with all but negligible probability). At a high level, FE is a type of encryption which enables generating functional keys, such that decryption of a ciphertext corresponding to a message m with a functional key for a circuit C results in $C(m)$. The hope is that if we can adapt the [16] transformation to our case, then we can attempt to transform the correct FE back to XiO.

From approximately correct XiO to correct FE. In [16], they first observe that by averaging and standard BPP-type amplification, their XiO scheme can be amplified to one which is correct with probability .9 *only over the inputs*. Then, they transform this XiO to a correct FE using an error-correcting code, as follows. To encrypt a message m, they obfuscate a circuit G_m which, on input i, outputs an encryption of (m, i) using a succinct functional encryption scheme sFE, that exists based on LWE [46]. Call the resulting obfuscated circuit $\widetilde{G}_m$. To generate a secret key for a circuit C, they generate an sFE secret key for a circuit C' that on input (m, i) outputs the ith bit of $\mathsf{ECC}(C(m))$, where ECC is an error-correcting code. To decrypt, they first evaluate the obfuscated circuit $\widetilde{G}_m$ on every input i to obtain a list of encryptions of (m, i) for all i. Then, they use the sFE secret key to decrypt each of these encryptions and finally, decode the result.

The reason why this is enough for [16] is that, first, by the BPP amplification, they obtain correct encryptions of (m, i) for a .9 fraction of i's, with all but negligible probability over the obfuscation. This lets them calculate $(\mathsf{ECC}(C(m))_i$ for a *large* $(\gg 3/4)$ fraction of the i's. Second, they rely on the error-correcting code which, given $(\mathsf{ECC}(C(m))_i$ for *many* $(\gg 3/4)$ i's, can recover $C(m)$.

In our case, a natural attempt would be to replicate their first step and then use an error-correcting code with better parameters for the second step. However, this approach fails: we are only guaranteed correctness with probability $(1/2 + 1/\mathsf{poly}(\lambda))$ over the obfuscation and the inputs, which is not enough for averaging and BPP-type amplification. Nevertheless, the framework of [16] is still a convenient starting point for us.

Our first challenge is to obtain every bit of the encryption of (m, i) for sufficiently many i's. One idea is to apply an error-correcting code to the output of G_m, so that for any index i for which G_m correctly outputs enough of the bits of the encryption of (m, i), we can decode successfully. While this is not possible for our regime of parameters using classical binary error-correcting codes, this is achievable with binary *list-decodable* codes, which output a list of possibilities upon decoding a codeword, rather than a unique decoding. Therefore, we modify the circuit G_m to output a list-decodable encoding of the encryption of (m, i), one bit at a time, which will be decoded at decryption time. This introduces the complication that list-decoding gives many possibilities for the encryption of (m, i) for each i. To address this, we employ a combination of NIZK proofs and commitments which enable us to uniquely decode from the decoded list. At a high level, we impose the requirement that in addition to the ciphertext of (m, i), the circuit G_m on input i must output a NIZK proof certifying that the ciphertext is correct. This ensures that we obtain sFE encryptions of (m, i) for a noticeable fraction of the inputs i. Thus, we have replaced the BPP-type amplification of [16] with list-decodable codes, NIZK proofs, and commitment schemes.

After this change, we have that for a noticeable (but small, say 1%) fraction of the i's, we obtain a correct encryption of (m, i). If we decrypt this with the sFE secret key of [16], we would hope to obtain $(\mathsf{ECC}(C(m)))_i$ for enough

i's such that ECC can successfully decode to $C(m)$, but this does not quite work because we only have a very small fraction of correct encryptions. Indeed, no (binary) error-correcting code can recover from more than 50% error! To overcome this, we notice that we have additional information (thanks to the NIZK) – we know exactly for which i's we obtained correct sFE encryptions of (m, i). Therefore, we replace the error-correcting code in the [16] construction with a code that can recover from a high fraction (say 99%) of *erasures*. To obtain optimal parameters, this requires us to have sFE output alphabet symbols rather than bits, but this does not impact the correctness of the scheme. Combining these two steps, we obtain an FE scheme with amplified correctness. As far as we know, this combination of list-decodable codes and erasure-correcting codes is novel to this work.

These techniques nearly work, with the caveat that our first step only gives us the correct encryptions of enough (m, i) when the obfuscator uses "good" random coins. Nevertheless, this can be remedied by using BPP-type amplification and leveraging the fact that our FE scheme always decrypts to $\bot$ or to the correct output, $C(m)$. Therefore, this results in an FE scheme which is correct for all inputs with all but negligible probability.

From correct FE to correct XiO. The only remaining step is to transform the FE back to XiO. The FE scheme we obtain from the above transformations is *weakly sublinear compact*, a weak notion of compactness which does not suffice for known transformations to XiO without assuming sub-exponential security. FE with weak sublinear compactness has the property that while the encryption time is proportional to the circuit size of circuits supported by the scheme, the ciphertext lengths are compact. We take advantage of this by having an obfuscation consist of many "short" encryptions, which exactly captures the requirement that the obfuscator has a long running time but a nontrivial output length.

To obfuscate a circuit C, we encrypt a circuit C_x for each $x \in \{0,1\}^{n/2}$, where $C_x(\cdot) = C(x\|\cdot)$. Then, we generate a functional key sk for a circuit T, which, given a circuit on $n/2$ bits, outputs its truth table. The ciphertexts and functional key serve as our obfuscation, which gives the desired efficiency for XiO exactly because of the weak compactness of FE. To evaluate the obfuscation on an input $x = x_1\|x_2$, we use FE to decrypt C_{x_1} with sk, and select the element of the truth table corresponding to x_2. This transformation yields a correct and secure XiO scheme, in which for any circuit C and every input x, it holds that the obfuscation of C at the point x agrees with $C(x)$ with all but negligible probability.

In the technical section, we present the full construction in a more streamlined manner. In particular, we compose the XiO to FE transformation with the FE to XiO transformation described above, which yields a transformation from approximately correct XiO to XiO that is correct on any input with all but negligible probability over the randomness of the obfuscator.

Given an XiO which is correct on any input with all but negligible probability, we can then apply another BPP-style transformation (this time we apply parallel repetitions and then take the majority vote) to get an obfuscator that for all but negligible fraction of the obfuscations the obfuscated circuit completely agrees with the input circuit. To conclude our correctness amplification, we observe that the running time for XiO allows the obfuscator to compute the truth table of the circuit it obfuscates. Therefore, we modify the obfuscator to check if an obfuscation $\widetilde{C}$ of a circuit C is correct by running over all inputs. If $\widetilde{C}$ agrees with C, then $\widetilde{C}$ is used as the obfuscation, and if not, we simply output C in the clear. This takes advantage of the running time of XiO, and incurs only a negligible loss in security, thus resulting in a perfectly correct XiO.

A universal construction. An important application of correctness amplification is a universal construction. We show a universal construction for XiO (resp. SXiO) by combining our correctness amplification with the results of [1].

A universal construction for a primitive can be obtained via a *robust combiner* for that primitive, which is a transformation that takes several candidate constructions of the primitive and outputs one construction that is as good as any of the input constructions. It is robust in the sense that it should work even if some of the candidates have weak correctness guarantees, have bad running times, etc. A universal construction is then acquired by enumerating over all possible candidates while making sure not to be "fooled" by bad faulty candidates so that we end up with a correct candidate. Thus, it is guaranteed that the resulting candidate is correct and secure.

We observe that a *combiner* (i.e., a secure candidate assuming one exists) for XiO (resp. SXiO) can be obtained by adapting the construction for iO of Ananth et al. [1] which further relied on LWE. In the case of iO, their construction, on input circuit C, obfuscates a variant of C that has the same input domain as C. In the security proof, they go "input-by-input" over this obfuscated circuit which results in a sub-exponential security loss. We notice that, in the case of XiO (resp. SXiO), the number of inputs in the above obfuscated circuit is at most logarithmic, so the very same proof can be carried out, losing only a polynomial term. Then, to make the combiner robust we use our correctness amplification procedure. This results in a universal construction of perfect XiO (resp. imperfect SXiO), assuming the existence of XiO (resp. SXiO) with very weak correctness.

2.2   Impossibility of Key-Agreement

To illustrate the difference between the power of compressing obfuscation and iO, we revisit one of the primary applications of iO—transforming a private-key scheme into a public-key one. In the context of iO, this transformation is performed by obfuscating the encryption circuit of a private-key encryption scheme, while embedding the symmetric secret key into the circuit. The public key is then simply the obfuscated circuit. In order to encrypt a message m, one has to choose randomness r and run the obfuscated circuit on (m, r) to obtain the ciphertext c. An important property of this construction is the ability to obfuscate circuits with "hardwired cryptography", e.g., the evaluation circuit of a pseudorandom function with a hardwired PRF key.

Since XiO is efficient only when obfuscating circuits with logarithmic size input, one cannot use the above approach with XiO even when the message space is limited to a single bit. Given the public key, the adversary can learn the entire truth table of the obfuscated circuit by enumerating over all inputs, thereby breaking the secrecy of the underlying message. Our proof formalizes this intuition, and shows that other attempts to make such a transformation cannot succeed. We formalize this using a black-box separation, showing that no perfectly complete bit-agreement protocol can be constructed from perfectly correct XiO and one-way functions.

Modeling non-black-box constructions. Constructions that are based on indistinguishability obfuscation are almost always *non-black-box* in the underlying primitives. In the example above, the circuit being obfuscated is the encryption algorithm of a private-key encryption scheme and thus contains a specific circuit representation of the underlying one-way function as a sub-circuit. We follow the framework of Asharov and Segev [9,10] that captures such constructions by enabling the obfuscator to run on *oracle-aided* circuits, i.e., circuits that might contain oracle gates. We refer to [9,10] for details regarding this model (see also [15]), and for examples of how it capture common techniques such as the punctured programming technique of Sahai and Waters [70] and its variants.

The oracle. Our result is obtained by presenting an oracle Γ relative to which the following properties hold: (1) there exists a one-way function f; (2) there exists a perfectly-correct, exponentially-secure XiO scheme for all oracle-aided circuits C^f; (3) for any perfectly complete bit-agreement protocol between two parties, there exists an eavesdropping adversary that makes polynomially many queries to the oracle Γ and succeeds to recover the bit from the transcript of the interaction. Our oracle consists of three functions, similar to that of [10]: (1) a random function f that will serve as the one-way function; (2) a random length-increasing function $\mathcal{O}$ that will serve as the obfuscator (an obfuscation of an oracle-aided circuit C is a "handle" $\widehat{C} = \mathcal{O}(C, r)$ for a random string r), and (3) a function $\mathcal{E}$ that enables evaluations of obfuscated circuits: given some obfuscated circuit $\widehat{C}$ and an input x, the function $\mathcal{E}$ looks for the lexicographically first pair (C, r) for which $\mathcal{O}(C, r) = \widehat{C}$ and returns $C^f(x)$.

The main difference between our oracle and the oracle of [10] is the expansion factor of the oracle $\mathcal{O}$. In order to capture compressing obfuscation, the expansion factor that we use is (sub-)exponential in the input size of the circuit C. While this modification is somewhat minor in syntax, it has a major effect – if the expansion factor is "small" then it is possible to construct a *polynomial time* key-agreement protocol relative to such an oracle (following the construction of Sahai and Waters [70]), whereas for a larger expansion factor this becomes impossible. As for the existence of one-way functions and indistinguishability of obfuscated circuits, we derive these almost for free from [10].

In what follows, we first discuss how to break a perfectly complete key-agreement protocol relative to a random oracle as a warmup. We then discuss the challenges when dealing with our (more structured) oracle, and discuss why our approach does not work for iO.

Separating key-agreement from a random oracle. As a warmup, we first present an overview of the result of Impagliazzo and Rudich [51] and Brakerski et al. [26], who show that for any two polynomial time oracle-aided algorithms $\mathcal{A}$ and $\mathcal{B}$, if $\langle \mathcal{A}^f, \mathcal{B}^f \rangle$ implements a perfectly-correct bit-agreement protocol for all functions f, then there exists an oracle-aided algorithm E such that for any function f learns the agreed bit with probability 1 by making only a polynomial number of oracle queries to f. The adversary E is given a transcript T which is a result of an interaction of $\mathcal{A}$ and $\mathcal{B}$ relative to some oracle f, and is required to find the key $k^\star$ that $\mathcal{A}$ and $\mathcal{B}$ agreed on. Denote by $r_{\mathcal{A}}^\star$ (resp. $r_{\mathcal{B}}^\star$) the randomness used by $\mathcal{A}$ (resp. $\mathcal{B}$) in the real interaction that produced T. The adversary E initializes a set of queries/answers Q, which will contain the actual queries made by E to the true oracle f. It also initializes a multiset $K = \emptyset$, and repeats the following polynomially many times:

- **Simulation:** E simulates an oracle f' that is consistent with Q (i.e., $f'(w) = f(w)$ for every $w \in Q$), and randomness $r'_{\mathcal{A}}, r'_{\mathcal{B}}$ such that the interaction $\langle \mathcal{A}^{f'}(r'_{\mathcal{A}}), \mathcal{B}^{f'}(r'_{\mathcal{B}}) \rangle$ (i.e., running the protocol with respect to the function f' with randomness $r'_{\mathcal{A}}$ for $\mathcal{A}$ and $r'_{\mathcal{B}}$ for $\mathcal{B}$) results in the transcript T and key k'. E adds k' to K.
- **Update:** E asks f for all queries in f' that are not in Q, and updates the set Q.

At the end of the attack, E outputs the majority value in K. The proof then relies on the following observation: In each iteration, either (1) in the update phase, E finds at least one new query that is also made by either $\mathcal{A}$ or $\mathcal{B}$ during the real interaction with the function f that produced the transcript T; or (2) E adds the real key $k^\star$ to K.

Intuitively, if (1) does not hold, then the perfect correctness of the bit-agreement protocol guarantees that (2) holds. In particular, in that case it is possible to construct a "hybrid" oracle $\tilde{f}$ that behaves like f in the real execution of $\mathcal{A}$, i.e., $\mathcal{A}^f(r_{\mathcal{A}}^\star)$, and behaves like f' in the simulated evaluation of $\mathcal{B}$, i.e., $\mathcal{B}^{f'}(r'_{\mathcal{B}})$. According to this hybrid oracle, an execution of $\mathcal{A}$ with randomness $r_{\mathcal{A}}^\star$ and an execution of $\mathcal{B}$ with randomness $r'_{\mathcal{B}}$ would result in the transcript T, $\mathcal{A}$ would output $k^\star$ (as in the real execution) and $\mathcal{B}$ would output k' (as in the simulation). Perfect correctness then tells us that $k^\star = k'$. This hybrid oracle can be constructed since the simulated execution and the real execution have no intersection queries in addition to the queries which are already in Q, and therefore there are no contradicting queries (i.e., queries w that appear in both executions for which $f(w) \neq f'(w)$). As the number of oracle queries $\mathcal{A}$ and $\mathcal{B}$

make during the execution of the protocol is some polynomial q, the majority value in K is guaranteed to be the correct key after $2q + 1$ iterations.

Attacking key-agreement relative to our oracle. We extend the attack described above relative to our oracle Γ, which is a significantly more structured than a random oracle and therefore raises several challenges. Recall that our oracle Γ consists of a three functions f, $\mathcal{O}$, and $\mathcal{E}$, that are dependent. Following the above template, we construct an adversary that simulates an execution that produces the transcript T with some simulated oracle $\Gamma' = (f', \mathcal{O}', \mathcal{E}')$. There are two main challenges with this approach. The first is to show that $\mathcal{A}$ and $\mathcal{B}$ cannot gain "extra" information from oracle queries that are not in the intersection of their query sets. In particular, in the case of a random oracle, the shared information between $\mathcal{A}$ and $\mathcal{B}$ can be recovered completely from their shared oracle queries and the transcript T. In our setting, since the oracles f, $\mathcal{O}$, and $\mathcal{E}$ have dependence, this may not be the case.

The second challenge is to show that a hybrid oracle $\widetilde{\Gamma} = (\widetilde{f}, \widetilde{\mathcal{O}}, \widetilde{\mathcal{E}})$ can be constructed from the two sets of queries, i.e., from the simulated execution and the real execution.

As an example, suppose there is a query $\mathcal{E}(\widehat{C}, x)$ that is performed in the real execution and a different query $\mathcal{E}'(\widehat{C}, y)$ that appears in the simulated execution. Such two queries raise a challenge for constructing a hybrid oracle $\widetilde{\mathcal{E}}$ which is consistent with these two queries simultaneously. In order to see this, suppose that in the real execution, the lexicographically first pair (C, r) for which $\mathcal{O}(C, r) = \widehat{C}$ is some pair (C_1, r_1), and in the simulated execution the lexicographically first pair (C, r) for which $\mathcal{O}'(C, r) = \widehat{C}$ is some pair $(C_2, r_2) \neq (C_1, r_1)$. As a result, $\mathcal{E}(\widehat{C}, x)$ in the real execution is mapped to $C_1^f(x)$, whereas $\mathcal{E}'(\widehat{C}, y)$ is mapped to $C_2^{f'}(y)$, but $C_1 \neq C_2$.

We solve the first challenge by adding additional oracle queries to the set of real queries that the parties make, which makes the dependence between the oracles more explicit. As for the second challenge, interestingly, our proof does not completely solve it, and we do not fully control to which one of the two circuits C_1 or C_2 the hybrid oracle $\widetilde{\mathcal{E}}$ maps $\widehat{C}$. Nevertheless, we design the adversary such that, whenever there is such a contradicting scenario between the real execution and the simulated execution, it must hold that C_1 and C_2 are functionally equivalent with respect to the hybrid oracle $\widetilde{\Gamma}$. Otherwise, i.e., when there is some input for which C_1 and C_2 do not agree, we claim that the adversary learns a new query that is associated with the real execution. As a consequence, E learns the entire truth table of any obfuscated circuit $\widehat{C}$ that is associated in the real execution, which is possible due to the fact that querying the oracle Γ on all inputs of $\widehat{C}$ results in polynomially many queries. Notably, for a different expansion factor of the oracle $\mathcal{O}$ (which results in iO and not XiO), this becomes an exponential number of queries, and the above attack fails.

2.3   Statistically Secure Compressing Obfuscation

This set of results is composed of two main parts. One is positive results showing that for small classes of circuits compressing obfuscation exists unconditionally. The other complements the constructions and shows that improvements in the above obfuscator, either in the compression factor or in the circuit class, will imply some nontrivial speedup for protocols solving SAT or UNSAT. We have positive and negative results both for the case of perfect correctness and for the case of approximate correctness.

Negative results. First, we show that approximately correct and statistically secure 2^{n^ϵ}-SXiO cannot exist unless $\mathsf{coNP} \subseteq \mathsf{AM}[2^{n^\epsilon}]$ for $\epsilon > 0$. Here, we follow on the approach of [24] from the world of iO. There, they show how to use iO and puncturable PRFs to create two circuits that differ at a single point but their obfuscations (as random variables) are statistically far. Then, they use an algorithm that can distinguish these two distributions to solve Unique-SAT which then implies that $\mathsf{coNP} \subseteq \mathsf{AM}$ by a result of Mahmoody and Xiao [64]. We modify the argument to work with compressing obfuscation by making the two circuits receive only short inputs, and observe that the proof still goes through, but then solving Unique-SAT on short inputs (say of poly-logarithmic size). We then apply the result of Mahmoody and Xiao and finally obtain our result by scaling the parameters.

Second, we show that perfectly correct and statistically secure $2^{n(1-\epsilon)}$-SXiO cannot exist unless $\mathsf{coNP} \subseteq \mathsf{AM}[2^{(1-\epsilon)n}]$ (with large enough $0 < \epsilon < 1$). For this, we construct an $\mathsf{SZK}[2^{(1-\epsilon)n}]$ protocol for all NP. In this protocol, the verifier, given $x \in L$ for a language L, chooses a bit b uniformly at random and obfuscates a circuit that gets a witness w as input, checks whether it is a valid witness for x and if so, it outputs b (otherwise it outputs $\perp$). This protocol can be shown to be honest-verifier statistical zero-knowledge with a verifier that runs in time $2^{(1-\epsilon)n}$ for L. This argument is reminiscent to the argument of [47,53] in the context of iO. We then carefully apply the transformation of Okamoto [67] to translate this protocol into an (honest-verifier) SZK protocol for every language in coNP. This implies that $\mathsf{coNP} \subseteq \mathsf{AM}[2^{(1-\epsilon)n}]$.

Positive results. We show that compressing obfuscators exists unconditionally for restricted classes of circuits such as AC^0 (the class of all constant-depth circuits) and Mon (the class of all monotone functions). We again construct compressing obfuscators with perfect correctness and approximate correctness. The approximately correct obfuscators are obtained by running a classical (PAC) learning algorithm [71] on the given circuit and outputting the hypothesis. Using the most efficient learning algorithms for AC^0 and Mon, we obtain compressing obfuscators for these classes. This construction is aligned with the above impossibility that says that we are unlikely to be able to get such an obfuscator for classes that contain a (puncturable) PRF.

In the perfect correctness case, we use a different tool called a *circuit compression* algorithm [31]. In circuit compression one is given the truth table of a

Boolean function f computable by some *unknown* circuit from a known class of circuits, and the goal is to find in time $\mathsf{poly}(2^n)$ a circuit C (not necessarily from the aforementioned family) computing f so that the size of C is less than the trivial circuit size $\approx 2^n$. We apply such an algorithm on circuits in AC^0 and get an obfuscator with small compression.

3   Correctness Amplification

In this section, we present a correctness amplification procedure for XiO. We show that assuming the existence of an XiO scheme with very weak correctness, there exists an XiO construction with a very strong correctness guarantee.

Theorem 3.1. *Let $p(\cdot)$ be any polynomial. Let xiO be an XiO scheme for $\mathsf{P}^{\log}$ that is $\left(\frac{1}{2} + \frac{1}{p(\lambda)}\right)$-approximately correct. Assuming LWE and the existence of NIZKs, there exists a perfectly correct XiO scheme for $\mathsf{P}^{\log}$.*

The correctness amplification proceeds in three phases. First, we transform an approximately-correct XiO scheme to a $(1/\mathsf{poly}(\lambda) - \mathsf{negl}(\lambda))$-worst-case correct XiO scheme. Then, we transform the resulting scheme to a $(1 - \mathsf{negl}(\lambda))$-worst-case correct XiO scheme. Then, we transform the resulting scheme to a perfectly correct XiO scheme.

The main technical contribution of this section is the first step, transforming an approximately-correct XiO scheme to a $1/\mathsf{poly}(\lambda)$-worst-case correct XiO scheme. Therefore, in Sect. 3.1, we present the construction for this step. The full proof of Theorem 3.1 appears in the full version.

3.1   From Approximately-Correct XiO to Worst-Case Correct XiO

Fix any class of circuits $\mathcal{C}^{s,n} \in \mathsf{P}^{\log}$. Throughout this section, we let $s = s(\lambda)$ and $n = n(\lambda)$. Our transformation relies on the following primitives as building blocks:

- $\mathsf{xiO} = (\mathsf{xiO.Obf}, \mathsf{xiO.Eval})$ is a $(1/2 + \gamma)$-approximately correct XiO scheme for $\mathsf{P}^{\log}$, where $\gamma = 1/p(\lambda)$ for some polynomial p.
- ECC is a Reed-Solomon $\left(\frac{8 \cdot 2^{\frac{n}{d}}}{\gamma\lambda}, \frac{2^{\frac{n}{d}}}{\lambda}, \frac{8 \cdot 2^{\frac{n}{d}}}{\gamma\lambda} - \frac{2^{\frac{n}{d}}}{\lambda} + 1\right)_{2^\lambda}$ erasure correcting code that can correct up to a $(1 - \frac{\gamma}{8})$-fraction of erasures using the algorithm ECC.Dec, where $|\mathsf{ECC}|$ is a polynomial of degree $d - 1$ in its input length. We assume that all inputs to ECC are padded to size $2^{\frac{n}{d}}$ bits. We let $\ell_1 = O(\log(\lambda)) + \frac{n}{d}$ be the length of the output of ECC.
- LDC is a binary error-correcting code that is $(\frac{1}{2} - \frac{\gamma}{4}, \mathsf{poly})$-list decodable using the algorithm LDC.Dec. We let $\ell_2 = O(\log(\lambda) + \log(s) + \log(n))$ be the output length of LDC when run on inputs of size $\mathsf{poly}(\lambda, s, n)$.
- $\mathsf{IFE} = (\mathsf{IFE.Setup}, \mathsf{IFE.Enc}, \mathsf{IFE.Keygen}, \mathsf{IFE.Dec})$ is a λ-output succinct FE scheme for the class $\mathcal{C}^{s',n'} \in \mathsf{P}$ where $s' = \left(s \cdot 2^{\frac{n}{d}}\right)^{d-1} \cdot \mathsf{poly}(\lambda)$ and $n' = s \cdot \mathsf{poly}(\lambda, n)$.

- PRF = (PRF.Key, PRF.Punc, PRF.Eval) is a puncturable PRF.
- C = (C.Commit, C.Open) is a commitment scheme.
- NIZK = (NIZK.Gen, NIZK.P, NIZK.V) is a Multi-NIZK proof system for the NP language L given by $L = \Big\{(\mathsf{ct}, i, \mathsf{com}_C, \mathsf{com}_0, \mathsf{pk}) :$ either
  1. $\exists r_0, r_1, C$ such that ct encrypts (C, i) and com_C is a commitment to C, that is, $\mathsf{ct} = \mathsf{IFE.Enc}(\mathsf{pk}, (C, i); r_0) \wedge \mathsf{com}_C = \mathsf{C.Commit}(C, r_1)$, or
  2. $\exists r$ s.t. $\mathsf{com}_0 = \mathsf{C.Commit}(1, r)\Big\}$,

We let $t = t(\lambda) = \mathsf{poly}(\lambda, s, n)$ denote the upper bound on the length of statements and witnesses in L when instantiated with security parameter λ (with parameters as used in the following scheme).

In what follows, we denote by $C_{x_1 \cdots x_k}$ the circuit C with the first k bits hardwired to $x_1 \cdots x_k$. We let T denote a circuit in $\mathcal{C}^{s \cdot 2^{\frac{n}{d}}, s}$ that receives as input a circuit and outputs its truth table. The transformation is as follows.

Worst-case correct XiO scheme xi$\mathcal{O}'$:

- $\widetilde{C} \leftarrow \mathsf{xi}\mathcal{O}'.\mathsf{Obf}(1^\lambda, C)$:
  1. Sample $(\mathsf{msk}, \mathsf{pk}) \leftarrow \mathsf{IFE.Setup}(1^\lambda)$.
  2. Generate a key $\mathsf{sk}_\mathcal{U} \leftarrow \mathsf{IFE.Keygen}(\mathsf{msk}, \mathcal{U})$ for the circuit $\mathcal{U}$ such that

$$\mathcal{U}(D, i) = \mathsf{ECC}(T(D))[i],$$

  for any input circuit D, where $\mathsf{ECC}(T(D))[i]$ denotes the ith block of length λ of $\mathsf{ECC}(T(D))$.
  3. For every $x \in \{0,1\}^{n-\frac{n}{d}}$:
     (a) Sample $K_0^x, K_1^x \leftarrow \mathsf{PRF.Key}(1^\lambda)$, and $\sigma^x \leftarrow \mathsf{NIZK.Gen}(1^\lambda, 1^t)$.
     (b) Create commitments $\mathsf{com}_{C_x}^x = \mathsf{C.Commit}(C_x, r_0^x)$ to C_x and $\mathsf{com}_0^x = \mathsf{C.Commit}(0, r_1^x)$ to 0 using randomness $r_0^x \leftarrow \{0,1\}^\lambda$ and $r_1^x \leftarrow \{0,1\}^\lambda$.
     (c) Generate the circuit $G^x = G^x[C_x, \mathsf{pk}, K_0^x, K_1^x, \mathsf{com}_{C_x}^x, \mathsf{com}_0^x, r_0^x, \sigma^x]$ such that on input (i, j) does the following:
        i. Let $\mathsf{ct} \leftarrow \mathsf{IFE.Enc}(\mathsf{pk}, (C_x, i); \mathsf{PRF.Eval}(K_0^x, i))$.
        ii. Construct a NIZK proof $\pi = \mathsf{NIZK.P}(\sigma^x, v, w; \mathsf{PRF.Eval}(K_1, i))$ for the statement $v = (\mathsf{ct}, i, \mathsf{com}_{C_x}^x, \mathsf{com}_0^x, \mathsf{pk})$ using the witness $w = (C_x, \mathsf{PRF.Eval}(K_0^x, i), r_0^x)$.
        iii. Output the jth bit of $\mathsf{LDC}(\mathsf{ct}, \pi)$, denoted by $(\mathsf{LDC}(\mathsf{ct}, \pi))_j$.
     (d) Let $\widetilde{G}^x \leftarrow \mathsf{xi}\mathcal{O}.\mathsf{Obf}(1^\lambda, G^x)$ and let $\widetilde{C}^x = (\widetilde{G}^x, \sigma^x, \mathsf{com}_{C_x}^x, \mathsf{com}_0^x)$.
  4. Output $\widetilde{C} = \Big(\big\{\widetilde{C}^x\big\}_{x \in \{0,1\}^{n-\frac{n}{d}}}, \mathsf{sk}_\mathcal{U}, \mathsf{pk}\Big)$.
- $y' \leftarrow \mathsf{xi}\mathcal{O}'.\mathsf{Eval}(\widetilde{C}, x)$:
  1. Let $x = x_1 \| x_2$ where $|x_1| = n - \frac{n}{d}$.
  2. For every $i \in [2^{\ell_1}]$:
     (a) For every $j \in [2^{\ell_2}]$, let $c_{ij} = \mathsf{xi}\mathcal{O}.\mathsf{Eval}(\widetilde{G}^{x_1}, (i, j))$.
     (b) Run $\mathsf{LDC.Dec}(c_{i1} c_{i2} \cdots c_{i 2^{\ell_2}})$ to obtain a list of possible decodings, where the kth element of the list is $(\mathsf{ct}_i^k, \pi_i^k)$.

(c) Let $k^\star$ be the first index k such that $\mathsf{NIZK.V}(\sigma, v_i^k, \pi_i^k) = 1$ where $v_i^k = (\mathsf{ct}_i^k, i, \mathsf{com}_{C_{x_1}}^{x_1}, \mathsf{com}_0^{x_1}, \mathsf{pk})$. Set $\mathsf{ct}_i = \mathsf{ct}_i^{k^\star}$ if $k^\star$ exists and otherwise set $\mathsf{ct}_i = \bot$.

(d) Run $y_i \leftarrow \mathsf{IFE.Dec}(\mathsf{sk}_\mathcal{U}, \mathsf{ct}_i)$.

3. If there are at least $\frac{\gamma}{8} \cdot 2^{\ell_1}$ indices i for which $\mathsf{ct}_i \neq \bot$, let $y = y_1 y_2 \cdots y_{2^{\ell_1}}$ and run $\mathsf{ECC.Dec}(y)$ and output the element corresponding to x_2. Otherwise, output $\bot$.

Theorem 3.2. *Assume that* PRF *is a puncturable PRF,* IFE *is a selectively-secure λ-output succinct FE scheme for* $C^{s',n'}$, C *is a commitment scheme, and* NIZK *is a Multi-NIZK for* L. *Fix any class of circuits* $\mathcal{C}^{s,n} \in \mathsf{P}^{\log}$. *Let* $p(\cdot)$ *be any polynomial. Then, if* xiO *is a* $(1/2 + 1/p(\lambda))$*-approximately-correct XiO scheme for* $\mathsf{P}^{\log}$, *then* xiO' *is a* $\left(\frac{\gamma}{16} - \mathsf{negl}(\lambda)\right)$*-worst-case correct XiO scheme for* $\mathcal{C}^{s,n}$, *for a negligible function* negl.

The proof of this theorem appears in the full version.

4   On Key-Agreement from XIO and OWFs

In this section, we show a separation from compressing obfuscation and one-way functions to key-agreement. This separation is largely based on [9,10], and in particular follows the framework of black-box separations presented in [51].

We refer to the full version for important preliminaries, including the class of reductions that our proof captures. Throughout this section, for ease of notation, we denote both the security parameter and the size of circuits by s. While these could be distinguished, it is natural to combine them in this way, as everything can be thought of as a function of the circuit size in question. Hereafter, we say that an oracle-aided algorithm $M(1^s)$ with oracle access to Γ is a q-query algorithm if for every $s \in \mathbb{N}$, the algorithm $M(1^s)$ makes at most $q(s)$ queries, and each of its queries have size at most $q(s)$.

We show the separation by presenting a distribution over oracles Γ relative to which the following properties hold: (1) there does not exist a perfectly correct key-agreement protocol, (2) there exists an (exponentially) secure one-way function, and (3) there exists an (exponentially) secure XiO.

Let ℓ be a 2-ary function with $\ell(s, n) > s$. The distribution $\mathfrak{S}_\ell$ over oracles $\Gamma = (f, \mathcal{O}, \mathcal{E})$ is defined as follows:

- **The function** $f = \{f_s\}_{s \in \mathbb{N}}$. For every $s \in \mathbb{N}$, the function $f_s : \{0, 1\}^s \rightarrow \{0, 1\}^s$ is a uniformly chosen function. We will use f to implement a one-way function.
- **The function** $\mathcal{O} = \{\mathcal{O}_{s,n}\}_{s,n \in \mathbb{N}}$. For every $s, n \in \mathbb{N}$, the function $\mathcal{O}_{s,n} : \{0, 1\}^{2s} \rightarrow \{0, 1\}^{10\ell(s,n)}$ is a uniformly chosen function. Intuitively, $\mathcal{O}_{s,n}$ will receive a description of a circuit with size s and input length n, as well as a string of length s (which represents the randomness of the obfuscator), and will increase this to a uniformly chosen string of length $10\ell(s, n)$. This will be used to implement the obfuscator for xiO. Note that $\ell(s, n) > s$, and therefore the output of $\mathcal{O}_{s,n}$ is at least $10sn$.

- **The function** $\mathcal{E}^{f,\mathcal{O}} = \{\mathcal{E}_{s,n}^{f,\mathcal{O}}\}_{s\in\mathbb{N},n\in\mathbb{N}}$. For every $s,n \in \mathbb{N}$, the function $\mathcal{E}_{s,n}^{f,\mathcal{O}} : \{0,1\}^{10\ell(s,n)} \times \{0,1\}^n \rightarrow \{0,1\}^*$ is defined as follows. On input $(y,x) \in \{0,1\}^{10\ell(s,n)} \times \{0,1\}^n$, the function $\mathcal{E}_{s,n}^{f,\mathcal{O}}$ finds the lexicographically first oracle-aided circuit C of size s and input size n, and a string $r \in \{0,1\}^s$ such that $\mathcal{O}_{s,n}(C,r) = y$, and outputs $C^f(x)$. If no such (C,r) exists, it outputs $\perp$. Looking ahead, the oracle $\mathcal{E}^{f,\mathcal{O}}$ will be used to implement the evaluator for $\text{xi}\mathcal{O}$.

When $\ell(s,n) = 2^{n(1-\epsilon)} \cdot \text{poly}(s)$ for a constant $\epsilon > 0$ and a polynomial poly, relative to this oracle there exists a one-way function f and perfectly correct XiO scheme. The construction of XiO is natural: Given some circuit C of size s and input length n, the obfuscator chooses a random $r \leftarrow \{0,1\}^s$ and evaluates $\widehat{C} = \mathcal{O}_{s,n}(C,r)$. Then, it checks that the resulting handle $\widehat{C}$ agrees with the input circuit C: it runs over all inputs $x \in \{0,1\}^n$ and checks that $\mathcal{E}_{s,n}(\widehat{C},x) = C^f(x)$. If this holds for every input, it outputs $(0,\widehat{C})$. Otherwise, it outputs $(1,C)$. The evaluator on input circuit $(0,\widehat{C})$ and input x returns $\mathcal{E}_{s,n}(\widehat{C},x) = C^f(x)$, whereas on input circuit $(1,C)$ and input x evaluates $C^f(x)$.[5] The following holds, and is discussed in the full version:

Theorem 4.1. *Let* $\ell(s,n) = 2^{n\epsilon} \cdot \text{poly}(s)$ *for some constant* $0 \leq \epsilon < 1$ *and polynomial* poly *and let* $\Gamma \leftarrow \mathfrak{S}_\ell$ *with* $\Gamma = (f,\mathcal{O},\mathcal{E})$. *Then, for any oracle-aided* q-*query algorithm* $\mathcal{A}$ *with* $q(s) < 2^{s/4}$, *it holds that*

$$\Pr_{x\leftarrow\{0,1\}^s,\Gamma}\left[\mathcal{A}^\Gamma(f_s(x)) \in f_s^{-1}(f_s(x))\right] \leq 2^{-s/2}.$$

Moreover, for any class of circuits $\mathcal{C}$ *with* f-*gates, there exists an XiO scheme* $\text{xi}\mathcal{O}$ *relative to* Γ *for the circuit class* $\mathcal{C}$ *such that*

$$\left|\Pr_{r,\Gamma}\left[\text{Exp}_{\Gamma,\text{xi}\mathcal{O},\mathcal{D},\mathcal{C}}^{XiO}(\lambda;r) = 1\right] - \frac{1}{2}\right| < 2^{-s/4},$$

for any q-*query distinguisher* $\mathcal{D}$ *that makes at most* $q(s) < 2^{s/4}$ *queries.*[6]

The main technical difficulty is showing that there is no key-agreement protocol relative to Γ.

Theorem 4.2. *Let* $\ell(s,n) = 2^{n\epsilon} \cdot \text{poly}(s)$ *for a constant* $0 \leq \epsilon < 1$ *and a polynomial* poly. *Then, for any perfectly correct oracle-aided bit agreement protocol*

---

[5] We note that the technique of enumerating over all inputs is similar to that used in our correctness amplification, and takes advantage of the ability of XiO to compute the truth table of the obfuscated circuit.

[6] The game $\text{Exp}_{\Gamma,\text{xi}\mathcal{O},\mathcal{D},\mathcal{C}}^{XiO}(\lambda;r)$ is the indistinguishability experiment for XiO, defined as follows: (1) $b \leftarrow \{0,1\}$; (2) $(C_0,C_1,\text{state}) \leftarrow \mathcal{D}_1^\Gamma(1^s)$ where $|C_0| = |C_1| = s$ and $C_0^\Gamma \equiv C_1^\Gamma$. (3) $\widehat{C} \leftarrow \text{Obf}^\Gamma(1^s,C_b)$. (4) $b' \leftarrow \mathcal{D}_2^\Gamma(\text{state},\widehat{C})$. (5) If $b' = b$ then output 1. Otherwise, output 0.

$\langle \mathcal{A}(1^s), \mathcal{B}(1^s) \rangle$ *in which* $\mathcal{A}$ *and* $\mathcal{B}$ *run in time at most* $q(s)$, *there exists an oracle-aided adversary* E *that makes* $q(s)^{O(1)+1/\epsilon}$ *oracle queries such that*

$$\left| \Pr\left[\mathsf{Exp}^{\mathsf{KA}}_{\Gamma,(\mathcal{A},\mathcal{B}),E}(\lambda) = 1 \right] - \frac{1}{2} \right| \geq \frac{7}{16},$$

where the probability is over $\Gamma \leftarrow \mathfrak{S}_\ell$, *and the randomness of* $\mathcal{A}$, $\mathcal{B}$, *and* E.[7] *Moreover, the algorithm* E *can be implemented in polynomial time given access to a* PSPACE-*complete oracle.*

The full proof of this theorem appears in the full version. Here, we give a high level overview. We start by defining some notation.

Notation. Let $Q_\mathcal{A}$, $Q_\mathcal{B}$, and Q_E denote the set of oracle queries made by $\mathcal{A}$, $\mathcal{B}$, and E, respectively. Let $[O(x) = y] \in Q_p$ denote that a party p queried an oracle O on x and received y. For example, to denote that $\mathcal{A}$ queried $\mathcal{O}$ on C and received $\widetilde{C}$, we write $[\mathcal{O}(C) = \widetilde{C}] \in Q_\mathcal{A}$. Let $Q_{\mathcal{A}\mathcal{B}} = Q_\mathcal{A} \cup Q_\mathcal{B}$ be the set of oracle queries in the real protocol.

For a PPT oracle-aided key-agreement protocol $\langle \mathcal{A}^\Gamma(1^s), \mathcal{B}^\Gamma(1^s) \rangle$, we let $q = q(s)$ denote an upper bound on the running time of $\mathcal{A}$ and $\mathcal{B}$ for any oracle Γ. Since $\mathcal{A}$ and $\mathcal{B}$ are run in time at most q, this also bounds the space that the algorithms consume and their number of oracle queries. As a result, all $\mathcal{O}_{s,n}$ and $\mathcal{E}_{s,n}$ queries satisfy $s \leq q$ and $2^{\epsilon n} \cdot \mathsf{poly}(s) \leq q$. This implies that $n \leq \frac{1}{\epsilon} \log q$. We will use this bound on n to show that $\mathcal{A}$ and $\mathcal{B}$ can only query $\mathcal{O}$ on circuits with logarithmic size input, and thus the adversary can learn the truth table of any circuit queried this way by only making a polynomial number of queries.

We now define an extended set of queries for any query/answer set Q. Intuitively, this captures queries that are "known" to an algorithm that makes the queries in Q. For example, suppose an algorithm M queries $\mathcal{O}_{s,n}$ on some (C, r) and obtains $\widetilde{C}$, and queries f on all queries in the evaluation of $C^f(x)$. Then, intuitively M knows that $\mathcal{E}_{s,n}(\widetilde{C}, x) = C^f(x)$ (up to the probability of $\mathcal{O}$ being injective), even without making any $\mathcal{E}$ query. The following definition captures this dependence between the oracles, and will be helpful in our separation.

Definition 4.3. *Given a set of queries* Q *and an oracle* Γ, *the augmented set of queries* $\mathsf{Aug}(Q)$ *with respect to* Γ *is defined as follows:*

1. *Every query and answer in* Q *is also in* $\mathsf{Aug}(Q)$.
2. *For every query* $[\mathcal{O}_{s,n}(C, r) = \widetilde{C}] \in \mathsf{Aug}(Q)$, *the set* $\mathsf{Aug}(Q)$ *contains queries* $\mathcal{E}_{s,n}(\widetilde{C}, x)$ *for all* $x \in \{0,1\}^n$.
3. *For every query* $[\mathcal{E}_{s,n}(\widetilde{C}, x) = y] \in \mathsf{Aug}(Q)$ *with* $y \neq \bot$, *the set* $\mathsf{Aug}(Q)$ *contains the query* $\mathcal{O}_{s,n}(C, r) = \widetilde{C}$, *and all* f-*queries made in the evaluation of* $C^f(x) = y$. *where* (C, r) *is the lexicographically first pre-image of* $\widetilde{C}$ *under* $\mathcal{O}_{s,n}$.

---

[7] The game $\mathsf{Exp}^{\mathsf{KA}}_{\Gamma,\Pi,E}(\lambda)$ is defined as follows: (1) $(k_\mathcal{A}, k_\mathcal{B}, T) \leftarrow \langle \mathcal{A}^\Gamma(1^s), \mathcal{B}^\Gamma(1^s) \rangle$, (2) $k' \leftarrow E^\Gamma(1^s, T)$, (3) If $k' = k_\mathcal{A}$ then output 1, otherwise output 0.

For a given set Q with $|Q| < q$, we bound the size of the set $|\mathsf{Aug}(Q)|$, and recall that this implies that $s < q$ and $n < \frac{1}{\epsilon} \log q$. For every query to $\mathcal{O}_{s,n}$ in Q, there are at most 2^n corresponding $\mathcal{E}_{s,n}$ queries in $\mathsf{Aug}(Q)$, each implies at most s queries to f in $\mathsf{Aug}(Q)$. Likewise for any $\mathcal{E}_{s,n}$ query in Q might imply at most $2^n \cdot s$ queries in $\mathsf{Aug}(Q)$. Therefore, we have

$$|\mathsf{Aug}(Q)| \leq q \cdot s \cdot 2^n \leq q^2 \cdot q^{1/\epsilon}.$$

We are now ready to define the adversary E.

The adversary E.

- **Input:** A transcript T of an execution $\langle \mathcal{A}^\Gamma(1^s; r_A^\star), \mathcal{B}^\Gamma(1^s; r_B^\star) \rangle$.
- **Oracle Access:** $\Gamma = (f, \mathcal{O}, \mathcal{E})$.
- **Algorithm:**
  1. Initialize $Q_E = \emptyset$ and $K = \emptyset$.
  2. Repeat the following $2q + 1$ times:
     (a) **Simulation phase:** Find a valid oracle $\Gamma' = (f', \mathcal{O}', \mathcal{E}')$ and random strings r_A', r_B' such that the following holds:
        i. Every query in Q_E is answered the same way in Γ' as in Q_E.
        ii. $\mathcal{O}'_{s,n}$ is injective for all $s, n \in \mathbb{N}$.
        iii. The transcript T' outputted by $\langle \mathcal{A}^{\Gamma'}(1^s; r_A'), \mathcal{B}^{\Gamma'}(1^s, r_B') \rangle$ is the same as T.
        Abort if no such Γ', r_A', r_B' exist. Let k_A' be the key outputted by $\mathcal{A}$ in this simulation, and add k_A' to K.
     (b) **Update phase:** Let Q_{Sim} be the queries made by $\mathcal{A}$ and $\mathcal{B}$ in the execution $\langle \mathcal{A}^{\Gamma'}(1^s; r_A'), \mathcal{B}^{\Gamma'}(1^s, r_B') \rangle$, and consider the set $\mathsf{Aug}(Q_{\mathsf{Sim}})$ with respect to Γ'. Query Γ with all queries in $\mathsf{Aug}(Q_{\mathsf{Sim}}) \setminus Q_E$ and update Q_E with these queries.
- **Output:** The majority key k in K.

Observe that in each iteration, $|Q_{\mathsf{Sim}}| < q$ and E makes at most $|\mathsf{Aug}(Q_{\mathsf{Sim}})|$ queries to Γ. Therefore, the total number of queries that E makes is bounded by $(2q + 1) \cdot q^2 \cdot q^{1/\epsilon} \in q^{O(1)+1/\epsilon}$.

To complete the proof of Theorem 4.2, the main technical difficulty is in showing that the adversary E always succeeds to find the key computed in the real key agreement protocol, assuming that $\mathcal{O}$ is an injective function. We denote this event by injective$^{\Gamma,\ell}$ and in the full version, we show that the probability that $\neg$injective$^{\Gamma,\ell}$ occurs is bounded by 2^{-4}. We then show the following lemma.

Lemma 4.4. *Let $k^\star$ denote the key computed by $\mathcal{A}$ and $\mathcal{B}$ in the real execution of the protocol. If* injective$^{\Gamma,\ell}$ *holds, then E does not abort, and in each iteration either (1) E adds a query in $\mathsf{Aug}(Q_{\mathcal{AB}})$ to Q_E, or (2) E adds $k^\star$ to K.*

Proof Sketch. At a high level, the proof is as follows. First, assuming injective$^{\Gamma,\ell}$ holds, we show that E does not abort. This follows from the fact that the real oracle Γ and random strings $r_A^\star$ and $r_B^\star$ satisfy the properties needed to form the simulated oracle Γ' and random strings r'_A and r'_B. Thus, there exists at least one valid oracle and pair of random strings and therefore E does not abort.

Then, we show that in each iteration, either (1) E adds a query in $\mathsf{Aug}(Q_{AB})$ to Q_E, or (2) E adds k^* to K. Consider some iteration in which (1) does not hold. Let Γ', r'_A, r'_B be the oracle and random strings chosen by E in this iteration. By definition, the transcript of this execution is T. Let k' be the key outputted by $\langle \mathcal{A}^{\Gamma'}(1^s; r'_A), \mathcal{B}^{\Gamma'}(1^s; r'_B) \rangle$. Assuming that (1) does not hold, we show that there exists a hybrid oracle $\widetilde{\Gamma}$ for which $(k', k^*, T) \leftarrow \langle \mathcal{A}^{\widetilde{\Gamma}}(1^s; r'_A), \mathcal{B}^{\widetilde{\Gamma}}(1^s; r^*_B) \rangle$. That is, we show an oracle $\widetilde{\Gamma}$ such that when $\mathcal{A}$ uses the randomness of the simulation and $\mathcal{B}$ uses the randomness of the real protocol and both run with respect to $\widetilde{\Gamma}$, $\mathcal{A}$ outputs k' (as in the simulation) while $\mathcal{B}$ outputs k^* (as in the real), and the execution produces the transcript T (as in both the real and simulated protocols). We form this oracle by incorporating all queries in $\mathsf{Aug}(Q_{AB})$ and $\mathsf{Aug}(Q_{\mathrm{Sim}})$ into $\widetilde{\Gamma}$. Because (1) does not hold, E does not learn any new query in $\mathsf{Aug}(Q_{AB})$, and thus $\mathsf{Aug}(Q_{AB})$ and $\mathsf{Aug}(Q_{\mathrm{Sim}})$ agree on all queries and answers. In the full version, we show that this implies that $\widetilde{\Gamma}$ agrees with all queries in $\mathsf{Aug}(Q_{AB}) \cup \mathsf{Aug}(Q_{\mathrm{Sim}})$, and that this suffices for the result. Given the existence of such an oracle, by the perfect correctness, it must hold that $k' = k^*$, and therefore, since E adds $k' = k^*$ to K, the claim follows.   □

5   On Statistical Security

In this section we study the possibility for compressing obfuscation with perfect (information-theoretic) security. We will distinguish between approximately correct and perfectly correct compressing obfuscators and show almost tight results.

For approximately correct obfuscators, one the one hand, we show that there exists a statistically secure compressing obfuscator for the class of bounded depth circuits. On the other hand, we show that this is almost tight as any class that contains a (puncturable) PRF cannot be obfuscated with statistical secure (under complexity theoretic conjectures). See Theorems 5.4 and 5.5 for the precise parameters.

For perfectly correct obfuscators, on the one hand, we show that there exists a statistically secure compressing obfuscator for the class of bounded depth circuits, but the compression factor will be very weak (the obfuscation time is $\mathsf{poly}(2^n)$). On the other hand, we show that even for depth two circuits, better compression with better running time is implausible. See Theorems 5.2 and 5.7 for the precise parameters. Due to lack of space, all proofs from this section appear in the full version.

5.1   Negative Results

We show that it is unlikely that there is a statistically secure compressing obfuscator with good enough compression.

Our first result says that if such an obfuscator exists with strong enough compression, namely a $(2^{\epsilon n}, 2^{\epsilon n})$-compressing obfuscator with statistical security and perfect correctness, then $\overline{\mathsf{SAT}}$ (the problem of deciding whether a SAT formula is unsatisfiable) has an AM protocol in which the verifier's running time is bounded by $2^{\epsilon n}$. This is not believed to be likely for small enough values of $\epsilon > 0$, according to the best of our knowledge. Note that for this result we only need an obfuscator for depth-2 circuits. This argument relies on ideas from [53] and can be seen as an extension of an argument from [47].

Definition 5.1. *We denote by* $\mathsf{AM}[t, \ell]$ *the class of all languages on instances of size n that have an AM protocol in which the running time of the verifier is at most $t(n)$ and its messages size is at most $\ell(n)$. The class $\mathsf{coAM}[t, \ell]$ is defined, analogously, to be the class that contains all the complement languages. In case that $t = \ell$, we will write $\mathsf{AM}[t]$ to denote $\mathsf{AM}[t, t]$ and $\mathsf{coAM}[t]$ to denote $\mathsf{coAM}[t, t]$.*

Theorem 5.2. *There exists a universal constant $c > 0$ such that the following holds. If there is $0 < \epsilon < 1$ and a statistically secure and perfectly correct $(2^{\epsilon n}, 2^{\epsilon n})$-compressing obfuscation for depth-2 circuits, then $\overline{\mathsf{SAT}} \in \mathsf{AM}[2^{c\epsilon n}]$.*

The conclusion in Theorem 5.2 can be stated more generally as a conjecture that is interesting on its own right. This conjecture is parametrized by an $0 < \epsilon < 1$ and it says that $\overline{\mathsf{SAT}}$ is not in $\mathsf{AM}[2^{\epsilon n}]$.

Definition 5.3 (Conjecture). *There exist $\epsilon > 0$ for which $\overline{\mathsf{SAT}} \notin \mathsf{AM}[2^{\epsilon n}]$.*

It is known that the conjecture is *false* for $\epsilon = 1/2$ by the recent result of Williams [72] who showed that $\overline{\mathsf{SAT}} \in \mathsf{AM}[\tilde{O}(2^{n/2})]$. However, for smaller values of ϵ it is still unknown. The conjecture is particularly appealing in the case that ϵ is sub-constant (some $o(1)$).

Additionally, we give evidence that a compressing obfuscator with statistical security and only *approximate* correctness cannot exist for classes of functions that contain a (puncturable) PRF. This argument relies on and extends the proof of [24].

Theorem 5.4 *[Restatement of Theorem 1.2, part II]. There exists a universal constant $c > 0$ such that the following holds. If there is $0 < \epsilon < 1$ and a statistically secure and approximately correct $(2^{n^\epsilon}, 2^{n^\epsilon})$-compressing obfuscation for all circuits, then $\overline{\mathsf{SAT}} \in \mathsf{AM}[2^{n^\epsilon}]$.*

5.2   Positive Results

We show that for small classes of circuits there is a compressing obfuscation with perfect security. We start with the constructions that give approximate correctness.

Theorem 5.5 [*Restatement of Theorem 1.2, part I*]. *There exist constants* $0 <$ $\alpha < 1$ *and* $0 < \beta < 1$ *such that there exists a* $(1 - s/2^{n^{\beta}})$-*approximately correct* $(2^{n^{\alpha}}, 2^{n^{\alpha}})$-*compressing obfuscator with perfect security for the class of polynomial-size constant-depth n-input Boolean circuits.*

Theorem 5.6. *There exists a polynomial* $p(\cdot)$ *and a constant* $\alpha > 0$ *such that there exists a* $(1 - 1/p(n))$-*approximately correct* $(2^{(1-\alpha)n}, 2^{(1-\alpha)n})$-*compressing obfuscator with perfect security for the class of monotone n-input Boolean functions.*

We show that the class of bounded-depth circuits above can also be obfuscated with perfect correctness, while still resulting with a compressing obfuscator. However, the resulting compression is very weak (in particular, such compression, even for compressing obfuscation for all circuits is not known to imply full-fledged obfuscation).

Theorem 5.7 [*Restatement of Theorem 1.3*]. *There exists a perfectly correct* $(\text{poly}(2^n), 2^{n-n/O(\log s)^{d-1}})$-*obfuscator with perfect security for the class of size s depth d, n-input Boolean circuits.*

All of the obfuscators above treat their input circuit as a black box and run a classical *learning* or *compression* algorithm on it. We introduce these tasks next.

Preliminaries on PAC learning. We begin by introducing the concept of PAC learning. The Probably Approximately Correct (PAC) learning model, introduced by Valiant [71], is one of the most central definitions in the learning community and in computer science in general. We focus on PAC learning over the uniform distribution with membership queries. In this setting the learner may query the oracle at any point x and get back the value of the oracle at that point.

Definition 5.8 (PAC learning over the uniform distribution with membership queries). *Let* $\mathcal{F}$ *be a class of Boolean functions over* n *inputs. The class* $\mathcal{F}$ *is* (ϵ, δ)-*PAC learnable if there exists an algorithm* $\mathcal{A}$ *that gets as input two parameters* $\epsilon, \delta > 0$, *has membership query access to a function* $f \in \mathcal{F}$, *and outputs with probability* $1 - \delta$ *(over its internal randomness) a circuit* C *that agrees with* f *on all but an* ϵ-*fraction of the inputs. That is,*

$$\Pr_{\mathcal{A}}\left[C \leftarrow \mathcal{A}^f(\epsilon, \delta); \Pr_{x \leftarrow \{0,1\}^n}[C(x) \neq f(x)] \leq \epsilon\right] \geq 1 - \delta.$$

The running time of A is measures as a function of $n, 1/\epsilon, 1/\delta$, *and the circuit size of* f.

There has been a tremendous amount of work on obtaining efficient algorithms for PAC learning various classes of functions. It is known that no $\text{poly}(n)$-time algorithm can learn arbitrary Boolean functions $f \colon \{0,1\}^n \to \{0,1\}$ to accuracy non-negligibly better than $1/2$, but many positive results are known for restricted classes of functions. We fix $\delta = 2/3$, and note that this choice is somewhat arbitrary and enough for all of our applications. We thus say that a class is ϵ-PAC learnable if it is $(\epsilon, 2/3)$-PAC learnable.

One well known example is the quasi-polynomial time algorithm of Linial, Mansour, and Nisan [60] for the class of functions computed by AC^0 circuits (constant depth circuits with AND, OR, and NOT gates of unbounded fan-in and fan-out).

Theorem 5.9 (Learning bounded-depth circuits [60]). *The class of size-s depth-d circuits is ϵ-PAC learnable within $n^{O(\log^{d-1}(s/\epsilon))}$ queries.*[8]

Another notable example that is relevant for us is the algorithm of Bshouty and Tamon [28] for learning arbitrary monotone functions.

Theorem 5.10 (Learning monotone functions [28]). *The class of monotone functions is ϵ-PAC learnable within $n^{O(\sqrt{n}/\epsilon)}$ queries.*

A more recent result of Carmosino et al. [29] showed a (quasi-polynomial-time) learner for $AC^0[p]$, the class of Boolean constant depth circuits with unbounded fan-in and fan-out with AND, OR, NOT, and MOD-p gates.[9] Their result follows by a generic implication from natural properties to (randomized) algorithms for learning. More elaborately, [29] showed that any circuit lower bound proved through the very general natural proofs paradigm of Razborov and Rudich [69] yields algorithms for learning and compression. They then apply this result with the natural lower bound of Razborovand Smolenskyfor the class $AC^0[p]$. Informally, a "natural" lower bound for a circuit class $\mathcal{C}$ consists of an efficient algorithm that recognized some property that distinguishes between the truth tables functions in $\mathcal{C}$ and those of random Boolean functions.

Theorem 5.11 (Learning bounded-depth circuits with mod gates [29]). *For every prime $p > 1$, the class of $AC^0[p]$ circuits of size s is ϵ-PAC learnable within $2^{\text{poly} \log(ns/\epsilon)}$ queries.*

Tightness of the Approach. The approach of constructing obfuscators via learning algorithms is inherently limited. As observed by Valiant [71], any class that contains a pseudorandom function cannot be learned with nontrivial savings. Moreover, this approach, as shown above, gives the very strong notion of perfect security, which does not exist for all functions (even the computational version, known as virtual black-box, does not exist for circuits that contain a PRF [14]). Thus, to get an obfuscator (that satisfies only indistinguishability obfuscation) for a larger class of functions, one has to use the fact that the obfuscator has access to a circuit rather than treating it as a black-box.

---

[8] In Theorems 5.9 and 5.10 it is enough that the labels are for uniformly random inputs (i.e., random examples).

[9] We note that recently Carmosino et al. [30] generalized their result to get an implication from "tolerant" natural proofs to agnostic learning [52]. In agnostic learning, is the same as in PAC learning except that the learner is only guaranteed that f is close to the concept class $\mathcal{C}$ (rather than assuming it belongs to it).

Preliminaries on Circuit Compression. In the problem of circuit compression, studied by Chen et al. [31], one is given the truth table of a Boolean function f computable by some *unknown* circuit from a known class of circuits, and the goal is to find in time $\text{poly}(2^n)$ a circuit C (not necessarily from the aforementioned family) computing f so that the size of C is less than the trivial circuit size $\approx 2^n$. For general functions this is impossible as there are functions that require this size, so the focus is on restricted classes.

Definition 5.12 ($\mathcal{C}$-compression). *Given the truth table of an n-variate Boolean function $f \in \mathcal{C}$, find a Boolean circuit of size $< 2^n/n$ that is functionally equivalent to f.*

As mentioned in [31], compression of Boolean functions is related to the setting of exact learning with membership and equivalence queries [6]. In this learning setting, the size of the hypothesis produced by the learning algorithm is upper-bounded by the running time of the algorithm. In the circuit compression setting, the hypothesis (compressed image) size and the running time of the learning (compression) algorithm are decoupled: we allow more running time, but ask for a small-size compression. This may enable improvements in the class of circuits that we can handle. Concretely, exact learning is strictly stronger as any result in exact learning yields a compression algorithm for the corresponding class of functions, but the opposite direction is not known.

We notice that in general good enough compression implies compressing obfuscation where the output size is nontrivial but the running time can be large enough to read the truth table of the function (i.e., as in XiO). However, the other direction is not known since in XO one is given a witness (i.e., a circuit rather than the truth table). The most relevant circuit compression result that is relevant for us is stated next.

Theorem 5.13 ([31]). *If a Boolean n-variate function is computed by an AC^0 circuit of size s and depth d, then it is compressible to a circuit of size at most $2^{n-n/O(\log s)^{d-1}}$.*

As in the case of learning algorithms, the above compression algorithms directly imply perfectly correct compressing obfuscators satisfying perfect security.

We note that, as in the case of learning, it is impossible to compress a class of circuits that contains a PRF. For this, consider a PRF with key size n^2 and input size n which is exponentially secure (namely, secure for adversaries running in time $2^{\Omega(n^2)}$).[10] In this case, the PRF-or-Random adversary is allowed to query the oracle at all 2^n inputs and yet it still cannot distinguish PRF from random. The impossibility of compression for such a family of circuits now follows from the fact that random functions cannot be compressed.

Acknowledgments. We thank Zvika Brakerski for discussions about the possibility of SXiO and XiO with statistical security.

---

[10] The argument works even with sub-exponential security by increasing the size of the key.

References

1. Ananth, P., Jain, A., Naor, M., Sahai, A., Yogev, E.: Universal constructions and robust combiners for indistinguishability obfuscation and witness encryption. In: Robshaw, M., Katz, J. (eds.) CRYPTO 2016. LNCS, vol. 9815, pp. 491–520. Springer, Heidelberg (2016). https://doi.org/10.1007/978-3-662-53008-5_17
2. Ananth, P., Jain, A., Sahai, A.: Robust transforming combiners from indistinguishability obfuscation to functional encryption. In: Coron, J.-S., Nielsen, J.B. (eds.) EUROCRYPT 2017. LNCS, vol. 10210, pp. 91–121. Springer, Cham (2017). https://doi.org/10.1007/978-3-319-56620-7_4
3. Ananth, P., Jain, A.: Indistinguishability obfuscation from compact functional encryption. In: Gennaro, R., Robshaw, M. (eds.) CRYPTO 2015. LNCS, vol. 9215, pp. 308–326. Springer, Heidelberg (2015). https://doi.org/10.1007/978-3-662-47989-6_15
4. Ananth, P., Sahai, A.: Projective arithmetic functional encryption and indistinguishability obfuscation from degree-5 multilinear maps. In: Coron, J.-S., Nielsen, J.B. (eds.) EUROCRYPT 2017. LNCS, vol. 10210, pp. 152–181. Springer, Cham (2017). https://doi.org/10.1007/978-3-319-56620-7_6
5. Ananth, P.V., Gupta, D., Ishai, Y., Sahai, A.: Optimizing obfuscation: avoiding Barrington's theorem. In: Proceedings of the 2014 ACM SIGSAC Conference on Computer and Communications Security, pp. 646–658 (2014)
6. Angluin, D.: Queries and concept learning. Mach. Learn. 2(4), 319–342 (1987)
7. Apon, D., Döttling, N., Garg, S., Mukherjee, P.: Cryptanalysis of indistinguishability obfuscations of circuits over GGH13. In: 44th International Colloquium on Automata, Languages, and Programming, ICALP 2017, pp. 38:1–38:16 (2017)
8. Applebaum, B., Brakerski, Z.: Obfuscating circuits via composite-order graded encoding. In: Dodis, Y., Nielsen, J.B. (eds.) TCC 2015. LNCS, vol. 9015, pp. 528–556. Springer, Heidelberg (2015). https://doi.org/10.1007/978-3-662-46497-7_21
9. Asharov, G., Segev, G.: Limits on the power of indistinguishability obfuscation and functional encryption. SIAM J. Comput. 45(6), 2117–2176 (2016)
10. Asharov, G., Segev, G.: On constructing one-way permutations from indistinguishability obfuscation. In: Kushilevitz, E., Malkin, T. (eds.) TCC 2016. LNCS, vol. 9563, pp. 512–541. Springer, Heidelberg (2016). https://doi.org/10.1007/978-3-662-49099-0_19
11. Asmuth, C., Blakley, G.: An efficient algorithm for constructing a cryptosystem which is harder to break than two other cryptosystems. Comput. Math. Appl. 7(6), 447–450 (1981)
12. Barak, B., Brakerski, Z., Komargodski, I., Kothari, P.K.: Limits on low-degree pseudorandom generators (or: sum-of-squares meets program obfuscation). IACR Cryptology ePrint Archive 2017, 312 (2017)
13. Barak, B., Garg, S., Kalai, Y.T., Paneth, O., Sahai, A.: Protecting obfuscation against algebraic attacks. In: Nguyen, P.Q., Oswald, E. (eds.) EUROCRYPT 2014. LNCS, vol. 8441, pp. 221–238. Springer, Heidelberg (2014). https://doi.org/10.1007/978-3-642-55220-5_13
14. Barak, B., Goldreich, O., Impagliazzo, R., Rudich, S., Sahai, A., Vadhan, S.P., Yang, K.: On the (im)possibility of obfuscating programs. J. ACM 59(2), 6:1–6:48 (2012)
15. Bitansky, N., Degwekar, A., Vaikuntanathan, V.: Structure vs. hardness through the obfuscation lens. In: Katz, J., Shacham, H. (eds.) CRYPTO 2017. LNCS, vol. 10401, pp. 696–723. Springer, Cham (2017). https://doi.org/10.1007/978-3-319-63688-7_23

16. Bitansky, N., Lin, H., Paneth, O.: On removing graded encodings from functional encryption. In: Coron, J.-S., Nielsen, J.B. (eds.) EUROCRYPT 2017, Proceedings, Part II. LNCS, vol. 10211, pp. 3–29. Springer, Cham (2017). https://doi.org/10.1007/978-3-319-56614-6_1

17. Bitansky, N., Nishimaki, R., Passelègue, A., Wichs, D.: From cryptomania to obfustopia through secret-key functional encryption. In: Theory of Cryptography Conference, pp. 391–418 (2016)

18. Bitansky, N., Paneth, O.: ZAPs and non-interactive witness indistinguishability from indistinguishability obfuscation. In: Dodis, Y., Nielsen, J.B. (eds.) TCC 2015. LNCS, vol. 9015, pp. 401–427. Springer, Heidelberg (2015). https://doi.org/10.1007/978-3-662-46497-7_16

19. Bitansky, N., Paneth, O., Wichs, D.: Perfect structure on the edge of chaos - trapdoor permutations from indistinguishability obfuscation. In: Kushilevitz, E., Malkin, T. (eds.) TCC 2016. LNCS, vol. 9562, pp. 474–502. Springer, Heidelberg (2016). https://doi.org/10.1007/978-3-662-49096-9_20

20. Bitansky, N., Vaikuntanathan, V.: Indistinguishability obfuscation from functional encryption. In: IEEE 56th Annual Symposium on Foundations of Computer Science, FOCS 2015, pp. 171–190 (2015)

21. Bitansky, N., Vaikuntanathan, V.: Indistinguishability obfuscation: from approximate to exact. In: Kushilevitz, E., Malkin, T. (eds.) TCC 2016. LNCS, vol. 9562, pp. 67–95. Springer, Heidelberg (2016). https://doi.org/10.1007/978-3-662-49096-9_4

22. Bitansky, N., Vaikuntanathan, V.: A note on perfect correctness by derandomization. In: Coron, J.-S., Nielsen, J.B. (eds.) EUROCRYPT 2017. LNCS, vol. 10211, pp. 592–606. Springer, Cham (2017). https://doi.org/10.1007/978-3-319-56614-6_20

23. Boneh, D., Wu, D.J., Zimmerman, J.: Immunizing multilinear maps against zeroizing attacks. IACR Cryptology ePrint Archive 2014, 930 (2014)

24. Brakerski, Z., Brzuska, C., Fleischhacker, N.: On statistically secure obfuscation with approximate correctness. In: Robshaw, M., Katz, J. (eds.) CRYPTO 2016. LNCS, vol. 9815, pp. 551–578. Springer, Heidelberg (2016). https://doi.org/10.1007/978-3-662-53008-5_19

25. Brakerski, Z., Jain, A., Komargodski, I., Passelègue, A., Wichs, D.: Non-trivial witness encryption and null-iO from standard assumptions. IACR Cryptology ePrint Archive 2017, 874 (2017)

26. Brakerski, Z., Katz, J., Segev, G., Yerukhimovich, A.: Limits on the power of zero-knowledge proofs in cryptographic constructions. In: Ishai, Y. (ed.) TCC 2011. LNCS, vol. 6597, pp. 559–578. Springer, Heidelberg (2011). https://doi.org/10.1007/978-3-642-19571-6_34

27. Brakerski, Z., Rothblum, G.N.: Virtual black-box obfuscation for all circuits via generic graded encoding. In: Lindell, Y. (ed.) TCC 2014. LNCS, vol. 8349, pp. 1–25. Springer, Heidelberg (2014). https://doi.org/10.1007/978-3-642-54242-8_1

28. Bshouty, N.H., Tamon, C.: On the fourier spectrum of monotone functions. J. ACM **43**(4), 747–770 (1996)

29. Carmosino, M.L., Impagliazzo, R., Kabanets, V., Kolokolova, A.: Learning algorithms from natural proofs. In: 31st Conference on Computational Complexity, CCC 2016, vol. 50, pp. 10:1–10:24 (2016)

30. Carmosino, M.L., Impagliazzo, R., Kabanets, V., Kolokolova, A.: Agnostic learning from tolerant natural proofs. In: Approximation, Randomization, and Combinatorial Optimization. Algorithms and Techniques, APPROX/RANDOM 2017, vol. 81, pp. 35:1–35:19 (2017)

31. Chen, R., Kabanets, V., Kolokolova, A., Shaltiel, R., Zuckerman, D.: Mining circuit lower bound proofs for meta-algorithms. Comput. Complex. **24**(2), 333–392 (2015)
32. Chen, Y., Gentry, C., Halevi, S.: Cryptanalyses of candidate branching program obfuscators. In: Coron, J.-S., Nielsen, J.B. (eds.) EUROCRYPT 2017. LNCS, vol. 10212, pp. 278–307. Springer, Cham (2017). https://doi.org/10.1007/978-3-319-56617-7_10
33. Cheon, J.H., Han, K., Lee, C., Ryu, H., Stehlé, D.: Cryptanalysis of the multilinear map over the integers. In: Oswald, E., Fischlin, M. (eds.) EUROCRYPT 2015. LNCS, vol. 9056, pp. 3–12. Springer, Heidelberg (2015). https://doi.org/10.1007/978-3-662-46800-5_1
34. Coron, J.-S., Lepoint, T., Tibouchi, M.: Practical multilinear maps over the integers. In: Canetti, R., Garay, J.A. (eds.) CRYPTO 2013. LNCS, vol. 8042, pp. 476–493. Springer, Heidelberg (2013). https://doi.org/10.1007/978-3-642-40041-4_26
35. Coron, J.-S., Lepoint, T., Tibouchi, M.: New multilinear maps over the integers. In: Gennaro, R., Robshaw, M. (eds.) CRYPTO 2015. LNCS, vol. 9215, pp. 267–286. Springer, Heidelberg (2015). https://doi.org/10.1007/978-3-662-47989-6_13
36. Diffie, W., Hellman, M.E.: Multiuser cryptographic techniques. In: American Federation of Information Processing Societies, pp. 109–112 (1976)
37. Fischlin, M., Herzberg, A., Bin-Noon, H., Shulman, H.: Obfuscation combiners. In: Robshaw, M., Katz, J. (eds.) CRYPTO 2016. LNCS, vol. 9815, pp. 521–550. Springer, Heidelberg (2016). https://doi.org/10.1007/978-3-662-53008-5_18
38. Garg, S., Gentry, C., Halevi, S., Raykova, M., Sahai, A., Waters, B.: Candidate indistinguishability obfuscation and functional encryption for all circuits. In: 54th Annual IEEE Symposium on Foundations of Computer Science, FOCS 2013, pp. 40–49. IEEE Computer Society (2013)
39. Garg, S., Gentry, C., Sahai, A., Waters, B.: Witness encryption and its applications. In: Symposium on Theory of Computing Conference, STOC 2013, pp. 467–476 (2013)
40. Garg, S., Mahmoody, M., Mohammed, A.: Lower bounds on obfuscation from all-or-nothing encryption primitives. In: Katz, J., Shacham, H. (eds.) CRYPTO 2017. LNCS, vol. 10401, pp. 661–695. Springer, Cham (2017). https://doi.org/10.1007/978-3-319-63688-7_22
41. Garg, S., Mahmoody, M., Mohammed, A.: When does functional encryption imply obfuscation? In: Kalai, Y., Reyzin, L. (eds.) TCC 2017. LNCS, vol. 10677, pp. 82–115. Springer, Cham (2017). https://doi.org/10.1007/978-3-319-70500-2_4
42. Gentry, C., Gorbunov, S., Halevi, S.: Graph-induced multilinear maps from lattices. In: Dodis, Y., Nielsen, J.B. (eds.) TCC 2015. LNCS, vol. 9015, pp. 498–527. Springer, Heidelberg (2015). https://doi.org/10.1007/978-3-662-46497-7_20
43. Gentry, C., Halevi, S., Maji, H.K., Sahai, A.: Zeroizing without zeroes: cryptanalyzing multilinear maps without encodings of zero. IACR Cryptology ePrint Archive 2014, 929 (2014)
44. Gentry, C., Lewko, A.B., Sahai, A., Waters, B.: Indistinguishability obfuscation from the multilinear subgroup elimination assumption. In: IEEE 56th Annual Symposium on Foundations of Computer Science, FOCS 2015, pp. 151–170 (2015)
45. Goldwasser, S., et al.: Multi-input functional encryption. In: Nguyen, P.Q., Oswald, E. (eds.) EUROCRYPT 2014. LNCS, vol. 8441, pp. 578–602. Springer, Heidelberg (2014). https://doi.org/10.1007/978-3-642-55220-5_32
46. Goldwasser, S., Kalai, Y.T., Popa, R.A., Vaikuntanathan, V., Zeldovich, N.: Reusable garbled circuits and succinct functional encryption. In: Symposium on Theory of Computing Conference, STOC 2013, pp. 555–564 (2013)

47. Goldwasser, S., Rothblum, G.N.: On best-possible obfuscation. In: Vadhan, S.P. (ed.) TCC 2007. LNCS, vol. 4392, pp. 194–213. Springer, Heidelberg (2007). https://doi.org/10.1007/978-3-540-70936-7_11
48. Harnik, D., Kilian, J., Naor, M., Reingold, O., Rosen, A.: On robust combiners for oblivious transfer and other primitives. In: Cramer, R. (ed.) EUROCRYPT 2005. LNCS, vol. 3494, pp. 96–113. Springer, Heidelberg (2005). https://doi.org/10.1007/11426639_6
49. Herzberg, A.: On tolerant cryptographic constructions. In: Menezes, A. (ed.) CT-RSA 2005. LNCS, vol. 3376, pp. 172–190. Springer, Heidelberg (2005). https://doi.org/10.1007/978-3-540-30574-3_13
50. Herzberg, A.: Folklore, practice and theory of robust combiners. J. Comput. Secur. **17**(2), 159–189 (2009)
51. Impagliazzo, R., Rudich, S.: Limits on the provable consequences of one-way permutations. In: Proceedings of the Twenty-First Annual ACM Symposium on Theory of Computing, pp. 44–61. ACM (1989)
52. Kearns, M.J., Schapire, R.E., Sellie, L.: Toward efficient agnostic learning. Mach. Learn. **17**(2–3), 115–141 (1994)
53. Komargodski, I., Moran, T., Naor, M., Pass, R., Rosen, A., Yogev, E.: One-way functions and (im)perfect obfuscation. In: 55th IEEE Annual Symposium on Foundations of Computer Science, FOCS 2014, pp. 374–383 (2014)
54. Levin, L.A.: One-way functions and pseudorandom generators. Combinatorica **7**(4), 357–363 (1987)
55. Lin, H.: Indistinguishability obfuscation from constant-degree graded encoding schemes. In: Fischlin, M., Coron, J.-S. (eds.) EUROCRYPT 2016. LNCS, vol. 9665, pp. 28–57. Springer, Heidelberg (2016). https://doi.org/10.1007/978-3-662-49890-3_2
56. Lin, H.: Indistinguishability obfuscation from SXDH on 5-linear maps and locality-5 PRGs. In: Katz, J., Shacham, H. (eds.) CRYPTO 2017. LNCS, vol. 10401, pp. 599–629. Springer, Cham (2017). https://doi.org/10.1007/978-3-319-63688-7_20
57. Lin, H., Pass, R., Seth, K., Telang, S.: Indistinguishability obfuscation with nontrivial efficiency. In: Cheng, C.-M., Chung, K.-M., Persiano, G., Yang, B.-Y. (eds.) PKC 2016. LNCS, vol. 9615, pp. 447–462. Springer, Heidelberg (2016). https://doi.org/10.1007/978-3-662-49387-8_17
58. Lin, H., Pass, R., Seth, K., Telang, S.: Output-compressing randomized encodings and applications. In: Kushilevitz, E., Malkin, T. (eds.) TCC 2016-A. LNCS, vol. 9562, pp. 96–124. Springer, Heidelberg (2016). https://doi.org/10.1007/978-3-662-49096-9_5
59. Lin, H., Vaikuntanathan, V.: Indistinguishability obfuscation from DDH-like assumptions on constant-degree graded encodings. In: IEEE 57th Annual Symposium on Foundations of Computer Science, FOCS 2016, pp. 11–20 (2016)
60. Linial, N., Mansour, Y., Nisan, N.: Constant depth circuits, Fourier transform, and learnability. In: 30th Annual Symposium on Foundations of Computer Science, pp. 574–579 (1989)
61. Liu, Q., Zhandry, M.: Decomposable obfuscation: a framework for building applications of obfuscation from polynomial hardness. In: Kalai, Y., Reyzin, L. (eds.) TCC 2017. LNCS, vol. 10677, pp. 138–169. Springer, Cham (2017). https://doi.org/10.1007/978-3-319-70500-2_6
62. Lombardi, A., Vaikuntanathan, V.: Limits on the locality of pseudorandom generators and applications to indistinguishability obfuscation. In: Kalai, Y., Reyzin, L. (eds.) TCC 2017. LNCS, vol. 10677, pp. 119–137. Springer, Cham (2017). https://doi.org/10.1007/978-3-319-70500-2_5

63. Mahmoody, M., Mohammed, A., Nematihaji, S., Pass, R., Shelat, A.: Lower bounds on assumptions behind indistinguishability obfuscation. In: Kushilevitz, E., Malkin, T. (eds.) TCC 2016-A. LNCS, vol. 9562, pp. 49–66. Springer, Heidelberg (2016). https://doi.org/10.1007/978-3-662-49096-9_3
64. Mahmoody, M., Xiao, D.: On the power of randomized reductions and the checkability of SAT. In: CCC 2010, pp. 64–75. IEEE Computer Society (2010)
65. Micali, S., Peikert, C., Sudan, M., Wilson, D.A.: Optimal error correction against computationally bounded noise. In: Kilian, J. (ed.) TCC 2005. LNCS, vol. 3378, pp. 1–16. Springer, Heidelberg (2005). https://doi.org/10.1007/978-3-540-30576-7_1
66. Miles, E., Sahai, A., Zhandry, M.: Annihilation attacks for multilinear maps: cryptanalysis of indistinguishability obfuscation over GGH13. In: Robshaw, M., Katz, J. (eds.) CRYPTO 2016. LNCS, vol. 9815, pp. 629–658. Springer, Heidelberg (2016). https://doi.org/10.1007/978-3-662-53008-5_22
67. Okamoto, T.: On relationships between statistical zero-knowledge proofs. J. Comput. Syst. Sci. **60**(1), 47–108 (2000)
68. Pass, R., Seth, K., Telang, S.: Indistinguishability obfuscation from semantically-secure multilinear encodings. In: Garay, J.A., Gennaro, R. (eds.) CRYPTO 2014. LNCS, vol. 8616, pp. 500–517. Springer, Heidelberg (2014). https://doi.org/10.1007/978-3-662-44371-2_28
69. Razborov, A.A., Rudich, S.: Natural proofs. J. Comput. Syst. Sci. **55**(1), 24–35 (1997)
70. Sahai, A., Waters, B.: How to use indistinguishability obfuscation: deniable encryption, and more. In: Symposium on Theory of Computing, STOC 2014, pp. 475–484 (2014)
71. Valiant, L.G.: A theory of the learnable. Commun. ACM **27**(11), 1134–1142 (1984)
72. Williams, R.R.: Strong ETH breaks with Merlin and Arthur: short non-interactive proofs of batch evaluation. In: CCC, vol. 50, pp. 2:1–2:17 (2016)
73. Zimmerman, J.: How to obfuscate programs directly. In: Oswald, E., Fischlin, M. (eds.) EUROCRYPT 2015. LNCS, vol. 9057, pp. 439–467. Springer, Heidelberg (2015). https://doi.org/10.1007/978-3-662-46803-6_15

Author Index

Printed in the United States
By Bookmasters